Computing and Systems Technology Division

Core Programming Topic at the 2012 AIChE Annual Meeting

Pittsburgh, Pennsylvania, USA
28 October - 2 November 2012

Volume 1 of 2

ISBN: 978-1-62276-722-9

Printed from e-media with permission by:

Curran Associates, Inc.
57 Morehouse Lane
Red Hook, NY 12571

Some format issues inherent in the e-media version may also appear in this print version.

Printed by Curran Associates, Inc. (2013)

For permission requests, please contact AIChE
at the address below.

AIChE
3 Park Avenue
New York, NY 10016-5991

Phone: (203) 702-7660
Fax: (203) 775-5177

www.aiche.org

Additional copies of this publication are available from:

Curran Associates, Inc.
57 Morehouse Lane
Red Hook, NY 12571 USA
Phone: 845-758-0400
Fax: 845-758-2634
Email: curran@proceedings.com
Web: www.proceedings.com

TABLE OF CONTENTS

VOLUME 1

VOLUME 2

(16a) Overview of CAST Activities and Programming

B. Wayne Bequette, Chemical and Biological Engineering, Rensselaer Polytechnic Institute, Troy, NY and Richard D. Braatz, Department of Chemical Engineering, Massachusetts Institute of Technology, Cambridge, MA

An overview of activities of the Computing and Systems Technology (CAST) division, including conference technical programming, will be presented. The CAST Executive Committee consists of Richard Braatz (Chair), Mayuresh Kothare (1st Vice-Chair), Marianthhi Ierapetritou (2nd Vice-Chair), Karl Schnelle (Secretary-Treasurer), and the following Directors: Ray Adomaitis (2010-2012), Tom Badgwell (2010-2012), Matt Bassett (2011-2013), Bob Parker (2011-2013), Christos Maravelias (2012-2014), John Congalidis (2012-2014). The CAST Division Programming Chair is Wayne Bequette (2010-2013), and the Programming Vice-Chair is Nick Sahinidis (Programming Chair for 2014-2017).

(16b) Process Synthesis, Design and Intensification. an Integrated Approach

Deenesh K. Babi, Technical University of Denmark, CAPEC-Department of Chemical and Biochemical Engineering, DK-2800 Lyngby, Denmark, John Woodley, Center for Process Engineering and Technology, Department of Chemical and Biochemical Engineering, Technical University of Denmark, Lyngby, Denmark, Rafiqul Gani, CAPEC, Department of Chemical and Biochemical Engineering, Technical University of Denmark (DTU), Kgs. Lyngby, Denmark, Jeffrey J. Siirola, Purdue University and Carnegie Mellon University, Kingsport, TN and Fabrício Rodrigues, Department of Chemical Engineering, Federal University of Uberlândia, Uberlândia, Brazil

Process Synthesis, Design and Intensification. An integrated approach

Deenesh K. Babi[1], John M. Woodley[2], Rafiqul Gani[1], Jeffrey J. Siirola[3]

dkbabi@kt.dtu.dk, jw@kt.dtu.dk, rag@kt.dtu.dk, jjsiirola@gmail.com

[1] ***CAPEC**-Department of Chemical and Biochemical Engineering, Technical University of Denmark (DTU), DK-2800 Kongens Lyngby, Denmark*

[2] ***PROCESS**-Department of Chemical and Biochemical Engineering, Technical University of Denmark (DTU), DK-2800 Kongens Lyngby, Denmark*

[3] *Department of Chemical Engineering, Carnegie Mellon University, Pittsburgh, PA 15213-3890 USA*

Abstract

Process design has four main objectives: process synthesis or generation of process alternatives; process analysis; process evaluation; and the selection of the best process to achieve a desired goal [1]. According to Westerberg [2] the overall improvement of a process, new or existing, may be achieved by considering the original design as the base case, performing flowsheet optimization to further improve the design and through analysis of the real time process operation. Possible improvements are identified and are generally, where economically feasible, implemented during plant retrofits. Process intensification (PI) is a means by which processes, whether conceptual or existing, can be designed or redesigned to achieve a more efficient and sustainable process through the improvement of key process parameters, for example energy efficiency and waste generation. PI is defined as the improvement of a process at the phenomena level which ultimately has an impact at the higher levels of a process, for example the functional and unit operations levels. More specifically it is the enhancement achieved through the integration of unit operations, functions and/or phenomena. Therefore incentives exist for the inclusion of PI within the overall concept of process design and in particular as part of the process synthesis and design methodology.

With the inclusion of PI into process design, not only can a process be synthesised, analysed and evaluated but it can also be further analysed for the potential use of intensified equipment within that process, whether novel or mature. In order for this paradigm shift to occur the intensified process should be better than the original design. An example of PI was achieved by Eastman Chemical [3] which in 1983 intensified a process for the manufacture of methyl acetate by replacing with one single reactive-extractive distillation column a multi-step process which had consisted of a reactor, extractor, decanter, several normal and azeotropic distillation columns, and two mass separation agents.

For achieving PI via a systematic approach different methods exist, for example, the task-based means-ends analysis developed by Siirola [1]. However, there is a need for not only a more systematic, efficient and flexible PI method covering a wider range of applications, but also for one that is able to find truly innovative and predictive solutions, using not only the knowledge of existing methodologies at the unit operations level but also at a lower phenomena level, where new unit operations can be designed.

The objective of this work is twofold: (1) The further development of a phenomena based PI synthesis and design algorithm (PBS algorithm) and (2) To incorporate the PBS algorithm as part of an overall process synthesis/design framework that includes both task-based means-ends analysis and thermodynamic insights [4]. Within the overall synthesis/design framework this PBS algorithm is employed to generate, screen and analyse many possible alternatives. A first version of the PBS algorithm has been developed and consists of a six steps.

The starting point of the PBS algorithm is the base case design, which can either be the final design based on the current state of the art of process design or the design of an existing process. In Step 1 the problem is defined together with the objective function. In Step 2 a decomposition approach is applied where the process is represented in terms of tasks and process phenomena respectively and the process is then analysed using thermodynamic insights for example analysis of pure component and mixture properties. In Step 3 the limitations/bottlenecks (LBs) of the process can then be identified together with the desirable and accompanying phenomena to overcome these LBs. In Step 4 phenomena are connected to form simultaneous phenomena building blocks (SPBs) which are screened for the most feasible connections using for example connectivity rules. An example of a connectivity rule is heating and cooling cannot exist within the same SPBs. The SPB's are then connected to form operations which are then connected to form flowsheets. In Step 5 these flowsheets are first screened using for example logical constraints and performance metrics. In Step 6 the feasible flowhseets from Step 5 are optimized with respect to the objective function defined in Step 1 with the end result being the identification of the best intensified process candidate. The emerging flowhseet consists of either novel or existing equipment. Through the combined synthesis and design framework, the results of the means-ends analysis and thermodynamic insights methodologies are used in defining the problem (Step 1), in decomposing the problem (Step 2), and in evaluating the results from Step 4.

In the proposed presentation the overall process synthesis and design framework together with its supporting algorithms and tools will be illustrated by a re-examination of the methyl acetate process. It will be shown that the framework is able to generate a wider range of solutions, including those that have been reported earlier.

References:

[1] J. J. Siirola. Strategic process synthesis: Advances in the hierarchical approach. Comput. Chem. Eng. 1996, Supplement 2 (20) S1637-S1643

[2] A. W. Westerberg. A retrospective on design and process synthesis. Comput. Chem. Eng. 2004 (28) 447–458

[3] V. H. Agreda, L. R. Partin & W. H. Heise . High-purity methyl acetate via reactive distillation. Chem. Eng. Prog. 1986 (2) 40–46

[4] C. Jaksland, R. Gani, K. M. Lien. Separation process design and synthesis based on thermodynamic insights. Chem Eng Sci, 1995 (50) 511-530

(16c) Isolation and Handling of Sensor Faults in Nonlinear Process Systems

Miao Du, Chemical Engineering, McMaster University, Hamilton, ON, Canada and Prashant Mhaskar, Department of Chemical Engineering, McMaster University, Hamilton, ON, Canada

Automatic control technologies have been widely used in industrial processes, leading to improved efficiency and profitability. Their successful implementation is based on the reliability of actuating and sensing equipment, which are subject to faults that take place due to various reasons. The equipment abnormalities can lead to failures to meet product quality specifications, safety hazards to facilities and personnel, and damages to the environment in the absence of appropriately designed monitoring and fault-handling systems. Therefore, an accurate and timely detection of a fault and identification of the location of the faulty equipment is of great importance to reducing the effect of faults through the implementation of corrective control action. This realization has motivated significant research efforts to the development of fault detection and isolation (FDI) techniques.

Existing results on FDI include those that mainly use information embodied in plant data or a process model. In the data-based approach, abnormal behavior is detected using statistical monitoring techniques (see, e.g., [1, 2]). In particular, faults are reported when detection statistics exceed the limits for normal operation. Faults then can be isolated by using contribution plots [3, 4] or through comparison with additional data on past faults (see, e.g., [5]). The other approach to FDI is based on an explicit use of first-principles or identification models, where the analytic redundancy contained in the process model is used to generate residuals as fault indicators. A fault is reported when a residual breaches its threshold, and isolated according to certain isolation logic. The problem of FDI has been studied extensively for linear systems (see [6] for a survey).

For nonlinear process systems, there has been a significant body of work on detection and isolation of actuator faults. In a geometric approach, a nonlinear FDI filter is designed to solve the fundamental problem of residual generation (see [7]). Dedicated residuals are generated for fault isolation by exploiting the structure of the system (see [8]). Adaptive estimation techniques are used to handle unstructured but bounded modeling uncertainty for FDI of a class of Lipschitz nonlinear systems (see, e.g., [9]). To enhance data-based fault isolation, a feedback control law has recently been designed to enforce a closed-loop structure by decoupling the dependency between certain process variables (see, e.g., [10]). More recently, a method that compares the estimate of the actual input to its prescribed value has been proposed, where the estimate is obtained with the consideration of bounded parametric uncertainty (see [11]).

Compared to actuator faults, there exist limited results on detection and isolation of sensor faults. This problem has been studied for Lipschitz nonlinear systems (see, e.g., [12, 13]). In [12], a nonlinear state observer is designed to generate state estimates by using a single sensor. The fault isolation logic, however, is only limited to systems with three or more outputs. The method developed in [13] utilizes adaptive estimation techniques to deal with unstructured but bounded uncertainty for FDI. A sliding mode observer is designed to reconstruct or estimate faults by transforming sensor faults into pseudo-actuator faults (see [14]). This approach, however, requires a special system structure, and there is a limitation on system nonlinearity that can be handled. While a bank of observers are used to isolate sensor faults in [15], the observer gain is obtained through the first order approximation of the nonlinear dynamics. Therefore, the performance of the FDI design is subject to the type of nonlinearities. In summary, the problem of sensor FDI stands to gain from further results on designs that explicitly consider system nonlinearity in the FDI filter design.

As sensor faults are concerned, observers are typically required to fully or partly recover the system state. The design of observers, however, is a challenging problem for nonlinear process systems. In the context of output feedback control, high-gain observers are known to have good convergence properties (see, e.g., [16, 17]). Most existing results, however, rely on a special system structure for the system to be in to begin with, or after an appropriate transformation. To generalize the application of this type of observers, a model predictive control formulation has been studied in [18], where the discrete nature of the control implementation is exploited to relax the system structure required in the standard high-gain observer design. This generalization, however, is developed under the assumption of the locally Lipschitz continuity of the control input in the system state. Note that this assumption is hard to verify, especially under controllers such as model predictive control where the control law is not explicit but results from the solution to an optimization problem.

Motivated by the above considerations, this work considers the problem of isolation and handling of sensor faults in nonlinear process systems subject to input constraints. To this end, a high-gain observer design is first presented, which expands the class of nonlinear systems to which this type of observers can be applied. The convergence property of the observer and practical stability of the closed-loop system in the absence of faults are established under standard assumptions without assuming the locally Lipschitz continuity of the control input in the system state. Exploiting the enhanced applicability of the observer design, a sensor fault isolation scheme is then proposed, which uses a bank of high-gain observers. Each observer is driven by a subset of the measured outputs, for which residuals are generated that capture the differences between a state estimate and its expected trajectory. A fault is isolated when all the residuals breach their thresholds except for the one generated without using the erroneous measurements. After a fault is isolated, the state estimate generated using measurements from the healthy sensors is used by the controller to continue nominal operation. The implementation of the proposed fault isolation and handling framework subject to uncertainty and measurement noise is illustrated using a chemical reactor example.

References

[1] J. V. Kresta, J. F. MacGregor, and T. E. Marlin. Multivariate statistical monitoring of process operating performance. Can. J. Chem. Eng., 69:35–47, 1991.

[2] S. J. Qin. Statistical process monitoring: Basics and beyond. J. Chemometrics, 17:480–502, 2003.

[3] J. F. MacGregor, C. Jaeckle, C. Kiparissides, and M. Koutoudi. Process monitoring and diagnosis by multiblock PLS methods. AIChE J., 40:826–838, 1994.

[4] J. Yu. A nonlinear kernel Gaussian mixture model based inferential monitoring approach for fault detection and diagnosis of chemical processes. Chem. Eng. Sci., 68:506–519, 2012.

[5] S. Yoon and J. F. MacGregor. Fault diagnosis with multivariate statistical models part I: Using steady state fault signatures. J. Proc. Contr., 11:387–400, 2001.

[6] P. M. Frank. Fault diagnosis in dynamic systems using analytical and knowledge-based redundancy: A survey and some new results. Automatica, 26:459–474, 1990.

[7] C. De Persis and A. Isidori. A geometric approach to nonlinear fault detection and isolation. IEEE Trans. Automat. Contr., 46:853–865, 2001.

[8] P. Mhaskar, C. McFall, A. Gani, P. D. Christofides, and J. F. Davis. Isolation and handling of actuator faults in nonlinear systems. Automatica, 44:53–62, 2008.

[9] X. Zhang, M. M. Polycarpou, and T. Parisini. Fault diagnosis of a class of nonlinear uncertain systems with Lipschitz nonlinearities using adaptive estimation. Automatica, 46:290–299, 2010.

[10] B. J. Ohran, D. Munoz de la Pena, J. F. Davis, and P. D. Christofides. Enhancing data-based fault isolation through nonlinear control. AIChE J., 54:223–241, 2008.

[11] M. Du, J. Nease, and P. Mhaskar. An integrated fault diagnosis and safe-parking framework for fault-tolerant control of nonlinear systems. Int. J. Rob. & Non. Contr., 22:105–122, 2012.

[12] R. Rajamani and A. Ganguli. Sensor fault diagnostics for a class of non-linear systems using linear matrix inequalities. Int. J. Contr., 77:920–930, 2004.

[13] X. Zhang. Sensor bias fault detection and isolation in a class of nonlinear uncertain systems using adaptive

estimation. IEEE Trans. Automat. Contr., 56:1220–1226, 2011.

[14] X.-G. Yan and C. Edwards. Sensor fault detection and isolation for nonlinear systems based on a sliding mode observer. Int. J. Adapt. Contr. & Sign. Process., 21:657–673, 2007.

[15] M. Mattei, G. Paviglianiti, and V. Scordamaglia. Nonlinear observers with H_∞performance for sensor fault detection and isolation: a linear matrix inequality design procedure. Contr. Eng. Prac., 13:1271–1281, 2005.

[16] A. N. Atassi and H. K. Khalil. A separation principle for the stabilization of a class of nonlinear systems. IEEE Trans. Autom. Contr., 44:1672–1687, 1999.

[17] N. H. El-Farra, P. Mhaskar, and P. D. Christofides. Output feedback control of switched nonlinear systems using multiple Lyapunov functions. Syst. & Contr. Lett., 54:1163–1182, 2005.

[18] R. Findeisen, L. Imsland, F. Allgöwer, and B. A. Foss. Output feedback stabilization of constrained systems with nonlinear predictive control. Int. J. Rob. & Non. Contr., 13:211–227, 2003.

(16d) Multi-Agent Systems for Modeling Vascularization and Tissue Growth within Porous Biodegradable Scaffolds

Hamidreza Mehdizadeh[1], Elif S. Bayrak[1], Eric M. Brey[2] and Ali Cinar[1], (1)Chemical and Biological Engineering, Illinois Institute of Technology, Chicago, IL, (2)Biomedical Engineering, Illinois Institute of Technology, Chicago, IL

Biological and biomedical phenomena such as tissue growth are highly complex processes in which tissue constructs self-assemble as a result of the collective behaviors of individually acting cells. Such complex phenomena involve a combination of biodegradable natural or synthetic scaffold materials, numerous signaling molecules (growth factors, proteins, etc) and various cell types, integrated together in a well-orchestrated system that is designed to stimulate, promote, and at the same time control the formation of mature vascularized tissues.

To design functional replacement tissues, tissue engineers must be able to understand how a combination of chemical, mechanical, and physical factors and environments influences the coordinated growth and differentiation of multiple cell types. Ongoing scientific advancements have increased our ability to better control the properties of engineered tissues, however tissue engineers still have limited ability to predict how tissues grow over time.

While experimental studies provide in-depth insight into this process, it is well understood that experimentation alone would not be sufficient to address challenges ahead. Developing a computational model that can realistically simulate the behavior of such systems over time would be invaluable for many different theoretical and practical purposes. These purposes include, but are not limited to, better understanding the mechanisms involved, manipulating or measuring the important variables of the system and enabling us to qualitatively or quantitatively predict their effect on system level behaviors, and ideally in longer term, providing us with a tool for investigating ways of intervening in such systems.

Agent-based systems (ABS) are naturally suitable for modeling biological systems as they are comprised of individual discrete micro-scale constituents (e.g. cells) that interact with each other and their environment (e.g. extracellular microstructure) to form non-homogeneous macro-scale bodies (e.g. tissues and organs). The central idea in agent-based modeling (ABM) is to define software agents that can represent the building blocks of a system and to develop rules that regulate their interactions. The rules originate from the vast available knowledge gained through many years of studying the individual components of these biological systems.

In this paper, a multi-layer multi-agent ABM framework is developed to model the process of tissue growth within biodegradable non-vascularized porous scaffolds. As scaffolds have dimensions in the order of hundred micrometers, tissue cells require functional blood vessel networks to provide them with required nutrients and hence it is necessary to consider vascularization of the scaffold at the same time as tissue growth. As a result, the model includes separate layers to simulate the scaffold structure, the developing blood vessel network (that invades inside the scaffold), and the tissue cells that grow, migrate, and increase in number. Depending on the cell type, cell differentiation may also be considered. Different layers in the code dynamically interact together and result in a complex system.

The developed framework is implemented in Java, using Repast (Recursive Porous Agent Simulation Toolkit) which is an open-source agent-based modeling and simulation platform with features such as event scheduling procedures and visualization tools. Two different types of software agents are developed to represent tissue cells and blood vessels in the context of a spatially heterogeneous multi-cell tissue environment. These agents interact together and with their micro-environment, leading to tissue growth and formation of new blood vessel capillaries within scaffold structure. Tissue cell agents can perform various actions including sensing their environment, growth, migration and proliferation. Depending on cell type, cell differentiation or apoptosis may also be considered.

This computational framework couples two different cell types and cellular and tissue level cell scales. Using this simulation framework, we can quantify the dynamic impact of cellular level events and actions on a complex multi-cellular system. Also challenges in modeling systems with a large number of rapidly increasing interacting agents are addressed. This framework is then used to study tissue patterning processes which are an important step during tissue regeneration.

(16e) Fault Diagnosis of a Benchmark Fermentation Process Based On Probabilistic Qualitative Analysis of Trends

Kris Villez, School of Chemical Engineering, Purdue University, West Lafayette, IN and Venkat Venkatasubramanian, Chemical Engineering, Columbia University, New York, NY

Fault Diagnosis of a benchmark fermentation process based on Probabilistic Qualitative Analysis of Trends

K. Villez, V. Venkatasubramanian

Qualitative analysis of trends consists of the task to segment a time series into contiguous time windows in which the first and/or second derivative of the trend underlying to the (noisy) data is considered to be unique. As such, one identifies so called episodes in which the trend is of a selected type: A – convex antitonic (monotone decrease); B – convex isotonic (monotone increase); C – concave isotonic; D – concave antitonic; etc. The resulting segmentation is referred to as a qualitative representation.

Such an analysis has been proposed frequently for the purpose of fault diagnosis of batch and continuous processes. Indeed, because process knowledge is often formulated or obtained easily in qualitative terms by communication with plant operators. However, recent research (Villez et al., 2011) points out that at least some of these existing methods are hardly robust to realistic noise levels. All existing methods are either based on recursive schemes or the application of heuristic rules which offer computational advantages at the loss of accuracy. For this reason, a spline based method has recently been proposed for improved accuracy (Villez et al., 2012a). In this line of research, focus is given to accuracy and statistical validity, while ignoring computational efforts as of yet. In essence, this spline method fits a spline subject to a supposed qualitative representation. This is a convex Second Order Cone Problem (SOCP) as long as the type and location of the episodes are known in advance and the spline function is of order four (cubic splines) or less (Nesterov, 2000; Papp, 2011). In Villez et al. (2012a) it is proved and demonstrated that the location of the episodes can be identified in a globally optimal fashion, given a qualitative sequence. Such qualitative sequence is the ordered series of episode types.

Given this method, the remaining problem is to identify the proper qualitative sequence. In Villez et al. (2012b), a probabilistic strategy is developed. To this end, each considered fault is associated with a unique qualitative sequence and a prior likelihood. Given a time series, the conditional likelihood (the likelihood of the data conditional to a qualitative sequence) is approximated with the maximum likelihood. This maximum likelihood follows from maximization of the spline function fit by searching the optimal locations of the episodes as given by the proposed qualitative sequences. By use of Bayes' rule, fault diagnosis boils down to the selection of the maximum a posteriori (MAP) likelihood qualitative sequence. This means that the qualitative sequence leading to the MAP likelihood, following global optimizition of the episode locations, is selected.

The provided method is applied to the simulated benchmark simulation model of [Birol]. This model was set up for benchmarking of fault diagnosis methods, amongst others. In Villez et al. (2011), noisy Penicillin concentration profiles were analyzed with an existing, wavelet-based method. Profiles of three conditions were selected, each of which have distinct qualitative representations based on visual inspection. In Figure 1, one can see that Normal Operational Conditions (NOC) correspond to a BC sequence. Fault 2 correspond to BCBC sequence. Fault 3 correspond to BCDA a sequence. The results in Villez et al. (2011) indicated extreme sensitivity to noise. The shown results are the initial results obtained with the newly proposed spline-based method and are obtained in the contex of larger, comparitive study of techniques for qualitative analysis.

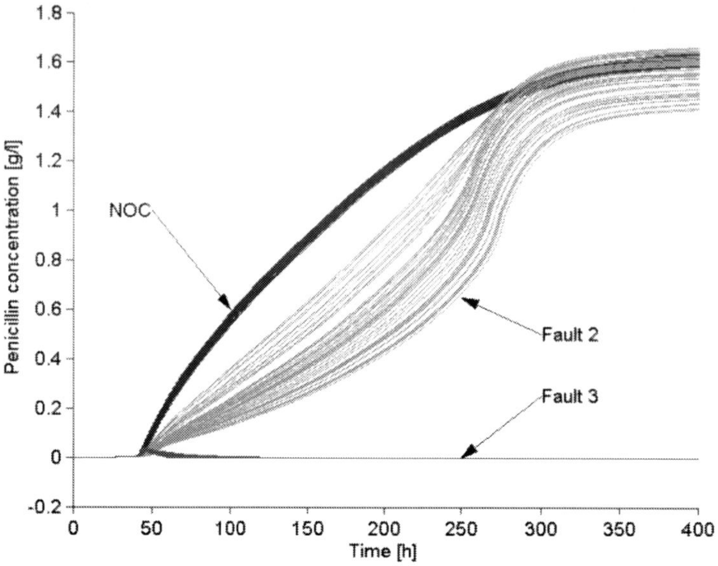

Figure 1: Simulated contration profiles for different conditions in the benchmark fermentation model.

Figure 2 shows the maximum likelihood fits for one batch with Fault 2 and for each of the proposed qualitative sequences. One can see that the spline fits for each shape constraints can hardly be distinguished.

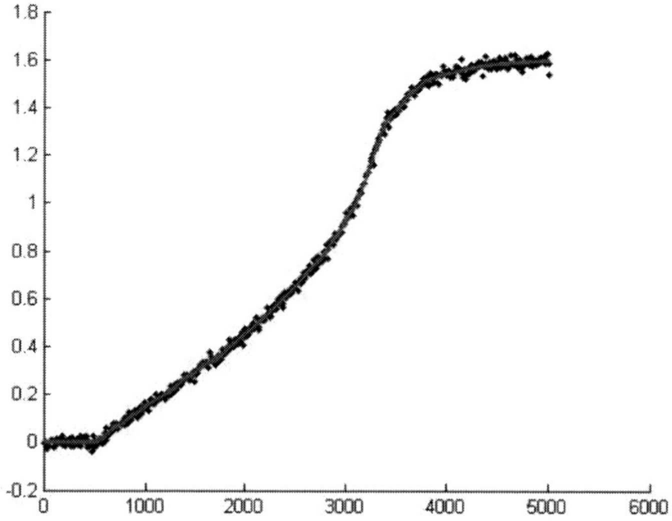

Figure 2: Maximum likelihood fits for a batch with Fault 2. The fits of different sequences are hard to distinguish.

Figure 3 shows the a posteriori likelihoods (normalized) which are obtained for a batch with Fault 2 and for three qualitative sequences (BC/BCBC/BCDA). Prior likelihoods for each of the qualitative sequences were set equal, implying complete ignorance. One can see that the maximum a posteriori (MAP) selection of a qualitative sequence is correct. Indeed, the BCBC sequence is selected. Note that the difference in likelihood is rather subtle as would be expected from the almost equal fits.

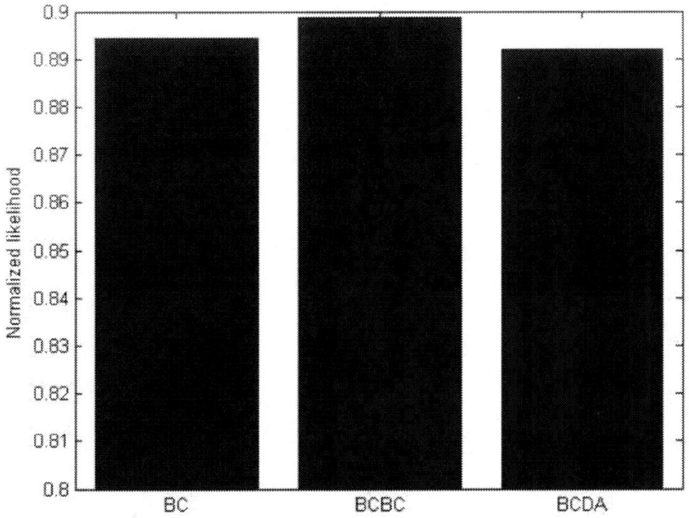

Figure 3: A posteriori likelihoods for three batches and three qualitative sequences. The maximum a posteriori (MAP) likelihoods lead to a correct matching.

REFERENCES

Birol, G., Undey, C., Cinar, A., "A modular simulation package for fed-batch fermentation: penicillin production", *Comput. Chem. Eng.,* 2002, 26:1553-1565.

Villez, K., Rosén, C., Duchesne, C., Anctil, F., Vanrolleghem, P. A., "Qualitative representation of trends (QRT): Extended method for proper inflection point recognition." *Comput. Chem. Eng.,* 2011 (Submitted)

Villez, K., Rengaswamy, R., Venkatasubramanian, V., "Generalized qualitative shape constrained spline fitting", *Technometrics,* 2012 (Submitted)

Nesterov, Y., "Squared functional systems and optimization problems", In: *Frenk, H., Roos, K., Terlaky, T., Zhang, S. (eds.) High performance optimization, applied optimization, vol. 33, pp. 405–440,* Kluwer Academic Publishers, Dordrecht, 2000.

Papp, D., "Optimization models for shape-constrained function estimation problems involving nonnegative polynomials and their restrictions", *M.Sc. thesis,* Rutgers University, 2011.

(43a) Looking to the Future in Continuous Pharmaceutical Manufacturing

Richard D. Braatz, Chemical Engineering, Massachusetts Institute of Technology, Cambridge, MA

This presentation describes the motivations, challenges, and opportunities in continuous pharmaceutical manufacturing. Some of the motivations are particularly intriguing, such as semicontinuous and continuous-flow unit operations that have the potential to have order-of-magnitude higher controllability and/or are much easier to scale up than batch and semibatch processes. Theoretical, computational, and experimental results are provided as evidence of the potential opportunities of continuous-flow pharmaceutical manufacturing operation. One of the conclusions of reviewing such unit operations is that they can look and operate very different from the batch and semibatch unit operations currently used in the pharmaceutical industry.

Example unit operations are drawn from the published literature, which include both journal papers and patents from various academic and industrial research groups. The presentation will also review and comment on the application of Quality-by-Design methodologies to pharmaceutical manufacturing. Of particular interest is the generation of design spaces, and the pros and cons of different approaches. In particular, it is shown that many of the methods for the construction of design spaces that have been proposed in the literature are useless from a control or operations point of view.

(43b) New Challenges and Opportunities for Pharmaceutical Manufacturing Science

Vincent Vilker, Office of Testing and Research, FDA, Chevy Chase, MD and Mansoor A. Khan, Cder/Ops/Otr/Dpqr, Food and Drug Administratin, Silver Spring, MD

New Challenges and Opportunities for Pharmaceutical Manufacturing Science

Vincent L Vilker and Mansoor A Khan, CDER/FDA

The increasing sophistication of modern drug products is leading to unprecedented challenges and opportunities for pharmaceutical manufacturing. At the same time, scientific and technological advancements in analytical instrumentation, coupled with implementation of the principles embedded in quality by design, enable the improved process monitoring and control that lead to better and more consistent product quality. ICH Q8(R2), the first of three International Conference on Harmonization (ICH) Guidance for Industry that spell out the concepts of Quality by Design, states that critical formulation attributes and process parameters should be identified through an assessment of the extent to which their variation can have impact on the quality of the drug product. In this presentation, examples will be given where modern instrumentation and manufacturing practices are required to maintain the critical formulation attributes of several drug products that have unique physicochemical and/or biological properties. Examples may include properties such as very low solubility, polymorphic solid states, poorly defined macromolecular size distributions, or where there is a tendency to become unstable in the formulated product.

() Recent Advances in the Molecular Modeling Foundation of Systems-Based Pharmaceutics

Constantinos C. Pantelides, Department of Chemical Engineering, Imperial College London,Center for Process Systems Engineering, London, United Kingdom

There are potentially significant benefits from the adoption of a systems-based approach to pharmaceutical science and engineering based on detailed mathematical first-principles models. However, the success of such an approach is predicated on the accurate description of the thermodynamic, kinetic and mechanical properties of the materials involved.

In this presentation, we review some recent advances in the description of the solid and fluid phases arising from work by the Molecular Systems Engineering group at Imperial College London. In particular, we review the state of the art for the *ab initio* prediction of crystal structure and polymorphism, and the prediction of vapour/liquid and liquid/solid phase equilibria using the SAFT-gamma group-contribution based equation of state. Significant progress has been achieved in all these areas over the last few years, but there are also significant challenges that still need to be addressed.

(43d) Experience with Regulatory Issues for Biomedical Devices and Vision for the Use of Advanced Control In Pharmaceutical Operations

Francis J. Doyle III, Chemical Engineering, UCSB, Santa Barbara, CA, Eyal Dassau, UCSB, Santa Barbara, CA and Howard Zisser, Sansum Diabetes Research Institute, Santa Barbara, CA

The landscape for clinical testing of biomedical devices has evolved rapidly over the last 5-10 years. In particular, the use of advanced control algorithms, such as model predictive control, are now being reviewed by the FDA as components of closed-loop devices. In our work with the Artificial Pancreas (AP), we have developed master files and other supporting materials to support IDE (Investigational Device Exemption) studies. In this talk, I will share some of the challenges in developing algorithms for regulatory review, including the role of simulation studies, communication architectures, algorithmic design, and IDE studies. I will extrapolate from these experiences to project the opportunities for advanced control in pharmaceutical operations.

Chemical Engineering & Biopharmaceutics: Linking the Process with the Ultimate Quality Attribute

Ravi M. Shanker

The discovery and development of heterocyclic, small molecule new drug entities is a highly sophisticated, multi-disciplinary, scientifically advanced, collaborative, highly competitive, extremely expensive and multi-year endeavor. The heavily R&D dependent pharmaceutical industry utilizes cutting edge scientific and technological tools and techniques encompassing the broad disciplines of chemistry, biology and medicine. To provide significantly improved therapeutic treatments for various acute and chronic diseases and to be able to meet the requirements and expectations of the regulatory agencies, the patients and the society is a constant challenge for the innovative pharmaceutical industry. The challenges forge deeper dives in science among the wide array of disciplines that constitute the pharmaceutical industry and consequently have provided tremendous opportunities for the advancement of computational and predictive tools within each of these disciplines and sub-disciplines. A key segment of R&D and industrial scale manufacture is pharmaceutical sciences. It is an amalgamation of applied disciplines of chemical engineering, physical and physical organic chemistry, synthetic organic chemistry, process chemistry, analytical chemistry, polymer chemistry, material sciences, crystallization science, drug product design and manufacture and evaluation of performance of drug product in the patients. Pharmaceutical sciences have also embraced the development and use of predictive computational tools among each of its areas of specialization and disciplines to ensure the highest quality of drug product being delivered to the patient. The multi-disciplinary components of pharmaceutical sciences are operationally integrated to provide the highest quality of drug product to meet the regulatory requirements in this highly regulated industry. However, there is a lack of integration of such predictive tools across disciplines to enable the assessment of local perturbations in quality during any segment of work flow on the ultimate quality and performance of the drug product. The ultimate quality attribute of a drug product is it's performance in patients, however, linking that performance to the upstream attributes of drug substance and drug product via engineering control during manufacture with emphasis on the principles of chemical engineering continues to be unrealized and unfulfilled possibility. The vision to provide such integrated approach of existing computational and predictive tools within disciplines and sub-disciplines in pharmaceutical sciences and to provide opportunities for further development of advanced integrated tools as a holistic system is the underlying concept for the creation of Systems Based Pharmaceutics.

Using Models to leverage process understanding; Case studies in O2 sensitivity of a Pd catalyzed CN bond forming reaction and isobutylene release in a acid catalyzed tertbutyl deprotection reaction.

Jeremy M. Merritt[1], Jonas Buser[2], Neil Kallman[1], Mark Pietz[1], Kevin Sullivan[1], Shekhar Viswanath[1], and Jared Fennell[1]

[1]Chemical Product R&D

[2]Analytical Sciences R&D

Eli Lilly and Co.

Abstract:

Oxygen sensitivity was observed in a Pd catalyzed amination reaction employed in the synthesis of an API. Dissolved O2 probes, 31P NMR, and kinetic modeling were used to help elucidate the oxidation mechanism and kinetics, to aid in process understanding and realization of the risks associated with scale-up. Elsewhere in the synthesis, an acid catalyzed tertbutyl deprotection reaction is employed. The largest observed API related impurity formed during the reaction results from re-attack of the tertbutyl group onto another part of the molecule. Several other side reactions were observed including reaction of the tertbutyl with the solvent system (HCl / water / isobutanol) to form tertbutanol, tertbutyl chloride, and isobutyl tertbutyl ether. Due to evolution of gaseous isobutylene during the reaction, there was concern over scale variability due to different gas-liquid mass transfer rates (k_la). A model was developed to predict the reaction rates and equilibria of the main and various side reactions and was used as an argument for scale invariability of the chemistry which was important in defining the impurity control strategy.

(99b) Modeling the Dynamics of the Film Coating Operation for Pharmaceutical Tablets

Salvador García-Muñoz, Process Modeling & Engineering Technology, Pfizer Worldwide Research & Development, Groton, CT, Mariana Moreno, Chemical Engineering, Tec de Monterrey, Monterrey, Mexico, Alejandro Cano, Process Systems Enterprise, Inc., Cedar Knolls, NJ and Simon Leyland, Process Systems Enterprise Inc, Cedar Knolls, NJ

Previous work in understanding the thermodynamics of the film coating operation has focused mainly around the pseudo-steady state reached between the drying air and the cooling effect from the evaporation of the spraying solution [1]. The work presented here aims to mathematically describe the time-varying state of the operation and provide a structured procedure with the optimal experimental approach to estimate the heat and mass transfer coefficients across the system.
The ultimate objective of this work is to deliver a model able to predict the temperature of the tablets and the amount of water in the tablets during (and at the end of) the operation. Such a model will aid the design of the operating cycle for the film coating process if there were constraints on the maximum temperature/water activity that the tablet can be exposed to, or the final amount of water in the tablet core. Such constraints are not uncommon in water sensitive products since this can accelerate degradation mechanism and limit the shelf life of the product.
To obtain a model that is accurate enough (given the uncertainties in the measurements); mass and energy balances were calculated around: drying air, nozzle air, liquids and solids in the solution, tablet cores and equipment. The model is implemented and all parameter estimations done using the gPROMS (Process Systems Enterprise Ltd, London, UK) modeling platform.
A key component to the model is the phenomena occurring at the surface of the tablet; that determines whether the water available in the impacting droplet is absorbed by the tablet or evaporated into the air. This mechanism is modeled as a function the thermodynamic state of the droplet as well as the hygroscopicity of the formulation, mass and surface area of the tablet, the vapour transmission properties of the coating solution and the extent of coating coverage in the tablet. The thermodynamic state of the droplet at the point of impact with the tablet is made dependent on the initial droplet size (calculated using rheological properties, nozzle geometrics, air pressure and the model by Aliseda et al [2]), the chemistry of the coating solution, the thermal conditions in the chamber and the distance traveled by the droplet (gun to bed distance).
The heat capacity of the equipment and the heat transfer from and to the equipment plays an important role to effectively model the energy transfer across the system (which in turn affects the prediction of temperatures). This is due to the fact that equipment becomes an energy sink that can absorb or emit energy in a non-linear way. These equations are established and coupled with the rest of the energy and mass transfer equations.
The model parameters are divided into those that characterize the equipment, the coating solution, the tablets and formulation, the tablet-gas mass and energy transfer and the tablet/liquid mass and energy transfer. Model-guided experiments are carried out in sequence to independently estimate each of the different interactions. The model is finally verified with 21 coating runs in an LDCS 20 coater using tablets of different sizes at multiple thermal conditions and extents of water exposure to the cores. The model is able to provide predictions well inside the experimental accuracy.

References

[1] M.T.am Ende and A.Berchielli, A Thermodynamic Model for Organic and Aqueous Tablet Film Coating, Pharmaceutical Development and Technology, 10 (2005) 47-58.
[2] A.Aliseda, E.J.Hopfinger, J.C.Lasheras, D.Kremer, A.Berchielli, and E.K.Connolly, Atomization of viscous and non-newtonian liquids by a coaxial, high-speed gas jet. Experiments and droplet size modeling, International Journal of Multiphase Flow, 34 (2008) 161-175.

(99c) Modeling of Fluid Bed Granulation Process for Aqueous Systems

Rajan Gupta, Pharmaceutical Commercialization Technology, Merck, Summit, NJ

This presentation discusses the development of a fluid bed granulation model and how it can be applied in the context of QbD. The model uses mass and heat balance to predict a complete profile of bed humidity, product temperature, and exhaust conditions during granulation and drying steps. It also incorporates empirical relations built from actual product data to account for heat losses, tortuosity aspects (particularly important during falling rate period), and sub-saturated conditions. The only inputs required in the model are typical batch parameters (such as spray rate, inlet air temperature / airflow, etc.), thus removing the need to know equipment (e.g., heat transfer coefficients) and formulation related factors (e.g., particle size) in advance.

The finalized model was tested on over 40 historical and new batches, and good agreement was observed between the theoretical predictions and the actual values. The model developed in the current work can be used as a valuable tool in feasibility studies (choosing appropriate parameters as starting points), process scaling-up (lab to commercial), and design space studies.

(99d) Plant-Wide Hybrid Model Predictive Control of a Continuous Pharmaceutical Tablet Manufacturing Process Via Direct Compaction

Ravendra Singh, Marianthi Ierapetritou and Rohit Ramachandran, Department of Chemical and Biochemical Engineering, Rutgers, The State University of New Jersey, Piscataway, NJ

QbD and PAT requires the next generation of pharmaceutical products be manufactured through continuous processing integrated with online monitoring tools coupled with efficient automatic model-based feedback control system to achieve the precise control of process variables so that the predefined end product quality can be achieved consistently.

The direct compaction process considered in this study is highly interactive, nonlinear, and has time delays due to sensor placements, that indicate that a simple feedback regulatory control system (e.g. PI(D)) by itself may not be sufficient to achieve the tight process control as desired by regulatory authorities (e.g. FDA). In previous work, we have demonstrated that under certain conditions, such a control system may not be adequate [1]. Advanced model predictive control (MPC [2]) is the desired control strategy for this complex system especially for systems with multi-variate interactions. As a closed loop optimal control method based on the explicit use of a process model, model predictive control has proven to be a very effective controller design strategy over the last twenty years and has been widely used in process industry such as oil refining and chemical engineering. However, because of different level of complexity involved (e.g. solid handling, irregular flow) in addition to the unavailability of the suitable process model, the design and implementation of MPC in the pharmaceutical manufacturing involving the solid dosage form is still state of the art. In comparison to the regulatory control strategy previously proposed [1], the MPC is more complex and requires the development of higher fidelity models and robust optimization strategies. Depending on the process dynamics either linear MPC (LMPC) that required linear model and less computational power or non-linear (NMPC) MPC that required first principle model and more computational power can be used. However, with the state of the art computational power, the NMPC is not practical for the plant-wide control of complex multi variable system but it could be used efficiently to control the part of a plant which is highly nonlinear. Similarly, LMPC cannot be efficient for the control of the part of the plant which is highly non-linear. Therefore, a hybrid control strategy integrating the regulatory control system, LMPC and NMPC is highly desired for plant-wide control of a process comprising of fast, delayed, linear and non-linear dynamics.

In the work reported here, a hybrid control system integrating PID, LMPC and NMPC for plant-wide control of a continuous tablet manufacturing process has been developed. The continuous manufacturing process involved the state of the arts continuous unit operations including blending, tablet pressing and coating. Based on the analysis of the process dynamics, linearity and non-linearity analysis, the control variables has been placed into three groups: (i) the variables for which the PID controller is sufficient, (ii) the variables for which LMPC is required, (iii) the variables for which NMPC is essential. These variables have been then controlled with the corresponding control strategies. The categorization of these variables ensures the better control performance with optimal resources and computation power. The first principle model as well as the linear model of the continuous tablet manufacturing process has also been developed. The performance of the hybrid control system has been evaluated and compared with the regulatory control strategy. The proposed hybrid strategy is currently being implemented in a continuous pilot plant at Rutgers University.

The objective of this presentation is two-fold: first to highlight the design of hybrid control system for a continuous tablet manufacturing process, and second to demonstrate the performance of the designed hybrid control system *in silico*.

References

[1] Ramachandran, R., Arjunan, J., Chaudhury, A, Ierapetritou, M. (2012). Model-Based Control Loop Performance Assessment of a Continuous Direct Compaction Pharmaceutical Processes. J. Pharm. Innov., 6(3), 249-263.

[2] Glaser, T., Sanders, C. F. W, Wang, F. Y., Cameron, I. T., Lister, J. D., Poon, J. M.-H., Ramachandran, R., Immanuel, C. D., Doyle III, F. J. (2009). Model predictive control of continuous drum granulation. Journal of process control, 19, 615-622.

(99e) PAT for Pharmaceutical Extrusion Monitoring and Supervisory Control

Patrick Wahl[1], Daniel Markl[1], Daniel Treffer[2], Jose C. Menezes[3], Gerold Koscher[1], Eva Roblegg[4] and Johannes G. Khinast[1,2], (1)Research Center Pharmaceutical Engineering GmbH, Graz, Austria, (2)Institute for Process and Particle Engineering, Graz University of Technology, Graz, Austria, (3)Research Center for Pharmaceutical Engineering, Graz, Austria, (4)IPS/Department of Pharmaceutical Technology, University of Graz, Graz, Austria

Introduction

Hot Melt Extrusion (HME) is increasingly attracting interest in pharmaceutical manufacturing, for two main reasons: (1) it allows to produce controlled release formulations by embedding API crystals in polymers, lipids or sugars (Kleinebudde, 2011) and (2) it significantly enhances the bioavailability of APIs with low solubility. This can be achieved with a solid solution, in which the API and the matrix are molecularly dispersed. Estimations predict that up to 90% of future new chemical entities may be classified as poorly soluble. Therefore, enhancing the bioavailability is a significant part for successful drug development. Further dosage forms like transdermal patches and implants among others can be produced by HME as well (Breitenbach, 2002).

HME also has the potential to implement continuous production in a straightforward way. One way is to use a hot-strand cutter as a downstream processing unit. The (hot) extrudate is cut directly at the die face, thereby forming nearly spherical pellets. If necessary, pellets with different APIs or release characteristics can be mixed and further processed by a capsule filling machine or a tablet press. This production line is in development at RCPE.

At present, many Process Analytical Technology (PAT) tools are available for establishing physicochemical product properties, such as the chemical composition. Multivariate data analysis (MVDA) combined with spectroscopic techniques, including near infrared (NIR) and Raman, allows quantitative and non-invasive process monitoring (Koller et al., 2011). Applying multivariate statistical process control (MSPC) reduces real-time process data streams to a convenient control chart. Clearly, process analysis calls for an IT infrastructure that can aggregate real-time process data from multiple unit operations, raw material data, PAT data and equipment status. SIPAT is a commercial PAT software solution that was created to meet the above requirements.

In this study the PAT concept consisted of: (1) In-line, real-time monitoring of extruder parameters and collection of NIR spectra of the extrudate close to the die; (2) Analysis of the spectrum with multivariate data analysis (MVDA) tools; (3) Integration and implementation of this PAT concept in SIPAT.

In this presentation we intend to show the setup of the HME system, Principal Component Analysis (PCA) to reduce the dimensionality of the spectra and a Partial Least Squares (PLS) model to predict the response of the extrusion process (the API content at the die) to step changes of the feed rates.

Experimental Methods

Materials

Paracetamol, donated by G.L. Pharma GmbH, Lannach, Austria, was used as an API (volume mean particle size 139.2 µm). The matrix carrier system was calcium stearate (Werba-Chem GmbH, Vienna, Austria; mean particle size 16.62 µm) (Roblegg et al., 2011). The paracetamol crystals were embedded in CaSt, resulting in a solid dispersion.

Extruder setup

Extrusions were performed on a co-rotating twin-screw extruder (ZSK 18, Coperion GmbH, Stuttgart, Germany) with 18mm screw diameter. The barrel temperatures and the screw setup was chosen in such way that the melt temperature at the die was 130°C. Two twin-screw gravimetric feeders (KT20, K-Tron, Pitman, USA) were filled with a premix. Mixtures with a mass fraction of paracetamol from 0% to 60% can be obtained by adjusting the feed rates of the two feeders. In the experiments the throughput of the extruder was 0.6 kg/h. To monitor, control and store process parameters of the extruder, NIR spectra and results of calculations (e.g. API content prediction) the software SIPAT 3.1.1 (Siemens AG, Brussel, Belgium) was used.

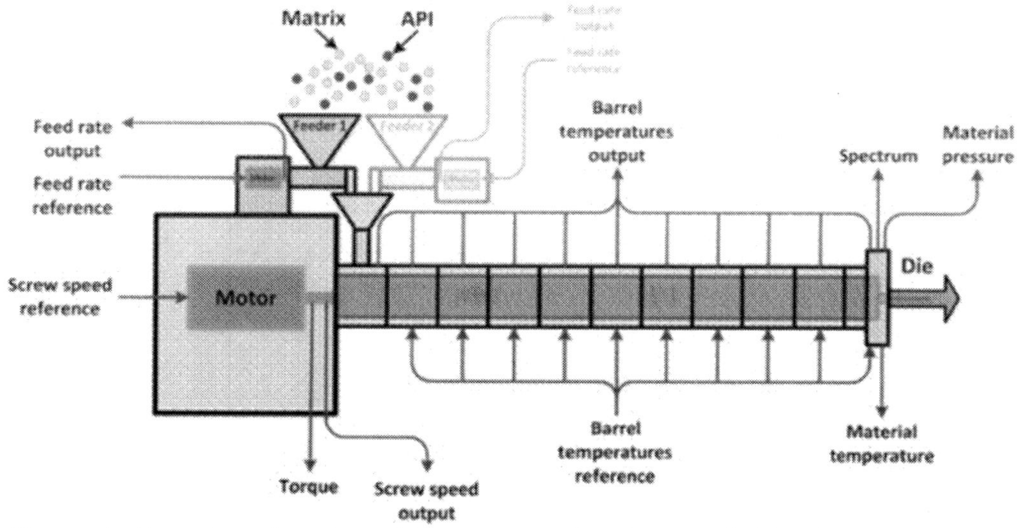

Figure 1: Overview of monitored (green, output) and controlled (red, input) parameters of the extruder and the NIR spectrum, which was taken close to the die.

NIR spectroscopy

NIR spectra were collected in transflexion mode with a process spectrometer SentroPAT FO (Sentronic GmbH, Dresden, Germany). The spectrometer has a spectral range of 1100 nm - 2200 nm and 2 nm resolution. A fibre-optic Dynisco NIR probe was mounted close to the extrusion die, after the screws, for in-line monitoring of the melt. For each measurement 120 spectra were averaged, with an integration time of 0.014s per spectrum. However, significant modifications of the die were necessary to obtain spectra which reflect the current bulk composition of the melt.

Chemometric models

Around 100 spectra of the melt of each step of API content were collected to develop a PCA and a PLS model. Principal component analysis applied to the observed data may be used to classify the main events affecting the process. The PLS model enables the direct prediction of the paracetamol content. The chemometric model was built with the software Simca P+ 12.0 (Umetrics, Umeå, Sweden). Simca-Q 12.0.1 (Umetrics, Umeå, Sweden) is used as an external calculation engine in SIPAT, which has the advantage of directly integrate the models developed in Simca P+ in SIPAT.

Results

Selecting a good NIR measurement position

Ensuring a good sample presentation to the Dynisco NIR probe is crucial for correct API predictions. The main reason for improper sample presentation is a too slow exchange of melt in the field of view of the NIR probe. First, if the probe is not perfectly in line with the barrel surface, melt will fill the dead volume once. The shear forces are not high enough to exchange the material. This is likely for measurements above the screws. A second possibility is a too large volume of melt directly before exiting the die. Here a funnel flow might build up, which causes slow melt exchange at the barrel walls, where the probe is situated. Although NIR penetrates a transparent melt, in the mentioned cases the measurement could be biased. Thus, a special probe set-up was designed and tested to represent the actual melt composition.

Monitoring the process

The feeding dynamics are the most important factor for content uniformity of the melt, especially for pharmaceutical powders, which are often not free flowing. Despite the mixing elements of the screw, the extruder has only a limited capability to compensate lower frequency feeding variations by its backmixing ability.

Monitoring the process in the reduced PC space is a straightforward way of implementing real-time process analysis. Adjusting the feed rates of the feeders yielded different API concentrations in the extrudate that appeared in the score plot as clusters of observations.

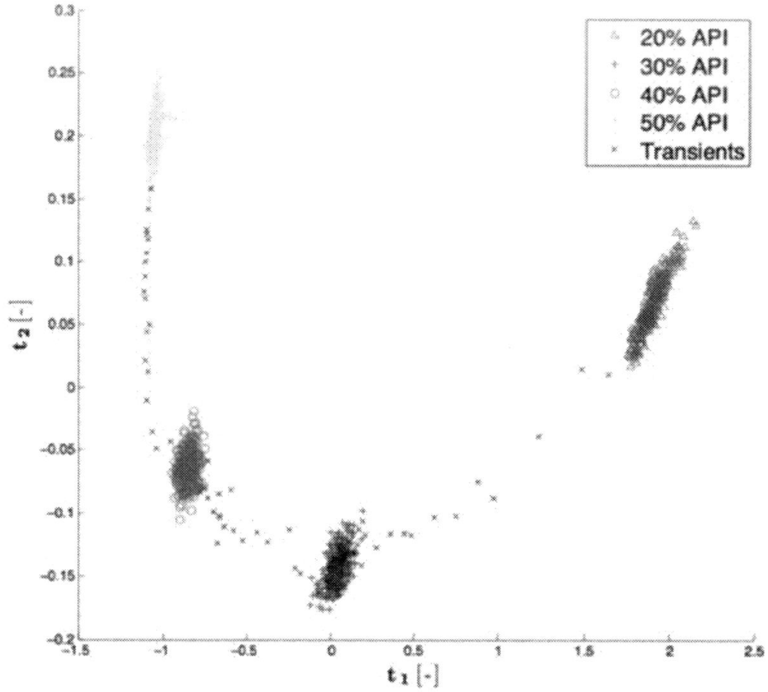

Figure 2: Different API concentrations (20%, 30%, 40% and 50%) result in the clustering of samples in the score plot (t_1 versus t_2).

Moreover, the prediction of the API content by the PLS model enables another way of monitoring the process in a convenient control chart.

Using Simca P+ models or models defined in MATLAB, the observations can be processed and visualized in SIPAT in real-time. The integration of the PCA and PLS models facilitates different ways to detect in real-time abnormalities in the process.

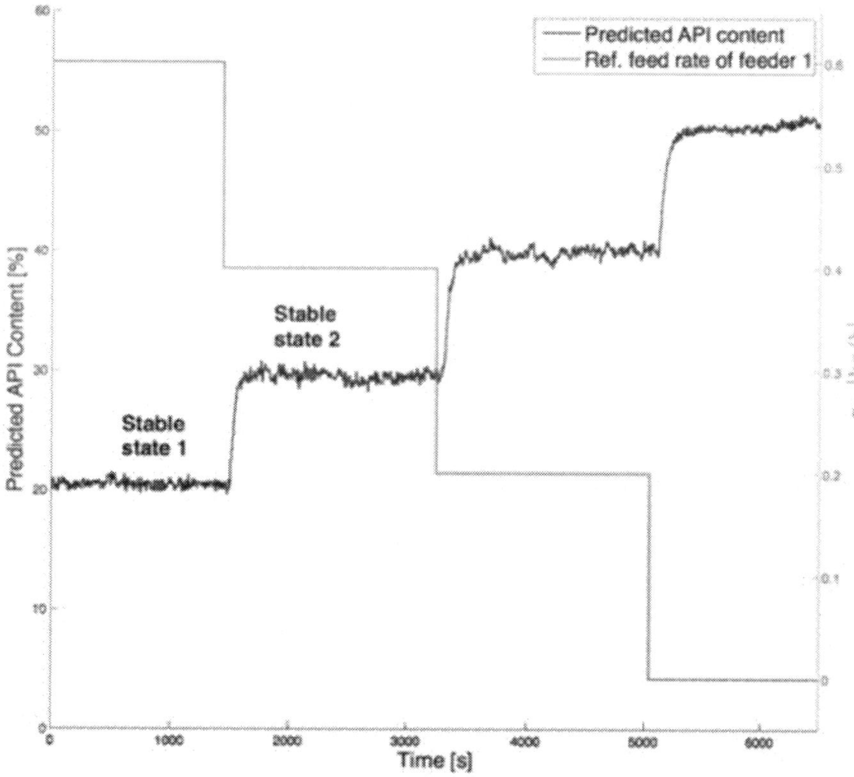

Figure 3: Predicted paracetamol content (blue). Variations induced by the feeders are noticeable. For the chosen screw configuration the time delay between feeder settings (magenta) and predicted API content at the extrusion die is 55s.

Conclusion

The presented hot melt extrusion system with an integrated PAT concept is capable of monitoring the critical process parameters. Special attention was paid to ensure a good NIR measurement position. However, the MVDA models that are developed in this process can be applied to only one formulation. In a typical campaign-driven manufacturing environment (which even continuous manufacturing of the future will be) formulations and/or products may change weekly or monthly. Automatic model development can be performed based on the setup including the HME process, SIPAT and Simca-Q. Using several models at the same time opens up the opportunity to validate different PCA or PLS models in real-time.

References

[1] Breitenbach J (2002) Melt extrusion: from process to drug delivery technology. *Eur. J. Pharm. Biopharm.* **54**:107-117.

[2] Kleinebudde P (2011) Pharmaceutical Product Design: Tailored Dissolution of Drugs by Different Extrusion Techniques. *Chem-Ing-Tech.* **83**:589-597.

[3] Roblegg E, Jäger E, Hodzic A, Koscher G, Mohr S, Zimmer A and Khinast J (2011) Development of sustained-release lipophilic calcium stearate pellets via hot melt extrusion. *Eur. J. Pharm. Biopharm.* **79**:635-645.

[4] Koller D M, Posch A, Hörl G, Voura C, Radl S, Urbanetz N, et al. (2011) Continuous quantitative monitoring of

powder mixing dynamics by near-infrared spectroscopy. *Powder Technology*. **205:** 87-96

2012 AIChE Annual Meeting QbD Topical Conference
"Model Based QbD in Drug Substance or Drug Product Development" Session
Pittsburgh, PA, Oct. 28-Nov. 2, 2012
Paper 99f

Scientific Validation of THz Spectroscopic Measurement and its Pharmaceutical Process Analytical Technology (PAT) Applications: Case Studies of First Principles Modeling Approach

Huiquan Wu* and Mansoor Khan

Division of Product Quality Research (DPQR, HFD-940)
OTR/OPS/CDER/FDA
FDA White Oak Research Center
10903 New Hampshire Ave
Silver Spring, MD 20993

Abstract: As an emerging technology, THz spectroscopy has gained increasing attention in the pharmaceutical area during the last decade. This attention is due to the fact that (1) it provides a promising alternative approach for in-depth understanding of both intermolecular interaction among pharmaceutical molecules and pharmaceutical product quality attributes; (2) it provides a promising alternative approach for enhanced process understanding of certain pharmaceutical manufacturing processes; and (3) the FDA pharmaceutical quality initiatives, most noticeably, the Process Analytical Technology (PAT) initiative. In this work, the current status and progress made so far on using THz spectroscopy for pharmaceutical development and pharmaceutical PAT applications, including FDA-NIST collaboration on THz spectroscopy pharmaceutical characterization and PAT applications, are discussed. In the spirit of demonstrating the utility of first principles modeling approach for addressing model validation challenge and reducing unnecessary model validation "burden" for facilitating THz pharmaceutical PAT applications, two scientific case studies are created and discussed. Furthermore, other technical challenges and opportunities associated with adapting THz spectroscopy as a pharmaceutical PAT tool are highlighted.

Keywords: Terahertz Spectroscopy, Pharmaceutical Product Quality Control, Process Analytical Technology (PAT), Chemometrics, Pharmaceutical Manufacturing Process, First Principal Modeling, Process Understanding.

Disclaimer: The views and opinions expressed in this paper are only of authors, and do not necessarily reflect the view or policies of the FDA.

*Corresponding author: Phone 301-796-0022; Fax: 301-796-9816; Email: Huiquan.wu@fda.hhs.gov

1. Introduction

The FDA recently released a strategic plan about advancing regulatory science at FDA [1]. One of the eight science priority areas identified in that strategic plan is "support new approaches to improve product manufacturing and quality". For implementation strategy, "FDA will support the application of novel technologies to product development and innovative analytical approaches to improve product manufacturing and quality…" by "enabling development and evaluation of novel and improved manufacturing methods", including "prompting two state-of-the-art manufacturing strategies-Process Analytical Technology, and Quality-by-Design approaches—for impact on manufacturers' ability to maintain consistent quality." Spectroscopic techniques and material characterization techniques have been playing a critical role for gaining pharmaceutical product and process understanding, ensuring pharmaceutical product quality [2] under the umbrella of the FDA's Pharmaceutical Quality Initiative for the 21st century [3], including Process Analytical Technology (PAT) [4] and Quality-by-Design (QbD) [5] approaches. It has been well recognized that process validation in the PAT domain could be challenging, as discussed in the FDA's new draft guidance on process validation [6].

Terahertz (THz) spectroscopy has been widely used to study the spectroscopic characteristics of a variety of materials in the spectral region spanning from 0.1 to 10 THz. The majority of crystalline solids have distinct and characteristic lattice vibrations within the THz frequency range, making THz spectroscopy a powerful analytical tool for probing the intermolecular interactions of molecular crystals. As a matter of fact, THz spectroscopy technique shows promising for the analysis of pharmaceutical products throughout the drug development process [7-8]. Most noticeably, THz spectroscopy as an emerging technology has been gained increasing attention in the area of pharmaceutical product and process understanding over the last several years. In this work, current status and progress on using THz spectroscopy for pharmaceutical product quality control and pharmaceutical Process Analytical Technology (PAT) [2] applications were reviewed. Two case studies of using first-principle approach to validate THz spectroscopic measurements were discussed. Challenges and opportunities associated with adapting THz spectroscopy as a pharmaceutical PAT tool were highlighted.

2. THz spectroscopy for pharmaceutical product characterization and pharmaceutical development

In this regard, THz spectroscopy has been applied to certain important areas, which includes but is non limited to:
(1) Crystalline pharmaceuticals, polymorph screening, and solid-state assessment [9-16];
(2) Assessing weak interaction (inter-molecular vibrations) and intra-molecular interaction [17-20];
(3) Mapping chemical compositions and tablet density [21-22];
(4) QA and QC for coating and controlled release dosage form manufacturing, dissolution performance, tablet thickness
Details could be found in the literatures cited.

3. THz spectroscopy for pharmaceutical PAT applications

According to the FDA's PAT Guidance for the Industry [2], PAT is a system for designing, analyzing, and controlling manufacturing through timely measurements (i.e., during processing) of critical quality and performance attributes of raw and in-process materials and processes, with the goal of ensuring final product quality. An ultimate goal of PAT is to enhance understanding and control the manufacturing process. Consequently, the tools and principles described in this guidance should be used for gaining process understanding and can also be used to meet the regulatory requirements for validating and controlling the manufacturing process. Since the FDA launched its PAT initiative in 2002, broad interests have been stimulated for the pharmaceutical community to adopt many analytical and characterization techniques for at-line, on-line, and in-line pharmaceutical PAT applications. Among these innovative spectroscopic pharmaceutical applications, THz spectroscopy has found some important yet promising applications due to its unique advantages. Some of the THz spectroscopy pharmaceutical PAT-oriented applications, most for small molecule drugs and several for biopharmaceuticals, are discussed below.

3.1. Film coating process of pharmaceutical tablets

Tablet coating thickness is a critical quality attribute (CQA) for modified (extended) release (MR or ER) or controlled release (CR) tablet dosage forms, as it directly dictating the dissolution rate [23] and dissolution performance of the MR/ER/CR tablet dosage form, thus linking to the pharmacological effect (AUC, t_{max}, etc.) and bioavailability. Traditionally, the tablet coating thickness is calculated based on the tablet weight gain obtained during the coating process. Direct measurement methods, such as scanning electron microscopy (SEM), need to cut the tablet to get the photographs of cut-through cross-section, thus are destructive and prone to error. In recent years, spectroscopy techniques have provided fast and non-destructive alternate approaches to evaluate the tablet coating and coat performances.

3.1.1 THz Pulsed Imaging for nondestructive analysis of tablet coating thickness

Few nondestructive techniques are capable of examining tablet coatings. THz Pulsed Imaging (TPI) is a fast (a waveform can be generated in less then 20 ms) and nondestructive technique. It was reported that TPI can distinguish two brands of over-the-counter ibuprofen tablets [24]. Good agreement was obtained between THz maps obtained via TPI and photographs of cutting-through sections of tablets. It is nondestructive, noninvasive, safe, fast, and provides information on factors influence bioavailability. Technically it can probe any position on a tablet in less than 20 ms, making it possible for rapid online assessment. Furthermore, the variation of internal coating layers can be observed which would not be readily apparent from separated and isolated point measurements. However, lateral resolution is limited by diffraction of the THz focus to about 150 μm at 3 THz, whereas the axial resolution is limited by the THz pulse duration (which is <200 fs) to about 30 μm. Therefore, more development work is to be done before TPI can provide reliable results on thin coating layers, especially for thickness <30 μm.

3.1.2. THz Pulsed Imaging (TPI) for assessing sustained-release tablet film coating quality including coating uniformity, dissolution performance; Critical assessment

Using TPI, Ho et al [25-27] conducted a thorough study to examine sustained-release tablet film coating quality including coating uniformity and how they impact dissolution performance of the coated tablets. The relationship between THz parameters (THz electric field peak strength (TEFPS) and coating layer thickness) and product performance (the Mean dissolution time, MDT) using SR tablets obtained during a scale-up process was examined. It was shown that weight gain (the traditional non-destructive indicator for coating quality in the dry-state) failed to reflect the functional performance of the product (with R^2 value of 0.09). However, THz parameters could be correlated with the actual product performance (with R^2 value around 0.84), thus showed potential to predict the dissolution performance of SR tablets. The proof-of-concept was proved. However, as shown in their Figures 3A-C, the R^2 values were based on two clusters of data points which are located at two ends of the regression lines. This situation will negatively impact the prediction power of the regression equation. Ideally, those data points should be evenly distributed along the regression line. A more systematic study to provide sufficient statistical power to further understand the extent of the relationship between the dry-state THz parameters and product performance is warranted. They also used TPI to determine film coating thickness and TEFPS for analysis of SR coated pellets [26]. Their results suggested that drug release for pellets studies was not predominately governed by drug diffusion through the polymeric film coating, but probably to a large extent by drug solubility. TPI's capability to detect non-homogeneities in the drug layer and polymeric film coating was demonstrated. Most recently, Ho et al. [27] using TPI to investigate dissolution performance critical areas on coated tablets, i.e., formation and location of film coating "weak spots" or film coating defects. TPI analysis showed that for the film coated tablets studied, the tablet central band exhibited the thinnest film coating, lowest coating density, and highest surface roughness, thus was the performance limiting area of the film coating. Dissolution tests confirmed these TPI observations: the film coating on the tablet central band has a faster MDT of 7.4 hr in comparison to 10.4 hr for the convex top/bottom surface. Therefore, TPI has potential to be adopted as a PAT tool to probe film coating defects during coating development and to evaluate the effect on subsequent dissolution performance.

3.1.3. Comparison of various non-destructive methods for quantifying tablet coating thickness

Zeitler and Gladden [28] reviewed *in-vitro* tomography and non-destructive imaging at depth of pharmaceutical solid dosage form. Four most promising techniques, X-ray computed microtomography, magnetic resonance imaging, THz imaging, and optical coherence tomography, were discussed to introduce the concept of tomography in a pharmaceutical setting. Their operating principles as well as their specific strengths and limitations were discussed. A recent study [29] compared non-destructive methods for quantifying tablet coatings thickness, including optical microscopy, micrometer, X-ray fluorescence (XRF), Raman, and Near-infrared (NIR) spectroscopy. Their conclusion was, direct thickness measurements were time-consuming but provided assured coat thickness data. On the other hand, XRF, Raman, and NIR spectroscopy methods were viable alternatives to provide complementary information for the tablet coatings studied. As discussed previously, THz spectroscopy also provides an important alternative to quantify the coating thickness.

3.2. Biologic molecules and biopharmaceutical applications, medical imaging modality

Johnson et al. [30] reported their study on simultaneous recording the mid-IR and far-IR (THz) spectra of the sporulated form of five *Bacillus* bacteria. Their results indicated that any room-temperature THz absorption features due to spores are at least 28 times weaker (based on p-p noise) than the corresponding mid-IR amide I band, thus limited the utility of the THz domain for the room temperature detection of spores. Tedy's group [31] studied THz absorption spectra of the similarly structured amino acids *L*-serine and *L*-cysteine in the solid phase at 77 and 298K. The THz spectra were compared to isolated molecule and solid-state infrared vibrational spectral calculations using force field and density functional theory. These molecules differ only by a single atom but the low frequency spectra exhibit very different absorption features, which can be attributed to changes of the internal atomic masses and hydrogen-bonding character of the O-H vs. S-H groups within the solid phase unit cell. These differences lead to different intermolecular interactions.

Brandt et al. [32] studied THz time-domain and Raman spectroscopy of the sulfur-containing cystein-based peptide dimmers in the region of low-frequency infrared vibrations at room temperature. The differences between Raman and THz absorption spectra can be attributed to different intermolecular interactions, and thus, the lowest frequencies reflect different lattice vibrations. Therefore, the low-frequency Raman and THz absorption lines reflect both inter- and intra-molecular motions. Typically, low-frequency vibrational modes of biological molecule consist of both intramolecular modes and intermolecular modes. The former is dependent on the molecule as a whole; the latter arises from hydrogen-bonding interactions and van der Waals forces. Therefore, vibrational modes contain information about confirmation dynamics of biological molecules and can be used for identification, as being used for small molecules discussed previously. However, quite often it is challenging to distinguish the broad and overlapping features generated by conventional FTIR spectroscopy and THz time-domain spectroscopy (THz-TDS). Laman et al [33] reported their development of a high-resolution waveguide THZ spectroscopy of biological molecules, which resulting sharper THz features. This technique has been successfully demonstrated with seven small biological molecules, thymine, deoxycytidine, adenosine, D-glucose, tryptophan, glycine, and L-alanine, which shows the possibilities and promise for future studying vibrational modes of large biological molecules. George et al. [34] reported THz time-domain spectroscopy of bovine serum albumin in aqueous solution using Zeonor microfluidic channels. The measured results agree well with those previously reported using high-power THz free-electron lasers. Their work demonstrated the feasibility of performing high sensitivity THz spectroscopy of biomolecules in aqueous solutions with detectable molecular quantities as small as 10 picomoles using microfluidic devices.

Liu et al [35] described sensing minute changes on biological cell monolayers (bovine lung microvessel endothelial cells in response to vascular endothedial growth factor (VEGF)) with THz differential time-domain spectroscopy (THz-DYDS). These changes were reflected by alterations in THz wave attenuations and THz dielectric properties of the treated cells. They revealed that the VEGF-induced THZ attenuations of cell monolayers correlated well with changes in transendothelial resistance measured by electrical cell-substrate impedance sensing (ECIS). However, the morphological differences causing these changes were not observed with standard optical phase contrast microscopy. Thus, THz-DTDS could be a complementary technique for an optical microscopy or ECIS to measure properties of live cells, including cell growth, cell volume, and morphological change over time. Its limitation is that cell samples need to be extracted from the medium for free-space measurements, in order to avoid bulk water absorption.

A thorough review by Pickwell-MacPherson and Wallace [36] discussed the progress on THz Pulsed Imaging (TPI) in its application to biomedical problems with the possibility of developing it into viable medical imaging modality in the future. Several challenges are to be tackled if THz imaging is to become an established medical imaging modality: (1) The origin of image contrast needs to be better understood. Presumably, the changes in optical properties of the tissues observed due to increased water content and structural changes due to the presence of disease are responsible for the contrast. The question is how can we distinguish them from benign conditions? (2) The technology needs to improve such that *in vivo* imaging can be easily performed on a variety of sites within the body and not just on the skin. Given the low penetration depth, THz imaging during surgery or via an endoscope could be possibilities.

3.3. Chemometrics aspects

Like any other spectroscopic technique, for successful development of THz spectroscopy pharmaceutical PAT applications, chemometrics is an enabling tool for extracting critical process and product information from THz data to build up a model. Depending on which type of model is established, such a chemometrics model can be used for different purposes, such as process monitoring and process supervision, quantification, process control, etc. To establish a good model based on THz spectral data for pharmaceutical PAT applications, it is important to recognize and address the following aspects during the life cycle of THz PAT model: (1) the impact of THz data quality and spectral data preprocessing algorithms on predictive power and robustness of the model; (2) the role of Design of Experiments (DOE) in PAT modeling; and (3) model validation, maintenance, and update.

3.3.1. Impact of THz data quality and spectral data preprocessing algorithms on predictive power and robustness of the THz PAT model

Based on the discussion in previous subsections, it is recognized that THz absorption spectral values of pharmaceutical materials and pharmaceutical dosage form are a complex function of a series variables, such as (1) properties, compositions, and structure of the pharmaceutical materials and pharmaceutical dosage form; (2) the measurement conditions; and (3) THz instrumental variables, etc. Our research results [37] demonstrated the effects of compositional and instrumental variables on THz spectral data quality for THz pharmaceutical PAT application. In addition, the influences of spectral data preprocessing algorithms on the predictive power and robustness of the multivariate statistical models have been recognized [23, 38-41]. It is well known that spectral data are information rich, often capturing both chemical and physical information of the substance. It is always challenging to separate the chemical information and physical information embedded with spectral data. If use right, spectral data preprocessing algorithms could be useful to filter certain irrelevant information in the data, thus allows establishing more realistic correlation between the independent variables and responsible variables. As an integral part of chemometrics modeling, the objective of the pre-processing is to remove physical phenomena in the spectra so as to improve the subsequent multivariate regression, classification model or exploratory analysis. There are two types of pre-processing methods which are most frequently used: scatter-correction methods and spectral derivatives. Multiplicative Scatter Correction (MSC), de-trending, Standard Normal Variate (SNV), and normalization are belong to the first group. Savitzky-Golay (S-G) polynomial derivative filter and Norris-Williams (NW) derivative are representatives of the 2nd group. Those data preprocessing algorithms are available in several commercial software packages specialized in handling spectral data for multivariate data analysis and chemometrics modeling.

3.3.2. Role of Design of Experiments (DOE) in PAT modeling

One essential element for pharmaceutical PAT is process understanding. According to the FDA's PAT Guidance, "A process is generally considered well understood when (1) all critical sources of variability are identified and explained; (2) variability is managed by the process; and, (3) product quality attributes can be accurately and reliably predicted over the design space established for materials used, process parameters, manufacturing, environmental, and other conditions. The ability to predict reflects a high degree of process understanding." Design of Experiments (DOE) provides a rationale approach to identify the variability and possible interactions. Integration DOE with PAT real time process monitoring could provide a powerful tool to understand the variability occurred with the process, and more importantly, to reveal how the process variability are translated into product variability. Most importantly, such integration is essential to achieve a high degree of process understanding due to the facts (1) the variability is examined systematically via DOE; and (2) the process information is acquired over the entire time course of the process via PAT real time process monitoring, which allows us to evaluate the process kinetics and process dynamics in a detailed fashion. Therefore, integration DOE and PAT real time process monitoring is essential to PAT modeling, product and process understanding, and design space development. Several FDA internal research projects have discussed those aspects in details [42-47].

3.3.3. Model validation, maintenance, and update

Model validation, maintenance, and update are important aspects for any process models associated with pharmaceutical PAT applications, including THz pharmaceutical PAT applications, if the model developed is to be used for quantification or process control. To the best of our knowledge, currently there is no standard or regulatory guidance that can be specifically address all of issued associated with THz pharmaceutical PAT applications in the aspect of model validation, maintenance, and update. However, relevant references including ASTM standards and EMEA NIR Guideline [48-52] may be adapted to address those issues, at least providing a good starting point.

In author's scientific opinion, the validation requirements for different types of models are different. For first-principles models and mechanistic models, we do need to examine the conditions and assumptions made during the modeling exercise to make sure those conditions and assumptions are still valid during the applications of those first-principles models and mechanistic models. For chemometric models or data-driven empirical models, a significant validation effort is mandatory, including but is not limited to cross-validation as the first safe guard, external validation via independent data set that has not been used previously for model calibration exercise. For this chemometric model, model maintenance and update are important steps to ensure the accuracy, representativeness, and robustness of the calibration model established during the pharmaceutical PAT application, in the event unaccounted sources of variability occurs.

4. Case Studies on First principle modeling approach to validate THz Spectroscopic Measurement

4.1. *In situ* monitoring technique for examining solvent diffusion in polymers: first-principles modeling case study

The ingress of liquid solvent molecules into solid polymers is of importance to certain applications, such as coating of pharmaceutical tablets (for controlled release/modified release dosage form design) and pharmaceutical devices where solvent transport is desirable, and packaging where solvent transport causes degradation and contamination of the polymer. Therefore, accurate measurement of the liquid solvent ingress position within the polymer film is critical for purposes such as validation of the models of the process, improved understanding of the liquid penetration process, and potentially the provision of on-line or at-line measurement during processing or storage as advocated by the FDA's PAT Guidance. Obradovic et al [53] reported a novel application of THz time-domain reflection spectroscopy to measure the penetrant front position of acetone in various polymers (polycarbonate and PVC) *in situ*, and thus as a function of time and temperature. In their work, the reflection of a short THz pulse off a penetrant front was observed. The data of depth of solvent penetration derived from *in situ* THz time-domain reflection spectra were then validated by detailed comparison with Magnetic Resonance Imaging (MRI) measurements on the same systems. The schematic of the THz reflection measurement is shown in Figure 1. The first-principles which govern the measurement process are expressed mathematically below.

The path length differences in polymer x_2 and in air x_1 can be written as:

$$x_2 = \frac{2d_a}{\cos\beta} \qquad (1)$$

$$x_1 = 2d_a \tan\beta \sin\alpha \qquad (2)$$

Therefore, the overall path length difference Δx is:

$$\Delta x = x_2 - x_1 = \frac{2d_a}{\cos\beta}(1 - \sin\beta \sin\alpha) \qquad (3)$$

By substituting Shell's law [54], $\frac{\sin\alpha}{\sin\beta} = \frac{n_2}{n_1}$, we have:

$$\Delta x = \frac{2d_a}{\cos\beta}(1 - \frac{n_1}{n_2}\sin^2\alpha) \qquad (4)$$

The time delay Δt at the detector introduced by the additional path length Δx is given by:

$$\Delta t = t_2 - t_1 = \frac{x_2 n_2}{c} - \frac{x_1 n_1}{c} = \frac{2d_a(t)}{c\cos\beta}(n_2 - n_1 \sin\beta \sin\alpha) \qquad (5)$$

Clearly, the THz reflection measurement results were expressed by first-principles models. Furthermore, good agreement with respect to the actone's penetration front position was obtained for both polycarbonate and PVC, when compared with corresponding MRI data. Therefore, the THz reflection measurement was validated by independent MRI measurement. In author's opinion, this work provides an excellent scientific case study which demonstrated how first-principles models play a critical role in the PAT and QbD implementation.

4.2. *Ex-situ* monitoring technique for examining moisture content in test disk: first-principles modeling case study

Previously [55], we used THz spectroscopy to identify water peak and derive the disk drying process kinetics. Pharmaceutical tablet (Theophylline as the drug with several excipients) was crushed into fine powder, then mixed with polyethylene. The mixture was then compressed into disks of thickness 2 mm in a 13 mm die at approximately 200 psi pressure. All disks were made under lab humidity conditions (relative humidity (RH)= 40 ±1 %). All disks especially the subsurface portions would contain a small amount of water from either moisture absorbed by components (Hancock and Shamblin, 1998) [56] or residual water in the tablets, from the mass balance point of view. Consequently, when the disk is placed into the spectrometer sample compartment for dried-air (RH=0%) purging, any subsurface water would diffuse out and evaporate into the gas phase in the compartment, and then quickly removed by the dried-air purging process, due to the gradient of water levels across the subsurface of disk and the gas phase according to the inter-phase mass-transfer theory [57]. THz measurement was done for various disk drying time points. Good agreement was obtained between the THz experimental results of disk drying, previously identified THz peaks for water, and theoretical mass-transfer analysis of the disk drying process, as discussed below.

Previously [58-61] (Rahman and Stillinger, 1971; Hasted et al, 1985; Thrane et al, 1995; Castner and Chang, 1995) two absorption bands centered at 50 cm^{-1} and 200 cm^{-1} for liquid water were identified in far-infrared measurements. The band at 200 cm^{-1} was broad and corresponding to the stretching mode of the O—H...O unit of the water molecules. The exact position and width could vary slightly, depending on the chemical and physical environment. Our study shows that there are two bands with peak maxima located around 54 cm^{-1} and 210 cm^{-1}on the tablet's corrected THz absorption spectra, which obviously are not associated with characteristic peaks of excipients in the tablet formulations. Meanwhile, two prominent peaks for the theophylline component around 52.1 cm^{-1} and 210 cm^{-1} were observed. Given that theophylline is stable under our experimental conditions, presumably we could assume these two prominent theophylline bands are superimposed on the water bands in their vicinity around 50 cm^{-1} and 200 cm^{-1}. The following derived drying process kinetics analysis further supported this rational.

According to Beer's law, the spectral absorbance (A) is proportional to the concentration (c) of species being detected:

$$A = \varepsilon \cdot l \cdot c \qquad (6)$$

Where ε is the molar absorbiptivity, l is the light path length. As a first appropriation, both ε and l can be assumed constants. Thus, the absorbance difference at the peak maxima can be related to the mass change associated with the disk drying process over time. Furthermore, the absorbance difference divided by the drying time difference can be a measure of the drying rate, if we make a reasonable assumption that the disk surface area remains unchanged during this relative short period of drying time. Figure 2(a)-(d) illustrates the relationship between this derived disk drying rate vs. disk drying time. For the bands centered around 54 cm^{-1} and 210 cm^{-1}, most WG tablets and DC tablets exhibit a distinct two-stage drying process: (1) at the 1st stage, the absolute value of the drying rate decreases very quickly when t_{disk} increased from 5 minutes to 30 minutes; and (2) at the 2nd stage, a nearly constant drying rate (almost zero) was exhibited when t_{disk} increased further to 60 minutes. This two-stage drying model appears to qualitatively agree with the typical drying curve of materials at constant drying conditions, and could be well explained by the interphase mass-transfer theory [57]. Therefore, the bands located at 54 cm^{-1} and 210 cm^{-1} are interpreted as combinations of THz absorption features of theophylline component and moisture. Furthermore, the absorbance difference observed over disk drying time was primarily attributed to residual water associated with the disk due to the high sensitivity of THz spectroscopy to water, although the amount of residual water does not appear to be significant.

This work provides a scientific case study which illustrated how first-principles models (in this case mass-transfer and heat-transfer) were applied to analyze THz test disk drying data to derive disk drying kinetics. The benefits of this approach lie in: (1) identification characteristic THz peaks for water in THz absorption spectra; (2) enabling optimal THz testing conditions to ensure THz data quality for THz pharmaceutical PAT application.

5. Challenges and Opportunities for adopting THz spectroscopy for a pharmaceutical PAT tool

From the public literatures, so far the majority of THz spectroscopy pharmaceutical PAT applications have been focused on real time process monitoring of tablet coating processes. Proof-of-concept has been proved. Preliminary success has been achieved for relative thick coating thickness. However, limited sensitivity to low coating thickness has been a significant challenge for the applicability of this development. On the other hand, from quantitative modeling perspective, a reference method which can directly measure coating thickness in an accurate and reliable manner is much needed. Destructive method (cutting-trough the film) is still most dominant method of directly measuring coating thickness at the moment, in spite of its possibility of introducing additional variability and error. As discussed above, TPI provides an alternate method to detect defects in the film, which might be linked to the dissolution performance critical areas, if the linkage could be established and confirmed by direct and microscopic site evidence.

The ultimate goal for THz spectroscopy pharmaceutical PAT application is to gain in-depth pharmaceutical process understanding, establish a process model, and hopefully use this model to control the pharmaceutical process. To achieve this ultimate goal, such a model must be representative (to the situations examined) and robust (to certain level of noises and/or process disturbance). As most of the works published so far are at laboratory scale, significant challenges and opportunities are coexisting when move the laboratory scale work at R&D to pilot scale and manufacturing scale, as discussed previously [62] and highlighted in the FDA's pharmaceutical quality initiatives in the 21st century [63-64].

6. Concluding Remarks

It can be easily appreciated that THz spectroscopy has found interesting and promising applications in the following areas. (1) pharmaceutical product characterization and pharmaceutical product quality control; (2) pharmaceutical PAT, especially real-time process monitoring of certain pharmaceutical manufacturing processes, etc. Due to its unique capability of probing the intermolecular interaction and intra-molecular interactions, THz spectroscopy provides a novel approach to study pharmaceutical polymorphism, solid-state properties, crystallinity, film coating process, and even biological and biopharmaceutical applications, etc. While direct validation of THz spectroscopy measurement result has been a significant challenge, first-principles models can provide a reliable approach to validate THz spectroscopic measurement, as demonstrated in the two case studies created in this work. Both scientific and regulatory challenges in adopting THz spectroscopy for pharmaceutical PAT applications were discussed in the context of THz measurement science and PAT chemometrics modeling.

Acknowledgement

The author greatly acknowledges scientific discussions on THz technology with Dr. Edwin Heilweil at the National Institute of Science and Technology (NIST), Gaithersburg, Maryland, USA. The authors gratefully acknowledge the original FDA PAT Policy Team, the FDA PAT Review and Inspection Team, the FDA PAT Research Team, and FDA CDER Senior Management Team. FDA CDER Regulatory Science and Review (RSR) Funding Support (RSR-04-16) and Interagency Agreement between the FDA and NIST are greatly acknowledged. Great contributions from the THz field from various technology vendors and research groups are appreciated. The scientific discussion on validation of new technology with Dr. Vincent Vilker, Director, Office of Testing and Research, CDER, FDA, is acknowledged.

References

[1] FDA, August 2011. A Strategic Plan: Advancing Regulatory Science at FDA. Available at: http://www.fda.gov/downloads/ScienceResearch/SpecialTopics/RegulatoryScience/UCM268225.pdf (accessed on 09/24/2012)

[2] J. Woodcock, 2004. The Concept of Pharmaceutical Quality, *American Pharmaceutical Review*, **7** (6), pg. 10-15.

[3] FDA, 2004. Pharmaceutical CGMPs for the 21st Century: A Risk-Based Approach. Final Report. Available at:

http://www.fda.gov/downloads/Drugs/DevelopmentApprovalProcess/Manufacturing/QuestionsandAnswersonCurrentGoodManufacturingPracticescGMPforDrugs/UCM176374.pdf (accessed on 09/24/2012)
[4] FDA, 2004. "Guidance for Industry. PAT—A Framework for Innovative Pharmaceutical Development, Manufacturing, and Quality Assurance". Available at: http://www.fda.gov/downloads/Drugs/GuidanceComplianceRegulatoryInformation/Guidances/UCM070305.pdf (accessed on 09/24/2012)
[5] FDA/ICH, 2009. Guidance for Industry, Q8(R2) Pharmaceutical Development. available at: http://www.fda.gov/downloads/Drugs/GuidanceComplianceRegulatoryInformation/Guidances/UCM073507.pdf (accessed on 09/24/2012)
[6] FDA, 2011. "Guidance for Industry. Process Validation: General Principles and Practices. Rev 1". Available at: http://www.fda.gov/downloads/Drugs/GuidanceComplianceRegulatoryInformation/Guidances/UCM070336.pdf (accessed on 09/24/2012)
[7] J. Aaltonen, K.C. Gordon, C.J. Strachan, T. Rades, 2008. Perspectives in the use of spectroscopy to characterise pharmaceutical solids. International Journal of Pharmaceutics 364(2008)159-169.
[8] P.F. Taday, 2004. Applications of terahertz spectroscopy to pharmaceutical sciences. Trans. R. Soc. Lond. A (2004) 362: 351-364.
[9] P.F. Taday, I.V. Bradley, D.D. Arnone, M. Pepper, 2003. Using Terahertz Pulse spectroscopy to Study the Crystalline Structure of a Drug: A Case Study of the Polymorphs of Ranitidine Hydrochloride. Journal of Pharmaceutical Sciences 92(4): 831-838.
[10] C.J. Strachan, T. Rades, D.A. Newnham, K.C. Gordon, M. Pepper, P.F. Taday, 2004. Using terahertz pulsed spectroscopy to study crystallinity of pharmaceutical materials. Chemical Physics Letters 390: 20-24.
[11] C. J. Strachan, P.F. Taday, D.A. Newnham, K.C. Gordon, J.A. Zeitler, M. Pepper, T. Rades, 2005. Using Terahertz Pulsed Spectroscopy to Quantify Pharmaceutical Polymorphism and Crystallinity. Journal of Pharmaceutical Sciences 94(4):837-846.
[12] Y. Ikeda, Y. Ishihara, T. Moriwaki, E. Kato, and K. Terada, 2010. A Novel Analytical Method for Pharmaceutical Polymorphs by Terahertz Spectroscopy and Optimization of Crystal Form at the Discovery Stage. Chem. Pharm. Bull. 58(1) 76-81.
[13] H-B. Liu, Y. Chen, X.-C. Zhang, 2007. Characterization of Anhydrous and Hydrated Pharmaceutical Materials with THz Time-Domain Spectroscopy. Journal of Pharmaceutical Sciences 96(4): 927-934.
[14] H.B. Liu, X.-C. Zhang, 2006. Dehydration kinetics of D-glucose monohydrate studied using THz time-domain spectroscopy. Chemical Physics Letters 429:229-233.
[15] M. D. King, W. D. Buchanan, T. M. Korter, 2011. Understanding the Terahertz Spectra of Crystalline Pharmaceuticals: Terahertz Spectroscopy and Solid-State Density Functional Theory Study of (S)-(+)0Ibuprofen and (RS)-Ibuprofen. Journal of Pharmaceutical Sciences 100(3):1116-1129.
[16] D.G. Allis, A.M. Fedor, T.M. Korter, J.E. Bjarnason, E.R. Brown, 2007. Assignment of the lowest-lying THz absorption signatures in biotin and lactose monohydrate by solid-state density functional theory. Chemical Physics Letters 440: 203-209)

[17] W. Limwikrant, K. Higashi, K. Yamamoto, K. Moribe, 2009. Characterization of ofloxacin-oxalic acid complex by PXRD, NMR, and THz spectroscopy. International Journal of Pharmaceutics 382: 50-55.
[18] M. Takahashi, Y. Kawazoe, Y. Ishikawa, H. Ito, 2006. Low-frequency vibrations of crystalline α, α-trehalose dehydrate. Chemical Physics Letters 429: 371-377.
[19] N. Nagai, R. Kumazawa, R. Fukasawa, 2005. Direct evidence of inter-molecular vibrations by THz spectroscopy. Chemical Physics Letters 413: 495-500.
[20] J. Shen, G. Wang, D. Jiang, L. Liang, X. Xu, 2010. Terahertz spectroscopic investigations of caffeine and 3-acetylmorphine. Optik 121 :1712-1716.
[21] Y. Watanabe, K. Kawase, T. Ikari, H. Ito, Y. Ishikawa, H. Minamide, 2004. Component anslysis of chemical mixtures using terahertz spectroscopic imaging. Optics Communication 234 : 125-129.
[22] R. Palermo, R. P. Cogdill, S. M. Short, J. K. Drennen III, P. F. Taday, 2008. Density mapping and chemical component calibration development of four-component compacts via terahertz pulsed imaging. Journal of Pharmaceutical and Biomedical Analysis 46 : 36-44.
[23] H. Wu, R.C. Lyon, A.S. Hussain, J.D. Drennen III, *et al.*. Application of Principal Component Analysis in Assessing Pharmaceutical Formulation Design: Exploring the Casual Links between the Tablet Processing Conditions and Drug Dissolution Rate. *Proceedings of Symposium on Innovations in Pharmaceutical and Biotechnology Development and Manufacturing, AIChE Annual Meeting* [TE001], pg. 635-642. AIChE Pub. No. 187, ISBN 0-8169-0929-6. San Francisco, Nov. 17-21, 2003.
[24] A.J. Fitzgerald, B.E. Cole, P.F. Taday, 2005. Nondestructive Analysis of Tablet Coating Thickness Using Terahertz Pulsed Imaging. Journal of Pharmaceutical Sciences 94(1):177-183.
[25] L. Ho, R. Müller, K.C. Gordon, P. Kleinebudde, M. Pepper, T. Rades, Y. Shen, P.F. Taday, J. A. Zeitler, 2008. Applications of terahertz pulsed imaging to sustained-release tablet film coating quality assessment and dissolution performance. Journal of Controlled Release, 127: 79-87.
[26] L. Ho, Y. Cuppok, S. Muschert, K.C. Gordon, M. Pepper, Y. Shen, F. Siepmann, J. Siepmann, P.F. Taday, T. Rades, 2009. Effects of film coating thickness and drug layer uniformity on in vitro drug release from sustained-release coated pellets: A case study using terahertz pulsed imaging. International Journal of Pharmaceutics, 382:151-159.
[27] L. Ho, R. Müller, C. Krüger, K.C. Gordon, P. Kleinebudde, M. Pepper, T. Rades, Y. Shen, P.F. Taday, J. A. Zeitler, 2010. Investigating Dissolution Performance Critical Areas on Coated Tablets: A Case Study Using Terahertz Pulsed Imaging. Journal of Pharmaceutical Sciences 99(1): 392-402
[28] J.A. Zeitler, L.F. Gladden, 2008. *In-vitro* tomography and non-destructive imaging at depth of pharmaceutical solid dosage forms. European Journal of Pharmaceutics and Biopharmaceutics, 71: 2-22
[29] C. Cahyadi, A.D. Karande, L.W. Chan, P.W.S. Heng, 2010. Comparative study of non-destructive methods to quantify thickness of tablet coatings. International Journal of Pharmaceutics, 398: 39-49
[30] T.J. Johnson, N.B. Valentine, S.W. Sharpe, 2005. Mid-infrared versus far-infrared (THz) relative intensities of room temperature *Bacillus* spores. Chemical Physics Letters 403:152-157.

[31] T.M. Korter, R. Balu, M.B. Campbell, M.C. Beard, S.K. Gregurick, E.J. Heilweil, 2006. Terahertz spectroscopy of solid serine and cysteine. Chemical Physics Letters 418: 65-70.
[32] N.N. Brandt, A.Yu. Chikishev, A.V. Kargovsky, M.M. Nazarov, O.D. Parashchuk, D.A. Sapozhnikov, I.N. Smirnova, A.P. Shkurinov, N.V. Sumbatyan, 2008. Terahertz time-domain and Raman spectroscopy of the sulfur-containing peptide dimmers: Low-frequency markers of disulfide bridges. Vibrational Spectroscopy 47: 53-58.
[33] N. Laman, S.Sree Harsha, D. Grischkowsky, and J. Melinger, 2008. High-Resolution Waveguide THz Spectroscopy of Biological Molecules. Biophysical Journal 94: 1010-1020.
[34] P.A. George, W. Hui, F. Rana, B.G. Hawkins, A.E. Smith, B.J. Kirby, 2008. Microfluidic devices for terahertz spectroscopy of biomolecules. Optics Express 16(3): 1577-1582.
[35] H.-B. Liu, G. Plopper, S. Earley, Y. Chen, B. Ferguson, X.-C. Zhang, 2007. Sensing minute changes in biological cell monolayers with THz differential time-domain spectroscopy. Biosensors and Bioelectronics 22:1075-1080.
[36]. E. Pickwell-MacPherson and V.P. Wallace, 2009. terahertz pulsed imaging-A potential medical imaging modality? Photodiagnosis and Photodynamics Therapy 6:128-134.
[37] H. Wu, E.J. Heilweil, A.S. Hussain, M.A. Khan, 2008. Process Analytical Technology (PAT): Quantification Approaches in Terahertz Spectroscopy for Pharmaceutical Applications. Journal of Pharmaceutical Sciences, 97(2): 970-984.
[38] P. Chalus, Y. Roggo, S. Walter, M. Ulmschneider, 2005. Near-infrared determination of active substance content in intact low-dosage tablets. Talanta 66:1294-1302.
[39] S.R.-Torres, J.D.P.-Ramos, K. R. Morris, E.R. Grant, 2005. Raman spectroscopic measurement of tablet-to-tablet coating variability. Journal of Pharmaceutical and Biomedical Analysis 38:270-274.
[40] C.F.Pereira, M.F. Pimentel, R.K.H. Galvao, F.A. Honorato, L. Stragevitch, M. N. Martins, 2008. A comparative study of calibration transfer methods for determination of gasoline quality parameters in three different near infrared spectrometers. Analytica Chimica Acta 611:41-47.
[41] Å. Rinnan, F. v. d. Berg, S.B. Engelsen, 2009. Review of the most common pre-processing techniques for near-infrared spectra. Trends in Analytical Chemistry 28 (10):1201-1222.
[42] Xie, L.,Wu, H., Shen, M., Augsburger, L., Lyon, R.C., Khan, M.A., Hussain, A.S.,and Hoag, S.W., 2008. Quality-by-Design (QbD): Effects of Testing Parameters and Formulation Variables on the Segregation Tendency of Pharmaceutical Powder Measured by the ASTM D 6940-04 Segregation Test. *Journal of Pharmaceutical Sciences* 97(10):4485-97.
[43] H. Wu and M. Khan, 2009. Quality-by-Design (QbD): An Integrated Approach for Evaluation of Powder Blending Process Kinetics and Determination of Powder Blending End-Point. *Journal of Pharmaceutical Sciences* 98(8):2784-2798.
[44] H. Wu, M. Tawakkul, M. White, and M. Khan. Quality-by-Design (QbD): An Integrated Multivariate Approach for the Component Quantification in Powder Blends. *International Journal of Pharmaceutics* 372(1-2):39-48, 2009.

[45] H. Wu and M. Khan, 2010. Quality-by-Design (QbD): An Integrated Process Analytical Technology (PAT) Approach for Real-Time Monitoring and Mapping the State of a Pharmaceutical Co-precipitation Process. *Journal of Pharmaceutical Sciences* 99(3): 1516-1534.
[46] H. Wu, M. White, M. Khan, 2011. Quality-by-Design (QbD): An Integrated Process Analytical Technology (PAT) Approach for a Dynamic Pharmaceutical Co-precipitation Process Characterization and Process Design Space Development. *International Journal of Pharmaceutics*, 405:63-78.
[47] H. Wu and M. Khan, 2011. Quality-by-Design (QbD): An Integrated Process Analytical Technology (PAT) Approach to Determine the Nucleation and Growth Mechanisms during a Dynamic Pharmaceutical Co-precipitation Process. *Journal of Pharmaceutical Sciences* 100(5): 1969-1986.
[48] ASTM Standard E1655-00. Standard Practices for Infrared Multivariate Quantitative Analysis.
[49] ASTM Standard D 3764-01: Standard Practice for Validation of Process Stream Analyzer Systems
[50] EMEA CPMP and CVMP Draft: Note for Guidance on the use of near infrared spectroscopy by the pharmaceutical industry and the data to be forwarded in part II of the dossier for a market authorisation. London, 14 Nov 2001, EMEA/CVMP/961/01; CPMP/QWP/3309/01 draft.
[51] R.D. Snee, 1977. Validation of regression models: Methods and examples, Technometrics 19:415-428.
[52] J. Michaelsen, 1987. Cross-validation in statistical climate forecast models, J. of Climate and Applied Meteorology 26:1589-1600.
[53] J. Obradovic, J. H.P. Collins, O. Hirsch, M. D. Mantle, M. L. Johns, L.F. Gladden, 2007. The use of THz time-domain reflection measurements to investigate solvent diffusion in polymers. Polymer 48: 3494-3503.
[54] Dykaar DR, Greene BI, Federici JF, Levi AFJ, Pfeiffer LN, Kopf RF., Appl Phys Lett 1991 59(3):262-4.
[55] H. Wu, E.J. Heilweil, A.S. Hussain, M.A. Khan, 2007. Process Analytical Technology (PAT): Effects of instrumental and compositional variables on terahertz spectral data quality to characterize pharmaceutical materials and tablets. International Journal of Pharmaceutics, 343: 148-158.
[56] B. Hancock, S. Shamblin, 1998. Water vapor sorption by pharmaceutical sugars. Pharmaceutical Science Technology Today, 1:345-351.
[57] R. Bird, W. Stewart, E. Lightfoot, 1960. Transport Phenomena. John Wiley, New York.
[58] A. Rahman, F. Stillinger, 1971. Molecular dynamics of liquid water. J. Chem. Phys. 55: 3336-3359.
[59] J. Hasted, S. Husain, F. Frescura, J. Birch, 1985. Far-infrared absorption in liquid water. Chem. Phys. Lett. 118: 622-625.
[60] L. Thrane, R. Jacobsen, P. Uhd Josen, S. Keiding, 1995. THz reflection spectroscopy of liquid water. Chem. Phys. Lett. 240:330-333.
[61] E. Castner, Jr., and Y. Chang, 1995. The intermolecular dynamics of liquid water. J. Chem. Phys. 102:653-659.

[62] H. Wu, A.S. Hussain, and M.A. Khan, 2007. Process Control Perspective for Process Analytical Technology: Integration of Chemical Engineering Practice into Semiconductor and Pharmaceutical Industries. *Chemical Engineering. Communications*. **194**: 760-779.
[63] US Food and Drug Administration. 2004. Innovation or Stagnation: Challenge and Opportunity on the Critical Path to New Medical Products. http://www.fda.gov/oc/initiatives/criticalpath/whitepaper.html (accessed on 09/24/2012)
[64] J. Woodcock and R. Woosley, 2008. The FDA Critical Path Initiative and Its Influence on New Drug Development. *Annual Review of Medicine*, Vol. 59, DOI: 10.1146/annurev.med.59.09056.155819, Feb, 2008

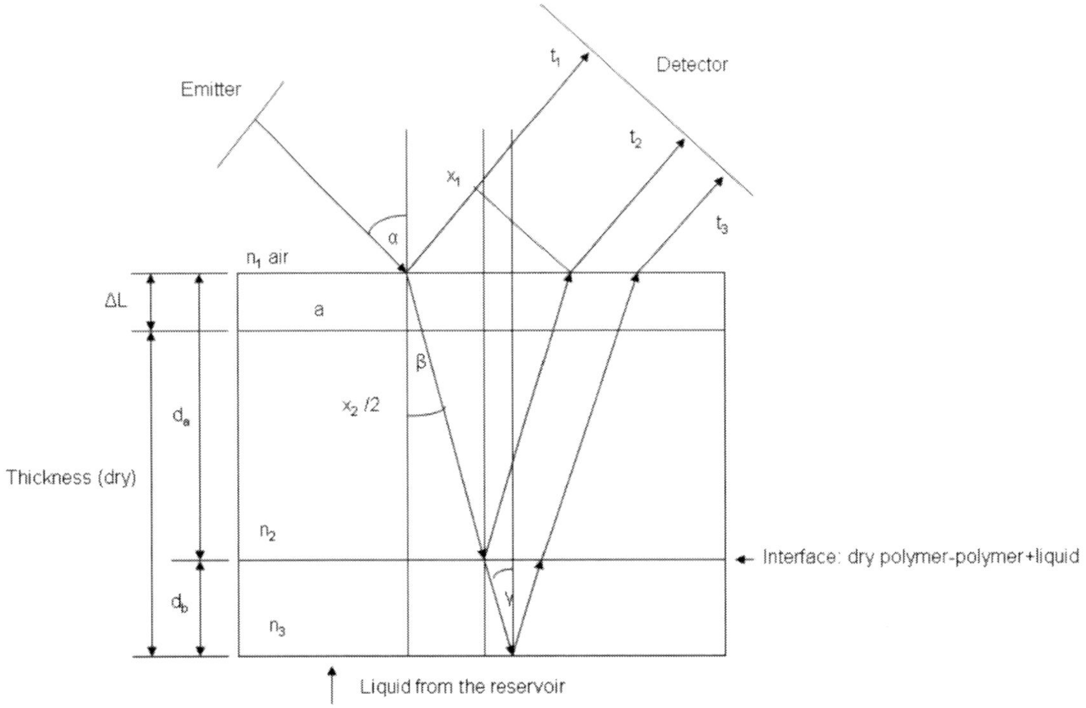

Figure 1. Figure 1 Schematic of the THz refection measurement from the polymer disc (adaption from Ref. [53] with kind permission from Elsevier)
n_1: index of refraction of air;
n_2: index of refraction of dry polymer;
n_3: index of refraction of the polymer penetrated by the liquid;
α: angle of incidence from air;
β: angle at which the THz beam is refracted when propagating from air into polymer disc;
a: shift of the reflection point;
x_1: path length difference in air;
x_2: path length difference in dry polymer;
d_a: thickness of the unpenetrated polymer layer;
d_b: thickness of the acetone-wet polymer layer;
ΔL: amount of swelling.

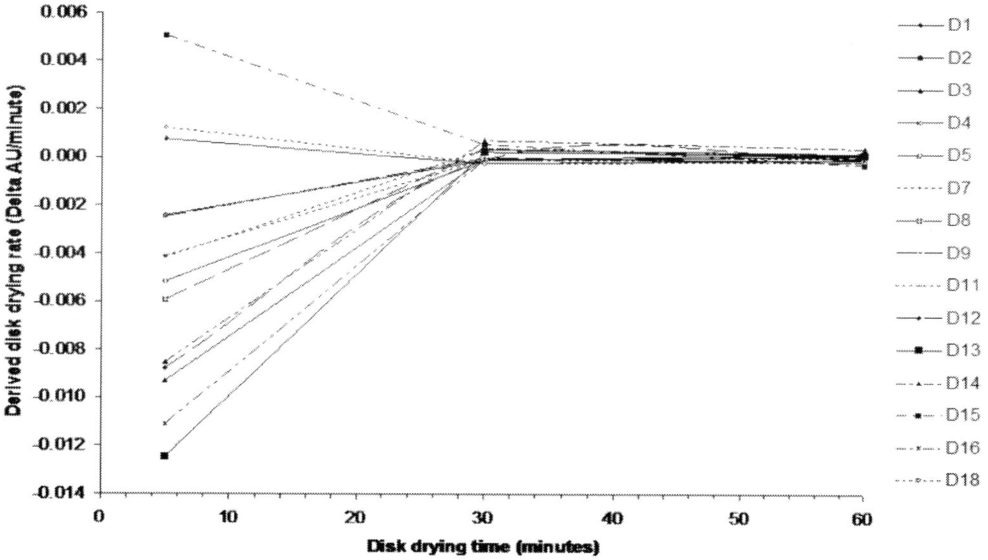

Figure 2(a)

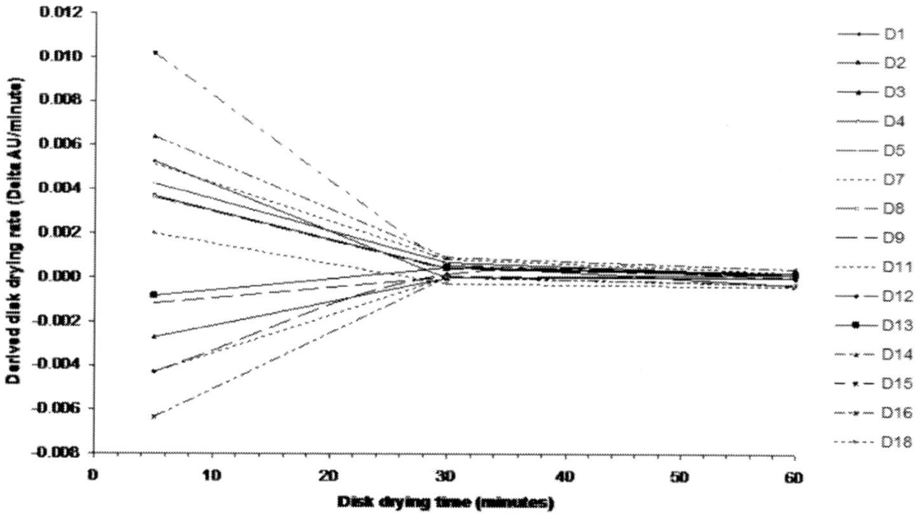

Figure 2(b)

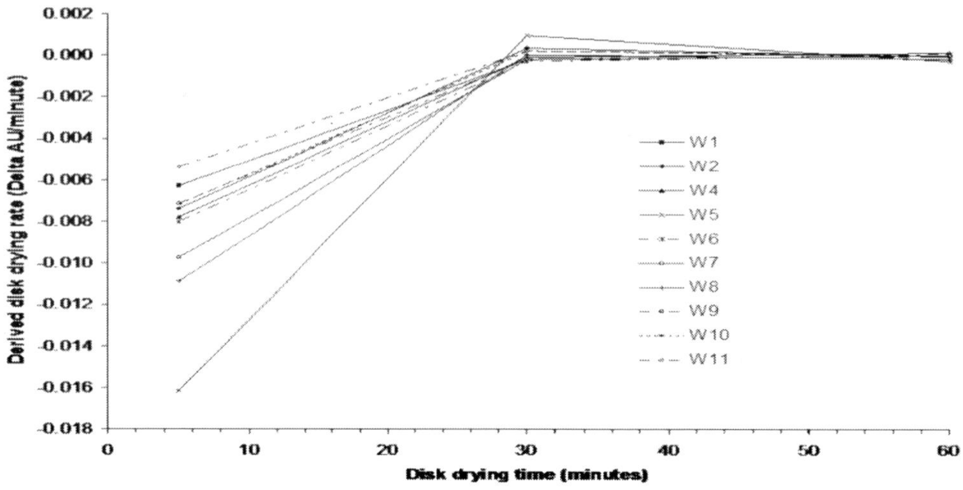

Figure 2(c)

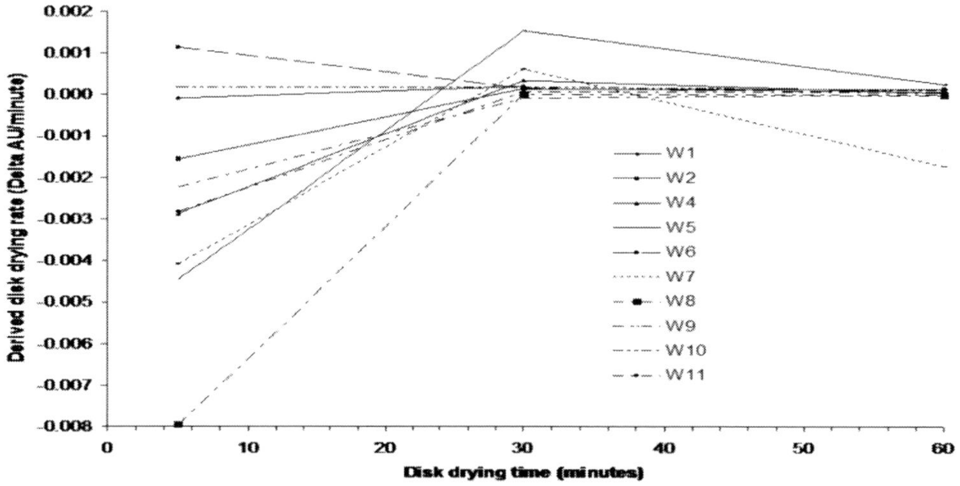

Figure 2(d)

Figure 2 Plots of derived disk drying rates vs. disk drying times based on absorbance data at THz bands around 54 cm^{-1} and 210 cm^{-1}. (a) Direct Compression (DC) tablets at 54 cm^{-1}; (b) DC tablets at 210 cm^{-1}; (c) Wet Granulation (WG) tablets at 54 cm^{-1}; (d) WG tablets at 210 cm^{-1}. In Figure 2(a)-(b), lengend D# stands for direct compression tablet #. In Figure 2(c)-(d), lengend W# stands for wet granulation tablet #. (adaption from Ref. [55] with kind permission from Elsevier)

(127b) Mobile Hydrogen Storage in CO2 Treated Ammonia Borane Complexes and Its Design Implications

Jae W. Lee, Department of Chemical Engineering, The City College of New York, New York, NY

Ammonium borane (AB, NH_3BH_3) is a promising on-board, solid-state hydrogen storage material (chemical hydride) with a high hydrogen content of 19.6 wt.%. Hydrogen extraction from AB has been carried out via hydrolysis, methanolysis, or pyrolysis (thermolysis or thermal decomposition) but hydrolysis and methanolysis accompany an additional component of H_2O and methanol. Thus, it lowers hydrogen storage capacities for vehicular applications. The thermolysis of AB is mostly employed to extract hydrogen from AB and involves three sequential steps occurring around 110, 150 and 1100 ^oC, with 6.5 wt. % of hydrogen liberated in each step. A primary research challenge for adopting AB as a hydrogen storage material is to enhance dehydrogenation rate at low temperatures around 85 ^oC that is the operating temperature for polymer electrolyte membrane fuel cells and also prevents the generation of byproducts (like borazine, harmful to electrodes). Several approaches have been discussed to thermolytically enhance the dehydrogenation kinetics of AB such as compositing AB with mesoporous materials, dispersing AB in ionic liquids, and adding catalysts and promoters. One major drawback associated with the first two approaches is the low hydrogen storage capacity, as a result of the mass fraction of mesoporous materials and ionic liquids being very high. Using transition metals or promoters adds up the cost of hydrogen storage system. As a result, the research efforts have continuously made to find cheap and effective promoters for hydrogen release without negatively impacting the storage capacity. In this talk, I will discuss the promoting effect of CO_2 on the dehydrogenation of solid AB and PAB (polyaminoborane) at 85 ^oC. Beyond the enhanced dehydrogenation kinetics of AB, I will also introduce design implications of hydrogen storage systems using AB complexes and present an economically-viable conversion process of CO_2 to carbon materials using AB.

A Semantic Representation of Policy Goals
in the Modeling of Electricity Generation and Water Treatment Systems

Aidid Chee Tahir and René Bañares-Alcántara[1]*, University of Oxford, Oxford, UK*

Abstract

A modeling system that converts policy goals and targets to configure an optimization model is presented. The system integrates semantic representation, inference and optimization techniques. The results of two case studies are discussed as a proof of concept, one case study for the formulation of an electricity mix and another for water treatment systems.

Introduction

Optimization models are used to evaluate different scenarios during the formulation of energy and water policies, including those formulated to ensure electricity and water supply security and sustainability. The strength of these models lies in their theoretical foundations built on mathematical equations that process numerical (quantitative) data. Nevertheless, a complete consideration of energy and water policy issues also requires the evaluation of non-numerical (qualitative) data. For example, MARKAL [1], MESSAGE [2] and DNE-21 [3] are optimization tools that aid in the evaluation of energy policies, but their use requires the intervention of modelers to "translate" the policy goals set by the policy maker into equations and constraints that are solvable by mathematical procedures.

This document introduces a modeling environment that can be applied to policy analysis and evaluation. This new modeling system aims to support policy makers by automatically converting their policy goals and targets as parts of an optimization model consisting of a set of mathematical expressions. To facilitate the inclusion of these additional aspects in a computer model, a modeling system, which operates at a level that is above that of optimization models, is proposed.

The modeling system integrates semantic representation techniques and knowledge inference functionalities with an optimization model and its solver. The selected method for semantic representation is a formalism known as ontologies which encodes knowledge in a manner that can be communicated and shared between people and software tools. Ontologies add semantics through a collection of associated terms, their conceptual relations and a set of logical axioms. In addition, we have integrated this semantic representation to a suite of engineering models (mathematical equations).

The document is organized as follows: the next section sketches the architecture of the modeling system. The following two sections present some of the results associated with two case studies, one for electricity generation and the other for water treatment. The last section discusses some of the limitations and contributions of the prototypes.

Architecture of the prototype Electricity Generation Modeling System

During the policy modeling task a modeler is required to translate the intention of the policy maker, as defined by a set of policy goals and targets, into mathematical equivalents fit for input into the electricity generation model. One of the complexities of this task is that a comprehensive formulation of an electricity generation mix must include aspects associated with the triple bottom line of sustainability (social, environmental and economic criteria), an evaluation of which requires

[1] Corresponding author.

the consideration of a significant amount of diverse non-numerical information. Two prerequisites must be fulfilled for energy models to consider the whole spectrum of sustainability aspects. First, the information associated with sustainability in the context of energy policies must be identified and defined. Second, a new approach to optimization-based energy modeling, which considers both quantitative data and qualitative information, must be developed.

We have developed a prototype system that uses a semantic representation of energy policy knowledge containing not only quantitative data related to technical, economics and environmental aspects, but also qualitative information related to social and political issues. Goals associated with sustainability issues are translated, via the ontology, to equality and inequality constraints using description logic. Currently, goals can be set with regards to the following criteria: capital cost, life-cycle energy payback, CO_2 emissions reduction, land utilization, water conservation, public health risk, new jobs creation, social acceptability and energy supply security. The semantic representation is integrated into a prototype energy modeling system used to formulate an optimized electricity generation mix.

A scenario-based analysis has been chosen to address the uncertainties associated with the future. In addition, a bottom up approach was chosen to represent technology from an engineering point of view. Furthermore, optimization was selected as the most appropriate method for the allocation of energy resources.

The architecture of the prototype *Electricity Generation Modeling System* is shown in Figure 1 with more detail provided in [4]. It consist of:

- A *Graphical User Interface* for user input (quantitative and qualitative policy goals, and the scenario parameters) and output display; it is written in Java and uses the Swing widget toolkit JfreeChart API [5].
- A *Semantic Representation Module* containing the domain ontologies, a mathematical-based set of engineering models and access to data from the web. The ontologies are coded in the OWL Web Ontology Language (OWL) [6] using Protégé [7] and the engineering models are coded in Java.
- A *Knowledge Inference and Expression Builder* which derives inferences form the ontologies to construct a set of mathematical expressions by translating each policy goal into one or more logic queries using Java and applying the FACT++ logic reasoner [8] for the inference. The output is generated as commands for Mathematica^{TM}.
- A *Support Module* containing a web crawler, data clustering and graphing components, and other parts of the program enabling the interoperation of the modules.

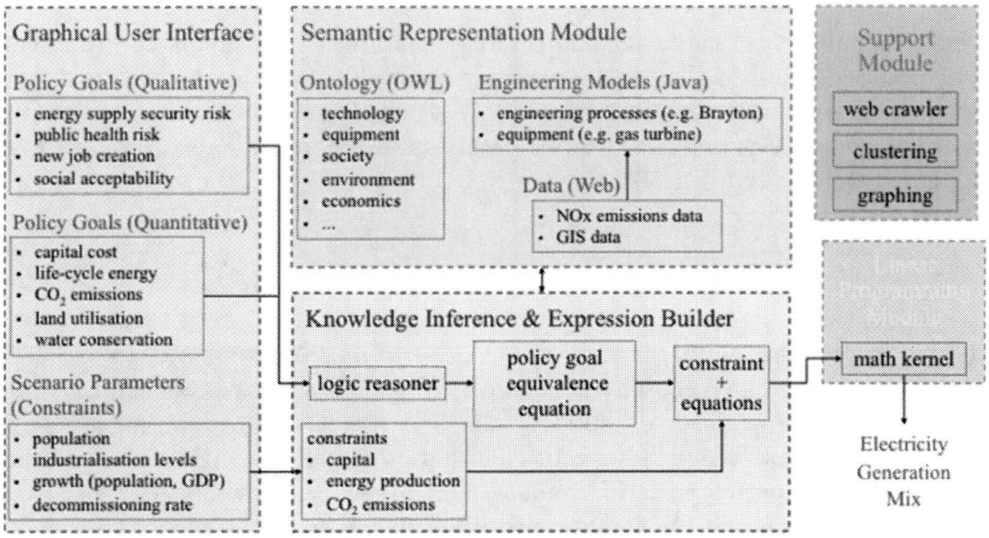

Figure 1. System architecture.

In the case of the prototype *Electricity Generation Modeling System* the semantic representation module consists of four ontologies and links to engineering models (e.g. mass and energy balances). The representation of the core knowledge is formed by the *Energy Policy Ontology* and the *Value Partition Ontology* with support from engineering models, while the *Scenarios Ontology* and *Equation Construction Ontology* constitute the representation of the application knowledge.

A simple example of the use of the ontologies structure and inference mechanisms in the generation of the optimization model is presented with the help of Figure 2.

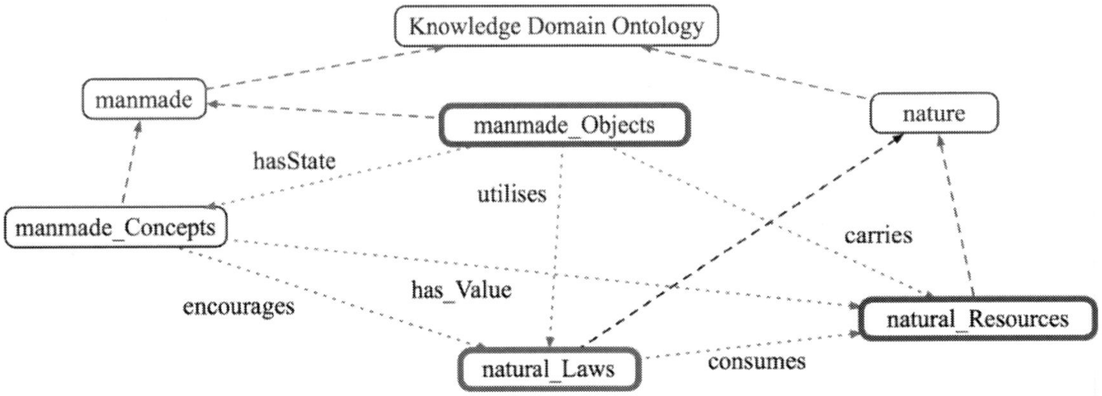

Figure 2. Top level classes in the Knowledge Domain Ontology.

For example, when a coal power plant (a `manmade_Object`) is considered as one of the alternative technologies, it will be established that one of its parts is a steam boiler (another `manmade_Object`) where a combustion process takes place (a `natural_Law`), and that the combustion process reacts coal and oxygen in the air to produce carbon dioxide (coal and air being `natural_Resources`), i.e.

`coal_Power_Plant` **consists_Of** `coal_steam_Boiler`
`coal_steam_Boiler` **utilises** `stoichiometric_Combustion`
`stoichiometric_Combustion` **consumes** `coal`
`stoichiometric_Combustion` **requires** `air`
`stoichiometric_Combustion` **emits** `CO2`
...

In other words, the prototype is able to establish that coal power plants emit CO_2 (but that this is not the case with photovoltaic or hydroelectric power generation); these inferences are directly relevant to policy goals and targets such as those related to CO_2 emission reduction. Additionally, the mass and energy balance equations associated to the combustion process are also retrieved from a library and are added to the optimization model.

Case study: electricity generation in Malaysia

To demonstrate the prototype *Electricity Generation Modeling System*, an example case study for the formulation of an electricity generation mix in Malaysia was conducted, which explores the influence of our extended set of policy goals and targets to the optimization of electricity generation mixes, i.e. the selection of type of electricity generation technologies and their quantity over a time horizon. The prototype can formulate mathematical expressions to be used as an input to the optimization of an electricity generation mix (the objective function, day and night electricity generation requirements, renewable energy generation limits due to intermittency, heat and power requirements, CO_2 production, etc.). It demonstrates that, for a set of scenario drivers (population size and growth, industrialization levels, GDP growth and decommissioning rates), an electricity generation mix that fulfils the constraints can be obtained.

The results shown in Figure 3 correspond to Scenario 1, a business-as-usual (BAU) case as outlined in the 10th Malaysian Plan (a plan that is mainly cost-driven with some concerns about energy supply dependence) [9]; data for solar radiation, wind velocity, maximum hydroelectric potential and biomass type and availability for Malaysia were taken from multiple sources and used to calibrate the electricity generation plan. The energy sources considered were (1) coal, (2) gas and oil, (3) nuclear, (4) renewables and (5) hydroelectric. Although the prototype system provides separate results for gas, oil and for the different types of renewable energy (wind, solar and biomass), the results shown in Figure 3 are consolidated to enable an easier comparison with the current plans of the Malaysian government as reported in [9]. The general trend moves form an initial situation in 2005 where most of the energy comes from fossil fuels (coal, gas and oil) to a situation in 2035 where the consumption of fossil fuels is maintained but the additional demand, due to population growth and industrialization, is met mainly by hydroelectric power, renewable sources and some nuclear power.

Policy Goals	Levels of Implementation or Concern		
	L	M	H
Cost			X
Life-Cycle Payback			X
CO_2 emission reduction	-	-	-
Land utilisation	-	-	-
Water consumption	-	-	-
Public Health Risk	X		
New Job Creation	X		
Social Acceptability	-	-	-
Energy Supply Risk		X	

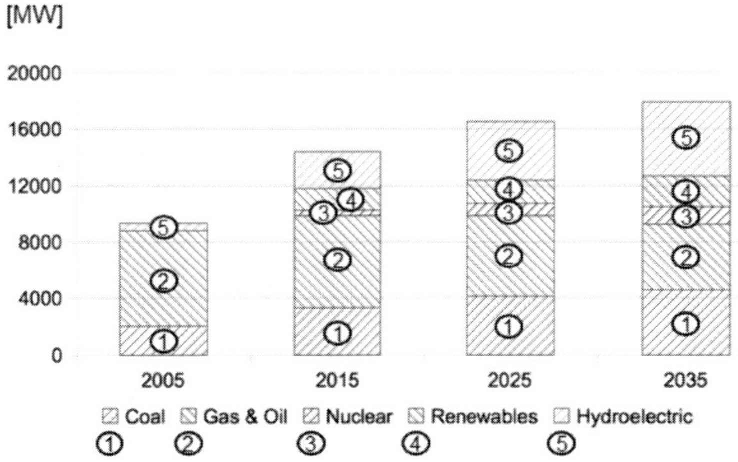

Figure 3. Proposed energy mix for Malaysia, Scenario 1 (BAU).

Despite the approximate nature of the models in the prototype there was a fair degree of similarity between the results generated from the *Energy Model* created by the prototype *Electricity Generation Modeling System* and the official Malaysian government plans. These results lend credence to the capability of the prototype to formulate electricity generation mixes that are acceptable and believable. The results from two other scenarios, i.e. EcoFriendly and Nuclear, are reported in [10].

Case study: water treatment alternatives for Penang, Malaysia

Similarly, optimization models can aid in the formulation of water policies, including those designed to ensure water supply security and sustainability. A prototype water modeling system based on a semantic representation of water policy has also been developed. The objective of this second application area was to demonstrate the generality of the overall architecture of the system: the *Energy Ontology* was substituted by a *Water Ontology*, the library of electricity generation engineering models was substituted by its equivalent for water treatment and supply, and the user interface was modified to reflect the new application (the new set of goals). The rest of the system, i.e. the *Knowledge Inference*, *Equation Builder* and *Linear Programming* modules were not modified. It only took approximately eight weeks to develop the *Prototype Water Modeling System*.

To demonstrate the concept of the second prototype, a case study for the state of Penang in Malaysia was developed. From a set of water policy goals, a water model is created automatically and subsequently optimized to generate plans for a water treatment and supply system for the region. In this case a single optimization period from 2000 to 2030 was used as some of the water supply options have a life span of 30 years or more.

Figure 4 shows the results of four scenarios based on six different water supply sources: (1) takes water directly from the Sungai Muda river to be used for agriculture, (2-3) both involve dam construction for fresh water catchment and its subsequent treatment by conventional means, (4) involves waste water or brackish water treatment using reverse osmosis, (5) waste water treatment by chemical and biological methods, and (6) brine or sea water desalination. The bars in Figure 4 represent the state of affairs in the year 2000 and three future scenarios for 2030.

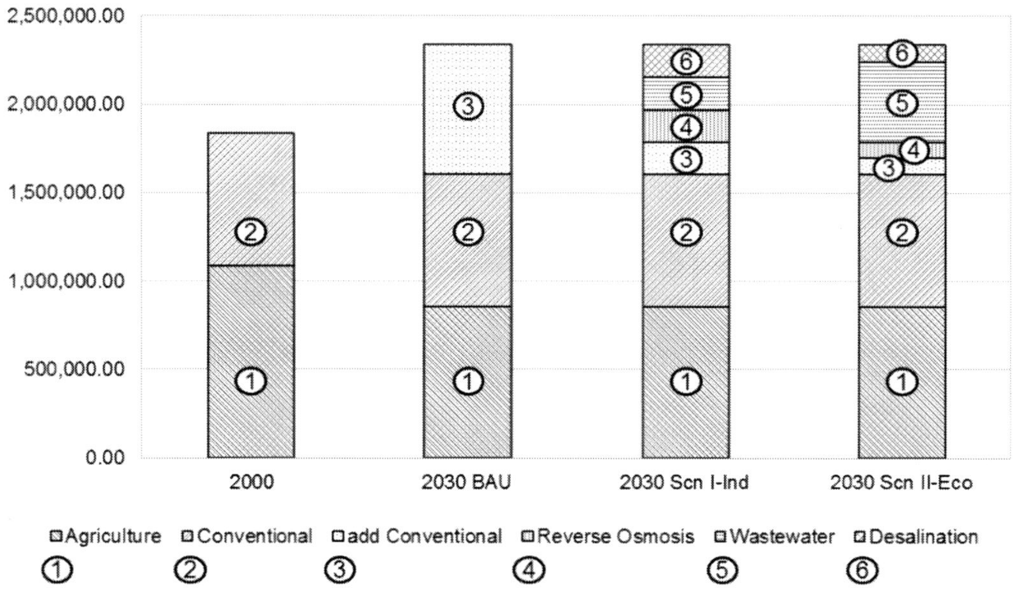

Figure 4. Proposed water treatment mix (m^3 / day) for agricultural and potable water supply for the state of Penang. Current situation (year 2000), BAU, Independence and Ecological scenarios (2030).

The '2000' bar in Figure 4 reflects the current situation whereby water for agriculture is diverted from the Sungai Muda river (coming from the State of Kedah) and the rest of the water requirements are met by fresh water catchment in existing reservoirs and their subsequent treatment [11].

The '2030 BAU' bar corresponds to the current government plans consisting in the construction of two new dams, Beris and Rui, to meet the additional water demand in 2030 [12].

The '2030 Scn I-Ind' and '2030 Scn II-Eco' bars explore the situation where either (a) Kedah's water requirements increase significantly and requires Penang to pay for the draw of water to compensate for economic loss to Kedah; and/or (b) possible draughts because of seasonal lack or rainfall or due to climate change, result in a reduction of the total downstream flow to the treatment plants. Both of these last scenarios shift away from the current practice of dam construction and reduce the amount of water withdrawn from the Sungai Muda river. In our estimation, based on published data, the Sungai Muda river currently has a spare capacity for an additional 300,000 m^3 of water per day, making it unnecessary to construct the Beris and Rui dams if other sources of water to meet future demand are found.

In particular, '2030 Scn I-Ind' addresses the case where Penang wishes to increase its water security and thus increase its water independence above any other considerations. This scenario shifts towards a balanced use of waste water treatment and desalination, technologies that are less dependent on water supply from a river or rain, thus increasing water security during the draught months of January, February, June and July. The '2030 Scn II-Eco' scenario is similar but it additionally involves more stringent limits to ecological and environmental impacts resulting in improved re-use and recycle of waste water.

In terms of both, financial and energy costs, the BAU scenario is the lowest, Scn I-Ind holds the middle ground and Scn II-Eco is the highest. This is expected as the installation of waste water treatment and of desalination facilities is more expensive than the treatment of fresh water from a river or rain. However, the additional costs would allow for a degree of water supply independence during the draught seasons and would alleviate water supply shortages due to river contamination episodes such as the ones that have been reported in local newspapers [13].

Discussion and Conclusions

The *Optimization Modeling System* presented in this paper, addresses the disconnect between the policy maker and the optimization model itself. We have sought to extend the boundaries of the formulation of optimization models used for policy evaluation and analysis with particular emphasis given to energy and water models. This has been achieved by semi-automating part of the function and the associated tasks of the modeler. This modeling system creates a *Model* by automatically formulating the necessary mathematical expressions for optimization based on both quantitative and qualitative inputs from the user. The modeling system achieves this task through the combined use of a semantic representation, knowledge inference, mathematical techniques and procedural programming.

Our modeling systems break new ground with regards to the use of ontologies. As far as can be ascertained, this is the first use of ontologies as a method to formulate mathematical expressions through inference in the knowledge domain of power generation and water treatment systems. This allows for a reduced dependence on manual programming of mathematical expressions where human intervention results in lack of uniformity and human error that can cause inaccurate results. From an application standpoint, the *Prototype Modeling Systems* and the *Models* formulated offer the advantage of a single evaluation platform, explicit model documentation, rapid model reconfiguration, ease of use, and an opportunity to explore many more alternative policy options within a very short time span.

Our modeling systems, however, do have several limitations. In particular, while the *Optimization Modeling System* accepts qualitative inputs, the analysis of these inputs remains limited. From an application standpoint, the ability of the *Optimization Modeling System* to process qualitative data is limited by the richness of the ontology, which in turn is limited by the available knowledge for codification. In addition, the *Prototype Modeling System* and the *Model* occasionally suffer from the vulnerability of an infeasible solution space for which only limited problem finding and solving functions have been incorporated.

Our proposed modeling system would benefit from additional research in the areas of semantic representation and its integration to linear programming. The four main areas of improvement are: (a) the expansion of the knowledge codification paths, (b) the improvement in the integration between the ontology and the engineering models, (c) the enhancement of the representation quality of the ontology, and (d) the adoption of a better multiobjective function optimization method.

In conclusion, we view the prototype modeling systems introduced in this paper as a proof-of-concept showing their potential use by policy makers in their efforts to formulate, analyze and evaluate sustainable and effective policies.

References

1. Seebregts A.J., G. Goldstein and K. Smekens (2001), "Energy/environmental modelling using the MARKAL family of model", In *Operations research proceedings* (OR2001), Duisburg, Germany; Sep 3–5.
2. Messner S., A. Golodnikov and A. Gritsevskii (1996), "A stochastic version of the dynamic linear programming model MESSAGE III", *Energy* 21:775–84.
3. Akimoto K., T. Tomoda , Y. Fujii and K. Yamaji (2004), "Assessment of global warming mitigation options with integrated assessment model DNE21", *Energy Econ* 26:635–53.
4. Chee Tahir, Aidid and René Bañares-Alcántara, (2012), "A knowledge representation model for the optimisation of electricity generation mixes", *Applied Energy* 97:77-83.
5. Object Refinery Limited (2005), "Welcome To JfreeChart!" [online] www.jfree.org/jfreechart
6. Bechhofer S., F. van Harmelen, J. Hendler, I. Horrocks, D.L. McGuinness, P.F. Patel-Schneider and L.A. Stein (2004), "OWL Web Ontology Language Reference", *W3C Recommendation 10 February 2004*, [online] http://www.w3.org/TR/owl-ref/
7. Knublauch H., R.W. Fergerson, N.F. Noy and M.A. Musen (2004), "The Protégé OWL Plugin: An Open Development Environment for Semantic Web Applications", In Mcllraith S.A., D. Plexousakis and F. Harmelen eds., *Proceedings of the Third International Semantic Web Conference*, Hiroshima, Japan 7-11 November.
8. Tsarkov D. and I. Horrocks (2006), "FaCT++ Description Logic Reasoner: System Description", *LNCS 4130/2006 Springer*, pp.292-297.
9. Economic Planning Unit (2010), "The Malaysian Economy In Figures 2010", *Putrajaya: Prime Minister's Department*.
10. Chee Tahir A. (2012), "A Framework for the Semantic Representation of Energy Policies related to Electricity Generation", DPhil thesis, *Department of Engineering Science, University of Oxford*.
11. Perbadanan Bekalan Air Pulau Pinang (2011), "Penang State Water Supply Infrastructure" [online] http://www.pba.com.my
12. Department of Irrigation and Drainage (2000), "National Water Resources Study 2000", *Putrajaya: Ministry of Natural Resources and the Environment*.
13. The Star (2003), "Checking troubled waters", *The Star Newspaper*. [online] http://www.ecologyasia.com/news-archives/2003/feb-03/thestar_20030208_3.htm

The "*Art*" of Modeling in Process Systems Engineering

Vicente Rico-Ramirez and Rasiel Toledo-Hernandez
Instituto Tecnologico de Celaya, Departamento de Ingenieria Quimica, Av. Tecnologico y Garcia Cubas S/N, Celaya, Guanajuato, Mexico 38010
vicente@iqcelaya.itc.mx, rth199@hotmail.com

Introduction

The first part of this paper focuses on the development of the ASCEND system as one of the results of the effort and leadership of Professor Art Westerberg for designing and implementing an efficient user-friendly environment for the creation and solving of complex engineering models. Then, the second part is concerned with the use of fractional calculus as an emerging modeling tool in process systems engineering. We will intend to demonstrate that the dynamics of some systems are described more accurately by using fractional order derivatives. In particular, we show that the kinetics of some reactive systems displaying atypical behavior, such as some biological processes, can be represented by fractional order differential equations (FDE). A simple case-study including fractional kinetics is used to describe the mathematical issues involved and the effect of the fractional order derivatives in the model formulation.

Keywords: ASCEND, fractional calculus, fractional kinetics

The Development of the Advanced System for Computations in ENgineering Design

The ASCEND program has been one of the main results of the significant effort and leadership of Professor Art Westerberg in the area of PSE, as well as the work of his graduate students at Carnegie Mellon University and many external collaborations that participated in the development of the system since 1978. ASCEND is both an equation-based object-oriented modeling environment and a system for solving large sets of linear, non linear and differential equations. Five versions of the system (ASCEND I, II, III, IIIc and IV) are clearly identified. The list of the ASCEND developers is long; among them: Dean Benjamin (1978-1982), Michael Locke (1981), Tom Berna, Larry Gaydos (1984), Peter Piela (1984), Tom Epperly, Karl Westerberg, Roy McKelvey, Kirk Abbott, Benjamin Allan (1992), Bob Huss, Boyd Safrit, Mark Thomas, Vicente Rico-Ramirez, Kenneth Tyner, Jennifer Perry, John Pye (2005) and Krishnan Chittur. The history behind the development of ASCEND, which was born and will remain as a free academic software, is a clear example of professor Westerberg´s philosophy, which inspires hard work, creativity and innovation. ASCEND IV provides the user with many useful tools for computations and modeling in engineering. The conceptual input of Professor Westerberg resulted in many features in ASCEND that have later been incorporated into commercial simulation programs. In ASCEND, models are written in a modeling language that combines declarative variables and equations with procedural methods; because of its object-oriented capabilities, complex models are constructed from simpler sub-models. The ASCEND tools allow one to structurally analyze the models, examine their behavior and work out how they can be solved. My talk will provide a personal opinion about the significance of the development of the ASCEND IV system from the perspective of a developer, user, and both a student and a teacher of the design course in the area of process system engineering.

An Emerging Modeling Tool: Fractional Calculus

Fractional calculus is a generalization of ordinary calculus which introduces derivatives and integrals of fractional order. Fundamental physical considerations favor the use of fractional models; in particular, fractional order derivatives and integrals allow the description of memory and hereditary effects. Several works report the fundamental theory of fractional calculus[1]. Recent literature reports have shown that fractional calculus is a powerful modeling tool to represent the behavior of a number of electrical, mechanical and chemical dynamic systems[2].

Fractional derivation of order α of the function *f(t)* can be given in terms of the Caputo definition:

$$ {}_0^C D_t^{\alpha} f(t) = \frac{1}{\Gamma(1-\alpha)} \int_0^t \left\{ (t-\tau)^{-\alpha} \frac{d}{d\tau} \left[f(\tau) \right] \right\} d\tau \tag{1} $$

where Γ is the gamma function. Such a definition includes a convolution integral between an ordinary derivative of the function and a power-law of time. The convolution integral is precisely the mathematical tool which accounts for the memory effects of the system. The definition given by (1) introduces the fractional operator *D*; the lower indexes represent the limits of the integral and the positive sign of the upper index α indicates derivation (as opposed to integration). Upper index *C* stands for Caputo. This work intends to illustrate the modeling capabilities of fractional calculus to represent the behavior of simple reactive systems where memory effects could be significant.

Fractional Models for Reactive Systems which Exhibit Atypical Kinetics

One of the goals of this work is to show how fractional calculus can be used as a modeling tool for representing the dynamics of reactive systems which exhibit atypical kinetics that do not necessary follow the classical mass-action form. On this area, previous works report successful applications of fractional calculus in drug release/dissolution processes[3]. We have studied the suitability of fractional models for several reactive biological systems. In this abstract we show the results only for the case of one fermentation process.

Illustrative Example

The case-study corresponds to the fermentation reaction involved in the Tequila production process. The experimental data and the conventional empirical model have been taken from the word of Herrera et al.[4] The kinetics of this kind of processes is generally represented by Michaelis-Menten-like expressions. The conventional mathematical model of the dynamics of the process is given as follows:

$$ \frac{dx_1}{dt} = (\mu - D) x_1 \tag{2} $$

$$ \frac{dx_2}{dt} = -\left(Y_{xs} - Y_{ps}\right) \mu x_1 - m_s x_1 - \left(-S_0 + x_2\right) D \tag{3} $$

$$\frac{dx_3}{dt} = \alpha\mu x_1 - Dx_3 \tag{4}$$

where x_1 is the biomass concentration, x_2 is the substrate concentration and x_3 is the product concentration, and μ is defined as:

$$\mu = \frac{\mu_{\max} x_2}{k_s + x_2 + x_2^2 k_i}\left(1 - x_3 k_p\right) \tag{5}$$

Table 1 provides the values of the rest of the parameters used to describe the dynamic behavior of the system.

Table 1. Parameters of the dynamic model for the tequila production process

Parameter	Value
μ^{max}	0.2735
k_s	9.2891
k_i	0.0029
k_p	0.0179
Y_{ks}	11.7
Y_{ps}	4.7125
m_s	0.0017
α	7.5498

We propose that, instead of Michaelis-Menten type of kinetics, fractional expressions can be used to represent the dynamic behavior of the system. To generate the alternative fractional model, we used a combination of non linear fitting tools provided in Matlab with an integration technique for fractional differential equations developed by Diethelm et al.[5]. Such technique is a modification of the Adams-Bashforth-Moulton method, which has also been coded in Matlab.

Figure 1 shows the results for the substrate concentration against time; the experimental data are plotted along with the results obtained by the conventional Michaelis-Menten kinetics and the fractional model. A clear advantage is the simplicity of the fractional kinetics expression as given by Equation (6) for the substrate concentration with α_2=0.872 and k_2= 0.0557.

$$\frac{d^{\alpha_2} x_2}{dt^{\alpha_2}} = -k_2 x_1 x_2 \tag{6}$$

A simple comparison between the simplicity of Equation (6) against Equation (3) shows that fractional order kinetics may indeed simplify the complexity of typical semi-empirical models.

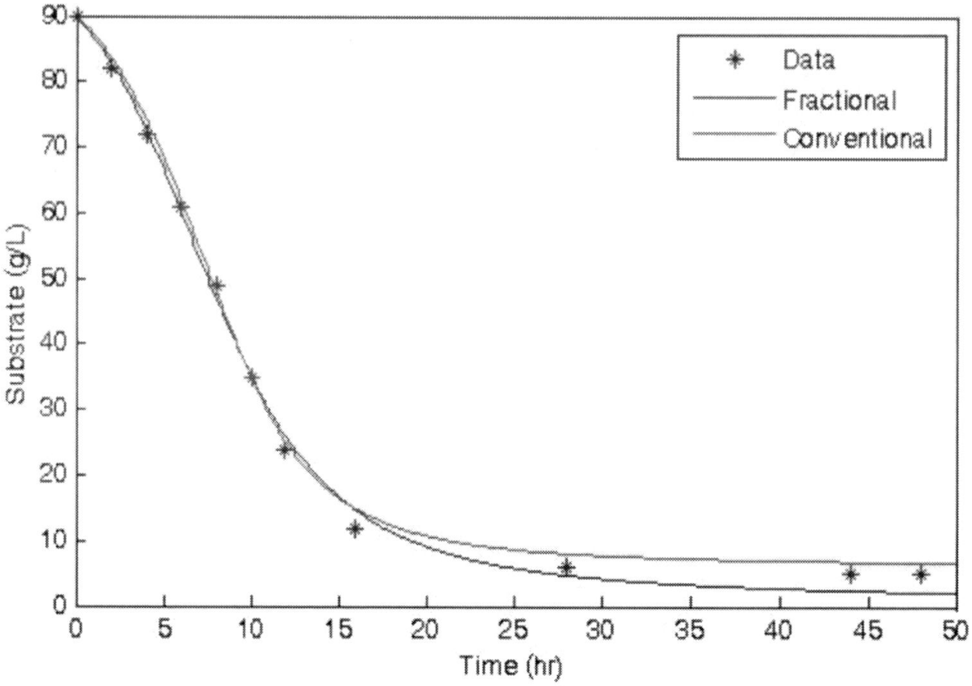

Figure 1. Substrate concentration against time

Discussion

Modeling is an "Art". This paper is concerned with modeling in two aspects. First, I provide a brief overview of the development of the ASCEND IV modeling environment. ASCEND IV is without a doubt one of the main examples of Art Westerberg´s legacy and leadership in the area of modeling and flowsheeting simulation in PSE. In the second part of the paper, our results have shown the potential of fractional calculus as a modeling tool for practical applications in the area of process systems engineering. In particular, the kinetics expressions of some biological reactive systems, which are typically represented by semi-empirical models, have been formulated through fractional differential equations.

References

1. Podlubny I., 1999, Fractional Differential Equations, Academic Press, San Diego, CA, USA.
2. Magin, R. L., 2006, Fractional Calculus in Bioengineering, Begell House Publishers, Redding, CT, USA.
3. Dokoumetzidis A., P. Macheras, 2009, Fractional Kinetics in Drug Absorption and Disposition Processes, *J. Pharmacokinet. Pharmacodyn.*, 36, 165-178.
4. Herrera, E. , G. Hernández, A. Moran, M. Flores, J. Cordova, R, Femat and D. Díaz-Montaño, 2009, Modelado Matematico del Cultivo en Lote y Continuo del Proceso Fermentativo del Tequila, in the proceedings of the 2009 Annual Conference of the Mexican Association of Automatic Control, Zacatecas, Mexico
5. Diethelm, K., N. J. Ford, A. D. Freed, Y. Luchko, 2005, Algorithms for Fractional Calculus: A Selection of Numerical methods, *Comput. Methods Appl. Mech. Engrg.*, 194, 743-773.

(127e) The Transition From Florida to Pittsburgh

Jigar Shah, Institute for Industrial Productivity, Washington, DC

This presentation recounts the journey of one of Art's graduate students in the field of energy efficiency - from operations research based
methods on reducing energy use in the late 70's to now working with industry and governments in China, India and the USA to help design policies and
programs for climate change mitigation.

Art Westerberg's Contributions in Design Research

Abstract ID# 288360

Steven Fenves, University Professor of Civil and Environmental Engineering Emeritus
Carnegie Mellon University

Abstract. This talk reviews Art Westerberg's contributions to multidisciplinary design research through his work at Carnegie Mellon University, particularly through the NSF Engineering Design Research Center.

The context

This section reviews the context in which Art Westerberg operated in the years 1974 to 1997.

In the beginning ... Herbert Simon, Nobel prize-winning economist, computer scientist and cognitive scientist, often walked between home and work with Herb Toor, a prominent chemical engineer and Dean of Carnegie Institute of Technology (CIT), the engineering college of Carnegie Mellon University. A frequent subject of conversation was the possibility of implementing the concepts of Simon's book, *The Sciences of the Artificial*, as a generic, teachable program of design, in contrast to the discipline-dependent, anecdotal teaching of "design" that prevailed at that time.

After I arrived at CIT in 1972 as Head of the Civil Engineering Department, Dean Toor informed me of my additional responsibility: initiating a program based on the two Herbs' conversations. It was a period of extreme fiscal restraint at Carnegie Mellon, and the idea of a Center for the Theory of Design, with three or four new faculty positions, was a non-starter. However, I found a few early allies, particularly Chuck Eastman, who held appointments in Architecture, Computer Science and the School of Urban and Public Affairs. More importantly, that first summer I met Art Westerberg and Stephen Director, both from Florida. I reported to Dean Toor that these two ought to be hired in the Chemical and Electrical Engineering departments, respectively, whether we have a Center or not. These appointments materialized, eventually augmented by the appointments of Gary Powers and Sarosh Talukdar, again in Chemical and Electrical Engineering, respectively.

The Design Research Center. With an initial critical mass at hand, the Design Research Center became a reality in 1974. Gary Powers served as the first director, followed by Art Westerberg, myself, and Sarosh Talukdar, in turn. The Dean's office provided a modest seed funding. A four-year grant from the Alfred P. Sloan Foundation, obtained primarily through Art's efforts, provided additional support. The center eventually comprised 22 faculty members with strong interests and excellent credentials in design, in all engineering disciplines and several related fields. The Center explored themes in engineering design of common interest, strove to penetrate beyond the jargon of the respective engineering disciplines, and attempted to transpose design methods developed in one discipline to other disciplines. Soon, interdisciplinary courses and joint interdepartmental research projects developed, followed by such trappings of a Center as a distinguished lecturer series, a seminar series, and a distribution service for the Center's research reports. In sum, in 12 years the EDRC did not make a breakthrough in a theory of design or even a science base in support of design, but gained the increase in maturity and cohesion that made possible a dramatic expansion of the Center's scope.

In 1984, the National Science Foundation (NSF) announced its new program of Engineering Research Centers (ERCs). It eventually received 160 proposals. At Carnegie Mellon (CMU), Dean James Williams decided to put the college's resources behind one proposal only, and held a mini-competition among competing faculty groups. DRC argued that 22 faculty members, about 15% of the college, were engaged in the field – a criterion in the original NSF report on the ERCs but not contained in the actual request for proposals. The argument carried, and CMU submitted a proposal for an Engineering Design Research Center. CMU was among the finalists but was not funded the first year. Based on encouraging reviews, we submitted a revised proposal the following year, and received the award in 1986.

The Engineering Design Research Center. The Engineering Design Research Center (EDRC) operated for 11 years, from 1986 to 1997. The transition from the small, entirely self-regulated, and minimally managed ERC to the large scale (up to $9M yearly budgets), closely regulated by NSF staffers, and tightly managed EDRC was much more difficult than anticipated. Art Westerberg was the first director, followed by Fritz Prinz and then Dan Siewiorek. An Associate Director and a full complement of administrative staff supported the director, and an Industrial Advisory Board, meeting twice a year, provided guidance. Yet, much of the collegial flavor of the DRC carried over to the EDRC. In particular, the Director's Committee - comprised of the director, the thrust area leaders, a few senior faculty members, and a representative of the research staff – planned the activities, prepared the reports and annual renewal requests to NSF, and hosted the NSF annual site visits, acting essentially as a "collective directorship."

EDRC's design research contributions are too numerous to list here. EDRC has created and improved numerous design tools for optimization and synthesis, environments and frameworks for collaborative and multidisciplinary design, shared geometric representations and databases, rapid prototyping environments, and some of the earliest applications of artificial intelligence methods to engineering design. EDRC faculty have taken a lead role in introducing design courses into the curriculum, including knowledge-based and optimization approaches for the synthesis of engineering systems; design process synthesis; interdisciplinary design and fabrication; and rapid prototyping. The last census of EDRC graduates shows that 258 students have completed graduate degrees while involved with the EDRC (116 MS/ME degrees and 142 PhD degrees). Of the PhD recipients, roughly half worked in industry, and the other half in tenure-track faculty positions.

After 11 years of existence, EDRC evolved into a new entity, the Institute for Complex Engineered Systems (ICES), chartered to define, seed, and sustain interdisciplinary research projects that build upon the existing strength of the College of Engineering.

Art Westerberg's contributions to design research

Art Westerberg has made seminal contributions to many aspects of design research. Many of these aspects overlap. The classification of the aspects of Art's design research presented here is the author's, as is the assignment of specific research papers to these aspects (Ignacio Grossmann kindly supplied the list of Art's design research-oriented papers). Art would undoubtedly provide an argument on both the classification and the paper assignments. The primary objective is to demonstrate the breadth and comprehensiveness of Art's design research contributions.

Design and synthesis concepts. (See references 1. – 6. in the Appendix) Art Westerberg never produced a "Theory of Design," nor did he have the ambition to do so. Rather, he made two strong arguments. First, that design needs to be defined in the broadest sense, from the perception of a need or opportunity to the disassembly and recycling of the artifact (what Art called "design in the large"), rather than the traditional view of engineering design, from user needs statement to specification of the artifact to be built ("design in the small"). Second, that synthesis is the key design activity, for the artifact as well as the design process itself; methods for the selection among synthesized alternatives are readily available, such as the application of utility theory.

Equation-based modeling. (See references 7. – 14.) The modeling of systems described by very large sets of sparse, structured equations has been a recurrent theme in Art's work starting with his doctoral dissertation. The ASCEND system is undoubtedly the most comprehensive and widely used toolkit in this area (references 10. – 12.). ASCEND features a language in which models are created, hierarchies are defined, and compatible parts are merged, with deferred binding throughout, allowing for the reuse of modules.

Expert systems and Artificial Intelligence applications. (See references 15. – 28.) The proximity of EDRC to the Carnegie Mellon Computer Science Department and frequent collaborations between the two faculties created a fertile ground for research in engineering applications of AI. The initial emphasis at EDRC, as elsewhere, was on knowledge-based expert systems. Westerberg and his students developed expert systems for the synthesis of separation systems, physical property prediction, catalyst selection, and separation process design. Invariably, the work involved methodological extensions of the expert systems' reasoning framework, as well as the collection and codification of domain knowledge. Subsequent work came closer to then-current research issues in Artificial Intelligence: informal modeling, shared memory, multiple knowledge sources, machine learning, and agent-based architectures.

Concurrence and social aspects of engineering. (See references 29. – 45.) Art's perennial use of project groups in all his research eventually made support of concurrent work and of the social aspects of engineering design the key concerns of his research. Shared environments, asynchronous communication, shareable information models, and similar facilities needed and used in collaborative design became themselves the foci of research. These concerns culminated in the development of the *n*-dim support environment designed to assist in capturing, organizing, and sharing information and in integrating diverse tools that can operate on this information.

Art's further contributions

The two preceding sections sketched the environment in which Art Westerberg operated in the years from 1974 to 1997, and Art's research program in design within that environment. To provide a full picture, the two brief sections below summarize Art's further contributions.

Art as an administrator, Art served as director of both DRC and EDRC, followed by a stint as Head of the Chemical Engineering Department. In all three of these positions, Art practiced "participatory democracy" to the maximum extent allowed by the environments of the college and the university. He kept a wide open door, put great emphasis on faculty recruitment and retention, fought for the recognition and reward of non-tenure track research faculty, and treated faculty, staff, and students as valuable friends. The success of DRC and EDRC in weathering the turmoils precipitated by the diverse participants from many disciplines and cultures is due largely to Art's equanimity in facing and extinguishing these flareups.

Art as an engineering educator, Art has always been at the forefront design education, interdisciplinary education, and the introduction of new educational technology. Art participated substantially in essentially all the DRC and EDRC courses mentioned above. He brought faculty from the Department of Design in the College of Fine Arts and the Graduate School of Industrial Administration into teaching a product development course.

Summary

Art Westerberg was a major force in shaping design research and design education at Carnegie Mellon University through his leadership positions in the Design Research Center (1974 – 1986) and the Engineering Design Research Center (1986 - 1997). In turn, Art took full advantage of the diversity of disciplines, professional views and attitudes, and individual research interests of the participants to refine and even redirect his own research activities. It has been a pleasure to share a close association for a quarter of a century.

Appendix. References.

Design and synthesis concepts.

1. Westerberg, A.W., Award Lecture: "Design Research. Both Theory and Strategy," Chem. Engng. Educ., 12-16, cont'd in winter (1982), 62-66, spring (1982).

2. Westerberg, A.W., "Synthesis in Engineering Design," Comput. Chem. Engng., Vol. 13, No. 4/5, pp 365-376 (1989).

3. Demes. G.H., S.J. Fenves, I.E. Grossmann, C.T. Hendrickson, T.M. Mitchell, F.B. Prinz, D.P. Siewiorek, E. Subrahmanian, S. Talukdar and A.W. Westerberg, "The Engineering Design Research Center of Carnegie Mellon University," Proc. IEEE, Vol. 81, No. 1, pp 10-24 (1993).

4. Wahnschafft, O.M., J.P. LeRudulier and A.W. Westerberg, "A Problem Decomposition Approach for the Synthesis of Complex Separation Processes," Ind. Eng. Chem. Res., Vol. 32, pp 1121-1141 (1993).

5. Cheng, L., E. Subrahmanian, A.W. Westerberg, "Design and Planning under Uncertainty: Issues on Problem Formulation and Solution," AIChE Annual Mtg, Nov (2001); Chem. Engng., Vol. 27, pp 781-801 (2003).

6. Westerberg, A.W., "A retrospective on design and process synthesis," Comp. Chem. Engng., Vol. 28, No. 4, pp 447-458 (2004).

Equation-based modeling.

7. Westerberg, A.W., and S.W. Director, "A Modified Least Squares Algorithm for Solving Sparse nxn Sets of Nonlinear Equations," Comp. Chem. Engng., Vol. 2, pp 77-81 (1978).

8. Director, S.W., A.W. Westerberg, and M.R. Lightner, "Simulation Procedures for Large-Scale Electronic Systems," J. Inst. Electronics and Telecom. Engrs, Vol. 24, No. 7, pp 271-285 (1978).

9. Chao, N.H., S.J. Fenves, and A.W. Westerberg, Applications of a Reduced Quadratic Programming Technique to Optimal Structural Design," in Atrek, A., R.H. Gallagher, K.M. Ragsdell, and O.C. Zienkiewicz, New Directions in Optimal Structural Design, Wiley, New York, pp 412-428 (1984).

10. Piela, P.C., T.G. Epperly, K.M. Westerberg, and A.W. Westerberg, "ASCEND: An Object Oriented Computer Environment for Modeling and Analysis. Part 1 - The Modeling Language," Comp. Chem. Engng, Vol. 15, No. 1, pp 53-72 (1991).

11. Krishnan, R., P. Piela and A.W. Westerberg, "Reusing Mathematical Models in ASCEND," in A. Whinston and C. Holsapple (ed.), Proc. NATO ASI on Decision Support Systems, Il Ciocco, Italy., Springer-Verlag (1991).

12. Piela, P.C., R.D. McKelvey, and A.W. Westerberg, "An Introduction to ASCEND: Its Language and Interactive Environment," J. Management Information Sci., Vol. 9, No. 3, pp 91-121, Winter (1992-93).

13. Abbott, K.A., B.A. Allan, and A.W. Westerberg, "Global Preordering for Newton Equations Using Model Hierarchy," AIChE J, Vol. 43, No. 12, pp 3193-3204 (1997).

14. Allan, B.A., and A.W. Westerberg, "Anonymous Class in Declarative Mathematical Modeling," I&EC Res., Vol. 38, pp 692-704 (1999).

Expert systems and Artificial Intelligence applications.

15. Banares-Alcantara, R., A.W. Westerberg and M.D. Rychener, "Development of an Expert System for Physical Property Prediction," Comput. Chem. Engng, Vol. 9, No. 2, pp 127-142 (1985).

16. Banares-Alcantara, R., Srivam, D., V. Venkatasubramanian, A.W. Westerberg and M. Rychener, "Knowledge-based Expert Systems: An Emerging Technology for CAD in Chemical Engineering," Chem. Eng. Prog., Vol. 81, No. 9, pp25-30, Sept (1985).

17. Banares-Alcantara, R., E.I. Ko, M.D. Rychener and A.W. Westerberg, "DECADE: A Hybrid Expert System for Catalyst Selection," Comput. Chem. Engng., Vol. 11, No. 3, pp 265-277 (1987).

18. Lien, K., G. Suzuki, and A.W. Westerberg, "The Role of Expert Systems Technology in Design," Chem. Engng Sci., Vol. 42, No. 5, pp 1049-1071 (1987).

19. Banares-Alcantara, R., E.I. Ko, A.W. Westerberg, and R.D. Rychener, "DECADE - A Hybrid Expert System for Catalyst Selection. Part II: Final Architecture and Results," Comput. Chem. Engng., Vol. 12, No. 9/10, pp 923-938 (1988).

20. Banares-Alcantara, R., A.W. Westerberg, E.I. Ko, and M.D. Rychener, "The DECADE Catalyst Selection System," in Rychener, M.D. (ed.), Expert Systems for Engineering Design, Academic Press, New York, pp 53-91 (1988).

21. Westerberg, A.W., I.E. Grossmann, S.N. Talukdar, F. Prinz, S. Fenves and M. L. Maher, "Applications of AI in Design Research at Carnegie Mellon University's EDRC," AI in Engineering Design, Vol. 5, No. 2, pp 110-124, April (1990).

22. Wahnschafft, O.M., T.P. Jurain and A.W. Westerberg, "SPLIT: a Separation Process Designer," Comput. Chem. Engng, Vol. 15, No. 8, pp 565-581 (1991).

23. Subrahmanian, E., S.L. Konda, S.N. Levy, Y. Reich, A.W. Westerberg and I. Monarh, "Equations Aren't Enough: Informal Modeling in Design," AIEdam, Vol. 7, No. 4, pp 257-274 (1993).

24. Subrahmanian, E., S.L. Konda, S.N. Levy, I.A. Monarch, Y, Reich, and A.W. Westerberg, "Computational Support for Shared Memory in Design," in Tzonis, A., and I. White (eds.), Automation Based Creative Design: Current Issues in Computers and Architecture, Elsevier Publishers, Amsterdam (1994).

25. Modi, A.K., A. Newell, D.M. Steier and A.W. Westerberg, "Building a Chemical Process Design System within Soar: Part 1 - Design Issues," Comput. Chem. Engng, Vol. 19, No. 1, pp 75-89 (1995).

26. Modi, A.K., A. Newell, D.M. Steier and A.W. Westerberg, "Building a Chemical Process Design System within Soar: Part2 - Learning Issues," Comput. Chem. Engng, Vol. 19, No. 3, pp 334-361 (1995).

27. Davis, J.G., E. Subrahmanian, and A.W. Westerberg, "SCOPE: A Blackboard Model-Based Decision Support System for Crude-oil Trading," Int. J. Intelligent Systems in Finance and Management, No. 8 (1999).

28. Siirola, J.D., S. Hauan and A.W. Westerberg, "Toward Agents-Based Process Systems Engineering: Proposed Agent Framework," Comp. Chem. Engng., Vol. 27, No. 12, pp 1801-1811 (2003).

Concurrent and social engineering.

29. Subrahmanian, E., G. Podnar, and A.W. Westerberg, "Toward A Shared Computational Environment for Concurrent Engineering," Proc. 2nd Nat'l. Symp. Concurrent Engng, CER Center, W.Va. Univ., Morgantown, WV, Feb. 7-9 (1990), in Sriram, D. and R. Logcher (eds.), Lecture Notes on Collaborative Product Development, Springer-Verlag, (1991).

30. Westerberg, A.W., "Human Aspect: Redefining the Role of Chemical Engineers," CEP, Vol. 89, No. 1, pp 60-66, (1993).

31. Subrahmanian, E., R. Coyne, S. Konda, S.N. Levy, R. Martin, I. Monarch, Y. Reich, and A. Westerberg, "Supporting Asynchronous Collaboration for Concurrent Engineering," Proc. Second Workshop on Enabling Technologies in Concurrent Engineering, UWVA, IEEE Press, July (1993).

32. Robertson, J.L., E. Subrahmanian, M.E. Thomas and A.W. Westerberg, "Management of the Design Process: The Impact of Information Modeling," in Biegler, L.T., and M.F. Doherty (eds.), Fourth International Conf. on Foundations of Computer Aided Process Design Conference, AIChE Symp. Series No. 304, 91, CACHE Corp. and AIChE, pp 154-165 (1995).

33. Westerberg, A.W. and the *n*-dim group, "Distributed and Collaborative Computer-aided Environments in Process Engineering Design," in Davis, J.F., G. Stephanopoulos and V. Venkatasubramanian (eds.), Intn'l Conf. Intelligent Systems for Process Engineering (ISP'95), AIChE Symp. Ser. 312, Vol. 92, CACHE, and AIChE, pp 184-194 (1996).

34. Westerberg, A.W., E. Subrahmanian, Y. Reich, S. Konda and the *n*-dim group, "Designing the Design Process," Computers Chem Engng, Vol. 21, Suppl., pp S1-S9 (1997). Presented as Invited Keynote Lecture, PSESCAPE 97 (1997).

35. Subrahmanian, E., Y. Reich, S.L. Konda, A. Dutoit, D. Cunningham, R. Patrick and A.W. Westerberg, "The n-dim Approach to Building Design Support Systems," ASME Design Theory and Methodology Conference, San Diego, Sept. (1997).

36. Cunningham, D., E. Subrahmanian, and A. Westerberg, "User-Centered Evolutionary Software Development Using Python and Java," Proc. Sixth International Python Conf., San Jose, CA, Oct (1997).

37. Reich, Y., S. Konda, E. Subrahmanian, D. Cunningham, A. Dutoit, R. Patrick, M. Thomas, A.W. Westerberg, "Being Agile when Building Design Information Systems with n-dim," Research in Engineering Design," Vol. 11, No. 2, pp 67-83 (1999).

38. Herder, P.M., A.L. Turk, E. Subrahmanian, and A.W. Westerberg, "Challenges for Process Systems Engineering in Infrastructure Design," Comp. Chem. Engng. Vol. 24, No. 2-7, pp 1775-1780 (2000).

39. Westerberg, A.W., and E. Subrahmanian, "Product Design," Invited Plenary Lecture, PSE2000, Keystone, CO, July 17-22 (2000), Comp. Chem. Engng, Vol. 24, No. 2-7, pp 959-966 (2000).

40. Davis, J.G., E. Subrahmanian, S.L.Konda, H. Granger, M. Collins, A.W. Westerberg, "Creating Shared information Spaces for Collaborative Engineering Design," Information Systems Frontier, Vol. 3, No. 3, pp 377-392 (2001).

41. Subrahmanian, E., A.W. Westerberg, S. Talukdar, J. Garrett, A. Jacobson, C. Paredis, C. Amon, P.M. Herder, A. Turk, "Integrating Social Aspects and Group Work Aspects in Engineering Design Education," International Journal of Engineering Education, Vol. 19, No. 1 (2003).

42. Subrahmanian, E., I. Monarch, S.L. Konda, H. Granger, M. Collins, R. Milliken,, A.W. Westerberg, A.W., "Boundary Objects and Prototypes at the Interfaces of Engineering Design," Intn'l J Computer Supported Co-operative Work, Vol. 12, No. 2, pp 185-203 (2003).

43. Davis, J.G., E. Subrahmanian, and A.W. Westerberg (eds.), Knowledge Management: Organizational and Technological Dimensions, Springer Verlag, Heidelberg (2004).

44. Davis, J.G., E. Subrahmanian, and A.W. Westerberg (eds.), "Knowledge Management: Conceptual Foundations, Emerging Directions," in Davis, J., E. Subrahmanian, and A.W. Westerberg (eds.), Knowledge Management: Organizational and Technological Dimensions, Springer Verlag, Heidelberg, pp 1-20 (2004).

45. Davis, J.G., E. Subrahmanian, and A.W. Westerberg, "The Global and the Local in Knowledge Management," J. Knowledge Management, Special Issue on Technology, Vol. 9, No. 1, pp 101-112 (2005).

(127g) Art Westerberg's Contributions to PSE

G. V. Reklaitis, Chemical Engineering, Purdue University, West Lafayette, IN

This talk gives an overview of the major contributions that Art Westerberg made in various areas of PSE, including equation-based modeling, large-scale flowsheeting and optimization, process synthesis, expert systems, and product design.

(146a) A Model-Based Approach for the Construction of Design Spaces in Quality-by-Design

Masako Kishida, University of Illinois at Urbana-Champaign, Urbana, IL and **Richard D. Braatz**, Chemical Engineering, Massachusetts Institute of Technology, Cambridge, MA

This presentation considers the characterization of the set of allowable real parameter uncertainties for the output of a nonlinear system to be located in a prespecified target range. This word "parameter" is used by FDA and in this presentation to refer to any variable held at a fixed value during a single experiment. For example, this could be a parameter in a model, an initial condition such as the initial concentrations in a semibatch chemical reactor, or an operational parameter such as the cooling rate in batch reactor.

This uncertainty analysis problem is motivated by various industrial initiatives for Quality by Design (QbD) (Juran 1992), which have the objective of designing a manufacturing process so as to ensure that the product achieves its desired quality qualifications despite perturbations in the manufacturing process. The QbD literature refers to the set of allowable parameters as the design space (Juran 1992), which in industrial practice is constructed from the data collected from a large number of time-consuming and expensive experiments (e.g., see Garcia-Munoz et al, 2010, for a description of the state-of-the-art in industrial practice and Togkalidou et al, 2001, for an example of experimental design for generating models for pharmaceutical processes). This presentation considers the alternative construction of the design space based on a first-principles, grey-box, or black-box model of the manufacturing process, as fewer experiments are required to construct a model than to directly construct a design space from experiments.

In the control literature, this design space is typically characterized in terms of a robustness margin (Safonov and Athans 1977). Numerous publications quantify the effects of a specified norm-bounded perturbation in a system model on a signal norm on the system output (Wang et al 1992; Zhou et al, 1995) or on outer bounds for state or output trajectories (Durieu et al 2001; Kishida and Braatz 2010, 2011; Kishida et al 2011; Scott and Barton, 2009). The structured singular value and quadratic stability margin quantify the amount of uncertainty allowed to maintain closed-loop stability or a signal norm characterizing output performance (Ma et al 2001, 2002; Hovd et al 2003; Zhou et al 1995; Boyd et al 1997), for linear time-invariant or time-varying systems or linear systems coupled with bounded nonlinear perturbations. This presentation considers nonlinear systems and defines allowable variations in parameters for instantaneous specified bounds on the states or outputs to be satisfied, as this characterization of output performance is used in QbD applications (Juran 1992). The overall approach applies to both open- and closed-loop systems. As the exact generation of a design space is a challenging computational problem for a large number of parameters (Braatz et al 1994), tight upper bounds are determined.

The system output is approximated by a polynomial function (Taylor series expansion) or a rational function (Pad´e approximant) followed by application of the skewed structured singular value to obtain a set of allowable real parameters that ensures that the system output will be in a target set. A target set described by upper and lower bounds on each output is considered here; the same analysis can be used for an ellipsoidal target set by slight modification of the skewed structured singular value analysis (Smith 1990). Both polynomial and rational approximants can be written in the form of a linear fractional transformation, which is used in the next step of the analysis of applying the skewed structured singular value to quantify the allowable uncertainty set. Algorithms are also discussed that generate the algebraic constructions needed by the method as well as model reductions to reduce computational cost (Russell et al 1997; Russell and Braatz 1998).

Several design spaces are constructed for models reported by pharmaceutical companies, including a crystallization operation and a nasal spray.

References:

S. Boyd, L. El Ghaoul, E. Feron, and V. Balakrishnan. Linear Matrix Inequalities in System and Control Theory.

SIAM Press, Philadelphia, PA, 1997.
R. D. Braatz, P. M. Young, J. C. Doyle, and M. Morari. IEEE T. Automat. Contr., 39:1000–1002, 1994.
C. Durieu, E.Walter, and B. Polyak. J. Optimiz. Theory App., 111:273–303, 2001.
S. Garcia-Munoz, S. Dolph, and H. W. Ward. Comput. Chem. Eng., 34:1098–1107, 2010.
M. Hovd, D. L. Ma, and R. D. Braatz. Ind. Eng. Chem. Res., 42:2183–2188, 2003.
J. M. Juran. Juran on Quality by Design: The New Steps for Planning Quality into Goods and Services. The Free Press, New York, 1992.
M. Kishida and R. D. Braatz. Opt. Contr. Appl. Met., 31:433–449, 2010.
M. Kishida and R. D. Braatz. In Proc. of IEEE CDC/ECC, pages 5671–5676, Orlando, FL, 2011.
M. Kishida, P. Rumschinski, R. Findeisen, and R. D. Braatz. Efficient polynomial-time outer bounds on state trajectories for uncertain polynomial systems using skewed structured singular values. Proc. of the Joint Symposium on Computer-Aided Control System Design and Systems with Uncertainty, Denver, Colorado, 216-221, 2011
D. L. Ma and R. D. Braatz. IEEE Trans. Control Syst. Tech., 9:766–774, 2001.
D. L. Ma, J. G. VanAntwerp, M. Hovd, and R.D. Braatz. IEE Proc. - Control Theory & Applications, 149:423–432, 2002.
E. L. Russell and R. D. Braatz. Comput. Chem. Eng., 22:913–926, 1998.
E. L. Russell, C. P. H. Power, and R. D. Braatz. Int. J. Robust Nonlin. Control, 7:113–125, 1997.
M. G. Safonov and M. Athans. IEEE TAC, 22:173–179, 1977.
J. K. Scott and P. I. Barton. Comput. Chem. Eng., 34:717–731, 2009.
R. S. R. Smith. Model Validation for Uncertain Systems. PhD thesis, California Institute of Technology, Pasadena, CA, 1990.
T. Togkalidou, R. D. Braatz, B. K. Johnson, O. Davidson, and A. Andrews. AIChE J., 47:160–168, 2001.
Y. Wang, L. Xie, and C. E. de Souza. Syst. Control Lett., 19:139–149, 1992.
K. Zhou, J. C. Doyle, and K. Glover. Robust and Optimal Control. Prentice Hall, 1995.

(146b) Fluid Bed in a Flash

Otto Scheibelhofer[1], Roland Hohl[1], Sharareh Salar-Behzadi[1], Detlev Haack[2], Kai Koch[3], Stephan Sacher[1], Jose C. Menezes[4] and Johannes G. Khinast[5,6], (1)Research Center Pharmaceutical Engineering, Graz, Austria, (2)Hermes Arzneimittel GmbH, Munich, Germany, (3)Innojet Herbert Hüttlin, Steinen, Germany, (4)Research Center for Pharmaceutical Engineering, Graz, Austria, (5)Institute for Process and Particle Engineering, Graz University of Technology, Graz, Austria, (6)Research Center Pharmaceutical Engineering GmbH, Graz, Austria

Fluid Bed in a Flash

O. Scheibelhofer[1], Roland Hohl[1], Sharareh Salar Behzadi[1], Detlev Haack[2], Kai Koch[3], Stefan Sacher[1], Jose C. Menezes[1], J. G. Khinast[1,4]

[1] Research Center Pharmaceutical Engineering GmbH, Graz, Austria

[2] Hermes Arzneimittel GmbH, Munich, Germany

[3] Innojet Herbert Hüttlin, Steinen, Germany

[4] Institute for Process and Particle Engineering, Graz University of Technology, Graz, Austria

Email for correspondence: otto.scheibelhofer@rcpe.at (O. Scheibelhofer), khinast@tugraz.at (J.G. Khinast)

Fluid Beds are a versatile and widely used unit operation in the pharmaceutical industry. Putting to use their property of very large particle to air contact surface, they are an advanced step for drying, coating and more. A great deal of interest has already been invested, concerning the properties of fluid beds, especially their geometry and flow pattern. Mostly, this is a field for simulations.

Nonetheless, the efficiency of fluid beds is often driven by the experience of their operators. This is a result of fluid beds being very sensitive to subtle changes, which can affect the flow pattern significantly, even dis-enabling the fluid bed process as a whole. Even more so, this increases the difficulty of placing sensors in such environment, making systematic experimental investigations an exhaustive work.

To overcome some of these difficulties, near-infrared spectroscopy seems a natural choice, for two reasons. First, the method is non-invasive, or even contactless, so to avoid the disturbance of the established flow pattern. Secondly, the method can be very fast and can therefore cope with the time-scale of flow inside the bed.

A single NIR probe is used for monitoring the fluid bed and following the trajectory of our process, for this a process NIR from Sentronic is used.

However, as there are different zones of flow and progress inside the bed, a single probe is not enough, to get an idea of the overall maturity inside the fluid bed.

Here we used a new spectrometer prototype provided by EVK, Raaba, Austria in order the monitor a coating process. The prototype is based on a push-broom principle, but instead of a line, a standardized port plate enables the mapping of different near-infrared probes, via a prism-grating-prism system onto the detection chip. As a consequence, several positions inside the fluid bed can be monitored simultaneously, with a frequency of several Hz. even more, the records of the probes can be directly compared, as they are guaranteed to be recorded at the same time, and there is no spectrometer drift in between probes, etc.

The probes are distributed inside the fluid bed vessel at carefully chosen positions, at the wall, in order to not disturb the flow pattern, but give at the same time the most relevant information.

The information gained is of value twofold. First, it tells about the process itself, and can be used to estimate necessary process time and quality. Second, it tells about the mechanisms inside the fluid bed, spatial differences, etc.; which is very valuable for final probe positioning, the task of up-scaling, and more.

Estimation of the sample positions and the process is done with risk analysis, and design of experiments is also used for generating an experimental setup pattern. Within the multiprobe measurement system, we will also evaluate the sample position of the single probe.

Figure 1: The Ventilus 2.5 fluid bed reactor is shown., with the Sentronic NIR probe attached.

(146c) From Experimental Data to Building Integrated Dynamic Flowsheet Models for Pharmaceutical Processes

Fani Boukouvala[1], Fernando J. Muzzio[2] and Marianthi Ierapetritou[2], (1)Chemical and Biochemical Engineering, Rutgers University, Piscataway, NJ, (2)Department of Chemical and Biochemical Engineering, Rutgers, The State University of New Jersey, Piscataway, NJ

FDA harmonization guidelines [1] aim to promote building quality into design of pharmaceutical processes through mechanistic understanding of powder process behavior. Based on this paradigm, the industry is moving with a fast pace towards a model-based process and product design, which will undoubtedly lead to competitive manufacturing and better product quality [2]. However, process modeling is highly dependent on experimental or computational data, either in the form of empirical correlations or during the stage of parameter estimation. At the same time, there has been a high interest in the population of a flowsheet model library for solids-based pharmaceutical processes. Flowsheet modeling has been an imperative process engineering tool for the design and optimization of liquid-based processes [3-4]. This work aims to summarize all of the identified challenges that originate from the incorporation of experimental and computational data into a dynamic flowsheet model library for pharmaceutical processes, in order to develop a generic model library which may be used for a wide range of raw materials and processes, integrated to form complete production lines.

Many challenges arise due to the nature of pharmaceutical materials and processes compared to other industries, such as increased variability of powder material properties and operating conditions, uncharacterized materials, lack of well-established correlations of material properties and process operation and oftentimes lack of predictive models for certain processing steps. However, during the recent years where there has been a strong focus in enhancing process understanding, large amounts of data for the characterization of processes or materials have been collected. In certain cases, these data have provided insight about the governing underlying mechanisms and have lead to the development of first-principle models, while in other cases data have been used to build empirical response surface models, or multivariate latent variable models in cases of high dimensionality. In fact, even in the well-established liquid-based flowsheet models, certain physical phenomena which are insufficiently understood, such as chemical reaction kinetics or heat and mass transfer effects are represented through empirical correlations which are incorporated into first-principle models to create hybrid or semi-empirical modules [5].

Moreover, the complex nature of powder based-processes has also lead to the development and use of high-fidelity spatially distributed simulation models that aim to capture the complex behavior of interacting particles in order to provide insight into detailed critical material attributes such as particle size distributions. This introduces another challenge towards the final goal of flowsheet model building, which is the incorporation of multiscale computational data, derived from very expensive computer models. Fortunately, however, through this work it will be shown that the handling of such computational data is very similar to that of experimental data.

Finally it should be mentioned that the link between data and models is always a closed-loop process, since even after the development of the necessary models, the need of experimental data is once again imperative for the estimation of the model parameters. In this case the experimental design will be guided by the models, in order to ensure that they contain the necessary information needed for the parameter estimation.

A great opportunity arises in order to efficiently incorporate the information obtained from large data-bases and build and/or refine the flowsheet model library following well-established techniques of the process systems engineering community. This work aims to discuss the state of the art in first-principle models used to simulate pharmaceutical manufacturing and their proposed combinations and modifications in order to maximize the captured information through incorporation of useful data-based correlations. Most importantly, the need to consider each process as part of a large integrated puzzle should be considered even from the early stages of experimental design, since due to the need for simulation of continuous integrated lines, the propagation of a common set of material properties of powder mixtures is essential. The challenges and proposed remedies will be shown through specific case studies for a variety of manufacturing scenarios of pharmaceutical tablets such as direct compaction, dry and wet granulation.

References:

1. *Q8 Pharmaceutical Development*. 2006; Available from: http://www.fda.gov/downloads/RegulatoryInformation/Guidances/ucm128029.pdf.

2. Oksanen, C.A. and S. García Muñoz, *Process modeling and control in drug development and manufacturing*. Computers and Chemical Engineering, 2010. 34(7): p. 1007-1008.

3. Boukouvala, F., et al., *An integrated approach for dynamic flowsheet modeling and sensitivity analysis of a continuous tablet manufacturing process.* Computers and Chemical Engineering, In Press, 2012.

4. Werther, J., C. Reimers, and G. Gruhn, *Design specifications in the flowsheet simulation of complex solids processes.* Powder Technology, 2009. 191(3): p. 260-271.

5. Kahrs, O. and W. Marquardt, *The validity domain of hybrid models and its application in process optimization.* Chemical Engineering and Processing: Process Intensification, 2007. 46(11): p. 1054-1066.

(146d) Modeling and Simulation of a Wurster Column for Particulate Coating Using Population Balance Moldels and Thermodynamics

Jianfeng Li, Chemical Engineering, Purdue University, West Lafayette, IN, **Salvador García-Muñoz**, Process Modeling & Engineering Technology, Pfizer Worldwide Research & Development, Groton, CT, William R. Ketterhagen, Pfizer Global Research and Development, Pfizer Inc., Groton, CT and James D. Litster, School of Chemical Engineering, Department of Industrial and Physical Pharmacy, Purdue University, West Lafayette, IN

The Wurster column coating process of pharmaceutical solids has been studied for many years and a considerable body of literature has been produced covering various aspects of the process. To achieve uniform active coating content is one of the key interests for the pharmaceutical manufacturing industry. Thus, a process model for predicting the evolution of a granule size distribution from first principles is desired.

In this work, a compartment based population balance model, combined with thermodynamic coating model has been developed to model a pharmaceutical coating process in the Wurster column coater. The model can predict certain key process parameters and the dynamic particle size distribution, and it can also be used to simulate responses of the system to a change in the working conditions.

In many industrial practices, the coating equipment has been identified into coating/non-spray regions. It is believed that both the variation of residence time in the spray region and the variation of cycle time lead to the non-uniform coating mass distribution among individual particles. Crites and Turton (2005) have developed a compartment based model to incorporate the cycle time distribution feature for the coating process in the Wurster column coater. The model was composed of fundamental flow types such as plug flow and well-mixed flow regions to describe phenomena including dead zone and side pass. Other instances include a series of well-mixed tanks model structure for a top-spray fluidized bed coater (Ronsse 2005), a twin region compartmental model for a rotating drum coater (Dennis et al. 2001).

In this work we use a population balance (PB) approach to model the particle size distribution of the coated particulates as a function of time. While the particle growth due to coating strongly relates to the number and duration particles passing through the spray region, a practical way to constructing PB models is based on the decomposition of the coater into spray/non-spray compartments. Thermodynamics of the spray are evaluated with a simple energy balance (see am Ende et al. 2005)

In terms of the flow of particles across the unite, the Wurster column coater is divided into three compartments including an ideal plug flow region, an ideal fountain CSTR, and a non-ideal annulus CSTR (in a similar way to that proposed by Crites and Turton (2005)).

1. The plug flow compartment refers to the region enclosed by the draft-tube, where particles are highly fluidized and heated by the upflow hot air. The coating solution is assumed to be delivered on to these particles from the spray nozzle centered in the middle of the distribute plate.

2. The ideal CSTR compartment represents the fountain region, in which particles enter into the chamber and continuously receive spray.

3. Particles are then settled down towards the bottom of the coater in the annulus region, described by the non-ideal CSTR. Particles spend most of the time in this region. The non-ideality refers to the factor that smaller particles have longer recycle time compared to larger particles (Liu and Litster 1993), so the characteristic time of the CSTR for the annulus region is a function of particle size.

Experimental data is used to determine the thermal properties of the column as well as growth kinetics. Predicted values of particle size as a function of time were found in good agreement with particle size distribution measurements (by laser diffraction) of samples taken throughout the process.

References

am Ende, M.T. and Berchielli, A., 2005. A thermodynamic model for organic and aqueous tablet film coating. Pharmaceutical Development and Technology 1, 47–58.

Crites, T and Turton, R., 2005. Mathematical model for the prediction of cycle-time distributions for the Wurster column-coating process. Ind. Eng. Chem. Res. 44, 5397-5402.

Denis, C., Hemati, A., Chulia, D., Lanne, J.Y., Buisson, B., Daste, G., Elbaz, F., 2003. A model of surface renewal with application to the coating of pharmaceutical tablets in rotary drums. Powder Technology 130,174-180.

Liu, L. X., Litster, J. D., 1993. Coating mass-distribution from a spouted bed seed coater - experimental and modeling studies. Powder Technology 74, 259-270.

Ronsse, F., Pieters, J.G., Dewettinck, K., 2007. Combined population balance and thermodynamic modelling of the batch top-spray fluidised bed coating process. Part I—Model development and validation. Journal of Food Engineering 78, 296–307.

(146e) Exceptional Event Managemnt Applied to Continuous Pharmaceutical Manufacturing

Anshu Gupta, Gintaras V. Reklaitis and Venkat Venkatasubramanian, Chemical Engineering, Purdue University, West Lafayette, IN

One of the important challenges in effective real time process management is the implementation of intelligent systems that can assist human operators in making supervisory control decisions, instead of simply sounding an alarm when process variables go out of range. Operator failures to exercise the appropriate mitigation actions often have an adverse effect on product quality, process safety, occupational health and environmental impact. The economic effect of such exceptional events is immense; an estimated $20billion/year losses in petrochemical industry have been reported. The challenges and opportunities for improvements are even larger in the pharmaceutical manufacturing domain because so much of the processing involves particulate and granular systems whose characteristics tends to be more problematic than that of fluids. The development and manufacturing of pharmaceutical products are governed by strict safety regulations but with the advent of Process Analytical Technology (PAT) initiative advanced by the FDA; gives the pharmaceutical industry an opportunity to apply various systems engineering tools. Early detection and diagnosis of process faults while the plant is still operating in a controllable region can help avoid abnormal event progression, production disruptions and productivity losses. An EEM framework has been developed to deal with fault detection, diagnosis and mitigation of conditions that result from process anomalies. The framework developed uses a combination of Principal Component Analysis (PCA), Signed Digraphs (SDG) and Qualitative Trend Analysis (QTA) in applications involving continuous pharmaceutical product manufacturing line. An ontological database has been created to maintain records of the signatures of a number of exceptional events typical of a dry granulation line and their associated mitigation strategies. In trials on pilot scale equipment we have found that the EEM framework was able to detect and diagnose several types of faults within a few seconds of their inception and to provide mitigation advisories to the operator.

(146f) Modelling of a Filter Drying Process

Giovanni Giorgio[1], Jason Mustakis[2], Mark A. Pinto[3] and Sean K. Bermingham[3], (1)Particle Generation and Control Engineering, GlaxoSmithKline Ltd., Stevenage, United Kingdom, (2)Pfizer Global Research & Development, Pfizer Inc., Groton, CT, (3)Process Systems Enterprise, London, United Kingdom

Filter drying, also referred to as vacuum contact drying, is a process widely used in the pharmaceutical industry to remove water or organic solvents. The advantage of this process over other drying operations is that it can be used to dry solids that are sensitive to oxygen and temperature.

The entire drying process is typically carried out in two stages in a vessel. Typically a cylindrical vessel is used although conical vessels are also encountered. In the first blowdown stage, pressure in the form of nitrogen is applied to remove excess solvent by convective drying. At the end of this stage, the wet filter cake typically has around 45% solvent on a wet basis.

The second, longer, period involves heating the walls of the filter cake while vacuum is applied to dry the cake to about 1% to 0.1%. During this stage, heat conduction from the wall to the interior of the cake is often the rate limiting step. While the process can be carried out without agitating the bed, this can lead to excessively long drying times and heterogeneity of the cake. On the other hand, continuous long agitation often results in agglomeration and/or breakage of the solid. As a compromise, intermittent periods of agitation are carried out to redistribute solvent and heat, resulting in faster drying times but without significant agglomeration and/or breakage of solid taking place.

A detailed model of the filter dryer unit operation has previously been published in the literature (Murru et al, 2011). The key features of the model are:

1. The second drying phase is modelled and drying is assumed to take place primarily due to heat conduction from the equipment walls. The blowdown phase where convective drying takes place is not accounted for.

2. A single solvent is considered. This allowed for process optimisation by allowing different solvents to be examined for the effects on total drying times.

The model was successfully used to predict drying times for various equipment sizes and solvents at both pilot plant and manufacturing scales.

In this work, the model is further extended to consider the following phemonena.

1. Convective drying is considered. This allows the modelling of the initial blowdown drying phase. The effects of the applied suction during the second drying phase are also explicitly accounted for. The application of hot nitrogen flow over the bed can significantly reduce the time of drying.

2. The model is extended to non-ideal mixtures of solvents. The presence of solvent mixtures in the drying operation has a significant effect on the drying time but most importantly lead to change of composition of the residual solvent during drying. This can lead to stability issues (hydrates / solvates) or can lead to particle bridging and agglomeration. Understanding the dynamics of the solvent composition is crucial to avoid these situations. The ability to handle multiple solvents simultaneously will allow for greater flexibility in selecting solvents for process optimisation.

3. Modelling of the vapour processing step is undertaken. This allows for data from downstream condensation and scrubbing unit operations to be used in model validation.

References

M Murru, G Giorgio, S Montomoli, F Ricard, F Stepanek (2011) Model-based scale-up of vacuum contact drying of pharmaceutical compounds. *Chemical Engineering Science* **66**, 5045-5054

(202a) AIChE/CACHE Mobile Device APP Competition

Robert P. Hesketh, Chemical Engineering, Rowan University, Glassboro, NJ

This presentation will report on the effort to start the first AIChE/CACHE mobile device App competition The objective of this competition is:

- To give students practice at developing their creative skills.
- To enhance and promote student's interest in new mobile device APPs for use in professional activities.
- To provide chemical engineering students with the opportunity to participate in a multidisciplinary team-oriented software design competition.
- To encourage students to become actively involved in their professional society.
- To show the connection between chemical engineering and society needs.
- To create useful mobile applications for the chemical engineering community.

National Competition

Undergraduate student teams will create an APP for a mobile device. This APP must be deemed to be useful for an AIChE member. Possible application areas include scientific,
engineering, teaching and professional meeting organizers. Teaching apps could include a game designed to teach chemical engineering concepts. The APP that is developed must be able to run on a mobile device such as a smartphone (iPhone, Android, Blackberry etc.) Student teams can consist of any discipline, but at least half of the team must be members of AIChE.
The competition at the AIChE annual meeting will consist of a poster competition where finalists will be selected to give a short presentation to a general audience. There will be two prizes, with monetary awards given to the student team's AIChE chapter:

1. \$500 from CACHE for the first place winner of the best mobile device APP
2. \$500 from a corporate sponsor for the best mobile device APP that addresses safety.
   Intellectual Property of the APP remains with the students who developed the APP.

Rules

- Student teams must consist of fulltime students
- Student teams can consist of any discipline, but at least half of the team must be members of the AIChE
- Student developed APPs can be constructed using existing software templates and components.
- The final APP must be an original creation of the team.
- Submissions must comply with intellectual property rules (e.g. copyright etc.)
- The student team must have at least one team member presenting at Annual Meeting.
- The App must be able to run on a mobile device that uses one of these operating Systems: Android, BlackBerry, iPhone, Palm, Symbian, or Windows Mobile.
- Timeline:
  - APP and Technical Documentation Submission 1 June 2012 through website upload (details to be announced)
  - Poster Competition – 10/28/12 Pittsburgh, PA

Judging of the overall competition will be based on:

- Creativity and uniqueness of the APP (15%)
- Usefulness of the APP (20%)
- Ease of use of the APP(15%)

- Professional and/or Societal Impact of APP (20%)
- Poster Presentation which includes demonstration of APP on a mobile device at AIChE Annual Meeting (30%)

Judging of Electronic Submission portion of the APP competition

- Submission deadline 1 June 2012; submission instructions will be announced at a later date
- Technical documentation containing
  - Description of the APP and equations used including screen shots
  - Sources of technical information and physical properties
  - Sources of software components that were used in APP
  - Flow chart illustrating how the APP works
  - Sufficient worked-out examples to verifying the utility of the APP
  - Broad impacts of chemical engineering on society
- APP Code

(202b) Integrating the Ipad and Iphone with Mass & Energy Balances

Jason E. Bara, John Patrick McLemore, Ashley M. Parker, Katie E. Jennings and Harvis J. Smith, Chemical & Biological Engineering, University of Alabama, Tuscaloosa, AL

The Chemical Engineering App was conceived to transform the information presented in a typical mass and energy balance course into a fully interactive suite of tools, spreadsheets, equations and data that will enhance students' learning experiences throughout their undergraduate curriculum yet also be highly useful to current chemical engineering professionals. While most of the relatively small number of existing apps geared to chemical engineers merely present basic mathematical formulas into which the user provides one or more inputs (e.g. the ideal gas equation) or perform routine unit conversions, we have taken a much more rigorous approach to implementing problem-solving tools in mobile devices. The major aspect of our strategy is to allow the user to directly access actual data within the calculations, providing a rapid and robust software package for tackling undergraduate course material. Distinct features of our app include thermophysical property data for hundreds of compounds and elements, steam tables, spreadsheet tools, graphing of VLE, Antoine Equation and heat capacity, unit converters, as well as a number of other functions and information encountered in a mass and energy balance course. We have also incorporated engines which can randomly generate an infinite number of mass and energy balance questions for a variety of unit operations as supplementary homework problems.

This presentation will discuss the history and motivation behind the Chemical Engineering App, the design philosophy, student feedback/response, and the impacts, challenges and opportunities in utilizing iPhones and iPads in teaching a mass and energy balance course.

(202c) Using an Operator Training Simulator in the Undergraduate Chemical Engineering Curriculum

Debangsu Bhattacharyya[1], **Richard Turton**[1] and Stephen E. Zitney[2], (1)Department of Chemical Engg., West Virginia University, Morgantown, WV, (2)AVESTAR Center, National Energy Technology Laboratory, Morgantown, WV

An operator training simulator (OTS) is to the chemical engineer what a flight simulator is to the aerospace engineer. The basis of an OTS is a high-fidelity dynamic model of a chemical process that allows an engineer to simulate start-up, shut-down, and normal operation. It can also be used to test the skill and ability of an engineer or operator to respond and control some unforeseen situation(s) through the use of programmed malfunctions.

West Virginia University (WVU) is a member of the National Energy Technology Laboratory's Regional University Alliance (NETL-RUA). Working through the NETL-RUA, the authors have spent the last four years collaborating on the development of a high-fidelity OTS for an Integrated Gasification Combined Cycle (IGCC) power plant with CO_2 capture that is the cornerstone of the AVESTAR™ (**A**dvanced **V**irtual **E**nergy **S**imulation **T**raining **A**nd **R**esearch) Center with sister facilities at NETL and WVU in Morgantown, WV. This OTS is capable of real-time dynamic simulation of IGCC plant operation, including start-up, shut-down, and power demand load following. The dynamic simulator and its human machine interfaces (HMIs) are based on the DYNSIM and InTouch software, respectively, from Invensys Operations Management.

The purpose of this presentation is to discuss the authors' experiences in using this sophisticated dynamic simulation-based OTS as a hands-on teaching tool in the undergraduate chemical engineering curriculum. At present, the OTS has been used in two separate courses: a new process simulation course and a traditional process control course.

In the process simulation course, concepts of steady-state and dynamic simulations were covered prior to exposing the students to the OTS. Moreover, digital logic and the concept of equipment requiring one or more permissive states to be enabled prior to successful operation were also covered. Students were briefed about start-up procedures and the importance of following a predetermined sequence of actions in order to start-up the plant successfully. Student experience with the dynamic simulator consisted of a six-hour training session in which the Claus sulfur capture unit of the IGCC plant was started up. The students were able to operate the simulator through the InTouch-based HMI displays and study and understand the underlying dynamic modeling approach used in the DYNSIM-based simulator. The concepts learned during the training sessions were further reinforced when students developed their own DYNSIM models for a chemical process and wrote a detailed start-up procedure.

In the process control course, students learned how the plant responds dynamically to changes in the manipulated inputs, as well as how the control system impacts plant performance, stability, robustness and disturbance rejection characteristics. The OTS provided the opportunity to study the dynamics of complicated, "real-life" process plants consisting of hundreds of pieces of equipment. Students implemented ideal forcing functions, tracked the time-delay through the entire plant, studied the response of open-loop unstable systems, and learned "good practices" in control system design by taking into account the real-world events where significant deviations from the "ideal" or "expected" response can occur. The theory of closed-loop stability was reinforced by implementing limiting proportional gain for stability limits of real plants. Finally, students were divided into several groups where each group was tasked to control a section of the plant within a set of operating limits in the face of disturbances and simulated process faults. At the end of this test, they suggested ways to improve the control system performance based on the theory they learned in class and the hands-on experience they earned while working on the OTS.

(202d) Improving Process Engineering Tools and Their Application in Chemical Engineering Curriculum

Ajay Lakshmanan, Aspen Technology, Inc., Burlington, MA, Boyd Gochenour, Aspen Technology, Inc., Burlington and George W. Huber, Department of Chemical Engineering, University of Massachusetts-Amherst, Amherst, MA

Accurate modeling can offer valuable insight into understanding the intricacies and trade-offs involved in a process plant. Over the past several decades different decision-support tools have been developed for specific business processes such as process modeling, design, simulation and optimization, control, production planning and scheduling. These tools are highly specialized and were focused on expert users. Advances in computing technology and devices have enabled educators and practitioners to introduce these tools to students and to other mainstream users earlier in the curriculum. To stay in sync with this new user community significant steps have been made in the past decade to improve these tools. This paper discusses how feedback from users in the universities and in the industry impacted the design of chemical engineering simulation and planning tools that are faster to learn and easier to use.

Specific topics that will be detailed include the design of a user workflow based interface for process engineering and planning tools that improves discoverability of "expert" features by the novice user; enhanced collaboration and search capabilities that enable users to locate relevant information within seconds, improved access to process engineering tools any time anywhere through mobile devices; and the joint work with universities to develop training modules for students.

The paper will also detail case studies and best practices at various universities on how professors have applied these tools in their university curriculum to further Chemical Engineering education.

Co-current Parameter Estimation and Model Refinement in Dynamical Systems

ID# 257668

Michael Elly, Chem. Eng. Dept., Ben-Gurion University, Beer-Sheva, Israel
Mordechai Shacham, Chem. Eng. Dept., Ben-Gurion University, Beer-Sheva, Israel
Jose C. Merchuk, Chem. Eng. Dept., Ben-Gurion University, Beer-Sheva, Israel

Introduction

Parameter estimation in dynamical systems has been used since the 1960s, mainly for the estimation of chemical reaction rate parameters. Recently there is a considerable interest in estimation of kinetic parameters in biological models. The computation of such parameters represents specific challenge (as pointed out, for example by Leppävuori et al, 2011) as the models often contain a large number of unknown kinetic parameters that cannot be measured and the number of parameters that can be reliably estimated, based on available experimental data, is often smaller than the total number of parameters. To reduce the number of equations and parameters the model is usually simplified (using, for example the pseudo steady state assumption, Ji and Luo, 2000) and some of the parameters are assigned a constant value a-priory. The presently used parameter estimation techniques (for a recent review see Michalic et al., 2009) may converge to a local minimum. However, even if they do converge to a global minimum the predicted values may differ significantly from the experimental data because of incorrectness of the assumptions that were associated with the model derivation and the assignment of constant, fixed values to some of the parameters.

To alleviate these difficulties we have developed an interactive tool that incorporates the human investigator in the parameter estimation and model refinement loop. This tool is basically a program which uses the, so called, "sequential approach" for parameter estimation. In an outer loop the weighted squared error between the experimental data set and the corresponding model predictions is minimized. In the inner loop an integration routine is used to determine the state variable values at time intervals where experimental data are available. The user can select the parameters that remain fixed and the ones that can be changed by the optimization algorithm.

The proposed interactive parameter estimation program has been implemented in POLYMATH 7.0 (POLYMATH is a product of Polymath Software, http://www.polymath-software.com). In this paper we report on the evaluation of the program using test problems for which the optimal solutions are available in the literature. A systematic procedure for simultaneous model improvement and initial guess variation for attaining a global optimum is also presented.

Basic Concepts

The parameter estimation problem involving an ODE model can be formulated as follows:

$$\min_{\theta} \Phi(\theta) = \sum_{\mu=1}^{n} \sum_{i=1}^{m} \left(x_{\mu,i} - x_{\mu,i}^{c} \right)^2 / W_i \qquad (1)$$

s.t.

$$\frac{d\mathbf{x}}{dt} = \mathbf{F}(\mathbf{x}, \theta, t) \qquad (2)$$
$$\mathbf{x}(0) = \mathbf{x}_0$$

where **F** is a system of *m* ordinary differential equations, **x** is a vector of *m* dynamic variables, **θ** is a vector of *p* parameters, $\mathbf{x}_\mu$ is a vector of *m* observed values at the μth data point, $\mathbf{x}^c_\mu$ is a vector of *m* calculated variable values at the μth data point and **W** is a vector of *m* weighting factors. In order to scale the variables in the $0 \le x \le 1$ interval the weighting factor:

$$W_i = \max_\mu |x_{\mu,i}| \qquad (3)$$

is used. The ODE parameter estimation problem is solved in POLYMATH 7.0 using a "sequential approach". Using this approach the minimization defined in Eq (1) is carried out in an outer loop while in the inner loop an integration routine is used to determine the state variable values at time intervals where experimental data are available. This approach is also called the feasible path approach because the estimated parameter values always satisfy the model. Presently only the Runge-Kutta-Fehlberg (RKF45, Forsythe et al, 1977) algorithm is used for the integration. This algorithm monitors the estimate of the integration error, and reduces or increases the step size of the integration in order to keep the error below a specified threshold. The accuracy requested is that both the relative and absolute (maximal) errors be less than the truncation error tolerance. The default value of this tolerance is 1.0E-6. The integration is carried out in a piecewise manner from the point t_μ to $t_{\mu+1}$. At this point the μth component of the objective function is calculated and added to the previous components. This process is continued until reaching the last set of data points where $t_{\mu+1} = t_n$.

The Levenberg-Marquardt (LM, see Press et al. 1992) algorithm is used for finding the optimal parameter values. This method is a modification of the Newton method where the second derivatives appearing in the Hessian matrix are neglected. Also, the algorithm switches to the steepest descent method when the Hessian matrix becomes nearly singular.

Graphical representation of the data and the calculated values as well as comparison of the optimal parameter values with the respective standard deviations (σ_K)are used for assessing the adequacy of the model and the parameters.

Example 1 – Catalytic cracking of gas oil

This example demonstrates the case where there is no need to modify the model and no user involvement is required to reach the global optimum. The example involves the modeling of the catalytic cracking of gas oil (*A*) to gasoline (*Q*) and other side products (*S*).

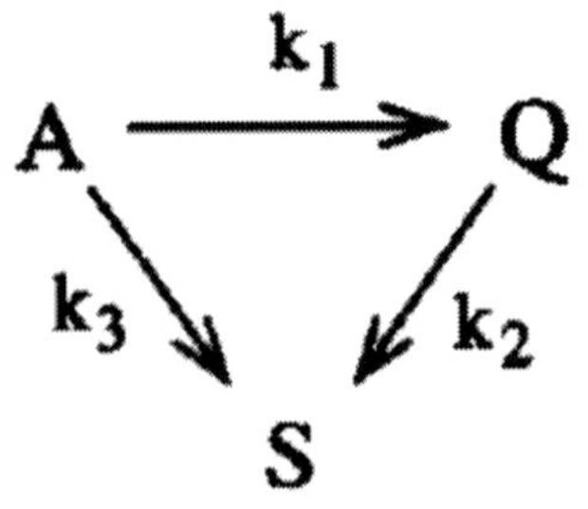

Only the concentrations of A (x_1) and Q (x_2) were measured and the concentration of S is not included in the model. This example was presented by Tjoa and Biegler (1991) and by Floudas et al. (1999). The models and data provided by Floudas et al. (1999) are used here.
The model equations of the catalytic cracking reactions are

$$\frac{dx_1}{dt} = -(\theta_1 + \theta_3)x_1^2$$
$$\frac{dx2}{dt} = \theta_1 x_1^2 - \theta_2 x_2 \qquad (4)$$

where θ_1, θ_2 and θ_3 are the rate constants. The parameter values are bounded $0 \leq \boldsymbol{\theta} \leq 20$.The experimentally observed values of x_1 and x_2 at times t_μ are shown in Table 1. Part of the Polymath input and the Polymath results are shown in Fig. 1. The Polymath input includes the observed values of the variables in a form of one dimensional arrays (only parts of the arrays are shown in Fig. 1), the model equations (the notation θ_1, θ_2 and θ_3 was changed to p1, p2 and p3) the parameter bounds and initial estimates for the parameter values. In this case the use of an arbitrary initial estimate $\theta_1 = \theta_2 = \theta_3 = 1$ within the parameter bounds was sufficient in order to reach the optimal solution.

The solution is reported in both numerical and graphical forms (Fig. 1). The plot of the calculated curves vs. the observed data points shows excellent fit for this problem. The numerical results for the parameter values and the sum of squares of the error at the optimal solution are compared with the results of Floudas et al. (1999) in Table 2. There are minor differences (in the 3rd or 4th decimal digits) in the solution values for the parameters and the sum of squares. These differences can result from the use of different error tolerances and/or different weighting factors in the objective function.

The standard deviations for θ_1, θ_2 are smaller by two orders of magnitude and for θ_3 by more than one order of magnitude than the parameter values. Thus, all the numerical and graphical indicators show that the global optimum has been found in this case without any need for user's intervention.

Example 2 – Conversion of methanol to hydrocarbons

This example demonstrate the case where the proposed reaction model includes redundant terms (parameters) that need to be removed by analysis of the ratio between the optimal parameter values and their standard deviations. This problem was suggested by Maria (1989) and Floudas et al. (1999). The model and data suggested by Floudas et al. (1999) are used here. The simplified kinetic model for conversion of methanol to various hydrocarbons is:

$$A \xrightarrow{k_1} B$$
$$A+B \xrightarrow{k_2} C$$
$$C+B \xrightarrow{k_3} P$$
$$A \xrightarrow{k_4} C$$
$$A \xrightarrow{k_5} P$$
$$A+B \xrightarrow{k_6} P$$

where *A* represents the oxygenates, *B* is the intermediate $\ddot{C}H_2$, *C* represents the olefins, and *P* represents the paraffins, aromatics and other products. The model of the reaction is formulated under the assumption of simple kinetics and quasi- steady state for *B*.

$$\begin{aligned}
\frac{dx_1}{dt} &= -(2\theta_1 - \frac{\theta_1 x_2}{(\theta_2+\theta_5)x_1 + x_2} + \theta_3 + \theta_4)x_1 \\
\frac{dx_2}{dt} &= \frac{\theta_1 x_1(\theta_2 x_1 - x_2)}{(\theta_2+\theta_5)x_1 + x_2} + \theta_3 x_1 \\
\frac{dx_3}{dt} &= \frac{\theta_1 x_1(x_2 - \theta_5 x_1)}{(\theta_2+\theta_5)x_1 + x_2} + \theta_4 x_1
\end{aligned} \qquad (5)$$

where x_1, x_2 and x_3 are the mole fractions of *A*, *B* and *C* respectively. The parameter values are bounded $0 \le \boldsymbol{\theta} \le 100$. The experimentally observed values of x_1, x_2 and x_3 at times t_μ are shown in Table 3.

The optimal parameter values were first sought for the model containing all five parameters starting at the initial estimate $\theta_k = 1$, for k = 1, 2, … 5. The results are shown in the first six rows of Table 4. Observe that for θ_4 the standard deviation is considerably larger than the parameter value (σ_4 = 6.831E-05 while θ_4 = 2.587E-07) meaning that the value of θ_4 can be set at zero. Rerunning the problem while setting θ_4 at a fixed value of $\theta_4 = 0$ yields the optimal results shown in Table 4, as the 2nd set of results. In this case the value of θ_5 becomes smaller than its standard deviation. The optimal parameter values for the case when both θ_4 and θ_5 set at the constant value of zero include the θ_3 value that is smaller than the associated standard deviation.

The final optimal model contains only the parameters θ_1 and θ_2 while the additional parameters should be set at zero. Thus the experimental data can justify only the simplified model presented in Eq. 6 for the methanol conversion reaction.

The optimal values for θ_1 and θ_2, obtained by Polymath (Table 4) are larger than the respective standard deviations by more than four orders of magnitude and they are very close to the values reported by Floudas et al. (1999). The model provides good representation of the data with regard to the products *B* and *C* (Figure 2), as for the oxygenates, there is still room for improvement of the model.

$$\frac{dx_1}{dt} = -(2\theta_1 - \frac{\theta_1 x_2}{\theta_2 x_1 + x_2})x_1$$
$$\frac{dx_2}{dt} = \frac{\theta_1 x_1(\theta_2 x_1 - x_2)}{\theta_2 x_1 + x_2} \qquad (6)$$
$$\frac{dx_3}{dt} = \frac{\theta_1 x_1 x_2}{\theta_2 x_1 + x_2}$$

Example 3 – The "Lotka-Volterra" predator-prey model

This example demonstrates the case where there is convergence to a local optimum from an arbitrary initial guess and user intervention is required to reach the global optimum. The "Lotka-Volterra" problem was originally studied by Luus (1998). This problem involves a two parameter predator-prey model similar to those used in ecology. The model determines the population over time of the two different species (the predators, and the prey) given parameters relating to the interaction between them. The state profiles at the solution are cyclic and out of phase with each other. The model and data presented by Floudas et al. (1999) are used here. The models for the predator (x_2) and the prey (x_1) populations are given by

$$\frac{dx_1}{dt} = \theta_1 x_1(1 - x_2)$$
$$\frac{dx2}{dt} = \theta_2 x_2(x_1 - 1) \qquad (7)$$

The parameters $\boldsymbol{\theta}$ are related to the birth, death, and interaction rates of the two species and they are bounded: $0 \le \boldsymbol{\theta} \le 10$. The data provided by Floudas et al. (1999) are shown in Table 5. We first attempted to find the optimum starting at the initial guess $\theta_1 = \theta_2 = 1$. The results are shown in Fig. 3 and in the first three rows of Table 6. Observe that in Fig. 3 both the data and the calculated curves exhibit oscillatory behavior, however the fit between the data and the calculated values is very poor. The relatively large standard deviation values (Table 6) also signal poor fit of the calculated values to the data. The similarity between the trend of the data and the calculated curve may indicate that the model is correct and the poor fit is caused by convergence to a local minimum. Changing the initial guess to $\theta_1 = 5$ and $\theta_2 = 1$ yields much better results. Fig. 4 shows that in this case there is a perfect match between the data and the calculated curves. The numerical results in Table 6 show that with the optimal parameter values found here, the sum of squares of errors and the standard deviations are reduced by at least two orders of magnitude in comparison to the previous case.

Conclusions

A new parameter estimation program for dynamic systems, intended for interactive model analysis and parameter estimation, has been developed. This program was evaluated by solving problems for which optimal solutions are available in the literature. Three examples were presented to demonstrate the use of the numerical and graphical feedbacks provided by the program for analyzing the model and attaining the global optimum in the parameter values. It has been demonstrated that a globally optimal solution for an adequate model is usually characterized by: 1. Similar trend and random residual distribution of the data around the associated calculated curve and 2. Small standard deviation/ abs(parameter value) ratios for all the parameters.

It was shown that if the absolute parameter value is smaller than its standard deviation this usually signals a redundant parameter. Consequently, this parameter and the associated term/s can be removed from the model. In cases where the trend of the data and the calculated curves are similar but there are large discrepancies between the data points and the associated curves this may signal convergence to a local optimum. In such cases variation of the initial guesses of the parameter values within their pre specified bounds is recommended.

Work is currently underway to apply the Polymath ODE parameter estimation program and the proposed interactive model analysis and parameter optimization procedure for identification of the model and parameters for a tobacco mosaic virus (TMV) proliferation problem (Enden et al., 2005). The results of this study will be reported elsewhere.

References

1. Enden, G., Y.H. Zhang, and J.C. Merchuk, A model of the dynamics of insect cell infection at low multiplicity of infection. Journal of Theoretical Biology, 2005. 237(3): p. 257-264.
2. Floudas, C. A., Pardalos, P. M. et al., Handbook of Test Problems in Local and Global Optimization, Kluwer, Dordrecht, The Netherlands, 1999
3. Forsythe, G.E., M.A. Malcolm, and C.B. Moler, Computer Methods for Mathematical Computation, Prentice-Hall, Englewood Cliffs, 1977.
4. Ji, F. and L. Luo, A hyper cycle theory of proliferation of viruses and resistance to the viruses of transgenic plant, Journal of Theoretical Biology, 2000, 204(3), 453-465.
5. Leppävuori, J. T.; Domach, M.M.; Biegler, L.T., Parameter Estimation in Batch Bioreactor Simulation Using Metabolic Models: Sequential Solution with Direct Sensitivities, Ind. Eng. Chem. Res. 2011, 50, 12080-12091
6. Luus, R., Parameter estimation of lotka-volterra problem by direct search optimization, Hung. J. Ind. Chem. 26, 1998
7. Maria, G., An adaptive strategy for solving kinetic model coccomitant estimation-reduction problems., Can. J. Chem. Eng. 67, 825(1989)
8. Michalik, C.; Chachuat, B.; Marquardt, W., Incremental Global Parameter Estimation in Dynamical Systems, Ind. Eng. Chem. Res. 2009, 48, 5489–5497.
9. Press, W.H., P.B. Flannery, S.A. Teukolsky and W.T. Vetterling, Numerical Recipes, 2nd Ed., Cambridge University Press, Cambridge, 1992.

Table 1. Data of observed values for Example 1 (Floudas et al. 1999).

Point No.: μ	t	$x_{\mu 1}$	$x_{\mu 2}$
0	0	1	0
1	0.025	0.7307	0.1954
2	0.05	0.5982	0.2808
3	0.075	0.4678	0.3175
4	0.1	0.4267	0.3047
5	0.125	0.3436	0.2991
6	0.15	0.3126	0.2619
7	0.175	0.2808	0.2391
8	0.2	0.2692	0.221
9	0.225	0.221	0.1898
10	0.25	0.2122	0.1801
11	0.3	0.1903	0.1503
12	0.35	0.1735	0.103
13	0.4	0.1615	0.0964
14	0.45	0.124	0.0581
15	0.5	0.119	0.0471
16	0.55	0.1109	0.0413
17	0.65	0.089	0.0367
18	0.75	0.082	0.0219
19	0.85	0.0745	0.0124
20	0.95	0.0639	0.0089

Table 2. Comparison of the optimal results from Floudas et al. (1999) with the Polymath results for Example 1.

Param. No.	*Value*[1]	σ	*Value*[2]
1	12.24071	0.11935	12.214
2	7.922492	0.0566524	7.9798
3	2.25226	0.1010523	2.2216
sum of squares	0.0026777	-	0.002656

[1]Polymath 7.0 results

[2]Optimal results from Floudas et al. (1999)

Table 3. Data of observed values for Example 2 (Floudas et al. 1999).

Point No.: μ	t	$x_{\mu 1}$	$x_{\mu 2}$	$x_{\mu 3}$
0	0	1	0	0
1	0.05	0.461	0.114	0.018
2	0.065	0.426	0.135	0.035
3	0.08	0.383	0.157	0.045
4	0.123	0.305	0.194	0.047
5	0.233	0.195	0.231	0.084
6	0.273	0.17	0.234	0.095
7	0.354	0.139	0.228	0.111
8	0.397	0.112	0.228	0.134
9	0.418	0.112	0.226	0.168
10	0.502	0.09	0.22	0.148
11	0.553	0.082	0.214	0.157
12	0.681	0.066	0.198	0.206
13	0.75	0.053	0.188	0.206
14	0.916	0.043	0.183	0.214
15	0.937	0.041	0.184	0.213
16	1.122	0.029	0.166	0.23

Table 4. Optimal results for Example 2 with various numbers of parameters

Param. No.	*Value*[1]	σ	*Value*[2]
1	3.998714	0.0002032	
2	0.9478271	8.686E-05	
3	0.760283	0.0001154	
4	**2.59E-07**	**6.83E-05**	
5	0.0685149	4.367E-05	
sum of squares	0.14928		
1	4.329012	0.0155159	
2	0.9410537	0.0064165	
3	0.3091262	0.0309942	
5	**2.55E-06**	**0.002628**	
sum of squares	0.12741		
1	6.122073	0.0004623	
2	1.905657	7.768E-05	
3	**3.39E-07**	**9.99E-05**	
sum of squares	0.16044		
1	5.052398	0.0004623	5.2407
2	1.220247	7.768E-05	1.2176
sum of squares	0.10721		0.10693

[1]Polymath 7.0 results

[2]Optimal results from Floudas et al. (1999)

Table 5. Data of observed values for Example 3 (Floudas et al. 1999).

Point No.: μ	t	$x_{\mu 1}$	$x_{\mu 2}$
0	0	1.2	1.1
1	1	0.799	1.0758
2	2	0.8731	0.8711
3	3	1.2487	0.9393
4	4	1.0362	1.1468
5	5	0.7483	1.0027
6	6	1.0024	0.8577
7	7	1.2816	1.0274
8	8	0.8944	1.1369
9	9	0.7852	0.9325
10	10	1.1527	0.9074

Table 6. Optimal results for Example 3 obtained using different initial guesses

Param. No.	***Initial guess***	***Value***[1]	σ	***Value***[2]
1	1	1.177802	0.1759329	
2	1	0.7200475	0.1198334	
sum of squares		0.79365	-	
1	1	3.192145	0.0038293	3.2424
2	1	0.9385447	0.0012158	0.9209
sum of squares		0.0013234	-	0.001249

[1]Polymath 7.0 results

[2]Optimal results from Floudas et al. (1999)

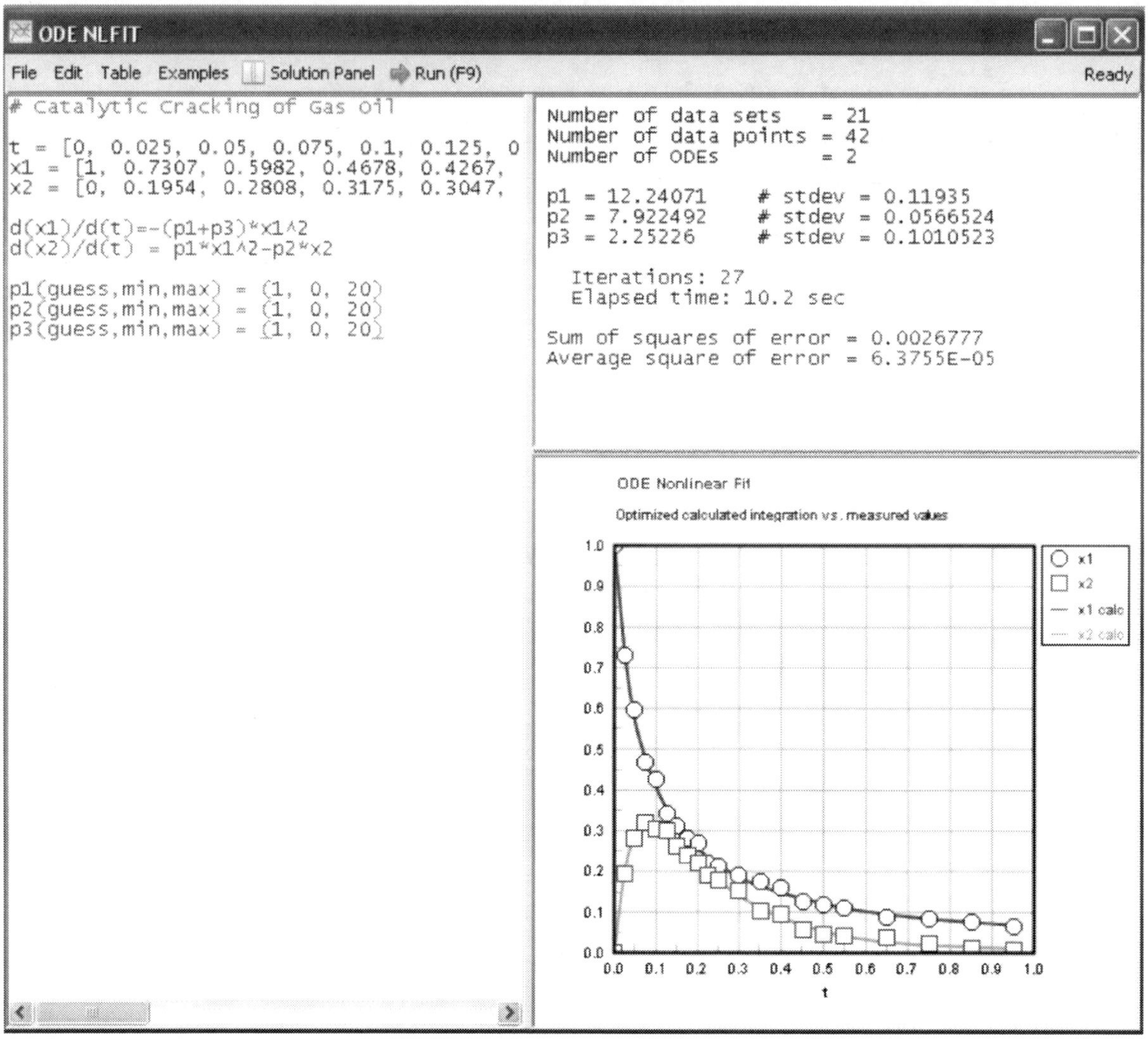

Figure 1. Polymath input and results for Example 1.

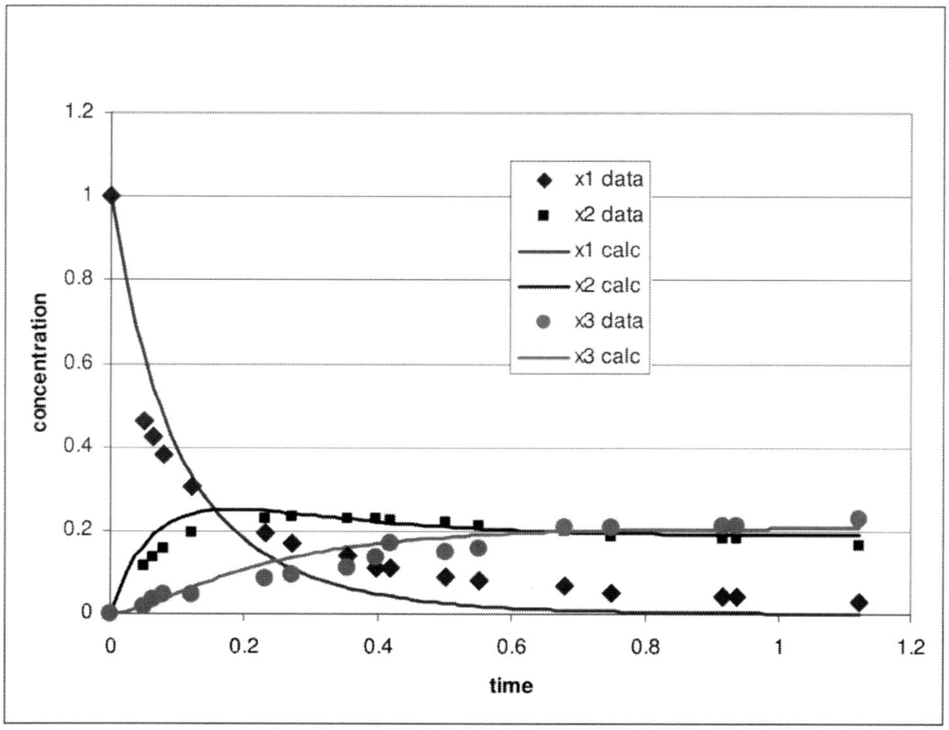

Figure 2. Plot of the calculated curves vs. the observed data points for the two parameter's optimal model for Example 2.

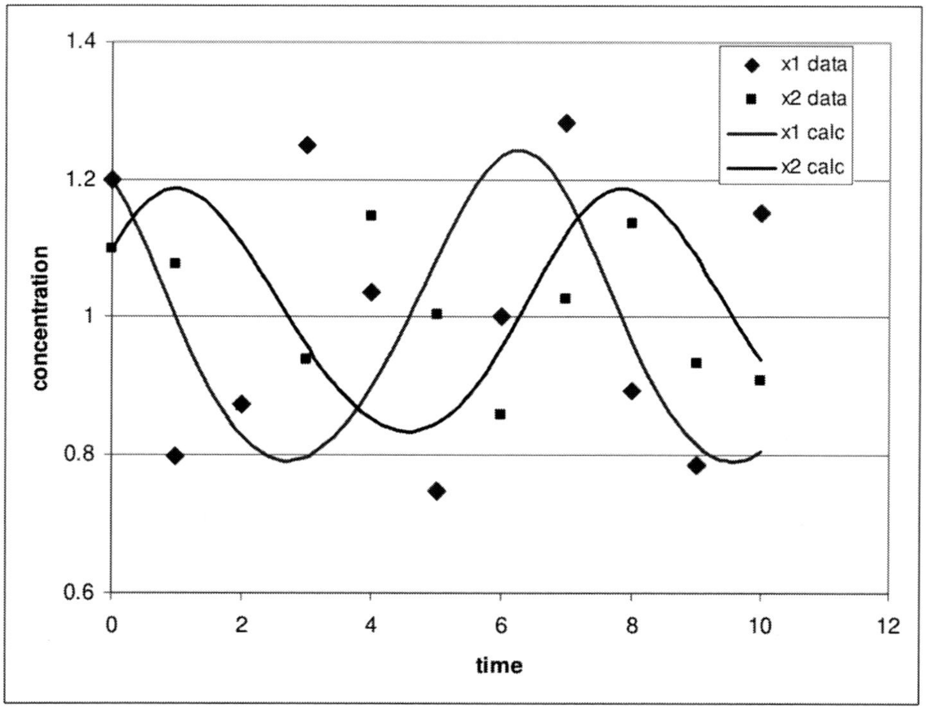

Figure 3. Plot of the calculated curves vs. the observed data points for the optimum found from the initial guess $\theta_1 = \theta_2 = 1$ for Example 3.

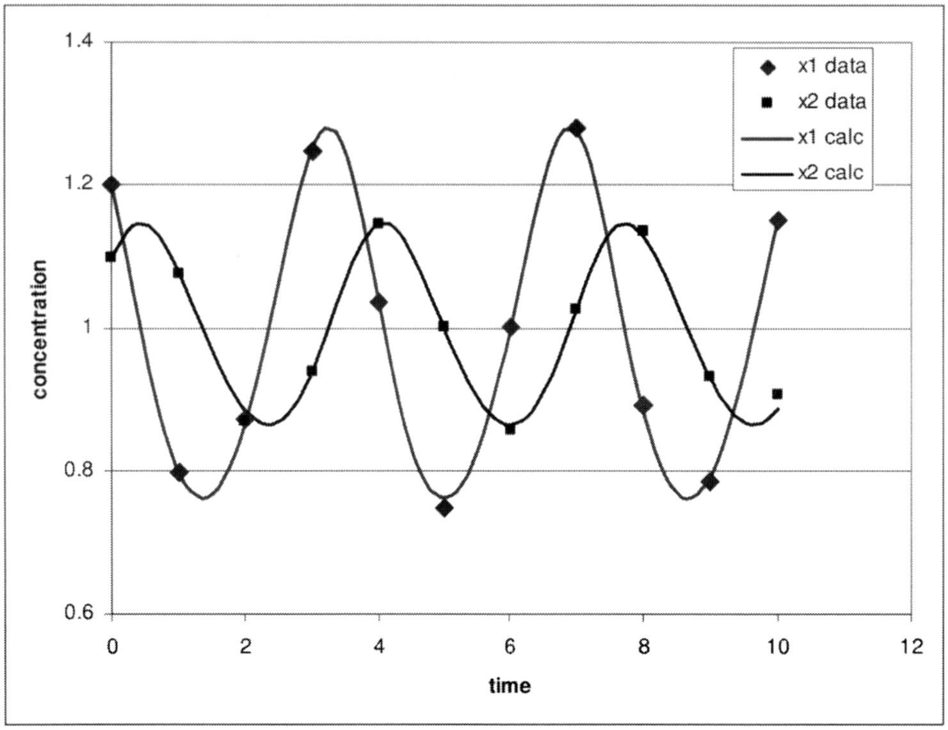

Figure 4. Plot of the calculated curves vs. the observed data points for the optimum found from the initial guess $\theta_1 = 5$, $\theta_2 = 1$ for Example 3.

(202f) A Process Systems Approach to Teaching Distillation

Kody Powell, Chemical Engineering, The University of Texas at Austin, Austin, TX and Thomas F. Edgar, Department of Chemical Engineering, The University of Texas at Austin, Austin, TX

First principles modeling is a valuable instructional tool to help undergraduate chemical engineering students understand how a process works. Distillation is perhaps the most difficult process for students to understand. In this work, a methodology where distillation is approached from a process systems point of view is presented. This approach was implemented in a unit operations distillation lab where students were required to build a first principles model of the process, which they would later use in optimization to maximize the revenue of a hypothetical chemical plant. From a pedagogical point-of-view, a systems engineering approach is beneficial to students as it requires them to gain an intimate understanding of the process through modeling and allows them to develop basic skills in optimization. Furthermore, a systems-engineering approach allows students to see the full scope of the process, while incorporating lab results through model parameter estimation and constraint identification.

The student's response to an open-ended versus a well-defined assignment is also studied. It is observed that, while the open-ended assignment is more difficult and time consuming, the understanding gained by this exercise is far greater. Additionally, this type of assignment compels the student to exercise more creativity and take a higher amount of ownership in the project, resulting in projects that are of higher quality and more diverse. In order for assignments of this nature to be successful, the proper balance must be established between keeping the assignment open-ended while giving students enough background information and guidance for the student to be successful.

In this work, evidence is presented comparing essentially the results from two semesters, one in which the assignment is predominantly instructor-directed, and another in which the assignment is largely student-directed. Although purely qualitative, evidence indicates that the student-directed methodology generally results in higher-quality work and a better learning experience.

(595a) The Computational Challenges of Simulating Atomic Layer Deposition (ALD) Process Dynamics

Curtisha D. Travis and Raymond A. Adomaitis, Department of Chemical & Biomolecular Engineering; Institute for Systems Research, University of Maryland, College Park, MD

The computational challenges of simulating atomic layer deposition (ALD) process dynamics

Curtisha D. Travis* and Raymond A. Adomaitis
Department of Chemical and Biomolecular Engineering
Institute for Systems Research
University of Maryland
College Park, MD 20742 USA

*Presenting author: cdtravis@umd.edu
Submitted to the 2012 AIChE Annual Meeting, Pittsburgh, PA

A physically based model of atomic layer deposition reaction kinetics is developed and applied to alumina ALD using water and trimethylaluminum precursors [1]. The ALD surface reaction models are treated as true dynamic systems, with continuous ALD reactor operation described by limit-cycle solutions, numerically computed using a polynomial collocation technique. We compute a solution by a full discretization of the differential equations followed by a Newton-Raphson method for the resulting nonlinear algebraic equations. Because the dynamics of the ALD system are relatively well-behaved, a low-order polynomial collocation scheme will work well for the discretization procedure.

To implement the collocation method, we first write the modeling equations over both half-cycles in vector form, subject to initial conditions for each full cycle equal to the final conditions at the end of the previous cycle and a condition requiring that the coverage be continuous between the two half-cycles. Using a simple polynomial Lobatto collocation technique over the unit interval with a set number of collocation points for each half cycle, we compute the discrete-ordinate formulation of the 1st-order differentiation array. We use an orthogonal polynomial collocation on finite elements discretization procedure for forced-periodic systems, where the resulting set of nonlinear algebraic equations are solved using a Newton-Raphson method. It was found that that convergence can be sensitive to the accuracy of the initial estimate of the solution, therefore, a predictor-corrector arc-length continuation procedure generally was used to compute solutions starting from sets of parameters corresponding to limit-cycle solutions for which accurate solution estimates could be made.

To compute the film growth per cycle (GPC), we compute the number of Al and O atoms deposited per unit area over one deposition cycle and by numerically integrating the reaction rates over each half-cycle using quadrature weights defined at the collocation points. Because the limit-cycle solutions represent continuous reactor operation, the ALD film composition then can be computed, which lends to calculation of the growth per cycle. A representative limit cycle solution computed using the collocation procedure is shown in Fig. 1 where the state of the surface during each each exposure cycle (TMA and water half-cycles denoted by A and B, respectively) is plotted as a function of time. An alternate view of this limit-cycle solution is shown in Fig. 2, where the changes in surface -CH3 and -OH groups during the TMA and water exposures are shown. To illustrate the dependency of GPC as a function of both precursor exposure levels for a fixed precursor pressure of 1 Torr, a GPC map generated by the limit cycle solutions is shown in Fig. 2 (right). Model predictions indicating optimal operating conditions will be discussed.

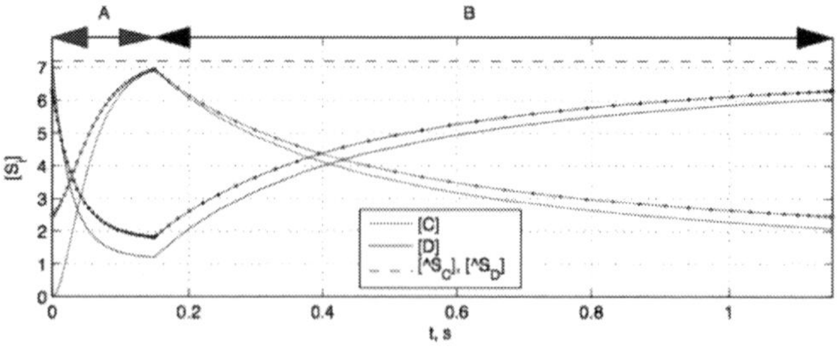

Figure 1: *TMA and water dose dynamics where the red solid curves correspond to surface -CH3 groups, the blue correspond to surface -OH during the first A-B cycle starting from a fully hydroxylated surface, and the dashed line indicates the close-packing limit of methyl groups on the growth surface. Dotted curves indicate the corresponding*

limit-cycle solutions with the dots showing the location of collocation points.

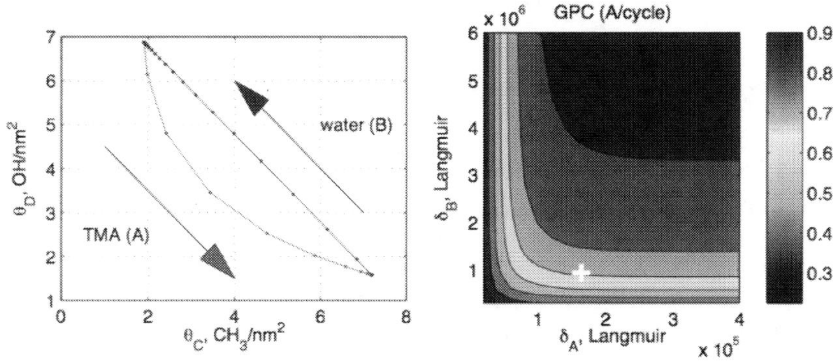

Figure 2: *Surface -CH3 and -OH limit-cycle coverage dynamics (left) for a representative set of alumina ALD operating conditions. The red curve corresponds to the TMA dose, the blue to water. Alumina GPC (A/cycle) map (right) as a function of each precursor exposure level, with limit-cycle conditions marked as +.*

Reference

[1] Puurunen, R. L., ``Surface chemistry of atomic later deposition: A case study of the trimethylaluminum/water system," Appl. Phys. Rev. 97 121301 (2005).

(595b) Multipoint Pade Method for Cyclic Adsorptions with Reactions in Slab, Cylinder and Sphere Catalysts

Jietae Lee, Chemical Engineering, Kyungpook National University, Taegu, South Korea

An accurate mathematical description for the dynamic mass transfer in a particle is the pore diffusion model represented by a parabolic partial differential equation. Practical separation processes such as packed bed adsorbers involve mass balance equations of the bulk flow through the adsorber in addition to this pore diffusion model equation. These coupled partial differential equations are rather complicate and the pore diffusion model equation need be approximated for a reduced computational load. The linear driving force (LDF) model is the simplest approximation. In pressure swing adsorption (PSA) for gas separation, the adsorbent particles in the bed are subject to a periodic concentration change. The LDF approximation has been used to simplify modeling and to reduce computation time in simulations of cyclic operations. As the frequency of the cyclic operation increases, it has been shown that the LDF approximation becomes invalid.

For fast cyclic operations, approximations derived for general concentration forcing should be modified. The LDF whose coefficients are varied as a function of the frequency has been suggested by several researchers. Kim (*Chem. Eng. Sci.* **1996**, 51, 4137-4144) proposed the first order model for which exact response is obtained for sinusoidal wave of the given frequency. For the cyclic operations, the first order model can also be obtained by dividing the particle into active and inactive zones and applying the one-point collocation method to the active zone. High-order approximations for general periodic concentration changes are available (Lee, J.; Kim, D. H., *Chem. Eng. Sci.*, **1998**, 53, 1209-1221).

Here the multipoint Pade method based on the continued fraction expansion is applied to obtain high order models for a fast cyclic operation of diffusion, adsorption and a first-order reaction in a slab, cylinder and sphere catalyst. Model parameters can be calculated easily. Trends of responses as the operation period decreases are obtained.

(595c) Operation Flexibility Study for Newly Designed Liquefied Natural Gas Receiving Terminals

Meiqian Wang, Jian Zhang and Qiang Xu, Dan F. Smith Department of Chemical Engineering, Lamar University, Beaumont, TX

Receiving terminal is one of the most important sectors along the whole liquefied natural gas (LNG) value chain. And the major concern at receiving terminal is that a great deal of heating duty will be consumed during LNG re-gasification. On the other hand, natural gas liquids (NGL) recovery process normally needs large quantity of cold energy for liquefaction and condensation. NGL recovery section usually proceeds prior to liquefaction and products. Moreover, a lot of shale gas plays distribute close to the receiving terminals in U.S. Currently, shale gas is usually burnt to generate electricity after simply treatment. Apparently, it would be much more ideal if required utility can be provided conveniently for shale gas to conduct NGL recovery, since NGL has more economic values than that as part of the natural gas stream. Hence, a novel design for receiving terminal has been proposed that integrates natural gas and shale gas NGL recovery process with LNG re-gasification process. Simulation of this newly designed receiving terminal indicates that the cold energy from LNG re-gasification could be efficiently recovered by shale gas liquefaction and NGL fractionators' condensers.

The problem about this design is how to get the optimal operation condition under the uncertainty of LNG and shale gas flow rates and compositions. However, the expected profit can be maximized under the stochastic optimization approach. With the analysis of historic data and prior information, the probability distribution of LNG and shale gas flow rates and compositions can be estimated. An MINLP model is developed with the integration of stochastic optimization approach. Heat integration is the main concern in this MINLP model so that minimum utility will be consumed, in another word, maximum profit could be obtained. Different operation conditions are calculated according to the changes on the flow rate and composition in order to attain the maximum average profit. Numerous simulations have been conducted for building the model and the MINLP model is solved in GAMS. The final results and analysis are also presented in this work.

OPTIMIZATION OF THE NITROXIDE MEDIATED COPOLYMERIZATION OF STYRENE AND ALPHA-METHYL STYRENE

Cecilia Fortunatti, PLAPIQUI (CONICET-UNS), Bahía Blanca, Argentina,
Claudia Sarmoria, PLAPIQUI (CONICET-UNS), Bahía Blanca, Argentina,
Adriana Brandolin, PLAPIQUI (CONICET - UNS), Bahía Blanca, Argentina, and
Mariano Asteasuain, PLAPIQUI (CONICET-UNS), Bahía Blanca, Argentina

Introduction

Free radical polymerization is widely used in industry. However, this synthesis technique presents the disadvantage of not allowing a good control of the molecular structure of the polymer, due to the high reactivity of the polymer radicals. As an alternative, controlled radical polymerization (CRP) has been developed since the mid-90s, offering enormous possibilities for creating materials with unique properties. The three best known and effective CRP techniques are atom transfer radical polymerization (ATRP), nitroxide mediated polymerization (NMP) and reversible addition-fragmentation transfer (RAFT). All of these CRP techniques are based on the idea of establishing a balance between active and inactive chains. The interactions between living chains are limited by the addition of an agent that can deactivate them reversibly. Thus, in a CRP the total effect of termination is reduced because the number of active chains is small. CRP allows obtaining (co) polymers with controlled molecular weight and polydispersity, as well as composition and chain architecture. In this way it is possible to improve the properties of products currently on the market, or create novel polymers with unique properties.[1] NMP is one of the most studied mechanisms of CRP. In this mechanism, the reaction is controlled through a reversible termination reaction with a stable radical. The propagation and termination rates are reduced by the balance between active and inactive species, which allows all chains to be initiated and grow at the same speed on average.[2]

Copolymers may be characterized by chain length, composition and sequence length distributions. It is known that the weight average molecular weights and the molecular weight distribution (MWD) are key properties that determine final properties of the copolymer. In addition, global and instantaneous compositions, as well as sequence length and distribution, have significant effects on the macroscopic properties of copolymers.[3,4] In order to produce tailor-made materials able to meet pre-specified applications, it is important to be able to control these molecular properties by an appropriate selection of the operating conditions. In this context, the modeling and optimization of CRP processes become critical.

Most articles about modeling of controlled radical copolymerization (CRcoP) processes aim to predict and control average properties such as composition, conversion, sequence lengths and molecular weights.[5-8] On the other hand, there are just a few reports about modeling full bivariate MWD for free-radical copolymerization systems.[9,10] To the best of our knowledge, no efforts have been made to obtain full CC-MWD in CRcoP systems and only a few articles have dealt with the optimization of these processes.[3,11]

In this work we study the NMP of styrene (St) and alpha-methyl styrene (AMS) using BPO (benzoyl peroxide) as initiator and TEMPO as mediator. The most widespread uses of poly(St-co-AMS) -or SAMS- are elastomer enhancers, adhesives, and dry inks for photocopiers and laser printers. Besides, small concentrations of AMS in the feed allow manufacturing low molecular weight polymers without the requirement for large amounts of chain transfer agents or initiator thanks to its low ceiling temperature (64ºC). [12,13] In previous works we presented a mathematical model of this process in tubular and semibatch reactors that is able to predict average molecular properties as well as the full bivariate MWD and the sequence length distribution of the copolymer.[14,15] In this work, this model is used for the optimization aiming at obtaining resins with pre-specified molecular properties. The optimization space included reactor design variables and operating policies, such as reactor configuration, initial charge, feeding policy and temperature profile. The performance and flexibility of two reactor types, semibatch and tubular with side feeds, is analyzed and compared. The optimization problem was formulated and solved in gPROMS (Process Systems Enterprise, Ltd.).

Methods

Mathematical Model

The mathematical model considers the following kinetic mechanism:

Initiation: $$I \xrightarrow{f\,k_d} 2\,I^\circ \;,\; I^\circ + M_j \rightarrow R^j_{2-j,j-1} \quad (1)$$

Thermal initiation of styrene: $$3\,M_1 \xrightarrow{k_{th}} R^1_{1,0} + R^1_{2,0} \quad (2)$$

Capping and uncapping: $$R^j_{n,m} + Te \underset{k_{uncap,j}}{\overset{k_{cap,j}}{\rightleftarrows}} D^j_{n,m} \quad (3)$$

Propagation and depropagation: $$R^i_{n,m} + M_j \underset{k_{p^{-1},22}}{\overset{k_{p,ij}}{\rightleftarrows}} R^j_{n+2,m+j-1} \quad (4)$$

Chain termination by coupling: $$R^i_{n,m} + R^j_{s,r} \xrightarrow{k_{t,ij}} P_{n+s,m+r} \quad (5)$$

Chain transfer to monomer: $$R^i_{n,m} + M_j \xrightarrow{k_{trm,ij}} P_{n,m} + R^j_{2-j,j-1} \quad (6)$$

The chemical species involved are: initiator BPO (I), monomers St (M_1) and AMS (M_2), TEMPO (Te), active radicals ending in monomer j with n units of M_1 and m units of M_2 ($R^j_{n,m}$), dormant (inactive) radicals ending in monomer j with n units of M_1 and m units of M_2 ($D^j_{n,m}$), and terminated (dead) copolymer chains with n units of M_1 and m units of M_2 ($P_{n,m}$).

Kinetic parameters were previously obtained by fitting experimental data. [14] Chain termination by disproportionation and disproportionation involving nitroxide were not included since their effects can be neglected. [16] The model takes into account the gel effect applying the correlation presented by Faliks et al. [17] for controlled radical polymerization of styrene.

The well-known moment technique with double index moments was used for modeling the average properties of the copolymer. This technique consists in transforming the mass balances of the polymeric species in order to obtain balances for the first moments of the copolymer MWD, from which the average properties can be computed.

Monomer sequences are the chain fragments in any copolymer chain composed by a unique monomer. A parallel kinetic mechanism was formulated considering these sequences as the reacting species in order to model the sequence length distribution (SLD). Mass balances for the monomer sequences are drawn from this kinetic mechanism that can be solved straightforwardly due to the short length of the monomer blocks. [14]

The full bivariate MWD was obtained using 2-D probability generating functions (pgf). This technique consists in transforming mass balances of polymeric species, characterized by the number of units of each comonomer, to the 2-D pgf domain. As a result, balances for the pgf transform of the bivariate MWD are obtained. The full bivariate MWD is recovered from the pgf applying an appropriate inversion method.[19] It is important to point out that this method does not require any a priori knowledge of the MWD shape and can deal with complex kinetic mechanisms.

Mass, moment and pgf balances for both semibatch and tubular reactors are shown in **Table A1** in the Appendix along with average properties equations.

Process Optimization

The optimization goal involved finding optimum design and operating conditions for obtaining copolymers of St and AMS with pre-specified molecular properties. Two reactor types were considered: semibatch and tubular with side feeds. Semibatch operation has the advantage that its flexibility for implementing different policies of reagents addition permits achieving good control on the product molecular properties. On the other hand, the continuous nature of tubular reactors allows improving productivity. At the same time, molecular properties can be controlled by tuning the position, number and flowrates of the side feeds. The process optimizations were performed aiming at synthesizing copolymers with a given uniform composition along chains or with a desired average length of St blocks. It should be remembered that due to AMS depropagation at the operating temperatures considered, only unit length blocks of AMS are possible for this copolymer.

The following process design and operating variables were included in the optimization runs: for the semibatch reactor, initial charge ($m_{i,0}, i = St, AMS, Te, BPO$), temperature profile (linear continuous), and reactants feeding policies ($f_i(t), i = St, AMS, Te, f_i(t)$ piecewise continuous); for the tubular reactor, main feed flowrates ($f_{i,0}, i = St, AMS, Te, BPO$), and lateral feeds positions ($z_j, j = 1,2,3$), and flowrates ($f_{i,j}, i = St, AMS, Te,\ j = 1,2,3$). The total length of the tubular reactor (z_f) is fixed, since it is assumed that an existing reactor is to be used. As a productivity requirement, the total residence time (t_f) in either reactor could not exceed 90 minutes. A minimum of 10% conversion was required while the final $\overline{Mn}$ should be between 10000 and 15000 g/mol to meet product specifications. It was also specified that the final fraction of dormant molecules, that is, active chains capable of continuing the polymerization reaction, should be no less than 0.9.

The mathematical formulations of the optimization problems for each reactor were posed as follows:

Semibatch Reactor

$$\min_{m_{i,0}, T_0, sT, f_{i,j}, t_j} FO_1 = \int_0^{t_f} (Inst\_comp_{St} - Desired\_comp_{St})^2 dt$$

or

$$\min_{m_{i,0}, T_0, sT, f_{i,j}, t_j} FO_2 = (\overline{Ln}_{Seq,St}(t_f) - Desired\_\overline{Ln}_{Seq,St})^2$$

s.t

Process model

$$\begin{aligned}
&\frac{dT}{dt} = sT,\ T(0) = T_0 && \text{(a)}\\
&t_f = \textstyle\sum_{j=1}^{4} t_j && \text{(b)}\\
&f_i(t) = f_{i,j}\ when\ \textstyle\sum_{k=1}^{j-1} t_k \le t \le \sum_{k=1}^{j} t_k && \text{(c)}\\
&Conv\ (t_f) \ge 10\% && \text{(d)}\\
&110°C \le T \le 155°C && \text{(e)}\\
&10000\ g/mol \le \overline{Mn}(t_f) \le 15000\ g/mol && \text{(f)}\\
&y_{Dormant}(t_f) \ge 0.9 && \text{(g)}\\
&Reaction\ volume \le 2\ dm^3 && \text{(h)}\\
&30\ min \le t_f \le 90\ min && \text{(i)}\\
&5\ min \le t_j \le 30\ min && \text{(j)}\\
&0 \le m_{St}(0) \le 1000\ g && \text{(k)}\\
&0 \le m_{AMS}(0) \le 1000\ g && \text{(l)}
\end{aligned} \qquad (7)$$

Tubular Reactor

$$\min_{f_{i,0}, T_0, sT, f_{i,j}, z_j} FO_1 = \int_0^{z_f} (Inst\_comp_{St} - Desired\_comp_{St})^2 dz$$

or

$$\min_{f_{i,0}, T_0, sT, f_{i,j}, z_j} FO_2 = (\overline{Ln}_{Seq,St}(z_f) - Desired\_\overline{Ln}_{Seq,St})^2$$

s.t

Process model

$$\begin{aligned}
&\frac{dT}{dz} = sT,\ T(0) = T_0 && \text{(a)}\\
&f_i(z_j) = f_{i,j}\ when\ z = z_j && \text{(b)}\\
&Conv\ (z_f) \ge 10\% && \text{(c)}\\
&110°C \le T \le 155°C && \text{(d)}\\
&10000\ g/mol \le \overline{Mn}(z_f) \le 15000\ g/mol && \text{(e)}\\
&y_{Dormant}(z_f) \ge 0.9 && \text{(f)}\\
&30\ min \le t_f \le 90\ min && \text{(g)}\\
&z_f = 63.15\ cm && \text{(h)}\\
&5\ cm \le z_j \le 30\ cm && \text{(i)}\\
&0 \le f_{St}(0) \le 3.5\ g/min && \text{(j)}\\
&0 \le f_{AMS}(0) \le 3.5\ g/min && \text{(k)}\\
&0.0001\ g/min \le f_{BPO}(0) \le 0.2\ g/min && \text{(l)}
\end{aligned} \qquad (8)$$

$0.5 \leq m_{BPO}(0) \leq 10\ g$ (m)
$0.5 \leq m_{TEMPO}(0) \leq 10\ g$ (n)
$0 \leq f_{St,j} \leq 3.5\ g/min$ (o)
$0 \leq f_{AMS,j} \leq 3.5\ g/min$ (p)
$0.0001\ g/min \leq f_{BPO,j} \leq 0.2\ g/min$ (q)
$0.0001\ g/min \leq f_{TEMPO,j} \leq 0.2\ g/min$ (r)

$0.0001\ g/min \leq f_{TEMPO}(0) \leq 0.2\ g/min$ (m)
$0 \leq f_{St,j} \leq 3.5\ g/min$ (n)
$0 \leq f_{AMS,j} \leq 3.5\ g/min$ (o)
$0.0001\ g/min \leq f_{BPO,j} \leq 0.2\ g/min$ (p)
$0.0001\ g/min \leq f_{TEMPO,j} \leq 0.2\ g/min$ (q)

The process optimization was performed in gPROMS (Process Systems Enterprise, Ltd.) in a standard PC.

Results and Discussion

Recipes to obtain desired copolymer composition in a semibatch reactor

Optimum operating policies were sought for obtaining copolymers with uniform compositions of styrene of 50% or 70%. **Table 1** shows the operating policy for the semibatch reactor for synthesizing a copolymer with a uniform 50% content of styrene. It can be observed that the amount of AMS predominates, both in the initial charge and in the feed flows. This is expected due to the depropagation of AMS, which makes a significant incorporation of this monomer into the copolymer difficult. **Figure 1** shows that the desired uniform composition of 50% St is achieved during the polymerization. It can also be observed in **Figure 1** that the requirements on the product properties and process variables are satisfied.

The optimum temperature profile in this case is isothermal operation at the maximum permitted temperature.

Table 1. Recipe to produce a poly(St-co-AMS) with 50% of St.

Desired Instantaneous Composition = 50% St					*Continuous flowrates (g/min)*		
$m_{St,0}$	62.615 g	*Time periods*	*Length*	*Start time*	$f_{St,j}$	$f_{AMS,j}$	$f_{TEMPO,j}$
$m_{AMS,0}$	466.487 g	t_1	25.0 min	0 min	0	0	0.0025
$m_{BPO,0}$	0.7698 g	t_2	8.4 min	25.0 min	0	0.0599	0
$m_{TEMPO,0}$	0.6112 g	t_3	9.6 min	33.4 min	0	0	0
T_0	155°C	t_4	11.3 min	43 min	0	0	0
sT	0 °C/min	***Total time 54.3 min***					

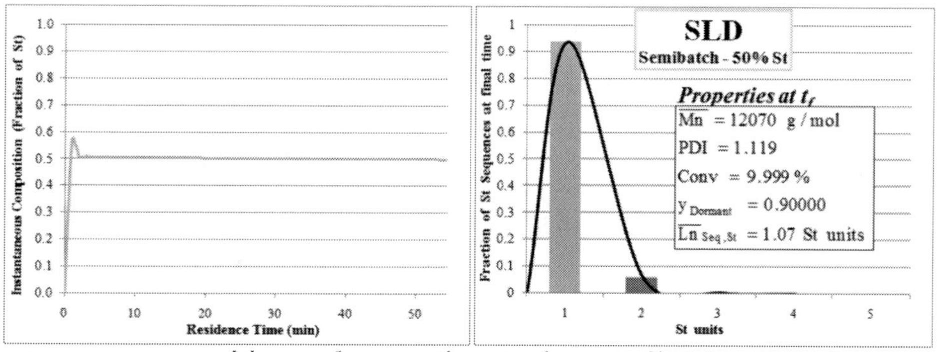

Figure 1. Instantaneous composition and properties at t_f in a semibatch reactor for a goal of 50% St in the copolymer.

As expected, it can be seen in **Table 2** that when the desired composition is 70% St, the mass of this comonomer initially charged and the mass added in the feed flows must be increased considerably. A small decrease in temperature is also necessary to obtain optimum product properties.

Figure 2 shows global and comonomer conversions for each case. The small amount of St fed in the first case reacts in a high proportion while AMS does not, even thought the concentration of this monomer in the reacting medium is considerably higher.

Table 2. Recipe to produce a poly(St-co-AMS) with 70% of St.

Desired Instantaneous Composition = 70% St					*Continuous flowrates (g/min)*		
$m_{St,0}$	430.363 g	*Time period*	*Length*	*Start time*	$f_{St,j}$	$f_{AMS,j}$	$f_{TEMPO,j}$
$m_{AMS,0}$	207.757 g	t_1	25.6 min	0 min	0.9310	0	0.0035
$m_{BPO,0}$	0.966 g	t_2	8.5 min	25.6 min	0.2369	0.2346	0
$m_{TEMPO,0}$	0.7735 g	t_3	10.8 min	34.1 min	0.1749	0.1677	0
T_0	155°C	t_4	12.4 min	44.9 min	0	0.0278	0
sT	-0.12 °C/min	***Total time 57.3 min***					

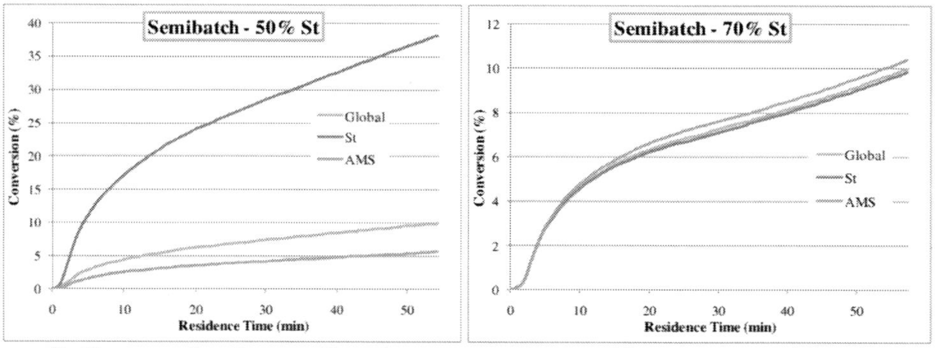

Figure 2. Conversion in semibatch reactor when uniform composition is sought.

A uniform composition of 70% St is achieved and the requirements on the product properties and process variables are satisfied (**Fig. 3**). The St sequences length distribution shows that the fraction of St blocks with more than 1 unit considerably increases with respect to the 50% case.

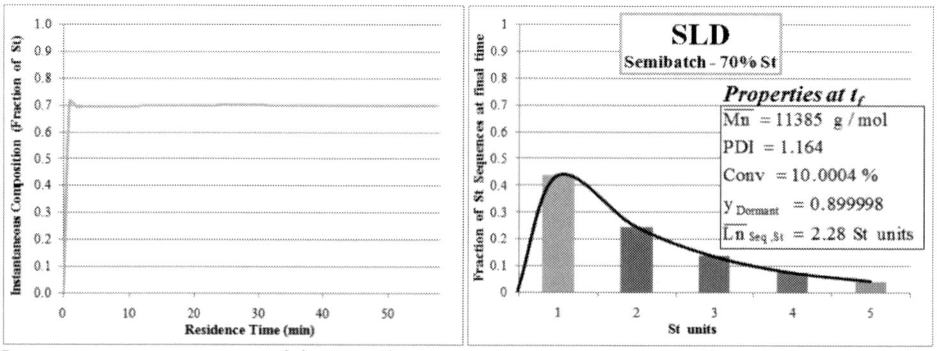

Figure 3. Instantaneous composition and properties at t_f in a semibatch reactor when uniform composition of 70% St is sought.

Recipes to obtain desired average length of styrene sequences in the same semibatch reactor

We set as a goal for the optimization obtaining unitary St sequences in the copolymer. Surprisingly, as can be observed in **Table 3** the optimum process operation that allows this involves appreciable feeds flows of this comonomer. This is simply due to the minute amount of St initially charged in the reactor. To maximize AMS incorporation into the copolymer chain it is necessary to maintain St concentration at a minimum.

Table 3. Recipe to produce a poly(St-co-AMS) with $\overline{Ln}_{seq,St}$ of 1.

Desired Average St Sequences Length = 1					*Continuous flowrates (g/min)*		
$m_{St,0}$	32.462 g	*Time periods*	*Length*	*Start time*	$f_{St,j}$	$f_{AMS,j}$	$f_{TEMPO,j}$
$m_{AMS,0}$	707.215 g	t_1	25.6 min	0 min	0.3452	0	0
$m_{BPO,0}$	1.3742 g	t_2	24.2 min	25.6 min	0.3257	0	0
$m_{TEMPO,0}$	1.1750 g	t_3	28.0 min	49.8 min	0.2011	0	0
T_0	155°C	t_4	12.2 min	77.8 min	0	0	0
sT	0 °C/min	***Total time 90 min***					

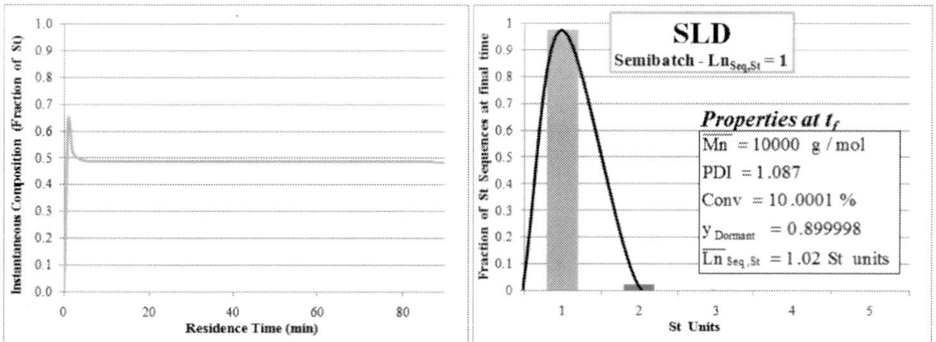

Figure 4. Instantaneous composition and properties at t_f with desired $\overline{Ln}_{seq,St}(t_f)$ of 1 in Semibatch reactor.

For comparison purposes, we set as a goal to obtain a copolymer with average length of St blocks of 3. **Table 4** shows the operating policy that will produce it. A lower temperature and small feed rates of TEMPO are required to maintain the reaction under control and the fraction of dormant chains sufficiently high.

The instantaneous composition needs to increase slightly in order to fulfill the requirements of this objective function (**Fig. 5**).

Table 4. Recipe to produce a poly(St-co-AMS) with $\overline{Ln}_{seq,St}$ of 3.

Desired Average St Sequences Length = 3					*Continuous flowrates (g/min)*		
$m_{St,0}$	883.004 g	*Time periods*	*Length*	*Start time*	$f_{St,j}$	$f_{AMS,j}$	$f_{TEMPO,j}$
$m_{AMS,0}$	305.418 g	t_1	16.8 min	0 min	1.7245	0.0000	0.0004
$m_{BPO,0}$	5.5017 g	t_2	19.9 min	16.8 min	4.8467	0.0014	0.0002
$m_{TEMPO,0}$	4.4625 g	t_3	17.9 min	36.7 min	1.5290	0.0008	0.0001
T_0	144.9 °C	t_4	17.8 min	54.6 min	1.4711	0.0025	0.0003
sT	-0.04 °C/min	***Total time 72.5 min***					

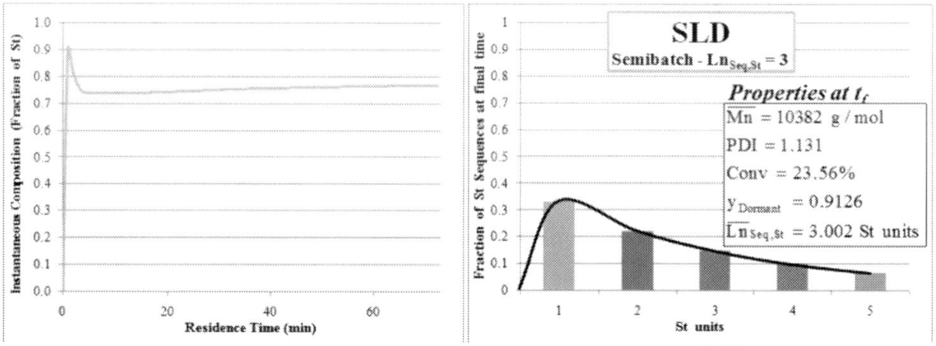

Figure 5. Instantaneous composition and properties at t_f with desired $\overline{Ln}_{seq,St}(t_f)$ of 3 in a Semibatch reactor.

Recipes to obtain desired copolymer composition in a tubular reactor

The operating policy to obtain a copolymer with uniform composition of 50% St in a tubular reactor is shown in **Table 5**.

Table 5. Recipe to produce a poly(St-co-AMS) with 50% of St.

Desired Instantaneous Composition = 50% St					*Injection flowrates (g/min)*		
$f_{St,0}$	0.0684 g/min	*Length extensions*	*Lateral feed positions*		$f_{St,j}$	$f_{AMS,j}$	$f_{TEMPO,j}$
$f_{AMS,0}$	1.2581 g/min	5.00 cm					
$f_{BPO,0}$	0.0045 g/min	25.98 cm	z_1	5.00 cm	0.1907	0.7099	0
$f_{TEMPO,0}$	0.0036 g/min	18.71 cm	z_2	30.98 cm	0.1552	0.1826	0
T_0	155°C	13.46 cm	z_3	46.69 cm	0	0.0171	0.0110
sT	-0.15 °C/cm	***Total residence time 76.8 min***					

As can be seen in **Figure 6**, the control over the copolymer composition is not as precise as the one observed for the semibatch reactor. Even so, almost the same molecular structure can be obtained in both the semibatch and tubular reactors.

Considering that the side feeds in the tubular reactor are point injections, it is expected that higher feed flows of AMS should be required for this reactor design than for the semibatch to obtain a copolymer with the same uniform composition of 50% St. Our results confirm this expectation.

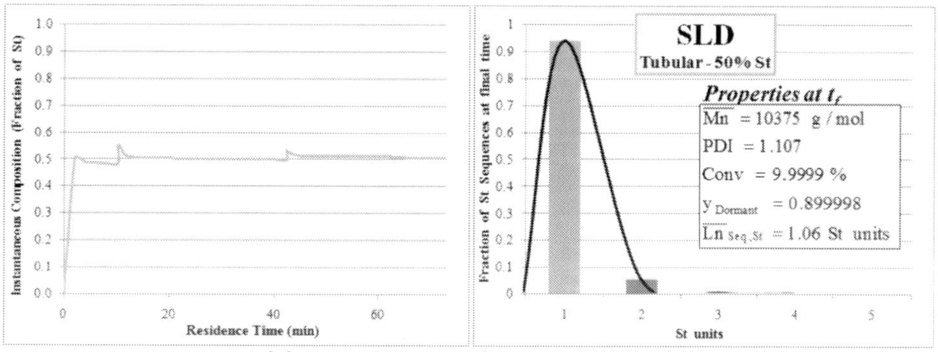

Figure 6. Instantaneous composition and properties at t_f in Tubular reactor for a goal of uniform composition of 50% St.

With respect to conversion, **Figure** 7 shows that in the tubular reactors occurs a similar phenomenon of that in the semibatch; the St fed in a much smaller proportion undergoes a higher conversion. The large drops are due the fresh monomer injections.

It is worth noting that due to the lack of flexibility of the tubular reactor the system often chooses to inject a sufficiently large amount of TEMPO to "numb" the reaction. In this way the reaction remains under control but it can be seen than the last section of the reactor is idle.

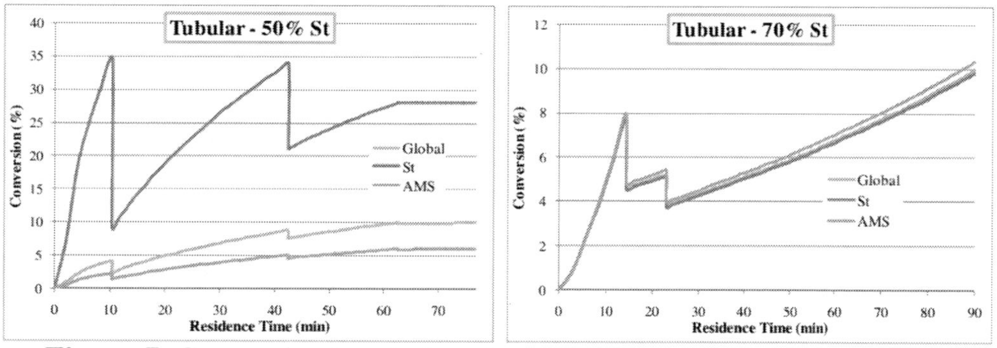

Figure 7. Conversion in tubular when uniform composition is desired.

Table 6 shows the recipe to produce a copolymer with 70% St. An important difference from the previous case are the lateral feeds positions. Flow injections to obtain a copolymer with 70% St are concentrated in the first half of the reactor while for 50% St the side feeds were more distributed.

Table 6. Recipe to produce a poly(St-co-AMS) with 70% of St.

Desired Instantaneous Composition = 70% St					*Injection flowrates (g/min)*		
$f_{St,0}$	0.6250 g/min	*Length extensions*	*Flow injection in*		$f_{St,j}$	$f_{AMS,j}$	$f_{TEMPO,j}$
$f_{AMS,0}$	0.2953 g/min	5.00 cm					
$f_{BPO,0}$	0.0029 g/min	5.01 cm	z_1	5.00 cm	0.4351	0.1952	0.0003
$f_{TEMPO,0}$	0.0024 g/min	23.20 cm	z_2	10.01 cm	0.3850	0.1763	0
T_0	155°C	29.94 cm	z_3	33.21 cm	0	0	0
sT	-0.13 °C/cm	***Total residence time 90 min***					

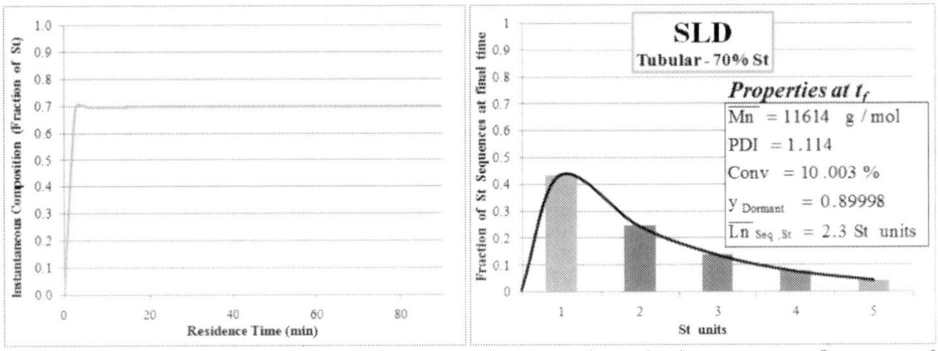

Figure 8. Instantaneous -composition and properties at t_f in tubular reactor for a goal of uniform composition of 70% St.

Recipes to obtain desired average length of styrene sequences in the same tubular reactor

We set as goals obtaining copolymers with St sequences of length 1 or 3. It can be seen in **Table 7** the operating policy to obtain a copolymer with unitary St blocks on average. Unlike the recipe to obtain a uniform composition of 50% St, it does not use any AMS or TEMPO lateral feeds.

Table 7. Recipe to produce a poly(St-co-AMS) with $\overline{Ln}_{Seq,St}$ of 1.

Desired Average St Sequences Length = 1					*Continuous flowrates (g/min)*		
$f_{St,0}$	0.0829 g/min	*Length extensions*	*injection position*		$f_{St,j}$	$f_{AMS,j}$	$f_{TEMPO,j}$
$f_{AMS,0}$	1.8143 g/min	5.86 cm					
$f_{BPO,0}$	0.0035 g/min	8.72 cm	z_1	5.86 cm	0.0128	0	0
$f_{TEMPO,0}$	0.0030 g/min	19.87 cm	z_2	14.58 cm	0.0182	0	0
T_0	155°C	28.71 cm	z_3	34.44 cm	0.0230	0	0
slope_T	0 °C/min	***Total residence time***	***90 min***				

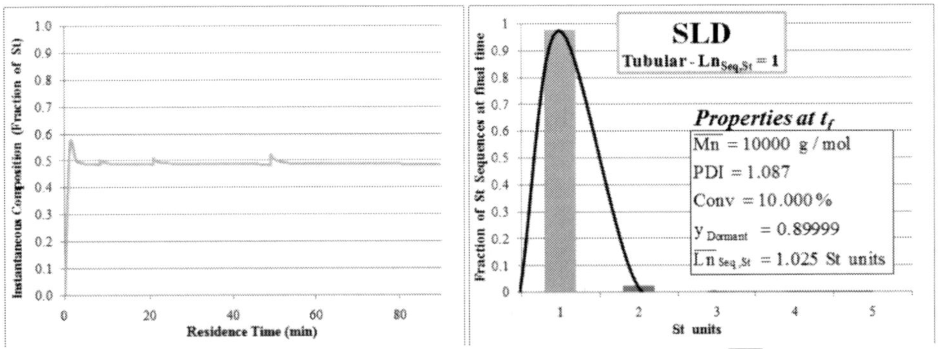

Figure 9. Instantaneous composition and properties at t_f with desired $\overline{Ln}_{Seq,St}(t_f)$ of 1 in a tubular reactor.

Table 8 shows how to operate a tubular reactor to obtain a copolymer with average length of St blocks of 3. A lower temperature is required, as it could be observed for the Semibatch reactor. This temperature decrease is due the higher reactivity of St. Since this comonomer is being fed in higher proportion a convenient way to control the reaction is to lower the temperature.

Table 8. Recipe to produce a poly(St-co-AMS) with $\overline{Ln}_{seq,St}$ of 3.

Desired Average St Sequences Length = 3					*Continuous flowrates (g/min)*		
$f_{St,0}$	0.7484 g/min	*Length extensions*	*injection position*		$f_{St,j}$	$f_{AMS,j}$	$f_{TEMPO,j}$
$f_{AMS,0}$	0.2438 g/min	6.53 cm					
$f_{BPO,0}$	0.0058 g/min	11.00 cm	z_1	6.53 cm	0.4109	0.1570	0.0000
$f_{TEMPO,0}$	0.0046 g/min	20.95 cm	z_2	17.53 cm	0.6786	0.1100	0.0000
T_0	141.8°C	24.67 cm	z_3	38.48 cm	0.3362	0	0.1777
sT	-0.08 °C/min	***Total residence time 83.6 min***					

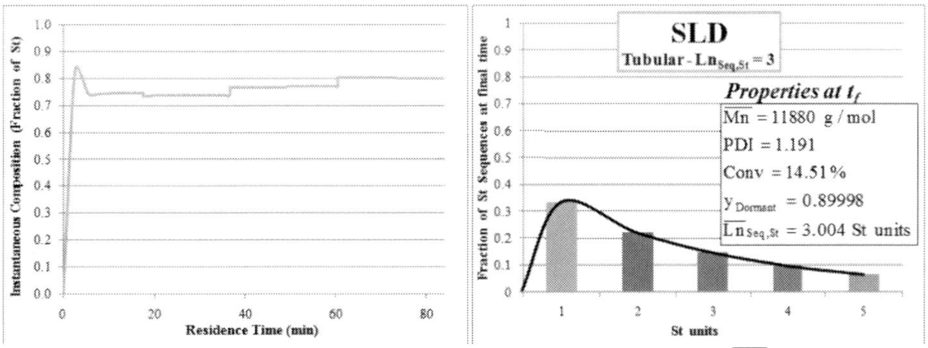

Figure 10. Instantaneous composition and properties at t_f with desired $\overline{Ln}_{seq,St}(t_f)$ of 3 in a tubular reactor.

Conclusions

The copolymerization of St and AMS by NMP has been successfully optimized for both a semibatch and a tubular reactor aiming at obtaining a resin with a pre-specified molecular structure. The results obtained are helpful to select the process and design conditions based on the required characteristics of the material to be produced and the product portfolio.

The major changes required to produce copolymer with such different molecular structure are observed in the mass or flowrate of each monomer initially fed and the feed flows at several times or positions. Isothermal operation at high temperature is the preferred policy.

Even though AMS depropagates at reaction temperature it is possible to obtain copolymers with a composition up to 50% of this comonomer.

The semibatch reactor showed a higher control over the reaction since more uniform materials are obtained. Although the tubular reactor is, by far, less flexible than a semibatch one, our results show that useful tailor-made copolymers can be produced in it. Furthermore, for both reactor types it is possible to produce resins with considerably different molecular properties in the same reactor. However, the design of the processes must be carefully thought in order to achieve an optimum control over the copolymer structure. If very different materials are to be produced in the same physical reactor, different sets of auxiliary equipment (such as pumps) may be needed to accommodate the widely different feed flowrates required. What is more, lateral feeds positions in the tubular reactor should be designed beforehand.

References

1. Matyjaszewski K. and Spanswick J., (2005), "Controlled/living radical polymerization", *Materials Today*, 8(3), pp. 26-33.
2. Braunecker W.A. and Matyjaszewski, K., (2007), "Controlled/Living Radical Polymerization: Features, Developments, and Perspectives", *Progress in Polymer Science (Oxford)*, 32 (1), pp. 93-146.
3. Wang R., Luo Y., Li B. G. and Zhu S., (2007), "Control of Gradient Copolymer Composition in ATRP Using Semibatch Feeding Policy", *AIChE Journal*, 53, pp. 174-186.
4. Lutz J. F., Pakula T. and Matyjaszewski K., (2003), "Synthesis and Properties of Copolymers with Tailored Sequence Distribution by Controlled/Living Radical Polymerization", *ACS Symposium Series, Vol. 854: Advances in Controlled/Living Radical Polymerization*, Chapter 19, pp 268–282.
5. Zhang M. and Ray W. H., (2002), "Modeling of "Living" Free-Radical Polymerization Processes. I. Batch, Semibatch, and Continuous Tank Reactors", *Journal of Applied Polymer Science*, 86, pp. 1630-1662.
6. Zhang M. and Ray W. H., (2002), "Modeling of "Living" Free-Radical Polymerization Processes. II. Tubular Reactors", *Journal of Applied Polymer Science*, 86, pp. 1047-1056.
7. Wang L. and Broadbelt L. J., (2009), "Explicit Sequence of Styrene/Methyl Methacrylate Gradient Copolymers Synthesized by Forced Gradient Copolymerization with Nitroxide-Mediated Controlled Radical Polymerization", *Macromolecules*, 42, pp. 7961-7968.
8. Wang L. and Broadbelt L. J., (2010), "Kinetics of Segment Formation in Nitroxide-Mediated Controlled Radical Polymerization: Comparison with Classic Theory", *Macromolecules*, 43, pp. 2228-2235.
9. Papavasiliou G. and Teymour F., (2003), "Reconstruction of the chain length distribution for vinyl–divinyl copolymerization using the numerical fractionation technique", *Macromolecular Theory and Simulations*, 12, pp. 543-548.
10. Krallis A., Meimaroglou D. and Kiparissides C., (2008), "Dynamic Prediction of the Bivariate Molecular Weight-Copolymer Composition Distribution using Sectional-Grid and Stochastic Numerical Methods", *Chemical Engineering Science*, 63, pp. 4342-4360.
11. Wang R., Luo Y., Li B. G., Sun X. and Zhu S., (2006), " Design and Control of Copolymer Composition Distribution in Living Radical Polymerization Using Semi-Batch Feeding Policies: A Model Simulation", *Macromolecular Theory and Simulations*, 15, pp. 356-368.
12. Kukulj D. and Davis T. P., (1998), "Average Propagation Rate Coefficients in the Free-Radical Copolymerization of Styrene and α-Methylstyrene Measured by Pulsed-Laser Polymerization", *Macromolecules*, 31, pp. 5668-5680.
13. Kukulj D., Davis T. P. and Gilbert R. G., (1998), "Chain Transfer to Monomer in the Free-Radical Polymerizations of Methyl Methacrylate, Styrene, and α-Methylstyrene", *Macromolecules*, 31, pp. 994-999.
14. Asteasuain M., Covan D., Sarmoria C., Brandolin A., Leite de Araujo C., and Pinto J.C. (2011), "Comprehensive Mathematical Modeling of Controlled Radical Copolymerization in Tubular Reactors", *Computer-Aided Chemical Engineering (21st European Symposium on Computer-Aided Process Engineering)*, 29, pp. 51-55.
15. Fortunatti C., Sarmoria C., Brandolin A. and Asteasuain M., (2012), "Simulation Study of the Nitroxide Mediated Copolymerization of Styrene and α-Methyl Styrene under Different Reactor Configurations", *Polymer Reaction Engineering VIII - ECI 2012*, Cancun, Mexico.
16. Asteasuain M., Matheus S., Lenzi M. K., Hutchinson R. A., Cunningham M., Brandolin A., Pinto J. C. and Sarmoria C., (2007), ""Living" Free radical Polymerization in Tubular Reactors. I. Modeling of the Complete Molecular Weight Distribution Using Probability Generating Functions", *Macromolecular Reaction Engineering*, 1, pp. 622-634.
17. Faliks A., Yetter R.A., Floudas C.A., Wei Y. and Rabitz H., (2000), "Optimization of living polymerization through distributed control of a nitroxide radical", *Polymer*, 42, pp. 2061–2065.
18. Asteasuain M. and Brandolin A., (2010), "Mathematical Modeling of Bivariate Polymer Property Distributions Using 2D Probability Generating Functions, 1 - Numerical Inversion Methods", *Macromolecular Theory and Simulations*, 19, pp. 342-359.

Appendix

Table A1. Mass, Moment and pgf balances and other equations.

Global mass balances for reactors
Balances for Tubular reactor: Global: $\frac{d(\rho_{mix}\cdot v)}{dz} = \frac{F_{lat}}{a\cdot\Delta z}$ Generic specie X$_j$: $\frac{d([X_j]\cdot v)}{dz} = r_{X_i} + F_{lat}[X_j]_{lat}$
Balances for Semibatch reactor: Global: $\frac{d(\rho_{mix}\cdot V)}{dt} = F_{lat}$ Generic specie X$_j$: $\frac{d([X_j]\cdot V)}{dt} = r_{X_i}\cdot V + F_{lat}[X_j]_{lat}$
Reaction Rates
Initiator: $r_I = -k_d[I]$ TEMPO: $r_{Te} = -k_{cap,1}[Te]\,\lambda^1_{0,0} + k_{uncap,1}\,\mu^1_{0,0} - k_{cap,2}[Te]\,\lambda^2_{0,0} + k_{uncap,2}\,\mu^2_{0,0}$
Monomer (j) - j= (1)-St or (2)-AMS/MMA: $r_{Mj} = -2f\,k_d[I]\left(\frac{[M_j]}{[M_j]+[M_i]}\right) - 3k_{th}[M_j]^3\delta_{j,1} - (k_{p,jj} + k_{trm,jj})[M_j]\,\lambda^j_{0,0} - (k_{p,ij} + k_{trm,ij})[M_j]\,\lambda^i_{0,0} + k_{p^{-1},jj}\,p_{22}\,\lambda^j_{0,0}\delta_{j,2}$
Reaction rates of polymeric species
Dormant copolymer with *n* units of monomer (1) and *m* units of monomer (2) ended in monomer (j), $D^j_{n,m}$: $r_{D^j_{n,m}} = k_{cap,j}[Te][R^j_{n,m}] - k_{uncap,j}[D^j_{n,m}]$
Macroradical chain with *n* units of monomer (1) and *m* units of monomer (2) ended in monomer (j), $R^j_{n,m}$: $r_{R^j_{n,m}} = \left\{2f\,k_d\,[I]\left(\frac{[M_j]}{[M_j]+[M_i]}\right) + (k_{trm,jj}\,\lambda^j_{0,0} + k_{trm,ij}\,\lambda^i_{0,0})[M_j]\right\}\delta_{n,1\delta_{j,1}}\delta_{m,1\delta_{j,2}} + \left(\frac{1}{2}\right)k_{th}[M_j]^3\,(\delta_{n,1}\delta_{m,0} + \delta_{n,2}\delta_{m,0})\delta_{j,1} + k_{uncap,j}\,[D^j_{n,m}] + k_{p^{-1},jj}\,p_{22}[R^j_{n,m+1}]\delta_{j,2} + k_{p,jj}[M_j]\left[R^j_{n-1\delta_{j,1},m-1\delta_{j,2}}\right] + k_{p,ij}[M_j]\left[R^i_{n-1\delta_{j,1},m-1\delta_{j,2}}\right] - \{k_{cap,j}[Te] + (k_{p,jj} + k_{trm,jj})[M_j] + (k_{p,ji} + k_{trm,ji})[M_i] + k_{t,jj}\,\lambda^j_{0,0} + k_{t,ji}\,\lambda^i_{0,0}\}\,[R^j_{n,m}]$
Dead copolymer with *n* units of monomer (1) and *m* units of monomer (2), $P_{n,m}$: $r_{P_{n,m}} = (k_{trm,11}[M_1] + k_{trm,12}[M_2])[R^1_{n,m}] + (k_{trm,21}[M_1] + k_{trm,22}[M_2])\,[R^2_{n,m}] + k_{t,12}\sum_{l=0}^{n-l}\sum_{r=0}^{m-r}[R^1_{n-l,m-r}]\,[R^2_{l,r}] + \left(\frac{1}{2}\right)k_{t,11}\sum_{l=0}^{n-l}\sum_{r=0}^{m-r}[R^1_{n-l,m-r}]\,[R^1_{l,r}] + \left(\frac{1}{2}\right)k_{t,22}\sum_{l=0}^{n-l}\sum_{r=0}^{m-r}[R^2_{n-l,m-r}]\,[R^2_{l,r}]$
Rates of ath, bth order moments for polymeric species (a=0, 1, 2 and b=0, 1, 2):
Moments of Dormant copolymer ended in monomer (j), $\mu^j_{a,b}$: $r_{\mu^j_{a,b}} = k_{cap,j}[Te]\,\lambda^j_{a,b} - k_{uncap,j}\,\mu^j_{a,b}$
Moments of Macroradical chain ended in monomer (j), $\lambda^1_{a,b}$: $r_{\lambda^j_{a,b}} = \left\{2f\,k_d\,[I]\left(\frac{[M_j]}{[M_j]+[M_i]}\right) + (k_{trm,jj}\,\lambda^j_{0,0} + k_{trm,ij}\,\lambda^i_{0,0})[M_j]\right\}(1^a\,0^b\delta_{j,1} + 0^a\,1^b\delta_{j,2}) + k_{th}[M_j]^3\,(1^a\,0^b + 2^a\,0^b)\delta_{j,1} + k_{uncap,j}\,\mu^j_{a,b} + k_{p,jj}[M_j]\left\{\left(\sum_{h=0}^{a}\binom{a}{h}\lambda^j_{a-h,b}\right)\delta_{j,1} + \left(\sum_{g=0}^{b}\binom{b}{g}\lambda^j_{a,b-g}\right)\delta_{j,2}\right\} + k_{p,ij}[M_j]\left\{\left(\sum_{h=0}^{a}\binom{a}{h}\lambda^i_{a-h,b}\right)\delta_{j,1} + \left(\sum_{g=0}^{b}\binom{b}{g}\lambda^i_{a,b-g}\right)\delta_{j,2}\right\} + k_{p^{-1},jj}\,p_{22}\sum_{g=0}^{b}\binom{b}{g}(-1)^g\,\lambda^j_{a,b-g}\delta_{j,2} - \{k_{cap,j}[Te] + (k_{p,jj} +$ *ktrm,jjMj+kp,ij+ktrm,ijMi+kt,jj λ0,0j+kt,ji λ0,0i+kp−1,jj p22δj,2 λa,bj*
Moments of Dead copolymer, $\varepsilon_{a,b}$: $r_{\varepsilon_{a,b}} = (k_{trm,11}[M_1] + k_{trm,12}[M_2])\,\lambda^1_{a,b} + (k_{trm,21}[M_1] + k_{trm,22}[M_2])\,\lambda^2_{a,b} + k_{t,12}\sum_{j=0}^{a}\sum_{g=0}^{b}\binom{a}{j}\binom{b}{g}\lambda^1_{a-j,b-g}\lambda^2_{j,g} + \left(\frac{1}{2}\right)k_{t,11}\sum_{j=0}^{a}\sum_{g=0}^{b}\binom{a}{j}\binom{b}{g}\lambda^1_{a-j,b-g}\lambda^1_{j,g} + \left(\frac{1}{2}\right)k_{t,22}\sum_{j=0}^{a}\sum_{g=0}^{b}\binom{a}{j}\binom{b}{g}\lambda^2_{a-j,b-g}\lambda^2_{j,g}$
Rates of ath, bth order pgf for polymeric species (a=0, 1, 2 and b=0, 1, 2):
pgf of Dormant copolymer ended in monomer (j), $\varphi^j_{a,b}(z_1, z_2)$: $r_{\mu^j_{a,b}\varphi^j_{a,b}} = k_{cap,j}[Te]\,\lambda^j_{a,b}\sigma^j_{a,b} - k_{uncap,j}\,\mu^j_{a,b}\varphi^j_{a,b}$

pgf of Macroradical chain ended in monomer (j), $\sigma^{j}_{a,b}(z_1, z_2)$:

$$r_{\lambda^{j}_{a,b}\sigma^{j}_{a,b}} = \left\{2f\ k_d\ [I]\left(\frac{[M_j]}{[M_j]+[M_i]}\right) + \left(k_{trm,jj}\ \lambda^{j}_{0,0} + k_{trm,ij}\ \lambda^{i}_{0,0}\right)[M_j]\right\}\left(z_1\ 1^a\ 0^b \delta_{j,1} + z_2\ 0^a\ 1^b \delta_{j,2}\right) + k_{th}[M_j]^3\ (z_1\ 1^a\ 0^b +$$

(z1)2 2a 0bδj,1+kuncap,j μa,bjφa,bj+kp−1,jj p22z2g=0bbg(−1)g λa,b−gjσa,b−gjδj,2+kp,jjMj z1 h=0aahλa−h,bjσa−h,bjδj,1+z2g=0bbgλa,b−gjσa,b−gjδj,2+kp,ijMj z1 h=0aahλa−h,biσa−h,biδj,1+z2g=0bbgλa,b−giσa,b−giδj,2−kcap,jTe+kp,jj+ktrm,jjMj+kp,ji+ktrm,jiMi+kt,jjλ0,0j+kt, ji λ0,0i+kp−1,jj p22δj,2 λa,bjσa,bj

pgf of Dead copolymer, $\vartheta_{a,b}(z_1, z_2)$:

$$r_{\varepsilon_{a,b}\vartheta_{a,b}} = \left(k_{trm,11}\left[M_1\right]+k_{trm,12}\left[M_2\right]\right)\lambda^{1}_{a,b}\sigma^{1}_{a,b} + \left(k_{trm,21}\left[M_1\right]+k_{trm,22}\left[M_2\right]\right)\lambda^{2}_{a,b}\sigma^{2}_{a,b} + k_{t,12}\sum_{j=0}^{a}\sum_{g=0}^{b}\binom{a}{j}\binom{b}{g}\lambda^{1}_{a-j,b-g}\sigma^{1}_{a-j,b-g}\ \lambda^{2}_{j,g}\sigma^{2}_{j,g}$$

$$+\left(\frac{1}{2}\right)k_{t,11}\sum_{j=0}^{a}\sum_{g=0}^{b}\binom{a}{j}\binom{b}{g}\lambda^{1}_{a-j,b-g}\sigma^{1}_{a-j,b-g}\ \lambda^{1}_{j,g}\sigma^{1}_{j,g} + \left(\frac{1}{2}\right)k_{t,22}\sum_{j=0}^{a}\sum_{g=0}^{b}\binom{a}{j}\binom{b}{g}\lambda^{2}_{a-j,b-g}\sigma^{2}_{a-j,b-g}\lambda^{2}_{j,g}\sigma^{2}_{j,g}$$

pgf of the overall copolymer, $\tau_{a,b}(z_1, z_2)$:

$$\tau_{a,b} = \frac{\lambda^{1}_{a,b}\ \sigma^{1}_{a,b} + \lambda^{2}_{a,b}\ \sigma^{2}_{a,b} + \mu^{1}_{a,b}\ \varphi^{1}_{a,b} + \mu^{2}_{a,b}\ \varphi^{2}_{a,b} + \varepsilon_{a,b}\vartheta_{a,b}}{\lambda^{1}_{a,b} + \lambda^{2}_{a,b} + \mu^{1}_{a,b} + \mu^{2}_{a,b} + \varepsilon_{a,b}}$$

Average Properties Equations

Instantaneous composition: $$Inst\_Comp = \frac{k_{p,11}\,[M_1]\,\lambda^{1}_{0,0} + k_{p,21}\,[M_1]\,\lambda^{2}_{0,0}}{k_{p,11}\,[M_1]\,\lambda^{1}_{0,0} + k_{p,21}\,[M_1]\,\lambda^{1}_{0,0} + k_{p,12}\,[M_2]\,\lambda^{1}_{0,0} + k_{p,22}\,[M_2]\,\lambda^{2}_{0,0} - k_{p-1,22}\,p_{22}\,\lambda^{2}_{0,0}}$$

Number Fractions of Dormant molecules: $$y_{Dormant} = \frac{\mu^{1}_{0,0}+\mu^{2}_{0,0}}{\mu^{1}_{0,0}+\mu^{2}_{0,0}+\lambda^{1}_{0,0}+\lambda^{2}_{0,0}+\varepsilon_{0,0}}$$

Number average St Sequence Length: $$\overline{Ln}_{Seq,St} = \frac{\lambda_{1}^{Seq1} + \mu_{1}^{Seq1} + \varepsilon_{1}^{Seq1}}{\lambda_{0}^{Seq1} + \mu_{0}^{Seq1} + \varepsilon_{0}^{Seq1}}$$

(595e) A Multi-Domain Read-Across Framework to Predict Aerobic Biodegradation of Organic Chemicals

Chandan Srivastava[1], Alberto Fernandez[1], Robert Rallo[2] and Francesc Giralt[3], (1)Enginyeria Química, Universitat Rovira i Virgili, Tarragona, Spain, (2)Departament d'Enginyeria Informatica i Matematiques, Universitat Rovira Virgili, Tarragona, Spain, (3)BIOcenit, Departament d'Enginyeria Química, Universitat Rovira i Virgili, Tarragona, Spain

The implementation of international regulations for chemicals (e.g., REACH in the EU) requires the integrated use of a variety of testing and non-testing methods to establish the PBT profiles of new and existing chemicals. Similarity-based methods such as read-across [1,2] provide, among the variety of techniques that can be used to implement non-testing strategies, a sound and cost-effective framework to predict physicochemical properties and biological activity of chemicals. Read-across methods strongly rely on both the type of information and the similarity metric used to identify the set of reference chemicals (families) used to generate the estimate. Accordingly, given the extense variety of molecular descriptors and properties available (i.e., data domains), finding the best information to characterize chemical families for a specific endpoint is a challenging task of crucial importance.

The present work introduces a multi-domain read-across framework, based on the integration of various types of chemical information, which is applied to predict the aerobic biodegradation of organic chemicals. The data used to demonstrate this approach were extracted from the Japanese Ministry of International Trade and Industry biodegradation database [3]. Compounds with insufficient structural information or with inconsistent biodegradation measures were filtered to produce a quality-controlled database containing experimental biodegradation measures (BOD) for 1063 chemicals. The intrinsic topology preservation properties of the Self-Organizing Map (SOM) were used to identify chemical families within each data domain (e.g., constitutional descriptors, functional group counts, atom center fragments). The quality and reliability of each chemical family was quantified using consensus clustering implemented *via*multiscale bootstrap methods [4]. Different approaches to compute read-across estimates (e.g., average, weighted average, value of the most similar chemical) for target compounds (i.e., not used during SOM training) were assessed. The best results were obtained from the similarity-weighted average of the chemicals assigned to the same cluster as the target compound. Finally, consensus techniques [5] were used to combine, in an optimal way, the output of each individual SOM to yield a final and more accurate read-across estimate. Results obtained using the new multi-domain read-across method outperformed results obtained via conventional read-across based on a particular type of information.

This work demonstrated that multi-domain read-across provides a suitable framework which is able to identify complementary chemical families with similarity features that remain hidden when only data from a single domain are used. Ongoing research work aims to include mechanistic information to enhance the characterization of the chemical families with the inclusion of biological information for the biodegrading organisms (e.g., activity of genes involved in biodegradation pathways).

References

[1] Schuurmann G., Ralf-Uwe E., Ralph K. (2011) Quantitative read-across for predicting the acute fish toxicity of organic compounds. *Environ. Sci. Technol*. 45, 4616–4622.

[2] Enoch SJ, Cronin MT, Schultz TW, Madden JC, (2008) Quantitative and mechanistic read across for predicting the skin sensitization potential of alkenes acting via Michael addition. Chem Res Toxicol, 21(2):513-20.

[3] CITI - Biodegradation and Bioaccumulation data of existing chemicals based on the CSCL Japan (1992). Chemicals Inspection and Testing Institute, Japan, ISBN 4-98074-101-1.

[4] Rallo R., France B., Liu R., Nair S., George S., Damoiseaux R., Giralt F., Nel A., Bradley K., Cohen Y. (2011)

Self-Organizing Map Analysis of Toxicity-Related Cell Signaling Pathways for Metal and Metal Oxide Nanoparticles. Environmental Science & Technology. 634-640.

[5] Fernández A; Lombardo A; Rallo R; Roncaglioni A; Giralt F; Benfenati E. (2012) Quantitative consensus of bioaccumulation models for integrated testing strategies. Environment International, (in press, doi: 10.1016/j.envint.2012.03.004).

Acknowledgement. This research was financially supported by the Spanish Ministry of Economy and Competitivity (CTM2011-24303).

(595f) Ozone Impacts of Flaring Emissions Considering Crosswind Effects

Ziyuan Wang, Jian Zhang, Qiang Xu and T. C. Ho, Dan F. Smith Department of Chemical Engineering, Lamar University, Beaumont, TX

Industrial flares emissions have been considered as a major part of point source emissions of CAMx model in a metropolitan region heavily concentrated with chemical plants, i.e., the Houston-Galveston region in Texas. Flare emissions depend heavily on a flare's destruction efficiency. Several studies have shown that crosswind may compromise the expected 98% combustion efficiency level of the flares. In this study, a CAMx simulation has been performed to predict the effect of crosswind on the formation of Ozone.

Both experimental flare data and computational fluid dynamics (CFD) simulation showed that the flare flames are prone to inefficient combustion under high crosswind velocity. In the CAMx simulations, the efficiencies of the flares in the Texas area vary with crosswind velocity. As the efficiency decreases, the emissions including VOCs and NO_X, both of which are ozone precursors, will change. The obtained flaring emissions are then incorporated into a CAMx air quality simulation model to simulate the effect of these VOCs and NO_X emissions on the regional ozone concentrations. The June 2006 ozone episode is selected to represent the meteorological conditions. The required emission inventories from different sources, i.e., point, mobile, area, and biogenic sources, are acquired from the Texas Commission on Environmental Quality (TCEQ). A model run has been conducted in which flare emissions are changed. At the same time, the potential source contribution function (PSCF)-based HRVOC reconciliation is removed. The results show that the model with crosswind effect can predict the ozone concentrations better.

(595g) Optimization of Agile Energy Network for System Design and Dispatch Under Emergency

Tianxing Cai and Qiang Xu, Dan F. Smith Department of Chemical Engineering, Lamar University, Beaumont, TX

The development of renewable technology has provided more potential sources to the energy network while it has added the difficulty in the mathematical modeling of design and dispatch of multiple energy types in a wide regional area. The optimal construction of energy network should consider the satisfactory coverage of variable demand, the advantage of local energy resources, the reasonable distribution of energy facility, the normal operation cost and the consequence of short-time energy shortage under extreme conditions. In local energy shortage, agile energy dispatching through an effective energy transportation network, targeting the minimum energy recovery time, should be a top priority, while the economic factor should be taken as the objective in routine energy utilization.

In this paper, a novel methodology is developed for energy network optimization for the capability of feasible energy dispatch under emergency of local energy shortage and the objective of minimal operation cost for routine operation under the constraint of demand requirement. This has included five stages of work. First, potential energy network needs to be characterized, where the capacity, quantity, and availability of various energy sources are put into the quantification. Second, the energy initial situation under emergency conditions needs to be identified. Next step, the energy dispatch requirement is represented as constraint based on a developed MILP (mixed-integer linear programming) model in the third stage. Then, mathematical modeling with the objective function of minimal operation cost is constructed. Finally, the sensitivity of the minimum dispatch time with respect to uncertainty parameters of energy loss and the construction cost with respect to energy dispatch capability is characterized by partitioning the entire space of parameters into multiple subspaces. The efficacy of the developed methodology is demonstrated via a case study with in-depth discussions.

(595h) Integration of Process Model and Life Cycle Assessment to Study Incineration and Gasification of Solid Waste

Zac Coventry[1], Amirhossein Mehrkesh[2], Ronald Tize[3] and **Arunprakash T. Karunanithi**[2], (1)Department of Civil Engineering, University of Coloraod Denver, Denver, CO, (2)Department of Civil Engineering, University of Colorado Denver, Denver, CO, (3)MWH, La Hulpe, Belgium

The focus of this work is to model waste to energy technologies, incineration & gasification, and study their environmental impact. Detailed process simulation of these technologies is constructed to derive mass and energy balances with municipal solid waste as feedstock. Material and energy balances for flue gas cleanup (incineration) and syngas cleanup (gasification) is also incorporated. For the incineration case, the energy recovery in the form of heat and conversion to electricity is modeled while for the gasification case syngas production and its conversion to electricity in an internal combustion engines is considered. Using the energy balances derived from the process model, we will examine the environmental impacts of these technologies through life-cycle assessment methods. We will present the process model results as well as the environmental life cycle assessment results of the two technologies: 1) advanced thermal recycling (incineration) with conversion to electricity; and 2) fluidized bed gasification with conversion to electricity. Life cycle environmental impacts are evaluated using the Tool for the Reduction and Assessment of Chemical and other environmental Impacts – TRACI.

(597a) Fire Incident Anomaly Detection and Monitoring System Based On Mining of National Fire Data

Dongil Peter Shin, Dept. of Chemical Engineering, Myongji University, Yongin, Kyunggido, South Korea and Kidon Joo, Dept. of Chemical Engineering, Myongji University, Yongin, South Korea

A real-time anomaly detection and monitoring system of fire incidents has been developed and is on test based on data mining of national fire data. U.S., Korea and most developed countries keep their records of fire incidents and provide basic statistics on the occurrence of fires. However, these massive and valuable data have not been used to proactively manage and reduce the number of fire occurrences or the loss from fire.

Like U.S. National Fire Incident Reporting System (NFIRS), The National Fire Data System (NFDS) of Korea has two objectives: to help national and local governments develop fire reporting and analysis capability for their own use, and to obtain data that can be used to more accurately assess and subsequently combat the fire problem at a national level. To meet these objectives, the NEMA has developed a standard NFDS package that includes incident and casualty forms, a coding structure for data processing purposes, manuals, computer software and procedures, documentation and a training course for utilizing the system.

(598a) Operability Analysis for Grade Transitions in the High-Density Polyethylene Process

Zhi Xia[1], Jianfeng Zhu[2] and Jinsong Zhao[2], (1)Institute of Process Systems Engineering, Department of Chemical Engineering, Tsinghua University, Beijing, China, (2)Institute of Process Systems Engineering, Department of Chemical Engineering, Tsinghua University, Beijing100084, China

According to customer's various and complicated demands, polymer plants have to produce many polymer grades with frequent grade transition operations. Numerous studies have demonstrated that polymerization processes always exhibit highly nonlinear dynamic behaviors, including input/output multiplicities, limit cycles, sustained oscillations, hysteresis, and chaos. These nonlinearities may adversely affect process operation and process controller performance and may lead to unsafe operations[1,2]. The grade transitions operation, resulting in changes of operating/design parameters or topology structure (e.g. parallel mode and series mode), will alter and complicate the overall nonlinearity and internal dynamic behavior of the polymerization process. Therefore, it is of critical importance to assess the effects of the nonlinear dynamics of the plant-wide process and the operability (e.g. stability and controllability) of feasible operating regions and different topology structures for producing required polymer grades.

The major objective of this paper is to analyze the operability for grade transitions in a high-density polyethylene process (HDPE) from a network perspective. The framework contains two parts. First, a plant-wide model of an HDPE process is established for steady-state simulation. Second, based on open-loop stability/instability and minimum/nonminimumphase behavior, the operability of the process for producing different polymer grades can be efficiently probed. An attempt is made with the help of this article to illustrate how to clarify the roots of the poor operability that arise in the grade transition to provide helpful guideline for design of the HDPE processes operation parameters and the control system.

References

1. Yuan ZH, Chen BZ, Zhao JS. Phase Behavior Analysis for Industrial Polymerization Reactors. *AIChE J.* 2011; 57: 2795-2087.
2. Seider WD, Brengel DD, Widagdo S. Nonlinear dynamics found in polymerization processes — a review. *Chem. Eng. Sci.* 2000; 55:275-290.

(598aa) Kinetic Studies of Calcium Oxide Carbonation At Different CO2 Environments

Lele Zhang[1], Yong-Hua Duan[2], Hui Li[3] and Yong Min[3], (1)Xi'an University of Architecture &Technology, Institute of Powder Engineering, Xi'an, China, xian, China, (2)Material Institute, Xi'an University of Architecture and Technology,, Xi'an, China, (3)Xi'an University of Architecture &Technology, Institute of Powder Engineering, Xi'an, China

Kinetic Studies of Calcium Oxide Carbonation at Different CO2 Environments

Zhang Lele, Duan Yonghua, Li Hui, Min Yong

1. Xi'an University of Architecture and Technology; Xi'an Shaanxi 710055; China;

Abstract:

The paper investigates the kinetics analysis pure CaO carbonation by utilizing thermogravimetric analyzer (TGA) at different heating rate (2.5 K/min, 7.5 K/min, 15 K/min, 20 K/min, 30 K/min) and different concentration of CO_2 atmosphere (0%, 30%, 50%, 80%, 100%). Double extrapolation method (Flynn-Wall-Ozawa) was used to obtain the apparent activation energy when conversion rate tends zero; Then use Coats-Redfern integral equation, push the heating rate to zero to get the activation energy of the sample in the thermal equilibrium state; Finally by comparing E-Bata with E-Alpha , confirmed the most probable function of the CaO carbonation in different atmosphere , thus get the dynamic reaction model , exponential factor A and reaction rate constant k, and the changing rule of the activation energy E-alpha of CaO carbonation with different atmosphere

Key words: CaO carbonation; thermal analysis kinetics; double extrapolation method the apparent activation energy E; dynamic reaction model.

(598ab) On-Line Fault Detection and Diagnosis of Pipeline in Water Distribution Network Systems

Shin Je Lee[1], Go Bong Choi[2], Dae Shik Kim[3], Jung Chul Suh[4], Gi Baek Lee[5] and Jong Min Lee[1], (1)School of Chemical and Biological Engineering, Seoul National University, Seoul, South Korea, (2)School of Chemical and Biological Engineering, Seoul National University, Seoul, 151744, South Korea, (3)Seoul National University, (4)Samchully, Seoul, South Korea, (5)Korea National University of Transportation, Chungju-si, South Korea

Water pipe networks are generally installed underground. Once they are equipped, it is difficult to recognize the state of pipes when faults such as leak or burst happen. Accordingly, post management is often delayed, thus making the loss of pipelines increase. Therefore, a systematic fault management system of water pipe network is required to prevent accidents and minimize the loss. The conventional fault detecting methods include listening, noise logging, tracer gas, and GPR. Those methods are easy to implement, but they have a limitation that considerable time, labor, and finances are required to apply those methods. Therefore, a novel approach for detecting and diagnosing faults of water pipelines is required.

In this work, we develop online fault detection and diagnosis system of water pipe networks that detects faults and diagnoses the location and cause of the faults using online data such as pressure or flow rate. The data are measured and gathered into communication devices in lab scales. The proposed approach takes the following steps. First, we decide the locations of sensors for measuring pressure or flow rate and the locations of fault tests for leak or burst. Genetic algorithm (GA) and inverse transient analysis (ITA) are applied to find optimal sensor locations and fault test locations, respectively. A modified scale space filtering (SSF) technique is applied to filter out noise included in the collected data. The proposed SSF approach reduces computational complexity and addresses the end-effect. Fault detection and diagnosis schemes are then applied to the filtered data. Cumulative sum (CUSUM), Fast Fourier transform (FFT), and Discrete wavelet transform (DWT) are applied for effective diagnosis. The proposed approach will be illustrated on a practically-sized pipe network with field test data.

(598b) The Use of a Time Variant IMC Controller for Robust Performance Improvement

Jonathan Whitlow and Latifat Adebayo, Chemical Engineering, Florida Institute of Technology, Melbourne, FL

The design of the IMC controller requires that the filter time constant be specified as a single tunable parameter with an increase resulting generally in lower system performance but increased system robustness. In this work the effect of using a variable filter time constant for implementation in an Internal Model Control (IMC) system is presented. During an initial closed loop response a high performance filter constant can be employed with it changing during the course of the response to a final value resulting in a robust filter constant. This work will demonstrate the potential performance improvement that can be obtained by this approach for various types of plant model structures. The internal models used will be examined with and without plant model mismatch and the performance improvement quantified. Implementation details of the discrete IMC controller and rules for how to adjust the filter constant over time as a function of the model parameters and model structure will also be presented.

A Stable Direct Inverse Neural Network Control of an Isothermal Continuous Stirred Tank Reactor with Input Multiplicities

G. Prabhaker Reddy[*,1] *and B. C. Eranna*[*]

[*]*Dept. of Chemical Engineering, University College of Technology, Osmania University, Hyderabad, India, Phone : 91-9866619899, Fax: 91-4027098472,*

[*,1] *email: gpreddy_ouct@yahoo.com, and *chinna_eranna@rediffmail.com*

Paper Number: 257658

Abstract

In this paper, a stable Direct Inverse Neural Network Controller (DINNC) is designed for an Isothermal continuous stirred tank reactor (CSTR) with input multiplicities. The Continuous Stirred Tank Reactor exhibits input multiplicities in space velocity on concentration. i.e., two values of space velocity will give the same value of concentration. Processes with input multiplicities may experience unstable response, oscillatory and switch over to lower stable input with conventional PID controller. In this work, the direct inverse NN controller is designed separately at lower and higher inputs and the DINNC controller output which is nearer to the operating is selected for the implementation on process. The Performance of direct inverse neural network controller has been evaluated through simulations at lower and higher input space velocities. The response of DINNC is found to be faster than conventional PI controller. Also, DINNC controller gives stable response whereas the PI controller designed for lower input, become unstable response at higher input space velocity. Thus, direct inverse neural network control is found to overcome the control problems due to the input multiplicities.

Keywords: Direct inverse neural network control, CSTR, Input Multiplicities, lower input space velocity and higher input space velocity.

Introduction

Generally in a single Input and Single Output (SISO) process, more than one value of input variable producing same value of output is known as input multiplicity. Input multiplicities occur due to the competing effects in the processes. Dynamic and steady state behavior of the process with input multiplicity will remain distinct at different input values for the same output. Processes with multiple reactions, multi reactors or recycle structures are shown to exhibit input multiplicities (Koppel, L. B. ,1982 &1983). Conventional linear PI controller will have control problems like instability, oscillatory and less economical (Dash, S.K. and Koppel, L.B.(1989)) due to input multiplicities in the process. In the last two decades, a new direction to control has gained considerable attention. This new approach to control is called 'Intelligent control'. The term 'intelligent control' addresses to more general control problems. It may refer to systems, which cannot be adequately described by differential equations framework. There are three basic approaches to intelligent control: knowledge-based experts systems, fuzzy logic and neural networks. The term 'conventional control' refers to theories and methods that are employed to control dynamic systems whose behavior is primarily described by differential and difference equations. Among these intelligent controllers, data based direct inverse neural control has become popular tool for control of dynamic process, demonstrating the ability of handling non linearity. Many neural network controllers are the data based type where the controller's output response is described by a series of data generation, training and validation of the control. The procedure used to perform the learning process is called a learning algorithm, the function of which is to modify the synaptic weights of the network in an orderly fashion to attain a desired design objective. Neural networks Neural network derives its computing power through first it's massively distributed structure and seconds its ability to learn and therefore generalize. Generalization refers to the neural network producing reasonable output for inputs not encountered during training. These two information processing capabilities make it possible for neural networks to solve complex problems that are currently intractable. However neural network cannot provide the solution by working individually. Rather they need to be integrated into a consistent system engineering approach.

There are many advantages of neural networks like Information is distributed over a field of nodes, Neural networks have the ability to learn, Neural networks allow extensive knowledge indexing, Knowledge indexing is the ability to store a large amount of information and access it easily, The network stores knowledge in two forms a) the connection between the nodes b) the weight factors of these connections, Neural networks are better suited for processing noisy, incomplete, or inconsistent data and Neural networks mimic human learning processes.

In this work, the design and evaluation of direct inverse neural network controller for an isothermal CSTR is presented to overcome the control problems associated with conventional PI controller due to input multiplicities.

Description of CSTR with input multiplicities

We consider here a continuous stirred tank reactor (CSTR) with the following isothermal series and parallel reactions (Van de Vusse, 1964);

$$A \xrightarrow{k_1} B \xrightarrow{k_2} C \qquad (1)$$

$$2A \xrightarrow{k_3} D \qquad (2)$$

The product B is the desired one. The mass balance equations for A and B are given by (Kravaris,1990):

$$dx_1/dt = -k_1x_1 - k_3x_1^2 + (C_{A,0} - x_1)u \qquad (3)$$

$$dx_2/dt = k_1x_1 - k_1x_2 + x_2u \qquad (4)$$

Where,

$$x_1 = C_A,\ x_2 = C_B,\ u = F/V \qquad (5)$$

and F is the flow rate (l/min), C_A and C_B are the concentration of A and B in the reactor (mol/l) and $C_{A,0}$ is the feed concentration of A (mol/l). The parameters considered for the present work are given by $k_1 = 0.8333$ (min^{-1}), $k_2 = 1.6667$ (min^{-1}), $k_3 = 0.16667$ ($l.mol^{-1}min^{-1}$),$C_{A,0} = 10$ (mol/l). The values of x_2 versus u_s is shown in Fig. 1 (Chidambaram M. and G.P.Reddy, 1995). It shows input multiplicities in u_s on the product concentration ($x_{2,s}$). That is two same values of $x_{2,s}$ for example, $x_2 = 1.117$ can be obtained at $u_s = 0.5714$ and also at u_s=2.8746. The gain is +0.5848 at $u_s = 0.5714$ min^{-1} where as the gain is -0.1208 at u_s= 2.8746 min^{-1} .

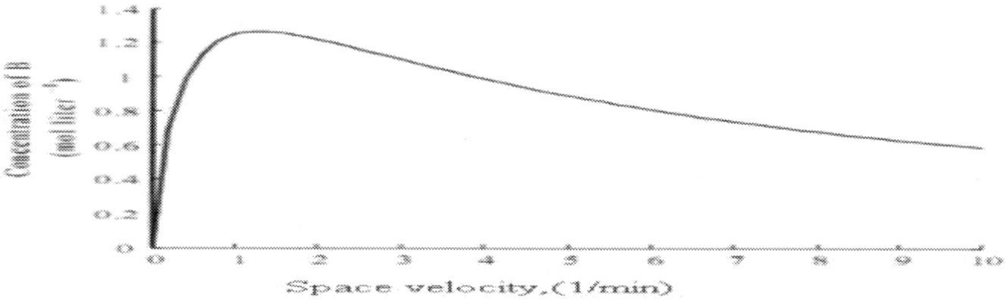

Fig.1. Steady-state behavior of CSTR with input multiplicities

Design of a Direct Inverse Neural Network Controller

The various steps of direct inverse neural network model based controller (Mohamed Azlan Hussain et al 2001) for the CSTR process are presented here. The Input-Output data are generated at lower and higher input space velocities with pseudo random signal (PRS). The process input-out data are shown in Figs.2 & 3 for lower input space velocity (u) of the CSTR system and respective input-output data for higher input space velocity are given in Figs. 4 & 5. They are generated using the simulink model of the process with a sampling time of 0.2s for about 1000 samples.

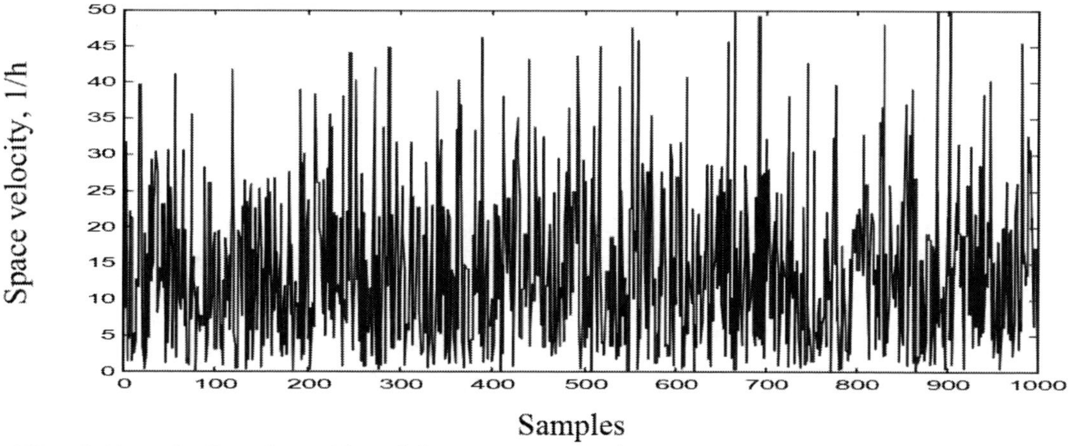

Fig. 2. Pseudo Random Signal for Space velocity, u (process input) to CSTR process at lower input space velocity.

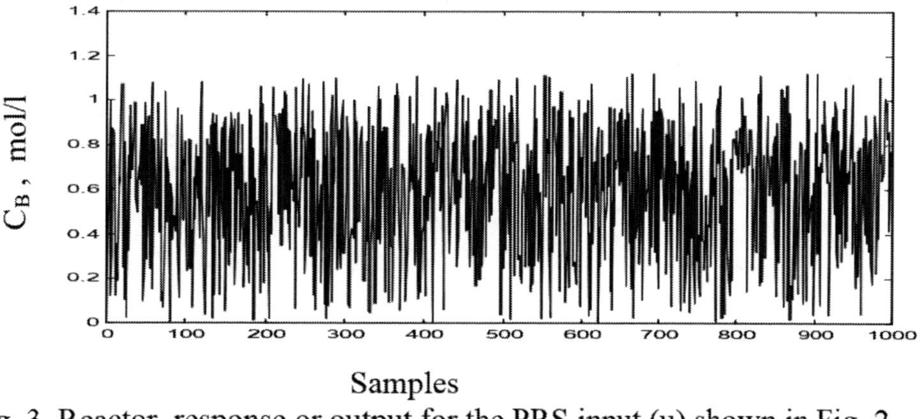

Fig. 3. Reactor response or output for the PRS input (u) shown in Fig. 2

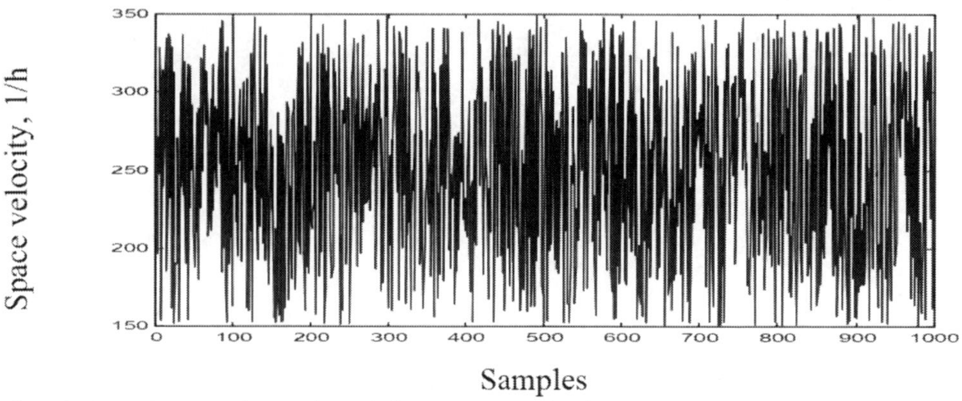

Fig. 4 Pseudo Random Signal for Space velocity, u (process input) to CSTR process at higher input space velocity.

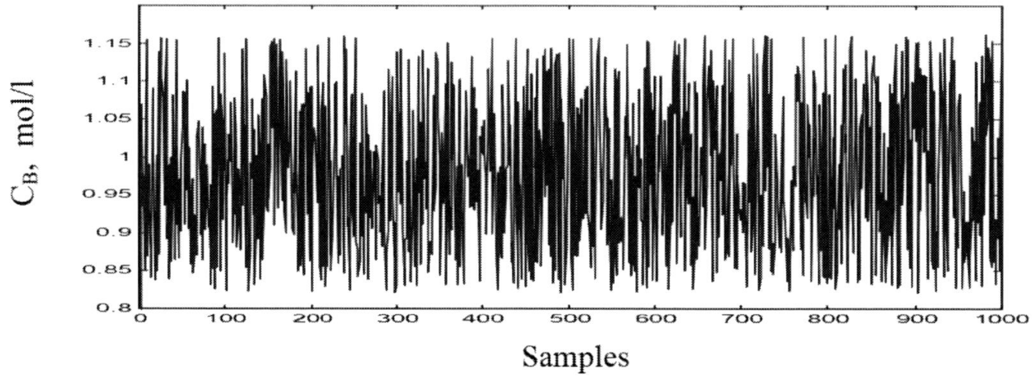

Fig. 5. Reactor response or output for the PRS input (u) shown in Fig. 4

The direct inverse neural network model shown in Fig.6 is basically the neural network structure representing the inverse of the system dynamics at the completion of training. The training procedure in this case is called inverse modeling. Here the network is fed with past inputs, past outputs, present outputs. The network then predicts the controller output, u (t) to make the output to reach the setpoint.

The final network representation of the inverse is given by

$$u(t) = f^{-1}[\, y(t+1), y(t), y(t-1), u(t-1)\ldots] \tag{6}$$

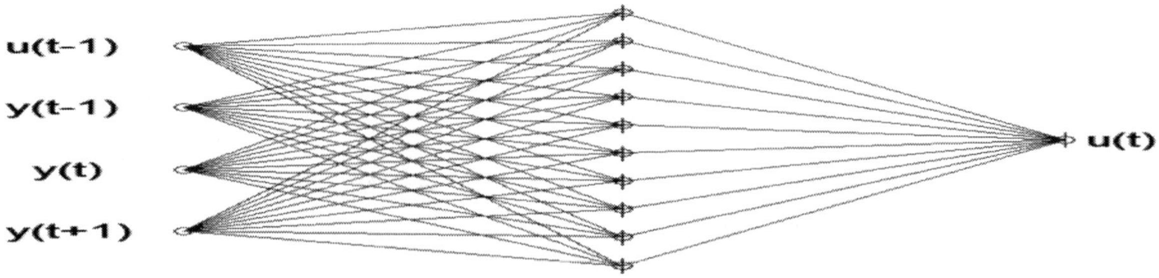

Fig. 6. Structure of inverse neural network model

In this NN design, the Levenberg-Marquardt training method(Baughman and Y. Liu, 1995) is used This method uses the past values of input, u & output, y, the control signal required for producing the desired output is found. The difference between expected u and the neural model output uN is the error eN which can be utilized for network learning. The input-output data obtained is divided into two parts each containing 500 data. The first 500 data are taken for training. Weights are initialized from input to hidden layer & hidden to output layer. Weight matrices:W11&W12 contains weights from input to hidden layer & weight matrices: W21&W22 contains weights from the hidden to output layer. The input matrix is chosen such that it contains the values of past input & output.

Results and Discussion

The performance of proposed direct inverse neural network controller and conventional PI controller to the CSTR with input multiplicities at lower and higher space velocities is evaluated using the closed loop block diagrams as shown in Fig.7. During the identification and control tasks, the NNSYSID (M. Nørgaard, 1996) and NNCTRL (Hunt, K.J., D. Sbarbaro, R. Zbikowski and P.J. Gawthrop 1992) Toolboxes for MATLAB are used. The simulation studies for servo i.e., setpoint tracking and regulatory problem i.e., maintenance of setpoint in the presence of disturbances have been presented below at lower and higher space velocities. The parameters of conventional PI controller used in the simulation studies are, K_c=1.25, τ_I =0.5848 min (Chidambaram M. and G.P.Reddy)

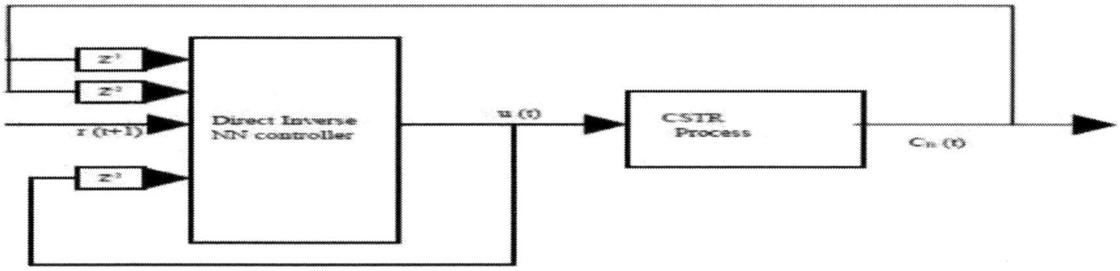

Fig. 7 Closed loop block diagram of Direct Inverse Neural Network Control of a CSTR

Servo problem

The servo response has been studied by giving a step change in set point of concentration of B (C_B) with direct inverse neural network controller (DINNC) and PI controller. At lower space velocity , the setpoint is changed from 1.117 to 1.22 and are shown in Fig.8. The Direct inverse controller shows faster response and reaches new setpoint in about 3 min but it gives some offset, whereas the PI controller response is slower and reaches setpoint in 6 min. To overcome the offset with DINNC, an integral action (Ti=1.25 min) is added. It works like a hybrid controller. The DINNC+I controller response is shown in Fig.9 and the corresponding control action in terms of space velocity is shown in Fig.10

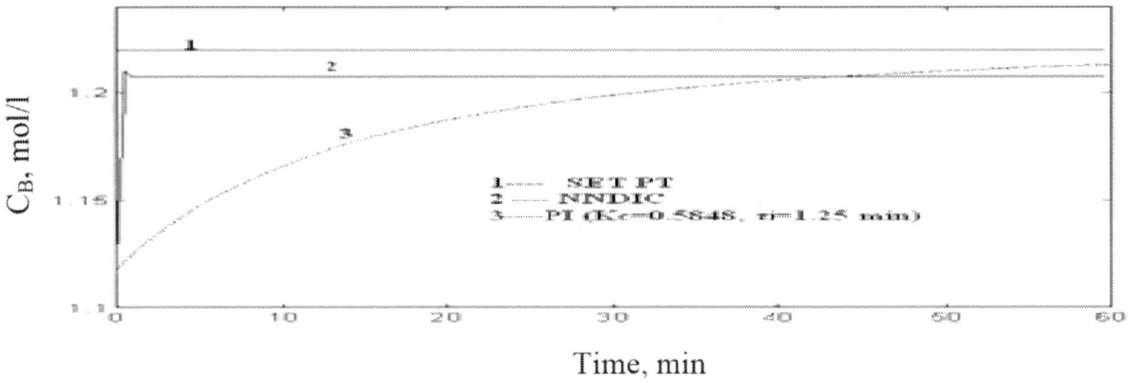

Fig.8.Reactor response for step change in setpoint from 1.117 to 1.22 at lower input space velocity.

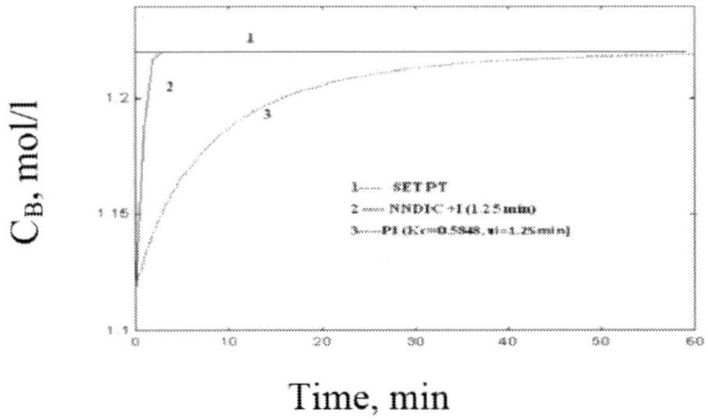

Fig.9. Reactor response for setpoint change from 1.117 to 1.22 at lower input space velocity.

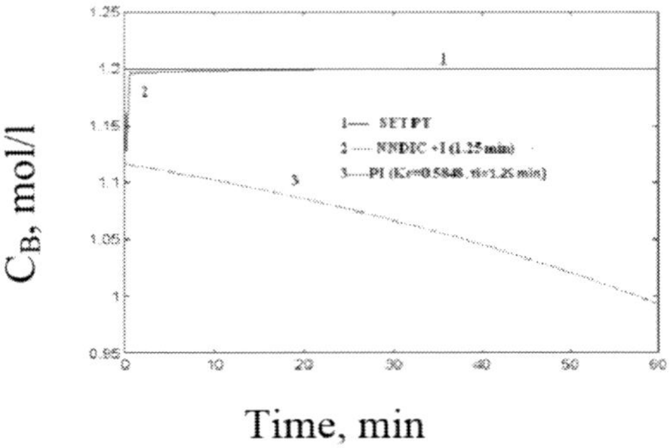

Fig.11 Reactor response for a setpoint change from 1.117 to 1.22 at higher input space velocity.

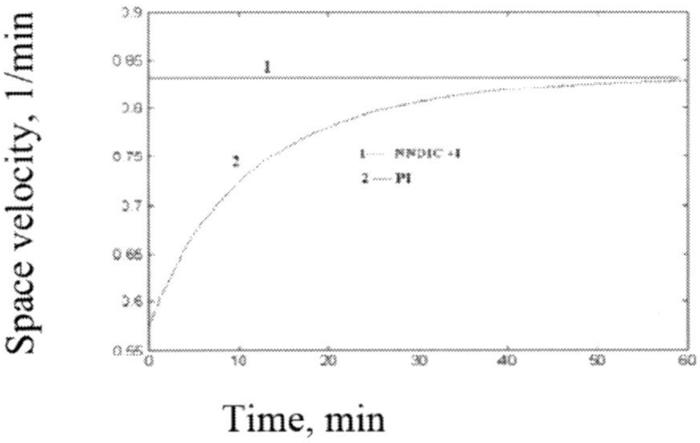

Fig.10 Control action for the response shown in Fig.9

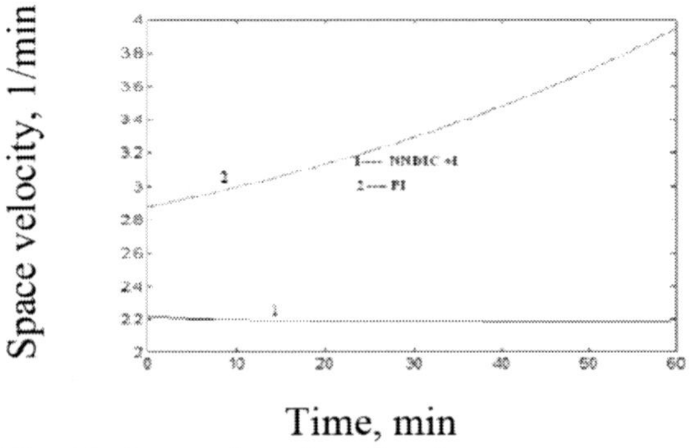

Fig.12 Control action for the response shown in Fig.11

At higher input space velocity, the closed responses of DINNC+I and PI are given in Fig.11 for setpoint change from 1.117 to 1.2. The Direct inverse control with integral action shows stable and offset free response and it reaches the setpoint at about 35 min whereas PI shows unstable response. The corresponding control action in terms of space velocity is shown in Fig. 12. Thus the PI controller designed for lower input space velocity become unstable at higher input whereas the proposed DINNC+I controller gives a stable response at higher input space velocity.

Regulatory problem

At lower input space velocity, the regulatory response of concentration B of direct inverse neural network controller and conventional PI for a step change in disturbance of feed concentration from 10 to 11(+10%) is given in Fig.13. As expected, It shows that the response of the Direct inverse neural network controller is faster than that of the linear PI. The Proposed neural network controller gives maximum deviation less than 1% whereas conventional PI controller has a maximum deviation of about 8%. Direct inverse neural network controller has lower settling time than the PI controller. The corresponding control action for manipulated variable in terms of space velocity versus time behavior is shown in Fig. 14. The regulatory response in concentration B of direct inverse neural network and conventional PI controller for a disturbance in feed concentration from 10 to 11(+10%) mol/l is given in Fig.15 at higher input space velocity. This result shows that the response of the Direct inverse neural network controller is faster and stable, whereas conventional PI controller results unstable response. The corresponding control actions in space velocity versus time behavior are given in Fig. 16.

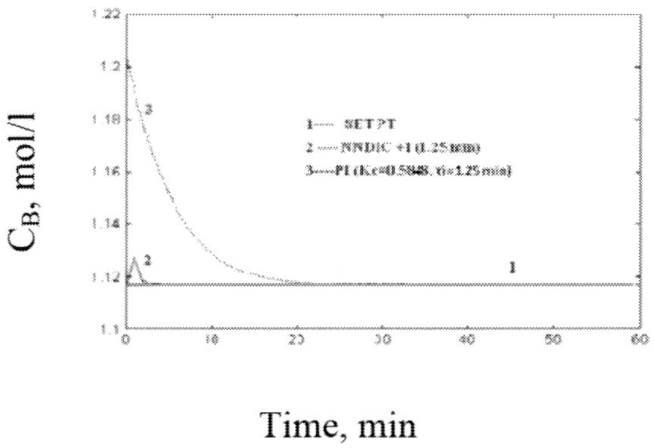

Fig.13 Regulatory response for a disturbance in feed concentration 10 to 11 mol/l at lower input space velocity.

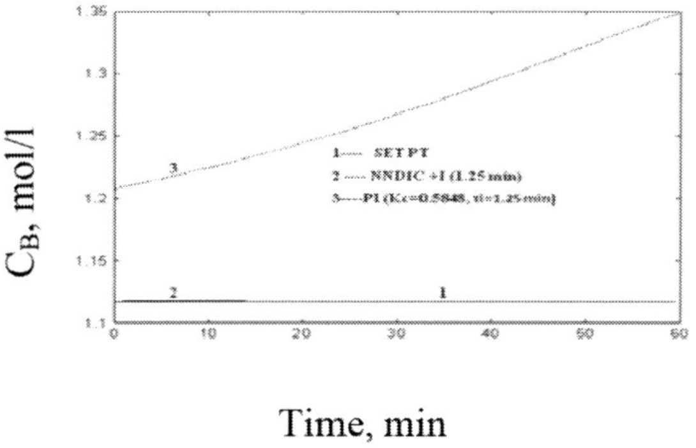

Fig.15 Regulatory response for a disturbance in feed concentration 10 to 11mol/l at higher input space velocity.

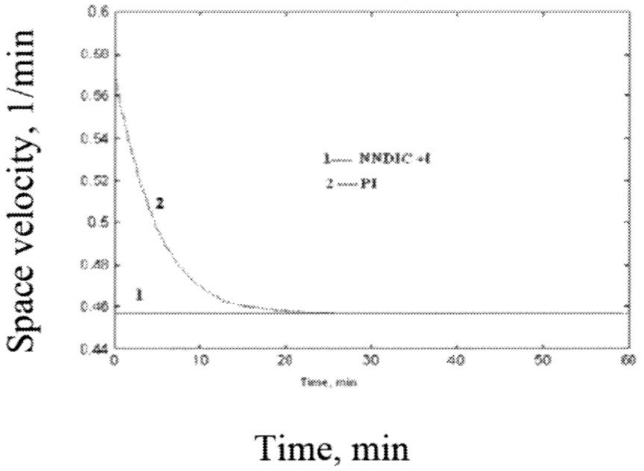

Fig. 14. Control action for the response shown in Fig.13

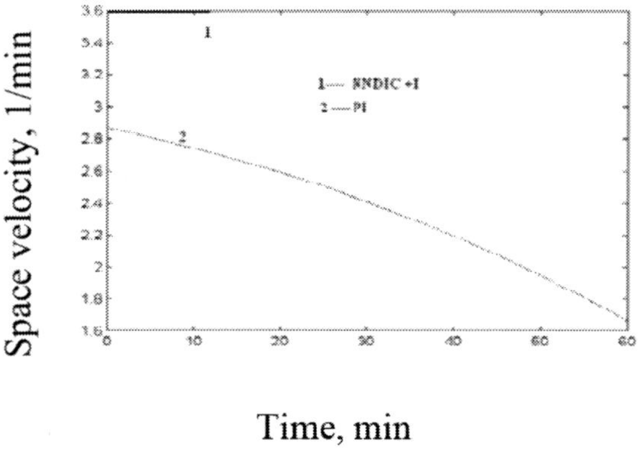

Fig. 16. Control action for the response shown in Fig.15

Conclusions

For a continuous stirred tank reactor with input multiplicities in space velocity, the performance of proposed direct inverse neural network controller plus integral at lower and higher input space velocities is found to superior to that of the conventional PI controller. At higher space velocity, the direct inverse neural network controller gives satisfactory responses, where as the linear PI controller (designed at lower space velocity) will give unstable responses at higher input space velocity. Thus, the proposed direct inverse neural network controller plus integral is stable and superior than linear PI controller at lower and higher space velocities and can overcome the control problems due to the input multiplicities.

References

1. Baughman,D.R. and Y. Liu, (1995), "Neural Networks in Bio processing and Chemical Engineering", Academic press.
2. Chidambaram, M and G.P.Reddy, (1995), "Nonlinear control of systems with input Multiplicities", *Computers and Chemical Engineering*, **19** pp249-252.
3. Dash, S.K. and L.B.Koppel, (1989) "Sudden destabilization of controlled chemical Processes ", *Chemical Engineering Communications*, **84** , pp 129-157.
4. Hunt, K.J., D. Sbarbaro, R. Zbikowski and P.J. Gawthrop, 'Neural Networks for Control System - A Survey,' Automatica, **28**, 1083-1112 (1992).
5. Koppel, L. B. (1982)," Input multiplicities in nonlinear multivariable control systems" *AIChE Journal*. **28** pp935 -945.
6. Koppel, L.B.(1983) "Input multiplicities in process control, Chemical Engineering Education, pp58-63, & 89-92.
7. Mohamed Azlan Hussain, Paisan kittisupokorn and Wachira Daosud (2001), "Implementation of Neural Network based inverse model control strategies on an exothermic reactor", *Science Asia* **27**, pp. 41- 50.
8. Norgaan, M.(1996)" Neural Network based control system design toolkit for use with MATLAB version 1.1 Technical Repot"Tech.Rep.,Technical University of Denmark.
9. Van de Vusse, (1964), " Plug flow type reactor versus tank reactor",*Chem.Eng.Sci.*,**16**, pp994-997.

(598e) Dual Unscented Kalman Filter for Leak Detection and Location in Natural Gas Pipelines

Weiting Tang, Chemical Engineering, Texas Tech University, Lubbock, TX and M. Nazmul Karim, Chemical Engineering Department, Texas Tech University, Lubbock, TX

Pipelines are efficient and economic ways for transporting and distributing water, natural gas, and petroleum products; these are extensively used all over the world. Monitoring the integrity of the pipeline systems has become an important issue for the safety and economical operation, and environmental protection. In this paper, an efficient and reliable leak detection and isolation method has been proposed for on-line monitoring of a natural gas pipeline system. Dual unscented Kalman filter (DUKF) was employed for the on-line estimation of the states and parameters, simultaneously. Small leaks, which are usually masked by the background noise, pose a challenge to all existing leak detection methods. The proposed leak detection method can adjust the parameters online according to the changes in the operating conditions, and thus was able to detect small leaks accurately, even with a normal transient state and with noisy measurement data. Simulation results demonstrated the capability of the proposed methodology to detect and locate the leak quickly, and accurately. Comparison studies of the extended Kalman filter (EKF), unscented Kalman filter (UKF), and DUKF show the superiority of DUKF over EKF and UKF for leak detection and location of the leaks in natural gas pipelines. Industrial data are used to validate the results.

(598f) Continuous Stirred Tank Reactor (CSTR) Temperature Control Via Multifractional PID Controller

Ruben D. Vargas and Watson L. Vargas, Department of Chemical Engineering, Universidad de los Andes, Bogotá, Colombia

Fractional calculus is the branch of mathematics dedicated to the extension of integral and derivative operations to the real domain (including complex), therefore allowing the possibility of calculating derivatives of , , orders among others. There has been an important number of works (Podlubny, 1999 and the references there in) devoted to applications of this mathematical tool in recent years. One of the most relevant areas of the so-called fractional applied analysis is the implementation of fractional order PID controllers.

These types of controllers allows the possibility of incorporating memory in the derivatives by changing the system order which results in changes in the stability regions (Tavazoei and Haeri, 2008). In this work this strategy is implemented on a second order reversible reaction occurring in a CSTR with the purpose of stabilizing the temperature that maximizes the reaction rate velocity. A classical and very important problem in the context of Chemical Reaction Engineering. The controller parameters are tune up via nonlinear optimization methods and comparison are made with implementations based on a classical PID controller.

References:

1. I. Podlubny, Fractional Differential Equations, Academic Press, San Diego (1999).
2. M.S. Tavazoei and M. Haeri, Chaos control via simple fractional-order controller, Physics Letters A 372 (2008) 798-807.

(598g) Use of Computer Modelling for Increase Efficienely of Methanol Synthesis Process

Uliana Osipenko, Chemical and biotechnology, Saint Petersburg State Institute of Technology, Saint Petersburg, Russia

A computer model of methanol synthesis was designed with the purpose of analysis and optimization in this work. Data for this work was taken from an actual industrial facility.

Methanol synthesis takes place on four shelves of reactor with bypassing cold gas, then methanol with water are separated from other gases. In the modeling, the reactor was presented in the form of four series-connected reactors, and implemented in the program HYSYS.

Performance analysis and improvement of technology of any chemical process require a mathematical description (mathematical model of the object). In this work, it was chosen the most suitable kinetic model of methanol synthesis. Based on the model implemented with the program HYSYS, we obtained dependencies for the thermal effects of reaction on pressure and temperature, made more exact estimation for the dependence for heat capacity of components on temperature, obtained equation for calculating the volatility of components in relation to the "pressure-temperature" working conditions.

Using this mathematical description of the process, the developed model of chemical-technological system for methanol synthesis was implemented using Borland Delphi programming environment.

The developed software allows you to perform these tasks: calculation of the circulating system of methanol synthesis, conducting computational studies to evaluate the effect of various parameters on the system as a whole, and tracking the system response to changes in these parameters, as well as calculation of the coefficient of catalyst activity.

(598h) Application of Computer Simulation and Optimization for the Increase of Xylenes Production Efficiency

Elena Manuylova, Resource-Saving Technologies Department, Saint-Petersburg State Institute of Technology (Technical University), St. Petersburg, Russia

This work is devoted to the study of the process of orthoxylene and paraxylene production at an operating oil refinery. Xylenes are valuable products used as raw materials for production of polymeric materials. Demand for xylenes increases by 6,5% annually, that's why the task of increasing their output is topical.

The aim of this work is to develop a virtual analyzer that allows to predict the composition of products and to calculate optimal parameters of the process.

The process under study was simulated in Aspen Hysys software in order to analyze it. The model was validated by the comparison of calculated and actual parameters.

A statistic model of the process was developed to predict the composition of products and to make optimization calculations. The software (virtual analyzer) was developed based on the results of process optimization.

Work of developed virtual analyzer was verified on an operational plant, yielding an increase in orthoxylene and paraxylene output by 8%.

(598i) Decentralized Multi-Loop PID Controller Design Directly From Plant Data

Jyh-Cheng Jeng, Department of Chemical Engineering and Biotechnology, National Taipei University of Technology, Taipei, Taiwan and Wan-Ling Tseng, National Taipei University of Technology, Taipei, Taiwan

Proportional-integral-derivative (PID) controllers are the most widely used controllers in the chemical process industries. As a result, PID controller design methods remain to be an active research topic in the past several decades. In particular, various model-based PID design methods can be found in the literature. There are two steps in the model-based PID designs: an empirical (low-order) model of the process is identified first, which is subsequently used together with a prespecified tuning algorithm to design a PID controller. Although these methods can give good PID design when the underlying process dynamics are reasonably described by the low-order models, the effectiveness of these methods would degrade for higher-order process dynamics owing to the inevitable modeling error. In addition, it is often difficult to collect input-output data for the identification of a controlled plant that has already been set up and is in full operation.

To alleviate these drawbacks, it is an attractive alternative to design PID controllers directly based on a set of process input and output data without resorting to a process model. Toward this end, several mode-free or data-based controller design methods were developed in the literature, such as the iterative feedback tuning (IFT) method, the virtual reference feedback tuning (VRFT) method, the fictitious reference iterative tuning (FRIT) method, and their variants. However, these previous works only discussed the PID controller design for single-loop feedback systems, and no application of the data-based methods to the controller design for multi-loop control systems is reported in the literature. In fact, multivariable systems are frequently encountered in the chemical process industries, and multi-loop PID controllers are still much more favored in most commercial process controls. Application of the data-based controller design method in multi-loop control systems is even more attractive, because the model identification for multivariable systems is more complicated and time-consuming than that for single-loop systems. This motivates our research to extend the data-based controller design method to multi-loop control systems with the specific aim of designing decentralized PID controllers directly from the plant data collected under open-loop or closed-loop operation. Thus, this method can design multi-loop PID controllers without resorting to the availability of process models.

The proposed direct PID controller design approximately solves a model-reference problem, and the design goal of the proposed method is to obtain PID parameters such that the corresponding multi-loop control system behaves as closely as possible to the prespecified reference models for each equivalent single loop. The selection of the reference models considers an appropriate tradeoff between the bandwidth of main loop and the resonant peaks of interaction loops. The optimization problems pertaining to the proposed design are derived, and the associated design issues are addressed. Extensive simulation results show that the proposed multi-loop PID design gives better or comparable control performance than those attained by the model-based PID designs. Consequently, the proposed design is an attractive alternative to the model-based decentralized PID design methods.

(598j) Online Estimation of Glucose Concentration and Oil Contents in Microalgal Biodiesel Production System

Se-Kyu Oh[1], Sung Jin Yoo[1] and Jong Min Lee[2], (1)School of chemical and biological engineering, Seoul National University, Seoul, South Korea, (2)School of Chemical and Biological Engineering, Seoul National University, Seoul, South Korea

Online estimation of glucose concentration and oil contents in microalgal biodiesel production system

Se-Kyu Oh, Sung Jin Yoo and Jong Min Lee*

School of Chemical and Biological Engineering, Seoul National University

Microalgae cultivation process has recently attracted much attention as a process for biodiesel production because of the potential of microalgae for lipid production. In this process, real-time measurements are important for the optimal control to maximize oil productivity. Unlike typical chemical processes, most of important variables are concentrations, which are difficult to measure. Pre-treatments and biological analysis equipments are required to measure the concentrations, and these tasks can take from 2 hours to 2 days. Although the concentrations are obtained offline, the measurements cannot be applied for feedback control because of the significant time delay between sampling and measurement.

In this study, we will present an integrated framework to predict glucose concentration and oil contents using Raman spectroscopy without any knowledge of spectroscopic analysis and biology. Raman spectroscopy is suitable for in-situ on-line monitoring in bioprocess. Raman spectrum has weak water interference peak and sample preparation is not required. In addition, sample or microalgae culture in bioreactor is not destroyed by Raman laser. Spectra are taken from microalgae culture using Raman spectroscopy, and Savitzky-Golay smoothing filter is used to reduce noise effect of the spectra. Fluorescence effects that inevitably exist in biological spectra are removed by Rolling-Circle Filter (RCF). Another challenge is that underlying relationship including linearity in correlations between the prediction variables and spectra is difficult to know. Hence, concentrations are estimated by multivariate analysis. This study compares different multivariate tools including Principal Component Regression (PCR), Partial Least Squares (PLS), Kernel PLS, Wide-kernel PLS and Support Vector Regression (SVR) to evaluate prediction performance and proposes the most suitable technique for online estimation applications.

(598k) Robustness Anlysis in Pipe Prediction Model of Water Distribution Network

Go Bong Choi[1], Shin Je Lee[2], Jong Min Lee[2], Gi Baek Lee[3] and Jung Chul Suh[4], (1)School of Chemical and Biological Engineering, Seoul National University, Seoul, 151744, South Korea, (2)School of Chemical and Biological Engineering, Seoul National University, Seoul, South Korea, (3)Korea National University of Transportation, Chungju-si, South Korea, (4)Samchully, Seoul, South Korea

Robustness anlysis in pipe prediction model of water distribution network

Gobong Choi, Shinje Lee, Dongwhi Jeong, Jong Min Lee*, Gibaek Lee, and Jungchul Seo

School of Chemical and Biological Engineering, Seoul National University,

1 Gwanak-ro, Gwanak-gu, Seoul, 151-744, Korea

Department of Chemical and Biological Engineering, Korea National University of Transportation

50 Daehak-ro Chungju-si, Chungbuk 380-702, Republic of Korea

Samchully, Samchully Bldg 35-6, Youido-dong, Yongdungpo-gu, Seoul 150-885 Republic of Korea

Water distribution network is a system that contains several components for supplying consumer with enough water of good quality. This distribution network can detorirate with not only enviromental conditions (soil type, acidity, etc.), but operational conditions (e.g., operational pressure), resulting in an serious economic burden and social impacts. Nevertheless, it is difficult to assess condition of the buried infrastrucure. Aging of pipe is used as a sole and primary criterion for replacement or rehabiliatation of pipe. It is also common that old water pipes can satisfactorily function while new pipes may show poor performance; this is because there are many other factors leading to pipe failures, including environmental conditions, physical characteristics, and historical factors. Hence, it can be costly to make decisions for only based on the age of pipe. There are two kinds of approaches for prediction of water main condition: first-principle and statistical approaches [1]. The first-principle approach constructs a mechanistic model to describe the physical mechanism leading to pipe breakage. It evalautes pipe failure due to the stesse applied on pipe under various conditions. Failure prediction of water main using this model is limited due the model complexity and expensive data requirement for reliable parameter estimation. On the other hand, statistical methods utilize historical data including recorded historical failure and enviromental factors and relates these to estimate pipe condition. While these models are easy to apply, obtaining such data sets is time-consuming for constructing reliable model. Hence, it is necessary to select a relevant set of variables before collecting data and determining a model structure. Towards this, this study first proposes a scoring scheme to rank the ease of data acquisition for each variable. Furthermore, robustness of existing statistical models for pipe condition assessment is examined using global sensitivity analysis and Monte-Carlo simulations. Incorporating these two schemes will aid in selecting the most sensitive factors in constructing statistical empirical model.

References

[1] Kleiner, Y. AND Rajani, B., "Comprehensive review of structural deterioration of water mains: statistical models", Urban Water, 3, pp. 131-150, May 2001

(598l) Model Predictive Control of Solar-Thermal Reactors

Elizabeth Saade Saade, Department of Chemical and Biological Engineering, University of Colorado, Boulder, CO, David E. Clough, Chemical and Biological Engineering, University of Colorado, Boulder, CO and Alan W. Weimer, Department of Chemical and Biological Engineering, University of Colorado at Boulder, Boulder, CO

A transport tube reactor uses concentrated solar irradiation in order to drive the endothermic biomass gasification reaction for the production of synthesis gas (H_2, CO and CO_2). This process is extremely dependent on the available sunlight and on the absence of clouds. Transient clouds constitute a significant disturbance for the steady operation of the reactor, causing unnecessary shut-downs and start-ups as well as complications in the separation processes downstream of the reactor. In order to make the process more feasible, flow manipulation is required to ensure steady conversion and a more efficient operation. Thus, a robust control system that allows continuous high performance operation of the reactor is required.

A model predictive control system was developed for the process. A digital charge coupled device (CCD) camera captures images of the sky at a sampling interval of one minute. These images are analyzed and used to anticipate the incoming solar irradiation at future instances. The predicted solar irradiation is used as an input for the model predictive controller (MPC), which ensures steady conversion despite transients in solar irradiation through the manipulation of the flow rates into the reactor. The control algorithm was developed based on linear time series models for different operating points of the reactor. These linear models were fitted to simulation results of a dynamic model of the process that was developed based on unsteady mass and energy balances. The control system was tested using SIMULINK and it was implemented at the High Flux Solar Furnace (HFSF) at the National Renewable Energy Laboratory (NREL). This poster describes the development of the control algorithm and presents the results obtained after its implementation.

(598m) Optimal Control of a System of Wastewater Treatment Ponds

María Paz Ochoa, Chemical Engineering, PLAPIQUI - UNS, Bahía Blanca, Argentina, Facundo Iturmendi, Chemical Engineering, Planta Piloto de Ingenieria Quimica (PLAPIQUI), Universidad Nacional del Sur, Bahia Blanca, Argentina, Vanina Estrada, Chemical Engineering, Planta Piloto de Ingenieria Quimica (PLAPIQUI), Universidad Nacional del Sur - CONICET, Bahia Blanca, Argentina and **Patricia M. Hoch**, Chemical Engineering Department, Plapiqui - UNS - Conicet, Bahía Blanca, Argentina

Wastewater treatment processes constitute a main issue to minimize major adverse impacts on freshwater and coastal ecosystems asssociated to urban and industrial growth. Activated sludge processes are currently the most widely used biological processes. Stabilization ponds, in turn, are large lagoons where wastewater is stored for long periods to allow a wide range of microorganisms to break down organic matter and sludge is not returned. In these ponds, there are aerobic, facultative and anaerobic zones, and different chemical and biochemical processes take place within the different zones, including mutualistic relationships between microalgae, heterotrophic bacteria and fungi that greatly influence the pond efficiency in biological wastewater treatment. A few models have been proposed for waste stabilization ponds describing hydrodynamic (Manga et al., 2004, Shilton and Mara, 2005) or biochemical processes (Kayombo et al., 2000; Dochain et al., 2003). However, the different pollutant removal processes have yet not been entirely understood.

We address the formulation of a detailed mechanistic model for a system of stabilization ponds (aerobic and facultative) for control purposes within a control vector parameterization framework (PSEnterprise, 2011). Dynamic mass balances are formulated for bacteria, yeast, main groups of phytoplankton, nitrogen, phosphorus, dissolved oxygen and biochemical demand of oxygen. Main model parameters have been estimated in previous work (Iturmendi et al., 2011), with data colleted from a juice plant in Argentina. In this work, we formulate an optimal control problem considering electrical motor power for mixers and nutrient addition rate as control variables. As the specification on biochemical demand of oxygen in the outlet stream is far from the target one, the objective is to minimize the offset between the desired value and the current one, along a time horizon of six months. Numerical results provide useful information on the complex relationship among microorganisms, nutrients and organic matter concentration, as well as optimal management of the system of ponds.

References

Dochain, D., Gregoire, S., Pauss, A., Schaegger, M., 2003. Dynamic modelling of a waste stabilization pond. Bioprocess and Biosystems Engineering 26, 19–26.

Iturmendi, F., V. Estrada, P. Ochoa, P.M. Hoch, M.S. Diaz, Biological Wastewater Treatment: Dynamic Global Sensitivity Analysis and Parameter Estimation in a System of Waste Stabilization Ponds, ESCAPE 22, 22th European Symposium on Computer Aided Engineering, London, UK,17-20 June, 2012

Kayombo, S., Mbwette, T.S.A., Mayo, A.W., Katima, J., Jorgensen, S.E., 2000. Modelling diurnal variation of dissolved oxygen in waste stabilization ponds. Ecol.Modelling 127, 21–31.

Manga, J.G., Molinares Nelson, R., Orlando Soto, E., Arrieta, J., Escaf Germa, J., Hernandez Gustavo, A., 2004. Influence of inlet-outlet structures on the flow pattern of a waste stabilization pond. In: Proceedings 6th International Conference of Waste Stabilization Ponds. Avignon, France.

PSEnterprise, 2011, gPROMS User guide.

Shilton, A.N., Mara, D.D., 2005. CFD (computational fluid dynamics) modelling of baffles for optimizing tropical waste stabilization pond system. Water Science &Technology 51,103–106.

(598n) Steady-State Analysis of Reactive Distillation for Simultaneous Etherification of Methanol and Hydrolysis of Methyl Acetate

Jer-Ming Hshieh, Chemical and Materials Engineering, Chang Gung University, Tao-Yuan, Taiwan and Gow-Bin Wang, Department of Chemical and Materials Engineering, Chang Gung University, Tao-Yuan, Taiwan

Till now distillation is the most widely used method of separation in the chemical process industry. Reactive distillation (RD) is well known as a process that combines reaction and separation in a single unit. Recently RD has drawn considerable attention due to its potential advantages over traditional processes.

It is known a large amount of methyl acetate (MA) is produced as a by-product from purified terephthalic acid (PTA) plants and polyvinyl alcohol (PVA) plants. Since the added value of this kind of MA is low, it is beneficial for PTA and PVA industries to effectively convert MA to more valuable products. Dimetyl ether (DME) has wide application, such as solvent, propellant, chemical intermediate, refrigerant replacement and transportation fuel. Recently, DME has attracted attention as an alternative fuel for diesel and LPG substitute due to less emission of pollutants such as CO, NO_x and particulate. Thus DME will be much used in fuel industries.

In industrial processes, DME is traditionally produced by dehydration of methanol using solid acidic porous catalyst in a fixed-bed reactor, followed by several distillation columns. MA is in general hydrolyzed in a fixed-bed reactor packed with cation exchange resin, also followed by a series of distillation columns. In recent years, the above mentioned process for methanol etherification and process for the hydrolysis of methyl acetate have been investigated and simulated in a reactive distillation (RD) process, respectively.

This work tries to develop alternative reactive distillation processes for simultaneously treating the etherification of methanol and the hydrolysis of methyl acetate to form valuable and high pure dimetyl ether and acetic acid. In our previous studies, this two-step reaction scheme with separation can be simulated in a single RD column which built on commercial process simulator platform. Valuable products with high purity can be formed under suitable feed ratio of methyl acetate to methanol.

Keywords: Reactive distillation, Etherification, Hydrolysis, Clean fuel, Process simulation

(598o) The Economic Potential of Process Stoichiometries in Incomplete Markets

Fernando Garcia[1], Fanghui Fan[2] and Jeffrey Kantor[1], (1)Chemical and Biomolecular Engineering, University of Notre Dame, Notre Dame, IN, (2)Chemical & Biomolecular Engineering, University of Notre Dame, Notre Dame, IN

This paper explores fundamental aspects of process economics and valuation using concepts established in modern quantitative finance. The objective is to understand process economics based on the underlying re- actions, stoichiometry, mass and energy balances and a minimal set of assumptions regarding stochastic price dynamics in the commodity markets. The goal of this work is to establish a modern stochastic interpretation of 'economic potential' previously developed for a deterministic framework by Douglas [3], Shinnar [6, 5], Hildebrandt and Glasser[2, 1, 4], among others.

Following Hildebrandt and Glasser [2, 1, 4], a 'simple process' is formulated using basic reaction stoichiometry, mass and energy balances, and second law feasibility. We show the dynamics of a self-financing portfolio consisting of the underlying commodities and risk-free assets cannot be spanned by the dynamics of the commodities alone, so in a technical sense the market is incomplete [7] for the process operator. The consequence of the incomplete market is, unlike for deterministic commodity processes or for standard financial options in a complete market, a unique price does not exist. Without modeling an owner's preferences for financial risk, the best one can do is establish a lower bound – the buyer's price – which any risk averse buyer would agree to buy a lease on the process, and an upper bound – the seller's price – for which any risk averse seller would agree to sell a lease. These prices are depend on process stoichiometry. Both of these bounds have special significance to investors and plant owners alike as they determine the fair price for the lease, or what in finance is known as bid-ask price. Tighter bounds are established by introducing second or- der stochastic dominance – the natural generalization of conditional value at risk – to measure preferences of risk-averse operators.

A crucial idea for the attainment of these bounds is that these types of processes occur in an incomplete market. These markets are characterized for subjecting their participants to some degree of financial risk, which in turn may represent monetary losses. A key point of this paper is to demonstrate efficient ways to hedge this risk.

References

1. David Glasser Bilal Patel, Diane Hildebrandt and Brendon Hausberger. Thermodynamics analysis of processes. 1. implications of work integration. Industrial & Engineering Chemistry Research, 44:3529– 3537, 2005.

2. Brendon Hausberger, Bilal Patel, Diane Hildebrandt, David Glasser and Benjamin J. Glasser. Producing transportation fuels with less work. Science, 323(5922):1680–1681, Mar 2009.

3. James M. Douglas. Conceptual Design of Chemical Processes. McGraw-Hill, 1988.

4. Bilal Patel, Diane Hildebrandt, David Glasser, and Brendon Haus- berger. Synthesis and integration of chemical processes from a mass, energy, and entropy perspective. Industrial & Engineering Chemistry Research, 46(25):8756–8766, 11 2007.

5. Reuel Shinnar. Thermodynamic analysis in chemical process and reactor design. Chemical Engineering Science, 43(8):2303–2318, 1988.

6. Reuel Shinnar, Dan Shapira, and Shlomlth Zakal. Thermochemical and hybrid cycles for hydrogen production. a differential economic comparison with electrolysis. Industrial and Engineering Chemical Process Design & Development, 20:581–593, 1981.

7. Aleš Černý. Mathematical Techniques in Finance: Tools for Incomplete Markets. Princeton University Press, 2nd edition, 2010.

(598p) Predictive Control and Analysis of a Sustainable Hybrid Energy System

Uday Guntupalli, Electrical, Missouri University Of Science & technology, Rolla, MO, Joseph D. Smith, Chemical and Biological Engineering, Missouri University of Science and Technology, Rolla, MO and Kyle Buchheit, Chemical & Biological Engineering, Missouri University Of Science & technology, Rolla, MO

Limited by the environmental regulations and the number of potential sources, the amount and rate of consumption of energy is largely constrained. Hybrid Energy Systems offer an effective and sustainable route to meet the exponential growth in energy demand. Due to the considerable carbon footprint of fossil based power and societies overwhelming dependence on this source, fossil based power generation can be hybridized with clean energy sources as a potential solution. The proposed hybrid energy system combines a Pressurized Circulating Fluid Bed combustor burning PRB coal together with a small modular nuclear reactor and renewable wind power.

Effective control of the proposed hybrid energy system poses a significant challenge given the number of control variables and the non-linear interaction between processes having a large disparity in the characteristic time scales. Model Predictive Control (MPC) has proven effective in decoupling interactions between control variables. In the present work, MPC is applied to the proposed hybrid energy system with its conflicting operating constraints to achieve better process efficiency and improved economic performance. The proposed hybrid energy system involves many conflicting hard constraints associated with intermittent energy source (i.e., wind) coupled with baseload energy (i.e., nuclear and/or coal). This work presents an effective control strategy capable of accommodating the proposed hybrid energy system.

(598q) IGCC Power Plant Design with Dispatch Capabilities

Benjamin P. Omell, Chemical and Biological Engineering, Illinois Institute of Technology, Chicago, IL and Donald J. Chmielewski, Illinois Institute of Technology, Chicago, IL

The Integrated Gasification Combined Cycle (IGCC) possesses a number of benefits over traditional power generation plants, ranging from increased energy conversion efficiency to flex-fuel and carbon capture opportunities. A less known benefit of the IGCC configuration is the ability to load track electricity market demands. The idea being that material storage can be used to time-shift power production away from periods of low power value to periods of high power value. In this work, we present a supervisory control scheme, known as Market Responsive Control (MRC), which has the objective of maximizing revenue for a given level of plant dispatchablity. The design of this controller is achieved by a fairly simple convex optimization problem. The MRC method is then embedded within an equipment design algorithm. This extended algorithm, known as MRC Embedded Equipment Design (MRC-EED), has the objective maximizing Net Present Value (NPV) in an effort to size the storage units as well as the required energy conversion devices. The design options considered in this study focus on hydrogen and compressed air storage as well as ramping abilities of the gasification block. Under the economic assumptions of this study, it is concluded that the NPV of the hydrogen storage option is significant, while that of compressed air is much less favorable. Finally, no scenario could be found in which ramping of the gasification block provided a positive NPV.

(598r) Infinite-Horizon Economic MPC for Building HVAC Systems

David Mendoza-Serrano, Chemical and Biological Engineering, Illinois Institute of Technology, Chicago, IL and Donald J. Chmielewski, Illinois Institute of Technology, Chicago, IL

Energy consumption by HVAC systems is usually heaviest when electricity prices are at their highest. The method of Economic Model Predictive Control (EMPC) can be used in conjunction with Thermal Energy Storage (TES) to time-shift power consumption away from periods of high demand to periods of low energy cost. In addition to enormous computational costs, implementation of such algorithms can result in unexpected and sometimes pathological closed-loop behavior, including inventory creep and bang-bang actuation. This paper will present an infinite-horizon formulation of the EMPC problem. While the design of this controller is achieved by a fairly simple convex optimization problem, it will be shown to alleviate many of the pathological behaviors observed in the finite-horizon case as well as significantly reduce the computational effort required for implementation. The method is illustrated on a simple building example using active TES. Implementation of the IH-EMPC policy resulted in a 99.987% reduction in computational effort, while only sacrificing 3.75% in operational cost. The new policy also eliminated much of the spiky behavior typically observed in EMPC implementations.

(598s) Development of a Dynamic Simulator for a Natural Gas Combined Cycle (NGCC) Power Plant with Post-Combustion Carbon Capture

Eric A. Liese, AVESTAR Center, National Energy Technology Laboratory, Morgantown, WV and Stephen E. Zitney, U.S. Department of Energy, National Energy Technology Laboratory, Morgantown, WV

The AVESTAR Center located at the U.S. Department of Energy's National Energy Technology Laboratory and West Virginia University is a world-class research and training environment dedicated to using dynamic process simulation as a tool for advancing the safe, efficient and reliable operation of clean energy plants with CO2 capture. The AVESTAR Center was launched with a high-fidelity dynamic simulator for an Integrated Gasification Combined Cycle (IGCC) power plant with pre-combustion carbon capture. The IGCC dynamic simulator offers full-scope Operator Training Simulator (OTS) Human Machine Interface (HMI) graphics for realistic, real-time control room operation and is integrated with a 3D virtual Immersive Training Simulator (ITS), thus allowing joint control room and field operator training. The IGCC OTS/ITS solution combines a "gasification with CO_2 capture" process simulator with a "combined cycle" power simulator into a single high-performance dynamic simulation framework.

This presentation will describe progress on the development of a natural gas combined cycle (NGCC) dynamic simulator based on the syngas-fired combined cycle portion of AVESTAR's IGCC dynamic simulator. The 574 MW gross NGCC power plant design consisting of two advanced F-class gas turbines, two heat recovery steam generators (HRSGs), and a steam turbine in a multi-shaft 2x2x1 configuration will be reviewed. Plans for integrating a post-combustion carbon capture system will also be discussed.

AVESTAR - Advanced Virtual Energy Simulation, Training, and Research (www.netl.doe.gov/avestar)

(598t) A Semi-Empirical Approach to Modeling and Simulation of Gasification Process

Fatemeh Rahmaninejad and Javad Abbasian, Chemical and Biological Engineering Department, Illinois Institute of Technology, Chicago, IL

There has been growing interest in understanding the coal gasification processes as a means of producing gaseous fuel. The inherent characteristics of coal gasification are complex and due to the high costs associated with the experimental investigations, modeling and simulation tools are becoming increasingly popular to assist with the design, analysis and optimization of this complex process. Most of the efforts to implement a gasifier model have been focused on two different approaches of euilibrium modeling and kinetic modeling.

The chemical equilibrium hypothesis assumes the gasification reactions have sufficiently fast kinetics to reach chemical reaction equilibria, which leads to over-prediction of the extent of the gasification reactions [1, 2, 3]. On the other hand, the kinetic based models taking into account transport and diffusion phenomena occurring in the reactor, which need detailed hydrodynamic and reaction kinetics for the heterogeneous and homogeneous chemical reactions [4, 5].

In this study, a versatile and user friendly gasifier simulation model was developed which incorporates a MS Excel interface aiming to provide a guide to gasification process analysis, evaluation and improvement. The model uses a set of semi-empirical correlations for the reaction conversions, which are based on experimental data, to predict the syngas composition, flow rate, and performance of the gasifier. The results of the parametric studies performed for the assessment of the effect of operating conditions, a parametric study are discussed and presented.

References:

1- Li, X., Grace, J.R., Watkinson, A.P., Lim, C.J., Ergudenler , A., "Equilibrium modeling of gasification: a free energy minimization approach and its application to a circulating fluidized bed coal gasifier", *Fuel* 80 (2001): 195-207

2- Huang, H.J., Ramaswamy, S., "Modeling Biomass Gasification Using Thermodynamic Equilibrium Approach", *Appl Biochem Biotechnol* 154 (2009):14-25

3- Emuna, F., Gadallaa, M., Majozib, T., Boer, D., "Integrated gasification combined cycle (IGCC) process simulation and optimization", *Computers and Chemical Engineering* 34 (2010) 331–338

4- Monaghan R.F.D, Ghoniem, A.F., "A dynamic reduced order model for simulating entrained flow gasifiers: Part I: Model development and description", *Fuel* 91 (2012) 61–80

5- Kim, Y.J., Lee, J.M., Kim, S.D., "Modeling of coal gasification in an internally circulating fluidized bed reactor with draught tube", *Fuel* 79 (2000) 69–77

(598u) Steady State and Dynamic Modeling and Simulation of Post-Combustion CO2 Capture Technologies

Qiang Zhang[1], Debangsu Bhattacharyya[2] and Richard Turton[2], (1)Department of Chemical Engineering, West Virginia University, Morgantown, WV, (2)Department of Chemical Engg., West Virginia University, Morgantown, WV

Steady State and Dynamic Modeling and Simulation of Post-Combustion CO2 Capture Technologies

Qiang Zhang, Debangsu Bhattacharyya, and Richard Turton Department of Chemical Engineering, West Virginia University, Morgantown WV, 26506, USA

The process of CO2 capture and storage has proven to be a promising and effective way to control the emission of CO2 from large-scale, stationary sources such as conventional, pulverized-coal, power plants. In this work, a liquid MEA-based, post- combustion CO2 capture and compression system, was simulated in Aspen Plus and Aspen Plus Dynamics. The process was also integrated with the steam system of a 550 MWe coal-fired power plant. The steady state simulation was carried out in Aspen Plus using rate-based separation processes for the absorber and stripper columns. Under the constraint of 90% capture efficiency, the effect of column height, flue gas temperature, lean-solvent temperature, stripper pressure, MEA wt%, and lean-solvent loading of CO2 were investigated in order to reduce the energy requirement for rich-solvent regeneration. With the optimized parameters of lean CO 2 loading of 0.208, a 30 wt.% MEA solution, 24 m of packed height in the absorber and a stripper operating pressure of 2.1 bar, the thermal energy requirement for CO2 recovery was 3.075 GJ/ton CO2 and the net power efficiency was 29.2%, which was more efficient than the values reported in the base case study of 4.048 GJ/ton CO2 and 26.55%, respectively. The steady state model developed in Aspen Plus was converted to a dynamic model in Aspen Plus Dynamics with appropriate modifications in the plant configuration for a pressure-driven dynamic simulation and by providing the required dynamic data such as equipment sizes, equipment heat transfer options, etc. The CO2 capture process should satisfy the overall target of CO2 capture in the face of typical disturbances such as changes in the flue gas flowrate and composition. However, for post-combustion CO2 capture in typical power plants, a number of CO2 capture trains are coupled together. Due to the pressure-flow dynamics, a precise distribution of the flue gas between the trains is a challenge for this highly interactive system. In addition, the control of liquid inventory and the regulation of the stripper efficiency pose further challenges. The presentation will include results from a number of innovative control strategies, both model-free and model-based, for maintaining the desired CO2 capture in the face of typical disturbances for this highly interactive process. .

(598v) Optimal Experimental Design of Chemical-Looping Combustion of Syngas in Fixed Bed Reactors

Lu Han, Ari Fischer and George M. Bollas, Department of Chemical, Materials & Biomolecular Engineering, University of Connecticut, Storrs, CT

Optimal experimental design of chemical-looping combustion of syngas in fixed bed reactors

Lu Han, Ari Fischer, George M. Bollas

Department of Chemical, Materials & Biomolecular Engineering, University of Connecticut, Storrs, CT

Abstract

The objective of this work is to evaluate and explore current processes for chemical-looping combustion of synthesis gas and propose optimal experimental design (OED) algorithms to assist in reactor design. Chemical-looping combustion (CLC) is a novel technology for efficient and low cost CO_2 separation. The basic concept of the process involves two reactors, a Reducer and an Oxidizer with an oxygen carrier (a metal oxide) circulating between the two (Figure 1). In the Reducer, synthesis gas (H_2, CO) reacts with metal oxide, according to the Reactions 1 and 2. The relevant catalytic reactions are methanation (Reaction 3), water gas shift (Reaction 4), carbon gasification by CO_2 (Reaction 5) or steam (Reaction 6), and dry reforming (Reaction 7). The reduced metal oxide, M_yO_{x-1}, is transferred to the Oxidizer, where it is oxidized back to M_yO_x with air, Reaction 8. The metal oxide is then returned to the Reducer and begins a new cycle of reactions. The flue gas leaving the Oxidizer contains N_2 and unreacted O_2. The exit gases from the Reducer contain CO_2 and H_2O, which are inherently separated from the rest of the flue gas. After condensation of the water almost pure CO_2 is obtained, without significant energy penalties for separation.

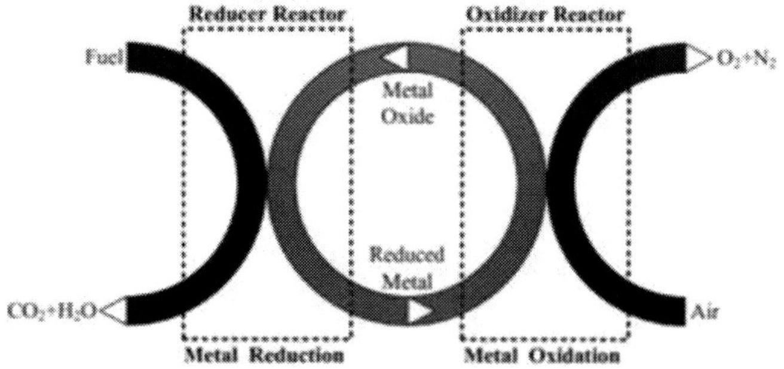

Figure 1: Chemical-looping combustion.

Reducer: $M_yO_x + H_2 \leftrightarrow M_yO_{x-1} + H_2O$ (1)

$$M_yO_x + CO \leftrightarrow M_yO_{x-1} + CO_2 \quad (2)$$

$$CO + 3H_2 \overset{Ni}{\leftrightarrow} CH_4 + H_2O \quad (3)$$

$$CO + H_2O \overset{Ni}{\leftrightarrow} CO_2 + H_2 \quad (4)$$

$$C + CO_2 \overset{Ni}{\leftrightarrow} 2CO \quad (5)$$

$$C + H_2O \overset{Ni}{\leftrightarrow} CO + H_2 \quad (6)$$

$$CH_4 + CO_2 \overset{Ni}{\leftrightarrow} 2CO + 2H_2 \quad (7)$$

Oxidizer: M_yO_{x-1} + ½ O_2 à M_yO_x (8)

Application of CLC on the oxidation of various fuels is being researched in studies of the effect of experimental parameters including pressure, temperature, and oxygen carrier reactivity, selectivity and stability. CLC with coal-derived synthesis gas is of interest due to the high CO_2 emissions from coal combustion. The effect of reactor pressure in syngas CLC is of interest due to the elevated pressure at which coal is typically gasified and the planned pressure for CO_2 sequestration. Therefore, the effect of pressure on reaction kinetics is a concern to the development of CLC technology. A literature review of this effect on CLC reactions in fixed bed reactors using natural gas, solid coal, and coal derived syngas reveals inconsistencies in experimental observations. Positive effects of increased pressure are noted by Xiao et al. [1], who observed an increase in reactor efficiency with pressures up to 0.5MPa and Zhang et al. [2] who observed intensified reactions of coal gasification products with the oxygen carrier at 0.5MPa. On the other hand, negative effects are reported by Jin et al. [3] who observed lower reduction rates at elevated pressures due to pressure restrictions on the reforming reactions and Tian et al. [4] who noted reduced reduction rates due to slower bulk diffusion. Utilizing previous experimental studies and current modeling of CLC reaction kinetics at the University of Connecticut, we are designing and constructing a bench-scale fixed bed reactor for oxidation of syngas with metal oxides. This will enable a study for measuring metal oxidation and reduction kinetics at high pressures, furthering the research of pressure effects on CLC technology. Results will be translated into reaction mechanisms and kinetic constants to be used in modeling and scale-up studies. Overall, the effects of pressure on CLC reactors are undetermined and require further investigation to understand the reaction kinetics. This presentation will focus on the peculiarities of high-pressure chemical-looping combustion and showcase a novel method for optimal design of a bench-scale experimental apparatus to be used for the identification of the effect of operating conditions (*e.g.*, reactor pressure) on the accuracy and consistency of experimental measurements.

References

1. Rui Xiao, Qilei Song, Shuai Zhang, Wenguang Zheng, Yichao Yang *Energy & Fuels 2010 24 (2), 1449-1463*

2. Shuai Zhang, Chiranjib Saha, Yichao Yang, Sankar Bhattacharya, Rui Xiao *Energy & Fuels 2011 25 (10), 4357-4366*

3. Hongguang Jin and Masaru Ishida *Industrial & Engineering Chemistry Research 2002 41 (16), 4004-4007*

4. Hanjing Tian, Karuna Chaudhari, Thomas Simonyi, James Poston, Tengfei Liu, Tom Sanders, Götz Veser, and Ranjani Siriwardane *Energy & Fuels 2008 22 (6), 3744-3755*

(598w) Modeling and Control Studies for the Elevated Pressure Air Separation Unit in an IGCC Power Plant

Kosan Roh, **Hyojin Lee** and Jay H. Lee, Chemical and Biomolecular Engineering, Korea Advanced Institute of Science and Technology, Daejeon, South Korea

Korea is one of the highest energy consuming countries, being ranked at the 8th in the world in terms of total energy consumption. Unfortunately, it is extremely resource-poor and 97% of the total energy resource is currently being imported from abroad. Therefore, the Korean government is intent on improving the energy supply and demand situation for the reason of national energy security. Recently, IGCC (Integrated Gasification Combined Cycle) power plant is attracting lots of attention as an alternative system of generating electricity from coal in an eco-friendly way. The coal is gasified with oxygen to generate the syn-gas, which is composed of carbon monoxide and hydrogen. After going through several cleaning steps, the syngas is sent to the combined cycle unit for generating the electricity. IGCC gives a higher thermal efficiency than the conventional pulverized coal power plant. Also, it is easier to implement CCS(Carbon Capture and Storage) technology in an IGCC power plant for slowing down the climate change.

However, the overall configuration is relatively complicated and also the level of integration among the units is very tight. The level of availability is often very low for an IGCC power plant because the operational know-hows and control strategies for IGCC power plants are yet to mature. Therefore, the plant-wide control study for an IGCC power plant is needed for its stable and efficient operation.

Air Separation Unit (ASU) is a part of the IGCC power plant. It separates air into oxygen and nitrogen and sends them to the gasifier and the gas turbine. The ASU consumes about 10% of the gross power output and also accounts for about 15% of the total operating cost. Also, it introduces the largest time lag during the operation. Therefore, the operation of ASU should be optimized in order to achieve good control of the overall IGCC plant.

In this research, Elevated Pressure Air Separation Unit (EP ASU) in an IGCC power plant is studied from the viewpoint of dynamic systems analysis. EP ASU, which is operated at a high pressure, is an improvement from the general ASU and gives significant energy savings. Firstly, equation-based modeling of the cryogenic rectification column in EP ASU has been carried out using the software platform of gPROMS. Also, the transient behavior in response to various disturbances is studied through the dynamic simulation. At the end, based on the dynamic analysis, PID controllers are applied to EP ASU for the product purity and flow rate control.

Acknowledgement

This research was supported by Basic Science Research Program through the National Research Foundation of Korea(NRF) funded by the Ministry of Education, Science and Technology(2011-0006839).

PERFORMING SOLUTION OF MARKET-DRIVEN OPTIMIZATION FOR LARGE-SCALE CORPORATES

Eng. Matteo D'Isanto, Politecnico di Milano, Milano, Italy
Eng. Flavio Manenti, Politecnico di Milano, Italy
Eng. Mariagrazia Grottoli, Politecnico di Milano, Italy
Prof. Sauro Pierucci, Politecnico di Milano, Italy
Prof. Guido Buzzi-Ferraris, Politecnico di Milano, Italy
Eng. Marcello Altavilla, Linde Gas Italia Production Division, Terni, Italy
Dr. Ornella Martinelli, Linde Gas Italia Headquarter, Arluno (MI), Italy
Davide Jurissevich, Linde Gas Italia, Logistics Division, Trieste, Italy
Eng. Roberto Di Marco, Linde Gas Italia Headquarter, Arluno (MI), Italy

Abstract

Many projects to rationalize and optimize the supply chain of complex corporates are ongoing in United States and Europe. The leitmotif is the high potential margins that may come from a high level optimization of decentralized productions and distribution networks, but also the need to raise the decision-making level from plantwide's purpose to the enterprise wide level. Actually, the global crisis of these last years is strongly pushing towards global optimization since the single production sites, although optimized, present relevant economic losses and the net operating margins are very small with respect to several years ago, especially in chemical and process industries. Thus, the corporate optimum, which almost never corresponds to the optimum of the single production sites, is a promising way to significantly improve the profits as well as to debottleneck inefficiencies highlighting and exploiting hidden potentialities.

The main problem is that such an approach unavoidably means to optimize in an integrated way all the operations of the corporate and, hence, to account simultaneously for billions of variables and to elaborate a solution in real-time for the optimal decision support.

Science fiction for the large industries, at least until computers will not be enough powerful, the parallel computing enough exploited, and the algorithms enough robust and efficient at the same time.

Introduction

Operations research is an important topic for a long time and it is of further interest in the last years since many production fields are undergoing a long period of reduced market demand due to the ongoing worldwide crisis. Actually, the current scenario is causing relevant economic losses as many processes lose efficiency when they are forced to operate far from their nominal conditions. It is worth saying that many production plants were engineered and constructed years ago, when the main objective was to maximize the production (production-driven scenario) in spite of the plant flexibility and operability, whereas these features should be particularly useful in the current market-driven scenario.

The industrial gases supply chains are particularly involved in this scenario and they have the need to handle a current persistent low-demand condition: contrary to many other production fields, facing a significant reduction in the market demand (hence in profits), industrial gases supply chains have not a corresponding decrease in costs since their raw material (air) is free. Moreover, the overall energy

efficiency of the energy-intensive plants such as air separation units (ASUs) dramatically drops down when the production is low (nonlinear chemical-physical relationships behind their behavior).
There are no process optimizations at the plant-wide level that can ensure a net operating margin to the ASUs when they are forced to operate below a certain competitive threshold. Notwithstanding, there is the possibility to exploit the typical decentralized-production structure of the industrial gases companies (i.e. Air Products, Linde, Air Liquide) not only to practically limit the losses, but even to obtain positive margins: the decision-making process of the industrial gases production must be raised from the plant-wide (or business-wide) level to the enterprise-wide (or corporate) level.
In fact, it is well-known that the production of nitrogen is larger than the demand and, in these last years of low-demand conditions, also the oxygen demand is so reduced that ASUs must vent sometimes also this product, therefore, the low-demand scenario must be tackled from a corporate point of view, accounting also for connections among production sites, logistics, and geographical distribution of the production and storage capacities. From this perspective, the supply chain planning at the corporate level is an appealing approach to provide an economic solution if combined with Boolean logic, since it has the possibility to ponder the temporary shutdown of certain ASUs and, at the same time, to force some others to operate close to their nominal conditions and thus with good plant efficiencies. As it is easy to understand, the corporate optimum imposed to the ASUs may be significantly far from the ones dictated by the optimization of the single ASU.
The present paper discusses (and applies by the field – Linde's air separation units) certain novel methodologies to solve separately several problems behind the real-time market-driven corporate optimization, looking forwards to their integration for a performing solution of the overall problem.
Specifically, the following topics will be discussed:

- The analysis of very large data sets. New methods for the analysis of very large (industrial) data sets have been developed and validated. Such methods are able to promptly detect the outliers that could affect the data acquired by the field and the DCS.
- Prevision of the future market demand. A combination of weighted averages based on the historian and linear short-term extrapolations can be successful to foresee the incoming market demand for each final customer.
- Allocation of available commodities to match the demand. A Gaussian-based approach can be useful and very performing to assign the right source of commodity to the final customer so as to satisfy it by accounting for the distance, the availability of cryogenic trucks, and the availability of product.
- KPIs for final customers supply. Each society monitors its production and distribution by means of specific key performance indicators (KPIs). Some of the most important ones will be introduced and the item of the cryogenic truck is optimized basing on them and clustering the customers.
- Just in time production and minimum inventory. Being one of the most relevant cost items, the liquid levels of the cryogenic tanks must be minimized. To do so, it is necessary to foresee the market demand with a certain reliability and to change the operational sets of the production plants accordingly. If the former bullets are mainly related to mixed-integer linear programming, this point is strongly related to nonlinear programming.
- Optimal production. Basing on nonlinear models for air separation units, the production and the energy consumptions are both optimized in a predictive way.

Supply Chain Management

The supply chain optimization is defined as the ability to capture the synergy among vertical administration levels and horizontal production hierarchy of a company, by considering interactions between their suppliers and customers and incorporating them into a well-defined decision-making process . The modern supply chain involves the entire corporate network and every single area, from the production to the market. The logistics, for its intrinsically enterprise-wide nature, played a determinant role in these last years for every type of company also for strategic meanings heavily concurring to increase the entire corporate business profitability, whereas the production is still optimized plant-by-plant usually bringing to a difficulties in plant conducting. Nevertheless, the recent economic crisis and the increasing market competitiveness emphasize the need for an overall approach that involves the entire production network. Thus to survive, each plant of decentralized productions cannot be seen as a single entity anymore, but all of them have to be settled up like a constellation where they all take part in the same objective: the customer satisfaction. From this perspective, the supply chain management can read the specific needs of chemical processes and their chemical-physical nature, down to their molecular scale and can optimize complex production networks with the aim to maximize not only the production but also the economic yield. This is nowadays possible using certain well-established detailed models for the chemical unit operations together with the characterization of commercial and economic phenomena through a kind of multiscale methodology for the optimization of complex production networks (Figure 1).

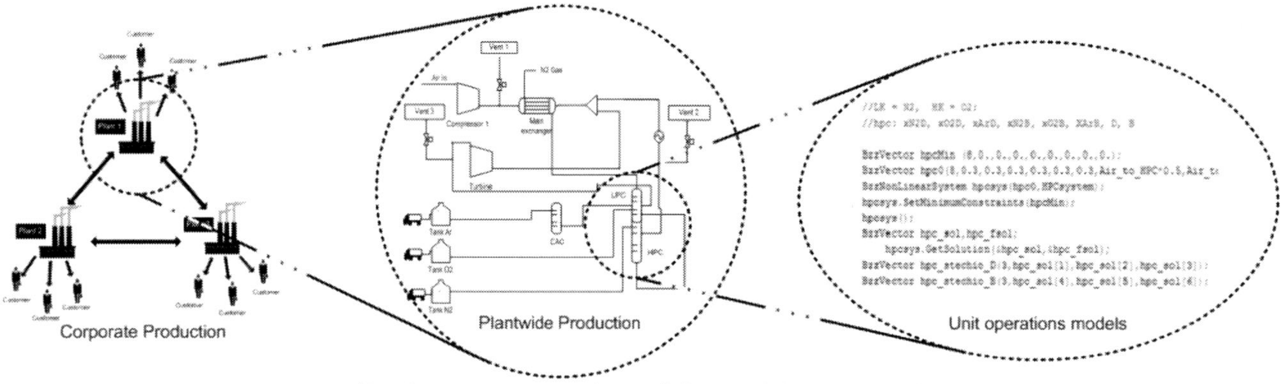

Figure 1. Qualitative representation of the multi-scale optimization problem.

Supply chain contextualization to industrial gases manufacturing

Industrial gases market and enterprises are subject to specific conditions that make their mixed-integer optimization different from other applications:

- The air is the raw material: the raw material supply is free and without any limitations, contrary to the majority of the industrial applications. Thus, the only external expenses are represented by electrical energy to be supplied to each ASU.
- The power supply is regulated by complex agreements and contracts with the National Electric Energy Supplier. These contracts ensure the electrical supply to the ASU, but they also define possible penalties and large discount opportunities in the monthly/yearly energy bill. The energy cost is also very difficult to predict because of its variability and high market volatility. In the present work night (20:00-8:00) and day (8:00-20:00) energy price variability has been taken into consideration according to the Italian energy market manager. The electric energy

price is referred to the Italian situation: there are several rules for the energy deregulated market but they are slightly different from other countries: the deregulation (like "take or pay" contracts, or cost-effective contracts) can be applied up to a certain threshold of power consumption (i.e. about 2 MW but it depends from the different zones). ASU's are over this threshold and have to pay the PUN's price (PUN = Unique National Price) stated by the Energy Supplier and it is the price that you can see in Fig. 2Some ASUs may be directly connected via pipelines to customers like steel mills or oil refineries. The connection is usually related only to one of the products of the ASU and all the other products derived by air liquefaction have a totally different demand, causing certain unbalanced situations.

- Cryogenic storages of industrial gases manufacturing are very expensive and, nowadays, the increasing market competitiveness is pressing companies to move towards the so-called just in time production policy, characterized by minimum stored volumes and real-time inventory scheduling.

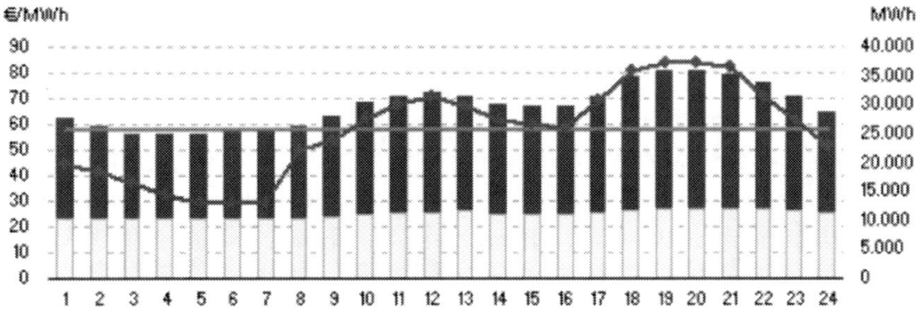

Figure 2. Energy price variability in 24 hours for Italy (http://www.mercatoelettrico.org)

Enterprise-wide mathematical models

The enterprise-wide production planning is a complex optimization problem since it involves nonlinearities of chemical/industrial processes, discrete variables to support the decision-making, large dimension of the resulting systems due to the mathematical models of production plants, and uncertainty due to market/demand fluctuations and volatilities. To be effective and preserve the connection with the production field, the enterprise-wide production planning should be based on detailed physical-chemical models that properly characterize the relevant phenomena behind the behavior of each ASU and the connections among ASUs as well (pipelines, storages, trucks, railways…). This unavoidably means to account for certain nonlinearities typical of this kind of processes such as air fractionators, polytrophic compression, cryogenic storages, etc.. The intrinsic nonlinearity of ASUs makes the enterprise-wide production planning problem particularly cumbersome: we need to solve a usually large-scale, mixed-integer, constrained, non-deterministic, real-time, nonlinear optimization problem:

- It is large-scale since it involves several plants with production trains each and many processes and unit operations. Thus, the number of state variables, decision variables, and degrees of freedom increases exponentially, leading to a large-scale problem.
- It is mixed-integer since it includes certain discrete (Boolean or integer) variables to model the decision-making process (i.e. 0 means off and 1 means on for a plant). The enterprise-wide production planning is a strategic optimization usually aimed at assigning most profitable production load to each plant so as to maximize the net operating margin. The final purpose changes when the demand is low, since the corporate production planning is aimed at defining which plant should produce and which other should not, by assigning a Boolean variable to each plant. Thus, decision variables are of discrete/Boolean nature and they transform the process optimization into a mixed-integer nonlinear programming (MINLP) problem.
- It is constrained since the optimization is subject to mathematical constraints such as the mass and energy balances governing the ASUs and liquid inventories or the take-or-pay contracts regulating the sales and purchase tasks.
- It is non-deterministic since there is the need to predict the future behavior of the enterprise-wide and the prices/costs to obtain an effective production planning. This means that we need to work with production and market uncertainties, which increase with longer prediction horizons. To reduce the effect of uncertainty, the so-called rolling horizon methodology must be adopted.
- It is real-time since the problem must be solved promptly to have effective support to the decision-making.
- As mentioned, it is nonlinear for the nonlinear nature of industrial gases plant.

The resulting problem can be solved using several tools: the most common one is the GAMS environment. We selected the MS Visual C++ programming language and the numerical methods belonging to the BzzMath library (www.chem.polimi.it/homes/gbuzzi/BzzMath).

Case study

The selected case is an enterprise for industrial gases production. One of the most efficient technologies for large production capacities is the air separation via cryogenic process. A typical ASU is an energy-intensive plant, which separates the air components into gaseous and liquid oxygen (GOX and LOX), gaseous and liquid nitrogen (GAN and LIN) and liquid argon (LAR). The gaseous argon is not required by any customer in the selected industrial case-study. The first part of this paragraph deals with the modeling of a single ASU, taking into consideration the key-unit operations and investigating the main issues at the plant-wide level. The second part of the paragraph deals with the additional enterprise-wide constraints.

The objective function consists of the evaluation of revenues and costs for each plant $i = 1,...,NP = 3$ and sampling time (half a day each) $j = 1,...,NT = 60$ of the selected time:

$$\min_{\mathbf{x},\mathbf{b}} \Phi = -\sum_{j=1}^{NT=60} \left[REV_j(\mathbf{x},\mathbf{b}) - \sum_{i=1}^{NP=3} Cost_{i,j}(\mathbf{x},\mathbf{b}) \right]$$

$s.t.:$

$$\begin{cases} f(\mathbf{x},\mathbf{b}) = 0 \\ g(\mathbf{x},\mathbf{b}) \ge 0 \end{cases}$$

Given the demand for every single product, products and energy costs, the optimizer determines the best corporate configuration in order to satisfy all the constraints.

Market and enterprise-wide selected scenarios

Two different scenarios are taken into consideration to study the applicability and the advantages of the new concept of supply chain management described in this paper and to compare the differences in decisions between the conventional plant-wide optimization and the global (enterprise-wide) optimization. The first scenario is characterized by relevant, initial, liquid product storages; this is the typical condition of the operating plants. The second scenario is characterized by null liquid storages; this is the typical condition of cold plant start-ups (start-up after relatively long shutdown period). Both the scenarios accounts for a monthly production planning problem under low-demand conditions.

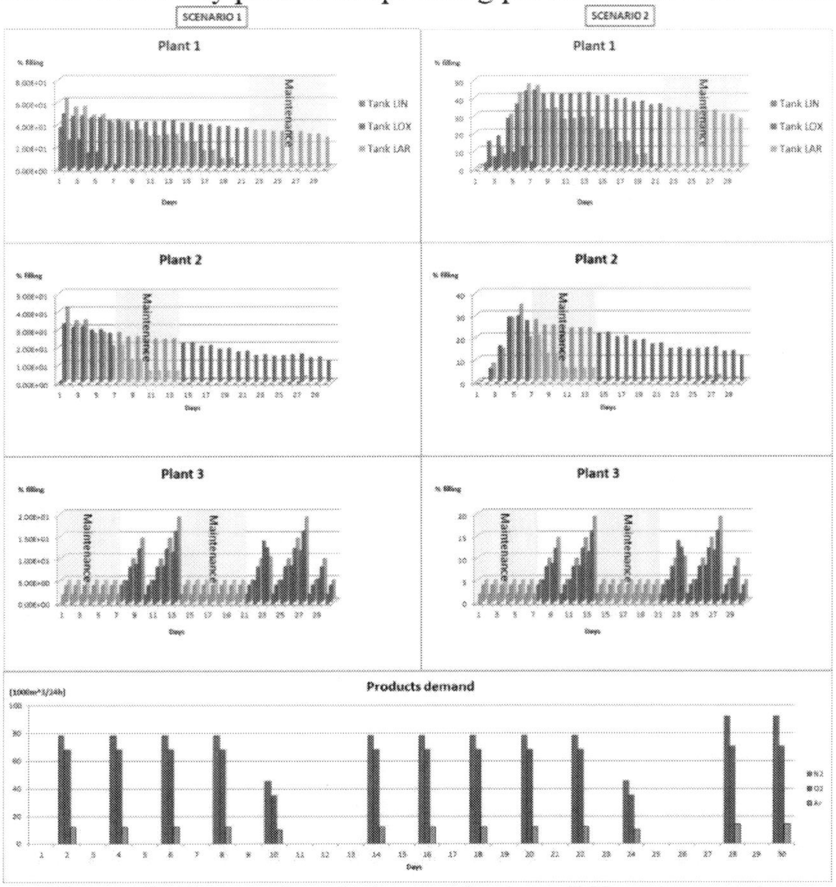

Figure 3. Comparison between the two different scenarios

Conclusions and next development

The paper shows certain tangible benefits of the corporate production planning to tackle low-demand conditions. The proposed approach is fully developed in Microsoft Visual C++ and based on the scientific BzzMath library (freely downloadable numerical tool for scientific computing, Politecnico di Milano). The novelty of the proposed methodology is related to two key-points:

- The possibility to insert certain reasonably detailed physical-chemical models for the production plants into the optimization procedure of the enterprise-wide production planning.

– The possibility to introduce a Boolean variable for the production plant (0 = off; 1 = on) so as to higher the decision-making from the plant-wide level to the enterprise-wide level.

The results have demonstrated that the methodology is particularly appealing to reduce the losses under worldwide economic crisis and to plan the production of complex networks. Nevertheless, by doing so, certain additional social aspects deriving from enterprise-wide decisions (due to plant turnover) must be taken into consideration. Moreover, the instability conditions of the plant-wide optimization with respect to the proposed approach have been demonstrated. The proposed methodology allows enterprises in general to deal with massive economic losses deriving from the impossibility to merge plant and operational aspects to economic and logistics ones helping the manager to make right choices from both points of view avoiding high products losses for excessive production.

References

1. Ierapetritou, M. G., D. Wu, et al. (2002). "Cost minimization in an energy-intensive plant using mathematical programming approaches, Industrial & Engineering Chemistry Research 41(21): 5262-5277.

2. Buzzi-Ferraris G, Manenti F. "Fundamentals and Linear Algebra for the Chemical Engineer: Solving Numerical Problems", Wiley-VCH, Weinheim, Germany. 2010.

3. Buzzi-Ferraris G, Manenti F. "Interpolation and Regression Models for the Chemical Engineer: Solving Numerical Problems", Wiley-VCH, Weinheim, Germany. 2010.

4. D'Isanto M, Manenti F, Lima NMN, Linan LZ. "Corporate Production Planning for Industrial Gas Supply Chains under Low-Demand Conditions", In: Pistikopoulos ENGMCKAC, ed. 21st European Symposium on Computer Aided Process Engineering. Vol 292011:965-969.

5. Mitra S, Grossmann IE, Pinto JM, Arora N., "Optimal production planning under time-sensitive electricity prices for continuous power-intensive processes", *Computers & Chemical Engineering*. Mar 5 2012;38:171-184.

6. Neiro SMS, Pinto JM., "Multiperiod optimization for production planning of petroleum refineries" Chemical Engineering Communications. Jan 2005;192(1):62-88.

7. Floudas CA, Gounaris CE., "A review of recent advances in global optimization", J. Glob. Optim. Sep 2009;45(1):3-38.

8. Buzzi-Ferraris G., "Linear Programming with the Attic Method", Industrial & Engineering Chemistry Research, May 4 2011;50(9):4858-4878.

(599aa) Developing Distributions for Biomass Feedstock Supplies

Joseph Amundson, University of Kentucky, Lexington, KY, Erdinc Oner, Department of Industrial Systems Engineering, Izmir University of Economics, Izmır, Turkey, Jeffrey Seay, Chemical and Materials Engineering, University of Kentucky, Lexington, KY and Fazleena Badurdeen, Institute for Sustainable Manufacturing, Lexington, KY; Mechanical Engineering, University of Kentucky, Lexington, KY

Biomass as an alternative for fossil fuels as a resource to create fuels for transportation intrigues many researchers largely due to its widespread availability and applicability as a potentially sustainable resource. Toward this end, progress has been made in recent years with regards to modeling integrated biorefineries and the supply chains necessary to provide them with raw materials. Decision support tools will provide potential investors, agriculture workers, policy makers, and other stakeholders with a means to judge the suitability and sustainability of nascent industries in specific regions because of this research in optimization and simulation.

Not least among the challenges of biomass as an alternative to fossil fuels is the seasonal and highly variable supply that is inherent to agricultural products. Due to a relatively low energy density of biomass feedstocks, large quantities of material may be required for conversion. Reasonable estimations of regional biomass that incorporate the stochastic nature of the supply become key to having any type of meaningful decision support tool.

This presentation focuses on the development of distributions for corn stover availability in the Jackson Purchase Region of Western Kentucky. At the current resolution (mostly monthly or yearly) the supply information cannot be used directly in any models, particularly those looking to simulate performance of corn stover based bioenergy systems. However, combinations of the available historical data have been used to develop distributions for each county in the region and for each month during a typical corn-harvesting season in Western Kentucky. Arena Input Analyzer and Minitab were used to develop the distributions and comparisons are drawn. These distributions provide the resolution necessary to begin region specific modeling.

(599ab) Use Dynamic Simulation to Minimize Flare Emissions During Ethylene Plant Shutdown

Jian Zhang, Meiqian Wang, Jie Fu, Chuanyu Zhao, Ha Dinh, Shujing Zhang, Karthik Krishnadevarajan, Qiang Xu and Kuyen Li, Dan F. Smith Department of Chemical Engineering, Lamar University, Beaumont, TX

There are large amount of emissions from flare during startup, shutdown and upset of an ethylene plant. Literature survey shows a shutdown procedure of a 300 kt ethylene emits about 150 ton of hydrocarbon through flare. Based on the 98% combustion efficiency that EPA recommended, it emits 1.2 ton of carbon monoxide, 0.22 ton of nitrogen oxides (NOx) and 3.4 ton of volatile organic compounds (VOCs). Since flare emissions happen within a short time, they will cause big impact on ambient air quality.

In this paper, dynamic simulation was used for shutdown procedure of an ethylene plant. Rigorous dynamic models were built, which included charge gas compressor (CGC), chilling train, demethanizer, deethanizer, depropanizer, hydrogenation reactor, ethylene splitter, propylene splitter, debutanizer and depentanizer. The shutdown procedure included many stages, and each stage had one or more dynamic models. Through dynamic simulation, a new shutdown procedure was proposed, which would decrease flare emissions by 50%. The new procedure was accepted by the ethylene plant and will be used for their next shutdown.

(599ac) Simultaneous Hoist Scheduling and 2D Production Line Design for Multi-Stage Material Handling

Chuanyu Zhao and Qiang Xu, Dan F. Smith Department of Chemical Engineering, Lamar University, Beaumont, TX

Multi-stage material handling (MSMH) processes are broadly used in industries for manufacturing massive amount of products/workpieces (jobs). A hoist is a controlled device for manufacturing by following a preset movement schedule, including job lifting, releasing, and moving along a production line. In previous works, how to feasibly and timely organize hoist schedules has been broadly studied. However, the productivity of an MSMH process not only depends on hoist movement schedule, but also substantially relies on the production line arrangement, i.e., the design for spatial allocations of various processing units. This is because a hoist needs to travel among processing units to pick up or drop different jobs for different processing purposes, and such travelling time greatly depends on unit spatial allocations, which affects both job processing and change-over time. Thus, the production line arrangement and hoist scheduling are better to be considered simultaneously for the best performance of a hoist-employed multi-stage material handling process.

In this paper, an MILP model has been developed to couple cyclic hoist scheduling and 2D production line arrangement for the maximum productivity of a general MSMH process. The major advantage of this study is that the spatial allocation of processing units and hoist movements of a 2D production line have more options than those of a traditional 1D production line, so that its productivity has better chance to be improved significantly. The efficacy of the proposed methodology is demonstrated by a case study. The comparisons between 1D and 2D production line arrangements have also been conducted.

(599ad) Simultaneous Design and Scheduling of a Material Handling System

Chuanyu Zhao, Jie Fu and Qiang Xu, Dan F. Smith Department of Chemical Engineering, Lamar University, Beaumont, TX

In multi-stage material handling processes, such as electroplating and polymeric coating, the design of production line is very important. A typical production line includes a series of units/tanks for different processing purposes; each unit may be built with different capacities; and there is a hoist above the production line to transport jobs among different processing units. In previous research, hoist scheduling problems have been broadly studied in order to maximize the productivity of a material handling system. But the design of the production line draws little attention. Actually, the design of the production line affects the productivity critically. Because the design contains the allocation of each unit, which inherently determines the distance, and hence the travelling time between each pair of units. The design also contains the capacity of each unit, which determines the system processing capability, and the job processing time in each unit. If the production line in a material handling system is designed optimally, with a corresponding superior hoist scheduling, the productivity of the system will be even boosted.

In this paper, an MILP model has been developed to design an optimized production line; and through the same model, a corresponding optimal hoist scheduling can also be provided. The objective for the model is to maximize the productivity of a material handling system, and various constraints need to be met. For instance, there is a budget limit for the design, and the hoist scheduling has to meet all the production requirements of various jobs. The major merit in this work is that through this model, an optimal design and an optimal hoist scheduling can be generated simultaneously. The efficacy of the proposed methodology is demonstrated by a case study. The comparisons between an optimal design and a heuristic design have also been conducted.

(599ae) A Comparison of Sampling Approaches Used in a Sequential Sampling Approach for Stochastic Programming Problems with Continuous Uncertain Parameters

Aroonsri Nuchitprasittichai, Chemical Engineering, The University of Tulsa, Tulsa, OK and Selen Cremaschi, Department of Chemical Engineering, University of Tulsa, Tulsa, OK

The size of a multi-stage stochastic program grows exponentially with the number of scenarios, which increases quickly for uncertain parameters represented with continuous distributions. Therefore, it is necessary to determine the minimum number of scenarios to reach a preset confidence in the solution to reduce the computational cost of solving stochastic programming with continuously distributed random variables. This talk presents a comparison of sampling approaches used in a sequential sampling procedure. [1] This procedure is used to determine an appropriate number of scenarios required to cover the overall uncertain space in a multi-stage stochastic optimization problem with uncertain parameters that can be represented by continuous distributions. In the procedure, the sampling approach is used to discretize the continuous probability distributions for different numbers of scenarios. The studied sampling approaches are as follows: Monte Carlo method which relies on the repeated random sampling; Stratified sampling which divides an overall population into subpopulations and samples each subpopulation independently; Low-discrepancy sequences which consist of a sequence of points with the property that each subsequence ($x_1, \ldots, x_N$) for all N points has a low discrepancy; Univariate dimemsion reduction method (UDR) which estimates a multivariate function with multiple univariate functions.[2] The different sampling approaches are compared in term of the necessary number of scenarios required to reach the preset optimality gap estimator defined in [1].

For each number of scenarios (n), the optimality gap estimator is calculated as the difference between the optimum solution of the stochastic program with n scenarios and the candidate solution, which is the optimum solution of the stochastic program with $2n$ scenarios. The procedure terminates when the estimation of the optimality gap is sufficiently small. Otherwise, the number of scenarios is increased.

A multi-stage stochastic program of the biomass to commodity chemicals (BTTC) investment planning, i.e., ethylene production from biomass, is used as the case study. The uncertainty parameters are the availability of the raw materials and the demand for the product. The objective is to minimize the expected total cost, which includes the raw material cost, the capital cost, and the R&D expenditure. The problem was formulated in GAMS 23.5 and solved using BARON. Low-discrepancy sequences modeling approaches required smaller number of scenarios to achieve the preset solution accuracy.

References:

[1] G. Bayraksan, D. P. Morton. A sequential sampling procedure for stochastic programming. *Operations Research*. 59. 898-913. 2011.

[2] S. H. Lee, W. Chen. A comparative study of uncertainty propagation methods for black-box-type problems. *Struct Multidisc Optim*. 37. 239-253. 2009.

(599ag) Uncertainty Relationship Analysis for Multi-Parametric Programming in Modeling and Optimization

Tianxing Cai and Qiang Xu, Dan F. Smith Department of Chemical Engineering, Lamar University, Beaumont, TX

In the progress of development of modeling and optimization for process design and operations problems under uncertainty, the process system engineering community has put more attention in the multi-parametric programming from the field of system theory and optimal control. This will help to achieve the full characterization of the entire range of solution space and provides a complete map of the various alternatives in the face of uncertainty.

Uncertainty and variability exists in all levels of the industry operation and manufacturing because they are the inherent characterization of any process system. Up to now, all the research that handles with multi-parametric programming (mp-LP, mp-QP, mp-NLP, mp-MILP, mp-MINLP) treats the uncertainty parameters to be linearly independent with each other while actually there exist some kinds of quantitative relationship between each other. This area is lack of research while the relationship between these uncertainty parameters can help to reduce dimension of uncertainty space and provide the solution for multi-parametric programming with nonlinear uncertainty items appear in the constraints.

This paper presents the definition of multiple types of quantitative relationship between the uncertainty parameters which can be generalized into two classifications of strong relationship and weak relationship. The strong relationship can be used to reduce the dimension of uncertainty space while the weak relationship is used to reduce the region of search uncertainty space. With the combination of the above relationships, the scenarios will be provided to solve different kinds of multi-parametric programming which cannot be solved directly with historical research result.

An Integrated Platform for Optimization of Heat and Water Use in Multipurpose Batch Plants

Esmael R. Seid[1], Omobolanle Adekola[1], Jane D. Stamp[1] and Thokozani Majozi[1,2]*

[1]Department of Chemical Engineering, University of Pretoria, Lynnwood Road, Pretoria, 0002, South Africa

[2] Modelling and Digital Science, CSIR, Meiring Naude Road, Pretoria, 0002, South Africa

**Corresponding author:*

Email: thoko.majozi@up.ac.za; Tel: +27 12 420 4130; Fax: +27 12 362 5173

Abstract

Water and utilities are major contributors to the overall running costs of multipurpose batch plants such as breweries, dairies, biochemical plants and agrochemical facilities. Water is essential for process equipment cleaning, due to the sharing of equipment by different tasks. To ensure that the amount of freshwater consumed and the amount of wastewater generated is minimized, efficient use of water is pursued. Similarly, heating and cooling are unavoidable aspects of many plant facilities, with operations where heat is generated and others where heat is required. It is because of this occurrence that heat integration becomes a possibility.

Heat optimization and wastewater minimization have mainly been treated as separate problems in literature. The batch production schedules resulting from the separate formulations do not guarantee that the plant is operated optimally. Consequently, it is required to develop a formulation that caters for opportunities that exist for both wastewater minimization and heat integration. This may result in production schedules that improve the operation of the batch plant when compared to optimizing water and heat separately.

Presented in this contribution is a formulation that addresses the optimization of both water and heat, while simultaneously optimizing the batch process schedule. The scheduling framework used in this study is based on the formulation by *Seid and Majozi* (2012). This formulation was shown to result in a reduction of computational time, an improvement of the objective function and led to fewer time points. Opportunities for direct water reuse were explored for wastewater minimization. With respect to heat optimization, direct heat integration was explored for reducing external utility requirements. The objective is to improve the profitability of the plant by minimizing wastewater generation and utility usage.

Keywords: Wastewater minimization, heat integration, multipurpose

1. Introduction

In recent years, batch processes have been getting more attention due to their suitability for the production of small volume, high value added products. The flexibility of batch plants allows the production of different products within the same facility. Despite the advantage of batch plants being flexible, the discreteness of tasks poses a challenge.

A considerable advancement has been made for scheduling of batch operations as compared to minimising energy and wastewater. However, greater public awareness of the impact of industrial pollution, more stringent environmental regulations, and escalating raw material costs, energy, and waste treatment costs have now motivated energy and water saving measures for more sustainable operations (Halim and Srinivasan, 2011). Since scheduling, energy and wastewater minimization go hand in hand we review work that has been done in those areas.

1.1. Scheduling of batch plants

Shaik et al. (2009) reviewed the current scheduling techniques based on different time representations and the associated challenges. In their review, the different models are classified as slot based, event based and precedence based (sequence-based) time representation. They concluded that due to heterogeneous location of events across the unit, the unit specific event based models require fewer events and perform better when compared to both the global event based and slot based models. Seid and Majozi (2012) presented an MILP formulation based on state sequence network and unit specific time points that can handle the proper sequence of tasks and FIS policy. The model results in reduction of binary variables and event or time points required. As a result, the model gives better computational performance in terms of objective value and CPU time required.

1.2. Energy integration in batch plants

Increasing energy costs and environmental considerations have made heat integration in batch plants more important. Energy savings in batch plants were neglected in the past as these plants are generally smaller in comparison to continuous plants. It was believed that energy use in batch plants was less significant. This is, however, untrue for some batch plants, such as the food industry, dairies, brewies, biochemical and agrochemical facilities. Halim and Srinivasan (2009) developed a sequential model for considering direct heat integration in batch plants. Chen and Chang (2009) proposed a cyclic model based on the resource task network representation. Majozi (2009) considered the addition of heat storage for indirect heat integration in batch plants with a flexible schedule, while Stamp and Majozi (2011) improved on this formulation and optimised the heat storage capacity and initial heat storage temperature.

1.3. Wastewater minimization in batch plants

Wastewater reuse in the form of direct reuse, indirect reuse using storage vessels and regeneration reuse have been explored in literature to minimize wastewater generation. Liu et al. (2009) considered the use of a central regenerator using a predefined schedule. Majozi and Gouws (2009) addressed wastewater minimization problems with multiple contaminants. In this method, the batch production schedule and the minimum wastewater target were obtained simultaneously. Adekola and Majozi (2011) extended the concept of reuse to include regeneration for multiple contaminant problems.

Wastewater minimization and heat integration in batch plants have largely been addressed separately. Halim and Srinivasan (2011) addressed this gap using a sequential methodology. The same problem is examined in this contribution by solving energy integration and wastewater minimization simultaneously within the same scheduling framework.

2. Problem statement

Given:

(1) The production recipe (STN or SSN representation), (2) The capacity of units and the type of tasks each unit can perform, (3) The maximum storage capacity for each material, (4) The task processing times, (5) Hot duties for tasks requiring heating and cold duties for tasks that require cooling, (6) Operating temperatures of heat sources and heat sinks, (7) Minimum allowable temperature differences, (8) The material heat capacities, (9) The units' washing times, (10) The mass load of each contaminant, (11) The concentration limits of each contaminant, (12) The costs of raw materials, utilities, freshwater, effluent treatment and selling price of products and (13) The scheduling horizon.

Determine:

a) The production schedule (i.e. allocation of tasks to units, timing of all tasks, and batch sizes),

b) The heat exchanger configuration, and

c) The direct water-reuse network.

3. Mathematical formulation

The scheduling model by Seid and Majozi (2012) was used since it has proven to result in fewer binary variables, shorter CPU time and a better optimal objective value compared to other scheduling models. Below some of the constraints in the model is discussed.

3.1 Heat integration model

Constraints (1) and (2) are active simultaneously and ensure that one hot unit will be integrated with one cold unit when direct heat integration takes place, in order to simplify operation of the process.

$$\sum_{s_{injc}} x(s_{injc}, s_{injh}, p, pp) \le y(s_{injh}, p), \quad \forall\ p \in P,\ \ s_{injh} \in S_{inJh},\ s_{injc} \in S_{inJc} \tag{1}$$

$$\sum_{s_{injh}} x(s_{injc}, s_{injh}, p, pp) \le y(s_{injc}, p), \quad \forall\ p \in P,\ \ s_{injh} \in S_{inJh},\ s_{injc} \in S_{inJc} \tag{2}$$

Constraints 3 and 4 describe the average heat flow for the hot and cold unit, respectively during the processing time which is the same as time average (TAM) model to address the energy balance during heat integration properly.

$$cl(s_{injh}, p) = avcl(s_{injh}, p)(tp(s_{injh}, p) - tu(s_{injh}, p)), \forall\ p \in P,\ \ s_{injh} \in S_{inJh} \tag{3}$$

$$hl(s_{injc}, p) = avhl(s_{injc}, p)(tp(s_{injc}, p) - tu(s_{injc}, p)), \forall\ p \in P,\ \ s_{injc} \in S_{inJc} \tag{4}$$

Constraints 5 and 6 define the heat load at time point *p* and extended time point *pp* for the hot and cold unit.

$$hlp(s_{injc}, p, pp) = avhl(s_{injc}, p)(tpp(s_{injc}, p, pp) - tuu(s_{injc}, p, pp)), \forall\ p, pp \in P,\ \ s_{injc} \in S_{inJc} \tag{5}$$

$$clp(s_{injh}, p, pp) = avcl(s_{injh}, p)(tpp(s_{injh}, p, pp) - tuu(s_{injh}, p, pp)), \forall\ p, pp \in P,\ \ s_{injh} \in S_{inJh} \tag{6}$$

Constraints 7 and 8 are used to calculate the temperature of the hot and cold unit at the intervals.

$$hlp(s_{injh}, p, pp) = mu(s_{injh}, p)cp(s_{injh})(T^{in}(s_{injh}, p, pp) - T^{out}(s_{injh}, p, pp)), \forall\ p, pp \in P,\ \ s_{injh} \in S_{inJh} \tag{7}$$

$$clp(s_{injc}, p, pp) = mu(s_{injc}, p)cp(s_{injc})(T^{in}(s_{injc}, p, pp) - T^{out}(s_{injc}, p, pp)), \forall\ p, pp \in P,\ \ s_{injc} \in S_{inJc} \tag{8}$$

Constraints 9 and 10 ensure that the minimum thermal driving forces are obeyed when there is direct heat integration between a hot and a cold unit.

$$T^{in}(s_{injh}, p, pp) - T^{out}(s_{injc}, p, pp) \ge \Delta T - \Delta T^{Max}(1 - x(s_{injh}, s_{injc}, p, pp)), \\ \forall\ p, pp \in P,\ \ s_{injh} \in S_{inJh},\ s_{injc} \in S_{inJc} \tag{9}$$

$$T^{out}(s_{injh}, p, pp) - T^{in}(s_{injc}, p, pp) \ge \Delta T - \Delta T^{Max}(1 - x(s_{injh}, s_{injc}, p, pp)), \\ \forall\ p, pp \in P,\ \ s_{injh} \in S_{inJh},\ s_{injc} \in S_{inJc} \tag{10}$$

Constraints (11) to (14) ensure that the times at which units are active are synchronized when direct heat integration takes place.

$$tuu(s_{injh}, p, pp) \ge tuu(s_{injc}, p, pp) - M(1 - x(s_{injh}, s_{injc}, p, pp)), \\ \forall\ p, pp \in P,\ \ s_{injh} \in S_{inJh},\ s_{injc} \in S_{inJc} \tag{11}$$

$$tuu(s_{injh}, p, pp) \le tuu(s_{injc}, p, pp) + M(1 - x(s_{injh}, s_{injc}, p, pp)), \\ \forall\ p, pp \in P,\ \ s_{injh} \in S_{inJh},\ s_{injc} \in S_{inJc} \tag{12}$$

$$tpp(s_{injh}, p, pp) \ge tpp(s_{injc}, p, pp) - M(1 - x(s_{injh}, s_{injc}, p, pp)), \\ \forall\ p, pp \in P,\ \ s_{injh} \in S_{inJh},\ s_{injc} \in S_{inJc} \tag{13}$$

$$tpp(s_{injh}, p, pp) \le tpp(s_{injc}, p, pp) + M(1 - x(s_{injh}, s_{injc}, p, pp)), \\ \forall\ p, pp \in P,\ \ s_{injh} \in S_{inJh},\ s_{injc} \in S_{inJc} \tag{14}$$

Constraint (15) states that the cooling of a hot unit will be satisfied by direct heat integration and external cooling utility, if required.

$$cl(s_{injh},p) = cw(s_{injh},p) + \sum_{s_{injc}} Qex(s_{injh},s_{injc},p,pp), \forall\ p,pp \in P,\ \ s_{injh} \in S_{inJh},\ s_{injc} \in S_{inJc} \qquad (15)$$

Constraint (16) states that the heating of a cold unit will be satisfied by direct heat integration and external heating utility, if required.

$$hl(s_{injc},p) = st(s_{injh},p) + \sum_{s_{injh}} Qex(s_{injh},s_{injc},p,pp), \forall\ p,pp \in P,\ \ s_{injh} \in S_{inJh},\ s_{injc} \in S_{inJc} \qquad (16)$$

3.2 Wastewater minimization model

Constraint 17 defines the amount of water entering the unit as the sum of freshwater and reuse water from other units.

$$mwin(s_{inj},p) = mfw(s_{inj},p) + \sum_{s_{inj'}} mrw(s_{inj'},s_{inj},p),\ \forall\ p \in P,\ \ s_{inj},s_{inj'} \in S_{inJ} \qquad (17)$$

Constraint 18 states that the amount of water leaving the unit is equal to the sum of reuse water sent to other units and water sent to effluent treatment.

$$mwout(s_{inj},p) = \sum_{s_{inj'}} mrw(s_{inj},s_{inj'},p) + mew(s_{inj},p), \forall\ p \in P,\ \ s_{inj},s_{inj'} \in S_{inJ} \qquad (18)$$

Constraint 19 is the water balance around the unit and states that the amount of water entering the unit equals the amount of water leaving the unit.

$$mwin(s_{inj},p) = mwout(s_{inj},p), \forall\ p \in P,\ \ s_{inj} \in S_{inJ} \qquad (19)$$

Constraint 20 defines the inlet contaminant load as the mass of contaminant, entering with reuse water.

$$cin(s_{inj},\Psi,p)mwin(s_{inj},p) = \sum_{s_{inj'}} cout(s_{inj'},\Psi,p)mrw(s_{inj'},s_{inj},p), \forall\ p \in P,\ \ s_{inj},s_{inj'} \in S_{inJ} \qquad (20)$$

Constraint 21 states that the amount of contaminant leaving the unit equals the mass of contaminant entering into the unit and the mass of contaminant removed from the process.

$$mwout(s_{inj},p)cout(s_{inj},\Psi,p) = SMC(s_{inj})mu(s_{inj},p) + cin(s_{inj},\Psi,p)mwin(s_{inj},p), \forall\ p \in P,\ \ s_{inj} \in S_{inJ} \qquad (21)$$

Constraint 22 ensures that the amount of reuse water from unit j to other units does not exceed the maximum allowable water in the receiving units. It also indicates whether water from unit j is reused or not.

$$mrw(s_{inj},s_{inj'},p) \leq W_{in}^{\max}(s_{inj'})yre(s_{inj},s_{inj'},p), \forall\ p \in P,\ \ s_{inj},s_{inj'} \in S_{inJ} \qquad (22)$$

Constraint 23 ensures that the reuse of water from unit j in other units can occur only if the receiving units are active.

$$yre(s_{inj},s_{inj'},p) \leq y(s_{inj'},p), \forall\ p \in P,\ \ s_{inj},s_{inj'} \in S_{inJ} \qquad (23)$$

Constraint 24 gives the upper bound on the water entering into unit j. It also ensures that water enters into the unit only if it is active.

$$mwin(s_{inj},p) \leq W_{in}^{\max}(s_{inj})y(s_{inj},p), \forall\ p \in P,\ \ s_{inj} \in S_{inJ} \qquad (24)$$

In Constraints 25 and 26, wastewater can only be directly reused if the finishing time of the unit producing wastewater and the starting time of the unit receiving wastewater coincide.

$$tuw(s_{inj},p) \geq tpw(s_{inj},p) - M^{*}\, yre(s_{inj},s_{inj'},p), \forall\ p \in P,\ \ s_{inj},s_{inj'} \in S_{inJ} \qquad (25)$$

$$tuw(s_{inj},p) \leq tpw(s_{inj},p) + M^{*}\, yre(s_{inj},s_{inj'},p), \forall\ p \in P,\ \ s_{inj},s_{inj'} \in S_{inJ} \qquad (26)$$

Constraint 27 defines the finishing time of the washing operation as the starting time of the washing operation added to the duration of washing.

$$tpw(s_{inj},p) \geq tuw(s_{inj},p) + \tau w(s_{inj})y(s_{inj},p), \forall\ p \in P,\ \ s_{inj} \in S_{inJ} \qquad (27)$$

Constraints 28 and 29 ensure that the inlet and outlet concentrations do not exceed the maximum allowable contaminant concentrations.

$$cin(s_{inj},\Psi,p) \leq cin^{\max}(s_{inj},\Psi), \forall\ p \in P,\ \ s_{inj} \in S_{inJ} \qquad (28)$$

$$cout(s_{inj},\Psi,p) \leq cout^{\max}(s_{inj},\Psi), \forall\ p \in P,\ \ s_{inj} \in S_{inJ} \tag{29}$$

Constraint 30 is the objective function in terms of profit maximization, with profit defined as the difference between revenue from product, cost of utility, freshwater cost and effluent treatment cost.

$$\max\left\{\begin{array}{l}\sum_{p}\sum_{s^p} price(s^p)d(s^p) - \sum_{p}\sum_{s_{injh}} costcw * cw(s_{injh},p) - \\ \sum_{p}\sum_{s_{injc}} costst * st(s_{injc},p) - \sum_{p}\sum_{s_{inj}} costfw * mfw(s_{inj},p) - \sum_{p}\sum_{s_{inj}} costew * mew(s_{inj},p)\end{array}\right\}, \tag{30}$$

$\forall\ p,pp \in Ps_{injh} \in S_{inJh},\ \ s_{injc} \in S_{inJc},\ \ s_{inj} \in S_{inJ}$

4. Case Study

This case study is a simple sequential process. A single product is produced via three tasks in five units. Figure 1 shows the recipe representation of the process. The necessary scheduling data as well as the heat integration and wastewater minimization data may be obtained from literature (Halim & Srinivasan, 2011). Steam is available for heating, cooling water is available for cooling and freshwater is available for washing. The objective is to maximise profit over a 12 h time horizon, while minimising freshwater usage and utility requirements. Direct heat integration and direct water reuse may be used.

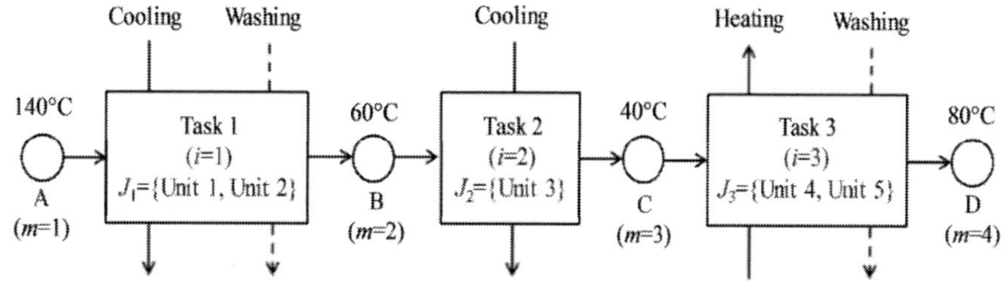

Figure 1: recipe representation of simple case study (Halim & Srinivasan, 2011).

The results from the proposed model were obtained using CPLEX 9 as MILP solver and CONOPT 3 as NLP solver in DICOPT interface of GAMS 22.0. The problem was solved using a 2.4 GHz, 4 GB of RAM, Acer TravelMate 5740G computer. Table 1 compares the results obtained using the proposed formulation to the results achieved by Halim and Srinivasan (2011).

As can be observed from Table 1 column 3, a better profit overall was obtained with the proposed formulation. The amounts for steam, cooling water, freshwater and effluent were lower for the proposed formulation. Similar to the method by Halim and Srinivasan (2011), the washing of a unit may not necessarily occur immediately after the end of a processing task, but can be delayed to improve reuse opportunities with other units.

Figure 2 illustrates energy integration between the cold and hot unit and the time intervals the energy integration has occurred. The energy requirement by unit *j2* and unit *j4* during the interval of 3.2 h to 5 h is taken to discuss the application of the proposed model in calculating and satisfying the energy requirements. The cooling demand of unit *j2* during interval 3.2 h to 4.9 h was 32 MJ. This is partly satisfied by energy integration with unit *j4* in the same time interval; resulting in external cooling requirement of 26.8 MJ rather than 32 MJ if operated in standalone mode. At the beginning of the operation of unit *j2* from time 3.2 h to 4.25 h the cooling requirement is 19.76 MJ obtained using the time average model by multiplying the duration (4.25 h – 3.2 h) and the energy demand per hour (32 MJ/1.7 h (total duration of the task) = 18.823 MJ/h). This cooling requirement is fully satisfied by external cooling. For the rest of the processing time between 4.25 h to 4.9 h, the cooling requirement is 12.24 MJ and is satisfied by energy integration (5.2 MJ) and external cooling (7.03 MJ). The heating requirement of unit *j4* during time interval 4.25 h to 5 h was 6 MJ. During the interval from 4.25 h to 4.9 h, the steam requirement is 5.2 MJ obtained from the time average model. The heating requirement in this interval is fully satisfied by integrating with hot unit *j2*. The rest of the heating required during its operation between 4.9 h to 5 h, 0.8 MJ, is satisfied by external steam.

Table 1: Comparison of results for case study

	Halim & Srinivasan (2011)	This formulation
Profit ($)	4 764.1	4 808.2
Steam (MJ)	43.9	39
Cooling water (MJ)	313.9	309
Total freshwater (kg)	1 238.4	977.7
Revenue from product ($)	5 000	5 000
Cost of steam ($)	43.9	39
Cost of cooling water ($)	6.3	6.2
Cost of freshwater ($)	123.8	997.7
Cost of wastewater ($)	61.9	48.9
CPU time (s)	not reported	3000

Figure 3 shows the Gantt chart with the resulting production schedule for the case study. The clear horizontal blocks represent the tasks associated with production processes. The amount of material processed in each unit is labelled within the clear blocks. The cross-hatched blocks represent the washing of a unit. Water reuse between units has been depicted with arrows and has been clearly labelled.

Figure 4 shows the water use network in detail which includes freshwater usage, amount of water reused and wastewater produced in the washing of each unit. For example, the washing operation of unit *j2* between 4.9 h and 5.2 h requires 200 kg of freshwater to remove a contaminant load of 20 g. At the end of washing, water with a concentration of 100 ppm is produced. Part of this water, 50 kg is used in unit *j4* to remove a contaminant load of 10 g. This is possible because the outlet concentration from unit *j2* (100 ppm) is lower than the maximum inlet contaminant concentration (150 ppm) for unit *j4.* From Figure 4 the total amount of reused water is 358.23 kg, minimising the water usage from 1105 kg (without water integration) to 977.7 kg (with water integration) resulting in a 11.5 % saving of freshwater usage and wastewater produced.

5. Conclusions

In the presented method, wastewater minimization and heat integration are both embedded within the scheduling framework and solved simultaneously. Results from a case study shows that profit maximization together with heat integration and wastewater minimization gave a much better overall economic performance when compared to a sequential based method.

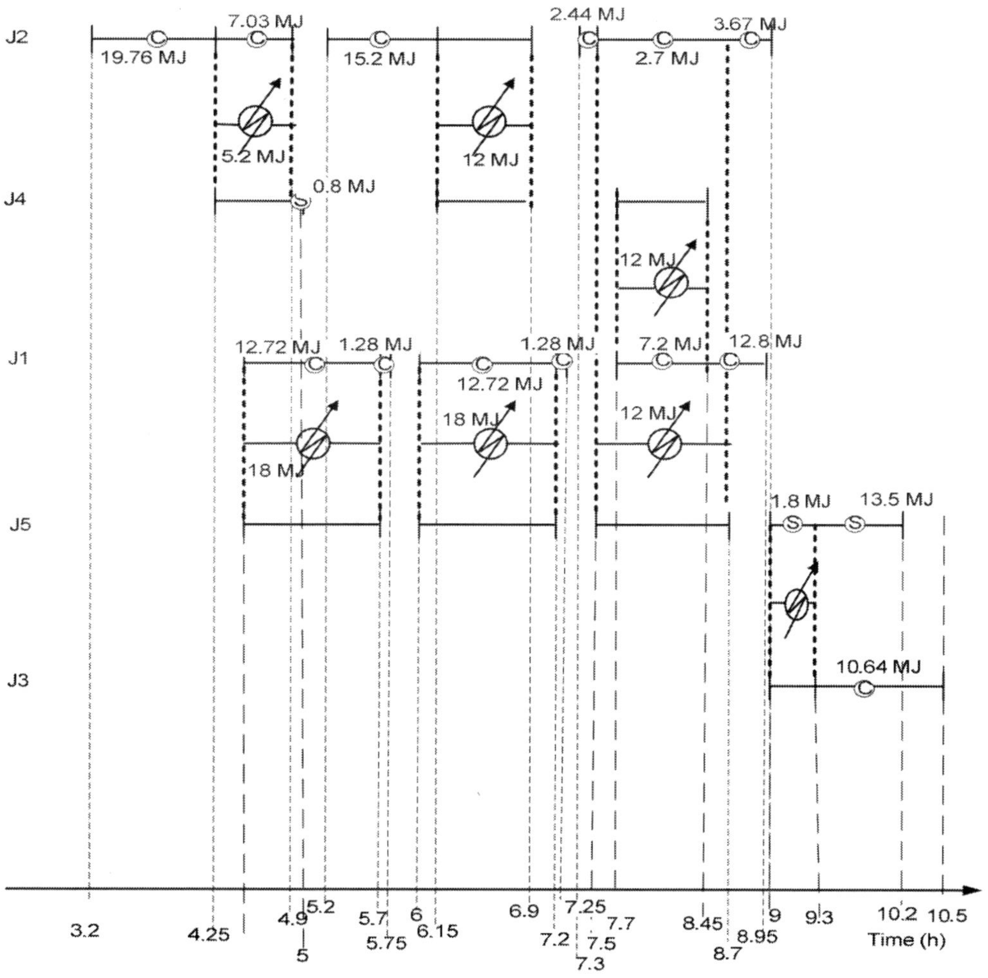

Figure 2. Energy integration within the time horizon of 12 h

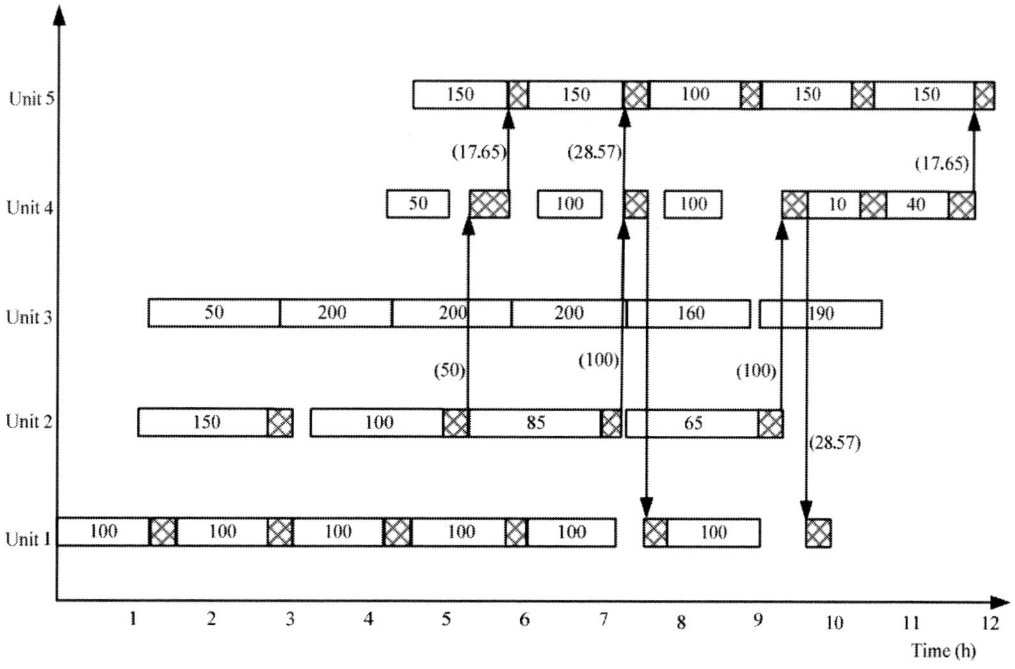

Figure 3. Gantt chart for the time horizon of 12 h.

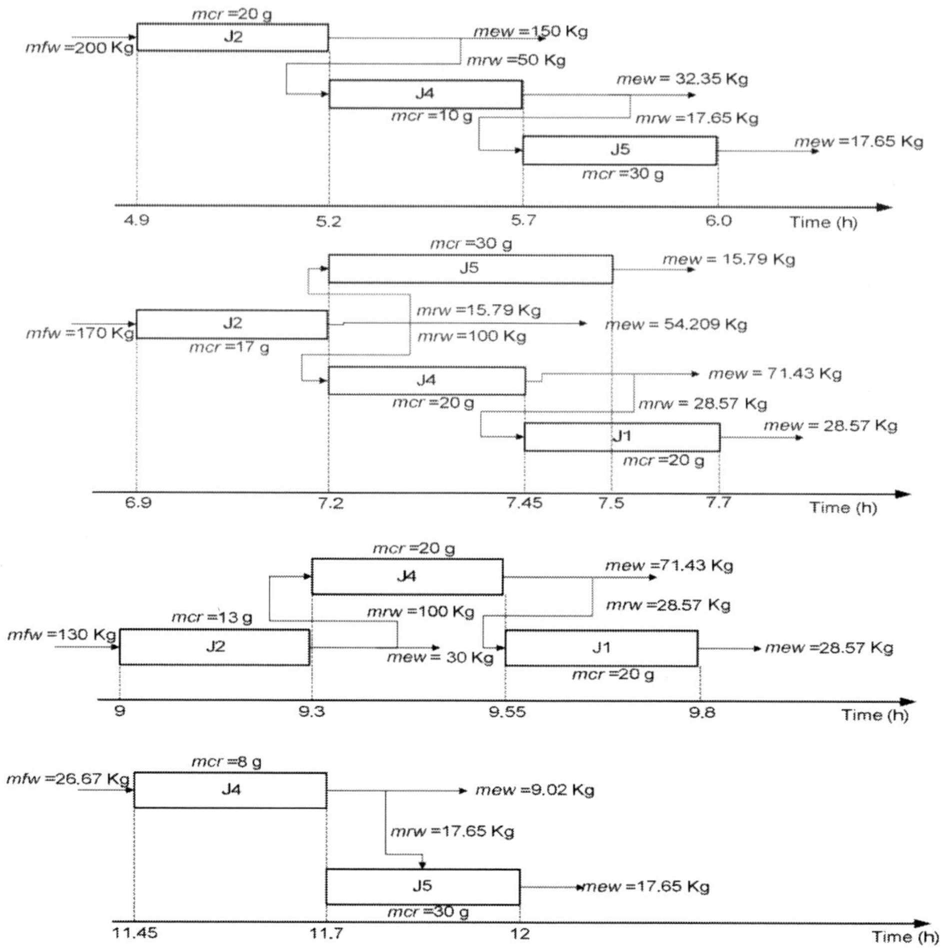

Figure 4. Water network for minimum water use in a 12 h time horizon

References

Adekola O.; Majozi T. Wastewater minimization in multipurpose batch plants with a regeneration unit: Multiple contaminants. Comput. Chem. Eng. 2011, 35, 2824-2836.

Chen, CL.; Chang, CY. A resource-task network approach for optimal short-term/periodic scheduling and heat integration in multipurpose batch plants. Appl. Therm. Eng. 2009, 29, 1195–208.

Halim I.; Srinivasan R. Sequential methodology for integrated optimization of energy and water use during batch process scheduling, Comput. Chem. Eng. 2011, 35, 1575-1597.

Liu, Y.; Li, G.; Wang, L.; Zhang, J.; & Shams, K. Optimal design of an integrated discontinuous waterusing network coordinating with a central regeneration unit. Ind. Eng. Chem. Res. 2009, 48, 10924–10940.

Majozi, T. Minimization of energy use in multipurpose batch plants using heat storage: an aspect of cleaner production. J. Clean. Prod. 2009, 17, 945–50.

Majozi, T.; & Gouws, J. F. A mathematical optimisation approach for wastewater minimisation in multipurpose batch plants: Multiple contaminants. Comput. Chem. Eng. 2009, 33, 1826–1840.

Seid, R.; Majozi, T. A. Robust mathematical formulation for multipurpose batch plants. Chem. Eng. Sci. 2012, 68, 36–53.

Shaik, M.A.; Floudas, C.A. Novel unified modeling approach for short term scheduling. Ind. Eng. Chem. Res. 2009 48, 2947-2964.

Stamp J.; Majozi T. Optimum heat storage design for heat integrated multipurpose batch plants. Energy. 2011, 36, 5119-5131.

(599ai) Proactive Abnormal Situations Management Using Anticipatory Alarms

Shichao Xu, Institute of Chemical and Engineering Sciences, Singapore , Singapore, Singapore and Rajagopalan Srinivasan, Department of Chemical and Biomolecular Engineering, National University of Singapore, Singapore, Singapore

ABSTRACT

Modern chemical plants are built with large amount of integrated and interlinked process units. Such complexity ensures that these plants are operated in an optimized and profitable fashion. When an abnormal situation occurs, the automation systems in these plants alert the operators through alarms and help orient them to the new state. However, due to the coupling nature of chemical processes, the alarming system may generate many simultaneous alarms resulting in an alarm flood[1]. Such a deluge of alarms may disorient and cause information (or data) overload on the part of the operator. To make matter worse, for inexperienced operators managing the plant this would lead to more confusion in the control room, delaying or preventing timely sense-making and eventually resulting in an emergency shut-down in the best case, or sometimes worse.

To address this issue, many alarm flood management techniques has been developed. For example, an alarm prioritization framework was introduced to ease the burden of operators with meaningless or false alarms[2] and an intelligent alarm management system was proposed to suppress nuisance alarms in a chemical refinery to reduce the number of alarms to a more manageable level[3,4]. Other techniques also include the development of a refinery robot to assist operators in managing alarms in refineries during abnormal situations[5] and the use of an intelligent operator decision support system, called Op-Aide, to assist the operator in quantitative diagnosis and assessment of alarms during abnormal situations[6]. While the above techniques have shown to improve alarm management especially during an alarm flood, none of them has proposed strategies to look at the sense-making facilities offer by the alarm system. As result, operators may still be disorient and overloaded with information caused by the number alarms during an alarm flood. In this paper, we introduce a new type of alarms, called anticipatory alarms, that seeks to improve the sense-making facilities offered by the alarm system in the face of alarm floods.

The idea of *anticipatory alarms*is to help plant operators, especially those that are less experienced, to manage alarms during an abnormal situation in a more proactive and organized manner. In most chemical plants, process variables are measured and monitored in real-time by tens of thousands of sensors. To ensure safe operation of the plant, thousands of alarms are configured on these process variables. In the event when a large disturbance or abnormal situation occurs in the process, operators' attention is abducted by the occurring alarms which are also highlighted in the DCS schematic. Other variables that are moving towards their alarm limits but have not yet reached their alarm threshold would appear to be in a normal state. In the absence of trend information in today's DCS schematics, the operator does not have the complete information required for inferring the true state of the plant or the process units. Anticipatory alarms seek to overcome this handicap of partial information offered by modern-day alarm systems. This type of alarms provide operators with anticipatory information on incipient alarms that are about to occur with a certain time-window. By having this information, operators can obtain a complete sense of the true state of the process which would help them, localize and rectify the problem early before the plant reaches a dangerous operating state. The availability of the anticipatory alarms in advance of their actual occurrence reduces the confusion in the control room caused by alarm floods and allows efficient handling of abnormal situations.

The proposed anticipatory alarms are built around an alarm anticipation algorithm, which utilizes dynamic models of the process to anticipate the alarms. These dynamic models, called *anticipatory alarming models*, are different to typical first-principle or data-driven models as they are able to (1) represent the dynamics of the process during an abnormal situation and (2) account for structural and parametric changes to the nominal model of the process so that they do not get less and less reliable during abnormal situations when the process deviates further from the nominal operating point. The alarm anticipation algorithm utilizes the anticipatory alarming models to predict the rate-of-change of process variables which are translated into predictions of time horizons for occurrence of various alarms. The time-prediction of alarms is then passed on to the operators through an anticipatory alarms display. The effectiveness of the anticipatory alarms is demonstrated using a case study of a depropanizer unit subjected to different abnormal situations.

References

[1] J. Errington, D. Vernon Reising and C. Burns, "ASM consortium guidelines: Effective alarm management practices," ASM Consoritum, Phoenix, 2009.

[2] O.M. Foong, S.B. Sulaiman, D.R.B. Awang Rambli and N.S.B. Abdullah, "ALAP: Alarm prioritization system for oil refinery" In proceedings of the World Congress on Engineering and Computer Science (ISBN:978-988-18210-2-7), San Francisco, 2009.

[3] J. Liu, K.W. Lim, W.K. Ho, K.C. Tan, R. Srinivasan and A. Tay, "The intelligent alarm management system", IEEE Software, vol. 20, pp. 66-71, 2003.

[4] J. Liu, K.W. Lim, R. Srinivasan, K.C. Tan and W.J. Ho, "The intelligent alarm management system", Hydrocarbon Processing , pp. 47-53, 2004.

[5] K. Krebsbach, and D. Musliner, "A refinery immobot for abnormal situation management," Automated Reasoning Group, Honeywell Technology Center, 1998.

[6] H. Vedam, S. Dash and V. Venkatasubramanian, "An intelligent operator decision support system for abnormal situation management", Computers adn Chemical Engineering, vol. 23, pp. 577 – 580, 1999.

(599ak) New Approach to Gas Hydraulics Calculations

Mesude Ozturk[1], Lutfiye Hacioglu[2] and Miguel Bagajewicz[1], (1)Chemical Engineering, University of Oklahoma, Norman, OK, (2)Chemical Engineering, University of Oklahoma, Norman

New Approach to Gas Hydraulics Calculations

Mesude Ozturk, Lutfiye Hacioglu and *Miguel Bagajewicz

School of Chemical Engineering and Material Science, University of Oklahoma

100 East Boyd Street, T-335 – Norman, OK 73019-0628 USA

*corresponding author

Abstract

The methods used by engineers to evaluate gas-hydraulics in pipelines have not changed in the last few decades. Indeed, several books recommends the usage of equations developed several years ago, when computers were not available, like Modified Colebrook-White, AGA, Panhandle A and B, Weymouth, etc. With the advance of computers, piping gas hydraulics can in principle be computed very accurately integrating numerically differential forms of the Bernoulli equation, thus rendering all approximate existing formulas for pressure drop rather inadequate. In fact, when accurate calculations are compared to the results rendered by these corelations large deviations are observed. In this paper we use an accurate implementation of pressure drop to determine the error of the different existing formulas. We also propose a way of constructing a more accurate analytical formula to replace existing correlations that can also be useful in operations optimization, pipeline design using mathematical programming, leak detection, among other uses. Finally, we report the errors of our formula and compare with others.

An Optimization Strategy for Batch Process Scheduling Under Processing Time Uncertainty

Jun-hyung Ryu†, Taeyoung Lee*, In-Beum Lee*

*Dept. of Chem. Eng., POSTECH, Pohang, Gyungbuk, Korea, 790-784

Dept of Nuclear and Energy System, Dongguk University, Gyeongju, Gyungbuk, Korea, 780-714

Abstract

Various events can occur during the operation of processes. Process scheduling thus should reflect the presence of such uncertainty while seeking for maximizing the performance in terms of economics and process stability. There are many areas scheduling can make its impact including short term, medium term and long term horizon of interest. While short term scheduling is similar to control of specific process and long term scheduling is more associated with planning problem, medium term scheduling should take into account both factors. In this paper, scheduling problem of batch processes with parallel units is considered. We aim to allocate units for serial multiproduct multistage scheduling problem. In order to take into account the processing time uncertainty, the concept of buffer is introduced to address the complex processing time uncertainty. The resulting scheduling model is constructed using State-Task-Network (STN) representation. For the sake of computing the solution of such challenging problem, several techniques have been developed in this paper such as decomposing the overall problem into subproblems, simultaneously considering mass balance and time duration. The concept of complex task is proposed in this paper reflecting input, output state, conversion factor, batch size, processing time, cycle time and combination of available units. Numerical examples with computational statistics have been presented with some remarks to illustrate the applicability of the proposed modeling methodology.

Key words: Scheduling, uncertainty, time buffer, State-Task-Network

---

† corresponding author, email: jhryu@dongguk.ac.kr

1. Introduction

Scheduling is concerned with determining the allocation of resources into process units over time. The role of scheduling is to decide the completion time of individual tasks with their detailed equipment. There is a lot of works on scheduling in process systems engineering community over the last decades in terms of various issues such as deterministic and stochastic, time representation, multiproduct, multipurpose processes, batch and continuous processes, et al. The objective of scheduling take many forms such as minimizing the time required to complete all the tasks, so called makespan, minimizing the number of orders completed after their committed due dates, maximizing customer satisfaction by completing orders in a timely fashion, maximizing plant throughput, maximizing profit or minimizing production costs, etc. A good schedule allows more to be done in the fixed and save the consumption of resources. Without attention on scheduling, a process can be still operated. However with the good and manipulated schedule, we can accomplish much by saving resources and increase profitability. Therefore the attention on scheduling is fully motivating in terms of its financial contribution. The rigorous scheduling modeling may be of little help in the occurrence of unexpected happening. The events such as delayed processing times beyond expectation, sudden breakdown of tool or malfunction of utility systems enforces the entire process schedule out of the previously arranged operations. Due to the presence of such uncertainty, the process operation using the schedule that is based upon the deterministic expectation may turn out to be worse than using the initially ill-planned schedule.

Uncertainty is thereby a very important concern in production scheduling since it can cause infeasibilities and production disturbances. Thus scheduling under uncertainty has received a lot of attention in the open literature in recent years from chemical engineering and operations research communities. Therefore it is necessary to consider the presence of uncertainty in the work of scheduling. The potential schedule computing methodology should incorporate the uncertainty. Many works have been made and most widely employed approach is to state task network and resource task network.

2. Process Variability

Many petro-chemical processes are operated through multiple unit operations. These operations are very complex and their performances are subject to variations. The presence of variation may make the operation in a stage hold or delayed. We should put some buffer.

The uncertainty is in the form of processing time variations that is continuous time representation format. Equipment breakdown is a discrete type of uncertainty. Uncertainty in the scheduling problem has been mainly addressed in terms of (1) bounded form, (2) probability description, and (3) fuzzy description. (Li and Ierapetritou, 2008).

This paper investigates the processing time variation proposes a new methodology in terms of reformulating the scheduling model by proposing a concept of buffer. On one hand the presence of buffer make the resources of the overall processes wasted. Otherwise the insertion of the buffer allows us to have a schedule that is similar to the actual happening. To include the description of uncertain parameters within the optimization model of the scheduling problem, several methods have been used in this paper.

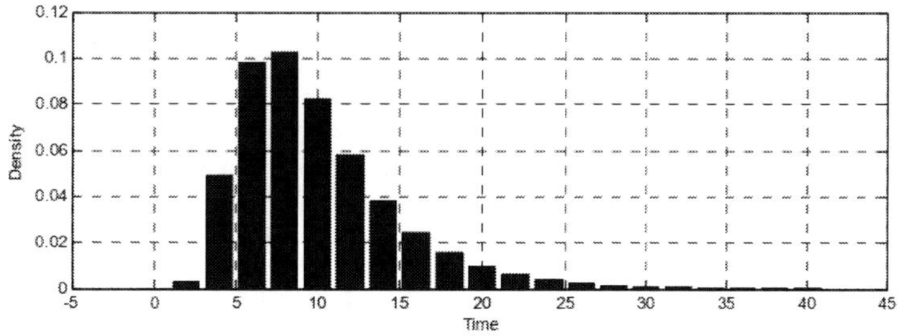

Figure 1 An illustration of Processing time variation (Log-normal Distribution with μ = 10 , σ = 5)

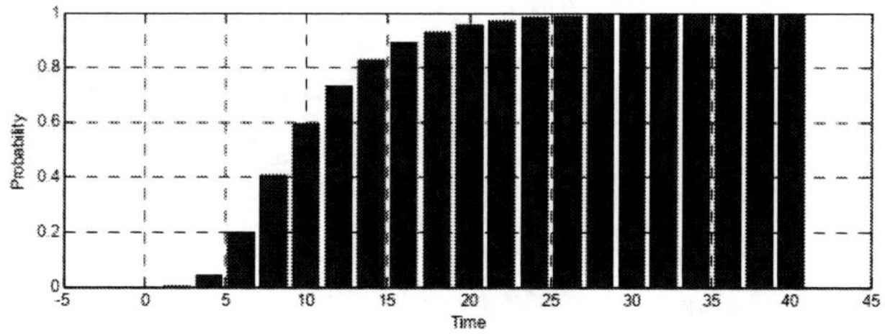

Figure 2 Log-normal Distribution with μ = 10 , σ = 5

How to manage variability in processing time is one of the most important issues in process systems. This paper proposes the concept of time buffer to address the scheduling problem under uncertainty. Considering the variation of processing time, it means that the probability of successful completion increases significantly. For example, we may increase the chance of successful completion to 90 % by adding extra time 6.39

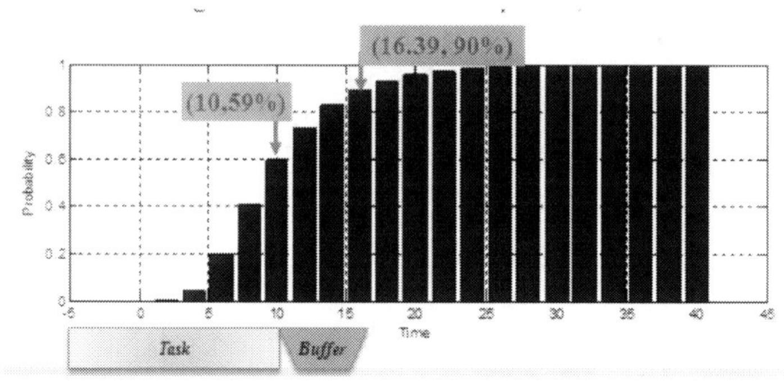

Figure 3 Log-normal Distribution with $\mu = 10$, $\sigma = 5$

The advantages of the proposed campaign mode operation are (1) reduction of # of setup and clean-up time and (2) reduction of contamination of units and (3) consequently, improvement of qualities of products.

3. Case Study

This paper investigates a serial multiproduct batch processes with multistage. There exist nonidentical parallel units as shown in Figure 4. The following constraints have been considered: (1) mass balance, (2) time duration to satisfy the required amount of final products. Individual subproblems are optimized with the objective function of minimizing the makespan. We aim to determine (1) the overall processing time of each complex task and (2) time gap of each complex task and its previous complex task.

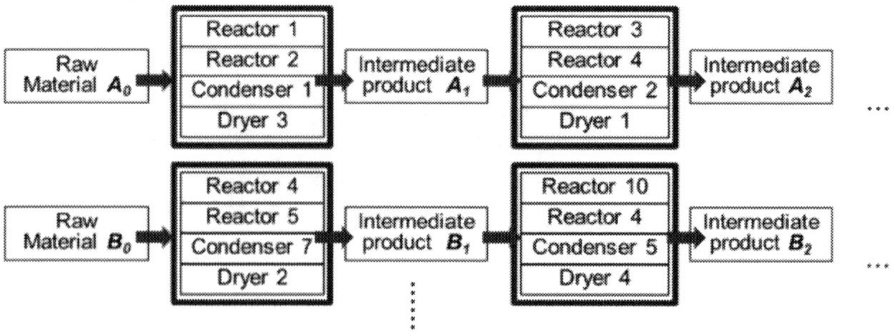

Figure 4 Illustration of The multi stage processes

Table 1 Objective function values according the proposed methods

Method	Original	Method 1	Method 2
Horizon(h)	801.0	1281.19	1087.68

4. Concluding remarks

We proposed a robust approach for optimizing batch scheduling considering a serial multiproduct / multistage batch plant. In order to increase the chance of successful completion of tasks, the concept of buffer time is incorporated with the state-task-network formulation.

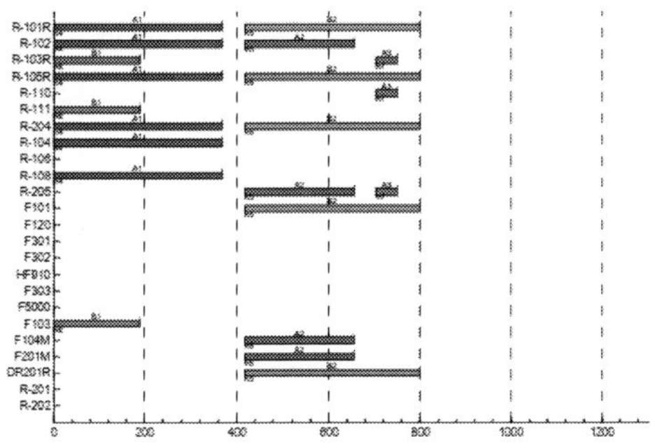

Figure 5 Gantt Chart of Initial Schedule using STN

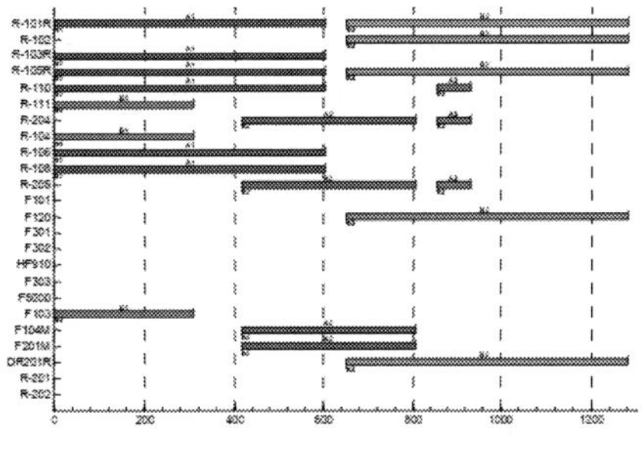

Figure 6 Gantt Chart using the proposed method 1

5. References

Li, Z., Ierapetritou, M. Process scheduling under uncertainty: Review and challenges, Computers and Chemical Engineering, 32 (2008) 715–727

6. Acknowledgement

This research was supported by Basic Science Research Program through the Nation al Research Foundation of Korea (NRF) funded by the Ministry of Education, Science and Technology (2010-0023678).

(599am) Process Systems Engineering Approach to Modeling and Configuration Optimization of Anti-Aircraft Guns

Dongil Peter Shin, Dept. of Chemical Engineering, Myongji University, Yongin, Kyunggido, South Korea and Hwan Il Kang, Myongji University, Yongin, South Korea

Process systems engineering approach is quite useful in systematic modeling, analysis and optimization of military systems. In this study, we investigated contributing factors of each subcomponent of anti-aircraft guns based on first-principles modeling and data-based study. Then the optimal configurations of anti-aircraft guns suitable for designated purposes have been produced out of mathematical programming.

An anti-aircraft gun is fired from the ground or shipboard in defense against aerial attack. Antiaircraft weapons development began as early as 1910, when the airplane first became an effective weapon.There are many types of anti-aircraft guns, including K-9.

Fire distribution is the key point in the operational process of integrated anti-aircraft gun and missile system. Aimed at the economically efficient quality and characteristics of air-defense missile system and anti-aircraft gun system, the criterion of optimum fire distribution is established on the basis of analyzing the main factors, which are killer area and kill probability of the system in fire distribution (Fan et al., 2004).

(599an) Dynamic Modeling and Parameter Estimation for Unit Operations in Lignocellulosic Bioethanol Production

Federico Andersen, Susana Moreno and **Maria Soledad Diaz**, Process Systems Engineering, Planta Piloto de Ingenieria Quimica (PLAPIQUI), CONICET - Universidad Nacional del Sur, Bahia Blanca, Argentina

Lignocellulosic biomass in nature is by far the most abundant and low-cost feedstock for the production of the so-called second generation biofuels. These biofuels are likely to replace first generation ones in the future to avoid the controversies resulting from the use of food crops for transport fuel production.

Lignocellulosic ethanol is one of the major second-generation biofuels and a considerable amount of research is being made in order to develop appropriate technologies for achieving a large conversion through biological processes (Martín & Grossmann, 2012; Andersen et al., 2011). Direct bioconversion of lignocellulosic materials to ethanol needs a pre-treatment stage aimed to remove lignin and hydrolyze hemicelluloses. This operation makes cellulose more accessible allowing the liberation of fermentable sugars from their carbohydrate polymers in a further hydrolysis stage. In the following microbial fermentation, the sugars extracted from cellulose and hemicellulose are converted to ethanol.

Although different pre-treatment techniques have been proposed, the dilute acid hydrolysis is preferred for industrial applications due to its simplicity and high sugar yield from hemicelluloses. However, it has some disadvantages such as generation of toxic compounds, like furfural and hydroxymethyl furfural, which can be inhibitory to microorganisms in the downstream process. Therefore, an additional operation of detoxification is required after this pre-treatment where the toxic materials are converted to less inhibitory components.

A number of kinetic models for the hydrolysis, detoxification, and fermentation operations involved in the production of bioethanol have been reported in the literature. However, they have not included process variables such as operating temperatures, concentrations of inhibitory compounds, etc., that would allow evaluating their influence when they are used in process simulation and optimization studies.

Consequently, the focus in this study is on the parameter estimation of kinetic models considering the main process variables for the unit operations of hydrolysis, detoxification, and fermentation in a bioethanol production plant. More specifically, we aim at identifying realistic dynamic models for the dilute acid hydrolysis, the detoxification by addition of $Ca(OH)_2$(overliming) and the simultaneous fermentation of mixtures of glucose and xylose (co-fermentation).

Particularly, for the diluted acid hydrolysis, the model postulated by Lavarack et al. (2002) has been adopted. They proposed several schemes for the production of xylose, glucose, furfural, and soluble lignin, considering the influence of temperature, acid concentration, and mass ratio of solid to liquid. In this work, we have extended original model with the inclusion of acetic acid generation. Regarding the overliming process, the model published by Purwadi et al. (2004) was selected. In this model the sugars or furans generate transient complexes with calcium ions which are degraded to a certain product. Here, we have proposed that the reaction rate constants are represented by the Arrhenius equation to account for the influence of the temperature in the process. Furthermore, a new expression for determining the initial concentration of calcium cation related to pH level was proposed. Finally, for the co-fermentation operation we have modified the unstructured model developed by Leksawasdi et al. (2001) by including a new factor that accounts for the furfural toxicity. Also, a first order kinetics with respect to furfural and biomass concentration is proposed for representing the conversion of furfural by the fermentative microorganism. The original high number of estimated parameters was significantly reduced in this study.

In this work we have formulated the dynamic parameter estimation problem for the above mentioned models within a simultaneous approach. In this approach, each differential algebraic equation (DAE) model is discretized using orthogonal collocation on finite elements (Biegler & Zabala, 2008, Estrada et al., 2009). Thus, the DAE constrained optimization problems are transformed into large-scale nonlinear programming (NLP) problems using a weighted least-squares objective function.

In order to accomplish parameter estimation, the experimental data of Cassales et al. (2011), Purwadi et al. (2004), and Gutiérrez-Padilla & Karim (2005) are used for hydrolysis, overliming, and fermentation processes, respectively. Main unknown kinetic parameters such as activation energies, pre-exponential factors in the hydrolysis and detoxification operations and the maximum overall specific growth rate and furfural inhibition coefficient for the fermentation process have been obtained. Results show reasonable agreement with the reported experimental data.

References

Andersen, Moreno & Diaz (2011) Model Based Optimization of Bioethanol Production from Lignocellulosic Biomass. 2011 AIChE Annual Meeting, Minneapolis.

Cassales, Rech, Souza-Cruz & Ayub (2011) Optimization of soybean hull acid hydrolysis and its characterization as a potential substrate for bioprocessing. Biomass and Bioenergy 35, 4675 – 4683.

Estrada, Parodi & Diaz (2009) Determination of biogeochemical parameters in eutrophication models with simultaneous dynamic optimization approaches. Computers and Chemical Engineering 33, 1760–1769.

Gutiérrez-Padilla & Karim (2005) Influence of Furfural on the Recombinant Zymomonas mobilis Strain CP4 (pZB5) for Ethanol Production. The Journal of American Science 1, 24-27.

Lavarack, Griffin & Rodman (2002) The acid hydrolysis of sugarcane bagasse hemicelluloses to produce xylose, arabinose, glucose and other products. Biomass and Bioenergy 23, 367 – 380.

Leksawasdi, Joachimsthal & Rogers (2001) Mathematical modeling of ethanol production from glucose/xylose mixtures by recombinant *Zymomonas mobilis*. Biotechnology Letters 23, 1087 – 1093.

Martín & Grossmann (2012) Energy Optimization of Bioethanol Production Via Hydrolysis of Switchgrass. AIChE Journal 58, 1538-1549.

Purwadi, Niklasson& Taherzadeh (2004) Kinetic study of detoxification of dilute-acid hydrolyzates by $Ca(OH)_2$. Journal of Biotechnology 114, 187 – 198.

Biegler & Zavala (2008) Large-scale nonlinear programming using IPOPT: An integrating framework for enterprise-wide dynamic optimization. Computers and Chemical Engineering 33, 575–582.

(599ao) Simulation-Optimization Approaches to the Solve Strategic and Tactical Problems of Biomass Utilization for Energy and Commodity Chemicals

Ismail Fahmi, Chemical Engineering, The University of Tulsa, Tulsa, OK and Selen Cremaschi, Department of Chemical Engineering, University of Tulsa, Tulsa, OK

Simulation-Optimization Approaches to the Solve Strategic and Tactical Problems of Biomass Utilization for Energy and Commodity Chemicals

Ismail Fahmi, Selen Cremaschi

Department of Chemical Engineering, The University of Tulsa, 800 South Tucker Drive, Tulsa, Oklahoma 74104, USA

10C08 Poster Session: Computers In Operations and Information Processing

Abstract

As the fossil-based material reserves deplete, but its demand by energy and commodity chemicals are increasing, alternative feedstocks need to be discovered. One alternative feedstock is the renewable, locally available, and abundant biomass. Incorporating biomass as a feedstock to energy and commodity chemicals industry requires a significant amount of investments in capacity expansions and in the research and development (R&D). In addition, for each selected technology, the optimal configuration and operating conditions of unit operations should be identified. As a result, there are two investigation levels in the problem of incorporating biomass to both energy and commodity chemicals industries: (1) Strategic level, which covers all possible technologies supporting the overall biomass processing system and the optimal decision will answer in which technologies, when, and how much to invest, (2) Tactical level, which covers all production alternatives for a selected processing technology and the solution will identify the optimal process flowsheet and operating conditions. In this poster, two different simulation-optimization approaches are presented to address the strategic and tactical problems of incorporating biomass to energy and commodity chemicals production.

The first type of simulation-optimization approach, also known as the surrogate-based optimization, uses simulation to generate surrogate models to accurately represent the original first-order principal relationship. The surrogate models are then used in a deterministic optimization problem formulation, which is solved to yield the optimum solution. In this work, surrogate-based optimization is utilized to solve the process synthesis problems, which has been demonstrated in [1] with biodiesel production as the case study. The three process alternatives considered were using super critical methanol as reactant, using base catalyst, and using acid catalyst. In [1], it was shown that using surrogate models to replace the original first-order principal relationship of each and every unit operation in the optimization formulation can reduce the computational costs to obtain a solution.

The second type of simulation-optimization approach uses deterministic optimization with stochastic simulation in an integrated fashion to study the impact of uncertainties on the system performance. In the second simulation-optimization approach, deterministic optimization is used to produce the optimum decision set of the system at its current state. Stochastic simulation is used to predict the system performance under uncertainties. In a single simulation-optimization run, the simulation proceeds through the time steps of a timeline based on the decisions set determined by the deterministic optimization. Because of the accumulated differences between the expected and the realized behavior of the system, the system performance under these decisions may no longer be applicable or feasible at one point in time. This situation is called a trigger event. In case of a trigger event, the simulation is halted and deterministic optimization is recalled to produce a new set of optimal decision for the remainder of the timeline. The control is passed back to simulation along with the new set of decisions. The iteration between the deterministic optimization and stochastic simulation continues until the final time step. After the completion of the final step, a unique controlled evolution of the system performance under uncertainties (i.e., one unique timeline) is generated. Multiple unique timelines are produced because of different realizations of the uncertain parameters. The statistically significant number of timelines can be used to study the system performance under uncertainties. In this work, the second type of simulation-optimization approach is used to address the problem of strategic investment planning to shift from fossil-based feedstocks to biomass feedstock for the chemical process industry (CPI). This problem is referred to as the Biomass to Commodity Chemicals (BTCC) investment planning problem. A case study of ethylene production from biomass and naphtha has been presented to demonstrate the performance of the second type of the simulation-optimization [2]. However, generating a single timeline may require significant amounts of computational costs. The major sources of the computational costs are the number of timelines and the computational resources necessary to solve the deterministic optimization problem, which may be solved many times during a single timeline. To address the former problem, a systematic analysis on different sampling methods to cover the uncertain parameters space of the BTCC investment planning problem is presented. To address the latter problem, a study of the development of a specialized global solver to solve the optimization problem is presented.

Keywords: simulation-based optimization, biomass feedstock incorporation, strategic investment planning and tactical process synthesis problem

References:

[1] I. Fahmi and S. Cremaschi, "Process Synthesis of Biodiesel Production Plant using Artificial Neural Networks as the Surrogate Models," *Computers and Chemical Engineering (in review)*, 2011.

[2] I. Fahmi and S. Cremaschi, "A Prototype Simulation-based Optimization Approach to Model Feedstock Development for Chemical Process Industry," *22nd European Symposium on Computer Aided Process Engineering*, 2012.

(599ap) Modeling and Simulation of a FT Reactor for the Process Scale-up for Gtl-FPSO Development

Dongil Peter Shin, Dept. of Chemical Engineering, Myongji University, Yongin, Kyunggido, South Korea and Gi Hoon Hong, Myongji University, Yongin, South Korea

GTL processes are emerging as a form of alternative energy sources. Korean Government is sponsoring research on several GTL processes, and GTL-FPSO aims to enable an FPSO system producing synthetic fuel out of stranded gas wells.

One of the most challenging parts of reactor and reaction systems is parameter estimations in reaction models minimizing differences between predicted and experimental values. Due to the dynamic complexity, combinatorial and highly nonlinear property of the models, conventional optimization theories used to fail. In this work, particle swarm optimization (PSO) method as a novel metaheuristic optimization is adopted for estimations of model parameters in reaction systems. Additionally, modifications of the essential model parameters in PSO are brought in. The new method takes advantages of nonlinearly changing adaptive strategy (aPSO) and defines a new changing combination of inertia weight and study factors. The proposed aPSO improves the performances of the basic version greatly and succeeds in parameter estimations in the three study cases of highly nonlinear (bio)chemical systems.

(599aq) Simultaneous Design of Ionic Liquids and CO2 Recovery Processes

Qi Chen[1], Brianna Christian[1], Brock C. Roughton[2], Igor Mitrofanov[3], Kyle V. Camarda[1] and Rafiqul Gani[3], (1)Chemical and Petroleum Engineering, University of Kansas, Lawrence, KS, (2)Bioengineering Graduate Program, University of Kansas, Lawrence, KS, (3)CAPEC, Department of Chemical and Biochemical Engineering, Technical University of Denmark, Kgs. Lyngby, Denmark

A methodology and tool set for the simultaneous design of ionic liquids and separation processes is presented. By manipulating the structure of the cation and anion, the properties of an ionic liquid can be adjusted for specific applications, such as azeotropic separation or CO_2 recovery. To estimate key properties, several group contribution models available in the literature have been used, along with newly developed group contribution models. For a given set of operational constraints, an ionic liquid is designed using a computer-aided molecular design (CAMD) method and the UNIFAC-IL model is used to screen design candidates or known ionic liquids.

The developed tool set allows users to select target objectives and input and output constraints while optimally designing both the ionic liquid and the separation process. Alternatively, the software package can simultaneously select from a database of ionic liquids with experimentally verified properties while optimizing the separation process. New group contribution models have been developed to predict ionic liquid viscosity and CO_2 solubility. Viscosity prediction allows computation of pumping costs, providing more accurate estimation of energy requirements for a given process. As an example, an absorption system for CO_2 capture is designed, involving both the design of the IL solvent itself as well as the separation scheme to remove the CO_2 from the ionic liquid. Simulations were used to determine the overall energy usage for the entire process, such that the optimal solvent structure and process design is determined.

(599ar) Parameter Estimation of Bioethanol Production Model by a Genetic Engineered Cyanobacterium

Vanina Estrada[1], Rebeca Vidal[2], Francisco Javier Florencio[2], Miguel García Guerrero[2] and **María Soledad Diaz**[1], (1)Chemical Engineering, Planta Piloto de Ingenieria Quimica (PLAPIQUI), Universidad Nacional del Sur - CONICET, Bahia Blanca, Argentina,, Bahia Blanca, Argentina, (2)Instituto de Bioquímica Vegetal y Fotosíntesis, CSIC-Universidad de Sevilla, Sevilla, Spain

In the last years biofuels are being investigated as alternative to reduce the dependence on fossil fuels. Ethanol derived from sugar cane and corn grain is the most common renewable fuel. Now, it is possible obtain ethanol as a third generation biofuel through microalgae and cyanobacteria, nevertheless current reported ethanol yields through this process still require improvement to make this technology economically attractive.

In this work we present a dynamic model for the production of ethanol by the engineered metabolic strain of the cyanobacterium *Synechocystis* sp. PCC 6803 (SGE9) constructed by Vidal (2009). Vidal (2009) has constructed a novel strain of the cyanobacterium *Synechocystis* sp. PCC 6803 able to produce ethanol from CO_2 through the inclusion of the genes *pdc* and *adhB* of the ethanologenic bacterium *Zymomonas mobilis* under the control of the gene *PetE*promoter, which is a different strategy to that used by Deng & Coleman, 1999 and Dexter & Fu, 2009.

The model we propose for ethanol production from this cyanobacterium includes mass balances for biomass, ethanol, nitrate and phosphate. Biomass equation includes a growth limiting function for light and nutrients (NO_3 and PO_4). The biomass equation also includes a term that takes into account the kinetics of growth inhibition for ethanol toxicity. The decrease in the available light for photosynthesis associated to the increase in cell concentration is also described by the model.

We formulate a parameter estimation problem with a weighted least-squares objective function subject to mass balances equations (Estrada *et al.*, 2009). The dynamic parameter estimation problem is solved in GAMS through a simultaneous optimization approach by prior transformation it into a nonlinear programming (NLP) problem by discretizing state variables by orthogonal collocation over finite elements. The NLP is solved with an Interior Point algorithm with Successive Quadratic Programming techniques (Biegler *et al.*, 2002).

Data set for parameter estimation were obtained in an experiment performed over 73 hours for SGE9 and wild type strains of *Synechocystis* grown in batch liquid cultures and include profiles for biomass (Chlorophyll a and total organic carbon), ethanol, NO_3 and PO_4. Numerical results provide useful insights on ethanol production by the genetic modified cyanobacteria, reproducing a 0.025% (v/v) ethanol yield.

References

Biegler L.T., Cervantes, A. & Waechter, A. (2002). Advances in simultaneous strategies for dynamic process optimization. Chemical Engineering Science, 57: 575-593.

Deng, M. D. & Coleman, J. R. (1999). Ethanol synthesis by genetic engineering in cyanobacteria. Applied Environmental Microbiology, 65: 523-528.

Dexter, J. & Fu, J. (2009). Metabolic engineering of cyanobacteria for ethanol production. Energy Environmental Science, 2: 857-864.

Estrada V., Parodi, E. & Diaz, M. S. (2009). Determination of biogeochemical parameters in eutrophication models as large scale dynamic parameter estimation problems, Computers and Chemical Engineering: 33, 1760-1769.

Vidal, R. (2009). Producción fotosintética de etanol por la cianobacteria Synechocystis sp. PCC 6803. Tesis doctoral, Universidad Nacional de Sevilla.

(599as) A Mathematical Model for a Biofuel Production Process From Macroalgae

Claudio Delpino, Chemical Engineering, Planta Piloto de Ingeniería Química (PLAPIQUI), Universidad Nacional del Sur-CONICET, Bahía Blanca, Argentina, Vanina Estrada, Chemical Engineering, Planta Piloto de Ingenieria Quimica (PLAPIQUI), Universidad Nacional del Sur - CONICET, Bahia Blanca, Argentina and **María Soledad Diaz**, Chemical Engineering, Planta Piloto de Ingenieria Quimica (PLAPIQUI), Universidad Nacional del Sur - CONICET, Bahia Blanca, Argentina,, Bahia Blanca, Argentina

In this paper, we review published data and past efforts towards the production of ethanol using macroalgae as raw materials. When analyzing macroalgae biomass as a source of polysaccharides, which can be directly fermented by special organisms or hydrolized into simple sugars that are fermentable by well known ethanol producing, a key fact to have into account is the absence of lignin. In the group known as "brown" algae, these polysaccharides are represented by storage molecules like laminarin and mannitol, and alginates and cellulose, which conform the amorphous and fibrillar part of the cell wall, respectively. When it comes to the so called "red" algae, the fermentable polysaccharides include cellulose and galactans (which are nowadays used for agar production), which could amount to higher ethanol yields. A biorefinery scheme is proposed and mathematically modeled based on published data and efforts from other sources ([1], [2], [3]). The proposed biorefinery main route includes cultivation, harvesting and processing of macroalgae. Cultivation can be carried out in offshore or nearshore "farms" for which several technologies have been proposed, or in onshore installations, which offer an easier and more precise control of the cultivation, with higher yields, at the expense of higher capital and operating costs. Processing of macroalgae includes succesive acid and enzymatic hydrolisis of polysaccharides and their posterior fermentation.

We formulate a mathematical model for integrated macroalgal-based biorefineries, allowing for cost, energy and water need calculations. The model also includes the possibility to allow for several macroalgae species and origins, diverse processing routes, and several products and by-products. This allows comparison of the different alternatives, and optimization of the economical and energetic performance of the process using mathematical programming techniques. The optimization framework is specially useful for the macroalgae context, where the integration with current algal high-value products facilities, or even the production of this high-value byproducts (i. e. agar) in the integrated biorefinery, is essential for the process economic sustainability. The model has been implemented in GAMS as a Mixed Integer Nonlinear Programming problem. Numerical results provide quantitative information and give useful insights on macroalgal-based biorefineries. At the same time, the local potential for a given processing environment can be easily evaluated. We accomplish these for several options within Argentina

References

[1] G. Roesijadi et al., "Techno-Economic Feasibility Analysis of Offshore Seaweed Farming for Bioenergy and Biobased Products," 2008.
[2] T. Burton et al., "A review of the potential of marine algae as a source of biofuel in Ireland," 2009.
[3] M. Yanagisawa et al., "Production of high concentrations of bioethanol from seaweeds that contain easily hydrolyzable polysaccharides," Process Biochemistry, vol. 46, no. 11, pp. 2111-2116, Nov. 2011.

(599at) A Reliability-P-Median Formulation for Optimization of Gas Detector Layout in Process Facilities

Alberto Benavides-Serrano[1,2], Sean Legg[1,2], M.Sam Mannan[1,2] and Carl D. Laird[1], (1)Chemical Engineering, Texas A&M University, College Station, TX, (2)Texas A&M University, Mary Kay O'Connor Process Safety Center , College Station, TX

A Reliability-P-Median formulation for optimization of gas detector layout in process facilities

Alberto Benavides-Serrano, Sean Legg, Sam Mannan, Carl Laird

A large number of variables influence the risk associated with gas leaks in process facilities. These variables include leak conditions, fluid properties and dispersion characteristics, process equipment geometry, detection equipment, environmental factors, and safety considerations. Given this large number of variables, the task of gas detector layout in the process industries is challenging. Mixed-integer linear programming (MILP) has been proposed as a quantitative approach for numerical optimization of gas detector layout. Legg et al (2012c) proposed a stochastic programming formulation that seeks a sensor placement that minimizes the expected time to detection across any number of leak scenarios. Extensions to this MILP formulation were proposed to improve the resilience of the solution placement to unforeseen scenarios (Legg et al, 2012a) and the tail-behavior of the distributions of detection times (Legg et al 2012b).

However, this previous work assumed the use of perfect gas sensors; in reality gas sensors are prone to false-positives and false-negatives. In the process industries, two solutions are usually implemented. First, additional confirmation from other detectors may be required before emergency actions are triggered, and several voting logic schemes are used. Second, the Probability of Failure on Demand (PFD) of the detectors should be considered in the placement strategy. In this work, we present an MILP that performs optimal gas detector placement while considering sensor failure and voting. This problem formulation is closely related to the Reliability-P-Median Problem (RPMP) proposed by Snyder and Daskin (2005) for the facility location problem.

Here, we show the relationship of our stochastic programming formulation to the RPMP formulation. Scenario data for this problem is generated with rigorous CFD simulations of a real process geometry using FLACS with different leak locations and conditions (provided by GexCon). The effectiveness of placement results are analyzed and compared with the previous formulation that ignores sensor reliability and voting.

References

Legg, S., Wang, C., Benavides-Serrano, A.J., Laird, C. (2012a). Optimal Gas Detector Placement Under Uncertainty Considering Conditional-Value-At-Risk. Submitted to the Journal of Loss Prevention in the Process Industries, 2012.

Legg, S., Benavides-Serrano, A.J., Siirola, J., Watson, J.P., Davis, S., Bratteteig, A., Laird, C. (2012b). A Stochastic Programming Approach for Gas Detector Placement Using CFD-Based Dispersion Simulations. Submitted to Computers & Chemical Engineering, 2012.

Legg, S., Siirola, J., Watson, J.P., Davis, S., Bratteteig, A., Laird, C. (2012c). A Stochastic Programming Approach for Gas Detector Placement in Process Facilities. Proceedings of the 2012 FOCAPO Conference, Savannah, January, 2012.

L. Snyder and M. Daskin. Reliability models for facility location: The expected failure cost case. Transportation Science, 39(3):400–416, August, 2005.

Efficient Object-Oriented Dynamic Modeling Framework with Source Code Generation for Nonlinear MPC

João Rui Leal, CIEPQPF/Chemical Engineering Dep., University of Coimbra, Coimbra, Portugal
Lino O. Santos, CIEPQPF/Chemical Engineering Dep., University of Coimbra, Coimbra, Portugal
Andrey Romanenko, Ciengis, Coimbra, Portugal

Abstract

This work describes a novel software framework designed to streamline and accelerate the model development process in the Plantegrity nonlinear model predictive control system. The first-principle based dynamic modeling library is written in C++ based on general object-oriented programming concepts as well as on the ForeSee modeling system[1] design considerations that employ four key modeling components: containments, core models, connectors, and coordinators. Some additional concepts, such as equilibrium, are included in the proposed framework. Owing to its dimensional analysis and conversion capabilities, this library allows the modeler to specify exogenous variables using the most convenient dimensions and units without the need to modify the model. Additionally, it is interfaced with several automatic differentiation (AD) packages, such as ADOL-C[2] and CppAD[3]. A new extension of the CppAD library was also developed in order to produce online efficient C code for the model and its derivative evaluations up to the second-order. A performance comparison of several AD packages was carried out in the context of NMPC and state estimation (using an Unscented Kalman filter). In addition, some models are presented in order to illustrate the modeling principles and work-flow.

Introduction

There is an increasing demand in chemical process plants to reach higher levels of performance due to several factors, such as increasing global market competition and stricter safety and environmental regulations. Consequently, there has been a trend to increase the level of process integration in order to reduce energy consumption, to improve material usage, to tighten the process constraints in order to comply with regulations, and to make more frequent changes to process operations in order to adapt to market, raw materials, and energy cost fluctuations.

A simple regulatory control level, which usually consists of a Distributed Control System (DCS) with many Proportional Integral Derivative (PID) controllers (or variants), may have difficulties in meeting these objectives. In some cases, Model Predictive Control (MPC), one of advanced process control (APC) algorithms, may achieve an improved control performance by exploiting the available operational freedom and considering process interactions and constraints. MPC implements a feedback control strategy based on the online solution of a moving horizon Optimal Control Problem (OCP) which uses a model to predict the system's dynamic behavior. MPC technology is currently present in a large variety of application areas, where refining, pretrochemicals, and chemicals take special predominance[4].

Nonlinear Model Predictive Control (NMPC), which typically requires sophisticated algorithms due to extra model complexity, has received significant contributions in recent years. More details on NMPC methods and applications may be found in relevant review works[5,6,7].

Plantegrity

Plantegrity is a modular NMPC framework derived from NEWCON[8] which contains several independent components that communicate through a database, as illustrated in Figure 1. The database

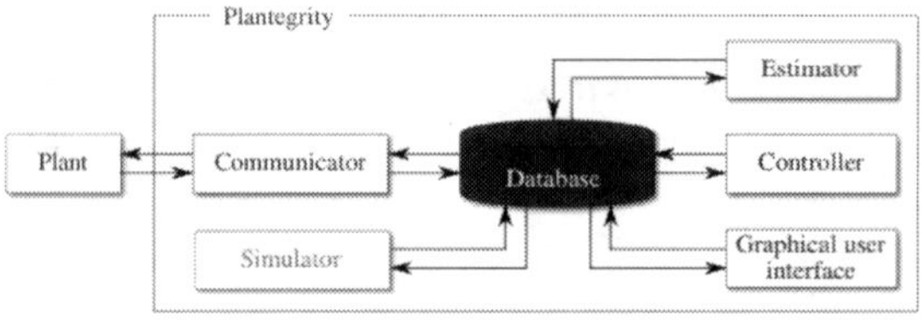

Figure 1: Module interaction diagram of the Plantegrity NMPC framework.

provides a standard interface between the modules. It is used for data exchange synchronization and for firing notifications to the appropriate components when it is updated. Additionally, it stores historical process data. The estimator module implements an Unscented Kalman Filter while the controller implements both a multiple-shooting strategy and an orthogonal collocation based simultaneous strategy. NLPs are solved using one of two Newton-type optimizers: a custom SQP solver or a primal-dual Interior Point solver (Ipopt). The simulator, estimator and controller modules are implemented under the C++ programming language and require the development of a model for each particular system or plant. Even if the system is the same, it is possible for each module to use slightly different models.

Since Plantegrity requires derivative information from the model in order to perform integrations and optimizations, the implementation was performed so that the use of Automatic Differentiation (AD) tools could be possible. It currently supports three AD packages: ADOL-C[2], CppAD[3], and CppADCodegen.

Both ADOL-C and CppAD make use of the operator overloading capabilities of C++ and are able to register all operations in the model equations into a tape. All model related information can then be retrieved, not directly from the model, but rather from the tape. Therefore, the model dependent variable values are determined by following the operations registered in that tape, while the derivative information is calculated by applying the differentiation rules throughout the tape. Since the number of dependent variables in the model (state time derivatives and algebraic equation residuals) is commonly much lower than the independent variables (states and controls), the reverse AD mode is employed.

The newly-created CppADCodegen uses CppAD for the differentiation rules and produces highly optimized C source-code that can be dynamically compiled and linked with Plantegrity.

Modeling

The derivation of a nonlinear model for NMPC and state estimation is a crucial step that can define the effectiveness of the controller. Furthermore, unlike traditional control, the development and identification of a suitable model can be one of the most time-consuming tasks during commissioning, taking about 80% of the total time and expense[9]. Several authors have noted that there is a lack of a systematic approach for the development of dynamic nonlinear models for NMPC[10,11]. It has been pointed out as being one of the factors preventing the adoption of NMPC in many industrial processes[5].

Unlike in LMPC, in NMPC both fundamental and empirical models are widely-used in the industry[4] and academia[12]. All these types of models come with fairly significant advantages and disadvantages and their selection may be difficult. A very brief comparison is presented in Table 1.

First-principle based models can be seen as facilitators for changing the operation conditions to a new and better operating window that may have never been previously explored. In some situations, a dynamic mechanistic model may already have been developed and validated for other purposes other than NMPC, and consequently its extension and use for NMPC may substantially reduce development time[13]. Unfortunately, there are few freely available modeling tools to support the development of models that can be easily applied to NMPC.

Plantegrity Modeling Framework

Most model equations in chemical engineering can be divided into four categories. These include

Table 1: Comparison between fundamental and empirical nonlinear dynamic modeling.

	Empirical	Fundamental
Required process knowledge	low	high
Selection of model form	difficult	intrinsic
Parameter identification	costly and prolonged plant tests	lab experiments and routine operating data
Extrapolation	unadvised	possible
Model complexity	lower/restricted	higher/may require Model Reduction
New units/plants model creation	after start-up	before and after start-up
Reuse in similar units/plants	harder	easier
Adaptation to process changes	harder	easier

equations for conservation laws (based on mass, energy and momentum), for physical and chemical equilibrium among species and phases, constitutive equations describing the rates of chemical transformation or transport of mass and energy, and constraint equations associated with the behavior of geometric variables. The *ForeSee* dynamic modeling system[1] explicitly contains most of these concepts translated into an Object Oriented (OO) framework. The modeling library presented here will be based on some of the concepts implemented in this modeling system. Unlike *ForeSee*, however, Plantegrity Modeling Framework (PMF) is implemented in C++, uses Automatic Differentiation (AD), explicitly allows the usage of measurement units, and can generate highly-efficient C source-code. *ForeSee* was implemented with a graphical user interface, while PMF was implemented solely as a programming library which adds additional freedom in the definition of custom process behavior. One the most significant conceptual differences, however, is the explicit consideration of equilibrium equations in PMF. *ForeSee* did not support such feature due to implementation difficulties associated with solving DAEs. Equilibrium does not, however, necessarily introduce algebraic equations into the model.

PMF was implemented so that it would be relatively easy to use by a chemical engineer with some modeling and programming background. It performs several checks and in many situations it does not allow to run physically or conceptually incorrect models.

Basic concepts

A substance, in a given physical state, is defined by the "AbstractPhaseInfo" class, which may contain several chemical components. In this class, these components' composition is only defined by intensive variables (e.g. mass fraction, molar concentration). Other intensive properties can be defined, such as fugacity and activity coefficients for each component, mixture vaporization enthalpy, etc. This class acts mostly as an interface, as shown in Figure 3, and can not be directly created by the user.

A "Phase" class is used to represent a mixture of chemical components, with a single physical state, which can be characterized by extensive quantities. In addition to the intensive variables used to describe the composition in its superclass, the composition may also be defined by the total amount (number of moles) or mass for each component. The mixture's total amount of material may be defined using either the total mass, total amount or total volume. In order to determine the quantity of material of a specific component, either an intensive composition variable of the component and a total mixture quantity or an extensive composition variable of the component must be defined. PMF can handle many possibilities for the computation of a composition depending on the available information. By default, it is not possible for the user to over-define a component's composition.

As in the *ForeSee* system, PMF uses streams and substreams. A substream is considered to be a flow of material with a single physical state, while a stream can contain one or more substreams in a temperature and pressure equilibrium. In many parts of the library, however, both stream and substream objects can be easily handled by using the "BaseStream" as a uniform interface (see Figure 2). Similar to the "Phase" class, in addition to intensive quantities, chemical component compositions can be defined as extensive partial flows (either mass or molar flow). The total flow of a substream can be specified in mass, amount (number of moles) or volume. Like in the "Phase" class, many auto-

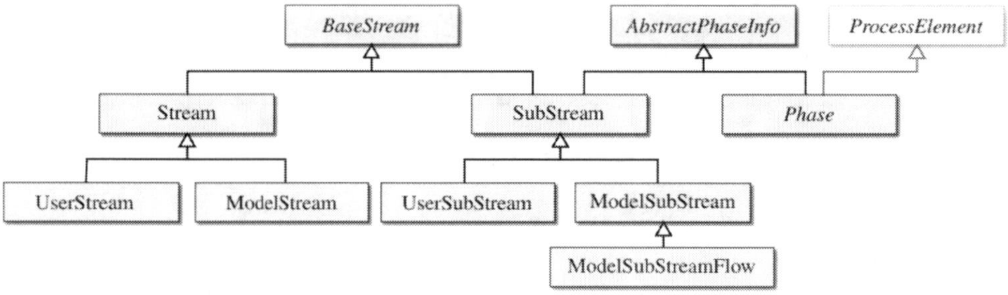

Figure 2: Relationship between Streams, SubStreams and Phase classes.

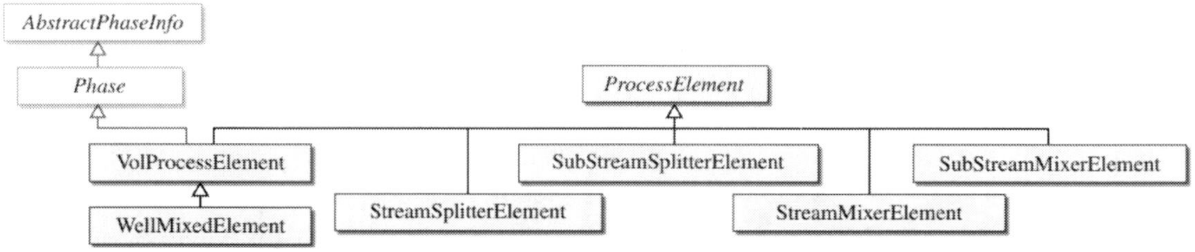

Figure 3: Process elements in the Plantegrity Modeling Framework.

matic conversions between different composition definitions is possible. Additionally, PMF is able to determine the chemical components present in each substream and process element by using an "event-listener" strategy for tracking component definition changes. PMF attempts to avoid unnecessary calculations by evaluating intermediate values only if they are considered as a dependency to a user requested value. Furthermore, it attempts to avoid unrequired conversions between different composition dimensions by using a data request system that does not only calculate upstream compositions for a specific component but also for a specific dimension.

Each model's implementation requires the creation of a single object of the class "Model". This class does not contain the model itself, but is rather referenced by all process elements and substreams so that it is possible to perform global model analysis. In PMF, several precoded process units, or part of them, may be available through the implementation of a "SubModel" class, just like in block-oriented modeling. Unlike the "Model" class, however, a "SubModel" contains the implementation. PMF contains many additional support classes associated, for instance, with thermodynamic properties, component and mixture properties, proportional controllers, etc.

Modeling Components

Each model equation type can be represented by a distinct fundamental modeling component. These are process elements, containers, connectors, coordinators, and equilibrium. Process elements in PMF are mostly associated with conservative equations and can be used to determine the dynamic behavior for a specific physical state. Unlike *ForeSee*, PMF not only allows intensive attributes but also extensive variables to be defined in process elements. Some process elements are presented in Figure 3. Connectors are used to transfer heat or mass to/from phases or to transform the composition within a single phase. Some examples include mass transport connectors – that provide mass transfer fluxes between phases–, chemical reaction connectors – that determine chemical reaction rates within a phase–, and heat transport connectors. Containers describe the physical features of an equipment item, such as a distillation column tray or a reactor, and may contain one or more phases where dynamic process changes occur. Coordinators are used to describe the dynamic behavior of geometric variables of a given phase, such as volume or surface area. Finally, equilibrium is used to account for extremely fast dynamic responses between two or more phases.

Equilibrium

Three different equilibria where considerer: temperature, pressure, and phase equilibrium.

A temperature equilibrium can be assumed when the heat transfer rate between phases is so high that it is considered to be instantaneous. When a temperature equilibrium is for two or more phases, a single energy balance is performed globally on all phases, instead of each individual phase.

Another type of equilibrium is the pressure equilibrium which behaves very similarly to the temperature equilibrium. It can be used, for instance, when two phases are in direct contact. These two types of equilibria typically do not require the addition of algebraic equations to the model and can, therefore, be modeled using ODEs.

The phase equilibrium can be used when two or more phases transfer mass with each other at a rate fast enough to reach a chemical potential equilibrium almost instantaneously. It typically also implies temperature and pressure equilibria. By specifying a phase equilibrium between a set of phases, only the chemical composition of a single phase should be defined since other compositions can be determined through the equilibrium. This type of equilibrium may require the addition of algebraic equations to the model. As an alternative, Newton methods can be employed to iteratively solve these equations within the model (e.g. solving bubble and dew points), without requiring the inclusion of algebraic equations in the model. PMF supports both strategies.

CppADCodeGen

One of the key stages of most NMPC implementations is the evaluation of model values and derivative information. An open-source extension to CppAD, CppADCodeGen, was developed in order to allow faster and accurate evaluation of such information. It can generate and compile source-code at runtime into a dynamic library that can be loaded during execution. This library can be reused for future executions, thus avoiding a lengthly model recompilation step.

CppADCodeGen uses a special template class, which supports most common arithmetic operations and math functions, instead of a numerical data type. This class is used to register the differentiation operations performed by CppAD into an operation graph that can later be used to produce source-code. Many unnecessary operations performed by CppAD can be completely eliminated, such as multiplications by 0 or 1, thus producing a better operation graph. Moreover, during source-code creation for the C language, only the absolutely needed temporary variables (variables used in more than one operation) are used, being reused as soon as they are no longer required. Additionally, this extension allows to explicitly consider "constant independent variables", such as the model parameters, for which no differential information is required but for which values can be changed. Typically, the resulting compiled model library is significantly faster than the original tape evaluation process. An additional benefit from generating source-code, resulting from using the several PMF classes, is the possibility of being able to visually inspect the model equations (including equations for derivative evaluation). Even though only the C language is implemented, the operation graph's independence and the source-code generation task allow CppADCodeGen to easily create code for different languages.

Dimension Analysis

When one of CppAD or CppADCodeGen AD libraries is used, PMF also performs dimensional analysis through the "boost::units" library. The most widely-used dimensions, units and quantities are defined for AD variable and constant data types, which allows to write:

```
Qa_pressure p1 = 1.0 * bar;
Qa_pressure p2 = 100000.0 * pascal;
```

Here, the variables *p1* and *p2* could be used interchangeably without any further model modification.

The Plantegrity configuration file allows to define each variable's units and dimension. If these properties are defined, both the dependent and independent variables in the model are declared as

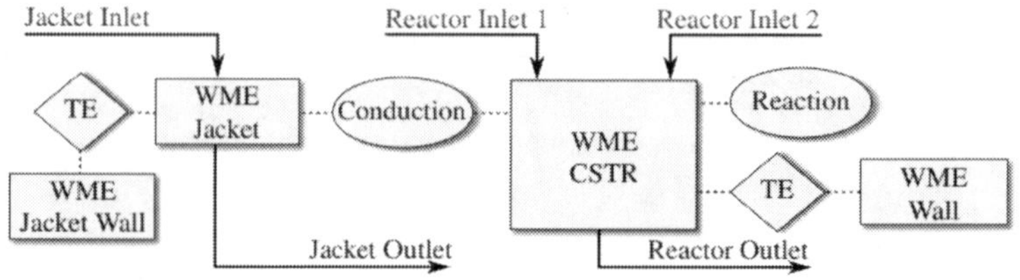

Figure 4: A representation of the usage of PMF to create a model for a CSTR heated by jacket. "TE" represents a temperature equilibrium and "WME" represents a Well Mixed Element.

quantities and appropriate C++ header and source files for the model are generated. This avoids the error prone task of unit conversion and dimensional checking. Unfortunately, ADOL-C could not be used with the "boost::units" library due to an implementation incompatibility among the two libraries that causes memory corruption. Nevertheless, when developing a model using ADOL-C, most of the units are defined as constant values so that it is easy and quick to change between AD packages without any changes to the model. Dimensional analysis is not, however, performed while using ADOL-C. Several benchmarks were performed to measure the impact of the usage of quantities in the framework, under the gcc compiler. As expected, in most tests no clear performance degradation was verified since all the dimensional checking is performed at compile time.

Examples

CSTR

A simple Continuously Stirred Tank Reactor (CSTR) can be used to demonstrate how to create a model with the framework. In this model, two aqueous liquid streams feed a reactor while one exits. The flow of the outlet stream of the reactor is manipulated by a pump. Since the streams are all in the liquid state, each stream has a single substream. A zero order reaction occurs inside the reactor where a chemical component A is transformed into B. A jacket is used to control the reactor's temperature. The masses of the metallic reactor wall and the jacket wall are significant and, consequently, the walls are considered in the model dynamics. The reactor wall is in a temperature equilibrium with the contents of the reactor while the jacket wall is also in a temperature equilibrium with the jacket fluid. Figure 4 shows how PMF can be used to model this system.

As can be seen from the execution measurements displayed in Figure 6 (a), CppADCodeGen can have a significant impact in the overall performance of the system (real time). It is relatively easy to parallelize a multiple-shooting strategy and run several model integrations concurrently. Although the total CPU time should ideally remain constant with the increase of the number of threads, an increase is observed most likely due to memory synchronization overheads. Since ADOL-C does not support concurrent evaluation of multiple tapes, only single-threaded ADOL-C tests were performed. ADOL-C does, nonetheless, have multi-threading support for the evaluation of a single Jacobian or Hessian from a single tape, which was not tested. The 4-threaded version of CppADCodeGen runs take 78% of ADOL-C's time and 92% of the 4-threaded CppAD's time. Improvements are not significant since the model is very simple and does not dominate the total evaluation effort.

Disitillation Column

In PMF, a multicomponent distillation column with equilibrium stages can be easily modeled into an ODE system. Unlike in non-equilibrium stage column models, mass and energy transport equations are not required. Figure 5 presents a schematic representation of a stage with uniformly-mixed liquid and vapor. In order to avoid difficult high-index DAE system, the bubble point computations are implemented within the model using a Newton method with a fixed number of iterations, thus originating an ODE system. Currently, activity coefficients can be considered ideal or estimated using a group contribution method, such as UNIFAC or modified UNIFAC (Dortmund).

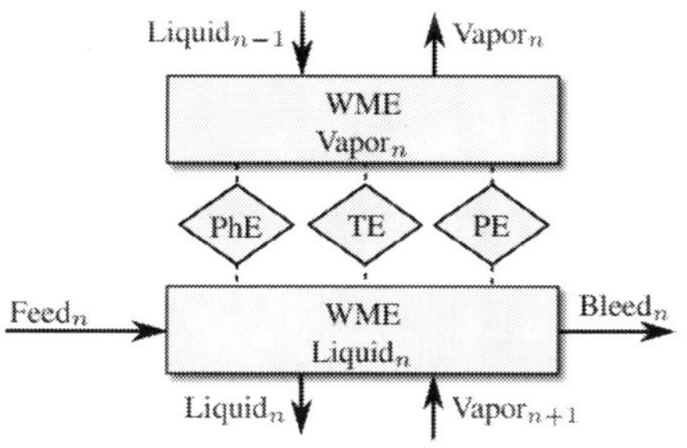

Figure 5: A representation of the usage of PMF to create a model for a distillation column stage. "PhE", "PE", "TE", "WME" represents a phase equilibrium, a pressure equilibrium, a temperature equilibrium and a Well Mixed Element, respectively.

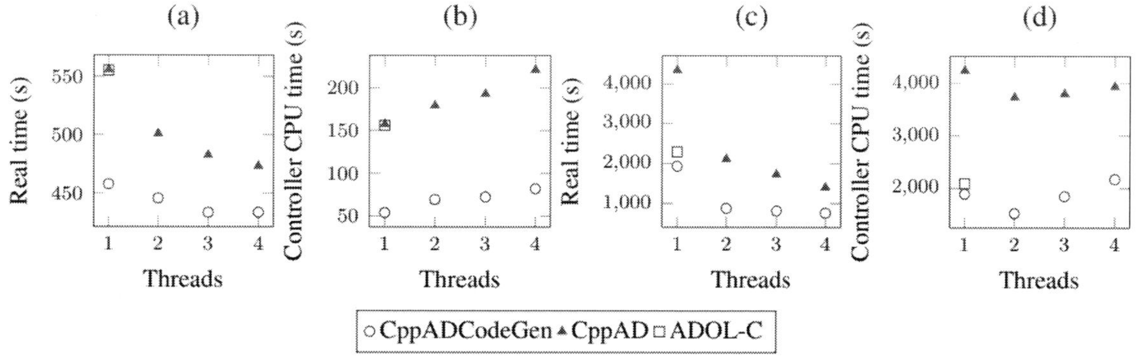

Figure 6: NMPC simulation using a multiple-shooting approach with an SQP solver. The system is an Intel i7-2600K CPU with 16 GiB of RAM running under GNU Linux (Kubuntu 12.04). Real time includes the time taken by 4 modules: simulator, estimator, controller and database. (a) (b) Times for the CSTR model after 1000 NMPC iterations with 40 prediction and 30 control horizon intervals of 30s. (c) (d) Times for the distillation model after 100 iterations with 10 prediction and 6 control horizon intervals of 10min.

Figures 6 (c) and (d) present some execution statistics for a distillation column model with 4 components, 18 stages, a reboiler and a condenser, negligible vapor holdup, constant pressure drop, ideal liquid and vapor behavior, and 100% stage efficiency. Liquid flow rates are determined through the Weir equation while the vapor flow rates are determined considering negligible changes in liquid molar enthalpy. The execution results show that the 4-threaded version of the CppADCodeGen implementation takes only 33% of ADOL-C's time and 54% of the 4-threaded CppAD's time. The generated model equations are significantly more complex than in the CSTR case and, hence, CppADCodeGen's impact in the overall execution time is much higher.

Final Remarks and Future Work

PMF is a dynamic modeling library written in C++ that takes profit of the OO programming paradigm and associates equation types to specific modeling components.

It is a relatively simple library to use, with a clear structure that makes it easy to map real-world concepts into a model. Several classes of the PMF can be applied to completely different models since each class translates relatively simple and basic engineering laws.

The modeling framework explicitly considers units of measurement and performs dimensional analysis checks at compile time.

The newly developed hybrid AD library, which combines operator overloading with source-code generation, clearly outperforms both ADOL-C and CppAD in the Plantegrity NMPC system.

Furthermore, the internal operation graph will allow the future implementation of automatic index reduction algorithms so that more complex DAEs can be handled. Several operation graph optimizations can still be implemented with a symbolic post-processing analysis.

Acknowledgements

João Rui Leal would like to acknowledge the Ph.D. Grant SFRH/BDE/51185/2010 from Fundação para a Ciência e Tecnologia (FCT), Portugal, and the financial support and cooperation from Ciengis.

References

[1] Tu, Hong and Rinard, Irven H.: (2006). ForeSee—A hierarchical dynamic modeling and simulation system of complex processes. *Computers & Chemical Engineering*, 30(9):1324 – 1345.

[2] Walther, A. and Griewank, A.: (2012). Getting started with ADOL-C. In *Combinatorial Scientific Computing*, Naumann, U. and Schenk, O., eds., chap. 7. Chapman-Hall CRC Computational Science, pp. 181–202.

[3] Bell, Bradley M.: (2012). CppAD: a package for C++ algorithmic differentiation. *Computational Infrastructure for Operations Research*, COIN-OR(`http://www.coin-or.org/CppAD`).

[4] Qin, S.Joe and Badgwell, Thomas A.: (2003). A survey of industrial model predictive control technology. *Control Engineering Practice*, 11(7):733 – 764.

[5] Qin, S. J. and Badgwell, T. A.: (2000). An overview of nonlinear model predictive control applications. In *Nonlinear Model Predictive Control*, Allgower, F and Zheng, A, ed., vol. 26 of *Progress in Systems and Control Theory*. Basel, Switzerland: Birkhauser Verlag AG, pp. 369–392.

[6] Camacho, Eduardo F. and Bordons, Carlos: (2007). Nonlinear model predictive control: An introductory review. In *Assessment and Future Directions of Nonlinear Model Predictive Control*, Findeisen, R and Allgower, F and Biegler, LT, ed., vol. 358 of *Lecture Notes in Control and Information Sciences*. Springer-Verlag Berlin, pp. 1–16.

[7] Diehl, Moritz, Ferreau, Hans Joachim and Haverbeke, Niels: (2009). Efficient Numerical Methods for Nonlinear MPC and Moving Horizon Estimation. In *Nonlinear Model Predictive Control: Towards New Challenging Applications*, Magni, L and Raimondo, DM and Allgower, F, ed., vol. 384 of *Lecture Notes in Control and Information Sciences*. Berlin, Germany: Springer-Verlag Berlin, pp. 391–417.

[8] Romanenko, A., Pedrosa, N., Leal, J. and Santos, L.: (2007). A Linux Based Nonlinear Model Predictive Control Framework. In *II Seminario de Aplicaciones Industriales de Control Avanzado*, J. F. Ardanuy, P. B. Escalante and Sánchez, J. M. M., eds. Madrid, Spain, p. 229.

[9] Agachi, P.S., Nagy, Z.K., Cristea, M.V. and Imre-Lucaci, Á.: (2007). *Model Based Control: Case Studies in Process Engineering*. John Wiley & Sons.

[10] Lee, Jay H.: (2000). Modeling and Identification for NonlinearModel Predictive Control: Requirements,Current Status and Future Research Needs. In *Nonlinear Model Predictive Control*, Allgöwer, Frank, Zheng, Alex and Byrnes, Christopher I., eds., vol. 26 of *Progress in Systems and Control Theory*. Birkhäuser Basel, pp. 269–293.

[11] Pearson, R.K.: (2003). Selecting nonlinear model structures for computer control. *Journal of Process Control*, 13(1):1 – 26.

[12] Kumar, A. S. and Ahmad, Z.: (2012). Model Predictive Control (MPC) and Its Current Issues in Chemical Engineering. *Chemical Engineering Communications*, 199(4):472–511.

[13] Foss, Bjarne A. and Schei, Tor S.: (2007). Putting nonlinear model predictive control into use. In *Assessment and Future Directions of Nonlinear Model Predictive Control*, Findeisen, R and Allgower, F and Biegler, LT, ed., vol. 358 of *Lecture Notes in Control and Information Sciences*. Berlin, Germany: Springer-Verlag Berlin, pp. 407–417.

(599c) Network-Induced Supervised Learning: Balancing Interpretation and Prediction Ability in Classification and Regression Tasks

Marco Reis, Department of Chemical Engineering, University of Coimbra, Coimbra, Portugal

Current supervised learning approaches are strongly focused on optimizing estimation accuracy metrics, leaving to a secondary concern the interpretation of the results produced (Hastie *et al.*, 2001; Reis & Saraiva, 2004, 2005, 2010). However, in the analysis of complex systems, one of the main interests is precisely the induction of relevant associations, in order to understand or clarify the way systems operate. The importance of this issue is not new and has motivated the development of approaches that try to improve the interpretation of results, while maintaining the quality of predictions, such as O-PLS (Trygg & Wold, 2002).

In this communication, we present two new frameworks for addressing supervised learning problems (classification and regression), that incorporate interpretational-oriented analysis features right from the onset of the analysis. These features constrain the predictive space, in order to introduce interpretable elements in the final model. After extensive testing, we verified that such constraints do not usually compromise the methods' performance, when compared to their unconstrained versions, and sometimes even led to improvements in prediction ability, due to the use of more parsimonious and robust models.

The frameworks, called Network-Induced Classification (NI-C) and Network-Induced Regression (NI-R), share a common methodological backbone, based on a variable clustering algorithm that takes into consideration the local associations among variables, in order to identify meaningful clusters of functionally related variables. These groups of variables constitute the basic predictive atoms that would be selectively requested at a second stage, for forming the predictive models. In the proposed communication, the methods will be described in detail, and results will be presented for several real world data sets.

References

Hastie, T., Tibshirani, R. & Friedman, J. (2001). *The elements of statistical learning*. NY: Springer.

Reis, M. S. & Saraiva, P. M. (2004). A comparative study of linear regression methods in noisy environments. *Journal of Chemometrics, 18*(12), 526-536.

Reis, M. S. & Saraiva, P. M. (2005). Integration of data uncertainty in linear regression and process optimization. *AIChE Journal, 51*(11), 3007-3019.

Reis, M. S. & Saraiva, P. M. (2010). Analysis and classification of the paper surface. *Industrial & Engineering Chemistry Research, 49*(5), 2493-2502.

Trygg, J. & Wold, S. (2002). Orthogonal projections to latent structures (o-pls). *Journal of Chemometrics, 16*(3), 119-128.

(599d) The Process of Dehydrogenating N-Paraffins Into Mono-Olefins: Simulation and Optimization

Natalia Gourko, Resource-Saving Technologies, Saint-Petersburg State Institute of Technology (Technical University), St Petersburg, Russia

Nowadays, detergents form one of the most developing markets in the world; note that linear alkylbenzene is the main component of these agents.
The major stage of linear alkylbenzenes (LABs) production is dehydrogenating the 10–13 n- paraffins into mono-olefins, performed by means of Pacol-Define assembly.
Conversion of paraffins into olefins in Pacol reactor makes 8–10%, in order to ensure high level of selectivity in mono-olefins production. Due to relatively low rate of n-paraffins generation in Pacol reactor, significant recycle of the paraffins from alkylation block to Pacol block is provided.
The aim of work is to investigate feasibility of improving the performance of the existing production process by simulation and optimization.
To develop the reactor's model, kinetic data on the process evolution from certain sources have been analyzed.
The estimation results have motivated the authors to suggest redesigning the reactor, i.e., implementing the process in two serially connected reactors (instead of the only one) with the catalyst volumes satisfying the ratio of 1:0.15.
Two-reactor scheme of dehydrogenating the n-para ns has been simulated in UniSim; it has been studied in the context of the following problems.
(1) increasing the level of conversion in the target reaction;
(2) increasing the rate of specific LAB production subject to 1-kg catalyst (under the same level of conversion).
In work model of complex industrial process of dehydrogenating the n-paraffins was improved and verified; principal feasibility of 15%-increase in conversion of 10–13 fractions in Pacol reactor has been demonstrated under the modified temperature mode in the reactor scheme(i.e., involving low-temperature and high-temperature reactors instead of the single reactor). Simulation model within the proposed scheme of reactor operation has been developed. It allows for solving the problems of structural and parametric synthesis of optimal chemical-engineering system for the dehydrogenating process. Simulation algorithm and relevant recommendations regarding the redesign of the reactor scheme used in Pacol-Define assembly have been provided.

(599e) Multivariate SPC of Dynamic Processes Using Integrated Dynamic Principal Components Analysis and Missing Data Methods (DPCA-MD)

Marco Reis and Tiago Rato, Department of Chemical Engineering, University of Coimbra, Coimbra, Portugal

Classical univariate and multivariate SPC charts are usually based on *iid* assumptions. For the multivariate case, the statistic commonly used in the monitoring of industrial processes is the Hotelling's T^2 [1] either applied to the original variables or to the latent scores after a principal component analysis (PCA) decomposition, usually complemented with the Q statistic (also known as square predicted error, SPE), in order to monitor that portion of variability not captured by the PCA model (PCA-MSPC, [2-4]). These SPC methodologies face some important practical limitations, as they assume variables to be independent along time, i.e., not autocorrelated, an hypothesis that is often not met in practice, specially with the high sampling rates currently achieved with modern instrumentation and data acquisition systems. In order to address this issue, Ku *et al.* [5] proposed an SPC procedure based on dynamic principal component analysis (DPCA), which is an extension of PCA that includes time shifted variables, in order to accommodate and tacitly model the dynamic behavior of variables within the same PCA framework. However, one can easily verify that the direct implementation of such method still leads to autocorrelated statistics, raising problems in its implementation.

Therefore, we have developed and tested alternatives procedures to cope with this problem. In this work, we present the results from the application of several statistics, some of them new, based on the PCA Hotelling's T^2 and Q statistics. These statistics use a combination of DPCA, ARMA models and missing data estimation methods, allowing the simultaneous reduction of the data dimensionality (correlation structure) while capturing their dynamic behavior, therefore handling the autocorrelation effects. Furthermore, a detailed procedure to select the number of lags for each variable in DPCA is also proposed, which improves the methodology proposed initially by Ku *et al.* [5].

In a first stage, the performance of the control chart procedures was assessed by their Average Run Length (ARL) on several systems, such as the Wood and Berry column. From this analysis we selected the proposed methodologies that presented the best overall performance and subsequently compared them with the traditional PCA and DPCA approaches, when applied to the Tennessee Eastman process.

The Wood and Berry column model [6] represents an approximation of the dynamical behavior of a binary distillation column separating methanol from water. In this dynamic model, the distillate and bottom methanol weight fraction are expressed as a function of the reflux and reboiler steam flow rate. The system was subject to a set of step perturbations in the sensor measurements, and the corresponding ARL were determined, for each perturbation. The upper control limits (UCL) for all statistics were previously adjusted, by trial and error, in order to enforce an equal in-control Average Run Length (ARL_0) of 370 for all of them.

In this system, we observed that there was no significant difference between the traditional static and the dynamic versions of PCA. However, even with DPCA, the resulting statistics still present some autocorrelation. In the case of the new proposed statistics, the one that incorporates an implicit prediction methodology, namely through the missing data imputation (MD) approach to estimate future values, presented the best results. In fact the DPCA-MD monitoring statistics, not only improves the control chart performance, but also reduces the statistics autocorrelation, overcaming all the studied statistics.

This analysis was also conducted on the multivariate AR(l) process presented by Ku *et al.* [5] and on a Continuous Stirred-Tank Reactor (CSTR) system with a heating jacket [7]. In all these systems, the DPCA-MD based statistics presented consistently superior performances and lower autocorrelation. Therefore, the DPCA-MD based statistics, were applied, at a second stage, to the Tennessee Eastman process, and compared with the traditional PCA and DPCA approaches.

The Tennessee Eastman process was developed by Downs and Vogel [8] and has been widely used by the process monitoring community as a source of data for comparing various approaches. The simulation model has 41 measurements (XMEAS), 12 manipulated (XMV) variables and allows for 21 process upsets.

In this study we used the data provided by Braatz in [9] where the control system reported by Lyman and Georgakis [10] was implemented to generate the closed loop simulated process data. Each data set contains 960 observations with a sample interval of 3 min. The fault was introduced 8 hours after the simulation start. All the manipulated and measurement variables were collected, with the exception of the agitation speed of the reactor's stirrer, giving a total of 52 variables.

The data sets, without faults, were used to construct the PCA, DPCA and DPCA-MD models and to determine their upper control limits (UCL). The UCL's were adjusted by trial and error so that all the monitoring statistics present the same false alarm rate of 1%. All of the studied statistics on this system failed to detect faults number 3 and 9 and had a low capability to detect faults number 15 and 21. On the remaining 17 faults, the DPCA-MD statistics had the highest fault detection rates, while the PCA and DPCA statistics were only capable to perform well on 4 faults each. The DPCA-MD statistics also presented a consistent out of control state during faulty periods.

The performance of the proposed method (DPCA-MD) was also assessed with a paired t-test. From this analysis, we concluded that, with a 5% significance level, the DPCA-MD statistics are significantly better than those from PCA and DPCA. Another advantage of the DPCA-MD statistics is its lower autocorrelation, which makes the DPCA-MD statistics more reliable and easy to implement in practice.

To sum up, in this work we addressed the problem of monitoring processes with correlated dynamical data. We have studied several statistics based on a combination of methods (including PCA, DPCA, PLS, Time Series and Missing Data) and conclude that those derived from missing data methodologies show, in general, better performances. However we would like to point out that such statistics do require a suitable method to estimate the number of lags needed to construct the DPCA model, an issue that was also treated. The best results were achieved with the DPCA-MD statistics, that also presented lower autocorrelation and consistent out of control detection, making it a very interesting and viable alternative to the current ones based strictly on PCA and DPCA.

References:

1. Hotelling, H., *The Generalization of Student's Ratio.* The Annals of Mathematical Statistics, 1931. **2**(3): p. 360-378.

2. Jackson, J.E., *Quality Control Methods for Several Related Variables.* Technometrics, 1959. **1**(4): p. 359-377.

3. Kresta, J.V., J.F. MacGregor, and T.E. Marlin, *Multivariate Statistical Monitoring of Process Operating Performance.* The Canadian Journal of Chemical Engineering, 1991. **69**: p. 35-47.

4. Reis, M.S. and P.M. Saraiva, *Multivariate and Multiscale Data Analysis*, in *Statistical Practice in Business and Industry*, S. Coleman, et al., Editors. 2008, Wiley: Chichester. p. 337-370.

5. Ku, W., R.H. Storer, and C. Georgakis, *Disturbance detection and isolation by dynamic principal component analysis.* Chemometrics and Intelligent Laboratory Systems, 1995. **30**(1): p. 179-196.

6. Wood, R.K. and M.W. Berry, *Terminal composition control of a binary distillation column.* Chemical Engineering Science 1973. **28**(9): p. 1707-1717.

7. Santos, L., *Simulação dinâmica de um sistema de constituído por um CSTR.* 2009.

8. Downs, J.J. and E.F. Vogel, *A plant-wide industrial process control problem.* Computers and Chemical Engineering, 1993. **17**(3): p. 245-255.

9. Braatz, R.D. *Multiscale Systems Research Laboratory.* 2002 [cited; Available from: http://brahms.scs.uiuc.edu.

10. Lyman, P.R. and C. Georgakis, *Plant-wide control of the Tennessee Eastman problem.* Computers and Chemical Engineering, 1995. **19**(3): p. 321-331.

(599f) Uncertainties and Stochastic Optimal Control in Batch Crystallization for Different Types of Objective Functions

Kirti Maheshkumar Yenkie, Bioengineering, University of Illinois, Chicago, Chicago, IL; VRI-CUSTOM, Chicago, IL and Urmila Diwekar, Vishwamitra Research Institute, Center for Uncertain Systems: Tools for Optimization and Management, Clarendon Hills, IL

Uncertainties and Stochastic Optimal Control in Batch Crystallization for different types of Objective Functions

Kirti M. Yenkie[1,2] *and Urmila Diwekar* [1,2]

[1]*Department of Bio Engineering, University of Illinois, Chicago, IL 60607 - USA*

[2]*Center for uncertain Systems: Tools for Optimization & Management (CUSTOM),*

Vishwamitra Research Institute, Clarendon Hills, IL 60514 – USA

Abstract:

Batch crystallization is an industrial separation and purification process. It is dependent on parameters like temperature, supersaturation, agitation, solution behavior, etc. The fundamental process parameters like solubility and crystal structure can have physical uncertainties, while the process can have uncertainties in engineering design parameters and operating conditions. These uncertainties were generally ignored while predicting operating policies and hence, the predicted outcomes were not fully in accordance with the actual outcomes. To overcome these shifts in the desired results uncertainties are taken into account while predicting the crystallization operating policies.

Uncertainties in the system are characterized as a special class of stochastic processes, known as Ito processess if they follow certain properties. These uncertainties on propagation into the process model yields dynamic uncertainties within the process states. Using the theory of stochastic modeling and Ito's calculus, these uncertainties in the process states are represented in the mathematical form. The modified process model involves the state equations with the stochastic term representing uncertainty. The objective functions in batch crystallization are dependent on the desired outcome and can be classified into two categories late growth and early growth functions. Thus, we consider several objectives to see the effect of uncertainty on both the categories of functions. This results into stochastic optimal control problems and they are solved using Ito's lemma, stochastic calculus and stochastic optimal control theory.

The aim is to study all possible sources of uncertainty in batch crystallization and model them as Ito processes based on their behavior. Observe the effects of uncertainty inclusion into the process model and their effects on the operating policies for the early and late growth categories of objective functions. Thus, providing operating policies which shall mimic the actual behavior of the system and yield results with better predictive value.

Keywords: uncertainty, operating policy, stochastic processes, Ito's calculus, optimal control

Case study 1: Optimal control while considering kinetic parameter uncertainty

1. Kinetic parameter uncertainty:

The kinetic parameters are generally empirical constants determined by fitting experimental data to the model, and hence are a source of uncertainty within the system. In batch crystallization kinetics, the growth and nucleation expressions have empirical constants shown in table 3, they can be assumed to follow a Gaussian distribution, with the fitted values from previous experiments to be the mean. The values are assumed to deviate around ±5% about the mean.

Table 1: Kinetic parameter uncertainty in batch crystallization model

Uncertainty	Kinetic constants	Value from experiments/ model fitting	Range of Values
G	k_g	1.44×10^8 μm s^{-1}	$1.368 \times 10^8 - 1.512 \times 10^8$
	E_g/R	4859 K	4616.05 – 5101.95
	g	1.5	1.425 – 1.575
B	k_b	285 (s $\mu m^3)^{-1}$	270.75 – 299.25
	E_b/R	7517 K	7141.15 – 7892.85
	b	1.45	1.3775 – 1.5225

We consider a 95 % confidence interval and hence, the kinetic parameters lie within two standard deviations of their values ($\mu \pm 2\sigma^{\mu \pm 2\sigma}$). Thus, we evaluate the standard deviation for all of them using the extreme deviations as minimum and maximum values.

ü The sampling operation for multivariable uncertain parameter domain is performed using the Monte Carlo sampling technique

ü 100 sample values for each of the kinetic parameter are generated using inverse transformation over cumulative probability distribution.

ü After generating 100 samples for the kinetic parameter data, the model is simulated using each set.

ü The resulting dynamic uncertainty in the state variables due to static uncertainty in the kinetic parameters is observed in the plots of the dynamic uncertainty as shown in figure 1.

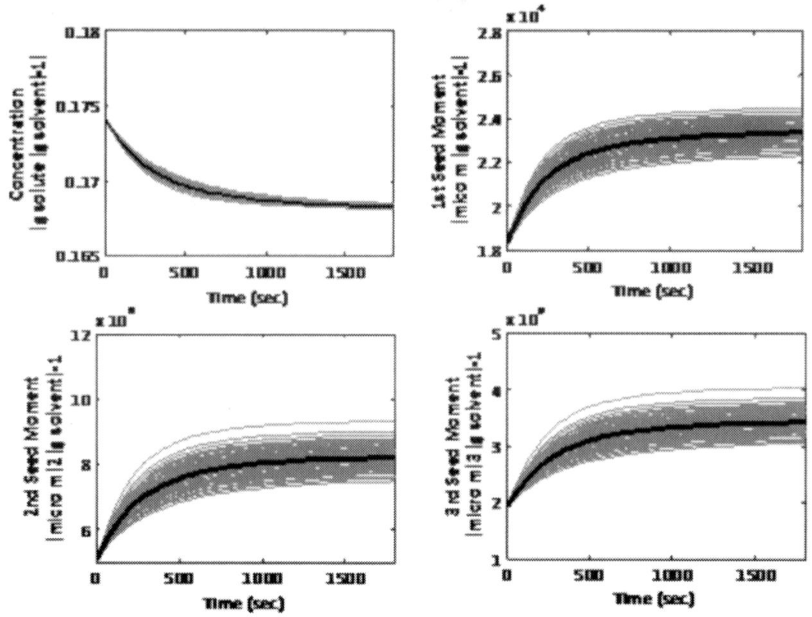

Figure 1 The dynamic uncertainties in state variables C, $\mu 1s$, $\mu 2s$, $\mu 3s$

2. Modeling uncertainties as Ito processes

By studying the dynamic uncertainty plots of the process variables and their correlation to Ito processes, it has been observed that the uncertainties can be best modeled with a simple Ito process known as Brownian motion with drift.

It is defined as: *dy=ay,tdt+by,tdz*

$$dy = a(y,t)dt + b(y,t)dz \quad (1)$$

where dz is the increment of the Wiener process equal to $\varepsilon_t\sqrt{\Delta t}$, and $a(y,t)$ and $b(y,t)$ are known functions. The random value ε_t has a unit normal distribution with zero mean and standard deviation of one. In this paper, a simplification of the above equation is used to represent the time dependent uncertainties in the concentration, seed and nucleation moments:

dYi=fYi,tDt+giεiDt

$$dY_i = f(\bar{Y}_i,t)\Delta t + g_i\varepsilon_i\sqrt{\Delta t} \quad (2)$$

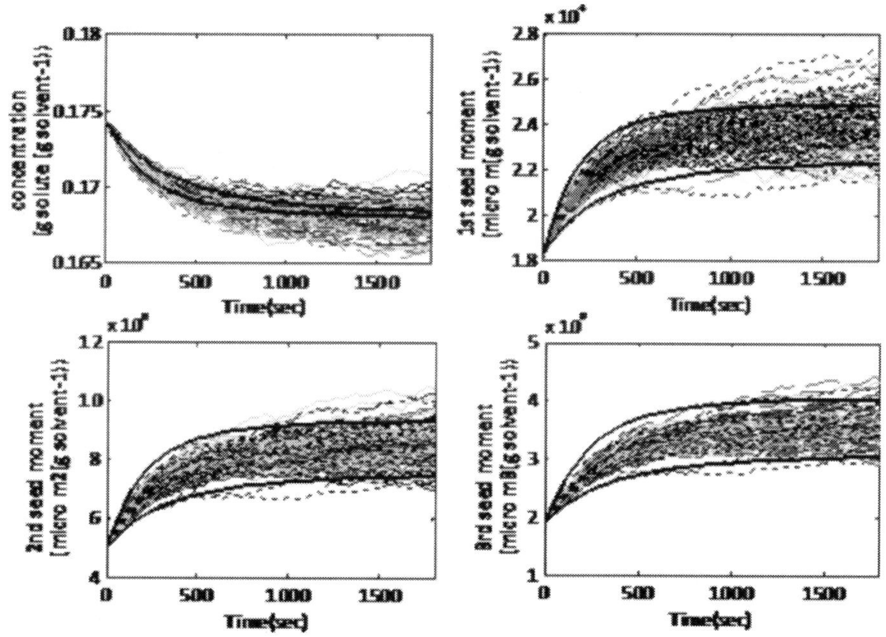

Figure 2. State variables modeled as Ito Processes

3. Stochastic Maximum Principle:

We use the stochastic maximum principle similar to the method illustrated in Ramirez and Diwekar. The objective is to maximize the expected value of mass of seeded crystals and minimize the expected value of mass of fines, considering the uncertainties associated with the concentration and moments of batch crystallization, finding the best operating temperature profile for the process.

The objective function for the stochastic formulation can be written as equation 3a or 3b where E is the expected value.

MaximizeT L=E0tdμ3sdt- dμ3ndtdt $\text{Maximize}_T\ L = E\left[\int_0^{t}\left(\frac{d\mu_3^s}{dt} - \frac{d\mu_3^n}{dt}\right)dt\right]$ (3a)

MaximizeT L=Eμ3s(tf)-μ3n(tf) $\text{Maximize}_T\ L = E[\mu_3^s(t_f) - \mu_3^n(t_f)]$ (3b)

Subject to seeded batch crystallization state variables modeled as Ito processes, initial conditions, constraints for supersaturation condition maintenance. The state equations in general can be represented as shown in eq 4.

dyi=fyi,tDt+giεiDt $dy_i = f(\bar{y}_i, t)\Delta t + g_i\varepsilon_i\sqrt{\Delta t}$ (4)

where, *yi=[C μ0s μ1s μ2s μ3s μ0n μ1n μ2n μ3n]* $y_i = [C\ \mu_0^s\ \mu_1^s\ \mu_2^s\ \mu_3^s\ \mu_0^n\ \mu_1^n\ \mu_2^n\ \mu_3^n]$

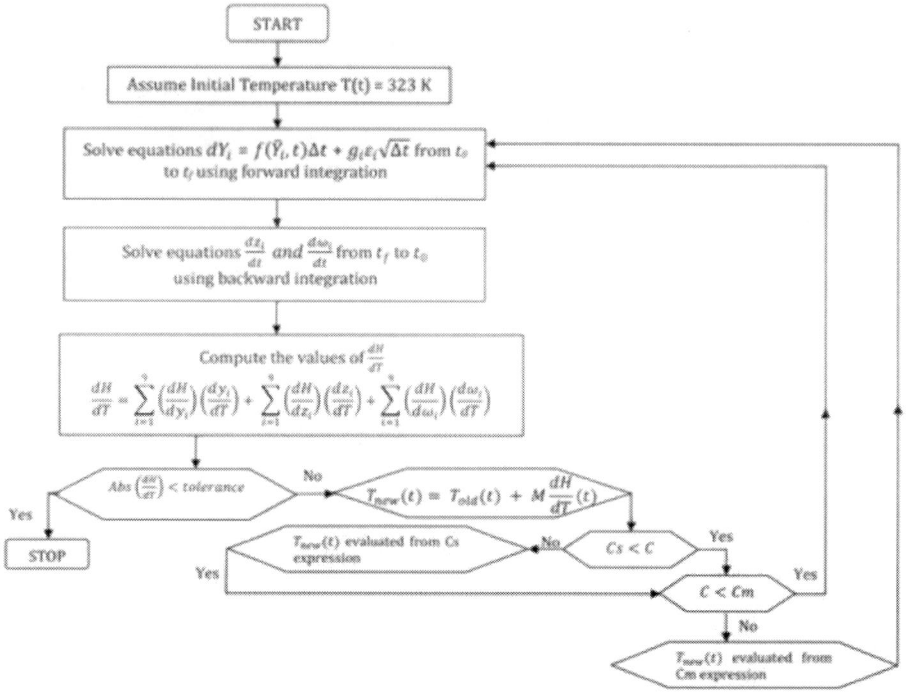

Figure 3. Flowchart for optimal temperature profile evaluation using stochastic maximum principle (active constraint strategy)

4. Results

Figure 4 summarizes the optimal temperature profiles evaluated by the deterministic method and stochastic method. The results for the linear cooling profile, which is most commonly used is also compared against the optimal trajectories. Figure 5 compares the final size distributions predicted from the moment values, and it can be seen that the stochastic case yields a much narrower probability density function for the distribution in case of seed as well as nucleated crystals.

Table 2: Results of Comparison between linear, deterministic and stochastic cases

Case	Objective function	% Increment in stochastic case
Linear	8.53×10^9	5.88
Deterministic	8.79×10^9	2.73
Stochastic	9.03×10^9	N.A.

5. Conclusion

The static uncertainty in the kinetics result in dynamic uncertainties within the state variables. These uncertainties could be modeled as Ito processes, since they followed certain properties and had a behavioral pattern. The most important aspect was to solve the stochastic optimal control problem involving Ito processes and application of stochastic calculus, Ito's Lemma and stochastic maximum principle.

6. Future work

We would like to consider several kinds of objective functions used by researchers for predicting the best optimal policy. The uncertainties associated with these functions would be incorporated in the study and a similar strategy to study and compare the determinstic and stochastic cases will be adopted. Other sources of uncertainty along with kinetic parameter uncertainty would also be taken into account to make the operating policies more robust.

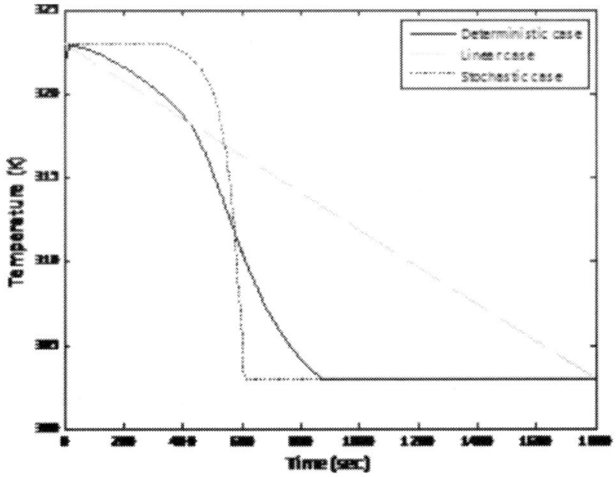

Figure 4. Comparison of Temperature profiles

(Linear, Deterministic and Stochastic case)

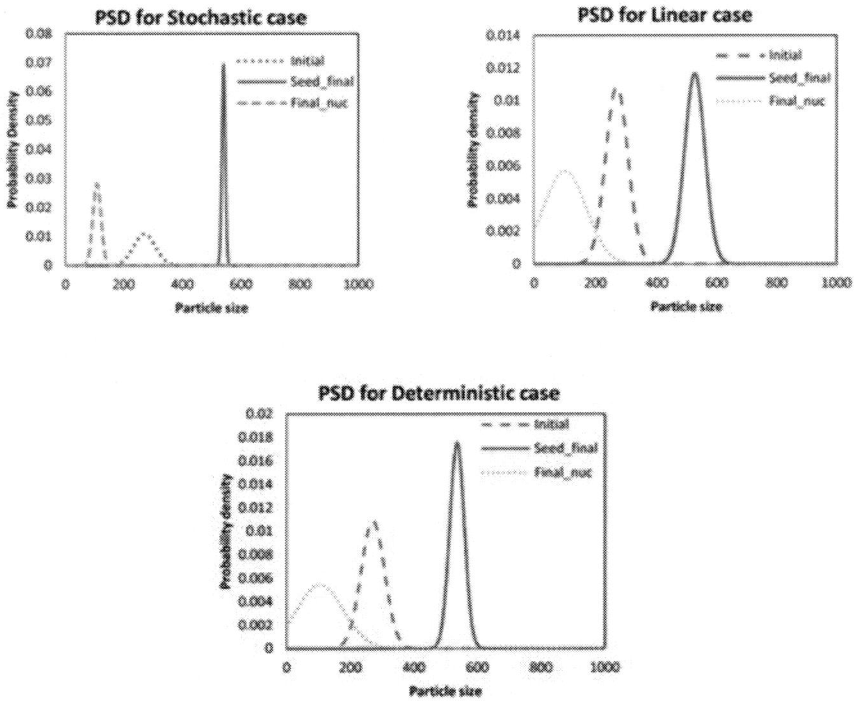

Figure 5. Predicted particle size distributions (PSD) from moment values for 3 cases.

Dynamic Simulation of an Ethylene Recycle Compressor System

*Dr. Martin Häfele** *Martin Kamann** *Felix Knödler**
*Dr. Andreas Kröner** *Dr. Martin Pottmann**
**Linde Engineering Division, Pullach, Germany*

1 Introduction

The Engineering Division of The Linde Group plans and delivers large scale chemical process plants for air separation, liquefaction of natural gas, separation of gases and for the production of olefins and synthesis gas. Compressors and turbines are major components in such plants. Therefore, safe operation of those machines over extended periods of time has to be ensured. This comprises not only steady state operation but also transient conditions associated with performance control and start-up or shutdown [2]. In particular, the preparation of the initial start-up has to be considered.

Linde uses dynamic simulation for both design and analysis of machine protection systems (MPS). The main purpose of a MPS is to ensure safe, i.e. non-surging operation during transient conditions such as normal stop or emergency shutdown (ESD) events. Such investigations are necessary e.g. for new process topologies. Moreover, these investigations are enforced by customers in order to prove adequate sizing of the MPS. Adequate sizing comprises both topological layout and dimensioning of bypass valves and bypass lines.

In this contribution, we will outline a dynamic simulation project of an Ethylene Recycle Compressor (ERC) system for the design and analysis of such an MPS.

2 Ethylene Recycle Compressor System

2.1 Process Topology

Fig. 1 provides the main process topology of the Ethylene Recycle Compressor (ERC) system. Ethylene is fed from the process at three different pressure levels, low, intermediate and high. Ethylene at low pressure is fed to the suction drum of the first stage KC401_1, ethylene at intermediate pressure is mixed with the outlet of the first stage and fed to the second stage KC401_2. Afterwards the outlet of the second stage is mixed with the feed at high pressure and compressed to the final pressure. It is cooled in the heat exchanger EE404AB and split into two streams. One is recycled to the process via valve MV45801, the other is compressed in the fourth stage KC401_4 to the product pressure and leaves the process as product. All four stages are driven by a steam turbine and have a common shaft. The bypasses of the first three stages are withdrawn from the pressure node U44D10 downstream of heat exchanger EE404AB. They tie in on the suction side of the corresponding compressor stage. The forth stage is bypassed by a pipe from pressure node U44D06 that binds in at pressure node U44D05, which is downstream from the check valve. Additionally, the first stage is equipped with a second bypass line with valve PV45504B for control purposes, as anti-surge valve FV45501 has only limited controllability

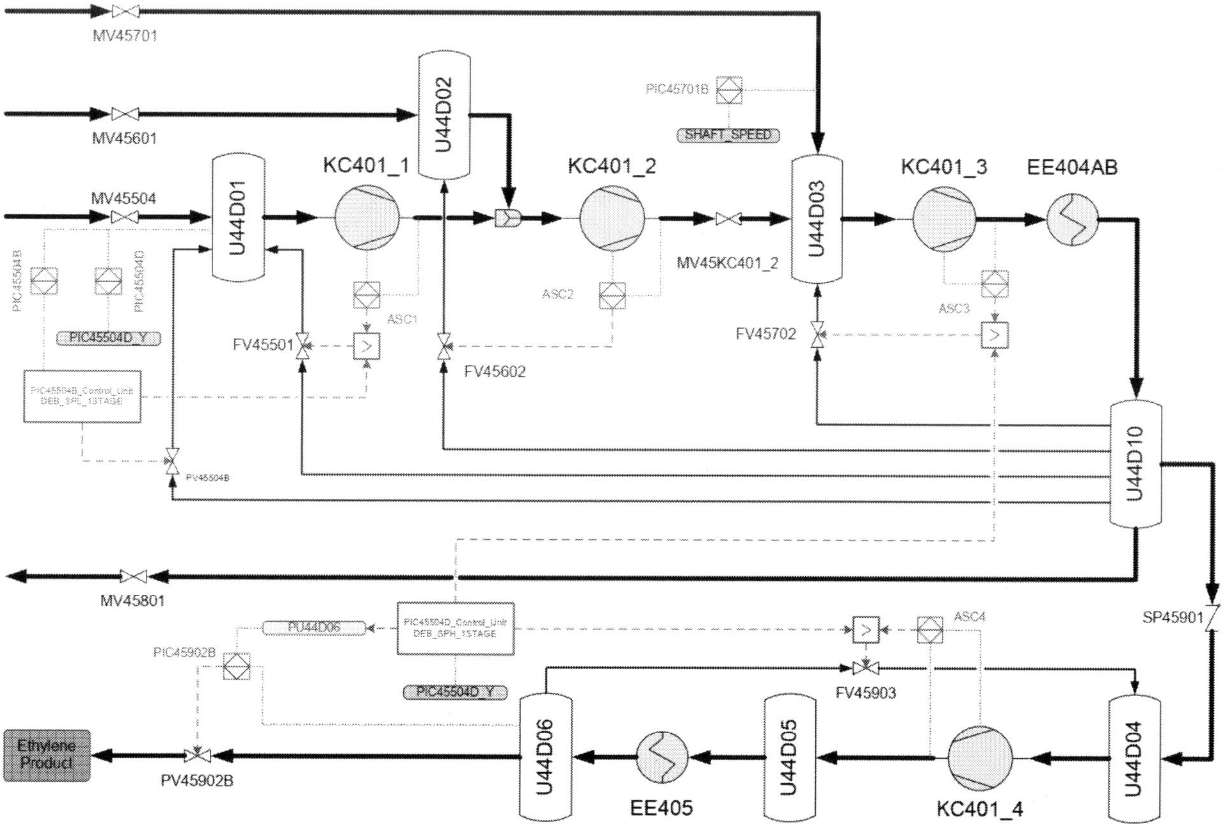

Figure 1: The flow-sheet of the ERC system of this study

for small streams. This system is analyzed for an adequate topology and sizing of the MPS. Additionally, the controllability of the process is investigated.

2.2 Process Modeling

Linde uses its proprietary process simulator OPTISIM® [3, 4, 6, 7] for steady state design, rating and dynamic simulations as well as for steady state [5, 8] and dynamic optimization.

In the steady state plant design an optimal process topology is identified and generates the transferred heat and outlet conditions of the specific process units. In a rating model, the process topology, inlet conditions and dimensions of a process unit are used to calculate the outlet variables. In OPTISIM®, a specific process unit is derived from first principles and is described by energy and material balance equations as well as by the phase equilibrium. The process units are connected via material, energy and signal streams. During a dynamic simulation, the energy and material balances are time dependent. Additional non-linear algebraic equations are used, e.g. for the evaluation of performance maps of machines. These model equations are transformed into a system of differential and algebraic equations, which is solved simultaneously with appropiate numerical methods. For the solution, analytical Jacobians are used, that are included into the process model of the specific process unit.

2.3 Compressor Modeling

As compressors are one major part of a highly complex plant, special care has been taken to include most accurate models of such rotating equipment. In particular, the model equations conform to the ASME PTC-10 Standard [1]. Additionally, performance maps from the machine

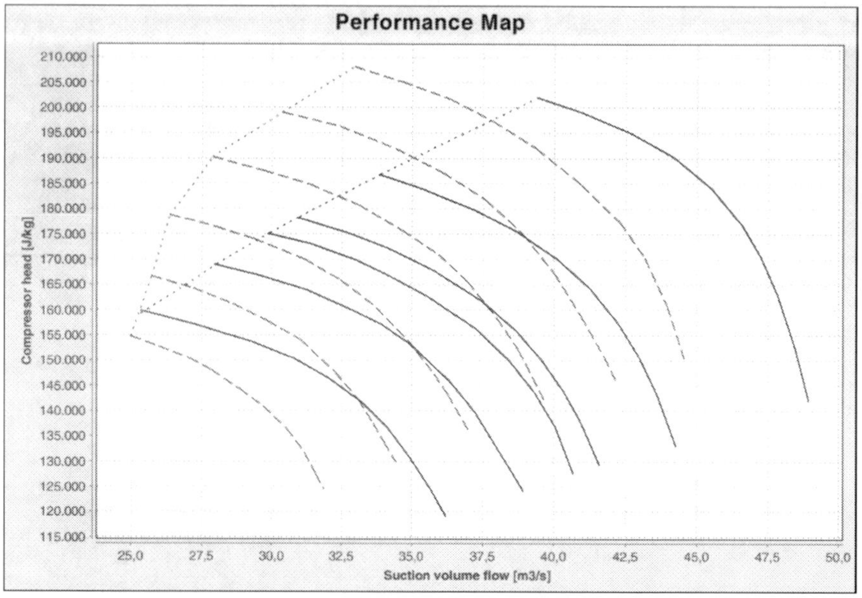

Figure 2: The vendor given performance map (blue solid line) and the map transformed to suction conditions at 40 K higher (red dashed line). For both maps the surge limit line is shown as dotted line in the corresponding color.

vendor are used. Such a performance map is the graphical representation that correlates shaft speed and volumetric flow rate on the suction side with the outlet pressure of the compressor. This performance map is only valid for the specified suction conditions. However, during rating or dynamic simulation, these suction conditions vary which has a strong influence on the obtained outlet pressure.

In order to be able to evaluate the performance map quickly - together with the analytical derivatives - a nonlinear monotonous function has been identified that contains very few parameters p which need to be fitted to the vendor data. This procedure is done automatically at the start of the calculations. Thereafter, this function calculates the head H_d of the compressor given volumetric flow rate on the suction side $\dot{V}$ and the machine Mach number M_{u2}, i.e.

$$H_d = f(\dot{V}, M_{u2}, p) \tag{1}$$

For $M_{u2} < M_{u2,min}$ or $M_{u2} > M_{u2,max}$ with $M_{u2,min}$ as minimum and $M_{u2,max}$ as maximum machine Mach number, the performance map is extrapolated according to fan law, i.e.

$$\dot{V} \sim M_{u2} \qquad \text{and} \qquad H_d \sim M_{u2}^2. \tag{2}$$

For $\dot{V} < \dot{V}_{surge}$ or $\dot{V} > \dot{V}_{st\_wall}$ the functional approach is extrapolated by an analytical continuation that is continuous in $\mathcal{C}^1$, i.e. continuously differentiable.

Using nonlinear similarity transformations for $\dot{V}$, M_{u2} and H_d, the outlet pressure is corrected in order to match operating conditions other than the ones from the performance map, as shown in Fig. 2. There, the vendor given performance map is printed in blue solid lines. The performance map transformed to a higher suction temperature is shown in red dashed lines. The difference between the two suction temperatures is 40 K. Such a difference may occur during plant operation if a bypass valve is opened and gas at ambient conditions from the discharge side of a compressor is fed to the cold suction side. From Fig. 2 follows, that a very large error would be made in the outlet pressure prediction, if the map was not adjusted for other suction conditions.

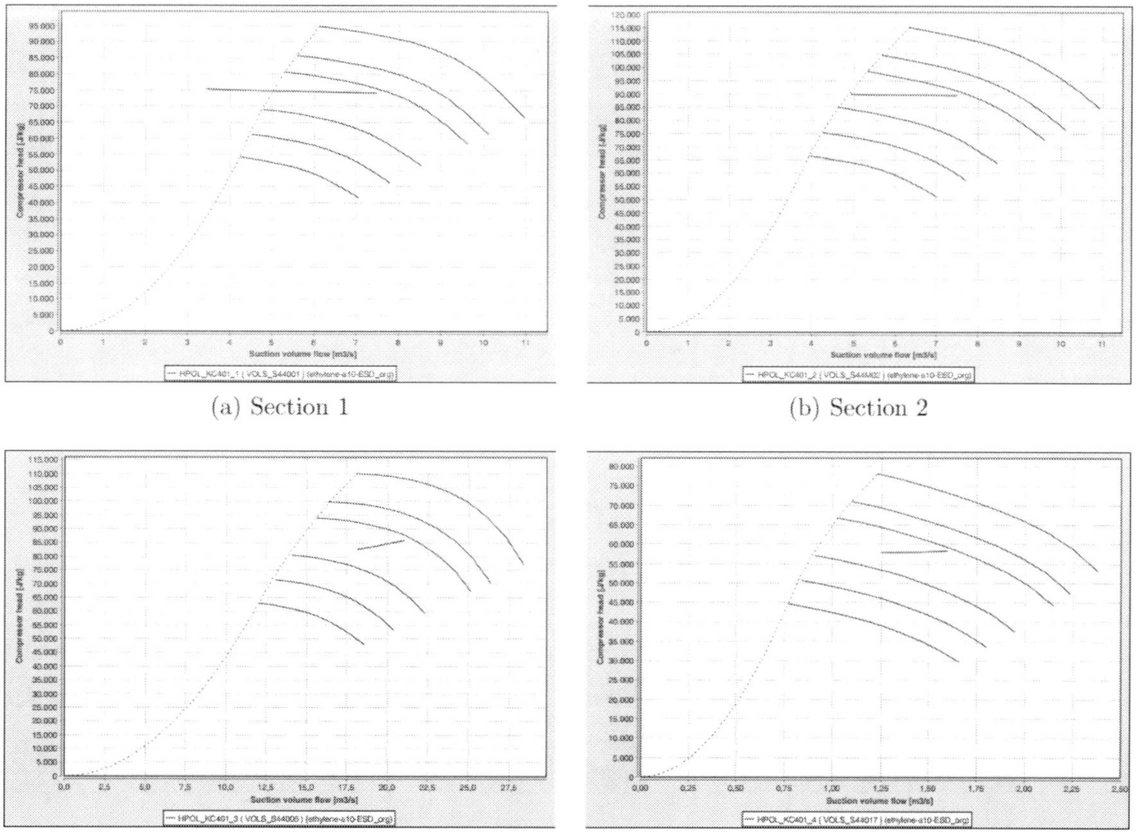

(a) Section 1 (b) Section 2

(c) Section 3 (d) Section 4

Figure 3: Rundown curves of all four stages of the ERC for an ESD.

Both the performance map fit and the transformation to other suction conditions have been verified using real plant data and it can be concluded, that the mathematical model of the compressor reliably predicts the outlet conditions.

3 Simulation Results

3.1 Emergency Shutdown

For the investigation whether the suggested MPS is correctly sized, an Emergency Shutdown (ESD) is simulated. This is the worst case condition for the design of a MPS.

In this case, the drive is stopped instantaneously and, while detecting this event, the bypass valves are forced to open. In the simulation model, the driving power is ramped down to zero within 0.5 s and the anti-surge valves are opened instantaneously afterwards taking into account the valve actuator dynamics. Opening the bypass valves should immediately lead to a significant pressure relaxation of the machine.

The simulation result with the original bypass configuration and anti-surge valve sizing is shown in Fig. 3. The head of the compressor stages are plotted against the suction volumetric flow rate of the considered compressor stage. In these diagrams, the rundown curves of all four stages are shown as red solid lines together with the iso speed curves (blue lines) of the performance map. As one can see, the operating point moves rapidly across the surge limit line, indicated by the blue dotted line.

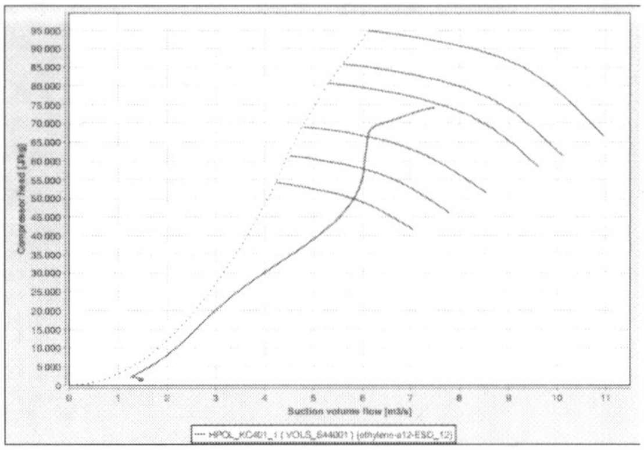

(a) Section 1

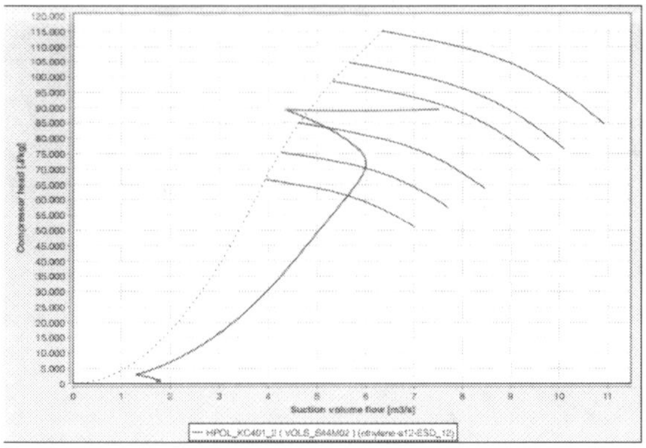

(b) Section 2

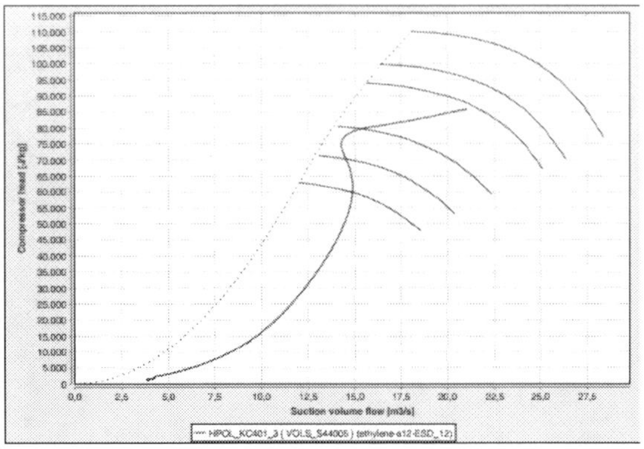

(c) Section 3

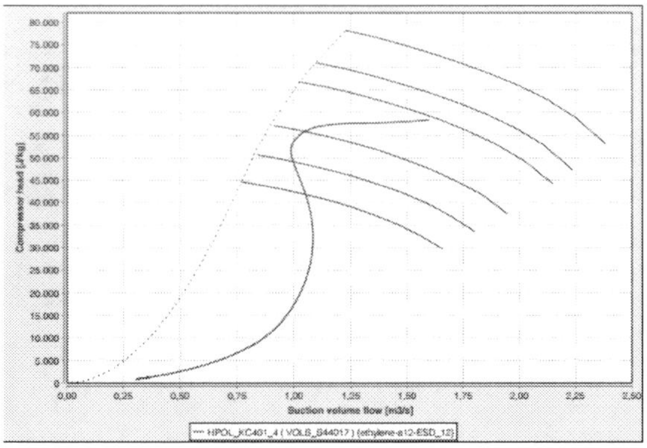

(d) Section 4

Figure 4: Rundown curves of all four stages of the ERC for an ESD with additional bypass.

This means, that this configuration does not guarantee an adequate MPS. As an alternative, an additional bypass has been installed (note: this bypass line is not depicted in Fig. 1), and this turned out to be a very robust solution (see Fig. 4). With this solution, the original anti-surge valves - which are specified by the machine vendor - can be used.

3.2 Control Study

For the ERC, a dynamic simulation study of the pressure control strategy is performed. Scope of the study is to evaluate the expected performance of the pressure control loops under various disturbances.

First the controller pairing is checked using a Relative Gain Analysis (RGA). The RGA is executed for different cases, i.e. (i) normal operating conditions (NOC), (ii) first stage pressure high via fourth stage discharge pressure control, (iii) first stage pressure high via stage 3 plus 4 bypass valves and (iv) first stage pressure low via stage 1 bypass valve. Here, we restrict the analysis to the NOC. In this configuration, the pressure high PIC4505B and pressure low PIC4505D controllers are saturated and do not contribute to the overall system dynamics. The RGA therefore only needs to be performed for the suction pressure control of the third stage PIC5701B and discharge pressure control of the fourth stage PIC45902B. The model is linearized using the input and output variables of those controllers. As one can see from the dynamic RGA in Fig. 5, the interactions between the two controllers are negligible. Over a wide frequency range, the shaft speed only affects the suction pressure of the third stage and the fourth stage discharge valve position only affects the fourth stage discharge pressure.

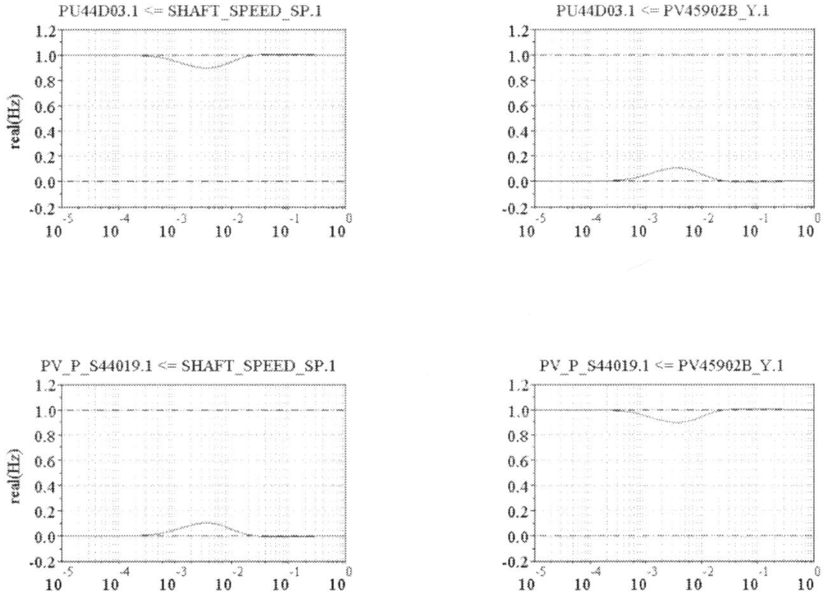

Figure 5: The dynamic RGA for NOC.

However, other RGA's suggested an alternative control approach, i.e. control of the first stage suction pressure via the compressor speed and control the third stage suction pressure via the fourth stage discharge valve and use the third stage and fourth stage anti-surge valves in a split range configuration. The original (Fig. 6a) and alternative (Fig. 6b) control approaches are compared in Fig. 6 which shows the response to a sinusoidal stimulation of the feed flow rate in stage 1. In Fig. 6, the reaction of the suction pressure of the first stage and the discharge pressure of the fourth stage on the disturbance are shown. It can be seen, that the alternative approach performs slightly better than the original control scheme, however, both can cope with the disturbance.

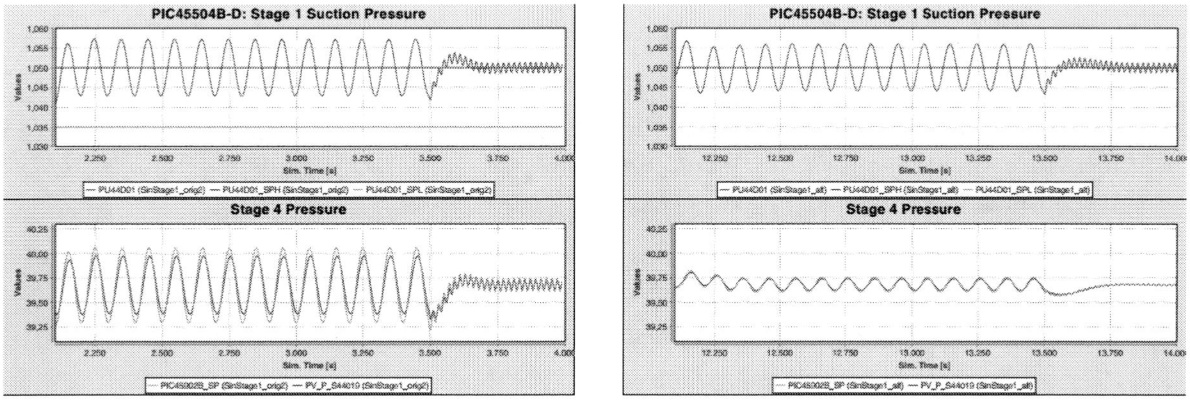

(a) Original control scheme

(b) Alternative control scheme

Figure 6: Original (on the left) and alternative (on the right) control scheme.

4 Conclusions

From the previous sections, one has seen, that dynamic simulations intensify the knowledge of the process and give deeper insight into the plant behavior under transient conditions, e.g. ESD or disturbances. However, the main challange for an engineering company that delivers turnkey chemical process plants is to learn from such simulations and directly include the result into current *and* new projects. Additionally, this approach allows a faster execution of simulation projects as they can be initiated earlier during the project execution.

It is obvious, that even the most accurate steady-state design does not automatically cover all aspects that are necessary for a safe plant operation, as most plants are not always operated at steady state. The dynamic simulations are complex but offer additional benefits, such as a theoretical verification of control strategies by linearizing the model or a verification by dynamic simulation.

References

[1] ASME PTC-10. *Performance Test Code on Compressors and Exhausters.* American Society of Mechanical Engineers, New York, 1997.

[2] K. K. Botros and S. T. Ganesan. Dynamic instabilities in industrial compression systems with centrifugal compressors. In *37th Turbomachinery Symposium*, pages 119–132, Houston, 2008.

[3] E. Eich, P. Burr, A. Kröner, and P. Lory. *Mathematik in der Praxis*, chapter Modellierung und numerische Simulation in der Chemischen Verfahrenstechnik, pages 61–85. Springer, Berlin, 1995.

[4] G. Engl and A. Kröner. Success Factors for CAPE in the Engineering Practice of a Process Plant Contractor. In W. Marquardt and C. Pantelidis, editors, *16th European Symposium on Computer Aided Process Engineering and 9th International Symposium on Process Systems Engineering*, pages 763–768. Elsevier, 2006.

[5] G. Engl and H. Schmidt. The optimization of natural gas liquefaction processes. In M. Brøns, M. P. Bendsøe, and M. P. Sørensen, editors, *Progress in Industrial Mathematics at ECMI 96*, pages 356–363, Stuttgart, 1997. B. G. Teubner.

[6] A. Kröner. Engineering Company's Approach to Filling "CAPE Gaps" in Process Simulation. In W. Marquardt and C. Pantelidis, editors, *16th European Symposium on Computer Aided Process Engineering and 9th International Symposium on Process Systems Engineering*, pages 781–786. Elsevier, 2006.

[7] I. Thomas. A Process Unit Modeling Framework Within a Heterogeneous Simulation Environment. In E. N. Pistikopoulos, M. C. Georgiadis, and A. Kokossis, editors, *21st European Symposium on Computer Aided Engineering*, pages 31–35. Elsevier, 2011.

[8] I. Thomas and A. Kröner. Mixed-Integer Optimization of Distillation Column Tray Positions in Industrial Practice. In W. Marquardt and C. Pantelidis, editors, *16th European Symposium on Computer Aided Process Engineering and 9th International Symposium on Process Systems Engineering*, pages 1015–1020. Elsevier, 2006.

(599h) Gas Pipeline Leak Detection Using Rigorous Hydraulics and Global Optimization

Mesude Ozturk, Sung Young Kim and Miguel Bagajewicz, Chemical Engineering, University of Oklahoma, Norman, OK

Gas Pipeline Leak Detection using Rigorous Hydraulics and Global Optimization

Mesude Ozturk, Sung Young Kim and *Miguel Bagajewicz

School of Chemical Engineering and Material Science, University of Oklahoma

100 East Boyd Street, T-335 – Norman, OK 73019-0628 USA

*** corresponding author**

Abstract

In this work, we use the Generalized Likelihood Ratio method to detect leaks and biases in pipelines. We assume that in addition to flows, pressures are also measured, so that gas hydraulic equations can be used to improve the power of the test. Instead of using the common gas hydraulics equations (Panhandle, Weymouth, etc), we resort to use a better correlation that we developed.

The GLR method requires solving an optimization problem. Because the gas hydraulic equations are nonlinear, the model has no longer the traditional analytical solutions of linear data reconcilation. The nonlinearities and non-convexities arise from bilinear terms corresponding to flow rates, inlet pressures, and outlet pressure in each pipeline segment. Thus, we resort to use our recently developed Bound Contractiong method (Faria and Bagajewicz, AICHE J., 2011). Because of the inclusion of pressure measurements, our leak detection method allows for the determination of the location of the leak in addition to determning its size.

(599i) Application of Global Solution Algorithm, Interval Analysis, in Pareto Optimal and Pareto Frontier Solving Strategy

Gang (Gary) Xu, R&D, Invensys/SimSci-Esscor, Lake Forest, CA

Pareto optimal (or Pareto set, Pareto frontier) is a concept in economics with applications especially in engineering and social sciences. The term is named after Vilfredo Pareto (1848–1923), an Italian economist who practiced the concept in his studies of economics.

Solving Pareto optimal and Pareto frontier involve solution strategy of multi-objective optimization where solutions come as "set". One approach extensively applied is the weighted sum approach, which had been studied with generic algorithm. Due to the fact that Pareto optimal set is a NP-hard problem, effort of obtaining the "whole set" is time consuming. Part of the "set" could even be missing if one (or more) of the functions within the multi-object optimization group is concave. Thus, more complicated approaches other than weighted sum have to be applied, which, if solved by generic algorithm, guarantee of discovering the "whole set" is still a mystery.

In our study, interval analysis, one of the global solution algorithms, had been applied to solve Pareto optimal and frontier. To avoid NP-hard, scheduling of the weighted sum factors had been applied with sufficient coverage. The ability of solving convex and concave optimal functions by global solution algorithm, such as interval analysis, enabled the weighted sum approach to discover the "whole set" of Pareto optimal solution with concave functions involved. The only problem by applying global solution algorithm seems to be the definition of the scheduling for the weighted sum factors. However, proper restricts under engineering consideration may provide reasonable suggestion on the coverage of the weighted sum factors, so that "whole set" is discovered with finite solutions.

This is the first attempt of applying deterministic solution algorithm to solve Pareto optimal and frontier. However, one has to apply proper restricts under engineering and economic considerations to avoid the NP-hard issue, as well as the ability of solving concave functions such as global solution algorithms.

(599j) Evaluation of Classification Algorithms for Automatic Inspection of TFT-LCD Glass Substrates

Ali Yousefian, Pukyong National University, Busan, South Korea and **J. Jay Liu**, Department of Chemical Engineering, Pukyong National University, Busan, South Korea

This paper presents an industrial case study of automatic inspection methodology to manufacturing of TFT-LCD glass substrates. We perform a three-way comparison of classification accuracy involving classification & regression tree (CART), multi-layer perceptron (MLP) and support vector machine (SVM). Decision making was performed in four stages: feature extraction by computation of wavelet co-occurrence signature, handling imbalanced data problem through the synthetic minority over-sampling technique (SMOTE), dimensional reduction with principal component analysis (PCA), and classification by using three different classification methods (CART, MLP and SVM optimized via simulated annealing algorithm (SA)).

The more SMOTE we used, the higher the accuracy we obtained by each three models. Optimized SVM and MLP models classified TFT-LCD glass substrates more accurately than the CART model. The SVM model was better in classification than the MLP model. The accuracy for the classification using CART and optimized MLP were 62.9% and 87.9%, respectively while a higher accuracy of 89.5% was observed for the optimized SVM model. This study shows possibility of automation of glass inspection in glass substrates manufacturing industry.

(599k) Advanced Unit-Specific Event Based Modeling Approach for Short-Term Scheduling of Batch Plants

Ramsagar Vooradi and Munawar A. Shaik, Department of Chemical Engineering, Indian Institute of Technology (IIT) Delhi, New Delhi, India

Scheduling of multipurpose batch plants has received considerable attention during the past two decades. Different modeling approaches have appeared in the literature. Most of the proposed approaches can be classified into two main groups based on time representation: discrete-time models and continuous-time models. Extensive reviews were written by Floudas and Lin [1], Mendez *et al*. [2], and Shaik *et al.*[3] who presented comparisons of different models and discussed associated challenges. Continuous-time models in the literature are further classified into slot-based, global-event-based, unit-specific event-based, and precedence-based formulations.

Unit-specific event-based models define events on a unit basis, allowing tasks corresponding to the same event point but in different units to take place at different times. Due to heterogeneous locations of event points used, the unit-specific event based models always require less event points and exhibit favorable computational performance compared to both slot-based and global-event-based models [3]. Shaik and Floudas [4] proposed a novel unified model using three index binary and continuous variables based on unit-specific event based continuous time representation that efficiently allows tasks to span over multiple events. This model is applicable for problems with and without resources such as utilities and with different storage policies, and has been demonstrated to give superior computational performance among all the models considered in their study. Recently, Vooradi and Shaik [5] presented an improved model compared to Shaik and Floudas [4] by taking advantage of the three-index variables and effectively incorporating the concept of active task leading to fewer constraints and big-M terms. The effectiveness of their model was demonstrated through various examples drawn from literature including Westenberger-Kallrath challenging benchmark problem [6].

Despite the significant advancements in the modeling and solution approaches in the past two decades, it is still a challenging task to solve real industrial scheduling problems because of the issues arising in handling large-scale problems. Hence, there is a global concern and obligation to develop innovative methods to reduce the number of events and/or the problem size of a scheduling model to enable efficient solution. In this study, we investigate innovative methods to further reduce the number of events required to efficiently solve a given scheduling problem using unit-specific event based approaches.

All the conventional approaches for batch process scheduling in the literature are based on the assumption that the consumption task corresponding to a given material/state typically occurs at the next event/slot relative to the production task unless there is enough inventory for the consumption task to start at earlier events. If we allow production and consumption tasks to correspond to the same event then it is possible to achieve a significant reduction in the number of events. With this motivation, in this work we propose an advanced unit-specific event-based continuous-time modeling approach by exploiting the fact that the unit-specific events correspond to heterogeneous locations along the time axis, thus allowing different tasks to occur at different times and still correspond to the same event point. The basic framework is adapted from Vooradi and Shaik [5] model. The main novelty in this formulation is that the production and consumption tasks corresponding to a given state/material are allowed to occur in a sequential manner at the same event point. We present few benchmark examples from literature [4,5,7-9] involving short-term scheduling of batch plants and compare the computational performance of the proposed model with respect to earlier models. The proposed model significantly reduces the number of event points required leading to fewer constraints, variables and yielding better RMIP values for several problem instances. The proposed model can handle problems with batch splitting and mixing, and different storage polices.

Acknowledgements: The authors gratefully acknowledge financial support received from Department of Science and Technology (DST), India, under SERC scheme, grant no. SR/S3/CE/075/2009.

References:

[1] Floudas, C.A. and Lin, X. (2004). Continuous-time versus discrete-time approaches for scheduling of chemical processes: A review. *Comp. Chem. Eng.*, **28**, 2109-2129.

[2] Mendez, C.A., Cerda, J., Grossmann, I.E., Harjunkoski, I. and Fahl, M. (2006). State-of-the-art review of optimization methods for short-term scheduling of batch processes. Comp. Chem. Eng., **30**, 913-946.

[3] Shaik, M.A., Janak, S.L. and Floudas, C.A. (2006). Continuous-time models for short-term scheduling of multipurpose batch plants: A comparative study. *Ind. Eng. Chem. Res.*, **45**, 6190-6209.

[4] Shaik, M.A. and Floudas, C.A. (2009). Novel unified modeling approach for short-term scheduling. *Ind. Eng. Chem. Res.*, **48**, 2947-2964.

[5] Vooradi, R. and Shaik, M.A. (2012). Improved three-index unit-specific event based model for short-term scheduling of batch plants. *Comp. Chem. Eng., http://dx.doi.org/10.1016 /j.compchemeng.2012.03.014*

[6] Kallrath, J. (2002). Planning and scheduling in the process industry, *OR Spectrum*, **24**, 219.

[7] Ierapetritou, M.G. and Floudas, C.A. (1998). Effective continuous-time formulation for short-term scheduling: 1. Multipurpose batch processes. *Ind. Eng. Chem. Res.*, **37**, 4341-4359.

[8] Li, J., Susarla, N., Karimi, I.A., Shaik, M.A. and Floudas, C.A. (2010). An analysis of some unit-specific event-based models for the short-term scheduling of non continuous processes. *Ind. Eng. Chem. Res.*, **49**, 633-647.

[9] Susarla, N., Li, J., Karimi, I.A. (2010). A novel approach to scheduling multipurpose batch plants using unit slots. *AIChE J.*, **56**, 1859-1879.

5991 - Most Common Pitfalls in Process Simulation

Presentation ID# 271009

Authors: Richard Pelletier and Mike Donahue, Invensys Operations Management

Process simulation is a tricky skill to master. Even with years of experience, a typical user is constantly facing new problems that they need to address. There are many issues that can lead to unexpected and unnecessary errors with the use of process simulation tools. Despite what simulation program you are using, many of these issues are common from one program to another. In this presentation, we will discuss the "Most Common Pitfalls in Process Simulation". For each topic, we will discuss various techniques and methods to fix these troublesome problems and prevent their occurrence.

Topics to be discussed include:

1. Bad data entry
2. Simulation defaults
3. Bad plant data
4. Thermodynamic method selection
5. Simulation tolerances and convergence
6. Recycle identification and handling
7. Model complexity
8. Difficult or impossible specifications
9. Assay pseudo-component development
10. Assay property blending

These topics are based on decades of experience and feedback from end users to technical support representatives for process simulators. They often represent the most commonly asked questions or areas that most users find to be problematic. While the specific examples highlighted in each topic come from SimSci-Esscor software, the intent of this presentation is to be broad based and should be applicable for both new and experienced users of process simulation programs. Also, no topic is more or less significant than another. The numbering of each topic is meant solely for organizational purposes, not for priority or importance.

#1: Bad Data Entry

Bad data entry is the easiest mistake to make when doing process simulation. Most often, units of measurement are the largest source of errors that occur. For example, in the specification below, we are incorrectly dividing a mass flowrate by a molar flowrate.

Specifications:

1 COL1SPEC1 - Column T1 Tray 17 Liquid Flowrate of All Components on a Wet basis in lb/hr / Stream FEED Flowrate of All Components on a Wet basis in lb-mol/hr = 0.030000 within the default tolerance

Sometimes these incorrect units of measurement are minor mistakes that play no impact on our modeling. More often than not however, a mistake in units of measure can be a major problem. A perfect example of this is the Mars Climate Orbiter (MCO)[1]. Launched in 1998 by NASA to study the climate, atmosphere, and surface changes on Mars, communication with the MCO was lost after the MCO passed behind Mars, and the result was a permanent loss of communication and $328 million. When investigated, it was found that the use of English units of measurement instead of Metric units of measurement caused the MCO to have an underestimated spacecraft trajectory.

To ensure that errors like this do not occur, we must use many different ways to help us ensure that our data entry is correct. As NASA's associate administrator for space science at the time of the MCO, Edward Weiler, once put it, "People sometimes make errors. The problem here was not the error, it was the failure of NASA's systems engineering, and the checks and balances in our processes to detect the error. That's why we lost the spacecraft."

Here are some ways that you can help ensure proper data entry in your process simulation files –

1. **Checks and balances:** Check the data that you entered after you entered it. Make sure that it matches what you intended to enter, paying close attention to the units of measurement.
2. **Organize your work:** Have all of your information ready for entry, in the units of measurement that you need for data entry. Keep everything in an organized manner so you can go through everything easily and double check as you go.
3. **Minimize distractions:** Try not to perform tasks like data entry without staying focused. Turn off distractions, like radios, TVs, etc. Close out of other windows, like web browsers or other pieces of work you were working on. Try to keep side conversations with co-workers, friends, or family to a minimum or eliminate them all together. Anything you can do to keep focused will help.
4. **Use "alternate forms" of data input:** Copy/paste information from one source to another. Import data from a data historian. Use data links, such as defines in Pro II, specifications, etc to ensure that the proper data is being entered into the data entry field at all times.
5. **Don't ignore warning messages:** While most warning messages are often there just to keep you informed in case it is something that you should be worried about, others (especially error messages) often tell you of a major problem in your simulation. Be sure to always review any messages, warnings, and/or errors the program gives you.

1 error(s), 0 warning(s), and 0 message(s) detected:

*** error *** Stream '1'-- assay data is invalid. The data must be supplied in ascending order.

[1] – Mars Climate Orbiter Official Website. http://mars.jpl.nasa.gov/msp98/orbiter/

#2: Defaults

Defaults are methods, values, and conditions that are automatically chosen for you by the program when you begin a new simulation model. These defaults will affect the results calculated by your model and every simulation program has them. Most defaults within simulation programs fall under these categories –

1. **Convergence Criteria:** How the program will converge (tolerances, calculation methods for various properties like pressures or flowrates, recycle handling conditions, etc)

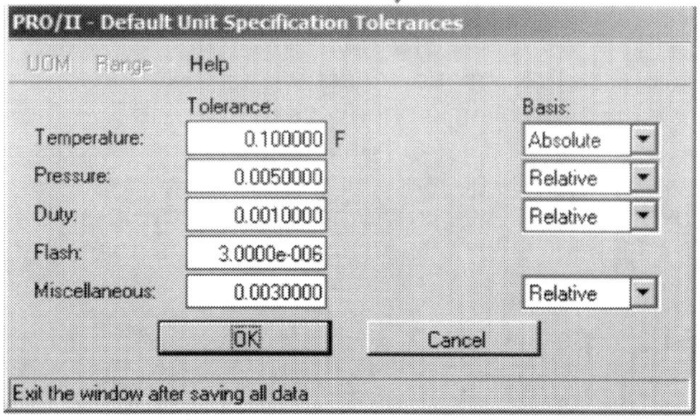

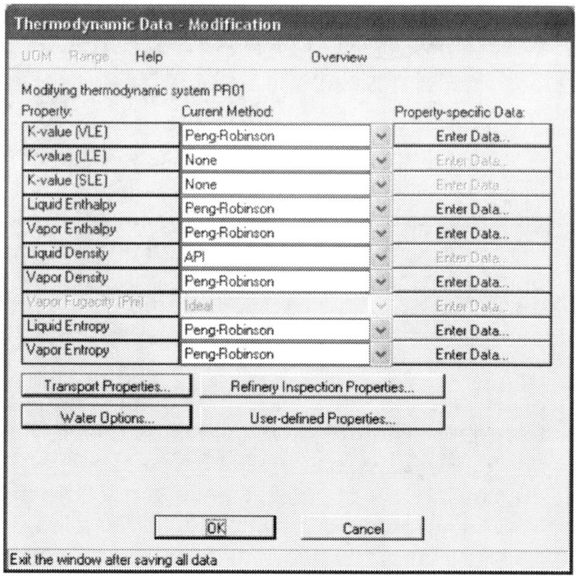

2. **Thermodynamic Method Selections:** When you select a thermo method, you are usually selected a grouping of defaults on how to calculate all of the various properties in the program.
3. **Units of measurement:** Most programs already have the units of measurement selected for you when you start a new model.

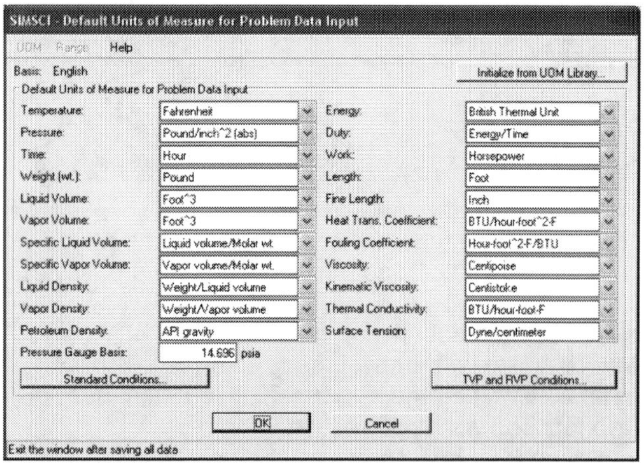

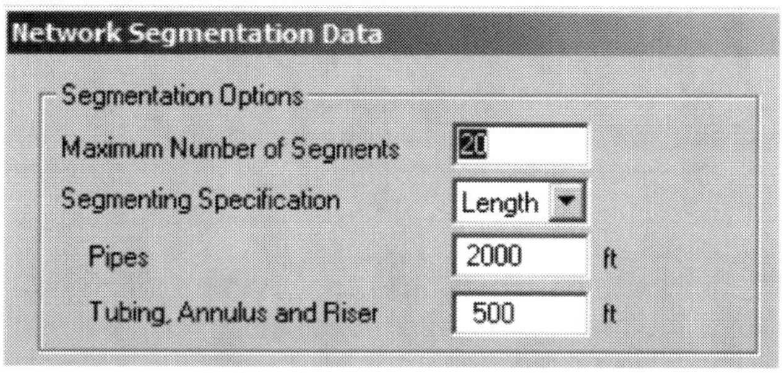

4. **Characterization Methods:** Information like assay processing methods or pipe segmentation data

Defaults within simulation programs are designed to make modeling quicker and easier. There is usually no reason for a user to have to specify every unit of measure for the model or all of the convergence criteria that the program will use for trying to solve the model every single time you begin a new model. Because of this, simulation programs come with defaults already selected to make your life easier.

The problem with defaults is when you didn't realize a default has had an unwanted affect on your model. Take the time when you begin working with a new simulation program to learn the defaults for the program and how those defaults will affect the results you will eventually obtain from the program. If you do not agree with the defaults selected by the program, most simulation programs will allow you to change the default to something else.

#3: Bad Plant Data

Plant data changes over time. What you enter into a simulation program when you initially build a model will not be the same information that is valid in the future. Your feed information may be different, especially in the case of assay data which always changes over time. Your unit operation's process conditions may also be different now than they were when you built the model. Not only does plant data change over time, but measurement information is often very inaccurate. This could be due to faulty or poorly calibrated measurement equipment, leaks in lines, sensor failures, and other issues. Because of these changes and inaccuracies over time, you always have to validate the information that is entered in your model to ensure that you are obtaining the proper results. A simple mass balance around the plant below would show that something does not add up properly!

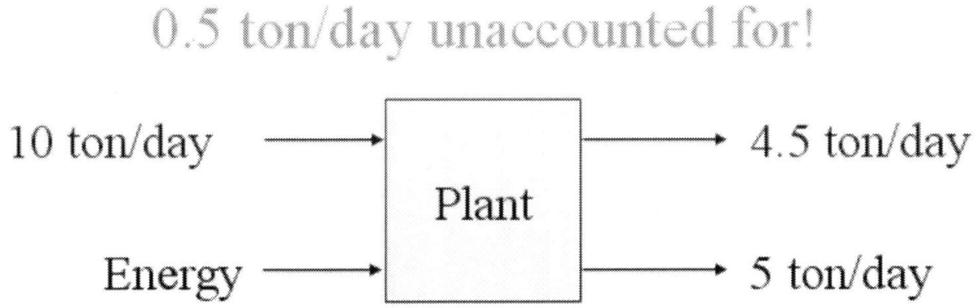

There are programs out there that can help with this data validation and reconciliation. ROMeo is SimSci-Esscor's online optimization program. With it, you can connect your model to your data historian to constantly pull up data measurements into your model and then reconcile those measurements to determine when problems arise within your plant and where those problems are most likely occurring. After running data reconciliation, you can manually transfer and use your reconciled plant data as feed information to our other software, like PRO/II and PIPEPHASE.

If you do not have a data reconciliation program to determine the quality of the plant data that you are given, you can use your engineering judgment to determine if the data is valid. For example, in the process below, depending on the mass balanced performed, we have 3 different flowrates for stream S4. If you look at just the measurement value, you will get 30.6. If you perform a mass balance around node C using the measured values for streams S3 and S5, you obtain a flowrate of 29.3. If you perform an overall mass balance for the process using the measured values given for S1, S5, and S6, you get a flowrate of 28. For the most complete evaluation, if you do reconciliation around all nodes (all streams with MS# shown; S1, S2, and S3 provide enough information to determine S7 = 59 and S8 = 37.9, which then allow you to measure S6 = 21.1), you get a flowrate of 29.3 for S4, same as the balance around node C. Four different mass balances providing different flowrate results for S4. Which is correct? As an engineer, it is up to you to decide, but knowing that this is a possibility and checking for this is how you help ensure good plant data instead of bad. In this situation, most likely 29.3 is correct and this would indicate that the plant data obtained from measurements MS6 and MS4 are incorrect.

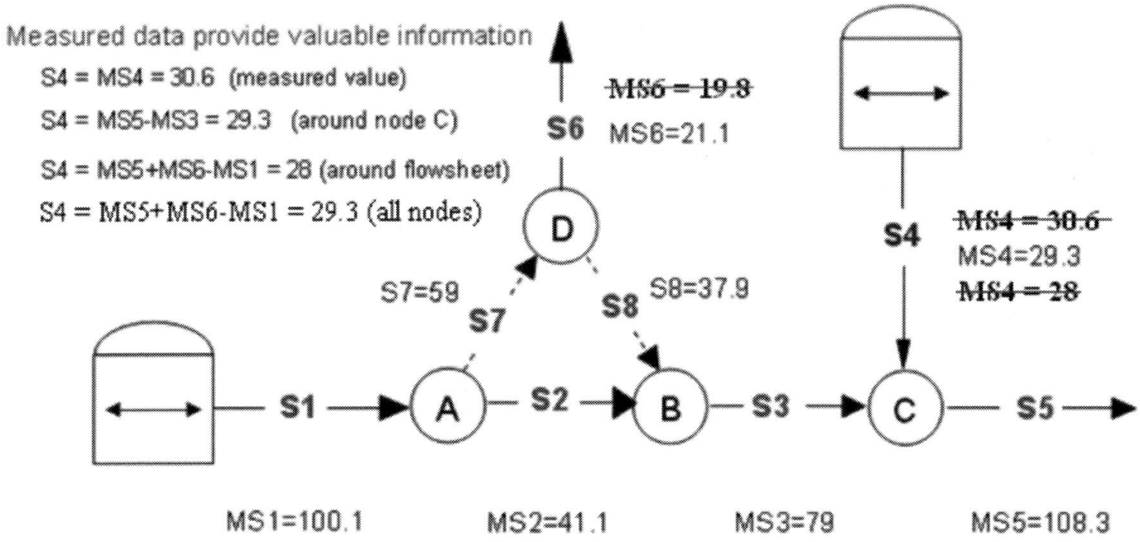

#4: Thermodynamic Method Selection

Proper selection of thermodynamic methods for your simulation is the most important decision that you make when modeling a process. The method you end up choosing could have a drastic change on your results. In the example below, using Peng-Robinson instead of Grayson-Streed for your column would result in a higher duty for your column's condenser and an increased reflux ratio. If you did not know that this was wrong, you could end up designing and purchasing a column much larger than you needed.

Thermodynamic System	Condenser Duty	Reflux/Feed Ratio
Peng-Robinson	-59.6	13.1
Grayson-Streed	-37.3	8.2

Normally, any number of thermodynamic methods are suitable for a given application. It is up to the user to determine the best method to use based on knowledge, personal experience, and best judgment. When making a decision, there are several factors that you will want to consider –

1. **What components are present in the simulation?** Take particular consideration for components that are permanent gases, hydrocarbons, water, and other chemical moieties present.
2. **How much of each component do you have?** Composition plays a large factor on selecting a thermo method. More or less of some components could make the difference between one thermo method and another.
3. **What are you operating conditions (temperature/pressure)?** Only use methods that apply to the temperature and pressure range you are operating at.

Oftentimes, the thermo method you have selected will probably still need some additional input from you as a user for it to most accurately model your system, even if you have selected a thermo system that most closely matches the type of process you are modeling. Most thermo methods do not contain all of the binary interaction parameters for all components in a process simulator. Any time a thermo method does not have the interaction parameters available for a specific pair of components, it treats those components as behaving ideally, which is probably not accurate. You will want to provide any missing interaction parameters to the software rather than letting the program treat these as ideal.

In instances where you have multiple thermo methods selected and used in your model, you are going to want to make sure that you use similar methods or use a thermo-reset unit operation to make sure that none of your mass or energy balances turn out unbalanced. PRO/II provides a thermo-reset unit operation through the SimSci Add-ons Unit Operation on the PFD palette.

Here are some ways to find help when selecting a thermo method –

1. **Manuals:** Most software comes with manuals that are installed on your computer when the software is installed. These manuals usually have guidelines on how to select the proper thermo method and a detailed description of technical equations and information used for each thermo method in the program. PRO/II for example has a Thermodynamic Keyword Manual and a Reference Manual (Volume 1) which both provide descriptions of different thermo and property methods.
2. **Application Library Examples and Other Examples:** Most software programs come with already built sample files that you can use as a reference for thermo selection. Most examples are named after the process that they model, making it easy to find one that is similar to your process. Also, if you have internal models available to you within your company, usually they will be the best way to determine an applicable thermo method for a new model.
3. **Training:** Many companies offer training on thermodynamic selection throughout the year in specific courses related to thermo and physical property calculations. In SimSci training courses, we also briefly discuss thermo selection in most of our basic Intro courses for our software.
4. **Hotline Support:** If you are still unsure of which thermo method to select, technical assistance is usually available for most companies. For SimSci technical support, please call 1-800-SIMSCI-1 (1-800-746-7241) or email support.simsci@invensys.com. We have a thermo expert on staff with over 30 years of thermodynamic experience who can assist you.

#5: Tolerances

Many of the calculations in process simulation are iterative. When these calculations iterate, the program needs to compare the results calculated from one iteration to the results calculated in the next iteration in order to determine whether it is converged to a solution and can stop calculating or if it needs to keep going. This way of comparing results is the basic formula for tolerance values, and simulation programs use tolerances to determine the stopping point. The diagram below shows how a typical tolerance calculation is performed, where N = the value of the current iteration number, N-1 = the value of the previous iteration number, and Ep is the tolerance value. This example is for a pressure calculation.

$$\left|\frac{P^{n} - P^{n-1}}{P^{n}}\right| \leq \varepsilon_{P}$$

There are tolerances for many different properties in simulation programs. Temperatures, pressures, duty, and mass balances are just a few examples of the type of calculations that are dictated by tolerance values. Not only do you have different properties, but different aspects of a model can be dictated by different tolerance values on the same calculation type. For example, you could have an overall temperature tolerance for your flowsheet equal to 0.1 F. You could then set an even tighter tolerance on recycle temperature calculations equal to 0.01 F. You could even go a step further and specify a 0.001 F tolerance for calculations within a given unit operation.

Tolerance values can come in two forms – relative or absolute. A relative tolerance is a percentage comparison (i.e. – 0.01 would be 1% difference between current and previous iteration values). An absolute tolerance means that the difference between the current and previous iteration values must be equal within an exact range (i.e. – 1 would mean that the current and previous iteration must be within a value of 1 from one another, like 101 and 100). Here is a mathematical representation of these, again using pressure as our property –

$$\left|\frac{P_{\text{Calculated}} - P_{\text{Specified}}}{P_{\text{Specified}}}\right| = \text{Relative Tolerance} \qquad P_{\text{Calculated}} - P_{\text{Specified}} = \text{Absolute Tolerance}$$

One of the major mistakes when modeling a process is not understanding or using the correct tolerance values. A looser tolerance may make the file easier and quicker to solve, but also reduces the accuracy of the model. A tighter tolerance may be more accurate, but might be impossible for the model to solve or increase the convergence time exponentially. As an engineer using process simulation, you need to investigate when you have reached the correct level of tolerance. A common suggestion is that when you have reached a converged solution with your model, tighten the tolerances within the model and solve it again. If you notice that your answers have changed significantly, you need to keep tightening the tolerance values until your results do not change much from one iteration to another. Then you will know that you have reached an accurate solution for your model.

Note: When tightening the tolerance values for calculations within your model, you will most likely also need to increase the number of iterations for the model to accommodate for the more iterative calculations that the program will need to perform in order to solve the model to the tighter tolerance.

#6: Recycles

By definition, recycles in process simulation are any time when you are returning to a previously calculated unit operation or operations to iteratively recalculate the unit due to a change made elsewhere in the model. They are very common and can often be difficult parts of a model to converge. When working with a model, it is important to identify and fully understand the recycles that are in your process. Recycles can come in many forms within process simulation. The most common form is the compositional recycle. A compositional recycle is a recycle that returns flow from one unit operation to a previously calculated one. The key identifier for a compositional recycle is the fact that the flow is mixing together with a previous flow and the compositions are changing due to this remixing. Another kind of flow recycle is a thermal recycle. In this case, flow returns to a previously calculated unit operation only to transfer heat and does not actually mix with a previously calculated flow. A good example of this would be a hot stream recycling back to feed the other side of a heat exchanger. In this case, it is transferring heat to the other stream, but there is no mixing of flow or changing of composition. The last type of recycle that you can have in a simulation is a controlled recycle. This non-flow recycle occurs when a unit operation, like a controller or calculator, makes a change to a process condition elsewhere in the model and causes the simulation to recalculate to achieve a specific purpose. While these recycles do not involve flow in their calculations, they can oftentimes be the most difficult and time consuming recycles to converge. Examples of all 3 types of recycles can be seen below –

*Compositional Recycle (**Blue**) and Thermal Recycle (Orange)*

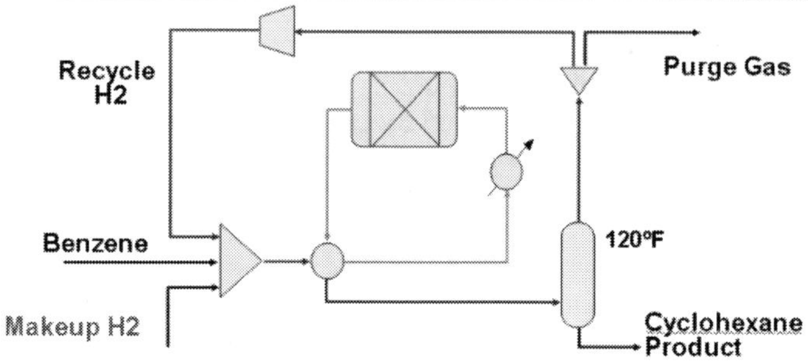

Controlled Recycle

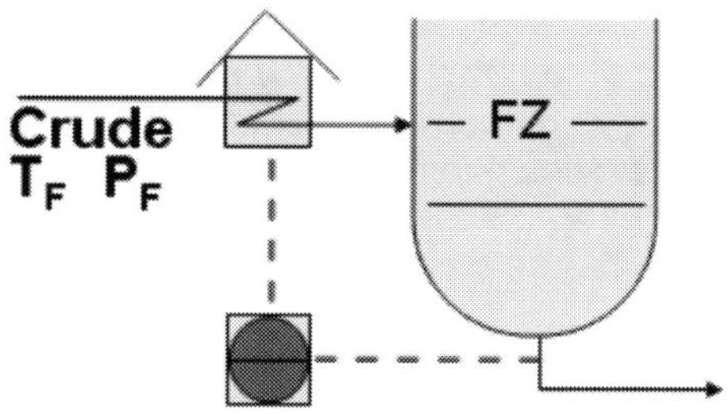

Once you have identified and understand your recycles and how they affect your model, you can begin troubleshooting their convergence. Here are some suggestions on troubleshooting recycle convergence –

1. Eliminate any recycles whenever possible. A thermal recycle is almost always avoidable if process conditions (temperature, pressure, flowrate) are known for the streams within the recycle.
2. Try solving your model without the recycles to see how it behaves, and then manually implement the changes necessary to get the model to where you want it to be with the recycle.
3. Avoid using complex unit operations within recycles. Move these calculations outside of the loop if possible and use reference streams to obtain their feed data.
4. Ensure all components have a way in and out of the recycle. Perform mass balances and use purge or makeup streams where necessary.
5. Use tighter tolerance values on unit operations within recycles than the total recycle tolerances themselves. Remember to increase the number of recycle iterations if you tighten the tolerance values.
6. Provide initial estimates to recycle streams.
7. Define your own calculation sequence and recycle loops.
8. If obtaining a converged solution takes a long time, use acceleration options (Wegstein and Broyden instead of direct substitution). These will help to achieve a solution faster. The chart below is a good example of this. Wegstein achieves convergence in 9 iterations and Broyden in 25.

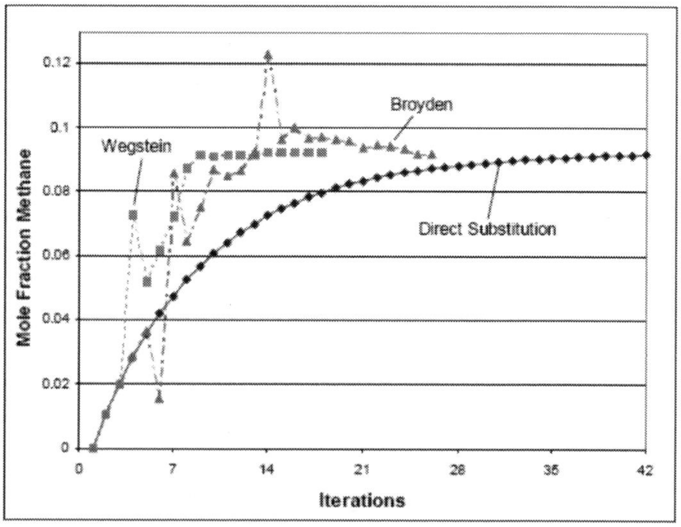

#7: Model Complexity

There are 2 ways to designing a process simulation model –

1. Make it so simple that there are **obviously no deficiencies** anywhere in the model.
2. Make it so complex that there are **no obvious deficiencies** anywhere in the model.

These 2 methods are polar opposites and could mean the difference between obtaining a converged model and saving significant amounts of time or not getting convergence and spending all of your time troubleshooting why it didn't converge. The preferred method is number 1. The general rule of thumb when modeling is to start simple and add complexity a little bit at a time as you go. Each time you add something, converge and save the model before moving on.

Make sure you justify any additions that you make to the model. You can most likely model what you are trying to accomplish without adding every single unit from your PFD into the simulation. When modeling, you should be thinking about what the goal of your simulation is and not about literally translating every real-world piece of equipment into your simulation model. Make the judgment call on what is most important and what the purpose of every unit you add is to your simulation. For example, if your real-life process shows a heat exchanger, a pump and a splitter, but the only thing that's really necessary is that the temperature, pressure and separation are achieved properly, consider using a flash drum to accomplish all three functions instead of using a separate unit operation for each. The flash drum can set the temperature, the pressure and perform your separation all at once.

Finally, think about whether you can use sub-flowsheets or multiple simulation files, if you can reduce or simplify recycles, and if you can eliminate controllers. These will all help reduce the level of complexity of your model.

In summary, don't start off doing this -

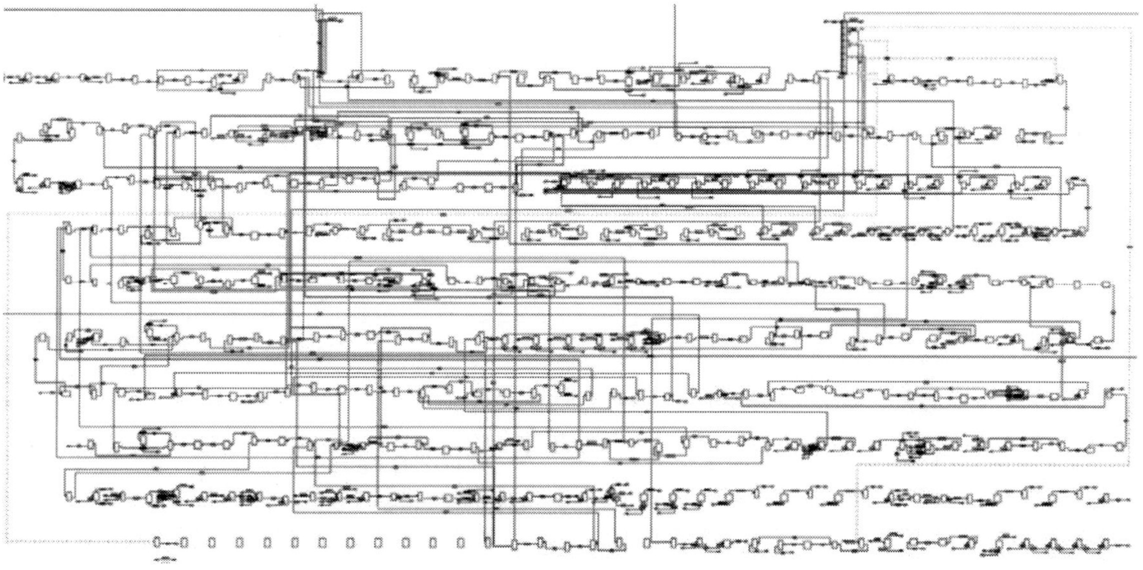

Build up to this a little bit at a time!

#8: Difficult or Impossible Specifications

Specifications within process simulation are everywhere. Every stream or unit operation you place in your model contains specifications. In the simplest form, a specification is a constraint that you place on the model in which the model needs to meet the constraint in order to successfully converge. For streams, these are usually temperatures, pressures, flowrates and compositions. For unit operations, they range from process conditions, like temperatures and pressures, to product conditions, like an overhead composition in a column.

Specifications can be categorized as either an absolute specification or a relative specification. An absolute specification means that the result calculated must be equal to the value that you specified. For example –

Column Overhead Methane Flowrate = 1000 lb/hr

In this example, the column must solve the overhead methane flowrate to be equal to 1000 lb/hr in order for the flowsheet to converge. This is a valid solution as long as there is at least 1000 lb/hr of methane coming in from the feed. When there is not 1000 lb/hr coming in from the feed however, this specification will always fail. A better solution here would be to use a relative specification.

A relative specification provides a little bit more flexibility in achieving a solution. With a relative specification, you tell the program to be within some range of the solution you are trying to achieve. For example –

(Column Overhead Methane Flowrate) / (Total Methane Feed Flowrate) = 99 wt%

In this example, the column has the flexibility to solve to a converged result where the overhead methane flowrate may be a different value every iteration depending on how much methane feed is coming into the column. With this kind of specification, the program should always be able to come to a converged solution no matter how much methane is coming in the feed.

As shown above, many absolute specifications have the ability to become infeasible over time. There are other pitfalls however that you can run into with specifications. Here are a few to consider –

1. **Infinite solutions:** Many times specifying 0 or 100% of something can lead to infinite solutions for the convergence algorithms and make it so the model does not converge.
2. **Azeotropes in columns:** Be careful not to specify a composition specification in a column which is above an azeotrope that you can not break.
3. **High purity specifications:** Very difficult to converge. Often better to provide impurity specifications instead.
4. **Multiple specifications may conflict with one another:** Be aware of what it is that you are specifying. Sometimes specifying one thing makes it extremely difficult or impossible to meet another specification. For example –
    a. Compositional and product flow specifications often clash with one another. Free one up and specify something else.
    b. Specifying all product flows constricts the material balance and gives no room to find a solution. Allow at least 1 product flowrate to be adjusted by the program.
    c. Two compositional specifications for the same product are difficult to converge. Also difficult is having the same component specification in different products.

#9: Assay Pseudo-Component Development

Petroleum assays are sets of distillation data that are used to characterize all of the components that reside in a given crude sample. When you enter the distillation data into a process simulator, the program needs to take this data and convert it into components that it can use when modeling your process. The components that are generated are called pseudo-components, and each pseudo-component represents all of the unknown components that would be within a given boiling point range in the crude.

In order to generate these pseudo-components, the program needs to do several things –

1. Convert the data that you have entered from the basis you entered it in (D86, D1160, or D2887) to TBP equivalent data.
2. Distribute the assay curve into separate cuts. Each cut represents a different pseudo-component.
3. Determine moles, mass, and volume of each cut.
4. Process any light ends
5. Determine the average normal boiling point, gravity, and molecular weight of each cut.
6. Characterize the rest of the component information for each cut.

Each step in this process is a possible point where errors could occur. Luckily, most process simulation programs provide you many options for processing methods and ways to enter the data into the program that can ensure that the pseudo-components that are developed are the most accurate pseudo-component possible for your assay. Here are some things you will want to ensure when properly developing your assay components –

1. **Did you enter your assay data correctly?** Always make sure to check that you used the correct distillation curve basis (TBP, D86, D1160, D2887), % distilled basis (weight or volume), entered the correct NBPs and gravity data vs % distilled, and so forth.
2. **Did you use enough pseudo-component cuts?** View the graph of your curve and see how your current cutpoint set distributes the curve into components. Notice any areas with large % distilled gaps? You probably need more cuts within the temperature range of that % distilled range.

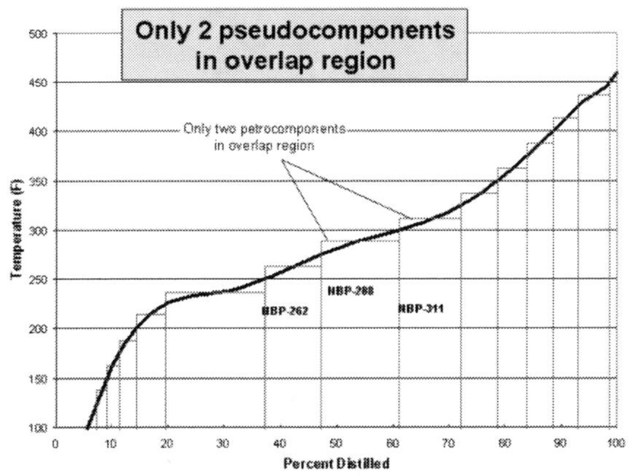

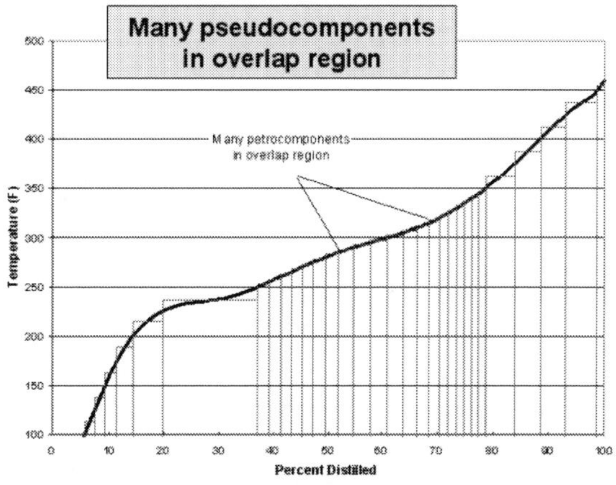

3. **Are you using the correct characterization methods?** There are different options for how the program fits a trend curve through the distillation curve data you enter, how it converts to and from TBP based on how you entered your data, how it calculates the gravity or molecular curves, how it calculates where to make the pseudo-component cuts, ways to help it fit the trend properly through end points, and more. Make sure you are using the most applicable methods for the type of assay you have, and if you are not sure, try switching up the methods to see if they give you a better fit or better results.

#10: Assay Property Blending

When you have more than 1 assay in a simulation model, you need to be concerned with assay blending and how it will affect the results of your file. Assay blending is the mixing of 2 or more groups of components (i.e. – sets of cuts/pseudo-components) in order to calculate averaged component properties for cuts that are in the same boiling point range as one another. Once these new averaged component properties are determined, the program will use these averages from there and on in the simulation model any time it needs to access component properties for a given pseudo-component. All component properties are blended, including normal boiling points, specific gravity, molecular weight and refinery inspection properties, and usually the blended component properties are calculated using a weighted average, like the equation below for normal boiling point –

$$NBP_{blend} = \sum_i NBP_i \cdot Volume\ Fraction_i$$

The below example demonstrates how this blending occurs –

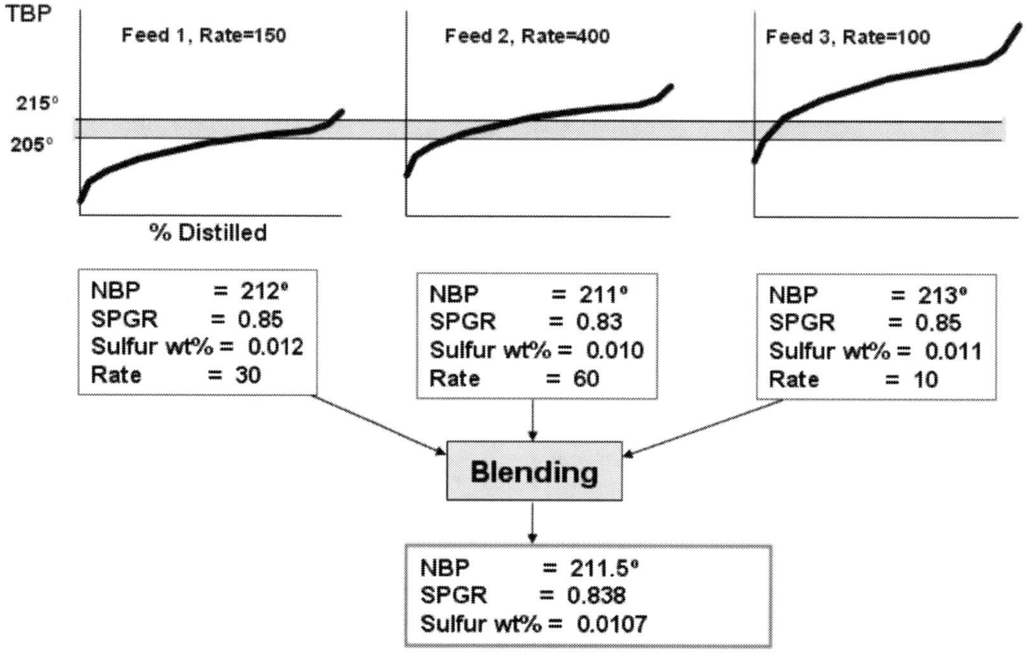

As you can see from the example, given the 3 separate assay curves above, for the boiling point range between 205 and 215 F, the component properties will be blended together and new component properties are calculated. Instead of using the original component property values for this NBP component, it will now use the average properties. This affects all 3 assays.

In order to retain the original assay curve properties, you will need to use different cut point sets for each curve. In PRO/II you would add more sets by going to the Assay Characterization window defining new sets. You can then go into each assay stream and choose that cut point set in the Flowrate and Assay data entry window, shown below. Choosing to use another cut point set will result in more total pseudo-components being generated for the model.

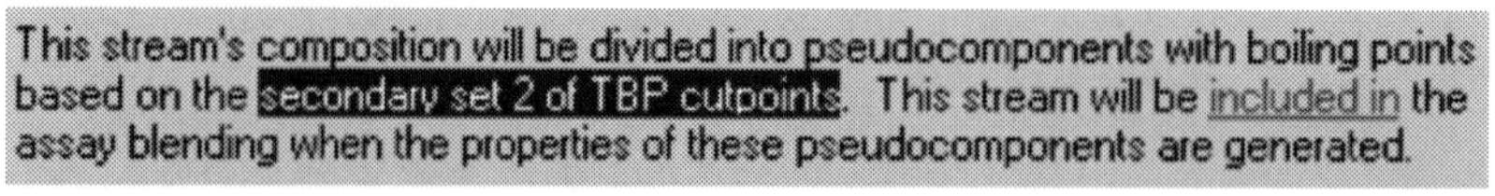
This stream's composition will be divided into pseudocomponents with boiling points based on the secondary set 2 of TBP cutpoints. This stream will be included in the assay blending when the properties of these pseudocomponents are generated.

There is a second option in this paragraph that also allows you to include or exclude the assay stream in the blending calculations. By choosing to exclude the assay from blending, the stream will not contribute its pseudo-component property information to the blending calculations, however, this does not necessarily mean that it will retain its individual pseudo-component properties. The stream itself will still be affected by the blending of other components within the same cutpoint set; it just won't contribute its own property data to the blending.

(599m) Simultaneous Optimization of Energy and Water in Multipurpose Batch Plants

Omobolanle Adekola[1], Jane D. Stamp[1] and Thokozani Majozi[1,2], (1)Chemical Engineering, University of Pretoria, Pretoria, South Africa, (2)Modelling and Digital Science, CSIR, Pretoria, South Africa

Utility and water requirements in many batch plants, such as in the food industry, breweries, dairies, biochemical plants and agrochemical facilities, contribute largely to their overall running costs. Process equipment cleaning is usually associated with large amounts of water, due to the inherent sharing of equipment by different tasks. Hence, efficient use of water for process equipment cleaning is pursued to ensure that the amount of freshwater consumed as well as the amount of wastewater generated is minimized. Similarly, heating and cooling are unavoidable aspects of many facilities, with operations where heat is generated and others where heat is required. It is because of this occurrence that heat integration becomes a possibility.

In the past, wastewater minimization and energy optimization in batch plants have mainly been treated as separate problems in literature. For instance, mathematical formulations that addressed wastewater minimization maximized revenue by minimizing wastewater generation while those developed for energy optimization maximized revenue by minimizing utility usage in the facility. The batch production schedules resulting from each of these formulations do not guarantee that the plant is operated optimally. Consequently, it is required to develop a formulation that caters for opportunities that exist for both wastewater minimization and heat integration. This may result in production schedules that improve the operation of the batch plant when compared to optimizing water and energy separately.

Presented in this study is a formulation that addresses wastewater minimization and energy optimization concurrently, while simultaneously optimizing the batch process schedule. With respect to wastewater minimization, opportunities for direct water reuse and indirect water reuse utilizing wastewater storage and wastewater regeneration are explored. With respect to energy optimization, both direct and indirect heat integration are explored for reducing external utility requirements. The objective is to improve the profitability of the plant by minimizing wastewater generation and utility usage.

(599n) Environmentally Conscious Optimization of Chemical Supply Chains Using Dimensionality Reduction Techniques

Andrei Kostin Sr.[1], Pedro J. Copado-Méndez Sr.[1], **Gonzalo Guillén**[2] and Laureano Jimenez[3], (1)Departament d'Enginyeria Química, Universitat Rovira i Virgili, Tarragona, Spain, (2)Chem. Enginering, Universitat Politécnica de Catalunya, Barcelona, Spain, (3)Chemical Engineering Department, University Rovira i Virgili, Tarragona, Spain

The growing consumer demand for "green" products and toughening environmental protection legislation have boosted research on how to decrease the negative environmental impacts caused by supply chains (SC). As a result, a variety of multi-objective optimization (MOO) models aiming to balance economic and environmental benefits have been proposed so far. The complexity of these models grows rapidly with the number of objectives. For this reason, most MOO models have included only one single environmental indicator, typically the amount of emitted greenhouse gases (GHGs) emitted. However, several researchers [6-8] have argued that this strategy is inadequate, mainly because reductions in GHG emissions might be achieved at the expense of increasing other negative environmental effects.

Aggregated environmental metrics [8-10] have been proposed to provide a holistic picture of the environmental performance while still keeping the model bi-objective, The computation of aggregated metrics involves two main steps: normalization and weighting, which are both applied to a set of single environmental indicators. The aim of normalization is to refer the original impact values to a common basis before being aggregated into a single metric. Weighting procedures range different indicators according to the views of a pannel of experts that reflect the interests of the society or a group of stakeholders. The weakness of aggregation is that it uses fixed normalization and weighting parameters that may not represent the decision-makers' interests. Moreover, when used in a MOO framework, aggregated metrics may change dominance structure of the problem in a manner such that some solutions may be left out of the analysis [11].

An alternative approach for the environmentally conscious MOO of SCs consists of constructing an approximated model where the key environmental objectives are kept and the redundant ones are omitted. So far, the elimination of objectives in environmental MOO problems has largely relied on the decision-makers' preferences, who typically select the most relevant criteria and drop the rest. This approach does not rely on any rigorous analysis, and for this reason may lead to large approximation errors. Objective reduction techniques arose in response to this situation. They allow transforming multi-objective problems with a large number of objectives into a meaningful equivalent with a reduced set of them. Ideally, the reduced representation should preserve the characteristics of the original problem, making it possible to identify the solutions of the original full space model by solving its simplified counterpart.

In this work, we propose an enhanced epsilon-constraint method that integrates an MILP-based strategy for dimensionality reduction, which computes the minimum subset of objectives that preserve the structure of a MOO model to the extent possible. To demonstrate the computational advantages of the proposed integrated strategy, we present two case studies of different complexity. These problems were solved *via* the proposed approach based on dimensionality reduction and the standalone epsilon-constraint method. By comparing the results of these two approaches, we conclude that our method outperforms the traditional epsilon constraint strategy in terms of Pareto quality (measured by the hypervolume indicator) and computational burden.

References

1. Hugo, A., Rutter, P., Pistikopoulos, S., Amorelli, A., Zoia, G., 2005. Hydrogen infrastructure strategic planning using multi-objective optimization. *International Journal of Hydrogen Energy* **30** (15), 1523–1534.

2. Zamboni, A., Murphy, R., Bezzo, F., Shah, N., 2011. Biofuels carbon footprints: Whole-systems optimisation for

GHG emissions reduction. *Bioresource Technology* **102** (16), 7457–7465.

3. Giarola, S., Zamboni, A., Bezzo, F., 2011. Spatially explicit multi-objective optimisation for design and planning of hybrid first and second generation biorefineries. *Computers & Chemical Engineering* **35** (9), 1782–1797.

4. Giarola, S., Shah, N., Bezzo, F., 2012. A comprehensive approach to the design of ethanol supply chains including carbon trading effects. *Bioresource Technology* **107**, 175–185.

5. Akgul, O., Shah, N., Papageorgiou, L., 2012. An optimisation framework for a hybrid first/second generation bioethanol supply chain. *Computers & Chemical Engineering*, In Press.

Scharlemann, J. P. W., Laurance, W. F., 2008. Environmental science — how green are biofuels? Science **319** (5859), 43–44.

6. Vries, S., Ven, G., Ittersum, M., Giller, K., 2010. Resource use efficiency and environmental performance of nine major biofuel crops, processed by first-generation conversion techniques. *Biomass & Bioenergy* **34** (5), 588–601.

7. Cooper, D., Sehlke, G., 2012. Sustainability and energy development: influences of greenhouse gas emission reduction options on water use in energy production. *Environmental Science & Technology* **46** (6), 3509–3518.

8. Guillén-Gosálbez, G., Grossmann, I., 2009. Optimal design and planning of sustainable chemical supply chains under uncertainty. *AIChE Journal* **55** (1), 99–121.

9. Guillén-Gosálbez, G., Mele, F. D., Grossmann, I. E., 2010. A bi-criterion optimization approach for the design and planning of hydrogen supply chains for vehicle use. *AIChE Journal* **56** (3), 650–667.

10. Bojarski, A., Laínez, J., Espuña, A., Puigjaner, L., 2009. Incorporating environmental impacts and regulations in a holistic supply chains modeling: An LCA approach. *Computers & Chemical Engineering* **33** (10), 1747–1759.

11. Brockhoff, D., Zitzler, E., 2010. New developments in multiple objective and goal programming. Vol. 638 of Lecture Notes in Economics and Mathematical Systems. Springer: Berlin, Ch. Automated aggregation and omission of objectives for tackling many-objective problems.

(599o) A Novel Integrated Model for Refinery Production and Pipeline Scheduling

Kailiang Tong and Gang Rong, Department of Control Science and Engineering, Institute of Cyber-System and Control, Zhejiang University, Hangzhou, China

The planning and scheduling in refinery operations attract much attention in recent years. Shah et al.[1] presented a comprehensive review for scheduling, planning and supply chain management in the petroleum refining industry, focusing on the integrated approach from the perspective of enterprise-wide optimization[2]. The key manufacturing operations in refinery include crude oil loading, unloading and mixing, production unit operations, product blending, and oil derivatives distribution. The goal of production scheduling and blending is to provide production profiles, material movements as well as inventory control under quantity and quality constraints. The pipeline system plays a key role in petroleum delivery business. As the annual operating costs are low compared to other transportation modes such as trains and trucks, scheduling of pipeline system has been widely developed. A real pipeline system transfers multiple products to multiple destinations. The main challenge is scheduling the optimal sequences, length and starting time of each pumping run from the refinery to the pipeline, together with the optimal timing of transferring these products from the pipeline to each depot. The complexity of this optimization process arises from the typical operational aspects of these systems such as the cost associated with contamination at product interface.

Many models and algorithms for refinery scheduling have been developed for decades. Neiro et al.[3] developed a general framework for operational planning for petroleum supply chain including production scheduling and product distribution. Shah et al.[4] presented a novel production scheduling model incorporating logistic details, such as minimum run length requirements, fill-draw-delay, one-flow out of blender, sequence-dependent switchovers, maximum heel quantity, and downgrading of better quality product to lower quality. Luo et al.[5] proposed an integrated framework iterating from upper planning optimization model and lower scheduling simulation model to achieve a better cooperation. On the other hands, many pipeline models have been proposed, such as discrete[6] and continuous[7] formulation, diverse pipeline types including single pipeline with several depots along the pipeline[6, 7], tree pipeline structure[8] and pipeline networks[9, 10].

In the pipeline scheduling model, the production profile at refinery is usually assumed to be given by the planning. As a result, a planning without detail pipeline scheduling may provide improper production plan. On the other hand, the product demand for refinery scheduling is assumed to be given, which is different from the one in depots because of the long distant pipeline transfer. As a result, integration of refinery production and pipeline scheduling is urgently needed to achieve a better performance. The integrated model can achieve better results while making sure there is no bottleneck along the refinery supply chain. Many researchers integrate some parts of the refinery scheduling[11, 12]. However, to the best of our knowledge, few consider the integration of production and pipeline scheduling. Tong et al.[13] first attempted to integrate production and pipeline scheduling in refinery. They used a discrete representation model for both time and product packages. The integrated model guarantees the feasibility and yields less cost compared to conventional sequential approaches, but it suffers from heavy computational burden as the large number of binary variables from discretization, especially when tiny time grids and package size are employed for model accuracy. Therefore, improvements are needed from both model and algorithm aspects.

In this work, we propose a novel integrated model including refinery production scheduling and pipeline scheduling. A continuous representation MILP model is considered for short time scheduling problem of production and blending as the upstream of the refinery supply chain. Similar to the work of Shah et al.[4], this model considers specific operations such as minimum run length requirements, fill draw delay and sequence-dependent switchovers. At the downstream of the refinery supply chain, a multi-product pipeline system using continuous MILP formulation is adopted for products delivering. This model is based on the work of Cafaro et al.[7, 14] Typical operations such as lot settling time for product quality control, and customer demands on daily basis are taken into account simultaneously. The goal of this paper is to explore the benefit of integrating production scheduling and pipeline scheduling. The novelty is that it is the first continuous representation for integrated model, which reduces the model size and is more accurate than the discrete one[13]. It also considers typical operations including minimum run length requirement, sequence-dependent changeover, and lot settling periods. The difficulties lie in the threefold. The first one is to determine what to model and the scope of the model. The proposed model considers the refinery production scheduling, product blending and oil derivatives distribution through a pipeline to several depots, whereas crude oil scheduling is omitted to maintain the problem size. Typical logistics in both refinery and depots are taken into account. Secondly, in order to reduce the gap between refinery scheduling and pipeline scheduling, the tank farm in refinery plays a key role communicating two activities. We present a novel modeling skill connecting the time slot and pump batch in tank farm modeling. The last one is related to the large problem size, especially for large number of binary variables in the discrete model. We propose a novel continuous representation model which employs the event slots in refinery and the product injection batches in pipeline to build the model. It explores the continuous manner in scheduling operations and reduced the problem size significantly.

The proposed model is demonstrated by a series of examples. The results show the integrated model not only guarantees feasibility but also reduces operation costs than the sequential approaches. Moreover, the continuous representation is much more efficient than the previous discrete one and obtains the solution within a reasonable time.

1. Shah, N. K.; Li, Z.; Ierapetritou, M. G., Petroleum Refining Operations: Key Issues, Advances, and Opportunities. *Industrial & Engineering Chemistry Research* **2010,** 50, (3), 1161-1170.

2. Grossmann, I., Enterprise-wide optimization: A new frontier in process systems engineering. *AIChE Journal* **2005,** 51, (7), 1846-1857.

3. Neiro, S. M. S.; Pinto, J. M., A general modeling framework for the operational planning of petroleum supply chains. *Computers & Chemical Engineering* **2004,** 28, (6-7), 871-896.

4. Shah, N. K.; Ierapetritou, M. G., Short-term scheduling of a large-scale oil-refinery operations: Incorporating logistics details. *AIChE Journal* **2011,** 57, (6), 1570-1584.

5. Luo, C.; Rong, G., Hierarchical Approach for Short-Term Scheduling in Refineries. *Industrial & Engineering Chemistry Research* **2007,** 46, (11), 3656-3668.

6. Rejowski, R.; Pinto, J. M., Scheduling of a multiproduct pipeline system. *Computers & Chemical Engineering* **2003,** 27, (8-9), 1229-1246.

7. Cafaro, D. C.; Cerdá, J., Optimal scheduling of multiproduct pipeline systems using a non-discrete MILP formulation. *Computers & Chemical Engineering* **2004,** 28, (10), 2053-2068.

8. MirHassani, S. A.; Fani Jahromi, H., Scheduling multi-product tree-structure pipelines. *Computers & Chemical Engineering* **2011,** 35, (1), 165-176.

9. Boschetto, S. N.; Magatão, L.; Brondani, W. M.; Neves-Jr, F. v.; Arruda, L. V. R.; Barbosa-Póvoa, A. P. F. D.; Relvas, S., An Operational Scheduling Model to Product Distribution through a Pipeline Network. *Industrial & Engineering Chemistry Research* **2010,** 49, (12), 5661-5682.

10. Cafaro, D. C.; Cerdá, J., Rigorous scheduling of mesh-structure refined petroleum pipeline networks. *Computers & Chemical Engineering* **2012,** 38, (0), 185-203.

11. Jia, Z.; Ierapetritou, M., Efficient short-term scheduling of refinery operations based on a continuous time formulation. *Computers & Chemical Engineering* **2004,** 28, (6-7), 1001-1019.

12. Guyonnet, P.; Grant, F. H.; Bagajewicz, M. J., Integrated Model for Refinery Planning, Oil Procuring, and Product Distribution. *Industrial & Engineering Chemistry Research* **2008,** 48, (1), 463-482.

13. Tong, K.; Yiping, F.; Gang, R. In *Integrated Model for Refinery Production and Pipeline System Scheduling*, 22th European Symposium of Computer Aided Process Engineering, London, UK, June 17-20, 2012; London, UK, 2012.

14. Cafaro, D. C.; Cerdá, J., Efficient Tool for the Scheduling of Multiproduct Pipelines and Terminal Operations. *Industrial & Engineering Chemistry Research* **2008,** 47, (24), 9941-9956.

(599p) Semantic Web Platform to Support and Automate the Process of Industrial Symbiosis

Franjo Cecelja, Process and Information Systems Engineering Research Centre, University of Surrey, Guildford, United Kingdom

Semantic Web Platform to Support and Automate the Process of Industrial Symbiosis

Franjo Cecelja, Antonis Kokossis, Tara Raafat, Nikolaos Trokanas, Aidong Yang

PRISE, FEPS, University of Surrey, Guildford, Surrey, GU2 7UB, U.K.,

f.cecelja@surrey.ac.uk

Abstract

Industrial Symbiosis is an innovative approach that brings together companies from different business sectors to process waste into energy and/or material with a view to generate revenue and allow sustainable management of industrial waste and by-products, as well as municipal waste [1]. Creating a synergy between these companies to form economically, environmentally and socially viable supply chains is a challenge that requires in-depth knowledge on waste and respective processing technologies [2]. Currently, the whole process of Industrial Symbiosis is practiced manually with performance which intuitively depends on practitioner's ability and hence is increasingly difficult to apply in rapidly elevating industrial diversity.

This paper introduces novel approach and respective design of a semantic web platform that fully automates selection of the Industrial Symbiosis synergies. The platform is based on the semantic formalisms for the waste and processing technology classifications which is further supported by the semantic web services technology in the process of establishing synergies.

Waste and technology classification is formalised through knowledge modelling for which we have selected ontologies that are proven to be flexible to implement, share and reuse, especially on the web. Ontologies are primarily used for knowledge representation in the form of a group of terms organised in a hierarchical structure, taxonomy or subsumption, and enhanced with properties/attributes and restrictions on these properties . We have developed ontologies which represent the knowledge in the domains of waste and respective processing technologies. As such, it is used to i) coordinate the whole process of collecting information from users (registration), and ii) formalising user's description for automated discovery, matching and validating relevance of symbiotic options. User descriptions are also in the form of ontology, service description ontology, hence enabling to establish semantic relevance between waste producers and technology providers by inferring and ontology matching. Final ranking of discovered synergies accounts for economic and environmental factors. Ontological formalism used for service description is based on universally recognized OWL-S framework. In addition, the developed waste and technology ontology is implemented in Ontology Web Language (OWL) which facilitates sharing and reasoning that can also automatically generate new knowledge.

The platform has currently been tested using general datasets available upon various waste producing industries and processing technology providers extracted from representative process industries. The models have also been validated against the same datasets and as part of the web platform.

Key words: Ontology, Semantic Web Services, Industrial Symbiosis

References:

1. Chertow, M., *Industrial Symbiosis*. Encyclopedia of Energy, 2004: p. 407-415.

2. Lehtoranta, S., et al., *Industrial symbiosis and the policy instruments of sustainable consumption and production.* Journal of Cleaner Production, 2011. **19**: p. 1865e1875.

(599q) On the Application of the Cascade Optimization Algorithm in Distributed Computer Networks and Grids

Franjo Cecelja, Process and Information Systems Engineering Research Centre, University of Surrey, Guildford, United Kingdom

On the application of the Cascade Optimization Algorithm in distributed computer networks and grids

Franjo Cecelja[a] , Antonis Kokossis[b], Du Du[a],

[a]*PRISE, FEPS, Unverisity of Surrey, Guildford, Surrey, GU2 7UB, U.K., f.cecelja@surrey.ac.uk*

[b]*School of Engineering, National Technical University of Athens, Zofrafou Campus, 9, GR-15780, Athens, Greece, akokossis@mail.ntua.gr*

Abstract

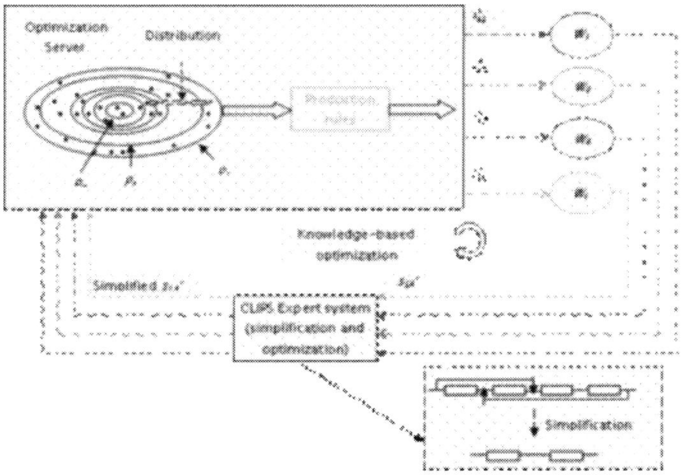

The paper builds on the Cascade Optimization Algorithm (COA) a recent approach with a strong potential to tackle complex problems using modular structures (pools). Figure 1 illustrates the pool structure, the coordination task, as well as flows of calculations in the course of optimization. COA deploys Markov processes by means of peer pools populated by clients propagating the Markov chains, and breaking down (decomposing) peer-to-peer communication through a (purely) parallel task (coordination task). Without any burden to the optimization, sophistication can be added to this task customizing the selection of chains and upgrading intermediate data into information and knowledge. Dotted lines in Figure 1 highlight exploitable information links with the coordination layer.

Figure 1

Using the distributable components of the Cascade Optimization Algorithm the paper presents evidence on the potential of the algorithm to couple with computer grids. Comparisons are drawn with the standalone algorithm as well as Tabu Search - an algorithm comparatively easy to implement on the grids - and does report noticeable improvements in computational performance. The algorithm is further tested on problems of industrial complexity as is the synthesis of chemical reactor networks offering evidence on biocatalytic reactions of several components and reaction mechanisms.

The paper further scopes for improvements in the cascading stages, also means to include self-supervised stages. As anticipated, the results show that the execution time to convergence reduces with increasing the number of CPUs with faster CPUs contributing more than slower ones. As for the self-supervised system, search directions have been intentionally biased by putting different weights on different patterns of intrinsic parameters for each solution according to on-time analytical results of these parameters. Results show that the optimization search converges more quickly by applying more parameters in the production rules in the model.

In conclusion, grid implementation of stochastic optimisation algorithm generally improves the performance in terms of execution time. Bringing in intelligent self-supervised and knowledge-based optimization, it is possible to improve the optimization performance, in particular convergence speed.

Kokossis, A., P. Linke and S. Yang, The Cascade Optimization Algorithm: A New Distributed Approach for the Stochastic Optimization of Engineering Applications, *Ind. Eng. Chem. Res*., 2011, 50 (9), pp 5266–5278

(599r) Logic Cuts for Strengthening the Relaxation Gaps of Network Planning Problems

Ali Fattahi, Industrial Engineering, Koc University, Istanbul, Turkey and **Metin Turkay**, College of Engineering, Koc University, Istanbul, Turkey

An important problem in solving mixed-integer programming problems is the closure of the relaxation gap to verify optimality of the solutions. It is well known that the initial relaxation gap is very critical in closing the relaxation gaps for hard problems using branch-and-bound and branch-and-cut algorithms. Different methods have been adopted for generic mixed-integer programming problems to strengthen the relaxation gaps in the literature such as Gomory cuts and mixed-integer rounding inequalities. These methods are shown to be useful in some cases; however they do not yield a much stronger relaxation for most of the hard problems due to the fact that they do not consider the special structure of the problem. In this talk, we present an approach to generate logic cuts that are specific to problem structure. We illustrate the details of the cut generation procedure for the network planning problems. We test the performance of the generated cuts on a large variety of test problems that are generated randomly. The results indicate that the initial relaxation gap can be reduced significantly resulting in several orders of magnitude reductions in the total CPU time to find the optimal solution.

(599s) Improved Pattern Discovery in Supply Chain Management with Graphic Processing Unit (GPU)

Mai Chan Lau, Chemical & Biomolecular Engineering, National University of Singapore, Singapore, Singapore and R. Srinivasan, Department of Chemical and Biomolecular Engineering, National University of Singapore, Singapore 117576, Singapore

Improved Pattern Discovery in Supply Chain Management with Graphic Processing Unit (GPU)

Lau Mai Chan[a], Rajagopalan Srinivasan[a,b]

[a] Department of Chemical and Biomolecular Engineering, National University of Singapore, Singapore, 10 Kent Ridge Crescent, Singapore 119260, Singapore

[b]Process Sciences and Modeling, Institute of Chemical and Engineering Sciences, Pesek Road, Jurong Island, Singapore 627833, Singapore

Abstract

While the use of supply chain management (SCM) is getting growing attention from all range of businesses, researchers are continue expanding the scope of supply chain management in numerous dimensions so as to achieve further improvement in organizational performance as well as to sustain the competitive position. Advanced supply chain management system attempts to embrace as many factors which are having potential impact on business decisions from various aspects, including operational, purchasing and delivery. The scope expansion is also going beyond the firm as the hierarchy of SCM has been observed growing externally which include indirect entities, for instance indirect suppliers which are one level higher than the direct suppliers. The adoption of supply chain management system entails business entities the competitive advantage only if meaningful and relevant knowledge are captured, which in turn enables reliable prediction to be made on business. However, the expansion of SCM scope has led to the generation of huge amount of data which makes the knowledge discovery process difficult. Owing to the limitation of computing resources, the conventional sequential data processing techniques are inefficient if not unable to handle the gigantic amount of data. As a result, parallel computing technologies evolve to be a potential solution for big data analysis. Among all the parallel computing resources such as supercomputers, grids and cluster, our group is particularly interested in harnessing the computing power from General Purpose Graphic Processing Unit (GPGPU) because of the relatively low cost and availability as commodity. On top of that, Graphic Processing Units consist of a number of Single Instruction Multiple Data (SIMD) multi-processors which map well with the implementation of many data analysis and pattern discovery methods. The aim of this work is to develop GPU parallel algorithm(s) which is (are) able to capture higher quality knowledge for business decision making purpose by having higher degree of efficiency. The quality and efficiency of the parallel algorithm will be examined with a simulated SCM system.

Keywords: Graphics Processing Unit (GPU) parallel computing, Supply Chain Management (SCM), Pattern Discovery

(599t) Comprehensive Investigation of Fault Detection Based On the Statistics Pattern Analysis Framework

Hector Galicia, Department of Chemical Engineering, Auburn University, Auburn, AL, Jin Wang, Auburn University, Auburn, AL and Qinghua He, Department of Chemical Engineering, Tuskegee University, Tuskegee, AL

The increasing demand for safer and more reliable systems in modern process operation leads to the rapid development of process monitoring techniques. With large number of variables measured and stored automatically by distributed control systems (DCS), multivariate statistical process monitoring approaches have been developed and successfully applied to monitor various industrial processes. Currently, Principal Component Analysis (PCA) based process monitoring methods have gained wide application in chemical and petrochemical industries [1-2]. PCA-based monitoring methods can easily handle high dimensional, noisy and highly correlated data generated from industrial processes [3-5], and provide superior performance compared to univariate methods such as Shewhart, CUSUM and EWMA charts. However, there are cases where PCA do not perform well due to the following reasons. First, PCA only considers the mean and variance-covariance of the process data, and lacks the capability of providing higher-order representation for non-Gaussian data. Second, the control limits of Hotelling's T^2 and the squared prediction error (SPE) charts are developed based on the assumption that the latent variables follow a multivariate Gaussian distribution. When the latent variables are non-Gaussian distributed due to the process nonlinearity or other reasons, using Hotelling's T^2 and SPE may be misleading.

In our previous work [6-9], a new multivariate statistical monitoring framework termed statistics pattern analysis (SPA) was developed. SPA addresses some challenges that cannot be readily handled by the commonly used multivariate statistical methods such as multiway-PCA for batch processes [6]. In addition, the SPA framework has also shown better monitoring performance in continuous processes compared to other linear and nonlinear monitoring methods using a moving window approach [7].

The major difference between the traditional multivariate monitoring methods such as PCA and SPA-based fault detection methods is that SPA monitors the statistics of the process variables (e.g. mean, variance, autocorrelation, cross-correlation, etc.) rather than the process variables themselves. In other words, in PCA singular value decomposition (SVD) is applied to the original process variables to build a model for normal operation, and the new measurements of the process variables are projected onto the PCA model to perform fault detection; while in SPA, SVD is applied to the computed statistics pattern matrix (i.e. various statistics of the process variables) of normal operating data to build the model, and the statistics calculated using the new measurements are projected onto the model to perform fault detection. In this way, different statistics that capture the different characteristics of the process can be selected to build the model for normal process operation, and higher order statistics can be utilized explicitly to cope with non-linearity and non-Gaussianity.

In this work, the key different aspects of fault detection for continuous processes, such as detection rate and false alarms rates are studied in detail using a simple benchmark simulation case study and a more complicated nonlinear quadruple-tank case study. Different faults with different characteristics are used to make a thorough analysis of the fundamental reasons of the performance improvement of SPA compared to other monitoring methods including PCA, dynamic PCA (DPCA), kernel PCA (KPCA), independent component analysis (ICA). In addition, since the application of SPA to monitor continuous processes is based on a window approach, detection delay may be a concern. However, our previous results [8-9] indicate that the detection delay associated with SPA is not significant compared with other linear and nonlinear monitoring methods. In this work, a comprehensive analysis of the insignificant detection delay associated with SPA is performed. Both simulated case studies with different types of faults are used to assess and comprehensively compare the different aspects associated with an accurate and prompt detection of faults using different monitoring methods including PCA, DPCA, KPCA, ICA and SPA.

[1] Jolliffe, I.T.,Principal Component Analysis. *Springer*, 2002.

[2] Wise B.M., et al. Principal component analysis for monitoring the West Valley liquid fed ceramic melter. *Waste Management'88 Proc*., 1988, 811-818

[3] Kourti, T. & MacGregor, J. F. Process analysis, monitoring and diagnosis, using multivariate projection methods. *Chemometrics and Intelligent Laboratory Systems*, 1995, 28, 3-21.

[4] Qin, S. J. Statistical process monitoring: basics and beyond. *Journal of Chemometrics*, 2003, 17, 480-502.

[5] Chiang, L. H.; Russell, E. L.; Braatz, R. D. Fault Detection and Diagnosis in Industrial Systems; *Springer: London*, 2001.

[6] He, Q. P. & Wang, J. Statistics pattern analysis: A new process monitoring framework and its application to semiconductor batch processes. *AIChE J*, 2011, 57, 107-121

[7] Wang, J. & He, Q. P. Multivariate Statistical Process Monitoring Based on Statistics Pattern Analysis. *Industrial & Engineering Chemistry Research,* 2010, 49(17), 7858-7869.

[8] Galicia, H.J., He Q.P., & Wang, J. Statistics pattern analysis - fault detection and diagnosis. *In Proceedings of the ISA Automation Week,* Mobile, AL, USA, 2011.

[9] Galicia, H.J. Q.P. He, and J. Wang. Statistics Pattern Analysis based fault detection and diagnosis. *CPC VIII Conference*, 55, 2012.

(599u) Capacity Planning in Novel Pharmaceutical Supply Chains Under Clinical Trials and Demand Uncertainties

Arul Sundaramoorthy, James M.B. Evans and P.I. Barton, Department of Chemical Engineering, Massachusetts Institute of Technology, Cambridge, MA

Upon successful completion of clinical trials and the FDA process, a new drug is introduced commercially (launched) into the market under patent protection. The launch phase includes capacity planning for meeting customer demands, and optimizing the production and inventory from launch to peak demands. Since time to market is the major driver in the pharmaceutical industry, decisions pertaining to the launch phase are usually made in parallel with the development process. Clearly, these decisions have to be made under the uncertainties of clinical trials and market demands. Thus, a holistic framework to facilitate the above strategic decision-making process in an integrated manner is highly desired (Shah, 2004).

Existing approaches for simultaneous drug development and capacity planning were developed for traditional pharmaceutical supply chains in which the primary and secondary production facilities are operated in geographically different locations (Maravelias and Grossmann, 2001; Shah, 2004). Furthermore, the batch mode of manufacturing has dominated the pharmaceutical industry for decades. Recently, the Novartis-MIT Center for Continuous Manufacturing has embarked on an innovative project to shift from the batch mode of manufacturing to the continuous mode, where raw materials through the active ingredients to the finished products are processed seamlessly in an integrated facility. Such an integrated end-to-end production scheme is very promising for several reasons: a) low costs of inventory, logistics, and production, b) short supply-chain cycle times, c) less exposure to supply-chain disruptions, and d) resilient supply chain. Thus, the option of continuous integrated manufacturing in the pharmaceutical industry brings in additional challenges and opportunities in the context of drug development and investment strategy. In this work, we develop a holistic optimization under uncertainty framework to address capacity planning for pharmaceutical supply chains that employ novel integrated production schemes. Since the outcomes of clinical trials as well as the market demands for products are uncertain, the problem naturally leads to a stochastic programming problem, which is formulated as a mixed-integer linear programming (MILP) model. However, the model size can grow dramatically with the number of scenarios, resulting in intractable large-scale MILP models. The underlying problem structure motivates a novel solution method to solve these large-scale models, and we illustrate the application of the proposed framework using several case studies.

References

Shah, N., 2004. Pharmaceutical supply chains: key issues and strategies for optimization. Comput. Chem. Eng. , 28, 929–941.

Maravelias, C.T., Grossmann, I.E., 2001. Simultaneous planning for new product development and batch manufacturing facilities. Ind. Eng. Chem. Res., 40, 6147–6164.

(599v) Fast Kernel Density Estimation Using Orthogonal Polynomials and Goertzel DFT

Sridhar Ungarala, Chemical and Biomedical Engineering, Cleveland State University, Cleveland, OH

Process monitoring methods to identify abnormal conditions in the operations of manufacturing processes are generally based on statistical principles. Statistical methods such as principal component analysis (PCA) and partial least squares (PLS) have gained widespread acceptance within process industries (Qin, 2003). The fundamental statistical measure to describe the behavior of process data is a probability density function (pdf) of the data. The problem of estimation of the pdf from data is relevant to routine data analysis problems in many disciplines of engineering, statistics and economics. Methods of statistical process control and process monitoring rely on the availability of a knowledge of the data pdf, which is generally a non-Gaussian pdf in many real world situations.

The central problem in statistical process monitoring may be posed as: given a set of N random numbers, estimate the underlying pdf to be evaluated on a chosen set of N support points of the density. A non-parametric estimate of the pdf is given by the kernel density estimator (KDE). The KDE is implemented as a normalized sum of kernel functions of a chosen width, known as bandwidth, each of which is centered on the process data points. The standard Gaussian function is a commonly employed kernel, although the exact form of the kernel is of less importance to the accuracy of the estimated pdf than the choice of the bandwidth. For example, Safavi et al. (1997) used wavelets as kernel functions and Johnston and Kramer (1994) used elliptical basis functions. Since these two publications, there is remarkably little attention given to improving the KDE among the process systems engineering community, given the importance of the pdf estimate for applications in routine process operations. Recently however, there is some interest in the use of KDE for state estimation by moving horizon estimator (MHE). The objective function of the MHE includes an arrival cost term summarizing the uncertainty at the beginning of each horizon, which is traditionally derived from a Gaussian pdf assumption. The use of KDE is suggested by Ungarala (2009) and Lopez-Negrete et al. (2011) for computing the arrival cost from a non-Gaussian pdf estimate.

The KDE procedure is a convolution of the data with the kernel, assumed here as the Gaussian function, requiring computation in O(NN). Silverman (1982) showed that the fast Fourier transform (FFT) can be used to implement KDE efficiently in O(NlogN) than the canonical discrete Fourier transform (DFT) in O(NN), and the FFT based KDE has become the standard method since. While it is computationally efficient, accurate estimation of the pdf by KDE is critically dependent on the bandwidth. Small bandwidth results in small bias but large variance and the estimate may suffer from high frequency flutter. Large bandwidth yields small variance while increasing the bias and the estimate may smooth out local detail due to dominant low frequency components (Silverman, 1986). The selection of the bandwidth according to various optimality criteria have been proposed (Jones et al., 1996). One of the many techniques is aimed to minimize the asymptotic mean integrated error (AIMSE) between the estimate and the true pdf. However, the minimizing bandwidth depends on a functional of the unknown pdf's second derivative. A KDE procedure for the estimation of the second derivative requires an optimal choice of its own bandwidth. In essence, the optimal choice of the bandwidth for a KDE estimate requires an estimate of the second derivative of the unknown density, and the optimal choice of the bandwidth of the kernel density second derivative requires the estimate of the fourth derivative of the unknown density and so on (Hardle et al., 1990).

In this paper a novel approach is proposed for a fast iterative method for the selection of the AIMSE minimizing bandwidth by estimating the pdf's second derivative and the implementation of the Fourier transform is sped up by an algorithm faster that the FFT under certain conditions. The FFT has several variants that have gained a formidable number of applications. However, a judicious selection of other incarnations of the DFT can prove to be more appropriate in particular situations. The mandated computation of Fourier coefficients at all allowable frequencies by FFT is by no means the most efficient pathway if a problem requires coefficients only at a ``few'' frequencies. In many pdf estimation problems, one often encounters pdfs that may be characterized by just a few low frequency harmonics, for example skewed Gaussian pdfs and dual or triple mode pdfs. In this context, a computationally efficient alternative to the FFT is the Goertzel algorithm for DFT (Goertzel, 1958). The algorithm computes a single point DFT at a selected frequency using a second order recursive impulse response (IIR) filter. The iteration is performed on real numbers and the output is calculated using a single complex multiplication. The algorithm for M nonzero frequencies requires computation in O(MN). However, it only requires about one quarter of real multiplications and about one half of real additions compared to the canonical DFT of O(MN) complex multiplications. For M<logN, the Goertzel algorithm is more efficient than the FFT.

The proposed fast algorithm for optimal bandwidth is based on orthogonal projections of the pdf on families of orthogonal polynomials such as Gegenbauer polynomials. There are certain inherent computational conveniences in employing a polynomial approximation for the pdf estimate. For a given number of terms in the sums, evaluation of the polynomial requires only real multiplications and additions compared to complex multiplications in trigonometric polynomials or the Fourier series. The differentiation and integration of polynomials is also fast and straightforward. In the proposed fast algorithm, the pdf is at first assumed to be represented by a truncated trigonometric polynomial or a partial Fourier sum by computing only the first few low frequency Fourier coefficients using the Goertzel algorithm for each coefficient or the FFT if necessary. The smoothing factors of the Fourier coefficients are initialized by a pilot bandwidth chosen according to the Silverman's rule of thumb, with the assumption that the unknown pdf belongs to the Gaussian family of pdfs. The resulting summation of the low frequency components of the density may or may not yield an adequate representation of the density. As a post processing step, the density is then assumed to be a partial Gegenbauer polynomial sum, whose coefficients are recovered as linear combinations of the previously computed Fourier coefficients. The relationship between the polynomial coefficients and Fourier coefficients is a standard result exploited by researchers focused on the resolution of Gibbs phenomenon (Gottlieb and Shu, 1997). The second derivative of the density polynomial is then used to estimate the density functional in the minimum AIMSE bandwidth formula to determine an updated value of the bandwidth. The whole procedure is iterated by recomputing the smoothing factors. As the iterations continue, the assumption of an underlying Gaussian density is progressively weakened.

In order to test the performance of the proposed algorithm, the battery of test pdfs published by Marron and Wand (1992) are used. These are fifteen pdfs expressed as Gaussian mixture pdfs to mimic frequently encountered pdfs in real life situations. In numerical experiments it is seen that all of these pdfs can be recovered accurately from the computation of just a few low frequency Fourier harmonics of the data, i.e. M<<N/2 for large N. Furthermore, it is seen that the fast algorithm handles data sets of size N=150,000 easily, incurring computation time in the order of milliseconds on today's computers. In conclusion, the proposed fast polynomial based KDE is a contribution towards improving the estimation of non-Gaussian pdfs from data sets, which may be employed for statistical process monitoring applications.

References:

Goertzel, G (1958) An algorithm for the evaluation of finite trigonometric series. Amer. Math. Month., 65, 33-35.

Gottieb, D. and C. Shu (1997) On the Gibbs phenomenon and its resolution. SIAM Rev., 39, 644-668.

Hardle, W., J. S. Marron and M. P. Wand (1990) Bandwidth choice for density derivatives. J. R. Statist. Soc. B, 52, 223-232.

Johnston, L. P. M. and M. A. Kramer (1994) Prabability density function using elliptical basis functions. AIChE J., 40, 1639.

Jones, M. C., J. S. Marron and S. J. Sheather (1996) A brief survey of bandwidth selection for density estimation. J. Amer. Statist. Assoc., 91, 401-407.

Lopez-Negrete, R., S. C. Patwardhan and L. Biegler (2011) Constrained particle filter approach to approximate the arrival cost in moving horizon estimation, J. Proc. Control, 21, 909-919.

Marron, J. S. and M. P. Wand (1992) Exact mean integrated square error. The Ann. Statist., 20, 712-736.

Qin, S. J. (2003) Statistical process monitoring: basics and beyond. J. Chemometrics, 17, 480-502.

Safavi, A. A., J. Chen and J. A. Romagnoli, Wavelet-based density estimation and application to process monitoring. AIChE J., 43, 1227-1241.

Silverman, B. W. (1982) Kernel density estimation using the fast Fourier transform. Stat. Algorithms, 65(1) 93-99.

Silverman, B. W. (1986) Density Estimation for Statistics and Data Analysis. Chapman & Hall, New York, NY.

Ungarala S. (2009) Computing arrival cost parameters in moving horizon estimation using sampling based filters. J. Proc. Control, 19, 1576-1588.

(599w) Optimal Placement of Gas Detectors in Process Facilities Using Conditional-Value-At-Risk

Sean Legg[1,2], Alberto Benavides-Serrano[2], M. Sam Mannan[1,2] and Carl D. Laird[2], (1)Mary Kay O'Connor Process Safety Center, Texas A&M University, College Station, TX, (2)Chemical Engineering, Texas A&M University, College Station, TX

Optimal Placement of Gas Detectors in Process Facilities Using Conditional-Value-at-Risk

Sean Legg, Alberto Benavides, Sam Mannan, Carl Laird

Gas detection systems are an integral part of modern process safety systems. These systems are reliant upon intelligent placement of the gas detectors to provide effective and timely response. A multi-scenario, mixed-integer programming (MILP) formulation for the optimal placement of gas detectors in petrochemical facilities is presented here. Early development of a basic MILP formulation for gas detector placement was presented by Legg et. al. (2012). The formulation was designed to minimize the expected detection time across the full set of scenarios. Additional constraints were added to enforce a minimum coverage area between sensors to improve solution resiliency in the face of unanticipated scenarios. In this work, we also discuss a modified formulation to improve tail-behavior in the distribution of detection times for the scenarios by considering Conditional-Value-at-Risk (CVaR). All data used in the problem scenarios were generated using the computational fluid dynamics software FLACS (GexCon 2011).

Here, three MILP formulations are presented: minimization of the expected detection time (SP), minimization of the expected detection time considering coverage (SPC), and minimization of the expected detection time considering CVaR (SP-CVaR). Results for each of these formulations are compared to a placement based solely on coverage (C). We include a comparison of all formulations on the entire scenario set and the results based on sampled subsets of the full scenario set. For the entire set, we show that each of the mixed-integer formulations greatly outperform the placement based solely on coverage. Additionally, results comparing the distribution of detection times from (SP-CVaR) and (SP) shows that the addition of a constraint on CVaR greatly improves the tail-behavior of the detection time distribution. By determining a placement with a subsample from the full scenario set, we can use the remaining scenarios to evaluate the resiliency of this solution to unanticipated scenarios. (SPC) shows significantly more resiliency to unknown scenarios than either (SP) or (C). Each of the results presented show that an optimization based approach to the placement of gas detectors shows significant promise in improving modern safety systems.

References:

GexCon. (2011). FLACS CFD Disperson Modeling and Explosion Software. http://gexcon.com/FLACSoverview.

Legg, S., Siirola, J., Watson, J.P., Davis, S., Bratteteig, A., Laird, C. (2012). A Stochastic Programming Approach for Gas Detector Placement in Process Facilities. Proceedings of the 2012 FOCAPO Conference, Savannah, January, 2012.

(599x) Energy Integration of Batch Processes Using a Robust Scheduling Platform

Esmael Reshid Seid and Thokozani Majozi, Chemical Engineering, University of Pretoria, Pretoria, South Africa

Utility requirements in many batch plants, such as in the food industry, pharmaceuticals, breweries, dairies, biochemical plants and agrochemical facilities, are a major contributor to overall running costs. Heating and cooling are unavoidable aspects of many chemical processing facilities, with operations where heat is generated and others where heat is required. It is because of this occurrence that heat integration becomes a possibility for reducing utility requirements.

Heat integration and scheduling have mainly been treated as separate problems in literature. The batch production schedules resulting from these formulations do not guarantee that the plant is operated optimally. Consequently, a formulation is required which integrates the task scheduling and heat recovery problems into a unified framework.

Presented in this contribution is a formulation that addresses the optimization of heat, while simultaneously optimizing the batch process schedule. In this work the novel scheduling framework of Seid and Majozi (2012) is used as a basis for capturing time. It has been proven that this scheduling framework has better computational efficiency and yields better objective value compared to previous models in literature. Both direct and indirect heat integration are considered as well as optimization of the heat storage size. Application of the proposed method to an agrochemical facility has shown a saving of more than 75% in external utility consumption.

Keywords: heat integration, heat storage, multipurpose batch plant

(599y) Fault Detection and Identification Based On Kalman Filtering Under Randomized Sensing Strategies

Kris Villez, School of Chemical Engineering, Purdue University, West Lafayette, IN, Venkat Venkatasubramanian, Chemical Engineering, Columbia University, New York, NY and Humberto E. Garcia, Idaho National Laboratory, Idaho Falls, ID

Fault Detection and Identification based on Kalman filtering under randomized sensing strategies

K. Villez, H. Garcia, V. Venkatasubramanian

Classic strategies for fault detection and identification (FDI) are focused on random and unintended faults in process systems. However, in an increasingly cyber-linked world, intended malicious attackes are an increasing reason for concern in safety-critical systems such as nuclear power generation systems, destillation plants and distribution systems for gas, oil and power. Recently, a probabilistic approach to sensor selection has been proposed to randomize the selected variables for which measurements are sent through a given channel. In doing so, one attempts to thwart cyber-attacks which take form in manipulating data so to mimic a non-existing situation. This can consist of emulating perfect operation when the actual system status is of a lower quality so to increase the probability of a dangerous event or in emulating an abnormal situation when there is none in reality so to promote an emergency shutdown and associated risks and costs. This is a promising approach to achieve resilience against cyber-attacks. However, this results in a time-varying nature of the obtained observations. In this contribution, an existing Kalman-filter based strategy is adjusted to account for this time-varying nature.

For a given linear state-space system, the equations with default monitoring strategies, i.e. with all states measured are written as follows:

$\mathbf{x}(k) = A \,.\, \mathbf{x}(k-1) + B \,.\, \mathbf{u}(k-1) + F \,.\, \mathbf{v}(k-1)$ (Eq. 1)

$\mathbf{y}(k) = C \,.\, \mathbf{y}(k) + D \,.\, \mathbf{w}(k)$ (Eq. 2)

In normal operational conditions, the input and output disturbances are distributed as white noise sequences:

$\mathbf{w} \sim N(0, Q)$ (Eq. 3)

$\mathbf{v} \sim N(0, R\,)$ (Eq. 4)

Equation (1) is modified as follows for an input fault as of the introduction of the fault start time, k_f:

bias type: $\mathbf{v}_f(k) = \mathbf{v}(k) + \mathbf{g} \,.\, b \mid k>= k_f$ (Eq. 5)

drift type: $\mathbf{v}_f(k) = \mathbf{v}(k) + \mathbf{g} \,.\, b \,.\, (k\text{-}k\_f) \mid k>= k_f$ (Eq. 6)

stuck type: $\mathbf{v}_f(k) = \mathbf{v}(k) + \mathbf{g} \,.\, (\mathbf{v}(k_f)\text{-}\mathbf{v}(k)) \mid k>= k_f$ (Eq. 7)

In these equations, b is a scalar value which represents the magnitude of the fault. The column vector g is called the signature vector and consists of zeros and ones and determines which input is affected. For instance for a fault in the first input: g = [1 0]' . With these equations, one adds a deterministic sequence to the zero-mean normal noise sequence. This modified disturbance input, vf(k), is used to replace $\mathbf{v}_k$ in Equation (1).

Similarly, the output faults are defined as follows:

bias type: $w_f(k) = w(k) + g * b \mid k >= k_f$ (Eq. 8)

drift type: $w_f(k) = w(k) + g * b * (k-k_f) \mid k >= k_f$ (Eq. 9)

stuck type: $w_f(k) = w(k) + g * (v(k_f)-v(k)) \mid k >= k_f$ (Eq. 10)

The scalar b and vector g are defined in the same way as for inputs. The modified disturbance input $\mathbf{w}_f$ is used to replace $\mathbf{w}$ in Equation (2).

The Fault Detection and Identification (FDI) scheme based on the Kalman filter consists of three major components. This includes (1) Kalman filtering, (2) Fault Detection Test and Fault Confirmation Test and (3) Fault Identification. We refer to [1-3] for details. In the presented work, we will elaborate on the effects of sensor selection probabilities, p_j, on the performance of the Fault Detection and Identification method.

1. Prakash, J., Patwardhan, S. C., & Narasimhan, S. (2002). A supervisory approach to fault-tolerant control of linear multivariable systems. Industrial & Engineering Chemistry Research, 41, 2270–2281.

2. Prakash, J., Narasimhan, S., & Patwardhan, S. C. (2005). Integrating model based fault diagnosis with model predictive control. Industrial & Engineering Chemistry Research, 44, 4344–4360.

3. Villez, K., Srinivasan, B., Rengaswamy, R., Narasimhan, S., Venkatasubramanian, V. (2011). Kalman-based strategies for Fault Detection and Identification (FDI): Extensions and critical evaluation for a buffer tank system. Computers and Chemical Engineering, 35, 806–816.

(604a) Synthesis of Biorefinery Systems for Lignocellulosic Biorenewables

Karthik Ramadoss[1], Wenkai Li[2], **Iftekar A. Karimi**[1] and Pui Kwan Wong[3], (1)Dept of Chemical & Biomolecular Engineering, National University of Singapore, Singapore, Singapore, (2)Graduate School of International Management, International University of Japan, Niigata, Japan, (3)Institute of Chemical and Engineering Sciences, Agency for Science, Technology and Research (A*STAR), Singapore, Singapore

Our society is actively looking for sustainable alternatives to meet the increasing energy and product demands. This is amid increasing concerns of global warming, fast depletion of petroleum resources, expected increase of the crude oil price due to rapid demand growth from China and India, and unstable oil supply due to turmoil in the Middle East. Although replacement of carbon-based economy is receiving a major research attention nowadays, mankind's reliance on carbon will continue for the foreseeable future. As the only carbon-rich material source, biorenewables processed in biorefineries is very likely to be the only viable alternative to fossil resources for the production of transportation fuels and chemicals.

The attractive future prospects for biorefineries are accompanied by several unique and challenging characteristics. Such challenges mainly include supply chain design and management (decentralized raw material sources etc.), process design and development (lack of models etc.) and intelligent decision making (uncertainty in markets and prices etc.). Furthermore, in most scenarios, successful process synthesis strategies for conventional refineries do not work for biorefineries. Thus, biorefinery synthesis using advanced systems engineering approaches require substantial research to design and manage economically viable, environmentally benign, and socially beneficial biorefineries.

Kamm et al. (2006) highlight existing works relevant to process, raw material, product, and technology options. However, a study that integrates all these components using a modeling and optimization approach based on mathematical programming does not exist. Here, we aim to bridge this knowledge gap by systematically and intelligently evaluating all possible alternatives. In this paper, we explore and evaluate a variety of plausible and promising options for the conversion of cellulose, lignin, and C5 sugars into a slate of potential value-added chemicals using different technologies. We propose a novel superstructure of various alternative configurations for a general biorefinery. For this, we develop a general mathematical model based on short-cut equations to obtain the optimal design of a biorefinery. In addition, we examine a number of alternatives in the superstructure and cost factors such as operation cost, utility cost, capital expenditures (investments, overheads etc.), taxes and inflation rates for the entire lifetime of the biorefinery. Then, based on market scenarios, availability of lignocellulosic biorenewables, and available technologies, we present a reduced and dedicated biorefinery superstructure for Singapore. Finally, we illustrate sensitivity analysis and debottlenecking opportunities in biorefineries using several case studies.

References:

U.S Department of Energy, 1997, Energy, environmental, and economics (E3) handbook - A resource tool to aid the office of industrial technologies, 1st ed.

National Research Council, 2000, Biobased industrial products, priorities for research and commercialization, National Academic Press, Washington, D.C.

Kamm, B., Kamm, M., 2004. Principles of biorefinery, Appl. Microbiol. Biotechnol. 64, 137–145.

(604b) Optimization of Product Formulations Using Multivariate Analysis and Property Clustering

Subin Hada[1], Robert Herring III[1], J. Colin Haser[1], Nishanth G. Chemmangattuvalappil[2] and Mario Richard Eden[1], (1)Department of Chemical Engineering, Auburn University, Auburn, AL, (2)Department of Chemical and Environmental Engineering, University of Nottingham - Malaysia, Semenyih, Malaysia

Mixture processes play an important role in today's manufacturing of value-added chemical products such as commodity chemicals, food, cosmetics, oil and pharmaceuticals. In classical non-mixture designs such as factorial and response surface designs, all the factors are orthogonal or independent. This means that it is possible to freely choose the level of a factor regardless of the other factors' levels. However, this freedom does not exist for mixture designs, because each component in a mixture is dependent upon the settings of the other component settings. In mixture design, the factors are interdependent and the effects of the factors on the responses are not separable. In addition, traditional mixture models are usually employed to investigate the relationships between the blend ratio matrix (**R**) and the final blend product property matrix (**Y**) only, given that the properties of the pure raw materials (**X**) and the process conditions used to manufacture them (**Z**) are already chosen. However, in the development of formulated products that meet target properties with minimum experiments and minimum total material cost, it is important to simultaneously account for all three degrees of freedom (**X**, **R**, and **Z**) available in the blending operations.

In this paper, projection methods like PCA and PLS are employed to extract and utilize the necessary information from all three degrees of freedom that are available to control the properties of any product manufactured in blending operation. This information is used to build models to simultaneously explore variations in the raw material properties data (**X**), their blend ratios (**R**), and process conditions (**Z**) and the resulting effect on the product quality (**Y**). These methods capture information about the variance and correlation structure within predictor and quality variables that were present during the production of previous product grades and that are important for consistency of the desired process conditions with past operating procedures. An integrated solution approach is employed that formulates two reverse problems, one for determination of design targets and another for matching of the design targets. The empirical models are used in a reverse problem formulation to ensure that a complete set of candidate materials and their mixtures are found which match the design target subject to the predictive power of the model. Mapping the design problem onto the lower dimensional cluster space utilizing property clustering algorithm facilitates visualization and simultaneous solution of the problem and also avoids combinatorial explosion when dealing with multi-component mixtures.

The method and concepts will be illustrated using a case study describing the design of polymer blending problems in a latent variable space.

(604c) Systematic Incorporation of 2D, 3D and Electrostatic Characterization in Molecular Design

J. Colin Haser[1], Robert Herring III[1], Subin Hada[1], Nishanth G. Chemmangattuvalappil[2] and Mario Richard Eden[1], (1)Department of Chemical Engineering, Auburn University, Auburn, AL, (2)Department of Chemical and Environmental Engineering, University of Nottingham - Malaysia, Semenyih, Malaysia

Computer-aided molecular design (CAMD) is emerging as a very powerful tool in chemical and pharmaceutical industries today. The ability to design products, which meet a set of specified target properties, exemplifies the applicability of CAMD. With the advent of new computing technologies, a more detailed and comprehensive description of molecular structures is possible. This growth in the ability to model and predict a wide range of molecular properties should parallel the increased stringency of regulations on environmental, health and safety standards. Thus, more tools must be developed which take advantage of relationships between molecular structure and the macro scale properties realized for these systems. It is well understood that most properties of interest are not solely a function of the topological arrangement and connectivity of the molecules under investigation. The spatial arrangement, e.g. steric hindrance or ligand binding, and electrostatic nature of these structures are also necessary to reveal more optimal candidate molecules. It has been suggested that searching the chemical space with a combination of 2D and 3D descriptors could prove more effective than employing single methods [1].

This contribution details a systematic methodology for doing just that, and it is based on the signature molecular descriptor [2]. One difficulty in the application of reverse molecular design using geometric based descriptors is that the minimum energy molecular conformation is a function of the complete structure. The most stable conformation is ultimately determined by interactions among subatomic particles through the force fields they generate, which can be accurately modeled by quantum mechanics. The effect of these force fields diminishes as a function of increasing inter-particle distance. Thus, a reasonable estimation of local geometry can be made by considering atoms extending radially, up to fixed distance, from a central atom. The signature descriptor is ideal for this approach as it details the extended valence sequence of an atom up to a pre-defined height or distance away from the central atom. Previous studies have proven the applicability and effectiveness of atomic signature descriptors utilized in molecular design with topological and topographical considerations [3-5]. These signatures can be geometry optimized through various quantum mechanical approaches, and this information is conveniently stored in the form of a geometry matrix. Also, the electron densities of these signature "fragments" can be generated through *ab initio* calculations, given the previously obtained geometry information, and recombined to produce complete molecular electron distributions. This approach, known as MEDLA (Molecular Electron Density Lego Approach) [6], utilizes a "fuzzy" electron density description so that boundary adjustments upon fragment recombination are unnecessary. The nature of these signature descriptors and boundary-less electron cloud fragments allow for a linear optimization approach and are consequently much less computationally demanding than other *ab initio* quality calculations. The framework and mathematical techniques necessary for formulation and solution of property or activity based molecular design problems, with 2D, 3D and electrostatic characterization, are presented here along with a case study which illustrates the benefits of such an approach.

[1] Nettles J.H., Jenkins J.L., Bender A., Deng Z., Davies J.W., and Glick M. (2006) Bridging Chemical and Biological Space: "Target Fishing" Using 2D and 3D Molecular Descriptors. *J. Med. Chem*., **49**, pp. 6802-6810

[2] Faulon J.-L., Visco D.P. Jr., and Pophale, R.S. (2003) The Signature Molecular Descriptor. 1. Using Extended Valence Sequences in QSAR and QSPR Studies. *Journal of Chemical Information and Computer Sciences,* **43**, pp. 707-720.

[3] Herring R.H., Rudolfs N., Chemmangattuvalappil N.G., Roberts C., and Eden M. (2012). Molecular Design Using Three-Dimensional Signature Descriptors. *Computer Aided Chemical Engineering*, **31**, (In Press).

[4] Chemmangattuvalappil N.G., Solvason C.C., Bommareddy S. and Eden M.R. (2009) Incorporating Molecular

Signature Descriptors in Reverse Problem Formulations. *Computer aided Chemical Engineering*, **27**(1), pp. 73-78.

[5] Chemmangattuvalappil N.G., Solvason C.C., Bommareddy S. and Eden M.R. (2009) Novel Molecular Design Technique using Property Operators based on Signature Descriptors. *Computer aided Chemical Engineering*, **27**(2), pp.897-902

[6] Walker P. D., Mezey P. G. (1993) Molecular Electron Density Lego Approach to Molecule Building. *J. Am. Chem. Soc.*, **115**, pp. 12423-12430.

(604d) A Systematic Approach for Conceptual and Sustainable Process Design: Production of Methylamines From Methanol and Ammonia

Seyed Soheil Mansouri, **Muhammad Imran Bin Ismail** and **Karim Almoor**, Chemical and Biochemical Engineering, Technical University of Denmark, Kongens Lyngby, Denmark

Methylamines are very important chemicals as they are commonly used as intermediates for a wide range of agricultural chemicals, animal nutrients, catalysts, electronics, explosives, fuel additives as well as gas and oil treatments.

A systematic method is applied to design a sustainable and environmentally acceptable plant for producing mono-methylamines, di-methylamines and tri-methylamines from methanol and ammonia. The systematic method divides the process design work into 12 sequential tasks that covers all stages of conceptual design, starting from the consideration of qualitative aspects of the process flowsheet and preliminary calculations to the detailed process simulations, equipment sizing, costing, an economic evaluation, and, sustainability and LCE assessment of the designed process. At the end of task-9, the base case design is obtained, which is then further refined and improved with respect to heat integration and process optimization. In the final task-12, a sustainability and LCA analysis is performed to assess the environmental impact of the process design. This is done through the following special software: SustainPro and LCASoft, that determine the key sustainability and LCA measures such as sustainability metrics for environment, economic and social; carbon footprint; safety index and many more. In addition, PRO/II is used for process simulation (for verification of design), ICAS for property prediction and analysis of design options; ECON for cost and economic analysis. This procedure can be applied to design and/or analysis for new or existing chemical or biochemical processes. The process design decisions are made considering the sustainability and LCA issues together with the process design specifications. The conceptual process design of methylamines producing plant is performed as part of a MSc-level course on Process Design at the Department of Chemical and Biochemical Engineering at DTU.

The annual profit of base case is estimated to be about 28 million US$ with a payback time of 2.5 years for production rates of 9792 metric tons/year, 33864 metric tons/year and 8976 metric tons/year of mono-methylamines, di-methylamines and tri-methylamines, respectively. The capital and operating costs are divided into the individual cost items. It is found that compressors are the most expensive equipment and the steam utility is the main contributor to the total operating cost. This information is then used as target for process improvement by heat integration and process optimization, thus increasing the annual profit and reducing the payback time. The environmental impact analysis identifies impact due to the release of chemicals and points to the need of better control mechanisms through the sustainable LCA analysis.

The design project was carried out under the supervision of Professor Rafiqul Gani.

(604e) A Systematic Approach for Conceptual and Sustainable Process Design: Production of Cumene

Sofie Thage Morthensen, Department of Chemical and Biochemical Engineering, Technical Univeristy of Denmark, Kgs. Lyngby, Denmark, Naia Wright, Department of Chemical and Biochemical Engineering, Technical University of Denmark, Kgs Lyngby, Denmark, Diogo Sebastiao, Department of Chemical and Biochemical Engineering, Technical University of Denmark, Kgs. Lyngby, Denmark and Deenesh K. Babi, Department of Chemical and Biochemical Engineering, Denmark Technical University, Lyngby, Denmark

A Systematic Approach for Conceptual and Sustainable Process Design: Production of cumene

Diogo M.B. Sebastiao[1], Naia R. Wright[1], Sofie T. Morthensen [1], Rafiqul Gani[1]

s111775@student.dtu.dk, s082422@student.dtu.dk, s082684@student.dtu.dk, *rag@kt.dtu.dk*

[1] Department of Chemical and Biochemical Engineering, Technical University of Denmark (DTU), DK-2800 Kongens Lyngby, Denmark

Abstract

Cumene (isopropylbenzene) is an important chemical since it is used as feed stock for acetone and phenol production, which are used in the organic chemical industry [1]. A systematic hierarchal decomposition method is applied to design a sustainable and environmentally friendly plant for producing cumene from the raw materials benzene and propylene [2]. The method consists of 12 sequential tasks which take into account all stages of conceptual design, starting from the consideration of qualitative aspects of the process flowsheet and preliminary calculations to the detailed process simulations, equipment sizing, costing, economic evaluation, sustainability and LCA assessment of the designed process. At the end of task-9 (the economic analysis), the base case design is investigated for improvements with respect to heat integration and process optimization. In the final task-12, a sustainability and LCA analysis is performed to assess the environmental impact of the process design. This is done using software: SustainPro and LCASoft that determine the key sustainability and LCA measures such as sustainability metrics for environment, economic and social; carbon footprint; safety index and many more. In addition, PRO/II is used for process simulation (for verification of design), ICAS for property prediction and analysis of design options; ECON for cost and economic analysis. This procedure can be applied to design and/or analysis for new or existing chemical or biochemical processes. The process design decisions are made considering the sustainability and LCA issues together with the process design specifications. The conceptual process design of the cumene producing plant is performed as part of a MSc-level course on Process Design at the Department of Chemical and Biochemical Engineering at DTU under the supervision of Professor Rafiqul Gani.

A production rate of 100,000 metric tons/year of cumene is considered. The highest annual profit of base case is obtained in the second year and is estimated to be 20 million USD with a payback time of 1-2 years depending on the type of depreciation. The capital and operating costs are divided into the individual cost items in order to analyze which equipment and utilities contribute significantly to these costs. It is found that the compressor is the most expensive equipment contributing to capital costs and the reboiler utility is the main contributor to the total operating cost. This information is then used as target for process improvement by heat integration and process optimization, thus increasing the annual profit and reducing the payback time. The environmental impact analysis identifies impact due to the release of chemicals and points to the need of better control mechanisms through the sustainable LCA analysis.

In this poster the following will be presented: Product characteristics and applications, process flow sheet, equipment sizing, costing, economic evaluation, optimization targets, sustainability and LCA assessment of the designed process.

References:

[1] F. Cavani, G. Centi, S. Perathoner, F. Trifiró. *Sustainable Industrial Chemistry*, John Wiley & Sons, 2009.

[2] L. T. Biegler, I. E. Grossmann and A. W. Westerberg. *Systematic methods of Chemical Process Design*, Prentice Hall, 1997.

(604f) A Systematic Approach for Conceptual and Sustainable Process Design: Production of Methanol

Morten Søe Jepsen, Asbjørn T. Pedersen and **Alexander Rubin**, Department of Chemical and Biochemical Engineering, Technical University of Denmark, Kgs. Lyngby, Denmark

Methanol is an important bulk chemical and is widely used in a variety of consumer products. Methanol produced from renewable resources furthermore has the potential to be used as an additive to transportation fuels in order to meet the requirement that 10% of the energy used for transportation within the European Union is supplied from renewable resources[1]. A systematic hierarchal decomposition method is applied to design a sustainable and environmentally plant for producing methanol from the raw materials biomass and hydrogen produced by electrolysis. The method consists of 12 sequential tasks which take into account all stages of conceptual design, starting from the consideration of qualitative aspects of the process flowsheet and preliminary calculations to the detailed process simulations, equipment sizing, costing, economic evaluation, sustainability and LCA assessment of the designed process. At the end of task-9 (the economic analysis), the base case design is investigated for improvements with respect to heat integration and process optimization. In the final task-12, a sustainability and LCA analysis is performed to assess the environmental impact of the process design. This is done using software: SustainPro and LCASoft, that determine the key sustainability and LCA measures such as sustainability metrics for environment, economic and social; carbon footprint; safety index and many more. In addition, PRO/II is used for process simulation (for verification of design), ICAS for property prediction and analysis of design options. This procedure can be applied to design and/or analysis for new or existing chemical or biochemical processes. The process design decisions are made considering the sustainability and LCA issues together with the process design specifications. The conceptual process design of methylamines producing plant is performed as part of a MSc-level course on Process Design at the Department of Chemical and Biochemical Engineering at DTU under the supervision of Professor Rafiqul Gani.

The base case is estimated to be loss about 14 million US$ annually for production rates of 100,000 metric tons/year of methanol, primarily due to the high hydrogen price. In order for the base case process to reach breakeven the methanol prices has to increase to 25 %. The capital and operating costs are divided into the individual cost items in order to obtain which equipment and utilities contribute significantly to these costs. It is found that the heat exchangers are the most expensive equipment contributing to capital costs and the hot utility is the main contributor to the total operating cost. This information is then used as target for process improvement by heat integration and process optimization, thus increasing the annual profit and reducing the payback time. The environmental impact analysis identifies impact due to the release of chemicals and points to the need of better control mechanisms through the sustainable LCA analysis.

This poster presentation will highlight the development of a process flowsheet, including heat and mass balances, for the production of methanol from biomass. The heat and mass integration performed in order to minimize the production cost are presented together with a LCA analysis to determine the overall sustainability of the process.

References:

[1] EU Directive (EC) 2009/28/EC of 23 April 2009, On the promotion of the use of energy from renewable sources

(604g) A Multidimensional Population Balance Model for Growth and Dissolution Identified From a Designed Temperature-Cycling Experiment

Mo Jiang[1], Xiaoxiang Zhu[1], Mark Molaro[1], Michael L. Rasche[2], Davide M. Raimondo[3], Kwang-Ki Kim[4], Haitao Zhang[1], Keith Chadwick[1], Lifang Zhou[1], Zhilong Zhu[2], Minhao Wong[2], Des O'Grady[5], Dominique Hebrault[5], John Tedesco[5] and Richard D. Braatz[6], (1)Chemical Engineering, Massachusetts Institute of Technology, Cambridge, MA, (2)Department of Chemical and Biomolecular Engineering, University of Illinois at Urbana-Champaign, Urbana, IL, (3)Università degli Studi di Pavia, Pavia, Italy, (4)Massachusetts Institute of Technology, Cambridge, MA, (5)METTLER TOLEDO, Columbia, MD, (6)Department of Chemical Engineering, Massachusetts Institute of Technology, Cambridge, MA

A large proportion of pharmaceutical crystallizations produce rod-like crystals, which can cause problems in process operations downstream from the crystallizer. Many recent efforts have been directed towards the construction of multidimensional population balance models suitable for the design of process operations to optimize crystal shape [1-11]. These models require the estimation of growth kinetics along multiple axes, with most methods based on sampling the slurry during crystallization, or on employing imaging technology that only applies for low crystal number densities or is not easy to implement in an industrial environment [12-21]. While it is not possible to construct an arbitrary two-dimensional crystal size distribution (2D CSD) from measurements of the chord length distribution (CLD) obtained by Focused Beam Reflectance Measurement (FBRM, which is a probe for in-situ laser backscattering measurements of particles), application to a specific solute-solvent system is simpler due to limited range of 2D CSDs that can occur for the allowed variation in seed mass and temperature profiles.

Previously, we have reported the application of the commonly available FBRM for the *in-situ* simultaneous measurement of crystal shape and size for rod-like organic crystals during a temperature cycling experiment, which resulted in substantial continual increases in aspect ratio, mean crystal length, and mean crystal width with cycle number. This presentation will describe the construction of a multidimensional population balance model for growth and dissolution rates along the two axes. The experimental observations motivated the incorporation of size-dependent dissolution into the multidimensional population balance model

$$\frac{\partial f}{\partial t}+\frac{\partial}{\partial L}(G_L f)+\frac{\partial}{\partial W}(G_W f)=0,\quad \text{for } S\geq 0$$
$$\frac{\partial f}{\partial t}-\frac{\partial}{\partial L}(D_L f)-\frac{\partial}{\partial W}(D_W f)=0,\quad \text{for } S<0 \qquad (1)$$

where f is the 2D population density, t is time, and the growth and dissolution rates in the length and width directions are 2D generalizations of the 1D equations [22], i.e.,

$$G_L = k_{GL0}\exp\left(-\frac{\Delta E}{RT}\right)S^{g_L}(1+\delta_{G,LL}L)$$
$$G_W = k_{GW0}\exp\left(-\frac{\Delta E}{RT}\right)S^{g_W}(1+\delta_{G,WW}W)$$
$$D_L = \frac{k_{DL}(-S)^{d_L}}{L^{\gamma}}$$
$$D_W = \frac{k_{DW}(-S)^{d_W}}{W^{\gamma}}$$
$$(2)$$

and the d_{ij} and γ define size dependencies. The size dependency broadens the crystal size distribution, which result in more crystals becoming dissolved in each cycle, as required to explain the experimentally observed systematic increase in both the average length and width of crystals with cycle number. A parameter estimation code (the numerical algorithms and software are similar to that used in [23] and [24]) employed a successive quadratic programming algorithm to minimize the sum-of-squared-deviations between the model predictions and experiments,

$$\min_{\underline{\theta} \leq \theta \leq \bar{\theta}} (1-\alpha)\sum_{i}\left[C_{measured}(t_i) - C_{model}(t_i;\theta)\right]^2 + \alpha\sum_{k}\left[L_{measured}(t_k) - L_{model}(t_k;\theta)\right]^2 + \alpha\sum_{k}\left[W_{measured}(t_k) - W_{model}(t_k;\theta)\right]^2 \quad (3)$$

where θ is the vector of model parameters and α weighs the confidence in each of the mean length and width measurements relative to the solute concentration measurements and takes into account the differing units and number of experimental data points. A Pareto-optimality plot of squared-error values was constructed with each point being associated with a value of α, which ensured that the value of α of 0.00008 reasonably weighed the sets of solute concentration and mean length/width data.

Compared to previous multidimensional population balance models (PBMs), this PBM includes the effects of dissolution rates that are estimated from experimental data. Although the PBM's modeling of the size dependency of dissolution is rather simplistic, the model fits the experimental data with limited number of parameters with a relatively small level of uncertainty in the parameters that was quantified by first-order multivariate statistical analysis. In addition, the model simulation at optimal kinetic parameters was able to predict the crystal size and shape distribution in a different temperature-cycling experiment reasonably well. These results suggest that growth and dissolution rates estimated in a single temperature-cycling experiment may be suitable for designing a temperature-cycling protocol for optimizing the mean length, mean width, mean aspect ratio, or some weighted combination of the three characteristics of the product crystals. These results also suggest that an automated protocol may be developed that can assess the potential usefulness of temperature cycling for removing or reducing undesirable crystal size and shape characteristics produced by more standard crystallization operations.

References

[1] D. B. Patience and J. B. Rawlings, "Particle-shape monitoring and control in crystallization processes," *AIChE J.*, vol. 47, pp. 2125-2130, 2001.

[2] R. Gunawan, D. L. Ma, M. Fujiwara and R. D. Braatz, "Identification of kinetic parameters in multidimensional crystallization processes," *Int. J. Mod. Phys. B, vol. 16, pp. 367-374, 2002.*

[3] K. Lee, J. H. Lee, M. Fujiwara, D. L. Ma and R. D. Braatz, "Run-to-run control of multidimensional crystal size distribution in a batch crystallizer," in *Proceedings of the American Control Conference,* pp. 1013-1018, 2002.

[4] F. Puel, G. Févotte and J. P. Klein, "Simulation and analysis of industrial crystallization processes through multidimensional population balance equations. Part 1: A resolution algorithm based on the method of classes," *Chemical Engineering Science,* vol. 58, pp. 3715-3727, 2003.

[5] H. Briesen, "Simulation of crystal size and shape by means of a reduced two-dimensional population balance model," *Chemical Engineering Science,* vol. 61, pp. 104-112, 2006.

[6] C. Y. Ma, X. Z. Wang and K. J. Roberts, "Multi-dimensional population balance modeling of the growth of rod-like L-glutamic acid crystals using growth rates estimated from in-process imaging," *Advanced Powder Technology,* vol. 18, pp. 707-723, 2007.

[7] C. Y. Ma, X. Z. Wang and K. J. Roberts, "Morphological population balance for modeling crystal growth in face

directions," *AIChE J.,* vol. 54, pp. 209-222, 2008.

[8] C. Y. Ma and X. Z. Wang, "Crystal growth rate dispersion modeling using morphological population balance," *AIChE J.,* vol. 54, pp. 2321-2334, 2008.

[9] J. Wan, X. Z. Wang and C. Y. Ma, "Particle shape manipulation and optimization in cooling crystallization involving multiple crystal morphological forms," *AIChE J.,* vol. 55, pp. 2049-2061, 2009.

[10] J. J. Liu, C. Y. Ma, Y. D. Hu and X. Z. Wang, "Effect of seed loading and cooling rate on crystal size and shape distributions in protein crystallization-A study using morphological population balance simulation," *Computers and Chemical Engineering,* vol. 34, pp. 1945-1952, 2010.

[11] J. J. Liu, C. Y. Ma, Y. D. Hu and X. Z. Wang, "Modelling protein crystallisation using morphological population balance models," *Chem. Eng. Res. Design,* vol. 88, pp. 437-446, 2010.

[12] F. Puel, P. Marchal and J. Klein, "Habit transient analysis in industrial crystallization using two dimensional crystal sizing technique," *Chem. Eng. Res. Design,* vol. 75, pp. 193-205, 1997.

[13] D. L. Ma and R. D. Braatz, "Worst-case analysis of finite-time control policies," *IEEE Trans. Control Syst. Technol.,* vol. 9, pp. 766-774, 2001.

[14] J. Calderon De Anda, X. Z. Wang, X. Lai and K. J. Roberts, "Classifying organic crystals via in-process image analysis and the use of monitoring charts to follow polymorphic and morphological changes," *J. Process Control,* vol. 15, pp. 785-797, 2005.

[15] J. De Calderon Anda, X. Z. Wang, X. Lai, K. J. Roberts, K. H. Jennings, M. J. Wilkinson, D. Watson and D. Roberts, "Real-time product morphology monitoring in crystallization using imaging technique," *AIChE J.,* vol. 51, pp. 1406-1414, 2005.

[16] J. De Calderon Anda, X. Z. Wang and K. J. Roberts, "Multi-scale segmentation image analysis for the in-process monitoring of particle shape with batch crystallisers," *Chemical Engineering Science,* vol. 60, pp. 1053-1065, 2005.

[17] P. A. Larsen, J. B. Rawlings and N. J. Ferrier, "An algorithm for analyzing noisy, in situ images of high-aspect-ratio crystals to monitor particle size distribution," *Chemical Engineering Science,* vol. 61, pp. 5236-5248, 2006.

[18] M. Oullion, F. Puel, G. Févotte, S. Righini and P. Carvin, "Industrial batch crystallization of a plate-like organic product. In situ monitoring and 2D-CSD modelling: Part 1: Experimental study," *Chemical Engineering Science,* vol. 62, pp. 820-832, 2007.

[19] J. Eggers, M. Kempkes and M. Mazzotti, "Measurement of size and shape distributions of particles through image analysis," *Chemical Engineering Science,* vol. 63, pp. 5513-5521, 2008.

[20] X. Z. Wang, J. Calderon De Anda and K. J. Roberts, "Real-time measurement of the growth rates of individual crystal facets using imaging and image analysis: A feasibility study on needle-shaped crystals of L-glutamic acid," *Chem. Eng. Res. Design,* vol. 85, pp. 921-927, 2007.

[21] X. Z. Wang, K. J. Roberts and C. Ma, "Crystal growth measurement using 2D and 3D imaging and the perspectives for shape control," *Chemical Engineering Science,* vol. 63, pp. 1173-1184, 2008.

[22] M. Shoji, T. Eto and H. Takiyama, "A kinetic study of the influence of modulated undersaturation operation on crystal size distribution in cooling-type batch crystallization," *J. Chem. Eng. Japan,* vol. 44, pp. 191-196, 2011.

[23] R. Gunawan, I. Fusman and R. D. Braatz, "High resolution algorithms for multidimensional population balance equations," *AIChE J.,* vol. 50, pp. 2738-2749, 11, 2004.

[24] T. Togkalidou, H. -. Tung, Y. Sun, A. T. Andrews and R. D. Braatz, "Parameter estimation and optimization of a

loosely bound aggregating pharmaceutical crystallization using in situ infrared and laser backscattering measurements," *Industrial and Engineering Chemistry Research,* vol. 43, pp. 6168-6181, 2004.

(604h) Surrogate-Based Optimization of CO_2 Capture Process with Aqueous Amines

Aroonsri Nuchitprasittichai, Chemical Engineering, The University of Tulsa, Tulsa, OK and Selen Cremaschi, Department of Chemical Engineering, University of Tulsa, Tulsa, OK

An increase in the level of greenhouse gases, mainly carbon dioxide, methane and nitrous oxide, in the atmosphere has become more severe in recent decades. The environmental and global warming concerns accelerated the research and development of carbon dioxide capture technologies to control and reduce the emissions of this primary greenhouse gas. As a result, there are three main technologies to capture CO_2 from the combustion of fossil fuels: pre-combustion, post-combustion, and oxy-combustion. In the post-combustion capture, CO_2 is recovered from the flue gas, whereas in pre-combustion capture, carbon is removed from the fuel prior to combustion. In oxy-combustion capture, the fuel is burned using an almost pure oxygen stream resulting in an almost pure CO_2 stream. We focus on the post-combustion capture technologies because they can be readily used to retrofit existing plants and have the highest potential to reduce the CO_2emissions in the short to medium term.

We developed surrogate-based optimization approaches to study the process synthesis and optimization of post-combustion carbon capture technologies, i.e., a conventional amine-based absorption. Surrogate- based optimization approaches can be categorized into two groups based on their implementations. In the first class, the optimization module uses local search space to estimate an appropriate direction to reduce the objective function (assuming minimization), e.g., response surface methodologies. In the second category, the simulation is used to build surrogate models, e.g., the artificial neural network (ANN), of the objective function and/or the constraint over the whole decision space, and the optimization is performed using these surrogate models.

In this work, the surrogate-based optimization frameworks using both categories were developed: response surface methodology and artificial neural network as the objective function. 1365 kg mole/h of gas turbine flue gases at a pressure of 1.1 bar and a temperature of 40 °C was processed in a conventional amine-based absorption plant for CO_2 recovery. The simulation was developed using Aspen HYSYS Version 7.1. The amines property package was used as the thermodynamic model. Li-Mather electrolyte model was selected as the equilibrium model.

In the response surface methodology, the process simulation was run that corresponds to the operating conditions and design variables generated by an experimental design. The combination of the decision variables as the inputs and the CO_2 capture cost as the outputs of the simulation was used to obtain the first order regression model for the objective function. We used the R^2 with the cutting value of 0.5 and adjusted-R^2 statistics in order to assess the fitness of the first and second order response models. If it was concluded that the first-order model represents the data set, the steepest descent was performed to move the system in the direction of maximum decrease in the objective function. On the other hand, if it was found that the data set cannot be represented accurately with the first order model, the second-order model was utilized to find the local minimum CO_2 capture cost and the corresponding design and operating conditions. The impact of different amine absorbents and their concentrations on the optimum CO_2 cost was analyzed. The results suggested that CO_2 recovery with 48 wt% DGA required the lowest CO_2 removal cost. Sensitivity analysis for the optimum amine-based CO_2 capture plant was used to study the impact of flue gas feed composition and utility cost fluctuations on the minimum CO_2 capture cost. The CO_2 capture cost showed an inversely proportional relationship with the amount of CO_2composition in flue gas while the utility cost did not show significant impact.

In the artificial neural network approach, the ANN was utilized to mimic the objective function behavior, cost of CO_2 removal, in terms of the decision variables. Feed forward neural network was used because of its mathematical simplicity. A back propagation (BP) algorithm was employed to train the network. We developed an algorithm to determine the appropriate sample size for constructing accurate artificial neural networks as surrogate models in optimization problems. In the algorithm, two model evaluation methods—cross-validation and/or bootstrapping—were used to estimate the performance of various networks constructed with different sample sizes. The CO_2 capture process with 48 wt% DGA solution resulted in the lowest CO_2 removal cost. The minimum CO_2 capture costs obtained from both approaches are similar for all solvents, except in the case of TEA, where the ANN approach is able to locate a lower CO_2 capture cost compared to the RSM algorithm results. This is because the RSM algorithm searches for the minimum CO_2 capture cost in the local region of large number of absorber stages, while the ANN approach was able to identify the region of lower CO_2 capture cost over the entire domain of the number of absorber stages for a better prediction of the minimum CO_2 capture cost.

(604i) Tailoring Ionic Liquids for Optimal CO2 Capture

Amirhossein Mehrkesh and Arunprakash T. Karunanithi, Department of Civil Engineering, University of Colorado Denver, Denver, CO

Due to increasing global climate action efforts, there is a great need to reduce concentration of greenhouse gases from our atmosphere. Approximately 86 % of greenhouse gas emissions arise from combustion of fossil fuels. This number is 83 % for United States. One strategy is to gradually shift towards renewable energy resources (e.g. solar, wind and biofuels). This approach has lot of benefits in the long run, both from the environment and energy security stand point. However, in the short to medium term a significant proportion of our energy portfolio is going to be based on non-renewable resources. This would necessitate initiatives to mitigate the environmental problems arising out of using fossil fuels. Carbon Capture and Sequestration (CCS) is a central strategy in these initiatives, which holds great promise as it offers us the ability to use fossil fuel while meeting global GHG emission targets. The idea behind this approach is to capture CO_2 from point sources (e.g. power plants) and sequester it in deep geological formations (e.g. depleted oil fields, saline formation).

Ionic liquids are a class of solvents that offer significant promise for CO_2capture. Ionic liquids are molten salts that exist as liquids at a relatively low temperature. Ionic liquids have shown great promise for CO2 absorption. CO2 solubility in different ionic liquids has been previously reported. In this paper we show the use of computer-aided ionic liquid design (CAILD) method to design optimal ionic liquids with enhanced CO_2 absorption capacities. We utilize ionic liquid structure-property computational models in conjunction with optimization algorithms to achieve this objective.

(604j) A Molecular Signature Based Approach for the Design of Reactive Systems

Nishanth G. Chemmangattuvalappil[1], Vikrant Dev[2] and Mario Richard Eden[2], (1)Department of Chemical and Environmental Engineering, University of Nottingham - Malaysia, Semenyih, Malaysia, (2)Department of Chemical Engineering, Auburn University, Auburn, AL

The simultaneous consideration of process and product design aspects is essential for the optimal design of a process. A variety of different algorithms have been published in the last decade that addresses this issue. However, the existing techniques for solving such problems are limited to non-reactive systems. In this work, we are introducing an algorithm that can be used for solving integrated process and molecular design problems of reactive systems on a property based platform. Property clustering is a well-known technique for a component independent design. Property clustering technique will be extended into identifying a set of target properties for the components in the system that provides the optimum process performance. Once the property targets have been identified, a molecular design problem can be formulated to identify the potential candidate molecules that meet the targets identified in the process design problem. The molecular design involves the identification of potential molecules possible from the specific types of reactions in the process. To design molecules, a recently introduced concept known as molecular signature descriptors have been used. The signature is a systematic coding system of atom types and the signature of a molecule can be obtained as a linear combination of its atomic signatures (Visco *et al* 2002). The molecular signatures can be tailored to track the change in molecular groups in a molecule resulting from different types of chemical reactions. This task is achieved by forming the signatures for both reactants and products and by considering them as un-connected graphs. The changes in the chemical structure can be correlated with the changes in the properties of the molecule. Therefore, the changes in the molecular structure due to reactions can be represented as a function of the property. The developed algorithm applies different qualitative structure activity/property relationships (QSAR/QSPR) to estimate properties from the molecular structure. QSAR relationships make use of different topological indices and it has been proved that a number of topological indices of molecules can be represented in terms of molecular signatures and it is possible to correlate the topological indices to the actual properties and biological activities (Faulon *et al* 2003). Here, the new algorithm utilizes molecular property operators based on signatures for solving the inverse problem of obtaining the molecular structures that satisfy the property targets estimated in the process design step. A new set of equations will be employed to ensure that the molecule meets the safety and environmental constraints as well. The principles in graph theory are incorporated to track signatures and to avoid the generation of infeasible molecular structures. This contribution will illustrate the developed methods and highlight their use through a case study.

(604k) Economic Optimization of Lipid Production in Microalgae Bioreactor Using Nonlinear Programming

Javad Abdollahi and Stevan Dubljevic, Chemical and Materials Engineering, University of Alberta, Edmonton, AB, Canada

Economic feasibility of biodiesel production from microalgae cultivation can be guaranteed by increasing the lipid production and minimizing nutrients consumption. A microalgae growth model can be used to mathematically calculate the feeding strategy to achieve these goals. This work is based on an existing heterotrophic microalgae (*Auxenochlorella* protothecoides) fed-batch model[1] which is developed to predict the microalgae growth and oil production. Microalgae grows on two nutrients, glucose and glycine, and the growth and lipid production can be manipulated by different feeding strategies. In addition, the mathematical model has highly nonlinear and time varying parameters such as growth, nitrogen uptake and oil production rates[2].

In this work, mathematical model of heterotrophic growth of microalgae is used to maximize lipid production. The most expensive feed stock in this model is glucose and the economic performance of the bio-reactor can be calculated based on glucose consumption and lipid production. The objective function is defined as the rate of lipid production per use of glucose feed rate. Maximizing this objective function with respect to feeding flow rates and also process time will maximize the economic profitability of microalgae bioreactor. The highly nonlinear rate functions and the constraints on the states and inputs make the conventional optimization tools too slow or not applicable. The optimization problem is formulated as a constrained nonlinear program and solved using Ipopt solver[3]. Ipopt is an open source software package for large-scale nonlinear optimization and the nonlinear program is formulated and passed to Ipopt through AMPL[4] interface. Finally, a feeding strategy is designed to maximize the lipid productivity of microalgae and keep the economic performance at the maximum possible value.

References:

[1] H. De la Hoz Siegler, A. Ben-Zvi, R.E. Burrell, W.C. McCaffrey, The dynamics of heterotrophic algal cultures, Bioresource Technology 102, 2011, 5764-5774

[2] H. De la Hoz Siegler, R.E. Burrell, W.C. McCaffrey, A. Ben-Zvi, Optimization of microalgal pro- ductivity using an adaptive, non-linear model based strategy, Bioresource Technology, 104, 2012, 537-546

[3] A. Wachter and L. T. Biegler, On the Implementation of a Primal-Dual Interior Point Filter Line Search Algorithm for Large-Scale Nonlinear Programming, Mathematical Programming 106(1), 2006, pp. 25-57

[4] R. Fourer, D. Gay, and B. Kernighan, AMPL: A Mathematical Programming Language, Duxbury Press; Second edition, 2002

(604l) Multi-Objective Optimization in Industrial Chemical Process Development by Navigation On Pareto Fronts

Andreas Scheithauer[1], Jakob Burger[1], Norbert Asprion[2], Sergej Blagov[2], Roger Böttcher[2], Oliver Ryll[2], Richard Welke[3], Agnes Dittel[3], Uwe Nowak[3], Michael Bortz[3], Karl-Heinz Küfer[3] and Hans Hasse[1], (1)Laboratory of Engineering Thermodynamics, University of Kaiserslautern, Kaiserslautern, Germany, (2)Chemical and Process Engineering, BASF SE, Ludwigshafen, Germany, (3)Fraunhofer Institute for Industrial Mathematics (ITWM), Kaiserslautern, Germany

Process development in the chemical industry is a multi-objective optimization problem. There are many objectives, e.g. product quality, operation and capital cost, energy efficiency or reliability. For a given process configuration and given design specification, typical parameters of this optimization problem are, e.g. reactor volumes, overall and feed stage numbers of distillation columns, temperatures, pressures, mass flows and compositions of internal streams. In industrial process design, the developer usually generates only a certain number of process variants by varying process parameters intuitively. The aim is to find a solution that is acceptable regarding all or at least most criteria, optimality can usually not be guaranteed.

In the present work a feasible alternative approach is presented. It is based on the concept of Pareto optimality. The Pareto front describes the set of solutions for which no improvement in one objective can be reached without a decline in at least one other objective. Hence the Pareto front describes all optimal trade-offs between objectives. The final choice among these alternatives is left to the engineer. In the present approach, tools were developed and implemented in BASF's in-house process simulation environment that a) allow determining the Pareto front for a given design problem and b) support the user in the exploration of that front by navigation. The depth of the underlying process model can be varied from "short-cut" to the industrial standard of "equilibrium-stage based". The navigation on the Pareto front enables the designer to efficiently find the best compromise between the objectives. Upon navigating, the designer gains insight in the solution space that allows him to rationally justify his final decision.

The Pareto front is approximated within a user defined accuracy by highly efficient optimization techniques; at the same time, a robust and accurate process simulation is used.

In the presentation the approach will be described and results of the application of the Pareto navigation for different processes will be presented. The examples include a comparison of a pressure swing distillation to an entrainer distillation as well as the optimization of an industrial reaction-distillation train system.

(604m) Fast Pyrolysis of Empty Fruit Bunches (EFB) for Bio-Oil Production in Fluidized-Bed

Truong Xuan Do, Chemical Engineering, Hankyong National University, Anseong, South Korea, **Young-Il Lim**, Integrated Biotechnology, Hankyong National University, Anseong, South Korea and **Young-Tai Choi**, High temperature processing R&D, Korea Institute of Industrial Technology (KITECH), Cheonan, South Korea

The Empty Fruit Bunch (EBF), a waste of the palm oil industry, is one of the most recent renewable energy resources, and promises high yield of bio-oil (about 50%) with a low of gas and char. The objective of this research is to develop a comprehensive process model for the fast pyrolysis of EBF in fluidized-bed using the ASPEN PLUS simulator that can predict process performance under various operating conditions. The proposed model integrates specific thermodynamic properties and reaction kinetic models. The ASPEN PLUS reactor models and the external FORTRAN subroutines nested in ASPEN PLUS are used to represent the fluidized-bed pyrolysis reactor. A five-step reaction kinetic model of fast pyrolysis is used to predict bio-oil formation. The five-step model gives reliable results over a wide range of heating rates, and proper trends for bio-oil formation even without considering physical parameters like heat transfer. The process model is validated with experimental data from fluidized-bed pyrolysis system in a pilot scale. The sensitive analyses are performed to investigate effects of operating conditions on product yields.

(604n) Economic Optimum Design and Control of the Monoisopropylamine Process

Ojasvi Ojasvi and **Nitin Kaistha**, Chemical Engineering, IIT Kanpur, Kanpur, India

In this work, the steady state economic optimum design and control of a process producing monoisopropyl amine (MIPA) via isopropyl alcohol (IPA) amination is studied. The principal reactions are the MIPA formation main reaction (IPA + NH_3 ® MIPA + H_2O) and the di-isopropylamine (DIPA) formation side reaction (IPA + MIPA ® DIPA + H_2O). The separation of MIPA, DIPA, H_2O, unreacted IPA and NH_3, by the presence of IPA-H_2O and IPA-DIPA homogeneous azeotropes is complicated.

A process flowsheet consisting of a high conversion packed bed reactor, 2 distillation columns, a decanter and a stripper along with two liquid recycles is shown to be capable of producing 99.5% pure MIPA. A high conversion reactor must be used to ensure the reactor effluent has sufficiently small amount of IPA. Post NH_3 recovery in the first column, the decanter then exploits ternary liquid-liquid phase split to obtain nearly pure water (some IPA) and DIPA (some water) phases. The water stream is stripped to recover IPA and nearly pure water is discharged. The DIPA rich stream along with the recovered IPA is recycled back to the reactor. The single pass reactor conversion is a crucial design variable affecting the feasibility of product separation as too much IPA (low conversion) causes the free reactor effluent composition to be outside the liquid-liquid phase envelope. Even as the economic optimum design attempts to minimize the reactor size to the extent that liquid-liquid phase split just occurs in the decanter. Process operability considerations dictate reactor size over design. With sufficient reactor overdesign the hard operating constraint at maximum achievable throughput are flooding limits for the two columns and not the decanter.

An effective plantwide regulatory control structure is synthesized for the overdesigned reactor flowsheet and shown to provide robust process regulations for large throughput changes. The case study provides an interesting example of the ever-present conflict between economic design and process operability.

(604o) Indirect Carbon (CO) PEM Fuel Cell

Gifty Osei-Prempeh[1], Asad Davari[2], Ali Manesh[3], Farshid Zabihian[4] and Kamran Rostami[4], (1)Department of Chemical Engineering, West Virginia University Institute of Technology, Montgomery, WV, (2)Department of Electrical and Computer Engineering, West Virginia University Institute of Technology, Montgomery, WV, (3)American Science and Technology, Chicago, IL, (4)Department of Mechanical Engineering, West Virginia University Institute of Technology, Montgomery, WV

The advancement of the use of hydrogen fuel cell in many applications has been delayed by the need for safe storage or easy on-site generation of hydrogen fuel. Coal is an abundant source of fuel in the US. However, the use of coal and other fossil fuels in energy generation is inefficient which results in burning high quantities of coal (fuel) and the subsequent release of significant amounts of CO_2 into the atmosphere. Hence the need for "Clean Coal Technologies" as an alternative path for future power generation. Direct carbon fuel cells (DCFC) present a good alternative to hydrogen fuel cells in the area of energy density and fuel storage. Also, the efficiency from DCFC is projected to be about 70% which is more than what can be achieved in a hydrogen fuel cell.

Nonetheless, due to the internal operation of a fuel cell, it is difficult to achieve a reasonable three-phase boundary where solid fuel (e.g. coal), electrolyte, and electro-catalyst meet. To avoid this problem, coal can be pre-processed by either gasification or partial oxidation to produce CO and H_2. Partial oxidation produces a fuel gas that is high in CO concentration. The product of this process can be used as the inlet fuel for fuel cells. This work utilizes partial oxidation to produce CO fuel for a polyelectrolyte membrane fuel cell (CO-PEMFC). At the anode, carbon monoxide reacts with water in a water gas-shift reaction, generating carbon dioxide, electrons and protons. Protons transport through the membrane and react with oxygen at the cathode to generate water. The generated electrons get transported through an external circuit from anode to cathode thus producing the desired product of the system, electricity. Unfortunately, carbon monoxide cannot be used as fuel in the conventional PEMFC. Traditionally, the electrocatalysts of PEMFC are made of platinum-based material, which is strongly poisoned in the presence of carbon monoxide. However, platinum alloys such as platinum-ruthenium, can be used as the electrocatalyst. In this work, Platinum-Ruthenium based electrocatalysts were made and used in standard PEM fuel cell. Power density from CO-PEMFC will be compared based on composition of electrocatalyst, CO fuel concentration and cell temperature.

(604p) Techno-Economic Analysis of Fast Pyrolysis and Upgrading Facilities Employing Two Depolymerization Pathways

Yanan Zhang, Mechanical Engineering, Iowa State University, Ames, IA, Tristan R. Brown, Bioeconomy Institute, Iowa State University, Ames, IA, Guiping Hu, Iowa State University, Ames, IA and Robert Brown, Center for Sustainable Environmental Technologies, Iowa State University, Ames, IA

We evaluate the economic feasibilities of fast pyrolysis and upgrading facilities employing two depolymerization pathways which are fluid catalytic cracking (FCC) with a zeolite catalyst and hydrocracking. Three scenarios are analyzed: two-stage hydrotreating followed by FCC of the aqueous phase of pyrolysis oil with hydrogen purchasing; two-stage hydrotreating followed by FCC of the aqueous phase of pyrolysis oil with hydrogen from steam methane reforming; one-stage hydrotreating followed by hydrocracking. The main products for the first two scenarios are the petrochemicals while transportation fuels and hydrogen are the main products for the third scenario. The two pathways are modeled using Aspen Plus® for a 2000 metric tons/day facility. Equipment sizing and cost calculations are based on Aspen Economic Evaluation® software.

The bio-oil yield is assumed to be 65% based on dry biomass and we calculate the internal rate of return (IRR) for each scenario as functions of the costs of feedstock, fixed capital investment (FCI), hydrogen and catalyst costs and projected revenues for the facility. The analysis indicates that a facility employing two-stage hydrotreating followed by FCC of the aqueous phase of pyrolysis oil with hydrogen from steam methane reforming is able to generate the highest IRR among the three scenarios. Facility IRRs are 7.6%, 13.3% and 9.8% for facilities employing the three scenarios, respectively. Sensitivity analysis demonstrates that product yield, FCI, and biomass cost have the greatest impacts on facility IRR. Monte-Carlo uncertainty analysis shows that two-stage hydrotreating followed by FCC of the aqueous phase of pyrolysis oil with hydrogen from steam methane reforming is of low risk for investment.

(604q) Process Modeling of a Coal, Wind and Nuclear Hybrid Energy System

Kyle Buchheit, Chemical & Biological Engineering, Missouri University Of Science & technology, Rolla, MO and **Joseph D. Smith**, Chemical and Biological Engineering, Missouri University of Science and Technology, Rolla, MO

Independent non-renewable traditional sources of energy are unsustainable, but purely renewable sources of energy are not yet economically viable. A proposed hybrid energy system combining new technologies utilizing traditional energy sources coupled with a renewable source of power is considered. A Pressurized Circulating Fluid Bed (PCFB) combustor is examined to provide clean energy from traditional coal. The hypothetical plant is also coupled with a small modular nuclear reactor with renewable energy generation coming from harnessing and storing wind power.

A steady-state process simulation program is used to construct and evaluate the hybrid energy system process model. The major issue facing the simulation is the time dependent nature of energy demand. Since air currents are typically not in sync with the energy demands from the grid, the sources of wind power must be stored for later consumption. A High Temperature Steam Electrolysis (HTSE) unit is incorporated into the model to generate and store hydrogen as a fuel source to offset the time differential from energy generated to energy demanded.

The proposed system, combining both traditional sources of energy and new renewable sources, is economically and environmentally sustainable. Effluent from the PCFB combustor is fed into a gas scrubber to remove emissions including carbon dioxide which is fed to the HTSE to produce oxygen for oxy-fired operation of the PCFB and hydrogen for subsequent synfuel production. Although the many components of the proposed Hybrid Energy System currently exists, the current work analyzes the integrated energy system using a detailed process model to assess thermal efficiency and economic sustainability.

(604r) Identification of Components of Petroleum Fractions by Molecular Weights for the Formation of Molecular Models

Tatiane Gercina de Vilas and Cláudio Oller, Chemical Engineering Department, University of São Paulo, São Paulo, Brazil

Molecular models are built based on the characterization of the composition of oil, which consists of detailed molecular information to know in order to identify the compounds present. Molecular weight is one of the most important parameters for the characterization of petroleum fractions and various physical properties can be calculated from it. This work aims to build a database for classifying constituent components of asphaltenes in molecular groups. The classification is done by the similarity of their molecular weights in groups of homologous series. This classification is done through the use of data from the analysis of mass spectrometry and a search algorithm in language VBA (Visual Basic for Applications) with the aim of identifying a molecule in a specific set of molecules known (database). Due to the fact that the database is generated from compounds whose composition is known, it is known that a given compound composition will have all the possible compositions. To generate molecules based on knowledge of the oil composition, it is used the general formulas of organic compounds which are formed by homologous series. Such series are generated and their molecular weights are calculated and stored together with their molecular formulas and can therefore generate the molecular groups with the support of a computing routine in VBA language.

(135c) Mini-Session On Gary Powers

Il Moon, Chemical and Biomolecular Engineering, Yonsei University, Seoul, South Korea

This session is in honor of Professor Gary Powers. Professor Powers was active in research centered on design and systems analysis. His work specifically concentrated on process synthesis, risk and reliability, and engineering education techniques.

Process Synthesis
The synthesis of operating procedures for continuous and batch processing facilities is an important task in plant design. Dr. Powers' work in computer programs tested such models and assisted designers in generating high-integrity procedures and flowsheets. Island-driven hierarchical planning was combined with functional models to unify synthesis of the procedures and flowsheet.

Chemical Process Risk and Reliability Assessment
Dr. Powers developed a strategy to assess process hazards and diagnose faults using real-time data using fault trees and digraph models to develop causal and probabilistic relationships between sensed variables annd process hazards.

Powers was also active in the development of a theory for the verification of real-time control systems by combining chemical engineering process models with software engineering techniques. Symbolic verification tests were developed for stages in the control system that range from the initial specification of the user's requirements to final control code and hardware.

(135b) Mini-Session On David Himmelblau

Thomas F. Edgar, Department of Chemical Engineering, The University of Texas at Austin, Austin, TX and Warren D. Seider, Chemical and Biomolecular Engineering, The University of Pennsylvania, Philadelphia, PA

This session will feature reflections on the life and career of David Himmelblau.

Professor Himmelblau taught at the university for 42 years. He authored 11 books and over 200 articles on process analysis, process optimization and fault detection using artificial neural networks.

His book "Basic Principles and Calculations in Chemical Engineering" was recognized as one of the most important books in the field by the American Institute of Chemical Engineers (AIChE).

"Dr. Himmelblau is one of the professors I remember most fondly from my years in the graduate program at UT Austin," said Victor Martinez. "I used his textbook in basic chemical engineering calculations as an undergraduate and it was such a great honor to meet him in person when I arrived for my doctorate. He was such a joy to talk to, always in a good mood and always willing to help."

Himmelblau was president of the Computer Aids for Chemical Engineering Education Corporation (CACHE) as well a director of AIChE during his career. He received numerous awards, including the CACHE Award from the Chemical Engineering Division of the American Society for Engineering Education; the Joe J. King Professional Engineering Achievement Award; and the AIChE Founders Award.

He received his bachelor's in chemical engineering from the Massachusetts Institute of Technology in 1947 after serving as an officer in the U.S. Army during World War II. In 1957 he earned a doctorate in chemical engineering from the University of Washington.

(184a) An Algorithm for Bound-Constrained Problems in Simulation Optimization

Satyajith Amaran, Carnegie Mellon University, Pittsburgh, PA, Nick Sahinidis, Chemical Engineering, Carnegie Mellon University, Pittsburgh, PA, Bikram Sharda, The Dow Chemical Company, Freeport, TX and Scott J. Bury, Engineering Sciences, The Dow Chemical Company, Midland, MI

The field of simulation optimization (SO), which deals with optimizing over simulations or experiments, is faced with challenges at every turn. Its use is of import in several engineering and business applications, including inventory management, urban traffic control, manufacturing plant design, and call center operations.

Apart from the lack of an explicit functional form for the objective function, and the steep cost of evaluating the objective at a point--the stumbling blocks in derivative-free optimization--SO also involves underlying simulations that are inherently stochastic in nature. Furthermore, the fundamental questions regarding the choice of an appropriate objective to include variability in function value and the characterization of optimal points add to its complexity.

We pose the problem as the local minimization of a statistic of a continuous black-box function, assuming that bounds are available on the control variables. Though the setup of the problem prohibits the direct use of traditional optimization, we draw heavily both from math programming and machine learning techniques to support our algorithmic framework. We present a deterministic algorithm that simultaneously looks for minima and performs relatively few function evaluations. The method does this by applying modern regression techniques while trading off performing simulations at new points against performing repeated simulations, and then embedding this in an iterative trust-region scheme.

A sketch of the fundamental ideas, a method for characterizing and comparing algorithms, and an evaluation of performance on test cases will be presented.

(184b) Globally Optimizing Mixed-Integer Quadratically-Constrained Quadratic Programs: Advances in Glomiqo

Christodoulos A. Floudas and Ruth Misener, Chemical and Biological Engineering, Princeton University, Princeton, NJ

Major applications of mixed-integer quadratically-constrained quadratic optimization problems (MIQCQP) include quality blending in process networks, separating objects in computational geometry, and portfolio optimization in finance [10]. Specific instantiations of MIQCQP in process networks optimization problems include: pooling problems, distillation sequences, wastewater treatment and total water systems, hybrid energy systems, heat exchanger networks, reactor-separator-recycle systems, separation systems, data reconciliation, batch processes, crude oil scheduling, and natural gas production. Computational geometry problems formulated as MIQCQP include: point packing, cutting convex shapes from rectangles, maximizing the area of a convex polygon, and chip layout and compaction. Portfolio optimization in financial engineering can also be formulated as MIQCQP.

We consider a general framework for deterministically addressing MIQCQP to epsilon-global optimality [9, 10]. Algorithmic components include: reformulating user input, detecting special mathematical structure, generating tight convex relaxations, dynamically adding separating cuts, partitioning the search space, bounding variables, and finding feasible solutions. We consider individual solution strategies and their practical interaction.

We discuss computational experience with the Global Mixed-Integer Quadratic Optimizer, GloMIQO. GloMIQO is validated against a MIQCQP test suite including the following problem classes: process networks; computational geometry; maximum clique; quadratic assignment. We also address standard test library examples from MINLPLib [4], GLOBALLib [5, 7], PerformanceLib, www.minlp.org, and the Bonmin test set [3].

New components in GloMIQO include integrating a validated interval arithmetic library, dynamically adding separating hyperplanes, addressing discrete/discrete and discrete/continuous products, selectively adding bilinear terms to the model for advanced application of the Reformulation-Linearization Technique (RLT) [6], eliminating bilinear terms based on knapsack constraint inferences, and removing RLT equations to expedite the search for feasible solutions. With respect to the cutting planes, we highlight the inherit complementarity between globally valid cuts derived from alphaBB convexification [1, 2] and locally valid cuts based on higher-order (4 - 7D) edge-concave expressions [8, 11, 12].

References

[1] C. S. Adjiman, I. P. Androulakis, and C. A. Floudas. A global optimization method, alphaBB, for general twice differentiable NLPs-II. Implementation and computional results. *Comput. Chem. Eng.*, 22:1159– 1179, 1998b.

[2] C. S. Adjiman, S. Dallwig, C. A. Floudas, and A. Neumaier. A global optimization method, alphaBB, for general twice differentiable NLPs-I. Theoretical advances. *Comput. Chem. Eng.*,22:1137 – 1158, 1998a.

[3] P. Bonami, L. T. Biegler, A. R. Conn, G. Cornuejols, I. E. Grossmann, C. D. Laird, J. Lee, A. Lodi, F. Margot, N. Sawaya, and A.Waechter. An algorithmic framework for convex mixed integer nonlinear programs. *Discrete Optimization*, 5(2):186 – 204, 2008.

[4] M. R. Bussieck, A. S. Drud, and A. Meeraus. MINLPLib - a collection of test models for mixed-integer nonlinear programming. *INFORMS J. Comput*,15(1), 2003.

[5] C. A. Floudas, P. M. Pardalos, C. S. Adjiman, W. R. Esposito, Z. H. Gumus, S. T. Harding, J. L. Klepeis, C. A.

Meyer, and C. A. Schweiger. Handbook of Test Problems in Local and Global Optimization. Kluwer Academic Publishers, 1999.

[6] L. Liberti and C. C. Pantelides. An exact reformulation algorithm for large nonconvex NLPs involving bilinear terms. *J. Glob. Optim.*, 36(2):161–189, 2006.

[7] A. Meeraus. GLOBALLib. http://www.gamsworld.org/global/globallib.htm.

[8] C. A. Meyer and C. A. Floudas. Convex envelopes for edge-concave functions. *Math. Program.,*103(2):207–224, 2005.

[9] R. Misener and C. A. Floudas. Global optimization of mixed-integer quadratically-constrained quadratic programs (MIQCQP) through piecewise-linear and edge-concave relaxations. *Math. Program. B.*Accepted for Publication; http://www.optimization-online.org/DB_HTML/2011/11/3240.html.

[10] R. Misener and C. A. Floudas. GloMIQO: Global Mixed-Integer Quadratic Optimizer. *J. Glob. Optim.* Accepted for Publication (DOI: 10.1007/s10898-012-9874-7).

[11] A. D. Rikun. A convex envelope formula for multilinear functions. *J. Glob. Optim.*, 10:425 – 437, 1997.

[12] F. Tardella. Existence and sum decomposition of vertex polyhedral convex envelopes. *Optim. Lett.*, 2:363–375, 2008.

(184c) Globally Optimal Parameter Identification for ODE Models with Discontinuities

Jeremy A. Conner, Chemical & Biomolecular Engineering, University of California, Los Angeles, Los Angeles, CA and Vasilios Manousiouthakis, Chemical & Biomolecular Engineering Department,, University of California Los Angeles, Los Angeles, Los Angeles, CA

This work applies a previously-developed globally optimal method of determining parameters from experimental data to ODE/DAE models in which the state variable trajectories can potentially exhibit discontinuities. Typical non-linear regression techniques used for ODE/DAE models with no analytical integral solution do not guarantee global optimality of the solution, and fail to ascertain validity of the proposed problem parameterization for the given data. To address these shortcomings, our method employs finite-dimensional global optimization techniques, and is guaranteed to identify the global optimum of the corresponding non-linear regression problem that employs a numerical approximation of the ODE model. The method employs two phases employing interval analysis, relaxation strategies for identifying efficient over and underestimators.

The aforementioned two-stage method will be employed in a case study involving a non-isothermal reaction/diffusion model which can potentially exhibit ignition/extinction phenomena. A comparison to the behavior of an isothermal reaction/diffusion model will be carried out. The identifiability of model parameters from gas concentration data will also be discussed.

(184d) Computing Elements of Generalized Jacobians for Nonsmooth Systems

Kamil A. Khan and Paul Barton, Process Systems Engineering Laboratory, Dept. of Chemical Engineering, Massachusetts Institute of Technology, Cambridge, MA

Nonsmoothness can enter chemical engineering models through intrinsic nonsmooth phenomena such as flow reversals, incorporation of check and relief valves, thermodynamic phase transitions, and flow regime changes, and through the use of nonsmooth physical property and chemical kinetics models. Conventional derivative-based methods for equation solving and optimization can fail when applied to nonsmooth models, since required derivative information may not be available, and convergence results may depend on differentiability assumptions.

Specialized methods for nonsmooth systems make use of the (Clarke) generalized Jacobian [1], which is a set-valued mapping containing slope information. In particular, semismooth Newton methods for equation solving [2] and bundle methods for local optimization [4] require computation of matrix-valued elements of the generalized Jacobian at each iteration. Computing these generalized Jacobian elements is difficult, however, because the generalized Jacobian does not obey calculus rules sharply.

In this work, a method is presented for evaluating generalized Jacobian elements for a nonsmooth function that is expressed as a finite composition of piecewise differentiable functions. The method is fully-automatable, and is guaranteed to be computationally tractable relative to the cost of a function evaluation. The method works by using a nonsmooth variant of the forward mode of automatic differentiation [3] to compute directional derivatives along certain basis directions. These basis directions are then perturbed in a systematic manner until they provide sufficient information to evaluate a generalized Jacobian element.

A developed implementation of the method in C++ is discussed, in which operator overloading is used to evaluate generalized Jacobian elements for nonsmooth functions that are provided as subroutines. The method is applied to several particular nonsmooth systems for illustration, including pipeline network models and flash models.

[1] F. H. Clarke, Optimization and Nonsmooth Analysis, John Wiley & Sons, Inc., New York, 1983.
[2] F. Facchinei, J.-S. Pang, Finite-dimensional variational inequalities and complementarity problems, Springer-Verlag New York, Inc., New York, 2003.
[3] A. Griewank, Automatic directional differentiation of nonsmooth composite functions. In:Recent Developments in Optimization, French-German Conference on Optimization. Dijon (1994)
[4] C. Lemarechal, et al., On a bundle algorithm for nonsmooth optimization. In: O.L. Mangasarian, R.R. Meyer, S.M. Robinson (eds.) Nonlinear Programming 4. Academic Press, New York, NY (1981)

(184e) Efficient Generation of Moment-Equations for Stochastic Simulation and Moment-Closure

Patrick Smadbeck, Chemical Engineering and Materials Science, University of Minnesota - Twin Cities, Minneapolis, MN and Yiannis Kaznessis, Department of Chemical Engineering and Materials Science, University of Minnesota, Minneapolis, MN

The use of stochastic mathematics has become common in chemical reaction simulation, especially for biological models, but current methods have severe computational drawbacks. These probability-based approaches are necessary when a system exists far from the thermodynamic limit, a common condition in biological systems. Unlike deterministic simulations, however, the most computationally effective way to produce simulation results is a kinetic Monte Carlo method which samples the underlying Chemical Master Equation (CME) dynamics. Unfortunately, such a method makes complete system analysis computationally expensive.

An alternative proposed by McQuarrie and others in the 1960s was a probability moment-based approach that reduces the CME to a set of linear ordinary differential equations (ODEs).[1] While this transform is effective in limited cases, such a reduction results in an infinite set of linear ODEs, or equivalently an infinite dimensional matrix, when 2nd-order reactions are involved. Despite this drawback the production of analytical expressions for the time-derivative of moments for an arbitrary chemical network is an active area of research. While these analytical expressions are useful for simple system, there is a need for a streamlined algorithmic approach when considering complex systems and the potential for moment-closure.

In the case where a moment-closure scheme can be applied it is obvious that all moments up to an order-M (for an M+1 moment closure) will be needed. Analytical approaches waste computation time in producing this potentially massive generate matrix. Additionally, the matrix is nearly lower-diagonal in nature and can require enormous memory allocation in the case of complex systems. The work presented focuses on the efficient production of all moment equations up to an order-M. The approach is based on applying differentials to the Z-transform (discrete Laplace transform) of the CME (Z-CME). It is shown that in all cases the modified basis set will conserve memory storage and produce a banded matrix rather than a nearly-lower diagonal matrix as is found with polynomial moments.

The advantages of this novel and efficient algorithm are that: 1) It is regressive in nature such that all moment equations are produced with little wasted computation; 2) It produces a banded matrix that necessitates the allocation of relatively little memory. A discussion of current work concerning moment-closure of arbitrary chemical networks, including the cumulant-zero techniques and System-Size Expansion, will be provided. Promising future paths will be presented in the context of applications utilizing the Z-CME moment generating algorithm.

[1] D. A. McQuarrie, C. J. Jachimowski, and M. E. Russell, Journal of Chemical Physics 40 (1964).

(184f) A Superbasin Kinetic Monte Carlo Method

Kristen Fichthorn, Dept. of Chemical Engineering, Pennsylvania State University, University Park, PA and Yangzheng Lin, Chemical Engineering, Penn State University, University Park, PA

A ubiquitous problem in atomic scale simulation of materials is the small barrier problem, in which the energy landscape presents "superbasins" with low intra-basin energy barriers relative to the inter-basin barriers. Rare-event simulation methods, such as kinetic Monte Carlo (KMC) and accelerated molecular dynamics, are inefficient for such systems because considerable effort is spent in simulating short-time, intra-basin motion without evolving the system significantly. We have developed a method for treating fast, intra-basin motion using a Master-Equation approach to obtain net, inter-basin escape rates. These escape rates are then incorporated into KMC simulations to reach significantly longer times than could be probed otherwise.

We present two different criteria for defining superbasins on the fly within a KMC simulation, then we discuss and apply our Master Equation solution strategy to model systems with superbasins that have various attributes. As the final application of our method, we use it to model Ga adatom diffusion on the beta 2(2x4) reconstruction of GaAs(001). The diffusion of Ga is a complex phenomenon involving many local minima. According to an analytic bond-order potential, up to 23 local minima can be involved in motion between two nearest-neighbor binding sites – although this number is less when first-principles density-functional theory is used to quantify the minima. We find that the diffusion of Ga involves climbing many small energy barriers to escape a superbasin and it is a classic example of the small-barrier problem. We simulate Ga adatom diffusion using our approach and achieve accelerations of up to ten orders of magnitude compared to a conventional KMC simulation.

(184g) A Re-Indexing Method for Performance Improvement of Gas-Solids Flow Simulations

Jean-Francois Dietiker, National Energy Technology Laboratory, Morgantown, WV; West Virginia University Research Corporation, Morgantown, WV

A Cartesian cut-cell method was recently implemented into the open-source multiphase flow solver MFIX (Multiphase Flow with Interphase Exchange). In the Eulerian/Eulerian approach, both gas and solids phases are treated as interpenetrating phases, and mass and momentum balance equation are solved for each phase. Momentum transfer between the solids and gas phase is achieved through an empirical drag law. MFIX employs a three-dimensional structured, staggered grid, which needs to be intersected with the geometry to identify fluid cells, boundary cells (called cut-cells), and blocked cells. During time advancement of the flow solution, a special treatment is required only for the cut-cells. For complex geometries, a large number of blocked cells can be generated, leading to a loss in performance. A new re-indexing method has been implemented to rearrange the data and loop only over active computational cells. The re-indexing method shows performance gain in the computations, in terms of memory management, and CPU time to solution. For parallel execution of the code, the data transfer is minimized and grid partition is optimized in terms of load imbalance. The re-indexing method is minimally invasive and timing results will be shown for gas-solids flow systems.

(186a) Regulating and Trapping an Ensemble of Brownian Particles by Broadcast the Stochastic Receding Horizon Control Policy

Gautam Kumar and Mayuresh V. Kothare, Chemical Engineering, Lehigh University, Bethlehem, PA

In the last decade, the attention of many control engineers has been focused in developing control strategies for regulating trajectories of micro and nano level objects in a specific medium. Emerging applications in this direction include but not limited to 1) the understanding of dynamical behaviors of biological objects and systems such as cellular interactions, DNA etc. in their natural environments for facilitating biomedical applications, 2) manipulation and regulation of nano and molecular structures for applications in drug delivery systems and laboratory on a chip technologies. The key challenge in facilitating these applications lies in the control over trajectories of these submicroscopic objects which are under continuous random fluctuations, also called Brownian fluctuations, induced by interactions of these objects with the medium in which they are placed. Thus the main question arises here is the design of optimal control strategies for suppressing or canceling these Brownian effects while achieving the desired performance of the system. Motived by this, in this work, we develop a "Broadcast Stochastic Receding Horizon Control (BSRHC)" strategy for regulating and trapping an ensemble of Brownian particles. The central idea of our BSRHC strategy lies in the integration of existing finite horizon based receding horizon control strategy with probabilistic tools such as the theory of supermartingale and the idea of broadcast of the same information to all particles in the ensemble. Further, the key feature of this new control architecture is the utilization of the minimum possible feedback information in designing optimal control inputs.

In the AICHE 2009 annual meeting, we presented our work on the broadcast model predictive control for multi-cellular systems using the aggregate system behavior as the measured feedback information [1]. We extended this work by developing a mathematical theory for regulating the behaviors of Brownian particles in one-dimensional space which was presented in the AICHE 2010 annual meeting [2]. Building on the same principles, here, we develop a novel "Broadcast Stochastic Receding Horizon Control" strategy to show the regulation and trapping of an ensemble of Brownian particles in one, two and three dimensional space by utilizing the measured position of a single particle in the ensemble as the only available feedback information.

We consider a system consisting of n particles. We assume that all particles in the ensemble make independent decisions i.e. the interactions among particles and the hydrodynamic effects of the medium on particles are neglected. The dynamical behaviors of particles are modeled using well known discrete time random walk models. Using the Donsker's theorem, first we show that the model is a reasonable approximation of continuous time Brownian model which shows the time-wise homogeneous phenomenon in a homogeneous medium. With this, we design the broadcast stochastic receding horizon control framework. The controller objective within this framework is to drive all particles towards a minimum trapping region and keep all particles in this region forever. For this, we formulate a finite horizon based stochastic receding horizon controller by developing the conditional expectation based non-homogeneous Markov model of the system. The transition probabilities of the model serve as the manipulated or control inputs. In the experimental designs, these probabilities are typically manipulated by applying external forces such as laser induced forces, magnetic fields, electro-kinetic fields etc. on particles placed in a given medium. At a given time step, these transition probabilities are designed over a control horizon by minimizing a conditional expectation based cost function over the finite prediction horizon under the system constraints. The first optimally computed transition probabilities are then broadcast among all particles. At each time step, the position of a particle closest to the origin which is outside the trapping region is measured. This information is used as the only available feedback in designing the next control moves.

In the presentation, we show the trapping of 100 Brownian particles simultaneously in one, two and three dimensional space by implementing the BSRHC strategy in a simulation environment. We show the trapping of all particles in the ensemble for an extended period of time. Finally, we guarantee convergence and stability of the controller with probability 1.

1. G. Kumar and M. V. Kothare. Broadcast Model Predictive Control of Multi-Cellular System, CAST Plenary Session, AICHE 2009.
2. G. Kumar and M. V. Kothare. A Mathematical Theory of Manipulating Suspended Multiple Brownian Particles Simultaneously in a Solution, AICHE 2010.

(186b) Improved State Estimation for High-Mix Semiconductor Manufacturing

Jin Wang, Auburn University, Auburn, AL and Q. Peter He, Chemical Engineering, Tuskegee University, Tuskegee, AL

High-mix manufacturing in semiconductor industry has driven the development of several non-threaded state estimation methods in the past a few years [1-7], which share information among different manufacturing context and avoid data segregation that threaded methods require. Assuming that the interaction among different contexts is linear, different algorithms such as linear regression [5-7] and the Kalman filter [1,2,6] have been applied to estimate the contributions from different variation sources. Among these methods, process disturbance is either modeled as a white noise disturbance [2,5,6] or as an integrated white noise disturbance [6,7]. However, it has been recognized that an integrated moving average (IMA) model provides much better approximation of the disturbances in industrial processes [8].

In this work, we extend the previously developed general framework [6] for non-threaded state estimation to cover a more realistic case, where the state disturbance is modeled as an integrated moving average process. Specifically, we first derive the state-space representation of the non-threaded state estimation problem which not only considers the integrated moving average disturbance, but also considers the fact that if a context item is not involved in a process run then its state does not change. Then the Kalman filter is applied to estimate the state. We use simulation examples to evaluate the performance of the proposed non-threaded state estimation method, and compared with an existing Kalman filter approach which considers integrated white noise disturbance. Three case studies are given to examine the performance and robustness of the proposed method, and we confirm that the proposed approach can provide satisfactory state estimation performance no matter the actually state disturbance is an IMA process or not; in addition, we found that the state estimation performance is not sensitive to the tuning parameter, which enables easy tuning for the proposed method. Finally, we found that by explicitly considering the fact that if a context item is not involved in the process run then its state should not evolve, we modified the Kalman filter based on the integrated white noise model. The modified version performs significantly better than the original one.

References

[1] A.J. Pasadyn and T.F. Edgar. Observability and state estimation for multiple product control in semiconductor manufacturing. IEEE Trans. Semic. Manuf., 18(4):592–604, 2005.

[2] C.K. Hanish. Run-to-run state estimation in systems with unobservable states. In Proc. AEC/APC Symp. XVII, Indian Wells, CA, 2005.

[3] E. Mos, V. Wang, A. Kisteman, L. Verstappen, and H. Megens. Generalized data-sharing and frequency domain optimization of overlay run-to-run feedback control. In The 6th European AEC/APC symposium, Dublin, Ireland, 2005.

[4] T. Levin, I. Geier, A. Zhivotovsky, n. Aframian, and H. Friedlander-Klar. Automated process control optimization to control low volume products based on high volume products data. In Proceedings of

the SPIE - data analysis and modeling for process control II, pages 145–146, vol. 5755, 2005.

[5] S.K. Firth, W.J. Campbell, A. Toprac, and T.F. Edgar. Just-in-time adaptive disturbance estimation for run-to-run control of semiconductor processes. IEEE Trans. Semicond. Manuf., 19(3):298–315, 2006.

[6] J. Wang, Q.P. He, and T.F. Edgar. State estimation in high-mix semiconductor manufacturing. J. Proc. Cont., 19:443–456, 2009.

[7] Amogh V Prabhu and Thomas F Edgar. A new state estimation method for high-mix semiconductor manufacturing processes. J. Proc. Cont., 19(7):1149–1161, 2009.

[8] G.E.P. Box, G.M. Jenkins, and J.F. MacGregor. Some recent advances in forecasting and control. Applied Statistics, 23:158–179, 1974.

Constant - yield control of continuous bioreactors

Georgios Savoglidis and Costas Kravaris

Process Control Laboratory, Department of Chemical Engineering,

University of Patras, 26500 Patras, Greece

Paper No. 186c, 2012 AIChE Annual Meeting, Pittsburgh, PA

Abstract: The process of anaerobic digestion is a common application in the treatment of industrial and communal wastewater treatment plants. The biochemical reactions taking place inside the fermenter can be described by a system of first order differential equations of the biomass and the limiting substrate. When the fermenter operates near optimal conditions that maximize production rate, the process is marginally stable. The intuitive idea of constant-yield control is known to improve the stability properties of the anaerobic digestion process by enlarging the stability region. In the present study, we develop a mathematical formulation of constant-yield control in the presence of biomass decay and endogenous metabolism, with main emphasis on stability analysis. In particular, we derive concrete formulas to estimate the size of the stability region of the closed loop system, and demonstrate the effectiveness of the proposed control strategy in enlarging the stability region.

Keywords: Constant – yield control, Output feedback control, Nonlinear control, Stability region estimate, Stabilization.

1. Introduction

Continuous stirred microbial bioreactors, often called chemostats, cover a wide range of applications; specialized "pure culture" biotechnological processes for the production of specialty chemicals (proteins, antibiotics etc.) as well as large-scale environmental technology processes of mixed cultures such as wastewater treatment. The dynamics of the chemostat is often adequately represented by a simple dynamic model involving two state variables, the microbial biomass x and the limiting organic substrate s.

A general model for chemostat dynamics is of the form (see e.g. Smith and Waltman, 1995, Chapter 1; Bailey and Ollis, 1986):

$$\begin{aligned} \frac{dx}{dt} &= -Dx + \mu(s)x - K_d x \\ \frac{ds}{dt} &= D(S_0 - s) - \frac{1}{Y_{x/s}}\mu(s)x - mx \end{aligned} \tag{1}$$

where D is the dilution rate, S_0 is the feed substrate concentration, $Y_{x/s}$ is a biomass yield factor, K_d is the biomass decay rate constant, m is the endogenous metabolism rate constant and $\mu(s)$ is the specific growth rate, a given function of s. The most widely used expressions for the specific growth rate are:

the Andrews or Haldane equation (Eq. 2a)

$$\mu(s) = \frac{\mu_{\max} s}{K_S + s + \dfrac{s^2}{K_I}} \tag{2a}$$

and the Monod equation (Eq. 2b)

$$\mu(s) = \frac{\mu_{\max} s}{K_S + s} \tag{2b}$$

where μ_{max} is the maximum specific growth rate, K_S the Monod kinetic constant and K_I the substrate inhibition kinetic constant, with the Monod kinetics being a special case of the Andrews $\left(\frac{1}{K_I}=0\right)$ for biological reactions where substrate inhibitory phenomena can be neglected.

The problem of the chemostat stabilization has received much attention in the literature (e.g. De Leenheer and Smith, 2003; Mailleret and Bernard, 2001; Syrou et al., 2004; Gouze and Robledo, 2006; Harmand et al., 2006; Karafyllis et al., 2008; Karafyllis et al., 2009). In the vast majority of the cases, the control of the chemostat is performed by adjusting the dilution rate D, which in most cases is the input of continuous reactors. Usually, the objective of control is to regulate the system at specific design conditions, which will net optimal performance.

One important class of applications of the chemostat is related to the anaerobic digestion process. In anaerobic digestion, the measured output of the system is the methane production rate

$$Q=Y_m\mu(s)x \tag{3}$$

where Y_m is the yield coefficient for methane production.

In the present study, we develop a mathematical formulation of constant-yield control for system (1) and we investigate the effect of the control strategy in enlarging the stability region of the continuous stirred biochemical reactor. Although the framework is developed around a widely used application, it can easily be generalized for a wider range of systems exhibiting similar dynamic behavior.

2. Open-loop system: Properties and optimal operating conditions

2.1 Hypothesis

Consider the dynamic system (1), with $\mu(s)$ given by (2a) or (2b), where the dilution rate D is the input variable of the system and $S_0, K_d, Y_{x/s}, \mu_{max}, K_S, K_I, m$ are constant parameters. The following hypothesis will be made throughout this paper:

$$-mY_{x/s} < K_d < \mu(S_0) \tag{H}$$

Assumption (H) guarantees that the steady states of the system exist and that they will always be positive; thus, within feasible bounds for the chemostat's states.

2.2 Equilibrium curve

The steady states of system (1) can be calculated from the set of equations:

$$\begin{aligned} x_s &= Y_{x/s}(S_0 - s_s)\frac{\mu(s_s) - K_d}{\mu(s_s) + mY_{x/s}} \\ D_s &= \mu(s_s) - K_d \end{aligned} \tag{4}$$

Hence the equilibrium curve is defined by

$$x_s = Y_{x/s}(S_0 - s_s)\frac{\mu(s_s) - K_d}{\mu(s_s) + mY_{x/s}} \quad \Leftrightarrow \quad x_s = (S_0 - s_s)\frac{1 - \dfrac{K_d}{\mu(s_s)}}{\dfrac{1}{Y_{x/s}} + \dfrac{m}{\mu(s_s)}} \tag{5}$$

Under assumption (H), the equilibrium curve (locus of points (x_s, s_s) with $x_s > 0, s_s > 0, D_s > 0$) has the shape shown in Figure 1.

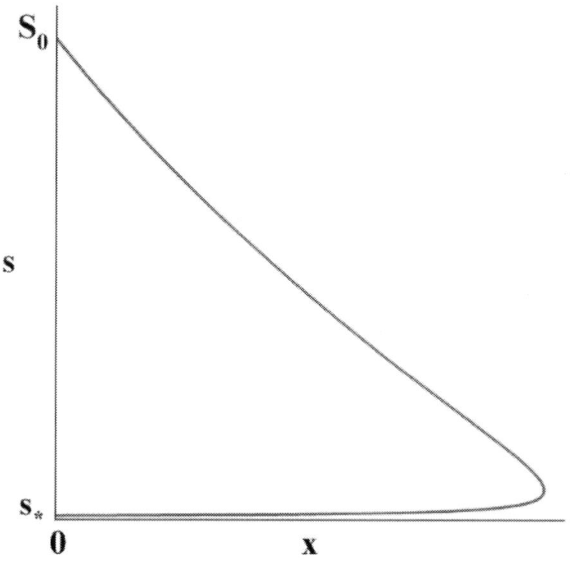

Figure 1.Equilibrium curve of the open-loop system

In Figure 1, s_* represents the smallest root of the equation $\mu(s) = K_d$ and is given by:

$$s_* = K_s \cdot \frac{2\dfrac{K_d}{\mu_{\max}}}{\left(1-\dfrac{K_d}{\mu_{\max}}\right)+\sqrt{\left(1-\dfrac{K_d}{\mu_{\max}}\right)^2-4\left(\dfrac{K_s}{K_I}\right)\left(\dfrac{K_d}{\mu_{\max}}\right)^2}} \tag{6a}$$

In the special case of Monod kinetics $\left(\dfrac{1}{K_I}=0\right)$,

$$s_* = K_s \cdot \frac{\dfrac{K_d}{\mu_{\max}}}{1-\dfrac{K_d}{\mu_{\max}}} \tag{6b}$$

is the unique root of $\mu(s) = K_d$.

Remark: For $\frac{1}{K_I} \neq 0$, the equation $\mu(s) = K_d$ has two roots, which are both real and positive; the smallest root s_* is given by (6a) and the largest root is

$$s^* = K_s \cdot \frac{2\frac{K_d}{\mu_{\max}}}{\left(1-\frac{K_d}{\mu_{\max}}\right) - \sqrt{\left(1-\frac{K_d}{\mu_{\max}}\right)^2 - 4\left(\frac{K_s}{K_I}\right)\left(\frac{K_d}{\mu_{\max}}\right)^2}} \tag{7}$$

Assumption (H) implies that $s_* < \min\left\{S_0, \sqrt{K_s \cdot K_I}\right\}$ and $s^* > \max\left\{S_0, \sqrt{K_s \cdot K_I}\right\}$.

In the special case of Monod kinetics $\left(\frac{1}{K_I} = 0\right)$, assumption (H) implies that the unique root s_* of $\mu(s) = K_d$ satisfies $s_* < S_0$.

Finally, it is important to point out that in practice, $\frac{K_d}{\mu_{\max}} \ll 1$ (usually significantly less than 0.1), therefore, in practice, s_* is a small fraction of K_s. Typically, s_* is two orders of magnitude smaller than S_0.

2.3 Optimal steady state maximizing methane production

For a given feed concentration, there is a value of the dilution rate that maximizes the methane production rate.

The steady state that corresponds to the maximization of methane production rate, i.e.

$$Q_s = Y_m \mu(s_s) x_s = Y_m Y_{x/s} \frac{\mu(s_s) - K_d}{\mu(s_s) + m Y_{x/s}} (S_0 - s_s) \mu(s_s) = Y_m \left(\mu(s_s) - K_d\right) \frac{(S_0 - s_s)}{\frac{1}{Y_{x/s}} + \frac{m}{\mu(s_s)}},$$

draws technical interest. The methane production rate is maximized when:

$$\frac{dQ_s}{ds}=0 \Leftrightarrow \frac{d\mu(s_s)}{ds}(S_0-s_s)=\frac{\mu(s_s)}{\dfrac{\mu(s_s)}{\mu(s_s)-K_d}+\dfrac{mY_{x/s}}{\mu(s_s)+mY_{x/s}}}$$

or equivalently

$$\frac{d\mu(s_s)}{ds}(S_0-s_s)=\frac{\mu(s_s)}{\dfrac{1}{1-\dfrac{K_d}{\mu(s_s)}}+\dfrac{\dfrac{mY_{x/s}}{\mu(s_s)}}{1+\dfrac{mY_{x/s}}{\mu(s_s)}}}. \tag{8}$$

Substituting $\mu(s)$ from (2a) to the above expression, leads to:

$$\frac{\mu_{\max}\left(K_s-\dfrac{s_s^2}{K_I}\right)}{\left(K_s+s_s+\dfrac{s_s^2}{K_I}\right)^2}(S_0-s_s)=\frac{\dfrac{\mu_{\max}s_s}{K_s+s_s+\dfrac{s_s^2}{K_I}}}{\dfrac{s_s}{s_s-\dfrac{K_d}{\mu_{\max}}\left(K_s+s_s+\dfrac{s_s^2}{K_I}\right)}+\dfrac{\dfrac{mY_{x/s}}{\mu_{\max}}\left(K_s+s_s+\dfrac{s_s^2}{K_I}\right)}{s_s+\dfrac{mY_{x/s}}{\mu_{\max}}\left(K_s+s_s+\dfrac{s_s^2}{K_I}\right)}}$$

which is, in general, a 6th-degree polynomial equation with respect to s_s. Only one of its roots corresponds to a positive steady state $(x_s>0,\ s_s>0, D_s>0)$ and it corresponds to the optimal steady state value s_s^{opt}.

For the following values of the parameters:

$$S_0=15\,g/\ell, Y_{x/s}=0.215\,g/g, K_d=0.031d^{-1}, m=0.1d^{-1}$$
$$\mu_{\max}=0.412\,d^{-1}, K_S=0.551\,g/\ell, K_I=5\,g/\ell$$

the optimal nontrivial, positive steady state calculated through (8) is $s_s^{opt}=1.354\,g/\ell$, which substituted into (5) gives the optimal steady state value for the biomass to be $x_s^{opt}=2.357\,g/\ell$. The optimal operating conditions are achieved for a dilution rate value of $D_s^{opt}=0.215\,d^{-1}$. With the exception of the value for the endogenous metabolism m, where a representative value has been used, all other parameter values come from parameter fitting of experimental data from a lab-scale anaerobic digester (Savoglidis et al., 2010).

2.4 Local asymptotic stability

The eigenvalues of the linearization of the open-loop system (1) are the roots of the quadratic polynomial

$$\lambda^2+\left(\mu(s_s)-K_d+\frac{\mu(s_s)-K_d}{\mu(s_s)+mY_{x/s}}(S_0-s_s)\frac{d\mu}{ds}(s_s)\right)\lambda+\left(\mu(s_s)-K_d\right)\left(S_0-s_s\right)\frac{d\mu}{ds}(s_s).$$

Hypothesis (H) guarantees that $\left\{\begin{matrix}\mu(s_s)>K_d\\ \mu(s_s)>-mY_{x/s}\\ s_s<S_0\end{matrix}\right\}$, therefore it is the sign of

$\frac{d\mu}{ds}(s_s)$ that determines its local stability characteristics. In particular, a positive equilibrium will be:

- locally asymptotically stable if $\frac{d\mu}{ds}(s_s)>0$
- unstable if $\frac{d\mu}{ds}(s_s)<0$.

Note that, because

$$\frac{d\mu(s_s)}{ds}(S_0-s_s)=\frac{\mu(s_s)}{\frac{\mu(s_s)}{\mu(s_s)-K_d}+\frac{mY_{x/s}}{\mu(s_s)+mY_{x/s}}}>0$$

the optimal steady state is always locally asymptotically stable.

2.5 The need for control

Figure 2A depicts the phase portrait of the system dynamics under constant dilution rate *D* and in particular, for the optimal value of $D_s^{opt}=0.215\,d^{-1}$

calculated in §2.3. For this value of dilution rate, the system has two steady states: one is the optimal steady state $\left(x_s^{opt} = 2.357\, g/\ell, s_s^{opt} = 1.354\, g/\ell\right)$, which is stable, and the other is an unstable steady state $\left(x_s^{unst} = 2.240\, g/\ell, s_s^{unst} = 2.034\, g/\ell\right)$. In the graph, the points in green and red solid circles represent the optimal and the unstable steady states respectively. Figure 2B is a detail of 2A close to the steady states, revealing the region of attraction of each steady state. Notice that the optimal steady state (green dot) is locally asymptotically stable but the stability region is very small, due to the fact that very close to it lies the unstable steady state (red dot). This makes the optimal operation of the chemostat very sensitive to disturbances, which may come in the form of parameter uncertainty or input perturbation.

The goal of the proposed control law is the stabilization of the system in the sense of enlargement of the stability region.

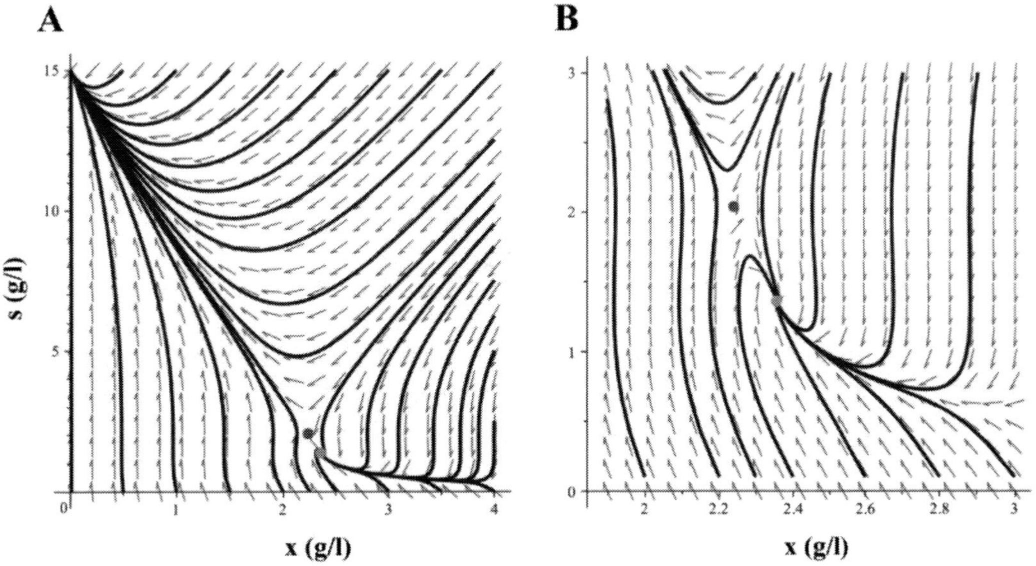

Figure 2.**(A)**Phase portrait under constant dilution rate and $K_d \neq 0, m \neq 0$. **(B)**Detail of the open loop phase portrait close to the steady states. Red dot: unstable steady state, green dot: stable steady state.

3. Closed-loop system: Output feedback control and estimation of the stability region

Consider the dynamic system (1), with measured output given by (3). The control objective is to stabilize the system at a given design steady state, e.g. the optimal steady state that corresponds to maximal methane production rate.

3.1 Constant Yield Control Law in the absence of biomass decay and endogenous metabolism $(K_d = 0, m = 0)$

For the control of anaerobic digesters, a successful practical approach is the so-called constant-yield control policy (Pullammanappallil et al., 1998), which is based on the intuitive idea that the control law must keep

$$\frac{\text{Methane production rate}}{\text{Organic substrate feed rate}} = \text{constant}\,.$$

In mathematical terms, this translates to

$$\frac{Q}{DS_0} = \text{constant} \quad \rightarrow \quad \frac{Q}{DS_0} = \frac{Q_s^{des}}{D_s^{des} S_0} = \frac{Y_m Y_{x/s}\left(S_0 - s_s^{des}\right)}{S_0},$$

which leads to the following control law:

$$D = \frac{1}{Y_m Y_{x/s}(S_0 - s_s^{des})} Q \qquad (9)$$

where s_s^{des} is the value of the substrate concentration at the design steady state, which is naturally chosen to be the optimal steady state that maximizes the methane production rate. The control law given by (9) is a proportional output feedback control law, with gain $\frac{1}{Y_m Y_{x/s}(S_0 - s_s^{des})}$, which depends upon the design steady state conditions.

The closed-loop stability properties under the feedback control law (9) are well established in the literature for the special case of $K_d = 0$, $m = 0$ (Mailleret and Bernard, 2001; Syrou et al., 2004; Karafyllis et al., 2008). In particular, in Syrou et al. (2004), a control-Lyapunov function approach was formulated, that led to exactly the same control law (9), and also it was shown that the resulting closed loop system is asymptotically stable, with stability region the entire open first quadrant. Moreover, it was shown that the control law (9) is robust with respect to bounded errors in S_0 (Karafyllis et al., 2008).

Figure 3 depicts the phase portrait of the closed loop system for of $K_d = 0$, $m = 0$ and all the other parameter values as in §2.3, with the design steady state being the optimal steady state. As it has been shown in Syrou et al.(2004), the trajectories of the closed loop system are straight lines, terminating at the optimal steady state.

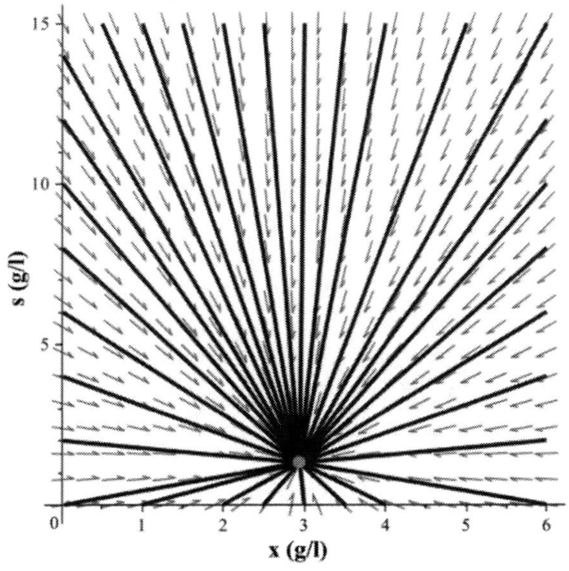

Figure 3. Phase portrait of closed-loop dynamics for $K_d = 0, m = 0$ leading to the optimal steady state. Green dot: stable steady state.

When $K_d \neq 0$ and/or $m \neq 0$, it is possible to generalize the control law. If the objective is to enforce closed loop stability over the entire first quadrant, this is still feasible, but the necessary control law is state feedback (Karafyllis et al., 2009). In the present paper, we generalize the concept of constant yield control in the presence of death rate of the biomass $K_d \neq 0$ and/or consumption of the substrate for endogenous metabolism $m \neq 0$, which will lead to an output feedback controller, and subsequently, the resulting closed loop stability region will be estimated. This point of view is preferable in terms of the significant practical advantages of output feedback. However, as we will see, the closed loop stability region will not cover the entire open first quadrant, even though the enlargement of the stability region will be significant.

3.2 Constant Yield Control Law in the presence of biomass decay and/or endogenous metabolism

Requesting that the control law must keep

$$\frac{\text{Methane production rate}}{\text{Organic substrate feed rate}} = \text{constant}$$

leadsto

$$\frac{Q}{DS_0} = \frac{Y_m \mu(s_s^{des}) x_s^{des}}{D_s^{des} S_0} = \frac{Y_m \mu(s_s^{des}) Y_{x/s}(S_0 - s_s^{des}) \dfrac{\mu(s_s^{des}) - K_d}{\mu(s_s^{des}) + mY_{x/s}}}{\left(\mu(s_s^{des}) - K_d\right) S_0} = \frac{Y_m (S_0 - s_s^{des})}{\left(\dfrac{1}{Y_{x/s}} + \dfrac{m}{\mu(s_s^{des})}\right) S_0}$$

from which, solving for the dilution rate, leads to the following control law:

$$D = \frac{\dfrac{1}{Y_{x/s}} + \dfrac{m}{\mu(s_s^{des})}}{Y_m (S_0 - s_s^{des})} Q \tag{10}$$

Control law (10) generalizes (9), while maintaining its attractive feature of being a simple linear proportional output feedback. The gain of the controller explicitly depends on the value of m, but in essence it depends on both m and K_d, since they both affect the selection of the design steady state s_s^{des}.

Under control law (10), or equivalently

$$D = \frac{\frac{1}{Y_{x/s}} + \frac{m}{\mu(s_s^{des})}}{(S_0 - s_s^{des})} \mu(s)x \tag{11}$$

the closed-loop system is

$$\begin{aligned} \frac{dx}{dt} &= \left(1 - \frac{\frac{1}{Y_{x/s}} + \frac{m}{\mu(s_s^{des})}}{S_0 - s_s^{des}} x \right) \mu(s)x - K_d x \\ \frac{ds}{dt} &= \left(\frac{\frac{1}{Y_{x/s}} + \frac{m}{\mu(s_s^{des})}}{S_0 - s_s^{des}} (S_0 - s) - \frac{1}{Y_{x/s}} \right) \mu(s)x - mx \end{aligned} \tag{12}$$

or equivalently

$$\begin{aligned} \frac{dx}{dt} &= -\left(\frac{\frac{1}{Y_{x/s}} + \frac{m}{\mu(s_s^{des})}}{S_0 - s_s^{des}} x - \left(1 - \frac{K_d}{\mu(s_s^{des})} \right) \right) \mu(s)x - K_d \left(\frac{1}{\mu(s)} - \frac{1}{\mu(s_s^{des})} \right) \mu(s)x \\ \frac{ds}{dt} &= -(s - s_s^{des}) \left(\frac{\frac{1}{Y_{x/s}} + \frac{m}{\mu(s_s^{des})}}{S_0 - s_s^{des}} + m \frac{\frac{1}{\mu(s)} - \frac{1}{\mu(s_s^{des})}}{s - s_s^{des}} \right) \mu(s)x \end{aligned} \tag{13}$$

from which it is transparent that $s_s = s_s^{des}$, $x_s = (S_0 - s_s^{des}) \frac{1 - \frac{K_d}{\mu(s_s^{des})}}{\frac{1}{Y_{x/s}} + \frac{m}{\mu(s_s^{des})}} = x_s^{des}$ is a steady state of the closed-loop system.

3.2.1 *Steady state(s) of the closed-loop system*:

From the substrate equation of (12), we see that s_s is a steady state of the closed-loop system if and only if it is a root of

$$\frac{\frac{1}{Y_{x/s}}+\frac{m}{\mu(s_s^{des})}}{S_0-s_s^{des}}(S_0-s_s)=\frac{1}{Y_{x/s}}+\frac{m}{\mu(s_s)} \tag{14}$$

When $m=0$, the above has only one solution $s_s=s_s^{des}$.

For $m\neq 0$ and $\mu(s)$ given by (2a) or (2b), (14) is a quadratic equation in s_s.

- One root is

$$s_s=s_s^{des}$$

  with corresponding biomass steady state being

$$x_s=(S_0-s_s^{des})\frac{1-\frac{K_d}{\mu(s_s^{des})}}{\frac{1}{Y_{x/s}}+\frac{m}{\mu(s_s^{des})}}=x_s^{des}$$

- and the other root is

$$s_s=\frac{S_0-s_s^{des}}{1+\left(1+\frac{S_0}{K_I}+\frac{\mu_{max}}{Y_{x/s}m}\right)\frac{s_s^{des}}{K_S}}=s_u$$

  with corresponding value of the biomass steady state being

$$x_s=\frac{(S_0-s_s^{des})\left(1-\frac{K_d}{\mu(s_u)}\right)}{\frac{1}{Y_{x/s}}+\frac{m}{\mu(s_s^{des})}}=x_s^{des}\frac{1-\frac{K_d}{\mu(s_u)}}{1-\frac{K_d}{\mu(s_s^{des})}}=x_u$$

  but it is not always positive.

The second steady state (x_u,s_u) will be in the first quadrant $(x_u>0 \text{ and } s_u>0)$ if and only if $\frac{K_d}{\mu(s_u)}<1$. Moreover, the condition $\frac{K_d}{\mu(s_u)}<1$

also guarantees positivity of the value of the dilution rate corresponding to (x_u, s_u).

Thus, we see that, for $m \neq 0$, the closed loop system may have one or two positive steady states, one always being the design steady state (x_s^{des}, s_s^{des}).

The other steady state (x_u, s_u) has the following important properties:

(a) As long as the design steady state is chosen to be the optimal steady state (see §2.3), $s_u < s_s^{des}$.

(b) As long as the design steady state is chosen to be the optimal steady state and (x_u, s_u) is in the 1st quadrant, (x_u, s_u) is unstable.

The proofs of the above properties are given in Appendix I.

Finally, it is important to note that, because $\mu(s)$ is a strictly increasing function over the interval $[0, s_s^{des}]$, the condition $\frac{K_d}{\mu(s_u)} < 1$ is equivalent to $s_* < s_u$.

Thus,

$$\begin{aligned}
&When \quad s_* < s_u \quad \Leftrightarrow \quad \frac{K_d}{\mu(s_u)} < 1 \quad \leftrightarrow \quad two\ positive\ steady\ states\ (x_s^{des}, s_s^{des})\ and\ (x_u, s_u) \\
&When \quad s_* > s_u \quad \Leftrightarrow \quad \frac{K_d}{\mu(s_u)} > 1 \quad \leftrightarrow \quad only\ one\ positive\ steady\ state\ (x_s^{des}, s_s^{des})
\end{aligned}$$

<u>Remark</u>: The steady states of the closed loop system can be alternatively calculated from (13) instead of (12). In this case, we see immediately from the substrate equation of (12) that $s = s_s^{des}$ is a steady state. The other steady state $s = s_u$ is the unique solution of the equation $\frac{\frac{1}{Y_{x/s}} + \frac{m}{\mu(s_s^{des})}}{S_0 - s_s^{des}} + m\frac{\frac{1}{\mu(s)} - \frac{1}{\mu(s_s^{des})}}{s - s_s^{des}} = 0$.

3.2.2 A lower limit on the region of stability

For any initial condition $(x(0), s(0))$ satisfying $s(0) = s_s$, where s_s is a root of (14), the solution of the differential equations of the closed loop system is

$$s(t) = s(0) = s_s$$

$$x(t) = \frac{1}{\dfrac{\dfrac{1}{Y_{x/s}} + \dfrac{m}{\mu(s_s^{des})}}{\left(S_0 - s_s^{des}\right)\left(1 - \dfrac{K_d}{\mu(s_s)}\right)} + \left(\dfrac{1}{x(0)} - \dfrac{\dfrac{1}{Y_{x/s}} + \dfrac{m}{\mu(s_s^{des})}}{\left(S_0 - s_s^{des}\right)\left(1 - \dfrac{K_d}{\mu(s_s)}\right)}\right) e^{(K_d - \mu(s_s))t}}$$

which proves that the straight lines

$$s = s_s^{des}$$

$$s = s_u = \frac{S_0 - s_s^{des}}{1 + \left(1 + \dfrac{S_0}{K_I} + \dfrac{1}{Y_{x/s}} \dfrac{\mu_{\max}}{m}\right) \dfrac{s_s^{des}}{K_S}}$$

are trajectories of the system.

- In the event that $K_d < \mu(s_u)$, in which case (x_u, s_u) is a positive steady state, the system trajectory lying on the straight line $s = s_u$ terminates at the point (x_u, s_u) since

$$x(t) = \frac{1}{\dfrac{\dfrac{1}{Y_{x/s}} + \dfrac{m}{\mu(s_s^{des})}}{\left(S_0 - s_s^{des}\right)\left(1 - \dfrac{K_d}{\mu(s_u)}\right)} + \left(\dfrac{1}{x(0)} - \dfrac{\dfrac{1}{Y_{x/s}} + \dfrac{m}{\mu(s_s^{des})}}{\left(S_0 - s_s^{des}\right)\left(1 - \dfrac{K_d}{\mu(s_u)}\right)}\right) e^{-(\mu(s_u) - K_d)t}} \rightarrow \frac{\left(S_0 - s_s^{des}\right)\left(1 - \dfrac{K_d}{\mu(s_u)}\right)}{\dfrac{1}{Y_{x/s}} + \dfrac{m}{\mu(s_s^{des})}} = x_u$$

- In the event that $K_d > \mu(s_u)$, in which case the closed-loop system does not have a second positive steady state (the design steady state is the only one), the system trajectory lying on the straight line $s = s_u$ terminates at the s-axis (at the point $(0, s_u)$) since

$$x(t)=\frac{1}{\dfrac{\dfrac{1}{Y_{x/s}}+\dfrac{m}{\mu(s_s^{des})}}{\left(S_0-s_s^{des}\right)\left(1-\dfrac{K_d}{\mu(s_u)}\right)}+\left(\dfrac{1}{x(0)}-\dfrac{\dfrac{1}{Y_{x/s}}+\dfrac{m}{\mu(s_s^{des})}}{\left(S_0-s_s^{des}\right)\left(1-\dfrac{K_d}{\mu(s_u)}\right)}\right)e^{(K_d-\mu(s_u))t}}\rightarrow 0$$

In either case, the line $s=s_u$ represents a lower limit on the region of stability of the design steady state.

Indeed, for every initial condition with $s(0)<s_u$, the system's trajectory will not be able to cross the line $s=s_u$, therefore it will not be able to reach the design steady state.

3.2.3 Qualitative aspects of the closed loop dynamics

In order to understand the qualitative behaviour of the closed loop dynamics and in particular the effect of the parameters K_d and m on the closed loop system's dynamic characteristics including the stability region, a series of numerical simulations have been performed. One of the key issues is to examinethe difference between having an additional unstable steady state of the closed loop system in the first quadrant, versus outside the first quadrant. Also, it is important to examine the limiting cases of zero death rate or zero rate of endogenous metabolism.

In the limiting case $m=0$, $K_d>0$, the closed-loop system has only one steady state: the design steady state. Figure 4A depicts the phase portrait of the closed-loop system for the case of $m=0$, with all the other parameters assuming the same values as in §2.3. The phase portraits for initial values of substrate close to $s_*=0.045\,g/\ell$ (figure 4B) and $s^*=61.407\,g/\ell$ (figure 4C), show that some trajectories terminate at the s-axis, leading to death of the microorganisms. These trajectories are undesirable and lie outside the region of stability of the design steady state.

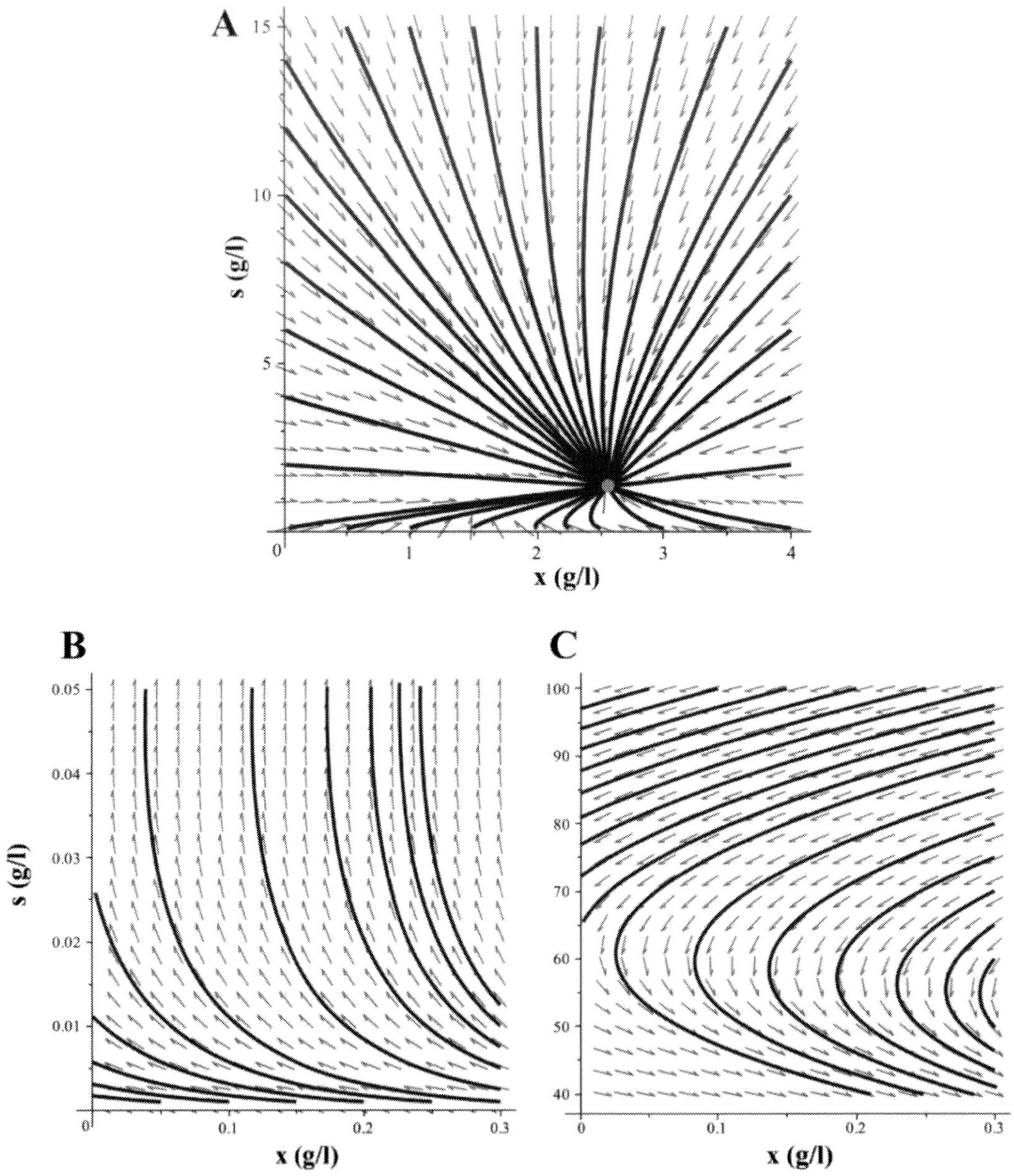

Figure 4.**(A)**Phase portrait of closed-loop dynamics for $K_d > 0, m = 0$ **(B)**detail for $s(0)$ close to s_* and **(C)**detail for $s(0)$ close to s^*. Green dot: stable steady state.

The phase portrait of figure 4indicates that there are two limiting trajectories that represent upper and lower bounds of the stability region. The upper bound is the one that terminates at the point $(0,s^*)$ (figure 5A) and the lower bound is the trajectory that terminates at the point $(0,s_*)$ (figure 5B).

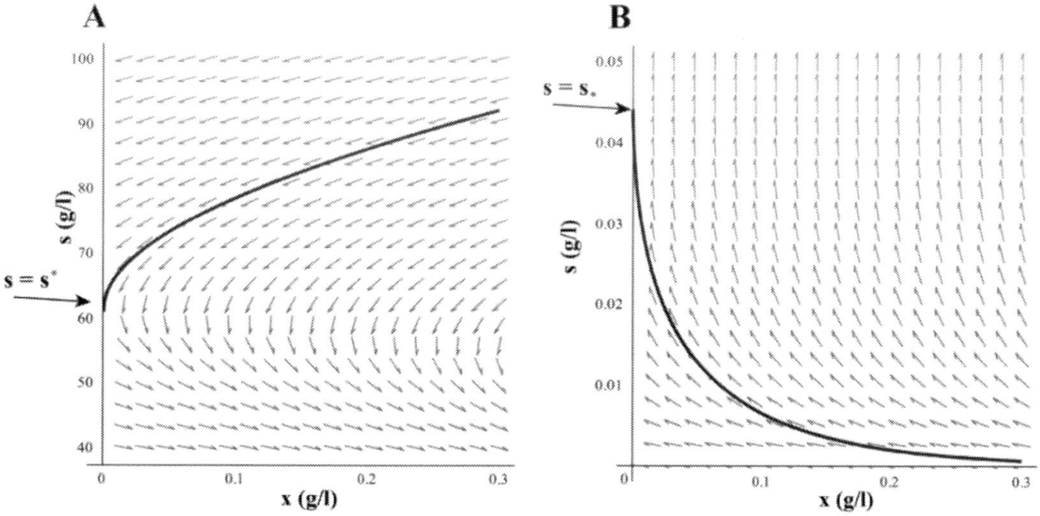

Figure 5. Trajectories defining the **(A)** upper bound of the stability region, ending in $s = s^*$and **(B)** lower bound of the stability region, ending in $s = s_*$.

Even though these exact bounds can be computed numerically, one can draw the following simple and important conclusion:

The strip

$$I = \left\{(x,s) \in \mathbb{R}^2 \mid x > 0,\ s_* < s < s^*\right\}$$

isa subset of the stability region of the design steady state. Because in practice s_* is very small and s^* is very large, the stability region essentially covers all practically meaningful initial conditions.

In the limiting case $K_d = 0$, $m > 0$, the closed-loop system has two positive steady states. One is the design steady state, which is stable, and the other is

$$\left\{\begin{aligned} x_u &= \frac{S_0 - s_s^{des}}{\dfrac{1}{Y_{x/s}} + \dfrac{m}{\mu_{\max}}\left(1 + \dfrac{K_S}{s_s^{des}} + \dfrac{s_s^{des}}{K_I}\right)} = x_s^{des} \\ s_u &= \frac{S_0 - s_s^{des}}{1 + \left(1 + \dfrac{S_0}{K_I} + \dfrac{\mu_{\max}}{Y_{x/s} m}\right)\dfrac{s_s^{des}}{K_S}} \end{aligned}\right\} \text{ which is unstable.}$$

Figure 6A depicts the closed loop phase portrait for $K_d = 0, m = 0.1\, d^{-1}$ and the other parameters assuming the same values as in §2.3. In the graph, the points in green and red solid circles represent the optimal and the unstable steady states respectively. Figure 6B is a focus of the phase portrait for low values of s close to $s_u = 0.241\, g/\ell$. It is clear from figure 6B that the stability region is limited below by the line $s = s_u$. For initial conditions $s(0)$ below s_u, system trajectories terminate at the x-axis, indicating depletion of the substrate, as a result of endogenous metabolism.

The stability region of the closed loop system is the set

$$I = \left\{(x,s) \in \mathbb{R}^2 \mid x > 0,\ s > s_u\right\}$$

Because s_u is typically small relative to s_s^{des}, the system has a large stability region.

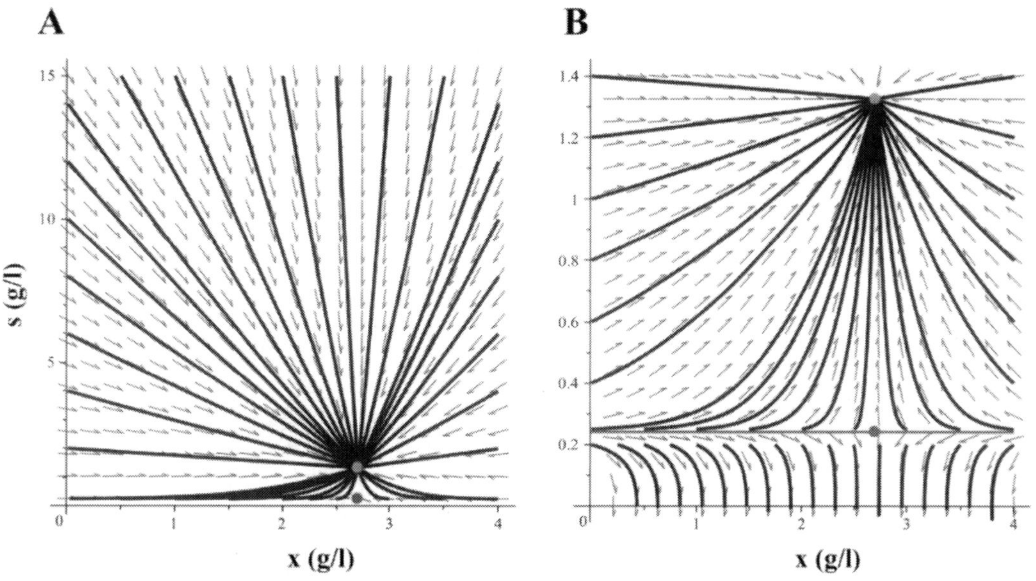

Figure 6.**(A)**Phase portrait of closed-loop dynamics for $K_d = 0$, $m > 0$ and **(B)**detail focus for concentrations of s close to s_u. Red dot: unstable steady state, green dot: stable steady state.

Consider now the situation where both K_d and m are positive. Then, the closed-loop system has two steady states $\left(x_s^{des}, s_s^{des}\right)$ and $\left(x_u, s_u\right)$, the latter being positive if and only if $K_d < \mu(s_u) \iff S_* < S_u$.

For the specific values $m = 0.1\, d^{-1}$, $K_d = 0.031\, d^{-1}$ that correspond to $K_d < \mu(s_u)$, the unstable steady state $\left\{\begin{matrix} x_u = 2.010\, g/\ell \\ s_u = 0.236\, g/\ell \end{matrix}\right\}$ lies in the first quadrant.

Figure 7A depicts the phase portrait of the closed loop system for the case under consideration. A detail of the phase portrait for values of s close to s_u (figure 7B) shows that there is a lower bound in the stability region: the line $s = s_u$. The second detail, for values of s close to s^* (figure 7C), reveals that there exist an upper bound trajectory (similar to figure 5A) ending on the s-axis at $\left(0,s^{*}\right)$.

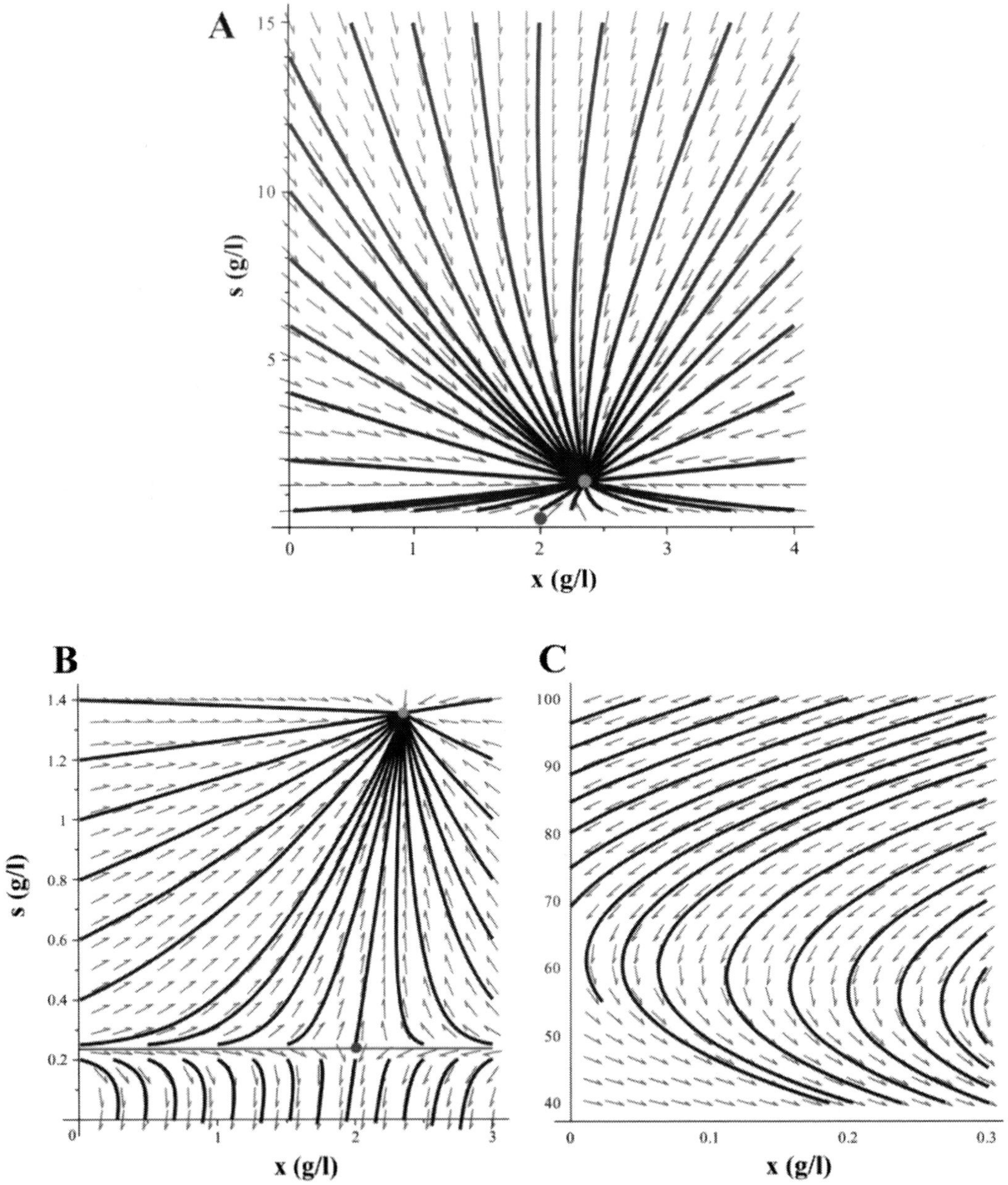

Figure 7.**(A)**Phase portrait of closed-loop dynamics for $K_{d}>0, m>0$ and $K_{d}<\mu\left(s_{u}\right)$, **(B)**detail for concentrations of s close to s_{u} and **(C)** detail for concentrations of s close to s^{*}. Red dot: unstable steady state, green dot: stable steady state.

On the other hand, for a smaller value of $m = 0.05\,d^{-1}$, and the other parameter values being the same, $K_d > \mu(s_u)$ the unstable steady state $\left\{\begin{matrix} x_u = -0.942\ g/\ell \\ s_u = 0.123\ g/\ell \end{matrix}\right\}$ does not lie in the first quadrant, the picture is somewhat different. The stability region is still limited by the line $s = s_u$, below which system is led to depletion of the substrate, and it is further limited by the value of s_* (which is larger than s_u), below which the system may be led to death of the biomass. In this case, the line $s = s_u$ can be considered to be a conservative lower limit of the stability region whereas the trajectory ending on the s-axis at the point $(0, s_*)$ is the exact lower bound.

Figure 8A depicts the phase portraits of the closed loop system, and two details of the same graph, for low values of s (figure 8B) and for low values of both x and s (figure 8C). In figures 8B and 8C the red horizontal line indicates, as in the previous case (figure 7), the line $s = s_u$. One can see that there exist trajectories above the red line that terminate on the s-axis and trajectories below the red line that terminate on the s-axis or on the x-axis.

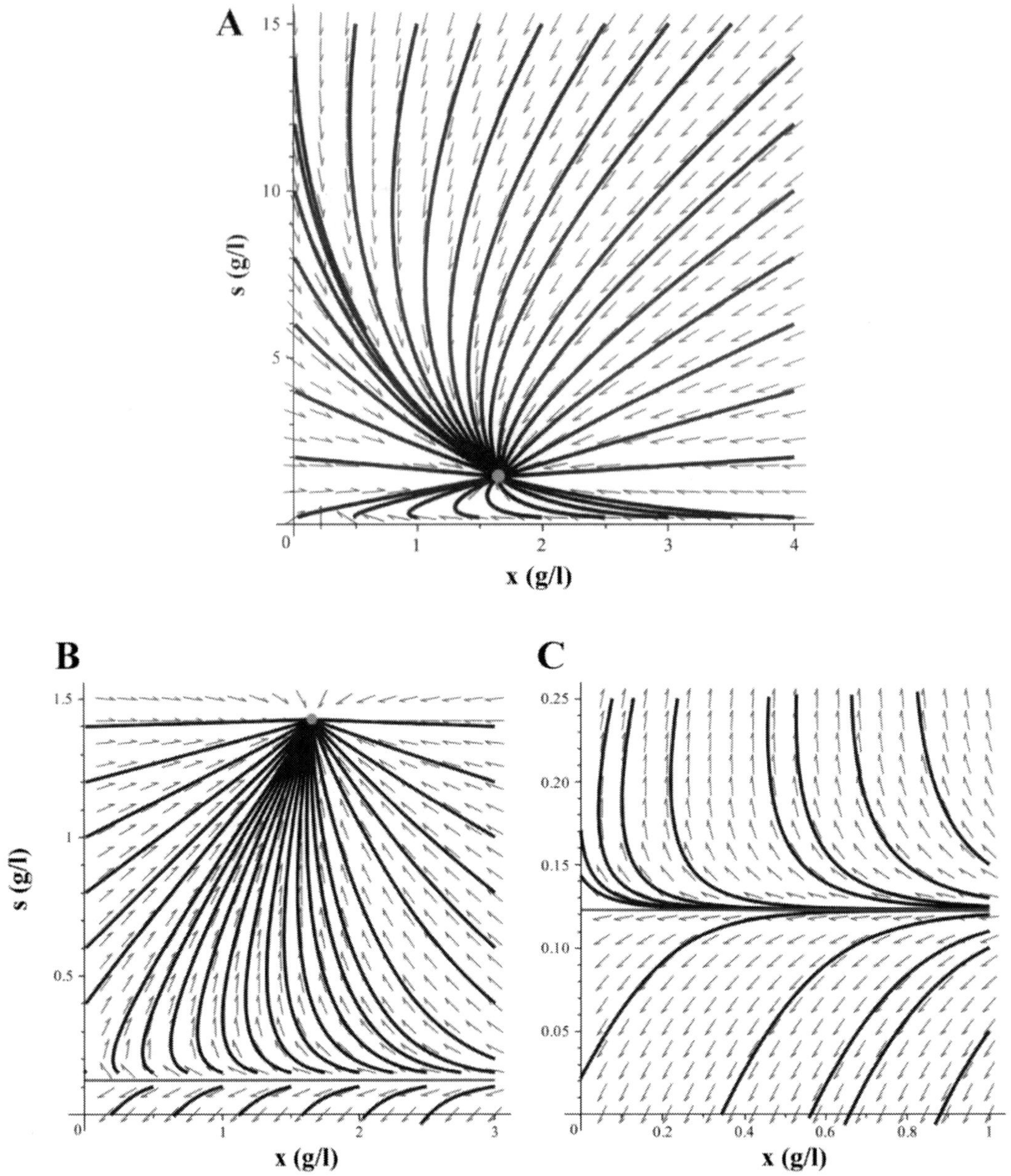

Figure 8.**(A)**Phase portrait of closed-loop dynamics for $K_d > 0, m > 0$ and $K_d > \mu(s_u)$,**(B)**detail for small concentrations of s and **(C)**detail for small values of x and values of s close to s_u. Green dot: stable steady state.

The table that follows summarizes the findings for the stability region in each case.

Table 1: Summary of the observations in the numerical simulations, depending on the relative sizes of death rate and endogenous metabolism

Case	Case
The stability region is the whole first quadrant $=\{(x,s)\in\mathbb{R}^2\}$. 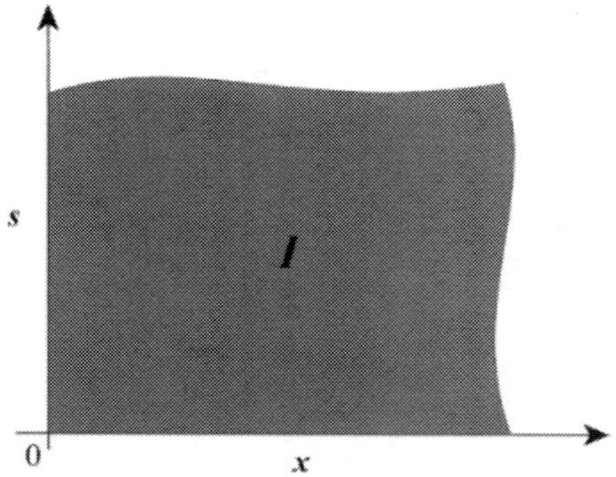	The strip is contained in the stability region. 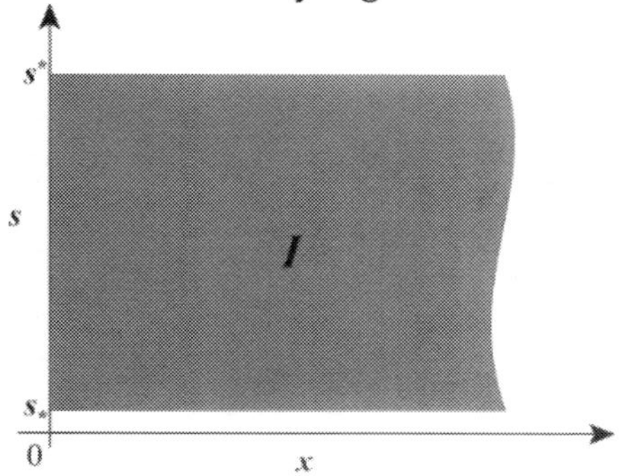

Case	Case	
The set is exactly the stability region 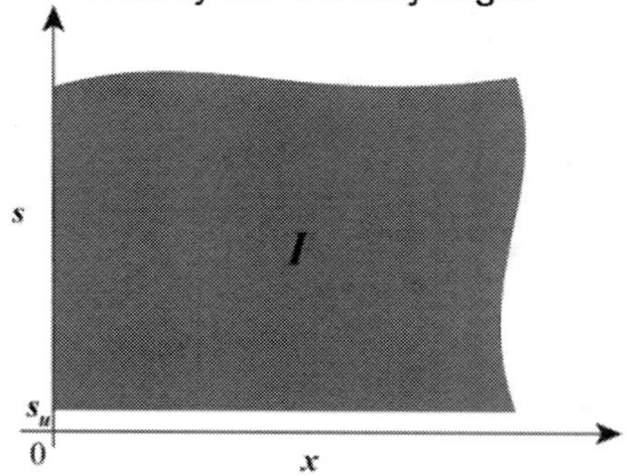	If the strip is contained in the stability region 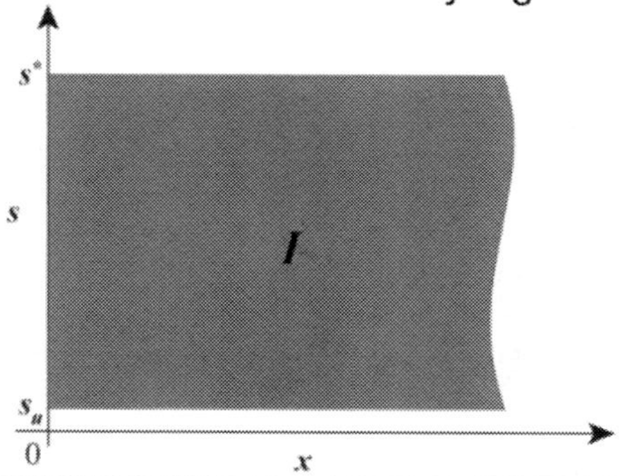	If the strip is contained in the stability region 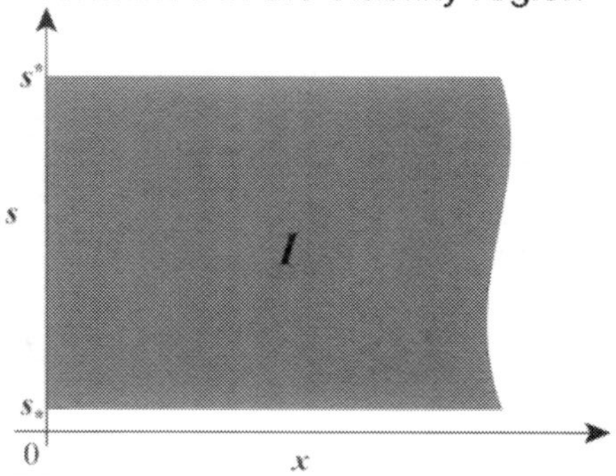

3.2.4 Estimation of the size of the stability region via a Lyapunov function approach

The conclusions drawn in the previous section are based on numerical simulations for different pairs of the parameters K_d and m. It is important to perform a rigorous Lyapunov-function stability analysis (Khalil, 1996) and actually prove bounds for the stability region.

The stability region of the chemostat under the output feedback control law (10) can be estimated using a quadratic Lyapunov function, accounting for the distance from the design steady state. In particular, using a Lyapunov function candidate of the form

$$V(x,s)=\frac{1}{2}\left[\left(\frac{\frac{1}{Y_{x/s}}+\frac{m}{\mu(s_s^{des})}}{S_0-s_s^{des}}x-\left(1-\frac{K_d}{\mu(s_s^{des})}\right)\right)^2+\eta\left(\frac{s}{s_s^{des}}-1\right)^2\right] \tag{15}$$

we prove the following proposition:

<u>Proposition</u>: Assume that $\mu(s)$ is given by (2a) or (2b) and $m \geq 0$. Also, assume that s_s^{des} is chosen to be the optimal steady state that corresponds to maximal value of Q on the equilibrium curve.

Then, the set

$$I=\left\{(x,s)\in\mathbb{R}^2 \,\middle|\, x>0,\ \mu(s)>K_d,\ \frac{\frac{1}{Y_{x/s}}+\frac{m}{\mu(s_s^{des})}}{S_0-s_s^{des}}+m\frac{\frac{1}{\mu(s)}-\frac{1}{\mu(s_s^{des})}}{s-s_s^{des}}>0\right\}$$

*is contained in the stability basin of the closed-loop system (**13**).*

In order to prove the Proposition, we need two Lemmas:

Lemma 1: Assume that $\mu(s)$ is given by (2a) or (2b) and $m \geq 0$. The positive definite quadratic function (15) with $\eta \geq \frac{1}{4}$, satisfies

$$\dot{V}(x,s) < 0 \quad \text{for every } (x,s) \in E - \left\{\left(x_s^{des}, s_s^{des}\right)\right\}$$

where

$$E = \left\{ (x,s) \in \mathbb{R}^2 \mid x > 0, \left[\left(m \frac{S_0 - s_s^{des}}{\frac{1}{Y_{x/s}} + \frac{m}{\mu(s_s^{des})}} \right) + \sqrt{\left(m \frac{S_0 - s_s^{des}}{\frac{1}{Y_{x/s}} + \frac{m}{\mu(s_s^{des})}} \right)^2 + \frac{1}{\eta}\left(K_d s_s^{des}\right)^2} \right] \frac{\frac{1}{\mu(s)} - \frac{1}{\mu(s_s^{des})}}{s - s_s^{des}} > -2 \right\}.$$

Equivalently, the set E can be represented as

$$E = \left\{ (x,s) \in \mathbb{R}^2 \mid x > 0, s > \tilde{s} \right\}$$

where

$$\tilde{s} = \frac{K_S}{\frac{s_s^{des}}{K_I} + \frac{2}{\left(\frac{\frac{m}{\mu_{\max}} \frac{S_0 - s_s^{des}}{s_s^{des}}}{\frac{1}{Y_{x/s}} + \frac{m}{\mu_{\max}}\left(1 + \frac{s_s^{des}}{K_I} + \frac{K_S}{s_s^{des}}\right)} \right) + \sqrt{\left(\frac{\frac{m}{\mu_{\max}} \frac{S_0 - s_s^{des}}{s_s^{des}}}{\frac{1}{Y_{x/s}} + \frac{m}{\mu_{\max}}\left(1 + \frac{s_s^{des}}{K_I} + \frac{K_S}{s_s^{des}}\right)} \right)^2 + \frac{1}{\eta}\left(\frac{K_d}{\mu_{\max}}\right)^2}}}.$$

Remark: When $K_d = 0$ and $m > 0$, $\tilde{s} = \dfrac{K_S}{\frac{s_s^{des}}{K_I} + \frac{\frac{1}{Y_{x/s}} + \frac{m}{\mu_{\max}}\left(1 + \frac{s_s^{des}}{K_I} + \frac{K_S}{s_s^{des}}\right)}{\frac{m}{\mu_{\max}} \frac{S_0 - s_s^{des}}{s_s^{des}}}} = s_u$, whereas $\tilde{s} > s_u$ when $K_d > 0$ and $m > 0$.

Lemma 2: Assume that $\mu(s)$ is given by (2a) or (2b) and $m \geq 0$. Also, assume that s_s^{des} is chosen to be the optimal steady state that corresponds to maximal value of Q on the equilibrium curve.

For every ε *with* $0<\varepsilon<\mu(s_s^{des})-K_d$ *and* $0<\varepsilon<\left[\dfrac{d\mu}{ds}(s_s^{des})\right](S_0-s_s^{des})$,

the set $I_\varepsilon=\left\{(x,s)\,\middle|\,x\geq\gamma\varepsilon,\ \mu(s)\geq K_d+\varepsilon,\ \dfrac{\dfrac{1}{Y_{x/s}}+\dfrac{m}{\mu(s_s^{des})}}{S_0-s_s^{des}}+m\dfrac{\dfrac{1}{\mu(s)}-\dfrac{1}{\mu(s_s^{des})}}{s-s_s^{des}}\geq\beta\varepsilon\right\}$,

where $\gamma=\dfrac{S_0-s_s^{des}}{\left(\dfrac{1}{Y_{x/s}}+\dfrac{m}{\mu(s_s^{des})}\right)\mu_{\max}}$, $\beta=\dfrac{\dfrac{1}{Y_{x/s}}+\dfrac{m}{\mu(s_s^{des})}}{\left(S_0-s_s^{des}\right)\left(\mu(s_s^{des})-K_d\right)}$, *is positively invariant.*

The methodology followed in order to derive the stability region can be summarized in the following steps:

<u>*Step 1*</u>: We consider a quadratic Lyapunov function candidate, and we identify the region over which its time derivative is guaranteed to be negative definite; this is done in Lemma 1.

<u>*Step 2*</u>: We identify a positively invariant set for the closed loop system; this is done in Lemma 2.

<u>*Step 3*</u>: We prove that the region defined in Lemma 1 is a superset of the positively invariant set of Lemma 2.

A similar methodology has been used by Mailleret and Bernard (2001) for the case $K_d=0$, $m=0$.

The proofs of the two Lemmas and the Proposition are given in Appendix II.

In what follows, we will see how the foregoing Proposition specializes in the cases (i) $K_d>0, m=0$, (ii) and (iii) $K_d>0, m>0$.

Case I: Negligible endogenous metabolism

When $K_d > 0$ and $m = 0$, the inequality $\dfrac{\dfrac{1}{Y_{x/s}}+\dfrac{m}{\mu(s_s^{des})}}{S_0 - s_s^{des}} + m\dfrac{\dfrac{1}{\mu(s)}-\dfrac{1}{\mu(s_s^{des})}}{s - s_s^{des}} > 0$ is automatically satisfied everywhere and thereforethe set I is

$$I = \left\{(x,s) \in \mathbb{R}^2 \mid x > 0,\ \mu(s) > K_d\right\}$$

or equivalently,

$$I = \left\{(x,s) \in \mathbb{R}^2 \mid x > 0,\ s_* < s < s^*\right\}$$

where s_*, s^* are the solutions of $\mu(s) = K_d$ given by (6a) and (7) respectively. This is a proper subset of the stability region; it is actually <u>the largest strip contained in the stability region</u>.

Case II: Negligible death rate of the biomass

When $m > 0$ and $K_d = 0$, the inequality $\mu(s) > K_d$ is obviously satisfied for all $s > 0$, hence

$$I = \left\{(x,s) \in \mathbb{R}^2 \mid x > 0,\ s > 0,\ \frac{\dfrac{1}{Y_{x/s}}+\dfrac{m}{\mu(s_s^{des})}}{S_0 - s_s^{des}} + m\frac{\dfrac{1}{\mu(s)}-\dfrac{1}{\mu(s_s^{des})}}{s - s_s^{des}} > 0\right\}$$

or equivalently,

$$I = \left\{(x,s) \in \mathbb{R}^2 \mid x > 0,\ s > s_u\right\}$$

In this case, I is exactly equal to the stability region of the closed loop system.

Case III:Non-negligible death rate and non-negligible endogenous metabolism

When both $K_d > 0$ and $m > 0$, the set I can be written equivalently as

$$I = \left\{ (x,s) \in \mathbb{R}^2 \mid x > 0,\ s^* > s > \max(s_*, s_u) \right\}$$

This is a proper subset of the stability region; it is the largest strip contained in the stability region.

The conclusion from the stability analysis is that all the main findings from the numerical simulations have been rigorously proven. The end result is a non-conservative estimate of the stability region, expressed in the form of very simple formulas for the upper and lower limits of the stability region.

4. Conclusion

In the present study, an output feedback control law that maintains constant the yield between feeding rate of the nutrients and the product in a chemostat simulating the process of anaerobic digestion was developed. The controller when applied to the open loop system manages to dramatically increase the stability region of the process and for all practical purposes avoid common phenomena for this type of process as washout of the biomass. The stability region was estimated using a Lyapunov function approach. The result of the stability analysis provides the largest strip contained in the stability region, whose upper and lower limits can be calculated through simple formulas.

5. Acknowledgement

The authors would like to thank Xakousti Skyllakou and Eleni Karnezi for their assistance at early stages of this work.

6. References

Savoglidis, G., C. Kravaris and G. Lyberatos, 2010, A robust Model-Based Control Policy Applied on a Glucose-Fed CSTR-Type Anaerobic Digester, Proceedings of the 7th European Congress of Chemical Engineering (ECCE-7), Prague, Czech Republic.

Smith, H. L., and P. Waltman, 1995, The theory of the chemostat: Dynamics of microbial competition, Cambridge University Press, Cambridge.

Bailey, J. E., and D. F. Ollis, 1986, Biochemical engineering fundamentals, 2nd edition, Mc-Graw-Hill, New York.

De Leenheer, P., and H. L. Smith, 2003, Feedback control for chemostat models, Journal of Mathematical Biology, 46, p. 48-70.

Gouze, J. L., and G. Robledo, 2006, Robust control for an uncertain chemostat model, International Journal of Robust & Nonlinear Control, 16(3), p. 133-155.

Harmand, J., A. Rapaport and F. Mazenc, 2006, Output tracking of continuous bioreactors through recirculation and by-pass, Automatica, 42, p. 1025-1032.

Karafyllis, I., C. Kravaris, L. Syrou and G. Lyberatos, 2008, A vector Lyapunov function characterization of input-to-state stability with application to robust global stabilization of the chemostat, European Journal of Control, 14(1), p. 47-61.

Pullammanappallil, P. C., S. A. Svoronos, D. P. Chynoweth, and G. Lyberatos, 1998, Expert system for control of anaerobic digesters, Biotechnology and Bioengineering, 58(1), p. 13-22.

Mailleret, L. and O. Bernard, 2001, A simple robust controller to stabilize an anaerobic digestion process, Proceedings 8th Conference on Computer Applications in Biotechnology, p. 213-218.

Syrou, L., I. Karafyllis, K. Stamatelatou, G. Lyberatos and C. Kravaris, 2004, Robust global stabilization of continuous bioreactors, Proceedings 7th International Conference on Dynamics and Control of Process Systems (DYCOPS 7), Boston, MA.

Karafyllis, I., C. Kravaris and N. Kalogerakis, 2009, Relaxed Lyapunov Criteria for Robust Global Stabilization of Nonlinear Systems, International Journal of Control, 82, p. 2077-2094.

Khalil H. K., 1996, Nonlinear Systems, 2nd edition, Macmillan, New York.

7. Appendix I: Properties of the additional steady state of the closed loop system

a)For $m \neq 0$,consider the function

$$\varphi(s) = \frac{(S_0 - s)}{\frac{1}{Y_{x/s}} + \frac{m}{\mu(s)}} \tag{A1}$$

This is a bell-shaped function of s (see figure A1) and equation (14) can be expressed as $\varphi(s) = \varphi(s_s^{des})$, with one root being s_s^{des} and the other s_u.

Now from the calculation of the optimal steady state (see §2.3), the condition $\frac{dQ}{ds}(s)=0$, can be expressed as $\frac{d\mu}{ds}(s)\varphi(s)+(\mu(s)-K_d)\frac{d\varphi}{ds}(s)=0$, from which

$$\frac{d\varphi}{ds}(s_s^{opt})=-\frac{\frac{d\mu}{ds}(s_s^{opt})\varphi(s_s^{opt})}{\mu(s_s^{opt})-K_d}=-\frac{\left[\frac{d\mu}{ds}(s_s^{opt})\right](S_0-s_s^{opt})}{\left(\mu(s_s^{opt})-K_d\right)\left(\frac{1}{Y_{x/s}}+\frac{m}{\mu(s_s^{opt})}\right)}<0. \qquad \text{(A2)}$$

This proves that the point $s=s_s^{des}=s_s^{opt}$ corresponds to the right branch of the graph of $\varphi(s)$, hence $s=s_u$ is at the right branch (see figure A1).

Hence $s_u<s_s^{des}$.

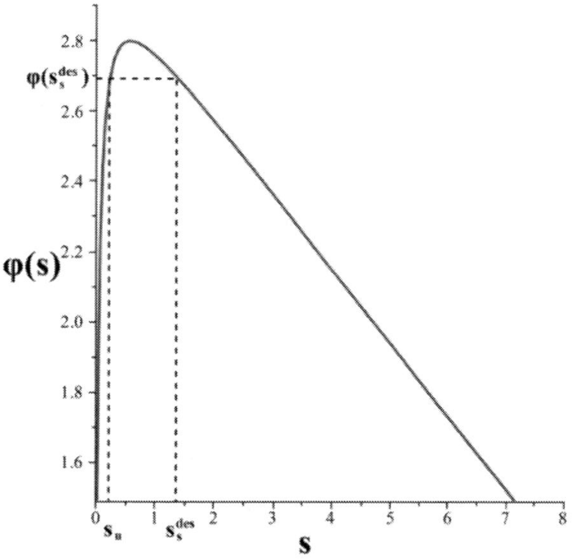

Figure A1. Function $\varphi(s)$ showing the two steady state values of the substrate.

b)Linearizing the closed loop system (12) or (13) around the steady state (x_u, s_u) and computing the corresponding eigenvalues, we find:

$$\lambda_1 = -\frac{\frac{1}{Y_{x/s}} + \frac{m}{\mu(s_s^{des})}}{S_0 - s_s^{des}} \mu(s_u) x_u$$

$$\lambda_2 = \left[-\frac{\frac{1}{Y_{x/s}} + \frac{m}{\mu(s_s^{des})}}{S_0 - s_s^{des}} + m \frac{\frac{d\mu}{ds}(s_u)}{[\mu(s_u)]^2} \right] \mu(s_u) x_u \quad . \tag{A3}$$

Taking into account that $\frac{\frac{1}{Y_{x/s}} + \frac{m}{\mu(s_s^{des})}}{S_0 - s_s^{des}} = \frac{\frac{1}{Y_{x/s}} + \frac{m}{\mu(s_u)}}{S_0 - s_u}$, eigenvalues (A3) can be expressed equivalently as

$$\lambda_1 = -(\mu(s_u) - K_d)$$

$$\lambda_2 = \left[-1 + m \frac{S_0 - s_u}{\frac{1}{Y_{x/s}} + \frac{m}{\mu(s_u)}} \frac{\frac{d\mu}{ds}(s_u)}{[\mu(s_u)]^2} \right] (\mu(s_u) - K_d) \tag{A4}$$

and we observe that, when $\mu(s_u) > K_d$, $\lambda_1 < 0$.

To determine the sign of the second eigenvalue, we consider again the function $\varphi(s)$, whose derivative is given by

$$\frac{d\varphi}{ds}(s) = \frac{1}{\frac{1}{Y_{x/s}} + \frac{m}{\mu(s)}} \left[-1 + m \frac{(S_0 - s)}{\left(\frac{1}{Y_{x/s}} + \frac{m}{\mu(s)} \right)} \frac{\frac{d\mu}{ds}(s)}{[\mu(s)]^2} \right]. \tag{A5}$$

Using (A5), the second eigenvalue can be expressed as

$$\lambda_2 = \frac{d\varphi}{ds}(s_u) \left(\frac{1}{Y_{x/s}} + \frac{m}{\mu(s_u)} \right) (\mu(s_u) - K_d) \tag{A6}$$

But $\frac{d\varphi}{ds}(s_u) > 0$ (see figure A1). Thus, when $\mu(s_u) > K_d$, the second eigenvalue is always positive $(\lambda_2 > 0)$. Therefore, the steady state (x_u, s_u) is unstable.

8. Appendix II: Proofs of the Lemmas and the Proposition in the stability analysis of the closed loop system

8.1 Proof of Lemma 1

Computing the time derivative of V along the trajectories of the closed-loop system (13), we find

$$\dot{V}(x,s) = -\left\{ \left(\frac{\frac{1}{Y_{x/s}} + \frac{m}{\mu(s_s^{des})}}{S_0 - s_s^{des}} x - \left(1 - \frac{K_d}{\mu(s_s^{des})}\right) \right)^2 + \left[K_d s_s^{des} \frac{\frac{1}{\mu(s)} - \frac{1}{\mu(s_s^{des})}}{s - s_s^{des}} \right] \left(\frac{\frac{1}{Y_{x/s}} + \frac{m}{\mu(s_s^{des})}}{S_0 - s_s^{des}} x - \left(1 - \frac{K_d}{\mu(s_s^{des})}\right) \right) \left(\frac{s}{s_s^{des}} - 1 \right) \right.$$
$$\left. + \eta \left[1 + m \frac{(S_0 - s_s^{des})}{\frac{1}{Y_{x/s}} + \frac{m}{\mu(s_s^{des})}} \frac{\frac{1}{\mu(s)} - \frac{1}{\mu(s_s^{des})}}{s - s_s^{des}} \right] \left(\frac{s}{s_s^{des}} - 1 \right)^2 \right\} \frac{\frac{1}{Y_{x/s}} + \frac{m}{\mu(s_s^{des})}}{S_0 - s_s^{des}} \mu(s) x \tag{A7}$$

Note that

$$\chi^2 + \left[K_d s_s^{des} \frac{\frac{1}{\mu(s)} - \frac{1}{\mu(s_s^{des})}}{s - s_s^{des}} \right] \chi\psi + \eta \left[1 + m \frac{(S_0 - s_s^{des})}{\frac{1}{Y_{x/s}} + \frac{m}{\mu(s_s^{des})}} \frac{\frac{1}{\mu(s)} - \frac{1}{\mu(s_s^{des})}}{s - s_s^{des}} \right] \psi^2 > 0 \quad \forall \chi, \psi \in \mathbb{R} - \{0\}$$

if s is such that, $\left[K_d s_s^{des} \frac{\frac{1}{\mu(s)} - \frac{1}{\mu(s_s^{des})}}{s - s_s^{des}} \right]^2 - 4\eta \left[1 + m \frac{(S_0 - s_s^{des})}{\frac{1}{Y_{x/s}} + \frac{m}{\mu(s_s^{des})}} \frac{\frac{1}{\mu(s)} - \frac{1}{\mu(s_s^{des})}}{s - s_s^{des}} \right] < 0$ or,

equivalently, after rearranging and regrouping terms, if

$$\left(\left(m \frac{S_0 - s_s^{des}}{\frac{1}{Y_{x/s}} + \frac{m}{\mu(s_s^{des})}} \right)^2 + \frac{1}{\eta} \left(K_d s_s^{des} \right)^2 \right) \left(\frac{\frac{1}{\mu(s)} - \frac{1}{\mu(s_s^{des})}}{s - s_s^{des}} \right)^2 - \left(2 + m \frac{(S_0 - s_s^{des})}{\frac{1}{Y_{x/s}} + \frac{m}{\mu(s_s^{des})}} \frac{\frac{1}{\mu(s)} - \frac{1}{\mu(s_s^{des})}}{s - s_s^{des}} \right)^2 < 0$$

or, further, after factorizing, if

$$\left[\left(-\left(m\frac{S_0-s_s^{des}}{\frac{1}{Y_{x/s}}+\frac{m}{\mu(s_s^{des})}}\right)+\sqrt{\left(m\frac{S_0-s_s^{des}}{\frac{1}{Y_{x/s}}+\frac{m}{\mu(s_s^{des})}}\right)^2+\frac{1}{\eta}\left(K_d s_s^{des}\right)^2}\right)\left(\frac{\frac{1}{\mu(s)}-\frac{1}{\mu(s_s^{des})}}{s-s_s^{des}}\right)-2\right]$$

$$\times\left[\left(\left(m\frac{S_0-s_s^{des}}{\frac{1}{Y_{x/s}}+\frac{m}{\mu(s_s^{des})}}\right)+\sqrt{\left(m\frac{S_0-s_s^{des}}{\frac{1}{Y_{x/s}}+\frac{m}{\mu(s_s^{des})}}\right)^2+\frac{1}{\eta}\left(K_d s_s^{des}\right)^2}\right)\left(\frac{\frac{1}{\mu(s)}-\frac{1}{\mu(s_s^{des})}}{s-s_s^{des}}\right)+2\right]<0$$

Considering now the first factor, we observe that:

a) because of the fact that $s\mu(s)$ is a strictly increasing function,

$$\frac{s\mu(s)-s_s^{des}\mu(s_s^{des})}{s-s_s^{des}}>0\Rightarrow\frac{\frac{s_s^{des}}{\mu(s)}-\frac{s}{\mu(s_s^{des})}}{s-s_s^{des}}<0\Rightarrow\frac{s_s^{des}}{s-s_s^{des}}\frac{1}{\mu(s)}-\frac{s_s^{des}}{s-s_s^{des}}\frac{1}{\mu(s_s^{des})}<\frac{1}{\mu(s_s^{des})}\Rightarrow$$

$$s_s^{des}\left(\frac{\frac{1}{\mu(s)}-\frac{1}{\mu(s_s^{des})}}{s-s_s^{des}}\right)<\frac{1}{\mu(s_s^{des})}$$

b) because $-\lambda+\sqrt{\lambda^2+\kappa^2}\leq\kappa$, $\forall\kappa,\lambda\geq 0$, it follows

$$-\left(m\frac{S_0-s_s^{des}}{\frac{1}{Y_{x/s}}+\frac{m}{\mu(s_s^{des})}}\right)+\sqrt{\left(m\frac{S_0-s_s^{des}}{\frac{1}{Y_{x/s}}+\frac{m}{\mu(s_s^{des})}}\right)^2+\frac{1}{\eta}\left(K_d s_s^{des}\right)^2}\leq\frac{K_d s_s^{des}}{\sqrt{\eta}}$$

and thus we conclude

$$\left(-\left(m\frac{S_0-s_s^{des}}{\frac{1}{Y_{x/s}}+\frac{m}{\mu(s_s^{des})}}\right)+\sqrt{\left(m\frac{S_0-s_s^{des}}{\frac{1}{Y_{x/s}}+\frac{m}{\mu(s_s^{des})}}\right)^2+\frac{1}{\eta}\left(K_d s_s^{des}\right)^2}\right)\left(\frac{\frac{1}{\mu(s)}-\frac{1}{\mu(s_s^{des})}}{s-s_s^{des}}\right)-2<$$

$$<\frac{K_d s_s^{des}}{\sqrt{\eta}}\cdot\frac{1}{s_s^{des}\mu(s_s^{des})}-2=\frac{1}{\sqrt{\eta}}\cdot\frac{K_d}{\mu(s_s^{des})}-2<\frac{1}{\sqrt{\eta}}-2$$

which means that the first factor of the inequality condition is necessarily negative,

and we are left with

$$\left(\left(m\frac{S_0 - s_s^{des}}{\frac{1}{Y_{x/s}}+\frac{m}{\mu(s_s^{des})}}\right)+\sqrt{\left(m\frac{S_0 - s_s^{des}}{\frac{1}{Y_{x/s}}+\frac{m}{\mu(s_s^{des})}}\right)^2+\frac{1}{\eta}\left(K_d s_s^{des}\right)^2}\right)\left(\frac{\frac{1}{\mu(s)}-\frac{1}{\mu(s_s^{des})}}{s-s_s^{des}}\right)+2>0$$

or equivalently

$$\left(\left(m\frac{S_0 - s_s^{des}}{\frac{1}{Y_{x/s}}+\frac{m}{\mu(s_s^{des})}}\right)+\sqrt{\left(m\frac{S_0 - s_s^{des}}{\frac{1}{Y_{x/s}}+\frac{m}{\mu(s_s^{des})}}\right)^2+\frac{1}{\eta}\left(K_d s_s^{des}\right)^2}\right)\left(\frac{\frac{1}{\mu(s)}-\frac{1}{\mu(s_s^{des})}}{s-s_s^{des}}\right)>-2.$$

As long as s is such that the above inequality holds and $x>0$, the time derivative of V along the trajectories of the system will be negative.

Finally, because

$$\frac{\frac{1}{\mu(s)}-\frac{1}{\mu(s_s^{des})}}{s-s_s^{des}}=\frac{1}{\mu_{\max}s_s^{des}}\left(\frac{s_s^{des}}{K_I}-\frac{K_S}{s}\right),$$

the above inequality can be easily solved for s, leading to

$$s>\tilde{s}=\frac{K_S}{\frac{s_s^{des}}{K_I}+\frac{2}{\left(\frac{\frac{m}{\mu_{\max}}\frac{S_0-s_s^{des}}{s_s^{des}}}{\frac{1}{Y_{x/s}}+\frac{m}{\mu_{\max}}\left(1+\frac{s_s^{des}}{K_I}+\frac{K_S}{s_s^{des}}\right)}\right)+\sqrt{\left(\frac{\frac{m}{\mu_{\max}}\frac{S_0-s_s^{des}}{s_s^{des}}}{\frac{1}{Y_{x/s}}+\frac{m}{\mu_{\max}}\left(1+\frac{s_s^{des}}{K_I}+\frac{K_S}{s_s^{des}}\right)}\right)^2+\frac{1}{\eta}\left(\frac{K_d}{\mu_{\max}}\right)^2}}}$$

8.2 Proof of Lemma 2

The proof will be made for $K_d>0$ and $m>0$. The limiting cases $K_d=0$, $m>0$ and $K_d>0$, $m=0$ can be handled similarly, mutatis mutandis.

Consider the flow on the boundaries of I_ε (see figure A2):

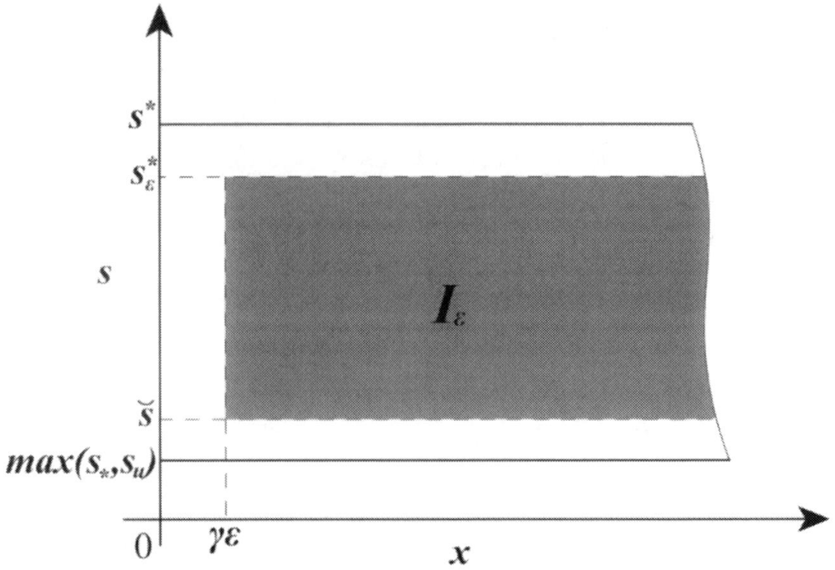

Figure A2. Boundaries of the stability region I_ε.

a) On the left boundary of I_ε, i.e. for $x = \gamma\varepsilon$, $\mu(s) \geq K_d + \varepsilon$,

$$\left.\frac{dx}{dt}\right)_{x=\gamma\varepsilon} = \left[\left(1 - \frac{\varepsilon}{\mu_{\max}}\right)\mu(s) - K_d\right]\gamma\varepsilon .$$

But since

$$\left(1 - \frac{\varepsilon}{\mu_{\max}}\right)\mu(s) - K_d \geq \left(1 - \frac{\varepsilon}{\mu_{\max}}\right)(K_d + \varepsilon) - K_d$$

$$= \varepsilon - \frac{\varepsilon}{\mu_{\max}}(K_d + \varepsilon) = \varepsilon\left(1 - \frac{K_d + \varepsilon}{\mu_{\max}}\right) > 0$$

it follows that $\left.\frac{dx}{dt}\right)_{x=\gamma\varepsilon} > 0$.

b) On the lower boundary of I_ε, $x \ge \gamma\varepsilon$, $\breve{s} = \max(s_{\varepsilon *}, \overline{s}_\varepsilon)$, where $s_{\varepsilon *}$ is the smallest root of $\mu(s) = K_d + \varepsilon$ and $\overline{s}_\varepsilon$ is the root of $\dfrac{\dfrac{1}{Y_{x/s}} + \dfrac{m}{\mu(s_s^{des})}}{S_0 - s_s^{des}} + m\dfrac{\dfrac{1}{\mu(s)} - \dfrac{1}{\mu(s_s^{des})}}{s - s_s^{des}} = \beta\varepsilon$, and we have:

$$\left.\frac{ds}{dt}\right)_{s=\breve{s}} = \left(s_s^{des} - \breve{s}\right)\left(\frac{\dfrac{1}{Y_{x/s}} + \dfrac{m}{\mu(s_s^{des})}}{S_0 - s_s^{des}} + m\frac{\dfrac{1}{\mu(\breve{s})} - \dfrac{1}{\mu(s_s^{des})}}{\breve{s} - s_s^{des}}\right)\mu(\breve{s})x\,.$$

We need to examine the sign of the right hand side and in particular the sign of the first two factors (the other two are positive).

At first, observe that the condition $\varepsilon < \left[\dfrac{d\mu}{ds}(s_s^{des})\right](S_0 - s_s^{des})$ can be rewritten as $\beta\varepsilon < \dfrac{\dfrac{1}{Y_{x/s}} + \dfrac{m}{\mu(s_s^{des})}}{\mu(s_s^{opt}) - K_d} \cdot \dfrac{d\mu}{ds}(s_s^{des})$. But because s_s^{des} is chosen to be the optimal steady state in the sense of corresponding to maximal Q (see §2.3), it has to satisfy the optimality condition (8), which can be rearranged to $\dfrac{\dfrac{1}{Y_{x/s}} + \dfrac{m}{\mu(s_s^{des})}}{S_0 - s_s^{des}} - m\dfrac{\dfrac{d\mu}{ds}(s_s^{des})}{\left[\mu(s_s^{des})\right]^2} = \dfrac{\dfrac{1}{Y_{x/s}} + \dfrac{m}{\mu(s_s^{des})}}{\mu(s_s^{opt}) - K_d} \cdot \dfrac{d\mu}{ds}(s_s^{des})$,

therefore $\beta\varepsilon < \dfrac{\dfrac{1}{Y_{x/s}} + \dfrac{m}{\mu(s_s^{des})}}{S_0 - s_s^{des}} - m\dfrac{\dfrac{d\mu}{ds}(s_s^{des})}{\left[\mu(s_s^{des})\right]^2}$.

Equivalently, the inequality condition can be expressed as

$$\frac{\dfrac{1}{Y_{x/s}} + \dfrac{m}{\mu(s_s^{des})}}{S_0 - s_s^{des}} + m\frac{\dfrac{1}{\mu(\overline{s}_\varepsilon)} - \dfrac{1}{\mu(s_s^{des})}}{\overline{s}_\varepsilon - s_s^{des}} < \frac{\dfrac{1}{Y_{x/s}} + \dfrac{m}{\mu(s_s^{des})}}{S_0 - s_s^{des}} + m\lim_{s \to s_s^{des}} \frac{\dfrac{1}{\mu(s)} - \dfrac{1}{\mu(s_s^{des})}}{s - s_s^{des}},$$

or further, as $\psi(\overline{s}_\varepsilon) < \psi(s_s^{des})$, where

$$\psi(s)=\frac{\frac{1}{Y_{x/s}}+\frac{m}{\mu(s_s^{des})}}{S_0-s_s^{des}}+m\frac{\frac{1}{\mu(s)}-\frac{1}{\mu(s_s^{des})}}{s-s_s^{des}}=\frac{\frac{1}{Y_{x/s}}+\frac{m}{\mu_{\max}}\left(1+\frac{s_s^{des}}{K_I}+\frac{K_S}{s_s^{des}}\right)}{S_0-s_s^{des}}+\frac{m}{\mu_{\max}}\frac{1}{s_s^{des}}\left(\frac{s_s^{des}}{K_I}-\frac{K_S}{s}\right).$$

Because $\psi(s)$ is a strictly increasing function, it follows that $\bar{s}_\varepsilon < s_s^{des}$.

On the other hand, the condition $\varepsilon < \mu(s_s^{des}) - K_d$ implies that $s_{\varepsilon *} < s_s^{des}$.

Therefore, $\breve{s} = \max(s_{\varepsilon *}, \bar{s}_\varepsilon) < s_s^{des}$, i.e. the first factor $\left(s_s^{des} - \breve{s}\right)$ is positive.

For the sign of the second factor, consider again the function

$$\psi(s)=\frac{\frac{1}{Y_{x/s}}+\frac{m}{\mu(s_s^{des})}}{S_0-s_s^{des}}+m\frac{\frac{1}{\mu(s)}-\frac{1}{\mu(s_s^{des})}}{s-s_s^{des}},$$ which is a strictly increasing function and, moreover, notice that $\psi(s)$ vanishes at $s = s_u$. Because $\psi(\bar{s}_\varepsilon) = \beta\varepsilon > 0$, we will have $\tilde{s}_\varepsilon > s_u$ and moreover, because $\breve{s} \geq \tilde{s}_\varepsilon$, it follows that $\psi(\breve{s}) > 0$, i.e. the second factor

$$\frac{\frac{1}{Y_{x/s}}+\frac{m}{\mu(s_s^{des})}}{S_0-s_s^{des}}+m\frac{\frac{1}{\mu(\breve{s})}-\frac{1}{\mu(s_s^{des})}}{\breve{s}-s_s^{des}}>0.$$

Thus, we conclude that $\left.\frac{ds}{dt}\right)_{s=\breve{s}} > 0$.

c) On the upper boundary of I_ε, $x \geq \gamma\varepsilon$, $s = s_\varepsilon^*$, where s_ε^* is the largest root of $\mu(s) = K_d + \varepsilon$, we have

$$\left.\frac{ds}{dt}\right)_{s=s_\varepsilon^*}=\left(s_S^{des}-s_\varepsilon^*\right)\left(\frac{\frac{1}{Y_{x/s}}+\frac{m}{\mu(s_s^{des})}}{S_0-s_s^{des}}+m\frac{\frac{1}{\mu(s_\varepsilon^*)}-\frac{1}{\mu(s_s^{des})}}{s_\varepsilon^*-s_s^{des}}\right)\mu(s_\varepsilon^*)x.$$

From the condition $\varepsilon < \mu(s_s^{des}) - K_d$, it follows that $s_\varepsilon^* > s_S^{des}$.

Moreover, since $s_s^{des} > \bar{s}_\varepsilon$, which implies $s_\varepsilon^* > \bar{s}_\varepsilon$,

$$\frac{\frac{1}{Y_{x/s}}+\frac{m}{\mu(s_s^{des})}}{S_0-s_s^{des}}+m\frac{\frac{1}{\mu(s_\varepsilon^*)}-\frac{1}{\mu(s_s^{des})}}{s_\varepsilon^*-s_s^{des}}=\psi(s_\varepsilon^*)>\psi(\overline{s}_\varepsilon)=\beta\varepsilon>0 .$$

Consequently $\left.\frac{ds}{dt}\right)_{s=s_\varepsilon^*}<0$.

8.3 Proof of the Proposition

We will prove that, given a sufficiently small, but otherwise arbitrary $\varepsilon>0$, the positively invariant set I_ε of Lemma 1 is a proper subset of the set E of Lemma 2, over which the time derivative of the positive definite quadratic function V is negative definite. We will conclude that I_ε is contained in the region of stability, therefore $\bigcup_{\varepsilon>0} I_\varepsilon = I$ is also contained in the region of stability.

- When $K_d=0, m>0$, we have

$$\tilde{s}=\frac{K_S}{\frac{s_s^{des}}{K_I}+\frac{K_S+s_s^{des}+\frac{(s_s^{des})^2}{K_I}+\frac{\mu_{\max}s_s^{des}}{mY_{x/s}}}{S_0-s_s^{des}}}=\frac{K_S(S_0-s_s^{des})}{S_0\frac{s_s^{des}}{K_I}+K_S+s_s^{des}+\frac{\mu_{\max}s_s^{des}}{mY_{x/s}}}=$$

$$\frac{K_S(S_0-s_s^{des})}{K_S+\left(1+\frac{S_0}{K_I}+\frac{\mu_{\max}}{mY_{x/s}}\right)s_s^{des}}=s_u$$

forany value of the parameter η. Therefore

$$E=\left\{(x,s)\in\mathbb{R}^2 \mid x>0,\ s>s_u\right\}=$$

$$\left\{(x,s)\in\mathbb{R}^2 \mid x>0,\ s>0,\ \frac{\frac{1}{Y_{x/s}}+\frac{m}{\mu(s_s^{des})}}{S_0-s_s^{des}}+m\frac{\frac{1}{\mu(s)}-\frac{1}{\mu(s_s^{des})}}{s-s_s^{des}}>0\right\}\supset I_\varepsilon .$$

- When $K_d>0, m=0$, and any choice $\eta\geq\frac{1}{4}$, we have

$$\tilde{s} = \frac{K_S}{\dfrac{s_s^{des}}{K_I} + \dfrac{2\sqrt{\eta}}{\dfrac{K_d}{\mu_{\max}}}} = K_S \frac{\dfrac{K_d}{\mu_{\max}}}{\dfrac{K_d}{\mu_{\max}}\dfrac{s_s^{des}}{K_I} + 2\sqrt{\eta}}.$$

But because

$$s_* = K_s \cdot \frac{2\dfrac{K_d}{\mu_{\max}}}{\left(1-\dfrac{K_d}{\mu_{\max}}\right) + \sqrt{\left(1-\dfrac{K_d}{\mu_{\max}}\right)^2 - 4\left(\dfrac{K_s}{K_I}\right)\left(\dfrac{K_d}{\mu_{\max}}\right)^2}} \geq$$

$$K_s \cdot \frac{\dfrac{K_d}{\mu_{\max}}}{1-\dfrac{K_d}{\mu_{\max}}} > K_s \cdot \frac{K_d}{\mu_{\max}} > K_s \frac{\dfrac{K_d}{\mu_{\max}}}{\dfrac{K_d}{\mu_{\max}}\dfrac{s_s^{des}}{K_I} + 2\sqrt{\eta}} = \tilde{s}$$

it follows that

$$E = \left\{(x,s) \in \mathbb{R}^2 \mid x > 0,\ s > \tilde{s}\right\} \supset \left\{(x,s) \in \mathbb{R}^2 \mid x > 0,\ s > s_*\right\} \supset I_\varepsilon .$$

(186d) Optimal Control System Design of an Acid Gas Removal (AGR) Unit for an IGCC Power Plant with CO2 Capture

Dustin D. Jones[1], Debangsu Bhattacharyya[2], Richard Turton[2] and Stephen E. Zitney[3], (1)Chemical Engineering, West Virginia University, Morgantown, WV, (2)Department of Chemical Engg., West Virginia University, Morgantown, WV, (3)AVESTAR Center, National Energy Technology Laboratory, Morgantown, WV

Optimal Control System Design of an Acid Gas Removal (AGR) Unit for an IGCC Power Plant with CO_2 Capture

Dustin Jones[a,b], Debangsu Bhattacharyya[a,b], Richard Turton[a,b], and Stephen E. Zitney[b]

[a] Department of Chemical Engineering, West Virginia University, Morgantown, WV, USA.
[b] AVESTAR™ Center, National Energy Technology Laboratory, Morgantown, WV, USA

Abstract:

Future IGCC plants with CO_2 capture should be operated optimally in the face of disturbances without violating operational and environmental constraints. To achieve this goal, a systematic approach is taken in this work to design the control system of a selective, dual-stage Selexol-based acid gas removal (AGR) unit for a commercial-scale integrated gasification combined cycle (IGCC) power plant with pre-combustion CO_2 capture. The control system design is performed in two stages with the objective of minimizing the auxiliary power while satisfying operational and environmental constraints in the presence of measured and unmeasured disturbances.

In the first stage of the control system design, a top-down analysis is used to analyze degrees of freedom, define an operational objective, identify important disturbances and operational/environmental constraints, and select the control variables. With the degrees of freedom, the process is optimized with relation to the operational objective at nominal operation as well as under the disturbances identified. Operational and environmental constraints active at all operations are chosen as control variables. From the results of the optimization studies, self-optimizing control variables are identified for further examination. Several methods are explored in this work for the selection of these self-optimizing control variables. Modifications made to the existing methods will be discussed in this presentation. Due to the very large number of candidate sets available for control variables and due to the complexity of the underlying optimization problem, solution of this problem is computationally expensive. For reducing the computation time, parallel computing is performed using the Distributed Computing Server (DCS®) and the Parallel Computing® toolbox from Mathworks®.

The second stage is a bottom-up design of the control layers used for the operation of the process. First, the regulatory control layer is designed followed by the supervisory control layer. Finally, an optimization layer is designed.

In this paper, the proposed two-stage control system design approach is applied to the AGR unit for an IGCC power plant with CO_2 capture. Aspen Plus Dynamics® is used to develop the dynamic AGR process model while MATLAB is used to perform the control system design and for implementation of model predictive control (MPC).

MPC tuning based on impact of modeling uncertainty on closed-loop performance

Quang N. Tran, Leyla Özkan, A.C.P.M. Backx
Eindhoven University of Technology
Den Dolech 2, 5612 AZ Eindhoven, The Netherlands
n.q.tran@tue.nl; l.ozkan@tue.nl; a.c.p.m.backx@tue.nl

I. Introduction

Model predictive control (MPC) represents a class of control algorithms, in which an optimization problem is solved online using a model of the plant. MPC has become a standard advanced controller in refining and petrochemical industries ([1]) due to its ability of dealing with constraints and operating the plant closely to its operational constraints. The basic idea of MPC is that at any given time, it solves an online open-loop optimal control problem over a finite horizon and only the first element of the optimal control sequence is actually implemented on the plant. The same procedure is implemented when the next measurement is available. The quality of MPC, like any other model-based operation support systems (such as Real-Time Optimization), is mainly determined by the accuracy and the maintained calibration of the model. If proper supervision is not performed, the performance of MPC degrades over time due to the model-plant mismatch. Hence, the influence of the modeling uncertainty on the performance of MPC is important.

The current tuning practice of these controllers is heuristic and there has been no standard way of tuning MPC that takes into account model-plant mismatch. In literature, there have been several studies that deal with modeling uncertainty in the context of controller tuning. In [6], the proposed tuning method consists of two steps: The first step is to tune the controller for nominal stability; the second step is to de-tune the controller by adjusting the Kalman filter gain and the covariance of the disturbance. The de-tuning method is based on the analysis of the sensitivity and complementary sensitivity functions in the frequency domain. Elsewhere, the robustness problem in MPC can also be tackled by matching a favorite robust controller ([4], [5]). When the parametric uncertainty is considered in MPC, a min-max algorithm is usually employed ([7]). The problem of robustness in MPC is reviewed in more depth in [8] and [9].

Despite the research efforts in tuning methods for MPC in literature, MPC tuning strategies that consider robustness in process industries often lead to a conservative tuning, which might be too far from the optimal trade-off between robustness and nominal performance. With this observation in mind, this research focuses on finding the optimal tuning which achieves this balance.

The paper is organized as follows. First, background information on MPC is given in Section II. The impact of uncertainty on the closed-loop performance of MPC initiated in [3] is described in Section III. Section IV presents the tuning method proposed based on the link between model uncertainty and closed-loop performance. In section V, the tuning method is implemented on a binary distillation column. Conclusions and future work are given in Section VI.

II. Preliminaries

Consider a linear time-invariant system described in the state space:

$$\begin{aligned} x(k+1) &= Ax(k) + B_1 e(k) + B_2 u(k) \\ y_1(k) &= C_1 x(k) + D_{11} e(k) + D_{12} u(k) \\ y_2(k) &= C_2 x(k) + D_{21} e(k) + D_{22} u(k) \end{aligned} \tag{1}$$

where $x(k) \in \mathrm{R}^{n_x \times 1}$, $u(k) \in \mathrm{R}^{n_u \times 1}$, $y_1(k) \in \mathrm{R}^{n_{y1} \times 1}$, $y_2(k) \in \mathrm{R}^{n_{y2} \times 1}$ and $e(k) \in \mathrm{R}^{d \times 1}$ are the system states, control inputs, controlled outputs, measured outputs and the unmeasured disturbance at time instant k, respectively. Assume that all steady-state values are already incorporated in the formulation above, the purpose of the controller is to steer all the states to zero.

The MPC uses the state-space model above to compute system predictions over a prediction horizon N as follows:

$$\begin{bmatrix} x(k+1) \\ x(k+2) \\ \vdots \\ x(k+N) \end{bmatrix} = \begin{bmatrix} A \\ A^2 \\ \vdots \\ A^N \end{bmatrix} x(k) + \begin{bmatrix} B_2 & 0 & \cdots & 0 \\ AB_2 & B_2 & \cdots & 0 \\ \vdots & \vdots & \ddots & \vdots \\ A^{N-1}B_2 & A^{N-2}B_2 & \cdots & B_2 \end{bmatrix} \begin{bmatrix} u(k) \\ u(k+1) \\ \vdots \\ u(k+N-1) \end{bmatrix}$$

$$\Leftrightarrow \quad X(\mathrm{k}) = \Psi x(k) + \Upsilon U(k)$$

where N is the prediction and control horizon of the system.

At each time instant, MPC solves an optimization problem to find the sequence of inputs that minimizes a cost function:

$$\begin{aligned} J(k) &= x'(k+N)Px(k+N) + \sum_{i=1}^{N-1} \left(x'(k+i)Qx(k+i) + u'(k+i-1)Ru(k+i-1) \right) \\ &= X'(k)\mathbf{Q}X(k) + U'(k)\mathbf{R}U(k) \\ &= \left(\Psi x(k) + \Upsilon U(k) \right)' \mathbf{Q} \left(\Psi x(k) + \Upsilon U(k) \right) + U'(k)\mathbf{R}U(k) \end{aligned}$$

where $\mathbf{Q}^{Nn_x \times Nn_x} = \begin{bmatrix} Q & 0 & \cdots & \cdots & 0 \\ 0 & Q & \ddots & & \vdots \\ \vdots & \ddots & \ddots & \ddots & \vdots \\ \vdots & & \ddots & Q & 0 \\ 0 & \cdots & \cdots & 0 & P \end{bmatrix}$ and $\mathbf{R}^{Nn_u \times Nn_u} = \begin{bmatrix} R & 0 & \cdots & 0 \\ 0 & R & \ddots & \vdots \\ \vdots & \ddots & \ddots & 0 \\ 0 & \cdots & 0 & R \end{bmatrix}$ are the weighting matrices, i.e. tuning parameters of the controller.

The unconstrained solution to the optimization problem is given by

$$U^*(k) = -\left(\Upsilon'\mathbf{Q}\Upsilon + \mathbf{R} \right)^{-1} \left(\Upsilon'\mathbf{Q}\Psi \right) x(k)$$

MPC applies the first input element of the sequence; therefore the actual control input is multiplied with a selection matrix $\Theta^{n_u \times Nn_u} = [I, 0, \ldots, 0]$:

$$u(k) = -\Theta\left(\Upsilon'\mathbf{Q}\Upsilon + \mathbf{R}\right)^{-1}\left(\Upsilon'\mathbf{Q}\Psi\right)x(k)$$
$$= -K_{MPC}x(k)$$

Hence, in the case of no active constraints, MPC can be considered as a state feedback controller and its sensitivity and complementary sensitivity functions are calculated to derive the impact of modeling uncertainty on closed-loop performance as shown in [3].

III. Impact of modeling uncertainty on closed-loop performance

MPC aims first to reduce the variance of the key performance variable and then pushes it towards the system constraints so that the system operates closely its economically optimal condition. Therefore, the variance of the key variable is a good indication of the performance of the closed-loop system. In addition, the closed loop performance, the tuning of controllers and the model accuracy are inter-related. This relationship has been extensively studied and presented in robust control theory ([2]) using frequency domain techniques. It was shown that the performance of the closed-loop system becomes sensitive to the modeling uncertainty at a certain bandwidth of frequency. Increasing the bandwidth further results in closed-loop performance deterioration.

As a starting point of the research, a similar analysis was done for MPC by representing the cost function as functions of sensitivity, complementary sensitivity functions and weighting matrices. The same impact of uncertainty in robust controllers was observed. In robust control theory, the sensitivity and complementary sensitivity functions are tuned to adjust the closed-loop bandwidth. In case of MPC, the closed-loop bandwidth is determined by the weighting matrices on controlled variables (CV) and manipulated variables (MV).

For the sake of simplicity and consistency in calculating the sensitivity and complementary sensitivity functions, consider $x(k)$ as controlled outputs of the system. In order to show the effect of the tuning parameters, the simple case $\mathbf{Q} = I$ and $\mathbf{R} = \rho I$ is considered. The transfer function from the disturbance *e(k)* to the state *x(k)* is given by

$$T_{e\to x} = \left(zI - A + B_2K\right)^{-1} B_1$$
$$= \left(I + (zI - A)^{-1} B_2 K\right)^{-1}\left(zI - A\right)^{-1} B_1$$
$$= S\left(zI - A\right)^{-1} B_1$$

where $T_{e\to x}$ denotes the transfer function from *e(k)* to *x(k)* and S denotes the sensitivity function of the system. The transfer function from *e(k)* to *u(k)* is given by:

$$
\begin{aligned}
T_{e\to u} &= K\left(zI - A + B_2K\right)^{-1} B_1 \\
&= K\left(I + (zI - A)^{-1} B_2K\right)^{-1}\left(zI - A\right)^{-1} B_1
\end{aligned}
$$

It follows that:

$$
\begin{aligned}
\left(zI - A\right)^{-1} B_2 T_{e\to u} &= \left(zI - A\right)^{-1} B_2 K\left(I + (zI - A)^{-1} B_2 K\right)^{-1}\left(zI - A\right)^{-1} B_1 \\
&= \left(I - \left(I + (zI - A)^{-1} B_2 K\right)^{-1}\right)\left(zI - A\right)^{-1} B_1 \\
&= \left(I - S\right)\left(zI - A\right)^{-1} B_1 \\
&= T\left(zI - A\right)^{-1} B_1
\end{aligned}
$$

From the relations above, it follows that increasing the bandwidth of $T_{e\to x}$ is increasing that of the sensitivity function and increasing the bandwidth of $T_{e\to u}$ is increasing that of the complementary sensitivity function. It implies directly that making both $T_{e\to x}$ and $T_{e\to u}$ small at every frequency range is impossible.

On the other hand, $T_{e\to x}$ and $T_{e\to u}$ reflect the state and input energy via the following equalities ([10]):

$$
\begin{aligned}
\left\|u(k)\right\|_2^2 &= \frac{1}{2\pi}\int_{-\infty}^{\infty} \mathrm{Trace}\left[T_{ue}(j\omega)^* T_{ue}(j\omega)\right] d\omega = \| T_{e\to u} \|_2^2 \\
\left\|x(k)\right\|_2^2 &= \frac{1}{2\pi}\int_{-\infty}^{\infty} \mathrm{Trace}\left[T_{xe}(j\omega)^* T_{xe}(j\omega)\right] d\omega = \| T_{e\to x} \|_2^2
\end{aligned}
$$

Therefore, the infinite-horizon MPC cost function can be written as

$$
\begin{aligned}
J &= \left\|x(k)\right\|_2^2 + \rho\left\|u(k)\right\|_2^2 \\
&= \left\|T_{e\to x}\right\|_2^2 + \rho\, \| T_{e\to u} \|_2^2
\end{aligned}
$$

From the equality above, it is obvious that tuning ρ directly affects the closed-loop bandwidth of the system. Increasing ρ penalizes the input energy, makes the complementary sensitivity function smaller and reduces its bandwidth, i.e. the bandwidth of $\| T_{e\to u} \|$. As a result, the bandwidth of the sensitivity function, i.e. that of $\left\|T_{e\to x}\right\|$ is also reduced. Hence, the variance of the state increases and that of the input decreases. Similarly, decreasing ρ increases the bandwidth and reduces the variance of the state.

The statement above holds for the case where no modeling errors are present. In the following, it is shown that with model inaccuracy, increasing the bandwidth from certain frequency does not decrease the state variance, on the contrary, it increases this variance. Assume that the transfer function from $u(t)$ to $x(t)$ is subject to an input uncertainty Δ:

$$P_{unc} = P(I + \Delta)$$

where P_{unc} is the true transfer function and P is the modeled one. The uncertain sensitivity function is given by:

$$\begin{aligned} S_{unc} &= \left(I + P\left(I + \Delta\right)K\right)^{-1} \\ &= \left(I + PK + P\Delta K\right)^{-1} \\ &= \left(I + PK\right)^{-1}\left(I + P\Delta K(I + PK)^{-1}\right)^{-1} \\ &= S\left(I + P\Delta KS\right)^{-1} \end{aligned}$$

It is well-known that a model is often inaccurate at high frequencies. Therefore, Δ is assumed to have a high-pass characteristic. When the bandwidth is increased up to the point Δ becomes significant, the difference between S_{unc} and S at that crossover frequency becomes substantial and this reduces the stability margin and may result in closed-loop instability. This explains the increased variance when the bandwidth is raised from certain frequency. The phenomenon is illustrated in Figure 1. In case of no modeling uncertainty (nominal scenario), a large bandwidth, which corresponds to small penalty on MV, leads to a low variance in output (blue dashed line). On the other hand, in case of model-plant mismatch (worst-case scenario), increasing the bandwidth further from some frequency results in an increase in the output variance.

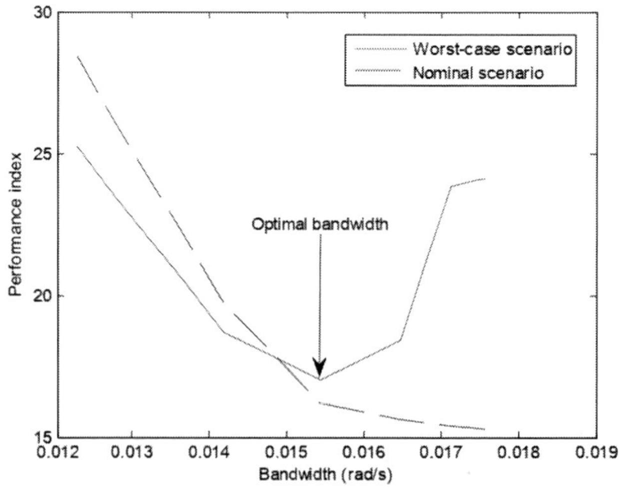

Figure 1. Relation between bandwidth and performance index, which is variance of key output

Hence, the limitation of bandwidth in the presence of modeling errors which is well-known in robust control is shown in MPC framework. This shows that there exists an optimal bandwidth where the balance between nominal performance and robustness is achieved. The tuning method proposed in the next section allows us to find this optimal bandwidth.

IV. Tuning method based on the impact of modeling uncertainty on closed-loop performance

The study of impact of model uncertainty on closed-loop performance points out that there exists an optimal bandwidth that gives the best trade-off between nominal performance and robustness. Motivated by this analysis, an approach where an H-infinity controller is designed for the desired bandwidth followed by a controller matching step, so that the resulting MPC matches the H-infinity controller, is proposed. Some methods of controller matching are introduced in [3], [4], [5] and [12]. In [3], an initial controller matching method was used to transform the tuning problem of MPC into the tuning of the frequency-domain-based input and output weights in an H-infinity controller. The method is based on the inverse optimality ([12]), which gives the weighting matrices $\mathbf{Q}$ and $\mathbf{R}$ so that the state feedback gain of the H-infinity controller $K_{H\infty}$ matches that of the infinite horizon state feedback gain K_{MPC} of MPC. The favorite H-infinity controller is designed based on the output weight $W_e(z)$ and input weight $W_u(z)$ as shown in the following scheme:

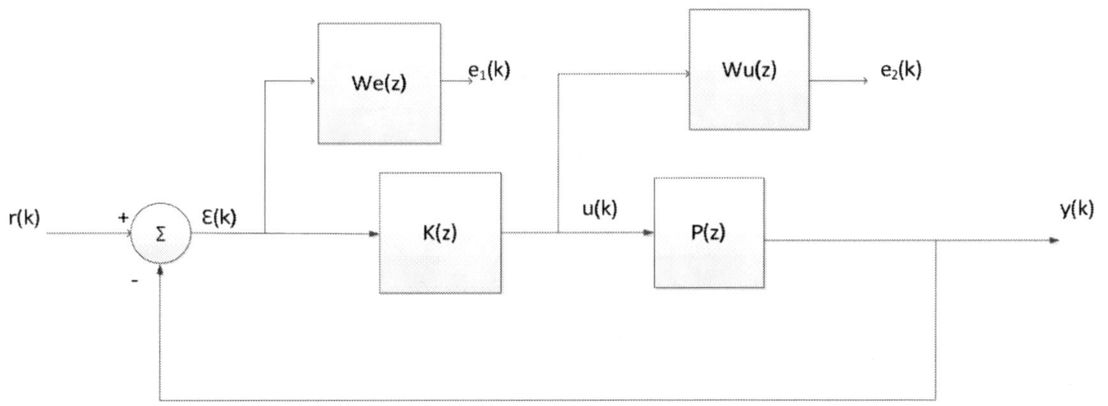

Figure 2. Scheme of an H-infinity controller.
Signals $e_1(k)$ and $e_2(k)$ are the filtered controlled outputs.

The MPC tuning method consists of the following steps:

1. Initialization: Design a favorite H-infinity controller: Fix the uncertainty weight (input weight) at a very low value. Design a low-bandwidth performance weight (output weight) with low-frequency and high-frequency gains satisfying specifications of the system. Find the corresponding MPC tuning parameters by inverse optimality. The control horizon and the prediction horizon are fixed at the same value and a terminal weight is used to obtain an infinite horizon performance.

2. Automation: Increase the bandwidth of the performance weight while monitoring the output variance. Find the corresponding MPC tuning parameters by inverse optimality.

3. Monitoring: Monitor the output variance. Compare it with the previous tuning. If the variance decreases, keep raising the bandwidth. If it increases, the previous tuning is assumed to be optimal.

V. Example

The procedure is applied to the binary distillation column benchmark problem [6]. A few major nominal parameters of the model are:

- NT = 110 (Number of trays)
- NF = 39 (Feed location)
- α= 1.35 (Relative volatility)
- F = 215 Kmol/min (Feed-rate)
- qF = 1 (Feed liquid fraction)
- zF = 0.65 (Feed composition)
- M = 30 Kmol (Molar holdup)

The control objective is to obtain the top composition of 0.95 [mole fraction] and the bottom composition of 0.05 [mole fraction]. The manipulated variables are the liquid flow L and vapor flow V. The feed-rate and feed composition are considered as disturbances and assumed to vary as sine wave signals:

$$\Delta F = 20\sin\left(\left(2\pi/750\right)t\right)$$

$$\Delta zF = 0.05\sin\left(\left(\pi/750\right)t\right)$$

where ΔF and ΔzF are the deviation of the feed-rate and feed composition from their nominal values of 215 and 0.65, respectively; and t is the simulation time going from 1 to 2000 minutes. Other parameters of the process are kept constant at their nominal values. The sampling period of the MPC and the process is T_s=5 minutes.

As the biggest problem of the distillation column is its directionality, an input uncertainty that changes the direction of the column may limit the achievable bandwidth of the controller as shown in [11]. Therefore, a rotation of the control input is used as the modeling error in the example, i.e.

$$\begin{pmatrix} L_{imp} \\ V_{imp} \end{pmatrix} = \begin{pmatrix} \cos\alpha & \sin\alpha \\ -\sin\alpha & \cos\alpha \end{pmatrix} \begin{pmatrix} L_{MPC} \\ V_{MPC} \end{pmatrix}$$

where L_{MPC}, V_{MPC} are the liquid flow and vapor flow computed by the MPC; L_{imp}, V_{imp} are the liquid flow and vapor flow implemented to the column and $\alpha \in [-\pi;\pi]$ is the rotation angle. In this example, α is chosen $-\pi/5$.

As initialization, the prediction and control horizons are both 15 samples, which cover the main dynamics of the open-loop system. The weights of the favorite H-infinity controller are diagonal:

$$W_e(z) = \begin{bmatrix} w_e(z) & 0 \\ 0 & w_e(z) \end{bmatrix}, \quad W_u(z) = \begin{bmatrix} w_u(z) & 0 \\ 0 & w_u(z) \end{bmatrix}$$

where $w_u(z)$ and $w_e(z)$ are scalar transfer functions with the following characteristics:

Weights	Low-frequency gain	Crossover frequency	High-frequency gain
$w_u(z)$	10^{-10}	$\pi / Ts = 0.627$	1.001
$w_e(z)$	10^4	0.1	0.5

The small low-frequency gain of the input weight $w_u(z)$ together with its high crossover frequency shows that a very exact model at low frequencies is considered. In high-frequency region, an uncertainty of 100% is set. The output weight $w_e(z)$ provides the performance specification: Steady-state offset less than 10^{-4}, closed-loop bandwidth of 0.1 rad/s and amplification of high-frequency noise less than 2.

At each tuning step, the bandwidth of $w_e(z)$ is increased by 0.05 rad/s and the output variance is computed after a simulation period of 2000 minutes. In Figure 3, the relation between the output variance and the bandwidth is provided. It is shown that 0.45 rad/s is the optimal bandwidth of the system. The corresponding outputs and inputs are given in Figure 4 and 5. It is shown that the compositions stay within acceptable ranges. Increasing or decreasing the bandwidth from this configuration raises the variance of the outputs.

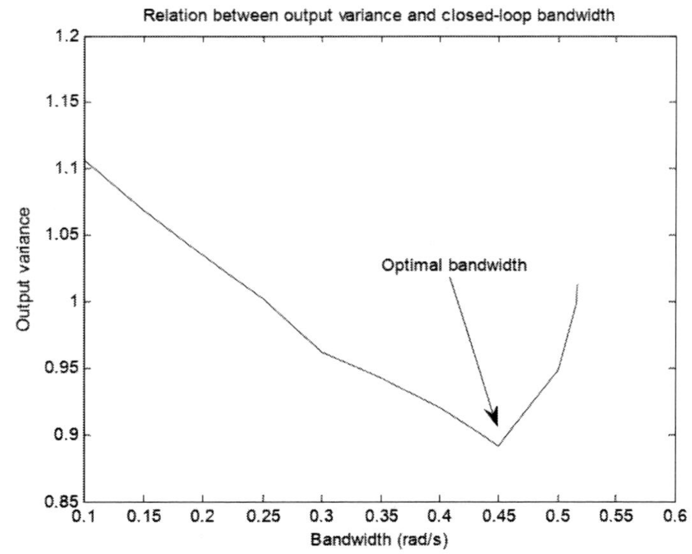

Figure 3. Relation between closed-loop bandwidth and output variance of the system

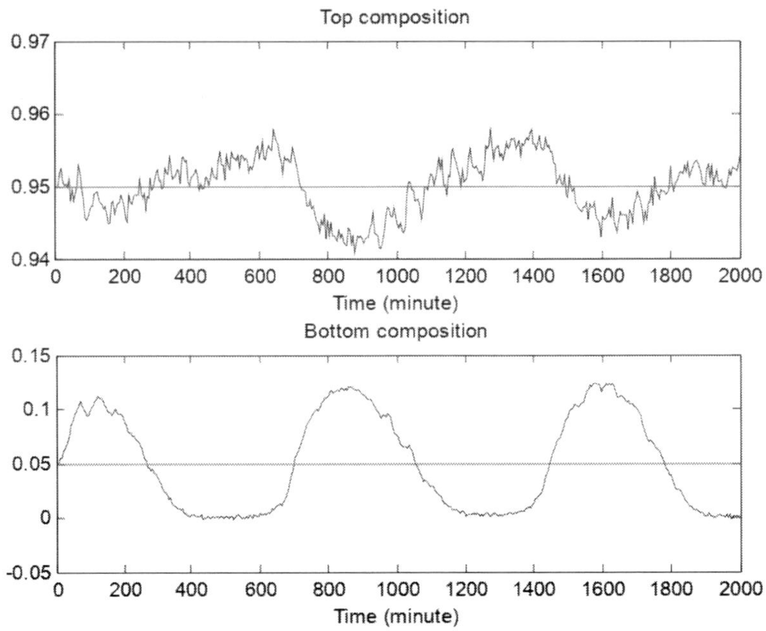

Figure 4. Top and bottom compositions corresponding to the optimal bandwidth

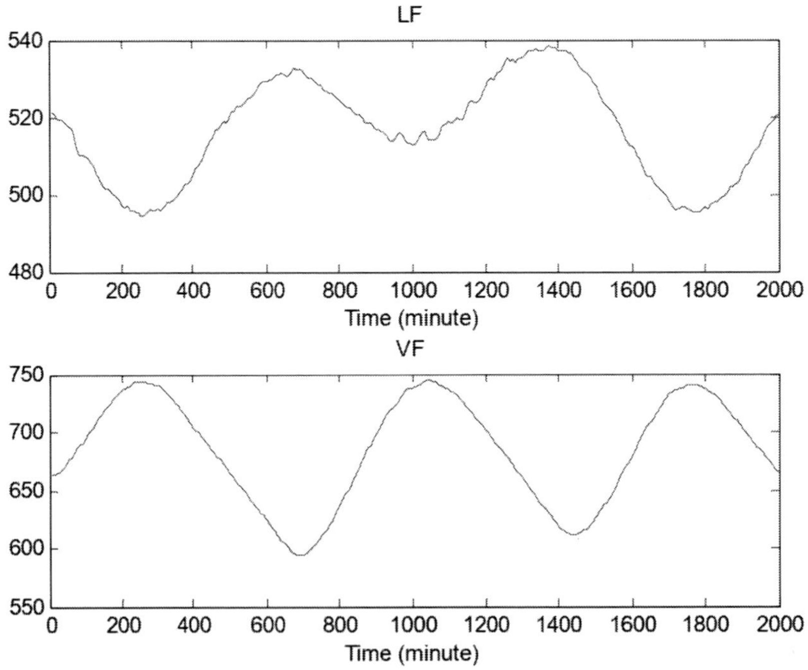

Figure 5. Control inputs of the system (liquid and vapor flows) corresponding to the optimal bandwidth

VI. Conclusion and future work

A tuning method based on the impact of model uncertainty on the closed-loop performance and the controller matching method is described. The method considers finding the optimal bandwidth the system can achieve in the presence of modeling errors. The tuning of MPC is indirectly done by the tuning of the frequency-domain weights of an H-infinity controller. Using a low-bandwidth configuration as a starting point, we increase the bandwidth until the optimal one, i.e. the best trade-off between nominal performance and robustness, is obtained.

The controller matching is based on the inverse optimality, which requires equal control and prediction horizon, as well as penalizing the states and the inputs in the cost function. However, equal control and prediction horizons are not generally considered in industrial implementation. In order to accommodate different control and prediction horizons, an optimization problem can be formulated to minimize the matching error between the feedback gain of the MPC and that of the favorite controller, as shown in [12].

The controller matching method proposed allows MPC to match a one-degree-of-freedom state-feedback H-infinity controller, which only considers unmeasured disturbances. Since MPC includes the prediction of measured disturbances, we plan to investigate two-degree-of-freedom H-infinity controller design in the proposed tuning methodology.

REFERENCE

[1] Qin, S.J. and Badgwell, T.A. (2003). A survey of industrial model predictive control technology. *Control Engineering Practice,* 11, 733-764.

[2] Skogestad, S. and Postlethwaite I. *Multivariable Feedback Control.* Wiley, 2nd edition, August 2005.

[3] Özkan , L., Meijs, J. and Backx, A.C.P.M. A frequency domain approach for MPC tuning. *Proceedings of the 11th International Symposium on Process Systems Engineering*, 15-19 July 2012, Singapore.

[4] Rowe, C. and Maciejowski, J. Robust finite horizon MPC without terminal constraints. *Proceedings of the 39th IEEE Conference on Decision and Control.* December 2000.

[5] Rowe, C. and Maciejowski, J. Tuning MPC using H_∞ loop shaping. *Proceedings of the American Control Conference,* June 2000.

[6] Lee, J.H. and Yu, Z.H. Tuning of model predictive controllers for robust performance. *Computers Chemical Engineering,* 18(1): 15-37, 1993.

[7] Han, K.; Zhao, J. and Qian, J. A novel robust tuning strategy for model predictive control. *Proceedings of the 6th World Congress on Intelligent Control and Automation*, pages 6404-6410, 2006.

[8] Bemporad, A. and Morari, M. Robust model predictive control: A survey. *Robustness in Identification and Control,* vol. 245, pages 207-226, 1999.

[9] Jalali, A.A. A survey on robust model predictive control from 1999 -2006. *International Conference on Computational Intelligence for Modeling, Control and Automation,* Australia, 2006.

[10] Zhou, K.; Doyle, J. and Glover, K. *Robust and Optimal Control.* Prentice Hall, New Jersey 07458, 1996.

[11] Skogestad ,S. and Morari, M. Effect of disturbance directions on closed-loop performance. *Ind.Eng.Chem.Res.* 26, pages 2029-2035, 1987.

[12] Di Cairano, S. and Bemporad, A. Model Predictive Control Tuning by Controller Matching. *IEEE Transactions on automatic control,* vol. 55, no. 1, January 2010.

(186f) State Estimation-Based Economic Model Predictive Control of Nonlinear Systems

Mohsen Heidarinejad[1], Jinfeng Liu[2] and Panagiotis D. Christofides[2], (1)Electric Engineering, University of California, Los Angeles, Los Angeles, CA, (2)Department of Chemical and Biomolecular Engineering, University of California, Los Angeles, Los Angeles, CA

Traditionally, optimal operation and control policies for chemical process systems are addressed via a two layer approach in which the upper layer carries steady-state process optimization to obtain economically optimal process operating points (steady-states) while the lower layer utilizes appropriate feedback control systems to steer the process state to an economically optimal steady-state. Model predictive control (MPC) is widely used in the lower layer to obtain optimal manipulated input values by minimizing a (typically) quadratic cost function which usually penalizes the deviation of the system state and manipulated inputs from their economically-optimal steady-state values subject to input and state constraints [1]; however, this two-layer approach usually limits process operation around a steady-state. Economic model predictive control (EMPC) framework deals with a reformulation of the conventional MPC quadratic cost function in which an economic (not necessarily quadratic) cost function is used directly as the cost in MPC, and thus, it may, in general, lead to time-varying process operation policies (instead of steady-state operation), which directly optimize process economics. In a previous work [2], we presented a two-mode Lyapunov-based economic MPC (LEMPC) design for nonlinear systems which is also capable of handling asynchronous and delayed measurements and extended it in the context of distributed MPC [3]. Currently, all economic MPC schemes have been developed under the assumption of state feedback. State estimation in certain classes of nonlinear systems can be carried out within the framework of high-gain observers, however, at this stage these estimation techniques have not been used in conjunction with economic MPC schemes. Motivated by this, in this work, we focus on a class of nonlinear systems and design an estimator-based EMPC system. Working with the class of full-state feedback linearizable nonlinear systems, we use a high-gain observer to estimate the nonlinear system state using output measurements and a Lyapunov-based approach to design an EMPC system that uses the observer state estimates. We prove, using singular perturbation arguments, that the closed-loop system is practically stable provided the observer gain is sufficiently large. We use a chemical process example to demonstrate the ability of the state-estimation based EMPC to achieve process time-varying operation that leads to a superior cost performance metric compared to steady-state operation. In the example, the high-gain observer is used to obtain estimates of the reactant concentration from temperature measurements; a meaningful case in process control practice.

[1] P. D. Christofides, J. Liu, and D. Munoz de la Pena. "Networked and Distributed Predictive Control: Methods and Nonlinear Process Network Applications", Advances in Induatrial Control Series. Springer-Verlag, London, England, 2011. [2] M. Heidarinejad, J. Liu, and P. D. Christofides. "Economic model predictive control of nonlinear process systems using Lyapunov techniques" ,AIChE Journal, 58:855–870, 2012. [3] X. Chen, M. Heidarinejad, J. Liu and P. D. Christofides, "Distributed Economic MPC: Application to a Nonlinear Chemical Process Network", Journal of Process Control, 22, 689-699, 2012.

(186g) PID Design for Optimal Closed-Loop Response with Specified Gain and Phase Margins for Second Order Systems

Keyu Li, Thermal Energy Engineering, China University of Petroleum-Beijing, Changping, Beijing, China

The Proportional-Integrative-Derivative(PID) control has been widely used in industry and large amount of research work has been done on PID design. Most modern PID tuning methods are model-based and there is always model mismatch. Thus, robustness plays an important role in PID design. In practice, gain and phase margins (GPM) serve as important indicators of system robustness. Intensive research has been done on designing PID to meet gain and phase margin specifications.

In general, the GPM based PI design process usually leads to 4 equations and 4 unknowns. Numerical methods can be used to directly solve the problem, or approximation on arctan function can be adopted to simplify the problem and then the analytical solution can be obtained. For PID, there are 5 unknowns for total 4 equations, so extra approximations on derivative term are usually made to reduce the number of unknowns. The extra degree of freedom can also be used to achieve another objective, such as the minimization of the integral square error (ISE). The internal model control (IMC) tuning algorithm can also be combined with GPM, which significantly simplifies the problem since there is only one tuning parameter for IMC. The analytical solution can be easily obtained and suitable for online application such as adaptive control. However, the down side is that there would be only 3 unknowns for 4 equations, thus gain margin and phase margin are not independent in such case and the robustness criterion can not be exactly met. So far, most research is based on first-order plus time-delay (FOPTD) model, and very few work has been done on second-order plus time-delay (SOPTD) system due to its complexity. Ho et al. explored the PID tuning method with specified GPM for the under-damped second-order system with approximations on natural frequency of the controller ω_c and the *arctan* function.

Besides robustness, good closed-loop controller performance is also required in PID tuning, and usually there is a trade-off between performance and robustness. The maximum value of closed-loop amplitude ratio M_T and bandwidth ω_b are important closed-loop performance criteria. A larger M_T leads to a faster response, but also implies a larger overshoot and a lower robustness level, so it should be properly bounded. The bandwidth ω_b, however, should be as large as possible, since larger bandwidth values result in faster closed-loop responses and smaller settling time. However, there is no report yet on directly applying performance criteria M_T and ω_b with robustness criteria GPM in PID tuning. Thus, the motivation of this work is to develop a novel robust PID tuning method by using M_T and ω_b as performance criteria combined with the robustness criteria GPM for SOPTD systems. In this work, M_T and ω_b are adopted as performance criteria combined with GPM as robustness criteria in PID tuning for the first time. Then the PID design for SOPTD systems is formulated into a nonlinear optimization problem to maximize the bandwidth with constraints on both GPM and M_T, so that the settling time is minimized, and criteria on robustness and closed-loop performance are both satisfied simultaneously.

In the theory part of this work, the open-loop amplitude ratio and the phase equations are explicitly given for SOPDT model and PID in parallel form without any approximation, so that gain and phase margin can be calculated. Further more, the closed-loop amplitude ratio equation is also derived to calculate M_T and ω_b. Then the PID tuning method based on nonlinear optimization is proposed and some simulation results would be discussed in the end of the paper.

(189a) Assessing Excipient Variability within QbD Drug Product Development

Joseph Kushner IV, Drug Product Design, Pfizer Worldwide Research and Development, Groton, CT

Understanding variability in excipient physico-chemical properties is becoming an important aspect of Quality-by-Design (QbD) drug product development. However, present experimental methods are only able to study a few physico-chemical properties on a few excipient lots due to time, cost, and sample gathering considerations. To improve the robustness of drug products to excipient variability, analysis methods are needed to examine a broader range of excipient physico-chemical properties. An alternative analysis method is proposed which uses quantitative physico-chemical property data gathered by excipient vendors to evaluate excipient lot-to-lot variability in a comprehensive, low cost manner. Microcrystalline cellulose, spray-dried lactose, and magnesium stearate were selected as commonly-used excipients for this demonstration. The proposed analysis method, which is based on multivariate approaches, offers drug product developers several advantages over present experimental methods, including the ability to: 1) examine excipient products for manufacturing site and/or year-to-year variations, 2) quantify a domain of prior experience for each excipient by determining the percentage of excipient lots contained within a multi-dimensional ellipsoid described by the excipient lots used during drug product development, and 3) rationally select excipient lots from the vendors inventory to maximize the domain of prior experience throughout the drug development process. Case studies will be reviewed illustrating successful application of this method supporting drug product robustness to critical excipient material properties. Leveraging prior knowledge from excipient vendors' experience with excipient lot-to-lot variability should be an integral part of a QbD drug product development program to understand the role of excipient variability on the robustness of drug product manufacturability and performance.

(189b) Inversion of Latent Variable Regression Models to Support Product and Process Development

Emanuele Tomba[1], Massimiliano Barolo[1] and **Salvador García-Muñoz**[2], (1)DII - Department of Industrial Engineering, University of Padova, Padova, Italy, (2)Process Modeling & Engineering Technology, Pfizer Worldwide Research & Development, Groton, CT

When designing a new product and its manufacturing process, there is the need to understand how the raw material properties and the process parameters interact and affect the product performances, in order to control them and optimize the manufacture. For these reasons, usually large amounts of experiments are carried out to independently excite all the driving forces acting on the system, in order to gain a deep understanding of the relations between inputs (raw materials, process parameters) and outputs (final product properties). Even if systematically designed, the number of experiments to be performed may become prohibitive, especially if a large number of candidate materials and permutations have to be considered, as for the formulation of a pharmaceutical product. If data on historically developed products were available, useful information could be drawn from these databases, to support the design and the optimization of new products and of their manufacturing processes, with the aim of accelerating the development cycle guiding the experimentation from the first steps.

Latent variable regression models (LVRMs) are tools specifically designed to analyze large datasets of highly correlated data reducing the vast information in them in a small number of variables (the *latent variables*) which indentify the underlying driving forces relating input data with system outputs. The driving forces identified from the available historical databases represented by the LVRM equations can then be used to support the design of new products: since the design activity aims at finding the best combination of inputs (raw material properties and process conditions) which ensure a desired product profile in output, the design objectives could be achieved through a LVRM inversion operation.

In this contribution we show how LVRM inversion [1,2] can be feasibly used to support pharmaceutical product and process development activities. Since the LVRM inversion can have several solutions depending on the rank of the historical datasets used to build the model, the inversion has to be posed as an optimization problem [3,4]. A framework is therefore proposed to manage the different constraints and objectives the design problem may have, in order to solve in each situation the most appropriate LVRM inversion problem.

The proposed procedure is tested on two different product design applications, related to a particle engineering problem for the design of a wet granulated product of desired properties and to the selection of the best formulation for a given API, in order to obtain blends suitable for direct compression. In both cases the LVRM inversion is shown to be very effective in estimating the best inputs to obtain the desired product in output, giving a systematic tool to guide experiments since the first stages of development for the establishment of the system design space.

References

[1] Jaeckle, C. M. and J. F. MacGregor (1998). Product design through multivariate statistical analysis of process data. *AIChE J.*, **44**, 1105-1118.

[2] Jaeckle, C. M. and J. F. MacGregor (2000). Industrial applications of product design through the inversion of latent variable models. *Chemom. Intell. Lab. Sys.*, **50**, 199-210.

[3] García-Muñoz, S., T. Kourti, J. F. MacGregor, F. Apruzzese and M. Champagne (2006). Optimization of batch operating policies. Part I. Handling multiple solutions. *Ind. Eng. Chem. Res.*, **45**, 7856-7866.

[4] García-Muñoz, S., J. F. MacGregor, D. Neogi, B. E. Letshaw and S. Mehta (2008). Optimization of batch operating policies. Part II. Incorporating process constraints and industrial applications. *Ind. Eng. Chem. Res.*, **47**, 4202-4208.

(189c) Enabling Predictive Catalyst and Reaction Design in Asymmetric Organocatalysis (Invited)

Donna G. Blackmond, Dept. of Chemistry, The Scripps Research Institute, La Jolla, CA

Over the past several decades, the burgeoning field of asymmetric organocatalysis has introduced novel solutions to a number of transformations that construct carbon-carbon and carbon-heteroatom bonds in organic synthesis directed toward medicinal chemistry and drug development. This work has produced a wealth of experimental observations – primarily concerning product selectivity – including the exciting concept of cascade reactions where consecutive reactions fix multiple stereocenters in one pot. The area is now poised to emerge from empirical discovery to prediction and design of reaction outcomes enabling practical application. This next stage requires an increased focus on the underlying reaction mechanisms that afford efficient and selective transformations. Understanding complexity in competitive reactivity patterns in these systems is key to the development of design tools for improving both reactivity and selectivity in known reactions as well as for enabling transformations that have eluded us to date.

Our goal is to enable predictive catalyst and reaction design by developing experimental kinetic and spectroscopic protocols to probe complex reaction pathways. Investigations to date have relied primarily on empirical optimization based on simple analysis of final reaction product outcomes, which may neglect the importance of reaction dynamics and the interplay between stability and reactivity of intermediates in competitive networks. The temporal monitoring of reaction and selectivity profiles combined with spectroscopic identification of dynamic intermediate species is an underutilized approach in organocatalysis. Its development will aid in the delineation, rationalization, and exploitation of unexpected modes of stereocontrol. In this work we explore the generality of a proposed Curtin-Hammett paradigm for different classes of organocatalysts, applying comprehensive kinetic and spectroscopic protocols to a series of key transformations in pyrrolidine-based organocatalytic systems involving enamine intermediates. The central theme is the incorporation of our pioneering approach to quantitative kinetic analysis into a range of mechanistic tools. Kinetic-assisted mechanistic analysis will be developed as an orthogonal tool for discovery and optimization by rational design based careful documentation of reactivity patterns. This approach will address unexplained features of existing reactions as well as enable design of new catalyst/substrate combinations for high efficiency and selectivity, including complex reaction networks with multiple sequential transformations that begin to resemble metabolic systems, whence the inspiration for organocatalysis originates.

(189d) Implementing a Recipe Based Framework to Support Product Lifecycle Management

Adam Fermier[1], John Cunningham[2], Thomas Difeo[3], John Dingerdissen[4], James Kenyon[3], Steve Mehrman[3], Terry Murphy[5], Eugene Schaefer[2], Tom Schultz[6], Brian Sherry[7], James Weber[3] and Paul McKenzie[7], (1)Janssen Pharmaceuticals, Spring House, PA, (2)PDMS, Malvern, (3)PDMS, Spring House, (4)PDMS, Radnor, (5)PDMS, Cork, Ireland, (6)Janssen, Titusville, (7)Janssen Supply Chain, Horsham

The pharmaceutical industry to date has not fully leveraged data analysis methods such as multivariate techniques, e.g., multivariate statistical process controls, as achieved in some other industries. It has been postulated that the lack of such analysis was in caused in part, by the ability of the pharmaceutical industry to absorb sub-optimal / inefficient processes. We contend it is due to a more systematic challenge in the industry related to streamlined and effective data management.

The data management challenge stems from the diversity of data generated across the lifecycle of a compound which needs to be linked and interpreted. For example, it is common for a compound developed internally to be transferred between 3-4 sites and up to 30 separate systems prior to commercialization. Each system has valuable information that helps to drive key design and development decisions. As more data and information is gathered across these source systems, it becomes exponentially more difficult to manage the isolated data streams resulting in very high barriers to effectively use the information to create valuable knowledge.

To overcome these barriers, we will introduce you to a recipe based framework that is being implemented to facilitate knowledge lifecycle management. The strategy will enable more routine application of multivariate techniques for which case studies will be demonstrated.

(189e) Variable Selection in Multivariate Modeling of Drug Product Formula and Process

Yong Cui[1], Xiling Song[2], King Chuang[1], Cadapakam Venkatramani[1], Sueanne Lee[3], Gregory Gallegos[4], Thirunellai Venkateshwaran[5] and Minli Xie[6], (1)Small Molecule Pharmaceutical Sciences, Genentech, Inc. , South San Francisco, CA, (2)Small Molecule Pharmaceutical Sciences, Genentech, Inc., South San Francisco, CA, (3)Samll Molecule Pharmaceutical Sciences, Genentech, Inc. , South San Francisco, CA, (4)Small Molecule Pharmaceutical Sciences, Genentech Inc., South San Francisco, CA, (5)Small Molecule Pharmaceutical Sciences, Genentech, Inc. , South san francisco, CA, (6)Small Molecule Pharmaceutical Sciences, Genentech, Inc. , south san francisco, CA

Multivariate data analysis methods such as partial least square (PLS) modeling have been increasingly applied to product development, particularly in the quality-by design paradigm. This study applied the PLS modeling to analyze a product development dataset combining a design of experiment study and historical batch data. Attention was paid in particular to the assessment of the importance of predictor variables, and subsequently the variable selection in the PLS modeling. The results indicate that in conducting PLS modeling irrelevant and collinear predictors could be extensively present in the initial model. Therefore, variable selection is an important step in the model optimization. The VIP and coefficient values can be employed to rank the importance of predictors and to help remove irrelevant predictors. To reduce collinear predictors, on the other hand, multiple rounds of PLS modeling on different combinations of predictors may be necessary. To this end, stepwise reduction of predictors based on their VIP/coefficient ranking was introduced and appeared to be an effective approach to identify and remove redundant collinear predictors.

(189f) A Novel QbD Tool for Statistical Data Analysis

Linas Mockus, Purdue University, West Lafayette, IN, José Miguel Laínez, School of Chemical Engineering, Purdue University, West Lafayette, IN and G. V. Reklaitis, Chemical Engineering, Purdue University, West Lafayette, IN

One of cornerstones of Quality by Design (QbD) is Design of Experiments (DoE). However, in addition to DoE, ICH Q8 (R2) recommends an enhanced approach to product and process development which is based on mechanistic first principle models. Such predictive models require parameters such as kinetic constants which are functions of Critical Process Parameters (CPPs) and determined via designed experiments. Therefore, a robust statistical analysis is a key step for the in-depth process understanding. In this paper we describe a novel technique which is both statistically sound and computationally efficient.

Markov chain Monte Carlo (MCMC) approacheshave provided the computational machinery that has brought Bayesian inference applications to a variety of fields. However, the computational burden that they may require for the inference of complex models involving a considerable number of parameters can be prohibitive. As an alternative, variational approximation methods have been proposed. This family of methods translates the inference problem into an optimization problem by substituting the posterior distribution by a functional form. A decomposition strategy has been developed for the variational approximate inference to deal with models described by differential-algebraic equations (Laínez et al. 2012). In this work we further extend this framework to handle hierarchical models whose number of parameters generally increases with the number of "submodels". The relationship between the hyperparameters and "sub-model"parameters is exploited in order to apply a Lagrangian based decomposition so as to reduce the computation expense of the optimization problem.

A pharmaceutical reaction kinetic case study is presented to demonstrate the advantages of the proposed approach against MCMC approaches for hierarchical models.

References

Laínez, J.M., L. Mockus, S. Orcun, G. Blau, and G.V. Reklaitis. "A decomposition strategy for the variational inference of complex systems", submitted to Technometrics (2012).

Acknowledgements

Support from the United States National Science Foundation(Grant NSF-CBET-0941302) is gratefully acknowledged.

(189g) Mixture Component Prediction Using Iterative Optimization Technology

Koji Muteki[1], Daniel O. Blackwood[1], Kyle R. Leeman[1], George L. Reid[1], Yong Zhou[1], Yang A. Liu[1], Brent Maranzano[1], Paul R. Gerst[1], Howard W. Ward[1], Andreas AM. Mühlenfeld[2] and Matthias Danner[2], (1)Pfizer Worldwide Research and Development, Groton, CT, (2)Pfizer GMS, Freiburg, Germany

Spectroscopy techniques such as near infrared (NIR), mid Infrared (MIR) and ultraviolet (UV) are commonly used to monitor various process operations such as blending, mixing and chemical reactions in pharmaceutical development. The calibration required for a quantitative measurement can be a labor-intensive and time-consuming to implement. The objective of this study is to predict/monitor/estimate the mixture component ratio without making calibration standards in advance (e.g., minimum effort) while keeping similar prediction ability as with calibration standards. Reference methods such as HPLC/GC are used just for the validation purpose.

Iterative Optimization Technology (IOT) involves the use of an optimization computation using only pure component spectra collected prior to a batch and mixture component spectra collected during a batch. The proposed approach has been successfully applied to many practical projects (solvent mixtures, chemical reactions, feed-frame process, blending process). It is expected that the use of IOT will enhance the adoption of PAT as a routine analytical tool in project labs, especially for process understanding and early development activities.

(234a) Alamo: Automatic Learning of Algebraic Models for Optimization

Alison Cozad[1], **Nick Sahinidis**[1] and David C. Miller[2], (1)Chemical Engineering, Carnegie Mellon University, Pittsburgh, PA, (2)U.S. Department Of Energy, National Energy Technology Laboratory, Morgantown, WV

We address a central problem in machine learning, namely that of learning an algebraic model from data obtained from simulations or experiments. The problem arises naturally in situations in which there is a need to replace a computationally intensive model with a cheaper to compute surrogate or reduced order model. The problem also arises in situations in which one is interested in developing a theoretical model from experimental measurements. We are interested in developing a technique that learns models that are (a) as accurate as possible and (b) as simple as possible. Requirement (a) is obvious, while requirement (b) is driven by our desire to utilize the developed model in further studies, for instance by embedding it in a larger multi-scale model for optimization, simulation, or analysis. Finally, we are interested in developing methodology that achieves the above goals without requiring too many simulations or experiments.

We present a methodology aiming to achieve the above requirements. The proposed approach begins by building a low-complexity surrogate model. The model is built using a best subset technique that leverages a mixed-integer linear programming formulation to allow for considering a very large number of possible functional components in the model. The model is then tested, exploited, and improved through the use of derivative-free optimization solvers to adaptively sample new simulation or experimental points.

Finally, we describe ALAMO, the computational implementation of the proposed methodology, along with extensive computational comparisons between ALAMO and a variety of machine learning techniques, including Latin hypercube sampling, simple least squares regression, and lasso.

(234b) Information Management for Environmental Impact Assessment in an Integrated Enterprise Structure Based On an Ontological Framework

Edrisi Muñoz[1,2], Elisabet Capon[1,3], José Miguel Laínez[4], Antonio Espuña[1] and Luis Puigjaner[1], (1)Chemical Engineering, Universitat Politecnica de Catalunya, Barcelona, Spain, (2)Software Engineering, Centro de Investigacion en Matematicas A.C., Guanajuato, Mexico, (3)Chemistry and Applied Biosciences, Institute for Chemical- and Bioengineering, Swiss Federal Institute of Technology, Zurich (ETHZ), Zurich, Switzerland, (4)School of Chemical Engineering, Purdue University, West Lafayette, IN

Ontologies stand for an excellent choice for building complex models maintaining a high level of flexibility, re-usability, usability and easiness of maintenance.

Enterprises are highly involved systems, in which decision-making becomes a highly challenging task, and decision processes are usually separated in several levels or subprocesses. Such levels share data and information, which requires an efficient system integration and communication structure.

Therefore, effective integration among the different hierarchical levels, by means of tools improving information sharing and communication, may play a crucial role for the enhanced enterprise operation, and consequently for fulfilling the enterprise's goals [1].

This work proposes the re-use of an ontological model for the integrated enterprises structure in order to include the environmental assessment function. The ontological framework provides a common modeling framework which facilitates integration among the different decision levels, and works as the mechanism for information and knowledge sharing for multiple applications in the enterprise [2]. The general semantic framework developed is applied to an enterprise supply chain network design-planning problem case study considering environmental issues.

References:

[1] Grossmann, I. Enterprise-wide optimization: A new frontier in process systems engineering Aiche Journal, 2005, 51, 1846-1857.

[2] Muñoz, E., Capon-Garcia, E., Lainez, J., Espuña, A., Puigjaner, L., 2011a. Ontological framework for the enterprise from a process perspective. In: SciTePress (Ed.), Proceedings of the International Conference on Knowledge Engineering and Ontology Development. pp. 538 – 546.

(234c) Ontology Based Data Integration with ISO 15926

Michael Wiedau, AVT - Process Systems Engineering, RWTH Aachen University, Aachen, Germany, Manfred Theißen, AixCAPE e.V. , Aachen, Germany and Wolfgang Marquardt, Process Systems Engineering, RWTH Aachen, Aachen, Germany

In the domain of Computer-Aided Process Engineering (CAPE), several approaches exist for the exchange and integration of plant data from different software tools. In the past, we have realized a prototype for data integration based on OntoCAPE [], a formal heavyweight ontology for the CAPE domain. However, OntoCAPE was never intended as a means to exchange detailed plant data between tools. The international standard ISO 15926[1] addresses such data exchange in a flexible way. In this presentation, we give first results of a current project in which our proven integration approach is applied to ISO 15926.

In the past, several software applications have been developed by our group to demonstrate OntoCAPE's potential and wide range of use [6]. In particular, a software prototype for the integration and reconciliation of – potentially inconsistent – data from distributed CAPE tools has been created in cooperation with partners from the chemical and software industries. To integrate a certain CAPE tool with the prototype, it must support a common data format, either directly or by means of a converter. However, this proprietary exchange format – an XML schema [2] – is tailored to the requirements of the industrial partner involved.

A current project in cooperation with BASF SE, Bayer Technology Services GmbH, and Evonik Industries AG focuses on data exchange with the international standard ISO 15926. Several software vendors are expected to support data import and export in their tools in a foreseeable future, which will hopefully solve the data exchange problem in an acceptable way. However, data integration is still an open issue. Thus, we examine the adaptation of our approach for data integration to the more widespread ISO 15926 instead of the restricted XML schema as before. From an application-oriented point of view, this would allow to integrate data from arbitrary tools that support ISO 15926.

In contrast to an XML schema, ISO 15926 allows for semantic annotations which can be exploited for data integration. However, the semantic community [5] has reported on several deficiencies of the underlying ontology of ISO 15926. We examine in how far concepts from OntoCAPE can be integrated with ISO 15926 in order to simplify data integration. Our approach will be evaluated and compared with other integration approaches [3]. In the presentation, also a practical example for data integration between different tools is given. References

[1] ISO 15926-1, 2004. Industrial automation systems and integration – Integration of life-cycle data for process plants including oil and gas production facilities – Part 1: Overview and fundamental principles.

[2] F. Anhäuser, H. Richert, and H. Temmen. PlantXML – integrierter Planungsprozess mit flexiblen Bausteinen. Automatisierungstechnische Praxis, 46(10):62–72, 2004.

[3] B. Bayer, R. Schneider, and W. Marquardt. Integration of data models for process design–first steps and experiences. Computers & Chemical Engineering, 24(2-7):599–605, 2000.

[4] W. Marquardt, J. Morbach, A. Wiesner, and A. Yang. OntoCAPE: a re-usable ontology for chemical process engineering. Springer Verlag, 2010.

[5] B. Smith. Against idiosyncrasy in ontology development. In Proceeding of the 2006 conference on Formal Ontology in Information Systems: Proceedings of the Fourth International Conference (FOIS 2006), pages 15–26. IOS Press, 2006.

[6] A. Wiesner, M. Wiedau, J. Morbach, W. Marquardt, H. Temmen, and J. Redmer. Ontology-based integration and management of distributed design data. In WCCE8, Montreal, Canada, August 23-27, 2009.

(238a) The Autocovariance Least-Squares Method for Batch Processes: Application to Experimental Chemical Systems

Franklin D. Rincon[1], Marcelo Esposito[2], Pedro H. H. Araújo[2], Claudia Sayer[2], Fernando V. Lima[3] and **Galo A. C. Le Roux**[1], (1)Chemical Engineering Department, Polytechnic School of the University of São Paulo, São Paulo - SP, Brazil, (2)Departamento de Engenharia Quimica e Alimentos, Universidade Federal de Santa Catarina, Santa Catarina, Brazil, (3)Department of Chemical Engineering and Materials Science, University of Minnesota, Minneapolis, MN

Heat flow calorimetry is frequently used to monitor and control batch and semibatch polymerization reactions[1–5]. This technique requires the determination of the overall heat transfer coefficient between the reaction medium and the jacket (UA), which is a function of not only the reactor setup (geometry and wall material) and operating conditions (stirring and cooling fluid flow rate), but also of the properties of the reaction medium that can change throughout the polymerizations, as the viscosity can increase drastically and fouling at the reactor wall can occur. Different approaches for the estimation of UA have been proposed in the literature, for example, temperature oscillation calorimetry (OC)[6], Luenberger observer (LO)[7], extended Kalman filter (EKF)[8], cascaded observers (CO)[9] and Unscented Kalman filter (UKF)[10]. These techniques have presented the following issues: difficulty to be extended to industrial applications (e.g., OC), disregard of the stochastic nature of the system (e.g., LO), complex calibration procedure (e.g., CO), and the presence of negative estimates during the process (e.g., EKF and UKF). On the other hand, Moving Horizon Estimator (MHE) is a more robust estimator that can incorporate restrictions to its states and handle nonlinear models[11]. Unlike a recursive-based estimator (e.g., EKF and UKF), the MHE is based on the solution of an optimization problem at each time step. To obtain accurate estimates, the covariance matrices of the process (Q) and of the measurement noises (R) must be provided to specify the statistics of such estimators. These covariances are typically determined by trial and error, but for real applications, the covariance calculation can be a challenging problem. The estimation of the covariances can be performed by several techniques for batch and semibatch processes, including the linearized and Monte Carlo approach[12], the direct optimization[13], the direct and sensitivity method[14]. Recently, the Autocovariance Least-Squares (ALS) method has emerged as a viable solution to estimate process and measurement noise covariances for continuous systems (linear and nonlinear) for simulated and experimental data[15–17]. This presentation will discuss the extension of the ALS technique to address batch processes and the application of this method to laboratory chemical systems.

First, some important aspects of covariance and state estimation will be introduced; then, three experimental systems will be presented: 1) the hydrolysis of acetic anhydride carried out with excess of water in three different non-adiabatic vessels: (a) a cylindrical plastic vessel, (b) a volumetric flask, and (c) a thermal bottle or Dewar flask[18]; 2) the in-line monitoring of a batch emulsion polymerization reactor (without initiator), heated to the reaction temperature by three different temperature trajectories (step, ramp and arc); 3) the vinyl acetate emulsion polymerization in a semibatch reactor, where the reactions were started as a conventional batch emulsion polymerization, and then after 20 min, monomer additions were performed with 10 min intervals[19]. Finally, the problem of the simultaneous online estimation of states and parameters in these systems, using EKF, UKF and MHE with statistics defined by ALS, will be discussed in terms of the accuracy of the estimated variables. Preliminary results associated with the hydrolysis of acetic anhydride indicated that the UKF is more effective than the EKF to monitor the reactors under different conditions with heuristic diagonal covariances[20]. This work will show that high-quality estimates for this and the other experimental systems in focus are obtained when the statistics of the state estimators are systematically determined by ALS, especially for the implementation of the MHE in systems under repeated perturbations.

References

[1] L. M. Gugliotta, M. Arot, J. R. Leiza, and J. M. Asuaf, "Estimation of conversion and copolymer composition in semicontinuous emulsion polymerization using calorimetric data," *Polymer*, vol. 36, no. 10, pp. 2019-2023, 1995.

[2] I. Sa, D. Buruaga, P. D. Armitage, R. Leiza, and M. Asua, "Nonlinear Control for Maximum Production Rate of Latexes of Well-Defined Polymer Composition," *Industrial & Engineering Chemistry Research*, vol. 36, no. 1992, pp. 4243-4254, 1997.

[3] C. Sayer, E. L. Lima, J. C. Pinto, G. Arzamendi, and J. M. Asua, "Molecular weight distribution in composition controlled emulsion copolymerization," *Journal of Polymer Science Part A: Polymer Chemistry*, vol. 38, no. 7, p. 1100, Apr. 2000.

[4] M. Esposito, C. Sayer, R. a. F. Machado, and P. H. H. Araújo, "Effect of Cooling Fluid Flow Rate on the Estimation of Conversion by Calorimetry in a Lab-Scale Reactor," *Macromolecular Symposia*, vol. 271, no. 1, pp. 38-47, Sep. 2008.

[5] F. B. Freire and R. Giudici, "Temperature oscillation calorimetry by means of a Kalman-like observer: the joint estimation of Qr and UA in a stirred tank polymerization reactor," *Macromolecular Symposia*, vol. 206, no. 1, pp. 15-28, Feb. 2004.

[6] R. Carloff, A. Prob, and K. H. Reichert, "Temperature oscillation calorimetry in stirred tank reactors with variable heat transfer," *Chemical Engineering & Technology*, vol. 17, no. 6, pp. 406-413, 1994.

[7] G. Fevotte, T. F. McKenna, S. Othman, and A. M. Santos, "A combined hardware/software sensing approach for on-line control of emulsion polymerisation processes," *Computers & Chemical Engineering*, vol. 22, no. 98, p. S443-S449, 1998.

[8] R. Gesthuisen, S. Krämer, G. Niggemann, J. R. Leiza, and J. M. Asua, "Determining the best reaction calorimetry technique: theoretical development," *Computers & Chemical Engineering*, vol. 29, no. 2, pp. 349-365, Jan. 2005.

[9] S. Benamor, D. Colombie, and T. Mckenna, "Online Reaction Calorimetry. Applications to the Monitoring of Emulsion Polymerization without Samples or Models of the Heat-Transfer Coefficient," *Industrial & Engineering Chemistry Research*, vol. 41, no. 17, pp. 4233-4241, 2002.

[10] F. D. Rincon, M. Esposito, P. H. H. Araújo, G. A. C. Le Roux, and C. Sayer, "Calorimetric estimation employing Unscented Kalman Filtering for a batch emulsion polymerization reactor," *In preparation*.

[11] C. V. Rao and J. B. Rawlings, "Constrained Process Monitoring : Moving-Horizon Approach," *AIChE Journal*, vol. 48, no. 1, pp. 97-109, 2002.

[12] J. Valappil and C. Georgakis, "Systematic Estimation of State Noise Statistics for Extended Kalman Filters," *AIChE Journal*, vol. 46, no. 2, pp. 292-308, 2000.

[13] D. I. Wilson and M. Agarwal, "Experiences implementing the extended Kalman filter on an industrial batch reactor," *Computers & Chemical Engineering*, vol. 22, no. 11, pp. 1653-1672, 1998.

[14] N. P. G. Salau, J. O. Trierweiler, A. R. Secchi, and W. Marquardt, "A new process noise covariance matrix tuning algorithm for Kalman based state estimators," *7th IFAC International Symposium on Advanced Control of Chemical Processes*, vol. 7, no. 1, p. 572-, 2009.

[15] M. R. Rajamani and J. B. Rawlings, "Estimation of the disturbance structure from data using semidefinite programming and optimal weighting," *Automatica*, vol. 45, no. 1, pp. 142-148, Jan. 2009.

[16] F. V. Lima and J. B. Rawlings, "Nonlinear Stochastic Modeling to Improve State Estimation in Process Monitoring and Control," *AIChE Journal*, vol. 57, no. 4, pp. 996-1007, 2011.

[17] F. V. Lima, M. R. Rajamani, T. A. Soderstrom, and J. B. Rawlings, "Covariance and state estimation of weakly observable systems: application to polymerization processes," *Accepted for publication in IEEE Trans. Ctl. Sys. Tech*, 2012.

[18] W. H. Hirota, R. B. Rodrigues, C. Sayer, and R. Giudici, "Hydrolysis of acetic anhydride: Non-adiabatic calorimetric determination of kinetics and heat exchange," *Chemical Engineering Science*, vol. 65, no. 12, pp. 3849-3858, Jun. 2010.

[19] M. Esposito, C. Sayer, and P. H. H. de Araújo, "In-Line Monitoring of Emulsion Polymerization Reactions Combining Heat Flow and Heat Balance Calorimetry," *Macromolecular Reaction Engineering*, vol. 4, no. 11-12, pp. 682-690, Aug. 2010.

[20] F. D. Rincon, W. H. Hirota, C. Sayer, G. A. C. Le Roux, and R. Giudici, "Parameter estimation via Kalman filters and Luenberger-like observer," *In preparation*.

(238b) Model Predictive Quality Control of Batch Processes

Siam Aumi[1], **Brandon Corbett**[1] and Prashant Mhaskar[2], (1)Chemical Engineering, McMaster University, Hamilton, ON, Canada, (2)Department of Chemical Engineering, McMaster University, Hamilton, ON, Canada

This work addresses the problem of driving a batch process to a specified product quality using model predictive control (MPC) with data-driven models. To address the problem of unavailability of online quality measurements, an inferential quality model, which relates the process conditions over the entire batch duration to the final quality, is required for this problem. At a given sampling instant, the accuracy of this type of quality model, however, is sensitive to the prediction of the future (unknown) batch behavior. That is, errors in the predicted future data are propagated to the quality prediction, adding uncertainty to any control action based on the predicted quality. To address this "missing data" problem, we integrate a previously developed data-driven modeling methodology, which combines multiple local linear models with an appropriate weighting function to describe nonlinearities, with the inferential model in a predictive control framework. The key benefit of this approach is that the causality and nonlinear relationship between the future inputs and outputs are accounted for in predicting the final quality, resulting in more effective control action. The efficacy of the proposed predictive control design is illustrated via closed-loop simulations of an industrially relevant nylon-6,6 batch polymerization process.

(238c) Dynamic Data Reconciliation Using Conventional Dynamic Simulation Software for LNG Terminal

Sangho Lee, Chemical and Biological Engineering, Seoul National University, Seoul, South Korea and Chonghun Han, Seoul National Univ., Seoul, South Korea

While the risk of nuclear power generation became real, the Liquefied Natural Gas(LNG) demand is growing higher because of relative safety and eco-friendliness. In order to supply LNG to consumers and power plant consistently, LNG transport terminal is necessary. LNG Terminal has several storage tanks those are huge enough to prevent LNG supply risk and long and complicate pipeline to send LNG from LNG carrier to storage tanks or from storage tanks to demands. In management of this facility, safety is the most important thing in many other issues such as economically optimal operation or environmental harm because LNG is one of the highly flammable fossil fuels and there were numbers of catastrophic accident which kill hundreds of people in history. For the safe operation, many and various type of operation data of LNG must be monitored and in general LNG terminals plenty of temperature, pressure, flow rate sensors are installed for this reason and they are printing the data out in real-time. However using the real-time data without any pre-process has risk of error included in raw data. Every sensor is exposed of noise and failure that can cause misunderstanding about terminal's status. In addition about this terminal case, the problem is more serious because it is intimately related in safety. For an example, if temperature of LNG is increased but it is not measured because of sensor failure, then without any notice BOG is formed and finally it can make serious problem such as BOG leakage due to pressure rise. In order to prevent the sensor error, data reconciliation is widely used in many chemical processes. From 1970s there have been many researches about the technique and recently dynamic data reconciliation is developed for rectification of time-varying variables. Though its technological progress, dynamic data reconciliation remains unable to be simultaneous by the real-time data because of its computation time.

In this research, a new data reconciliation algorithm in real-time with conventional dynamic process simulation program is proposed to solve these data problems. A well-developed dynamic simulation model built with HYSYS dynamics is utilized as first-principle equations and for minimizing sum of squared differences MISO controller acts as an optimizer. Using the controller prevents to fall into calculation loop and is able to guess the reconciled value in real-time which is difficult for other researches. These are shown through a case study targets LNG terminal unloading pipeline during operation mode changes. As a result of the case study, the algorithm makes quite satisfying reconciled data close enough to other dynamic data reconciliation methods and requires much less work than others.

(238d) Comparison of Model Approximation Methodologies for Online and Multi-Parametric/Explicit Model Predictive Control

Romain Lambert, Centre for Process Systems Engineering, Department of Chemical Engineering, Imperial College, LONDON, United Kingdom, Pedro Rivotti, Department of Chemical Engineering, Centre for Process Systems Engineering, Imperial College, London, United Kingdom, Alexandra Krieger, Chemical Engineering, Imperial College London, London, United Kingdom and Efstratios N. Pistikopoulos, Centre for Process Systems Engineering, Department of Chemical Engineering, Imperial College London, London, United Kingdom

Comparison of model approximation methodologies for online and multi-parametric/explicit model predictive control.

Romain S.C. Lambert, Alexandra Krieger, Pedro Rivotti, Efstratios N. Pistikopoulos

Model Approximation is very useful in the context of complex high fidelity modelling of systems for which it is very impractical or impossible to derive an accurate model directly from plant data. Despite powerful prediction capability, such complex models pose significant difficulties to model predictive control (MPC) applications. To address the challenge posed by complex models, a large number of model approximation techniques have been introduced [1] [4] [6]. These methods are diverse in nature and may entail very different control implementations.

In this work we attempt to establish a practical classification of model approximation as a decision tool for selection of the most appropriate strategy for online and multi-parametric model based control. Firstly we study the performance of online model approximation techniques. These model approximations techniques aim to reduce the complexity of the control problem by adaptively reducing the dynamical system online and reformulating the MPC problem. The main advantage of these approaches is the reduction in memory storage requirements. In this study we compare two approximation strategies consisting of online linearization and model order reduction. The first approach [1] consists of online linearization of the system followed by linear model order reduction. For comparison we suggest a second approach which is a combination of online linearization of the system subsequently to applying an empirical nonlinear model reduction [2]. It can be shown that the order at which linearization and dimensionality reduction are performed has an influence on the controller design. Secondly we compare different offline approximation techniques. These approximation techniques present the advantage of being compatible with multi-parametric/explicit model predictive control (mp-MPC) algorithms [3]. The type of approximation employed implies different control strategies. We survey the use of piece-wise affine systems which entails the use of mixed integer model predictive control strategies (essentially branch and bound), trajectory piecewise linearization which can be addressed through linear mp-MPC [6], and finally nonlinear model reduction which can be combined to nonlinear mp-MPC [5]. The different combinations of approaches are compared on benchmark biomedical and chemical systems. Criteria for evaluations will consist of: Close-loop performance, cost of extraction of the approximate models, memory storage and online computation speed.

References

[1] Astrid, P., (2004). Reduction of process simulation models: a proper orthogonal decomposition approach. PhD dissertation, Eindhoven University of Technology,

[2] Bao Z., Sun Y., (2008). *Control Theory Appl* 2008 6 (3) 305–310.

[3] Bonis I, Theodoropoulos C. Model Chemical Engineering Science. 2012; 69: 69-80

[4] Hahn, J.,. Edgar, T.F, and Marquardt, Journal of Process Control 13, No. 2, pp. 115-127 (2003)

[5] Pistikopoulos, E.N., Dua, V., Bozinis, N.A., Bemporad, A., Morari, (2000) A. On-line optimization via off-line parametric optimization tools. Computers & Chemical Engineering. Volume 24, Issues 2-7, 15 July 2000, pp 183-188.

[6] Rewieński M, White J (2003) IEEE Trans. Computer-aided Design of Integrated Circuits and Systems, 22, 155-170.

[7] Rivotti, P., Lambert, S.C., Dominguez, L., Pistikopoulos, E.N. (2011). 21st European symposium on computer Aided Process Engineering

[8] C. Seatzu, D. Corona, A. Giua, and A. Bemporad, *IEEE Trans. Automatic Control*, vol. 51, no. 5, pp. 726–741, 2006

[9] Xie W., Bonis I., Theodoropoulos C. Comput. Chem. Eng. (2011) 35: 750-757.

(238e) A Semi-Physical Valve Stiction Model and Its Application for Stiction Quantification

Q. Peter He, Chemical Engineering, Tuskegee University, Tuskegee, AL and Jin Wang, Chemical Engineering, Auburn University, Auburn, AL

A semi-physical valve stiction model and its application for stiction quantification

Q. Peter He[1] and Jin Wang[2]

[1]Department of Chemical Engineering, Tuskegee University, Tuskegee, AL 36088, USA

[2]Department of Chemical Engineering, Auburn University, Auburn, AL 36849, USA

Valve stiction is one of the most common equipment problems that can cause poor performance in control loops. Consequently, there is a strong need in the process industry for non-invasive methods that can not only detect but also quantify stiction [1].

To simulate valve stiction, both detailed physical models and empirical models have been developed. Physical models [2] describe the stiction phenomenon using force balances based on Newton's second law of motion. The main disadvantage of these models is that they require knowledge of several parameters such as the mass of the moving parts and different type of friction forces which cannot be easily measured and depend on the type of fluid and valve wear. On the other hand, empirical or data-driven models [3-5] use simple empirical relationships between valve input and output to describe valve stiction, with just a few parameters that can be determined from operating data. Due to their simplicity and easy implementation, data-driven models have gained tremendous research interest in recent years.

In this work, a semi-physical valve stiction model is derived based on the analysis of the dynamic response of a physical model of a pneumatic valve, which is derived based on the Newton's second law of motion [6]. Comparison of the semi-physical model with the three existing data-driven models [3-5] to the physical model [2] shows that only the semi-physical model can reproduce the same valve behavior described by the physical model. One advantage of the semi-physical model is that it contains only two parameters compared to eight parameters of the physical model. Another advantage of the semi-physical model is that its implementation is straightforward without the involvement of cumbersome numerical integration as in the physical model.

In addition, we propose a noninvasive valve stiction quantification method based on the semi-physical model using the routine operating data obtained from the process, i.e., process variable (PV) and controller output (OP). The algorithm is proposed to estimate the stiction parameters, namely static friction f_S and dynamic or kinetic friction f_D, without requiring the valve position (VP) signal. Identification is accomplished by using linear and nonlinear least-squares methods which are robust and easy to implement. Several simulation and industrial examples, including both self-regulating and integrating processes with different degrees of stiction, are used to demonstrate the effectiveness of the method.

Key words: valve modeling, stiction quantification; control valve; identification and estimation; fault diagnosis

1. Choudhury, M. S.; Shah, S.; Thornhill, N.; Shook, D. Automatic detection and quantification of stiction in control valves. *Control Engineering Practice* 2006, 14, 1395-1412.
2. Kayihan A, Doyle III FJ (2000) Friction compensation for a process control valve. Control Engineering Practice 8:799–812.
3. He QP,Wang J, PottmannM, Qin SJ (2007) A curve fitting method for detecting valve stiction in oscillating control loops. Industrial and Engineering Chemistry Research 46:4549–4560.

4. Choudhury MAAS, Thornhill NF, Shah SL (2005) Modeling valve stiction. Control Engineering Practice 13:641–658.
5. Kano, M., Maruta, H., Kugemoto, H., Shimizu, K. (2004). Practical model and detection algorithm for valve stiction. Proc IFAC DYCOPS, Cambridge, USA.
6. He, Q.P., Wang, J. & Qin, S.J. (2010), An alternative stiction modeling approach and comparison of different stiction models, in Detection and Diagnosis of Stiction in Control Loops – State of the Art and Advanced Methods, Jelali, M. and Huang, B. (Eds.), Springer, pp37-60

(238f) Sequential Monte Carlo Moving Horizon Estimation

Sridhar Ungarala, Chemical and Biomedical Engineering, Cleveland State University, Cleveland, OH

State estimation is a fundamental element of engineering operations because it provides the knowledge for implementing process monitoring, control, optimization and safety related tasks. State estimation must be solved in the presence of often limited and noisy measurements as well as incomplete and uncertain process models. Many engineering systems tend to be nonlinear in nature and the uncertainties may be characterized by non-Gaussian probability density functions (pdf). The general solution to the state estimation problem is given by the marginal conditional pdf of the state from which point estimates are drawn as representative of the unknown state. Most operational tasks such as control and optimization require a point estimate for decision-making. It may not be generally claimed that the conditional mean estimate or the conditional mode estimate is superior to the other. In cases such as the existence of multiple steady states in state space, the mean estimate may lie in region of low probability between multiple peaks. In problems where a binary loss function is chosen for estimation, the mode is the only available choice. The nature of the system and the utility of the estimate will determine the choice of the point estimator.

Recently there is a surge of interest in solving the nonlinear non-Gaussian conditional mean estimation using the sequential Monte Carlo approach (SMC). This implementation of the recursive Bayesian filter is also referred to as particle filters. There has been a flurry of research activity in process systems engineering community regarding nonlinear and constrained estimation using the Monte Carlo approach with various simplifications such as Gaussianity and statistical linearization. Much of the literature in particle filtering relates to the approximation of the marginal conditional pdf and the estimation of the conditional mean. The conditional mode is difficult to estimate with small particle sets, while the mean is more robust.

There is a considerable body of literature on maximum a posteriori (MAP) state estimation, based on the maximization of the joint pdf of the state trajectory conditioned on all available measurements. The estimation of the most probable state trajectory becomes a computationally intractable problem as more measurements are accumulated and the number of decision variables increases. In practice, moving horizon estimation (MHE) limits the MAP optimization problem to a moving horizon of measurements to solve for finite sequences of the modal trajectory. For each horizon, the joint conditional pdf includes the a priori state pdf, which summarizes the uncertainty about the state at the beginning of the horizon. Traditionally, the a priori pdf is assumed to be a Gaussian pdf, which is recursively updated using linearized system dynamics such as the extended Kalman filter. In recent years, more accurate sampling based filters are suggested for determining the a priori pdf, for example, the SMC particle filters and the unscented Kalman filter. This area of research is active in what is known as the arrival cost computation in MHE. Several researchers identified potential for improving the MHE implementation using the SMC particle filter and yet others saw the synergy as the MHE providing improvements to the SMC particle filter.

In this paper the concept of approximating unknown state pdfs with Monte Carlo samples and the concept of solving a finite moving horizon MAP problem are treated in the same framework of sequential Monte Carlo. The novelty of the proposed sequential Monte Carlo Moving horizon estimation (SMCMHE) lies in (1) implementing a dynamic programming approach using the sequentially evolving particles for solving the MAP problem, specifically the Viterbi algorithm combined with the iterated dynamic programming method, which has desirable convergence properties, (2) using kernel density estimation (KDE) with fast bandwidth selection for estimating the a priori state pdf, thus eliminating the need for a separate filter running in the background for updating the arrival cost and (3) imposing multivariate linear and nonlinear constraints into the iterated dynamic programming approach.

The evolving set of particles in sequential Monte Carlo methods form a family of potential discrete state trajectories. This family of trajectories naturally generates a non-uniform and randomized dynamic grid that is adaptive according to the evolution of the state conditional pdf, which conveniently identifies the feasible regions of the state space.

Dynamic programming can be effectively used to maximize the joint conditional density on this feasible grid. The dynamic randomized particle grid is computationally more favorable for dynamic programming compared to fixed grids that have to face the crippling effects of " the menace of the expanding grids." In this paper, the MAP trajectory identified by the Viterbi algorithm is used as an initial condition for the iterated dynamic programming method to converge to the globally optimum modal trajectory in continuous state space. The computational complexity is limited to appending computed values of the objective function with the latest measurement and dropping the earliest values of the horizon as it moves forward and the sweep searches of the Viterbi algorithm. Kernel density estimation is a nonparametric method to estimate the underlying density of a set of random numbers of particles. KDE is a convolution of the kernel function with the data, which is efficiently implemented by the fast Fourier transform (FFT). The width of the kernel, or bandwidth, must be optimally chosen for which a novel iterative method is proposed (subject of a companion paper submitted to the conference, Ungarala (2012)). The arrival cost of each horizon is computed via KDE for non-Gaussian a priori pdfs. Constraints are handled by generating importance densities over constraint surfaces in state space from which the particles are drawn. This random sampling approach is in contrast with a number of computationally demanding methods proposed in recent literature, where each particle is individually subject to constraints by further optimization methods.

The proposed approach is demonstrated with simulation examples taken from literature including benchmark nonlinear systems such as a pendulum subject to total energy conservation and a chemical reactor system subject to multivariate equality constraints on species compositions. The computational demands are compared with solutions obtained from using standard nonlinear programming code for the MAP optimization problem.

References:

Ungarala, S. (2012) Kernel density estimation using orthogonal polynomials and Goertzel DFT. Submitted to 10C07 Process Monitoring, Fault Detection and Diagnosis I, AIChE Annual Meeting, Pittsburgh, PA.

(243a) To Teach or Not to Teach, What Is the Answer?

Santiago Faucher, Technologies, Hatch, Mississauga, ON, Canada

The typical university graduate will have spent 17 of their 21 years being taught a prescribed education regimen. The retention of this education is tested through written exams which, in the best of cases, will permit the student to elaborate on the answer and outline the process they took to achieve it. In the worst of cases, the evaluation is reduced to a multiple choice format due to the limited capacity of the education system to do otherwise. In all cases, there is typically only one correct answer or approach thereto.

Fortunately the next 60 years of one's life and career are not so simple and much richer: there is no one right answer, but rather a multitude of possible acceptable solutions to choose from. For Chemical Engineers who have been subjected to problem-based and/or self directed learning, this realization may come early during their college education; for others, the capstone project may present the first opportunity toward this realization. Ultimately it is the professor's approach to learning that dictates when students begin to think more richly towards varied and novel solutions. Should learning be pushed onto the student through professing or be pulled by the student through the establishment of the appropriate problem and reward? Which of these approaches will lead to a deeper retention of the curriculum? As professor of a capstone course, I chose a problem-based learning approach thus taking on the role of mentor and client rather than professor. I responded to student questions with questions. Each question designed to guide but not direct; no answers were provided. My students' response to my unconventional approach fitted well with the model of Kübbler-Ross describing grief management, while their group dynamics nicely fit Tuckman's Group Development Model.

This talk will report on the pros and cons of a problem-based capstone project, the educator and student roles, student and group dynamics, mentoring mechanisms and effectiveness, and finally on an unexpected team performance. As is often the case, both in this course and in life, the question asked in this educational experiment has lead to learning and to a more profound question: should one teach or mentor the chemical engineers of the future?

(243b) Helping Students to Learn by Doing - Capstone Design Experience At Columbia University

Michael Hill and Stanley A. Leshaw, Chemical Engineering, Columbia University, New York, NY

Columbia has transformed its two-semester Capstone Design sequence from a traditional lecture-based format to one which imitates the real-world dynamics of industry. Students meet in small groups or teams for up to one hour each week with an advisor/technical manager/coach who critiques progress, offers suggestions for improvement, and provides direction for the following week. Student teams complete their design projects on a semi-independent basis, receiving individual guidance, but not complete answers, from their advisor. The format also plays to the strengths of adjunct faculty, who have found it easy to adopt. As a result, students "learn by doing" as they address three significant process or product design problems of increasing complexity.

Projects are selected by the instructors to provide the students with sufficient hands-on exposure to several important pedagogical issues in design, such as the interaction of components within a system and the impact of constraints on optimization in design. These projects are also chosen to reflect an awareness of broader societal issues, such as environmental relevance, green energy, and sufficient water.

Student feedback has expressed appreciation for the format, with students frequently citing Design as their favorite course.

(243c) Chemical Process Design and Projects Two Semester Sequence

Alan W. Weimer, Chemical and Biological Engineering, University of Colorado at Boulder, Boulder, CO

The Fall semester 3 Credit course, “Chemical Process Synthesis”, along with the Spring semester 2-Credit "Design Projects" course, is the *capstone* course sequence in chemical and biological engineering at the University of Colorado. The 1st semester utilizes a design text for approximately 8 weeks and ends with a "mini-design" team project which serves as the final exam. Topics covered from the text include: (1) an introduction to the design method; (2) health, safety, environmental, and ethics; (3) process creation/simulation, (4) process synthesis heuristics; (5) design methods reviewed for major unit operations; (6) heat and power integration; (7) ideal and non-ideal separation trains; and (8) process economics and profitability analysis (3 weeks of economics). Every class starts with 10 minutes dedicated to current events. The mini-design project is a well-defined problem that can be simulated and students are organized into typically 3-person teams. All teams work on the same design problem. Grading of the mini-design projects is taken very seriously (not hand-waved) in order to "set the bar" for the 2nd semester. For unsatisfactory mini-design projects, the teams must submit a satisfactory "retread" prior to the drop/add deadline of the 2nd semester. First semester grading is 30% homework, 30% two midterm exams, and 40% mini-design project, however, a passing grade is required on the mini-design project independent of all other grades in order to pass the 1st semester class. A peer-review counts as 20% of the mini-design project grade, but does not impact the requirement that the team project itself requires a passing grade - independent of the peer-review. The 2nd semester "Design Projects" course is supported by external typically industrial liaisons. The 3-person team projects are organized in the fall by the instructor and the projects are selected during the first week of class. Students work very closely with their external liaisons. Problem statements are typically very open-ended and may be only one or two paragraphs in length. Teams are required to research the project and to write a project proposal due within 3 weeks. The proposal needs to incluide all required "battery limits". The proposal is graded by the instructor with input by the external liaisons. Teams give two oral presentations during the remaining 8 weeks of class and are required to make an oral presentation at their liaison's facility, if local, during the last week of class, or during final exam week. GoToMeeting is used for non-local liaison presentations. Industry liaisons typically attend the oral presentations and comprise an external panel that asks quesions along with students in the class. Oral grading is done based on certain metrics with input from both the students and the external liaisons. Liaisons that cannot attend are engaged using GoToMeeting. A final written report is submitted and graded with input from the external liaisons. Organizing the external projects is challenging, but the instructor has been able to organize 36 different projects over each of the last two years. Student evaluations are typically excellent regarding the capstone sequence. Many of the industry liaisons are former students who want to contribute to the educational process. Many also recruit at the university and even hire students that they are engaged during the course. The instructor has spent half of his career in industry and half in academia and also directs a very active research program.

(243d) Senior Design At UC Irvine: Process Development and Economics Analysis with PRO/II

Gang (Gary) Xu, R&D, Invensys/SimSci-Esscor, Lake Forest, CA

Industrial standardized simulator PRO/II is applied in UC Irvine senior design class. Students are explore to real time simulation and project design assignments.

With fundamental knowledge of process design and economic analysis learnt from classroom, students get to develop projects by literature searching, PRO/II flow-sheet exercising, data reconciliation, and high-level process optimization from economic and environmental consideration.

Students were also stimulated by choosing their own projects reviewed by instructor. Interesting ones including generation of long-chain hydrocarbon (as fuel easy to store and transport) from natural gas, bio-energy generation from grass, supercritical CO2 decaffeination, sulfuric acid production from SO2, alcohol fermentation from potato (sugar) ... From which practice of fundamental engineering and economic understanding were encouraged.

With all the activities, popular topics such as

1. Azeotropic distillation

2. process optimization by controlling of design variables

3. economic model analysis such as Pareto optimal with engineering invest vs. environmental impact

were also emphasized as key portion of the design projects. Some of these topics, such as Pareto optimal, exceeded undergraduate level. However, basic knowledge of such topics were well understood with reasonable scale of exercise directly connected with projects chosen by the students.

(243e) Conduct Your Capstone Design Class As a Consulting Company

Richard L. Zollars, Voiland School of Chemical Engineering and Bioengineering, Washington State University, Pullman, WA

Capstone design classes are often taught in a similar manner – design a process (ammonia plant, methanol plant, etc.) to meet some production goal. By selecting an appropriate plant the instructor can be insured of covering almost any unit operation to which the students have been exposed at some time during their education. There are a number of shortcomings to such an approach, however. One of these is that the details of such plants, which unit operations are used, in what order, etc., are readily available in the literature. Students will repeat the standard design, go through a laborious series of calculations to size and cost equipment, perform an economic analysis and report their findings. In terms of promoting any in-depth thinking, there is little to offer. In addition, only a small fraction of students will ever be employed in such a situation.

A more common situation for students upon graduation is to be employed as a process or unit engineer. As such, in addition to keeping their unit operating, they have the responsibility of upgrading or debottlenecking their operation. This smaller scale, more focused, design work typically involves only a few unit operations. While few in number the unit operations that are involved are highly variable depending upon the unit in which the engineer is employed.

In an academic setting trying to operate a design class using this narrower focus is difficult without having the problems seem contrived. Thus, eleven years ago, we began operating our capstone design class in a different manner. The instructor for the capstone design class solicited regional industries for projects. With this set of projects available the design class is then conducted like an engineering consulting firm with teams of students vying for various projects. During the design process the students must keep in contact with their "clients" to insure that they are addressing the client's concerns. At the end the student groups must satisfy not only the class instructor, but also the industrial contact person who originally proposed the problem.

The capstone design class at Washington State University is a two semester sequence of classes. The fall semester course covers economic analysis, equipment sizing, safety and team work in addition to starting the design project. The spring semester class is devoted exclusively to the design project. In preparation for the academic year regional industries are contacted for proposed projects. For a variety of reasons (cost, laboratory facilities, safety) the projects are to be paper studies, not requiring work in a laboratory or shop. The projects should require approximately 500 person-hours to complete. In addition to developing such a project the industry is expected to provide part of a person's time to serve as the point of contact for the student design team as well as some financial remuneration to cover expenses for the project.

During the fall semester the students select the group in which they want to work. Ideally, each group is three people but no group will be larger than four people. Once the groups are formed one-page descriptions of the projects that are available for the year are given to each group. Each group then goes through a bidding process for the selection of the project on which they will work. At the completion of this bidding/selection process each group will visit the industrial site to meet with their contact as well as get a fuller understanding of the project. Prior to the end of the fall semester but after the initial site visit each group will conduct a brainstorming session to develop a slate of potential ideas for the solution of the problem they have been assigned.

During the spring semester the students work exclusively on the design project. They develop Gantt charts to help them monitor their progress. These are updated formally three times during the semester. They are expected to maintain regular communications with their industrial contact. A second site visit about half way through the semester is conducted to insure that both the students and industry contact are in agreement with the direction of the project. Final written reports are due after 12 weeks. This is followed by a day of formal oral presentations the following week. Because of the number of projects in any given semester the time for these presentations is short. Thus we have gone to a poster/oral presentation format. The student design groups participate in a poster session first followed by 20 to 25 minute oral presentations. Finally, before the end of the semester, each group is to hand-in a final revised written report which will be sent to the industrial contact.

To date a total of 58 different design projects have been conducted for 15 different industrial sponsors. The industries have ranged from major corporations, to government labs, to state government agencies, to small companies (5 – 10 employees). The range of topics is also quite broad with few projects covering similar processes. There have been some very traditional chemical engineering projects (heat exchanger optimization, hydrogen recovery in a refinery, debutanizer optimization), some others that are still definitely chemical engineering in nature (active metal reduction of nitrates, oxidation of uranium using Fenton's Reagent, purification of soy and palm oils, water treatment/recovery, and copper recovery) while a few others stretch the boundaries of the discipline (recycling of carbon fiber reinforced composites, recovery of neutraceuticals from fruit waste, and converting municipal and/or food waste into biofuels).

As measured by student involvement conducting the design course in this manner has been a success. Student evaluations of the course are uniformly high. More than one alumnus has reported that their experience in the design course was the closest thing they had to actual practice than any other course in the curriculum. The industries that have participated have also expressed their support. Not only have they continued to provide projects, manpower and financial support but also have hired students as a result of the student's work on a project supported by the industry.

Another way to assess the success of conducting the design course in this manner is the quality of the work performed by the students. The quality of the written reports has gone up, in part due to the fact that the reports are going to people who might be hiring the student. The oral/poster presentations have been excellent. More than one person from industry has commented that they meet or exceed the quality that they would expect from their engineers.

The quality of the designs is also outstanding. Approximately 20% of the student design suggestions have been implemented in some from by the company that supplied the project. The quality of the student's work was verified by a situation that occurred recently. One of the companies offered a project for which they had already signed a contract with an engineering consulting firm. They received both the professional firm's report at about the same time as they received the student's report. While the student's report lacked the detail of the professional report (types of welds, fasteners, support structures, etc.) the student design and the professional design contained almost identical unit operations, arranged in the same order, with equivalent material and energy balance calculations.

Integrating Active Research on CO2 Capture in Traditional Process Design

Omkar Namjoshi[1], Paul Nielsen[1], Matthew Walters[1], Bo Lu[1], Siyun Wang[1], Gary Rochelle[1]

[1]The University of Texas at Austin, Austin, Texas, USA

The typical capstone design course in most chemical engineering curricula integrates material from heat and material balances, unit operations, engineering economics, and reaction engineering. At the University of Texas at Austin, we also incorporate aspects from sustainable engineering and from our research program in the design class. Students who take this class are tasked with completing a preliminary design of a post-combustion carbon capture process (capturing CO_2 from a point source) using amine-based solvents or physical solvents.

Themes from sustainable engineering are incorporated at the beginning of the course and reinforced throughout the semester. Students are first given an introduction to climate change, how capturing CO_2 might help mitigate climate change, a brief overview on CO_2 storage sites. These themes are reinforced throughout the class as detailed design progresses and economics are finalized. Students must answer what price CO_2 must be "sold" for the plant to make a profit and also must discuss the environmental impacts of CO_2 capture facility. The link to sustainability gives students valuable insight regarding energy and environmental policy, which is not commonly encountered in a traditional "core" chemical engineering class.

Within the first three weeks of class, students are expected to form teams of three students each and submit a proposal outlining their choice of CO_2 source, capture solvent, and storage site, justifying their choices with quantitative arguments. The teams are expected to use as many resources as possible and are encouraged to get information and guidance from graduate students, research sponsors (which include large oil and gas companies, amine technology suppliers, and power plant operators), and old student reports (which are posted on the course website) to help strengthen their arguments for their proposal (as well as for general help, if it is needed). Students – past and present – have proposed a variety of CO_2 sources, including (but not limited to) coal-fired power plants, integrated combined-cycle power plants, fluidized catalytic crackers, and LNG treating plants. Example of capture solvents include aqueous amine solvents, such as monoethanolamine, concentrated piperazine, and activated MDEA. Students are encouraged to design plants using unconventional sources and solvents; those who do so are evaluated more on effort and a good initial, "first-draft" design. Students who have chosen solvents and sources that have been selected in prior years are evaluated on producing an optimized design that represents an improvement over the prior year's design.

The integration with our research program is achieved by incorporating thermodynamic models built using the AspenPlus simulation software / framework and developed by graduate students into the student projects. Our research group has developed rigorous models for activated MDEA and concentrated PZ and has developed preliminary frameworks for other amine systems; the use of these models by the undergraduate students is a helpful validation tool for the graduate students and helps in improving future revisions of the model.

Assignments are generally assigned each week, with a few exceptions for some of the lengthier and more involved assignments. Each assignment can be best described as "goal directed, mission oriented" – students will have a set number of deliverables for each assignment cycle and are expected to complete them (or explain why they were unable to) and are held accountable to these deliverables. Table 1 gives a list of assignments along with deliverables (note that the deliverables can change from year to year and are group specific) that are assigned to students. A few assignments are individual assignments, and the assessments from each individual assignment are used to differentiate student performance within the group.

The weekly assignment cycle ensures that students do not fall behind on their reports. The final report essentially consists of students putting all of their reports together and addressing any areas for improvement that TA feedback has identified. The grade for the final report is 50% of the overall course grade. This performance bias ensures that students who fall behind are able to catch up (should they choose to do so) and that students who earned good evaluations on prior reports will continue to maintain their performance through the end of the semester.

One unique aspect about course management is that highly motivated graduate students serve as teaching assistants (TA's) for the class. Each teaching assistant supervises three to four student design groups and is responsible for assigning deliverables to the students, mentoring student design groups and ensuring that they have the resources to be successful in completing the assigned deliverables, and holding the students accountable to those deliverables (addressing feedback). A head teaching assistant (generally the most experienced amongst the TA's) is also appointed and, in addition to having his/her own student groups to supervise, mentors and advises the other teaching assistant and also manages course coordination such as AspenPlus tutorial development and ensuring that student evaluation is completed in a way that is fair, consistent, and free from individual TA bias. The supervising professor (currently Dr Gary Rochelle) delivers the lectures during the classroom sessions, is responsible for supervising each TA, and holds a standing weekly "alignment" meeting to collectively set goals and deliverables for upcoming assignments with the TA's as well as address any TA concerns. The teaching assistant experience for this course is an excellent way for motivated graduate students to obtain "real-world" management skills that are useful in either a professional or academic setting. In recognition of this fact, UT's chemical engineering department grants three hours of Supervised Teaching course credit to each student who serves as a process design TA.

The emphasis on sustainability, the integration of research into the student design projects, coursework organization, and emphasis on graduate TA development make UT's design course unique.

Table 1. Plant Design Assignments at UT-Austin

Assignment / completion time	Individual or Group?	Deliverables
Proposal / 2 weeks	Individual	Pick a CO_2 source, capture solvent, and storage site
Proposal / 1 week	Group	Finalize source / solvent / storage site decision, submit one proposal, make improvements to proposal
PFD / 1 week	Group	Develop preliminary flow diagram, outline equipment required, model specifications (e.g. CO_2 compressor pressure)
Simple Heat and Material Balance / 1 week	Group	Develop basic heat and material balance, model CO_2-amine interactions based on an equilibrium model
Rigorous Heat and Material Balance / 2 weeks	Individual	Develop rigorous heat and material balance on one unit operation, model CO_2-amine interactions using rate-based models
Rigorous Heat and Material Balance / 1 week	Group	Combine reports, address feedback
Equipment Specifications / 1 week	Group	Size all equipment (absorption / regeneration columns, heat exchangers) and specify materials (e.g. stainless steel)
Equipment Costs / 1 week	Group	Cost all sized equipment using Aspen Economic Toolkit
Project Economics / 1 week	Group	Find price of CO2 required for plant to be profitable, develop project economics – cash flows, IRR, DCF
Process Safety and Environmental Assessment / 1 week	Group	Identify environmental impacts that the process might have
Draft Final Report / 1 week	Group	Combine all previous assignments into one report, address all feedback. No new information is needed beyond the environmental assessment section.
Final Final Report / 2 weeks	Group	Address feedback on draft final report, also present oral presentation on design to class

(243g) Incorporating Process Simulation Across the Chemical Engineering Curriculum to Improve Student Performance On the Capstone Design Project

David A. Rockstraw and Martha C. Mitchell, Chemical Engineering, New Mexico State University, Las Cruces, NM

The New Mexico State University Chemical Engineering Department recently made changes to the core curriculum to accommodate increased training in the use of chemical process simulation (Aspen Plus®) in parallel with the fundamental topics of the discipline. Over the course of five consecutive semesters, students learn aspects of the simulator, solving the same problems in simulation lab that were previously solved by hand calculation in their courses. The 2012 graduating seniors received twice the exposure hours to the simulator as previous graduating classes, and the 2013 graduating class will have received twice the hours of exposure of the 2012 graduating class. The increased level of understanding of how to properly use the simulator for process design has led to significant improvement in student performance on their capstone design projects in 2012. This presentation will describe the curriculum changes made that have permitted this improvement in student work, and the content and methodologies of the design courses in the sequence.

(245a) Closed-Loop Decision Making Under Uncertainty for Sustainability Enhancement of Industrial Systems

Zheng Liu and Yinlun Huang, Department of Chemical Engineering and Materials Science, Wayne State University, Detroit, MI

Sustainability performance of manufacturing systems could be significantly improved using carefully selected design, manufacturing, and management technologies. However, due to inherent uncertainty associated with accessible data and information relevant to technology candidates, how to identify most desirable technologies is always a challenge. This becomes more challenging for assuring future sustainability performance of the systems as future market conditions, environmental regulations, etc., could change along the time.

In this paper, we introduce a novel decision-making methodology for sustainability performance improvement under severe of uncertainty, where data uncertainty and scarcity leads to no data distribution information derivable. The introduced methodology is developed by resorting to sustainability principles, control science, and process systems engineering fundamentals. In development, we consider industrial sustainability problems as a general system control problem. In this context, a basic closed-loop feedback control scheme is used to characterize an industrial system under investigation, evaluate and analyze its sustainability performance, and identify most suitable technologies based on the preset sustainability goals, for the improvment of the system's sustainability under budget and other constraints. This sustainability decision making and system performance improvement process will continue as a response to the change of external situations that may be related to market, regulation, policy, etc. Since a large amount of information and data is uncertain, imprecise, and incomplete, an interval-parameter-based approach will be used in system characterization, assessment, and decision making. The effectiveness of the derived decisions and system stability will also be evaluated before and after implementation. The proposed decision making methodology has been successfully used to investigate a number of industrial problems. In this presentation, a comprehensive study on sustainable biodiesel manufacturing in a large geographical region will be illustrated.

(245b) Second Law of Thermodynamics and the Notion of Externalities of Macro-Economics: Two Pillars of Assessing the Sustainability of a Manufacturing System

L. T. Fan, Chemical Engineering, Kansas State University, Manhattan, KS and **Tengyan Zhang**, Western Research Institute, Laramie, WY

Sustainability engineering, an emerging discipline of engineering, can be regarded as an engineering discipline to craft frameworks, or establish paradigms, for attaining sustainability in any social or human endeavor including sustainable manufacturing. No discipline or branch of engineering can be enduring without theoretical principles underlying it or without a firm theoretical foundation. In this regard, two cornerstones of sustainability assessment of any system or process should be the second law of thermodynamics and the notion of externalities of macro-economics.

The inevitable entropy increase attributable to a process of concern manifests itself in exergy dissipation by the process, which can be evaluated by resorting to the available energy balance; this balance results from integrating the first and second laws of thermodynamics. In reality, however, the available energy balance also implicitly embodies the mass conservation law: It entails the detailed accounting of every material species involved in the process in light of the species' matter-energy.

Economic externalities attributable to a system or process of concern manifest themselves in the process' external costs. These external costs inherent to the system can be evaluated through cost accounting or estimation in conjunction with the process' internal costs, such as the labor and equipment costs.

Various methodologies are available for assessing the sustainability of manufacturing or process systems, which take into account exergy dissipation as well as cost, or can possibly take both into account. One of such methodologies is sustainability potential; this hierarchical approach is proposed to estimate the sustainability-potential of a system based on any of the alternative manufacturing or synthetic routes. This novel notion of sustainability-potential is a generalization of the notion of profit-potential, the estimation of which renders it possible to screen out unsustainable processes derived from the alternative synthetic routes at the earliest stage. The efficacy of sustainability potential is illustrated with the manufacture of vinyl chloride.

(245c) A Shortcut Graphical and Spreadsheet Technique for Identifying Mass Exchange Networks

Russell F. Dunn, Chemical and Biomolecular Engineering , Vanderbilt University, Nashville, TN and Kenneth A. Debelak, Chemical and Biomolecular Engineering, Vanderbilt University, Nashville, TN

Cost-effective waste minimization within chemical processes continues to be a critical component for future sustainability of large chemical and polymer manufacturing companies. Several methods have been developed for the synthesis of mass exchange networks (MENs) for waste minimization. In fact, several of the methods previously developed are based on a two stage targeting approach. The first stage uses the rich (waste) and lean (mass separating agents) composite curves to identify the maximum mass integration possible for an undesirable species to be transferred from a group of waste streams to a group of internal process streams that are used as mass separating agents. Furthermore, the second stage identifies the network of mass exchange technologies that meets the target identified in stage one. Several strategies have been employed for identifying the MEN in stage two including, but not limited to, mathematical optimization and state-space structural approaches.

This paper will present a new graphical/spreadsheet technique and design strategy for identifying the MEN in stage two. This technique allows the simultaneous evaluation of multiple diverse mass separation technologies. Advantages of this strategy will be highlighted and several case studies will be discussed that illustrate the effectiveness of this approach. Finally, experiences using this strategy in teaching undergraduate chemical engineering design courses will also be discussed.

(245d) Synthesis and Performance of Sodium 9,10-Dihydroxy Stearic Acid

Xiao Zhao, Shool of Chemical and material engineering, Jiangnan University, Wuxi, China and Yun Fang, School of Chemical and material engineering, Jiangnan University, Wuxi, China

9, 10-dihydroxy stearic acid (DHSA) and its derivatives are of high value due to their wide range of industrial application, including cosmetic, food, lubricant, plastic additives, personal care, pharmaceutial products and etc. It used to be synthesized by introducing hydroxy groups into carbon-carbon double bond of oleic acid with various catalysts. Almost all those process could not be industrialized for the more excessive co-oxidant and the boring separation process. Here we describe a more convenient method to obtain high purity DHSA by purified oleic acid and hydrogen peroxide with the catalyst of formic acid. The conversion of the oleic acid was over 97%. And the crude DHSA product was obtained by decomposition a small amount of excessive hydrogen peroxide over MnO_2, followed by evaporating water and a residue catalyst. The chemical structure of the recrystallised DHSA was characterized by IR, NMR and MS. Sodium 9,10-dihydroxy stearic acid (DHSA-Na) was prepared by neutralization of DHSA with NaOH. Surface tension, wetting property, lime soap dispersing power (LSDP), Krafft point, detergency and other performances of DHSA-Na were well determined. The results indicated a variety of applications in many fields for the excellent surface and interface chemical performance.

(245e) A Framework for Chemical Risk Management in Product Manufacturing and Its Application to Industrial Cleaning

Masahiko Hirao, Emi Kikuchi and Yasunori Kikuchi, Department of Chemical System Engineering, The University of Tokyo, Tokyo, Japan

For sustainable manufacturing, voluntary efforts are needed in industry to manage risks on human health and the environment associated with the use of chemicals. In this study, a practical framework for chemical risk management in product manufacturing based on simultaneous evaluation of functional, economic, and risk aspects is proposed. The framework is embodied through the case study of industrial cleaning processes, where open-top processes with chemical substances such as chlorinated organic solvents, inorganic compounds, and surfactants, are used.

Process and operation design aimed at risk reductions needs two major activities: evaluation of current and alternative processes, and decision-making on the selection of alternatives. Knowledge and tools, such as technical information, risk evaluation method, and process model, are necessary in order to evaluate risks of processes, as well as their functional and economic aspects. Technical information is systematically organized for cleaning processes, where processes have been designed on the basis of heuristics conventionally. Product defects associated with the careless process modifications aimed at risk reductions can be avoided by evaluating function of alternative processes simultaneously on the basis of the technical information. Risk analysis methods using LCA and risk assessment are applied to the evaluation in order to analyse risk trade-off relationships among alternative processes. A model of open-top cleaning processes is developed to estimate process data needed for evaluating alternative processes.

It is necessary to manage conflicts among evaluated functional, economic, and risk aspects in the decision-making phase. Engineers need to select appropriate alternatives for risk reductions, so that the improved process can achieve not only risk reduction targets but also functional requirements and economic viability. The proposed framework including concrete procedure of such multi-objective decision making is developed through the study on actual cases of process improvement in industrial cleaning. It is revealed that the flexibility assessment of constraints on product quality can be effective in managing conflicts between functional and risk aspects. Furthermore, the engagement of up- and downstream manufacturers in product supply chain can contribute to expanding the range of options of risk reduction measures.

(245f) Scheduling for Multiple Chemical Plant Turnarounds to Minimize Regional Air Quality Impacts

Tianxing Cai and Qiang Xu, Dan F. Smith Department of Chemical Engineering, Lamar University, Beaumont, TX

During the chemical manufacturing process, multiple chemicals are released into the atmosphere including benzene(C_6H_6),carbon monoxide(CO), nitrogen oxides(NO_x), sulfur dioxide(SO_2), and volatile organic compounds(VOC_s).Depending on meteorological conditions, emission quantity, and the fate and transport of chemical species, the environmental impact can affect air quality over a great area. This will cause the regional increased exposure to primary pollutants for neighboring populations with the summation of emission pollutant due to the temporal co-occurrence of start-ups and shutdowns among multiple chemical plants. Furthermore, these emission chemical will react with each other to form other secondary pollutants. The mechanism should follow the principle of reaction kinetics. Therefore, from the emission control point of view, the best practices for improving regional air quality should integrate both efforts on plant-wide emission minimization at every industrial point source, relationship and regional-wide multi-plant emission variability control.

In this paper, a scheduling methodology for coordinating multiple chemical plant start-ups is developed to minimize regional air quality impacts by the combination of reaction kinetics, atmosphere dispersion and scheduling will give the practical and feasible scheme to reduce primary and secondary pollutants in the region where multiple plants are co-located. The whole methodology will include two stages: proactive risk assessment and scheduling optimization. During the risk assessment, the focus is put on the changes in the weather conditions (stability class and wind condition), the pollutant and regional emissions from other sources, especially in the extreme weather or other emergent situations. The simulation result will indicate whether it needs to optimize the temporal sequence of multiple plants' activity. If necessary, the optimization modeling result will give an optimized temporal distribution of start-ups and shut-downs to reduce the chance of emission pairing and the concentration of primary and secondary pollutant. The development provide quantitative decision support for multiple stake holders, including government environment agencies, chemical industry, and local communities.

(246a) New Global Optimization Algorithm for Water-Using Networks

Pedro M. Castro and João Teles, Modeling and Optimization of Energy Systems, LNEG, Lisbon, Portugal

Design approaches for water networks can be classified into pinch analysis and mathematical optimization techniques. Pinch technology provides a comprehensive understanding of the problem at hand through simple graphical methods but is difficult to apply to large problems featuring multiple contaminants or to optimization goals besides flowrate minimization. On the other hand, rigorous optimization approaches through mathematical programming models start with the generation of a complex superstructure involving all design possibilities. A major advantage is their wider scope, which is why this work focuses on mathematical models.

Water networks (Bagajewicz, 2000; Jezowski, 2010) are a special type of process network problems (Quesada & Grossmann, 1995) that can be formulated as bilinear programs. Another example is the pooling problem (Haverly, 1978; Meyer & Floudas, 2006; Misener & Floudas, 2011). The non-convex bilinear terms arise in the mixing of the streams with different properties and are known to give rise to a multiplicity of local optima, which prevent gradient based solvers from certifying optimality of the nonlinear program (NLP). Further complexity will arise from binary decisions associated to alternative process units and/or connecting pipelines, leading to a mixed-integer nonlinear program (MINLP).

The most common global optimization algorithms are based on spatial branch-and-bound (Wicaksono and Karimi, 2008) and involve multiple formulations of a lower bounding problem that is a relaxation of the original bilinear problem. The relaxations are frequently based on the standard McCormick (1976) envelopes, leading to a linear problem (LP), or on piecewise envelopes (Bergamini et al. 2005; Karuppiah & Grossmann 2006). In the latter, the domain of the variables is divided ab initio into a given number of partitions, with the purpose of generating multiple McCormick envelopes that will provide a tighter relaxation. The optimal set of partitions is identified through binary variables leading to a mixed-integer linear programming problem (MILP).

We have recently proposed an alternative global optimization approach called multiparametric disaggregation (Teles et al. 2011) and applied it to the water-using networks (WUN) design problem (Teles et al. 2012). Through the discretization of the domain of one of the variables of the bilinear term, the original problem can be approximated to any desired accuracy. The discretization is based on the concept of significant digits and can be performed with respect to different numeric representation systems (e.g. decimal, binary). An important aspect is that the number of binary variables scales linearly, as opposed to exponentially (piecewise McCormick), with increasing accuracy.

In this paper, the multiparametric disaggregation (MDT) approach is upgraded by proposing a rigorous relaxation to the bilinear problem arising in the optimal design of WUNs, enabling us to generate a lower rather than an upper bound. The ability of becoming increasingly tight with the increase in accuracy is kept, providing the natural iterations of a novel global optimization algorithm. Through the solution of several test cases from the literature, we show that the new algorithm terminates with lower optimality gaps than the ones determined by the commercial solver BARON in 89% of the cases. Since MDT works by discretizing half of the variables appearing in bilinear terms, we test the two alternatives of parameterizing the concentrations and flowrates, with the results favoring the former.

References:

-Bagajewicz, M. (2000). A review of recent design procedures for water networks in refineries and process plants. Computers & Chemical Engineering 24, 2093.

-Bergamini, M.L., Aguirre, P., Grossmann, I.E. (2005). Logic-based outer approximation for globally optimal synthesis of process networks. Computers and Chemical Engineering 29, 1914-1933.

-Haverly, C. A. (1978). Studies of the behavior of recursion for the pooling problem. SIGMAP Bull. 25, 19.

-Jezowski, J. (2010). Review of Water Network Design Methods with Literature Annotations. Ind. Eng. Chem. Res. 49, 4475.

-Karuppiah, R., Grossmann, I.E. (2006). Global optimization for the synthesis of integrated water systems in chemical processes. Computers and Chemical Engineering 30, 650-673.

-McCormick, G.P. (1976). Computability of global solutions to factorable nonconvex programs. Part I. Convex underestimating problems. Mathematical Programming 10, 146.

-Meyer, C.A., Floudas, C.A. (2006). Global Optimization of a Combinatorially Complex Generalized Pooling Problem. AIChE J. 52 (3), 1027.

-Misener, R., Thompson, J.P., Floudas, C.A. (2011). APOGEE: Global Optimization of Standard, Generalized, and extended Pooling Problems via Linear and Logarithmic Partitioning Schemes. Computers & Chemical Engineering, 35, 876.

-Quesada, I., Grossmann, I.E. (1995). Global optimization of bilinear process networks with multicomponent flows. Computers & Chemical Engineering, 19, 1219.

-Teles, J.P., Castro, P.M., Matos, H.A. (2011). Multiparametric disaggregation technique for global optimization of polynomial programming problems. Journal of Global Optimization DOI: 10.1007/s10898-011-9809-8.

-Teles, J.P., Castro, P.M., Matos, H.A. (2012). Global Optimization of Water Networks Design using Multiparametric Disaggregation. Computers and Chemical Engineering 40, 132-147.

-Wicaksono, D.N., Karimi, I.A. (2008). Piecewise MILP Under- and Overestimators for Global Optimization of Bilinear Programs. AIChE J., 54, 991.

(246b) Optimal Synthesis for the Feed-Water-Heater Network of a Pulverized Coal (PC) Power to Minimize Water Consumption

Juan M. Salazar and Urmila Diwekar, Vishwamitra Research Institute, Center for Uncertain Systems: Tools for Optimization and Management, Clarendon Hills, IL

1. Introduction

Coal-fired power plants contribute to almost 50% of the United States' total electric power production. At the same time, pulverized coal (PC) power plants are large water consumers to the point that construction and operability of PC power plants have started to be constrained by water availability in some regions of the country. Two process systems engineering approaches have been studied to minimize the water consumption of power plants. First, a better optimization of nonlinear uncertain systems (BONUS) algorithm has provided solutions to the minimization of water consumption under uncertain atmospheric conditions. The calculated conditions for a 550 MW PC plant predicted reductions of 6.4%, 3.2%, 3.8% and 15.4% in the average water consumption for the four different seasons from fall to summer respectively. A second approach pursues the reduction of the residual heat in the steam cycle of the PC process by formulating an optimal design of the feed water heat exchange network (HEN). The proposed methodology uses Aspen Energy Analyzer (AEA) to determine the mass flowrates of the bleeding streams while generating alternative designs that can potentially reduce the water consumption by reducing the total cooling requirement. The optimization approach to process synthesis involves the selection of an optimal solution from the superstructure with a simulated annealing (SA) capability built in Aspen Plus. Results show at least a 5% reduction in water consumption for the PC power plant via this enhanced HEN synthesis technique

2. Process Description

The PC power plant model studied in this paper is based on Case 11 referenced by the

DOE/NETL's report on cost and performance of fossil energy plants [1] . This power plant is a re-heating cycle with feed-water heating. The process comprises a boiler section where coal is burned and the combustion heat is transferred to water generating steam, and a steam section where the high pressure steam is expanded through a train of high, intermediate and low pressure turbines. Steam is initially generated in the boiler and expanded in the high pressure (HP) turbine; then, it is re-heated at the boiler for later expansion at the intermediate (IP) and low pressure (LP) turbines. Then, exhausted steam is condensed with liquid water and returned to the boiler while heated with bleeding streams of steam from the turbines (feed-water heater). Cooling water is sent to the cooling tower where heat from the cycle is ultimately rejected to the environment by evaporative cooling.

3. Optimization under uncertainty

3.1. Stochastic Simulation

As it was previously said, water consumption is strongly affected by weather conditions like air temperature and humidity. Detailed information about air conditions for different locations within the US can be obtained from the website (http://www1.eere.energy.gov). Histograms of the frequency distributions were fitted to the appropriate probability density functions to represent the air conditions variability of an average U. S. Midwestern urban center during each of the four seasons [7] . Stochastic simulation is carried out by efficiently sampling these distributions to generate a set of scenarios under which the plant models are evaluated and a corresponding probability distribution of water consumption (output variable) is calculated.

3.2. Stochastic Optimization

Minimization of water consumption in power plants is a stochastic non-linear programming (SNLP) problem where one of the moments (expected value, standard deviation, etc.) of the water consumption's probability distribution is the objective function, the model is the set of constraints and model parameters are the decision variables. BONUS algorithm [4] is a "here and now" approach to solve this problem [8] determining the process conditions under which the expected value of the water consumption can be reduced compared to the base case originally reported, as shown in REF _Ref319918871 \h Table 1 08D0C9EA79F9BACE118C8200AA004BA90B02000000080000000E0000005F0052006500660033003100390039003100380038003700310000000 [9] . It can be seen that the reduction on the consumption strongly depends on the season consumption during drastic weather conditions as summer can be reduced by changing some of the parameters of the process.

Table SEQ Table * ARABIC 1 Minimization of average water consumption under uncertain air conditions for a 550 MW PC power plant located at Midwestern US for four different seasons [9] .

Season	Optimal values of decision variables					Savings %
	Par 1 (%)	Par 2 ^{0}F	Par 3 Mass	Par 4 pressure ratio	Par 5 pressure ratio	

			fraction			
Fall	38.9	1160.8	0.31	0.49	0.61	6.4
Summer	48.9	1174	0.42	0.49	0.66	15.4
Spring	35.5	1096.5	0.22	0.36	0.61	3.8
Winter	19.0	1141.9	0.30	0.49	0.79	3.2
	Base case values of decision variables					
All	20	1157	0.3	0.365	0.637	

4. Heat Exchange Network synthesis for water management

4.1. Modified HEN synthesis in AEA

Conventional methodology for the heat exchange network synthesis of the feed water heating system is based on equal enthalpy change on the liquid stream for each of the heaters. Mass flow rates of the bleeding streams are calculated based on such heat load distribution. AEA generates alternative designs with a mixed integer linear programming (MILP) algorithm. The solution of the MILP problem is based on the thermodynamic characteristics of the involved streams and on fixed mass flowrate of the bleeding streams. This approach was modified based on the early work by Linhoff [5] that proposes the employment of pinch technology to define mass flowrates from the bleeding streams for maximum heat recovery and improved cycle efficiency. This cycle efficiency is directly associated to the PC process water consumption through the amount of heat rejected by the cycle.

The main idea of this work is employing the AEA MILP algorithm to determine the mass flowrates of the bleeding streams while generating alternative designs that can potentially reduce the water consumption by reducing the total cooling requirement. For this purpose, the bleeding streams have been treated as utilities (instead of process streams) whose mass flowrates are determined by the calculated heat load assigned to them by the MILP algorithm.

4.2. Cost modification for alternative utility streams

Optimization algorithms based on process costs (as the one used by AEA) yield designs that employ mostly hot streams because they minimize heat transfer area leaving low temperature streams unused. To avoid this situation a cost penalty was assigned to bleeding (utility) streams. The approach assumes that feed water heating is expensive for the process when using high pressure steam and using low pressure steam is inexpensive to the point that some income can be generated. The main reason is that employing large amounts of high pressure steam may decrease the process productivity and preheating the feed water with low pressure steam will increase the process efficiency. The drain streams are mixed and also are used to transfer sensible heat to the feed water. The costs assigned to these drains are naturally related to utility streams. On the other hand the remaining drain streams are mixtures of two or three drains. Therefore their costs have been assumed to be a weighted average of their bleeding constituents having more weight the cost of the bleed that is more expensive. This formulation of the problem and its solution with AEA yielded interesting alternative designs whose cold utility consumption is lower than that of the base case as shown in REF _Ref314440794 \h Table 2 08D0C9EA79F9BACE118C8200AA004BA90B02000000080000000E0000005F005200650066003300310034003400340030003700390034000000 and it is expected that the water consumption will also be lower.

Table SEQ Table * ARABIC 2 Cooling requirements for designs that consider the bleed streams as utilities

Design	Cooling requirement X 10 7 BTU/h
base case	1.024
Design1	0.627
Design2	0.627

Design3	0.627

5. Conclusion

This paper presents a framework for integrated water management in power systems. It has been found that water consumption depends on the weather conditions and the uncertainties in weather affect the consumption considerably. We presented an optimization under uncertainty problem for savings water consumption. The second approach is heat integration. An algorithmic framework (Figure 1) based on simulated annealing (and OA/ER/AP MINLP) for discrete decisions, BONUS algorithm for the stochastic NLP, and Hammersley sequence sampling for the sampling saved 97% of computational time to solve this problem. Optimization under uncertainty allowed us to save up to 15% in the expected value of water consumption in a PC plant and a pinch technology approach to the heat exchange network synthesis allowed a 38% in cooling requirement which will be translated in water savings as well.

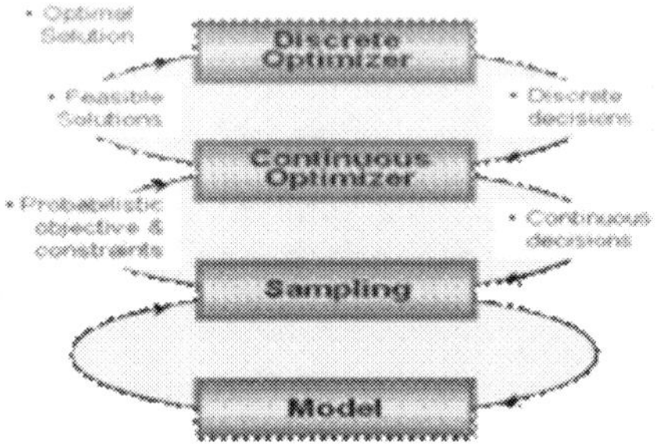

Figure 1: Algorithmic Framework

References

[1]DOE/NETL, Cost and performance baseline for fossil energy plants, (2007). DOE/NETL-2007/1281.

[2]DOE/NETL, Power plant water usage and loss study, (2007).

[3] U. Diwekar and E.S. Rubin, Comput. Chem. Eng. 15 (1991) 105-114.

[4] K. Sahin and U. Diwekar, Annals of Operations Research 132 (2004) 47-68.

[5] B. Linhoff and F.J. Alanis, ASME Advanced Energy Systems 85 (1989) 10-15.

[6] J.M. Salazar, U.M. Diwekar and S.E. Zitney, Comput. Chem. Eng. 35 (2011) 1863-1875.

[7] J.M. Salazar, U.M. Diwekar and S.E. Zitney, Energy Fuels 24 (2010) 4961-4970.

[8] U.M. Diwekar, Introduction to Applied Optimization, 2nd ed., SpringerLink, New York, 2008.

[9] J. Salazar and U. Diwekar. Energy Systems 2 (2011) 263-279.

(246c) Superstructure-Based Industrial Wastewater Treatment System Synthesis

Linlin Yang, Chemical Engineering, Carnegie Mellon University, Pittsburgh, PA, Raquel Salcedo, Carnegie Mellon University, Pittsburgh, PA; Chemical Engineering, Universidad de Alicante, Alicante, Spain and Ignacio E. Grossmann, Chemical Engineering Department, Carnegie Mellon University, Pittsburgh, PA

Stricter environmental effluent standards have forced industries to target increased water-use efficiency and water reuse. In order to meet this demand, systematic methods based on superstructure optimization for WN synthesis have been developed. Mathematical programming approaches are taken to optimize superstructures to generate the optimal water network for the system using NLP or MINLP models[2,6]. Water is supplied to water-using process units, and then wastewater streams generated from these processes are treated in various treatment units. In the typical superstructure-based models[7], process units are defined by their allowable contaminant concentration level, whereas wastewater treatment units are usually defined based on their contaminant removal ratio. This versatile superstructure considers systematic alternatives for water reuse, recycle, and recycle-reuse to minimize freshwater consumption or total network cost subject to discharge limit imposed. However, these models greatly simplify wastewater treatment units by using fixed recoveries, creating a gap for their applicability to industrial processes. It is clear that further improvements can be obtained by considering more accurate treatment models in the optimization of these water networks.

This presentation describes a unifying work combining various technologies capable of removing all the major types of contaminants through the use realistic models for the treatment units. Improvements over the simplified model are as follows. First, unit-specific short-cut models are developed in place of the fixed contaminant removal model to describe contaminant mass transfer in wastewater treatment units. Only a few works in the literature have considered more detailed models of treatment processes and these approaches are limited to a single treatment technology [4,8]. In this work, multiple types of treatment units and appropriate modeling equations that can satisfactorily predict unit performance with reasonable computational complexity are considered. For example, the rapid sand filter model and the Ergun equation are used to describe filtration units, and the osmotic pressure model is used to calculate the transmembrane pressure and the membrane area needed for reverse osmosis units. For units that do not demand complex mass transfer correlations (e.g. settling tank, filter press), new removal efficiency relations are developed to account for the dependency of the removal ratios on the presence of other contaminants[5]and the temperature of the unit for a more rigorous representation of these treatment units.

Furthermore, the architecture of the superstructure is modified to accommodate realistic potential structures. Different types of contaminants (suspended solids, dissolved solids) present in the system are removed by considering the Best Available Technology (BAT)[1], which provides the standards for discharge of priority pollutants. Since there are multiple treatment technologies for the removal of each type of pollutants, the modified superstructure allows for the selection of a subset of BAT treatment technology through the use of disjunction in the MINLP formulation. Another drawback in the original WN superstructure is that freshwater supplied to the system is assumed to be of high purity and therefore can satisfy the contaminant concentration requirement at the inlet of the process units. Yet this is not always the case in industrial applications, especially when freshwater is serving the purpose of feedwater to boilers. This work accounts for raw water treatment subsystem, which is placed prior to the water-using subsystem and wastewater treatment subsystem to accommodate more interactions in the overall WN superstructure[3]. Accordingly, the set of raw water pretreatment units can be used to process water streams with higher purity than the set of wastewater treatment units used to treat wastewater streams.

Finally, short-cut wastewater treatment cost functions (O&M cost and construction cost)[9]in the form of polynomial functions are incorporated into the model. The conventional network cost function usually consists of an operating cost term (linear with flowrate) and a capital cost term (exponential) for the sake of convenience and ease in computation.

The use of more complex objective in this more rigorous model enables the capability to design different WNs that allow for trade-offs that better meet the need of their respective decision criteria. The resulting superstructure, which is modeled as an MINLP, is illustrated through several industrial applications in metal finishing, and pulp and paper industries.

References:

[1] Integrated pollution prevention and control reference document on available techniques in common waste water and waste gas treatment/management systems in the chemical sector, Tech. Rep., European Commission, BREF document, Current Draft, 2009.

[2] M. J. Bagajewicz, "A review of recent design procedures for water networks in refineries and process plants," *Computers & Chemical Engineering*, vol. 24, no. 9-10, 2000, pp. 2093 – 2113.

[3] M. J. Bagajewicz and D. C. Faria, "On the appropriate architecture of the water/wastewater allocation problem in process plants," *Computer Aided Chemical Engineering*, vol. 26, 2009, pp. 1-20.

[4] E. Bringas, R. Karuppiah, M. F., San Roman, I. Ortiz, and I.E. Grossmann, "Optimal groundwater remediation network design using selective membranes," *Ind. Eng. Chem. Res.*, vol. 46, 2007, pp. 5555-5569.

[5] B. Galan and I. E. Grossmann, "Optimal design of distributed wastewater treatment networks," *Industrial & Engineering Chemistry Research*, vol. 37, no. 10, 1998, pp. 4036–4048.

[6] J. Jezowski, "Review of water network design methods with literature annotations," *Industrial & Engineering Chemistry Research*, vol. 49, no. 10, 2010, pp. 4475–4516.

[7] R. Karuppiah and I. E. Grossmann, "Global optimization for the synthesis of integrated water systems in chemical processes," *Computers & Chemical Engineering*, vol. 30, no. 4, 2006, pp. 650 – 673.

[8] Y. Lua, Y. Hua, X. Zhang, L. Wu, and Q. Liu, "Optimum design of reverse osmosis system under different feed concentration and product specification," *Journal of Membrane Science*, vol. 287, 2007, pp. 219-229.

[9] J. Sharma, Development of a preliminary cost estimation method for water treatment plants, master's thesis, University of Texas at Arlington, 2010.

(246d) Multi-Scale Design of Interplant Water-Allocation and Heat-Exchange Network with Fixed Flow-Rate and Fixed Contaminant-Load Processes

Ruijie Zhou, Dept. of Chemical and Biochemical Engineering, Rutgers University, Piscataway, NJ, Lijuan Li, Dept. of Chemical and Biological Engineering, Rensselaer Polytechnic Institute, Troy, NY and **Hong-Guang Dong**, Chemical Engineering, Dalian University of Technology, Dalian, China

Water and energy are two of the most essential resources in process industries. Large amount of water is needed not only as an effective mass-separating agent, but also as heat carriers in the utility systems. In aforementioned usages, both wastewater-treatment units and large amount of energy in the form of cooling and heating utilities are needed to satisfy both the concentration and temperature requirements. At present, escalating costs of freshwater, effluent treatment and energy have motivated integrated water-allocation and heat-exchange network (WAHEN) designs, which facilitate optimal distribution of water and energy resources simultaneously to satisfy process demands as well as environmental regulations at minimum total cost.

In the past few years, the tasks of designing single WAHEN have been examined in several works [1-3]. However, since industrial practices have necessitated the management of both quality and temperature of water across plants, there is therefore a need to consider both WANs and HENs for conceptual designs on an interplant scale simultaneously. In this work, the multi-scale state-space superstructure, a framework that considers all intra- and inter-plant configurations, are presented for interplant WAHEN design for both fixed flow rate (FF) and fixed contaminant-load (FC) processes. In particular, the proposed superstructure is capable of capturing all the water reuse, recycle and heat recovery opportunities at both intra- and inter-plant levels [4]. Additionally, in this work, both direct integration by process streams and indirect integration with interplant transfer network (ITN), which consists of junctions, treatment units and heat exchangers, are introduced for interplant WAHEN designs. To the best knowledge of the authors, this is the first time that the aforementioned problems have been addressed in its entirety. Finally, considering the costs of (1) utilities, (2) heat exchangers and water treatment units and (3) ITN, an mixed-integer nonlinear programming (MINLP) model has been developed for the interplant WAHEN model. It can be shown that better network design with 10-50% improvement in total annualized cost (TAC) can be obtained in all example studies.

References

(1) Savulescu, L. E.; Kim, J. K.; Smith, R. Studies on simultaneous energy and water minimization-Part II: systems with maximum re-use of water. *Chem. Eng. Sci.* **2005**, *60*, 3291.

(2) Dong, H. G.; Lin, C. Y.; Chang, C. T. Simultaneous Optimization Approach for Integrated Water-Allocation and Heat-Exchange Networks. *Chem. Eng. Sci.* **2008**, *63*, 3664.

(3) Bogataj, M.; Bagajewicz, M. Synthesis of non-isothermal heat integrated water networks in chemical processes. *Comput. Chem. Eng.* **2008**, *32*, 3130.

(4) Zhou, R. J.; Li, L. J.; Dong, H. G.; Grossmann, I. E. Synthesis of Interplant Water-Allocation and Heat-Exchange Networks. Part 1: Fixed Flow Rate Processes. *Ind. Eng. Chem. Res.* In Press

(246e) A Systematic Approach for Optimized Water Allocation Through Solution of Large Scale Water/Wastewater Networks Problems

Alberto Quaglia[1], Alessandra Pennati[2], Gürkan Sin[3] and Rafiqul Gani[1], (1)CAPEC, Department of Chemical and Biochemical Engineering, Technical University of Denmark, Kgs. Lyngby, Denmark, (2)Chemical and Biochemical Engineering, Technical University of Denmark, Lyngby, Denmark, (3)Department of Chemical and Biochemical Engineering, Technical University of Denmark, Kgs. Lyngby, Denmark

A systematic approach for optimized water allocation through solution of large scale water/wastewater networks problems

A. Quaglia, A. Pennati, G. Sin,* R. Gani

CAPEC, Department of Chemical and Biochemical Engineering, Technical University of Denmark, DK-2800 Kgs. Lyngby, Denmark; email: aq@kt.dtu.dk. *email:gsi@kt.dtu.dk.

Water is a resource of great relevance for the mankind, and its preservation and reuse will be a key priority for the coming decades. As water will become scarcer, optimization of its purification and use is of key importance. One example of water optimization problem is the water/wastewater network problem. Given a set of water sources with different levels of contamination, a set of water users requiring different levels of purity and a list of available water treatment technologies, the water/wastewater network problem involves allocating the sources to the users, while identifying the optimal water treatment network with respect to a defined objective function (e.g. minimization of clean water consumption, minimization of costs, etc.).

Process Systems Engineering (PSE) has the potential to contribute to water preservation, through the development of methods and tools needed to solve the water/wastewater network problem (Karuppiah et al., 2006; Foo 2008). However, due to problem complexity, most of the water networks proposed in the literature are limited to few contaminants and water treatment options, and often do not employ a rigorous model of the processing unit considered. Thus, the previous studies about the water networks are arguably not able to manage the complexity of a real industrial case with multi-contaminants (COD, N, SS, heavy metals, etc) and requiring a selection amongst a myriad of wastewater and water treatment technologies.

In this work, a systematic framework for the formulation and solution of water networks problems is proposed. The optimization problem is formulated as a Mixed Integer Non Linear Program (MINLP), which is solved to identify the best wastewater treatment process among a large set of predefined alternatives, according to selected optimum criteria (Quaglia et al, 2012). The key novelty of the proposed framework is to employ methods and tools for the analysis of the formulated problem and identification of the best solution strategy, in order to allow the solution of large scale problems, as it is required when dealing with real industrial cases. This is an exact (or full) method, which employs incremental decomposition-based solution strategies and includes all strategies which can be employed to solve any network synthesis problem, without affecting the quality of the results. The alternative, called here the pragmatic method, aims at formulating a simplified problem, which can be solved in order to obtain an approximate solution, which is then checked against the original problem to guarantee its feasibility. The simplification strategy is selected in order to minimize the loss of quality of the solution, with the help of analysis tools such as sensitivity analysis. Solutions obtained by pragmatic methods cannot be guaranteed as global optimum of the original problem, and therefore the use of such methods should be restricted to cases which cannot be treated with exact methods. Nevertheless, they represent a valuable approach to obtain approximate solutions of very large and complex problems, which cannot be solved otherwise.

In order to demonstrate and highlight the features of the tool, a case study dealing with refinery wastewater purification is presented.The case study considers 4 water sources (3 wastewater streams - caustic, sour and oily wastewater - at different level of contamination and a clean water stream) and 4 water sinks (3 process water users – boiler feed water, cooling tower make-up and desalter - requiring different level of purities and a water effluent stream). A list of 23 water treatment technologies is organized into a superstructure of possible network configuration, leading to more than 1.4 million possible process networks. The optimization problem is formulated and solved for different scenarios, both with respect to problem data (e.g. contamination levels, emission limits), and to objective function selection (e.g. minimization of treatment costs, minimization of fresh water use, etc.), and the results are qualitatively validated and discussed.

With respect to the water networks presented in literature, the one developed here accommodates a wider range of contaminants and allows for the use of engineering insights in the modeling of the treatment units. This approach constitutes a useful and flexible tool to provide a guideline for the process design of the treatment of different types of wastewater sources, as well as to design wastewater treatment systems for a new plant or retrofit or expansion of existing plants. The tool also allows for the evaluation of the effect of different scenarios e.g. different optimum criteria, wastewater sources, environmental regulations on water discharge, contaminant loads on purified water to recycle etc.

References:

R.Karuppiah, I.Grossmann, 2006, Global optimization for the synthesis of integrated water systems in chemical processes, Computers & Chemical Engineering (30) 650-673.

D. Foo, 2008, Flowrate targeting for threshold problems and plant-wide integration for water network synthesis, Journal of Environmental Management (88), 253-274.

A.Quaglia, B.Sarup, G.Sin, R.Gani, 2012, Integrated business and engineering framework for synthesis and design of enterprise-wide processing networks, Computers & Chemical Engineering (38) 213-223.

Optimization of Water Storage and Distribution Systems for Cities

Fabricio Nápoles-Rivera,[1] *Medardo Serna-González*[1], *Mahmoud M. El-Halwagi,*[2,3] *José María Ponce-Ortega*[1]

[(1)] *Chemical Engineering Department, Universidad Michoacana de San Nicolás de Hidalgo, Morelia, Mich., 58060, México.*
[(2)] *Chemical Engineering Department, Texas A&M University, College Station, TX, 77843, USA*
[(3)] *Adjunct Faculty at the Chemical and Materials Engineering Department, King Abdulaziz University, Jeddah, Saudi Arabia*

Introduction

Water scarcity in several regions of the world, along with the continuous increase in the water demands in the cities due to population growth, is currently one of the major concerns for the governments. In consequence, developing strategies for the appropriate use of water resources has become of paramount interest. One of the main concerns consists in the depletion of water resources. Thus, options such as the use of desalinated seawater and the collection of precipitation in storage tanks, artificial ponds and aquifers, which in the past were considered unpractical and expensive, are now being considered as attractive alternatives for common use and replenishment of water bodies. An additional concern is the appropriate distribution for domestic and agricultural use, which commonly is based solely in immediate demands, and not in projected demands. This causes water droughts and the reduction in water supply in some cases. Therefore, it is important to establish systematic procedures for the appropriate planning and distribution of the available resources. In this regard, this paper presents a mathematical programming approach for the optimal water distribution planning in a macroscopic water system. The problem is based on the representation presented in Figure 1. The proposed approach considers natural bodies as water sources and the possibility of installing storage tanks and artificial ponds for the collection of precipitation in a given location. The main sinks are domestic and agricultural users for which an optimal water distribution schedule can be devised satisfying demand constraints. A case study from the city of Morelia in México is solved to prove the applicability of the proposed approach; the results show the potential advantages of using a systematic procedure for the water distribution planning in a city considering alternative water sources.

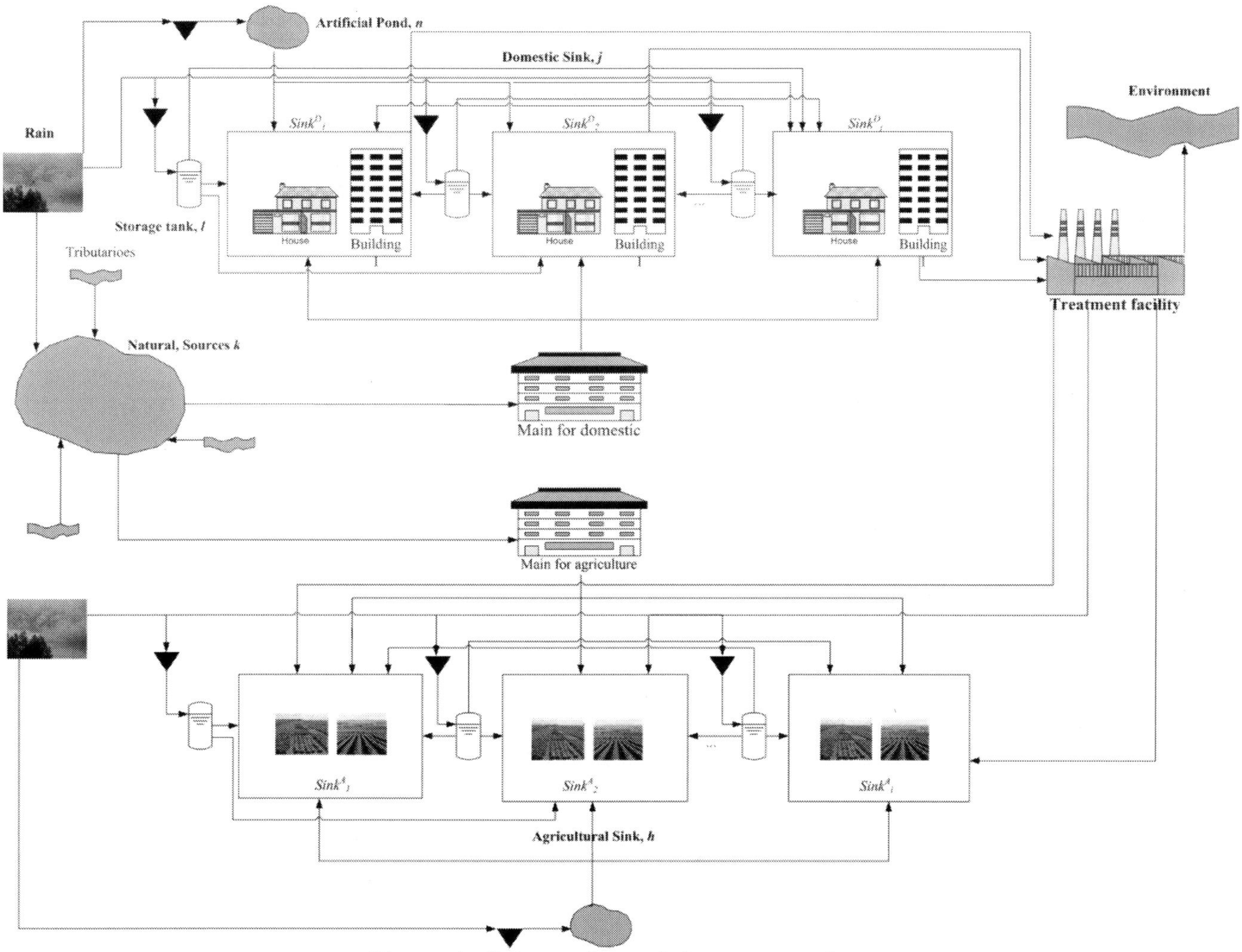

Figure 1. Schematic model representation.

Model Formulation

Balances for the natural sources of water

The available water in any given period of time in the natural bodies depends on the inventory in the previous time and the inlets and outlets in the same period.

$$G_{k,t} - G_{k,t-1} = \sum_{m \in M} r_{m,k,t} + p^{g}_{k,t} - g^{d}_{k,t} - g^{a}_{k,t} - v^{g}_{k,t}\,, \qquad k \in K, t \in T$$

Balance for precipitation in artificial water storages

The rainwater available for storage is a function of the area for collection, a runoff coefficient and the precipitation in a given location

$$P = f\left(Area, Ce, Precipitation\right)$$

The precipitation water can be collected either in storage tanks for which:

$$P_{l,t}^{s} = P_t \cdot A_l^{s} \cdot Ce^{s}, \qquad l \in L, t \in T$$

Or in artificial ponds, in such case

$$P_{n,t}^{a} = P_t \cdot A_n^{a} \cdot Ce^{a}, \qquad n \in N, t \in T$$

Due to capacity issues some water can be stored and the rest has to be discarded, the equations that describe this situation are:

$$P_{l,t}^{s} = s_{l,t}^{in} + v_{l,t}^{s}, \qquad l \in L, t \in T$$

$$P_{n,t}^{a} = a_{n,t}^{in} + v_{n,t}^{a}, \qquad n \in N, t \in T$$

Balance in storage tanks

The balance in the storage tanks states that the stored water in any given period is a function of the inventory in the storage tank, and the inlets and outlets of the tank

$$S_{l,t} - S_{l,t-1} = s_{l,t}^{in} - \sum_{j \in J} s_{l,j,t}^{out,d} - \sum_{h \in H} s_{l,h,t}^{out,a}, \qquad l \in L, t \in T$$

Balance in artificial ponds

Similarly for the artificial ponds the balance is:

$$A_{n,t} - A_{n,t-1} = a_{n,t}^{in} - \sum_{j \in J} a_{n,j,t}^{out,d} - \sum_{h \in H} a_{n,h,t}^{out,a}, \qquad n \in N, t \in T$$

Balance in main for domestic use

Natural water sources are sent to the main for domestic, from the main for domestic the water is sent to the domestic sinks:

$$\sum_{k} g_{k,t}^{d} = \sum_{j} f_{j,t}, \qquad t \in T$$

Balance in main for agricultural use

From the main for agriculture the water is distributed to agricultural sinks

$$\sum_{k \in K} g_{k,t}^{a} = \sum_{h \in H} r_{h,t}, \qquad t \in T$$

Balance in domestic sinks

The domestic sinks receive water from the domestic main, form storage tanks or from artificial ponds.

$$f_{j,t} + \sum_{l \in L} s_{l,j,t}^{out,d} + \sum_{n \in N} a_{n,j,t}^{out,d} = D_{j,t}^{ds}, \qquad j \in J, t \in T$$

Some of the water is used or consumed in the domestic sinks, the rest of the water is sent to a treatment plant

$$D_{j,t}^{ds} = cw_{j,t}^{d} + int_{j,t}^{in}, \qquad j \in J, t \in T$$

Treatment plant balance

In the treatment plant the water can be treated either to be reused in the agricultural sinks or for final disposal

$$\sum_{j} int_{j,t}^{in} = int_{t}^{out} + cw_{t}^{tp}, \in \qquad t \quad T$$

The treated water is then distributed to the agricultural sinks

$$int_{t}^{out} = \sum_{h} int_{h,t}^{out,ag}, \qquad t \in T$$

Balance in agricultural sinks

The demand in the agricultural sinks can be satisfied with water from the natural sources, from storage tanks, artificial ponds or with water from the treatment plant

$$r_{h,t} + \sum_{l \in L} s_{l,h,t}^{out,a} + \sum_{n \in N} a_{n,h,t}^{out,a} + int_{h,t}^{out,ag} = D_{h,t}^{as}, \qquad h \in H, t \in T$$

Disjunctions for installing or not storage tanks or an artificial pond

The following disjunctions are used to select either to install or not the storage tanks or artificial ponds in a given location

$$\begin{bmatrix} z_l^s \\ S_l^{\max} \geq \delta_l^{s,\min} \\ S_l^{\max} \leq \delta_l^{s,\max} \\ Cost_l^s = A + B\left(S_l^{\max}\right)^{\alpha} \end{bmatrix} \vee \begin{bmatrix} z_l^s \\ S_l^{\max} = 0 \\ Cost_l^s = 0 \end{bmatrix}, \qquad l \in L$$

$$\begin{bmatrix} z_n^a \\ A_n^{max} \geq \delta_n^{a,min} \\ A_n^{max} \leq \delta_n^{a,max} \\ Cost_n^a = C + D\left(A_n^{max}\right)^{\alpha} \end{bmatrix} \vee \begin{bmatrix} z_n^a \\ A_n^{max} = 0 \\ Cost_n^a = 0 \end{bmatrix}, \qquad \in n \quad N$$

In the previous disjunctions the cost functions include a nonlinear term, in order to avoid the nonlinearities these equations are linearized. Then the disjunctions are reformulated using a Big M formulation (Veccietti et al., 2003)

Objective function

The objective function consists in maximizing the gross profit for the sales of water minus the treatment, storage and distribution cost.

$$TAR = revenue - treatment\,cost - storage\,cost - pp\,cost$$

Case Study

The selected case study is the water distribution in the city of Morelia, México. The water consumption in Morelia for public uses is 90,168,374.77 m^3/year (CONAGUA, 2011), whereas for agricultural activities 21,348,208.28 m^3/year are required for a grand total of 111,516,583.05 m^3/year, which are provided by 105 deep wells (43.93%), springs (33.41%) and the Cointzio dam (22.66%) (Avila-Olivera and Garduño-Monroy, 2007). The objective is to find an optimal storage and distribution schedule that satisfies the demands in domestic and agricultural sinks, and at the same time maintain sustainable levels of water in the natural sources.

Results

The model was coded in the commercial software GAMS® and solved without numerical complications. In the optimal scenario 8.6% of the water used to satisfy the demands (domestic and agricultural) is provided by harvested rainwater while treated water provides 8.5%. This means 17.1% is provided by alternative water sources. The profit for the solution is $26,375,936/year, when compared to the traditional operation (i.e. no alternative water sources are used) this represents an increment of 7% in the annual revenue. The obtained solution does not only satisfy the water demands in the city, but it also maintains sustainable levels of water in the natural bodies. An example distribution scheme for the month of August is shown in Figure 2. During this month 0.172Mm^3 for agricultural uses is taken from the treatment plant, while 0.106Mm^3 is taken form storage tanks. The domestic sinks use 0.856Mm^3 of water from artificial ponds; this is a 10% of the domestic demand during the month this way reducing the amount of water taken form natural water bodies.

Conclusions

A model for optimal the management, storage and distribution of water in a city has been presented. The obtained solutions indicate that the use of alternative water sources in a systematic scheme might help reduce the pressure over natural water bodies and still be economically attractive, yielding at least the same or better solutions (from the economic point of view) and maintaining acceptable levels of the water bodies.

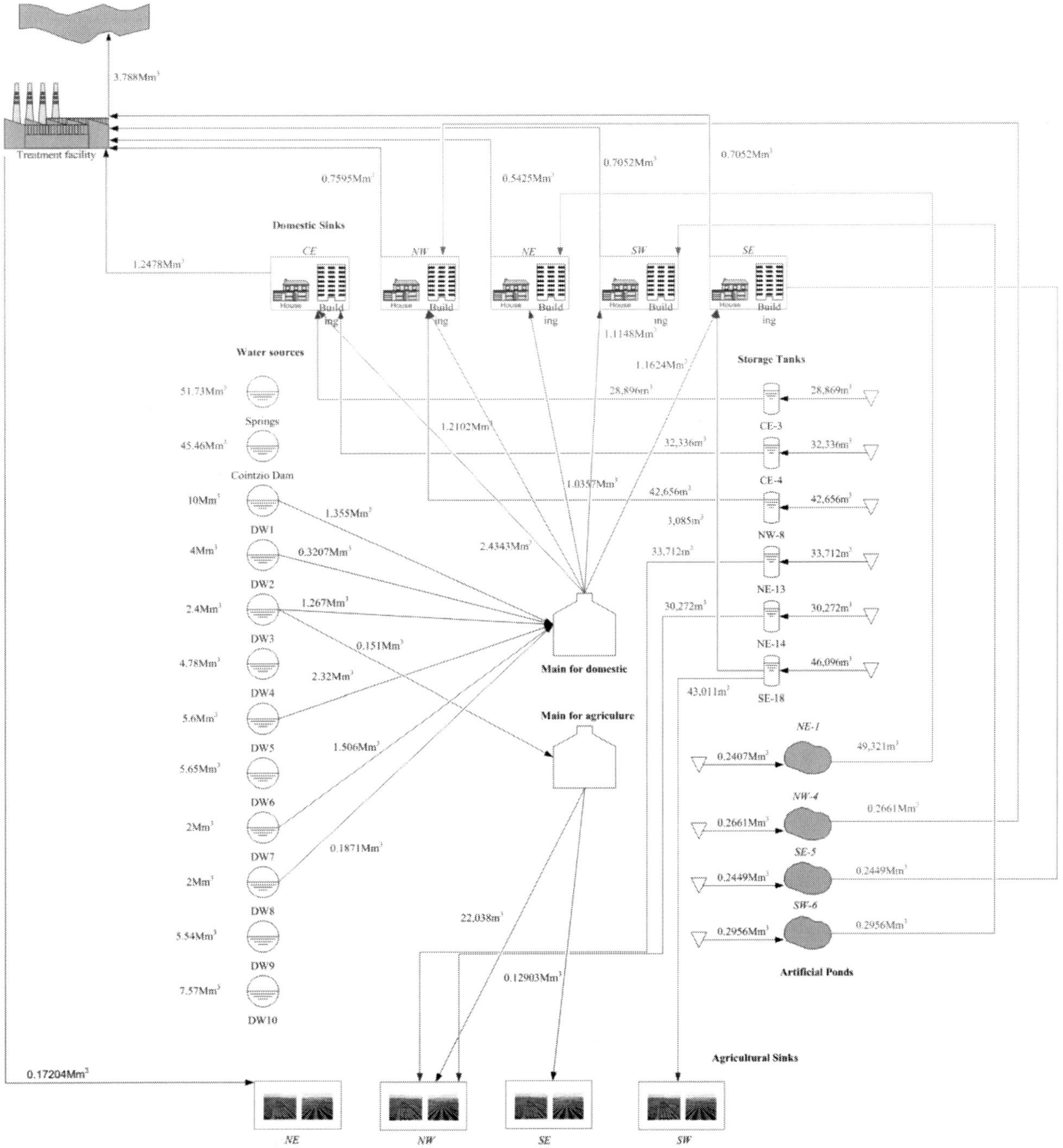

Figure 2 Water distribution during the month of August

References

- Avila-Olivera, J. A., Garduño-Monroy, V. H., 2007. Analysis for the decrement of the available water in the aquifers of Morelia. *Ciencia Nicolaita.* 46, 105-122.

- CONAGUA, 2011. National Council of Water, Statistic of the water in México, Chapter 3 Appendix. Available at http://www.conagua.gob.mx/CONAGUA07/Contenido/Documentos/SINA/Capitulo3Usosdelagua.zip. Accessed August 2012
- Vecchietti, A., Lee, S., Grossmann, I. E., 2003. Modeling of discrete/continuous optimization problems: Characterization and formulation of disjunctions and their relaxations. *Computers and Chemical Engineering*. 27(3), 433–448.

(258a) Modeling of Drug Delivery From PLGA Microspheres Using Reaction-Diffusion Equations with Hindered Diffusion

Ashlee N. Ford Versypt[1,2], Daniel W. Pack[2] and Richard D. Braatz[3], (1)Chemical Engineering, Massachusetts Institute of Technology, Cambridge, MA, (2)Chemical & Biomolecular Engineering, University of Illinois Urbana-Champaign, Urbana, IL, (3)Department of Chemical Engineering, Massachusetts Institute of Technology, Cambridge, MA

Controlled-release drug delivery systems provide alternatives to conventional medical drug therapy regimens which require frequent dosages due to short pharmaceutical in vivo half-life and poor oral bioavailability. Controlled-release systems have the potential to enhance control of drug concentrations, reduce side effects, and improve compliance as compared to conventional regimens.

Poly(D,L-lactic-co-glycolic acid) (PLGA) microspheres are the controlled-release drug delivery system modeled in this work. Biodegradable PLGA microspheres have been extensively studied as drug delivery devices [1-3]. Modeling of the drug release from the devices must consider the interdependent phenomena that contribute to drug release. The microspheres encapsulate drug molecules dispersed throughout the polymer. The polymer undergoes autocatalytic hydrolysis, breaking the polymer bonds and generating smaller polymer chains with acidic end groups that catalyze further hydrolysis of the degradation products. Autocatalytic hydrolysis is more significant in the interior of large microspheres where the diffusion of degradation products is more limited [4-6]. Sufficiently small oligomers produced by the degradation are water-soluble and can diffuse out of the polymer microspheres through water-filled pores. The resulting polymer mass loss increases the pore volume in the microspheres. Encapsulated drug molecules diffuse through the aqueous pores in the polymer by hindered diffusion [7]. As the pore network in the microspheres evolves, the effective diffusivity of the drug increases.

A mathematical model has been developed for the simultaneous treatment of PLGA degradation and erosion and diffusive drug release with autocatalytic effects and nonconstant effective diffusivity of the drug [8-9]. A mechanistic reaction-diffusion model with pore evolution coupled to hydrolysis and related to the effective diffusivity through hindered diffusion theory is proposed. The unique contribution of the modeling work is that it combines autocatalytic PLGA degradation mechanisms [10-11] with hindered diffusion in aqueous pores [12] having variable pore sizes. The model performance for the case of drug release from microspheres of different sizes is presented to highlight the capability of the model for predicting size-dependent, autocatalytic effects on the polymer and the release of drug.

References

[1] J. M. Anderson and M. S. Shive. Biodegradation and Biocompatibility of PLA and PLGA Microspheres. *Advanced Drug Delivery Reviews*, 28(1):5-24, 1997.

[2] N. K. Varde and D. W. Pack. Microspheres for Controlled Release Drug Delivery. *Expert Opinions on Biological Therapy*, 4(1):35-51, 2004.

[3] S. Freiberg and X. X. Zhu. Polymer Microspheres for Controlled Drug Release. *International Journal of Pharmaceutics*, 282(1-2):1-18, 2004.

[4] J. Siepmann, K. Elkharraz, F. Siepmann, and D. Klose. How Autocatalysis Accelerates Drug Release from PLGA-Based Microparticles: A Quantitative Treatment. *Biomacromolecules*, 6(4):2312-2319, 2005.

[5] C. Berkland, E. Pollauf, C. Raman, R. Silverman, K. Kim, and D. W. Pack. Macromolecule Release from Monodisperse PLG Microspheres: Control of Release Rates and Investigation of Release Mechanism. *Journal of*

Pharmaceutical Sciences, 96(5):1176-1191, 2007.

[6] D. Klose, F. Siepmann, K. Elkharraz, and J. Siepmann. PLGA-Based Drug Delivery Systems: Importance of the Type of Drug and Device Geometry. *International Journal of Pharmaceutics*, 354(1-2): 95-103, 2008.

[7] A. Zhao, S. K. Hunter, and V. G. J. Rodgers. Theoretical Prediction of Induction Period from Transient Pore Evolvement in Polyester-Based Microparticles. *Journal of Pharmaceutical Sciences*, 99(11):4477-4487, 2010.

[8] A. N. Ford, D. W. Pack, and R. D. Braatz. Multi-Scale Modeling of PLGA Microparticle Drug Delivery Systems. In E. N. Pistikopoulos, M. C. Georgiadis, and A. C. Kokossis, editors, *21st European Symposium on Computer Aided Process Engineering: Part B*, pages 1475–1479. Interscience, New York, 2011.

[9] A. N. Ford Versypt. *Modeling of Controlled-Release Drug Delivery from Autocatalytically Degrading Polymer Microspheres*. Ph.D. Dissertation, University of Illinois at Urbana-Champaign, Urbana, IL, 2012.

[10] H. Antheunis, J. C. van der Meer, M. de Geus, W. Kingma, and C. E. Koning. Improved Mathematical Model for the Hydrolytic Degradation of Aliphatic Polyesters. *Macromolecules*, 42(7):2462-2471, 2009.

[11] S. Lyu, R. Sparer, and D. Untereker. Analytical Solutions to Mathematical Models of the Surface and Bulk Erosion of Solid Polymers. *Journal of Polymer Science Part B: Polymer Physics*, 43(4):383-397, 2005.

[12] P. Dechadilok and W. M. Deen. Hindrance Factors for Diffusion and Convection in Pores. *Industrial and Engineering Chemistry Research*, 45(21):6953-6959, 2006.

(258b) Multi-Scale Dynamic Simulation and Optimization for Simultaneous Chemical Plant Emission Reduction and Regional Air Quality Improvement

Qiang Xu and Jian Zhang, Dan F. Smith Department of Chemical Engineering, Lamar University, Beaumont, TX

Chemical industrial zones heavily populated by chemical plants often experience highly localized and transient air pollution events, with elevated concentrations of multiple air pollutants, and ozone concentrations that violate the National Ambient Air Quality Standards (NAAQS). One identified major source related to these pollution events is air emissions of highly reactive volatile organic compounds (HRVOCs) from chemical industry flaring activities. Flaring is crucial to the chemical plant safety. However, excessive flaring, especially the intensive flaring during chemical plant turnarounds, emits huge amount of volatile organic compounds (VOCs) and HRVOCs and also results in tremendous industrial material and energy loss. Thus, the flare emission must be minimized by chemical plants. However, flare minimization at plant-wide scale is not enough to deal with regional air quality problems, because two problems are still insufficiently addressed or neglected today, which include i) what are the exact impacts of air emissions from chemical plant flaring? ii) are there any cost-effective ways available for chemical plants to help improve regional quality, instead of being simply enforced to reduce emissions continuously and aggressively.

In this paper, we propose a systematic methodology on multi-scale dynamic simulation and optimization for simultaneous chemical plant emission reduction and regional air quality improvement. It couples plant-wide dynamic simulation for flare minimization and emission quantification (i.e., hourly dynamic emissions from plant flaring activities), as well as regional scale air quality simulation for flare emission impact analysis. Based on such multi-scale dynamic simulations, optimization strategies for chemical plants operations will be proposed, which benefits regional air quality improvement, meanwhile provides cost-effective ways for chemical plant operations.

(258c) Simulation of Nanoparticle Transport In Airways Using Petrov-Galerkin Finite Element Methods

Jeffrey J. Heys, Chemical and Biological Engineering, Montana State University, Bozeman, MT and Prathish Rajaraman, Chemical and Biological Engineering, Montana State University, BOZEMAN, MT

The simulation of nanoparticle transport and deposition in human airways requires the solving of multiple differential equations. The bulk air phase is modeled using the Navier-Stokes equations in the laminar flow regime. Transport in the particle phase is dominated by advection with diffusion being less important but non-negligible for nanoparticles (d < 1 micron). The nanoparticle phase is typically modeled using multiple advection-diffusion equations with the diffusivity being a function of particle size and density. The numerical challenge here is two-fold: (1) a higher-order approximation is desirable due to the large boundary layer gradients, and (2) the larger magnitude of advection relative to diffusion requires upwinding or stabilization for most numerical approaches. Our focus is on Petrov-Galerkin finite element methods because they enable straightforward higher-order discretizations and provide stabilization for advection dominated flows. Various Petrov-Galerkin approaches are compared, including a novel combined approach, for transport down a straight cylinder. Using the results from straight cylinder study, the methods are then applied to multiple generations (typical 3-5) of realistic human airway geometries. The objective is to predict optimal nanoparticle size and density for targeted drug delivery to specific regions of the airways.

(258d) Using a CFD-DEM Modeling Approach to Predict Drill Bit Dynamics

Jeremiah N. Mpagazehe and C. Fred Higgs III, Mechanical Engineering, Carnegie Mellon University, Pittsburgh, PA

The increasing demand for energy has resulted in the need to drill for oil and natural gas at greater depths under the earth's surface. The extreme heat and pressure at such depths produces changes in the operation of traditional drilling components. Correspondingly, more research is being conducted in the optimization of drilling materials and parameters such as the weight-on-bit (WOB), angular velocity and the drilling fluid properties to better perform in such conditions. Along with lab-scale experiments and industrial testing, modeling is a valuable tool to help in this effort. Specifically, modeling can be used to explore the design space in these extreme environments in an economical fashion. In the current work, computational fluid dynamics (CFD) is used to model the drilling fluid while discrete element modeling (DEM) is used to model the porous rock formation. Physics-based algorithms are employed to couple the model drill bit, the rock formation and the drilling fluid. The model drill bit is rotated and pressed into the rock formation and the rate-of-penetration (ROP) under various conditions is studied.

(258e) Experimental Validation of High Reynolds Number CFD Simulations of Heat Transfer in a Pilot-Scale Fixed Bed Tube

Anthony G. Dixon[1], Greg Walls[2], Helen Stanness[3], Michiel Nijemeisland[3] and E. Hugh Stitt[3], (1)Worcester Polytechnic Institute, Worcester, MA, (2)East Coast Metrology, (3)Johnson Matthey Catalysts, Billingham, United Kingdom

The simulation of detailed flow and transport inside the interstitial space of packed tubes using computational fluid dynamics (CFD) is becoming a standard approach to obtaining insight into fixed bed transport phenomena. At the high flow rates of industrial practice, flow and heat transfer are highly turbulent and complex, and comparisons to experiment are necessary. In this study, CFD simulations of heat transfer in fixed beds of spheres were validated by comparison to experimental measurements in a pilot-scale rig. The comparisons were made for particle Reynolds numbers in the range $2200 < Re < 27000$ for tube-to-particle diameter ratio $N = 5.45$, and in the range $1600 < Re < 5600$ for $N = 7.44$. Radial temperature profiles were obtained at four axial positions in the heated bed spaced 0.16 m apart.

CFD models of a 0.72 m long tube containing 1000 spheres ($N = 5.45$) and a 0.35 m long tube of 1250 spheres ($N = 7.44$) were solved to obtain well-developed flow fields. These provided inlet velocity profiles to reduced models of 0.20 m heated packed length, consisting of 304 spheres for $N = 5.45$ and 722 spheres for $N = 7.44$. The measured temperature profiles were used as inlet boundary conditions and profiles 0.16 m downstream were computed. Verification with a medium and a fine mesh showed negligible difference between the profiles.

Contact points were treated by one of three methods: global reduction of sphere size resulting in gaps, or local insertion of bridges at particle-wall contact points with either surface flattening or bridges at the particle-particle contact points. The gaps method gave slightly poorer results; the two bridges methods were better and indistinguishable from each other. CFD simulations compared well to the experimental data: trends with Re, N and bed depth were captured, and quantitative agreement of temperature profiles was reasonable.

(258f) An Integrated Computational Fluid Dynamics Model to Simulate the Productivity of Industrial Scale Algae Raceway Ponds

Stephen Y. Park, Food, Agricultural and Biological Engineering, The Ohio State University - OARDC, Wooster, OH and Yebo Li, Department of Food, Agricultural and Biological Engineering, , The Ohio State University-OARDC, Wooster, OH

Minimizing the issues attachment due to poor flow conditions can have significantly negative impacts associated with the scale-up of microalgal growth systems is a necessary part of making microalgal biodiesel production a sustainable process. Self-shading caused by high cell concentration and biomass during commercial production. Through the simulation of algae growth with a computational fluid dynamics (CFD) model, process optimization can be performed while reducing the costs and efforts in operating an actual commercial growth system. The aim of this study is to integrate physical, thermochemical, and biological effects to predict the space- and time- dependent algal growth in industrial scale raceway ponds. A commercial CFD software, ANSYS-Fluent is used to solve the proposed model in regards to fluid flow, heat transfer, and also nutrient balance. Simulated results are compared with the distributions of light intensity, temperature, nutrient concentration, and algal biomass acquired from four industrial scale raceway ponds constructed for the growth of *Nannochloropsis salina*.

Proceedings of the
American Institute of Chemical Engineering
AIChE 2012, October 28 - November 2, 2012
Pittsburgh Convention Center
Pittsburgh, PA

VALIDATING THE PREDICTION CAPACILITIES OF COMMERCIAL ANSYS CFD SOFTWARE IN THE STUDY OF HYDROCARBON-CATAYST HYDRODYNAMICS OF FLUID CATALYTIC CRACKING (FCC) RISER REACTORS

M. N. Idris[1,3*], T. Mahmud[1,] and B. M. Gibbs[2]
[1]Institute of Particle Science and Engineering
[2]Energy and Resources Research Institute
School of Process, Environmental and Materials Engineering
The University of Leeds
Leeds LS2 9JT, UK
[3]Process Refining Technology, University of Maiduguri, Borno State, Nigeria

The application of the ANSYS commercial computational fluid dynamics (CFD) usefulness to study the hydrodynamics of hydrocarbon-catalyst flow in FCC is a very useful and complex phenomenon. In order to establish reliable, precise and reasonable predictions, a choice of runs for time-average comparison for solid (catalyst) velocities and concentrations to measure fluxes and densities along the riser axes of flow regime. The model structure were carried-out successfully and compared with experimental data on a FCC riser length of 15.1m and 0.10 m in diameter, which incorporates a two-fluid framework with kinetic theory of granular flows to simulate fully developed hydrocarbon catalyst flows in the vertical risers. The preliminary studies were based on the simulation of gas-solid hydrodynamics and flow behaviour in a riser circulating fluidised bed (CFB) reactor using the Eulerian-Eulerian two-fluid modelling approach, incorporating a kinetic theory constitutive model for dilute assemblies of FCC particulate solid. The interaction between gas and particles was modelled using particle interphase momentum transfer model with Gidaspow and Schiller Naumann drag model, using turbulent k-ε models. The CFD predictions were reasonably compared with established experimental data from the literature. The overall flow patterns within the FCC riser reactors were predicted well by the model. The results of this study assist in a better understanding of the hydrodynamics of gas-solid flow in CFB/FCC real life applications, which could help in areas of pressure regulations, process flow control, and energy efficiency and utilisation.

Keywords: CFD, FCC, kinetic theory, hydrodynamic, simulation,

1. INTRODUCTION

Circulating fluidised bed (CFB) reactors are widely used in the industry to carry out a variety of multiphase gas-solid catalytic and non-catalytic reactions including coal combustion and gasification, fluid catalytic cracking (FCC) of heavy oil fractions for the production of gasoline, LPG and middle distillates, and Fischer-Tropsch synthesis of hydrocarbons. In the *riser* of a CFB, gas flows through granular solids at a high velocity which entrains the solids and transports them out of the riser. Three distinct fluidisation regimes have been identified in the riser (Levenspiel, 1999), namely turbulent fluidisation, fast fluidisation and pneumatic conveying. However, most risers in industrial CFB reactors operate in the turbulent and fast fluidisation regimes (Grace, 2000; Jiradilok, et al., 2006,), which are not fully understood (Du, et al., 2003; Jiradilok, et al., 2006) and phenomenological flow models are not well developed (Levenspiel, 1999). Consequently, uncertainty exists in the design, scale-up and performance prediction of such reactors. In recent

* Corresponding Author: M. N. Idris,
E-mail: muhinu@hotmail.co.uk

years, there have been significant advances in computational fluid dynamics (CFD) modelling of multiphase flow, which can be used to provide an improved understanding of the riser hydrodynamics and to facilitate the reactor design and the selection of appropriate operating conditions to achieve the desired fluidisation regime. However, multiphase CFD models require validation against data from large-scale experimental rigs before they can be used with confidence for the above mentioned purposes.

A large number of CFD modelling studies of gas-solid flows through the riser of CFB reactors have been reported in the literature (see reviews in Ranade, 2002; Almuttahar and Taghipour, 2007). However, it appears that no systematic approach has been used to select models for the gas phase (such as laminar or turbulent) and for the solid phase (such as empirical constitutive relations or models based on the kinetic theory of granular flow) and the parameters of the constitutive equations (such as the values of restitution and specularity coefficients, and interphase drag coefficients) to simulate gas-solid flow in riser systems (Ranade, 2002; Almuttahar and Taghipour, 2007). Although multiphase flows in riser reactors generally occur at high Reynolds numbers associated with high levels of turbulence, many previous CFD studies have treated flows as laminar, in some cases (see Almuttahar and Taghipour, 2007) with satisfactory agreements between the predictions and experimental data. The riser flows are also three-dimensional, yet in many studies the flow has been treated as two-dimensional in order to reduce the computational demand. This has lead to discrepancies between the predictions and measurements.

In the present study, gas-solid flow behaviour in the riser section of a pilot-scale CFB investigated experimentally by Huang et al. (2007) has been simulated using a proprietary CFD code, ANSYS FLUENT. Calculations were carried out using an Eulerian-Eulerian multiphase flow model. The solids pressure, shear and bulk viscosity were obtained in terms of the granular temperature using the constitutive equations derived from the kinetic theory of granular flow (KTGF). The granular temperature was determined using both the algebraic and differential transport models. The gas phase turbulence was represented by the k ε model. The CFD predictions are compared with the measured radial distributions of particle mean axial velocity and concentration at different locations along the height of the riser.

2. MATHEMATICAL MODEL

An outline of the multiphase CFD modelling methodology for the gas-solid turbulent flows through the riser is given here; full details may be found in Idris (2010). An Eulerian-Eulerian multiphase flow modelling approach with the KTGF was used. A set of the mass and momentum equations was solved for each phase, where the momentum equations were coupled by an interphase exchange term.

The Reynolds-averaged continuity and momentum equations for the gas phase are given by (ANSYS CFD, 2006):

$$\frac{}{t}\left(\ _g\ _g\right)+\ .\left(\ _g\ _g\vec{v}_g\right)=0 \qquad (1)$$

$$\frac{}{t}\left(\ _g\ _g\vec{v}_g\right)+\ \left(\ _g\ _g\vec{v}_g\vec{v}_g\right)=\ \ _g\ p+\ \overset{=}{}_g+\ _g\ _g\vec{g}+K_{gs}\left(\vec{v}_s\ \ \vec{v}_g\right) \qquad (2)$$

The continuity and momentum equations for the solid phase are:

$$\frac{}{t}\left(\ _s\ _s\right)+\ .\left(\ _s\ _s\vec{v}_s\right)=0 \qquad (3)$$

$$\frac{}{t}\left(\ _s\ _s\vec{v}_s\right)+\ \left(\ _s\ _s\vec{v}_s\vec{v}_s\right)=\ \ _s\ p\ \ p_s+\ \overset{=}{}_s+\ _s\ _s\vec{g}+K_{gs}\left(\vec{v}_g\ \ \vec{v}_s\right) \qquad (4)$$

where $\vec{v}_g$ and $\vec{v}_s$ are the gas and solid velocity vector, respectively; p and p_s are the fluid and solids pressure, respectively; g is the gravitational acceleration, K_{gs} is the gas-solid momentum exchange coefficient which is obtained from the drag model of Gidaspow et al. (1992); $\overset{=}{}_g$ and $\overset{=}{}_s$ are the gas and solid phase stress tensors, respectively; $_g$ and $_s$ are the gas and solid density, respectively; and ε_g and ε_s are the volume fraction of gas and solid. In addition, the volume fractions add up to unity

$$\varepsilon_g+\varepsilon_s=1 \qquad (5)$$

The gas phase turbulence is represented by the widely used k ε turbulence model. The turbulence quantities of the solid phase are predicted using the dispersed turbulence modelling approach (ANSYS CFD, 2006). The solids stress tensor is given by:

$$\overset{=}{}_s=\ _s\ _s\left(\ .\vec{v}_s+\ .\vec{v}_s^{\,T}\right)+\ _s\left(\ _s\ \frac{2}{3}\ _s\right)\ .\vec{v}_s\overset{=}{I} \qquad (6)$$

Constitutive equations (ANSYS CFD, 2006) for the solids pressure (p_s), solids bulk ($_s$) and shear ($_s$) viscosity are derived from the KTGF (for details, see Gidaspow, 1994) in terms of the granular temperature defined as:

$$\Theta_s = \frac{1}{3}\overline{u_s'^2} \qquad (7)$$

The granular temperature, which is a measure of the kinetic energy due to the fluctuating velocities of the particles (u_s'), can be determined by solving a differential transport equation or from an algebraic equation resulting from neglecting the convection and diffusion terms in the transport equation.

3. APPLICATION OF THE MODEL

3.1 The Experimental Case

In this study, the experiments carried out by Huang et al. (2007) in a pilot-scale CFB system were simulated. Fig. 1 is a schematic diagram of the riser section of the CFB consisting of a cylindrical vessel with 0.10 m in diameter and 15.1 m in height, which provided a sufficiently long distance for flow development. The solids used in the riser were FCC particles of 67 μm mean diameter having a density of 1500 kg/m^3. The solid mass flow rate (G_s) was varied between 50 and 200 kg/m^{2}s. Air at ambient conditions was used in the experiments with a superficial velocity (U_s) varying between 3.5 and 8.0 m/s. Measurements of the solid concentration and mean axial velocity distributions along the radius of the riser at eight locations along the height, at z = 0.95, 2.59, 4.51, 6.34, 8.16, 10.0, 12.28 and 14.08 m (where z is the distance from the riser inlet), were reported. The operating conditions of the experimental cases simulated in this study are given in Table 1.

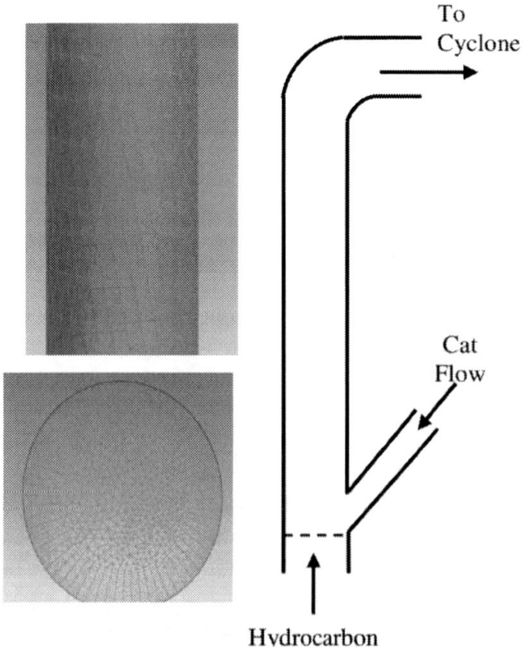

Fig. 1: Schematic diagram of the riser section of CFB and the computational mesh

Table 1 Experimental condition

Operating conditions	**Case 1**	**Case 2**	**Case 3**	**Case 4**
G_s (kg/m^{2}s)	**100**	50	**100**	200
U_g (m/s)	**3.5**	5.5	**5.5**	8.1

3.2 Computational Details

Three-dimensional, transient calculations were performed on an unstructured mesh consisting of 5.78 10^{5} cells. Four mesh sizes ranging from 1.34 10^{5} to 7.38 10^{5} cells were tested in order to examine the influence of the mesh size and its distribution on the predictions. The test calculations show that the selected mesh size is adequate to provide acceptable mesh independent solutions. The input parameters for the flow simulations are summarised in Table 2. The conservation equations were discretised using the finite volume method. A second order upwind discretisation scheme was used for the convection terms.

At the inlet of the riser, the velocities and volume fractions of gas and solid were specified. At the outlet, the pressure was assumed atmospheric. At the walls, the no-slip boundary condition was specified for the gas phase. The wall boundary condition developed by Johnson and Jackson (1987) was used for the solid phase. The discretised momentum and pressure correction equations, together with the equations for turbulence model quantities, were solved iteratively using FLUENT CFD code. Transient computations were performed using zero initialisation for a real time of 5 s, with a time step of 1 10^{4} s. The target value of the normalised residual for each variable was 1 10^{4}. For a converged solution, the overall mass conservations for the gas and solid phases were typically within 0.002 and 0.15%, respectively. The typical computational time for the transient simulation was 10-13 days on a 4 GHz processor.

Table 2 Input parameters for the simulation

Parameters	**Values**
Density of hydrocarbon gas ($_g$)	1.20 kg/m^3
Density of catalyst ($_s$)	1550 kg/m^3
Viscosity of gas ($_g$)	1.831 10^{5} kg/m.s
Particle diameter	67 μm
Particle-particle restitution coefficient	0.99
Particle-wall restitution coefficient	0.60
Specularity coefficient	0.60

4. RESULTS AND DISCUSSION

In order to establish the potentials and limitations of the multiphase CFD modelling methodology, the model was evaluated for all the experimental conditions given in Table 1 ranging from a low to a high density flow. However, simulation results for the experimental case 1 (G_s = 100 kg/m^{2}s and U_s = 3.5 m/s) are presented here as page limitation precluded coverage of all the cases.

The comparison between the measured and predicted solid velocity profiles using the differential transport equation (referred to as the transport-KTGF) and the algebraic equation (referred to as the algebraic-KTGF) modelling approaches for the granular temperature are shown in Fig. 2 only at z = 2.59, 8.16 and 14.08 m due to the space limitation. A scrutiny of the simulation results at all the measurement locations reveals that the general measured trends are reasonably well reproduced in the predictions obtained using both the modelling methods for the granular temperature. However, there are discrepancies between the predictions and measurements and also between the two sets of predictions. The solid velocity at and around the riser axis is underpredicted at all the locations except at z = 0.95 (not shown here). The algebraic-KTGF model for the granular temperature consistently under-predicts solid velocity in this region compared with the transport-KTGF model. However, as can be seen in Fig. 2, the quality of the predictions improves and the difference between the two sets of predictions reduces significantly in the region further away from the axis. An important feature of the solid flow in the riser observed in previous experimental studies including that of Huang et al. (2007) is the down flow of solids along the wall. This is evident at z = 8.16 and 14.08 m in Fig. 2. In fact, the experimental data show the reverse flow of solids close to the wall at all measurement locations between z = 4.51 and 14.08 m. This feature of the solid flow is well captured qualitatively in the predictions obtained using the transport-KTGF model, although the maximum reverse velocity of solids is underpredicted. However, in the case of the algebraic-KTGF model, the down flow of solids was predicted only near the top region of the riser at z = 10.0 and 14.08 m.

Fig. 3 shows the comparison between the measured and predicted (using the algebraic-KTGF and transport-KTGF models) volume fraction of solids at the corresponding locations. Again, the measured trends are generally well reproduced in the predictions. As can be seen in the figure, the solid concentration is low in the central region and high near the wall of the riser at all measurement locations conforming *core-annulus* flow behaviour observed in previous studies. Such a significant deviation of catalyst distribution from a plug-flow profile in a FCC riser reactor will adversely affect the cracking reaction process. As for the quality of the predictions, both the models for the granular temperature accurately predict the solid concentrations in the central region of the riser. In the region near the wall the transport-KTGF model generate better predictions compared with the algebraic-KTGF model at z = 2.59 and 8.16 m. However, both models significantly overpredict the solid concentration near the exit at z = 14.08 m.

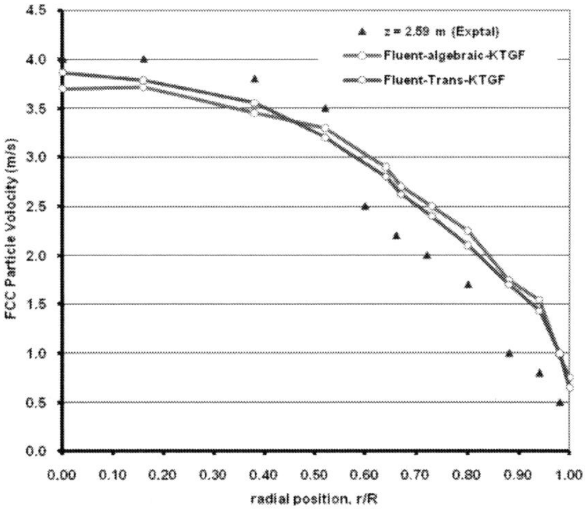

Fig 2: Velocity profile at z = 2.59 m of the riser

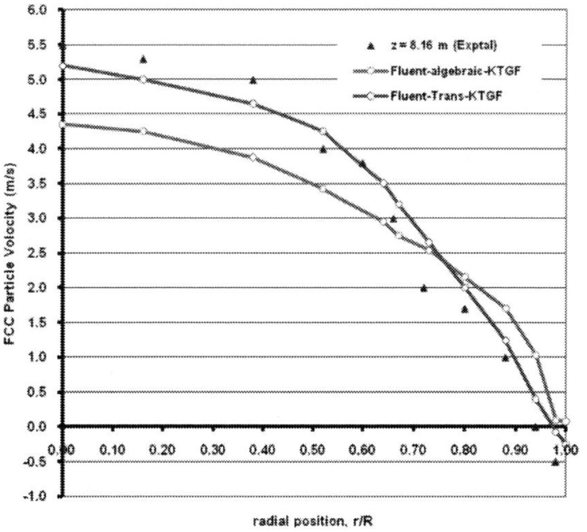

Fig 3: Velocity profile at z = 8.16 m of the riser

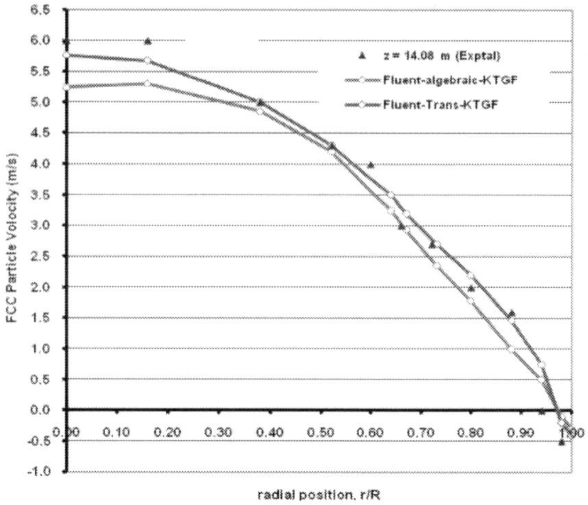

Fig 4: Velocity profile at z = 14.08 m of the riser

Figures Predicted and measured (Huang et al., 2007) solid velocity distributions

Profiles in the central region of the riser and in the solid concentration profiles near the wall. The down flow of particles along the wall is correctly predicted by the transport-KTGF models; whereas the algebraic-KTGF model is able to capture this feature of the flow close to exit. Overall, the former model for the granular temperature performs better than the latter model.

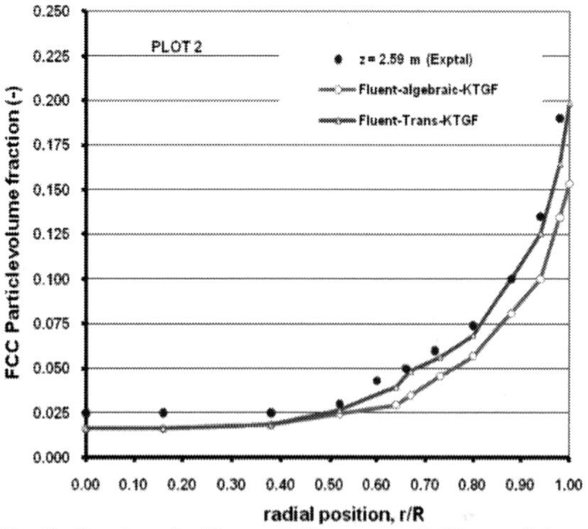

Fig 5: Catalyst holdup profile at z = 2.59 m of the riser

In a similar description, the catalyst solid holdup along the FCC riser were predicted between the axial length of z = 2.59, 8.16 and 14.08 m. From these predictions, it is generally seen that the modeling results shown on the various axial flow were in good agreement with the literatures. Refer to figures 5, 6 and 7.

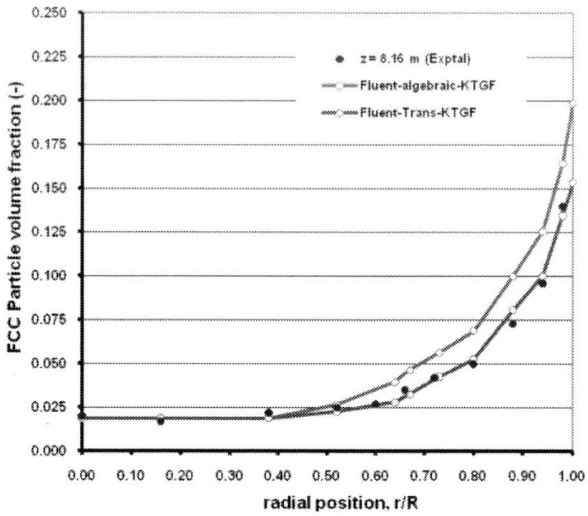

Fig 6: Catalyst holdup profile at z = 8.16 m of the riser

Figure 5 – 7 represents the predicted and measured (Huang et al., 2007) solid concentration distributions

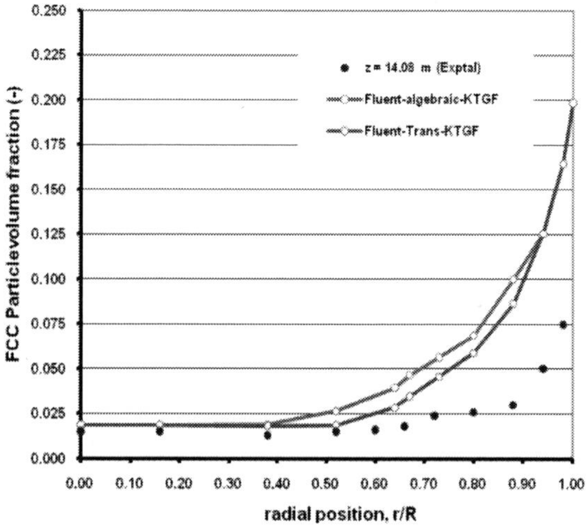

Fig 7: Catalyst holdup profile at z = 14.08 m of the riser

Figures 8 – 11 were used to establish the capabilities of the ANSYS CFD used in these studies. Various cases of the experimental setup were run to determine the solver strengths and weaknesses. This is the only way we can conclude on the CFD effectiveness in achieving the desired results with negligible error due to computation.

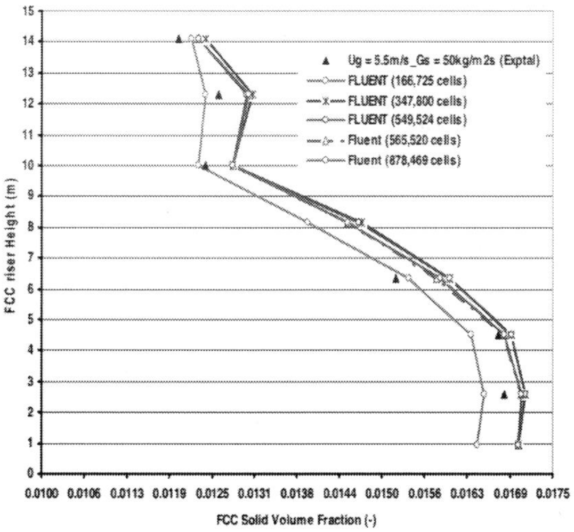

Fig 8: Grid test for concentration at 5.5m/s and Gs = $50kg/m^2s$

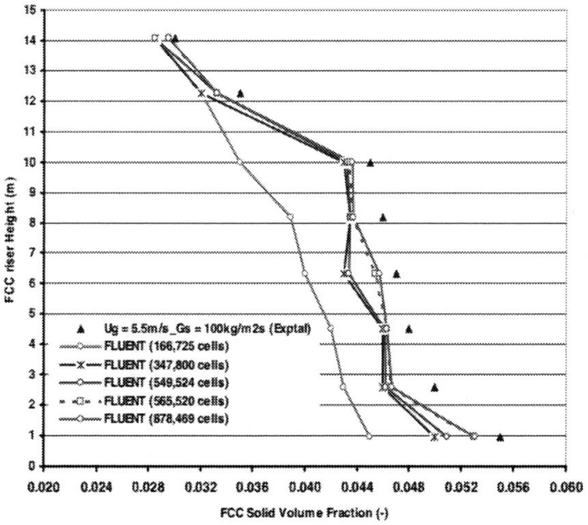

Fig 9: Grid test for concentration at 5.5m/s and Gs = $100kg/m^2s$

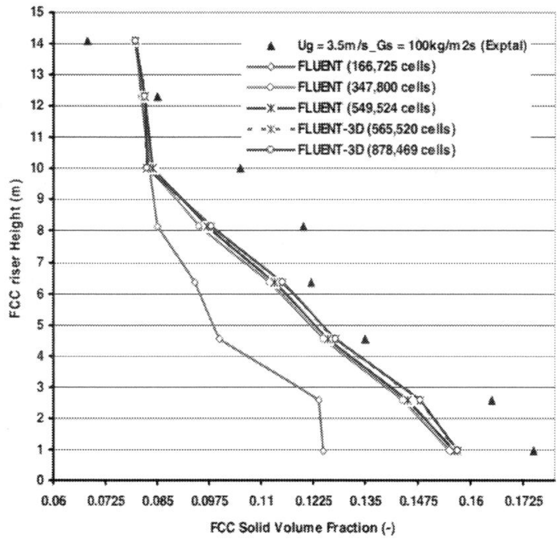

Fig 10: Grid test for concentration at 3.5m/s and Gs = $100kg/m^2s$

From Figure 10, that is, the grid test for concentration at 3.5m/s and Gs = $100kg/m^2s$ show some deviations with the various grids used for test. This figure is the main proof that was used to establish the choice of using grid size 738,890 cells because it gives a better agreement with the experimental data.

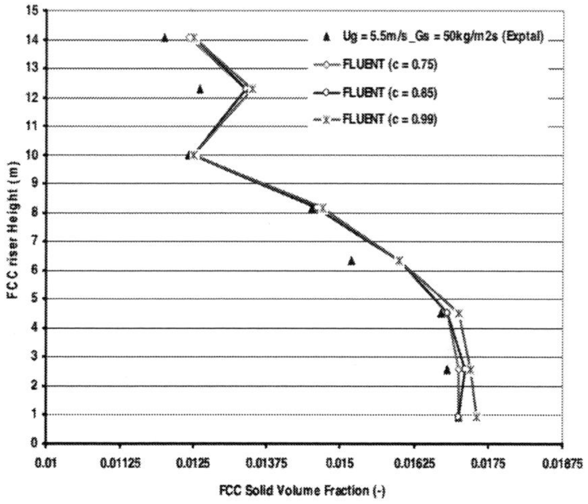

Fig 11: Grid test for restitution coefficient at 5.5m/s and Gs = $50kg/m^2s$

5. CONCLUSIONS

The CFD predictions were reasonably compared with established experimental data from the literature. The overall flow patterns within the FCC riser reactors were predicted well by the model. The results of this study assist in a better understanding of the hydrodynamics of gas-solid flow in CFB/FCC real life applications, which could help in areas of pressure regulations, process flow control, and energy efficiency and utilisation. An Eulerian-Eulerian multiphase CFD modelling methodology imbedded in FLUENT was used to simulate isothermal gas-solid flows in the riser section of a pilot-scale CFB reactor (Huang et al., 2007). The predicted solid velocity and concentration profiles obtained using the two approaches to calculate the granular temperature, namely the differential transport equation model and the algebraic equation model, are compared with measurements at different locations along the height of the riser. The overall agreement between the predictions and the reported experimental data is reasonably good. However, there are differences in the predicted and measured solid velocity.

6. REFERENCES

1. Almuttahar, A. and Taghipour, F. (2007), Computational fluid dynamics of high density circulating fluidized bed riser: Study of modeling parameters, *Powder Technology*, **185**, pp. 11-23
2. ANSYS Fluent (2006), User's Guide, ANSYS Europe Ltd.
3. Du, B., Warsito, W., Fan, L. A. and Bed, S. (2003), Nonhomogeneity in turbulent gas-solid fluidization, *A.I.Ch.E Journal*, **48**, pp. 1896-1909
4. Gidaspow, D., Bezburuah, R. and Ding. J. (1992), Hydrodynamics of circulating fluidized beds, kinetic theory approach. *Proceedings of 7th Engineering Foundation Conference on Fluidization*, pp. 75-82.
5. Gidaspow, D. (1994), Multiphase flow and fluidization: continuum and kinetic theory description, Academic Press, Boston.
6. Grace, J. R. (2000), Reflections on turbulent fluidization and dense suspension upflow, *Powder Technology*, **113**, pp. 242-248
7. Huang, W., Yan, A., and Zhu, J. (2007), Hydrodynamics and flow development in a 15.1 m circulating fluidised bed riser, *Chemical Engineering Technology*, **30(4)**, pp. 460-466
8. Idris M. N. (2010), Hydrodynamics and Process Modelling of Fluid Catalytic Cracking Reactors, PhD thesis, University of Leeds, United Kingdom.
9. Johnson, P. C. and Jackson, R. (1987), Frictional-collision constitutive relations for granular materials, with application to plane shearing, *Journal of Fluid Mechanics*, **176**, pp. 67-93
10. Jiradilok, V., Gidaspow, D., Damronglerd, S., Koves, W. J. and Mostofi, R. (2006), Kinetic theory based CFD simulation of turbulent fluidization of FCC particles in riser, *Chemical Engineering Science*, **61**, pp. 5544-5559
11. Levenspiel, O. (1999), Chemical Reaction Engineering, 3rd Edition, John Wiley & Sons
12. Ranade, V. (2002), computational flow modeling for chemical reactor engineering, Academic Press, Boston.
13. Idris, M. N. (2010), Hydrodynamics and Modelling of Fluid Catalytic Cracking (FCC) Risers, a PhD Thesis submitted to the University of Leeds, United Kingdom

(263a) CAPE-OPEN Laboratories Network Status Report

Michel Pons, CO-LaN, Lyon, France

The CAPE-OPEN Laboratories Network (CO-LaN) is the not-for-profit organization maintaining and developing the CAPE-OPEN (CO) standards providing interoperability between process simulation software. The objective of CO-LaN is ***to expand process modelling capability through software interoperability standards***. The CO-LaN is a user driven organization (http://www.colan.org/) and capitalizes on further opportunities that can be gained from open standard interfaces for process simulation.It has been in operation since 2001, growing in size and scope. Benefits of CO standard implementation include developing a broader market for simulation software vendors, giving users access to best in class process models, and allowing software component suppliers to support the full range of potential users.

CO-LaN will report on its current membership as well as on its managerial organization.

Five missions have been defined for CO-LaN:

- user priorities for CAPE-OPEN standards
- dissemination actions
- maintenance of interface specification standards
- testing and interoperability facilitation
- training facilitation

Reporting will be provided on activities related to each of these missions over the last 12 months. Activities of the Special Interest Groups (SIGs) within CO-LaN will be highlighted.

(263b) CAPE-OPEN Models for Hg Speciation and Partitioning Flowsheet Development

Megan Evans and Lloyd Brown, ConocoPhillips, Houston, TX

Mercury exists in a variety of forms in hydrocarbon production. CAPE-OPEN models were used to develop a flowsheet to accurately predict the distribution and speciation of elemental mercury (Hg^0) and meta-cinnabar (HgS) throughout process equipment in an oil and gas processing facility. The speciation of total mercury between Hg^0 and HgS is critical to reducing the risk of projects, since the equipment required to remove each species is entirely different and the HgS can convert back to Hg^0 at high temperatures. Mercury distribution and contamination of the gas, condensate, and water export fluids impacts both commercial and regulatory feasibility for the facility process selection. HgS has minimal risk if defined and intercepted properly in the production process design, but failure to properly handle the HgS risks conversion of a benign compound to elemental Hg by the production process. Appropriate definition and distribution prediction of HgS and Hg is thus critical to managing overall mercury risk on projects. The use of CAPE-OPEN models within COFE allowed for simulation of multiple phases (including mercury and solids) along with an appropriate property package to manage hydrocarbons, water and alcohols simultaneously.

(263c) Thermodynamic Models As CAPE OPEN Equilibrium Calculators

César Pernalete and Jimmy Contreras, Refining and Industrialization Management, PDVSA Intevep, Los Teques, Venezuela

Thermodynamic models as CAPE OPEN equilibrium calculators

C. G. Pernalete [1], J. M. Contreras [1], R. E. López [1,2], A. Nagel [1,3]

[1]Gerencia Técnica de Refinación. PDVSA Intevep. Venezuela.

[2] Departamento de Ingeniería Química. Vice-Rectorado de Barquisimeto. Universidad Nacional Experimental Politécnica "Antonio José de Sucre". Venezuela.

[3] Departamento de Ingeniería Química. Universidad Metropolitana. Venezuela.

Abstract

Nowadays, software technology for process simulation has been developed in such an important way that they have become a fundamental tool for the correct operation of several processes, specially in the oil field. The use of this technology had reached a level that enhances human work or indeed substitutes it, improving its efficiency and decreasing the time and effort to carry on activities like process monitoring, control and optimization. In a daily basis, the amount of commercial proposals available in the market for tools to process simulation is increasing.

One of the main strengths of most commercial process simulators is their flexibility to make different kinds of thermodynamic calculations. In the oil industry the equation-of-state approach (EOS) has had an important role in thermodynamic modeling of vapor - liquid equilibrium (VLE) for hydrocarbons. However, these models do not offer good results under certain conditions, like under the presence of hydrogen or at high temperature or pressure conditions, making the system far from an ideal situation. In these cases it is necessary to use special models like the correlation proposed by Chao and Seader in 1961, and the modification proposed by Grayson and Streed in 1963 or the modification made by Chorng H. Twu in 1995 to the Peng-Robinson equation of state.

In spite of all the benefits and advantages that offer the use of commercial software technology for chemical process simulation, from a practical point of view there are some drawbacks. Commercial proposals for process simulation require a license that in most cases is expensive and must be renewed every year. If there is no renewal or purchase of this license it is not possible to develop new simulation projects or to have access to models previously built. This situation generates a problem of technological dependence.

It is important to point out that, chemical process simulators are software tools that provide a set of functions such as physical and thermodynamic property databases, thermodynamic calculations, numerical methods, unit operation and kinetic models. At the beginning, process simulators combined these functions with a programming paradigm where there were not well defined boundaries among these functions. The current trend is to have a process simulator formed as a group of software components that interact or interoperate simultaneously, where each component have well defined duties.

Since 2001, the CO-LaN organization has promoted the creation of the CAPE-OPEN standards, where a set of free-to-use non-proprietary rules and interfaces that allow CAPE (Computer-Aided Process Engineering) applications or components to interoperate, are defined. Currently, most process simulators are somehow compliant with this standard; however, COCO (CAPE OPEN to CAPE OPEN) simulator is the CAPE OPEN compliant reference tool. COCO simulator is a non-commercial software for process simulation that was fully designed inspired by this standard.

According to the philosophy proposed in the CAPE OPEN standards, in the architecture for process modeling tools in steady state, there must be a "Process Modeling Executive" (PME), which is the component responsible for the building of the entire model and execute the overall calculations to solve the system of equations. In order to meet this goal, the PME must communicate with the other components that describe each unit operation or other calculations like thermodynamic equilibrium, specialized algorithms for solving the entire system or physical properties determination. These components are called "Process Modeling Components" (PMC) since they "wrap" specific tasks in a well defined software unit. PMC's can be implemented as independent software components but able to communicate among themselves.

In this work, we have implemented some thermodynamic methods, as CAPE OPEN compliant components: process modeling components following the aforementioned strategy. The idea is to have these components available for using in non-commercial process simulation tools in order to solve specific problems for the oil industry. We have developed three components that implement the following methods: Chao-Seader (1961), Grayson Streed (1963) and Peng Robinson – Twu (1995). Specifically, we have developed them as equilibrium calculators that could be used in a property package as described in the 1.1 version of the standard for physical and thermodynamic properties.

In this study the component TEA (Thermodynamic for engineering application) was used as the property package. TEA is a component conceived as a property package manager that comes with the distribution of COCO simulator. TEA is the only property package available as non-commercial software on the network that is able to deal with external CAPE OPEN equilibrium calculator components. Otherwise it would have been necessary to create a property package and this study would have considered the calculation of chemical-physical properties in addition to equilibrium calculations.

In that sense, as a first step we described the original thermodynamic models considered in this work, including the temperature, pressure and compositions constraints associated to each model. After that we designed the algorithms of each model using UML (Unified Modeling Language). The implementation of the algorithms was made using C++ as programming language taking into account the interoperability standards. As a result, we achieved three interoperable thermodynamic models as CAPE OPEN equilibrium calculators that could be used in commercial and non-commercial process simulation tools, even though the simulator implements the standard.

For testing and validation of the developed components we used COCO simulator process modeling environment. In first place, we tested the validity of the models through the comparison of its performance in predicting Vapor-Liquid Equilibriums (VLE) in systems composed only by pure components. Secondly, it was selected an experimental system composed by petroleum fractions and whose modeling is made through the use of pseudo compounds. It is important to clarify that temperature and pressure conditions of those systems were within the constraints established for the original models. Finally, a comparison between the results obtained with the developed models and the results showed by a commercial simulation tool that consider the models in study, was made.

We found that the developed models are able to predict with good agreement experimental data about VLE used in the validation step, both in case of pure compounds and in case of petroleum fractions. Moreover, we found that the error prediction of the commercial software were always similar or larger than the reported by the in-house models.

Finally, we noticed a robust performance of each one of the software components of COCO simulator, specially TEA which was responsible for establishing the communication with the equilibrium calculators developed.

(263d) Streamlining CAPE-OPEN Interoperability with Promax®

Michael W. Hlavinka, Jorge Martinis, Hugh Docherty, Mauricio Lopez and Christopher Skowlund, Bryan Research & Engineering, Inc., Bryan, TX

Streamlining CAPE-OPEN interoperability with ProMax®

Christopher Skowlund, Mauricio Lopez, Jorge M. Martinis, Hugh Docherty, Michael Hlavinka

**Bryan Research & Engineering, Inc. P.O.Box 4747, Bryan, Texas 77805*

As CAPE-OPEN (CO) becomes a *de facto* standard for process simulation software interoperability, a recent implementation of the Thermodynamic and Unit Operation interface specifications in ProMax® has been tested. The new implementation, which supports version 1.0 and 1.1 of the CO standard, extends the ability of ProMax® to consume foreign CO compliant components in the form of unit operations, and provides users with a seamless mechanism to host ProMax® powerful thermodynamics in their environment of choice for process simulation. In the present work, a case study case is presented to illustrate the interoperability of ProMax® simulation environment with CO unit operations from ProMax® and various other vendors. Furthermore, a second case study featuring ProMax® thermodynamic models interoperating with COCO simulation environment using a similar unit operation layout is presented. In both cases, the impact of the CO interface layer overhead on performance and the strategies to overcome current limitations in the CO standard for multithreading execution have been addressed. As a result, a series of recommendations to increase the generality and flexibility of the CO standard have been proposed.

REFINERY PETROLEUM PROPERTIES VIA CAPE-OPEN: FROM PROTOTYPE TO COMMERCIAL APPLICATION

Richard Baur, Ashok Dewan, Kevin Keyser – Shell Global Solutions[1]
Jasper van Baten – AmsterCHEM
David Jerome, Penukonda KrishnaMurthy – Invensys SimSci-Esscor

With tight refinery margins and stringent environmental regulations dictating refining operations today, any advantage that can be derived from optimal rigorous process modeling is gaining increasing importance and attention. To achieve this objective better in-house process models that capture the physical and chemical changes in a given refinery operation (e.g., desulfurization, cat cracking) using robust algorithms, in-house pilot plant data, *and* an accepted standardized method to link those process models to a given process flowsheeter/optimizer tool seamlessly, are needed. Furthermore, as several flowsheeter/optimizer vendors "accept" the standardized way to link up the in-house process models to the vendors' process tools, it adds the potential to conduct process simulation case studies for refinery process unit performance monitoring, real time optimization, catalyst cycle management, unit revamp, refinery wide optimization and/or operator training, using the same set of in-house process models in different process tools.

Within Shell, the CAPE OPEN framework was selected to provide the standardized look and feel to a process model, in a consistent and efficient manner across a variety of refinery conversion process applications. The calling application (or, process tool such as PRO/II) sets up and modifies the refinery application process model as needed, using the standardized interface. Thus the end-user needs to get acquainted with only one standardized interface, based upon the CAPE-OPEN framework, to integrate diverse refinery process units such as a hydrocracker, catcracker or reformer into a flowsheeter or optimizer. Alternatively, the same refinery application process model should be capable of being linked with multiple process tools that perform different process engineering case studies.

The first step is to link all the refinery application process models that are supported by the standardized framework together in a Process Modeling Environment (e.g., a flowsheet application, such as PRO/II). The CAPE-OPEN Unit Operation standard provides an ideal starting point. However, the data types, compound definitions, and property definitions required by the Shell process models are unique to Shell. Furthermore, the Shell process models require refinery petroleum properties, which so far had been missing from the CAPE-OPEN Unit Operation. This technical presentation focuses upon the interface between CAPE-OPEN based Unit Operation models and the Process Modeling Environment (PME) – in this case, PRO/II - that deals with the refinery petroleum properties.

---

[1] Shell Global Solutions is a network of independent technology companies in the Shell Group. In this publication the expression 'Shell' or 'Shell Global Solutions' is sometimes used for convenience where reference is made to these companies in general, or where no useful purpose is served by identifying a particular company.

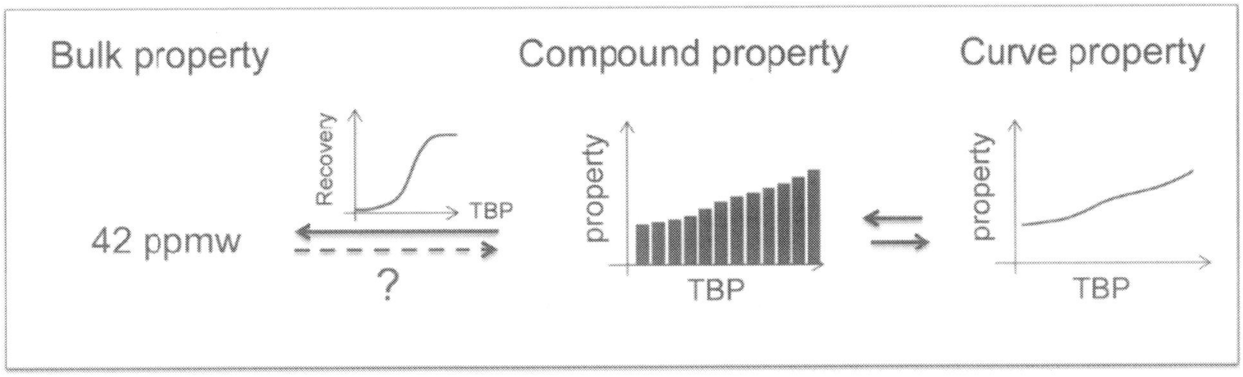

Figure 1. Bulk compound and curve properties

Following the work of the former CAPE-OPEN Petroleum Properties Special Interest Group (SIG), curve-, bulk- and compound properties need to be distinguished. In petroleum processing, generally there are real and pseudo compounds. The pseudo compounds are in fact a lumped range of real compounds that boil between certain temperatures. Such pseudo compounds have refinery physical properties that are not constant, as opposed to real compounds. Some of these properties affect the thermodynamics of the mixture, such as the pseudo critical temperature. Some do not affect the thermodynamics but are required to keep track of the content of the process stream. An example of a property that does not affect the thermodynamics is the sulphur content. The variable sulphur content of the pseudo compounds, in combination with a blend and the fixed sulphur content of real compounds and composition allows for calculating the bulk sulphur content. The calculation of pseudo compound sulphur from bulk sulphur however is not unique. Properties can also be expressed as a curve, commonly the property value vs. the normal boiling point temperature.

The standardized Unit Operation may set values of particular petroleum properties, and choose to not set other properties. In a separation unit, for example, only flow rate and composition are modified between feeds and products, neither of which is specific to petroleum properties. Still, access to petroleum properties such as boiling range may be required. On the other hand, a desulphurization unit will likely change the sulphur content, molecular weight, boiling range and other properties of the pseudo compounds. A scheme is therefore required such that the PME (calling application like PRO/II) can allow for changing certain petroleum properties while keeping others equal to values at a particular feed stream. In addition, changing certain petroleum properties may affect other properties. For example, changing either pseudo compound sulphur content or composition will affect the bulk sulphur content, and changing the molecular weight of a pseudo compound may affect all properties that are predicted by an underlying equation of state. Hence, the PME needs to know which properties apply, in case a petroleum property is not modified, and when the Unit Operation is done with modifying properties, so that the PME can update all values that are affected. The resulting workflow is that the standardized Unit Operation must define a stream from a reference to a feed stream, after which it can change petroleum and regular properties. Finally the standardized Unit Operation must indicate it is done with setting properties before the standardized Unit Operation can perform any calculation on the stream, such as phase equilibrium calculations (required for all product streams of a CAPE-OPEN compliant unit operation) or property calculations (e.g. enthalpy to close a heat balance).

The prototype interface presented herein is implemented in a selected set of Unit Operations in a Shell-customized test platform using the SimSci-Esscor PRO/II v9.2 simulator. The presentation will highlight some issues and solutions that were implemented. The interface has been tested and validated within Shell for hydro-processing conversion process models. The combination of the "CAPE-OPEN standard" Shell Unit Operations in PRO/II v9.2 is now at "production" stage, and will be validated next year for other Shell conversion process models. On completion of the validations, the prototype will be moved to "Production" and deployed within global Shell.

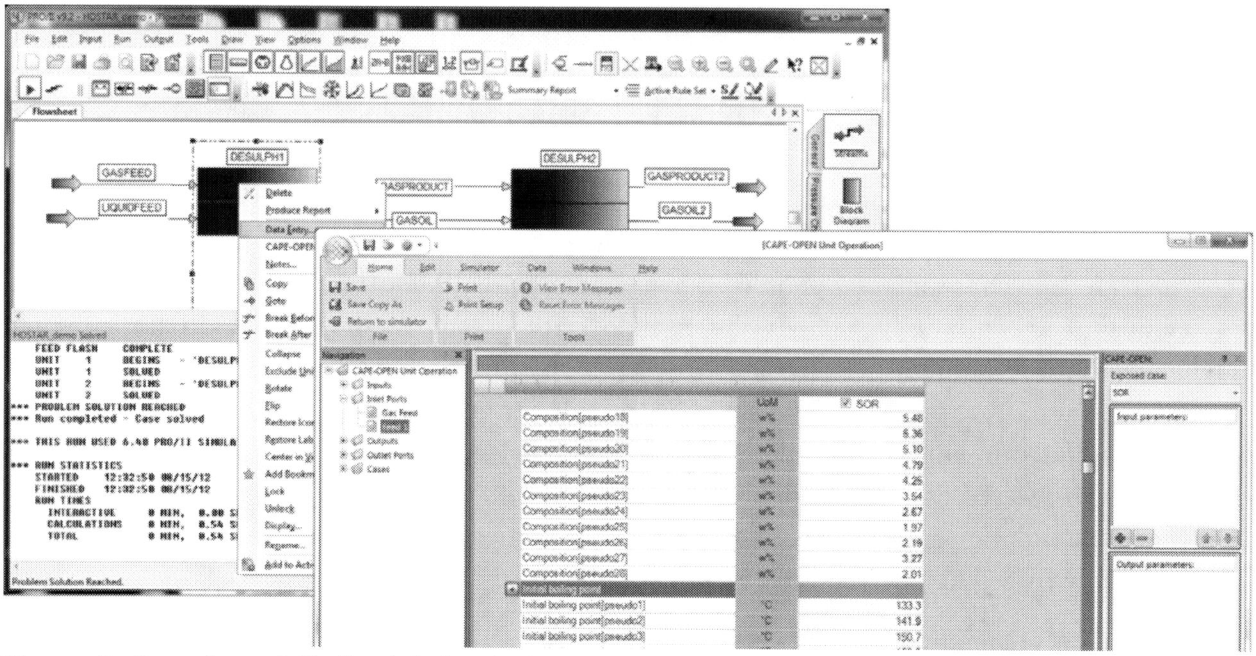

Figure 2. Snapshot of Shell Global Solution Unit Operation in SimSci Pro/II 9.2

A CAPE-OPEN Unit Operation Software Component for Solving CHEMKIN Reactor Networks Inside Process Modeling Simulations

T.J. Garratt, C.-P. Chou, D. Hodgson, and E. Meeks

Reaction Design *5930 Cornerstone Court West Suite 230, San Diego, Ca. 92121*

Process modeling simulators are used to model and simulate the physicochemical processes occurring in chemical plants and power-generation facilities. Examples of process simulators include COFE [13] and AspenPLUS [1], which represent the chemical processes as a *flowsheet diagram*, consisting of a number of single *Unit Operations (UO)* connected to each other by flow *streams*. Each simulator usually is coupled with its own library of UO models such as mixers, splitters, pumps, valves and reactors. In addition to built-in model options, process simulators also often provide one or more interfaces that allow users to insert their own models as UOs within the flowsheet. *CAPE-OPEN (CO)* [2] is a standard interface designed for interfacing UO components into process simulators. The interface provides methods to encapsulate a model based on *COM* and *CORBA* protocols.

For the reactor models employed in AspenPLUS and other process modeling environments, the chemical reactions occurring within the reactor are usually treated in a very simplified manner. Typically, global reactions are used, with just a few chemical steps. Reaction Design's CHEMKIN models, on the other hand, allow very detailed representation of the reaction kinetics within a reactor system. CHEMKIN is particularly well established for accurate modeling of gas-phase homogeneous combustion processes and of gas-surface interactions for catalyzed processes, in combustion simulations, reaction mechanisms typically used in CHEMKIN include 100s of chemical species and 1000s of chemical reactions. To bring this level of chemical detail into the flowsheet simulation environment, we have used the CO UO interface to create a new CO UO component that encapsulates a generalized CHEMKIN model so that a much more accurate chemistry model can be included in the reactor-model unit within the flowsheet.

The CO UO component allows the unit operation to be represented by a single PSR, a cluster of PSRs, or a more generalized Equivalent Reactor Networks (ERNs) that consists of perfectly stirred reactors (PSRs) and plug-flow reactors (PFRs). The PSRs and PFRs are CHEMKIN-PRO models, with which any level of gas-phase or gas-surface kinetics can be included. The CO UO was developed specifically to target gas-turbine combustors and coal gasifiers, which are essential for the design of integrated gasifier/gas-turbine combined-cycle systems (IGCC). The idea is to replace much more simplified coal gasifier models with an ERN that can both represent the complex flow within the gasifier and the complex kinetics, while still allowing fast flowsheet simulation times. This approach allows more accurate predictions of the gasifier operation and outlet composition using the detail kinetics models.

Researchers have long established that an Equivalent Reactor Network (ERN) can represent a reacting flowfield in a cost-effective and accurate way to account for detailed kinetics and to predict the behavior of a combustor or chemical process, including trace-species emissions. Gas-turbine manufacturers, for example, routinely use perfectly-stirred-reactor (PSR) and plug-flow-reactor (PFR) networks to model gas-turbine combustor behavior, particularly for predictions of CO and NO_x emission levels, where detailed kinetics are required [3-5]. Kendrick, and coworkers at UTRC [3], for example, showed that networks of perfectly stirred reactors (PSRs) with detailed chemistry can accurately simulate an industrial combustor. Using just 10 reactors in the network, they were able to quantitatively predict the correct NO_x levels in scale-up studies of the system and were also able to correctly predict NO_x emissions over a wide range of fuel/air ratios, combustion efficiency, and the effects of non-mixed regions. In recent years, such approaches have also been applied to industrial-scale coal combustors, furnaces, and boilers [6-9] as well as to bioreactors for multi-phase batch-reactor chemical production [10].

Recognizing the importance and power of the equivalent-reactor-network approach, Reaction Design developed a product specifically designed to generate, solve, and visualize the results of ERNs, using CHEMKIN-PRO [11] as the solution engine. This product, the ENERGICO® Simulation Package [12], provides flexibility to expert users, while allowing a turn-key approach for non-experts using established algorithms for extracting reactor networks. The reactor-network simulation is run in CHEMKIN-PRO using state-of-the-art solution techniques for stiff chemistry systems accommodating large reaction mechanisms with 100s of species and 1000s of reactions. The reactor network simulation within CHEMKIN-PRO allows for complex connectivity between idealized reactors (PSRs and PFRs), making use of a tear-stream solution algorithm. A schematic of the work-flow facilitated by ENERGICO is shown in Figure 1.

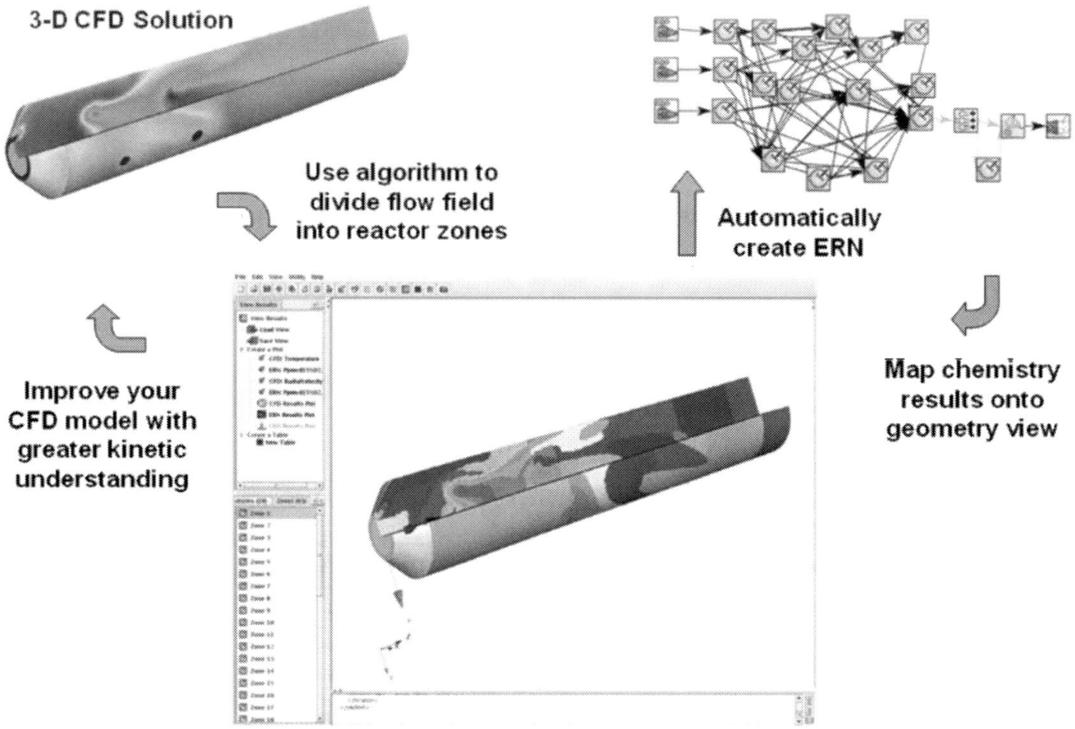

Figure 1. Workflow for creating a reactor network from a CFD solution within the ENERGICO Simulation Package.

This approach has proven to be very effective in predicting ppm-level NO_x and CO emissions from gas-turbine combustors [14], using high-fidelity models of the combustor geometry in CFD and high-fidelity chemistry models in the ERN solution for the gas-phase combustion and emissions production. Furthermore, once the ERN is established, studies of fuel loading effects on emissions can be performed efficiently and accurately without repeated simulation of the computationally expensive CFD solution.

Our CO ERN component uses the same CHEMKIN core reactor and solver components, together with a new and more efficient "in-memory" tearing implementation, to enable the solution of ERNs using detailed chemistry via the CO UI interface. This enables the ERN to be used within a process flowsheet simulation to more accurately model the whole process.

To validate our component, we solved a complex ERN representing a coal gasifier. This reactor-network model was created using ENERGICO, where a sample FLUENT model of a coal gasifier provided the CFD flow-solution input to ENERGICO. The case studied is a 3D coal combustor tutorial provided by FLUENT. Pulverized coal and air are injected separately into the combustor in which the coal particles will first undergo the devolatilization process (essentially the same as a gasifier) and then the remaining char particles are burned (as a charcoal combustor). The ERN is created according to the DPM (dispersed phase) variables, temperature, and gas velocity in the CFD solution and results in 28 zones2. These zones are then translated into a 28-PSR, multi-Cluster ERN with tearing and initialized tear streams, as shown in Figure 2.

The resulting ERN was then exported from ENERIGO via a CKCAPE file format, which is then read into our CO UO module which was instantiated in COFE flowsheet with identical inlet conditions. Reaction Design's C_2_NO_x detailed chemistry mechanism [15] was used, which has 99 species and 693 reactions. We solved the ERN in CHEMKIN and also in COFE using the CO UO software component. In both cases, an identical number of tear iterations (5 and 11) were required to achieve convergence of the tear loops using the default relative and absolute tolerances and the direct substitution algorithm. Table 1 compares the solution in CHEMKIN-PRO stand-alone and in COFE using the CO UO module. The values listed are for the species mole fractions and temperature in the COFE outlet. We note that lumping of the sum of trace-species components and assigning that fraction to N_2 is currently used by the CO UO component to allow matching of the chemical species used in the CHEMKIN mechanism species to the usually much smaller set of major species that the
flowsheet handles. It can be seen that the solutions are identical within solution tolerances, thus validating the CO UO ERN component.

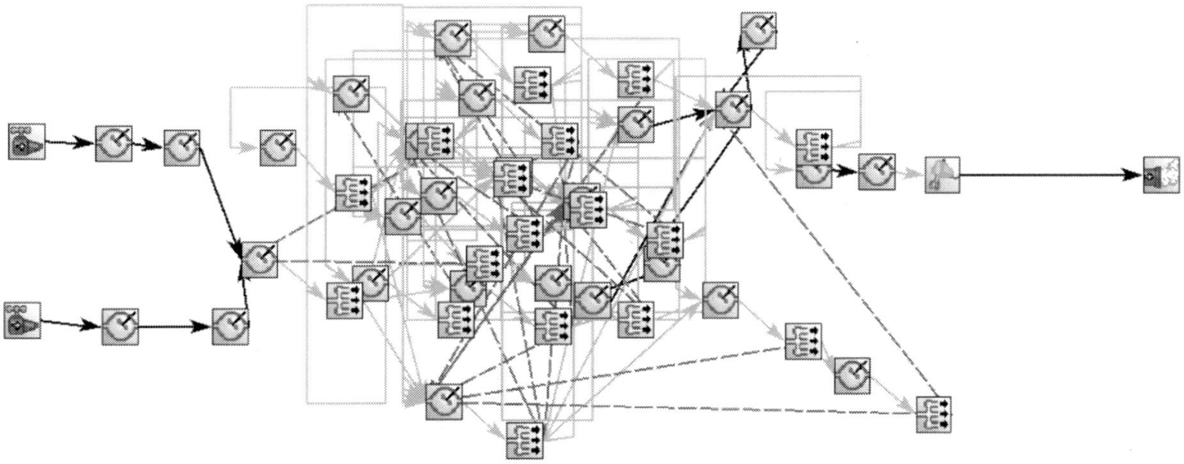

Figure 2: 28-PSR cluster ERN of goal gasifier.

Variable in Outlet Stream	CHEMKIN	COFE
Temperature (K)	1365	1365
Methane	5.489E-15	5.490E-15
Oxygen	3.554E-2	3.553E-2
Nitrogen	7.628E-1	7.638E-1
Carbon monoxide	4.558E-7	4.559E-7
Carbon dioxide	1.455E-1	1.454E-1
Hydrogen	8.072E-8	8.073E-8
Ethylene	9.519E-7	9.519E-7
Water	5.512E-1	5.512E-1
Hydrogen cyanide	9.555eE-12	9.555E-12

Table 1: Comparison of CHEMKIN and COFE outlets (all species units are mole fractions)

1 References

1. AspenPLUS, Cambridge, MA: AspenTech, Inc., 2009.
2. "CO-Lan (The CAPE [Computer Aided Process Engineering] Laboratories Network)," http://www.colan.org/,
3. D. W. Kendrick, A. Bhargava, M. B. Colket, W. A. Sowa, D. J. Maloney, and K. H. Casleton, "NOx Scaling Characteristics for Industrial Gas Turbine Fuel Injectors," ASME Turbo Expo, 2000.
4. I. V. Novosselov, P. C. Malte, S. Yuan, R. Srinivasan, and J. C. Y. Lee, "CHEMICAL REACTOR NETWORK APPLICATION TO EMISSIONS PREDICTION FOR INDUSTIAL DLE GAS TURBINE," Proceedings of GT2006, 2006-90282, 2006.
5. C. Russo, G. Mori, V. V. Anisimov, and J. Parente, "Micro Gas Turbine Combustor Emissions Evaluation Using the Chemical Reactor Modelling Approach," Proceedings of ASME Turbo Expo, GT2007-27687, 2007.
6. D. Benedetto, S. Pasini, M. Malcitelli, C. La Marca, and L. Tognotti, "NOX Emission Prediction from 3-D Complete Modelling to Reactor Network Analysis," Combustion Science and Technology, 153: 279-294, 2000.
7. M. Falcitelli, S. Pasini, N. Rossi, and L. Tognotti, "CFD+reactor network analysis: an integrated methodology for the modeling and optimization of industrial systems for energy saving and pollution reduction ." Applied Thermal Engineering, 22 971-979, 2002.
8. T. Faravelli, L. Bua, A. Frassoldati, A. Antifora, L. Tognotti, and E. Ranzi, "A New Procedure for Predicting NOx Emissions from Furnaces," Computers and Chemical Engineering, 25: 613-618, 2001.
9. G. Liu and S. Niksa, AFRC International Symposium, 2003.
10. F. Bezzo, S. Macchietto, and C. C. Pantelides, "General Hybrid Multizonal/CFD Approach for Bioreactor Modeling," AIChE J., 49: 2133-2148, 2003.
11. CHEMKIN-PRO Release 13082: Reaction Design, 2008.
12. ENERGICO Release 30081: Reaction Design, 2008.
13. "COCO CAPE-OPEN to CAPE-OPEN Simulation Environment," http://www.cocosimulator.org/, April 2010.
14. S.A. Drennan, C.-P. Chou, A. F. Shelburn, D. W. Hodgson, C. Wang, C. V. Naik, E. Meeks, and H. Karim, "Flowfield-Derived Equivalent Reactor Networks for Accurate Chemistry Simulation in Gas-Turbine Combustors", Proceedings of ASME Turbo Expo, GT2009-59861, 2009.
15 C. V. Naik, E. Meeks, and S. Drennan, "Detailed Chemical Kinetic Mechanism for Methane and Ethane Combuston including NOx Emissions for Gas Turbines Applications", in Fall Technical Meeting of the Western States Section of the Combustion Institute. Irvine, CA, 2009.

(271a) Optimal Simultaneous Production of Hydrogen and Liquid Fuels From Glycerol: Integrating the Use of Biodiesel Byproducts

Mariano Martín, Chemical Engineering, University of Salamanca, Salamanca, Spain and Ignacio E. Grossmann, Chemical Engineering Department, Carnegie Mellon University, Pittsburgh, PA

Biodiesel is one of the most important alternatives to fossil fuels due to its compatibility with the diesel engines and its supply chain. The economic feasibility and the net energy balance of the biodiesel are linked to the use of its byproducts. Among them, the most important one is glycerol. Glycerol has been used so far in cosmetics, drugs, personal care, detergents, food industry (Pagliaro & Rossi, 2010), and has kept its price at levels that can support the economy of the biodiesel. However, the expected increase in the production of biodiesel will eventually saturate the current glycerol market, and thus its price will drop affecting the production cost of biodiesel which can increase around 25%-50% with respect to the current production cost presented in previous papers (Martín & Grossmann 2012, Severson et al. 2012) for the expected future price of glycerol (Ahmed & Papadias 2010). Therefore, new markets for glycerol should be considered. Since the biodiesel production requires energy, around 2 to 4 MJ/gal (Martín & Grossmann 2012), there is an option to increase the production of fuels from glycerol and improve the energy balance of biodiesel.

In this paper we present the optimization of the simultaneous production of hydrogen and liquid fuels from glycerol. We propose a limited superstructure embedding the various process units involved in the production of hydrogen and FT-fuels. The process starts with the reforming of the glycerol. Three different alternatives are evaluated: aqueous phase reforming, which uses liquid water in large quantities but operates at low temperature; steam reforming, which has a higher yield to hydrogen but is an endothermic process; and autoreforming which has a lower yield to hydrogen but operates close to adiabatic conditions. Next, the raw syngas obtained is cleaned up and its composition is adjusted in terms of the ratio CO / H_2 using three possible technologies (bypass, pressure swing adsorption, PSA, and water gas shift reaction, WGSR) allowing the production of pure hydrogen. Later, the removal of CO_2 is performed by means of PSA. Once the syngas is prepared, the Fischer - Tropsch reaction is conducted and the products separated. Hydrocracking of the heavy products is also considered to increase the yield towards diesel. Reduced order models for the key units such as reformer, FT reactor, WGSR or hydrocraking, are developed based on experimental data and/or rules of thumb from the literature (Choi et al, 2003; Shabaker et al, 2004; Douette et al, 2007; Adhikari et al, 2007; Jones & Pujado, 2006; Bezergianni et al, 2009). Our objective is to optimize the superstructure and the operating conditions minimizing the energy consumption while maximizing the yield and determining the optimal distribution of synthetic liquid fuels. The optimization of the process is formulated as a Mixed Integer Non-linear Programming (MINLP) problem in GAMS.

The problem is solved first for the optimal production of Hydrogen alone and later for the simultaneous production of green gasoline, FT-diesel and hydrogen. It is shown that the production of hydrogen is competitive with that obtained from lignocellulosic biomass (Martín & Grossmann, 2011) as long as the glycerol price is below $0.06/lb, which above the target of the DOE (Ahmed & Papadias 2010), using aqueous phase reforming followed by WGSR. However, the production of hydrogen discards the use of glycerol as a carbon source. On the other hand, the simultaneous production of hydrogen and liquid fuels is attractive versus the use of switchgrass (Martín & Grossmann 2011b) only if the price of glycerol is below $0.05/lb, DOE target. For that option the selected technologies are auto reforming and bypassing the raw syngas partially to PSA for the purification of the hydrogen.

References:

Adhikari, S.; Fernando, Sandim, Gwaltney, S.R.; Filip To., S.D.; Bricka, R.M.; Steele, P.H.; Haryanto, A. (2007) Int. J. Hydrogen Energy., 32, 2875-2880

Ahmed and Papalias, (2010) Hydrogen from Glycerol: A Feasibility Study Presented at the 2010 Hydrogen Program Annual Merit Review Meeting Washington DC, June 8, 2010

Bezergianni, S., Kalogianni, A., Vasalos, I. A. (2009) Bioresour.Technol. 100 (2009) 3036–3042

Choi, Y., Stenger, H. G. (2003) Journal of Power Sources 124, 432–439

Douette, A.M.D.; Turn, S.Q.; Wang, W.; Keffer, V.I. (2007). Energ. Fuel., 21, 3499-3504

J.W. Shabaker, G.W. Huber, and J.A. Dumesic (2004) J. Catal. 222,180–191

Jones, D.S.J., Pujadó, P.R. (2006) Handbook of petroleum processing. Springer.

Pagliaro,M.; Rossi, M. (2010) Future of Glycerol. 2nd Edition The royal society of Chemistry. Cambridge

Martín, M., Grossmann, I.E. (2012) Process optimization bioDiesel production from cooking oil and Algae. Submitted to Ind. Eng. Chem Res

Severson, K., Martín, M., Grossmann, I.E. (2012) Process optimization bioDiesel production using bioethanol. Submitted AICHE J.

Martín, M., Grossmann, I.E. (2011a) Energy optimization of Hydrogen production from biomass. Computers and chemical engineering, 35, 9, 1798-1806

Martín, M., Grossmann, I.E. (2011b) Process optimization of FT- Diesel production from biomass. Ind. Eng. Chem Res, 50 (23),13485–13499

(271b) Coal Oxycombustion Flowsheet Optimization

Alexander W. Dowling[1], Lorenz T. Biegler[1] and David C. Miller[2], (1)Chemical Engineering, Carnegie Mellon University, Pittsburgh, PA, (2)U.S. Department Of Energy, National Energy Technology Laboratory, Morgantown, WV

Increasing atmospheric concentrations of carbon dioxide have unquestionable sparked scientific interest in developing fuel sources with reduced carbon footprints. Many researchers have begun developing technologies to enable carbon capture, utilization and sequestration with traditional fossil fuels. All of these new powerplant flowsheets require some separation to purify CO2 to high concentrations for utilization or sequestration at high pressures. In a coal oxycombustion process the additional separation happens at beginning. Oxygen is concentrated from air, and the coal is burned in a very nitrogen light environment. Thus the flue gas leaving the boiler contains mainly water and CO2. Unlike some other coal carbon capture schemes, large separation units are not required post-combustion. There is also some potential to retrofit existing coal power plants for oxycombustion.

Coal oxycombustion on a large scale is a very immature technology; heuristic methods tailored to design near high performance oxycombustion plants do not exist. Furthermore the oxyfuel combustion flowsheet is highly integrated; changing one part of the flowsheet can dramatically impact the performance of other units. This makes design and optimization especially challenging. However this also means the oxycombustion design problem an excellent test-bed to develop new flowsheet optimization technologies.

An example oxycombustion design dilemma is centered on the massive compressor train used to compress CO2 to 150+ atm for sequestration. Some members of the oxycombustion research community are advocating for adiabatic CO2 compression with only a few stages. Although this strategy requires additional mechanical work, it also heats the high pressure CO2 stream. Some of this energy can then be recovered through heat integration with the steam cycle. Other members of the community advocate for using streams in the cryogenic air separation unit for interstage compressor cooling, thus reducing the mechanical work for compression. Due to the integrated nature of the flowsheet both approaches have the potential to impact the design of every major unit. Without a framework to optimize the flowsheet, selecting the best compression strategy remains difficult and un-systematic task.

In this talk we present our efforts to answer the compressor design question through use of advance flowsheet optimization techniques. More specifically we employ a fully equation oriented optimization approach. All model equations, including thermodynamics, and their derivatives are open to the optimization algorithm. This allows us to utilize start of the art large scale nonlinear programming solvers. We present our models for each unit, along with our experiences incorporating cubic equations of state (non-ideal thermodynamics) and heat integration into the optimization problem. We conclude by presenting our plans to extend these developments to the entire oxycombustion flowsheet.

(271c) Computer-Aided Synthesis and Design of Working Fluid Mixtures for Organic Rankine Cycles

Athanasios I. Papadopoulos[1], Mirko Z. Stijepovic[2], Patrick Linke[2], Panos Seferlis[3] and Spyros S. Voutetakis[4], (1)Chemical Process Engineering Research Institute, Centre for Research and Technology-Hellas, Thermi-Thessaloniki, Greece, (2)Chemical Engineering, Texas A&M University at Qatar, Doha, Qatar, (3)Mechanical Engineering Department, Aristotle University of Thessaloniki, Thessaloniki, Greece, (4)Chemical Process Engineering Research Institute, Centre for Research and Technology - Hellas, Thessaloniki, Greece

Organic Rankine Cycle (ORC) systems are receiving increasing attention world-wide as they enable power generation from widely available resources of low thermal content (e.g. industrial waste heat streams, geothermal fields etc.). Their operation is based on the vaporization of a single or blended working fluid to drive a turbine, hence there are two major characteristics affecting the thermodynamic and economic cycle performance for a given heat source, namely a) the physical and chemical features of the working fluid and b) the design and operating system configuration. ORCs commonly utilize single working fluids to facilitate heat extraction, although reports [1] indicate that mixtures enable a significant increase of the ORC thermodynamic performance, while decreasing the produced electricity cost and improving the overall environmental and safety system features.

Despite the enormous importance of both single and blended working fluids for ORCs, their selection is based largely on trial-and-error approaches, applied in empirically compiled repositories containing conventional options. Such a small set is extremely limiting in view of the vast number of molecules/mixtures that could be considered as candidate ORC fluids, hampering opportunities for innovation. The authors have been the first to address such limitations through the use of a Computer-Aided Molecular Design (CAMD) method applied in the design and selection of single ORC working fluids [2, 3]. The considerable economic, operating, environmental and safety advantages of the designed fluids compared to conventional choices constitute a significant motivation for the development and implementation of a CAMD-based method for the design of ORC mixtures. The challenges associated with such an effort are due to the need to simultaneously determine the optimum mixture composition (chemical structure of all participating working fluids) and concentration (amount of each fluid in the mixture). While CAMD-based mixture design has yet to be considered in ORC research, it has only been reported for few other applications involving solvent-based separations and refrigeration, leaving significant opportunities for improvements.

This work presents a method for the simultaneous determination of the composition and concentration of binary ORC working fluid mixtures using CAMD-based optimization, implemented in two interacting stages. The first stage aims to explore and identify the highest possible economic, operating, environmental and safety performance limits of a wide set of mixtures in an ORC system. This is approached by searching for chemically feasible molecular structures only for one of the two components of a binary mixture, while emulating the mixture behaviour of the remaining component within a much wider structural design space by removing the chemical feasibility constraints. The identification of multiple optimum mixture candidates is accomplished through a multi-objective formulation of the CAMD-optimization problem, treating multiple ORC performance measures simultaneously and resulting in a comprehensive Pareto front revealing useful structural and property trade-offs among mixture components. The second stage serves to determine the optimum and chemically feasible structure of the second component for each one of the molecules already obtained in the first stage, together with the optimum mixture composition. In this stage, the mixture performance limits identified in the first stage are used as a reference point to efficiently avoid sub-optimal choices. The proposed approach avoids the need for immediate targeting of a single mixture with optimum features which often requires enormous computational effort. Instead, the effort is reduced as the user is allowed to review, interpret and analyse rich intermediate insights prior to exploiting meaningful conclusions between design stages. Optimum solutions are identified in a Pareto sense, enabling the exploitation of the often conflicting design objectives.

The merits of the proposed approach are illustrated through a case study on ORC systems. The considered design indices employed in CAMD reflect important ORC performance measures such as thermodynamic efficiency and exergy through an ORC mathematical model utilized in the course of CAMD-optimization. Important mixture properties such as flammability, toxicity and azeotropic mixture behaviour are also considered.

[1] Angelino, G., Di Paliano, P. C., Multicomponent working fluids for organic rankine cycles (ORCs), Energy 1998; 23(6): 449.

[2] Papadopoulos A.I., Stijepovic M and Linke P., On the systematic design and selection of optimal working fluids for Organic Rankine Cycles, Applied Thermal Engineering 2010, 30, 760.

[3] Stijepovic M.Z., Linke P., Papadopoulos A.I., Grujic A.S., On the role of working fluid properties in Organic Rankine Cycle performance, Applied Thermal Engineering 2012, 36, 406-413

(271d) Optimization-Based Process Synthesis Framework for Thermochemical Conversion of Natural Gas to Liquid Transportation Fuels (GTL) Systems

Josephine A. Elia, Richard C. Baliban and Christodoulos A. Floudas, Chemical and Biological Engineering, Princeton University, Princeton, NJ

A significant portion of natural gas resources in the United States is categorized as stranded natural gas since the wells are inaccessible from current pipeline infrastructure. Additionally, the associated gas production from oil operations is either vented, re-injected, or flared. There is a need to monetize these stranded gas resources to channel more fuels to the market as well as mitigate the environmental damage from gas flaring. Natural gas to liquid (GTL) systems can utilize these gas resources to add to the fuel market. A recent review of hybrid and single feedstock energy processes can be found in [1].

We investigate GTL systems through an optimization-based process synthesis approach to identify an optimal plant design for small-scale GTL systems out of a superstructure of conversion pathways [2-9]. Multiple natural gas conversion technologies are included, namely autothermal reforming, direct conversion to methanol, and direct conversion to ethylene. These processes are followed by the hydrocarbon production section via either a combination of Fischer-Tropsch units operating at multiple temperature levels using iron or cobalt catalysts, a methanol synthesis reactor to produce intermediate methanol from synthesis gas, a combination of methanol to gasoline or methanol to diesel and kerosene units, and an oligomerization unit to convert ethylene to a gasoline product. The upgrading of the fuels take place over ZSM-5 catalyst or a series of hydrocracker, hydrotreater, isomerizer, and alkylation units to produce gasoline, diesel, and kerosene. A hydrogen/oxygen production system that utilizes both carbon and non-carbon sources is included, using a combination of an air separation unit, a pressure-swing absorption unit, and an electrolyzer. The whole process is heat, power, and water integrated to minimize the utility consumption.

Case studies on the different compositions of liquid fuels produced by the GTL system are performed. A range of mixtures for gasoline, diesel, and kerosene products are imposed to elucidate the economic and environmental tradeoffs between technological routes in the superstructure, and identify the topology that will minimize the cost of fuel production.

[1] C.A. Floudas, J. A. Elia, R. C. Baliban (2012) Hybrid and Single Feedstock Energy Processes for Liquid Transportation Fuels: A Critical Review. Comp. Chem. Eng. 41:24-51.

[2] R. C. Baliban, J. A. Elia, C. A. Floudas (2010) Toward novel biomass, coal, and natural gas processes for satisfying current transportation fuel demands, 1: Process alternatives, gasification modeling, process simulation, and economic analysis. Ind. Eng. Chem. Res. 49:7343-7370.

[3] J. A. Elia, R. C. Baliban, C. A. Floudas (2010) Toward novel biomass, coal, and natural gas processes for satisfying current transportation fuel demands, 2: Simultaneous heat and power integration. Ind. Eng. Chem. Res. 49:7371-7388.

[4] R. C. Baliban, J. A. Elia, C. A. Floudas (2011) Optimization framework for the simultaneous process synthesis, heat and power integration of a thermochemical hybrid biomass, coal, and natural gas facility. Comp. Chem. Eng. 35:1647-1690.

[5] J. A. Elia, R. C. Baliban, X. Xiao, C. A. Floudas (2011) Optimal energy supply network determination and life cycle analysis for hybrid coal, biomass, and natural gas to liquid (CBGTL) plants using carbon-based hydrogen production. Comp. Chem. Eng. 35:1399-1430.

[6] R. C. Baliban, J. A. Elia, V. Weekman, C. A. Floudas (2012) Process synthesis of hybrid coal, biomass, and natural gas to liquids via Fischer-Tropsch synthesis, ZSM-5 catalytic conversion, methanol synthesis, methanol-to-gasoline, and methanol-to-olefins/distillate technologies. Submitted for publication.

[7] R. C. Baliban, J. A. Elia, C. A. Floudas (2012) Simultaneous process synthesis, heat, power, and water integration of thermochemical hybrid biomass, coal, and natural gas facilities. Comp. Chem. Eng. 37:297-327.

[8] R. C. Baliban, J. A. Elia, R. Misener, C. A. Floudas (2012) Global Optimization of a MINLP Process Synthesis Model for Thermochemical Based Conversion of Hybrid Coal, Biomass, and Natural Gas to Liquid Fuels. Comp. Chem. Eng. In press. DOI:10.1016/j.compchemeng.2012.03.008

[9] J. A. Elia, R. C. Baliban, C. A. Floudas (2012) Nationwide Supply Chain Analysis for Hybrid Feedstock Energy Processes with Significant CO2 Emissions Reduction. AIChE Journal. Submitted for publication.

(271e) An Optimization-Based Technology Assessment Framework for Biofuel Production Strategies

Sercan Murat Sen, Chemical and Biological Engineering, University of Wisconsin-Madison, Madison, WI, Jiyong Kim, University of Wisconsin, Madison, WI and Christos T. Maravelias, Department of Chemical and Biological Engineering; DOE Great Lakes Bioenergy Research Center, University of Wisconsin-Madison, Madison, WI

In the course of the past few decades, various technologies have been developed to replace fossil-based routes for the production of chemicals and fuels with sustainable and environmentally friendly biomass-based strategies. There are many alternative ways to utilize a certain biomass feedstock as a carbon source to produce a certain fuel and trade-offs between these alternative technologies make the selection of a production strategy a challenging task. Therefore, systematic efforts are required to analyze biomass-based production in terms of cost, energy efficiency, environmental impact, and productivity. Towards this aim, we propose an optimization-based framework to evaluate biomass-to-fuels strategies. We use this framework to identify limitations in existing approaches as well as identify new promising strategies.

First, we identify the major reactants and products of each technology, and by interconnecting these technologies we generate a "technology superstructure" that contains multiple strategies. Second, we gather from the literature or evaluate based on similar technologies the required technical and economic parameters: conversion coefficients, unit production cost, energy requirement, feedstock supply availability, and final product demand. Third, we develop a linear programming (LP) model to represent the underlying network structure. The model consists of: (i) material balances for all compounds, (ii) equations describing the consumption/production of compounds by technologies, (iii) production capacity constraints, and (iv) feedstock availability and/or product demand satisfaction constraints. With minor modifications in this model, we can address various types of questions (e.g., which feedstock/strategy is best for a given product, how a given feedstock can be utilized) using different types of criteria (e.g., economic, environmental). The model yields the optimal strategy (i.e., selection of technologies) for a given question and objective function. Fourth, we develop a mixed-integer programming (MIP) model to identify alternative strategies. Finally, we assess the uncertainty in the parameters of the various technologies based on their levels of maturity and complexity. Using these indicators, we assign uncertainty levels to each technology, which are subsequently used to determine a range for the unit production cost of each alternative strategy.

We illustrate the capabilities of the proposed framework with a case study. Ethanol can be produced from hardwood via hydrolysis (using dilute acid, ammonia fiber expansion, or hot water pretreatment), direct or indirect gasification, or pyrolysis. We find that, although gasification-based strategies have higher capital and operating costs, their unit production costs are lower than fermentation-based strategies mainly due to their high ethanol yields. Moreover, the byproduct (acetic acid) credit obtained in the gasification-based strategies significantly decreases the production cost (reduction of 24.6%), whereas the electricity credit in fermentation-based strategies lowers production costs by only 5.8~7.3%. Finally, sensitivity analyses reveal that the economics of gasification-based strategies can be improved primarily through system modifications (e.g. cheaper catalysts), while the cost of fermentation-based strategies can be lowered if cheaper feedstock is used since feedstock cost is the major cost driver for these strategies.

Sustainable Design of Integrated Absorption Refrigeration Systems

Luis Fernando Lira-Barragán,[a] *José María Ortega-Ponce,*[a] *Medardo Serna-González,*[a] *Mahmoud M. El-Halwagi*[b,c]

[a] *Universidad Michoacana de San Nicolás de Hidalgo, Morelia, Mich., México*

[b] *Chemical Engineering Department, Texas A&M University, College Station, TX, USA*

[c] *Adjunct Faculty at the Chemical and Materials Engineering Department, King Abdulaziz University, Jeddah, Saudi Arabia*

Introduction

Refrigeration is commonly used in many industrial processes that require cooling below room temperature. Conventional refrigeration cycles typically consume large amounts of energy and consequently yield enormous quantities of greenhouse gas emissions. Refrigeration duties may be reduced through heat integration and the use of absorption refrigeration which uses excess process heat to drive the cycle. Process integration through heat exchanger networks (HENs) has been extensively studied (see the review by Morar and Agachi, 2010). Additionally, the absorption refrigeration (AR) systems are driven by heat and are an alternative to compression refrigeration. Some studies have optimized absorption refrigeration systems without considering the integration with the process streams.

Recently, Ponce-Ortega et al. (2011) reported a methodology to address the optimization of refrigeration systems while accounting for economic and environmental objectives. Nonetheless, this methodology has not considered simultaneously the overall refrigeration system with the configuration synthesis of the heat exchange network including the transfer of heat from hot process streams to the absorption refrigeration system. This limits the applicability because without the HEN structure is not possible to determine the capital costs associated to the process. Achieving utility targets without accounting for fixed cost may not correspond to the cost-optimum solution when the associated capital cost is included after the network synthesis. Furthermore, to implement clean energies it is necessary to include social criteria in addition to the economic and environmental aspects. An important social criterion is the generation of jobs due to the use of clean energies. This approach allows the consideration of non-conventional design options such as the use of biofuels, because these involve job creation in rural areas that impact society. Therefore, to overcome the drawbacks of the previously reported methodologies, this work proposes an optimization approach for synthesizing integrated refrigeration systems considering explicitly the network configuration, accounting for the capital costs of the units, and simultaneously addressing economic, environmental, and social objectives. The approach deals with this multi-objective optimization problem to allow the decision maker to select the best solution that trades off these three objectives.

Model Formulation

The proposed approach is represented in **Figure 3**. As it can be seen, the hot process streams can transfer heat to the stripper located inside the absorption refrigeration system, which requires a minimum temperature of 80°C for its performance. After the units *AR1*, the heat integration between hot process streams and cold process streams is allowed. This process can be carried out in one or several stages to ensure any possible match between process streams. The next stage is a first cooling stage for hot process streams using cooling water (*CW*) available at 30°C. Finally, for the hot process streams that require cooling below room temperature, a set of coolers is placed at the cold-end (*AR2*) of the system to reach the target temperature of each hot process stream. For the cold process streams the heating required is provided in two stages: the first stage consists in the matching between hot and cold process streams to enhance heat recovery; then, at the hot-end of the system, the cold process streams that still have not reached their target temperatures will be processed through a set of heaters that use low pressure steam (*LPS*) as hot utility.

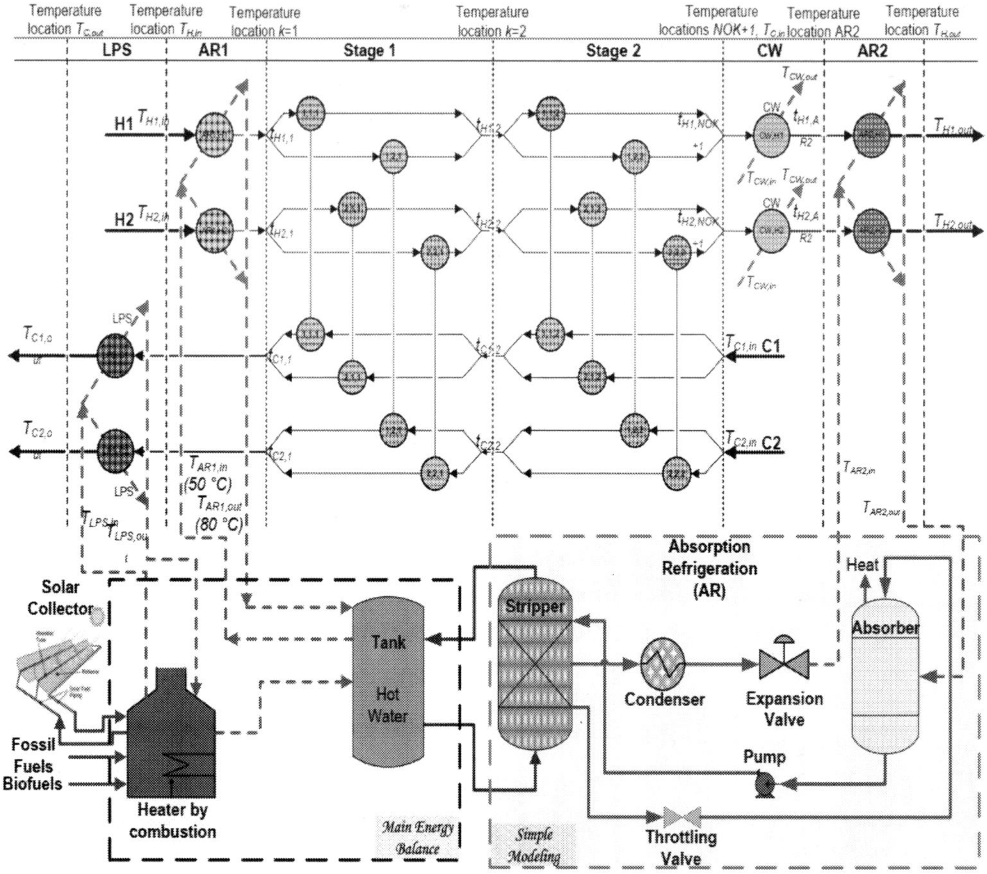

Figure 1. Proposed superstructure for integrated refrigeration system.

Model for the Heat Exchanger Network

Total energy balances for the process streams.

$$(T_{IN_i} - T_{OUT_i})FCp_i = q_i^{ar1} + \sum_{k \in ST} \sum_{j \in CPS} q_{i,j,k} + q_i^{cw} + q_i^{ar2}, \qquad i \in HPS \tag{1}$$

$$(T_{OUT_j} - T_{IN_j})FCp_j = \sum_{k \in ST} \sum_{i \in HPS} q_{i,j,k} + q_j^{LPS}, \qquad j \in CPS \tag{2}$$

Energy balance for each inner stage of the superstructure.

$$(t_{i,k} - t_{i,k+1})FCp_i = \sum_{j \in CPS} q_{i,j,k}, \quad i \in HPS, k \in ST \tag{3}$$

$$(t_{j,k} - t_{j,k+1})FCp_j = \sum_{i \in HPS} q_{i,j,k}, \quad j \in CPS, k \in ST \tag{4}$$

Energy balances for the stage of the hot process streams and the absorption refrigeration system.

$$(T_{IN_i} - t_{i,1})FCp_i = q_i^{ar1}, \ i \in HPS \tag{5}$$

$$(t_i^{ar2} - T_{OUT_i})FCp_i = q_i^{ar2}, \qquad i \in HPS \tag{6}$$

Balances for steam and cooling water utilities.

$$(t_{i,NOK+1} - t_i^{ar2})FCp_i = q_i^{cw}, \qquad i \in HPS \tag{7}$$

$$(T_{OUT_j} - t_{j,1})FCp_j = q_j^{LPS}, \qquad j \in HPS \tag{8}$$

Assignment of the temperatures for the extreme borders of the superstructure.

$$T_{IN_j} = t_{j,NOK+1}, \quad j \in CPS \tag{9}$$

Constraints for the feasibility of the temperatures in the superstructure.

$$T_{IN_i} \geq t_{i,1}, \quad i \in HPST_{IN_i} \geq t_{i,1}, \quad i \in HPS \tag{10}$$

$$t_{i,k} \geq t_{i,k+1}, \quad i \in HPS, k \in ST \tag{11}$$

$$t_{i,NOK+1} \geq t_i^{ar2}, \quad i \in HPS \tag{12}$$

$$t_i^{ar2} \geq T_{OUT_i}, \quad i \in HPS \tag{13}$$

$$T_{OUT_j} \geq t_{j,1}, \quad j \in CPS \tag{14}$$

$$t_{j,k} \geq t_{j,k+1}, \quad j \in CPS, k \in ST \tag{15}$$

Existence for the heat exchanger units.

$$q_{i,j,k} - Q_{i,j}^{\max} z_{i,j,k} \leq 0, \quad i \in HPS, j \in CPS, k \in ST \tag{16}$$

$$q_i^{ar1} - Q_i^{\max} z_i^{ar1} \leq 0, \quad i \in HPS \tag{17}$$

$$q_i^{cw} - Q_i^{\max} z_i^{cw} \leq 0, \quad i \in HPS \tag{18}$$

$$q_i^{ar2} - Q_i^{\max} z_i^{ar2} \leq 0, \quad i \in HPS \tag{19}$$

$$q_j^{LPS} - Q_j^{\max} z_j^{LPS} \leq 0, \quad j \in CPS \tag{20}$$

Feasibilities for the temperature differences.

$$dt_{i,j,k} \leq t_{i,k} - t_{j,k} + \Delta T_{i,j}^{\max}(1 - z_{i,j,k}), \quad i \in HPS, j \in CPS, k \in ST \tag{21}$$

$$dt_{i,j,k+1} \leq t_{i,k+1} - t_{j,k+1} + \Delta T_{i,j}^{\max}(1 - z_{i,j,k}), \quad i \in HPS, j \in CPS, k \in ST \tag{22}$$

$$dt_i^{ar1-2} \leq t_{i,1} - T_{IN}^{ar1} + \Delta T_i^{ar1\max}(1 - z_i^{ar1}), \quad i \in HPS \tag{23}$$

$$dt_i^{cw1} \leq t_{i,NOK+1} - T_{OUT}^{cw} + \Delta T_i^{cw\max}(1 - z_i^{cw}), \quad i \in HPS \tag{24}$$

$$dt_i^{cw2} \leq t_i^{ar2} - T_{IN}^{cw} + \Delta T_i^{cw\max}(1 - z_i^{cw}), \quad i \in HPS \tag{25}$$

$$dt_i^{ar2-1} \leq t_i^{ar2} - T_{OUT}^{ar2} + \Delta T_i^{ar2\max}(1 - z_i^{ar2}), \quad i \in HPS \tag{26}$$

$$dt_j^{LPS-2} \leq T_{OUT}^{LPS} - t_{j,1} + \Delta T_j^{LPS\max}(1 - z_j^{LPS}), \quad j \in CPS \tag{27}$$

$$\Delta T_{\min} \leq dt_{i,j,k}, \quad i \in HPS, j \in CPS, k \in ST \tag{28}$$

$$\Delta T_{\min} \leq dt_i^{ar1-2}, \quad i \in HPS \tag{29}$$

$$\Delta T_{\min} \leq dt_i^{cw1}, \quad i \in HPS \tag{30}$$

$$\Delta T_{\min} \leq dt_i^{cw2}, \quad i \in HPS \tag{31}$$

$$\Delta T_{\min} \leq dt_i^{ar2-1}, \quad i \in HPS \tag{32}$$

$$\Delta T_{\min} \leq dt_j^{LPS-2}, \quad j \in CPS \tag{33}$$

Absorption Refrigeration System.

Energy balance for the absorption refrigeration system.

$$\frac{\sum_{j \in CPS} q_j^{LPS}}{COP^{LPS}} + \frac{\sum_{i \in HPS} q_i^{ar2}}{COP^{ar2}} = \sum_{i \in HPS} q_i^{ar1} + \sum_{f \in F} Q_{f,t}^{Fossil} + \sum_{b \in B} Q_{b,t}^{Biofuel} + Q_t^{Solar}, \quad t \in T \tag{34}$$

Model for the solar collector.

$$y^{Solar} = \sum_{s \in S} z_s^{Solar} \tag{35}$$

$$z_s^{Solar} = \sum_{r \in R} w_{s,r}^{Solar}, \quad \forall s \in S \tag{36}$$

$$Q_t^{Solar} = \sum_{s \in S} q_{t,s}^{Solar}, \quad \forall t \in T \tag{37}$$

$$A_c^{Solar} = \sum_{s \in S} a_{c_s}^{Solar} \tag{38}$$

$$a_{c_s}^{Solar} = \sum_{r \in R} a_{c_{s,r}}^{dis}, \ \forall s \in S \tag{39}$$

$$C_{op}^{Solar} = \sum_{s \in S} c_{op_s}^{Solar} \tag{40}$$

$$C_{cap}^{Solar} = \sum_{s \in S} c_{cap_s}^{Solar} \tag{41}$$

$$c_{cap_s}^{Solar} = \sum_{r \in R} c_{cap_{s,r}}^{dis}, \ \forall s \in S \tag{42}$$

$$C^{Tank} = FC^{Tank} y^{Solar} + VC^{Tank} \left(V^{Tank}\right) \tag{43}$$

$$q_{t,s}^{Solar} \leq Q_{s,t}^{Useful\_Solar} a_{c_s}^{Solar} \frac{1}{D_t}, \quad \forall t \in T, \forall s \in S \tag{44}$$

$$c_{op_s}^{Solar} = Cu_s^{Solar} \sum_{t \in T} \left(q_{t,s}^{Solar} D_t\right), \ \ \forall s \in S \tag{45}$$

$$A_{s,r} w_{s,r}^{Solar} \leq a_{c_{s,r}}^{dis} \leq A_{s,r} w_{s,r}^{Solar}, \ \forall r \in R, \ \forall s \in S \tag{46}$$

$$c_{cap_{s,r}}^{dis} = FC_{s,r}^{Solar} w_{s,r}^{Solar} + VC_{s,r}^{Solar} \left(a_{c_{s,r}}^{dis}\right), \ \forall r \in R, \ \forall s \in S \tag{47}$$

$$q_{t,s}^{Solar} \leq Q_{Max}^{Solar} z_s^{Solar}, \quad \forall t \in T, \forall s \in T \tag{48}$$

$$a_{c_s}^{Solar} \leq A_{c_s}^{Solar\_Max} z_s^{Solar}, \quad \forall s \in S \tag{49}$$

$$c_{op_s}^{Solar} \leq C_{op_s}^{Max} z_s^{Solar}, \ \ \forall s \in S \tag{50}$$

$$c_{cap_s}^{Solar} \leq C_{cap_s}^{Max} z_s^{Solar}, \ \forall s \in S \tag{51}$$

$$c_{cap_{s,r}}^{dis} \leq C_{cap_s}^{Max} w_{s,r}^{Solar}, \ \forall s \in S, \forall r \in R \tag{52}$$

Maximum Availability for the Fuels.

$$Q_{b,t}^{Biofuel} \leq \frac{Heating_b^{Power} \ Avail_{b,t}^{Max}}{D_t}, \quad \forall b \in B, \ \forall t \in T \tag{53}$$

$$Q_{f,t}^{Fossil} \leq \frac{Heating_f^{Power} \ Avail_{f,t}^{Max}}{D_t}, \quad \forall f \in F, \ \forall t \in T \tag{54}$$

Objective Functions

Economic Objective Function

The economic objective function consists in minimizing the total annual cost, which is composed by the heat exchanger network cost and the absorption refrigeration system cost, in addition to the positive gains obtained for the tax credits due to the reduction in the greenhouse gas emissions achieved for the use of solar collectors, biofuels and some fossil fuels.

$$
\begin{aligned}
Min\ TAC = & \sum_{i\in HPS} C_{cw} q_i^{cw} + k_f \left[C_{CAP}^{Solar} + C^{Tank} \right] \\
& + k_f \Bigg[\sum_{i\in HPS}\sum_{j\in CPS}\sum_{k\in ST} C_{F_{i,j}}^{exc} z_{i,j,k} + \sum_{i\in HPS} C_{F_i}^{ar1} z_i^{ar1} + \sum_{i\in HPS} C_{F_i}^{cw} z_i^{cw} + \sum_{i\in HPS} C_{F_i}^{ar2} z_i^{ar2} + \sum_{j\in CPS} C_{F_j}^{LPS} z_j^{LPS} \\
& + \sum_{i\in HPS}\sum_{j\in CPS}\sum_{k\in ST} C_{i,j}^{exc} \left\{ \frac{ q_{i,j,k}\left(\frac{1}{h_i}+\frac{1}{h_j}\right) }{ \left[(dt_{i,j,k})(dt_{i,j,k+1})\left(\frac{dt_{i,j,k}+dt_{i,j,k+1}}{2}\right) \right]^{1/3} + \delta } \right\}^{\beta} \\
& + \sum_{i\in HPS} C_i^{ar1} \left\{ \frac{ q_i^{ar1}\left(\frac{1}{h_i}+\frac{1}{h_{ar1}}\right) }{ \left[\left(T_{IN_i} - T_{OUT}^{ar1}\right)\left(dt_i^{ar1-2}\right)\left(\frac{\left(T_{IN_i}-T_{OUT}^{ar1}\right)+dt_i^{ar1-2}}{2}\right) \right]^{1/3} + \delta } \right\}^{\beta} + \sum_{i\in HPS} C_i^{cw} \left\{ \frac{ q_i^{cw}\left(\frac{1}{h_i}+\frac{1}{h_{cw}}\right) }{ \left[\left(dt_i^{cw1}\right)\left(dt_i^{cw2}\right)\left(\frac{dt_i^{cw1}+dt_i^{cw2}}{2}\right) \right]^{1/3} + \delta } \right\}^{\beta} \\
& + \sum_{i\in HPS} C_i^{ar2} \left\{ \frac{ q_i^{ar2}\left(\frac{1}{h_i}+\frac{1}{h_{ar2}}\right) }{ \left[\left(dt_i^{ar2-1}\right)\left(T_{OUT_i} - T_{IN}^{ar2}\right)\left(\frac{dt_i^{ar2-1}+\left(T_{OUT_i}-T_{IN}^{ar2}\right)}{2}\right) \right]^{1/3} + \delta } \right\}^{\beta} + \sum_{j\in CPS} C_j^{LPS} \left\{ \frac{ q_j^{LPS}\left(\frac{1}{h_j}+\frac{1}{h_{LPS}}\right) }{ \left[\left(T_{IN}^{LPS} - T_{OUT_j}\right)\left(dt_j^{LPS-2}\right)\left(\frac{\left(T_{IN}^{LPS}-T_{OUT_j}\right)+dt_j^{LPS-2}}{2}\right) \right]^{1/3} + \delta } \right\}^{\beta} \Bigg] \\
& + H_Y \left\{ \sum_{f\in F}\left[C_f^{Fossil} \sum_{t\in T}\left(Q_{f,t}^{Fossil} D_t\right)\right] + \sum_{b\in B}\left[C_b^{Biofuel} \sum_{t\in T}\left(Q_{b,t}^{Biofuel} D_t\right)\right] + C_{op}^{Solar} - R^{Solar}\left[\sum_{t\in T}\left(Q_t^{Solar} D_t\right)\right] - \sum_{t\in T}\sum_{b\in B}\left[R_b^{Biofuel} Q_{b,t}^{Biofuel} D_t\right] - \sum_{t\in T}\sum_{f\in F}\left[R_f^{Fossil} Q_{f,t}^{Fossil} D_t\right] \right\}
\end{aligned}
\quad (55)
$$

Environmental Objective Function

The environmental objective function is an indirect environmental impact assessment through the overall quantification of the greenhouse gas emissions (*GHGE*).

$$Min\ NGHGE^{Overall} = \sum_{t\in T}\sum_{f\in B}\left[GHGE_f^{Fossil} Q_{f,t}^{Fossil} D_t\right] + \sum_{t\in T}\sum_{b\in B}\left[GHGE_b^{Biofuel} Q_{b,t}^{Biofuel} D_t\right] \quad (56)$$

Social Objective Function

This work considers the jobs generated for the energy sources. In this way, it is considered the social impact associated to the generation of energy sources and, therefore, one of the most important issues to implement the use of clean energies in industrial processes.

$$Max\ NJOBS^{Overall} = \sum_{t\in T}\sum_{f\in B}\left[NJOB_f^{Fossil} Q_{f,t}^{Fossil} D_t\right] + \sum_{t\in T}\sum_{b\in B}\left[NJOB_b^{Biofuel} Q_{b,t}^{Biofuel} D_t\right] + \sum_{t\in T}\left[NJOB^{Solar} Q_t^{Solar} D_t\right] \quad (57)$$

Results and Discussion

A pharmaceutical plant that has three hot process streams and two cold process streams to carry out the energy integration is considered in this example problem. **Table 1** shows the data for the process streams for this case. The minimum temperature difference for this example problem is 10°C. For this case, the low pressure steam is available at 255°C and the refrigerant of the cooling system has a temperature of -5°C. This example has been previously addressed by Ponce-Ortega et al. (2011); El-Halwagi (1996) and Tora et al. (2010).

Table1. Streams data.

Stream	Inlet temperature [°C]	Outlet temperature [°C]	*FCp* [kW/°C]
H1	125	65	38.75
H2	85	25	18.75
H3	45	25	43.75
C1	35	240	20
C2	35	55	36.25

Two solar collectors are considered (Parabolic trough solar collector –PTSC and Evacuated tube solar collector –ETSC). **Table 2** shows the useful collected energy per month per unit area of solar collector for PTSC and ETSC. The fuels and biofuels taken into account are presented in **Table 3** as well as the heating power, the overall *GHGE*, the cost and the number of jobs created by each fuel considered.

Table 2. Useful collected energy per month for different solar collectors.

Month/Type of Solar Collector	PTSC [kJ/(m^2 month)]	ETSC [kJ/(m^2 month)]
January	409,293	245,576
February	443,016	265,810
March	577,530	346,518
April	571,860	343,116
May	555,768	333,461
June	454,410	272,646
July	443,610	266,166
August	439,425	263,655
September	394,470	236,682
October	410,967	246,580
November	407,430	244,458
December	522,288	235,030

Table 3. Data for the fossil and biofuels considered in the examples presented.

#	Fuel	Heating power [kJ/kg]	Overall *GHGE* [ton CO_2 eq/kJ]	Cost [$/mm kJ]	Generation of Jobs [Jobs/kJ]
			Fossil Fuels		
1.	Coal	35,000	2.21357×10^{-7}	1.5559	1.06281×10^{-11}
2.	Oil	45,200	8.05408×10^{-8}	18.2447	1.81677×10^{-11}
3.	Natural Gas	54,000	7.90892×10^{-8}	5.8349	5.25431×10^{-11}
			Biofuels		
1.	Biomass	17,200	2.44307×10^{-8}	2.0303	6.6964×10^{-8}
2.	Biogas	52,000	2.68216×10^{-8}	8.5388	5.25431×10^{-7}
3.	Softwood	20,400	3.3482×10^{-8}	2.5332	1.46691×10^{-8}
4.	Hardwood	18,400	3.3482×10^{-8}	2.8975	5.43641×10^{-8}
5.	Biodiesel	40,200	5.13283×10^{-8}	31.3092	2.46582×10^{-6}
6.	Bioethanol	29,600	5.8436×10^{-8}	14.4212	2.87453×10^{-6}

The proposed methodology was applied with a tax credit of \$5/ton $CO_{2\ eq.}$ obtaining the results shown in **Figure 2**. As it can be seen, the results are represented through the Pareto curve obtained comparing the *TAC* and the overall *GHGE*, evaluating the number of jobs for each point. In the Pareto curve, points A, B, C and D can be identified, where point A represents the optimal solution for the lowest overall *GHGE*, being zero for this concrete solution (achieved due to the total required energy is supplied by a solar collector). However, at the same time, this solution A represents the worst economic solution with the highest cost (\$2.13x10^6/year) and the number of jobs is the lowest respect to all the illustrated solutions. Otherwise, point B is the best economic solution with a *TAC* of \$313,459/year and the worst environmental solution with a value of overall *GHGE* of 24,285 tons $CO_{2eq.}$/year and the corresponding number of generated jobs is 724, which is not the best solution in this aspect.

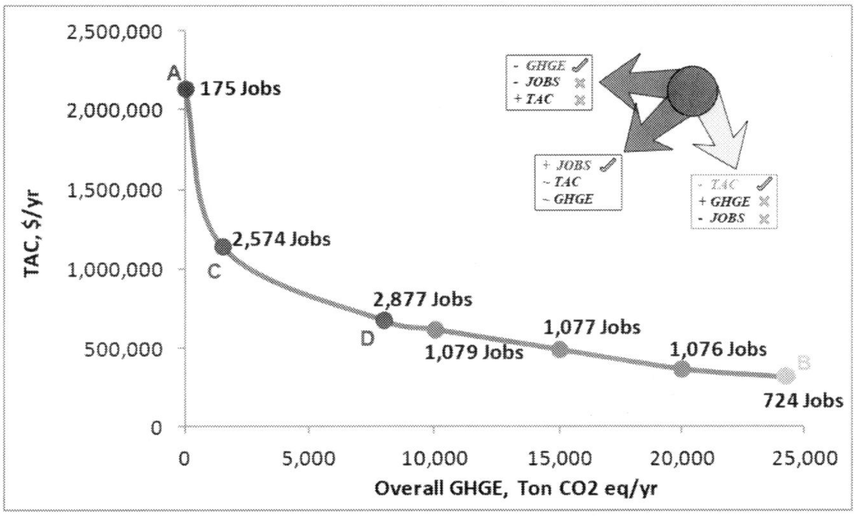

Figure 2. Pareto curve for a tax credit of \$5/ton $CO_{2\ eq}$.

Table 4 presents the detailed costs for solutions A, B, C, and D, as well as their characteristics for the three objective functions (*TAC*, *GHGE* and *NJOBS*) to make a comparison of the different scenarios, where solution A (minimum *GHGE*), solution B (minimum *TAC*), and other two feasible solutions are included.

Table 4. Identified solutions for a tax credit of \$5/ton $CO_{2\ eq}$.

Concept/Case	Point A (minimum *GHGE*)	Point B (minimum *TAC*)	Point C (environmental preference)	Point D (economic preference)
Heat exchanger network cost	\$112,543.87/yr	\$131,706.23/yr	\$142,267.05/yr	\$142,267.05/yr
Fossil fuel cost	0	\$168,568.79/yr	\$66,033.82/yr	\$537,117.94/yr
Biofuel cost	0	\$24,819.12/yr	\$67,907.02/yr	\$80,541.86/yr
Solar collector cost	\$2,215,342.34/yr	0	\$981,243.32/yr	0
Credit *GHGE* reduction	\$194,515.24/yr	\$11,635.03/yr	\$122,063.11/yr	\$89,563.11/yr
TAC	\$2,133,370.97/yr	\$313,459.13/yr	\$1,135,388.10/yr	\$670,363.74/yr
Overall *GHGE*	0	24,285 ton CO2 equiv/yr	1,500 ton CO2 equiv/yr	8,000 ton CO2 equiv/yr
Number of jobs	175	724	2,574	2,877

Figure 3 illustrates the configuration obtained for the point C, in which there are two heat exchanger units between process streams and cooling water is not required. The energy necessary to run the absorption refrigeration system (AR) is supplied by the solar collector, fossil fuels and biofuels. **Figure 4** shows the type of energy source required per month for solution C. Then, solution C considers a solar collector of the type PTSC with an area of 15,321 m^2, which almost satisfies the heat requirements in April.

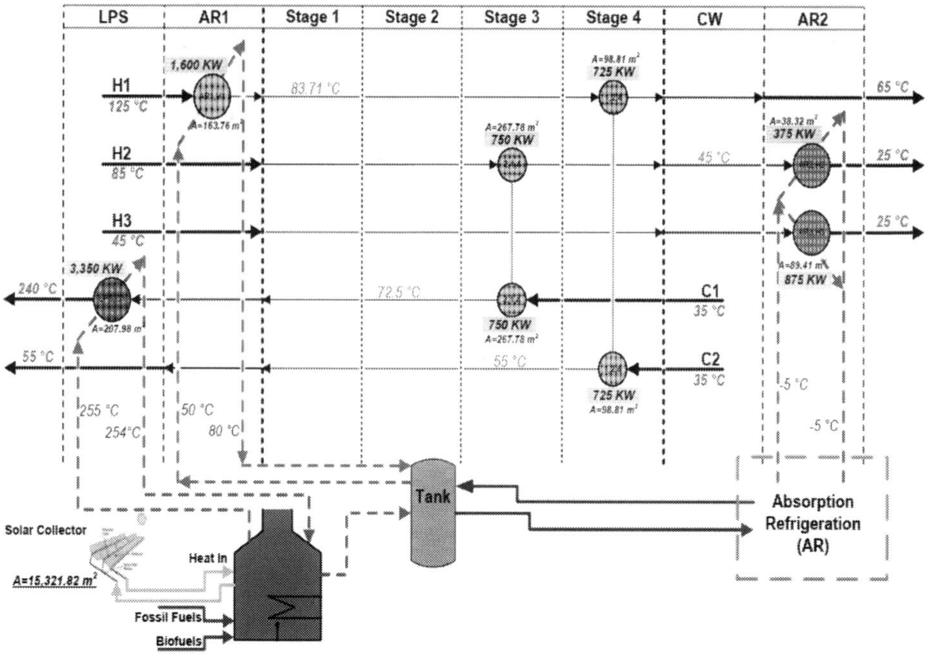

Figure 3. Configuration of the solution C of the example.

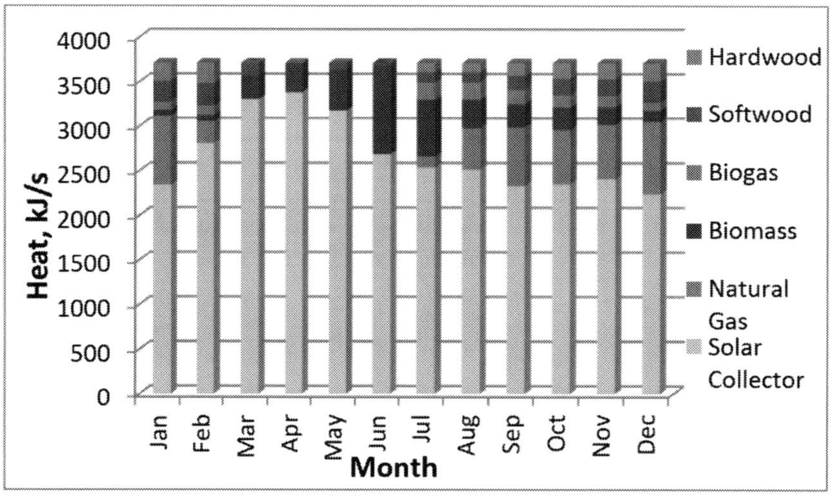

Figure 4. Energy sources required for each month for solution C of the example.

Conclusions

This work has presented a new mathematical programming model for the optimal integration of the process streams with absorption refrigeration systems considering simultaneously the minimization of the total annual cost, the minimization of the overall greenhouse gas emissions, and the maximization of the number of jobs that can be created for the use of the different types of energy sources. The heat-integration configuration is explicitly optimized. This enables the incorporation of the fixed cost along with the operating cost in the economic function. The approach also determines the optimal usage and distribution of external energy sources (solar, fossil, and biofuels). Pareto solutions are produced to enable tradeoffs among the economic, environmental, and social objectives. Two scenarios can be identified in Pareto curves: one corresponding to low emissions but high costs with a low number of jobs (using mostly solar energy) and one corresponding to low costs but high emissions and moderate job creation (using mainly fossil fuels). An intermediate scenario was also observed where the solutions correspond to a high number of jobs while the emissions and costs are low (consuming mainly biofuels).

References

1. Morar, M. and Agachi, P.S. (2010) Review: "Important contributions in development and improvement of the heat integration techniques," *Computers and Chemical Engineering* 34 (8) 1171-1179.
2. Ponce-Ortega, J.M.; Tora, E.A.; González-Campos, J.B.; and El-Halwagi, M.M. (2011) "Integration of renewable energy with industrial absorption refrigeration systems: systematic design and operation with technical, economic and environmental objectives," *Industrial and Engineering Chemistry Research* 50 (16) 9667-9684.
3. El-Halwagi, M.M. (1996) "Pollution prevention through process integration: Systematic design tools," Academic Press: San Diego, CA.
4. Tora, E.A. and El-Halwagi, M.M. (2010) "Integration of solar energy into absorption refrigerators and industrial process," *Chemical Engineering Technologies* 33 (9) 1495-1505.

(293a) Optimal Temperature Control Using Latent Energy Storage

Siyun Wang, Chemical Engineering, University of Texas at Austin, Austin, TX and Michael Baldea, University of Texas

Optimal Temperature Control Using Latent Energy Storage

Siyun Wang and Michael Baldea

Department of Chemical Engineering

The University of Texas at Austin, 1 University Station C0400, Austin, TX 78712

email: mbaldea@che.utexas.edu

An ever increasing number and broader range of mobile devices in industrial and consumer applications (e.g., wireless sensors, cellular phones, mobile computers, electric vehicles), depend on battery power to fulfill their function. Their utility depends strongly on the useful battery life - on the amount of time the device can operate until the battery must be recharged. Useful battery life can, evidently, be increased by increasing the stored energy (that is, by increasing battery capacity), which, at present, can entail a significant cost and carry an unacceptable weight penalty (Armand & Tarascon, 2008). The alternative lies in improving the way the stored energy is used, i.e., improving system energy management.

The operation of mobile devices is typically intermittent, ranging from a low-energy "standby" mode, to a high-intensity operation when the system operates at full capacity. The latter is often associated with a high rate of heat generation (Chen, Wu, & Hwang, 2008) (for example, heat output from central/graphics processors in a portable computer increases dramatically under full load; likewise, batteries installed in electric or hybrid vehicles require significant cooling under hard acceleration or during charging). High intensity operation is typically accompanied by the activation of cooling devices, typically fans, to maintain the device temperature within an acceptable range. In turn, active cooling reduces the battery charge and the total operation time of the mobile device.

In this work, we present a novel approach for temperature control and diminishing active cooling needs via latent energy storage. We consider an energy storage medium located in direct contact with the heat source of the system (e.g., processor, battery), which acts as a buffer that absorbs –at constant temperature – the heat generated during high-intensity operation. The heat is subsequently dissipated to the environment through a combination of natural and forced convection. Using a simple linear analysis, we demonstrate that, in the general case, this concept requires considerably less energy to maintain the temperature of the system at a desired value than the traditional approach based exclusively on forced-convective cooling.

Subsequently, we focus on the use of Phase Change Materials (PCMs) as thermal energy storage media. PCMs present the advantage of a relatively high latent heat and thus an increased storage density (Sharma, Tyagi, Chen, & Buddhi, 2009). In order to increase specific area (and, consequently, to decrease the time required for the PCM to undergo melting/solidification cycles), we consider a particular class of materials, where the PCM is encapsulated in a thermally conductive matrix. We develop a mathematical model capturing the melt dynamics of matrix PCM materials in the most general case, assuming that the PCM particles are polydisperse. Subsequently, we construct a model that describes the behavior of the ensemble consisting of the heat source, the temperature controlled system, the PCM thermal buffer, and a cooling fan, and employ this model to develop an optimal sizing method for the thermal buffer. To this end, we formulate a novel stochastic optimization method, based on a dynamic optimization where disturbances (e.g., heat generation patterns) are described as multi-level random signals.

Finally, we provide a numerical case study concerning the control of the CPU temperature in a portable computer. We show that the proposed latent energy storage temperature control approach yields significant energy savings by reducing fan use. Further, we demonstrate that the optimal size of the required storage buffer (using commercially available PCMs) is sufficiently small to encourage the implementation of this approach in portable devices. Reference

Armand, M., & Tarascon, J. (2008). Building better batteries. *Nature*, pp. 652-657.

Chen, C.-T., Wu, C.-K., & Hwang, C. (2008). Optimial Design and Control of CPU Heat Sink Processes. *IEEE Transactions on Components and Packaging Technologies*, pp. 184-195.

Sharma, A., Tyagi, V., Chen, C., & Buddhi, D. (2009). Review on thermal energy storage with phase change materials and applications. *Reneewable and Sustainable Energy Reviews*, 318 - 345.

(293b) Advanced Regulatory Control and Coordinated Plant-Wide Control Strategies for IGCC Targeted towards Improving Power Ramp-Rates

Priyadarshi Mahapatra and Stephen E. Zitney, AVESTAR Center, National Energy Technology Laboratory, Morgantown, WV

As part of ongoing R&D activities at the National Energy Technology Laboratory's (NETL) Advanced Virtual Energy Simulation Training & Research (AVESTAR™) Center[1,2], this paper highlights strategies for enhancing low-level regulatory control and system-wide coordinated control strategies implemented in a high-fidelity dynamic simulator for an Integrated Gasification Combined Cycle (IGCC) power plant with carbon capture. The underlying IGCC plant dynamic model contains 20 major process areas, each of which is tightly integrated with the rest of the power plant, making individual functionally-independent processes prone to routine disturbances. Single-loop feedback control although adequate to meet the primary control objective for most processes, does not take into account in advance the effect of these disturbances, making the entire power plant undergo large offshoots and/or oscillations before the feedback action has an opportunity to impact control performance. In this paper, controller enhancements ranging from retuning feedback control loops, multiplicative feed-forward control and other control techniques such as split-range control, feedback trim and dynamic compensation, applicable on various subsections of the integrated IGCC plant, have been highlighted and improvements in control responses have been given. Compared to using classical feedback-based control structure, the enhanced IGCC regulatory control architecture reduces plant settling time and peak offshoots, achieves faster disturbance rejection, and promotes higher power ramp-rates. In addition, improvements in IGCC coordinated plant-wide control strategies for "Gasifier-Lead", "GT-Lead" and "Plantwide" operation modes have been proposed and their responses compared. The paper is concluded with a brief discussion on the potential IGCC controller improvements resulting from using advanced process control, including model predictive control (MPC), as a supervisory control layer.

References:

1. http://www.netl.doe.gov/avestar/
2. Zitney, S.E., E.A. Liese, P. Mahapatra, R. Turton, D. Bhattacharyya, and G. Provost "Advanced Virtual Energy Simulation, Training, and Research: IGCC with CO2 Capture Power Plant," *Proc. of the 28th Annual International Pittsburgh Coal Conference*, Pittsburgh, PA, September 12-15 (2011)

(293c) Offset-Free Model Predictive Control of a Heat Pump

Matt Wallace[1], Prashant Mhaskar[1], John House[2] and Tim Salsbury[2], (1)Department of Chemical Engineering, McMaster University, Hamilton, ON, Canada, (2)Johnson Controls, Milwaukee, WI

Building comfort control is a complex task, consisting of not only conditioning but ventilation and transportation of a medium to/from regions requiring comfort maintenance. Such a task, is completed by various mechanical and electrical equipment usually interacting in a hierarchical manner, all in an attempt, which in certain cases, comes with an unknown degree of certainty, to simultaneously meet comfort and safety objectives. Improving such a process can be too vast when considering all interacting components at once. As a result, this work focuses on removing the uncertainty in the operation and control of the lowest level component within the hierarchical operation, which we consider as the heating, ventilating and air-conditioning (HVAC) unit.

One common HVAC unit commonly used to condition (by supplying both heating and cooling) building air is that of a heat pump. A heat pump is identical component-wise to that of a vapor compression cycle (VCC), with only one additional component being a reversing valve which allows for the refrigerant flow within the cycle to be reversed. Because of this additional valve, the heat pump can provide conditioning at any time in the year due to it's bimodal dynamics (i.e. reversing dynamics can be considered a discrete effect). As a result of the heat pump being operational in both summer and winter months, it, and it's respective control strategy, must be robust to fluctuations common in both seasonal periods, while also ensuring it's able to handle the usual fluctuation in demand common to the building occupants. The former requirements are taken care of during the system/component design and outside the scope of this work as our focus is on designing a control strategy which can account for such fluctuations while doing so in the most energy efficient manner. Our previous work has considered the problem of improving the local control approach in one mode of the system (i.e. cooling), however, by considering the bimodal nature of the heat pump, an opportunity arises to present the potential energy savings and tracking improvements associated with both it's modes of operation (i.e. cooling in summer months; heating in winter months).

Current practice fails to take into account interactions within a system and also handles constraints in a way that is separate and often damaging to control performance. Energy use is also not usually featured as an objective in the control performance and is handled instead by setting set-points in an ad-hoc way based on experience. Model predictive control (MPC) is a control methodology that has reached a certain level of maturity in several industries but has not yet been applied to buildings at any substantial scale. MPC is designed to handle many of the shortfalls of existing control mentioned earlier including interaction between variables, constraints, and the incorporation of higher level objectives such as energy use. The work reported here involves applying MPC methods to a heat pump system used as a roof-top conditioning unit. An MPC strategy is developed and tested for manipulating the electronic expansion valve and also the speed of a variable speed compressor in both winter and summer conditions. During both seasonal periods, the control objective is to regulate the air temperature being blown over the evaporator subject to constraints on the superheat. An optimization objective is formulated to minimize the energy used by the compressor while meeting control objectives and constraints. A linear model is used for the air and superheat temperature predictions with different models being active depending on the mode of operation which the heat pump is operating at. The linear MPC framework incorporates an offset-free mechanism in the form of an augmented model and state estimator in order for the controller to achieve offset-free tracking. This offset-free mechanism is able to account for variations in the air and load conditions outside the heat pump. A detailed simulation model was used to evaluate and compare the performance of the MPC strategy with that of a traditional control strategy. Results will be presented showing both improved temperature regulation and energy savings when regulating the heat pump using the MPC strategy.

(293d) Model Predictive Control of an Entrained-Flow Gasifier for an IGCC Power Plant

Hyojin Lee and Jay H. Lee, Chemical and Biomolecular Engineering, Korea Advanced Institute of Science and Technology, Daejeon, South Korea

Integrated Gasification Combined Cycle (IGCC) is a promising technology that can produce electricity from coal in an environmentally friendly way. Korea is a resource-poor country importing 97% of its energy resources; 42% of the electricity is generated from coal. Therefore, it is important to build up an optimal mix of energy sources by introducing new thermal power generation systems based on coal and other fossil fuels. IGCC is considered to be an excellent choice for Korea due to its environment-friendliness and high efficiency. In Korea, a 300MW coal-based IGCC power plant is currently being constructed and it is expected to generate electricity from year 2017. Two more plants are in the planning stage. The expected thermal efficiency of the demonstration plant being constructed is 42% (HHV, net) and the targeted availability level is 85%.

Throughout the world, 17 IGCC power plants are currently in operation. Among them, 5 plants use coal as the main feedstock (Gasification Data Base 2010, NETL). The coal-based IGCC power plants show lower availability figures than the plants using other types of feedstock. Moreover, it is reported that a newly commissioned IGCC plant suffers from low availability for a number of years before achieving a normal level of availability. The forced outages by equipment failure are caused mainly by the gas turbine and the gasifier. The coal gasifier is perhaps the most important unit not only because it is prone to failures but also because its dynamic responses propagate through the entire plant and impact the performances of other units like the gas turbine. Therefore it is necessary to develop good understandings of the gasifier's dynamic behavior in order to improve the whole plant's availability.

In this study, a detailed dynamic simulation model of an entrained-flow gasifier has been developed and implemented on the commercial software of g-PROMS and MATLAB. We test the MPC algorithm on the simulated plant in order to develop effective control strategies to fight against various disturbances and load changes that can occur during a typical operation. The Shell gasifier is chosen to be modeled as a typical entrained-flow gasifier.

• Reference

[1] NETL, "Gasification DataBase," 2010.

[2] J. Phillips "CoalFleet RD&D Augmentation Plan for IGCC power plants", EPRI report, 2007

[3] Bo Sun , Yongwen Liu, Xi Chen, Qulan Zhou and Ming Su, "Dynamic modeling and simulation of shell gasifier in IGCC", Fuel Processing Technolcogy, vol.92, Iss.8, pp. 1418-1425, Aug. 2011

[4] Y.T. Kim, "Overview of IGCC Demonstration Project in Korea", coal gasification symposium at KIChE, 2012

• Acknowledgement

1) This research was supported by Basic Science Research Program through the National Research Foundation of Korea(NRF) funded by the Ministry of Education, Science and Technology(2011-0006839)

2) This treatise was supported by the project of Global Ph.D. Fellowship which National Research Foundation of Korea conducts from 2011.

(293e) Nonlinear Control of a Water Gas Shift Membrane Reactor for Hydrogen Production

Dimitrios Georgis[1], Fernando V. Lima[2], Ali Almansoori[3] and Prodromos Daoutidis[1], (1)Chemical Engineering and Materials Science, University of Minnesota, Minneapolis, MN, (2)Department of Chemical Engineering and Materials Science, University of Minnesota, Minneapolis, MN, (3)Department of Chemical Engineering, The Petroleum Institute, Abu Dhabi, United Arab Emirates

Hydrogen is a promising solution for carbon free energy production in fuel cells and gas turbines. The majority of the worldwide hydrogen production is based on fossil fuels (e.g., coal, natural gas and oil) while, a small but increasing percentage of hydrogen production is based on renewable energy resources (e.g., biomass). Light and heavy hydrocarbons generate hydrogen through steam reforming, while coal- or biomass-derived syngas (a mixture of mainly carbon monoxide and hydrogen) produces hydrogen through the water gas shift (WGS) reaction.

The need for high-purity hydrogen production via WGS typically requires a packed-bed reactor operating at high conversions followed by a separation unit for hydrogen purification. However, the merging of the reaction and separation processes into one unit, also known as process intensification, increases process efficiency and reduces costs, despite the higher process complexity. This can be achieved by using membrane reactors to replace both unit operations.

In addition to the advantage of process intensification, membrane reactors enable the continuous removal of hydrogen from the reaction zone through the membrane, leading to higher conversions when compared to conventional packed-bed reactors under the same operating conditions. However, higher conversions (which correspond to more released heat for exothermic reactions) may cause steeper temperature gradients (hot spots) along the membrane reactors than in conventional reactors. Therefore, a proper control strategy should be developed to ensure a safe operation at a predefined temperature range and preclude catalyst deactivation during the transition between different operating points and under the effect of unexpected disturbances.

In this study, a WGS membrane reactor is considered consisting of three different compartments: the reaction zone (tube), where the moderately exothermic WGS reaction takes place, the permeation zone (shell), where the produced hydrogen is collected and the cooling zone (inner tube). A distributed model for the WGS membrane reactor is developed using a hydrogen-selective (zeolite-based) membrane and a high-temperature catalyst (Fe/Cr – based). This membrane reactor achieved higher carbon monoxide conversions than the ones achieved by conventional packed-bed reactors. A steady-state analysis under different flow configurations revealed the presence of hot spots along the reactor. Open-loop simulations demonstrated the necessity for a control strategy not only to suppress the hot spot but also to maintain the operation of the reactor under a specified temperature. A nonlinear controller is derived on the basis of the PDE model of the process which is a distributed analogue of nonlinear inversion-based geometric control methods for nonlinear ODEs. To assess the performance of the nonlinear controller, a case study is performed where a large step in the tube inlet volumetric flow rate (representing an increase in H_2 production) is imposed. The results revealed a superior performance of the nonlinear controller when compared to a linear PI controller in suppressing the hot spots and attaining an operating temperature under the maximum accepted temperature constraint.

(293f) Thermal Management of Lithium-Ion Batteries and Control Strategies

James C. Ng, Chemical & Materials Engineering, University of Alberta, Edmonton, AB, Canada and Stevan Dubljevic, Chemical and Materials Engineering, University of Alberta, Edmonton, AB, Canada

Lithium-ion batteries continue to be the predominant energy storage device for hand-held and portable electronics due to their high energy density and specific energy relative to other rechargeable battery devices which are based on lead-acid, nickel metal hydride, and other storage cell technologies. In recent years, there has also been a growing interest by automotive manufactures in the use of lithium-ion batteries as potential sources of power for electric vehicles (EVs) and hybrid electric vehicles (HEVs) [1, 3, 6]. The demand for higher performance batteries for applications within both of the electronics and automotive industries has placed pressure on battery manufacturers to improve on existing lithium-ion battery technology, and has resulted in a large amount of research in design, fabrication, and implementation.

One of the main challenges faced by both manufacturers and industries which utilize lithium-ion batteries, is the thermal effects on battery performance [1, 6]. Detrimental thermal effects on lithium-ion batteries, which includes self-discharge, capacity and power fade, cycle performance degradation, and thermal runaway, are further magnified in scale-up cell designs for energy intensive applications. In particular, the high discharge rates and rapid charge and discharge cycling in high-performance electronic devices, and required of in EVs and HEVs, lead to distributed temperature fluctuations exceeding optimal or permissible levels, and present major operational and safety concerns [5]. The temperature dynamics are due to complex interactions between electrochemical heat generating reactions, charge distributions, and temperature fields within individual cells and between multiple cells in the case of multi-cell battery packs. There have been many thermal models developed to describe and characterize the origins of heat generation and the overall transport phenomena, which range from lumped parameter models, empirical models, systems of partial differential equations derived from mass and energy balance relations, as well as combinations of these aforementioned approaches [1, 2, 4, 7].

In this work, we report on model based control strategies in the thermal management of lithium-ion battery devices to mitigate the thermal effects on performance, and to ensure safety and reliability under various operating conditions. The practical use of passive control strategies (e.g. battery geometry design, materials, and ambient air/liquid flow) and also active control methodologies (model based feedback/feedforward control to determine air/liquid flow) will be discussed and accompanied by numerical simulations.

References:

[1] Todd M. Bandhauer, Srinivas Garimella, and Thomas F. Fuller. A critical review of thermal issues in lithium-ion batteries. Journal of The Electrochemical Society, 158(3):R1–R25, 2011.

[2] D. Bernardi, E. Pawlikowski, and J. Newman. A general energy balance for battery systems. Journal of The Electrochemical Society, 132(1):5–12, 1985.

[3] N.A. Chaturvedi, R. Klein, J. Christensen, J. Ahmed, and A. Kojic. Algorithms for advanced battery-management systems. Control Systems, IEEE, 30(3):49 –68, june 2010.

[4] S.C. Chen, C.C. Wan, and Y.Y. Wang. Thermal analysis of lithium-ion batteries. Journal of Power Sources, 140(1):111 – 124, 2005.

[5] Kazuo Onda, Takamasa Ohshima, Masato Nakayama, Kenichi Fukuda, and Takuto Araki. Thermal behavior of small lithium-ion battery during rapid charge and discharge cycles. Journal of Power Sources, 158(1):535 – 542, 2006.

[6] Zhonghao Rao and Shuangfeng Wang. A review of power battery thermal energy management. Renewable and Sustainable Energy Reviews, 15(9):4554 – 4571, 2011.

[7] Venkat Srinivasan and C. Y. Wang. Analysis of electrochemical and thermal behavior of li-ion cells. Journal of The Electrochemical Society, 150(1):A98–A106, 2003.

(293g) Sensitivity Analysis and State Estimation for Microalgal Photobioreactor Systems

Sung Jin Yoo, Se-Kyu Oh and Jong Min Lee, School of Chemical and Biological Engineering, Seoul National University, Seoul, South Korea

Microalgae are photosynthetic microorganisms, which have an ability to produce large amounts of oil that can be used directly as a high value bioactives or be used to synthesize biodiesel. The oil content in microalgae ranges from 15% to 77% depending on species and culture conditions. Although the oil production rate in microalgae is strain dependent, it has several advantages as a feedstock for biodiesel because of high growth rate and the ability of producing large amounts of lipid. However, biodiesel from microalgae is not economically competitive compared to biodiesel from conventional plant sources or petrodiesel. For the economic competitivenesss, optimal model-based control strategy is required to increase the rate of microalgae growth and the amount of stored oil.

However, there are some difficulties in applying model-based control to microalgal bioreactor systems. Microalgal systems are highly nonlinear and some of the parameters and states are difficult to measure directly or estimate. Furthermore, metabolism inside the cells makes process response very slow and its effect on process is difficult to understand.

In this study, a first principles ODE model for microalgae growth and neutral lipid synthesis related with photo effect is investigated for the purpose of maximizing the rate of microalgae growth and the amount of neutral lipid. The model follows the assumption of Droop model which explains the growth as a two-step phenomenon; the uptake of nutrients is first occurred in the cell, and then use of intracellular nutrient to support cell's growth. In practice, microalgal bioreactor systems are complex and highly nonlinear. For this case, which has uncertainties in the model and has model-plant mismatches, robust model-based control using particle filter is studied. For this purpose, sensitivity analysis is performed to determine which parameters have a negligible effect on the model predictions. For state estimation, state variables are divided into frequent but inaccurate or accurate but infrequent measurements for same quality variable and Bayesian approach is exploited to account for multiple-source observations. To enhance the robustness, a practical Bayesian fusion formulation with time-varying variances is proposed and observation validity is taken into account. The Bayesian model calibration strategy is finally implemented by using a sequential Monte Carlo sampling approach based particle filter for simultaneously dealing with systematic and nonsystematic errors (i.e. bias and noise).

(293h) Dynamic Optimization of Solar Thermal Systems with Storage

Kody M. Powell, Chemical Engineering, The University of Texas at Austin, Austin, TX, John D. Hedengren, University of Texas at Austin, Austin, TX and Thomas F. Edgar, Department of Chemical Engineering, The University of Texas at Austin, Austin, TX

Solar energy will continue to become an increasingly important part of the energy mix. Its intermittency, however, is a challenge that must be overcome for solar to become a true replacement for fossil fuels. Therefore, energy storage, combined with effective control and dynamic optimization techniques will be essential to successful integration of solar and other intermittent renewables into the current energy infrastructure.

In this work, dynamic optimization using a 24-hour solar radiation forecast is used to improve performance of a solar-powered steam generation plant with thermal energy storage. The work considers a hybrid plant where natural gas is used to supplement solar energy to maintain a steady power supply. Two control approaches are compared. In the first, a standard control approach is used to maintain a constant solar field outlet temperature and a constant power output from solar energy. When sufficient stored thermal energy is not available, the supplemental fuel source is used. The other approach uses dynamic optimization, where a 24-hour forecast of available solar energy is used to predict future performance. Instead of controlling to a constant temperature, the optimal collection temperature is computed as a function of time. Hybrid operation, where the power output can be a combination of solar and fossil energy proves to be an essential part of operation. The ability to bypass storage is also beneficial as the optimal solution indicates that fluid stored at one temperature should not be mixed with fluid at another temperature due to the entropy that would be generated.

The solution to the dynamic optimization problem is obtained using a simultaneous method which converts the system of differential algebraic equations into a system of algebraic equations via orthogonal collocation on finite elements. The problem then becomes a nonlinear programming problem which is solved using an interior point algorithm.

Results indicate that the dynamic optimization approach is most effective on cloudy days, improving the solar share (the amount of energy delivered via solar energy) by up to sixty percent. This is largely attributed to reduced radiative heat losses in the solar field by virtue of shorter residence times and lower temperatures. The technique also improves performance on sunny days by approximately eight percent. A real-time implementation methodology is also discussed where the optimization strategy acts as a supervisory controller by feeding optimal temperature and power set-points to lower-level regulatory controllers. The system updates at regularly, taking the most recent state and forecast information into account. This methodology will help the system overcome the uncertainty in the forecast.

(301a) On the Derivation of Continuous-Time Scheduling Formulations From Generalized Disjunctive Programming

Pedro M. Castro, Modeling and Optimization of Energy Systems, LNEG, Lisbon, Portugal and Ignacio E. Grossmann, Chemical Engineering Department, Carnegie Mellon University, Pittsburgh, PA

Process Systems Engineering has been responsible for a variety of scheduling models that are primarily classified according to process and time representation. While the unified frameworks (STN, RTN) provide a systematic way of converting the real plant entities (e.g. units, products, utilities) into virtual entities that can be used by a mathematical model, most scheduling models feature variables and constraints derived from intuition and trial and error so as to maximize computational performance. This aspect becomes particularly relevant when switching from a discrete to a continuous-time representation, which has more freedom from a modeling point of view.

Continuous-time approaches can be classified as precedence (Méndez et al. 2001; Jain and Grossmann 2001), single (Castro et al. 2004) or multiple time grid (Castro and Grossmann, 2006; Susarla and Karimi, 2010; Seid and Majozi, 2012) based (a.k.a unit-specific; Ierapetritou and Floudas, 1998). Some of the latter, when applied to the more general multipurpose production environment, have raised serious concerns over the years with respect to generality, suggesting that: (i) either the underlying time representation concept does not account for all possibilities; (ii) not all the required variables or constraints are effectively part of the model. It is thus highly desirable to have a systematic modeling framework that, starting with a simple concept for time representation, can generate the constraints to make it work.

Generalized Disjunctive Programming (GDP) provides a high level framework for modeling mixed-integer programs starting from a logic representation in which mixed-integer logic is represented through disjunctions and integer logic through propositions. In fact, different formulations can systematically be derived using for example big-M and convex hull reformulations, which have complementary strengths. The former has the advantage of being simpler and does not require the definition of additional continuous variables and constraints but has the drawback of being less tight.

In this work, we examine the relatively simple case of single stage scheduling problems with single and parallel units focusing on continuous-time representations and specifically on the immediate, general precedence and multiple time grid alternatives. Starting with the linear GDP models, we derive the equivalent mixed-integer linear programming models using the big-M and convex hull reformulations to show that well-known models from the literature trace back to a particular elementary time representation/reformulation pair. Their performance is then evaluated on a set of test problems with up to 30 orders and 5 parallel units, with the results showing that the multiple time grid model with the convex hull reformulation is the best performer. This is essentially the model proposed by Castro and Grossmann (2006). Other interesting result is that the two similar precedence concepts give opposite results with represent to the most efficient reformulation.

References:

Castro, P.M.; Barbosa-Póvoa, A.P.; Matos, H.A.; Novais, A.Q. (2004). Simple Continuous-time Formulation for Short-Term Scheduling of Batch and Continuous Processes. Ind. Eng. Chem. Res. 43, 105.

Castro, P.; Grossmann, I.E. (2006). An Efficient MILP Model for the Short-term Scheduling of Single Stage Batch Plants. Comput. Chem. Eng. 30, 1003.

Ierapetritou, M.G.; Floudas, C.A. (1998). Effective continuous-time formulation for short-term scheduling. 1. Multipurpose batch processes. Ind. Eng. Chem. Res. 37, 4341.

Jain, V.; Grossmann, I.E. (2001). Algorithms for Hybrid MILP/CP Models for a Class of Optimization Problems. Informs Journal on Computing, 13 (4), 258.

Méndez, C.A.; Henning, G.P.; Cerdá, J. (2001). An MILP continuous-time approach to short-term scheduling of resource-constrained multistage flowshop batch facilities. Comput. Chem. Eng. 25, 701.

Seid, R.; Majozi, T. (2012). A robust mathematical formulation for multipurpose batch plants. Chem. Eng. Sci. 68, 36.

Susarla, N.; Li, J.; Karimi, I.A. (2010). A Novel Approach to Scheduling Multipurpose Batch Plants Using Unit-Slots. AIChE J. 56, 1859.

(301b) Hybrid Bilevel-Lagrangean Decomposition Scheme for the Integration of Planning and Scheduling of a Network of Batch Plants

Bruno A. Calfa[1], Ignacio E. Grossmann[1], Anshul Agarwal[2] and John M. Wassick[2], (1)Chemical Engineering Department, Carnegie Mellon University, Pittsburgh, PA, (2)Dow Chemical Company, Midland, MI

The Process Systems Engineering (PSE) and Operations Research (OR) communities have contributed with models and solution strategies to tackle the decision-making on three different time scales in Enterprise-Wide Optimization (EWO) problems. The long-term decisions (years) belong to the strategic level and are usually related to investments, plant capacity expansion and retrofit; the medium-term decisions (months, years) pertain to planning and commonly refer to production planning and inventory control; the short-term decisions (days, weeks) represent the scheduling of operations, *i.e.* determining the detailed timing and sequencing of operations.

The focus of this work is on integrated planning and scheduling for batch processes. Excellent reviews for these problems are available in the literature, Kallrath (2002) and Méndez *et al.* (2006). However, when attempting to solve large-scale industrial problems, some authors have identified the need to decompose the problem in smaller and sometimes independent subproblems. For instance, Iyer & Grossmann (1998) proposed a Bilevel Decomposition algorithm, which involves solving an aggregate formulation in the upper level and a detailed formulation in the lower level with some decisions fixed by the upper level. Lagrangean Decomposition, a particular case of Lagrangean Relaxation (Guignard, 2003), has also been investigated by Terrazas-Moreno *et al.* (2011) who decomposed their planning problem temporally and spatially in order to solve large instances.

This work deals with the integration of planning and scheduling of a network of batch plants that arises in industrial practice. The network consists of single-stage, multiproduct batch plants located in different sites, which can exchange intermediate products in order to blend them to obtain final products. The time horizon is given and divided into multiple time periods, at the end of which the customer demands have to be exactly satisfied. The planning model is based on the detailed planning model proposed by Erdirik-Dogan & Grossmann (2007), which incorporates Traveling Salesman Problem constraints to predict the sequence-dependent changeovers between products, within and across time periods, without capturing the detailed timing of operations. The scheduling model is based on the unit-specific general precedence (USGP) model proposed by Kopanos, Laínez, & Puigjaner (2009) with two new features: (1) number of batches is allowed to be variable, so it becomes part of the optimization problem, and (2) sequence-dependent changeovers are allowed to occur across time periods.

The solution of the full-space problem, *i.e.* scheduling formulation, for large-scale industrial problems becomes impractical mainly due to the problem size. Therefore, we investigate the combination of two decomposition schemes. The Bilevel Decomposition (BD) framework comprises the solution of an upper-level planning problem, which is an aggregate formulation of the scheduling model, and the solution of a lower-level scheduling problem by fixing some decisions from the solution of the upper level problem. The main motivation behind this approach is to attempt to match the planning and scheduling solutions in a loop structure. For real world applications, the planning problem can still be challenging to solve; hence, we further decompose it using Temporal Lagrangean Decomposition (TLD) by taking advantage of the multiperiod aspect of the problem and dualizing the inventory and assignment constraints. This decomposition yields smaller and independent subproblems for each time period, which are solved in parallel. The TLD of the planning problem is solved in an inner loop of the BD loop. The proposed hybrid decomposition framework is illustrated with three examples of increasing size; the first is a motivating small example in which the optimal schedule obtained by solving the full-space problem is compared with the one obtained by solving the decomposed problem; the second and third examples are medium- and large-scale problems and the computational advantage of the decomposition approach is demonstrated against the solution of the respective full-space problems.

References

Erdirik-Dogan, M., & Grossmann, I. E. (2007). Planning Models for Parallel Batch Reactors with Sequence-Dependent Changeovers. *AIChE Journal*, *53*(9), 2284-2300.

Guignard, M. (2003). Lagrangean Relaxation. *Top*, *11*(2), 151-228.

Iyer, R. R., & Grossmann, I. E. (1998). A Bilevel Decomposition Algorithm for Long-Range Planning of Process Networks. *Industrial & Engineering Chemistry Research*, *37*(2), 474-481.

Kallrath, J. (2002). Planning and scheduling in the process industry. *OR Spectrum*, *24*(3), 219-250.

Kopanos, G. M., Laínez, M. J., & Puigjaner, L. (2009). An Efficient Mixed-Integer Linear Programming Scheduling Framework for Addressing Sequence-Dependent Setup Issues in Batch Plants. *Industrial & Engineering Chemistry Research*, *48*(13), 6346-6357.

Méndez, C. A., Cerdá, J., Grossmann, I. E., Harjunkoski, I., & Fahl, M. (2006). State-of-the-art review of optimization methods for short-term scheduling of batch processes. *Computers & Chemical Engineering*, *30*(6-7), 913-946.

Terrazas-Moreno, S., Trotter, P. A., & Grossmann, I. E. (2011). Temporal and spatial Lagrangean decompositions in multi-site, multi-period production planning problems with sequence-dependent changeovers. *Computers & Chemical Engineering*. http://dx.doi.org/10.1016/j.compchemeng.2011.01.004.

(301c) A Constraint Propagation Algorithm and Valid Inequalities for Chemical Production Scheduling MIP Models

Sara Zenner and Christos Maravelias, University of Wisconsin, Madison, WI

A Constraint Propagation Algorithm and Valid Inequalities for Chemical Production Scheduling MIP Models

Sara Zenner and Christos T. Maravelias

Department of Chemical and Biological Engineering, University of Wisconsin – Madison

1415 Engineering Dr., Madison, WI 53706, USA

In chemical production industries (e.g. pharmaceutical, food, and petroleum), processes consist of a set of tasks to produce finished goods. The equipment units used to process a task have a minimum and maximum capacity. Materials, such as feedstocks, intermediate chemicals, and final products, can be mixed together, split apart, and recycled. Each task consumes and produces materials according to a fixed recipe. Material balances enforce this recipe and keep track of the inventory for all materials. In general, scheduling decisions include (i) how many tasks to carry out, (ii) which piece of equipment to use for each task, (iii) when to carry out each task, and (iv) how much material to process. Chemical production scheduling problems can be modeled and solved using mixed integer programming (MIP) methods.

Unfortunately, solving MIP scheduling problems for practical instances remains computationally intractable. Previous work showed that adding valid inequalities based on material balances in lot-sizing problems results in a significantly tighter formulation (Pochet and Wolsey, 1993; Wolsey, 1997; Pochet and Wolsey, 2000; and Miller and Wolsey, 2003). However, these methods are not effective in chemical production scheduling problems due to the presence of minimum capacity constraints and recycle streams. To address this challenge, we propose two new types of valid inequalities based on material balances and customer orders for final products.

To construct these inequalities, we back-propagate demand for final products through the network to determine the minimum amount of each material that is needed and the minimum amount that each task must produce. Two things make the back-propagation for chemical production processes especially challenging. First, units have minimum capacities, which means that not all values for a task's minimum production are attainable. For example, if a task must produce at least 50kg and can take place in a single unit with a 30-40kg capacity, then the task will have to produce at least 60kg in at least two batches; we must use this improved value in the back-propagation. To account for this, we first determine the set of attainable production amounts for each task, and then, using this information, we adjust the minimum necessary amount for each material. Second, recycle loops are common in chemical industries, which makes back-propagating demand sequentially challenging. To overcome this challenge, we use a tear stream.

The demand propagation algorithm, which is run within a few seconds even for large scale process networks, provides us with a lower bound on the number of batches of each task and the amount of material that each task has to produce. These bounds are then used to formulate two types of tightening constraints. The first lower bounds the total number of times each task is run, while the second lower bounds the total production of each task. Our constraint propagation algorithm can be applied in process networks of arbitrary structure (including networks with nested recycle loops). Also, the proposed tightening constraints can be added to any time-indexed (discrete or continuous) MIP scheduling formulation.

To illustrate the effectiveness of our methods, we present results from 36 instances using the discrete-time model of Shah et al. (1993). Adding these inequalities reduces the average computational time for problems that are solved to optimality from 189s to 1.9s and increases the percent of problems that are solved to optimality within 30 minutes from 39% to 94%. Similar enhancements are observed for continuous-time formulations.

Shah, N.; Pantelides, C.C.; Sargent, R.W.H. A General Algorithm for Short-Term Scheduling of Batch Operations – II. Computational Issues. *Comput. Chem. Eng.*, **1993**, 17, 229-244.

Pochet, Y.; Wolsey, L.A. Lot-Sizing with Constant Batches: Formulation and Valid Inequalities. *Math. Oper. Res.*, **1993**, 18, 767-785.

Wolsey, L.A. MIP Modelling of Changeovers in production Planning and Scheduling Problems. *European Journal of Operational Research*, **1997**, 99, 154-165.

Miller, A.J.; Wolsey, L.A. Tight MIP Formulations for Multi-Item Discrete Lot-Sizing Problems. *Op. Res.*, **2003**, 51, 557-565.

Pochet, Y; Wolsey, L.A. Production Planning by Mixed Integer Programming. *Springer, New York,* **2000**.

(301d) A State-Space Model and Advanced Control Methods for Chemical Production Scheduling

Kaushik Subramanian, University of Wisconsin - Madison, Madison, WI, Christos T. Maravelias, Department of Chemical and Biological Engineering; DOE Great Lakes Bioenergy Research Center, University of Wisconsin-Madison, Madison, WI and James B. Rawlings, Chemical and Biological Engineering, University of Wisconsin-Madison, Madison, WI

A State-space Model and Advanced Control Methods for Chemical Production Scheduling Kaushik Subramanian, Christos T. Maravelias and James B. Rawlings Department of Chemical and Biological Engineering, University of Wisconsin Madison 1415 Engineering Dr., Madison, WI 53706, USA Traditionally, scheduling is considered as a static optimization problem. We optimize at the beginning of the scheduling horizon, implement the schedule, and optimization is performed again for the next scheduling horizon. In case of disturbances affecting the schedule, various heuristic and recipe based methods are employed to locally correct the schedule within the current scheduling horizon. In reality, however, scheduling is a dynamic process, in which the optimal schedule can drastically change on observing a disturbance. Advance model based controllers are a natural choice to deal with dynamic processes. Accordingly, to facilitate the use of advanced process control methods to scheduling, we first develop a state-space form of the standard state task network (STN) model and then discuss how concepts from model predictive control (MPC) can be used to effectively address scheduling problems. In developing the state-space STN model, we identify the inputs and introduce new state variables, via input lifting, to fully describe the state of the system. Then, we model typical scheduling disturbances that lead to rescheduling (e.g., shipment/delivery disturbances, unit breakdowns, and processing delays) using disturbance variables in the state-space model. We discuss how a wide range of scheduling models, with different types of decisions and processing constraints, can be expressed in state-space form. The state-space form of the STN model allows researchers to study scheduling models from a process control perspective, using existing theory (e.g., hybrid MPC) or developing new results to study the dynamics of the scheduling problem. Furthermore, we discuss how concepts from control can be used in scheduling. Specifically, we discuss: (i) if and what input/state trajectories can be used in scheduling; and (ii) what stability and operability mean in the context of scheduling problems. Finally, we discuss two methods to achieve recursive feasibility for the scheduling problem that is responding to a nominal demand scenario. We achieve recursive feasibility using terminal constraints, that is constraints at the end of the scheduling horizon, which is the standard method used to ensure recursive feasibility (and stability) for dynamic control problems. First, we develop periodic schedules and use them as the terminal constraint, thus, developing a sub-optimal infinite horizon schedule for the plant. Second, we develop a safety state (that includes constraints on the inventories in the STN as well as the tasks). We also present some ideas to develop safety stocks that ensure recursive feasibility and terminal conditions that are robust to multiple demand-scenarios.

(301e) Operational Planning in Residential Microcogeneration Energy Production Networks

Georgios M. Kopanos, Chemical Engineering Department, Imperial College London, Centre for Process Systems Engineering, London, United Kingdom, Michael C. Georgiadis, Chemical Engineering, Aristotle University of Thessaloniki, Thessaloniki, Greece and Efstratios N. Pistikopoulos, Chemical Engineering Department, Imperial College London, Centre for Process Systems Engineering, London, United Kingdom

Nowadays, it is evident that the classical Energy Supply Chain (ESC) is rapidly changing to an energy-efficient and low-carbon energy market economy by moving towards more decentralized energy production. There is clearly a distributed energy generation option which could play a vital role within the development of sustainable future energy systems, the energy *microgeneration*. Specifically, the most promising microgeneration technology involves the *cogeneration* (i.e., combined generation) of electrical energy and heat in small-scale energy generation units that can be directly embedded in the buildings wherein the heat and electricity are to be used; also known as *microcogeneration* (Pehnt et al., 2006).

Indeed, the domestic sector constitutes a key consumer of both electricity and heat, and could benefit from consolidation to meet these demands via micro Combined Heat and Power (microCHP) generators (Hawkes et al., 2006). Today, the principal microCHP technologies are based on Stirling engines or fuel cells. Stirling engines are already available in the market in competitive prices, while fuel cells constitute a still- experimental technology that features high initial capital costs. Also, there are financial incentives (such as the Feed-In-Tariff support scheme introduced by the UK Government) initiated by many countries in an attempt to encourage the uptake of small-scale microgeneration technologies.

ESCs based on microgeneration have been recently emerged, and they are expected tp grow rapidly. In this new energy environment, the electrical grid (macrogrid) is seen not as a primary power supplier but instead as a back-up of the microgeneration ESC network network. A residential energy microgrid (i.e., a microgeneration ESC) could be formed by connecting several domestic microCHP systems. Importantly, by constructing such a microgrid while integrating various energy alternatives in a residential scale, residents can trade and consume the power flexibly (and more efficiently) with each other. Noticeably, the planning and the overall management of microCHP-based ESC networks result into highly sophisticated decision making problems.

In this work, a mathematical programming framework has been developed for the operational planning of such ESC networks. The minimization of total costs (including microgeneration system's startup and operating costs as well as electricity production revenue, sales, and purchases), under full heat demand satisfaction, constitutes the objective function. Additionally, an alternative microgrid structure that allows the heat interchange within subgroups of the overall microgrid is proposed, and the initial mathematical programming formulation is extended to deal with this new aspect. Afterwards, a number of real-world size case studies, using real data provided from the UK Energy Research Centre, have been solved by the proposed approaches so as to shed light on the potential benefits of the residential microgeneration ESC networks suggested in this study. Finally, some concluding remarks are drawn and potential future research directions are identified.

Acknowledgments

The authors acknowledge the financial support from the Engineering and Physical Sciences Research Council (EPSRC) under the research project EP/G059071/1. Financial support from the European Union's Seventh Programme managed by REA-Research Executive Agency http://ec/europa.eu/rea (FP7/2007-2013) under grant agreement PIRSES-GA-2011-294987 (ESE project) is also gratefully acknowledged.

References

Hawkes, A. D., Aguiar, P., Hernandez-Aramburo, C. A., Leach, M. A., Brandon, N. P., Green, T. C., Adjiman, C. S., 2006. Techno-economic modelling of a solid oxide fuel cell stack for micro combined heat and power. Journal of Power Sources 156 (2), 321-333.

Pehnt, M., Cames, M., Fischer, C., Praetorius, B., Schneider, L., Schumacher, K., Voβ, J., 2006. Micro Cogeneration: Towards Decentralized Energy Systems.Springer, Berlin, Heidelberg, Germany.

UK Department of Energy & Climate Change, June 2011. Microgeneration strategy.

UK Energy Research Centre Energy Data Centre (UKERC-EDC), 2008. Milton Keynes Energy Park Dwellings 1990 dataset.

(301f) Optimal Scheduling of Industrial Combined Heat and Power Plants Under Time-Sensitive Electricity Prices

Sumit Mitra, Department of Chemical Engineering, Center of Advanced Process Decision-making, Carnegie Mellon University, Pittsburgh, PA, Lige Sun, RWTH Aachen University, Aachen, Germany and Ignacio E. Grossmann, Chemical Engineering Department, Carnegie Mellon University, Pittsburgh, PA

Combined heat and power (CHP) plants are widely used in industrial applications, e.g. in pulp and paper mills, aluminum plants, refineries and other chemical processes. In the aftermath of the recession, many of these processes are largely under-utilized, which puts chemical companies under pressure to remain competitive on a global basis.

Fortunately, under-utilization can be an opportunity for tighter interaction with the power grid, which is in transition to the so-called smart grid. Deregulation and an increasing share of renewable energies lead to more variability in time-sensitive electricity prices, offering potential incentives for industrial consumers if they are able to cope with the fluctuations (Samad and Kiliccote, 2012). Therefore, it is essential to characterize the flexibility of a CHP plant and use the obtained insights to determine how to dynamically respond to the time-sensitive electricity prices. An additional challenge is variability in steam demand (Velasco-Garcia et al., 2011).

In this work, we will answer three main questions related to the decision-making for CHP plants under time-sensitive electricity prices:

1. Is it technically and economically feasible to operate an under-utilized industrial CHP plant flexibly in interaction with the smart grid?

2. How large can the potential economic gains be, i.e. what is the optimal way to operate?

3. How can these plants be modeled such that the process flexibility is captured appropriately?

For this purpose, we describe a generalized mode model on a component basis that addresses the operational optimization of under-utilized industrial CHP plants. The mode formulation tracks the state of each plant component in a detailed manner and can account for different operating modes and transitional behavior. Different operating modes include fuel switching for boilers and supplementary firing for gas turbines with a heat recovery steam generator (see Cohen and Ostrowski, 1996; Liu et al., 2009). Transitional behavior such as warm start-ups, cold start-ups, shutdowns and pre-computed trajectories during start-ups is modeled with modes as well. The feasible region of operation for each component is described based on input-output relationships that are thermodynamically sound (Mavromatis and Kokossis, 1998; Varbanov et al., 2004; Aguilar et al., 2007). Furthermore, we emphasize the use of mathematically efficient logic constraints that allow solving the large-scale models fast (Hedman et al., 2009; Rajan and Takriti, 2005). We provide an industrial case study and investigate the impact of different scenarios for under-utilization, where we obtain profit improvements of up to 10%.

References

Aguilar, O.; Perry, S.J.; Kim, J.-K.; Smith, R. Design and Optimization of Flexible Utility Systems Subject to Variable Conditions Part 1: Modelling Framework. Chemical Engineering Research and Design, 85:1136–1148, 2007.

Cohen, A.I.; Ostrowski, G. Scheduling units with multiple operating modes in unit commitment. IEEE Transactions on Power Systems, 11:497–503, 1996.

Hedman, K.W.; O'Neill R.P.; Oren, S.S. Analyzing Valid Inequalities of the Generation Unit Commitment Problem.

Power Systems Conference and Exposition, 2009. PSCE '09. IEEE/PES, pages 1–6, 2009.

Liu, C.; Shahidehpour, M.; Li, Z.; Fotuhi-Firuzabad, M. Component and Mode Models for the Short-Term Scheduling of Combined Cycle Units. IEEE Transactions on Power Systems, 24:976–990, 2009.

Mavromatis, S.P.; Kokossis, A.C. Conceptual optimisation of utility networks and operational variations - I. Targets and level optimisation. Chemical Engineering Science, 53:1585–1608, 1998.

Rajan, D.; Takriti, S. Minimum Up/Down Polytopes of the Unit Com- mitment Problem with Start-Up Costs. Technical report, IBM Research Division, 2005.

Samad, T.; Kiliccote, S. Smart Grid Technologies and Applications for the Industrial Sector. Proceedings of FOCAPO, 2012.

Velasco-Garcia, P.; Varbanov, P.S.; Arellano-Garcia H.; Wozny, G. Utility systems operation: Optimisation-based decision making. Applied Thermal Engineering, 31:3196–3205, 2011.

Verbanov, P.S.; Doyle, S.; Smith, R. Modelling and Optimization of Utility Systems. Chemical Engineering Research and Design, 82:561– 578, 2004.

(301g) A Novel Approach to the Optimal Scheduling of Continuous Blending Operations

Stacy L. Janak, Aspen Technology, Houston, TX and Dimitrios Varvarezos, Aspen Technology, Inc.,, Houston, TX

Optimal product blending is a complex process that, to a large extent, defines the profitability of refining operations. The goal of product blending in a refinery is to meet all the shipment commitments on time and with the required specifications, while operating within the inventory and production constraints both for the blending components and the blended products. In addition, numerous other operational constraints dictated by the hardware infrastructure of the blending apparatus have to be observed. Lastly, this operation should be executed in an optimal way in terms of operational margin, quality giveaway, and operational efficacy. Continuous blending operations (often referred to as rundown blending) are a special class of refinery blending operations whereby there is one or more blending component that does not have intermediate storage tanks. This type of operation is common in many refineries, such as those in Eastern Europe and Asia, as well as North American refineries involved in blending diesel and fuel oil with more strict quality specifications.

In this paper, we present a novel modeling and optimization approach that determines the optimal sequence and timing of blend events, as well as rundown component tank switches, in order to handle blending of "hot" streams (coming directly off of a process unit without the benefit of intermediate tank storage) to a finished product tank. This system incorporates multiple blend headers and multiple blends in a multi-period, event-driven campaign, using open-equation based optimization and modeling technology. It produces the optimum schedule for multi-period blending along with optimum recipes and volumes for each blend, addressing the underlying inventory optimization problem. In addition, it manages all of the production blending operations including external component and product receipts and shipments, as well as relevant intra-refinery transfers. This new modeling formulation allows for the determination of the optimum split ratio for any rundown splitters, the duration and timing of the rundown components to the blends, and the amount of each static component for each blend. The resulting problem is a large MINLP that is solved using the Aspen XSLP Optimizer. The proposed model combines a continuous-time formulation with operationally dictated time periods. In this context, we also present an example problem based on an actual commercial blending operation. The solution of the example problem maximizes the profit while taking into account all of the economic factors including additives, rundown components, storage costs, and the robustness of the schedule.

The proposed approach is far superior to the current practice, which is typically based on trial-and-error methods in Excel or other heuristic tools and used primarily to obtain a feasible blend schedule including rundown operations. The optimal blend schedule provides for stream containment and ensures that all products meet their specifications. It also minimizes the use of slop tanks (used only for containment purposes) and the incidence of product giveaway. This approach is both comprehensive and practical as we are able to avoid these costly practices while finding a feasible and economically optimal blend schedule. Despite the large size of the resulting problem, which can be on the order of 50,000 continuous and several thousand discrete variables, the algorithm is very efficient and typically converges in less than a couple minutes even for complex refinery configurations.

(304a) Process Intensification of the Production of Di-Methyl Carbonate (DMC) Using a New Synthesis and Design Process Intensification Methodology

Deenesh K. Babi, Department of Chemical and Biochemical Engineering, Technical University of Denmark, Lyngby, Denmark, Johannes Holtbruegge, TU Dortmund University, Department of Biochemical and Chemical Engineering, Laboratory of Fluid Separations, D-44227 Dortmund, Germany, Philip Lutze, Department of Biochemical and Chemical Engineering, Laboratory of Fluid Separations, TU Dortmund University, D-44227 Dortmund, Germany, Andrzej Gorak, Dortmund University, Chair of Fluid Separation Processes, 44221 Dortmund, Germany, John Woodley, Center for Process Engineering and Technology, Department of Chemical and Biochemical Engineering, Technical University of Denmark, Lyngby, Denmark and Rafiqul Gani, CAPEC, Department of Chemical and Biochemical Engineering, Technical University of Denmark (DTU), Kgs. Lyngby, Denmark

Process Intensification of the production of Di-Methyl Carbonate (DMC) using a New Synthesis and Design Process Intensification Methodology

Deenesh K. Babi[1], Johannes Holtbruegge[2], Philip Lutze[2], Andrzej Górak[2], John M. Woodley[1], Rafiqul Gani[1]

dkbabi@kt.dtu.dk, *johannes.holtbruegge@bci.tu-dortmund.de,* philip.lutze@bci.tu-dortmund.de, *andrzej.gorak@bci.tu-dortmund.de* jw@kt.dtu.dk, rag@kt.dtu.dk

[1] *Department of Chemical and Biochemical Engineering, Technical University of Denmark (DTU), DK-2800 Kongens Lyngby, Denmark*

[2] *Department of Chemical and Biochemical Engineering, Technical University of Dortmund (TU Dortmund), D-44227 Emil-Figge-Str. 70 Dortmund*

Abstract

Process Intensification (PI) is a means by which processes, whether new or existing, can achieve a more efficient and sustainable chemical process through the improvement of, for example, energy efficiency and waste generation. PI can be defined as the improvement of a process at the phenomena level which ultimately has an impact at the higher levels of a process, that is the functional and unit operations (Unit-Ops) levels. More specifically PI is the enhancement of the involved phenomena through the integration of Unit-Ops, functions and phenomena. Except for reactive distillation and dividing wall columns, the implementation of PI still faces hurdles [1]. For achieving PI via a systematic approach different methodologies exist, of which two are mentioned here, the task-based means-ends analysis developed by Sirrola [2] and the Unit-Ops based synthesis and design method for achieving PI developed by Lutze et al. [1]. These methods, which work at the Unit-Ops level, are limited to PI equipment already in existence and are therefore able to generate new integrations/combinations of intensified existing equipment but may not generate new/innovative PI solutions.

The objective of this work is threefold: (1) the development of a new/innovative systematic PI synthesis and design methodology, which includes model based synthesis/design and experimental based validation; (2) the development of a phenomena based synthesis and design method (PBS) which in addition to the knowledge-based (KBS) and Unit-Ops based (UBS) methods (thereby extending further, the application range) is one of the methodologies within the new PI methodology; and (3) the application of the new PI methodology to a case study - the production of di-methyl carbonate (DMC).

The new PI methodology has the following characteristics: (a) flexible for a wide range of applications (b) finds innovative (new) and predictive (existing) solutions and (c) model and experimental based validation of intensified options. The new PI methodology follows a systematic decomposition approach starting with a base case as a reference point. In Step 1 a general problem definition is determined. In Step 2 system information is collected based on thermodynamic data and the type of chemical/process systems that are identified. Using this data, the process is analysed and in Step 3, it is possible to use one or two or all of the PI methods. The three synthesis and design methods considered are the KBS, the UBS and the PBS. The KBS and UBS methods use knowledge stored in a knowledge-based tool on implemented PI equipment used for overcoming certain limitations/bottlenecks (LB's), for example, the use of reactive distillation for overcoming unfavourable equilibrium in a reaction. The PBS methodology not only uses the knowledge of the existing methodologies at the Unit-Ops level but also operates at a lower level of aggregation (i.e phenomena level), where the apriori knowledge of existing Unit-Ops is needed to some degree but also where new Unit-Ops, in principle, can be designed. In Step 4 the feasible flowsheets obtained from Step 3 are validated/ verified either through a model or experimental based analysis. At the end of Step 4 the best PI process candidate and the other feasible options are used to update the systems information. It should be noted that the degree of complexity increase from the KBS to PBS while the possibility of obtaining/generating novel PI equipment decreases from PBS to KBS.

A preliminary version for a phenomena-based synthesis/design (PBS) methodology has been developed [3] and consists of a six step workflow: Problem and objective function definition→ Process analysis using thermodynamic insights→ Identification of LB's of the process together with the desirable and accompanying phenomena to overcome these LB's→ Phenomena building blocks are connected to form simultaneous phenomena building blocks (SPBs) which are screened for the most feasible connections using for example connectivity rules. The SPB's are then connected to form operations which are then connected to form flowsheets just as atoms are combined to form molecules→ Flowsheets are first screened using for example logical constraints and performance metrics to obtain feasible PI options→ The feasible flowsheets from are optimized with respect to the objective function with the end result being the identification of the best intensified process candidate. The emerging flowsheets consists of either novel or existing PI equipment. Defined phenomena are for example mixing, two-phase mixing, phase transition and phase separation. This preliminary version is now extended and implemented within the general PI synthesis design methodology.

In this presentation the PI synthesis and design methodology with focus on the PBS workflow will be highlighted together with the details of its application to the production of DMC. DMC is an important bulk chemical used as a solvent and fuel additive. It will be shown that the PBS methodology systematically generates, besides other options, the current intensified option proposed by the co-authors who have followed the UBS path of the combined methodology and is as follows: a reactive distillation column→ membrane separation →distillation.

References:

[1] P. Lutze, A. Román-Martinez, J. M. Woodley & R. Gani. A systematic synthesis and design methodology to achieve process intensification in (bio) chemical processes. Comput. Chem. Eng. 2012 (36) 189– 207

[2] Jeffrey J. Siirola. Strategic process synthesis: Advances in the hierarchical approach. Comput. Chem. Eng. 1996, Supplement 2 (20) S1637-S1643

[3] P. Lutze, R. Gani & J. M. Woodley. Phenomena-based Process Synthesis and Design to achieve Process Intensification. Comp. Aided Chem. Eng. 2011 (29) 221–225

A NOVEL MILP APPROACH TO THE SYNTHESIS OF THERMALLY COUPLED DISTILLATION SEQUENCES.

José A. Caballero; Ignacio E. Grossmann***
**Department of Chemical Engineering. University of Alicante. Ap. Correos 99. 03080 Alicante. Spain*
*** Department of Chemical Engineering. Carnegie Mellon University. Pittsburgh, PA. USA.*

Introduction

The distillation is still the most widely used separation technique in the chemical industry. It handles more than 90% of separations and purifications and this trend seems unlikely to change in the near future. The capital investment for these distillation systems was estimated in $8x10^9$ US$ [1]. Using data by Mix et al [2], Soave & Feliu [3] calculated that distillation accounts about the 3% of the total US energy consumption which is equivalent to $2.87x10^{18}$ J (2.87 million TJ) a year or to a power consumption of 91 GW or 54 million tons of crude oil.

A renew interest in Thermally Coupled Distillation (TCD) appeared in, say the last 15-20 years, due the important potential savings in energy: typical values around 10% to 40% (in some cases even larger) has been reported [4-9] when compared with conventional distillation sequences. An although the TCD can only be implemented in mixtures involving the separation of at least three components the reader can have an idea of the impact just looking at the numbers presented in the first paragraph.

The design of TCD sequences is much more complex than conventional distillation (each column with a condenser and a reboiler) by at least two reasons. First, the number of alternatives is much larger and second when we introduce a thermal couple we are introducing also two thermodynamically equivalent configurations (TEC) [10-12],. Therefore, the number of alternatives rapidly increases (i.e. in a five component mixture there are 203 basic configurations, more than 8000 if we consider internal heat exchangers and more than $2 \cdot 10^5$ if we take into account TEC). In order to avoid the degeneracy created by TEC it is convenient to use a task based approach instead of a column based approach. –All TEC shares the same sequence of separation tasks-.

As noted by Andrecovich and Westerberg [13] in the 80's of last century, when we consider conventional separation sequences, we can, a priori, optimize each possible column of the sequence, and then use a MILP approach to extract the optimal sequence. But, when we deal with non-sharp separations we cannot know a priori the distribution of no-key components and therefore we must simultaneously optimize the columns and the sequence which produces a non-convex MINLP even with shortcut methods [4, 5, 14]. However, taking into account the following two facts, it is possible to develop a strategy that allows decoupling the column and sequence optimization in TCD systems.

1. Any distillation sequence of zeotropic mixtures have a structure without recycles (A thermal couple can be simulated with an equivalent system without recycles).

2. In TCD sequences the flows of some column sections are not in the optimal operation conditions because they must be modified in order to accomplish with the mass balance. But, the introduction of heat exchangers allows decoupling these sections and avoids the necessity of a simultaneous optimization.

It is possible then generating a tree of possible separation tasks as follows:

1. First we consider, and optimize, all the initial separation tasks (those that have the initial mixture as feed).

2. For each of the products (distillate and bottoms of each column) we generate all the possible separations. At this point we consider that mixtures of the same components with different concentrations or different thermal states are different -explicitly takes into account the possibility of a heat exchanger or a thermal couple-.

3. Following recursively this approach we can optimize all possible separation tasks and, implicitly, generate a tree that includes all the separation tasks.

The number of possible separation tasks is large, but much smaller than the number of sequences. Using the Underwood-Fenske shortcut method the complete tree for a 5 component mixture can be generated in around 4-5 seconds of CPU time.

The final step consists of extracting the optimal separation sequence. However this is not straightforward. The simple connectivity, like in conventional columns, does not assure a feasible basic sequence:

a) It is necessary to include logical relationships accomplish with these task (a comprehensive description of these logical relationships can be found in the work by Caballero & Grossmann [5].

b) For some internal mixtures, there are more than one possible separation task that depends on their origin (which separation task generate it) and the thermal state (include or not a heat exchanger or a thermal couple). Therefore it is necessary to include a set of logical relationships to assure the feasibility and relates the tree of separations with the feasible alternatives.

c) Divided Wall Columns (DWC) are explicitly included. A DWC is generated by a subsystem of three fully thermally coupled tasks [15] and some logical relationships added to avoid only feasible configurations.

As all the data of all the separation tasks is calculated when each of the individual columns are calculated, the resulting problem is a MILP in where we select the best feasible sequence of tasks (i.e. the sequence with minimum Total Annualized Cost). Calculate the excess or defect of heat (or vapor flows) in the connectivity of each individual task and estimate the extra cost (energy).

Example.

As an example considers a mixture of 5 hydrocarbons (all relevant data and main results are showed in Table 1).

The optimal solution includes a DWC and a set of small heat exchangers in the final products that appears as a consequence of the differences in the flows in the different column sections –See Figure 1-. Alternatively it is possible to remove all these heat exchangers by adjusting the flows in the implied column sections. The net result will be an increase in the heats of reboiler (associated with product E) and condenser (product A) approximately in the same quantity of the heat removed in the intermediate products. This could have two opposite effects. In one side we remove three heat exchangers, but at the same time we lose the possibility of using utilities at intermediate temperatures.

It is possible to re-optimize the optimal sequence presented in Figure (1) taking now into account the possibility of removing the heat exchangers related to products B, C and D. Remove one of those heat exchangers save its investment cost but the energy consumed by the heat exchanger is transferred to the upper heat exchanger if the heat exchanger removed is a condenser, and to the lower heat exchanger if it is a reboiler. At the same time the internal flows in the corresponding column sections increases and then also the investment costs in the column. The optimal solution obtained varies in less than a 2% in TAC respect to that shown in Figure (1) but in this case, the reboiler C and the condenser B were removed. Removing reboiler D produce an important increase in the internal flows of the separation tasks (D/E) and then it is better to maintain the reboiler D.

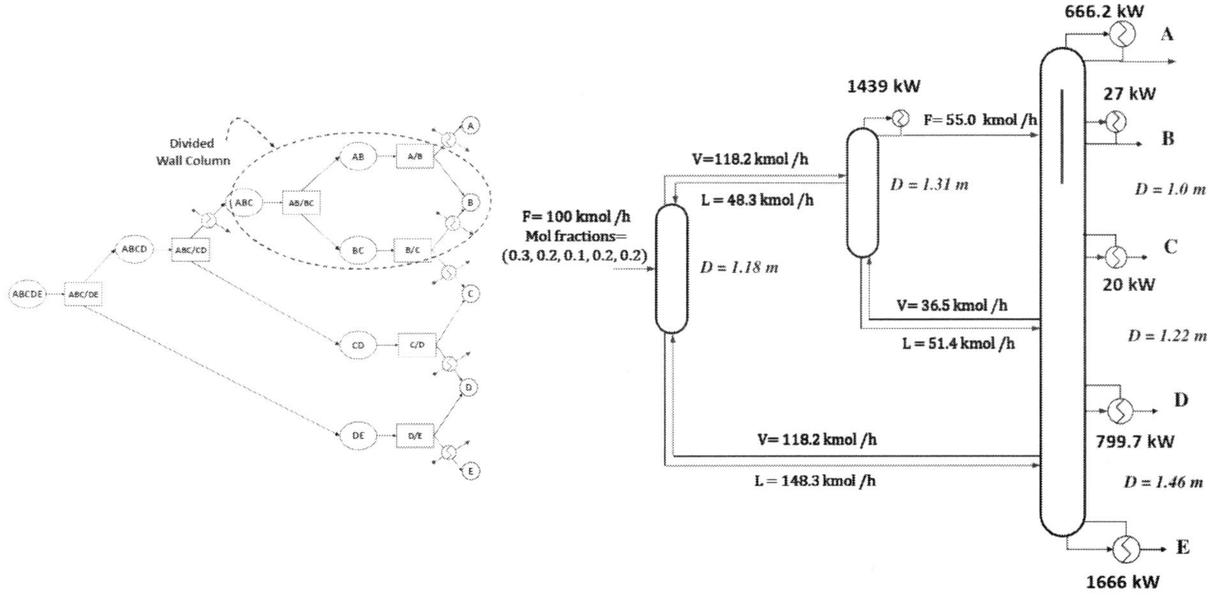

Figure 1. ***Sequence of separation tasks and a possible arrangement in actual columns.***

Table 1. Data and results for example 1.

Component	**Feed composition (mol fraction)**		
(A)- Benzene	0.3	Total Flow	100 kmol/h
(B)- Toluene	0.2	Pressure	1 atm
(C)- Ethylbenzene	0.1	Steam cost	5.02 $/GJ
(D)- Styrene	0.2	Cold water cost	0.19 $/GJ
(E)- α-methyl styrene	0.2	(8000 h/year)	
		Capital charge factor 0.18 (9% interest ; 8 years)	
Optimal sequence	ABC/DE - ABC/CD - AB/BC – A/B – B/C – C/D – D/E		
	Condensers: ABC; A; B		
	Reboilers: C; D; E		
Vessels cost (k$)	574.27		
Tray cost (k$)	256.51		
Heat Exchangers cost (k$)	599.5		
Total Investment (k$)	1430.03		
Total utilities cost (k$/year)	568.42		
TAC ((k$/year)	**825.8**		

Table 1 (cont). Data and results for example 1.

Model Statistics	
CPU time (s)	57
Nº equations	12064
Nº continuous variables	3142
Binary variables	2563

Acknowledgements

The authors wish to acknowledge support from the Spanish Ministry of Science and Innovation under project (CTQ2009-14420-C02).

References

1. Humphrey, J., *Separation processes: playing a critical role.* Chemical Engineering Progress, 1995. **91**(10): p. 43-54.
2. Mix, T., et al., *Energy conservation in distillation.* Chemical Engineering Progress, 1978. **74**(4).
3. Soave, G. and J.A. Feliu, *Saving energy in distillation towers by feed splitting.* Applied Thermal Engineering, 2002. **22**(8): p. 889.
4. Caballero, J.A. and I.E. Grossmann, *Design of distillation sequences: from conventional to fully thermally coupled distillation systems.* Computers & Chemical Engineering, 2004. **28**(11): p. 2307-2329.
5. Caballero, J.A. and I.E. Grossmann, *Structural considerations and modeling in the synthesis of heat-integrated-thermally coupled distillation sequences.* Industrial & Engineering Chemistry Research, 2006. **45**(25): p. 8454-8474.
6. Fidkowski, Z.T., *Distillation configurations and their energy requirements.* AIChE Journal, 2006. **52**(6): p. 2098-2106.
7. Fidkowski, Z.T. and R. Agrawal, *Multicomponent thermally coupled systems of distillation columns at minimum reflux.* AIChE Journal, 2001. **47**(12): p. 2713-2724.
8. Rudd, H., *Thermal Coupling For Energy Efficiency.* Chemical Engineer-London, 1992(525): p. S14-S15.
9. Triantafyllou, C. and R. Smith, *The Design And Optimization Of Fully Thermally Coupled Distillation-Columns.* Chemical Engineering Research & Design, 1992. **70**(2): p. 118-132.
10. Agrawal, R. and Z.T. Fidkowski, *More operable arrangements of fully thermally coupled distillation columns.* AIChE Journal, 1998. **44**(11): p. 2565-2568.
11. Caballero, J.A. and I.E. Grossmann, *Thermodynamically equivalent configurations for thermally coupled distillation.* AIChE Journal, 2003. **49**(11): p. 2864-2884.

12. Rong, B.G., A. Kraslawski, and I. Turunen, *Synthesis and optimal design of thermodynamically equivalent thermally coupled distillation systems.* Industrial & Engineering Chemistry Research, 2004. **43**(18): p. 5904-5915.
13. Andrecovich, M.J. and A.W. Westerberg, *An Milp Formulation For Heat-Integrated Distillation Sequence Synthesis.* AIChE Journal, 1985. **31**(9): p. 1461-1474.
14. Caballero, J.A. and I.E. Grossmann, *Generalized disjunctive programming model for the optimal synthesis of thermally linked distillation columns.* Industrial & Engineering Chemistry Research, 2001. **40**(10): p. 2260-2274.
15. Caballero, J.A. and I.E. Grossmann, *Logic-Sequential Approach to the Synthesis of Complex Thermally Coupled Distillation Systems*, in *Computer Aided Chemical Engineering*, M.C.G. E.N. Pistikopoulos and A.C. Kokossis, Editors. 2011, Elsevier. p. 211-215.

(304c) Process Intensification in Ethylene Oxide Production: Optimal Reactor Design From a Process Point of View

Hannsjörg Freund[1,2], Andreas Peschel[2] and Kai Sundmacher[2,3], (1)Chemical Reaction Engineering, University of Erlangen-Nürnberg, Erlangen, Germany, (2)Process Systems Engineering, Max Planck Institute for Dynamics of Complex Technical Systems, Magdeburg, Germany, (3)Process Systems Engineering, University of Magdeburg, Magdeburg, Germany

In this work, a new method for the identification of the best reaction concept from an overall process point of view is proposed. Using this method the optimal reaction route is determined independently of existing apparatuses and the recycle affecting the reactor is fully considered. The reaction concept and the design parameters of the process are simultaneously optimized. For this purpose, a detailed model of the reaction system including reaction kinetics, thermodynamic relationships, and intrinsic bounds such as explosion limits as well as a complete model of the downstream process is set up. The potential of different reaction concepts including innovative process intensification alternatives can easily be determined by comparison with an optimized reference case.

The example considered in this work is the oxygen based ethylene oxide process, which is one of the most important chemical bulk processes. Sophisticated cooling in combination with distributed oxygen dosing is identified as best technical reaction concept with a significant potential of reducing the running expenses compared to an optimized reference case. In addition, the utility consumption (e.g. the consumption of electricity) can be drastically reduced.

In conclusion, the proposed method enables the predictive determination of the best reaction concept from an overall process point of view taking the full interconnections between the reaction concept and the process into account and considering highly innovative process intensification options. Thereby, this method provides a key stone for the development of more economical and more sustainable chemical reactors and processes of the future.

(304d) Microwave Assisted Flow Processing: Coupling of Electromagnetic and Hydrodynamic Phenomena

Narendra Patil[1], Erik Esveld[2], Faysal Benaskar[1], Evgeny V. Rebrov[3], Lumbertus Hulshof[1], Jan Meuldijk[1], Volker Hessel[1] and Jaap C. Schouten[1], (1)Chemical Engineering and Chemistry, Eindhoven University of Technology, Eindhoven, Netherlands, (2)Food and Biobased Research, Wageningen UR , Wageningen, Netherlands, (3)School of Chemistry & Chemical Engineering, Queen's University Belfast, Belfast, United Kingdom

Microwave assisted flow processing: coupling of electromagnetic and hydrodynamic phenomena

Narendra G. Patil,[1] Erik Esveld,[2] Faysal Benaskar,[1] Evgeny V. Rebrov,[3] Lumbertus A. Hulshof,[1] Jan Meuldijk,[1] Volker Hessel[1] and Jaap C. Schouten[1]

[1]*Laboratory of Chemical Reactor Engineering, Eindhoven University of Technology, P.O. Box 513, 5600 MB Eindhoven, the Netherlands*

[2]*Wageningen UR Food and Biobased Research, P.O. Box 17, 6700 AA, Wageningen, The Netherlands*

[3]*School of Chemistry and Chemical Engineering, Queen's University Belfast, Stranmillis Road, Belfast, BT9 5AG, United Kingdom*

Introduction

Fast and volumetric heating behavior of microwaves is attracting attention for their application in continuous flow synthesis of specialty chemicals. However, the use of tuned microwave field even in small defined reactors does not always provide the expected temperature patterns.[1,2] This is majorly due to a lack of position specific information with respect to velocity profile of liquid. Additionally, the accuracy of these predictions in microwave heating is limited when no space distribution of volumetric heating source is considered.[1] Understanding the space distribution of microwave heating process with detailed modeling is a prerequisite for process development and control. In this work, we take a closer look at the influence of liquid velocity profiles on the axial and radial temperature profiles in a tubular microwave integrated millireactor combined with a heat exchanger (Figure 1a).

Results and discussion

Horizontal co-current flow of a reactant (ethanol) and a microwave transparent coolant (toluene) was established in a teflon supported quartz tube (i.d.: 3 mm, o.d.: 4 mm) and shell (i.d.: 7 mm, o.d.: 9 mm), respectively (Figure 1a). The co-current flow of coolant was assumed to avoid overheating of the reactant by inherently fast microwave heating. The reactor was inserted in a microwave cavity with a length of 47 mm. The physical model included electromagnetic interactions, fluid dynamics and heat transport (Figure 1b). The temperature and flow profiles were obtained in a 3-D domain using the COMSOL software. As the dielectric properties of the reactant were changing with temperature, it was expected that they should influence the heating rate. However, this effect was of minor importance (Figure 1c) and the remaining simulations were made in a 2-D domain. Effect of gravitation was observed due to horizontal arrangement of the reactor-heat exchanger. The effect of gravity was clearly pronounced in the coolant flow where a recirculation pattern was observed due to difference in density (Figure 2). The coolant flow rate had a minor effect on temperature profiles and energy dissipation at a high heating rate of 180 °C/s (Figure 3, a and b). Higher temperatures were observed at the wall (stagnant liquid films), leading to drastic drop in the energy dissipation (Figure 3, c and d). The laminar flow of ethanol was disturbed by a temperature probe (Figure 4). The probe was inserted either from the outlet (Figure 4a) or inlet (Figure 4b) side of the assembly. In the former case, measured temperature was lower as compared to the latter case (Figure 5). This effect could be explained by local hydrodynamics near the tip of the probe.

When the probe was inserted from the outlet, the highest velocities were found at the tip. Conversely, when the probe was inserted from the inlet, a stagnant film assured lowest velocity at the probe tip. These observations were then validated by experiments (data points, Figure 5). The temperature profiles obtained by the modeling study agreed with the experimental observations.

Conclusions

The stagnancy of the flow of the microwave absorbing fluid influenced the temperature distribution in a microwave integrated flow reactor. The buoyancy influence (as a result of gravitational forces) was already visible for horizontal arrangement of reactor-heat exchanger assembly at millimeter sizes. The stagnant layer formation caused by any insertion or at reactor walls yielded higher temperatures and lower energy dissipation regions. The coolant flow, unless used as a heated jacket to minimize heat losses, was found to be ineffective. Lastly, complex 3-D geometry modeling efforts can be eliminated at very early stages unless there are specific disturbances in electric field pattern.

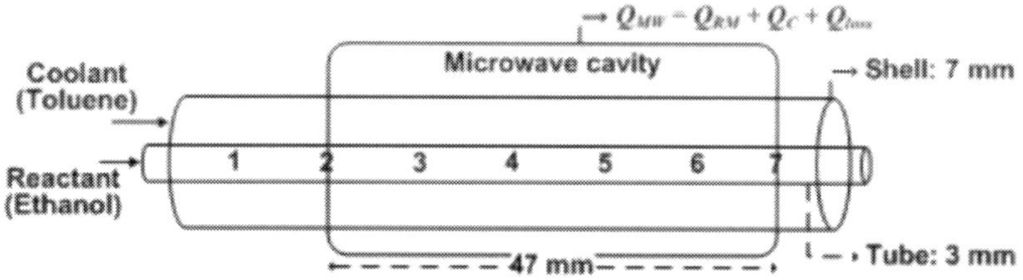

(a)

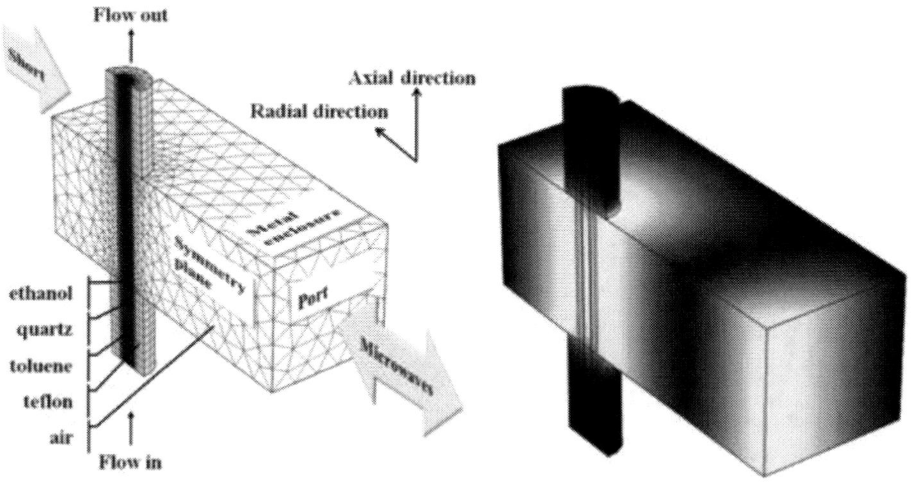

(b) (c)

Figure 1: *a)* Schematic view and process details of the microwave integrated reactor-heat exchanger assembly used for experimental validations, *b)* 3 D computational domain of the microwave cavity showing individual components of the assembly, *c)* Electric field intensity in and around the microwave integrated reactor-heat exchanger. Ethanol flow rate: 40 ml/min, toluene flow rate: 100 ml/min. Input microwave power: 55 W.

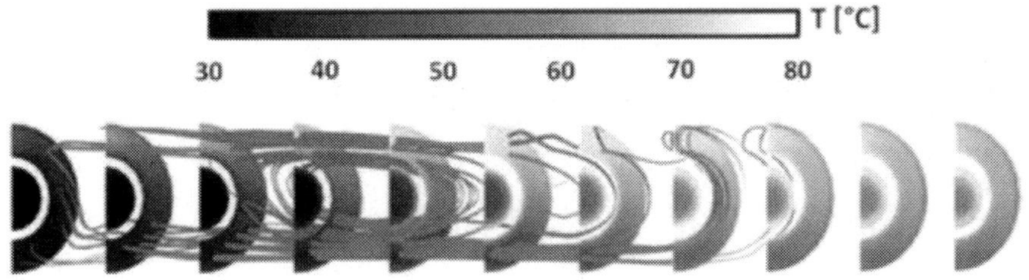

Figure 2: Effect of buoyancy in a horizontal arrangement of the microwave integrated reactor-heat exchanger assembly. Gray lines (with thickness indicating the strength) show velocity profiles of coolant. Axially sectioned colored pattern shows temperature in the reactant as well as the coolant section.

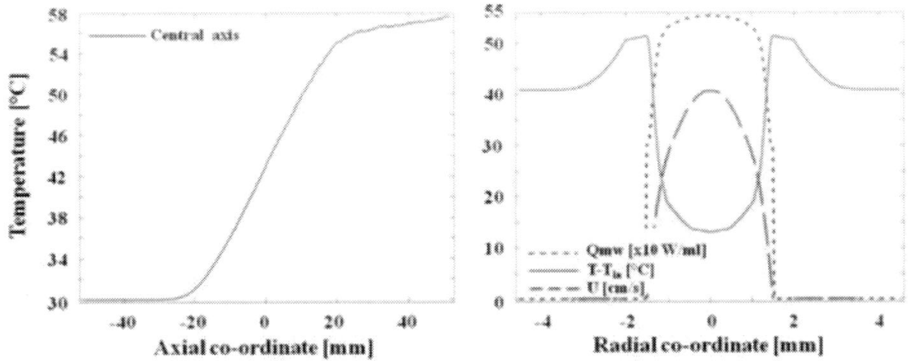

(a) (b)

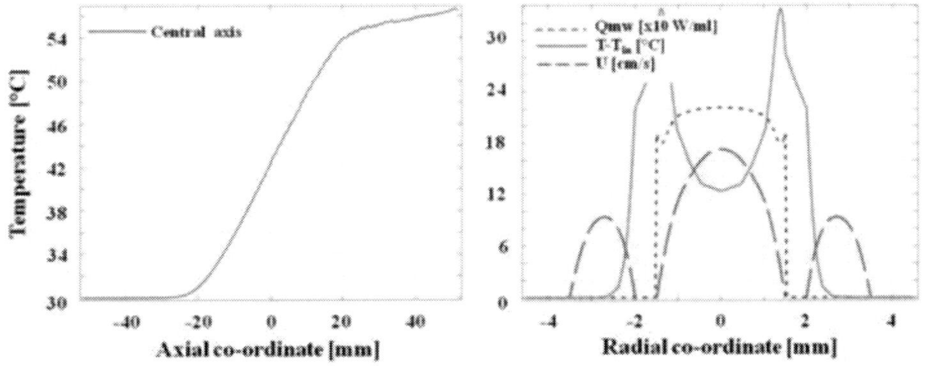

(c) (d)

Figure 3: Temperature profiles *(a,c)* as a function of axial co-ordinate, and *(b,d)* energy dissipation of microwaves (Q_{MW}, W/ml), temperature rise ($T\text{-}T_{in}$, °C) and velocity profile (U, cm/s) as a function of radial co-ordinate. *(a)* near stagnant coolant flow, F_{EtOH} = 100 ml/min, F_{TOL} = 0.5 ml/min. *(b)* high coolant flow, conditions are the same as those in Fig 1c.

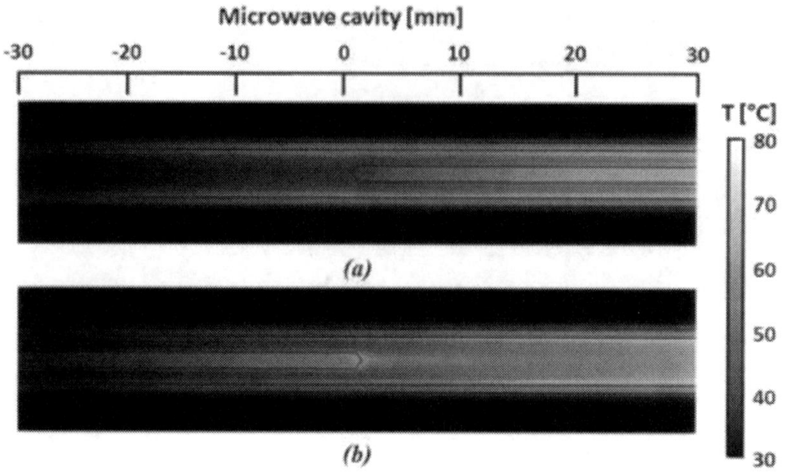

Figure 4: Influence of the direction of probe insertion on the temperature profiles obtained by modeling, a) probe inserted from the outlet, *b)* probe inserted from the inlet. Conditions are the same as those in Fig 1c.

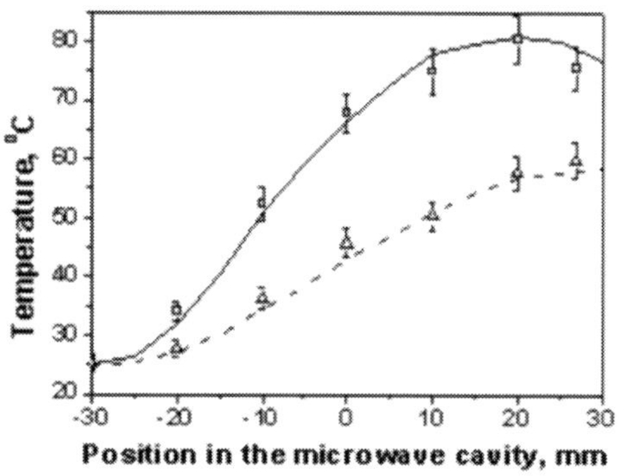

Figure 5: Influence of the direction of probe insertion on the temperature profiles obtained by modeling (lines) and experiments (data points). Solid line and squares: probe inserted from inlet, dotted line and triangles: probe inserted from outlet. Conditions are the same as those in Fig 1c.

Reference:

1. Patil, N. G.; Hermans, A. I. G.; Benaskar, F.; Rebrov, E. V.; Meuldijk, J.; Hulshof, L. A.; Hessel, V.; Schouten, J. C. Energy efficient and controlled flow processing under microwave heating by using a milli reactor-heat exchanger. *AIChE J.* **2011**, *Accepted*. DOI 10.1002/aic.13713.

2. Patil, N. G.; Benaskar, F.; Rebrov, E. V.; Meuldijk, J.; Hulshof, L. A.; Hessel, V.; Schouten, J. C. Continuous multi-tubular milli-reactor with a Cu thin film for microwave assisted fine-chemical synthesis. *Ind. Eng. Chem. Res.* **2012**, *submitted*.

Process Intensification for Hydrogen Generation using Micro-Channel Steam Reforming

Presentation ID# 281083

Peter Bossard, Jacques Mettes and Luis Breziner **Power & Energy, Inc.**
Benjamin A. Wilhite **Texas A&M University**

1. Abstract. Today, the advantages of micro-channel technology for next generation steam reforming have been demonstrated on a pilot scale. Modeling and experimental efforts are reported in the context of Power & Energy, Inc's (P&E) development of a radial tube-in-tube microreactor (TTμR) for efficient, scalable production of hydrogen (H_2) and/or synthesis gas (CO, H_2) from natural gas at competitive costs. The novel radial micro-channel reactor allows manipulating heat- and mass-transport phenomena in catalytic-wall heat-exchanger microreactors to achieve increased catalyst utilization, process intensification and reactor stability. Using similar time proven production methods as used in hydrogen purifier manufacturing, a successful commercialization of the novel microreactor technology is expected with the potential to achieve transformative breakthroughs in system cost, durability and thermal efficiency as compared to existing planar μreactor systems.

2. Introduction. The efficient catalytic conversion of hydrocarbons, including natural gas or methane, to synthesis gas remains a critical challenge for both emerging clean energy industries and existing petrochemical and natural gas markets.[1,2,3,4,5,6,7,8] Hydrogen gas represents a universal energy currency, in that it may be harvested from myriad hydrocarbon resources. Hydrogen can be utilized in electrochemical power systems capable of exceeding Carnot-Cycle thermodynamic limits on efficiency. Process intensification strategies have been employed in the design of next-generation methane reformers where closely coupling of endothermic steam reforming with exothermic combustion,[9,10] yields breakthroughs in thermal efficiency. Potential integration with gas purification via membrane[11,12] or adsorption[13,14] systems further enhances the reaction conversion process while providing a pure hydrogen product. Thanks to pioneering efforts of companies like the Battelle, Velocys and the Institut fur Mikrotechnik Mainz (IMM), micro-reactor technologies have been identified as a potentially transformative technology for realizing next-generation methane reformers for several markets.

Microchemical systems have been demonstrated as a promising technology for chemical processing, with (i) improved heat and mass transport owing to reduced characteristic lengths in microchannels, (ii) resulting enhanced control of reaction temperature and hot-spot formation, and (iii) their potential for integrating multiple processes within one compact unit for unrivalled process intensification and portability.[15] Enhancement of heat and mass transport rates within micro-channels has fueled significant research towards developing thermally-integrated μreactor systems capable of combining multiple endothermic and exothermic catalytic and/or physical processes within a compact, efficient device. To-date, several μreactor designs have been reported for integrating catalytic combustion with endothermic steam reforming of hydrocarbons [16, 17] and subsequently with water-gas-shift [18] and/or membrane separations[19,20] for further cleanup of hydrogen product.

Fabrication of μreactors traditionally involved machining two-dimensional patterns into individual silicon,[18] metal[7,21] or ceramic[22,23] plates via chemical etching or laser cutting, followed by sealing and packaging of a "stack" of such plates to achieve the desired total reactor volume.[24,15] This approach results in a near-linear cost of scale-up, [15] which has limited the market penetration of μreactors to the research and niche-chemical markets. Recently, the extension of mass-production techniques such as roll-embossing and laser cutting of individual metal plates has enabled the production of industrial-scale μreactors at competitive costs[25,26,7]

The planar μreactor design is a major achievement, it has demonstrated the potential of micro-channel technology, but it does not capture the true potential that micro-channels offer. It suffers inherent drawbacks in terms of thermal efficiency for heat-exchanger applications. In the planar μreactor design, the multitude of individual reaction channels are embedded in a single monolithic substrate, allowing significant thermal "cross-talk" between individual microchannels; more importantly, this means that each microchannel is provided with a direct conductive pathway to the outer surface of the micro-reactor. These drawbacks have been shown to play a significant role in reducing the effectiveness of micro-reactors.[27,28,29,30] Current solutions to this challenge involve the combustion of additional hydrogen and/or

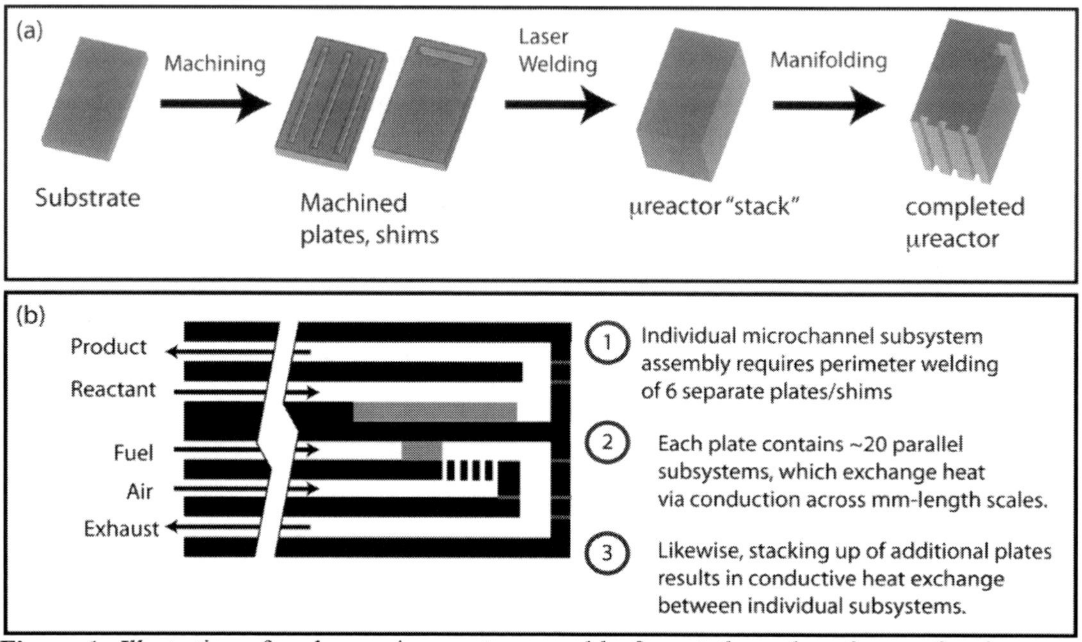

Figure 1: Illustration of a planar microreactor assembly for autothermal methane reforming to synthesis gas; (a) generalized process flow for microreactor construction, adapted from LeViness et al., 2011; (b) cross-section schematic a planar microchannel subsystem for integrated methane combustion and steam reforming, adapted from Tonkovich et al.

fuel to compensate for heat losses at the external surface of the micro-reactor system.[28] In addition to transverse or radial heat dispersion, solid-phase axial heat conduction further limits the thermal efficiency of planar microreactors. Analysis of microchannel heat-exchangers [31, 32] demonstrate the importance of maintaining significant axial thermal gradients within the solid phase to ensure maximum driving force for heat transfer between fluid volumes along the entire axial length of the exchanger. However, the high substrate-to-process volume ratios of existing planar systems result in significant axial dispersion of process heat by the solid phase, such that the lower limit of an isothermal slab is approached with commiserate losses in heat-exchanger efficiency.

	Planar System*	**Tubular System****
Subsystem Volume	75 cm^3	17.9 cm^3
Process Volume	10 cm^3	9.3 cm^3
Sealing Surface	53.3 cm^2	21.4 mm^2
Heat Transfer Surface	256 cm^2	260 cm^2
Surface Area-Process Volume Ratio	~2,500 m^{-1}	~2,800 m^{-1}
Sealing Area-Process Volume Ratio	~5,000 m^{-1}	~4.6 m^{-1}
Process Volume-System Volume Ratio	0.14	0.52
Volumetric Heat Transfer Flux	65 W/cm^3 @ 4ms	840 W/cm^3 @ 5ms

*values estimated based upon dimensions and performance reported by Tonkovich et al.[53] for a single microchannel subsystem depicted in Figure 1.

**values estimated based upon nominal tube dimensions discussed in Section 3 and preliminary results for methane steam reforming.

3. Tube-in-Tube Microreactors (TTμR): A Novel Approach to Microreactor Design

P&E has developed a novel manufacturing process for realizing compact, high-efficiency radial microchannel systems with significant advantages over their planar counterparts. The heart of this technology is the use of individual tube-in-tube assemblies to realize thermally isolated μchannel subsystems (Figure 2). Individual tube-in-tube assemblies are constructed by coaxially mounting an inner stainless-steel microtube (nominally 1.47 mm OD) inside an outer stainless-steel μtube (nominally 2.03 mm ID). The outer tube is closed on one end, such that two radially-symmetric microchannels are formed by the inner diameter of the inserted tube and the annulus formed between the inner and outer tubes. Manufacturing of the tube-in-tube assembly is achieved via automated 200W continuous laser-welding system developed by P&E capable of achieving high-quality sealing at a rate of ~3 seconds/tube. Mechanical stabilization is provided at the sealed end of the outer-most tube via insertion in a "slip plate," which minimizes transverse deflection of individual tubes while allowing axial expansion with changing temperature. The P&E TTμR design offers several unique industrial advantages over existing planar μreactor, including:

- **Reduction in manufacturing costs:** By utilizing mass-produced tubular substrates and conventional (automated laser-welding) sealing technologies, dramatic reduction in cost of scale can be achieved, as compared to planar μreactors requiring customized manufacturing of individually-patterned plates.
- **Increased Durability, Scalability:** By minimizing the packaging and sealing contact area, device durability is increased by reducing the occurrence of gas leaks, seal failure; likewise, placement of seals is such that each seal can be inspected and repaired if necessary prior to final μreactor assembly. Finally, scale-up is no longer limited by sealing challenges (discussed in Section 2), as individual tubes are not sealed directly to each other.
- **Increased Resistance to Thermal Stresses:** By anchoring the tubular μreactor at one end and allowing the other to "float", the resulting system is capable of absorbing significant thermal gradients without the risk of failure due to thermal strain; likewise, because individual tubes are not sealed directly to each other, thermal expansion does not compromise seals.
- **Improved Power and System Density:** By minimizing substrate volumes through the use of the tube-in-tube architecture, breakthroughs in system density, portability construction costs are achieved.
- **Better Handling of High Pressure Operation:** the hoop stress advantage of, particularly, small OD cylinders allows larger pressure drops for a given wall thickness compared to planar technology. Alternatively, walls can be thinner for similar pressure differences providing better heat transport. The ability to operate at higher pressures may be particularly relevant when sequestering CO_2.

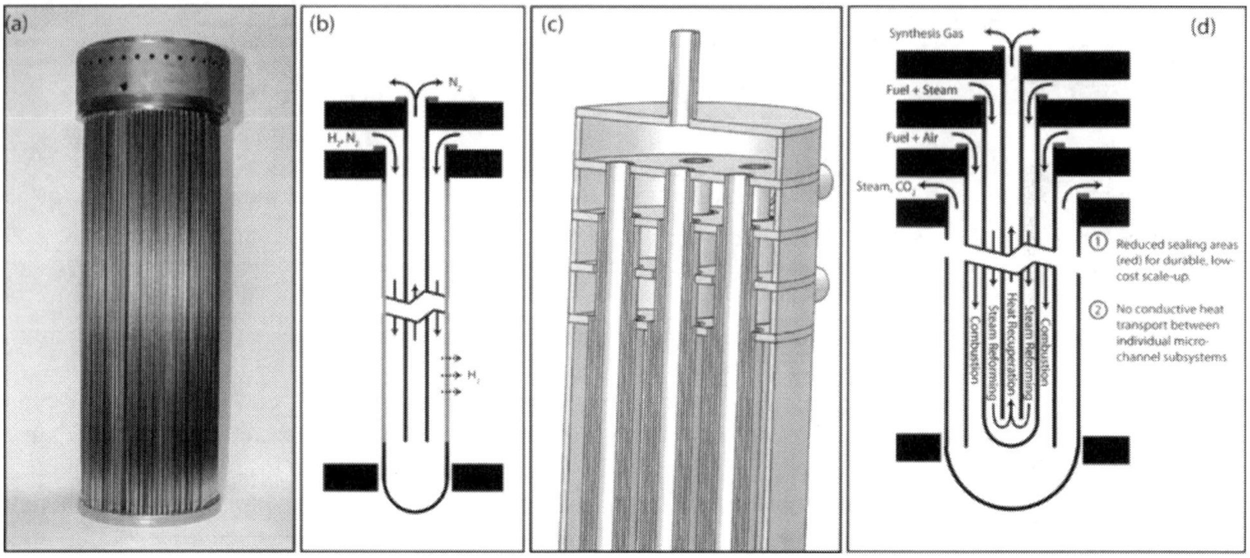

Figure 2: P&E tube-in-tube µchannel technology; (a) tube-in-tube Pd-µmembrane system, 6" x 12.5" containing 1400 individual µmembrane subsystems; (b) schematic of individual tube-in-tube µmembrane subsystem; (c) CAD drawing of proposed µreactor system showing packaging strategy; (d) schematic of planned µreactor system for coupling methane combustion with steam/autothermal reforming for compact production of hydrogen and/or synthesis gas.

4. Experimental

4.1. Manufacturing of TTµRs

A video will be shown of existing automated manufacturing methods currently employed in the production of commercial hydrogen Palladium membrane purifiers that is largely similar in the methods used to fabricate the novel tube-in-tube micro-channel devices. Hereunder, are various pictures showing tools and parts of the latter process.

Fig. 6, laser welding connecting tubes on one end only.

Fig. 7, large parallel tubes assemblies

Figure 8 Automated laser welding positioning with computer vision software.

Figure 9 traditional hand weld attempt.

Fig. 10, laser weld result.

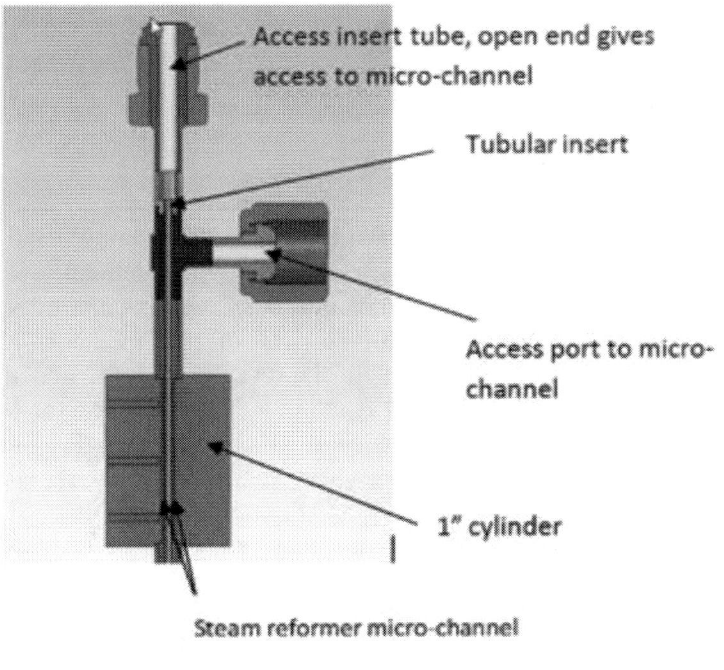

Figure 3: Experimental single cylindrical micro reactor systems developed by **P&E**, which allows the unique ability to precisely monitor heat flux and temperature in a single, isolated microchannel.

4.2. Prototype TTµR for Methane Steam Reforming

P&E's researchers have constructed a bench-scale TTµR system for evaluating heat transfer and reaction rates of individual micro-channel processes. This single TTµR system is comprised of a band heater attached to the outside of a 1" OD Aluchrom tube with bore diameter machined to match that of final tube-in-tube µreactor designs, see figure 3. The band heater allows accurate control of reactor temperature, to within +/- 0.1 K. The inner wall of the bore is coated with steam reforming catalyst and an insert tube is coaxially mounted within the bore to complete the TTμR geometry. Reactor temperatures are measured by thermocouples placed approx. 0.03" from the inner surface of the micro-channel. Heat flux from the band heater to the catalyst wall can then be accurately determined via geometry. Multiple cylindrical reactors are put in series with individual power control and samples can be extracted on either end of any of the reactors. This enables detailed and precise studies of reaction dynamics of the fuel, steam and catalyst in a test setup that has the essential internal features of TTμR based production systems. This apparatus provides, for the first time, operation of individual microchannels with precise monitoring and/or control of heat flux to the microchannel volume and inner-wall surface temperature while minimizing heat losses common to planar single-channel systems.

4.3. Two-Dimensional Simulations of P&E system using COMSOL

Detailed simulations of laboratory-scale single-tube μreactor described above were implemented using the COMSOL multiphysics programming environment. Gas velocity through individual μchannels was described using the Navier-Stokes equation assuming weakly compressible fluids in tandem with the continuity equation while mass transfer of individual species is described by Maxwell-Stefan multicomponent diffusion.[33] Fluid density (ρ) was calculated assuming ideal gas law and dynamic viscosity of the gas mixture is evaluated as a function of composition and temperature using the semiempirical formula developed by C.R.Wilke[34] with all values for Lennard-Jones parameters taken from Bird et al. (2006).[35] Fluid-phase heat transport was described by conduction and convection, assuming negligible radiative heat transport within the μchannel. Fluid-phase thermal properties were taken from physical property databases.[36] The resulting 2-dimensional, radially-symmetric computational fluid-dynamic (CFD) model is identical to models previously implemented by Vlachos' and Vacarro's research groups.[27,28,29,30,37]

Reaction, mass diffusion and heat conduction within the catalyst film was modeled assuming a combination of Darcy's Law (to describe convective transport in porous media) and Maxwell-Stefan diffusion and reaction. The resulting reaction-diffusion model has recently been shown to be capable of comparable accuracy to the more rigorous dusty-gas model of Jackson[38,39,40] for pore diameters > 500 μm.[41,42] Kinetics of methane pyrolysis and water-gas-shift provided in the literature for a $Ni/MgAl_2O_4$ catalyst by Xu and Froment[43] were employed. By imposing a constant temperature boundary condition at the catalyst-μreactor wall interface, i.e. the inner boundary of the catalyst layer, numerical solutions to the above system of partial differential equations predict extent of catalytic reaction as well as the total heat duty required to maintain the externally-imposed μreactor wall temperature.

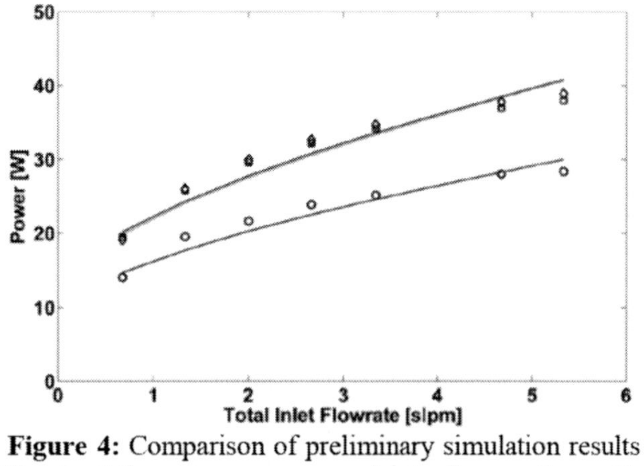

Figure 4: Comparison of preliminary simulation results (lines) of the net power entered into the reaction with experimental data points for single-tube microreactor provided by P&E; feed composition 25% CH_4:75%H_2O at 10bar and (o) 600°C, (□) 650°C and (◇) 700°C.

5. Preliminary Results

Experimental data was initially obtained using a proprietary methane steam reforming catalyst washcoated within a single 1" μchannel system (described in Section 4.2) over a range of inlet gas superficial velocities of 25% methane, 75% steam in terms of the total heat supply added to the reactor to maintain a designated reactor-wall temperature (Figure 4). For the single TTμR described here, heat duties approaching 40W correspond to a volumetric heat flux of > 800W/cm^3, more than an order-of-magnitude improvement over competing planar designs.

Simulations using the model described above showed good agreement between COMSOL predictions and experimental data provided by

P&E relating heat duty vs. process flowrate for the 1" single-channel μreformer (Figure 3). Validation of the COMSOL design model allows theoretical exploration of different catalyst and/or reactor configurations for further improvement in process intensification. Additional experimental and simulation results will be presented in this talk.

6. Conclusion/outlook

Today, the advantages of micro-channel technology for next generation steam reforming have been demonstrated on a pilot scale. A radial microreactor is presented for efficient, scalable production of hydrogen and/or synthesis gas from natural gas at competitive costs. Modeling and experimental tools are introduced that provide, for the first time, operation of individual microchannels with precise monitoring and/or control of heat flux to the microchannel volume and the inner-wall surface temperature while minimizing heat losses common to planar single-channel systems. Simulation, in the absence of any refinement of kinetics or transport properties, showed good agreement between COMSOL predictions and experimental data relating heat duty vs. process flowrate. The presentation shows some of the tools of commercial production methods used for micro-channel hydrogen purifier manufacturing. These tools are very similar to tools needed for radial microreactor manufacturing and successful commercialization of this technology has the potential to achieve transformative breakthroughs in system cost, durability and thermal efficiency as compared to existing planar microreactor systems.

Future work is planned to provide the possibilities and limitations of the radial micro-channel reactor such as: the exceptionally low operational stresses, the possibility of manipulate heat- and mass-transport phenomena in catalytic-wall heat-exchanger microreactors to achieve transformational breakthroughs in catalyst utilization, process intensification and reactor stability.

7. Acknowledgement Funding support for the Development in the form of contract N00014-11-C-0194 was provided by the Office of Naval Research, Washington D.C., with technical oversight provided by the Naval Surface Warfare Center, Philadelphia PA.

8. References

[1] A. Kirubakaran, S. Jain, R.K. Nema, A review on fuel cell technologies and power electronic interface, **Renewable and Sustainable Energy Reviews**, 13 (2009) 2430-2440

[2] F. de Bruijn, The current status of fuel cell technology for mobile and stationary applications, **Green Chemistry**, 7 (2005) 132-150.

[3] A. Zagoria, R. Huycke, Refinery hydrogen management – the big picture, **Hydrocarbon processing**, 82 (2003) 41-46.

[4] N. Hallale, F. Liu, Refinery hydrogen management for clean fuels production **Advances in Environmental Research**, 6 (2001) 81-98

[5] D.C. Elliott, Historical Developments in Hydroprocessing Bio-oils, **Energy & Fuels**, 21 (2007) 1792-1815.

[6] M.E. Dry. The Fischer-Tropsch Process – Commercial Aspects. **Catalysis Today**, 6 (1990) 183-206.

[7] S. LeViness, A.L. Tonkovich, K. Jarosch, S. Fitzgerald, B. Yang, J. McDaniel. Improved Fischer-Tropsch Economics Enabled by Microchannel Technology. White Paper generated by Velocys (2011) http://www.oxfordcatalysts.com/press/wp/wp110224_microchannel_FT_white_paper_24Feb11.pdf

[8] M. Ball, M. Wietschel. The future of hydrogen – opportunities and challenges. **Int J Hydrogen Energy**, 34 (2009), 615-627.

[9] G. Kolios, B. Glockler, A. Gritsch, A. Morillo, G. Eigenberger. Heat-integrated reactor concepts for hydrogen production by methane steam reforming. **Fuel Cells**, 5 (2005), 52-65.

[10] G.D. Stefanidis, D.G. Vlachos. Intensification of Steam Reforming of Natural Gas: Choosing Combustible Fuel and Reforming Catalyst. **Chem. Eng. Sci.,** 65 (2010), 398-404.

[11] F. Galluci, M. Van Sint Annaland, J.A.M. Kuipers. Autothermal Reforming of Methane with Integrated CO_2 Capture in a novel fluidized bed membrane reactor. Part 1: Experimental Demonstration. **Topics in Catalysis**, 51 (2008), 133-145.

[12] A. Vigneault, S.S.E.H. Elnashaie, J.R. Grace. Simulation of a Compact Multichannel Membrane Reactor for the Production of Pure Hydrogen via Steam Methane Reforming. **Chem. Eng. Tech.**, 35 (2012), 1520-1533.

[13] L. Barelli, G. Bidini, F. Gallorini, S. Servilli. Hydrogen Production through Sorption-Enhanced Steam Methane Reforming and membrane Technology: a Review. **Energy**, 33 (2008), 554-70.

[14] Balasubramanian B, Lopez Ortiz A, Kaytakoglu S, Harrison DP. Hydrogen from methane in a single-step process. **Chem Eng Sci** 54 (1999) 3543-3547.

[15] Jensen, K.F., "Microchemical Systems: Status, Challenges, and Opportunities," **AIChE Journal**, **45**(10), 2051-2054 (1999).
[16] Arana, L.R., S.B. Schaevitz, A.J. Franz, M.A. Schmidt and K.F. Jensen, "A Microfabricated Suspended-Tube Chemical Reactor for Thermally Efficient Fuel Processing," **J. Microelectromech. Syst**., 12(5), 600-612 (2003)
[17] Tonkovich, A., D. Kuhlmann, A. Rogers, J. McDaniel, S. Fitzgerald, R. Arora and T. Yuschak, "Microchannel Technology Scale-Up to Commercial Capacity," **Chem. Eng. Res. Des.,** 83(A6), 634-639 (2005).
[18] K. Shah, R.S. Besser. Key Issues in the Microchemical Systems-based Methanol Fuel Processor: Energy Density, Thermal Integration and Heat Loss Mechanisms. **J. Power Sources**, 2007 (166), 177-182.
[19] S.V. Karnik, M.K. Hatalis, M.V. Kothare. Towards a Palladium Micro-Membrane for the Water-Gas-Shift Reaction: Microfabrication Approach and Hydrogen Purification Results. **J. MEMS**, 12 (2003), 93-98.
[20] Wilhite, B.A., S.E. Weiss, J.Y. Ying, M.A. Schmidt and K.F. Jensen, "Demonstration of 23wt% Ag-Pd Micromembrane Employing 8:1 LaNi0.95Co0.05O3/Al2O3 Catalyst for High-Purity Hydrogen Generation," **Advanced Materials,** 18, 1701 (2006)
[21] E.R. Delsman, M.H.J.M. DeCroon, G.J. Kramer, P.D. Cobden, Ch. Hofmann, V. Cominos, J.C. Schouten. Experiments and Modeling of an Integrated Preferential Oxidation-heat Exchanger Microdevice. **Chem. Eng. J.,** 101 (2004), 123-129.
[22] A. Rouge, B. Spoetzl, K. Gebauer, R. Schenk, A. Renken. Microchannel Reactors for Fast Periodic Operation: The Catalytic Dehydrogenation of Isopropanol. **Chem. Eng. Sci.,** 56 (2001), 1419.
[23] R. Knitter, D. Gohring, P. Risthaus, J. Haubelt. Microfabrication of Ceramic Microreactors. **Microsyst. Technol.**, 7 (2001), 85
[24] K. Shah, X. Ouyang, R.S. Besser. Microreaction for Microfuel Processing: Challenges and Prospects. **Chem. Eng. Technol**., 28 (2005), 303.
[25] V .Hessel, B. Cortese, M.H.J.M. de Croon, Novel process windows – Concept, proposition and evaluation methodology, and intensified superheated processing. **Chem. Eng. Sci**., 66 (2011) 1426-1448.
[26] J.J. Brandner in "Micro Process Engineering – A Comprehensive Handbook, Volume 2", V. Hessel, A. Renken, J.C. Schouten, J. Yoshida (Eds.), WILEY-VCH Verlag GmbH & Co. KGaA, Weinheim, 2009. (Chapter 2 pp. 25-43).
[27] G.D. Stefanidis, D.G. Vlachos. High vs. Low Temperature Reforming for Hydrogen Production via Microtechnology. **Chem.Eng. Sci**., 64 (2009), 4856-4865.
[28] M.S. Mettler, G.D. Stefanidis, D.G. Vlachos. Enhancing Stability in Parallel Plate Microreactor Stacks for Syngas Production. **Chem. Eng.Sci.**, 66 (2011), 1051-1059.
[29] M.S. Mettler, G.D. Stefanidis, D.G. Vlachos. Scale-out of Microreactor Stacks for Portable and Distributed Processing: Coupling of Exothermic and Endothermic Processes for Syngas Production. **Ind. Eng. Chem. Res.,** 49 (2010), 10942-10955.
[30] G.D. Stefanidis, D.G. Vlachos. Intensification of Steam Reforming of Natural Gas: Choosing Combustible Fuel and Reforming Catalyst. **Chem. Eng. Sci**., 65 (2010), 398-404.
[31] A. Moreno, K.M. Murphy, B.A. Wilhite. Parametric Study of Solid-Phase Axial Heat Conduction in Thermally Integrated Microchannel Networks. **Ind. Eng. Chem. Res**., 47 (2008), 9040-9054.
[32] T. Steif, O.-U. Langer, K. Schubert. Numerical Investigations of Optimal Heat Conductivity in Micro Heat Exchangers. **Chem. Eng. Technol**. 22 (1999), 297.
[33] C.F. Curtiss, R.B. Bird, Multicomponent Diffusion, **Ind. Eng. Chem. Res.,** 38 (1999) 2515-2522.
[34] C.R. Wilke, A Viscosity Equation for Gas Mixtures, **The Journal of Chemical Physics**, 18 (1950) 517-519.
[35] B. Bird, W.E. Stewart, E. N.Lightfoot, Transport Phenomena, revised Second ed., 2006
[36] R.H. Perry, D.W. Green. Perry's Chemical Engineers' Handbook (7th Ed). McGraw-Hill, (1997).
[37] Vaccaro et al., Modelling of a Catalytic Micro-Reactor Coupling Endothermic Methane Reforming and Combustion, published Online: 10/04/2012, citation Information: International Journal of Chemical Reactor Engineering. Volume 8, Issue 1, Pages –, ISSN (Online) 1542-6580, DOI: 10.2202/1542-6580.2121, March 2010.
[38] R. Jackson, Mass Transport and Reaction in Porous Catalysts, Elsevier, Amsterdam, 1977.
[39] E.A. Mason, A.P. Malinauskas, Transport in Porous Media: The Dusty Gas Model, Elsevier, New York, 1983.
[40] J.W. Veldsink, R.M.J. van Damme, G.F. Versteeg, W.P.M. van Swaaij. The use of the dusty-gas model for the description of mass transport with chemical reaction in porous media. **Chem. Eng. J.,** 57 (1995), 115-125.
[41] R. Suwanwarangkul, E. Croiset, M.W. Fowler, P.L. Douglas, E.Entchev, M.A. Douglas. Performance comparison of Fick's, dusty-gas and Stefan-Maxwell models to predict the concentration overpotential of a SOFC anode. J. **Power. Sources**, 122 (2003), 9-18.
[42] J. Solsvik, H.A. Jakobsen. Multicomponent mass diffusion in porous pellets: effects of flux models on the pellet level and effects on the reactor level. Application to methanol synthesis. **Can. J. Chem. Eng.,** articles ASAP (2012).
[43] J. Xu, G.F. Froment, Methane Steam Reforming, Methanation and Water-Gas-Shift: I. Intrinsic Kinetics. **AIChE J., 35** (1989), 88-96.

AUTHOR INDEX

AUTHOR INDEX

AUTHOR INDEX

AUTHOR INDEX

AUTHOR INDEX

AUTHOR INDEX

AUTHOR INDEX

AUTHOR INDEX

AUTHOR INDEX

AUTHOR INDEX

Computing and Systems Technology Division

Core Programming Topic at the 2012 AIChE Annual Meeting

Pittsburgh, Pennsylvania, USA
28 October - 2 November 2012

Volume 2 of 2

ISBN: 978-1-62276-722-9

Printed from e-media with permission by:

Curran Associates, Inc.
57 Morehouse Lane
Red Hook, NY 12571

CURRAN ASSOCIATES INC. proceedings .com

Some format issues inherent in the e-media version may also appear in this print version.

Printed by Curran Associates, Inc. (2013)

AIChE
3 Park Avenue
New York, NY 10016-5991

Phone: (203) 702-7660
Fax: (203) 775-5177

www.aiche.org

Additional copies of this publication are available from:

Curran Associates, Inc.
57 Morehouse Lane
Red Hook, NY 12571 USA
Phone: 845-758-0400
Fax: 845-758-2634
Email: curran@proceedings.com
Web: www.proceedings.com

TABLE OF CONTENTS

VOLUME 1

VOLUME 2

(304f) Kinetic Modeling of Methanol to Olefins (MTO) On SAPO-34 Catalyst

Guozhen Qi, Siqing Zhong and Hongyuan Zong, Sinopec, Shanghai, China

Methanol to olefins (MTO) coupled with transformation of coal or natural gas to methanol gives an interesting and promising way to produce ethylene and propylene. The kinetics of methanol conversion to olefins on SAPO-34 catalyst was studied in an isothermal multiple fixed bed reactors in parallel. A reaction kinetic model including five lumps was proposed on the basis of Hydrocarbon Pool parallel reaction and parallel deactivation of catalyst. The influence of water and coke deposition was considered in the model. The simulation in fixed bed reactor using the obtained general kinetic equations was performed and also compared with the experimental value. The result showed a good match between the estimated and experimental values. The average activity of catalyst showed a corresponding evolution with the increasing coke formation. Some water could be effective to reduce the deactivation of catalyst and methanol conversion did not drop instantaneously but instead it remained constant for some time before deactivation breakthrough. An empirical model for the coke formation was obtained and made the coke content as a function of time on stream, temperature, WHSV, water/methanol in feed. The obtained model showed a very good linearity correlation.

A simulation using the bubble fluidized bed reactor model coupled with the obtained kinetic model to predict the performance of MTO was performed. The values calculated by the model were identical to the experimental ones. Besides, the probability application of riser reactor in MTO process was also investigated.

(304g) Numerical Investigation of Effect of Hollow Cross-Over Disk On Flow and Heat Transfer in Tube of Cracking Furnace

Yuan Zong, Chemcial Engineering, East China University of Science and Technology, Shanghai, China, Xiaogang Yang, Glyndŵr University , Wrexham, United Kingdom and Ling Zhao, East China University of Science and Technology, Shanghai, China

This paper presents the influence of main parameters of a novel hollow cross-over disk (HCD) on flow, heat transfer enhancement and flow resistance in tube of ethylene cracking furnace. The location of HCD in the tube, geometric size and shape of HCD are studied. Global analysis of thermal performance and local analysis of vorticity flux, turbulence mixing and friction factor are performed based on computational fluid dynamics (CFD). The numerical results indicated the overall Nusselt number of tube decreased when HCDs' location was far from the inlet and the space between HCD pair decreased. While the total pressure drop of the tube had no significant change with the variance of HCDs' location. The effects of the geometrical size of the HCD on heat transfer enhancement were analyzed from the view point of the field synergy principle and it was found that the increase of heat transfer enhancement was always accompanied with the decrease of field synergy angle between the velocity and temperature gradient when the geometrical parameters of HCD were changed. This confirms that the basic mechanism of heat transfer enhancement by HCD can be well described by the field synergy principle. This present results show that heat transfer enhancement can be achieved without much flow resistance using strategies of geometrical optimization and appropriate placement for HCDs, which is benefit to energy consumption decreasing and coke formation inhabitation for tubular cracking furnace.

(305a) Early Hypoglycemia Alarm Systems Based On Autoregressive and Recursive Partial Least Squares Models

Elif S. Bayrak[1], Kamuran Turksoy[2], Ali Cinar[1], Lauretta Quinn[3], Elizabeth Littlejohn[4] and Derrick Rollins[5], (1)Chemical and Biological Engineering, Illinois Institute of Technology, Chicago, IL, (2)Biomedical Engineering, Illinois Institute of Technology, Chicago, IL, (3)College of Nursing, University of Illinois at Chicago, Chicago, IL, (4)Section of Adult and Pediatric Endocrinology, Diabetes, & Metabolism, Institute for Endocrine Discovery and Clinical Care, Chicago, IL, (5)Iowa State University, Ames, IA

Hypoglycemia, having unusually low glucose concentration in blood (usually defined as ≤ 60 mg/dL), caused by intensive insulin therapy is a major challenge of artificial pancreas systems. Patients with diabetes regulate their glucose concentrations by insulin administration. Since patients with Type 1 diabetes (T1D) do not have natural means to reduce plasma insulin levels and their glucagon response is often impaired, they are often unable to prevent low blood glucose levels. Hypoglycemia is prevented or treated by eating and/or reducing or stopping insulin infusion. Hence, early detection and prevention of hypoglycemia is essential for insulin pump users and necessary for the implementation of fully automated closed-loop control. The frequency of hypoglycemia among people with T1DM is variable, and may occur as often as 2-3 times within a day. Hypoglycemia may lead to unconsciousness and seizures if untreated. Availability of continuous glucose monitoring sensors (CGMS) has enabled collection of frequent glucose concentration data and monitoring, prediction and control blood glucose in T1D.

Early alarms to warn the potential of hypoglycemia are essential and should provide enough time to take the necessary action to avoid hypoglycemia. Many alarm systems proposed in the literature are based on interpretation of recent trends in glucose values. In the present study, subject-specific linear models are introduced as a better alternative to capture glucose variations and predict future blood glucose concentrations. These models are then used in early hypoglycemia alarm systems that notify patients to take required action to prevent hypoglycemia before it happens.

An autoregressive and recursive partial least squares (ARPLS) modeling algorithm is proposed to analyze the frequent data from CGMS and predict future glucose values. Partial Least Squares (PLS) is a multivariate regression method, especially convenient for large number of highly correlated data set. PLS summaries the original data matrix to extract the most predictive information and maximize the covariance between input variables and response variables. Current or future glucose concentration can be expressed as a function of previous glucose measurement with autoregressive models. Here in this model, previous glucose concentrations are used as input to model and future glucose concentrations, response variables, are predicted.

In order to adapt the changes in the system, it is necessary to avoid accumulation of old data as new data become available. Autoregressive PLS models developed are recursively updated at each sampling step with a moving window size of one day. Different prediction horizons are analyzed to keep the prediction error minimum while providing a long enough horizon for early detection. An early hypoglycemia alarm algorithm using the developed models is proposed and evaluated.

ARPLS models are tested retrospectively using T1D subject data collected in University of Illinois at Chicago. The algorithm is also tested in *silico*prospectively using an FDA approved metabolic simulator. The data from 17 T1D subject is first filtered to avoid adverse effect of noise on prediction. Even though continuous glucose sensors have analog filters embedded, the output signal is still noisy and should be filtered in order to enhance their signal to noise ratio. A Savitzky-Golay filter is used to reduce noise in the CGMS data.

The alarm algorithm is developed based on the future glucose predictions from ARPLS algorithm. The algorithm first checks the current data and if the value (actual glucose reading) is under the hypoglycemic threshold, immediate hypoglycemia alarm is triggered. Then it checks further for future predictions to trigger early hypoglycemia alarms. When the n-step predicted value crosses the hypoglycemia threshold (70mg/dl), an early hypoglycemia alarm is issued. Only early hypoglycemia alarms are considered to evaluate our algorithm's performance; alarms held during the event are not counted as true positive as the continuous sensors are already equipped with immediate alarms for the current data point. Sensitivity, false positive ratio and time to detection are reported to determine alarm performance.

ARPLS models are evaluated in terms of root mean squared error (RMSE) and sum of squares of glucose prediction error (SSGPE) and compared with previously proposed time series algorithms. For a CGM sampling rate of 5 minutes, 4.45 RMSE and 4.04% SSGPE are obtained for 6 step-ahead glucose predictions for recursive ARPLS. For nonrecursive ARPLS, RMSE is 6.17 and SSPGE is 4.55%. For nonrecursive and recursive time series models (ARMA (3,1)) SSPGE's are reported as 10.32% and 5.56% respectively. Alarm systems developed are tested retrospectively both using the T1D patient data and in *silico* simulations. The performance of alarm systems based on autoregressive recursive PLS is evaluated. Out of 69 hypoglycemia events, 62 of them are detected with average detection time of 28.25 min. Sensitivity of 90% is reported for the early alarm based on 6 step-ahead predicted glucose values. False alarm rate is also reported as 0.36 false positive/day.

The recursive and autoregressive algorithm proposed enable the dynamic adaptation of the models to inter-/intra-subject variation and glycemic disturbances. The ARPLS models developed provide satisfactory glucose prediction with relatively less error than other proposed algorithms and are good candidates to detect and prevent hypoglycemia.

(305b) Multi-Rate Sampled-Data Fault Detection and Fault-Tolerant Control of Hybrid Process Systems

Ye Hu, Department of Chemical Engineering and Materials Science, University of California, Davis, Davis, CA and Nael H. El-Farra, Department of Chemical Engineering & Materials Science, University of California, Davis, Davis, CA

With the increasing emphasis on safety and reliability in industrial process operation and the susceptibility of process control systems to malfunctions in the control actuators and measurement sensors, the problems of fault detection and fault-tolerant control have become the focus of considerable attention in both the academic and industrial communities. While an extensive and growing body of research work already exists on these topics, the majority of existing methods are developed for purely continuous processes. For hybrid processes wherein discrete events are superimposed on the underlying continuous dynamics, however, results have been more limited, primarily due to the difficulty of handling the strong and inherent interactions between the continuous dynamics and discrete events. It is now well understood that these interactions need to be explicitly taken into account in the development of the fault detection and fault-tolerant control methods for hybrid systems. An effort to address this problem was initiated in [1] where an integrated approach for fault detection and monitoring of a class of nonlinear hybrid processes with control actuator faults, uncertain continuous dynamics and uncertain mode transitions was developed. A key idea was the design of a bank of dedicated mode observers using unknown input observer theory to identify the active mode at any given time and distinguish between faults, mode transitions and uncertainties.

In addition to the challenges posed by the combined discrete-continuous dynamics for fault detection in hybrid systems, there are a number of practical implementation issues that must also be addressed in the fault-tolerant control system design. One such issue is the sampled-data constraint which arises due to the increasing use of networked and/or digital control systems in process operations. The information is often collected and transmitted at a certain rate which is dependent on the capabilities of the particular digital control device, the measurement system, and/or the communication medium. The lack of continuous information transfer between the control system components restricts the ability of the control system to accurately monitor and regulate the process. In most practical applications, the discrete sampling constraint impacts not only the sensor-controller interface, but extends to the controller-actuator interface as well. Furthermore, the control system typically makes use of multiple outputs which may be subject to different sampling rates. For example, in chemical reactors, the composition and density measurements typically need several minutes of analysis, while the temperature can be measured at a relatively fast rate. Moreover, the importance of the measurement collected is another factor that can trigger the use of multi-rate sampling. It is reasonable to apply a fast sampling rate to the sensors placed at certain critical locations in the process (e.g., where frequent monitoring and tight control are required), while reducing the sampling rates of the other sensors in order to reduce cost and optimize energy resource consumption.

In this work, we develop a framework for fault detection and fault-tolerant control for hybrid process systems with sensor-controller and controller-actuator communication constraints and multi-rate sampling. To compensate for the lack of continuous measurements from the sensors, an inter-sample model predictor of each mode of the hybrid system is embedded within the corresponding feedback controller to provide estimates of the measurements between sampling instants, and the states of the model are updated by the actual measurements when they become available from the sensors. Owing to the different sampling rates of the different sensors, the updates are not performed in a synchronized fashion. A sample-and-hold scheme is used on the controller-actuator link, which means that the controller outputs are implemented for a certain period of time before the next controller outputs are transmitted. An augmented system is constructed using a switched system formulation, and its dynamics are analyzed to obtain explicit closed-loop stability conditions (in terms of sampling periods and model uncertainty) for each mode in the absence of faults. The fault-free behavior of each model is then used as the basis for the design of a time-varying alarm threshold for actuator fault detection. To ensure stability in the event of faults, the control system must be reconfigured by using another set of healthy fall-back actuators. The key consideration here is that the fall-back actuator configuration must be chosen with the stability of the future operating modes in mind. Specifically, the fall-back configuration should be available to the subsequent modes, and -- more importantly -- the stability conditions for the future operating modes with the fall-back actuator configuration being used must be satisfied given that the operative sampling periods and inter-sample model predictors are fixed. Finally, the fault detection and reconfiguration capabilities of the fault-tolerant hybrid control system are illustrated using a simulated model of a hybrid chemical process.

References:

[1] Hu, Y. and N. H. El-Farra, "Robust Fault Detection and Monitoring of Hybrid Process Systems with Uncertain Mode Transitions," *AIChE Journal*, 57, 2783–2794, 2011.

(305c) Data-Based Proportional-Integral Controller Assessment and Retuning

Tim Spinner[1], Babji Srinivasan[2] and Raghunathan Rengasamy[1], (1)Department of Chemical Engineering, Texas Tech University, Lubbock, TX, (2)Department Chemical Engineering, Columbia University, New York, NY

Proportional-integral (PI) control is the dominant control structure implemented within the process industries, often to less than acceptable effect. In this work, a new algorithm is presented for performance assessment of PI control of self-regulating processes within single-input single-output (SISO) control loops wherein disturbance rejection is the main concern. Two methods are used for diagnosis of sluggish or aggressive tuning: (i) pattern classification of the estimated closed-loop load response to disturbances, and (ii) determination of the process output's Hurst exponent through the method of detrended fluctuation analysis. The Hurst exponent gives a measure of long-standing positive or negative correlations in the system's output, which are indicative of poor controller performance. In an earlier work, the authors applied the method of detrended fluctuation analysis for assessment of a more general class of control loops; however by focusing on the simple PI structure, the present work allows specific recommendations on retuning of controller parameters to be made. The methods used herein rely solely on the measured process output, and therefore, *a priori* model information on the plant to be controlled is not required. The diagnosis and retuning techniques are demonstrated using results from several simulation examples.

(305d) Optimal Device Placement for Fault Tolerant Controller Design of Transport-Reaction Processes with Irregular Process Domain

Antonios Armaou, Chemical Engineering, The Pennsylvania State University, University Park, PA and Michael A. Demetriou, Aerospace Engineering, Worcester Polytechnic Institute, Worcester, MA

Actuator failures have too often plagued chemical processes, often leading to deteriorating product quality and potentially dangerous process operation, such as runaway conditions. Motivated by the importance of the aforementioned problem, the issue of fault tolerant and fault accommodating controller design has been an active research topic in the chemical engineering community. However, while there has been extensive research from the control community on fault detection and diagnosis of finite dimensional systems using model-based robust and adaptive control techniques, the issue of fault tolerant control for distributed processes has been investigated in the last five years.

In this presentation we address the issue of fault tolerance by explicitly considering the spatial variability that transport-reaction processes naturally enjoy, The underlying idea is to identify actuator locations that while being physically apart they have the same authority on the process states. This is only applicable to distributed parameter systems wherein two or more different locations within the spatial domain of definition can provide the same level of controllability and at the same time the same feedback gain can be applicable to all such locations. We capitalize on this property of spatially distributed parameter systems to find locations for different groups of actuators with similar controllability levels, and when a given actuator group fails, then simply deactivating the faulty actuator group and activating another actuator group constitutes the fault accommodation policy, while the control signal in this case is not changed.

Therefore the major conceptual contribution of this work is the efficient identification of ``optimal" actuator locations that have the same level of controllability and design a single feedback controller for the archetypal actuator group. A major hurdle in such a procedure is the complexity of the spatial domain which in general would render the approach computationally infeasible. In principle, the spatial operator eigenfunctions could be combined with the method of weighted residuals to identify an ordinary differential equation (ODE) based approximation of the process description. However, for processes with complex spatial domains such functions cannot be analytically derived. We circumvent this by developing a proper orthogonal decomposition based algorithm that is used to reduce the process description to the ODE form. The algorithm is specifically tailored to account for the statistical nature of the order reduction step and converge to the true optimal actuator locations. An on-line supervisor is subsequently built that includes a fault detection scheme that monitors for possible faults by comparing the plants performance with that of the nominal performance. When a deviation is observed, then a fault is declared. The fault accommodation scheme in this special case does not require a control reconfiguration, but simply the deactivation of the current actuator group and the activation of another healthy actuator group. The profound simplicity of the fault accommodation significantly reduces the costs associated with process supervision with obvious economic and performance savings.

(305e) Use of Parallel Coordinates to Monitor Absorption and Stripper Columns for Optimal CO_2 Removal

Ricardo Dunia, Chemical Engineering Department, The University of Texas at Austin, Austin, TX, Thomas F. Edgar, Department of Chemical Engineering, The University of Texas at Austin, Austin, TX, Gary Rochelle, Chemical Engineering, University of Texas at Austin, Austin, TX and Mark Nixon, Process Management, Emerson, Austin, TX

Parallel Coordinates is a well-recognized visualization technique introduced by Inselberg [1] [2] in which data points are represented with unlimited number of coordinates by overlaying parallel axes. It is a scalable way to represent data as any additional dimension can be visualized by adding an extra axis without affecting the rest of the graph. In parallel coordinates each dimension is drawn as a vertical (or horizontal) line, and a point is visualized as a poly-line that intercepts each axis/coordinate at the appropriate location.

The Figure below illustrates the use of parallel coordinates to demonstrate the temperature profile along an absorption column in a carbon dioxide removal process. Each of the seven poly-lines represent a different time sample with eleven temperature measurements. The time axis is the first parallel coordinate, and the temperature parallel axes are in the same order placed in the absorber (bottom to top). The high and low values and temperature scaling are made the same for all temperature axes. The use of parallel coordinates not only assists the visualization of the temperature profiles, but also allows the user to append additional axes without altering the existing configuration.

This work exploits the use of parallel coordinates to monitor the absorber and stripper columns in a CO_2 removal pilot plant. In particular, parallel coordinates are used in combination with principal components to determine optimal operating regions. This technique is called PC^2 [3] and consists of the visualization of scores or latent variables and the square prediction error in parallel coordinates. The insight provided by PC^2 assists the data visualization and estimates the necessary adjustments in order to improve operating conditions in the CO_2 removal columns. The results demonstrate the great potential that PC^2 can provide to monitor industrial processes.

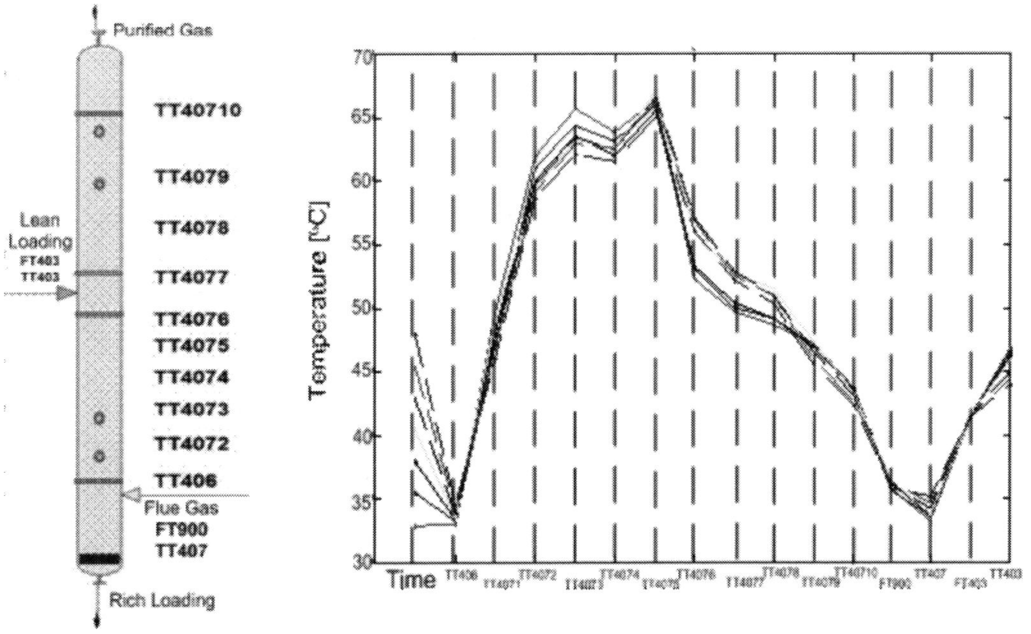

The measurements from the absorption column sensors (left) are used in the parallel coordinates plot (right) for process monitoring. Such a visualization tool can provide an important insight of the plant operations and multivariable statistical control of the process.

References

[1] A. Inselberg and B.Dimsdale, Parallel coordinates - a tool for visualizing multidimensional geometry. 1st IEEE Conf On Visualization (Visualization 90), San Francisco, CA, OCT 23-26, 1990; 361--378.

[2] A. Inselberg, Visual data mining with parallel coordinates, Computational Statistics. 1998; 13,(1) ; 47--63.

[3] R. Dunia, T. Edgar and M. Nixon, Process Monitoring using Principal Components in Parallel Coordinates. Paper accepted for Publication in AIChE Journal.

(305f) A Novel Dynamic Bayesian Network Approach for Networked Process Monitoring and Fault Propagation Diagnosis

Mudassir Rashid and Jie Yu, Department of Chemical Engineering, McMaster University, Hamilton, ON, Canada

Bayesian networks are graphical models for performing probabilistic inference that can be used for a large number of problems under systematic uncertainty such as reasoning, learning causality analysis, etc. Bayesian networks provide graphical structures that effectively characterize the complex casual relationships among different variables for evaluating system uncertainty using probabilistic inference. They are directed acyclic graphs consisting of a set of nodes that represent variables with a finite set of states. The arcs or edges connecting the various nodes represent the probabilistic causal dependence among the variables.

A novel networked process monitoring and fault propagation diagnosis method is developed in this study. Dynamic Bayesian networks (DBN) are used to monitor and diagnose abnormal events in chemical processes as well as to model the propagation of faults throughout the processes. Despite plenty of literature research on process monitoring, inadequate attention has been paid on diagnosing the root causes of abnormal events when multiple potential sources of faults exist. The proposed process monitoring scheme integrates preliminary process knowledge with operational data to identify the root causes and diagnose the propagation of process faults. The preliminary process knowledge is used to map process flow diagrams into the Bayesian network structure and the DBN is then trained using plant historical data. Due to the dynamic and stochastic natures of chemical processes, a probabilistic analysis technique based on Bayesian inference is developed to compute the likelihood of abnormality of process variables for fault identification. The trained network model is also used to determine the fault propagation throughout the processes due to intricate variable interactions and thus diagnose the root causes of process faults. The presence of uncertainty is further quantified by the node relationships and the adaptive model parameters. The identified fault propagation sequence provides a better understanding as to the abnormal process operation and the interactive effects of faults throughout the processes.

The proposed approach is demonstrated using the Tennessee Eastman Chemical process and the fault identification and diagnosis performance is evaluated against conventional principal component analysis (PCA) and independent component analysis (ICA) based variable contribution plots. The fault identification and fault propagation diagnosis capability of the proposed networked process monitoring approach is proven through accurate fault detection and root-cause diagnose of the abnormal events. Moreover, the propagation pathways of process faults provide a deep understanding regarding the effects of the abnormal event throughout the processes. It is shown that the proposed networked process monitoring and fault propagation diagnosis method accurately identifies the faulty process variables and effectively determines the fault propagation pathways to diagnose the root causes of process faults.

(310a) Optimal Design of Transient Microreactor Experiments

Spencer D. Schaber and Paul I. Barton, Chemical Engineering, Massachusetts Institute of Technology, Cambridge, MA

Development of synthetic routes for active pharmaceutical ingredient (API) synthesis is increasingly performed using microreactors, because their use can reduce experimental cost (both time and material use), make hazardous reactions safer, and enable more precise control [1,2]. While microreactor experiments can be designed to find conditions maximizing yield or selectivity by adjusting processing conditions, others can be designed to discriminate between reaction mechanisms and identify rate parameters as precisely as possible. The latter approach allows more flexibility because mechanisms and model parameters (such as kinetic and equilibrium constants) remain relevant for a range of reactor inlet conditions. This is useful during iterations in a full process design, allowing a mathematical model for a whole process to be optimized. With a dynamic process model in hand, the process analytical technology (PAT) and control systems can be designed as well. Model-based optimal experimental design techniques have been used to maximize the information content of experiments, allowing a single mechanism to be selected and its parameters to be estimated precisely, and the results used to scale up to a 500x larger flow reactor [3]. McMullen and Jensen [3] used a sequence of steady-state experiments in a fully-automated microreactor configuration to choose the best reaction mechanism, then fit a preexponential factor and activation energy. It used relatively little experimental time and material due to automation and the small size of the microreactor. Still, ever faster and cheaper process development is desired to scale up and certify processes, and enter the market as soon as possible. Here we present an evolution of the above concept, in which transient experiments are designed rather than steady-state ones. By modeling the behavior of the system at all times instead of only at steady state, any physically-relevant input functions (such as ramp functions for temperature and flow rate) can be used and measurements can be taken at any time (subject to the limitations of the measurement technology). This eliminates the waste of time and material which occurs between steady states in sequences of steady-state experiments, thus allowing more informative experiments to be performed at less cost. We present case studies with a pharmaceutically-relevant reaction to show for a fixed experimental cost, the degree to which transient experiments reduce uncertainty in (i) parameter estimates and (ii) product purity. We also show for a fixed parameter precision, the degree to which experimental resources can be reduced by optimally designing transient experiments.

[1] Roberge, D. M.; Ducry, L.; Bieler, N.; Cretton, P.; Zimmerman, B. Microreactor technology: a revolution for the fine chemical and pharmaceutical industries? Chemical Engineering and Technology 28 (3), 318–323, 2005.

[2] Ahmed-Omer, B.; Brandt, J. C.; Wirth, T. Advanced organic synthesis using microreactor technology. Organic and Biomolecular Chemistry 5, 733–740, 2007.

[3] McMullen, J. P.; Jensen, K. F. Rapid determination of reaction kinetics with an automated microfluidic system. Organic Process Research and Development 15, 398–407, 2011.

(310b) Building Process Robustness and Flexibility Through Implementation of QbD Methodologies

Kevin Sirk[1], Lorrie Berwick[2], Jason Dorwart[2], Rositza Petrova[3], Paul Fernandez[4] and Luke Schenck[5], (1)Chemical Process Development and Commercialization, Merck & Co., Inc., Rahway, NJ, (2)Analytical Development and Commercialization, Merck & Co., Inc., Rahway, NJ, (3)Center for Material Science and Engineering, Merck & Co., Inc., Rahway, NJ, (4)Chemical Process Development and Commerciliazation, Merck & Co., Inc., Rahway, NJ, (5)Chemical Process Development and Commercialization, Merck & Co, Inc., Rahway, NJ

Effective Quality by Design (QbD) achieves a thorough understanding of processes via a systematic set of development activities guided by a comprehensive risk assessment. The output of the thorough process understanding is a well-defined control strategy that achieves consistent delivery of a product's critical quality attributes (CQAs). Elements of this control strategy include raw material specifications, in process testing, and control via design space definition; moving away from the prior methodology of "testing in quality."

This talk focuses on the QbD development work related to a drug substance crystallization process. Factors from the risk assessment were investigated in a multi-factor Design of Experiments (DOE), with design space ranges selected such that the output of the DOE demonstrates the following: the process ensures all drug substance CQAs are achieved, the process is robust and well understood, and the process is flexible to accommodate the supply chain. Responses from the DOE were statistically analyzed to evaluate their impact on drug substance CQAs across the design space. The results will show that all drug substance CQAs were achieved across the entire design space, demonstrating a robust process that delivers quality API.

(310c) Scale-up Considerations for a Continuous Process with QbD in a Manufacturing Setting

Stephen B. Jeffery, Process Engineering, Eli Lilly and Company, Kinsale, Ireland

To be further developed.

A minimum of 150 words is required for the abstract text, and no more than 1500 words may be submitted. Each abstract is allowed to include up to 20 figures.

As part of a new process API introduction we undertake an equipment comparability study to justify equipment selection scale up decisions. This part of the technical transfer process requires that the knowledge developed at smaller scales is understood at the larger scale, in particular that process parameters, the basis of safety and operability issues are adequately considered for each unit operation and risks identified and controlled.

This talk will outline the issues arising when applying this procedure to continuous processes at manufacturing scale at Lilly's Kinsale, Ireland facility, taking QbD principles into account and translating these to a real manufacturing plant.

This talk will outline the issues arising when applying this procedure to continuous processes at manufacturing scale at Lilly's Kinsale, Ireland facility, taking QbD principles into account and translating these to a real manufacturing plant.

(310d) Application of the Discrete Element Method (DEM) to Understand a Powder Blending Process and De-Risk Scale-up

William R. Ketterhagen and Patrick Daugherity, Worldwide Research and Development, Pfizer Inc., Groton, CT

Scale-up of batch manufacturing operations in the pharmaceutical industry often presents technical challenges and various risks. In this work, a case study describing the use of a discrete element method (DEM) model to study a batch blending process is described. This model was developed to (1) gain insight into undesirable blend uniformity results observed during the manufacture of a recent drug product registration batch and (2) to assess the risk associated with scaling-up the process. During execution of the pilot-scale registration batch, the team had observed the presence of a sub-potent region of the blend that was insufficiently mixed in addition to variability in the potency of stratified samples. The model has confirmed that poor blending is to be expected in this particular bin and visually depicts the presence of a "dead zone" where mixing does not occur due to the particular processing conditions used in the operation of this blender. The model also revealed that, by employing certain modifications to the process, a uniform blend could be achieved in this equipment. To assess the risk in scaling-up this process, the blending performance in the commercial scale bin was predicted to be satisfactory. Results from the commercial scale batch confirm that a well-mixed blend is in fact achieved in this equipment. The use of process models such as this enable scientists and engineers to better understand the process and the associated risks which help to enable successful drug product development and scale-up.

(310e) Validation of a Scale-up Model for the Lubrication Blending Process From Material Sparing to Commercial Scale

Joseph Kushner IV[1], Holger Schlack[2], Margit Karle[2] and Deborah Booth-Schindler[2], (1)Drug Product Design, Pfizer Worldwide Research and Development, Groton, CT, (2)Pfizer PPD, Freiburg, Germany

Turbula bottle blenders are often used in material sparing scale experiments during early-stage pharmaceutical product development. However, applying knowledge gained with these blenders to larger scale diffusion mixers is limited by the lack of blending models that are validated with commercial scale equipment. The present investigation attempts to incorporate Turbula bottle blenders and commercial scale blenders into a recently-developed lubrication blending process scale-up model:

$$QA/QA_0 = (1-\beta) + \beta*\exp(-\gamma*L*F_{Headspace}*r)$$

where QA_0 is initial quality attribute value, β is sensitivity of the QA to lubrication, γ is formulation-specific lubrication rate constant, L is characteristic mixing length scale, $F_{Headspace}$ is the blender headspace fraction, and r is the number of blending cycles.

2:1 blends of microcrystalline cellulose and spray-dried lactose or dibasic calcium phosphate were mixed with 1% magnesium stearate using Turbula bottle blenders and pilot/commercial scale blenders, varying blender volume, V, (30-1250 ml; 100-2000 Liters); $F_{headspace}$(30, 70%); and r (24 to ~190,000 cycles). Tablets were manufactured using a Huxley-Bertram compaction simulator to evaluate the change in tablet tensile strength as a function of the extent of lubrication mixing. Other attributes of the blends/tablets were also evaluated including compressibility, bulk/tapped density, friability, and disintegration.

The validity of the proposed lubrication blending process scale-up model was shown for blenders from 30-ml to 2000-liter (an approximately five order of magnitude change in scale) for both placebo formulations. This validated scale-up model can be used to replace commercial-scale experiments with bench-scale, material-sparing experiments to support QbD lubrication blending process understanding and design space verification.

(310f) Quality by Design Development of a Film Coating Process Resulting in a Dimensionless Design Space

Alan R. Silverman[1], W. Mark Eickhoff[2], Fochive N. Njikam[1] and Manoharan Ramasamy[3], (1)Pharmaceutical Commercialization Technology, Merck & Co., Inc., West Point, PA, (2)Pharmaceutical Commercialization Technology, Merck & Co., Inc, West Point, PA, (3)Process Analytical Technologies, Merck & Co., Inc., West Point, PA

The Quality by Design development of a film coated tablet drug product began with a risk assessment which identified moisture sensitivity as a key product quality attribute risk. Additionally, tablet moisture was viewed as a key measure for film coated tablet elegance attributes. The film coating unit operation, where the drug product was sprayed with an aqueous based coating suspension, was a key focus of this development. A theory was devised that the tablet moisture could be described by overall thermodynamic factors, spray and tablet flux factors, and geometric factors. A design of experiments was executed at both lab and commercial film coating scales. The result of the design of experiments was the identification of the high, medium, and low risk process factors impacting the tablet moisture response via a statistical model in support of this theory. The design of experiments indicated the scale independence of the moisture response over the range of study. This allowed the description of the film coating process design space with a collection of dimensionless groups in the product New Drug Application. This film coating control strategy was accepted by regulatory agencies with the approval of the product and it has been implemented at the commercial site, first with process validation and then with subsequent commercial production.

(341a) Developing of Heat Integration Techniques in a Biorefinery for the Sustainable Production of Biodiesel and Succinic Acid

Anestis Vlysidis, Chemical Engineering and Analytical Science, University of Manchester, Manchester, United Kingdom, Dimitrios Kastritis, The University of Manchester, Simon J. Perry, Centre for Process Integration, The University of Manchester, Manchester M60 1QD, United Kingdom and **Constantinos Theodoropoulos**, School of Chemical Engineering and Analytical Science, University of Manchester, Manchester, United Kingdom

Recently, the concept of integrated biorefineries for the production of biofuels and chemicals is becoming very attractive as humanity searches for different methods to reduce dependency on petroleum. The major reason why there is a very small number of applications of this concept so far, is the elevated cost of the feedstocks and/or the processing technologies of biomass. Biofuel industry can become more economic efficient if it is upgraded to integrated biorefineries (Vlysidis et al., 2011). Techniques which are originally implemented in petroleum refineries, like heat integration (Smith R., 2005), need to be applied to biorefineries in order to improve their sustainability.

In this work, heat integration techniques are developed on a biodiesel biorefinery that co-produces bio-succinic acid. The latter is produced by microbial fermentation from glycerol, the main by-product of the biodiesel industry (Vlysidis et al., 2011). Succinic acid can be used as a building block for the production of many specialty and commodity chemicals and it is considered to be one of the top value added chemicals (Bozell J. J. and Petersen G. R., 2010). Industrial biotechnology, like the succinic acid production, contains processes that offer large energy recovery potentials such as the sterilization and the downstream recovery units. Here, the initial heat exchange network (HEN) is being optimized in terms of maximum energy recovery and minimum total annualized cost by using stochastic optimization methodologies. New HENs are designed and compared with the initial scheme in terms of energy recovery and total cost. The acquired environmental benefits (e.g. decrease of CO_2emissions due to the energy recovery) for each case are also assessed and compared with the primary process. Finally, a sensitivity analysis is carried out for the lifespan of the plant and the interest rate which were originally set to 20 years and 7%, respectively.

Bozell J. J., Petersen G. R., (2010) Technology development for the production of biobased products from biorefinery carbohydrates-the US Department of Energy's "Top 10" revisited. *Green Chemistry,* 12,539-554.

Smith R., (2005). Chemical Process Design and Integration. *Wiley.*

Vlysidis A., Binns M., Webb C. and Theodoropoulos C., (2011), A techno-economic analysis of biodiesel biorefineries: Assessment of integrated designs for the co-production of fuels and chemicals, Energy, 36, 4671-4683.

(341b) Optimal Planning of Drilling and Production in Oil Reservoirs

Mohammad Sadegh Tavallali, Department of Chemical & Biomolecular Engineering, National University of Singapore, Singapore, Singapore, Iftekar A. Karimi, Dept of Chemical & Biomolecular Engineering, National University of Singapore, Singapore, Singapore and Kwong Meng Teo, Department of Industrial & Systems Engineering, National University of Singapore, Singapore, Singapore

The oil demand is increasing and therefore oil companies have increased their drilling activities to meet this demand. However, oil reservoirs are limited and mostly are depleting. Hence, in order to reduce drilling costs and increase the efficiency of oil recovery, it is inevitable to focus on optimal well placement and optimal production planning. To do so, all three technical elements of upstream production should be considered in a unified manner. That includes multiphase flow (a) in the reservoir, (b) through the well and (c) into the surface facilities. The production efficiency strongly depends on all these three elements and economic constraints of the problem. In our previous work [1], we focused on the subsurface section as the first above element and the main source of the oil phase. We developed a non-convex MINLP model and modified an outer approximation algorithm to solve that. In this study, we extend our previous work to include the remaining elements (b) and (c). The main objective is to design a state-of-art model that locates the most promising drilling sites for the new/infill oil producers and decides the optimal number of wells. Additionally, it should decide the optimal injection and production flow rates to maximize the net present value (NPV) of the overall (drilling and production) project. Although there have been attempts to address similar problems using evolutionary methods [2,3] or derivative based algorithms [4], we use the power of mathematical programming, which enables us to rigorously model, analyse and solve the problem. In contrast, the first two approaches rely mostly on the numerical reservoir simulators as black box models.

Developing a mathematical model that thoroughly captures sub-surface, well string and surface elements of upstream production is very challenging. It is different from combining several software packages, each specialized for a specific task and phenomenon. The former is a complex task but it can provide much flexibility and has a wide scope; the latter is more straightforward, however it has limited scope. A holistic model should offer very versatile and general formulation that can accommodate different possible events. That is truly a complicated task. This complexity is due to several contributors including: the different nature of spatial grids with and without wells in the reservoir, highly combinatorial nature of the problem and strong nonlinearity of the multiphase flow in the porous media and then through the wells. We have attempted to overcome most of these challenges and we have considered all the three aforementioned elements in the current extended non-convex, dynamic and multi-period MINLP model. Since the problem is inherently very large in size, we have also ensured that the model remains as sparse as possible. That can help both the MILP and NLP solvers.

We tested our extended model using two case studies for two synthetic reservoirs. Both are infill-drilling problems with old producer and injector wells already active in the reservoir. It is targeted to drill up to 8 and 11 wells for the first and second case studies respectively. For comparison purpose, we also defined a base case where no new wells are drilled and old wells continue the production. Our method successfully located 5 wells for the first example and achieved NPV of \$ 5.96×10^8 (even by considering the drilling cost of 5 new wells) whereas the base case only obtained \$$4.50 \times 10^8$. It also suggested to drill 3 new producer wells for the second test to yield NPV of \$ 4.31×10^8. The base case NPV is \$ 3.69×10^8 where no new wells are drilled. In both examples, the optimizer has effectively determined and distributed the production and injection flow rate level of different wells. Moreover, their flow rate profile confirmed the effective responses from the injectors at the event of water breakthrough to maximize the NPV.

Acknowledgements

We would like to appreciate National University of Singapore and SINGA program for the financial support of this research. We are also thankful to Mr. David Baxendale from RPS Energy Limited and Professor Sh. Ayatollahi from

Shiraz University for their valuable industrial and academic insights. Finally, we would like to extend our gratitude to Schlumberger for the provision of the ECLIPSE software package to aid us in our work.

References

[1] M.S. Tavallali, I.A. Karimi, & K.M. Teo, 2011, Dynamic Optimal Well Placement in Oil Reservoirs, Annual Meeting of AICHE, Minneapolis, USA.

[2] B. Yeten, L.J. Durlofsky & A. Khalid, 2003, Optimization of Nonconventional Well Type, Location, and Trajectory, SPE Journal, Volume 8(3), Pages 200-210.

[3] Onwunalu JE, Durlofsky LJ, 2009, Application of a particle swarm optimization algorithm for determining optimum well location and type, Computational Geosciences:1-16

[4] Forouzanfar, F., G. Li, & A.C. Reynolds (2010). A Two-Stage Well Placement Optimization Method Based on Adjoint Gradient. SPE Annual Technical Conference and Exhibition. Florence, Italy.

* Corresponding author: Tel.: +65 6516-6359, Fax: +65 6779-1936, Email : cheiak@nus.edu.sg

(341c) Closed-Loop Design of Building HVAC Systems with Thermal Energy Storage

David Mendoza-Serrano, Chemical and Biological Engineering, Illinois Institute of Technology, Chicago, IL and Donald J. Chmielewski, Illinois Institute of Technology, Chicago, IL

Energy consumption by Heating Ventilation and Air Conditioning (HVAC) systems is usually heaviest when electricity prices are at their highest, presenting significant opportunities for the improvement of the underlying control algorithms. The idea being that thermal energy storage can be used to time-shift power consumption away from periods of high power demand to periods of low power cost. In this work, we present a supervisory control scheme, known as Market Responsive Control (MRC), which has the objective of minimizing the expenditure required for the cooling of a given building. The design of this controller is achieved by a fairly simple convex optimization problem. The MRC method is then embedded within an equipment design algorithm. This extended algorithm, known as MRC Embedded Equipment Design (MRC-EED), has the objective of maximizing Net Present Value (NPV) in an effort to size the storage unit and cooling device. Under the economic assumptions of this study, it is concluded that the NPV of installing Thermal Energy Storage (TES) can be significant if electric price variability is sufficiently large.

(341d) Integration of Solid-Oxide Fuel Cells and Compressed Air Energy Storage for Peaking Power with Zero Carbon Emissions

Jake Nease and Thomas A. Adams II, Chemical Engineering, McMaster University, Hamilton, ON, Canada

In recent years, sustainable production of environmentally friendly electricity has become a major topic of discussion and research worldwide. One promising technology is the solid-oxide fuel cell (SOFC) power generation system [1]-[4]. An SOFC utilizes a carbonaceous fuel gas (synthesis gas, for example) and an oxidant (typically air) to efficiently produce electrical power through an electrochemical reaction across a solid oxide barrier. There are key advantages to using SOFCs for power production. For example, the anode exhaust is mainly H_2O and CO_2, which are easily separated if the anode and cathode streams are not mixed downstream, and the cathode exhaust is mostly N_2, which can be used for additional power generation or heat recovery and then safely vented. However, a significant disadvantage of the SOFC system is that there are currently cost-prohibitive operational challenges associated with its dynamic use, which limits the usefulness of an SOFC system for following a typical diurnal demand profile for electrical power.

This work investigates a novel system that integrates a natural gas fuelled SOFC system for base load power production (Adams and Barton (2010)) with a compressed air energy storage (CAES) plant for load-following capabilities [1]. This new system takes advantage of the already hot and compressed cathode exhaust by temporarily storing it underground with a relatively low parasitic energy penalty. The SOFC/CAES plant may be switched from storage mode (where the CAES system consumes power to store compressed cathode exhaust) to expansion mode (where the CAES system generates power by releasing stored compressed exhaust through a turbine). The plant operates in storage or expansion mode depending on whether the current power demand is lower or higher than the base power output provided by the SOFC system, respectively. The cathode and anode exhaust streams of the SOFC system are not mixed, and thus 100% CO_2 capture from the anode exhaust can be maintained at all times. Simulations of the combined system under a variety of charging and discharging conditions are performed in Aspen-Plus and dynamic simulations of the load-following scenarios and dynamic mass balances on the storage cavern are executed in MATLAB.

Simulation results based on real scaled market demand and pricing data show promising results for the SOFC/CAES system with regard to both its ability to provide peaking power as well as improve gross revenues due to hourly variations in electricity pricing. Moreover, the addition of CAES turbomachinery allows for effective load-following capabilities to be added to the SOFC system with very minor reductions in overall plant efficiency (~1% HHV). Furthermore, unlike other standalone CAES plants, the SOFC/CAES hybrid plant does not require any additional air or fuel in order to operate the CAES section; all of the required electricity, heat, and compressed air can be obtained from the SOFC and its waste streams. Therefore, the SOFC/CAES system provides load-following power generation from fossil fuels with essentially zero CO_2 emissions at high efficiencies. A techno-economic analysis of the combined SOFC/CAES system is also performed, and its profitability is discussed and compared to other systems.

References

[1] Adams T.A. II, Barton, P.I. High-efficiency power production from natural gas with carbon capture. *Journal of Power Sources*. **2010**, *195*, 1971.

[2] Adams T.A. II, Barton P.I. High-efficiency power production from coal with carbon capture. *AIChE J.* **2010**, *56*, 3120.

[3] Duan, L.; Yang, Y.; He, B.; Xu, G. Study on a novel solid oxide fuel cell/gas turbine hybrid cycle system with CO_2

capture. *Int. Journ. of Energy Res.* 2012, 36, 139-152.

[4] Becker, W.L.; Braun, R.J.; Penev, M.; Melaina, M. Design and technoeconomic performance analysis of a 1 MW solid oxide fuel cell polygeneration system for combined production of heat, hydrogen, and power. *Journal of Power Sources*. 2012, 200, 34-44.

Optimization of Biofouling Control and Energy Supply Policies in thermally integrated plants

Fabricio Nápoles-Rivera,[1] Abdullah Bin-Mahfouz,[2] Arturo Jiménez-Gutiérrez[3], Mahmoud M. El-Halwagi,[2] José María Ponce-Ortega[1]

[(1)] *Universidad Michoacana de San Nicolás de Hidalgo, Morelia, 58060, México.*

[(2)] *Texas A&M University, College Station, TX, 77843, USA*

[(3)] *Instituto Tecnológico de Celaya, Celaya, Gto., México*

One of the most important aspects of a process economy is the proper use of energy resources; in this context heat integration strategies have been widely used in order to reduce the amount of external heating and cooling utilities. Although the degree of integration may vary from one process to another, there is typically the need for additional external utilities. Cooling water is the most commonly used cooling utility; however, low salinity resources are limited and, furthermore, not homogeneously distributed in the planet, which has driven the interest for the use of seawater as a cooling fluid. Two main issues associated with the use of seawater are the formation of biofilms and the need for additional refrigerant utilities when the required temperature of any hot process stream is below room temperature. Two design strategies are considered in this work for these types of situations. One is the addition of biocides to control biofilm growth, and the other one is the inclusion of an absorption refrigerant system when cooling water is not feasible as a cooling utility. An integrated scheme of heat exchanger networks together with the use of seawater as cooling utility and an absorption refrigeration system is considered in this paper, so that energy integration aspects are tied to proper biocide dosing scheduling in the search for an optimal implementation policy. A mathematical programming approach is developed to deal with this problem and obtain an optimal scheduling of biocide dosing, mechanical maintenance and energy supply for thermally integrated facilities. The resulting model is a multi objective mixed integer non-linear programming problem. A case study of an Acrylonitrile plant located in Musqat, Oman is used to show the applicability of the proposed approach. Pareto curves are obtained, which help devise attractive solutions considering technical, economic and environmental constraints for the selected process.

Biocide and maintenance scheduling model

The biocide dosing is based in the structure shown in Figure 1.

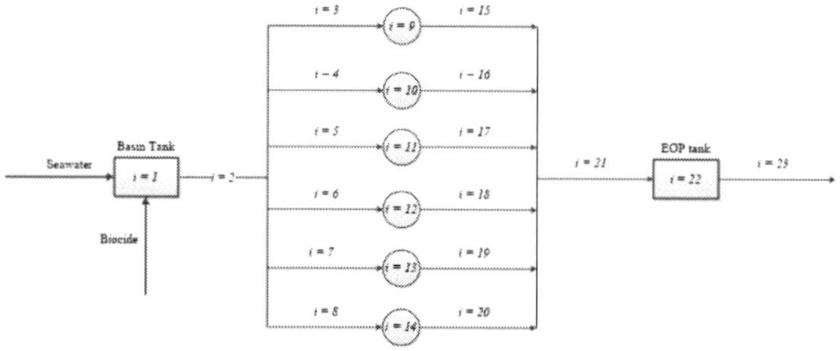

Figure 6. Discretized segments for the seawater cooled system

The concentration in the basin tank is obtained as follows

$$C_t^{basin} = \frac{F_t^B C^{Feed}}{F_{basin,t}}, \qquad \forall t \in NT$$

Biocide can be either produced in plant or purchased off site, the flowrate of biocide used is:

$$F_t^B = F_t^{B,purchased} + F_t^{B,produced}, \qquad \forall t \in NT$$

The residual concentration in each process segment is

$$C_{i,j,t} = C_{i-p,j,t} e^{-k_{i,t}\tau_{i,t}}, \qquad \forall i \in NI, \forall j \in NC, \forall t \in NT$$

The decay constant is proportional to the temperature T in process segment i at time t:

$$k_{i,t} = 6.8\times10^{-4}\left(\frac{1}{\min{}^\circ C}\right)\cdot T_{i,t}, \qquad \forall i \in NI, \forall t \in NT$$

From the energy balance, the following relationship is obtained to calculate the temperature in each process segment:

$$T_{i,t} = \frac{Q_{i,t}}{F_{i,t}\rho^{SW} Cp^{SW}} + T_{i-p,t}, \qquad \forall i \in NI, \forall t \in NT$$

In the EOP tank the residual concentration is calculated as follows:

$$C_{i,j,t} = \frac{C_{i-p,j,t}F_{i-p,t} - C^{Chem}F_t^{Chem} + \psi_t C^{Chem}}{F_{i,t}} e^{-k_{i,t}\tau_{i,t}}, \qquad \forall i = EOP, \forall j \in NC, \forall t \in NT$$

The relationship between the percentage of heat transfer reduction ($\%HTR_{i,t}$) and the biofilm thickness $d_{i,t}$ in process segment i at time t is given by:

$$\%HTR_{i,t} = 0.2153 \cdot d_{i,t}, \qquad \forall i \in NHE, \forall t \in NT$$

The actual heat transferred in process segment i at time t, is

$$Q_{i,t} = Q_{i,t}^{Req}\left(1 - \frac{\%HTR_{i,t}}{100}\right), \qquad \forall i \in NHE, \forall t \in NT$$

The net effect on the biofilm growth is described as:

$$d_{i,t} = 250\frac{\left[1-e^{-\beta_{i,t}\sigma_{i,t}}\right]}{e^{-\gamma_{i,t}\omega}}, \qquad \forall i \in NHE, \forall t \in NT$$

The biofilm growth rate constant $\beta_{i,t}$ for process segment i at time interval t is:

$$\beta_{i,t} = 2.909\times10^{-7}\left(\frac{1}{\min{}^{\circ}C}\right)T_{i,t}, \qquad \forall i \in NHE, \forall t \in NT$$

The biofilm destruction rate $\gamma_{i,t}$:

$$\gamma_{i,t} = 2.829\times10^{-4}\left(\frac{1}{\min(mg/L)}\right)C_{j,i,t-1}, \qquad \forall j = KC, \forall i \in NHE, \forall t \in NT$$

$\sigma_{i,t}$ can be described with the following disjunction:

$$\begin{bmatrix} Y_t \\ \sigma_{i,t+1} = \omega,\ \forall i \in NHE \end{bmatrix} \vee \begin{bmatrix} \neg Y_t \\ \sigma_{i,t+1} = \omega + \left(\frac{-1}{\beta_{i,t}}\right)\ln\left(1-\frac{d_{i,t}}{d^{\infty}}\right),\ \forall i \in NHE \end{bmatrix}, \qquad \forall t \in NT$$

The disjunction is reformulated with a big-M technique [1]. These equations select the day of the year in which maintenance will be scheduled. To give flexibility to the operation schedule of the plant, it is considered that the maximum allowed biofilm thickness can vary depending on whether maintenance is performed during a particular month or not, according to the following disjunction:

$$\begin{bmatrix} Y_M \\ d_{i,M} \le 70\mu m,\ \forall i \in NHE \\ d_{i,M-1} \le 60\mu m, \forall i \in NHE \end{bmatrix} \vee \begin{bmatrix} Y'_M \\ d_{i,M} \le 50\mu m,\ \forall i \in NHE \end{bmatrix}, \qquad \forall M \in NM$$

Therefore, the maximum allowed biofilm thickness is *50μm* if no maintenance is scheduled in two consecutive months, but if maintenance is scheduled during month M, the maximum thickness is *70μm* for that month and *60μm* for the previous month (i.e., month M-1). The disjunction is reformulated again using the Big-M

The maximum number of treatments that can be performed in a month is limited to one:

$$y_M \le 1, \qquad \forall M \in NM$$

The maximum number of treatments during a year is limited to n with the following constraint:

$$\sum_{M} y_M \leq n$$

The relationship between the binary variables y_t and y_M is given by:

$$\sum_{t \in M} y_t = y_M, \qquad \forall M \in NM$$

t is the time interval in days and M is the time interval in months.

Energy Balance

The energy balance for the stripper unit is given by:

$$Q^{Str} = \frac{Q^{\mathrm{Re}f}}{COP} = Q^{Excess\,Process} + \sum_{f \in F} Q_{f,M}^{Fossil,str} + \sum_{b \in B} Q_{b,M}^{Biofuel,str} + Q_M^{Solar,str}, \qquad \forall M \in NM$$

The energy balance for biocide production is:

$$Q^{biocide} = \frac{\left(F_M^{B,produced} C^{Feed}\right)(CBP)}{COB} = \sum_{f \in F} Q_{f,M}^{Fossil,biocide} + \sum_{b \in B} Q_{b,M}^{Biofuel,biocide} + Q_M^{Solar,biocide}, \qquad \forall M \in NM$$

The biocide produced during a given month is calculated with the following expression:

$$F_M^{biocide,produced} = \sum_t 24 \cdot F_t^{biocide,produced}, \qquad \forall M \in NM$$

The total heat provided by each energy source is the sum of the heat required in the stripper unit and the heat required to produce the biocide:

$$Q_{f,M}^{Fossil} = Q_{f,M}^{Fossil,str} + Q_{f,M}^{Fossil,biocide}, \qquad \forall f \in F, \forall M \in NM$$

$$Q_{b,M}^{Biofuel} = Q_{b,M}^{Biofuel,str} + Q_{b,M}^{Biofuel,biocide}, \qquad \forall b \in B, \forall M \in NM$$

$$Q_M^{Solar} = Q_{f,M}^{Solar,str} + Q_{f,M}^{Solar,biocide}, \qquad \forall M \in NM$$

Model for the Solar Collectors

Different types of solar collectors are available. Each collector has a given capacity to transform solar radiation into energy per unit area. In addition, each collector has a specific operational and capital cost per unit area. The selection of the solar collector (if it is required) is modeled with the following disjunction:

$$\left[\begin{array}{c} Y^{Solar} \\ C^{Tank} = FC^{Tank} + VC^{Tank}\left(Q^{Solar\_Max\{M\}}\right) \\ \underset{s\in S}{\vee}\left[\begin{array}{c} Z_s^{Solar} \\ Q_t^{Solar} \le Q_{s,M}^{Useful\_Solar}\, A_c^{Solar}\, \dfrac{1}{D_M}, \quad \forall M \in NM \\ C_{op}^{Solar} = Cu_s^{Solar} \sum_M \left(Q_M^{Solar}\, D_M\right) \\ \left[\begin{array}{c} W_{s,1}^{Solar} \\ A_{s,1} \le A_c^{Solar} \le A_{s,2} \\ C_{cap}^{Solar} = FC_{s,1}^{Solar} + VC_{s,1}^{Solar}\left(A_c^{Solar}\right) \end{array}\right] \vee \left[\begin{array}{c} W_{s,2}^{Solar} \\ A_{s,2} \le A_c^{Solar} \le A_{s,3} \\ C_{cap}^{Solar} = FC_{s,2}^{Solar} + VC_{s,2}^{Solar}\left(A_c^{Solar}\right) \end{array}\right] \vee \ldots \vee \left[\begin{array}{c} W_{s,R}^{Solar} \\ A_{s,R} \le A_c^{Solar} \le A_{s,R+1} \\ C_{cap}^{Solar} = FC_{s,R}^{Solar} + VC_{s,R}^{Solar}\left(A_c^{Solar}\right) \end{array}\right] \end{array}\right] \end{array}\right] \vee \left[\begin{array}{c} \neg Y^{Solar} \\ Q_t^{Solar} = 0 \\ A_c^{solar} = 0 \\ Q^{Solar\_Max\{t\}} \\ C_{op}^{Solar} = 0 \\ C_{cap}^{Solar} = 0 \\ C^{Tank} = 0 \end{array}\right]$$

These disjunctions are reformulated using a Convex Hull relaxation technique [2]

Fuels availability

The heat provided by fossil and biofuels depends on the availability of such resources

$$Q_{b,M}^{Biofuel} \le \frac{Heating_b^{Power}\ Avail_{b,M}^{Max}}{D_M}, \quad \forall b \in B,\ \forall M \in NM$$

$$Q_{f,M}^{Fossil} \le \frac{Heating_f^{Power}\ Avail_{f,M}^{Max}}{D_M}, \quad \forall f \in F,\ \forall M \in NM$$

Economic Objective Function

The economic objective is to minimize the operational and capital cost of the plant

$$\begin{aligned} Min\,TAC = & \sum_{t=1}^{Nt}\left[F_t^{B,purchased} \cdot \frac{\omega}{60\left[{}^{\min}/_{hr}\right]} \cdot BiocideCost + F_t^{Chem} \cdot \frac{\omega}{60\left[{}^{\min}/_{hr}\right]} \cdot ChemicalsCost + MaintenanceCost\ \ y_t\right] \\ & + H_Y \left\{\begin{array}{l} \sum_{f\in F}\left[C_f^{Fossil} \sum_{M\in NM}\left(Q_{f,M}^{Fossil}\, D_M\right)\right] + \sum_{b\in B}\left[C_b^{Biofuel} \sum_{M\in NM}\left(Q_{b,M}^{Biofuel}\, D_M\right)\right] + C_{op}^{Solar} \\ -R^{Solar}\left[\sum_{M\in NM}\left(Q_M^{Solar}\, D_M\right)\right] - \sum_{M\in NM}\sum_{b\in B}\left[R_b^{Biofuel}\, Q_{b,M}^{Biofuel}\, D_M\right] - \sum_{M\in NM}\sum_{f\in F}\left[R_M^{Fossil}\, Q_{f,M}^{Fossil}\, D_M\right] \end{array}\right\} \\ & + k_f\left\{C_{cap}^{Solar} + C^{Tank}\right\} \end{aligned}$$

Environmental Objective Function

The environmental objective is to minimize de GHGE emissions of the process

$$Min\,NGHGE^{Overall} = \sum_{M\in NM}\sum_{f\in B}\left[GHGE_f^{Fossil}\, Q_{f,M}^{Fossil}\, D_M\right] + \sum_{M\in NM}\sum_{b\in B}\left[GHGE_b^{Biofuel}\, Q_{b,M}^{Biofuel}\, D_M\right]$$

Multi-Objective Problem Solution

The constrained method [3] is implemented to solve the multi objective mixed integer nonlinear problem MO MINLP. The multi-objective function:

$$OF = Min\,TAC + Min\,NGHGE^{Overall}$$

is converted into a single objective constrained problem:

$$Min\,TAC$$

subject to:

$$NGHGE^{Overall} \le \varepsilon_i$$

$$Model\ Equations$$

Case study: Acrylonitrile plant

An acrylonitrile plant located in Musqat, Oman is analyzed. The plant has twelve hot process streams and eight cold process streams. The data for these streams is shown in Table 1.

Table 1. Hot and Cold process streams data

Hot stream	Inlet Temp.°C	Outlet Temp.°C	F*Cp kW/°C	Q kW	Cold stream	Inlet Temp. °C	Outlet Temp.°C	F*CpkW/°C	kW
h1	21.11	15	115.1	703.37	c1	37.78	55.56	39.56	703.3706
h2	15	4.44	111.06	1172.28	c2	37.78	55.56	65.94	1172.284
h3	470	287.78	88.46	16118.91	c3	87.78	287.78	80.59	16118.91
h4	287.78	260	126.61	3516.85	c4	22.22	26.67	263.76	1172.284
h5	65.56	51.67	1160.56	16118.91	c5	21.11	26.67	263.76	1465.355
h6	51.67	37.78	238.44	3311.7	c6	21.11	26.67	211.01	1172.284
h7	48.89	37.78	13.19	146.54	c7	87.78	104.44	1213.31	20221.9
h8	32.22	21.11	633.03	7033.71	c8	136.67	146.11	930.93	8792.132
h9	76.67	65.56	31.65	351.69					
h10	143.33	132.22	1503.45	16705.05					
h11	136.67	48.89	120.2	10550.56					
h12	43.33	21.11	112.1	2491.1					

These streams were thermally integrated using the model SYNHEAT [4]. This provided the minimum heating and cooling utilities. The resulting HEN requires 11,937.865 kW of heating utility and 39,349.95 kW of cooling utilities, from which 10,241.88 kW have to be removed using a refrigerant, 25,039.73 kW using seawater as cooling fluid and 4068.33 kW can be recovered by the stripper unit of the refrigerant system. From the total amount of heat removed with seawater and a heat balance the flow of seawater is calculated (4,397.45 m^3/h). The seawater-cooled system is discretized into i segments (divided into three categories: pipelines, tanks and heat exchangers), as shown in Figure 1. Other data required to run the model are the useful collected energy per month, and the cost for each of the considered solar collectors, the data and availability of fossil and biofuels and the cost of biocide and EOP chemical

Results

The problem was solved considering a governmental subsidy of \$20/ton CO_{2eq}. For this scenario the Pareto curve is shown in Figure 2. Notice that the higher cost for low GHGE is associated to the production of solar energy; this is due to the fact that this is the cleanest energy. For higher allowed emissions the use of fossil fuels is preferred. The use of biofuels is limited, not only for the cost but also for the availability of such resources in Musqat, Oman. Another important aspect to consider is that the existence of governmental subsidies makes feasible the use of alternative sources of energy even for scenarios where higher emissions are allowed (up to 40000 GHGE). Notice also that for lower emissions the cost of mechanical maintenance is the highest. This means that when lower emissions are required, it is preferred to purchase the biocide instead of producing it, in order to reduce the energy consumption in site. The downside of this scenario is that several plant shut downs must be programmed in order to maintain the operation of the plant in optimal conditions. For higher emissions it is preferred to produce the biocide in plant using the cheapest fossil fuel.

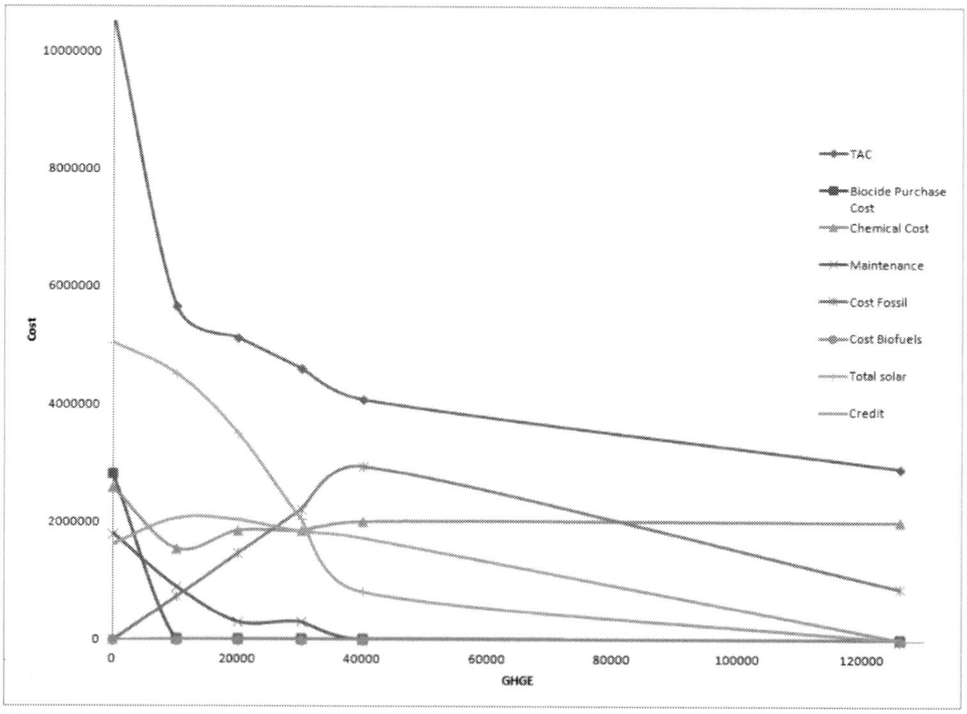

Figure 2 Pareto curve (Cost vs. GHGE) for \$20/ton CO_{2eq} tax credit reduction

Conclusions

A multi-objective MINLP model for the simultaneous optimization of energy management, biocide dosing and maintenance scheduling of a thermally integrated facility has been presented. A case study for an acrylonitrile plant was solved using the constrained method. A set of optimal Pareto curves was obtained, this curves might help devise attractive solutions taking into account economic, technical and environmental aspects.

References

[1] A. Vecchietti, S. Lee, I.E. Grossmann. Modeling of discrete/continuous optimization problems: characterization and formulation of disjunctions and their relaxations. *Computers and Chemical Engineering*. 27 (2003) 433-448.

[2] R. Raman, I.E. Grossmann. Modelling and computational techniques for logic based integer programming. *Computers and Chemical Engineering*. 18 (1994) 563-578.

[3] U. Diwekar. Introduction to applied optimization. Kluwer Academic Publishers Group, Netherlands, 2003.

[4] T.F. Yee, I.E. Grossmann. Simultaneous optimization models for heat integration—II. Heat exchanger network synthesis. *Computers and Chemical Engineering*. 14 (1990) 1165-1184.

(341f) Simultaneous Optimization and Heat Integration for the Production of Algae Based Biodiesel Using Bioethanol

Kristen Severson, Chemical Engineering, Carnegie Mellon University, Pittsburgh, PA, Mariano Martín, Chemical Engineering, University of Salamanca, Salamanca, Spain and Ignacio E. Grossmann, Chemical Engineering Department, Carnegie Mellon University, Pittsburgh, PA

Biodiesel, along with bioethanol, are the two most promising biofuels in today's market. Vegetable oils hold promise as alternative fuels for diesel engines. [1,2] However, using raw vegetable oils for diesel engines can cause numerous engine-related problems. [3] Thus, research has focused on developing transformation processes like pyrolysis, micro emulsion and transesterification. The process of transesterification has been most widely used. It removes glycerol from the triglycerides, and replaces it with radicals from the alcohol used for the conversion process. [4] As the source of oil we consider algae for several reasons such as the fact that microalgae convert sunlight, nutrients and CO_2 into proteins carbohydrates and lipids with a growth rate that doubles their biomass up to five times a day. Algae can grow not only on normal carbon sources, such as glucose, fructose, etc, but on waste from agriculture and food industries. It is also possible to use saline water or wastewater, eliminating the problem of water usage and lowering the cost of microalgae oil, algae growth can also be used to treat water in order to remove pollutants such as NH_4^+, NO_3^-, PO_4^{3-}. [5] Finally, the yield of biofuel for algae per area is between 10 and 100 times higher than conventional raw materials for biodiesel, and between 3 and 10 times higher than conventional raw materials for ethanol. Thus, it is expected that algae oil will be capable of meeting the US diesel need with 2-5% of the current US cropland. [6-8]

For quite some time the use of methanol in the transesterification process has been based on its lower cost and quicker reaction times compared to ethanol. [9,10] However, methanol is currently produced mainly from non renewable sources which implies that the production of biodiesel using methanol is highly dependent on fossil fuels. Current biorefineries are becoming petrochemical complexes where a number of different products including bioethanol and biodiesel are produced. Furthermore, lately it has been reported that similar yields of biodiesel can be obtained using either ethanol or methanol as the transesterification agent. Therefore, in order to avoid the dependence on fossil fuels, we investigate the integration of the use of bioethanol in the production of biodiesel because the use of bioethanol can result in a large economical benefit in the actual operation of the complex. [1,11]

In this paper we optimize the production of biodiesel applying simultaneous optimization and heat integration[12] followed by minimization of freshwater consumption. The proposed superstructure produces biodiesel from algae oil using bioethanol by defining the best transesterification technology and the operating conditions at the reactor including the products separation stages. We first grow the algae in ponds, dehydrate them and extract the oil. Next, we considered four different transesterification paths: alkali, enzymatic and heterogeneous catalysts and under supercritical conditions. The reactors are modelled using response surface methodology based on experimental results from the literature[13-18] to account for the effect of different variables such as operating temperature, excess of ethanol, catalyst load, time and pressure. These reactor models are implemented together with short-cut methods, experimental and rules of thumb based models for the recovery of ethanol, the separation of the polar and non polar phases and the final purification of the biodiesel and glycerol produced to formulate the problem as a superstructure of alternatives. The aim of this paper is to simultaneously optimize and heat[17] and water integrate the production of biodiesel using ethanol in terms of the reaction technology and the operating conditions.

Due to the simultaneous optimization and heat integration, the optimal conditions in the reactors differ from the ones traditionally used because our results take the separation stages into account. In terms of the best process, the alkali catalyzed process is the most profitable, while the enzymatic based one is even more promising due to the lower consumption of energy and water. However, it requires that the enzyme cost to be reduced. Finally, only for production costs of bioethanol below $0.5/gal will this process become economically competitive with that using methanol. On the one hand, his value is in the range of some of the values reported in the recent literature for bioethanol production.[19] Furthermore, this process does not depend on fossil fuels nor on fossil fuel based intermediates.

1 Scholl, KW.; S.C. Sorenson. SAE Paper N o. 930934. SAE, Warrendale, PA, **1993.**

2 Wagner, L.E.; Clark, S.J.; Schrock, M.D., SAE, **1984.,** Paper 841385.

3. Korus, R.A.; Mousetis, T.L.; Lloyd, L., ASAE Publication, **1982**, 4-82, Fargo, ND, USA, 218–223.

4 Kusy, P.F., ASAE, Publication 4-82, **1982**, Fargo, ND, USA, 127–137.

5 Mata, T. M.; Martins, A.A.; Nidia. S.; Caetano, N.S. *Renew. Sust. Energ. Revs,* **2010**, 14 , 217–232

6 Pimentel, D.; Patzek, T.W. *Nat. Resour. Res,* **2005,** 14, 1, 65-76

7 Zhang, W.; Wu, H., Zong, M., Optimal conditions for producing microalgal oil with high oleic acid content from *Chlorella vulgaris* LB 112 . **2009**

8 Chisti, Y., 2007,. *Biotechnol. Adv.*, **2007**, 25, 294-306.

9 Meneghetti, S.M.P., Meneghetti, M.R., Wolf, C.R., Silva, E.C., Lima, G.E.S., Coimbra, M.A., Soletti, J.I., Carvalho, S.H.V,. J. Am. Chem. Soc. **2006**; 83: 819–822.

10 Simoni M. Plentz Meneghetti, Mario R. Meneghetti, Carlos R. Wolf, Eid C. Silva, Gilvan E. S. Lima, Laelson de Lira Silva, Tatiana M. Serra, Fernanda Cauduro, Lenise G. de Oliveira Energy & Fuels. **2006;** *20:* 2262-2265

11 Kiss, A.A. Comp. Chem. Eng. **2010**; 34: 812–820

12 Duran, M.; Grossmann, I.E. *AIChE, J.,* **1986,** 32, 123-138

13Silva,N.L., Batistella, C.B., Filho, R.M., Maciel, M.R.W. Energy Fuels. **2009**; 23: 5636–5642

14 Joshi, H.C., Toler; J., Walker, T. J Am Oil Chem Soc. **2008**; 85:357-363.

15 Rodrigues, R.C., Volpato, G., Ayub, M.A.Z., Wada, K. *J Chem Technol Biotechnol, 2008,* 83:849–854

16 Li,E., Xu, Z.P., Rudolph, V. Appl. Cat. B: Environ. **2009**; 88: 42-49

17 Gui, M.M.; Lee, K.T.; Bhatia, S.; J. Supercritical Fluids. **2009**; 49: 286-292

18 Valle, P.; Velez, A.,; Hegel.; Mabe, G.; Brignole, E.A. J. Supercritical fluids. **2010**; 54: 61-70

19 Martin, M.; Grossmann, I. E. **2011** AICHE J 57 3408-3028

(359a) Modeling and Dynamic Optimization of Fuel-Grade Ethanol Fermentation Using Fed-Batch Process

Wei Dai, Chemical & Biological Engineering, Rensselaer Polytechnic Institute, Troy, NY, Daniel Word, Chemical Engineering, Texas A&M University, College Station, TX and Juergen Hahn, Biomedical Engineering and Chemical & Biological Engineering, Rensselaer Polytechnic Institute, Troy, NY

Ethanol produced from fermentation of biomass-derived sugar is increasingly being used as a fuel for transport, either neat or in petrol blends. The main advantage of fuel-grade ethanol is that it can be produced from plants such as maize, sugar cane and sweet sorghum, thereby balancing the amount of greenhouse gases produced during the combustion process by absorption during photosynthesis. Also, gasoline additives are not necessary and the combustion is more complete when fuel-grade ethanol is used [1].

As the world's largest biofuel producer, the U.S. will produce over 57 billion liters of ethanol in 2012, and 136 billion liters in 2022 [2]. Most of the ethanol in U.S. is produced from maize-based plant, and more than 90% of the plants make use of the dry mill process, which involves four steps: milling, liquification, simultaneous saccharification and fermentation (SSF), and distillation. The SSF process is the most important step, where dextrin is broken down into fermentable dextrose by glucoamylase and dextrose is converted into ethanol by yeast. The theoretical maximum yield of ethanol from maize starch indicates that the ethanol concentration after SSF process can be over 160 (g/L) in the batch fermenter [3]. Currently, the most advanced commercial plants can produce approximately 150 (g/L), however, the average yield from the plants using traditional techniques produce less than 140 (g/L). Therefore, improving the operations of the SSF process based on current facilities is undoubtedly the most economic approach to increase ethanol yield. However, optimization of SSF operations requires a good understanding of the process, usually represented by data and models.

Models of the SSF process usually consist of dynamic balances of components such as yeast, dextrose, and ethanol. Models of different complexities have been developed, ranging from relatively simple models [4] to quite detailed ones [5-7]. The main differences of the degree of complexity results from addition of intermediate steps in the saccharification process, use of more appropriate kinetics reflecting growth rate and inhibition term, and inclusion of correlations of the temperature influence on enzyme and yeast activity.

One significant challenge for the use of SSF models in commercial plants is the lack of data, since yeast, active yeast, and dead yeast cannot be measured independently. In addition, fed-batch fermenters are operated as a "black box" for the majority of the operating time. Dextrose and ethanol concentrations can be measured only every few hours and other intermediate substances are difficult to measure. The one process variable that is measured often in commercial plants is the process temperature, however, an energy balance is required if these temperature data are to be used for parameter estimation or monitoring other quantities. Unfortunately, none of the existing models for SSF include an energy-balance equation as SSF is a complex process and an energy balance includes a significant number of parameters and relationships to describe the thermo-physical properties.

Another challenge for operating ethanol plants is that ethanol fermentation is an exothermic process and temperature changes will significantly affect enzyme and yeast activity. Glucoamylase has an optimum activity at 140 in the saccharification process, while the optimal temperature for yeast growth and ethanol fermentation is in the range of 84-94 . These competing optimal process temperatures for the SSF highlight the complexity of selecting an optimal temperature profile over the course of a batch. It should be noted that most commercial plants hold the temperature constant between 86-90 .

The model developed in this work is based upon existing work [5-11], however, the model has been updated to address the challenges mentioned above: reformulate the flowrate equation which reduces the stiffness of the system and simplifies numerical solution of the model; substitute Monod kinetics expression with Haldane kinetics expression to enhance the synergistic substrate and product inhibition; expand the existing model to also include an energy balance with new parameters that are estimated from available data. Furthermore, this work addresses the above mentioned challenges by:

- Establishing the relationship between cooling water flowrate, temperature, and active yeast concentration, so that temperature data can be used to estimate unknown parameters arising from both mass-balance equations and energy-balance equations;
- Computing optimal input profiles for a fermenter used for ethanol fermentation by using the developed model and a simultaneous approach for solving the dynamic optimization problem.

The parameter estimation algorithm presented here is based on a Trust-Region approach with a fixed-step Runge-Kutta method. The simulation results fit the original experimental data well for both the training and testing set. In the dynamic optimization problem, both the input profile and the model are discretized using a three-point Radau collocation on finite elements [12]. Then, the optimization problem is implemented in AMPL [13], and is solved using nonlinear solver IPOPT [14]. It is shown that modifications of the input profiles for the cooling rate and the glucoamylase addition lead to an approximately 10% increase in ethanol yield. These are promising results, even though these findings will ultimately need to be validated during real plant operations.

Reference

1. Yan, L. and S. Tanaka, *Ethanol fermentation from biomass resources: current state and prospects.* Applied Microbiology & Biotechnology, 2006. **69**(6): p. 627-642.

2. Walker, G.M., *125th anniversary review: fuel alcohol: current production and future challenges.* Journal of the Institute of Brewing, 2011. **117**(1): p. 3-22.

3. Patzek, T.W., *A Statistical Analysis of the Theoretical Yield of Ethanol from Corn Starch.* Natural Resources Research, 2006. **15**(3): p. 205-212.

4. Fogler, H.S., *Elements of chemical reaction engineering, Fourth Edition, 2005.*

5. de Andrés-Toroa, B., Girón-Sierraa, J.M., López-Orozcoa, J.A., Fernández-Condea, C.,Peinadob, J.M., Garca-Ochoa, F., *A kinetic model for beer production under industrial operational conditions.* Mathematics and Computers in Simulation, 1998. **48**(1): p. 65-74.

6. Lee, C.G., C.H. Kim, and S.K. Rhee, *A kinetic model and simulation of starch saccharification and simultaneous ethanol fermentation by amyloglucosidase and Zymomonas mobilis.* Bioprocess and Biosystems Engineering, 1992. **7**(8): p. 335-341.

7. Ochoa, S., Yoo, A., Repke, JU.-U., Wozny, G., Yang, D.R., *Modeling and Parameter Identification of the Simultaneous Saccharification-Fermentation Process for Ethanol Production.* Biotechnology Progress, 2007. **23**(6): p. 1454-1462.

8. Roeva, O., *A genetic algorithms based approach for identification of Escherichia coli fed-batch fermentation.* Bioautomation, 2004. **1**(2): p. 30.

9. de Andres-Toro, B., Giron-Sierra, J., Lopez-Orozco, J., Fernandez-Conde, C., *Evolutionary optimization of an industrial batch fermentation process*, in *Proceedings of the Europan Control Conference*1997.

10. Iyer, M.S. and D.C. Wunsch, II, *Dynamic re-optimization of a fed-batch fermentor using adaptive critic designs.* Neural Networks, IEEE Transactions on, 2001. **12**(6): p. 1433-1444.

11. Mulchandani, A. and J.H.T. Luong, *Microbial inhibition kinetics revisited.* Enzyme and Microbial Technology, 1989. **11**(2): p. 66-73.

12. Zavala, V.M., *Computational strategies for the optimal operation of large-scale chemical processes*2008:

ProQuest.

13. Fourer, R., D. Gay, and B. Kernighan, *AMPL: A Modeling Language for Mathematical Programming*2002: {Duxbury Press}.

14. Wächter, A. and L.T. Biegler, *On the implementation of an interior-point filter line-search algorithm for large-scale nonlinear programming.* Mathematical Programming, 2006. **106**(1): p. 25-57.

(359b) Generic Modeling, Design and Operating Optimization of Hybrid Renewable Power Generation Flowsheets Including Hydrogen Storage

Damian Giaouris[1], **Athanasios I. Papadopoulos**[1], Chrysovalantou Ziogou[1], Dimitris Ipsakis[1], Panos Seferlis[2], Simira A. Papadopoulou[3] and Spyros S. Voutetakis[4], (1)Chemical Process Engineering Research Institute, Centre for Research and Technology-Hellas, Thermi-Thessaloniki, Greece, (2)Mechanical Engineering Department, Aristotle University of Thessaloniki, Thessaloniki, Greece, (3)Department of Automation, Alexander Technological Educational Institute of Thessaloniki, Thessaloniki, Greece, (4)Chemical Process Engineering Research Institute, Centre for Research and Technology-Hellas, Thessaloniki, Greece

To address the intermittent nature of largely unpredictable environmental phenomena, renewable energy sources (RES) are often transformed into dependable power flows by simultaneous utilization of different types of conversion equipment and storage media (e.g. PV panels, wind generators, chemical accumulators etc.). Hydrogen-based technologies are also receiving considerable attention as they enable flexible and long-term power storage, increasing the capacity provided by conventional accumulators. Such features result in hybrid infrastructures combining multiple sub-systems of heterogeneous characteristics that need to operate smoothly and efficiently within an integrated flowsheet. Their optimum design and operation involves significant complexities due to the existence of a very large number of interactions affecting the overall system performance. In addition to numerous structural options, the intense variability of renewable energy sources inflicts a varying temporal behaviour resulting in significant operating transitions among temporally different flowsheet topologies. Optimization can be used to systematically address such complexities by treating system features and interactions as decision options. However, the efficient consideration of all the available alternatives requires systematic models for the exhaustive representation of structural and temporal flowsheet options.

Recently reported works address the structural design of power and/or heating/cooling co-generation systems through generic modeling concepts developed around flow characteristics which are independent of specific equipment functions (e.g. energy hub and power node concepts [1], matrix-based models [2]). In a similar context, superstructure-based approaches have also been considered in the design of extensive polygeneration flowsheets, focusing mostly on chemical conversion processes for the co-generation of chemicals, fuels and power [3, 4]. Clearly, there is a lot of scope in exploiting the merits of both approaches for the efficient modeling and design of flowsheets integrating renewable power generation systems including hydrogen/power storage. On the other hand, models addressing the varying temporal behavior of such systems are often case-specific. They mostly utilize conditional rules in the form of empirical, tree-like representations to determine the appropriate time–instance for activation/deactivation of the desired equipment, ultimately aiming to maintain feasible, stable and prolonged system operation. Although useful, this practise requires the manual addition of terms and conditions in the event that new options (e.g. streams, equipment, rules etc.) need to be considered in the performed calculations. Whilst marginally feasible for reasonably small systems, the exhaustive consideration of all possible temporal combinations becomes impractical as the number of incorporated structural options increases.

This work addresses the systemic modeling, design and operating optimization of renewable power generation and hydrogen/power storage flowsheets based on generic representations integrating structural, temporal and logical features. Building on existing methods, an inclusive synthesis model is developed in the form of a superstructure to capture structural and operating characteristics of RES-based flowsheets, regardless of processing tasks or utilized equipment. Generic concepts are used to describe flows and tasks of conversion and accumulation, integrating streams utilized in multi-component material/energy or purely energy conversions and facilitating the representation of operations at different resolution within the same flowsheet.

The above are coupled with the use of logical propositions assembled around generic temporal features of power generation and storage systems to support the flexible development of simple and complex conditional statements representing operating requirements as constraints. Such features are based on the hysteresis phenomena dominating the operation of systems that switch between activation and de-activation, within or between subsequent time instances. They are linked with the proposed structural concepts by defining hysteresis as a zone developed around the capacity of the accumulators where the converters are allowed to operate. In this respect, operating policies are easily represented as logical schemes that maintain the generic features of the superstructure. The resulting schemes are subsequently incorporated in the synthesis models, supporting the temporal evolution of flowsheet topologies based on the conditions associated with the considered constraints. This is approached by appropriate transformations inflicted upon the superstructure model.

The proposed developments are illustrated through case studies on the optimum design of an extended power generation and hydrogen storage flowsheet, considering equipment such as PV panels, wind generators, batteries, fuel cells, electrolyzers etc. Important structural and operating characteristics such as the state of the accumulators at different time instances (e.g. capacity of batteries and hydrogen storage tanks at different pressures), the hysteresis zones and the types of equipment are considered as design variables, hence giving rise to an extensive number of feasible alternatives of utilizing and storing power at different forms.

[1] Heussen, K., Koch, S., Ulbig, A., Andersson, G., Unified system-level modeling of intermittent renewable energy sources and energy storage for power system operation, IEEE Systems Journal (2011), 10.1109/JSYST.2011.2163020

[2] Chicco, G., Mancarella P., Matrix modeling of small-scale tri-generation systems and application to operational optimization, Energy 2009, 34 (3), 261-273.

[3] Liu, P., Pistikopoulos, E.N. (2010), A multi-objective optimization approach to polygeneration energy systems design, AICHE Journal, 56(5), 1218-1234.

[4] Baliban, R.C., Elia, J.A., Floudas, C.A. Simultaneous process synthesis, heat, power, and water integration of thermochemical hybrid biomass, coal, and natural gas facilities, Computers and Chemical Engineering (2012), DOI: 10.1016/j.compchemeng.2011.10.002

(359c) Optimization-Based Power Management of a Hybrid Energy System

Milana Trifkovic[1], W. Alex Marvin[1], Mehdi Sheikhzadeh[2] and Prodromos Daoutidis[1], (1)Chemical Engineering and Materials Science, University of Minnesota, Minneapolis, MN, (2)Instrumentation and Control, Lambton College, Sarnia, ON, Canada

Future power generation systems will rely on utilization of renewable and non-renewable energy resources. These systems will vary in size and have a possibility of operating in autonomous or grid-connected configurations. Despite all the improvements in renewable energy technologies, the stochastic nature of renewable energy sources (RES) represents an obstacle in their utilization.

This study deals with the development of an optimization-based power management tool for a 10kW hybrid energy system built in Lambton College, Sarnia, ON. The system consists of photovoltaic (PV) arrays, wind turbine, electrolyzer, hydrogen storage tanks, and fuel cell.

In our previous work we developed a comprehensive dynamic model for each component and integrated the system. A hierarchical control strategy, consisting of a supervisory controller and a set of local controllers was designed. A supervisory, decision-based controller ensured power balance between the renewable energy sources and the load demand.[1]

In this study, we investigate the effectiveness of an on-line optimization framework as a supervisory controller. The optimization strategy incorporates weather and demand forecast information to determine the optimal set-points which are passed to a set of low level controllers. The hybrid system is designed and modeled in a modular approach for all components. We compare the use of the previously developed decision based supervisory power controller that neglects the future evolution of the weather and the load demand with the predictive optimization framework. The advantage of the dynamic optimization frameworks is that economic objectives and operational limits can be handled directly by the optimizer in a systematic manner. We demonstrate that incorporating forecasts in the control strategy can result in significant cost and energy savings through proof-of-concept simulation studies.

1. M. Trifkovic, M. Sheikhzadeh, K. Nigim, P. Daoutidis. *Hybrid energy system modeling and control*, submitted to IEEE Transactions on Control Systems Technology.

(359d) Carbon Dioxide Capture Processes: Sensitivity Analysis for Optimization and Control

Muhammad Zaman[1], Jay H. Lee[1] and Rafiqul Gani[2], (1)Chemical and Biomolecular Engineering, KAIST, Dajeon, South Korea, (2)CAPEC, Department of Chemical and Biochemical Engineering, Technical University of Denmark, Kgs. Lyngby, Denmark

Abstract

Carbon dioxide is responsible for 60 percent of the global warming caused by greenhouse gases (GHGs). Most of the CO_2 emissions are due to burning of fossil fuels, and power production in particular accounts for roughly 40 percent of the total CO_2 emissions, the main contributor of which is the coal based power generation. According to the International Energy Agency's roadmap, 20 percent of the total CO_2 emissions should be removed by carbon dioxide capture and sequestration (CCS) by year 2050. Hence the capture of CO_2 from power plants, mainly from coal based power generation is very significant for greenhouse gas reduction. The major difficulty in the post-combustion capture is the low concentrations of carbon dioxide in the flue gas and hence the larger volumes that need to be treated. Since this low concentration cannot be treated by physical absorption with reasonable absorption capacity, chemical absorption is commonly employed. Carbon dioxide capture by monoethanolamine (MEA) is a very mature technology for post-combustion capture. The major limitation with the amine-based absorption process is its parasitic power requirements including the steam requirements for the regeneration of the solvent, for CO_2 capture, and for compression. Other difficulties with MEA based CO_2 capture is the degradation of MEA and corrosion.

The main objective of this paper is to present a steady-state sensitivity analysis of the CO_2 process and to relate it with operability and control of the process to find the optimal design, operating conditions and control strategies. This design should be able to work at its optimal operating point in the presence of disturbances; this last item is to be verified through closed-loop dynamic simulations.

In this paper, an integrated process design and controller design methodology consisting of four steps including base case design, steady state sensitivity analysis, optimization problem & control structure selection, and optimization & control system validation, is presented. First a consistent simulation of the process verifying a base case design is performed, using an appropriate process simulator (Aspen Plus 7.2 is used). Thermodynamic properties play an important role in these simulations - an equilibrium based model with the electrolyte-NRTL model for the vapor liquid equilibrium and associated property models are used. Simulations are performed to investigate the sensitivity of the cost function to changes in the design variables and disturbances in the property model parameters. This sensitivity analysis using the base case design as a reference point helps to identify the cost function and the important variables, such as actuator (design) variables, control (process) variables, disturbance (input) variables, for future operability and control studies after developing the dynamic model of the process. The values for various cost functions, calculated in terms of the relative variations from their base case values for different values of the perturbed design variables and property model parameters are analyzed to identify the most sensitive design variables and model parameters.

Based on the sensitivity analysis, process optimization problems are defined and solved and, a preliminary control structure selection is made. The selected cost function is the cost of energy used by the process and the design variables affecting it most are the L/G ratio and the steam flowrates in the reboiler. The important process variable selected is the lean loading (moles of CO_2/moles of MEA in the lean solvent). The cost function (indirectly, the heat duties) and equipment parameters such as the stripper column diameter are also found to be sensitive to the stripper pressure, to positive variations of the vapor pressures as well as the flowrates of the hot utilities. The important disturbances in the process are found to be the flue gas flow rate, stripper pressure, and the lean solvent temperature. The formulated optimization problem therefore minimizes the cost of operation subject to selected design variables (L/G ratio and steam flowrates in the reboiler); fixed process specifications (lean loading, CO recovery goal, condenser temperature;

upper limit on solvent loss, etc.) and additional constraints (mass and energy conservation, i.e., the process model). Pairing the design variables (which are also actuators for the control problem) with the process specification variables (which are the measured variables in the control problem) helps to define control structure. According to the selected control structure, lean loading is controlled by manipulating steam flowrate to the reboiler of the stripper to ensure the optimum performance (reboiler duty). Alternatively, easily measured variable (e.g. stripper stage temperature) can be control variable, and both cases can be compared for better performance and control.

In the final step, the optimization problem is solved through the established steady state simulation model and the control structure is validated through a developed dynamic process model. As preliminary results, compared to the base case, the process is further improved by 7.3% in terms of operating costs and the control structure is shown to reject disturbances also at a minimum operating cost.

(359e) Modeling and Control of A Solar-Thermal Reactor for Synthesis Gas Production

Elizabeth Saade Saade[1], David E. Clough[1] and Alan W. Weimer[2], (1)Department of Chemical and Biological Engineering, University of Colorado, Boulder, CO, (2)Department of Chemical and Biological Engineering, University of Colorado at Boulder, Boulder, CO

A transport tube reactor uses concentrated solar irradiation in order to drive the endothermic biomass gasification reaction for the production of synthesis gas (H_2, CO and CO_2). This process is extremely dependent on the available sunlight and on the absence of clouds. Transient clouds constitute a significant disturbance for the steady operation of the reactor, causing unnecessary shut-downs and start-ups as well as complications in the separation processes downstream of the reactor. In order to make the process more feasible, flow manipulation is required to ensure steady conversion and a more efficient operation. Thus, a robust control system that allows continuous high performance operation of the reactor is required.

A model predictive control system was developed for the process. A digital charge coupled device (CCD) camera captures images of the sky at a sampling interval of one minute. These images are analyzed and used to anticipate the incoming solar irradiation at future instances. The predicted solar irradiation is used as an input for the model predictive controller (MPC), which ensures steady conversion despite transients in solar irradiation through the manipulation of the flow rates into the reactor. The control algorithm was developed based on linear time series models for different operating points of the reactor. These linear models were fitted to simulation results of a dynamic model of the process that was developed based on unsteady mass and energy balances. The control system was tested using SIMULINK and it was implemented at the High Flux Solar Furnace (HFSF) at the National Renewable Energy Laboratory (NREL). This paper describes the development of the control algorithm and presents the results obtained after its implementation.

(359f) Demand Response of Building HVAC Systems Using Infinite-Horizon Economic MPC

David Mendoza-Serrano, Chemical and Biological Engineering, Illinois Institute of Technology, Chicago, IL and Donald J. Chmielewski, Illinois Institute of Technology, Chicago, IL

Energy consumption by HVAC systems is usually heaviest when electricity prices are at their highest. The method of Economic Model Predictive Control (EMPC) can be used in conjunction with Thermal Energy Storage (TES) to time-shift power consumption away from periods of high demand to periods of low energy cost. In addition to enormous computational costs, implementation of such algorithms can result in unexpected and sometimes pathological closed-loop behavior, including inventory creep and bang-bang actuation. This paper will present an infinite-horizon formulation of the EMPC problem. While the design of this controller is achieved by a fairly simple convex optimization problem, it will be shown to alleviate many of the pathological behaviors observed in the finite-horizon case as well as significantly reduce the computational effort required for implementation. The method is illustrated on a simple building example using active TES. Implementation of the IH-EMPC policy resulted in a 99.987% reduction in computational effort, while only sacrificing 3.75% in operational cost. The new policy also eliminated much of the spiky behavior typically observed in EMPC implementations.

(359g) Modeling Interface Dynamics in Crystal Growth Processes: Applications to Silicon Wafer Manufacturing

German A. Oliveros, Chemical Engineering, Carnegie Mellon University, Pittsburgh, PA, Sridhar Seetharaman, Materials Science and Engineering, Carnegie Mellon University, Pittsburgh, PA and B. Erik Ydstie, Department of Chemical Engineering, Carnegie Mellon University, Pittsburgh, PA

Many industrial crystal growth processes require the careful management of the moving solid-liquid interface to avoid quality imperfections in the final product. In these processes it is necessary to control the formation of dendrites to guarantee a micro- and macro structurally good crystal. Dendrites are a consequence of a morphologically unstable interface; understanding how these instabilities form is the first step to addressing how they should be controlled. Our goal is to link the theory of stability of interfaces with the tools of systems engineering and apply them to a novel crystal growth process. Specifically, we incorporate the morphological stability analysis into the development and mathematical modeling of the Horizontal Ribbon Growth (HRG) process to manufacture silicon wafers. In this process a thin film of silicon is produced continuously from its melt making use of the fact that silicon floats on its melt just like ice floats on water. Our specific goal in this work is to analyze the effects of an applied perturbation to the growing solid-liquid interface of the HRG process.

The seminal investigation in the area of the stability of solid-liquid interfaces was carried out in the early '60s by Mullins and Sekerka [1]. They considered a binary alloy and analyzed the effect of applying a sinusoidal perturbation of infinitesimal amplitude to an interface moving at a constant velocity under purely diffusive transport. Under the steady state and infinite domain assumptions, they found an analytical expression for the rate of growth –or decay- of the perturbation as a function of the temperature gradients at each side of the interface, the concentration gradients in the melt, and the surface tension. A hill or valley of the sinusoidal wave ahead or behind the interface finds itself in a thermally unfavorable situation. If a tip of solid protruding into the liquid is surrounded by a domain that is at a temperature above the melting point, heat will flow from the liquid to the solid tip to equilibrate the thermal profile; this results in a stabilizing force. Surface tension offers resistance to deformation by minimizing the surface area of the system; hence the low free energy planar morphology is preferential over the high free energy curved interface. Concentration gradients, on the other hand, are a destabilizing force. The effect of melting point depression due to the presence of impurities causes an undercooling close to the front. Thus, when a sufficiently undercooled tip protrudes into the melt, it might not be immediately surrounded by a domain of a higher temperature since the undercooled melt lowers the temperature around the tip and counteracts the heat flow coming from the liquid. If the undercooled tip overcomes this "heat barrier", the interface breaks down resulting in the formation of cellular and dendritic morphologies.

In previous work we applied the Mullins-Sekerka stability criterion directly to an unperturbed silicon film under the assumption that the film is growing at a relatively low velocity [2]. In this work we integrate these two analyses and apply them to the growth of a silicon film in a bounded domain under radiative cooling. Aspects of the formulation that are reformulated are the use of proper boundary conditions and the incorporation of the initial transient effects of the growing film. The mathematical formulation of the new system yields a set of linear and nonlinear partial differential equations. The unperturbed and perturbed modes are independent of each other in both subdomains (solid and liquid) due to the linearity of the diffusion equations. The only coupling occurs at the solid-liquid interface, where the perturbation is applied. The system is modeled using standard numerical methods techniques.

[1] Mullins, W.W., Sekerka, R.F. Stability of a Planar Interface during solidification of a dilute binary alloy, Journal of Applied Physics, 35,444-451, 1964.

[2] Oliveros G.A., Liu R., Sridhar S., Ydstie B.E., Model development and control strategies for multicomponent alloy solidification processes and its application to solar cell manufacturing, Presented at the 2011 AIChE Annual Meeting, MN, USA.

(362a) Atomistic/Molecular Scale Integration of Polymer/Solid Surface Interaction Models

Robert L. Smith, Lorenz T. Biegler and Myung S. Jhon, Chemical Engineering, Carnegie Mellon University, Pittsburgh, PA

The physiochemical properties of oligomeric/polymeric materials are controlled by molecular level architecture, and the many polymer internal degrees of freedom span a range of time/length scales requiring an integrated multiscale description.[1] Many engineering applications involve polymeric fluids interacting with solid surfaces adding to the complexity of the theoretical description.[2] Typical applications in areas such as energy, tribology, and catalysis must meet the challenge of accurately and efficiently designing polymer melt/solid surface materials where performance is governed by multi-physical/multiscale phenomena.

A natural modeling path towards linking disparate scales in two-phase material design is a bottom-to-top approach. The essence of this multiscale integration approach is in constructing models with varying degrees of resolution while passing only the most relevant physical information from high resolution, highly complex models to lower resolution descriptions containing much less detail. This reduction in the degrees of freedom facilitates efficient computation of models approaching the macroscopic device scale for the purpose of technological advancement.

We utilize atomistic/ molecular scale modeling and integration techniques in modeling polymers with varied molecular architecture including linear di-block and random copolymers as well as branched/starlike polymers on a variety of solid surfaces. Case studies dealing with perfluorinated polymer nano films coating various allotropes of carbon are examined to illustrate our hierarchal multiscale approach. The bottom-level descriptions are established utilizing ab initio method based models of polymer-polymer and polymer-surface interactions and are linked with lower resolution molecular descriptions via coarse-graining procedures. The atomistic scale interactions are incorporated into coarse-grained molecular dynamics simulations of the polymer/solid surface system to quantify static and dynamic physiochemical properties. Thus, a correlation between atomistic/molecular architecture and macroscopic/mesoscale material performance is established.

1. P. S. Chung, M. S. Jhon, and L. T. Biegler, "Chapter 2- The Holistic Strategy in Multi-Scale Modeling", *Advances in Chemical Engineering,* Ed. G. B. Martin, New York: Academic Press 2011, 59-118.

2. P. S. Chung, R. Smith, S. H. Vemuri, Y. I. Jhon, K. Tak, I. Moon, L. T. Biegler, and M. S. Jhon, "Multi-scale/mult-physical modeling in head/disk interface of magnetic data storage," *J. Appl. Phys*. **111**, 07B712 (2012).

(362b) Multiscale Models for Characterizing the Pore Structure of Biochars

Hao Sun[1], Caroline A. Masiello[2] and **Kyriacos Zygourakis**[1], (1)Chemical and Biomolecular Engineering, Rice University, Houston, TX, (2)Earth Sciences, Rice University, Houston, TX

Biochar is charcoal generated for intentional soil amendment by pyrolyzing sustainable biomass feedstocks. Properly "engineered" charcoals can increase the water holding and cation exchange capacities of soils, improving the ability of plants to survive under drought conditions and reducing fertilizer runoff into watersheds. The environmental performance of biochars depends on their ability to absorb, retain and release water and nutrients. These biochar properties are controlled by their porosity and surface chemistry, which can vary widely depending on the composition of the biomass feedstocks and on the pyrolysis conditions employed during biochar production.

Biochars have a complicated pore structure consisting of multiple interconnected networks of micropores, mesopores and macropores that span multiple length scales: from sub-nanometer micropores to macropores with sizes of the order of 10 to 100 microns. Since biochars consist of a mixture of amorphous and crystalline phases, they usually exhibit two subpopulations of micropores: randomly distributed pores and domains of orderly arranged pores. The latter pores are the slits formed between the graphitic-like layers of aromatic carbon clusters that are turbostratically arranged in nanometer-size crystallites. Such structures have been confirmed with NMR and XRD measurements.

We report here the development of a new class of multiscale discrete models that can accurately characterize the complicated pore structure of biochars. The first step of our modeling approach is to generate a "nanoscale" model of the pore structure on a three-dimensional computational grid. Pores of various shapes and sizes are distributed on the grid in a random or orderly fashion to match the experimentally determined micro- and mesoporosities, as well as any available information about the shape of the pores or the size of crystalline domains. The generated porous solids are then eroded using rules that simulate a gas-solid, non-catalytic reaction that takes place in the kinetic control regime where the entire surface area attributed to micro- and mesopores is completely accessible to the reactant. In the case of very small slit pores, the model can be adjusted to consider that reaction takes place only at the pore mouths.

If the reaction takes place at high temperatures, however, strong diffusional resistances will appear first in the micropores and subsequently in the mesopores. As a result, progressively larger fractions of the micropore and mesopore structure will become inaccessible to the gaseous reactant. At sufficiently high temperatures, reaction will take place only on the micropore and mesopores "mouths," where they open up into the large macropore cavities identified in SEM or optical microscopy images. To model reaction in the diffusion control regime, we generate a "macroscale" model of the pore structure on a cubic computational grid. This model matches experimental macroporosity measurements and information obtained from SEM or optical microscopy images about the size and spatial arrangement of the large macropore cavities.

Brute-force computations with pores ranging from 0.5 nm to 10 μm or more would require arrays with more than 8×10^{15} computational cells. To overcome this limitation, we use grids with "porous" cells. Since the gaseous reactants can now diffuse into the cells, we have both external and internal (bulk) reaction: External reaction occurs on surfaces fully exposed to the gaseous reactant. Internal (bulk) reaction occurs on the surface of the pores that have been penetrated by the reactant.

Multiple realizations of a solid with the same pore structural properties are generated and reacted to estimate the average pore surface and to establish confidence intervals. For a given pore structure, this process will generate a family of curves that give the evolution of pore surface area with conversion as the intraparticle diffusional resistances increase. A comparison of these surface evolution patterns to the char reactivity patterns measured at different temperatures can provide important information about the complex pore structure of carbonaceous materials.

To validate our models, we compared simulation results with experimental data from the combustion of various biochars at several reaction temperatures. Chars produced from different biomass feedstocks exhibited different reactivity patterns in the kinetic control regime. These results were consistent with the presence of two pore subpopulations with relative volumes that strongly depend on the feedstock. In general, the first subpopulation consists of orderly arranged submicropores, while the second is formed by randomly distributed micropores. Biochars from different feedstocks, however, exhibited similar reactivity patterns when combusted with oxygen in the regime of strong diffusional limitations, when oxygen cannot penetrate deeply into the micro- and mesopores and the reaction takes place at the pore "mouths" where they open up into the large macropore cavities. These patterns were consistent with a macropore structure that consists of many large cavities separated by walls of similar thickness.

(362c) Modeling of Globular Protein Crystal Growth Via Kinetic Monte Carlo Simulations

Michael Nayhouse, Chemical and Biomolecular Engineering, UCLA, Los Angeles, CA, Panagiotis D. Christofides, Department of Chemical and Biomolecular Engineering, University of California, Los Angeles, Los Angeles, CA, Gerassimos Orkoulas, Chemical Engineering, University of California, Los Angeles, Los Angeles, CA and Sangil Kwon, Chemical Engineering, UCLA, Los Angeles, CA

The production of highly-ordered, high quality protein crystals through batch crystallization processes is vital in devising proteins for therapeutic purposes. Despite extensive experimental and theoretical work on understanding protein structure and function, there is lack of a systematic framework that relies on fundamental understanding of the nucleation and growth mechanisms of protein crystals at the microscopic level and utilizes such information to model and operate protein batch crystallization processes at the macroscopic level.

Our work focuses on investigating the growth of protein crystals via molecular dynamics and kinetic Monte Carlo simulations and using such knowledge to model and control batch protein crystallizers through population, mass and energy balances. The modeling and simulation work is tested against available experimental data.

The present work focuses on modeling and simulation of globular protein crystal growth. The model protein used for this work is tetragonal hen egg white lysozyme. Since crystal growth is a non-equilibrium process, kinetic Monte Carlo methods are developed. As is common practice in simulations of crystal growth, the solid-on-solid model is assumed. In this approximation, particles are deposited on the growing crystal lattice without voids or overhands, resulting in a highly compact crystal. In this work we consider three types of microscopic events: 1) molecular attachment, 2) detachment, and 3) migration events on the (101) and (110) faces of model protein. The implementation of the kinetic Monte Carlo methodology requires knowledge of the binding energies, the impingement rate, and the crystallization driving force. Previous work assigned a range of values to these parameters until satisfactory agreement between the calculated and the experimental growth rates was achieved.

Furthermore, the surface kinetics of the model protein is described for each microscopic event. For molecular attachment it is assumed that each lattice site is available for attachment, and thus the attachment rate is equal over the lattice. Conversely, for molecular detachment and migration events, the rates are dependent on the local environment. Each local environment comprises up to four nearest neighbors. This allows us to classify our local environment into five classes (zero to four nearest neighbors) in order to increase the computational efficiency when calculating the rates associated with executing a kinetic Monte Carlo event. The specific details of our simulation model and extensive simulation results covering the various aspects discussed above will be presented.

(362d) First Principles Modeling of Atomic Layer Deposition (ALD) Surface Reaction and Process Dynamics Using Absolute Reaction Rate Theory

Curtisha D. Travis and Raymond A. Adomaitis, Department of Chemical & Biomolecular Engineering; Institute for Systems Research, University of Maryland, College Park, MD

First principles modeling of atomic layer deposition (ALD) surface reaction and process dynamics using absolute reaction rate theory

Curtisha D. Travis* and Raymond A. Adomaitis
Department of Chemical and Biomolecular Engineering
Institute for Systems Research
University of Maryland
College Park, MD 20742 USA

*Presenting author: cdtravis@umd.edu
Submitted to the 2012 AIChE Annual Meeting, Pittsburgh, PA

A physically based model of atomic layer deposition reaction kinetics is developed and applied to alumina ALD using water and trimethylaluminum precursors [4]. The modeling strategy is based on computing the adsorbed adduct surface concentrations during each half reaction and the subsequent rate of the ligand exchange reactions using transition state theory. Upon precursor adsorption, barrierless formation of a stable Lewis acid-base adduct is facilitated through electron interactions between the aluminum and oxygen atoms [5]. An equilibrium is established between the adduct and transition states which is evaluated through a first principles analysis using statistical thermodynamics. Our derivations of the partition functions determining the reaction equilibrium relationships governing adsorbed surface species concentrations and reaction rates will be shown to produce tractable reaction kinetics models from which quantitative rate information can be extracted.

To complete the model, the resulting rate expressions are incorporated into time-dependent reaction material balance expressions accounting for the temporal evolution of the surface states during each exposure period. The key assumption in our model is that the adsorption, desorption, and processes responsible for formation of the two transition states instantly reach their equilibrium values relative to the changes that take place on the reaction surface due to the final irreversible ligand-exchange reaction. Elliott and Greer [3] demonstrate the TMA reaction takes place whether or not surface hydroxyl groups are present, and so the overall deposition process will be modulated by the extent to which the surface is covered by -CH3 ligands acting as an inert species or steric factor.

The ALD surface reaction models are treated as true dynamic systems, with continuous ALD reactor operation described by limit-cycle solutions, numerically computed using a polynomial collocation technique (see figure below). The linear nature of the water exposure dynamics is attributable to the ligand exchange reaction where one reacting water precursor molecule results directly in the production of one CH4 molecule desorbing to the gas phase. The curved nature of the TMA portion of the curve is due to the slower accumulation of -CH3 groups when the growth surface has a higher density of -OH groups, switching to a more rapid accumulation as -CH3 ligands are left by the TMA surface reactions. To illustrate the dependency of growth-per-cycle (GPC) as a function of both precursor exposure levels for a fixed precursor pressure of 1 Torr, a GPC map generated by the limit cycle solutions is shown. Under saturating conditions, we find GPCmax=0.925 A/cycle. While this value is less than the nominally observed value of 1.1 A/cycle GPC for the TMA/water ALD system, the prediction is significant given that the only limit to GPC in our model is the close-packing limit of -CH3 groups on the growth surface.

This work was motivated by the predictive capabilities of physically based models to decouple the effects of precursor pressure, exposure time, reactor temperature, and the dynamics of each exposure period on growth-per-cycle [2]. As such, our model demonstrates a good match between its predictions and observed GPC behavior in experimental alumina ALD studies. We note that the model has no adjustable parameters, eliminating the need for sticking coefficients. Model predictions indicating optimal operating conditions for most efficient ALD operation will be discussed. Work to couple these surface rate expressions to models of precursor transport in reactor-scale environments will also be discussed.

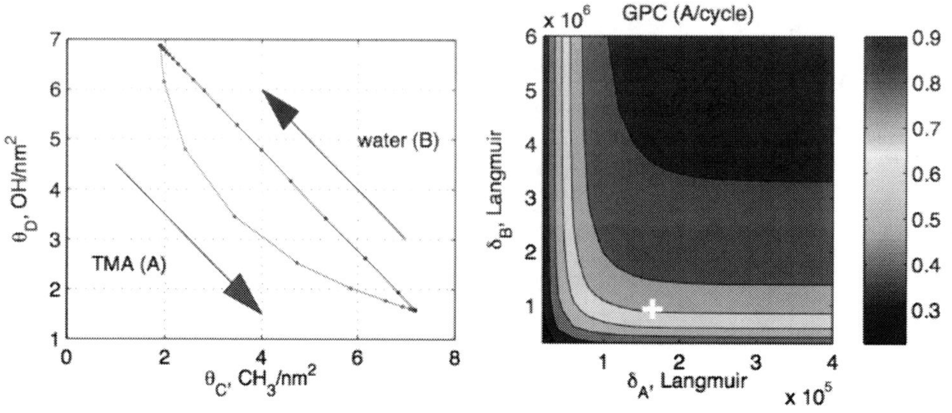

Figure 1: *Surface -CH3 and -OH limit-cycle coverage dynamics (left) for a representative set of alumina ALD operating conditions. The red curve corresponds to the TMA dose, the blue to water. Alumina GPC (A/cycle) map (right) as a function of each precursor exposure level, with limit-cycle conditions marked as +.*

References

[1] Adomaitis, R. A. ``A Ballistic Transport and Surface Reaction Model for Simulating Atomic Layer Deposition Processes in High Aspect-Ratio Nanopores," Chemical Vapor Deposition, 17 353-365 (2011).

[2] Deminski, M., A. Knizhnik, I. Belov, S. Umanskii, E. Rykova, A. Bagatur'yant, B. Potapkin, M. Stoker, and A. Korkin, ``Mechanism and kinetics of thin zirconium and hafnium oxide film growth in an ALD reactor," Surf. Sci. 549 67-86 (2004).

[3] Elliott, S. D. and J. C. Greer, ``Simulating the atomic layer deposition of alumina from first principles," J. Mater. Chem. 14 3246-3250 (2004).

[4] Puurunen, R. L., ``Surface chemistry of atomic later deposition: A case study of the trimethylaluminum/water system," Appl. Phys. Rev. 97 121301 (2005).

[5] Widjaja, Y. and C. B. Musgrave, ``Quantum chemical study of the mechanism of aluminum oxide atomic layer deposition," Appl. Phys. Lett. 80 3304-3306 (2002).

(362e) Multiscale Modelling of Backspillover Processes in Electrochemically Promoted Systems

Ioannis S. Fragkopoulos, Ioannis Bonis and Constantinos Theodoropoulos, School of Chemical Engineering and Analytical Science, University of Manchester, Manchester, United Kingdom

Backspillover in catalysis, denotes a migration of one or more species from the support to the catalytic surface. This metal-support interaction in some cases is followed by a significant alteration and enhancement of the catalytic activity increasing the surface reaction rate by up to 1500%. In the 1980s, it was first observed by Stoukides and Vayenas [1] that the catalytic activity of a metal catalyst deposited on a solid electrolyte can dramatically be enhanced by applying potential between the catalyst and a reference electrode. Later, it was found by Nicole and Comninellis [2] that this effect can be extended to an oxide catalyst, whose activity can be increased by up to a factor of 10 via anodic polarization. This phenomenon is known as Electrochemical Promotion of Catalysis (EPOC), also referred to as Non-Faradaic Electrochemical Modification of Catalytic Activity (NEMCA) and is of increasing interest nowadays in the fields of modern electrochemistry and surface science [3,4]. Although this technology exhibits a great potential, it has still not found industrial applications, partly because the underlying phenomenon is not fully understood yet and cannot be modelled to allow robust system design.

Here, we propose an accurate multi-scale model for such a system, which in conjunction with high fidelity, purpose-designed experiments currently underway, will illuminate electrochemical promotion. The various and complex phenomena taking place are described in their characteristic length-scales and any interactions are explicitly considered. The proposed model couples a macroscopic model based on a CFD software implementing the Finite Elements Method (COMSOL Multiphysics) and an in-house developed efficient implementation of the kinetic Monte Carlo method (kMC) [5].

The macroscopic model is based on partial differential equations (PDEs) and is employed for the simulation of charge transport throughout the cell as well as of the electrochemical phenomena taking place at the Triple Phase Boundaries (TPBs, where the gas phase is in contact with the catalyst and the support). The microscopic model is employed for the simulation of reaction-diffusion microprocesses on the catalytic lattice. The macro- and microscopic models are coupled via fluxes (from the support to the micro-lattice) of "backspillover" species at the anodic TPBs leading to an integrated multi-scale system which can efficiently be used for parameter estimation studies exploiting experimental data using appropriate optimisation techniques. The system considered comprises a Pt catalytic film (anode) deposited on an YSZ support and Au (cathode) reference/counter electrodes for the oxidation of carbon monoxide. The dimensions of both anodic and cathodic electrodes are in the order of nanometers, whereas the support is orders of magnitude larger.

References

[1] M. Stoukides, C.G. Vayenas, J. Catal. 1981; 70(1): 137-146.

[2] J. Nicole, C. Comninellis, Solid State Ionics. 2000; 136-137(1-2): 687-692.

[3] D. Poulidi, M.E. Rivas, I.S. Metcalfe, J. Catal. 2011; 281(1): 188-197.

[4] E.P.M. Leiva, C.Vazquez, M.I.Rojas, M.M. Mariscal. J.Appl.Electrochem. 2008; 38(8): 1065-1073.

[5] B. Hari, F. Goujon, C. Theodoropoulos. 2009; Integrated multi-scale models for microreactor simulation and design. In: S. Pierucci, ed. AIDIC Conference Series. pp 167-176 DOI: 10.3303/ACOS0909020

(362f) Accelerated Kinetic Monte Carlo (KMC) Algorithm for off-Lattice Particle- Based Reaction-Diffusion Systems

Vikram Thapar, Chemical Engineering, Cornell University, Ithaca, NY and Paulette Clancy, Chemical and Biomolecular Engineering, Cornell University, Ithaca, NY

Accelerated Kinetic Monte Carlo (KMC) Algorithm for Off-Lattice Particle- Based Reaction-Diffusion Systems

Vikram Thapar* and Paulette Clancy

School of Chemical and Biomolecular Engineering

Cornell University

Ithaca NY 14853

***vt87@cornell.edu**

We present an accelerated stochastic algorithm for simulating off-lattice/particle-based reaction-diffusion network. Off-lattice Kinetic Monte Carlo models are widely acknowledged to be valuable for a wide range of applications (especially in materials growth applications and cell biology) for which a fixed geometry lattice-based approach is restrictive. The most popular current particle-based models include ChemCell, MCell, Smoldyn and DADOS. The first three models (used largely for biological applications) make use of a fixed, or adaptive-based, time step, whereas the commercial package DADOS (used almost exclusively for Si-based semiconductor applications) uses a variable time step. DADOS is based on a Gillespie SSA-like that "fires" just one reaction or diffusion event in a time step. This, inevitably, makes it computationally inefficient and limits its scope to complex problems.

Our newly developed algorithm offers a computationally efficient, open source, code that makes use of a variable time step. The "tau-leaping" approach developed by Gillespie for "well-mixed" systems lays the foundation for our algorithm. The idea of "tau leaping" is based on firing *multiple* reaction-diffusion events in a time interval chosen such that the rates of those events will not change by a significant amount. Hence, the most critical and challenging aspect is to select an appropriate value of tau. The expression for the time step we have implemented is derived from concepts of how one event affects another. A traditional (single event, rejection-free) KMC method is implemented for use as a "gold standard" comparison to our accelerated simulations. Neighbor lists and data structures are used to increase computational efficiency. The use of constrained binomial random numbers to calculate the number of executions for each event takes care of the issue of overlapping of particles and negative populations. Two interesting reaction-diffusion examples – Fisher's Equation and the Lotka-Volterra (Predator-Prey) Equations -- are used to check the performance and accuracy of the algorithm. We highlight improvements in computational efficiency obtained by implementing this approach. Major challenges applying this technique to a general reaction-diffusion system are discussed, and possible solutions are presented.

(362g) Enforcing Elemental Mass and Energy Balances for Reduced Order Models Generated From CFD Simulations

Jinliang Ma[1], Christopher J. Montgomery[1], Khushbu Agarwal[2], Poorva Sharma[2], Yi-dong Lang[3], David E. Huckaby[4], Stephen E. Zitney[5], Ian Gorton[2], Deb Agarwal[6] and David C. Miller[7], (1)URS Corporation, Morgantown, WV, (2)Pacific Northwest National Laboratory, Richland, WA, (3)Department of Chemical Engineering, Carnegie Mellon University, Pittsburgh, PA, (4)Computational Science Division, National Energy technology Laboratory, Morgantown, WV, (5)AVESTAR Center, National Energy Technology Laboratory, Morgantown, WV, (6)Lawrence Berkeley National Laboratory, Berkeley, CA, (7)National Energy Technology Laboratory

Development of economically feasible gasification and carbon capture, utilization and storage (CCUS) technologies requires a variety of software tools to optimize the designs of not only the key devices involved (e., g., gasifier, CO_2 adsorber) but also the entire power generation system. High-fidelity models such as Computational Fluid Dynamics (CFD) models are capable of accurately simulating the detailed flow dynamics, heat transfer, and chemistry inside the key devices. However, the integration of CFD models within steady-state process simulators, and subsequent optimization of the integrated system, still presents significant challenges due to the scale differences in both time and length, as well the high computational cost. A reduced order model (ROM) generated from a high-fidelity model can serve as a bridge between the models of different scales. While high-fidelity models are built upon the principles of mass, momentum, and energy conservations, ROMs are usually developed based on regression-type equations and hence their predictions may violate the mass and energy conservation laws. A high-fidelity model may also have the mass and energy balance problem if it is not tightly converged. Conservations of mass and energy are important when a ROM is integrated to a flowsheet for the process simulation of the entire chemical or power generation system, especially when recycle streams are connected to the modeled device. As a part of the Carbon Capture Simulation Initiative (CCSI) project supported by the U.S. Department of Energy, we developed a software framework for generating ROMs from CFD simulations and integrating them with Process Modeling Environments (PMEs) for system-wide optimization. This paper presents a method to correct the results of a high-fidelity model or a ROM such that the elemental mass and energy are conserved perfectly. Correction factors for the flow rates of individual species in the product streams are solved using a minimization algorithm based on Lagrangian multiplier method. Enthalpies of product streams are also modified to enforce the energy balance. The approach is illustrated for two ROMs, one based on a CFD model of an entrained-flow gasifier and the other based on the CFD model of a multiphase CO_2 adsorber.

(368a) Advances in Robust Optimization: Improving the Solution Quality and Addressing Uncertainty Parameter Correlation in Planning and Scheduling

Zukui Li and Christodoulos A. Floudas, Chemical and Biological Engineering, Princeton University, Princeton, NJ

Among the various methods for addressing uncertainty in mathematical optimization problem, robust optimization is a type of rigorous yet efficient technique that can lead to solution robustness/reliability. Except the contribution from operations research community (e.g., [1-4]), robust optimization has also received lots of attention in the chemical process systems engineering community [5-6]. Specifically, Lin et al. [7] and Janak et al. [8] developed the theory of the robust optimization framework for general mixed-integer linear programming problems and considered both bounded and several known probability distributions. The robust optimization framework was later extended by Verderame and Floudas [9] for both continuous (general, bounded, uniform, normal) and discrete (general, binomial, Poisson) uncertainty distributions. The work was further compared with the conditional value-at-risk based method in Verderame and Floudas [10].

In our recent work [11-12], we systematically studied the set induced robust counterpart optimization technique for linear and mixed integer linear optimization problems. Different uncertainty sets are studied and their corresponding robust counterpart optimization formulations for both linear and mixed integer linear optimization problems are derived [11]. *A priori* and *a posteriori* probabilistic guarantees on constraint satisfaction for the robust solution of those different uncertainty set induced robust counterpart optimization models are developed, for both bounded and unbounded uncertainty, with and without detail probability distribution information [12].

Although significant progress has been made, there are several important issues which restrict the efficient application of robust optimization. First, the quality of the robust solution was often ignored in the past. In other words, while the desired solution feasibility (i.e., probability of constraint satisfaction) is met, how far is the solution from optimality? Can we find better robust solution while still satisfying the same probability requirements? Second, the issue of correlation between different uncertain parameters is often ignored in many applications.

To address the above issues, we make several advances for the robust optimization method. First, we propose a novel strategy for improving the robust solution using the *a priori* and *a posteriori* probability bounds developed in our recent work [11-12], where it has been shown that the *a posteriori* probability bound is in general tighter than the *a prior* probability bounds. Based on this fact, we developed an iterative solution framework which combines the advantage of robust optimization and the *a priori/a posteriori* probability bounds. The proposed method can essentially improve the solution quality of classical robust optimization framework without significant computational efforts. Comparing to the classical single-step robust optimization method, the quality of the robust solution can be improved while the same probabilistic guarantee on constraint satisfaction is still met. Second, we study the issue of correlation between different uncertain parameters in constraints, and developed robust counterpart optimization models for both linear and quadratic type correlations, which further provides a technique support for the application of the robust counterpart optimization in practical problems. The effectiveness of the proposed methods is illustrated through numerical examples and its application in planning and scheduling problems will be presented.

References:

1. Soyster, A.L., *Convex programming with set-inclusive constraints and applications to inexact linear programming.* Operations Research, 1973. **21**: p. 1154-1157.

2. Ben-Tal, A. and A. Nemirovski, *Robust solutions of uncertain linear programs.* Operations Research Letters, 1999. **25**(1): p. 1-13.

3. Ben-Tal, A. and A. Nemirovski, *Robust solutions of Linear Programming problems contaminated with uncertain data.* Mathematical Programming, 2000. **88**: p. 411-424.

4. Bertsimas, D. and M. Sim, *The price of robustness.* Operations Research, 2004. **52**(1): p. 35-53.

5. Verderame, P.M., et al., *Planning and Scheduling under Uncertainty: A Review Across Multiple Sectors.* Industrial & engineering chemistry research, 2010. **49**(9): p. 3993-4017.

6. Li, Z. and M.G. Ierapetritou, *Process scheduling under uncertainty: review and challenges.* Computers and Chemical Engineering, 2008. **32**: p. 715-727.

7. Lin, X., S.L. Janak, and C.A. Floudas, *A new robust optimization approach for scheduling under uncertainty: I. bounded uncertainty.* Computers and Chemical Engineering, 2004. **28**: p. 1069-1085.

8. Janak, S.L., X. Lin, and C.A. Floudas, *A new robust optimization approach for scheduling under uncertainty: II. Uncertainty with known probability distribution.* Computers and Chemical Engineering, 2007. **31**: p. 171-195.

9. Verderame, P.M. and C.A. Floudas, *Operational Planning of Large-Scale Industrial Batch Plants under Demand Due Date and Amount Uncertainty. I. Robust Optimization Framework.* Industrial & Engineering Chemistry Research, 2009. **48**(15): p. 7214-7231.

10. Verderame, P.M. and C.A. Floudas, *Operational Planning of Large-Scale Industrial Batch Plants under Demand Due Date and Amount Uncertainty: II. Conditional Value-at-Risk Framework.* Industrial & Engineering Chemistry Research, 2010. **49**(1): p. 260-275.

11. Li, Z., R. Ding, and C.A. Floudas, *A Comparative Theoretical and Computational Study on Robust Counterpart Optimization: I. Robust Linear and Robust Mixed Integer Linear Optimization.* Ind. Eng. Chem. Res, 2011. **50**: p. 10567-10603.

12. Li, Z. and C.A. Floudas, *A Comparative Theoretical and Computational Study on Robust Counterpart Optimization: II. Probabilistic Guarantees on Constraint Satisfaction.* Ind. Eng. Chem. Res., 2012: p. in press.

(368b) Online Integration of Scheduling and Control for Cyclic Production in CSTR

Yunfei Chu and Fengqi You, Department of Chemical and Biological Engineering, Northwestern University, Evanston, IL

Integration of scheduling and control achieves a better overall performance than the conventional sequential approaches because a good tradeoff can be made between the two problems [1-3]. The integrated problem is a mixed integer dynamic optimization (MIDO) problem [4, 5] where the differential equations describing the control problem are incorporated into the scheduling problem that includes discrete variables. The resulting MIDO problem might be quite computationally expensive, and as a consequence, most integration methods can only be applied offline. However, online implementation of the integration method is critical to deal with various uncertainties and disturbances in the production system. Yet the online solution is rather more challenging because there are stringent requirements on the computational efficiency. It is the goal of the paper to propose a novel framework for integrating scheduling and control and a fast computational strategy capable for online applications.

Different from most existing integration methods where the trajectory of the control variable (the process input) is determined directly in the integrated problem, a parameterized controller is considered in the proposed method. The controller designs, rather than the values of the control variables, are determined simultaneously with the scheduling decisions. The direct calculation of the control variable will incur the risk of open loop if the control variable cannot be updated immediately by solving the integrated problem in the sampling period. In contrast, the proposed integration structure is more robust since there is always a closed-loop controller working in real time even if the controller parameters are not updated by the integrated problem. Due to the existence of the real-time controller, the integrated problem does not need to be solved in real time, but can be solved in a relatively large time scale, e.g. once per production period. The most popular parameterized controller in the process industry is the PI controller [6, 7] and it is adopted in this work.

Efficient solution to the formulated MIDO problem is the key to the online implementation of the integrated method. Therefore, we propose an efficient computational strategy, which significantly reduces the model complexity by decomposing the dynamic optimization for the controller design from the scheduling model. For each potential transition, the range of the transition time is discretized by a number of sampling points and a controller is designed to minimize the transition cost at each transition time point. A set of controller candidates is then calculated and stored. Based on the controller candidates generated offline, the integrated MIDO problem is approximated by a simultaneous scheduling and controller selection problem, which is a mixed-integer nonlinear fractional programming problem. The objective function of this problem is the ratio of a convex quadratic function to a linear function and all constraints are linear, because the equations related to the dynamic system are all removed after the decomposition. The derived problem has a much smaller size than the original integrated problem and it can be globally optimized using a fast computational method based on the Dinkelbach's algorithm [8, 9].

The novelties in the proposed integration framework are summarized as

- A novel MIDO framework for scheduling and control which calculates the controller parameters in the integrated problem rather than the direct value of the control variable.
- A fast computational strategy based on the decomposition of control problems from the integrated problem where all dynamic optimization problems are solved offline.
- An efficient global optimization algorithm based on the model properties and Dinkelbach's algorithm.

The advantage of the proposed method is demonstrated through four case studies [10]. The proposed method based on the Dinkelbach's algorithm is compared with the solvers of BARON, SBB, and DICOPT. The results show that the proposed method globally optimizes the simultaneous scheduling and controller selection problem with the shortest computational time.

Comparisons with the traditional sequential method show that the proposed integration method can reduce the overall production cost rate because it provides a better balance between the inventory cost rate and the transition cost rate. Comparing with another integration method which solves the MIDO using the simultaneous approach, the proposed method is computationally much more efficient, especially for cases with a large number of products. The proposed method addresses the challenges in the online implementation of the integrated scheduling and control problems. The results also show that the online solution allows the production system to quickly respond to various uncertainties and disturbances.

Reference

[1] I. Grossmann, "Enterprise-wide optimization: A new frontier in process systems engineering," *AIChE Journal,* vol. 51, pp. 1846-1857, 2005.

[2] I. Harjunkoski, R. Nystrom, and A. Horch, "Integration of scheduling and control-Theory or practice?," *Computers & Chemical Engineering,* vol. 33, pp. 1909-1918, 2009.

[3] E. Munoz, E. Capon-Garcia, M. Moreno-Benito, A. Espuna, and L. Puigjaner, "Scheduling and control decision-making under an integrated information environment," *Computers & Chemical Engineering,* vol. 35, pp. 774-786, 2011.

[4] R. J. Allgor and P. I. Barton, "Mixed-integer dynamic optimization I: problem formulation," *Computers & Chemical Engineering,* vol. 23, pp. 567-584, 1999.

[5] V. Bansal, V. Sakizlis, R. Ross, J. D. Perkins, and E. N. Pistikopoulos, "New algorithms for mixed-integer dynamic optimization," *Computers & Chemical Engineering,* vol. 27, pp. 647-668, 2003.

[6] K. J. Astrom, H. Panagopoulos, and T. Hagglund, "Design of PI controllers based on non-convex optimization," *Automatica,* vol. 34, pp. 585-601, 1998.

[7] S. Skogestad, "Simple analytic rules for model reduction and PID controller tuning," *Journal of Process Control,* vol. 13, pp. 291-309, 2003.

[8] W. Dinkelbach, "On nonlinear fractional programming," *Management Science,* vol. 13, pp. 492-498, 1967.

[9] F. Q. You, P. M. Castro, and I. E. Grossmann, "Dinkelbach's algorithm as an efficient method to solve a class of MINLP models for large-scale cyclic scheduling problems," *Computers & Chemical Engineering,* vol. 33, pp. 1879-1889, 2009.

[10] Y. Chu and F. You, "Integration of scheduling and control with online closed-loop implementation: Fast computational strategy and large-scale global optimization algorithm," submited to *Computers & Chemical Engineering,* 2012.

(368c) Material Pipeline Scheduling and Inventory Planning Under Uncertainty

Pradeep Suresh[1], Jim Sturnfield[1] and John Wassick[2], (1)Dow Chemical Company, Freeport, TX, (2)Dow Chemical Company, Midland, MI

Inventories of raw materials for chemical manufacture are managed at various distributed sites within Dow based on supply and demand. Various sources of supply and demand exist at each of these sites, some dedicated and some shared. Multiproduct pipelines for material transport connect these different sites facilitating material movement to manage inventory. Various planning, scheduling and forecasting activities take place in this complex framework involving uncertainties around demand, supply and pipeline flows. In this work, we will present case studies where multiple optimization methods were applied to scheduling a multiproduct pipeline while simultaneously optimizing material inventories. Uncertainties around pipeline flows were factored in based on historical data. Results from stochastic optimization and deterministic optimization as well as a hybrid method will be compared. Additionally, results from these optimization methods will be compared to manually planned schedules. Certain practical caveats to applying scheduling optimization methods to this problem will also be discussed.

(368d) A Reactive Scheduling Framework for Batch Plants: Balancing Schedule Stability and Efficiency Performance Measures

Juan M. Novas, INTEC (UNL-CONICET), Santa Fe, Argentina and Gabriela P. Henning, Universidad Nacional del Litoral, CONICET, Santa Fe, Argentina

Predictive scheduling methodologies are aimed at addressing the generation of production agendas, while assuming that all the operational features considered at the beginning of the scheduling horizon do not change during such horizon. However, scheduling has to be considered as an ongoing process in which evolving and changing circumstances continually force reconsideration and revision of pre-established plans (Smith, 1995). Due to their intrinsic dynamics, real industrial environments frequently face unexpected events (new order arrivals, order cancellations/modifications, equipment failures, etc.), which disrupt in-progress schedules. Nevertheless, continuous and disturbing changes are not well accepted on the shop-floor since they lead to "shop nervousness". Consequently, not only predictive scheduling systems are required, but also reactive scheduling support ones, intended at coping with unforeseen events and preserving the main characteristics of the active agenda.

This paper presents new developments associated with a rescheduling support framework (Novas and Henning, 2010), aimed at addressing the repair-based reactive scheduling problem. This framework is based on a hybrid approach that combines an explicit object-oriented domain representation and a constraint programming (CP) methodology. Since the modeling of domain knowledge is a critical issue in any scheduling support system, the proposed framework has incorporated an explicit representation of the scheduling scenario, as well as a categorization of the disrupting events. This representation allows capturing the status of an in-progress agenda when an unexpected disrupting event occurs.

Based on the captured information, the reactive scheduling problem to be faced is specified. A rescheduling problem specification is defined as a set of associations between types of rescheduling actions and those processing tasks that comprise the reactive scheduling problem. The types of actions being considered are: task shift, swap, re-assign, among others. Each action type that is associated with a task does not define a particular rescheduling operation, but several possible ones. In fact, each association prescribes a category of rescheduling operation and a range of possibilities to apply it on the particular task. For instance, a task-shift action type defines a shift operation, as well as the interval in which the task is allowed to move. Similarly, a task-reassign action type specifies a reassignment operation and the possible units to which the task can be reallocated.

For a given specification, many solutions to the problem could exist. In order to find a good quality/optimal one, a constraint-based formulation is employed. Then, the second step of the solution methodology relies on the translation of the set of previously specified rescheduling action types into a constraint programming (CP) model. In this model, each rescheduling action type-task association is automatically translated into assignment, timing and/or precedence constraints. The solution of the model renders an updated schedule in which the repair action types that were suggested are instantiated.

Since the association between a task and an action type is not unique in many cases, different solution scenarios can be generated. Although some insights regarding the generation of alternative scenarios were discussed in Novas and Henning (2010), more elaborated techniques aimed at associating rescheduling action types with processing tasks are required. This paper focuses on this issue by considering different intervals along the scheduling horizon, as explained below.

The rescheduling framework allows identifying those tasks that are directly and indirectly affected by the disrupting event, as well as those non-affected tasks. Directly affected tasks are those that are associated with particular action types, which are mandatory and depend on the type of event (e.g., reassigning a task allocated to a broken unit). On the other hand, actions types to be associated with both indirectly and non-affected tasks cannot be prescribed beforehand, and are specified depending on the schedule status and the state of the plant at the rescheduling time point. These types of tasks are the ones that are expected to be less influenced by disruptions. Since maintaining the schedule stability is one of the main goals of the revision activity, scheduling decisions associated with these tasks should be directed towards maintaining as much as possible the current agenda. In consequence, tasks belonging to these two categories are not to be associated with very unsettling action types.

As it was pointed out by Musier and Evans (1990), whenever possible, batches which are about to be processed should not be moved to another unit or to another position in the sequence because their required raw materials have already been sent to the shop floor. Therefore, a possible scenario consists in freezing those non-affected tasks that have a start time close to the rescheduling time point, establishing a *Freezing Interval (FI)*. For those not frozen tasks, the less disturbing action type to be applied to them is the shift-jump action type. Thus, an interval located to the right of the freezing one is defined in order to identify tasks that can be shifted or can make small "jumps" (*Shift-Jump Interval, S-JI*). Finally, those tasks located further to the right and close to the end of the scheduling horizon can be subject to any type of rescheduling action, including the more disrupting ones, like the reassign action type. They give rise to a *Reassign Interval* (*RI*). The challenge is to set adequate limits of such time windows, because the length of the *FI*, *S-JI* and *RI* intervals will affect the stability of the updated schedule, as well as its quality in terms of time-related or cost performance measures. Progress on this matter is discussed in this paper. It is shown how the length of these intervals depends on the average task processing time, the time remaining before the end of the scheduling horizon, etc.

This paper presents the framework components explained above and describes how to cope with unexpected events, while trying to optimize a conventional performance measure (e.g. makespan, tardiness, etc.), and at the same time, attempting to preserve the schedule stability. The contribution put emphasis on the new work that is related to the proper specification of the *FI*, *S-JI* and *RI* intervals lengths.

The use of the framework and the new enhancements is illustrated by means of a case study involving a multiproduct multistage batch plant. Different scenarios, associated with various disruptive events, occurring at distinct time points, are considered. Efficiency performance measures and stability ones are both sought.

Novas, J.M., Henning G.P. (2010). Reactive scheduling framework based on domain knowledge and constraint programming. Computers and Chemical Engineering 34, 2129–2148.

Musier, R. F. H., Evans, L. (1990). Batch process management. Chemical Engineering Progress, 87, 66–77.

Smith, S. F. (1995). In: D. E. Brown, & W. T. Scherer (Eds.), Reactive scheduling systems. Intelligent Scheduling Systems. Kluwer Press.

(368e) Integrated Scheduling and Control for Multiproduct Continuous and Batch Processes

Jinjun Zhuge and Marianthi Ierapetritou, Department of Chemical and Biochemical Engineering, Rutgers-The State University of New Jersey, Piscataway, NJ

Simultaneous scheduling and control involves process scheduling and control problems with an integrated framework in which production time and task transitions are considered simultaneously. Using the simultaneous approach a Mixed Integer Dynamic Optimization (MIDO) problem is formed and then is discretized into Mixed Integer Nonlinear Programming (MINLP) through collocation point method [1]. Solving the resulting MINLP generates the optimal production time, product sequence and transition profiles between different products. However, the resulting MINLP is generally computational very expensive. In the literature in order to achieve better computational performance, Lagrangean decomposition was applied to the integrated model, resulting in a master scheduling problem and a primal control problem [2]. Other works that decoupled MIDO problem into dynamic optimization problem and MILP problem can be found in [3] where the solution was obtained using an iterative approach.

In our study, we explore the structure of the integrated optimization problem for continuous and batch processes and establish an efficient decomposition scheme based on the mathematical structure of the corresponding model. In continuous processes, we decompose the scheduling problem into an optimization model for the production time and a separate one for product sequence, and integrate the sequence consideration with the control problem. In this way, two sub-problems are formed. One is an MINLP dealing with the production sequence and control profile during transitions, and the other is an NLP dealing with the production time for each product considering the market demands. For batch processes, we decompose the integrated problem into an MINLP problem generating the optimal production sequence and operating conditions during process and an NLP yielding the optimal batch sizes to satisfy demands.

We prove the separability of the resulting two sub-problems analytically after decomposition. Unlike the Lagrangean decomposition approach, no iteration is needed in our approach. Moreover, the computational complexity is significantly reduced compared to the simultaneous scheduling and control problem under varying demand. The main contribution of this work is that there is no need to solve the entire integrated problem for different demand specifications because the only decision variables that need to be updated are the production times, and the batch sizes for continuous, and batch processes, respectively. Based on this finding we are investigating the integration of the planning level decisions moving towards enterprise wide optimization.

References:

(1) Flores-Tlacuahuac, A.; Grossmann, I. E., Simultaneous Cyclic Scheduling and Control of a Multiproduct CSTR. *Industrial & Engineering Chemistry Research* **2006,** 45, (20), 6698-6712.

(2) Terrazas-Moreno, S.; Flores-Tlacuahuac, A.; Grossmann, I. E., Lagrangean heuristic for the scheduling and control of polymerization reactors. *AIChE Journal* **2008,** 54, (1), 163-182.

(3) Nystrom, R. H.; Franke, R.; Harjunkoski, I.; Kroll, A., Production campaign planning including grade transition sequencing and dynamic optimization. *Computers & Chemical Engineering* **2005,** 29, (10), 2163-2179.

(368f) Tactical Planning and Scheduling of Flexible Process Networks Under Uncertainty with Stochastic Inventory

Dajun Yue and Fengqi You, Department of Chemical and Biological Engineering, Northwestern University, Evanston, IL

In order to take advantage of the synergies that exist between processes and also ensure maximum flexibility, the chemical process industry often constructs large production sites composed of many interconnected processes and chemicals [1]. But, on the other hand, the risks associated with demand uncertainty and supply disruptions may significantly affect tactical decisions made by a chemical complex. Although inventories can serve as buffers in dealing with demand and supply uncertainties, excessive inventory can be costly [2]. In addition, a chemical complex usually involves many chemical states, including feedstocks, intermediates, and final products. Thus, it is of significant importance to determine which chemicals should be stored and what is the optimal inventory level for each of them so as to achieve a certain service level and production target [3]. A body of research closely related to this work is the multi-echelon stochastic inventory theory, which is originally developed for the supply chain management. By using the guaranteed service approach, one is allowed to model the inventory allocation across the entire system [4-6]. However, the application of the stochastic inventory management policy is not limited to the optimization of supply chain configurations. You and Grossmann [3] integrated this approach with the tactical planning of chemical complexes under supply and demand uncertainty. Though problems like how to model chemical process networks involving recycle flows and how to model the propagation of uncertainties to quantify internal demands are addressed in their work, only a set of dedicated processes are considered. However flexible processes that are able to perform different production schemes are quite common in industries, because these processes exhibit flexibility with both raw materials and products, thus improving the robustness and economic performance of the entire chemical complex [1]. It is known that ignoring the effects of changeovers may lead to infeasible schedules or suboptimal solutions [7]. Thus, it is desirable to take into account the detailed scheduling when making tactical decisions. Furthermore, the planning and scheduling are closely related to the stochastic inventory, because the production timings would have great impacts on the order processing delays. This results in an aggregated problem involving the integration of planning, scheduling and stochastic inventory management. The simplest approach is to formulate a single simultaneous model, while how to solve the aggregated problem efficiently remains a challenge

In this work, we approach this challenge from the following aspects. To capture the stochastic nature of the demand variations and supply delays, we use the guaranteed service time approach. To address the detailed scheduling problem, we employ the cyclic scheduling policy. We propose a single MINLP model that simultaneously determines the scheme selections, production schedules, purchases, sales, production/consumption amounts and working and safety inventory levels. This model exhibits multiple tradeoffs among variables from all decision levels, thus seamlessly integrate the planning, scheduling and stochastic inventory management. While this basic model contains multi-linear and concave terms, which would be computationally intractable for large-size problems, we reformulate the model to an MINLP with only square root and linear terms by exploiting the problem properties and using general linearization methods. In order to obtain global optimal solutions with modest computational times, we further develop a tailored branch-and-refine algorithm based on successive piecewise linear approximations. Also, inspired by the work of Dogan and Grossmann [7], novel symmetry breaking cuts are added to eliminate the degeneracy of the assignment configurations, which in turn greatly accelerate the computation. Three examples with up to 16 processes and 25 chemicals are presented to illustrate the application of the model and its computational performance. To examine the responsiveness issues[8] of chemical complexes, we present approximated Pareto-optimal curves to reveal the trade-offs between the total cost and inventory versus the maximum guaranteed service times to the markets by solving a series of instances. The results show that the more responsive the process network needs to be, the higher cost and more inventory it will have.

References

[1] L. C. Norton and I. E. Grossmann, "Strategic-Planning Model for Complete Process Flexibility," *Industrial & Engineering Chemistry Research,* vol. 33, pp. 69-76, Jan 1994.

[2] F. Q. You and I. E. Grossmann, "Design of responsive supply chains under demand uncertainty," *Computers & Chemical Engineering,* vol. 32, pp. 3090-3111, Dec 2008.

[3] F. Q. You and I. E. Grossmann, "Stochastic Inventory Management for Tactical Process Planning Under Uncertainties: MINLP Models and Algorithms," *Aiche Journal,* vol. 57, pp. 1250-1277, May 2011.

[4] S. C. Graves and S. P. Willems, "Optimizing the Supply Chain Configuration for New Products," *Management Science,* vol. 51, pp. 1165-1180, Aug 2005.

[5] F. Q. You and I. E. Grossmann, "Integrated Multi-Echelon Supply Chain Design with Inventories Under Uncertainty: MINLP Models, Computational Strategies," *Aiche Journal,* vol. 56, pp. 419-440, Feb 2010.

[6] F. Q. You and I. E. Grossmann, "Mixed-Integer Nonlinear Programming Models and Algorithms for Large-Scale Supply Chain Design with Stochastic Inventory Management," *Industrial & Engineering Chemistry Research,* vol. 47, pp. 7802-7817, Oct 2008.

[7] M. Erdirik-Dogan and I. E. Grossmann, "A decomposition method for the simultaneous planning and scheduling of single-stage continuous multiproduct plants (vol 45, pg 299, 2006)," *Industrial & Engineering Chemistry Research,* vol. 46, pp. 5250-5250, Jul 2007.

[8] F. Q. You and I. E. Grossmann, "Balancing Responsiveness and Economics in Process Supply Chain Design with Multi-Echelon Stochastic Inventory," *Aiche Journal,* vol. 57, pp. 178-192, Jan 2011.

(368g) Short-Term Scheduling of Multiproduct Batch Plants: A Novel CP Approach

Franco Novara, Industrial Engineering, Universidad Nacional del Litoral, Santa Fe, Argentina, **Juan M. Novas**, INTEC (UNL-CONICET), Santa Fe, Argentina and Gabriela P. Henning, Universidad Nacional del Litoral, CONICET, Santa Fe, Argentina

Short-Term Scheduling of Multiproduct Batch Plants: A Novel CP Approach

Franco M. Novara#*, Juan M. Novas*, Gabriela P. Henning*

Facultad de Ingeniería Química, UNL, Santiago del Estero 2829, 3000 Santa Fe, Argentina

* INTEC (UNL, CONICET), Güemes 3450, 3000 Santa Fe, Argentina

This contribution addresses the short-term scheduling problem of multiproduct multistage batch plants by means of a novel Constraint Programming (CP) methodology. This problem has been extensively studied in the last decade by several authors, by resorting to various solution approaches (Méndez et al., 2001; Harjunkoski and Grossmann, 2002; Castro and Grossmann, 2005; Castro et al., 2006, Castro et al., 2009; Liu and Karimi, 2007a, 2007b, 2008; Marchetti and Cerdá, 2009; Zeballos et al., 2011). Most of the available solution procedures assume that the batching problem has been solved beforehand (Maravelias, 2012), and so does this work. This means that no batching decisions are made, and that the solution approach focuses on unit assignment, batch sequencing and timing decisions (as well as on the proper handling of some limited renewable resources), while optimizing a time-based or cost-based objective function.

The proposed approach is a totally new CP model, which is based on the ILOG-IBM OPL language and the CP Optimizer (ILOG-IBM, 2012). It takes advantage of some popular CP constraints, like cumulative functions, as well as other user-defined ones. The formulation can easily handle different features found in industrial environments, like production orders requiring several batches of each product, which are addressed in a campaign mode. When these campaigns are run, upper and lower bound limits on the number of batches per product campaign need to be complied with. The proposed CP model can also take into account the existence of dissimilar parallel equipment at each stage, topology constraints, forbidden job-equipment assignments, order release times, finite unit ready times, as well as limitations on renewable discrete resources, like heating and cooling utilities, manpower, etc. Regarding discrete resources, the approach can also address sequence-dependent changeover tasks that require limited cleaning resources. With respect to inter-stage storage and operational policies, the proposal considers UIS, NIS/ZW and NIS/UW, as well as mixed ones.

The approach has been extensively tested by means of several literature examples having various difficulty degrees. The formulation has rendered good computational results for a variety of inter-stage storage policies, campaign sizes and objective functions. Results that compare this approach with others are presented and discussed. Additionally, scalability studies are reported.

References

Castro, P.M., & Grossmann, I.E. (2005). New Continuous-Time MILP Model for the Short-Term Scheduling of Multistage Batch Plants. Ind. Eng. Chem. Res., 44, 9175-9190.

Castro, P.M., Grossmann, I.E., & Novais, A.Q. (2006). Two New Continuous-Time Models for the Scheduling of Multistage Batch Plants with Sequence Dependent Changeovers. Ind. Eng. Chem. Res., 45, 6210-6226.

Castro, P.M., Harjunkoski, I., & Grossmann I.E. (2009). Optimal Short-Term Scheduling of Large-Scale Multistage

Batch Plants. Ind. Eng. Chem. Res., 48, 11002–11016.

Harjunkoski, I., & Grossmann, I.E. (2002). Decomposition Techniques for Multistage Scheduling Problems Using Mixed-integer and Constraint Programming Methods. Computers and Chemical Engineering, 26, 1533-1552.

ILOG-IBM CP Optimizer, http://www-01.ibm.com/software/integration/optimization/cplex-cp-optimizer/

Liu, Y., & Karimi, I.A. (2007a). Scheduling multistage, multiproduct batch plants with non identical parallel units and unlimited intermediate storage, Chemical Engineering Science, 62, 1549-1566.

Liu, Y., & Karimi, I.A. (2007b). Novel continuous-time formulations for scheduling multi-stage batch plants with identical parallel units. Computers and Chemical Engineering, 31, 1671-1693.

Liu, Y., & Karimi, I.A. (2008). Scheduling multistage batch plants with parallel units and no interstage storage. Computers and Chemical Engineering, 32, 671-693.

Maravelias, C., (2012). General Framework and Modeling Approach Classification for Chemical Production Scheduling, AIChE Journal. In Press

Marchetti, P.A., & Cerdá, J. (2009). A general resource-constrained scheduling framework for multistage batch facilities with sequence-dependent changeovers. En: Computers and Chemical Engineering, 33, 871–886.

Méndez C.A., Henning, G.P., & Cerdá, J. (2001). An MILP continuous time approach to short-term scheduling of resource-constrained multistage flowshop batch facilities. Computers and Chemical Engineering, 25, 701–711.

Zeballos, L.J., Novas, J.M., & Henning, G.P. (2011). A CP formulation for scheduling multiproduct multistage batch plants. Computers and Chemical Engineering, 35, 2973– 2989.

Abstract Title:

A Method of Estimating the Probability of Rare Events with Application to Risk Assessment in Processing Plants

Authors:

Taha Mohseni[1], Masoud Soroush[1*], Warren D. Seider[2], Ulku Oktem[3], Darrin Feather[4], Jeffrey Arbogast[4], and Amogh Prabhu[4]

[1] Department of Chemical and Biological Engineering, Drexel University, Philadelphia, PA 19104
[2] Department of Chemical and Biomolecular Engineering, University of Pennsylvania Philadelphia, PA 19104
[3] Risk Management and Decision Processes Center, Wharton School, Philadelphia, PA 19104
[4] American Air Liquide, Newark, DE 19701

Keywords:
Rare events; probability estimation, Bayesian network modeling, risk assessment, processing plants

Abstract:
Since the introduction of a tractable inference technique for Bayesian Networks (BNs) by Spiegelhalter and Lauritzen [1] in 1988, BNs have become very popular in modeling real world problems. Although BN models are capable of handling incomplete and qualitative data, they become unreliable when there is no information on every possible scenario such as those regarded as rare events. Several solutions have been proposed to address this problem. They are referred to as sampling techniques. The solutions, however, have several weaknesses. First, many of them are not suitable for systems that lack an accurate quantitative model, and underestimate the probability of rare events. Second, commonly used sampling techniques such as importance sampling become unstable when system dimension increases significantly [2]. Third, some sampling techniques are only useful for constructing asymptotic probability distribution functions. Fourth, majority of sampling techniques have slow rate of convergence in estimating the probability distributions of rare events.

In this work, we present a method of generating historical data and estimating probability distributions in cases for which there is no accurate mathematical model or adequate historical data. In the framework of BN modeling, the problem arises when (a) there is no observation for a possible combination of states of parent variables (due to rarity of that instantiation) and (b) an underlying governing equation is not known. In such situations,

* Corresponding Author: soroushm@drexel.edu

it is impossible to estimate entries of conditional probability tables (CPTs) of child nodes (variables) accurately, which leads to unreliable inference. We have developed a method of estimating unknown entries of CPTs by "extrapolating" existing conditional probability distributions obtained from available historical data. We have successfully applied this method to a series of rare event case studies, and simulation results show good agreement with true values. This method is applicable to BNs with many nodes and complex interactions, and its convergence rate is considerably higher compared to conventional sampling techniques. An application of this method is in risk assessment of rare events, which are usually results of a sequence of rare causes. Available databases usually lack adequate information on rare events. The application and performance of the method are shown using a process example case study.

References:

[1] Lauritzen L., Spiegelhalter J., "Local computations with probabilities on graphical structures and their application to expert systems (with discussion)," *Journal of the Royal Statistical Society* **50**(2), 157-224 (1988).

[2] Gelman A., Meng X.-L., "Simulating normalizing constants: From importance sampling to bridgesampling to path sampling," *Statistical Science* **13**(2), 163–185 (1998).

(373b) An Efficient Method for Leak Detection and Isolation in Natural Gas Pipelines Using Dual Unscented Kalman Filter

Weiting Tang, Chemical Engineering, Texas Tech University, Lubbock, TX and M. Nazmul Karim, Chemical Engineering Department, Texas Tech University, Lubbock, TX

Pipelines are efficient and economic ways for transporting and distributing water, natural gas, and petroleum products; these are extensively used all over the world. Monitoring the integrity of the pipeline systems has become an important issue for the safety and economical operation, and environmental protection. In this paper, an efficient and reliable leak detection and isolation method has been proposed for on-line monitoring of a natural gas pipeline system. Dual unscented Kalman filter (DUKF) was employed for the on-line estimation of the states and parameters, simultaneously. Small leaks, which are usually masked by the background noise, pose a challenge to all existing leak detection methods. The proposed leak detection method can adjust the parameters online according to the changes in the operating conditions, and thus was able to detect small leaks accurately, even with a normal transient state and with noisy measurement data. Simulation results demonstrated the capability of the proposed methodology to detect and locate the leak quickly, and accurately. Comparison studies of the extended Kalman filter (EKF), unscented Kalman filter (UKF), and DUKF show the superiority of DUKF over EKF and UKF for leak detection and location of the leaks in natural gas pipelines. Industrial data are used to validate the results.

(373c) Development of a Sensor Placement Algorithm for Maximizing the Efficiency of an Acid Gas Removal (AGR) Unit for an IGCC Power Plant with CO2 Capture

Prokash Paul[1], Debangsu Bhattacharyya[1], Richard Turton[1] and Stephen E. Zitney[2], (1)Department of Chemical Engg., West Virginia University, Morgantown, WV, (2)AVESTAR Center, National Energy Technology Laboratory, Morgantown, WV

Future integrated gasification combined cycle (IGCC) power plants with CO_2 capture will face stricter operational and environmental constraints. Accurate values of relevant states/outputs/disturbances are needed to satisfy these constraints and to maximize the operational efficiency. Unfortunately, a number of these process variables cannot be measured while a number of them can be measured, but have low precision, reliability, or signal-to-noise ratio. In this work, a sensor placement (SP) algorithm is developed for optimal selection of sensor location, number, and type that can maximize the plant efficiency and result in a desired precision of the relevant measured/unmeasured states.

In this work, an SP algorithm is developed for an selective, dual-stage Selexol-based acid gas removal (AGR) unit for an IGCC plant with pre-combustion CO_2 capture. A comprehensive nonlinear dynamic model of the AGR unit is developed in Aspen Plus Dynamics® (APD) and used to generate a linear state-space model that is used in the SP algorithm. The SP algorithm is developed with the assumption that an optimal Kalman filter will be implemented in the plant for state and disturbance estimation. The algorithm is developed assuming steady-state Kalman filtering and steady-state operation of the plant. The control system is considered to operate based on the estimated states and thereby, captures the effects of the SP algorithm on the overall plant efficiency. The optimization problem is solved by Genetic Algorithm (GA) considering both linear and nonlinear equality and inequality constraints.

Due to the very large number of candidate sets available for sensor placement and because of the long time that it takes to solve the constrained optimization problem that includes more than 1000 states, solution of this problem is computationally expensive. For reducing the computation time, parallel computing is performed using the Distributed Computing Server (DCS®) and the Parallel Computing® toolbox from Mathworks®. In this presentation, we will share our experience in setting up parallel computing using GA in the MATLAB® environment and present the overall approach for achieving higher computational efficiency in this framework.

(373d) Intelligent Recursive Soft Sensor Adaptation Via Bayesian Outlier Detection and Classification

Hector Galicia, Department of Chemical Engineering, Auburn University, Auburn, AL, Qinghua He, Department of Chemical Engineering, Tuskegee University, Tuskegee, AL and Jin Wang, Auburn University, Auburn, AL

Data-driven soft sensors that predict the primary variables of a process by using the secondary measurements have drawn increased research interests recently. Among them, the partial least squares (PLS) based soft sensor is the most commonly used approach for industrial applications. As industrial processes often experience time-varying changes, it is desirable to update the soft sensor model with the new process data once the soft sensor is implemented online. In our previous work [1-2], the recursive reduced-order dynamic PLS (RO-DPLS) soft sensor is developed to provide quality estimates of primary process variables in the event of large transport delays and time-varying process conditions. Both simulated and industrial case studies of a continuous Kamyr digester showed that a recursive RO-DPLS soft sensor can provide accurate estimates of the extent of reaction (i.e., the Kappa number) and can cope with time-varying process behavior.

Since the focus in [2] was to investigate the properties of different recursive updating schemes and data scaling methods, the industrial datasets were pre-processed to remove all outliers before subjecting them to different experiments. This step was taken since static and recursive PLS algorithms are sensitive to outliers in the dataset [3]. Therefore, outlier detection and handling plays a critical role in the development of the PLS-based soft sensors. Although there exists extensive studies on outlier detection for off-line model building [3-7], outlier detection remains a challenging problem. In addition, the detection of outliers online poses some additional challenges. First, the resources dedicated to develop the soft sensor off-line (e.g. expert knowledge) are not available during the online phase. Second, for online adaptation of soft sensor models, if erroneous readings are used to update the soft sensor model, future predictions from the updated model may deteriorate significantly. Furthermore, the challenge increases since outliers online not only can be erroneous readings, but they could also be normal samples of new process states that represent a new process behavior. These samples are necessary to use to provide the updated model with accurate information of the current process behavior and thus improve the soft sensor adaptation.

In this work, a multivariate approach for online outlier detection based on the squared prediction error (SPE) statistic is developed to improve the automatic adaptation of the soft sensor model. In addition, to differentiate outliers caused by erroneous readings from those caused by process changes, a Bayesian supervisory approach is proposed to analyze and classify the detected outliers. A challenging simulation case study of a continuous Kamyr digester [8] is used to assess the performance of the recursive soft sensor with outlier detection and classification. In this case study a major process disturbance, a wood type change, is simulated. It is worth noting that after this major process change occurs, the process settles in a completely new state, and the normal monitoring indices (i.e., SPE indices) may switch to a different level. Therefore, the thresholds of the online monitoring SPE indices need to be updated as well. Otherwise, the performance of outlier detection will deteriorate considerably. In this work, a robust way to update the monitoring thresholds based on an exponentially moving average (EWMA) filter is proposed. The update strategy is shown in Eqn. (1)

$$L_{new} = \lambda(L_{old}) + (1-\lambda)L_{current} \quad (1)$$

where L_{old} denotes the previous monitoring threshold for outlier detection before the update; L_{new} the threshold after the update; $L_{current}$ the threshold estimated using the reconstructed SPE indices of new measurements. The initial thresholds are determined using historical data under normal operation condition. The parameter λ is a tuning parameter which controls how fast the thresholds are updated. In this work, the thresholds are updated for normal operation every 20 samples i.e., no outliers are detected. For this case a relatively conservative setting of λ is used (0.9 < γ < 0.7). For cases where a process change is detected, i.e., detected outliers are classified as part of a process change, a more conservative setting is used ($0.95 < \lambda < 0.99$). It is also worth noting that usually different settings are used for monitoring the independent and dependent variable spaces via SPEx and SPEy indices. This is necessary due to the differences in variability on their corresponding monitoring indices. Due to the limited number of samples for regular update (i.e. 20 samples) and for process change update (i.e. 5 samples), $L_{current}$ is estimated through an empirical way as shown in Eqn. (2).

$$L_{current} = \mu_{SPE} + a \cdot \sigma_{SPE} \qquad 2$$

where μ_{SPE} and σ_{SPE} are the mean and standard deviation of SPE of the samples used for update;   is a tuning parameter usually around 2~3.

The results obtained from the challenging simulation case study indicate that the recursive soft sensor with outlier detection, classification and threshold update, provides a robust way to address the time-varying nature of industrial processes and to provide an intelligent adaptation of the soft sensor model. In addition, the good performance observed in the simulation case study of the proposed approaches is confirmed by the application to a more challenging industrial case study of an industrial continuous Kamyr digester.

References:

[1] Galicia, H. J.; He, Q. P. & Wang, J. A reduced order soft sensor approach and its application to a continuous digester. *Journal of Process Control*, 2011, 21, 489-500.

[2] Galicia, H. J.; He, Q. P. & Wang, J. Comparison of the performance of a reduced-order dynamic PLS soft sensor with different updating schemes for digester control. *Control Engineering Practice*, 2012. Accepted

[3] Hubert, M. and Branden, K.V. Robust methods for partial least squares regression, *Journal of Chemometrics*, vol. 17, pp. 537-549, 2003.

[4] Hodge, V. and Austin, J. A survey of outlier detection methodologies. *Artificial Intelligence Review*, vol. 22, pp. 85-126, 2004.

[5] Pearson, R. K. Outliers in process modeling and identification. *Control Systems Technology, IEEE Transactions on*, vol. 10, pp. 55-63, 2002.

[6] Davies, L. and Gather, U. The identification of multiple outliers. *Journal of the American Statistical Association*, vol. 88, pp. 782-792, 1993.

[7] Jolliffe, I. T. Principal Component Analysis: *Springer*, 2002.

[8] Wisnewski, P. A., Doyle, F., Kayihan, F. (1997). Fundamental continuous-pulp-digester model for simulation and control. *AIChE Journal,* 43(12): 3175-3192.

(373e) Soft Sensor Models Based On a Process Variable and Dynamics Selection Method and Support Vector Regression

Hiromasa Kaneko and Kimito Funatsu, Department of Chemical System Engineering, The University of Tokyo, Tokyo, Japan

Plant operators have to monitor the operating conditions of industrial plants and control process variables, such as temperature, pressure, liquid level, and concentration of products. These variables, therefore, must be measured online. However, it is not easy to measure all variables online due to technological limitations, large measurement delays, and high investment costs. Thus, soft sensors have been widely used to estimate process variables that are difficult to measure online. An inferential model is constructed between those variables that are easy to measure online and those that are not, and an objective variable **y** is then estimated using that model. Through the use of soft sensors, the values of **y**can be estimated with a high degree of accuracy in real-time. In addition, soft sensors can give useful information in terms of fault detection by working with hardware sensors in parallel.

Soft sensor models predicting values of **y** should be constructed with only important explanatory variables **X** in terms of predictive ability, better interpretation of models and lower measurement costs. Besides, some **X**-variables can affect **y** with time-delays. Therefore, we previously proposed the method for selecting important **X**-variables and optimal time-delays of each variable simultaneously, by modifying the genetic algorithm-based wavelength selection method that is one of the wavelength selection methods in spectrum analysis. This proposed method was named as genetic algorithm-based process variable and dynamics selection (GAVDS). The GAVDS method can select time-regions of **X**-variables as a unit by using process data that includes process variables that are delayed in the range from zero to a set/given maximum value.

However, a nonlinear relationship between **X** and **y** cannot be modeled with GAVDS because a partial least squares (PLS) method is used as a modeling technique. In fact, nonlinearity between **X** and **y** has been reported in the fields of soft sensors. We therefore attempted to select appropriate **X**-variable regions and construct a highly predictive model in the case of a nonlinear relationship between **X** and **y**, and then, we proposed a new method for variable region selection based on GAVDS and support vector regression (SVR), which is a nonlinear regression method. The proposed method is named as GAVDS-SVR. Both the representation of nonlinearity between **X** and **y**and variable region selection could be achieved employing the GAVDS-SVR method.

To verify the effectiveness of the proposed method, we applied it to simulation data and soft sensor analysis. In the analysis of simulation data, we compared PLS, GAVDS, SVR, and GAVDS-SVR in the case of nonlinearity between **X**, in which variables have high correlation, and **y**. From the results of the analysis of the simulation data, it was found that although GAVDS could handle monotonically increasing and decreasing functions, appropriate variable selection was not accomplished by the GAVDS method when the nonlinear function is a quadratic function with a local minimal value. In this case, the proposed GAVDS-SVR method could select appropriate **X**-variable regions and construct a highly accurate and predictive model.

Afterward, we applied the GAVDS-SVR method to soft sensor analysis and used actual industrial data obtained during an industrial polymer process. The proposed method could construct models predicting polymer quality with predictive ability higher than that of models constructed by other methods. In addition, process variables and the dynamics of the variables could be selected by employing the GAVDS-SVR method.

The proposed method will be able to reduce the change in parameters and effort of maintenance of soft sensor models during model reconstruction, and probably contribute to the optimal positioning of sensors.

(373f) Estimation of Faults in Water Gas Shift (WGS) Reactor Using Extended Kalman Filter

Jeevan Maddala[1], **Parham Mobed**[1], Richard Turton[2], Debangsu Bhattacharyya[2] and Raghunathan Rengaswamy[1], (1)Chemical Engineering, Texas Tech University, Lubbock, TX, (2)Department of Chemical Engg., West Virginia University, Morgantown, WV

Efficient operation of the water gas shift reactor (WGSR) is very important for satisfying the CO_2 capture target in integrated gasification combined cycle (IGCC) plants with pre-combustion CO_2 capture. The operation of the WGSRs should not only satisfy the desired H_2/CO ratio at the inlet of the acid gas removal (AGR) plant, but also convert most of the COS content in the syngas. For pre-combustion CO_2 capture, a sulfur-tolerant catalyst is used in the WGSR for simultaneous hydrolysis of CO and COS. In these WGSRs, a number faults can occur. For example, the residual fly ash in the syngas can deposit on the catalyst reducing the number of active sites for reaction and decreasing the porosity of the catalyst bed. As the WGS reactions are exothermic, temperatures that are higher than that allowed by the catalyst manufacturer can cause micro-structural changes to the catalyst. Our current work focuses on estimation of these faults using an Extended Kalman Filter (EKF) framework.

The first-principles dynamic WGSR model used in this work is a nonlinear differential algebraic equations (DAE) system that poses unique challenge in estimation. Some of the algebraic equations have a process noise term while others have to be satisfied exactly. Further, some of the algebraic states are also a part of the measurement system. A recently proposed EKF approach for state estimation in DAE systems is modified to solve the WGSR state estimation problem. The fault variables are considered as additional states and estimated jointly with the other states in the proposed approach. The results show that a very accurate estimation of the faults in a WGSR can be obtained by using our proposed enhanced EKF. The presentation will include a discussion on the effect of number of measurements and their sampling rates on the accuracy of prediction of the time of occurrence of the faults.

(373g) Ontology Based Process Supervision System for Large Scale Chemical Plants

Sathish Natarajan and **Rajagopalan Srinivasan**, Department of Chemical and Biomolecular Engineering, National University of Singapore, Singapore, Singapore

Process supervision is a set of activities for determining the state or condition of the process and controlling it to ensure foremost that it remains in a safe regime. The first step in process supervision, determining the condition of the process, involves interpreting real-time information from the process primarily from online measurements. This interpretation requires a model of the process that describes the expected behavior of the process, for a particular state. The development of this model requires various forms of information such as fundamental knowledge (chemical engineering principles), process design related information (flowsheet, instrumentation, etc) or data and information from operating experience (operating procedures, historical data, etc). Implicit in the development of most process supervision methods is the assumption that the underlying information, or descriptors, is static. Changes in descriptors, for instance a change in the flowsheet configuration during maintenance activities and retrofits is not uncommon and has been widely recognized as affecting the safety profile of the plant. The root cause(s) of many major accidents such as the Flixborough, Piper Alpha and PetroBras have originated from significant changes in the process flowsheet, equipment specification, or other such primary descriptors. Alarm management systems too highlight the importance of reflecting changes in the process descriptors to process supervision systems; omissions would lead to spurious alarms and alarm floods that obfuscate abnormal situation management. In other words, while process supervision methods are usually developed for a specific process configuration and modes of operation, modern day plants change significantly throughout their lifecycle.

Process supervision methods therefore need to be cognizant of changes in the underlying process and adapt to them. In cases when the process or instrumentation structure changes during operation, the results from process supervision methods, developed assuming a static nature, would be unreliable. Such adaptation requires that each process supervision method is (1) explicitly aware of the information that has been used for developing its model(s), and (2) able to respond to the changes as necessary or become inactive so as not to generate misleading results. Specifically, since supervision models are dependent on disparate information drawn from distributed sources, an explicit linkage to each is essential. Shared semantics based on a common ontology offers a way to develop these linkages. Hence, an ontology, specifically suited for process supervision is developed and presented here. This ontology forms the basis for the development of a multi agent based process supervision system called ENCORE. The performance of this system under several scenarios is reported for the Kongsberg Generic Oil and Gas Production Simulator.

(377a) Flowsheet Monitoring Interface Specification

William M. Barrett, Sustainable Technology Division, United States Environmental Protection Agency, Cincinnati, OH and Jasper M. Van Baten, AmsterCHEM, Cuevas del Almanzora, Spain

This presentation proposes a standardized mechanism for obtaining process information directly from a process modelling environment (PME) using a strongly-integrated interface set, called Flowsheet Monitoring. The Flowsheet Monitoring interface provides read-only access to the unit operation collection and material, energy, and information stream collection within the process model, and can be used to obtain the material flow data directly from the simulation. This material flow data can then be used to for a variety of purposes to evaluate the process. The Flowsheet Monitoring interface has been proposed as an extension of the CAPE-OPEN chemical process simulation interface set.

An implementation of the US Environmental Protection Agency (USEPA) WAste Reduction (WAR) Algorithm is presented that uses Flowsheet Monitoring to access process flow information from a process modelled with AmsterCHEM's COFE (CAPE-OPEN Flowsheeting Environment). The WAR Algorithm add-in is included in the latest release of its COCO (CAPE-OPEN to CAPE-OPEN) Simulation Environment, available from http://www.cocosimulator.org/.

(377b) Proposal for a New CAPE-OPEN Object Model

Jasper M. Van Baten, AmsterCHEM, Cuevas del Almanzora, Spain, Michel Pons, CO-LaN, Lyon, France and William M. Barrett, US Environmental Protection Agency, Cincinnati, OH

Process simulation applications require the exchange of significant amounts of data between the flowsheet environment, unit operation model, and thermodynamic server. Packing and unpacking various data types and exchanging data using structured text-based architectures, including XML-based systems such as the OPC Unified Architecture (UA), impose significant processing overhead that greatly affect process simulation application performance. For this reason, process simulation middleware protocols need to utilize low-level native data calls to maintain performance.

When CAPE-OPEN was initially developed, it focused on two middleware architectures: Microsoft's Common Object Model (COM) and Common Object Request Broker Architecture (CORBA). The COM object model is tightly integrated with the Windows architecture, while CORBA provides nearly the same level of integration for Unix/Linux-based systems. At present, most CAPE-OPEN implementations are in-process and COM-based because COM's overhead is similar to C function calls. However, the limitations of COM and CORBA have limited the ability of CAPE-OPEN to expand the scope of process simulation into areas such as high-performance computing and the overall enterprise architecture of operating companies.

COM-based CAPE-OPEN has become a mature environment for developing process simulation applications. However, as information technology architectures continue to evolve, CAPE-OPEN needs to adapt to current programming paradigms. This paper presents a proposal for updating the CAPE-OPEN object model to incorporate modern capabilities such as .Net, JAVA, high performance computing and web-based interoperation while maintaining the functionality of the existing COM- and CORBA-based implementations. In addition, the proposed object model will allow for easier and less error-prone CAPE-OPEN software development.

(398a) Third Generation Branch-and-Reduce Algorithms and Software

Nick Sahinidis, Chemical Engineering, Carnegie Mellon University, Pittsburgh, PA and Aida Khajavirad, Carnegie Mellon University, Pittsburgh, PA

The development of the BARON algorithms and software for global optimization began in the early 1990s. Initially, the software integrated factorable nonlinear relaxations within a domain-reduction-based branch-and-bound algorithm. At that time, various feasibility- and optimality-based techniques were introduced for range reduction [1]. BARON became more widely used and recognized only when it entered its second phase of development, when nonlinear relaxations were replaced by polyhedral ones [2, 3] that increased its robustness and performance. The third generation of branch-and-reduce algorithms began to appear with the introduction of relaxation techniques for simultaneous convexification of several quadratic terms [4].

In this talk, we propose techniques that are crucial in extending BARON's simultaneous convexification techniques via polyhedral cutting planes to a variety of problems. Key in our development is the observation that facets of polyhedral envelopes for a variety of nonlinear functions and constraint sets can be obtained by solving a specific linear optimization problem (LP). When used as cutting planes, these facets can significantly enhance the quality of conventional relaxations in global solvers. However, in general, the size of this LP grows exponentially with the number of variables in these nonlinear functions. To cope with this growth, we propose a graph decomposition scheme that exploits the structure of the nonlinear function to decompose it to lower-dimensional components, for which the aforementioned LP can be solved very efficiently by employing a customized simplex algorithm.

We embed this cutting plane generation strategy at every node of the branch-and-reduce global solver BARON, and carry out an extensive computational study on quadratically constrained quadratic problems, multilinear problems, and polynomial optimization problems. Results show that the proposed cuts enable BARON to solve many more problems to global optimality and result in considerable speed-ups of the algorithm.

Finally, other recent advances in the branch-and-reduce algorithm are summarized and extensive computational results are presented for large classes of standard problem sets.

References cited:

1. Ryoo, H. S. and N. V. Sahinidis, Global optimization of nonconvex NLPs and MINLPs with applications in process design, Computers & Chemical Engineering, 19(5), 551-566, 1995.
2. Tawarmalani, M. and N. V. Sahinidis, Global optimization of mixed-integer nonlinear programs: A theoretical and computational study, Mathematical Programming, 99(3), 563-591, 2004.
3. Tawarmalani, M. and N. V. Sahinidis, A polyhedral branch-and-cut approach to global optimization, Mathematical Programming, 103, 225-249, 2005.
4. Bao, X., N. V. Sahinidis, and M. Tawarmalani, Multiterm polyhedral relaxations for nonconvex, quadratically-constrained quadratic programs, Optimization Methods and Software, 24, 485-504, 2009.

(398b) Global Optimization of Bilinear Programs with a Multiparametric Disaggregation Technique

Scott Kolodziej [1], **Pedro M. Castro** [2] and Ignacio E. Grossmann [1], (1)Chemical Engineering Department, Carnegie Mellon University, Pittsburgh, PA, (2)Modeling and Optimization of Energy Systems, LNEG, Lisbon, Portugal

The global optimization of bilinear programming problems is important in such areas as water networks and petroleum blending operations (e.g. Haverly, 1978; Bagajewicz, 2000; Misener et al, 2011). Nonconvex, bilinear constraints are required to model the mixing of various streams in these systems, and are in some cases the only nonlinearities in the models. The pooling problem, stemming from the original Haverly paper, contains these bilinear constraints and has received much attention in the literature. Recently, Misener & Floudas (2011) have demonstrated a novel logarithmic relaxation for modeling bilinear terms with piecewise McCormick envelopes while addressing various classes of pooling problems. Water network optimization problems containing bilinear terms have also received much attention in the literature (see Jezowski, 2010 for a review). The same blending constraints present in the pooling problem are present in water network problems, and thus numerous advances in solving bilinear programs have been made addressing these problems.

In this paper we present the derivation of the multiparametric disaggregation technique (MDT) by Teles et. al (2011) for solving nonconvex bilinear programs. The basic idea relies on a mixed-integer linear programming approximation that is based on a novel discretization scheme in which a discretized variable in the bilinear term is represented by a sum of digits over different powers. While commonly a base of 10 is used, the use of other bases is also possible (Teles et al. 2012), e.g. base 2. We derive both upper and lower bounding formulations corresponding to mixed-integer linear programs that are derived using disjunctive programming and exact linearizations. These are then incorporated into two global optimization algorithms that are used to solve bilinear programming problems. The extension to mixed-integer bilinear programs is also discussed.

First, a set of small problems is considered in order to compare the MDT with piece-wise McCormick convex envelopes. The results show that the relaxation derived using the MDT is shown to scale much more favorably (i.e. linearly vs. exponentially) than the relaxation that relies on piecewise McCormick envelopes, yielding smaller mixed-integer problems and faster solution times for similar optimality gaps. The proposed relaxation also compares well with general global optimization solvers on large problems.

Finally, we present results on both large-scale water network problems and multiperiod blending problems to show that the proposed MDT is competitive, and often more efficient that the direct use of general optimization solvers such as BARON and GloMIQO.

References

Bagajewicz, M.: A review of recent design procedures for water networks in refineries and process plants. Computers & Chemical Engineering 24(9–10), 2093-2113 (2000)

Haverly, C. A.: Studies of the behavior of recursion for the pooling problem. SIGMAP Bull. 25, 19-28 (1978)

Jezowski, J.: Review of Water Network Design Methods with Literature Annotations. Industrial & Engineering Chemistry Research 49 (10), 4475-4516 (2010)

Karuppiah, R., Grossmann, I.E.: Global optimization for the synthesis of integrated water systems in chemical processes. Computers and Chemical Engineering 30, 650-673 (2006).

Misener, R., Thompson, J.P., Floudas, C.A. APOGEE: Global optimization of standard, generalized, and extended pooling problems via linear and logarithmic partitioning schemes, Computers & Chemical Engineering, 35(5), 876-892

(2011)

Teles, J.P., Castro, P.M., Matos, H.A. Multiparametric disaggregation technique for global optimization of polynomial programming problems. Journal of Global Optimization doi: 10.1007/s10898-011-9809-8 (2011).

Teles, J.P., Castro, P.M., Matos, H.A.: Global Optimization of Water Networks Design using Multiparametric Disaggregation. Computers and Chemical Engineering 40, 132-147 (2012).

(398c) Globally Optimizing Mixed-Integer Signomial Programs

Christodoulos A. Floudas and Ruth Misener, Chemical and Biological Engineering, Princeton University, Princeton, NJ

We develop a computational framework for addressing mixed-integer signomial optimization problems (MISO) to epsilon-global optimality. Chemical engineering applications of MISO include heat exchanger network design, chemical reactor design, optimal condenser design, oxygen production, membrane separation process design, optimal design of cooling towers, chemical equilibrium problems, optimal control, batch plant modeling, and optimal location of hydrogen supply centers [13].

This work encompasses and extends previous development of the Global Mixed-Integer Quadratic Optimizer, GloMIQO [18, 19]. As in GloMIQO, our approach is based on (1) reformulating user input, (2) detecting special mathematical structure, and (3) branch-and-bound global optimization. All our solution strategies for quadratic programs are integrated into this work; this presentation highlights new components for signomial functions.

In the reformulation stage, we transform a factorable programming tree into a flattened expression tree. This automatic procedure effectively flattens nonconvex expressions from vertical, operator-based data structures (*e.g.*, [8, 21]) to horizontal, term-based data structures (*e.g.*, [1, 2, 12]).

In the detection stage, we elucidate special mathematical structure in the terms including convexity/concavity and edge-convexity/edge-concavity. The term-based data structures allow us easy access to specially developed tight convex underestimators for trilinear terms [15, 16], quadrilinear terms [6], odd-degree monomials [9], signomial terms [10, 11, 13], low-dimensional edge-concave terms [17, 22], products of convex functions [12], and general nonconvex terms [1, 2, 12]. In addition to discerning tight convex underestimators for individual terms, we interlink the terms using an implementation of the Reformulation-Linearization Technique that considers all possible combinations of products between variables, nonlinear terms, and equations [20].

In the global optimization stage, we use the special mathematical structure elucidated in the detection stage to: generate tight convex relaxations, dynamically add separating cuts, partition the search space, bound the variables, and find feasible solutions. Different underestimation strategies may have varying relative efficacy throughout the branch-and-bound tree [3, 8]; we generate candidate hyperplanes using an array of convexification strategies. Due to steep computational costs associated with generating the deepest possible cut, we temper our analysis of each cutting plane strategy with complexity analysis.

We extensively test and validate our work based on MINLPLib [5], GLOBALLib [7, 14], PerformanceLib, www.minlp.org, and the Bonmin test set [4].

References

[1] C. S. Adjiman, I. P. Androulakis, and C. A. Floudas. A global optimization method, aBB, for general twice differentiable NLPs-II. Implementation and computional results. *Comput. Chem. Eng.,*22:1159 – 1179, 1998b.

[2] C. S. Adjiman, S. Dallwig, C. A. Floudas, and A. Neumaier. A global optimization method, aBB, for general twice differentiable NLPs-I. Theoretical advances. *Comput. Chem. Eng.*, 22:1137 – 1158, 1998a.

[3] A. Bompadre and A. Mitsos. Convergence rate of McCormick relaxations. J. Glob. Optim., 52(1):1–28, 2011.

[4] P. Bonami, L. T. Biegler, A. R. Conn, G. Cornuejols, I. E. Grossmann, C. D. Laird, J. Lee, A. Lodi, F. Margot, N. Sawaya, and A.Waechter. An algorithmic framework for convex mixed integer nonlinear programs. *Discrete*

Optimization, 5(2):186 – 204, 2008.

[5] M. R. Bussieck, A. S. Drud, and A. Meeraus. MINLPLib - a collection of test models for mixed-integer nonlinear programming. *INFORMS J. Comput*, 15(1), 2003.

[6] S. Cafieri, J. Lee, and L. Liberti. On convex relaxations of quadrilinear terms. *Journal of Global Optimization*, 47:661–685, 2010.

[7] C. A. Floudas, P. M. Pardalos, C. S. Adjiman, W. R. Esposito, Z. H. Gumus, S. T. Harding, J. L. Klepeis, C. A. Meyer, and C. A. Schweiger. Handbook of Test Problems in Local and Global Optimization. Kluwer Academic Publishers, 1999.

[8] E. P. Gatzke, J. E. Tolsma, and P. I. Barton. Construction of convex relaxations using automated code generation techniques. *Optim. Eng.*, 3:305–326, 2002.

[9] L. Liberti and C. C. Pantelides. Convex envelopes of monomials of odd degree. *J. Glob. Optim.*, 25:157–168, 2003.

[10] A. Lundell, J. Westerlund, and T. Westerlund. Some transformation techniques with applications in global optimization. *J. Glob. Optim.*, 43:391–405, 2009.

[11] A. Lundell and T. Westerlund. Convex underestimation strategies for signomial functions. *Optim. Methods Software*, 24(4-5):505–522, 2009.

[12] C. D. Maranas and C. A. Floudas. Finding all solutions of nonlinearly constrained systems of equations. *J. Glob. Optim.*, 7(2):143–182, 1995.

[13] C. D. Maranas and C. A. Floudas. Global optimization in generalized geometric programming. *Comput. Chem. Eng.*, 21(4):351 – 369, 1997.

[14] A. Meeraus. GLOBALLib. http://www.gamsworld.org/global/globallib.htm.

[15] C. A. Meyer and C. A. Floudas. Trilinear monomials with positive or negative domains: Facets of the convex and concave envelopes. In C. A. Floudas and P. M. Pardalos, editors, Frontiers in Global Optimization, pages 327–352. Kluwer Academic Publishers, 2003.

[16] C. A. Meyer and C. A. Floudas. Trilinear monomials with mixed sign domains: Facets of the convex and concave envelopes. *J. Glob. Optim.*,29(2):125–155, 2004.

[17] C. A. Meyer and C. A. Floudas. Convex envelopes for edge-concave functions. *Math. Program.*,103(2):207–224, 2005.

[18] R. Misener and C. A. Floudas. Global optimization of mixed-integer quadratically-constrained quadratic programs (MIQCQP) through piecewise-linear and edge-concave relaxations. *Math. Program. B*. Accepted for Publication; http://www.optimization-online.org/DB_HTML/2011/11/3240.html.

[19] R. Misener and C. A. Floudas. GloMIQO: Global Mixed-Integer Quadratic Optimizer. *J. Glob. Optim*. Accepted for Publication (DOI: 10.1007/s10898-012-9874-7).

[20] H. D. Sherali and W. P. Adams. A Reformulation-Linearization Technique for Solving Discrete and Continuous Nonconvex Problems. Nonconvex Optimization and Its Applications. Kluwer Academic Publishers, Dordrecht, Netherlands, 1999.

[21] E. M. B. Smith and C. C. Pantelides. A symbolic reformulation/spatial branch-and-bound algorithm for the global optimisation of nonconvex MINLPs. *Computers & Chemical Engineering*, 23(4 - 5):457– 478, 1999.

[22] F. Tardella. Existence and sum decomposition of vertex polyhedral convex envelopes. *Optim. Lett.*, 2:363–375, 2008.

(398d) A Parametric Approach for the Global Optimization of Mixed-Integer Fractional Programs

Yunfei Chu and Fengqi You, Department of Chemical and Biological Engineering, Northwestern University, Evanston, IL

Mixed-integer fractional programming (MIFP) problems frequently occur in process engineering, e.g. cyclic scheduling [1], and integration scheduling and control [2, 3]. In this work, we specifically consider the MIFPs with a convex quadratic function in the numerator and a linear function in the denominator. This is a non-convex mixed integer nonlinear programming (MINLP) problem and we developed an efficient approach for its global optimization.

First, the relaxed MIFP problem is proved to be quasi-convex. Based on the property, the MINLP solver SBB [4] can guarantee the global optimality. Because each nonlinear programming (NLP) subproblem in the branch-and-bound is solved to the global optimum, the global solution to the MINLP problem can be obtained by a standard branch-and-bound algorithm. Another popular MINLP solver DICOPT [5] uses the outer-approximation method and it might cut off some feasible solution space during the iterations. Thus, DICOPT cannot guarantee the global optimality of the solution. The global optimal solution can be also obtained by the global MINLP solvers, such as BARON [6] and GloMIQO [7].

Second, the MIFP is transformed into an equivalent parametric mixed-integer programming (PMIP) problem which is the numerator minus the denominator multiplied by a parameter. The solution to the PMIP problem is a function of the parameter, called the optimal value function. A feasible solution to the MIFP problem is the global optimal solution if and only if the solution is also the global optimal solution to the PMIP problem with the parameter which is the root of the optimal value function [8-10]. The condition for the equivalence is very general, i.e. the denominator should be positive. There are no assumptions required on the types or the convexities of the functions in the numerator and the denominators. However, the equivalence is true only for the global minima of the two problems. For the problem we consider, the equivalent PMIP problem is a convex mixed-integer quadratic programming (MIQCQ) problem and it can be solved to the global minimum by state-of-the-art solvers, such as CPLEX, in an efficient way. To guarantee the equivalence, the PMIP problem must be solved at the parameter which is the root of the optimal value function.

Third, two approaches are applied to find the root of the optimal value function which has no explicit expression. They are the bisection method [8] and the generalized Newton's method [10]. The bisection method only uses the sign information of $F(q)$ in the iteration while the generalized Newton's method requires the information of the subgradient. By using more information of the function, the generalized Newton's method has a faster convergence rate than the bisection method. The Newton's method converges supper-linearly while the bisection algorithm converges linearly.

The PMIP solution approaches based on the bisection method and the Newton's method are demonstrated by a large-scale MIFP problem derived from an integrated scheduling and control problem [3]. The results are compared with those returned by the MINLP solvers, SBB, DICOPT, and BARON 9.3.1. The results showed that the proposed approaches can globally optimize the MIFPs within much shorter times than general-purpose MINLP methods.

References

[1] J. M. Pinto and I. E. Grossmann, "Optimal cyclic scheduling of multistage continuous multiproduct plants," *Computers & Chemical Engineering,* vol. 18, pp. 797-816, 1994.

[2] A. Flores-Tlacuahuac and I. E. Grossmann, "Simultaneous cyclic scheduling and control of a multiproduct CSTR," *Industrial & Engineering Chemistry Research,* vol. 45, pp. 6698-6712, 2006.

[3] Y. Chu and F. You, "Integration of scheduling and control with online closed-loop implementation: Fast computational strategy and large-scale global optimization algorithm," submitted to *Computers & Chemical*

Engineering, 2012.

[4] M. R. Bussieck and A. S. Drud. SBB: A new solver for mixed integer nonlinear programming [Online]. Available: http://www.gams.com/presentations/or01/sbb.pdf

[5] G. R. Kocis and I. E. Grossmann, "Computational experience with DICOPT solving MINLP problems in process systems engineering," *Computers & Chemical Engineering,* vol. 13, pp. 307-315, 1989.

[6] M. Tawarmalani and N. V. Sahinidis, "Global optimization of mixed-integer nonlinear programs: A theoretical and computational study," *Mathematical Programming,* vol. 99, pp. 563-591, 2004.

[7] R. Misener and C. A. Floudas, "GloMIQO: Global mixed-integer quadratic optimizer.," *Journal of Global Optimization,* in press, 2012.

[8] J. R. Bradley and B. C. Arntzen, "The simultaneous planning of production, capacity, and inventory in seasonal demand environments," *Operations Research,* vol. 47, pp. 795-806, 1999.

[9] W. Dinkelbach, "On Nonlinear Fractional Programming," *Management Science,* vol. 13, pp. 492-498, 1967.

[10] F. Q. You, P. M. Castro, and I. E. Grossmann, "Dinkelbach's algorithm as an efficient method to solve a class of MINLP models for large-scale cyclic scheduling problems," *Computers & Chemical Engineering,* vol. 33, pp. 1879-1889, 2009.

(398e) Convex Relaxation of Multivariate Compositions

Angelos Tsoukalas, Mechanical Engineering, MIT, Cambridge, MA and **Alexander Mitsos**, Department of Mechanical Engineering, Massachusetts Institute of Technology, Cambridge, MA

McCormick relaxations [1] provide the framework for the computation of convex relaxations of composite functions. McCormick's relaxations are clearly a very important tool, but they have the limitation of only allowing univariate composition. Although most functions can be decomposed in a way that only univariate functions are used as building blocks, this often results in weak relaxations.

We propose a reformulation of McCormick's composition theorem, which while equivalent to the original, suggests a straight forward generalization to multi-variate outer functions. In addition to extending the framework, the multi-variate McCormick relaxation is a useful tool for the proof of relaxations: by direct application to the product, division and minimum/maximum of two functions, we obtain improved relaxations when comparing with uni-variate McCormick.

Furthermore, we generalize the theory for the computation of subgradients [2] to the multi-variate case, envisioning practical methods that utilize the framework. Finally, our approach suggests that an algorithm operating directly on recursively obtained McCormick relaxations, can be seen as a decomposition method for the auxiliary variable method (AVM) [3,4]. This interpretation suggests hybrid methods, which can eliminate the disadvantage of the potentially weaker relaxations due to repeated terms in McCormick's framework with minimal increase to the dimension of the optimization space.

[1] G. P. McCormick. "Computability of global solutions to factorable nonconvex programs: Part I. Convex underestimating problems." Mathematical Programming, 10(1):147–175, 1976.

[2] A. Mitsos, B. Chachuat, and P. I. Barton. "McCormick-based relaxations of algorithms". SIAM Journal on Optimization, 20(2):573–601, 2009.

[3] E. Smith and C.C. Pantelides. "Global optimisation of nonconvex minlps." Computers & Chemical Engineering, 21:791–796, 1997.

[4] E. Smith and C. C. Pantelides. "A symbolic reformulation/spatial branch-and-bound algorithm for the global optimisation of nonconvex minlps." Computers & Chemical Engineering, 23(4-5):457–478, 1999.

(398f) Improved Relaxations for Global Optimization with ODEs and DAEs Embedded

Joseph K. Scott, Chemical Engineering, Massachusetts Institute of Technolgy, Cambridge, MA and Paul Barton, Process Systems Engineering Laboratory, Dept. of Chemical Engineering, Massachusetts Institute of Technology, Cambridge, MA

An improved method is presented for computing convex and concave relaxations of the parametric solutions of nonlinear ordinary differential equations (ODEs) and nonlinear semi-explicit index-one systems of differential-algebraic equations (DAEs). Such relaxations, termed state relaxations, are fundamental to deterministic global optimization algorithms for problems with ODEs and DAEs embedded, and their computation is the subject of several recent articles. Largely owing to the difficulty of this computation and the weaknesses of available methods, it remains an unfortunate fact that state-of-the-art deterministic methods for global dynamic optimization can only solve problems of modest size with reasonable computational effort, typically on the order of 5 state variables and 5 decisions. On the other hand, potential applications for such techniques are ubiquitous, including parameter estimation problems with dynamic models, optimal control of batch processes, safety verification problems, optimal catalyst blending , optimal drug scheduling, etc. Moreover, representative case studies in the literature suggest that these applications commonly lead to problems with multiple suboptimal local minima, especially when the embedded dynamic system involves a model of chemical reaction kinetics. Thus, the need for improved relaxation techniques is clear.

In this presentation, we study a class of state relaxation techniques in which the desired relaxations are computed as the solutions of an auxiliary dynamic system (either ODEs or DAEs) derived by relaxing the governing equations of the original system in some way. Methods of this type have the advantage that the state relaxations can be evaluated efficiently using a state-of-that-art numerical integration code. We begin by presenting a theoretical analysis of this class of methods that formalizes properties of the auxiliary system that are guaranteed to lead valid state relaxations. Next, two existing state relaxation methods for parametric ODEs are classified according to our analysis, and it is shown that each of these methods has certain undesirable properties that follow directly from this classification. Based on these observations, a new method is designed in order to avoid these deficiencies. By numerical experiments, it is shown that the proposed method outperforms the previous methods in terms of the tightness of the computed relaxations and the empirical rate of convergence. Finally, the new approach is extended to the class of semi-explicit index-one DAEs.

(398g) A Deterministic Global Optimization Algorithm, Branch-and-Sandwich, for Optimistic Bi-Level Programming Problems: Implementation and Computational Results

Polyxeni-M. Kleniati and Claire S. Adjiman, Centre for Process Systems Engineering, Department of Chemical Engineering, Imperial College London, London, United Kingdom

We describe an implementation of the Branch-and-Sandwich algorithm, introduced in (Kleniati and Adjiman, J. Global Optim., 2012), which can be used to solve optimistic bi-level programming problems, where the inner problem is nonconvex but satisfies a regularity condition, and the functions involved are twice continuously differentiable. The main components of the algorithm are (i) the partition of both the inner and the outer spaces without distinction, but at the same time the consideration of all inner (successively refined) subdomains, where global optima may lie, for successively refined outer subdomains; (ii) the generation of guaranteed upper and lower bounds on the inner optimal value function and on the outer objective value for a given region - the proposed bounding problems are nonconvex but do not grow in size with the number of iterations and are always obtained from the corresponding problems of the parent node; (iii) the appropriate tree management with auxiliary lists of nodes as well as inner and outer fathoming rules. The theoretical convergence properties have also been investigated in the aforementioned work and the Branch-and-Sandwich algorithm was proved to be epsilon-convergent, i.e. at termination an epsilon-optimal solution of the bi-level problem is computed. This result followed from exhaustiveness and the general theory concerning the convergence of branch-and-bounds methods developed in (Horst and Tuy, Springer-Verlag, Third ed., 1996). The Branch-and-Sandwich algorithm was tested on 33 small problems with promising numerical results. The full implementation of the algorithm and computational experience are the focus of the present work. Practical considerations, algorithmic options and performance on larger problems are explored. In particular, the efficiency of the algorithm is strongly dependent on the progress of the outer lower bound on the global optimal value of the bi-level problem, which in turn may often be dependent on the progress of the inner upper bound. In addition, what choices we make with respect to how the global optimization subproblems are tackled, e.g. via convexification versus solving them directly to global optimality, may affect the performance of the algorithm. The C++ implementation of the algorithm is a customization of the branch-and-bound platform provided by the MINOTAUR open-source toolkit (Leyffer, Mahajan, Munson, Linderoth and Luedtke).

Process Understanding Tools in PAT Implementations

João Machado, Tiago Robalo, Pedro Felizardo
4Tune Engineering Ltd - Technology Services & Products, Lisbon, Portugal

Abstract

For the past years the pharmaceutical industry has been trying to increase its process understanding in order to keep pace with regulatory demands and also to increase its competitiveness by being able to find improvement opportunities and increase efficiency. This is also a requirement from the new paradigm of Quality by Design.

For this increase in process understanding demand, the industry has started to collect increasingly amounts of new data from the process and the data itself is now of different types, particularly with the introduction of PAT tools. These tools that are able to collect complex data (NIRS, 2D-Fluorescence, dielectric spectroscopy, etc.) pose challenges for the data organization, structuring, accessibility, prioritization and knowledge extraction. In fact, without these challenges solved, the amount and type of data collected can reveal to be a considerable investment and irrelevant for practical use.

If data is not systematically and automatically organized (presuming that it is being automatically gathered), making sense of it can become an unreasonably time-consuming activity. Diagnostic and troubleshooting efforts are lost in the data matrix complexity and energy is spent in the organization of the data in the context of the production process. Also, this lack of structure and organization also poses barriers to the implementation of prediction and MSPC models across the entire cascade of process steps (from raw-materials to end product). To perform true whole-process analysis across the entire process and process life-cycle in the lights of the new guidelines and regulatory demands, this issue needs to be solve with new tools that enable the shift of energy from data organization to the more useful process understanding and process knowledge sphere with a true system engineering environment.

Therefore a new tool that provides a solution to the industry to tackle these problems is needed. This tool should allow the organization and storage (with proper structure) of different types of data: Process data (temperatures, pressures, flow rates, etc.), PAT fingerprint data (NIR. MIR, 2D-Fluorescence, etc.), LIMS data and in-line monitoring data. These data have to be added in real-time or be a part of an historical database of the process enabling fast monitoring and diagnosis of the process or historical analysis of previous runs. The necessary graphical instruments are also required in this tool to enable quick and intuitive visualization of the data according to its type (single plots, landscapes, etc.).

Comparison of data should also be possible in this graphical interface that takes the previous automatic organization of data and allows easy evaluation of different data types, from different process steps (ex: unit operations) at different time steps or from different runs. This should prove very useful when performing root-cause analysis and assessing cause & effect on the process.

Many of the data being gathered from pharmaceutical production process is used (or could be used) to predict production quality, control the process, diagnose performance, assist in real time release, etc. This is especially true since the introduction of PAT tools that can extract increasingly more informative data from the process. By applying modeling techniques to these data (PAT data, process data, etc.) more process understanding and control can be implemented. For these reasons, this tool should support an easy integration of inline prediction models, calibrations, MSPC, etc. from third-party software. In fact, by having this feature, the user has the chance to improve its models in accuracy and robustness by taking advantage of the organization and accessibility of all the data that ultimately leads to a greater process understanding.

The importance of fast sharing of information and findings is an increasing necessity in the industry when knowledge-based decisions often have to be taken with haste but also truthful. In this context, a new tool has to support mechanisms to share the information found (through normal channels such as email, PDF, etc.) and also to store it (with systems like dropbox, database storage or similar).

In order to exemplify the above-mentioned needs and the advantages of such tool in the pharmaceutical industry, it is intended to present iSEE™ (Systems Engineering Environment interface) in the context of a case study with data from a bio-production pharmaceutical process.

2012 AIChE Annual Meeting QbD Topical Conference
(Pittsburgh, PA, Oct 28-Nov 2 2012)
Session 401 "**Application of PAT in the Manufacturing Setting**"
Paper 401c

An Integrated Process Analytical Technology (PAT) Approach to Examine the Effect of Temperature on Nucleation Kinetics of a Dynamic Pharmaceutical Co-precipitation Process

Huiquan Wu*, Maury White, Ryan Foringer, and Mansoor Khan
Division of Product Quality Research (DPQR, HFD-940), OTR/OPS/CDER/FDA
FDA White Oak Research Center, 10903 New Hampshire Ave
Silver Spring, MD 20993

Abstract: Process Analytical Technology (PAT) has become an essential part for pharmaceutical process and product understanding, real-time process monitoring and control during the last several years. Previously, we reported the development of a novel real-time PAT-based approach to measure the nucleation induction time and elucidate the nucleation and growth mechanisms of a dynamic pharmaceutical co-precipitation process at 25^{0}C. The goal of this work is to understand the effect of temperature (15^{0}C, 25^{0}C, 35^{0}C) on nucleation induction time for a dynamic pharmaceutical co-precipitation process (naproxen-Eudragit L100-alcohol-water) at various ratios of drug/polymer. Real-time inline process monitoring and nucleation induction time (t_{ind}) measurement have been accomplished using FBRM and PVM simultaneously. It was shown that: (1) the plots of t_{ind} vs. the ratio of naproxen to polymer (Q) display two distinct linear segments; (2) the plots of $\ln(t_{ind})$ vs. $(\ln S)^{-2}$ (S is the maximum allowable super-saturation level) display two distinct linear segments which agree well with the classic nucleation theory (CNT); (3) Quantitative relationships based on CNT among the kinetic parameter A, dimensionless thermodynamic parameter B, and temperature (T) suggest that the nucleation and growth process is followed by the interface transfer mechanism. Therefore, at low S level, the nucleation is governed by primary heterogonous nucleation; while at high S level it is governed by primary homogeneous nucleation; after nucleation, the growth is controlled by the interface transfer mechanism. This work demonstrated the utility of real time process PAT monitoring on process understanding and process kinetic mechanism elucidating, thus could enhance the science-based pharmaceutical regulation in the Quality-by-Design (QbD) domain.

Key words: PAT, QbD, Nucleation and Growth Kinetics, Dynamic Co-precipitation Process, Nucleation and Precipitation Induction Time, Dimensionless Thermodyanmic Parameter and Kinetic Parameter

*Corresponding author: Huiquan Wu (Telephone: 001-301-796-0095; Fax: 001-301-796-9816; Email: Huiquan.wu@fda.hhs.gov)

Disclaimer: The views and opinions expressed in this work are only of the authors, and do not necessarily reflect the policy or statement of the US FDA.

1. PAT approach to examine the nucleation phenomena and crystallization onsets

Previously, an integrated PAT (FDA, 2004) approach based on real time Focused Beam Reflectance Measurement (FBRM), real time Process Near-infrared (NIR) Spectroscopy, and process chemometrics had been developed for measuring and validating the co-precipitation nucleation induction time, and elucidating the nucleation and growth mechanisms of a 4-component co-precipitation process at 25 ^{0}C (Wu and Khan, 2011).

Several experimental techniques have been used for examining the nucleation phenomena and crystallization onsets, such as polarized light microscopy (Varon et al, 2010), Focused Beam Reflectance Measurement (FBRM) and Particle Vision Microscopy (PVM) (Wu and Khan, 2011). Nucleation induction time (Söhnel and Mullin, 1988) (t_{ind}) is a quantitative parameter for characterizing the nucleation tendency for solution crystallization at isothermal condition. It is the time elapsed between super-saturation creation and new phase detection in a system. It is a function of the initial super-saturation and temperature. However, it is not a fundamental property of a system because the new phase detection is open to a variety of interpretations, such as the first appearance of crystals or the onset of a change of solution property. There are a number of experimental techniques used for new phase detection, such as differential scanning calorimetry (Yosjihashi et al., 2010), conductivity measurement (Van Der Leeden et al., 1992), turbidity measurement (Guo et al., 2006), intensity of transmitted or scattered light method (Wakita and Masuda, 1983), and combined process NIR and turbidity measurement (Wu and Khan, 2010).

From the transport process perspective, it is important to recognize that the nucleation and growth process involves multiple steps, including nucleation initiation, bulk diffusion, and interface transfer (between the nucleus and substrate). Depending on the process conditions and the crystallization system being studied, one or a combination of these could be the rate-limiting step for the nucleation and growth process. In theory, the impacts of critical process parameters (such as temperature and super-saturation level) on nucleation and growth kinetics are highly complicated. Based on our previous Design of Experiments (DOE) and ANOVA results (Wu et al.,2011), temperature is the most influential process variable with respect to its impact on several response variables for the co-precipitation process, including derived process rates during transition periods, mode and peak frequency at the end of steady state. In this work, the effect of temperature on nucleation kinetics and growth kinetics is characterized by the measurement and analysis of the nucleation induction time via an integrated real-time online PAT monitoring scheme, under the framework of Classical Nucleation Theory (CNT).

2. Experimental

Naproxen USP (lot No: NPX 368) was obtained from Albemarle Corporation (Orangeburg, SC). Eudragit L 100 (lot No: 1221203048) was obtained from Röhm America Inc. (Somerset, NJ). Solvent reagent alcohol (HPLC grade, lot No: 053546) was purchased from Fisher Scientific (Fair Lawn, NJ). Non-solvent DI water was obtained from a FDA in-house facility (Silver Spring, MD) and was kept refrigerated at 3^{0}C prior to use. All chemicals, solvent, and non-solvent were used without any further processing or purification prior to the use for this experimental work.

Experimental setup is similar to previous work, except that both FBRM and PVM are simultaneously monitoring the dynamic co-precipitation process in this work. Both FBRM and PVM data were acquired every two seconds.

To quantify the effect of temperature on nucleation and growth kinetics mechanisms for this pharmaceutical co-precipitation system, other process parameters were fixed throughout the study: stirring rate 200 rpm; DI water addition rate 1.8 ml/sec. Various ratios of API/Polymer (Naproxen)/Eudragit L100) were used for the study, although it was fixed for each individual experiment. Nucleation induction time was measured for each co-precipitation system based on real-time inline PAT monitoring results. As discussed in our previous work (Wu and Khan, 2011), the nucleation induction time measurement for this particular dynamic co-precipitation system has been quantitatively validated by an independent method, which is based on process trajectory and real-time online Near-infrared (NIR) spectroscopic monitoring results. In this work, the nucleation induction time measurement based on FBRM chord-length-distribution (CLD) was qualitatively validated by real time online PVM measurement results. Theoretical analysis of those nucleation induction times and corresponding super-saturation ratios was carried out based on the classical nucleation theory (CNT), as discussed below.

3. Classical Nucleation Theory (CNT)

Based on the classical primary nucleation theory (Nývlt 1968; Sőhnel and Mullin, 1988; Mullin 1972; Kashchiev et al., 1991), we have

$$\ln(t_{ind}) = \ln(\frac{N_c}{VA_{\hom,J}}) + \frac{16\pi\gamma^3 v^2}{3(kT)^3} \cdot \frac{1}{(\ln S)^2} \quad \text{for HON process;} \tag{3.1}$$

$$\ln(t_{ind}) = \ln(\frac{N_c}{VA_{het,J}}) + \frac{16\pi\gamma^3 v^2}{3(kT)^3} \cdot f(\theta) \cdot \frac{1}{(\ln S)^2} \quad \text{for HEN process.} \tag{3.2}$$

Eq. (3.1) and Eq.(3.2) suggests that, at a given temperature, the plot of $\ln(t_{ind})$ vs. $(\ln S)^{-2}$ should give a straight line. For a HON process, the slope and intercept are:

$$B_{\hom,t_{ind}} = \frac{16\pi\gamma^3 v^2}{3(kT)^3} \tag{3.3}$$

$$A_{\hom,t_{ind}} = \ln(\frac{N_c}{VA_{\hom,J}}) \tag{3.4}$$

For a HEN process, the slope and intercept are:

$$B_{het,t_{ind}} = \frac{16\pi\gamma^3 v^2}{3(kT)^3} \cdot f(\theta) \tag{3.5}$$

$$A_{het,t_{ind}} = \ln(\frac{N_c}{VA_{het,J}}) \tag{3.6}$$

For HON mechanism,

(1) If growth is volume-diffusion controlled, we have

$$A_{\hom,t_{ind}} = \ln(\frac{N_c}{V}) + \ln(\frac{1}{A_{\hom,j}}) = A_{\hom}^{diff} - 0.5\ln T \tag{3.7}$$

(2) If growth is interface transfer controlled, we have

$$A_{\hom,t_{ind}} = \ln(\frac{N_c}{V}) + \ln(\frac{1}{A_{\hom,j}}) = A_{\hom}^{surf} + 0.5\ln T \tag{3.8}$$

For HEN mechanism,

(1) If growth is volume-diffusion controlled, we have

$$A_{het,t_{ind}} = \ln(\frac{N_c}{V}) + \ln(\frac{1}{A_{het,j}}) = A_{het}^{diff} - 0.5\ln T \tag{3.9}$$

(2) If growth is interface transfer controlled, we have

$$A_{het,t_{ind}} = \ln(\frac{N_c}{V}) + \ln(\frac{1}{A_{het,j}}) = A_{het}^{surf} + 0.5\ln T \tag{3.10}$$

Therefore, for both HON and HEN mechanisms, the slope of $A_{\hom,t_{ind}}$ vs. $\ln T$ and the slope of $A_{het,t_{ind}}$ vs. $\ln T$ will be negative if the growth is volume-diffusion controlled, and positive if the growth is interface-transfer controlled.

4. Results and Discussion

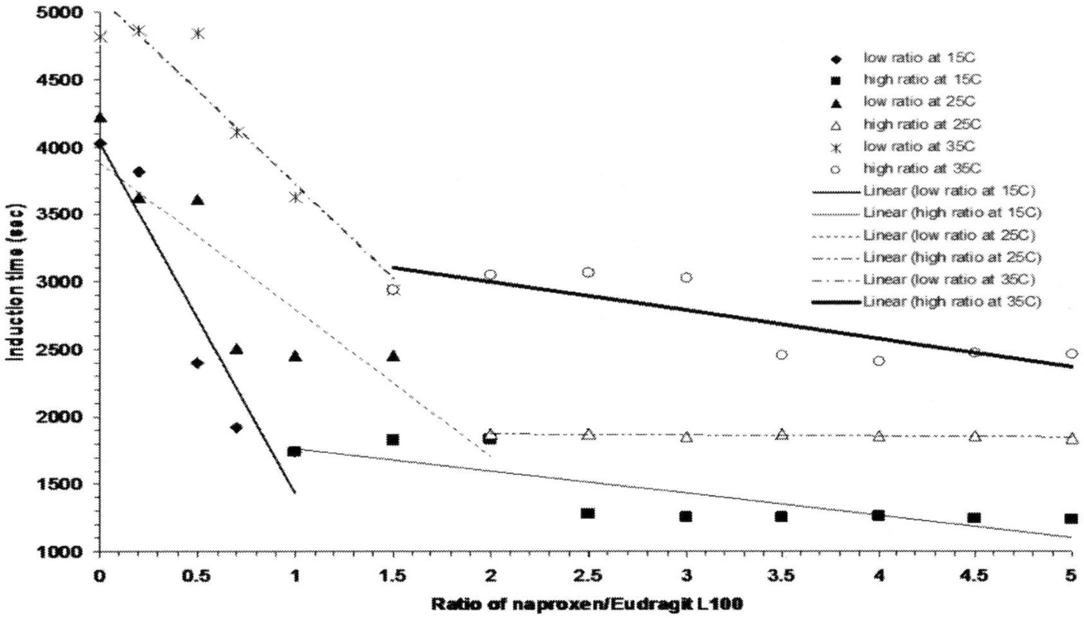

Figure 1. Plots and linear regression lines of nucleation induction time vs. drug/polymer ratio at15^{0}C, 25^{0}C, and 35^{0}C.

Figure 1 shows that there are two distinct segments for each temperature: at low drug/polymer ratio, the induction time decreases quickly with the increase of drug/polymer ratio; after the drug/polymer ratio reaches a threshold value which changes with the temperature, the decreasing rate of induction time vs. drug/polymer ratio becomes much smaller. Eventually, the induction time reaches a plateau value specific to the temperature.

Figure 2. Nucleation and/or precipitation induction and growth phenomena monitored in real time by FBRM and PVM simultaneously. PVM images and FBRM CLD profiles at 15 ^{0}C, with various *Q*.
(2a) *Q*=1.0 at time point of t_{ind}

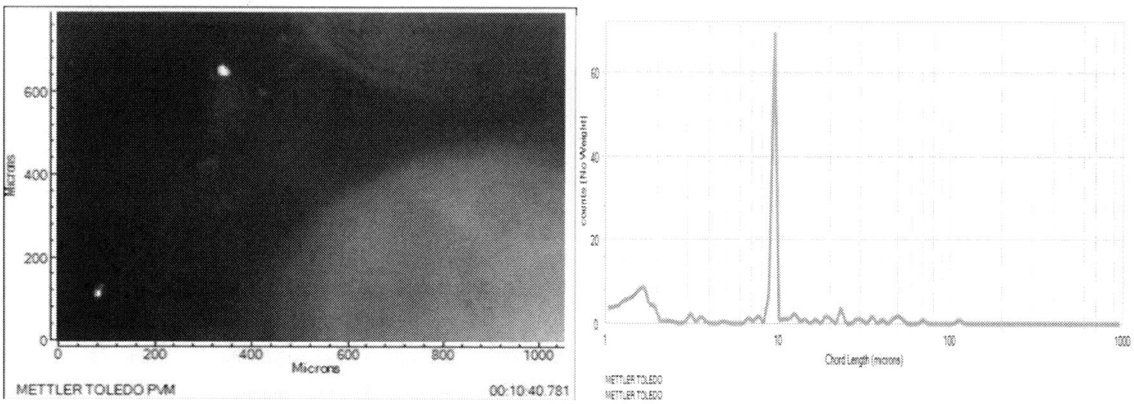

(2b) *Q*=1.0 at time point of t_{ind} +10 minutes

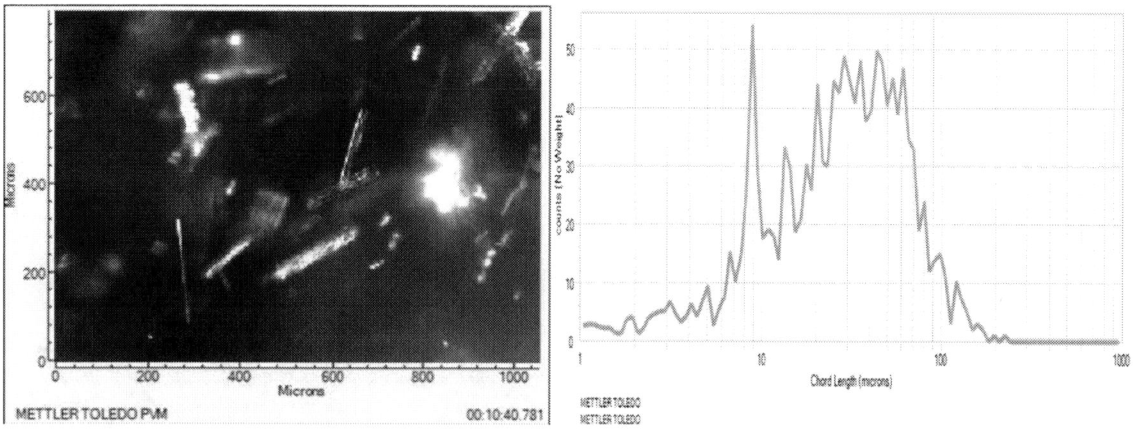

(2c) *Q*=5.0 at time point of t_{ind}

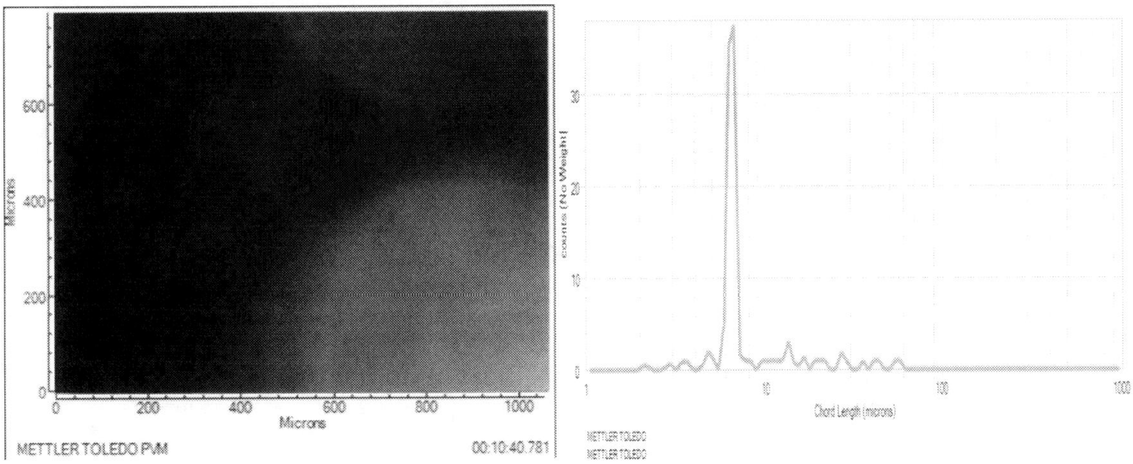

(2d) Q=5.0 at time point of t_{ind} +10 minutes

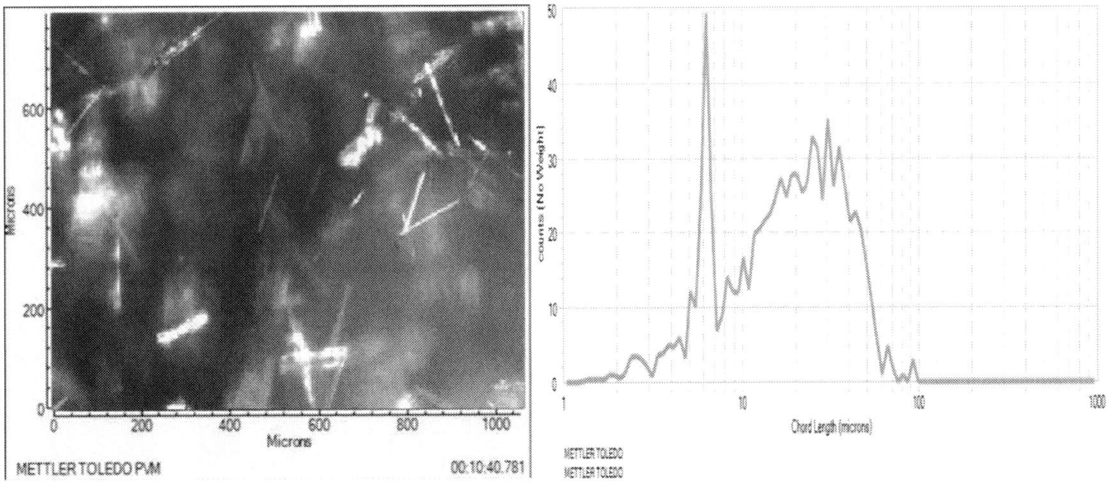

At time point of t_{ind}, PVM pictures show several small particles are detected, while FBRM shows that the counts/s is in the range of magnitude of 10. At time point of t_{ind} +10 minutes, the increased number of crystals and increased dimensions of particles are evident by PVM pictures, while the increased dimensions of particles and crystals are also evidenced by FBRM CLD.

Figure 3. Nucleation induction time as a function of super-saturation. Plots and regression lines of $ln(t_{ind})$ vs. $(lnS)^{-2}$ at15^{0}C.

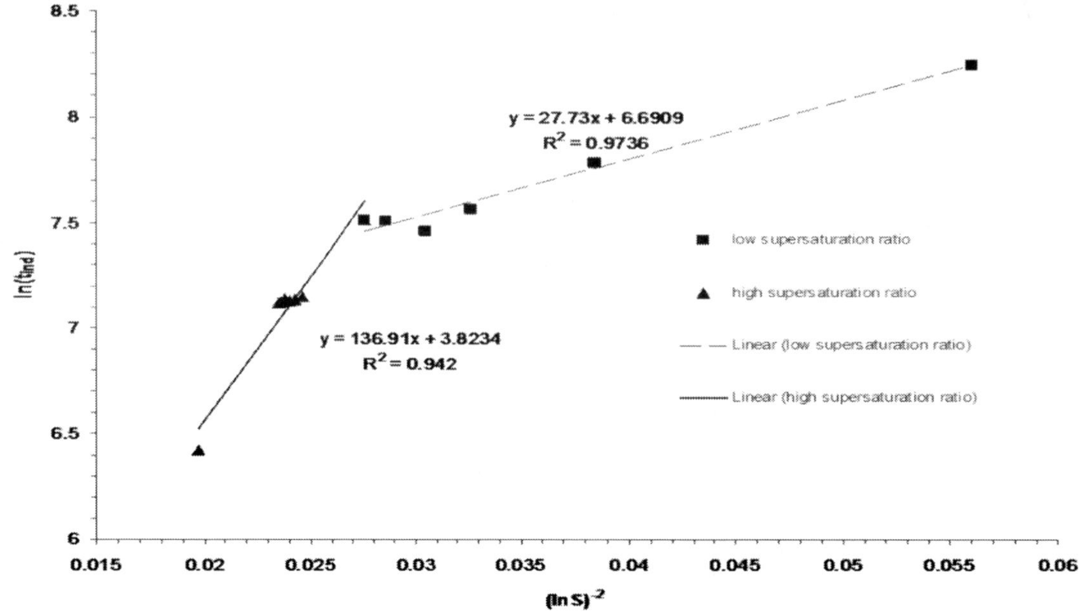

The different nucleation regimes are highlighted and the corresponding parameters are indicated. At low S level, the nucleation is governed by heterogeneous nucleation mechanism; at low S level, the nucleation is governed by homogeneous nucleation mechanism.

Figure 4a Plots and regression lines of the dimensionless thermodynamic parameter B vs. T^{-3} at both low and high S value.

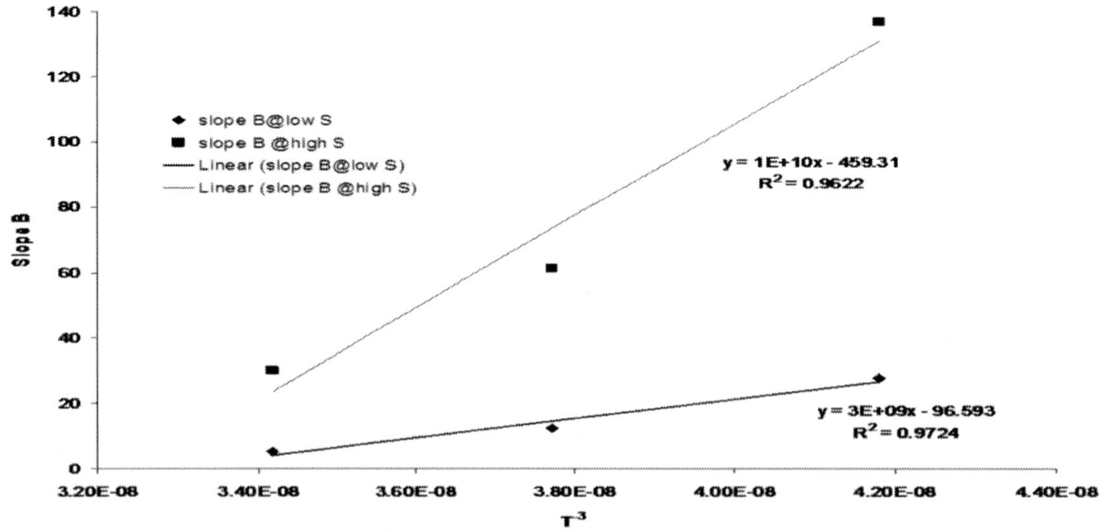

At low S segment, the HEN mechanism prevails; at high D segment, the HON mechanism is domaint.

Figure 4b Plots and regression lines of kinetic parameter *A* vs. ln*T* at both low and high S value.

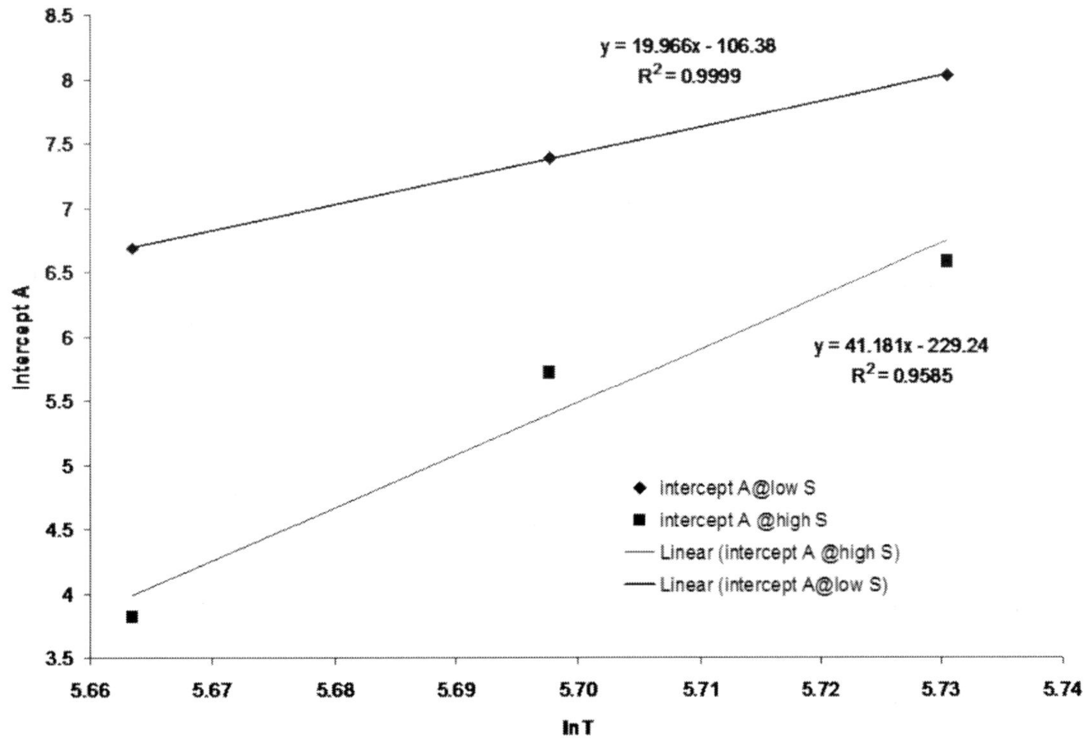

The intercept A changes with lnT in a linear fashion with a positive slope. This suggests that the growth is controlled by the interface transfer mechanism rather than the volume-diffusion mechanism.

5. Conclusions:

An integrated Process Analytical Technology (PAT) approach has been established to examine the effect of temperature on nucleation and growth kinetics of a 4-component dynamic pharmaceutical co-precipitation process. It reveals that at low S level, the nucleation is governed by primary heterogonous nucleation; while at high S level it is governed by primary homogeneous nucleation; after nucleation, the growth is controlled by the interface transfer mechanism.

Acknowledgements:
H.Wu sincerely acknowledge the technical guidance and supports from Dr. Vincent Vilker, Director, Office of Testing and Research (OTR), CDER, FDA (Silver Spring, MD), and Dr. San Kiang at BMS Research Institute (New Brunswick, NJ). The following persons and organizations are thankful: Mr. John Casamento from CDRH, FDA (Silver Spring, MD), Mr. Eric Dycus, Ms. Vaso Vlachos, Mr. Ben Smith from Mettler-Toledo AutoChem, Inc. (Columbia,MD), Mr. Steve Ware and Mr. Brendan Simon from ChemGlass Life Science (Vineland, NJ) for their assistance on instrumentations; Röhm America Inc. (Somerset, NJ) for technical assistance.

References:

1. FDA, 2004. "Guidance for Industry. PAT—A Framework for Innovative Pharmaceutical Development, Manufacturing, and Quality Assurance". Available at: http://www.fda.gov/cder/guidance/6419fnl.pdf (accessed on 10/01/2012)
2. Guo, Z, Jones, A.G., Li, N., 2006. Interpretation of the ultrasonic effect on induction time during $BaSO_4$ homogeneous nucleation by a cluster coagulation model. J. Colloid and Interface Science 297:190-198.
3. Kashchiev, D., Verdoes, D., and van Rosmalen, G.M., 1991. Induction time and metastability limit in new phase formation. J. Crsyatl Growth 110:373-380
4. Mullin, J.W. Crystallization. 2nd Edition. Butterworth-Heinemann Ltd., London, 1972, p.573.
5. Nývlt, J., 1968. Kinetics of nucleation in solutions. J. Cryst Growth 3,4:377-383.
6. Söhnel, O. and Mullin, J.W.1988. Interpretation of Crystallization Induction Periods. J. Colloid and Interface Science 123: 43-50.
7. Van Der Leeden, M.C., Kashchiev, D., and Van Rosmalen, G.M., 1992. Precipitation of Barium Sulfate: Induction Time and the Effect of an Additive on Nucleation and Growth. J. Colloid and Interface Science 152:338-350.
8. Varon, V., Bhugra, C., and Pikal, M.J., 2010. Prediction of Crystallization in Amorphous Pharmaceutical Systems: Phenobarbital, Nifedipine/PVP, and Phenobarbital/PVP. J. Pharm. Sci. DOI 10.1002/jps.22232.
9. Wakita, M.I.H. and Masuda, I., 1983. Analysis of precipitation processes of bis(dimethylglyoximato)Ni(II) and related complexes. J. Crystal Growth 61:377-382.
10. Wu, H. and Khan, M.A., 2010. Quality-by-Design (QbD): An Integrated Process Analytical Technology (PAT) Approach for Real-time Monitoring and Mapping the State of a Pharmaceutical Co-precipitation Process. Journal of Pharmaceutical Sciences 99(3): 1516-1534.
11. Wu, H. and Khan, M.A., 2011. Quality-by-Design (QbD): An Integrated Process Analytical Technology (PAT) Approach to Determine the Nucleation and Growth Mechanisms during a Dynamic Pharmaceutical Co-precipitation Process. Journal of Pharmaceutical Sciences 100(5): 1969-1986
12. Wu, H, White,M., and Khan, M.A., 2011. Quality-by-Design (QbD): An Integrated Process Analytical Technology (PAT) Approach for a Dynamic Pharmaceutical Co-precipitation Process Characterization and Process Design Space Development. *International Journal of Pharmaceutics*, 405:63-78.
13. Yosjihashi, Y., Yonemochi, E., Maeda, Y., Terada, K., 2010. Prediction of the induction period of crystallization of naproxen in solid dispersion using differential scanning calorimetry. J. Therm. Anal. Calorim. 99:15-19.

(401d) Making PAT an Integral and Sustained Part of Manufacturing Operations

John O'Reilly, Roche, Clarecastle, Ireland

To be further developed.

A minimum of 150 words is required for the abstract text, and no more than 1500 words may be submitted. Each abstract is allowed to include up to 20 figures.

Modern process analytical technology (PAT) can promise many innovative solutions in the manufacture of API. However, to implement such technologies in a sustainable way is often the greatest challenge.

This talk looks at some of the current PAT projects at Roche Clarecastle and in particular the innovative strategy we are developing to make PAT an integral and sustained part of our manufacturing operations.

The concepts of in-process data samples (IPDS) and PAT-QC are introduced. The use of PAT as a continuous real-time process verification/validation tool is also discussed with reference to the latest FDA process validation guideline.

The concepts of in-process data samples (IPDS) and PAT-QC are introduced. The use of PAT as a continuous real-time process verification/validation tool is also discussed with reference to the latest FDA process validation guideline.

(401e) Design and Implementation of An Efficient Control System in a Continuous Pharmaceutical Manufacturing Process Via Roller Compaction

Ravendra Singh, Marianthi Ierapetritou and Rohit Ramachandran, Department of Chemical and Biochemical Engineering, Rutgers, The State University of New Jersey, Piscataway, NJ

Pharmaceutical industries are facing enormous challenges to achieve the predefined end product quality consistently as well as to satisfy the regulatory constraints, flexible market demands, operational complexities and economical limitations [1]. A novel manufacturing strategy based on continuous processing integrated with online monitoring tools coupled with efficient automatic feedback control system is therefore highly desired [2]. However, because of different level of complexities involved (e.g. solid handling, irregular flow, multi-variate interactions, difficulty in real time measurement) in addition to the requirement of very precise product quality, the design and implementation of an efficient control system in the pharmaceutical manufacturing involving the solid dosage form is still a challenging task. The process involving the solid flow is not as simple to control as the process involving the liquid/gas flow therefore the conventional control strategy may not be efficient, although it could be adapted. The design and implementation of the process control system involves the identification of critical control variables, coupling of the control variables with suitable actuators, selection of suitable monitoring tools, selection of control strategy followed by controller parameters tuning, model-based closed-loop performance evaluation and finally implementation to the manufacturing plant through the available hardware (e.g. Delta V) and control interface (e.g. OPC).

In the work reported here, an efficient control system for an integrated continuous pharmaceutical tablet manufacturing process via roller compaction has been designed. The continuous manufacturing process consists of the following unit operations: blending, roller compaction, milling, tablet compression and coating. The roller compactor has been used as a means of dry granulation to avoid any API water sensitive issues involved. The designed control system consists of four cascade control loops and three single control loops giving 36 tuning parameters. An effective controller parameter tuning strategy involving ITAE methods coupled with an optimization strategy has also been proposed. The designed control system has been implemented in a first principle model-based flowsheet simulated in gPROMS (Process System Enterprise) to verify its performance and is being implemented in a continuous pilot plot at Rutgers University. The systematic methodology for pharmaceutical process control system design as well as supporting process models and knowledge/data base has also been developed.

The objective of this presentation is two-fold: first to highlight the design of control system for a continuous tablet manufacturing process with roller compaction, and second to demonstrate the performance of the designed control system through the first principle model as well as in a pilot plant.

References

[1] Singh, R., Gernaey, K. V., Gani, R., (2010). ICAS-PAT: A Software for Design, Analysis & Validation of PAT Systems. Computers & Chemical Engineering, 34(7), 1108-1136.

[2] Ramachandran, R., Arjunan, J., Chaudhury, A, Ierapetritou, M. (2012). Model-Based Control Loop Performance Assessment of a Continuous Direct Compaction Pharmaceutical Processes. J. Pharm. Innov., 6(3), 249-263.

(401f) Near Infrared Spectroscopy (NIRS): An Enabling Tool for Pharmaceutical Quality by Design (QbD)

K. Geoffrey Wu, Suhas Patankar, Lane V. Christensen, Robert A. Lionberger and Lawrence X. Yu, Office of Generic Drugs, Food and Drug Administration, Rockville, MD

Near Infrared Spectroscopy (NIRS): An Enabling Tool for Pharmaceutical Quality by Design (QbD)

K. Geoffrey Wu, Suhas Patankar, Lane V. Christensen, Robert A. Lionberger, Lawrence X. Yu

Office of Generic Drugs, Food and Drug Administration, Rockville, MD 20855

This presentation will provide an overview of the application of NIRS within the paradigm of pharmaceutical QbD and elaborate on the current thinking of the Office of Generic Drugs (OGD) regarding the deployment of such a useful application of process analytical technology (PAT).

Applying various types of NIRS methods for pharmaceutical development has been proposed by the generic industry to OGD. Case studies will be given to demonstrate some of the major technical issues observed during the review of these proposals. Such issues may be related to the NIR method development and validation, model maintenance, and parallel testing. Points to consider during the model transfer (e.g., from R&D to manufacturing phase) will also be discussed.

OGD encourages the use of NIRS and considers it as an important tool to facilitate the implementation of pharmaceutical QbD. From OGD's perspective, possible regulatory pathways to file the use of NIRS will be presented.

(412a) Chemical Transportation Network Optimization Under Extreme Weather Conditions

Tianxing Cai and Qiang Xu, Dan F. Smith Department of Chemical Engineering, Lamar University, Beaumont, TX

Chemical transportation is a segment with special risks during the whole life cycle of chemical processing. It often deals with toxic or hazardous chemicals, which presents a high potential of dangerous accidents such as explosion or toxic gas leakage. Therefore high attention should be paid to the design of transportation route in order to reduce the potential risk to the environment and densely populated areas. The previous studies have put focus on the optimization between population distribution, roadway safety and traffic congestion to give the recommendation. The requirement of chemical transportation route has already been included in the national wide or regional wide regulations for the safety management and emergency response. Despite this progress, many serious events have occurred during chemical transportation and have made severe impacts on environment. One major reason to aggravate its affect on the environment is the meteorological change. Therefore, the optimized route to cope with different weather condition is of great significance to all stake holders.

In this study, a methodology is developed for risk assessment and dynamic optimization for the transportation of chemicals and evaluation in the situations of weather condition change. In such case, the coincidence of heavy raining and passing through the bridge will add the risk to cause chemical release into the downstream of the river. The difference in the atmosphere stability will also have different pollutant dispersion in the surrounding area. The whole problem is set into a grid network. Afterwards, geographic and meteorological parameters have been assigned and a MINLP model will be set up to achieve the minimal potential impact of environment and populated zones.

(412b) Coarse-Graining the Dynamics of Network Evolution: An Equation-Free Approach and Perspectives Using Data Mining

Karthikeyan Rajendran[1], Constantinos I. Siettos[2] and Ioannis G. Kevrekidis[1], (1)Department of Chemical and Biological Engineering, Princeton University, Princeton, NJ, (2)School of Applied Mathematics and Physical Sciences, National Technical University of Athens, Athens, Greece

Complex dynamic systems, exhibiting emergent dynamics, often arise in the form of graphs (or networks): the internet, social networks, communication networks, chemical and biochemical reaction networks, and so on. We focus on the dynamics of evolving networks, i.e., networks whose connectivity changes dynamically. In such systems, coarse-graining approaches are essential in helping us understand the interplay between the structure and the dynamics of networks.

In this work, we propose a coarse-graining approach [1] based on the equation-free framework: short bursts of detailed network evolution simulations are coupled with lifting and restriction operators that translate between actual network realizations and their (appropriately chosen) coarse observables. This framework is used to accelerate temporal simulations (through coarse projective integration), and implement coarse-grained fixed point algorithms (through matrix-free Newton-Krylov GMRES) in a simple network evolution example. We also discuss the choice of good observables when the network is heterogeneous (its nodes are not identical).

A crucial step in our coarse-graining approach is a judicious selection of coarse variables for the problem at hand. This step is usually carried out heuristically. We explore the possibility of automating this step by using data mining procedures like Diffusion Maps [2] to "extract" the intrinsic parameters (coarse variables) of the system making use of data (graphs) collected either empirically or from a model simulating the dynamical process. We present such data mining results using illustrative families of graphs.

References

[1] Bold, K. A., Rajendran, K., Ráth, B. and Kevrekidis, I. G., An equation-free approach to coarse-graining the dynamics of networks, arXiv:1202.5618 (2012).
[2] Nadler, B., Lafon, S., Coifman, R. R. and Kevrekidis, I. G., Diffusion maps, spectral clustering and reaction coordinates of dynamical systems, Applied and Computational Harmonic Analysis, 21, 113 – 127 (2006).

(412c) On Closed-Loop Performance of Lyapunov-Based Economic Model Predictive Control of Nonlinear Systems

Mohsen Heidarinejad[1], Jinfeng Liu[2] and Panagiotis D. Christofides[2], (1)Electric Engineering, University of California, Los Angeles, Los Angeles, CA, (2)Department of Chemical and Biomolecular Engineering, University of California, Los Angeles, Los Angeles, CA

The development of optimal operation and control policies for chemical process systems aiming at optimizing process economics has always been an important research subject with major practical implications. Generally, economic considerations are addressed via a two layer approach in which the upper layer carries steady-state process optimization to obtain economically optimal process operating set points (steady-states) while the lower layer employs appropriate feedback control laws to steer the process state to an economically optimal steady-state operating point. Model predictive control (MPC) is widely utilized in the process control layer to provide optimal manipulated input values by minimizing a (typically) quadratic cost function which usually penalizes the deviation of the system state and manipulated inputs from their economically-optimal steady-state values subject to input and state constraints. This two-layer approach restricts process operation to seek a solution within the neighborhood of steady-state operation. In order to account for general economic optimization considerations, the quadratic cost function used in standard MPC should be replaced by an economics-based cost function which may result in a time-varying operation. Consequently, the standard MPC should be re-formulated in an appropriate way to guarantee closed-loop stability. At this point, there is limited work to ensure improvement of closed-loop performance through time-varying operation via economic MPC with respect to operation under conventional MPC in the context of finite time operation. In [1], we demonstrated through closed-loop simulations that there is improvement in economic closed-loop performance when we apply time-varying solution of economic MPC instead of steady-state solution of Lyapunov-based MPC.

Motivated by the lack of available methodologies to guarantee performance of economic MPC, the present work focuses on a Lyapunov-based economic model predictive control (LEMPC) scheme for nonlinear systems which is capable of optimizing closed-loop performance with respect to a general objective function that may directly address economic considerations. Unlike steady-state operation of conventional Lyapunov-based model predictive control (LMPC), LEMPC design through time varying operation guarantees to improve economic cost function value with respect to conventional LMPC by incorporating appropriate constraints in its formulation and solving an auxiliary LMPC problem at each sampling time. The proposed scheme takes advantage of a predefined Lyapunov-based feedback law through a two mode operation to characterize its stability region while maintaining the closed-loop system state in an invariant set. The first operation mode corresponds to the periods in which the cost function should be optimized (e.g., normal production periods); and in this operation mode, the MPC maintains the closed-loop system state within a pre-defined stability region and optimizes the cost function to its maximum extent. The second operation mode corresponds to operation in which the system is driven by the MPC to an appropriate steady-state. Theoretical results are demonstrated through a nonlinear chemical process example.

[1] M. Heidarinejad, J. Liu, and P. D. Christofides. Economic model predictive control of nonlinear process systems using Lyapunov techniques. AIChE Journal, 58:855-870, 2012.

(412d) Networked Model Predictive Control of Spatially Distributed Processes

Zhiyuan Yao, Department of Chemical Engineering and Materials Science, University of California, Davis, Davis, CA and Nael H. El-Farra, Department of Chemical Engineering & Materials Science, University of California, Davis, Davis, CA

Control over communication networks has attracted considerable attention in recent years in both the academic and industrial research communities. The use of wireless networks in process control systems, for example, allows for modular and flexible system design (e.g. distributed processing and inter-operability), simple and fast implementation, and reduced installation and maintenance costs. However, the integration of communication networks in general (and wireless networks in particular) introduces a number of challenges due to the intrinsic limitations on the collection, processing and transmission capabilities of the measurement system and the communication medium. These challenges have been a major driving force behind an extensive and growing body of research work on networked control of dynamic systems, particularly those modeled by lumped parameter systems. In recent years, a number of efforts have also been made towards the development of a methodology for networked control of spatially-distributed systems modeled by partial differential equations (PDEs) that arise naturally in the modeling of transport-reaction processes and fluid flows [1],[2]. The focus of these efforts has been primarily on achieving closed-loop stability with minimal network resource utilization. This is an appealing objective given that it reduces the susceptibility of the networked control system to communication disruptions. Beyond the stabilization objective, performance and optimality requirements, as well as the handling of constraints on the process states and inputs, are key objectives that need to be addressed in the design of any process control system. One of the few systematic methods suited for incorporating performance objectives and constraint handling is Model Predictive Control (MPC), which is based on iterative, finite-horizon optimization of a plant model. At this stage, systematic methods for networked predictive control of spatially distributed systems remain lacking and in need of development.

In this contribution, we develop and present a framework for networked model predictive control of spatially distributed processes modeled by parabolic PDEs with sensor measurements that are transmitted to the controller over a resource-constrained communication medium. The aim of this work is to enforce the desired stability and optimality properties with minimal sensor-controller information exchange. To this end, a finite-dimensional approximate model is initially obtained using model reduction techniques to capture the dominant dynamics of the infinite-dimensional system, and is then used to design a Lyapunov-based model predictive controller (LMPC) that enforces closed-loop stability for a given sensor-controller communication rate. The control action is generated by solving a model-based finite-horizon optimization problem based on an appropriate cost functional subject to constraints on the process dynamics, states and inputs. The controller guarantees the decay of the Lyapunov function over each sampling interval within a well-defined stability region. Unlike conventional LMPC formulations, however, the sensors do not transmit the measurements at a fixed rate, but instead communicate with the controller in an adaptive fashion. The key idea is to monitor the evolution of the Lyapunov function at the sampling times and suspend communication for periods when the prescribed stability threshold is satisfied. During such periods, the predictive controller relies on the available model to estimate the slow states of the infinite-dimensional system, and computes the control action using an open-loop optimization formulation. At times when the sampled state begins to breach the expected stability threshold, communication is restored and the controller switches back to using the sampled measurements to update the model predictions and repeat the optimization at each sampling time. Finally, the implementation of the finite-dimensional networked control structure on the infinite-dimensional system is analyzed, and the results are illustrated using a representative diffusion-reaction process example.

References:

[1] Sun, Y., S. Ghantasala and N. H. El-Farra, ``Networked Control of Spatially Distributed Processes with Sensor-Controller Communication Constraints," *Proceedings of American Control Conference*, pp. 2489-2494, St. Louis, MO, 2009.

[2] Yao, Z. and N. H. El-Farra, ``Resource-Aware Scheduled Control of Distributed Process Systems Over Wireless Sensor Networks," *Proceedings of American Control Conference*, pp. 4121-4126, Baltimore, MD, 2010.

(412e) Autocatalytic Reactions in a Spatially Segregated Environment

Swapnil Hegishte and **Satish J. Parulekar**, Chemical and Biological Engineering, Illinois Institute of Technology, Chicago, IL

Autocatalytic reactions and processes are commonly encountered in growth of all living cells, processes involving free radicals, polymerization processes, many inorganic and organic reactions, and crystallization processes. A comprehensive analysis of static and dynamic behavior of an autocatalytic reaction occurring in two coupled CSTRs is presented. The generation of the autocatalyst from a primary resource is followed by its decay. A single well-mixed reactor may operate at up to three steady states and can give rise to limit cycles. The reactors in the two-reactor system are identical only in terms of feed composition and reactor space time. The reactors need not have identical volumes and volumetric feed, effluent, and exchange rates. A two-reactor system may admit up to nine steady states at least at very low interaction rates. Three of these steady states, symmetric steady states, correspond to identical composition in the two reactors. The remaining steady states, the asymmetric steady states, correspond to different composition in the two reactors and are admissible over a range of interaction rates. The six asymmetric steady states exhibit four different pairing sets with two or more limit points. Numerical illustrations reveal the rich steady state structure of the reaction scheme in coupled reactors. The steady states are tracked systematically as the interaction rate is varied. Numerical illustrations for complex dynamics of the coupled reactors reveal phenomena such as extinction or synchronization of oscillations and chaotic operation through period doubling. The two-reactor system is operationally more flexible and more robust vis-a-vis single reactor. Emergence of additional steady states and higher order periodic states and aperiodic states at intermediate values of interaction rates reveals that the coupled reactors involving autocatalytic reactions are an example of a complex system.

(412f) Many-to-Many Routing of Vehicles in a City Network with Facility Interception Constraints

Zeeshan Pervaiz[1], **Kris Villez**[1] and Venkat Venkatasubramanian[2], (1)School of Chemical Engineering, Purdue University, West Lafayette, IN, (2)Chemical Engineering, Columbia University, New York, NY

Many-to-many routing of vehicles in a city network with facility interception constraints

Zeeshan A. Pervaiz, K. Villez, V. Venkatasubramanian

Providing infrastructures which enable a shift to a sustainable society is of special importance in the transportation sector. A particular problem concerns the optimal location of service stations for electric vehicles. In addition, optimal routing of vehicles to these stations is crucial given that current driving ranges are limited for many electrical vehicles. The many-to-many routing of vehicles to such service stations is solved by means of a Minimum Cost Multi Commodity Flow (MCMCF) problem formulation. Our results indicate that this is indeed possible for small to medium-scale settings.

The city of Alexandria, VA, is used as a case study to test our methodology. This case study is available in TRANSIMS and has been used in prior work (Shukla et al., 2011, Villez et al., 2011). Figure 2 shows the road network for this case study with five facility locations. TRANSIMS is an agent-based simulator developed by Los Alamos National Laboratory. This simulator allows to extract two essential inputs for our methodology which are (1) the topology of the traffic network including arcs (roads) and nodes (road junctions or terminal ends) and (2) the source-destination pairs and the average number of vehicles (flow) per day. The road network, excluding public transport infrastructures, has a total of 2573 nodes with 7214 edges. There are a total of 13892 unique source destination pairs (commodities), each of which have an associated commodity.

The third requirement is that one knows the location of the facilities. To this end, we use the optimized locations as found in Villez et al. (2011) for ten different scenarios, namely for the number of facilities ranging from one to ten.

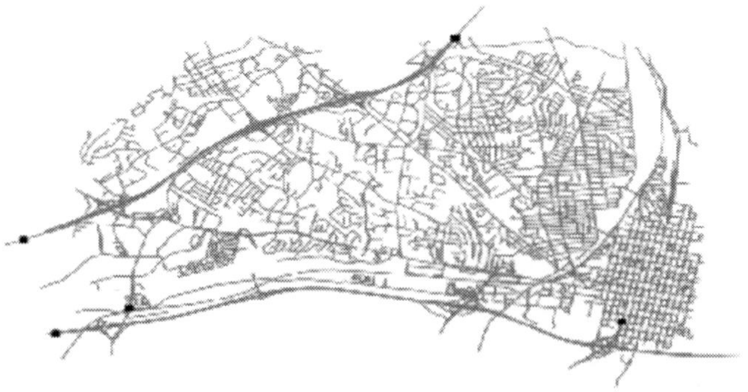

Figure 2. Road map of Alexandria, VA, with 5 service stations (squares), adopted from Villez et al. (2011).

For the formulation of the problem, consider a directed Graph G=(V,E) with nodes V=(1,….,n) and directed arcs $E=\{e_{ij}|(i,j)\in V\}$. Each arc e_{ij} has associated with it a cost c_{ij} and a capacity b_{ij}. A commodity k=(1,…..,a) is associated with each unique pair of source-destination nodes (s_k,t_k) such that s_k is the source while t_k is the destination for the commodity k. Also, let $X=\{x|x\in V\}$ be the set of nodes with facilities. Let f_{ij}^k be the flow on the edge e_{ij} for the commodity k. Then solving the MCMCF problem is equivalent to the minimization of the total cost:

$$\sum_{e_i \in E} \sum_{k=1}^{a} c_{ij} \cdot f_{ij}^{k} \quad \text{Eq. 1.1}$$

subject to the following constraints:

$$\sum_{j=1}^{n} f_{ij}^{k} - f_{ji}^{k} = \begin{cases} r_k & i = s_k \\ -r_k & i = t_k \\ 0 & \text{otherwise} \end{cases} \quad \text{Eq. 1.2}$$

$$\sum_{k=1}^{a} f_{ij}^{k} \leq b_{ij} \quad \text{Eq. 1.3}$$

$$\sum_{x \in X}^{a} f_{xj}^{k} = r^{k} \quad \text{Eq. 1.4}$$

Equations 1.2 represent the nodal balances in each node of the graph (junction of the road network). Equations 1.3 represent the capacity constraint in each edge of the graph (road in the network). Equations 1.1 to 1.3 represent the problem formulated in (Tomlin 1966). Equation 1.4 are new. These are added to ensure that all routes are constrained to go through at least one of the provided facilities.

The total load on the network as presented by all the commodities is 165263 units, while the loads presented by the selected pairs (commodities) are 6527, 13205, 27632 and 41728 units respectively. So the percentage load offered by the four sets of commodities is 1.7%, 3.43%, 7.18% and 10.85% respectively.

In Figure 4, the objective function values are plotted as function of the fraction of the load on the Alexandria city network represented by the selected source-destination pairs (commodities). As expected one can see that the objective function is increasing with the number of commodities. As can be seen, this increase is almost linear. More importantly, the result are largely the same for all considered sets of facility locations (1 to 10 locations). This indicates the benefit of additional facilities at this limited load fraction is very limited.

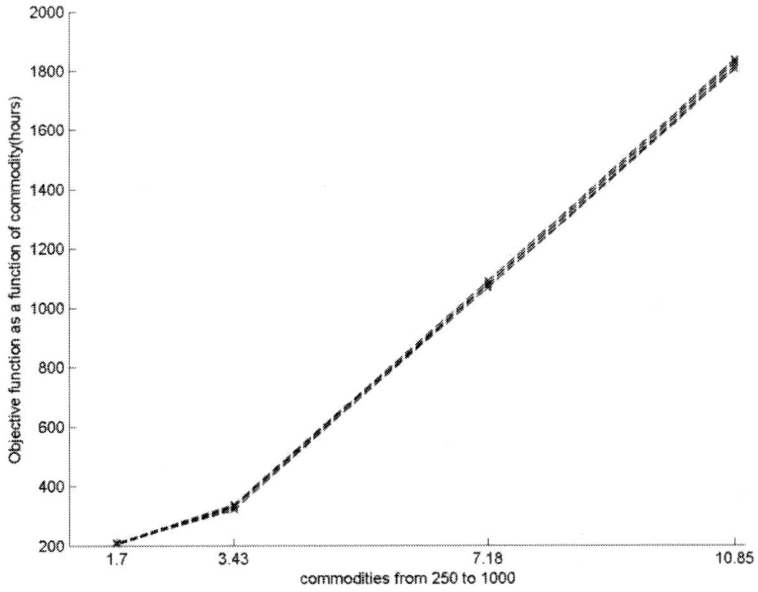

Figure 4. Objective function plotted as function of the fraction of total load (%) in the Alexandria city network. Ten lines are shown for each set facility locations with facility numbers ranging from 1 to 10.

REFERENCES

SHUKLA, A., PEKNY, J. & VENKATASUBRAMANIAN, V. 2011. An optimization framework for cost effective design of refunding station infrastructure for alternative fuel vehicles. *Computers and Chemical Engineering,* 35, 1431-1438.

TOMLIN, J. A. 1966. Minimum-Cost Multicommodity Network Flows. *Operations Research,* 14, 45-51.

VILLEZ, K., GUPTA, A., VENKATASUBRAMANIAN, V. & RIEGER, C. Resilient design of recharging station networks for electric transportation vehicles. 4th International Symposium on Resilient Control Systems, ISRCS 2011, August 9, 2011 - August 11, 2011, 2011 Boise, ID, United states. IEEE Computer Society, 55-60.

(429a) Recent Results in Chemical Reaction Network Theory

Martin Feinberg, Ohio State University, Columbus, OH

Reuel Shinnar was an early pioneer in the theory of reactor behavior. Reuel and others laid the foundations for a body of theory, largely driven by practical chemical engineering problems, that now permits one to make penetrating and far-from-obvious statements about highly complex chemical systems in a precise way. This talk will describe some of what is known about the behavior of intricate chemical reaction networks more than a half century after Reuel began his career.

Reuel Shinnar was an early pioneer in the theory of reactor behavior. Reuel and others laid the foundations for a body of theory, largely driven by practical chemical engineering problems, that now permits one to make penetrating and far-from-obvious statements about highly complex chemical systems in a precise way. This talk will describe some of what is known about the behavior of intricate chemical reaction networks more than a half century after Reuel began his career.

() The Mathematical Structure of Glycosylation in Living Cells

Frederick Krambeck, ReacTech Inc., Alexandria, VA

Glycans are highly variable and structurally diverse sugar chains that, when

attached to membrane proteins and lipids, are a dominant feature of the mammalian cell surface. In contrast to protein structures, glycan structures are not directly encoded in the genome. Instead, the structure of glycans is determined during their assembly in intracellular compartments through a sequence of enzymatic reactions. The many thousands of glycan structures found in cells result from the activities, concentrations, and localization of a much smaller number of enzymes in different types of cells. In this paper a previously developed mathematical model of the glycosylation process is used to explore the structural and quantitative features of this complex process.

(429b) The Impact of Acid Site Concentration, Crystal Size and Reaction Temperature in the Production of C9+ Aromatics in ZSM-5 Catalyzed Xylene Isomerization

Thomas F. Degnan, ExxonMobil Research and Engineering Company, Annandale, NJ

ZSM-5 is widely used as a commercial catalyst for xylene isomerization in the production of para-xylene for the ultimate synthesis of polyethylene terephthalate (PET). Previous studies have shown that the crystal structure (ZSM-11 vs. ZSM-5 vs. mordenite) is very important in determining the selectivity of the disproportionation products in the zeolite catalyzed isomerization of mixed xylene feeds (i.e., those containing ethylbenzene and mixed xylenes). In this study, we have examined the effects of ZSM-5 crystal size, crystal acid site concentration, and reaction temperature in the selective production of diethylbenzene and dimethylethylbenzene from a mixture of ethylbenzene (14 wt%) and mixed-xylenes (86 wt%). As expected from diffusivity calculations, all three parameters have a significant effect on DEB and DMEB selectivity. Reaction conditions used in he study were: 350 - 390oC, Hydrocarbon WHSV = 2 to 150 hr-1, Hydrogen / Hydrocarbon ratios of 1-2 and a reaction pressure of 200 psig. The findings are important in the design of shape selective zeolitic xylene isomerization catalysts.

(429c) Aerobic Reduction of NOx in the Sequential LNT-SCR Catalytic Reactor

Arun Kota, Dan Luss and Balakotaiah Vemuri, Department of Chemical and Biomolecular Engineering, University of Houston, Houston, TX

The effect of key parameters on the DeNO$_X$ performance of the sequential LNT-SCR catalyst under non-isothermal operation is investigated in detail. The model simulates the cyclic operation of the sequential catalyst with lean NO$_X$ storage, aerobic reduction of the stored NO$_X$ using H_2 in the LNT and selective reduction of NO$_X$ with NH_3 in the SCR. Global kinetics with fast and slow sites are used to account for the storage and regeneration reactions in the LNT and the reduction of NO$_X$ due to NH_3 in the SCR catalyst. H_2 oxidation and H_2-NO reactions are described using micro-kinetic mechanism to predict the experimentally observed selectivities. The transient simulations reveal traveling concentration and temperature fronts along the length of the catalyst. The effect of various parameters like the precious metal loading, space velocity, inlet and initial temperature of the catalyst, number of sequential bricks and the total cycle time on the formation and speed of the traveling fronts are studied to determine the optimal operating conditions of the sequential catalyst. H_2 oxidation exotherm resulted in enhanced reaction rates and selectivities during non-isothermal operation. Consequently significant DeNOx activity is observed even with the Pt loading as low as 0.1wt%. However, from the ignition point of view too low Pt loading resulted in decrease in the DeNOx activity. The performance of the sequential LNT-SCR catalyst under cold start conditions is studied and the results are compared to those observed in the literature.

(429d) Glycerol Selective Oxidation: A Reaction Engineering Study

Arvind Varma, School of Chemical Engineering, Purdue University, West Lafayette, IN

Increase in the production of biodiesel has resulted in an oversupply of glycerol, the primary byproduct which forms in a 1:10 ratio with biodiesel. An effective utilization of glycerol (currently ~\$0.02/lb) by conversion to the higher value chemical dihydroxyacetone (DHA; currently ~\$20/lb) will enhance the economics of biodiesel production. Towards this end, we have developed a Pt-Bi/C catalyst capable of selective oxidation of the secondary hydroxyl group of glycerol to produce DHA. The catalyst and operating conditions were optimized in a semi-batch reactor to achieve a high yield of DHA (>40%) [1]. A kinetic model for the full 10-step glycerol oxidation network and a simplified kinetic model were also developed [2]. To increase production, a continuous-flow trickle-bed reactor can be utilized. In this direction, the hydrodynamics of trickle-bed reactors have been characterized, particularly with respect to the material attributes of the activated carbon support for our optimal Pt-Bi catalyst. Detailed catalyst, kinetic, and reactor characterization provide a fundamental basis for the development of a continuous-flow process for selective conversion of glycerol to DHA, which in turn will increase the economic viability of biodiesel production.

[1] Hu, W., Knight, D., Lowry, B., Varma, A. Selective oxidation of glycerol to dihydroxyacetone over Pt-Bi/C catalyst: Optimization of catalyst and reaction conditions. *Ind. Eng. Chem. Res.* 2010, **49** (21), 10876-10882.

[2] Hu, W., Lowry, B., Varma, A. Kinetic study of glycerol oxidation over Pt-Bi/C catalyst. *Appl. Catal. B: Environmental* 2011, **106** (1-2), 123-132.

Large Scale Convection during Phase Separation of Liquid Mixtures

Roberto Mauri

Dipartimento di Ingegneria Chimica, DICCISM, Università di Pisa, 56126 Pisa, Italy.

Abstract

Complete phase segregation may occur on a *10* centimeter scale even in the absence of buoyancy, due to unidirectional, large scale rapid bulk flows driven by chemical potential gradients. Using a hexadecane-acetone nearly density-matched liquid mixture in a *20 cm*-long condenser tube with a *1 cm* diameter, we observed the rapid axial migration of the acetone-rich drops towards the warmer regions of the condenser. Conversely, the hexadecane-rich drops moved in the opposite direction, therefore ruling out thermocapillary effects as a possible explanation of the phenomenon. These flows lead to a complete phase segregation within *10* seconds, with the formation of a single interface perpendicular to the axial direction. Changing the temperature gradient along the tube from *0.25 °C/cm* to *1 °C/cm* no change was detected, with typical drop speeds up to *6 cm/s*, irrespectively of the distance of the drop from the wall, showing that the phenomenon is not due to a flow instability.

I. Introduction

When a binary mixture is quenched (or heated) from its single-phase region to a temperature below (above) the composition-dependent spinodal curve, it phase separates through a process called spinodal decomposition,[1] which is characterized by the spontaneous formation of single-phase domains, proceeding then to grow and coalesce. Unlike nucleation, which starts from a metastable state where activation energy is required to initiate the separation, spinodal decomposition starts from an unstable state and involves the growth of any fluctuations whose wavelength exceeds a critical value. In general, as the growth of the single-phase patches is due to the movement of one phase relative to the other, it can be driven by two mechanisms, namely diffusion and convection. Diffusion is induced by the velocity fluctuations of the molecules that compose the system at thermal equilibrium and consists of their incoherent, random motion, with no specific preferential direction. On the contrary, convection can exist only for a system far from equilibrium, as it consists of the collective, coherent motion of its molecules, and is therefore a much faster process than diffusion.

If the mixture is quenched to a temperature only by a few millikelvins below its critical value or for very viscous systems such as polymer mixtures, the process is driven by diffusion and therefore is very slow.[2-5] On the contrary, when low (i.e. water-like) viscosity, liquid mixtures are quenched to a temperature deeply below their critical point of miscibility, the process of phase separation is much faster and is driven by convection.[2,3,5,6] Based on dimensional analysis, such convection must be due to surface tension effects.[7] This is confirmed by a more detailed analysis through the so called coalescence-induced coalescence[8,9], where the strong gradient in curvature at the neck of each coalescing pair of droplets induces a flow which, in turn, causes further coalescence. These models, though, assume that the system is at local equilibrium, so that surface tension can be defined, while experimentally it has been shown[10] that convection starts well before the appearance of sharp interfaces. Such hypothesis of local equilibrium is removed in the more fundamental diffuse interface model,[11-13] where convection is induced by a (non equilibrium) body force that is proportional to the chemical potential gradients. At the late stages of phase separation, after the system has developed well-defined phase interfaces, this body force reduces to the more conventional surface tension,[14,15] so that, eventually, the diffuse interface model becomes a generalization of other surface tension-based models. Accordingly, the convective driving force can be seen as a non-equilibrium attractive capillary force among drops which enhances coalescence[16-18] until, eventually, the single-phase patches become large enough that buoyancy dominates surface tension effects and the mixture separates by gravity.[19]

Based on the diffuse interface model, we have simulated the phase separation of regular mixtures with water-like viscosity in isothermal conditions,[20-22] finding that the process is driven by convection, and not by diffusion neither by gravity, so that the typical drop size grows linearly with time, in agreement with many experimental results of late stage phase separation.[2,3,6,7,17,23] In addition, we saw that, in the absence of any preferential direction, convection occurs in random directions, producing effects that are very similar to turbulent transport. We also found that during the process of phase segregation the system is far from local equilibrium, explaining why we observe bulk motion even at late stages, when the system is composed of nucleating drops with sharp interfaces.[17,24] These predictions, though, could not be fully confirmed in our previous experimental works,[10,17,24] since we were using the narrow glass cell shown in figure 1a, where wall effects could never be ruled out. For example, quenching a critical, density-matched mixture in the cell of Figure 1a to a temperature of *20°C*, we showed that,[24] at the end of the phase separation process, one phase is distributed along the walls of the cell, while the other forms a large domain at the center of the cell, revealing that the non-equilibrium force that is responsible for phase separation is too small to overcome wall effects. In the present study, we either strengthen this driving force by increasing the depth of the temperature quench within the cell, or weaken all wall effects, replacing the narrow cell with the condenser tube of figure 1b, showing that phase segregation in the presence of weak temperature gradients may result in unidirectional, large scale convection.

II. Experimental Results

In our experiments we used the acetone-hexadecane liquid mixture that was used in previous works,[10,24] presenting a $T_c = 27°C$ critical temperature. At ambient temperature, this mixture separates into two phases, with a density difference $\Delta\rho = 6\ 10^{-4}\ g/cm^3$ that, although very small, is still sufficient to cause a slow separation of a droplet dispersion by settling and coalescence. Similar, so called, isopycnic mixtures have been also used by Guenoun *et al.*[6] In addition, *100 ppm* of crystal violet, a dye that absorbs preferentially in acetone, was added to the mixture to enhance the visualization of the process. When dissolved in such small percents, this dye does not change the phase diagram of the mixture, or the characteristics of the phase separation process. Two systems were used, the former having a *50%* acetone, *50%* hexadecane volumetric critical composition, in which the dispersed phase is acetone-rich, while the second system consisted of a *70%* acetone - *30%* hexadecane mixture, where the dispersed phase is hexadecane-rich.

In all our experiments, we started with the mixture in its phase-separated state at a *20 ℃* constant temperature. Then, the solution was first heated to *38 ℃* (i.e. well above its *27 ℃* critical temperature), mixed thoroughly and finally quenched back to *7 ℃*, with a quench rate of about *3 ℃/s*. This quenching was achieved by circulating cooling water at *7°C* through the outer chamber of the cell or of the condenser. Inserting three thermocouple within a cross section, we saw that there was no detectable temperature difference along the radial direction. Instead, this rapid quench resulted in a small temperature gradient along the axial direction, which played a crucial role in our results. In fact, we saw that the temperature profile along the axial direction of the condenser during phase separation was, at most, *0.28 ℃/cm*. Convection induced by temperature gradients alone, though, was ruled out, as no axial fluid flow was observed when the mixture was replaced with water. Similar results were obtained by Santonicola *et al.*[10]

In order to investigate the interplay between gravity and growth, we considered two sets of experiments. In the first, we used a temperature-regulated, *1 mm* thick, *10 mm* high, *40 mm* long sample cell, an optical microscope with Xenon illuminator and a high-resolution digital camera. The sample cell was kept horizontal and filled with the critical mixture, while its temperature was regulated by placing it into an *8 mm*-thick water jacket, where temperature-controlled water was circulated.

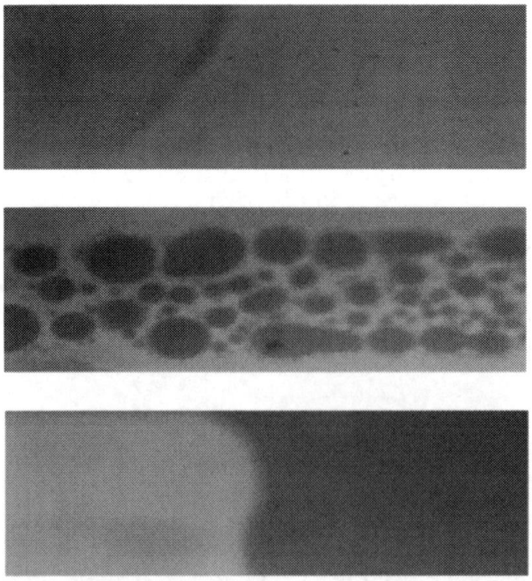

Figure 1. Phase segregation within the cell of a 50% acetone – 50% hexadecane critical mixture, *0*, *3* and *25* seconds after being quenched with water at 7°C.

The results of the visualization of the phase separation within the cell are shown in Figure 1, with the time $t=0$ corresponding to the moment when the temperature of the mixture crossed the miscibility curve. The separation process occurred in two stages. First, small droplets formed in the whole cell, which rapidly coalesced into larger droplets, as it was observed in previous experimental works.[17,23] Then, blue, acetone-rich droplets moved rapidly towards the warmer region of the cell along the longitudinal direction (i.e. to the right in Figure 1) and coalesced, until, eventually, the two phases were completed separated by a clear vertical interface, which remained stable even when complete phase equilibrium was achieved. When the flow of the cooling water was inverted, phase segregation took place in the same way, but the acetone-rich phase moved to the left and the hexadecane-rich phase to the right. Repeating the experiments in a vertical cell, we obtained identical results.

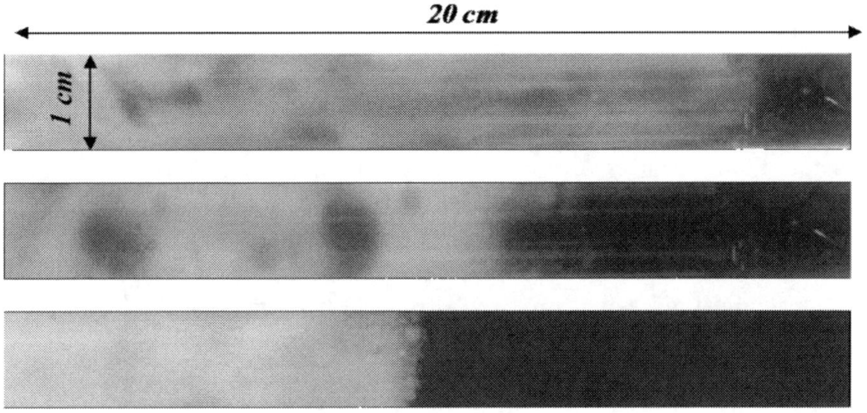

Figure 2. Phase segregation within the condenser tube of a 50% acetone – 50% hexadecane mixture, *2*, *5* and *10* seconds after being quenched with water at 7°C.

Naturally, one can suspect that within the small cell interfacial glass-liquid forces as well as capillary forces could play an important role. Furthermore since we were interested to obtain results that could be applied on industrial scales, we needed to see whether our results could be scaled up. Therefore, we designed and built a new experimental setup consisting in the condenser tube of figure 1b, *20 cm* long and *1 cm* in diameter, where individual droplets in a size range of *0.5 mm* of diameter and up could be visualized even without microscope.

In Figure 2 we see that the phase separation of the critical mixture in the condenser tube occurred in the much same way as in the cell: acetone-rich droplets formed in the whole tube and started to move along the axial direction, towards the warmer region, until a vertical interface appeared at the warmer end of the tube. Then, this interface moved towards the middle (i.e. the colder region) of the tube, until complete phase segregation was achieved at time $t=8s$, where, as before, $t=0$ corresponds to the moment when the temperature of the mixture crossed the miscibility curve. At the end, when $t=20s$, the small density difference made the system unstable, so that the heavier phase moved to the bottom part of the condenser, resulting in a horizontal interface. Again, when we reversed the temperature gradient by exchanging the inlet and the outlet of the cooling system, the flow was reversed as well. In addition, we observed that, initially, when the drops were small, a strong random motion was superimposed to their net drift, while as the drop size increased, such random motion weakened considerably.

We also noted that, as the separated acetone phase cooled further, secondary phase separation occurred, leading to the formation of droplets of the hexadecane-rich phase, flowing towards the colder region. To confirm this effect, we repeated the experiments using a *70%* acetone - *30%* hexadecane mixture, where the continuous phase was acetone-rich. In that case, we observed that the hexadecane-rich dispersed phase moved to the colder region, with similar motion of the droplets and morphology as in Figure 2. Again, once the interface formed at the warmer end of the tube, it moved towards the middle (i.e. the warmer region) of the tube.

At this point, we measured the velocity of individual drops, exploiting a feature of the Diagnostic Instruments Spot Camera (Model 3.2.1) digital camera, and saw that at $t=2s$ droplets were about *2 mm* in size and moved with a drift velocity of approximately *3 mm/s*, irrespectively of their size, with a weak random motion superimposed to it. Later, at $t=6s$, droplets were over *5 mm* in size and reached velocities up to *6 cm/sec*, with no random motion. At this point, though, larger drops moved faster than smaller drops, thereby enhancing coalescence, although the dependence of the drop speed on the drop size could not be determined (we can say, however, that this dependence is less than linear). To make sure that this effect was not caused by flow instabilities, we took pictures at different cross sections of the tube and measured the velocity of a few hundred droplets, finding that all drops moved in the same direction (i.e. with no back flow), both those near the wall as well as those at the center of the tube. In addition, when the flow rate of the cooling water was reduced, so that the temperature gradient increased from *0.25℃/cm* to *1℃/cm*, no change was reported, therefore confirming that drop migration here was not caused by thermocapillary forces. This indicates that temperature gradient is not responsible for the force acting on the droplets, although it acts as a switch for the direction of the flow.

As mentioned in the Introduction, previous experiments using density matched mixtures have not revealed this unidirectional convection because the temperature quench was not deep enough. In fact, in a previous work[24] we showed that, quenching our critical, quasi-isopycnic mixture in the cell of Figure 1a to a temperature of *20°C*, instead of *7°C*, at the end of the phase separation process the mixture was segregated so that the acetone-rich phase was distributed along the walls of the cell, while the hexadecane-rich phase formed a large domain at the center of the cell, showing that the non-equilibrium force that is responsible for the phase separation is too small, in this case, to overcome wall effects.

III. Discussion and Conclusions

In this paper we presented an unexpected result: when a low viscosity and nearly isopycnic binary mixture is quenched deeply and rapidly into its spinodal region, it develops unidirectional and large scale convective motion of the nucleating droplets along the direction of the temperature gradient. In fact, thanks to such large-scale flows, complete phase segregation may occur within *10* seconds on a *10* centimeter scale, even in the absence of buoyancy.

The results that we describe here are surprising and quite reproducible, yet they are not completely understood, although we can rule out thermo-capillary effects as a possible explanation of this phenomenon. In fact, thermocapillary velocity is proportional to the temperature gradient, while here we saw that this factor hardly influences the velocity of the drops. In addition, the direction of the thermo-capillary velocity of a drop is proportional to the gradient of its surface tension with respect to temperature. Accordingly, the thermocapillary velocity of acetone-rich drops has the same direction of that of hexadecane-rich drops, while in our experiments acetone-rich and hexadecane-rich drops moved in opposite directions.

The strong convection that we observed cannot be ascribed to flow instability either, because in that case the drops, at least initially, would follow closed trajectories, while here we saw explicitly that they all moved in the same direction, with no back flow.

A preliminary explanation of the phenomenon, instead, must be based on the fact that the system is far from equilibrium and therefore it is subjected to a force that is proportional to the local chemical potential gradient.[8] In fact, it has been observed experimentally[2,3] and confirmed numerically[9] that, in the absence of temperature gradients, drops move in random directions, with an instantaneous velocity that depends on the distance of the system from its equilibrium state and is therefore a function of the composition of the drops (and the composition of the continuous phase as well). As soon as a small temperature gradient is applied to the mixture, drops start moving coherently either parallel or antiparallel to the temperature gradient. Accordingly, the temperature gradient has merely the effect of orienting all the drop speeds along the same direction and that is why its magnitude does not influence the phenomenon. The temperature gradient, though, must lie within a certain range: it cannot be too small, otherwise it cannot orient the drop velocities, but it cannot be too large either, otherwise conventional flow instabilities and thermocapillary effects become relevant. Although this explanation is far from complete, it is consistent with previously obtained results,[20] where we studied the phase separation in a system with a preexisting concentration gradient. In addition, this explanation is consistent with the fact that no radial temperature gradient was observed, as the radial convection predicted by the model would strongly increase heat transport in that direction.

While the existence of flows driven by chemical potentials has been previously observed on a smaller scale, flows of this magnitude have never been reported. This new phenomenon is totally reproducible and controllable, therefore offering a challenging subject for theoretical and experimental investigations of separation processes, as well as for inducing flows in micro-gravity environments.

References

[1] Reviews on spinodal decomposition can be found in: J.S. Langer, in *Systems Far from Equilibrium*, L. Garrido, ed. (Springer Verlag, Berlin, 1980); J.D. Gunton, M. San Miguel, and P.S. Sahni, in *Phase Transition and Critical Phenomena*, Vol. 8, C. Domb and J.L. Lebowitz, eds. (Academic Press, London, 1983).

[2] Y.C. Chou and W.I. Goldburg, Phys. Rev. A **20**, 2105 (1979) and references therein.

[3] N.C. Wong and C. Knobler, Phys. Rev. A **24**, 3205 (1981) and references therein.

[4] X. Cumming, P. Wiltzius, F.S. Bates, and J.H. Rosedale, Phys. Rev. A, **45**, 885 (1992) and references therein.

[5] W.R. White and P. Wiltzius, Phys. Rev. Lett. **75**, 3012 (1995).

[6] P. Guenoun, R. Gastaud, F. Perrot, and D. Beysens, Phys. Rev. A **36**, 4876 (1987).

[7] E. Siggia, Phys. Rev. A 20, 595 (1979).

[8] K.S. McGuire, A. Laxminarayan, D.S. Martula and D.R. Lloyd, J. Colloid. Interface Sci. **182**, 46 (1996).

[9] D.S. Martula, T. Hasegawa, D.R. Lloyd and R.T. Bonnecaze, J. Colloid. Interface Sci. **232**, 241 (2000).

[10] G. Santonicola, R. Mauri, and R. Shinnar, Ind. Eng. Chem. Res. **40**, 2004 (2001).

[11] P.C. Hohenberg and B.I. Halperin, Rev. Mod. Phys. **49**, 435 (1977).

[12] D.M. Anderson, G.B. McFadden, and A.A. Wheeler, Annu. Rev. Fluid Mech. **30**, 139 (1998).

[13] J. Lowengrub and L. Truskinovsky, Proc. R. Soc. London, Ser. A **454**, 2617 (1998).

[14] D. Jasnow and J. Viñals, Phys. Fluids **8**, 660 (1996).

[15] D. Jacqmin,. Fluid Mech. **402**, 57 (2000).

[16] H. Tanaka, J. Chem. Phys. 105, 10099 (1996).

[17] R. Gupta, R. Mauri, and R. Shinnar, Ind. Eng. Chem. Res. **38**, 2418 (1999).

[18] N. Vladimirova, A. Malagoli, and R. Mauri, Phys. Rev. E **60**, 2037 (1999).

[19] H. Furukawa, Physica A, **204**, 237. (1994).

[20] N. Vladimirova, A. Malagoli, and R. Mauri, Phys. Rev. E **60**, 6968 (1999).

[21] N. Vladimirova, A. Malagoli, and R. Mauri, Chem. Engng. Sci. **55**, 6109 (2000).

[22] A.G. Lamorgese and R. Mauri, in *Non-Linear Dynamics and Control in Process Engineering – Recent Advances*, G. Continillo, S. Crescitelli and M. Giona eds. (Springer, Berlin, 2002), pp. 139-152.

[23] F. Califano and R. Mauri Ind. Eng. Chem. Res. **43**, 349 (2004).

[24] R. Mauri, F. Califano, E. Calvi, R. Gupta, and R. Shinnar, J. Chem. Phys. **118**, 8841 (2003).

(444a) Design of an IGCC Power Plant with Dispatch Capabilities

Ming-Wei Yang[1], **Benjamin P. Omell**[1] and Donald J. Chmielewski[2], (1)Chemical and Biological Engineering, Illinois Institute of Technology, Chicago, IL, (2)Illinois Institute of Technology, Chicago, IL

The Integrated Gasification Combined Cycle (IGCC) possesses a number of benefits over traditional power generation plants, ranging from increased energy conversion efficiency to flex-fuel and carbon capture opportunities. A less known benefit of the IGCC configuration is the ability to load track electricity market demands. The idea being that material storage can be used to time-shift power production away from periods of low power value to periods of high power value. In this work, we present a supervisory control scheme, known as Market Responsive Control (MRC), which has the objective of maximizing revenue for a given level of plant dispatchablity. The design of this controller is achieved by a fairly simple convex optimization problem. The MRC method is then embedded within an equipment design algorithm. This extended algorithm, known as MRC Embedded Equipment Design (MRC-EED), has the objective maximizing Net Present Value (NPV) in an effort to size the storage units as well as the required energy conversion devices. The design options considered in this study focus on hydrogen and compressed air storage as well as ramping abilities of the gasification block. Under the economic assumptions of this study, it is concluded that the NPV of the hydrogen storage option is significant, while that of compressed air is much less favorable. Finally, no scenario could be found in which ramping of the gasification block provided a positive NPV.

(444b) Real-Time Optimization of an Industrial Hydrocracking Plant

Hasan Sildir[1], Yaman Arkun[2], Berna Cakal[3], Dila Gokce[3] and Emre Kuzu[3], (1)Chemical and Biological Engineering, Koc University, Istanbul, Turkey, (2)Chemical and Biological Engineering, Koc University, İstanbul, Turkey, (3)TUPRAS Refineries, Kocaeli, Turkey

Hydrocracking is a catalytic chemical process which converts high–boiling hydrocarbons to more valuable lighter products like diesel, kerosene, naphtha and LPG. Environmental concerns and increased demand for low-sulphur diesel and high smoke point jet fuel have played a major role in the increased utilization of the hydrocracking technology. Hydrocracking is carried out in the presence of hydrogen at high temperatures (260 –495°C) and pressures (35 – 200 bar). The main reactions are cracking and hydrogenation which take place in multiple catalytic beds. The overall system is highly exothermic and hydrogen quench is used for interstage cooling. Due to hydrogenation, sulfur and nitrogen compounds are removed in the process to yield low–impurity reactor product. In the fractionation part of the plant, the reactor effluent is separated into the final valuable products.

Our research covers the real time plant-wide optimization of the hydrocracking reactors and the fractionation columns of TUPRAS refineries. We have developed a hierarchical control structure which is based on the spatial decomposition of the plant into reactor and fractionation subsystems and temporal decomposition of the control tasks based on the characterization of disturbances, availability of measurements and frequency of control actions. The hierarchical control structure consists of a cascade of MPCs operating in tandem with economic optimization as shown in Fig.1. The highest level in the hierarchy is the management layer which considers the demand for the products by the market and different units in the refinery. A refinery-wide linear programming is performed to compute the prices of feed, hydrogen and products. In addition product specifications and operating constraints for the hydrocracking plant are also defined. The next layer below in the hierarchy is steady-state economic optimization. Based on prices, product specifications and operating constraints set by the management, steady–state optimization maximizes profit by calculating the optimal product distribution (i.e. the relative amounts of light ends, light naphta, heavy naphta, diesel, kerosene, and bottoms). The 95% boiling point temperatures (or cut-points) of the products are also computed. Next optimal operating conditions found by the steady-state optimizer are realized by a cascade of MPCs. In this cascade, inner MPCs consist of the individual reactor and fractionator MPCs. Reactor MPC controls the bed exit temperatures by changing the set-points of hydrogen quench PID controllers that regulate the bed inlet temperatures. In Fig.1 these PID controllers are within the reactor block. Fractionator MPC adjusts the set-points of the product withdrawals to control the 95% boiling point temperatures. Similar to the reactor MPC, the PID flow controllers that regulate the product withdrawals are within the fractionator block in Fig.1. The 95% boiling point temperatures are estimated by an on-line soft sensor that uses an empirical model that predicts the 95% BPs from tray temperatures and the column pressure readings. The model has been built using historical data and is updated on-line when the laboratory distillation assay measurements become available.

Outer MPC's task is to control the optimal product disribution at values determined by steady-state optimization. It accomplishes this by manipulating the set-points of inner MPCs. Specifically it adjusts the reactor bed temperature set-points and the 95% boiling point temperature set-points of the fractionater. Reactor MPC uses the dynamic model developed in our earlier work [1]. This model can predict the reactor bed temperatures and the product disribution. Steady-state optimization makes use of the steady-state predictions of this nonlinear model.

Steady-state optimization and control cycle is implemented in a moving horizon pseudo-steady state fashion to handle potential plant-model mismatch and unaccounted disturbances. At the end of each steady-state optimization only a fraction of optimal set-point changes is implemented by MPCs. Using feedback measurements the models are updated and steady-state optimization-control cycle is repeated.

1. H. Sildir, Y. Arkun, B. Cakal, D. Gokce, E. Kuzu A, “Dynamic Non-Isothermal Model for a Hydrocracking

Reactor: Model Development by the Method of Continuous Lumping and Application to an Industrial Unit" J. of Proc. Control (in review).

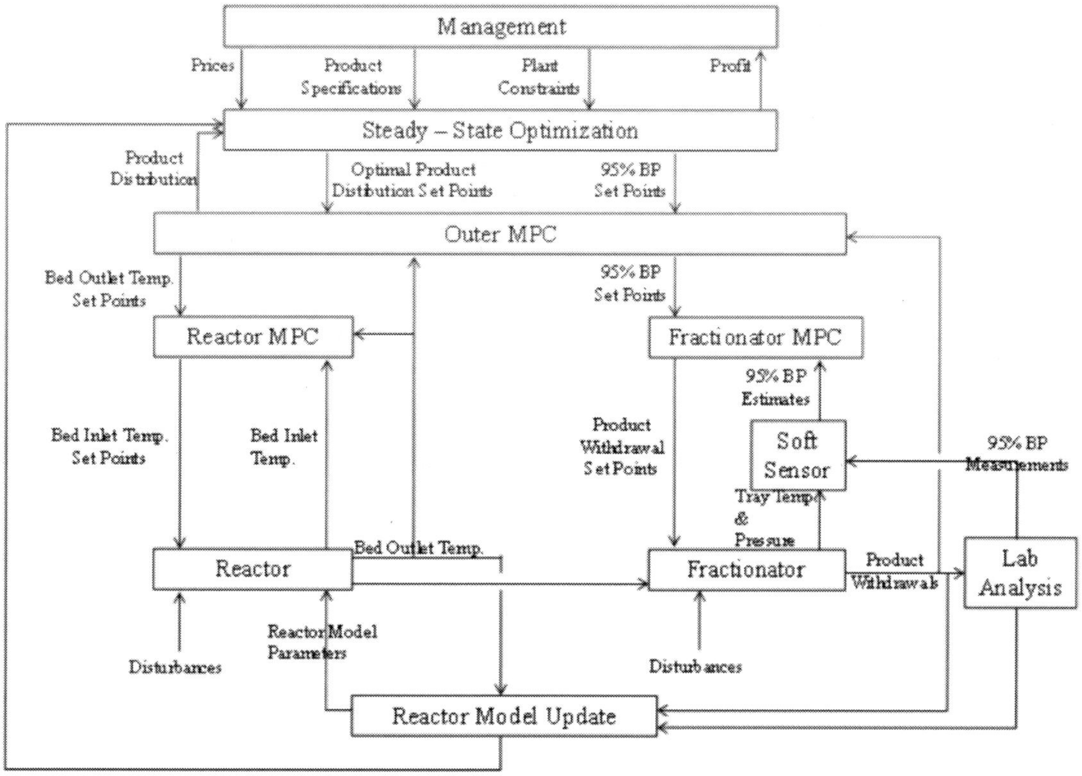

Fig.1. Hierarchical Control Structure for the Hydrocracking Plant.

(444c) New Control Structure for Fehe/Reactor Systems

William L. Luyben, Department of Chemical Engineering, Lehigh University, Bethlehem, PA

Feed-effluent heat exchangers (FEHE) are widely used in high-temperature exothermic adiabatic tubular reactor systems to conserve energy. The hot reactor effluent is recycled back to a feed preheater to provide all or a portion of the energy required to preheat the reactor feed to the optimum reactor inlet temperature. Several alternative flowsheets have been presented in the literature for this type of configuration. Some use bypassing of cold material around the FEHE, some use a furnace after the FEHE and before the reactor and some use a cooler (steam generator) after the reactor.

A number of control structures have been proposed to control the reactor inlet temperature, which is critical for the steady-state performance (achieving the desired conversion and staying below some maximum temperature). It is also critical for dynamic performance since these systems can be openloop unstable as a result of the positive feedback of energy from the reactor back to the preheat system. The control structure must prevent quenching to low temperatures ("blowout") and temperature runaways ("blowup"). The most commonly suggested control structure recommends controlling the temperature of the mixed stream after the FEHE by manipulating the bypass flow and controlling the reactor inlet temperature by manipulating fuel to the furnace.

This paper presents an alternative control configuration that reduces furnace energy consumption while maintaining good dynamic control in the face of very large disturbances.

(444d) Effect of Tray Locations On Temperature Based Control Structures of Ideal Reactive Distillation Columns

Baris Demirel and Devrim B. Kaymak, Department of Chemical Engineering, Istanbul Technical University, Istanbul, Turkey

Process intensification has been an area of growing interest in recent years. Reactive distillation columns are one of the well-known applications of process intensification in the chemical industry. The control structure selection is an important problem of reactive distillation columns. In the literature, dual temperature control structures are used for the inferential control of reactive distillation columns. Selection of the tray locations to temperature control have been performed using sensitivity analysis and singular value decomposition (SVD) method in the past decade. Recently, Kookos has presented a new methodology for the solution of tray selection problems, and applied it to a three temperature control structure of an ideal reactive distillation column with two reactants and two products [1].

The aim of this study is making quantitative comparisons between dual and triple temperature control structures of reactive distillation columns using different tray selection approaches. The comparison will be made for two different types of reactive columns: (i) a two-reactant and two-product column, and (ii) a two-reactant, one product and one inert column. The control structures will be tested for disturbances in the production rate and feed stream compositions.

[1] Kookos, I.K., Control Structure Selection of an Ideal Reactive Distillation Column, Ind. Eng. Chem. Res., 2011, 50, 11193-11200.

(444e) Studying Various Optimal Control Problems in Biodiesel Production in a Batch Reactor Under Uncertainty

Pahola T. Benavides, VRI- Custom, Chicago, IL and Urmila Diwekar, Vishwamitra Research Institute, Center for Uncertain Systems: Tools for Optimization and Management, Clarendon Hills, IL

Abstract

Optimal control problems encountered in biodiesel production can be formulated using various performance indices like maximum concentration, minimum time, and maximum profit. The problems involve determining optimal temperature profile so as to maximize these performance indices. This paper presents these three formulations and analyzes the solutions in biodiesel production. We also present the maximum profit problem where the variability and uncertainties in the feed composition of soybean are considered.

Introduction

Optimal control problems are defined in the time domain and their solution requires establishing a performance index for the system. Because of the dynamic nature, optimal control problems are much more difficult to solve compared to normal optimization. In this paper, we are proposing an alternative approach that avoids the use of large-scales NLP solvers. Moreover, we study the three optimal control problems on batch reactors: maximum concentration problem (MCP) of methyl ester, the minimum time problem (MTP), and the maximum profit problem (MPP) for biodiesel production. The MCP and MTP are solved using the maximum principle, and the approach is based on the Steepest Ascent of Hamiltonian. On the other hand, the MPP is solved using an algorithm that combines the maximum principle and NLP techniques. This algorithm is an efficient approach which avoids the solution of the two-point boundary value problem that results in the pure maximum principle or in the solution of the partial differential equations for the pure dynamic programming formulation. Biodiesel is one of the most well-known examples for alternative energy and is also environmentally friendly emission profile [1]. Here we present the optimal control problem where optimal temperature profile is derived using optimal control theory.

Optimal control problems become more challenging when variability in any parameter or variable is included. In biodiesel production, there are inherent uncertainties that have a significant impact on the process. One of the most influential uncertainties in this process is the feed composition since the percentage and type of triglycerides in biodiesel composition varies considerable [2]. This uncertainty can be modeled using probabilistic techniques, and can be propagated using stochastic modeling iterative procedures [3]. Therefore, at the end of this paper we propose a stochastic maximum profit problem (SMPP) that regards the variability in the feed composition to observe how this uncertainty affects the process economic in the batch reactor.

Optimal Control Problems

Table 1 summarizes the optimization problems presented in this paper.

Table 1 Optimal Control problems

Problem	Concentration	Time	Objective
MCP	Free	Fixed	Maximize C_E

MTP	Fixed	Free	Minimize t_f
MPP	Free	Free	Maximize profit

The kinetic model presented here is based on [4].

F1=dCTGdt=-k1CTGCA+k2CDGCE $F_1 = \frac{dC_{TG}}{dt} = -k_1 C_{TG} C_A + k_2 C_{DG} C_E$ (1)

F2 =dCDGdt=k1CTGCA-k2CDGCE-k3CDGCA+k4CMGCE $F_2 = \frac{dC_{DG}}{dt} = k_1 C_{TG} C_A - k_2 C_{DG} C_E - k_3 C_{DG} C_A + k_4 C_{MG} C_E$ (2)

F3 =dCMGdt=k3CDGCA-k4CMGCE-k5CMGCA+k6CGLCE $F_3 = \frac{dC_{MG}}{dt} = k_3 C_{DG} C_A - k_4 C_{MG} C_E - k_5 C_{MG} C_A + k_6 C_{GL} C_E$(3)

F4 =dCEdt=k1CTGCA-k2CDGCE+k3CDGCA-k4CMGCE+k5CMGCA-k6CGLCE (4)

$$F_4 = \frac{dC_E}{dt} = k_1 C_{TG} C_A - k_2 C_{DG} C_E + k_3 C_{DG} C_A - k_4 C_{MG} C_E + k_5 C_{MG} C_A - k_6 C_{GL} C_E \quad (4)$$

F5=dCAdt=-dCE dt $F_5 = \frac{dC_A}{dt} = -\frac{dC_E}{dt}$(5)

F6=dCGLdt=k5CMGCA-k6CGLCE $F_6 = \frac{dC_{GL}}{dt} = k_5 C_{MG} C_A - k_6 C_{GL} C_E$ (6)

where C_{TG}, C_{DG}, C_{MG}, C_E, C_A, and C_{GL} are the state variables and represent: triglycerides, diglycerides, monoglycerides, biodiesel, methanol, and glycerol, respectively. The initial conditions are: $C_i(t_0) = [C_{TG}; 0; 0; 0; C_A; 0]$ [mol/L] and k_i is the reaction constant.

The objective function for the MPP is represented by Eq.7 [5]

maxJ"=MEPr-BoCot+ts $\max J'' = \frac{M_E P_r - B_o C_o}{t + t_s}$(7)

where M_E is the amount if product (kg), P_r is the sales value of the product ($/kg), B_o is the amount of feed F (kg), C_o is the cost of feed ($/kg), t is batch time (minutes) and t_s is the setup time for each batch. However, this equation can be converted as:

maxJ"= max MEPr-BoCot+ts $\max J'' = \frac{(\max M_E) P_r - B_o C_o}{t + t_s}$(8)

Table 2 shows the information for profit function calculation. The amount of feed involves the quantity of methanol and triglycerides at the beginning of the reaction while the amount of product is the final concentration of methyl ester which is maximized by finding a temperature profile as a control variable.

Table 2 Information for maximum profit problem

Item	Data
Soy bean oil (Triglycerides) [a]	$0.62/kg
Methanol	$0.320/kg

Biodiesel (methyl ester) $ 3/gallon=$ 0.9/kg

Setup time (t_s) 10 min

[a] (www.icis.com). [2]

For the MCP and MTP, maximum principle is used to solve the optimal control problem. On the other hand, a combination of maximum principle and nonlinear optimization (NLP) technique based on SQP algorithm is used to solve the MPP.

In the SMPP, the objective function is subject to fluctuations due to the uncertainty arising in the feedstock content. In a previous work of our group [6], we showed the uncertainty characterization and the stochastic simulation for the feed stock composition of soybean oil. The interest of these types of problems is to determine the expected value of the maximum profit. Then, the objective function for the SMPP is shown in Eq. 9.

maxJ"=E max MEPr-BoCot+ts $\max J'' = E\left[\frac{(\max M_E)P_r - B_oC_o}{t+t_s}\right]$(9)

Result and discussion

Figure 1 shows the concentration profiles obtained for the three optimal control problems and two base cases (base case 1: 315K and 2: 323K). To start with, consider the concentration profile of methyl ester for the MCP. Here, we are comparing the concentration values at constant temperature with the values calculated at optimal temperature profile. It can be seen that at 100 minutes, the concentration of biodiesel using optimal control reaches its maximum value, 0.7944mol/L; while at constant temperature, the maximum concentration is 0.7324mol/L and 0.7829mol/L, respectively. This change represents an increase of 8.46 % (base case 1) and 1.47% (base case 2) on the concentration of methyl ester. The increment for the second base case is not significant since the constant profile at 323K belongs to the constant optimal profiles reported in the literature [7]. Moreover, if we fix the concentration at 0.7324mol/L (concentration reached in base case1), the reaction time needed is 30.5 minutes which represents 69.5% less than it was at the beginning (100 minutes) after using optimal control. Compare with base case 2, the reduction on time represents 46% of the original time. This improvement does not affect the behavior of the other components because after 50 minutes their concentrations remain constant. For the MTP, after fixing the concentration of methyl ester to 0.7324mol/L the minimum time reached is 30.6 minutes when optimal control is applied, while for base case 1 and 2 their minimum time is reached at 100 and 54 minutes, respectively. Although, the optimal control profiles shown in Figure 2 for the two optimal control problems are significantly different, their results are similar showing that this problem have multiple solutions.

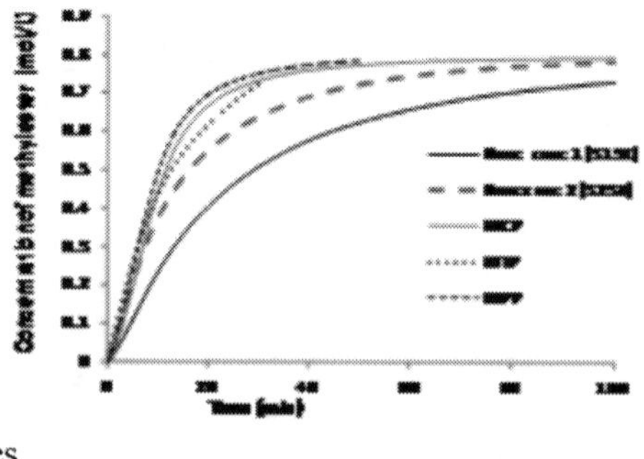

Figure 1 Concentration Profiles.

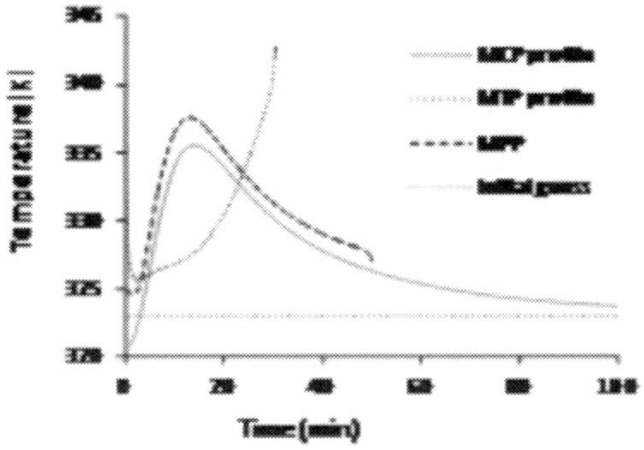

Figure 2 Optimal Temperature Profiles

For the case of MCP, it can be seen that at 50 of minimum time there is an increase of Biodiesel concentration of 25.56% (case 1) and 8.50% (case 2). Moreover, if we compute the profit in the MCP and MTP using Eq. 9 and compare these values with the profit found in the MPP, there is an increment of 45.32% and 355.58%, respectively.

Table 3 Comparison of the optimal control problems

Parameter	Maximum Concentration	Minimum Time	Maximum Profit
Biodiesel concentration (mol/L)	0.7944	0.7324	0.7802
Time (minutes)	100	30.5	50
Profit ($/hr)	103.1005	32.8868	149.8260

Table 4 shows results of these problems regarding uncertainty. As shown, in stochastic case there is an improvement of 6.68% compared to the deterministic case and very significant improvement compare with the two base cases. In other words, the SMPP gives 9.99$/hr more than MPP (deterministic) and 148.163$/hr and 76.951$/hr more than base case 1 and 2, respectively.

Table 4 Comparison deterministic and stochastic cases.

Parameter	(315K)	(323K)	Deterministic	Stochastic
Biodiesel concentration (mol/L)	0.7316	0.7767	0.7799	0.7786
Time (minutes)	100	92.653	50	44.912
Profit ($/hr)	11.197	82.409	149.368	159.360

Conclusions

The problems presented in this article involved determining optimal temperature profile so as to maximize or minimize three performance indices: concentration, time, and profit for biodiesel production. MCP and MTP were solved using maximum principle and steepest ascent of the Hamiltonian. The solution of these two problems results in similar equations for maximum principle. In both cases reaction time was around 30.5 to 30.6 minutes and concentration of biodiesel was the same. While in MPP, the technique used was based on combining the maximum principle and NLP techniques. Applying optimal control under uncertainty it resulted in better time to produce the same amount of biodiesel which improved the profit value of the problem.

References

1. Zhang, Y.; Dube, M.A.; McLean, D.D.; Kates, M. Biodiesel production from waste cooking oil: 1. Process design and technological assessment. Bioresour. Technol. 89 (2003) 1.

2. Linstromberg WW. Organic chemistry. MA: DC Heath and Co. Lexington. (1970) 129.

3. Diwekar U, Rubin ES. Stochastic modelling of chemical processes. Comput. Chem. Eng. 15 (1991) 105

4. Noureddini H, Zhu D. Kinetic of transesterification of soybean oil. J Am Oil Chem Soc. 74 (1997) 1457.

5. Kerkhof, L H. J.; Vissers, J. M. On the profit of optimum control in batch distillation. Chem. Eng. Sci. 33 (1978) 961.

6. Benavides, P.; Diwekar, U. Optimal Control of Biodiesel Production in a Batch Reactor Part II: Stochastic Control. Fuel. 94 (2012) 218.

7. Leung, D, Guo, Y. Transesterification of neat and used frying oil: optimization for biodiesel production. Fuel Process Technol. **87** (2006) 883.

(444f) Toward Glucose Control without Hypoglycemia in Critical Care

Stanislaw Gawel[1], Gilles Clermont[2], Balaji Yegneswaran[3] and Robert S. Parker[1], (1)Chemical and Petroleum Engineering, University of Pittsburgh, Pittsburgh, PA, (2)Critical Care Medicine, University of Pittsburgh School of Medicine, Pittsburgh, PA, (3)Critical Care Medicine, University of Pittsburgh, Pittsburgh, PA

In the United States alone, over 5 million patients are admitted into an intensive care unit (ICU) every year. US critical care medicine costs in 2005 were $81.7 billion; this is 13.4% of hospital costs, 4.1% of national health expenditures, and 0.66% of the gross domestic product [1]. These patients are prone to a condition known as "stress hyperglycemia" or "diabetes of injury", where they are unable to properly regulate blood glucose, which results in elevated blood glucose levels (hyperglycemia). Studies have shown that intensively managing patients' blood glucose levels with insulin can result in higher survival rates and lower healthcare costs [2,3]. While these studies have shown a substantial benefit of targeted glucose control, a large recent observational study has reported targeted glucose to result in increased mortality rates due to the much higher incidence rates of hypoglycemia (low blood glucose) and large glucose variability [4]. Severe hypoglycemia can result in coma or death, but new evidence has shown that even one incidence of mild hypoglycemia results in increased mortality rates and worse outcomes [5].

We have developed a model predictive controller (MPC) that will provide tight glucose control without hypoglycemia in critical care patients. Similar to the "artificial pancreas" for ambulatory diabetics, this system will measure glucose concentration, calculate (via MPC) the insulin and glucose needs of the patient to maintain normoglycemia, and administer these agents, all at 5 min intervals. To prevent hypoglycemia, glucose infusion was added to the MPC in an output-regulator (OR) formulation. In order to successfully manage glucose control across patients with marked inter-individual variability, as well as time-varying behavior in their dynamic response during recovery, parameters within the MPC need to be updated in real-time. For this purpose, we embed a moving horizon estimator (MHE) within the MPC algorithm. A state-space MPC with MHE was built in MATLAB (©2012, The MathWorks, Natick, MA).

The MPC+MHE+OR algorithm was compared against the MPC-Only algorithm (insulin as only manipulated input), which could only administer insulin, *in silico*. When a hypoglycemia-inducing challenge of approximately 67 mU/min of insulin was given to the patient for 2 hours, the MPC+MHE+OR algorithm had a maximum hypoglycemic excursion of 1 mg/dL from the 89 mg/dL reference. In contrast, the MPC-only algorithm could not avoid hypoglycemia, returning a minimum blood glucose of 45 mg/dL. The MPC-MHE+OR algorithm was also able to track a 2-hour slow insulin sensitivity parameter drift, with only a slight lag in the estimated value resulting from the ongoing rampwise drift.

In order to evaluate the two algorithms under more realistic conditions, a virtual patient was created from 17 days of ICU-patient data. This virtual patient was constructed by fitting two parameters in a nonlinear ODE model of glucose-insulin-fatty-acid system dynamics [6] to the patient data such that the simulation of the model under actual clinical inputs reproduced the blood glucose measurements of the actual patient. The MPC+MHE+OR algorithm was then compared against the MPC-only algorithm *in silico* with the virtual patient as the simulated patient. Both algorithms could account for measured and unmeasured disturbances and both performed better than the actual patient in the clinic. Quantitatively, the sum of squared error from the glucose reference was 31% better for the MPC+MHE+OR algorithm than the MPC-only algorithm. Furthermore, the minimum value of blood glucose for the MPC+MHE+OR algorithm was 82.5 mg/dL over the entire 17 day window, while the minimum blood glucose for the MPC-only algorithm was 28.3 mg/dL and the actual minimum observed in the real patient was 52 mg/dL.

The MPC+MHE+OR algorithm is able to provide targeted glucose control without hypoglycemia in ideal (nominal case) and non-ideal (patient-model mismatch case) simulated patients. The MHE improves the performance of the algorithm in the presence of parametric variations and patient-model mismatch by computing the time-dependent, patient-specific values of the model parameter, insulin action on glucose uptake. This action allows the algorithm to match model predictions to measurements so it remains accurate despite changing patient dynamics. By providing a patient-specific solution to targeted glucose control without hypoglycemia, the MPC+MHE+OR algorithm has the potential to improve the safety and robustness of further randomized clinical trials of targeted glucose. Moreover, it can form the basis of closed-loop approaches to glucose homeostasis in ICU patients.

List of References

[1] Halpern et al. "Critical care medicine in the United States 2000–2005: An analysis of bed numbers, occupancy rates, payer mix, and costs." Critical Care Medicine 38.1 (2010): 65-71.

[2] Van den Berghe, G et al. "Intensive Insulin Therapy in Critically Ill Patients." New England Journal of Medicine 345. (2001): 1359-1367.

[3] Krinsley, James Stephen. "Effect Of An Intensive Glucose Management Protocol On The Mortality Of Critically Ill Adult Patients." Mayo Clinic Proceedings 79.8 (2004): 992-1000.

[4] Meynaar IA et al. "Blood glucose amplitude variability as predictor for mortality in surgical and medical intensive care unit patients: a multicenter cohort study." J Crit Care (2012). S0883-9441(11)00486-2 [pii];10.1016/j.jcrc.2011.11.004 [doi].

[5] Krinsley, James S. et al. "Mild hypoglycemia is independently associated with increased mortality in the critically ill." Critical Care 15:R173 (2011)[6]. Roy, Anirban. *Dynamic Modeling of Free Fatty Acid, Glucose, and Insulin during Rest and Exercise in Insulin Dependent Diabetes Mellitus Patients*. Diss. University of Pittsburgh, 2008.

(445a) Economic Plantwide Control Design Procedure

Rahul Jagtap, Chemical Engineering, Indian Institute of Technology Kanpur, Kanpur, India, Nitin Kaistha, Chemical Engineering, IIT Kanpur, Kanpur, India and Sigurd Skogestad, Department of Chemical Engineering, Norwegian University of Science and Technology, Trondheim, Norway

A general systematic top-down procedure for an effective plantwide control system design for economically optimal operation of a process over a large throughput range is synthesized. As throughput is increased, operating constraints become active one by one at the optimum steady state till maximum plant throughput where maximum number of constraints is active. The crucial step in the systematic procedure is to design a control system for economic process operation at maximum throughput. In step 1 of the procedure control loops for tight control of active constraints and economically dominant CVs are implemented, then in Step 2 inventory control loops are set followed. Step 3 involves control loop configurations for throughput manipulation so that the setpoints that becomes unconstrained at lower throughput can take up additional close-by economic CV control for throughput manipulation. Step 4 involves control structure reconfiguration to eliminate fragile / unconventional inventory control loops resulting from above 3 steps. The procedure is applied on a realistic chemical process.

(445b) Systematic Synthesis of Augmented Biomass-to-Liquid Fuel Processes

Dharik S. Mallapragada[1], Mohit Tawarmalani[2], Fabio H. Ribeiro[1], W. Nicholas Delgass[1] and Rakesh Agrawal[1], (1)School of Chemical Engineering, Purdue University, West Lafayette, IN, (2)Krannert School of Management, Purdue University, West Lafayette, IN

Liquid fuels from biomass potentially offer a sustainable option for supplying high volumetric energy density fuel for the transportation sector in a future solar energy driven economy. However, there are limits on the amount of sustainably available (SA) biomass that can be recovered annually without using additional agricultural land. Since the continued use of liquid fuel for transportation seems likely, conversion processes that maximize the liquid fuel output from the limited SA biomass resource are worthy of exploration.

Here, we present systematic synthesis of augmented biomass-to-liquid fuel processes, which maximize biomass carbon conversion to liquid fuel by using supplemental energy derived from sunlight in the form of H_2, heat, and electricity. While standalone biomass-to-liquid fuel processes are limited to $<50\%$ biomass carbon recovered as liquid fuel, augmented processes overcome this barrier by utilizing biomass as a carbon source combined with supplemental solar energy use. The emphasis is on identifying augmented processes that require the least amount of supplemental solar energy input corresponding to a given biomass carbon recovery.

We focus on two promising biomass to liquid fuel thermochemical approaches of heat-assisted biomass gasification followed by fuel synthesis and biomass fast-hydropyrolysis/hydrodeoxygenation (HDO). A superstructure of all possible process configurations for the aforementioned thermochemical routes is developed. Subsequently, an equation-oriented nonlinear optimization model is solved to identify the optimal process configuration for different target carbon recovery levels. The novel features of the process synthesis include: 1) a simplified fast-hydropyrolysis/HDO model derived from experimental literature, 2) simultaneous process heat, power and mass integration while allowing for co-generation and 3) flexibility to design for different carbon recovery levels. The presentation will also briefly discuss the progress in developing a theoretical guarantee of the global optimality of the solutions presented.

(445c) Process Synthesis of Hybrid Coal, Biomass, and Natural Gas to Liquids (CBGTL) Via Fischer-Tropsch Synthesis, ZSM-5 Catalytic Conversion, Methanol Synthesis, Methanol-to-Gasoline, and Methanol-to-Olefins/Distillate Technologies

Richard C. Baliban, **Josephine A. Elia**, Vern W. Weekman Jr. and Christodoulos A. Floudas, Chemical and Biological Engineering, Princeton University, Princeton, NJ

Thermochemical conversion of biomass or coal to liquid hydrocarbons can proceed via syngas intermediate, followed by either the Fischer-Tropsch (FT) process or via methanol intermediate. Thermochemical conversion of natural gas generally occurs through indirect liquefaction using an autothermal reactor. A recent review of hybrid and single feedstock energy processes can be found in [1]. A coal, biomass, and natural gas to liquids (CBGTL) thermochemical superstructure was recently developed [2-9] that incorporates various technologies, unit operating conditions, and stream interconnections. A process synthesis mathematical model was proposed to determine the process topology that can produce fuels at the lowest levelized cost [4, 6-8]. The hydrocarbon conversion route was primarily based on a combination of Fischer-Tropsch processes operating at multiple temperature levels, using either iron or cobalt catalysts.

This superstructure is expanded to include conversion from syngas to fuels via methanol intermediate, followed by the methanol-to-gasoline and methanol-to-olefins/distillate technologies to produce gasoline, diesel, and kerosene. The hydrocarbons can be upgraded to fuel quality in a standard upgrading section akin to the operation of petroleum refineries, or through catalytic conversion over ZSM-5 catalysts. Formulated as a mixed-integer nonlinear optimization problem (MINLP), the process synthesis problem is solved with simultaneous heat, power, and water integration.

Case studies are presented to investigate the trade-offs between the Fischer-Tropsch and methanol routes, along with detailed parametric analysis on the effect of plant capacities on the technology selection and optimal topologies. These examples elucidate the process topological differences and their effect of the overall system cost, the process material/energy balances, and the well-to-wheel greenhouse gas emissions.

[1] C.A. Floudas, J. A. Elia, R. C. Baliban (2012) Hybrid and Single Feedstock Energy Processes for Liquid Transportation Fuels: A Critical Review. Comp. Chem. Eng. 41:24-51.

[2] R. C. Baliban, J. A. Elia, C. A. Floudas (2010) Toward novel biomass, coal, and natural gas processes for satisfying current transportation fuel demands, 1: Process alternatives, gasification modeling, process simulation, and economic analysis. Ind. Eng. Chem. Res. 49:7343-7370.

[3] J. A. Elia, R. C. Baliban, C. A. Floudas (2010) Toward novel biomass, coal, and natural gas processes for satisfying current transportation fuel demands, 2: Simultaneous heat and power integration. Ind. Eng. Chem. Res. 49:7371-7388.

[4] R. C. Baliban, J. A. Elia, C. A. Floudas (2011) Optimization framework for the simultaneous process synthesis, heat and power integration of a thermochemical hybrid biomass, coal, and natural gas facility. Comp. Chem. Eng. 35:1647-1690.

[5] J. A. Elia, R. C. Baliban, X. Xiao, C. A. Floudas (2011) Optimal energy supply network determination and life cycle analysis for hybrid coal, biomass, and natural gas to liquid (CBGTL) plants using carbon-based hydrogen production. Comp. Chem. Eng. 35:1399-1430.

[6] R. C. Baliban, J. A. Elia, V. Weekman, C. A. Floudas (2012) Process synthesis of hybrid coal, biomass, and natural gas to liquids via Fischer-Tropsch synthesis, ZSM-5 catalytic conversion, methanol synthesis, methanol-to-gasoline, and methanol-to-olefins/distillate technologies. Submitted for publication.

[7] R. C. Baliban, J. A. Elia, C. A. Floudas (2012) Simultaneous process synthesis, heat, power, and water integration of thermochemical hybrid biomass, coal, and natural gas facilities. Comp. Chem. Eng. 37:297-327.

[8] R. C. Baliban, J. A. Elia, R. Misener, C. A. Floudas (2012) Global Optimization of a MINLP Process Synthesis Model for Thermochemical Based Conversion of Hybrid Coal, Biomass, and Natural Gas to Liquid Fuels. Comp. Chem. Eng. In press. DOI:10.1016/j.compchemeng.2012.03.008

[9] J. A. Elia, R. C. Baliban, C. A. Floudas (2012) Nationwide Supply Chain Analysis for Hybrid Feedstock Energy Processes with Significant CO_2 Emissions Reduction. AIChE Journal. Submitted for publication.

(445d) Ab-Initio and Computer-Aided Molecular Design for the Identification of Optimal Solvents for Reactions

Eirini Siougkrou, Chemical Engineering, Imperial College London, Centre for Process Systems Engineering, London, United Kingdom, Amparo Galindo, Department of Chemical Engineering, Centre for Process Systems Engineering, Imperial College London, London, United Kingdom and Claire S. Adjiman, Centre for Process Systems Engineering, Imperial College London, London, United Kingdom

A solvent is frequently used as a medium for liquid-phase reactions and its role is significant, especially in the control of reaction rates and temperature. The prediction of the effects of solvents on a given reaction and the rational choice of the most appropriate solvent for the reaction are complex challenges. Strübing et al.[1] have proposed an algorithm for the Computer-Aided Molecular Design (CAMD) of solvents for chemical reactions, where quantum mechanical (QM) calculations and a surrogate (approximate) model are used for the prediction of the reaction rate constant in various solvents. This approach has been successfully applied to the design of an improved solvent for a Menschutkin reaction, which was subsequently tested experimentally and shown to lead to a 40% increase in the measured rate constant, relative to the best initial solvent. Here we develop the approach further by removing the need for a surrogate model, which can be shown to introduce modeling error in the overall methodology. An optimisation CAMD formulation is used to predict the solvent that maximises the reaction rate. A design of several thousand potential molecules can be explored in this way. The reaction rate constants in the solvents tested during the course of the algorithm are obtained from QM calculations, using conventional transition state theory (CTST). The rate constant, according to CTST, is a function of the standard state free energy of solvation of the reactants and transition state. The free energy of solvation is calculated in this work from the SMD solvation model[2]. The proposed approach is compared to the CAMD algorithm of Strübing et al.[1] through application to a Menschutkin S_N2 reaction and the numerical performance and accuracy of both approaches is discussed.

References

[1] H. Strübing, P. G. Karamertzanis, E. N. Pistikopoulos, A. Galindo, C. S. Adjiman, Solvent design for a Menschutkin reaction by using CAMD and DFT calculations, Comp. Aided Chem. Eng., **2010**, 28, 1291-1296

[2] A. V. Marenich, C. J. Cramer, D. G. Truhlar, Universal solvation model based on solute electron density and on a continuum model of the solvent defined by the bulk dielectric constant and atomic surface tensions, J. Phys. Chem. B, **2009**, 113, 6378-6396

(445e) Dynamic Simulation and Load-Following Control of an IGCC Power Plant with CO2 Capture

Debangsu Bhattacharyya[1], Richard Turton[1] and Stephen E. Zitney[2], (1)Department of Chemical Engg., West Virginia University, Morgantown, WV, (2)AVESTAR Center, National Energy Technology Laboratory, Morgantown, WV

Load-following control of future integrated gasification combined cycle (IGCC) plants with pre-combustion CO_2 capture is expected to be far more challenging as electricity produced by renewable energy is connected to the grid and strict environmental limits become mandatory requirements. To study control performance during load following, a plant-wide dynamic simulation of a coal-fed IGCC plant with CO_2 capture has been developed. The slurry-fed gasifier is a single-stage, downward-fired, oxygen-blown, entrained-flow type with a radiant syngas cooler (RSC). The syngas from the outlet of the RSC goes to a scrubber followed by a two-stage sour shift process with inter-stage cooling. The acid gas removal (AGR) process is a dual-stage physical solvent-based process for selective removal of H_2S in the first stage and CO_2 in the second stage. Sulfur is recovered using a Claus unit with tail gas recycle to the AGR. The recovered CO_2 is compressed by a split-shaft multistage compressor and sent for sequestration after being treated in an absorber with triethylene glycol for dehydration. The clean syngas is sent to two advanced "F"-class gas turbines (GTs) partially integrated with an elevated-pressure air separation unit. A subcritical steam cycle is used for heat recovery steam generation. A treatment unit for the sour water strips off the acid gases for utilization in the Claus unit. The steady-state model developed in Aspen Plus® is converted to an Aspen Plus Dynamics® simulation and integrated with MATLAB® for control studies. The results from the plant-wide dynamic model are compared qualitatively with the data from a commercial plant having different configuration, operating condition, and feed quality than what has been considered in this work.

For load-following control, the GT-lead with gasifier-follow control strategy is considered. A modified proportional–integral–derivative (PID) control is considered for the syngas pressure control. For maintaining the desired CO_2 capture rate while load-following, a linear model predictive controller (LMPC) is implemented in MATLAB®. A combined process and disturbance model is identified by considering a number of model forms and choosing the final model based on an information-theoretic criterion. The performance of the LMPC is found to be superior to the conventional PID control for maintaining CO_2 capture rates in an IGCC power plant while load following.

(445f) Design of Transesterification Processes

Cory Silva[1], Leonard Fabiano[1] and **Warren D. Seider**[2], (1)Chemical and Biomolecular Engineering, University of Pennsylvania, Philadelphia, PA, (2)Chemical and Biomolecular Engineering, The University of Pennsylvania, Philadelphia, PA

DESIGN OF TRANSESTERIFICATION PROCESSES

by

Cory Silva, Leonard A. Fabiano, and Warren D. Seider

Department of Chemical and Biomolecular Engineering

University of Pennsylvania

Philadelphia, PA 19104-6393

Abstract

Biofuels, derived from organic oils, have received considerable attention during the last decade; however, most of the research effort has concentrated upon bench-scale experiments, with little or no attention given to the large-scale industrial applications of the various conversion methods. Despite the numerous techniques available,[1,2]current process research has focused mainly upon the use of aqueous alkaline catalysts[3], with some attention being given to non-catalytic, supercritical transesterification[4]. The narrow focus is likely due to the scarcity of kinetic data, which is almost nonexistent for non-aqueous catalysts.

In this paper, an algae-oil transesterification process is presented, using a proprietary solid-acid catalyst. Kinetic parameters are regressed from bench-scale experimental data for oil harvested from *Nanochloropsis Salina* algae. The process model is simulated using ASPEN PLUS, and is then heat integrated to reduce the consumption of utilities. The equipment is sized and costed using ASPEN's Process Economic Evaluator to establish an overall project cost. Finally, our model and costs are compared with previous studies for algae-derived oils[5] and vegetable oils[3]to assess the viability of a process using heterogeneous catalysis.

References: 1. Vyas A. P., Verma J.L., Subrahmanyam N., 2010. A review on FAME production processes, *Fuel*, **89**, 1–9. 2. Demirbas A., 2008. Comparison of transesterification methods for production of biodiesel from vegetable oils and fats, *Energy Conversion and Management*, **49**, 125–130. 3. Chang A., Liu Y.A., 2010. Integrated process modeling and product design of biodiesel manufacturing, *Ind. Eng. Chem. Res*., **49**, 1197-1213. 4. van Kasteren J.M.N., Nisworo A.P., 2007. A process model to estimate the cost of industrial scale biodiesel production from waste cooking oil by supercritical transesterification, *Conservation and*

Recycling, **50**, 442–458. 5. Pokoo-Aikins G., Nadim A., El-Halwagi M.M., Mahalec V., 2010. Design and analysis of biodiesel production from algae grown through carbon sequestration, *Clean Techn. Environ.*

Policy, **12**, 239-254.

(445g) Comparison of Extractive Distillation and Pressure-Swing Distillation for Methanol-Chloroform System

Eda Hosgor and **Devrim B. Kaymak**, Department of Chemical Engineering, Istanbul Technical University, Istanbul, Turkey

Two of the most common methods for separating a binary homogeneous azeotrope are extractive distillation and pressure-swing distilllation. The first method is effective if a suitable entrainer can be found, while the second one is effective if the composition of the azeotrope changes significantly with pressure. In this study, the steady-state designs of these methods will be developed for a methanol-chloroform binary system which is an effective solvent for extraction of bioactive substances from biological sources. For this binary system extractive distillation is feasible using water as the entrainer. Also the pressure-swing distillation is feasible since the minimum-boiling azeotrope at 1 atm contains 63,32 mol % chloroform at 327 K and at 10 atm the azeotropic composition is 40,83 mol % chloroform at 404 K. For all the process simulations the feed mixture contains 50 mol % chloroform 50 mol % methanol with a flowrate of 100 kmol/h and the purities of the two products are set at 99.5 mol %. Wilson model is used as thermodynamic model. For the extractive distillation process, the design variables to be optimized are; entrainer flowrate, extractive distillation total number of stages, fresh feed and entrainer feed locations, entrainer recovery column total number of stages, feed location of entrainer recovery column. For the pressure swing distillation process, the design variables to be optimized are; total number of stages and feed locations for both columns and recycle flowrate. The aim of this study is the comparison of steady-state designs of these methods by investigating the effects of design variables and total annual costs.

(446a) Fault Detection and Isolation and Fault Tolerant Control of a Catalytic Alkylation of Benzene Process

David Chilin[1], Jinfeng Liu[2], Xianzhong Chen[3] and Panagiotis D. Christofides[2], (1)Chemical and Biomolecular Engineering, Univ. of California, Los Angeles, Los Angeles, CA, (2)Department of Chemical and Biomolecular Engineering, University of California, Los Angeles, Los Angeles, CA, (3)Chemical and Biomolecular Engineering, Univ. of California, Los Angeles, Los Angeles

Fault-tolerant control is an essential component in modern process industries as abnormal situations account for over $20 billion in lost annual revenue in the US alone. Traditionally, control systems rely on centralized control architectures utilizing dedicated wired links for measurement sensors and control actuators to operate a plant at desired conditions and a separate monitoring component for detecting faults. While this paradigm to process monitoring and control has been successful, modern chemical plants that rely on highly automated processes to maintain robust operations and efficient production are vulnerable to abnormal events. This issue has prompted significant research efforts in the integration with and application of fault-tolerant control methods to existing legacy control systems including chemical processes controlled by distributed model predictive control (DMPC) systems (e.g., [1]).

The focus of this work is on the application of an integrated fault detection and isolation and fault-tolerant control (FDIFTC) framework to a catalytic alkylation of benzene process which is controlled by a DMPC system and is subjected to unknown, persistent control actuator faults. The FDIFTC system uses measurements of process variables like temperature and concentrations. To design the fault detection and isolation (FDI) system, we take advantage of recent results on FDI that address both FDI filter design, residual generation and residual threshold computation using fault-free process data [2]. After isolation of an actuator fault, the FDIFTC system estimates the fault magnitude, recalculates a new optimal operating point, and ultimately reconfigures the DMPC system to maintain stability of the process in an optimal manner. Extensive simulations are carried out to demonstrate the performance of the FDIFTC system from stability and performance points of view.

1. D. Chilin, J. Liu, D. Muñoz de la Peña, P.D. Christofides, J. F. Davis, Detection, isolation and handling of actuator faults in distributed model predictive control systems, Journal of Process Control 20 (2010) 1059--1075.

2. D. Chilin, J. Liu, J. F. Davis, P. D. Christofides, Data-based monitoring and reconfiguration of a distributed model predictive control system, International Journal of Robust and Nonlinear Control 22 (2012) 68--88.

(446b) Industrial Experiences of On-Line Fault Detection and Process Monitoring

Zdravko Stefanov and **Leo H. Chiang**, The Dow Chemical Company, Freeport, TX

Process chemometrics (chemometrics and multivariate statistical modeling methods applied to chemical process data and manufacturing quality data) is a technology that has gained increasing significance in Dow Chemical's plant troubleshooting and proactive quality improvement efforts. Process upsets and product quality issues often result in loss of asset capability and/or customer concerns. When these unfavorable conditions happen, it is important to effectively and promptly identify the root cause(s) to recover to normal operating conditions. Solving these problems has lead to multimillion dollar savings for the company.

This presentation will show several examples of successful application of On-line Fault Detection and Process Monitoring using multivariate techniques such as Principal Component Analysis (PCA) and Partial Least Squares (PLS). The selected examples come from batch and continuous processes and show the versatility of the technology. Practical issues such as variable selection, outlier detection, on-line model implementation, and operation training will be discussed as well.

(446c) Non-Gaussian Dynamic Process Monitoring Using A New Dissimilarity Method Based On Independent Component Analysis and Multidimensional Mutual Information

Mudassir Rashid and Jie Yu, Department of Chemical Engineering, McMaster University, Hamilton, ON, Canada

Traditional multivariate statistical processes monitoring (MSPM) techniques like principal component analysis (PCA) and partial least squares (PLS) are not well-suited in monitoring non-Gaussian processes because the derivation of monitoring indices requires the approximate multivariate Gaussian distribution of the process data. To address the challenges of traditional MSPM methods, a novel pattern analysis driven dissimilarity approach is developed by integrating multidimensional mutual information (MMI) with independent component analysis (ICA) in order to quantitatively evaluate the statistical dependency between the independent component subspaces of the normal benchmark and monitored data sets.

In this study, ICA is integrated with MMI to measure the statistical independency between two independent component subspaces, which represents the subspace dissimilarity. The presented approach uses entropy based mutual information to assess the dissimilarity between the multidimensional independent component subspaces of the benchmark and monitored sets. The new MMI based ICA dissimilarity index is derived from the higher-order statistics so that the non-Gaussian process features can be extracted efficiently. The proposed approach does not require any assumptions regarding the relationships among measurement variables and can deal with the process data following an arbitrary probability density distribution. In contrast to the angle based dissimilarity factors, the new multidimensional mutual information based dissimilarity takes into account the higher-order statistics and thus can capture the non-Gaussian features of the process data well. Furthermore, a rolling window that moves over the monitored segment incrementally ensures that the time-varying process dynamics are accounted for in the MMI based dissimilarity approach. The higher-order statistics underlying the dissimilarity index can help extract the non-Gaussian features and quantify the statistical dependency between IC subspaces on a moving-window basis. Thus the non-Gaussian dynamic processes can be effectively monitored through the continuous comparison between the normal benchmark and monitored data sets.

The multidimensional mutual information based ICA dissimilarity method is applied to the Tennessee Eastman Chemical process. The comparison of monitoring results demonstrates that the new mutual information based dissimilarity index is superior to regular PCA, angle based PCA dissimilarity, conventional ICA and angle based ICA dissimilarity methods with the most reliable fault detection capability.

(446d) Fault Detection and Accommodation in Particulate Processes Using Multi-Rate Sampled and Delayed Measurements

Trina Napasindayao, Department of Chemical Engineering and Materials Science, University of California, Davis, Davis, CA and Nael H. El-Farra, Department of Chemical Engineering & Materials Science, University of California, Davis, Davis, CA

Fault detection and accommodation are fundamental problems at the interface of process control and operations. Agricultural, chemical, food, mineral, and pharmaceutical process industries place a huge importance on product quality which may deteriorate when faults are not properly diagnosed and handled once they occur. Many of the underlying processes in these industries are classified as particulate processes which, in addition to the continuous phase, also include a dispersed phase composed of particulates. The dispersed phase is described using a particle size distribution which is of significance in controlling the process since it links the product quality indices to product characteristics. Although there have been numerous studies on control of particulate processes, results on fault detection and accommodation are limited at present. Fundamental issues that arise in the design of model-based fault-tolerant control systems for particulate processes include the infinite-dimensional nature of the process models, as the complex nonlinear and uncertain dynamics of particulate processes. Efforts to address these problems were initiated in [1], [2] where a unified framework for the robust detection, isolation and handling of control actuator faults in particulate processes was developed using low-order models.

In addition to the presence of nonlinearities and uncertainty, there is a host of practical implementation issues that need to be accounted for in the design of the fault diagnosis and fault-tolerant control systems. Among these issues is the delayed arrival of discrete multi-rate measurements from the sensors. In particulate processes, the measurements of the discrete (e.g., principal moments of the particle size distribution) and continuous phase variables (e.g., solute concentration and temperature in a crystallizer) are typically available from the sensors at discrete time instances and are delayed. The control system may also make use of multiple outputs which may be subject to different sampling rates. For instance, the dispersed phase properties may be collected using light scattering techniques whereas properties of the solute concentration in the continuous phase may be obtained from a refractometer. The frequency and times at which the measurements are received by the control system are constrained by the inherent limitations on the data collection, processing and transmission capabilities of the measurement sensors. These limit the controller implementation as well as accuracy of process monitoring. Ignoring these factors in process monitoring and controller design may erode the fault diagnosis and fault handling capabilities of the fault-tolerant control system.

Motivated by these considerations, we develop in this work a model-based framework for component fault detection and accommodation in particulate processes modeled by population balance equations with delayed, discretely-sampled, multi-rate sensor measurements. Initially, model reduction techniques are used to derive an approximate finite-dimensional system that captures the dominant dynamics of the particulate process. An observer-based output feedback controller is then designed based on this system to stabilize the fault-free process. To compensate for the lack of continuous measurements, an inter-sample model predictor is included within the control system to provide the observer with an estimate of the process outputs when measurements are unavailable from the sensors. The model state is then updated when measurements are received at discrete times. Because of the dissimilar sampling rates of the process outputs, this update is performed using different combinations of outputs at a given sampling time depending on which measurements are available from the sensors. To compensate for the measurement delays, the control system includes a propagation unit that uses the low-order model together with the past values of the control input to calculate an estimate of the current output from the delayed measurements received. This estimate is then used to update the inter-sample model predictor which, together with the controller, generates the control signal for the process.

Specific conditions for fault-free closed-loop stability are obtained by formulating the closed-loop system as a hybrid discrete-continuous system that exhibits variable update patterns generated by the dissimilar sensor sampling rates. This condition is expressed in terms of the plant-model mismatch, the different sampling rates, the controller and observer design parameters, and the size of the measurement delay. The characteristic fault-free closed-loop behavior obtained from this analysis is used as the basis for deriving appropriate rules for fault detection and accommodation. The idea is to use the state observer as a fault detection filter and compare its output with the estimate of the current plant output generated by the propagation unit at the sampling times. The discrepancy is used as a residual and compared against a time-varying alarm threshold obtained from the stability analysis to determine the fault or health status of the process components. Once a fault is detected its accommodated by adjusting the controller and observer design parameters to preserve stability and minimize performance deterioration. Finally, the results are illustrated using a simulated model of a continuous crystallizer.

References:

[1] El-Farra, N. H. and A. Giridhar, "Detection and Management of Actuator Faults in Controlled Particulate Processes Using Population Balance Models," *Chem. Eng. Sci.*, 63:1185-1204, 2008.

[2] Giridhar, A. and N. H. El-Farra, "A Unified Framework for Robust Fault Detection, Isolation and Compensation in Uncertain Particulate Processes," *Chem. Eng. Sci.*, 64:2963-2977, 2009.

(446e) Optimal Sensor Placement for Fault Diagnosis in an Acid Gas Removal (AGR) Unit for an Integrated Gasification Combined Cycle (IGCC) Plant with CO2 Capture

M. R. Islam[1], Jeevan Maddala[1], Richard Turton[2], Debangsu Bhattacharyya[2] and Raghunathan Rengaswamy[1], (1)Chemical Engineering, Texas Tech University, Lubbock, TX, (2)Department of Chemical Engg., West Virginia University, Morgantown, WV

Future integrated gasification combined cycle (IGCC) power plants with CO_2 capture will face stricter environmental emission targets. The acid gas removal (AGR) unit for an IGCC plant plays a key role in satisfying the environmental constraints. However, a number of faults can occur in an AGR unit that can result in violation of the environmental constraints. For examples, fly ash present in the syngas can deposit on the trays of the absorber(s) and stripper(s). Heat exchangers may leak causing internal mixing of shell and tube side materials. An early detection of these faults can help to satisfy the emission targets, and to ensure safe operation of the unit in the presence of faults. It can also help to plan for maintenance in advance and thereby reduce the overall down-time.

In this work, a comprehensive model of a selective, dual-stage, Selexol-based acid gas removal (AGR) unit is developed in Aspen Engineering Suite (AES). Using a chilled solvent, the AGR unit captures about 90% CO_2 and more than 99.9% H2S present in the syngas. The captured CO_2 is recovered at three pressure levels and is sent for compression. The H_2S that is thermally stripped off from the loaded solvent is sent to the Claus unit.

The effect of faults gets manifested in process variables. However, it is not obvious which variables to measure for diagnosing a fault. It is neither economically viable nor practically possible to measure every process variable. A sensor placement (SP) algorithm can help to identify the number, type, and location of the sensors required for fault diagnosis.

In this work, first, a sensor placement (SP) algorithm is developed for the desired observability of the faults. For this, a set cover problem is formulated based on the Directed Graph (DG) representation of the fault propagation. The algorithm is further extended for a desired fault resolution by using a Signed Directed Graph (SDG) representation. The candidate sensors considered in this algorithm are those that can be measured with current state-of-the-art technology. Sensors with different levels of precision are considered in this algorithm. In this presentation, we will also discuss the impact of the budget for sensors on the overall fault resolution and the trade-off between precision and fault resolution. The proposed approach shows great promise for system level sensor placement for fault diagnosis in plants with a large number of possible sensor locations.

Fault Detection based on Kernel Mean Discrepancy Test

Jiusun Zeng,[†] Jingjing Sha,[‡] and Lei Xie[*,‡]

College of Metrology & Measurement Engineering, China Jiliang University, Hangzhou 310018, China, and National Laboratory of Industrial Control Technology, Zhejiang University, Hangzhou 310027, China

E-mail: leix@iipc.zju.edu.cn
Phone: +86 130-7184-7893. Fax: +86 571-8795-2279

Abstract

A fault detection method based on kernel mean discrepancy test is proposed, which deals with the fault detection problem from the aspect of non-parametric statistical test. The basic idea of kernel mean discrepancy is to project the training data and test data into the reproducing kernel Hilbert space, in which the mean discrepancy test is performed. Based on the kernel mean discrepancy test, a fault detection strategy is developed and a quantitative analysis on the sensitivity of the test statistic is presented. The confidence limit is obtained by a random sampling method and a moving window approach is used to detect process faults. Simulation and application results show that the method is more sensitive to process faults than methods based on support vector data description.

1 Introduction

Thanks to the advances in sensor technology and computer technology, large amounts of data is collected and accumulated during the operation of industrial systems, leading to a rapid development in data-driven fault detection and diagnosis (FDD) methods. Due to the capability to handle data co-linearity, multivariate statistical methods have received significant attention, e.g., techniques based on principal component analysis (PCA),[1] partial least squares (PLS)[2] and independent component analysis (ICA).[3] The general procedure of fault detection using multivariate statistical methods consists of a feature extraction step and a subsequent step of constructing monitoring statistics for process monitoring. If the data is Gaussian distributed, T^2 and SPE statistics are appropriate. However, if the variables are non-Gaussian distributed, T^2 and SPE may produce significant number of type 1 and type 2 errors. To deal with non-Gaussian processes, different methods have been developed to design monitoring statistics, such as kernel density estimation,[4] Gaussian mixture models,[5] methods based on ICA[6–8] etc. It has been reported that these methods can be problematic if the data distribution is sparse or clustered,[4] facing singularity problems[5] or causing disturbance of the monitoring performance.[7]

Provided the probability density function (PDF) of the normal data can be obtained, the task of process monitoring reduces to inspecting whether the test data are generated from the PDF of the normal data. However, direct estimation of PDF from data is computationally intensive, available methods like kernel density estimation cannot handle higher dimensional data sets. As an alternative, different kinds of two-sample test approaches have been developed in the field of machine learning and statistics to test whether two sets of samples

*To whom correspondence should be addressed

†College of Metrology & Measurement Engineering, China Jiliang University, Hangzhou 310018, China

‡National Laboratory of Industrial Control Technology, Zhejiang University, Hangzhou 310027, China

are sampled from the same distribution, where kernel mean discrepancy[9] is one of the state-in-art approach. The basic idea of kernel mean discrepancy is to project the normal and test data into the reproducing kernel Hilbert space (RKHS), in which the mean discrepancy test is performed. If the test data shares the same distribution as the normal data, the mean discrepancy would be very close to zero. Hence the mean discrepancy in the kernel space can be used as a test statistic to decide whether the normal and test data set are sampled from the same distribution.

In this article, a fault detection method based on kernel mean discrepancy test is proposed, which measures the difference between the distribution of normal and test data set. A moving window approach is firstly used to divide the test data into consecutive subsets. And the kernel mean discrepancy between the subset and the normal data set is computed. If the discrepancy is large, it is more likely that the subset contains faulty data. Hence, the key step would be to determine the critical value/ confidence limit of the mean discrepancy in the RKHS. To determine the confidence limit, a random sampling approach is performed based on the normal data set and the 95% or 99% percentile of the test statistic is used as the confidence limit. With the confidence limit obtained, a fault detection strategy is then proposed. A quantitative analysis on the sensitivity of the test statistic is further presented. The proposed monitoring strategy is nonparametric and distribution free; it is applicable to both Gaussian and non-Gaussian distributed data. Simulation and application study to the blast furnace ironmaking process show that the proposed method is more sensitive to process faults than the monitoring strategy based on support vector data description (SVDD).[10]

2 Kernel mean discrepancy as a test statistic

Kernel mean discrepancy was first proposed to deal with the two-sample test problem,[11] which tests whether two samples are drawn from the same distribution. Assume p,q to be two probability distributions defined in a compact metric space $\mathbb{R}^d$ (closed and bounded), $\mathbf{X} = \{\mathbf{x}_i\}_{i=1}^N, \mathbf{x}_i \in \mathbb{R}^d, i = 1,2,\dots,N$ and $\mathbf{Y} = \{\mathbf{y}_j\}_{j=1}^M, \mathbf{y}_j \in \mathbb{R}^d, j = 1,2,\dots,M$ to be observations drawn from p and q, the kernel mean discrepancy investigates the difference between p and q by exploiting the difference between the expectations of the samples mapped into a universal Reproducing Kernel Hilbert Space (RKHS) $\mathscr{H}$. The rationale is that, for a compact metric space $\mathbb{R}^d$, $p = q$ holds if and only if $\mathbb{E}_p[f(\mathbf{x})] = \mathbb{E}_q[f(\mathbf{y})]$ for all continuous bounded functions in $\mathbb{R}^d$. As a result, the difference between mean function values in the RKHS $\mathscr{H}$ can be used to test whether $p = q$ without direct estimation of density functions, hence significantly simplifying the task of distribution test. When the maximum difference of mean function is large, the two samples are likely to be drawn from different distributions.

More formally, let $\mathscr{F}$ be a class of functions $f : \mathscr{X} \to \mathbb{R}$, then the maximum mean discrepancy (MMD) is defined as

$$\mathrm{MMD}(p,q,\mathscr{F}) = \sup_{f\in\mathscr{F}} \left(\mathbb{E}_p[f(\mathbf{x})] - \mathbb{E}_q[f(\mathbf{y})]\right) \quad (1)$$

The empirical estimate of MMD is

$$\mathrm{MMD}(\mathbf{X},\mathbf{Y},\mathscr{F}) = \sup_{f\in\mathscr{F}} \left(\frac{\sum_{i=1}^{N} f(\mathbf{x}_i)}{N} - \frac{\sum_{j=1}^{M} f(\mathbf{y}_j)}{M} \right) \quad (2)$$

where the supremum operator denotes the least upper bound of a sets of values. From Eqn (1) and Eqn (2) it can be seen that the quantity of MMD heavily relies on the class of functions $\mathscr{F}$ it employed. Theoretically, any continuous bounded function in $\mathbb{R}^d$ can be used and allows $p = q$ to be identified uniquely. However, such a rich function class is unpractical to work with in the case of finite samples. In practice, the function class should be rich enough to uniquely identify whether $p = q$, yet restrictive enough to give useful finite sample estimates.

An ideal choice of the function class $\mathscr{F}$ is the unit ball in a RKHS $\mathscr{H}$ with associated kernel $k(\cdot,\cdot)$. With this kind of function class, the empirical MMD can be computed efficiently and it is proved that $\mathrm{MMD}(\mathscr{F},p,q) = 0$ if and only if $p = q$.[9] A common choice of the kernel function

$k(\cdot,\cdot)$ is the Gaussian kernel

$$k\left(\mathbf{x}_i,\mathbf{x}_j\right)=\exp\left(\frac{-\left(\mathbf{x}_i-\mathbf{x}_j\right)^T\left(\mathbf{x}_i-\mathbf{x}_j\right)}{\sigma}\right) \quad (3)$$

where $\sigma>0$ is the kernel parameter.

The RKHS is a complete, normed vector space endowed with a dot product $\langle\cdot,\cdot\rangle$, with its norm defined as $\|\mathbf{x}\|=\sqrt{(\mathbf{x},\mathbf{x})}$. In RKHS, the function can be written in the form of dot product as $f(x)=\langle k(\mathbf{x},\cdot),f\rangle$.[12] Hence MMD can be re-written as

$$\begin{aligned}\mathrm{MMD}(p,q,\mathscr{F})&=\sup_{\|f\|\le 1}\langle\mu_p-\mu_q,f\rangle\\&=\sup\left\|\mu_p-\mu_q\right\|_{\mathscr{H}}\end{aligned} \quad (4)$$

where $\mu_p=\mathbb{E}_p[k(\mathbf{x},\cdot)]$ is the expectation of $k(\mathbf{x},\cdot)$, $\mathbb{E}_p[f(\mathbf{x})]=\mathbb{E}_p[\langle k(\mathbf{x},\cdot),f\rangle]=\langle\mu_p,f\rangle$. In practice, the following squared form of MMD is commonly employed

$$\begin{aligned}\sup\left\|\mu_p-\mu_q\right\|_{\mathscr{H}}^2=\langle\mu_p,\mu_p\rangle_{\mathscr{H}}-2\langle\mu_p,\mu_q\rangle_{\mathscr{H}}\\+\langle\mu_q,\mu_q\rangle_{\mathscr{H}}\end{aligned} \quad (5)$$

For two sets of samples $\mathbf{X}=\{x_i\}_{i=1}^N$ and $\mathbf{Y}=\{y_j\}_{j=1}^M$ drawn from p and q, the mean function values can be determined as

$$\mu_{\mathbf{X}}=\frac{\sum_{i=1}^{N}k(x_i,\cdot)}{N} \quad (6)$$

$$\mu_{\mathbf{Y}}=\frac{\sum_{i=1}^{M}k(y_i,\cdot)}{M} \quad (7)$$

And the squared form of MMD can be determined by its empirical estimate as

$$\begin{aligned}\mathrm{MMD}^2(\mathbf{X},\mathbf{Y},F)=\sup\frac{\sum_{i,j=1}^{N}k(\mathbf{x}_i,\mathbf{x}_j)}{N^2}\\-\frac{2\sum_{i=1}^{N}\sum_{j=1}^{M}k(\mathbf{x}_i,\mathbf{y}_j)}{MN}+\frac{\sum_{i,j=1}^{M}k(\mathbf{y}_i,\mathbf{y}_j)}{M^2}\end{aligned} \quad (8)$$

In two-sample test problem, MMD^2 is used as the test statistic to determine whether $\mathbf{X}$ and $\mathbf{Y}$ are sampled from the same distribution. It is shown in ref.[9] that under the assumption that $M=N$ the empirical estimate in Eqn (8) is unbiased. If $M\neq N$, the estimate is biased, however, with an upper bound on the deviation between the true value and the expected value of empirical estimate. Since the MMD statistic involves kernelization, we call it kernel mean discrepancy (KMD) here. The advantages of kernelization are obvious, in the RKHS even some linear statistics can capture the potential nonlinear relationship in original space.[13] The KMD test directly compares the data points; it is applicable to arbitrary data distributions since no density estimation is involved. Furthermore, KMD has a light computation load, with a computation complexity of $O(N^2)$.

3 Fault detection strategy based on kernel mean discrepancy

Considering the advantages of KMD in comparing data distributions, we now develop a monitoring strategy by using KMD as a test statistic. The monitoring strategy is consisted of two steps, i.e., developing the normal operation condition model and online monitoring.

3.1 Developing the normal operation condition (NOC) model

Let $\mathbf{X}=\{\mathbf{x}_i\}_{i=1}^N$ denote the normal sample set collected and stored in the database. Assume we have a test sample set with M samples which is denoted as $\mathbf{Y}=\{\mathbf{y}_j\}_{j=1}^M$, the KMD statistic can be easily computed using Eqn (8). Theoretically, the value of KMD would be zero if and only if $\mathbf{X}$ and $\mathbf{Y}$ are sampled from the same distribution. If KMD is not equal to zero, they are from different distributions. This can be translated into the following hypothesis test

$$\begin{aligned}&H_0:\ \mathbf{X}\text{ and }\mathbf{Y}\text{ are sampled from}\\&\text{the same distribution, such that }\mathrm{MMD}^2=0;\\&H_1:\ \mathbf{X}\text{ and }\mathbf{Y}\text{ are sampled from}\\&\text{different distributions, such that }\mathrm{MMD}^2\neq 0.\end{aligned} \quad (9)$$

However, in practice, the KMD would not be equal to zero even if $\mathbf{X}$ and $\mathbf{Y}$ are drawn from the same distribution due to the effect of sampling. Hence a confidence limit should be determined before using the hypothesis test for fault detection. Ref.[9]

gave the confidence limits for the case of $M = N$ under the assumption that the data are Gaussian. The confidence limit for the case of $M \neq N$, however, cannot be determined. Here we propose a random sampling method to obtain the confidence limit from the normal data set; the confidence limit is applicable to both Gaussian and non-Gaussian data. The random sampling method determines the distribution of the test statistic under the null hypothesis by sampling from the normal sample set, which is performed as follows.

1. Randomly select $\frac{N}{2}$ samples from the normal data set $\mathbf{X} = \{\mathbf{x}_i\}_{i=1}^{N}$ as the reference set; denote the reference set as $\widehat{\mathbf{X}}$ and the remaining samples as $\tilde{\mathbf{X}}$;
2. Randomly select M samples from $\tilde{\mathbf{X}}$ and denote it to be $\tilde{\mathbf{Y}}$; compute the MMD^2 statistic between $\widehat{\mathbf{X}}$ and $\tilde{\mathbf{Y}}$ using Eqn (8);
3. Repeat step 2 T times until T values of MMD^2 are obtained;
4. Determine the critical value CV_δ as the $100 \times (1-\delta)$ percentile of the T values of the MMD^2 statistic, where δ is commonly selected as 95% or 99%.

3.2 Online monitoring

In the context of fault detection, a large number of normal samples can be collected and stored whilst the test sample is collected online sequentially, i.e., one by one. Ideally, it is desired that each sample collected online can be compared with the normal set to determine whether it is a normal or faulty sample by checking the KMD statistic. However, this is not feasible since KMD deals with the two sample test problem by comparing the distributions of two sample sets, which requires both the normal and test sample set having enough samples. In this article, we use a window-based approach to construct the test set by including the latest sample and and its past $M-1$ samples. The KMD statistic can then be computed through Eqn (8) and compared with the confidence limit; if MMD^2 is greater than the confidence limit, the test set contains faulty samples. More specifically, assume the confidence limit has been obtained from the normal sample set $\mathbf{X} = \{\mathbf{x}_i\}_{i=1}^{N}$, the online monitoring consists the following steps

1. Collect the new sample $\mathbf{x}_{N+1}$ and construct the test set corresponding to $\mathbf{x}_{N+1}$, denote it to be $\mathbf{X}_{N+1} = \{\mathbf{x}_k\}_{k=N-M+2}^{N+1}$;
2. Compute the KMD statistic between $\widehat{\mathbf{X}}$ and $\mathbf{X}_{N+1} = \{\mathbf{x}_k\}_{k=N-M+2}^{N+1}$ using Eqn (8) and compare with the confidence limit CV_δ. If the statistic is larger than CV_δ, then faulty condition arises;
3. Collect the sample at the next time instance, construct the corresponding test set by including the latest sample and removing the earliest sample;
4. Compute the KMD statistic and compare with the confidence;
5. Repeat step 3 through step 4;

The number of samples in the test set (window length) should be great enough to ensure good empirical estimation; on the other hand, it should not be too great to avoid large delay in the process of fault detection.

4 Simulation Example

In this section, a simulation example involving a non-Gaussian process is considered. For the non-Gaussian process, the support vector data description (SVDD) based monitoring strategy[10] is used for comparison. Consider a process with 5 variables, which are linear combination of 3 non-Gaussian source signals. The 3 source signals are generated as follows

$$\begin{cases} s_1(k) = 0.2\cos(0.08k)\sin(0.006k) \\ s_2(k) = 0.8\cos(0.1k) - 0.6\sin(0.05k) \\ s_3(k) \sim \mathcal{N}(0,\ 0.25) \end{cases} \quad (10)$$

where k is the sampling time. Based on the 3 source signals 5 process variables are generated as $\mathbf{z} = \mathbf{A}\mathbf{s}$, $\mathbf{s} = [\mathbf{s}_1, \mathbf{s}_2, \mathbf{s}_3]$. The mixing matrix $\mathbf{A}$ is as follows

$$
A = \begin{bmatrix} 0.86 & 0.79 & 0.67 \\ -0.55 & 0.65 & 0.46 \\ 0.17 & 0.32 & -0.28 \\ -0.33 & 0.12 & 0.27 \\ 0.89 & -0.97 & -0.74 \end{bmatrix} \quad (11)
$$

The process data is contaminated by a set of Gaussian-distributed noise $\hat{\mathbf{z}} = \mathbf{As} + \varepsilon$, with $\mathbb{E}(\varepsilon) = 0$ and $\mathbb{E}(\varepsilon\varepsilon^T) = 0.0025\mathbf{I}$. Using the above model, 2000 samples representing normal operation conditions are generated. For the non-Gaussian process, ICA is performed to extract 3 non-Gaussian components. The NOC model is then developed for both the extracted non-Gaussian components and residual components using the KMD statistic. The KMD statistic relies on the choice of kernel function and hence the kernel parameter in Eqn (3). A heuristic method by setting σ as the median distance between samples is proposed and used in ref.[9] Recently, an idea of using σ yielding the maximum KMD statistic is proposed by ref.[14] and confirmed by ref.[15] yielding a better result than the heuristic method. In this paper, we used a grid search approach to find the σ that maximizing the average KMD. A window length of $M = 50$ is set and the 99% percentile is used as the confidence limit. To test the performance of the proposed monitoring strategy on the non-Gaussian process, a test set with 2000 samples are generated and tested, with the first 1000 samples from normal condition and the remaining 1000 samples representing the fault conditions. The faulty condition of a sensor bias with the magnitude of 0.4 is considered.

For the sensor bias, Figure 1 shows the monitoring results of the extracted non-Gaussian components under the sensor fault. It is shown in Figure 1 that the fault is successfully detected by the MD^2 statistic. No false alarms is observed for the first 1000 samples and very significant violations are observed for the 1000 faulty samples. In contrast, Figure 2 gives the monitoring results using SVDD. For SVDD, the kernel width is set as $\sigma = 1000$ by trial and error and the significance level is also set as 99%. It is shown in Figure 2 that the sensor fault is also detected by SVDD. However, several false alarms can be observed and a large

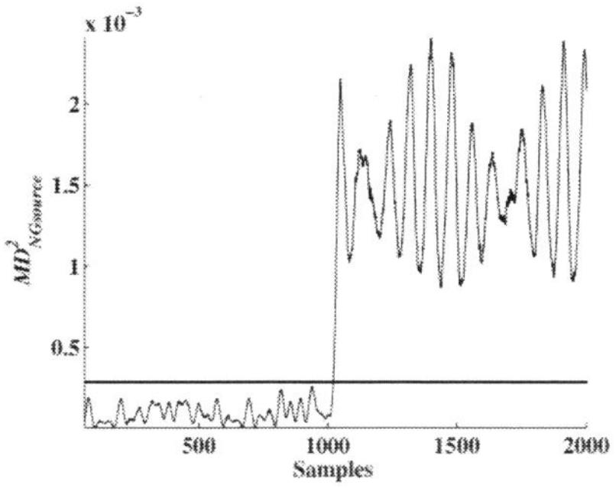

Figure 1: Monitoring results for the non-Gaussian components using KMD

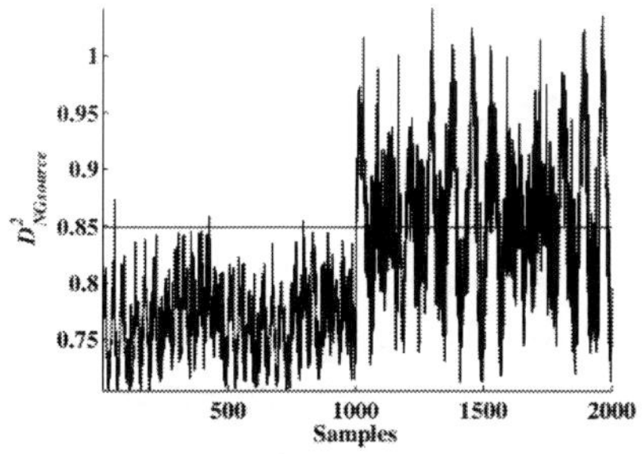

Figure 2: Monitoring results for the non-Gaussian components using SVDD

number of missing detection can be observed. For monitoring of the residual components, the KMD statistic as well as the SPE statistic are tested and compared, as is shown in Figures 8 and 4. Figures 8 and 4 show a similar picture. Again, there is no false alarm for the first 1000 normal samples and significant violations can be observed for the faulty samples. On the other hand, the SPE statistic presents several false alarms and many missing detections. This clearly shows that the KMD outperforms the SVDD based monitoring strategy

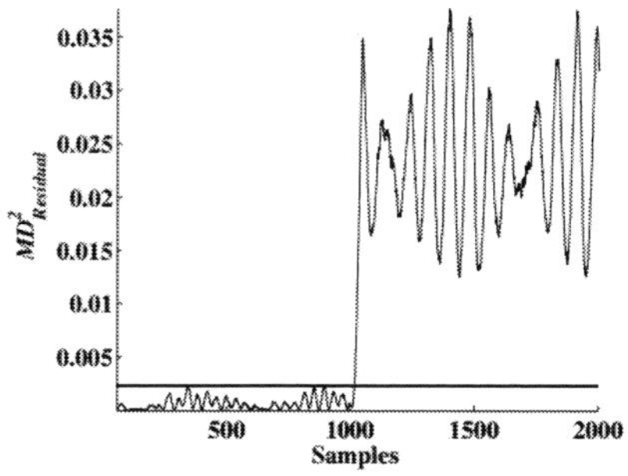

Figure 3: Monitoring results for the residual components using KMD

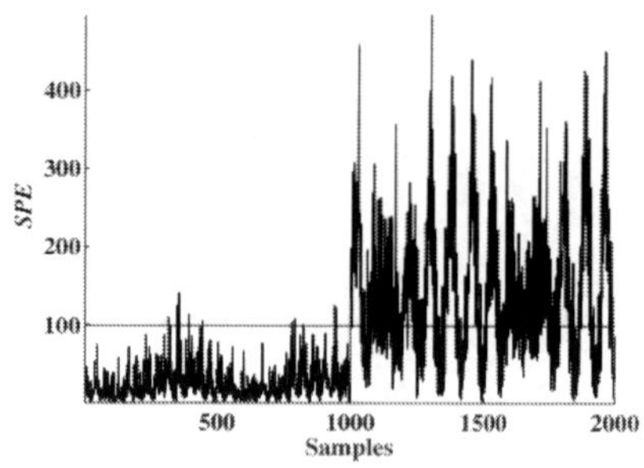

Figure 4: Monitoring results for the residual components using the SPE statistic

with regard to less false alarms and higher detection sensitivity.

5 Application Study

This section examines the performance of the proposed monitoring strategy by applying it to fault detection of the blast furnace ironmaking process. The blast furnace is a kind of shaft furnace used to produce molten iron from iron ore. In a blast furnace, raw materials like iron ore and coke are dumped from the top of the blast furnace layer by layer, while hot air (sometimes with oxygen enrichment) and coal power are blasted into the furnace from the bottom. The raw materials are heated during the downward movement and a series of chemical reactions take place throughout the furnace. The end products include molten iron and slag tapped from the bottom and the flue gases exiting from the top.

During the operation of the blast furnace ironmaking process, it is important to keep stable gas flows to facilitate smooth production of molten iron. A typical fault involving the gas flow is the hanging fault. The hanging fault usually results when the upward gas is restricted by excessive fines or the melted iron ore causes the bosh to bridge over.[16] If the hang is permitted and uncorrected, the top pressure and top heat of the furnace would increase. Eventually, the top structure can be seriously damaged. In this paper, we consider a hanging fault happening in a real blast furnace with an inner volume of 2500 m^3 in China. A total of 8 variables are considered and listed in Table 1. During the hanging fault, the quantity of blast and H_2 concentration in top gas decrease, the top pressure, CO_2 and CO concentration in top gas increase. For the purpose of fault detection, a training set with 2000 samples are collected under normal operation and the test set with 1080 samples are considered, with the last 580 samples consisting of faulty samples. Similar to the simulation example, ICA is first used to extract 5 non-Gaussian sources from the process data. The NOC model is then constructed for both the source signals and the residuals. Again, the kernel width is determined by a grid search and the window length is set as $M = 50$ and 99% percentile is used as the confidence limit. Figure 5 shows the monitoring results for the extracted non-Gaussian components. It can be seen that for the first 500 normal samples, no violation of the confidence limit is observed; while for the last 580 faulty samples, significant number of violations are observed, hence the fault is successfully detected.

Table 1: Selected process variables for monitoring of the hanging fault.

Process variables	Variable description
u_1	quantity of blast
u_2	blast temperature
u_3	blast pressure
u_4	top pressure
u_5	oxygen enrichment
u_6	CO_2 concentration in top gas
u_7	CO concentration in top gas
u_8	H_2 concentration in top gas

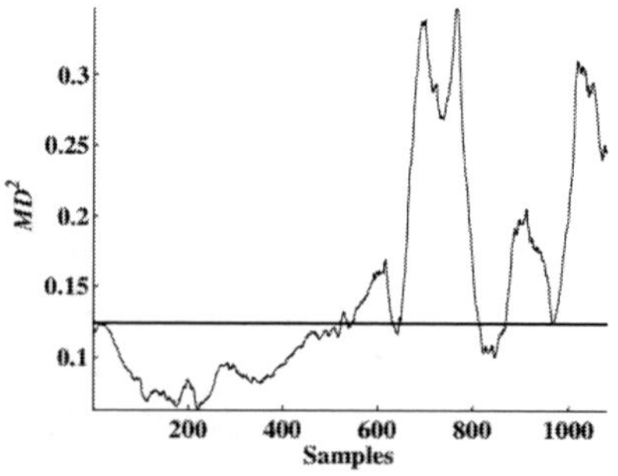

Figure 5: Monitoring results for the non-Gaussian components using KMD for the hanging fault

On the other hand, Figure 6 presents the monitoring results using SVDD. For SVDD, the kernel width is determined as $\sigma = 10$ by trial and error and the confidence level is set as 99%. Figure 6 shows that the hanging fault is also successfully detected by SVDD, however, with significant number of missing alarms. Comparing Figures 5 and 6 it can be seen the proposed KMD based monitoring strategy has higher sensitivity than SVDD.

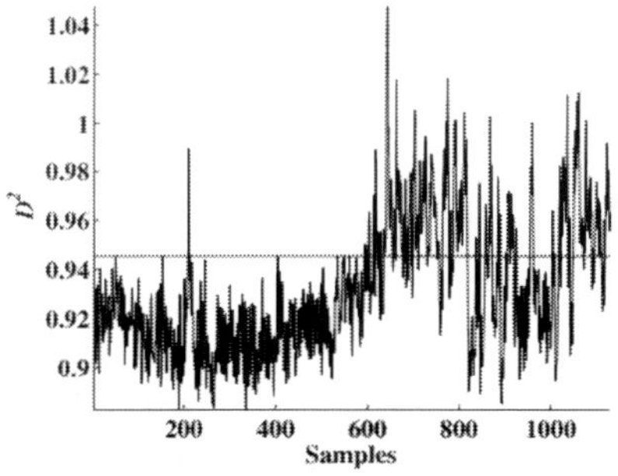

Figure 6: Monitoring results for the non-Gaussian components using SVDD for the hanging fault

Figure 7 and Figure 8 present a similar picture. It can be seen that the fault is successfully detected by KMD and significant number of violations can be observed. On the other hand, SVDD can also detect the fault, however, with lower sensitivity and many missing alarms can be observed. This further proves the effectiveness of the proposed monitoring strategy.

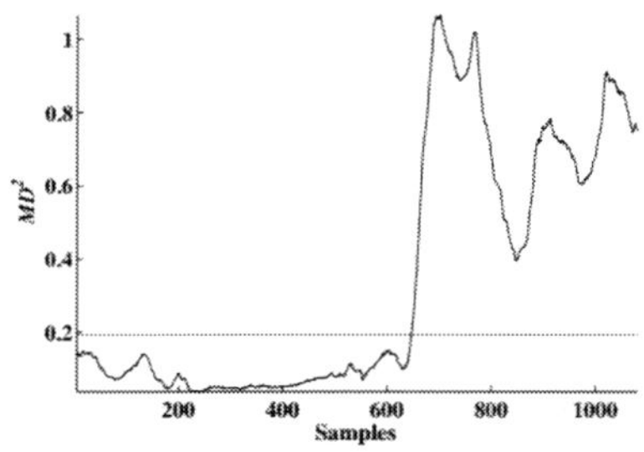

Figure 7: Monitoring results for the residual components using KMD for the hanging fault

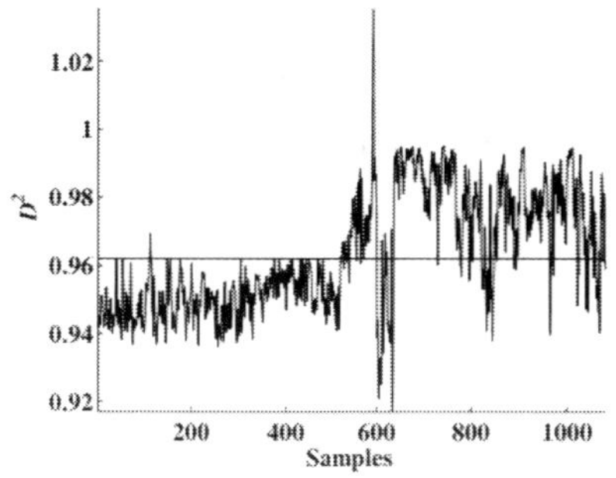

Figure 8: Monitoring results for the residual components using SVDD for the hanging fault

6 Concluding summary

This paper proposes to monitor process faults using the kernel mean discrepancy approach, which compares the test samples and normal samples from the aspect of PDF. The normal and test samples are projected into the RKHS and the mean functions are then compared. If the difference in the mean functions is large, the test samples are likely to contain faulty samples. A quantitative analysis on the sensitivity of the KMD statistic is then presented. A random sampling approach is developed to obtain the confidence limit of the statistic. Based on the confidence limit, a moving window approach is then developed for fault detection. The monitoring statistic does not rely on any distribution assumption and hence can be applied to both Gaussian and non-Gaussian processes. Application studies involving a simulation example and a blast furnace fault are then used to test the performance of the proposed monitoring strategy. The results shows that the proposed approach has higher sensitivity than the SVDD based approach.

References

[1] Tharrault, Y.; Mourot, G.; Ragot, J.; Maquin, D. *International Journal of Applied Mathematics and Computater Science* **2008**, *18*, 429–442.

[2] Kruger, U.; Dimitriadis, G. *AIChE Journal* **2008**, *54*, 2581–2596.

[3] Ge, Z. Q.; Song, Z. H. *Industrial and Engineering Chemistry Research* **2007**, *46*, 2054–2063.

[4] Chen, Q.; Kruger, U.; Leung, A. Y. T. *Control Engineering Practice* **2004**, *12*, 267–274.

[5] Yu, J.; Qin, S. J. *AIChE Journal* **2008**, *54*, 1811–1829.

[6] Kano, S.; Tanaka, S.; Hasebe, S.; Hashimoto, I.; Ohno, H. *AIChE Journal* **2003**, *49*, 969–976.

[7] Lee, J. M.; Yoo, C. K.; Lee, I. B. *Journal of Process Control* **2004**, *14*, 467–485.

[8] Lee, J. M.; Qin, S. J.; Lee, I. B. *AIChE Journal* **2006**, *52*, 3501–3514.

[9] Gretton, A.; Borgwardt, K. M.; Rasch, M. J.; Scholkopf, B.; Smola, A. *Journal of Machine Learning Research* **2012**, *13*, 723–773.

[10] Liu, X. Q.; Xie, L.; Kruger, U.; Littler, T.; Wang, S. Q. *AIChE Journal* **2008**, *54*, 2379–2391.

[11] Devore, J. L. *Probability and statistics for engineering and the sciences, 8th edition*; Brooks/Cole Publishing: Pacific Grove, 2011.

[12] Steinwart, I. *Journal of Machine Learning Research* **2001**, *2*, 67–93.

[13] Joachims, T. *Support vector and kernel methods*; 2003.

[14] Sriperumbudur, B.; Fukumizu, K.; Gretton, A.; Lanckriet, G.; Scholkopf, B. Kernel choise and classifiability or rkhs embeddings of probability distributions. 2009.

[15] Sugiyama, M.; Suzuki, T.; Itoh, Y.; Kanamori, T.; Kimura, M. *Neural Networks* **2011**, *24*, 735–751.

[16] Strassburger, J. H. *Blast furnace theory and practice*; Taylor and Francis: London, 1969.

(461a) Proactive Fault-Tolerant Model Predictive Control

Liangfeng Lao, Chemical Engineering, UCLA, Los Angeles, CA, Panagiotis D. Christofides, Department of Chemical and Biomolecular Engineering, University of California, Los Angeles, Los Angeles, CA and Jim Davis, Chemical Engineering, University of California - Los Angeles, Los Angeles, CA

Fault-tolerant control methods have been extensively researched over the last ten years in the context of chemical process applications and provide a natural framework for integrating process monitoring and control aspects in a way that not only fault detection and isolation but also control system reconfiguration is achieved in the event of a process or actuator fault. The motivation for this broad research direction is provided by the strong industrial need to handle process and control system faults effectively before they turn into major accidents. Most of the work in the area of fault tolerant control in the last few years is in the context of active fault tolerant control where a fault tolerant control system is designed to detect and isolate the fault and then automatically reconfigure the control system to maintain operation as close as possible to optimal conditions with the remaining control actuators.

In this paper, we embark into a new direction in the area of fault tolerant control; namely, we try develop algorithms that calculate in real-time from process data and manufacturer guidelines the probability of a process control system component to fail, and in the event this probability exceeds a certain threshold we proactively determine how to reconfigure the control system to guide the process state in an operational mode that allows to fix or replace the component that is about to fail. We formulate and solve this problem in the context of nonlinear chemical process models and use model predictive control as the feedback design technique. We use results from stochastic stability to assess the stability of the closed-loop system. We will use chemical process examples to evaluate the effectiveness of the approach using Monte-Carlo simulations.

(461b) Hierarchical Activity Model Using Integrated Definition Language (IDEF0) for Generation of Green Retrofit Alternatives

Alireza Banimostafa[1], Yasunori Kikuchi[2], Stavros Papadokonstantakis[1], Hirokazu Sugiyama[1], Masahiko Hirao[2] and Konrad Hungerbühler[1], (1)Institute for Chemical and Bioengineering, Swiss Federal Institute of Technology, Zurich (ETHZ), Zurich, Switzerland, (2)Department of Chemical System Engineering, The University of Tokyo, Tokyo, Japan

In chemical industry owing to its capital intensive nature, retrofit design is an important key strategic decision. However, each sector of industry tends to act upon alternative generation by different approaches, therefore a standardized procedure as a business model is vital for retrofit design. Besides, due to high level of complexity, for practical implementation it is also advantageous to clarify the actual procedure of alternative generation by information flows. The objective of this paper is to address retrofitting actions to enhance both monetary and non-monetary aspects of chemical process design. This is done by presenting a general template of type zero method of Integrated Definition Language (IDEF0) for identification of key activities. The developed IDEF0 method was based on a systematic Path Flow Decomposition (PFD) method for generation of design alternatives. In this study a set of Environmental, Health, and Safety (EHS) hazard assessment, Life Cycle Impact Assessment (LCIA) and economical indicators were used in order to exploit both monetary and non-monetary aspects of process design.

The above set of indicators has been associated with a table of heuristics based on EHS, LCIA, and cost as well as rules of thumb and established principles of green chemistry and engineering. The IDEF0 method is a useful framework to connect and visualize the aforementioned activities with information flows with a focus on green process alternative generation. Therefore, several activities were decomposed to sub-activities in a hierarchical manner in order to clarify exact activity definition, information, models, tools, and constraints. These step by step activities systematized the generation of retrofit alternatives, and customized and supported the appropriate implementation of alternatives in reality. Based on incoming information, resources, and constraints the outcome of alternative generation activity was a set of feasible options for improving process design. Furthermore, screening design alternatives as the next activity could bring sufficient understanding regarding constraints and alternative candidates for the multi-objective decision making.

The developed methodology was applied to an existent batch production plant from pharmaceutical industry for production of an essential intermediate product. In pharmaceutical industry generating alternative design is getting especially important as a response to studies drawing attention to strong contribution of solvents for API manufacturing. Currently, the plant is running under solvent incineration, however, besides its EHS hazard potentials; the incineration of solvents is the main contribution of the fine chemical industry to the greenhouse effects. The IDEF0 activity model paved the way for green alternative generation by PFD methodology and mainly preserved for reduction in solvent use, seeking for novel recovery scenarios, use of alternative green solvents with reduced emissions and hazard potentials (both for reaction and separation system) for overall process improvements compared to base case of incineration.

(461c) Use of Workflows in Managing Experimental Data

Girish Joglekar[1], Arun V. Giridhar[1], Linas Mockus[2], Juan G. Osorio[3], Fernando J. Muzzio[3] and Gintaras V. Reklaitis[4], (1)School of Chemical Engineering, Purdue University, West Lafayette, IN, (2)Purdue University, West Lafayette, IN, (3)Chemical and Biochemical Engineering, Rutgers, The State University of New Jersey, Piscataway, NJ, (4)Chemical Engineering, Purdue University, West Lafayette, IN

Systematic capture and storage of experimental data is critical in any research activity. Even today, experimental data are laboriously recorded by hand in lab notebooks. Sophisticated e-lab notebooks, and subsequently, laboratory information management systems (LIMS), have evolved over the past 10 years, but are not wholeheartedly embraced by the research community due to a variety of reasons, price of commercial software packages and resistance to change being the main ones. Due to the persistence of old techniques, the recorded data are either not machine readable (if entered in a notebook), or the data provenance is not properly recorded. This significantly jeopardizes the long term use of the data, for either reproducibility or verification, as well as the ability to share it among peers or do data mining at a later date. More systematic and affordable techniques are needed to capture all the information about experimental data in a semantically rich format.

An interactive data entry and retrieval system was developed at Purdue for managing the information generated during the dry powder blending experiments performed at Rutgers. The key component of the system is the workflow for the associated experiment. A workflow captures the operational aspect of a work procedure: how tasks are structured, who performs them, what their relative order is, how they are synchronized, how information flows to support the tasks and how tasks are being tracked. A workflow model provided the reference points for defining the data to be recorded, and designing the data entry forms. Each recorded experiment is an instance of its associated workflow. All the instances of a workflow provide the handles for all data retrieval and processing functions.

The experimental data is stored on the *pharmaHUB* server hosted at Purdue in normalized tables in a MySQL database, and can be retrieved using a collection of views. A graphical user interface provides a concise, pictorial representation of the underlying standard procedure, with data entry done through icon specific forms. In this paper, we will discuss the details of the system and its use at Rutgers.

Conceptual Design of Biorefineries using Agrowastes, its Decision Support and Design Rationale

Arturo Sánchez[1], René Bañares-Alcántara[2], Julian Hunt[2], José de Jesús Rodríguez[1] and Gabriela Magaña[1]; [1]CINVESTAV, Zapopan, México; [2] University of Oxford, Oxford, UK.

Introduction

This work explores the role of OUTDO, a decision support system, in the conceptual design of multipurpose-multiproduct plants producing lignocellulose-based biofuels and bioproducts. These plants are gaining importance as part of a paradigm change in the energy and process industries. In particular, the work addresses the process of designing in an academic environment a class of plants known as biorefineries, which employ agro-wastes and enzyme-based technologies to produce ethanol and bioproducts. OUTDO provides the basis to structure the information used and generated during the design process. This includes the design strategy and the documentation of the design process and of decision making, e.g. the selection of a model. A design step evolving a multiproduct plant into a multiproduct facility is employed to illustrate OUTDO's functionalities. The decision step is rather simple and has the purpose of showing the potential use of OUTDO as a design support and rationale documentation tool.

The paper introduces in the next section the context within which the conceptual design is taking place and the motivations to use OUTDO. The following section presents a brief introduction to OUTDO, highlighting those functionalities of interest for the conceptual design of chemical plants. Of particular relevance is the definition of *artefact*, comprising the relevant information of a design. Next, the use of OUTDO is shown within a design step evolving a multiproduct plant into a multipurpose facility. Decisions for choosing a design were made only on economic grounds. Finally, the last section closes with a discussion of some relevant issues to be considered in future work in order to employ OUTDO as a tool for design support and rationale documentation in this project. In particular, the inclusion of decision criteria based on environmental and social aspects as part of the Life Cycle Assessment (LCA).

Conceptual Design of Biorefineries for Mid-Size Economies

The conceptual design of this type of biorefineries is part of an inter-institutional research project currently under execution in Mexico. The project objective is to develop the concept of a biorefinery using agro/industrial wastes as feedstock within the context of a mid-size economy. Ten geographically distributed research groups participate in the project executing work packages related to each processing stage of a biorefinery and its design. They include pretreatment, saccharification, fermentation, separation, biogas, hydrogen and electricity co-production, enzyme enhancement, synthesis of lignine-based products, plant design and Life Cycle Assessment (LCA) of the biorefinery concept. As part of the conceptual plant design work package, several designs have been completed in the past four years using standard techniques and tools. Most of the process technologies employed can be found in the open literature and many of the process conditions for the feedstock conditioning, pre-treatment, hydrolysis and fermentation steps have been corroborated at a laboratory scale by the corresponding groups participating in the project. Each one of the conceptual designs has been modeled and analyzed: steady–state models were solved using commercial simulators, the economic assessment was carried out according to standard profitability analysis procedures, and the environmental impact assessment is being currently carried out with LCA employing conventional assessment methods that involve the use of heterogeneous criteria (qualitative and quantitative).

During the execution of the project new data has been produced that either modifies the existing process functionality or introduces new aspects to be considered. New stakeholders have been also incorporated along the duration of the project. Thus, conducting the conceptual design of the biorefinery has become a highly dynamic and *sui generis* activity. It is for these reasons that the need of a flexible environment to keep track of the technical issues associated with the design and the design process itself was identified.

In the context of this work an artefact is the set of information that describes a biorefinery conceptual design. Table 1 presents the description of the artefact components. The information contained in an artefact is stored in multiple files of different types (spreadsheets, word processor files, image files, etc.).

Table 1. Artefact components.

Component	Definition	Type of information
Chemical process flowsheet	- General process flowsheet - Detailed flowsheets	- Graphic files (JPG, PDF, etc.)
Production process description	- Brief description of the main purposes of each process stage	- Text files
Mass balance information	- Equations - Equipment design and operation parameters - Mass balance results	- Design package files (Super Pro, Aspen Plus, etc.) - Spreadsheet files (Excel)
Energy balance information	- Energy requirements - Production capacity of electricity or thermal energy	- Design package files (Super Pro, Aspen Plus, etc.). - Spreadsheet files (Excel)
Economic information	- Investment and operation costs - Economic and financial information - Results of economic studies	- Spreadsheet files (Excel)
Environmental information	- Indicators and metrics	- Spreadsheet files (Excel)
Miscellaneous	- Any other relevant information	- Graphic files (PDF, JPG, etc.)

OUTDO, A Decision Support System for Conceptual Design

OUTDO (Oxford University Tool for Decision Organisation) provides several functionalities related to decision making and to information and knowledge management that can be used to support the rationale and documentation of the conceptual design process. These functionalities include:

a) access to a library of multi-criteria decision analysis methods (MCDA) to improve decisions;
b) maintenance of the design history for future analysis and re-use;
c) explicit representation of the decision rationale, thus improving communication amongst design participants and allowing the exploration of the impact of changes in the technological, environmental, economic and social conditions on previous decisions (or even the impact of the change of a given decision on other decisions!);

d) integration of probabilistic forecasting methods with the MCDA methods to evaluate alternatives at different points in time (present and future);
e) operation as a repository of design models, data and documents, i.e. acting as a "glue" binding disparate pieces of information.

OUTDO is based on Compendium v.1.5.2 (a hypertext dialog mapping tool). It uses a decision tree structure to represent the design rationale [1]. Some of the concepts used in the representation are given in Table 2. In its graphical representation, *issues* are depicted as question mark icons, *alternatives* as light bulb icons and *recommended alternatives* as a light bulb with a check mark. Figure 3 shows the design rationale of the example considered in this paper. OUTDO provides mechanisms to define quantitative criteria to support the selection of an alternative amongst competing ones. Multiple criteria can be taken into account through the use of weighted sum models (WSM) or other more complex methods (e.g. ELECTRE). *Global Parameters* are OUTDO internal variables or expressions with numerical or logical values that can be declared locally or imported from external sources; they are used to adapt dynamically the decision process to varying parameters, for example, to analyse how changes in oil prices would affect the whole decision process. Though OUTDO provides recommendations, a decision can only be taken by the user.

Table 2. Terms used in OUTDO for the decision rationale.

Term	Definition
Issue	A question to be answered in a decision process
Argument	Constructs behind a decision and their relations (supports, refutes, qualifies)
Alternative or Option	Possible solutions to an issue. Individual alternatives and their relations (component of, incompatible, specializes)
Criteria	Judgment elements used to evaluate the alternatives
Evaluation	Rank of alternatives based on the degree of satisfaction of the criteria
Artefact	Anything created by a human that does not exist in nature, including abstract things such as plans

Design Rationale of Biorefineries using OUTDO

The first design in the process is referred as Artefact 1; the solid line part of Figure 1 shows its block flow diagram. The rest of the artefact components, as established in the previous section, are not included due to space restrictions. The process model was created in SuperPro Designer v8.5. The interested reader can find a detailed description of the process in [2]. Currently, the environmental information has not yet been included and the LCA is under way. It is important to mention that Artefact 1 did not include an energy balance in the first instance; the assumption was that the energy generated internally was enough to fulfil its energy requirements.

According to Figure 1, the first stage is a three-step continuous thermo-chemical pre-treatment (5:1 H_2SO_4 0.75% v/v aqueous solution:feedstock) that soaks and cooks the feedstock. Inhibitors in the liquid hidrolisates are eliminated by overliming. A 92% conversion of hemicelluloses is achieved (80% to xylose, 12% to arabinose) and only 3% from cellulose to glucose. The output stream is fed to the saccharification-fermentation stage in which hemicelluloses and celluloses are transformed enzymatically into sugars and then fermented. A 2%-alcohol stream is sent to the separation stage producing 99.5% ethanol. Biogas is produced as a by-product in the water treatment stage. Cogeneration is considered, by burning the lignin residues.

Artefact 2, which adds a Dark Fermentation block to Artefact 1 (Figure 1), represents a coproduction scheme of ethanol, hydrogen as well as acetic and butyric acids (i.e. a multipurpose plant). The pre-treatment stage was modified to use a batch pressure cooking process (again, 5:1 H_2SO_4 0.75% v/v aqueous solution:feedstock). The resulting liquid hidrolisates are sent to a dark fermentation stage (thermophilic bacteria, 60 °C, 25 hours residence time) that produces hydrogen and CO_2 (2:1 molar ratio) (2.23 mol H_2/ mol pentose). Acetic and butyric acids are the co-products sent to the water treatment stage for bio-gas production. A six-fold increment of biogas is achieved compared against the ethanol-only production scheme. The anaerobic reactor converts 93% of the organic contents to biogas (CH_4 and CO_2, 3:1 molar ratio) with mesophilic bacteria, at 25-45 °C and 6 hours residence time. The residual stream goes to the aerobic reactor where almost all the organic content is removed. Details of this plant are presented in [3].

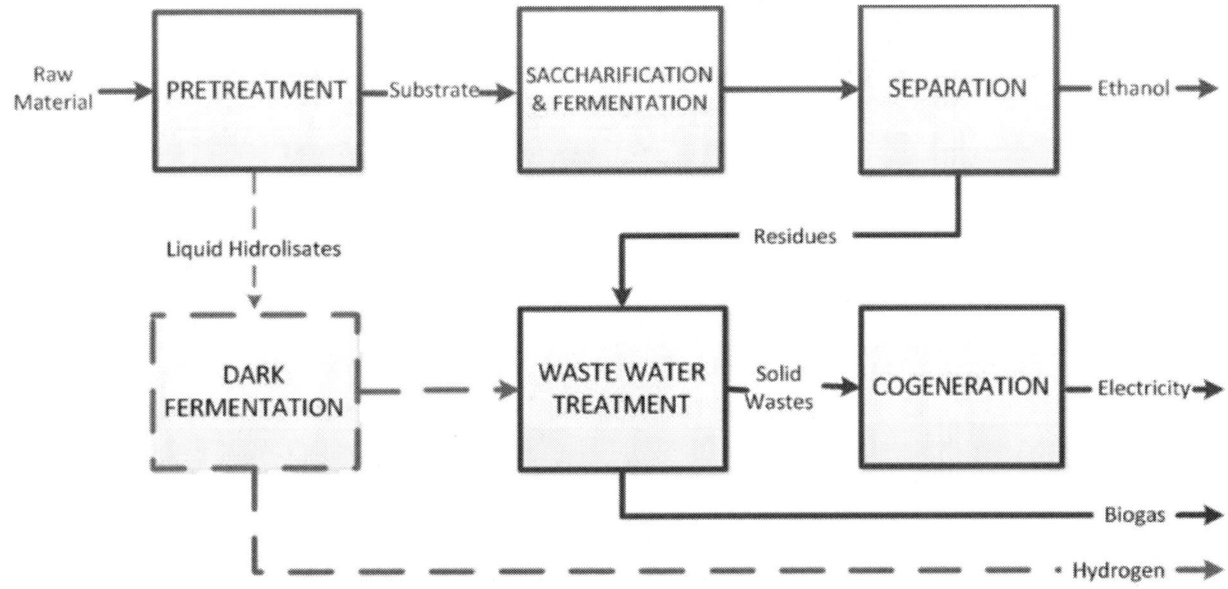

Figure 1. Biorefinery block flow diagram. Artefact 1 in solid lines; Artefact 2 adds the Dark Fermentation process.

The main reason for changing the pre-treatment technology was that the continuous technology is subject to patent royalties. Testing and validating the operating conditions was not possible due to the lack of equipment at laboratory scale. The introduction of the hydrogen production stage was made because of a collaboration agreement with a research group that developed novel genetically modified microorganisms enhancing the production of hydrogen and biogas production promoters together with the pilot scale reactors demonstrating the technical feasibility of this process.

Super Pro Designer 8.5 was employed to work out the plant mass balances, equipment sizes and capital investment. An economic comparison of these two artefacts was made calculating capital investment, cost contributions per stage and total production cost of ethanol as a function of plant capacity and feedstock price. Polysaccharide concentration of feedstock was established as a positive monotonic function of price and was adjusted to local market conditions. Typical results are shown in Figure 2 and Table 3. Note that the lowest production costs were obtained for coproduction schemes using cheap feedstock (i.e. 0.42 $/l EtOH @2,100 ton/day DB (dry basis)). However, for capacities below 1,000 ton/day DB using the ethanol production scheme may be more attractive from an economic point of view under the established conditions. Capital investment and patent royalties had a significant impact on the decision to discard the use of the continuous pretreatment. However, expected improvements in technology (not to mention the additional consideration of other criteria such as environmental, financial or political) may modify the economic conditions thus shifting this decision.

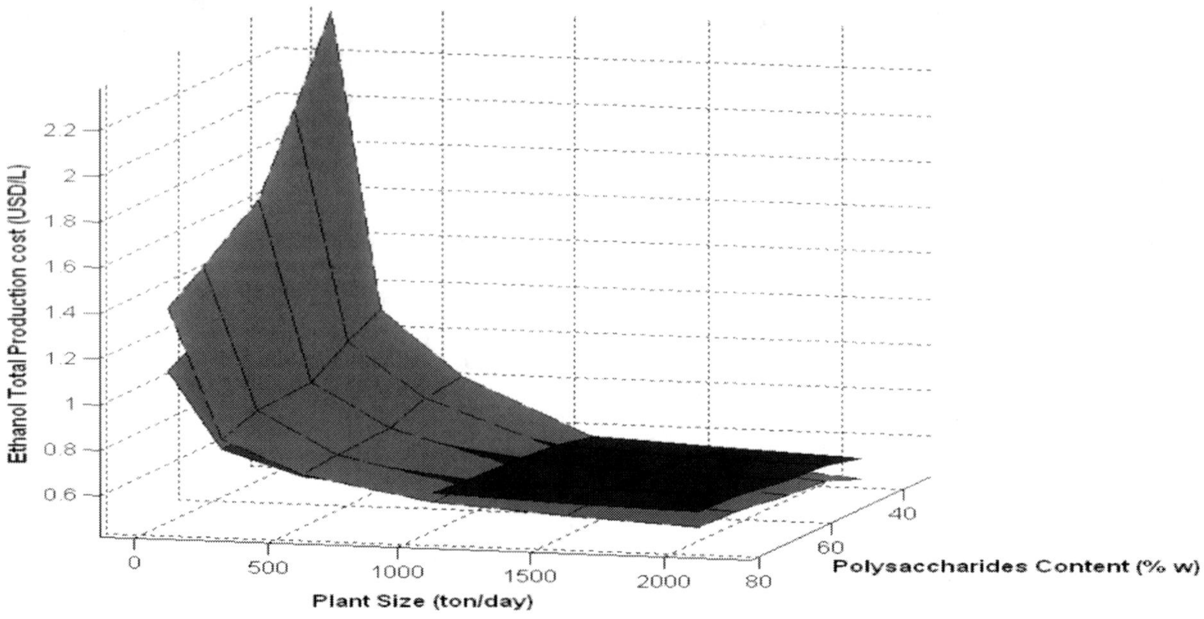

Figure 2. Ethanol total production costs. Enhanced coproduction (i.e. ethanol, biogas and hydrogen) in green and ethanol-biogas in blue.

Table 3. Boundary ethanol total production cost. These correspond to the Global Parameter values in OUTDO.

Plant capacity (ton/day)	Feedstock composition (polysaccharide content)	Production cost (US $/l EtOH)	
		Ethanol only plant (A1)	Ethanol, H_2 and biogas plant (A2)
100	35%	1.80	2.22
	80%	1.14	1.34
2,100	35%	0.51	0.42
	80%	0.61	0.54

These preliminary results have been included in OUTDO with the intent to record the design rationale and to further analyse the design process with MCDA methods. Part of the design rationale, depicted in Figure 3, was conceived as a three-step process. The first issue under consideration was the impact of plant capacity and feedstock polysaccharide concentration on ethanol production price. A sensitivity analysis was carried out taking as parameters plant capacity and polysaccharide concentration (i.e. purchasing price) of a feedstock available in central Mexico in sufficient amounts. The selected polysaccharide content for the feedstock was 35%. Finally, based on this information, the selection of plant type gave preference to the production of ethanol, biogas and hydrogen (a multipurpose-multiproduct plant).

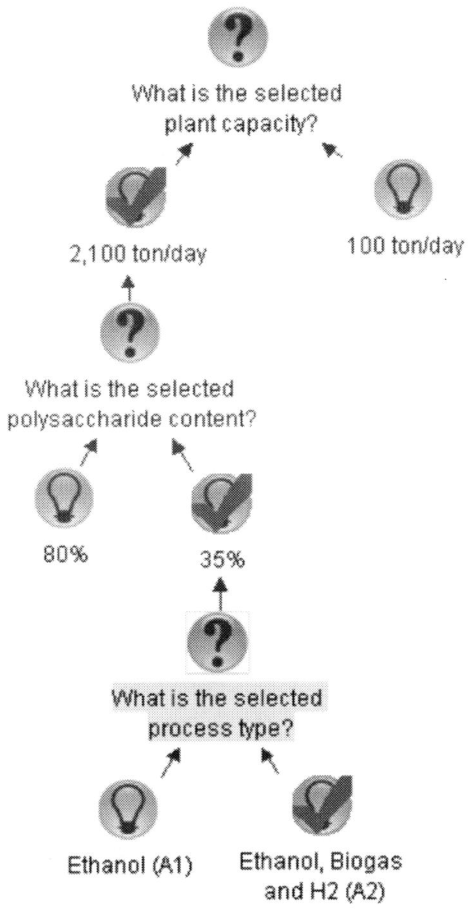

Figure 3. Design rationale for artefact selection.

In this paper, the only criterion used in the selection of process type is the cost of ethanol. However, there are various other criteria that are important for the selection between an *Ethanol* producing plant and an *Ethanol, Biogas and* H_2 producing plant. Thus, a *Decision Matrix* for the Issue node highlighted in yellow in Figure 3 was created and is shown in Figure 4. In this analysis, *Ethanol cost ($/L)* is the only criterion that contributes to the decision, as it is the only one with the weight of criterion different than 0. The weight of criterion for *Ethanol cost ($/L)* is negative because the smaller the cost of ethanol the better it is for the decision maker. The other criteria include other economical, environmental and social aspects involved in the decision. For example, *Economical Impact 1* measures the production flexibility of the plants, *Environmental Impact 1* compares the CO_2 emissions, *Environmental Impact 2* looks at the water requirements, *Social Impact 1* includes the contribution to local employment and, finally, *Social Impact 2* analyzes how the two alternatives would affect the regional energy prices.

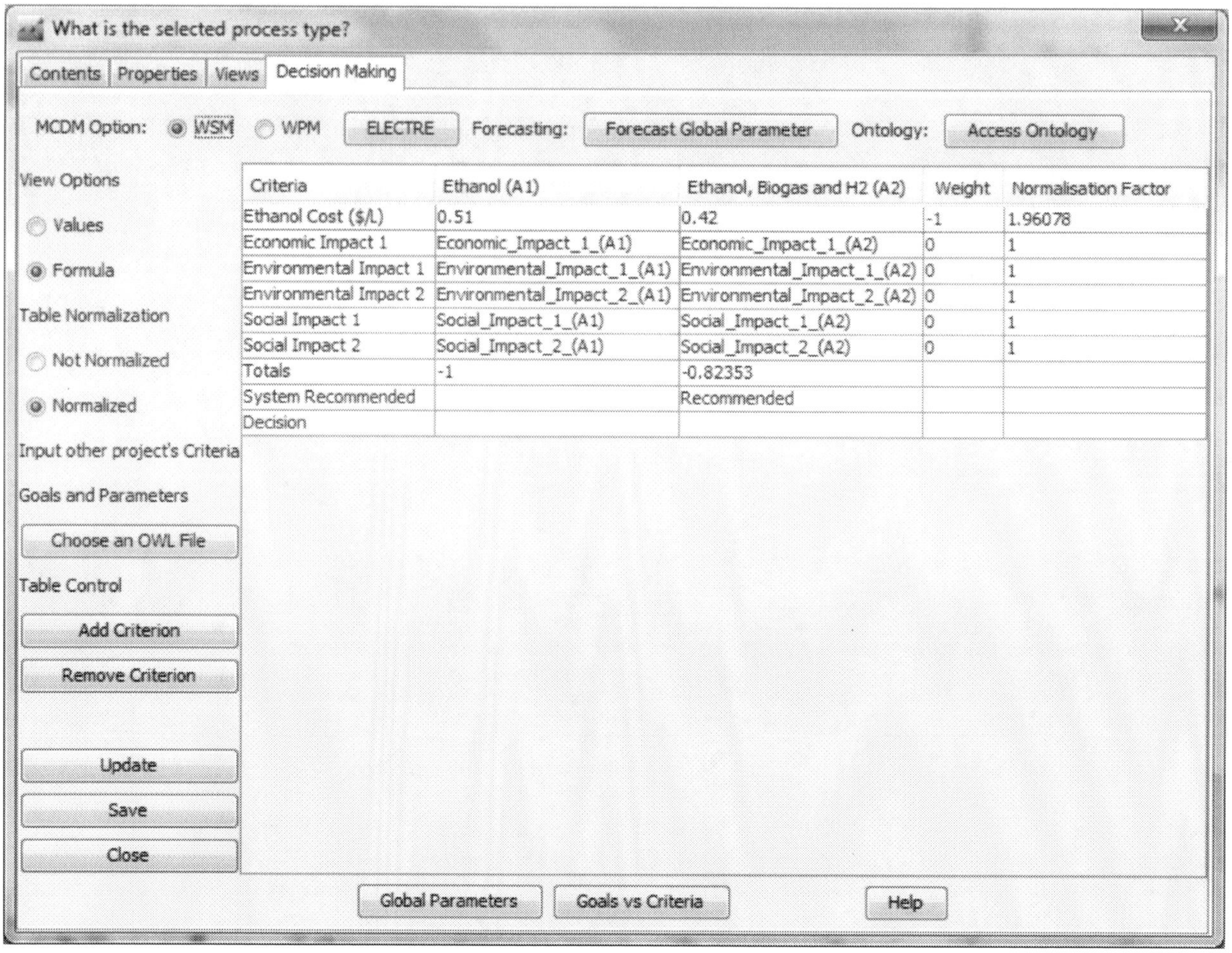

Figure 4: Multi-Criteria Decision Analysis example in OUTDO.

Conclusions

The use of OUTDO is a recent event in the biorefinery design project. This exercise is exposing some decisions that should be re-evaluated and is improving the organization of the accumulated information. More importantly, OUTDO is being used within the project to maintain its undergoing design history, improve current decision processes (through the use of its MCDA toolkit and the forecasting mechanisms) and foster discussion around the values and preferences of the design participants. It is expected that as the project unfolds, the recorded rationale will allow swift identification of those decisions affected by changes to various design variables (costs, operation conditions, targets, regulation limits, etc.) and the opportunity to backtrack and revise them.

References

1. Hunt J., R. Bañares-Alcántara and D. Hanbury (2012), "A new integrated tool for complex decision making", submitted to *Decision Support Systems*.
2. Sánchez A., V. Sevilla-Guitrón, L. Gutiérrez and G. Magaña (2012), "Conceptual Design and Process Economics of a 2G-Ethanol Production Plant in Medium-Scale Agriculture Sectors", submitted to *Industrial Engineering Chemistry*.

3. Sánchez A., V. Sevilla-Guitrón, G. Magaña, P. Melgoza and H. Hernández (2011), "Co-production of ethanol, hydrogen and biogas using agro-wastes. Conceptual plant design and NPV analysis for mid-size agriculture sectors", *Computer-Aided Chemical Engineering* 29:1884-1889.

(461e) Kpred and Vdss: Free Tools On the Pharmahub for Exploring Drug Tissue Partitioning and Distribution

Stephen D. Stamatis[1], Michael J. McLennan[2] and Lee E. Kirsch[1], (1)Division of Pharmaceutics and Translational Therapeutics, The University of Iowa, Iowa City, IA, (2)ITaP, Purdue University, West Lafayette, IN

Translating cutting edge simulation models to the next generation of pharmaceutical professionals and students is more than a problem of information management. Papers published with complex mathematical models often include summarizing remarks and discuss general trends, but the real utility of such models comes from being able to challenge them with so called "what-if" scenarios. In the case of many simple linear models, these "what if" scenarios can be done via inspection, but not always. Many nonlinear models common to pharmacokinetics do not have obvious trends, and the impact of a trend may change in different regions of parameter space. For students to explore these types of relationships, software implementation of the models with easy to use interfaces are required.

In this work, we will briefly review the PharmaHub (pharmahub.org) as an online community for exploring all aspects of pharmaceutical research from fundamental pharmacokinetics to manufacturing. We then focus on two modules in the phubpk suite of tools for pharmacokinetic modeling. At the heart of both tools is a 13-20 compartment whole body physiologically based pharmacokinetic model where each compartment corresponds to a different physiological tissue.

Kpred is a tool to explore a model for predicting the blood tissue partition coefficients for a given drug molecule (Rodgers and Rowland. Pharm Res. 24, 5 2007). In this highly nonlinear model, physicochemical descriptors of the drug molecule are coupled with information about how the drug distributes in the blood and plasma to estimate the partition coefficient. Users of the tool are presented with a simple interface to select drugs with known input parameters or the user can enter custom values to consider many different cases or scenarios. The Kpred tool also allows users to do Monte Carlo sampling to understand the sensitivity of the input parameters on the partition coefficients.

To gain more physical insight into the effect of different partition coefficient values mean for drug distribution, Vdss is a complimentary tool that allows the user to couple the partition coefficients with physiological tissue volumes to compute the steady state volume of distribution. A user is allowed to select a standard 70kg subject with nominal tissue volumes, or they may define a custom individual with nonstandard tissue distribution, i.e. an obese individual. The user can also use Monte Carlo sampling to understand the effect of the uncertainty in the partition coefficients and/or tissue volumes on the predicted steady state volume of distribution.

These tools serve as an example of translating complex models to audiences who would otherwise not have the resources to explore them.

(461f) Best Practices in Process Unit Health Monitoring

Kyle P. Kostroski, Refining & Logistics Technology, BP, Naperville, IL

The primary job function of an operating engineer at a refinery or chemical plant is to ensure safe and reliable operations of his or her process unit. It is therefore imperative that the operating engineer employs regular and rigorous unit health monitoring practices. One impediment to doing so is the number of data and information systems the engineer has to interface with (e.g. process data, laboratory data, operator-rounds data, production planning data, etc.). A common work process is to pull large amounts of data from these various systems into a common, easy-to-use platform where the data can be conditioned, used in calculations, and trended. Many times, this work process requires the engineer to spend more time doing "data jockeying" than analyzing the data.

This talk will focus on a new systems approach (part work process, part computer tool) that automates the data jockeying aspect of unit health monitoring and gives the engineer time to do engineering.

(461g) A Pragmatic Approach to Creating a Digital Information Hub for Plant Asset Information

Philip Simon, Global Business Consulting, AVEVA, Houston, TX

Title: A Pragmatic Approach to Creating a Digital Information Hub for Plant Asset Information

PTP Abstract:

Information management used to be and in some cases still is synonymous with document management. This is understandable because documents have always been used to create, manage, and store information in context. However, documents are limited as an information store in their ability to serve up relevant and selective data. With the advent of databases and other data management technologies, it has become possible to create and manage information down to the specific plant component and attribute value. This has created a new paradigm for information management that has enormous value over document-based information sources because information presentation can be selective and purpose-driven. As a result, companies are slowly warming up to the ideal where all the information needed for their work is available in context, accurate, and presented in a format relevant to the specific user, at the click of a button. This goal has significant business value associated with it, including but not limited to safe and efficient operations and effective configuration control.

However, amidst all the latest information management technologies, a fundamental challenge has also emerged, ironically fueled by information technology itself. On the one hand, all attempts to access current, trusted and relevant information implicitly assumes the existence or knowledge of all such information under one umbrella or centralized system and not in disparate sources and formats. On the other hand, with the increase in new technologies and best-of-breed discipline-specific applications, the number of disparate sources and formats of information is increasing rather than decreasing. In addition, in an almost instinctive manner, every system attempts to become the de-facto central system for all information simply because they recognize the importance of access to other information for their end user's work efficiency and presumes that their discipline is clearly the center of the information hub. In this tug-of-war of applications that are best of breed in one discipline vying to be also the central source of information for the rest of the information, two facts emerge: one is that information will continue to be created in multiple, disparate sources, and the other is that there is no simple way to get them all together with context in any one system. The second fact poses a serious challenge that manifests itself in the process required to create the data standard and to transform the data to be compliant to the standard in the information hub. We will look at this challenge in more detail and identify a solution to overcome it.

The process to create the data standard requires the concerted knowledge, attention, and time from subject matter experts from multiple disciplines, away from their day-jobs. Most significantly, this often is done as a theoretical exercise in a vacuum without visibility to the actual data in the field. Faced with this challenge, vocalized by one Facilities Integrity Manager as “I don't know where to begin to get such a system defined and populated…”, many companies either postpone this lofty goal for a ‘later time' or go to their shareholders and request a major company undertaking. The challenge may have been less immense if we had the standards such as proposed by the Semantic Web or the industry organizations such as Fiatech, POSC-Caeser, ISO, EPRI, etc. The reality though is that even with the progress made by these bodies, we don't have any comprehensive decisive standard available to propose for adoption. To add to the complexity, very often in the early phases of a capital project, these standards are either yet to be defined or cannot be adhered to due to engineering uncertainty at the early stages. Inconsistencies are expected and are intended to be resolved as the project progresses.

If and when the standard has been defined and approved, there is considerable delay in getting sufficient information validated and loaded into the system. Typically the information that is not compliant with the standard is simply rejected for further analysis and evaluation by humans to rectify or transform the data. Depending on the closeness of the standard to real information, as much as 80% of the information may get rejected especially if the standard was developed in a vacuum. A partially loaded system remains a work in progress and is not useful in the daily operations. Several iterations of adjusting the standard and transforming the data will need to occur before a critical mass of information becomes available in the centralized system. These activities will also require the input and attention from the very same subject matter experts, whose time is generally at a premium. All of these factors, including the need to make hard-and-fast decisions without visibility to the impact to real data, go toward making this undertaking time consuming and expensive, often holding up regular work and delaying the rollout of the system.

One solution to this problem is to have a system that can accommodate significant uncertainty as it takes a set of inconsistent, disparate sources from a state of relative chaos (whether it be due to project maturity state or due to poor data quality) to a state of order without holding up the work or the business or impacting existing systems. This system should provide visibility to the chaos so that appropriate and comprehensive standards can be developed relevant to the real data and not in a vacuum. It must then be able to progressively enforce those standards and guide the reconciliation of the data. The progressive nature of the process is critical in that it allows the development of the standard and the cleansing of data to be iterative. A setup of this nature will provide immediate returns for the investment while reducing the risk of the undertaking through multiple interative steps, slowly growing the quality of the information even as the knowledge and the standard matures.

In this presentation, we will compare and contrast this pragmatic approach against other current approaches and also discuss the key ingredients of such a system that can deliver value from the start.

(463a) A Branch and Bound Algorithm for the Solution of MIP Models Using Parallel Computing

Christos Maravelias and Sara Zenner, University of Wisconsin, Madison, WI

A Branch and bound Algorithm for the Solution of MIP Models Using Parallel Computing

Sara Zenner and Christos T. Maravelias

Department of Chemical and Biological Engineering, University of Wisconsin – Madison

1415 Engineering Dr., Madison, WI 53706, USA

Exploiting multiple cores in a computer or on a grid of computers can reduce the time required to solve a mixed-integer programming (MIP). The original MIP can be decomposed into several smaller problems, provided that the union of the feasible regions of the subproblems is the same as the feasible region of the original problem so that no solutions are removed. If such a partition is available, then each subproblem can be solved independently on a single core of a computer. However, for many classes of MIP model, simply decomposing and solving using multiple cores is not effective due to the combinatorial nature of these problems. Instead, problem specific information should be used to generate the subproblems (Ferris et al., 2009).

Accordingly, we develop a method for decomposing and solving hard MIP models using domain specific knowlege. The technique for decomposing and solving an MIP is based on a customized branch-and-bound method. We start by dividing the original problem into several problems by fixing or bounding some key integer variables that are chosen using knowledge about the problem. Next, we solve each of these MIP subproblems in parallel for a fixed amount of time. As in the branch-and-bound algorithm, the solutions to these problems provide bounds on the objective value, and problems are pruned when their best bound is worse than the best objective value found so far. Problems that are not solved to optimality and cannot be pruned are dynamically divided further. We use the GAMS grid computing facility to submit multiple problems to solve simultaneously and collect the solutions of solved problems. We also use the GAMS Branch-and-Cut-and-Heuristic (BCH) facility to share integer solutions among all subproblems as soon as they are found, which allows problems to be pruned sooner.

We apply the decomposition method to a chemical production scheduling problem using the discrete-time model of Shah et al. (1993). We consider a process consisting of a series of tasks that produce a set of final products. The optimization decisions include when to carry out each task, which piece of equipment to use, and how much material to process in order to satisfy customer orders at the lowest cost. In each level of the decomposition tree, we fix or bound the total number of batches for a particular task in a particular unit. The LP-relaxation of the original problem gives an estimate of the number of times a job will run in a good solution; in each level of decomposition, we create several subproblems with the number of batches fixed near this value. The order in which the jobs are fixed is based on specific information about the problem. For example, we can estimate how much time each unit is in use from the solution to the LP-relaxation and fix the number of batches in the bottleneck units first, since those will likely be the most difficult to schedule. We explore how different options for the algorithm and for decomposing the problem affect the overall solution time. Using this method, we are able to solve several problems in under 10 minutes that were not solved within an hour either by using a single core or with the CPLEX parallel option.

M.C. Ferris, C.T. Maravelias, and A. Sundaramoorthy, Simultaneous Batching and Scheduling Using Dynamic Decomposition on a Grid, *INFORMS Journal on Computing*, 2009, 21, 398-410.

N. Shah, C.C. Pantelides, and R.W.H. Sargent, A General Algorithm for short-term scheduling of batch operations--II. Computational Issues, *Computers & Chemical Engineering*, 1993, 17, 229-244.

(463b) Logic-Based Outer-Approximation Algorithm for Solving Discrete-Continuous Dynamic Optimization Problems

Ruben Ruiz-Femenia, Chemical Engineering Department, University of Alicante, Alicante, Spain, **Antonio Flores-Tlacuahuac**, Chemical Engineering Dept., Universidad Iberoamericana, Mexico City, Mexico and **Ignacio E. Grossmann**, Chemical Engineering Department, Carnegie Mellon University, Pittsburgh, PA

LOGIC-BASED OUTER-APPROXIMATION ALGORITHM FOR SOLVING DISCRETE-CONTINUOUS DYNAMIC OPTIMIZATION PROBLEMS

Many chemical process systems of practical interest are subject to changes that cause discontinuities in their dynamics. For instance, thermodynamic phases appear and disappear, flow regimes switch from laminar to turbulent and vice versa and fluids in pipes change direction of motion. Dynamic models are also required for the transient phase of continuous processes such as the start-up, shut-down and changeover procedures.

Optimization of discrete-continuous dynamic problems (also referred as hybrid systems) requires the treatment of non-smooth conditions within the problem formulation. These problems can be formulated as mixed integer nonlinear programing (MINLP) problems, that allow to handle logic conditions that lead to non-smoothness [1]. However, the associated computational expense may be high for large systems with many discrete decisions. This is often the case in hybrid systems with switches at any point in time. Complementary formulations also offer an alternative for some classes of disjunctive problems [2]. They can be embedded within a standard nonlinear programming (NLP), as they model disjunctions without the use of binary variables, but the introduction of complements introduces an inherent non-convexity.

Raman and Grossmann [3] generalized the Balas' disjuntive programming theory [4] and developed the Generalized Disjunctive Programming (GDP), as an alternative modeling framework to the traditional mixed-integer formulations for the discrete-continuous optimization problems. While the mixed-integer programming is based entirely on algebraic equations and inequalities, GDP involves Boolean and continuous variables that are specified in algebraic constraints, disjunctions and logic propositions. This alternative mathematical representation of discrete-continuous optimization problems, not only facilitates the modeling, but also keeps the underlying logical structure of the problem, which can be exploited to the development of customized algorithms. Indeed, two methods have been proposed for the case of convex nonlinear GDP, namely, the disjunctive branch and bound method [5] and the logic-based outer-approximation method [6]. The latter method is based on the idea of solving iteratively a NLP subproblem in reduced space that will provide an upper bound, and a master problem reformulated as an MILP by the convex hull of the linearization of the nonlinear inequalities, which will supply a lower bound and new values for the integer variables. In addition, the logic outer approximation algorithm requires solving several NLP subproblems to produce at least one linear approximation for each term in all the disjunctions. This initialization step implies to solve a set covering problem, which is of small size and easy to solve.

In this paper we address the optimization of discrete-continuous dynamic optimization problems that arise in chemical processes. We formulate these problems as dynamic GDP problems for which we examine an extension of the logic-based OA for static system. The logic-based OA leads to a substantial reduction in the size of the problem compared to the traditional OA algorithm, due to the fact that only the constraints that belong to the active terms in the disjunction (i.e., the associated Boolean variable is true) are imposed. The inactive terms are disregarded. Furthermore, the logic-based OA avoids solutions of the NLP subproblems at trivial solutions (e.g. zero flows) and performs the outer approximations of the nonlinear constraints at the optimal values of continuous variables for the corresponding subproblem.

We assess the performance of the proposed approach by solving several examples of increasing complexity: a problem based on an the example presented by Barton and Lee [7]; the diver problem [8]; and the cascading tank problem [9]. We approach the dynamic optimization problem as a multiperiod dynamic optimization problem without an embedded DAE solver. These periods are represented by finite elements in time, with piecewise polynomial representations of the state and controls in each element [2]. It is shown that the logic-based proposed approach leads to significantly reduced computational times compared to those offer by formulating the hybrid problem as a mixed-integer dynamic optimization problem. Furthermore, it is also shown that the computations for the optimization are much more robust than the full-space MINLP counterpart. Thus, this work has shown that the fundamental ideas from logic-based optimization can be applied to the optimization problems with disjunctive dynamic model constraints.

References

1. Avraam, M.P., N. Shah, and C.C. Pantelides, *Modelling and optimisation of general hybrid systems in the continuous time domain.* Computers and Chemical Engineering, 1998. **22**(SUPPL.1): p. S221-S228.

2. Baumrucker, B.T. and L.T. Biegler, *MPEC strategies for optimization of a class of hybrid dynamic systems.* Journal of Process Control, 2009. **19**(8): p. 1248-1256.

3. Raman, R. and I.E. Grossmann, *Modelling and computational techniques for logic based integer programming.* Computers and Chemical Engineering, 1994. **18**(7): p. 563-578.

4. Balas, E., *Disjunctive Programming.* Annals of Discrete Mathematics, 1979. **5**: p. 3-51.

5. Lee, S. and I.E. Grossmann, *New algorithms for nonlinear generalized disjunctive programming.* Computers and Chemical Engineering, 2000. **24**(9-10): p. 2125-2141.

6. Türkay, M. and I.E. Grossmann, *Logic-based MINLP algorithms for the optimal synthesis of process networks.* Computers & Chemical Engineering, 1996. **20**(8): p. 959-978.

7. Barton, P.I. and C.K. Lee, *Design of process operations using hybrid dynamic optimization.* Computers and Chemical Engineering, 2004. **28**(6-7): p. 955-969.

8. Galán, S. and P.I. Barton, *Dynamic optimization of hybrid systems.* Computers and Chemical Engineering, 1998. **22**(SUPPL.1): p. S183-S190.

9. Till, J., et al., *Applied hybrid system optimization: An empirical investigation of complexity.* Control Engineering Practice, 2004. **12**(10): p. 1291-1303.

(463c) An Algorithm for Preprocessing and Tightening Generalized Disjunctive Programming Models Through the Application of Basic Steps

Francisco Trespalacios, Chemical Engineering, Carnegie Mellon University, Pittsburgh, PA and Ignacio E. Grossmann, Chemical Engineering Department, Carnegie Mellon University, Pittsburgh, PA

An Algorithm for Preprocessing and Tightening Generalized Disjunctive Programming Models through the Application of Basic Steps

Francisco Trespalacios and Ignacio Grossmann

Many optimization problems are represented with algebraic equations, using continuous and discrete variables, giving rise to Mixed (Linear or Nonlinear) Integer Programs (MILP/ MINLP). An alternative way to represent this type of problems is Generalized Disjunctive Programming (GDP) proposed by Raman and Grossmann[2] that involves not only algebraic equations, but also disjunctions and logic propositions[1]. This higher level representation, which is of special importance in Engineering, allows us to exploit the logic structure of the problem to obtain better formulations.

GDP can be regarded as an extension of well-known disjunctive programming paradigm developed by Balas[3]. A particular operation in disjunctive programming that helps to generate a tighter formulation of the problem, using the logic structure of disjunctions, is the so-called Basic Step[4]. The basic steps can be applied sequentially to generate a hierarchy of relaxations, where the limiting case is given by the convex hull of the original GDP problem. The application of Basic Steps has proven to improve the tightness in linear GDP formulations, as shown by Sawaya and Grossmann[5], and in general convex GDP formulations, as shown by Ruiz and Grossmann[6]. The drawback in the application of this operation is the exponential growth of continuous variables and number of constraints (although the number of binary variables can be kept constant).

In this work we propose an algorithmic approach to sequentially apply basic steps to linear and nonlinear convex GDP problems that will, at most double the number of continuous variables, and will potentially decrease the number of binary variables in the MILP/MINLP reformulation. There are three main stages that will allow this: (a) Feasibility check in individual disjunctions through the solution of continuous programs; (b) application of basic steps in specific disjunctions; (c) logic transformation after reducing the number of disjunctive terms into a GDP.

For the first stage we analyze the feasibility of the binary variables, which have a one to one correspondence with each of the disjunctive terms. Through a pre-solving analysis, we discard those binary variables that make the problem infeasible when taking a value of 1. In this way we reformulate the GDP excluding the disjunctive terms that were proven to be infeasible for a corresponding "True" value. Through this first stage we can rigorously reduce the size of the GDP problem.

For the second stage we start from the observation that the growth in continuous variables is mainly driven by the increase of the number of terms in the disjunctions as we apply basic steps. Additionally, the number of terms in a disjunction (after the application of a basic step) is defined by the multiplication of the number of terms in the two disjunctions to which we applied the basic step. For example, if we apply a basic step between two disjunctions with two terms each, we obtain a new single disjunction with four terms; if one has 4, and the other 3, the resulting disjunction will have 12 terms. In this stage we apply basic steps between two disjunctions with two terms in each, resulting in 4-term disjunctions. In this way the number of terms in the disjunctions of the problem is kept constant. This stage, in combination with the feasibility check, allows us to sequentially apply basic steps without increasing the number of disjunctions.

The third stage relies on the fact that there is a one to one correspondence with binary variables and the disjunctive terms in a GDP. When we eliminate some of the disjunctive terms of a GDP, through the feasibility check, this results into a new GDP problem with fewer disjunctive terms (even if a basic step is applied in an earlier stage). With this in mind, we can assign a new binary variable to the disjunctive terms in the new GDP, and relate them to the logic propositions as needed, and in that way reduce the number of binary variables.

These three stages can be applied iteratively for a finite number of times. This will then lead to a tighter GDP, with potentially fewer binary variables and limited growth in continuous variables and constraints that can be solved more efficiently. The resulting GDP problems can then be either reformulated as an MILP or MINLP using the hull relaxation, or else be solved through a cutting plane algorithm.

We illustrate the application of this algorithm with several linear and convex GDP examples, including strip packing, reactor selection and layout problems. The results show the improvement in their relaxed solution, reduction in the number of binary variables, and relatively small growth in number of continuous variables and constraints.

References:

[1] Williams H.P., *"Model Building in Mathematical Programming"*, Wiley, 1985.

[2] Raman R. and Grossmann I.E., *"Modelling and Computational Techniques for Logic-Based Integer Programming"*, Computers and Chemical Engineering, 18, 563, 1994.

[3] Balas E., *"Disjunctive Programming: Properties of the Convex Hull of Feasible Points"*, MSRR #348, Carnegie Mellon University, Pittsburgh, PA, July 1974.

[4] Balas E., *"Disjunctive Programming and a Hierarchy of Relaxations for Discrete Continuous Optimization Problems"*, SIAM J. Alg. Disc. Meth., Vol. 6, No. 3, 1985.

[5] Nicolas Sawaya, Ignacio Grossmann, *"Reformulations, Relaxations and Cutting Planes for Linear Generalized Disjunctive Programming"*, Carnegie Mellon University, Department of Chemical Engineering, Pittsburgh, PA, U.S.A, 15201.

[6] Juan P. Ruiz, Ignacio E. Grossmann, *"A hierarchy of relaxations for nonlinear convex generalized disjunctive programming"*, Carnegie Mellon University, Department of Chemical Engineering, Pittsburgh, PA, U.S.A, 15201

(463d) Convergence Rate of Taylor Model-Based Estimators in Global Optimization

Agustín Bompadre, Graduate Aerospace Laboratories, California Institute of Technology, Pasadena, CA, Alexander Mitsos, Department of Mechanical Engineering, Massachusetts Institute of Technology, Cambridge, MA and **Benoit Chachuat**, Centre for Process Systems Engineering, Department of Chemical Engineering, Imperial College London, London, United Kingdom

We analyze the convergence of Taylor and McCormick-Taylor model estimators, which have been used, e.g., in enclosing the solutions of nonlinear differential equations [1,2]. Both estimators propagate a multivariate polynomial part symbolically that matches the Taylor expansion of the estimated function up to a specified order, together with an estimator of the Taylor remainder. The latter consists of an interval estimator in a Taylor model and of a pair of convex/concave McCormick relaxations [3] in a McCormick-Taylor model.

The concept of convergence order of an estimator finds its origins in interval extensions and compares the rates of convergence of the estimation error and of the range of the estimated function. Recently, Bompadre and Mitsos [4] applied and extended this concept to McCormick relaxations. Building upon this work (and refining some of their results), we determine how the convergence orders of the remainder estimators propagate through addition, multiplication and composition operations. It is proved that the convergence orders of both qth-order Taylor models and qth-order McCormick-Taylor models are at least $q+1$, under relatively mild assumptions. Moreover, it is verified through simple numerical examples that these bounds are sharp. A consequence of this analysis is that, unlike McCormick relaxations over natural interval extensions, McCormick-Taylor models do not result in increased order of convergence over Taylor models in general. As demonstrated by the numerical case studies however, McCormick-Taylor models can provide sharper bounds or even result in a higher convergence rate.

References:

1. Lin Y, Stadtherr MA (2007) Validated solutions of initial value problems for parametric ODEs. Appl Numer Math 57(10):1145–1162
2. Sahlodin AM, Chachuat B (2011) Convex/concave relaxations of parametric ODEs using Taylor models. Comput Chem Eng 35(5):844–857
3. McCormick GP (1976) Computability of global solutions to factorable nonconvex programs: Part I – Convex underestimating problems. Math Program 10:147–175
4. Bompadre A, Mitsos A (2012) Convergence rate of McCormick relaxations. J Global Optim 52(1):1-28

(463e) An Augmented Lagrangian Approach for Parallel Solution of NLP Problems On Graphics Processing Units

Yankai Cao and Carl D. Laird, Chemical Engineering, Texas A&M University, College Station, TX

Large-scale nonlinear programming (NLP) has proven to be an effective framework for optimal process design and operation. As problem sizes continue to grow, the capabilities of a single workstation can become insufficient. This, coupled with the emergence of new concurrent computing architectures, drives the need for effective parallel algorithms.

Graphics processing units (GPUs) are emerging as massively parallel systems that offer a large degree of parallelism at a relatively low cost. However, compared with classic multiple-instruction-multiple-data (MIMD) the single-instruction-multiple-thread (SIMT) architecture of the modern GPU limits its direct application in many numerical applications. This architecture contains a large number of processing cores, however, groups of these cores must be concurrently executing the same instruction. This makes the architecture inappropriate for parallel direct factorization – the primary numerical workhorse of most nonlinear programming solvers.

In this paper, we propose an Augmented Lagrangian approach for parallel solution of large-scale nonlinear programming problems on a GPU. The algorithm is iterative at three levels. The first level replaces the original problem by a sequence of bound-constrained optimization problems using an Augmented Lagrangian method. Each of these bound-constrained problems is solved using a nonlinear interior-point method. Inside the interior-point method, the barrier subproblems are solved using a variation of Newton's method, where the linear system is solved using a preconditioned conjugate gradient method (PCG). If the PCG approach detects negative curvature, a diagonal modification of the Hessian is made to ensure positive definiteness. The primary advantage of this algorithm is that is allows use of the PCG method, instead of using direct factorization. The PCG method can be implemented efficiently on the GPU in parallel.

Finally, numerical experiments are reported using a set of problems from COPS test collection. We compare the computational results of our serial and parallel versions with the IPOPT nonlinear optimization package, and show that significant speedup (approximately an order of magnitude) is possible.

(463f) Efficient Surrogate Model Generation with Adaptive Sequential Sampling

John P. Eason and Selen Cremaschi, Department of Chemical Engineering, University of Tulsa, Tulsa, OK

By constructing a simpler approximation of the full system, surrogate-based optimization allows for traditional derivative-based optimization techniques to be applied in more complex problems. However, the construction of a surrogate model requires the execution of the original model many times in order to gather the data that will be used to construct it. Depending on the complexity of the original model, this step could become cost prohibitive. For example, if the original model is a computational fluid dynamics simulation of a packed two-phase reactor, single simulation run might take anywhere from days to weeks to solve. Therefore, it is important to determine the number and location of the samples to minimize the upfront computational costs to generate accurate surrogate models. In this work, three adaptive sequential sampling algorithms are developed for surrogate model construction. Their performances are compared on the basis of number of samples needed to generate surrogate models with 5% accuracy for three challenge functions: the Shekel function, Ackley function, and Beale function. In this work, artificial neural networks are used as surrogate models.

All three algorithms are sequential design techniques, i.e., they begin with an initial sample set, train the networks, evaluate the performance of the trained networks (we used K-fold cross-validation as the model evaluation technique), select n new data points where n is a given fraction of the current sample size, and train the networks again with the new full data set. This procedure is repeated until a stopping criterion is satisfied.

The first algorithm begins the process of selecting new points by generating a large Latin hypercube sample of proposed points. The set of neural networks, each trained by the data of a single fold, is used to predict the output variable at each proposed point. The variance of the predictions at a given point provides an estimate of the surrogate-model variance at that point. The points with the highest variance estimates are where the models have the highest uncertain predictions and hence these points are selected to run the original model. The neural networks are then trained with the overall data set. This process is repeated until the maximum variance estimate decreases below a target value. The second algorithm is similar in its execution steps; however, it uses a performance metric combining the estimated variance and the distance to the nearest-neighbor sample point to select the sequential sample points rather than the variance alone. This algorithm terminates when the performance metric is below a target value for all proposed points. The addition of the nearest-neighbor distance makes this algorithm both space-filling and adaptive, to both locate and fully model fluctuations in the objective function. The final algorithm uses incremental Latin hypercube sampling (or iLHS, as presented in Nuchitprasittichai and Cremaschi 2012) until the mean squared error stabilizes, and then switches to the first algorithm to add more samples in areas of insufficient information.

Algorithm 1 is the simplest and fastest algorithm, but requires that the initial sample size be sufficient to locate all regions of interest in the objective function. The space-filling criterion in algorithm 2 overcomes this problem but at an additional computation cost to calculate the nearest neighbor distance. Algorithm 3 uses iLHS to create sufficiently space-filling samples before beginning the adaptive sampling, but the iLHS process is costly because the algorithm maintains a Latin hypercube sample for each iteration, eliminating previously sampled points if necessary. Although many past sampling algorithms were designed for a specific type of surrogate model, all three algorithms presented here generalize to any type of surrogate model. These algorithms also scale well to problems with higher dimensionality as are common in chemical engineering.

(463g) Simulation-Based Derivative-Free Optimization for Computationally Expensive Functions

Fani Boukouvala and Marianthi Ierapetritou, Chemical and Biochemical Engineering, Rutgers University, Piscataway, NJ

Nowadays, the need to develop predictive models for complex processes along with advances in computational tools and capabilities have lead to the development of a plethora of detailed complex mathematical methods. However, as the complexity and level of detail increases, so does the necessary computational cost which prohibits their use in any application that requires multiple simulations- such as optimization- and limits them to simple qualitative assessment tools. For this reason, there has been a lot of interest on simulation based optimization techniques which aim to minimize the necessary function evaluations by not relying on analytic derivatives of the objective function for optimization, which are costly to evaluate and oftentimes noisy [1]. In the literature, many gradient-free optimization algorithms have been proposed, which can be divided into direct-search or model-based, local or global and deterministic or stochastic based on their search criteria, search space and the nature of their search steps respectively [2]. In addition, one of the major challenges of gradient-free optimization is the handling of complicated constraints. In this work, a gradient-free algorithm will be proposed which employs a hybrid global/local search, is deterministic and model-based and can handle black-box constraints through the concept of black-box feasibility mapping [3].

Since the first significant direct search gradient-free methods ([4],[5]), a large number of methods have been proposed which aim to optimize models without the need for the calculation of derivatives [6]. The history of gradient-free optimization has transitioned from direct-search to surrogate-based optimization, which has been most popular recently based on the hypothesis that through the use of surrogate models, the search will be guided in a more intelligent way and thus function calls will be minimized. Local model-based methods focus within a small subregion of the domain, for which the appropriate amount of samples is collected to fit the chosen response surface, an optimizer is found within the trust region and the trust region radius is increased, decreased, or relocated based on a set of rules proven to lead to convergence. This procedure is repeated until the trust region radius is sufficiently small. On the other hand, many global methods have been proposed which differ in terms of the type of response surface (response surface methodology, kriging, radial basis functions etc) or the strategy that identifies the next promising sample (expected improvement, bumpiness, entropy etc)[7-10]. The basic ideas however are identical: collect an initial sample from the entire region, fit a global model, optimize it, find the next promising sample based on a search criterion and evaluate it, update response surface and re-optimize until expected improvement is low.

Driven by the need to optimize extremely expensive and noisy simulations, this work combines the main advantages of well-established global and local search algorithms with new concepts of feasibility space mapping within a framework that aims to optimize such models, while minimizing sampling without sacrificing quality of the solution. The proposed algorithm initially employs a global search, collecting an initial sample from the entire investigated region based on which not only an approximate global model is built, but also a mapping of the feasible space of the problem is performed. This first step serves two purposes since it provides promising initial starting points for the proceeding local search of the algorithm, and it allows the investigation of the feasible region without the need of any assumptions about its nature (i.e. linear, convex, continuous). Next, the algorithm switches to local-search mode starting from more than one in parallel (if applicable) warm-start locations, taking advantage of the fact that surrogate model approximations are more accurate within smaller trust-regions. The algorithm proceeds based on a set of rules for updating the location or size of the trust region, until the trust region radius is sufficiently small. During this procedure, the global model as well as the feasibility mapping are updated when new samples are collected while the size of the trust-region is continuously adapted based on the local information about the feasibility space, in order to avoid sampling in infeasible regions. The performance of the proposed approach is evaluated through a set of benchmark examples and an actual case study of an expensive simulation from the pharmaceutical industry.

References:

1. Kolda, T.G., R.M. Lewis, and V. Torczon, *Optimization by Direct Search : New Perspectives on Some Classical and Modern Methods.* 2003. **45**(3): p. 385-482.

2. Rios, L.M. and N.V. Sahinidis, *Derivative-free optimization: A review of algorithms and comparison of software implementation.* AICHE Annual Conference, Nashville, TN, 2009.

3. Boukouvala, F. and M.G. Ierapetritou, *Feasibility analysis of black-box processes using an adaptive sampling Kriging-based method.* Computers & Chemical Engineering, 2012. **36**(0): p. 358-368.

4. Hooke, R. and T.A. Jeeves, *"Direct Search" Solution of Numerical and Statistical Problems.* J. ACM, 1961. **8**(2): p. 212-229.

5. Nelder, J.A. and R. Mead, *A Simplex Method for Function Minimization.* The Computer Journal, 1965. **7**(4): p. 308-313.

6. Conn, A.R., K. Scheinberg, and L.N. Vicente, *Introduction to derivative-free optimization*. MPS-SIAM series on optimization. 2009, Philadelphia: Society for Industrial and Applied Mathematics/Mathematical Programming Society. xii, 277 p.

7. Huang, D. *Experimental planning and sequential kriging optimization using variable fidelity data*. 2005; Available from: http://etd.ohiolink.edu/view.cgi?acc_num=osu1110297243.

8. Jakobsson, S., et al., *A method for simulation based optimization using radial basis functions.* Optimization and Engineering, 2010. **11**(4): p. 501-532.

9. Jones, D.R., M. Schonlau, and W.J. Welch, *Efficient Global Optimization of Expensive Black-Box Functions.* Journal of Global Optimization, 1998. **13**(4): p. 455-492.

10. Villemonteix, J., E. Vazquez, and E. Walter, *An informational approach to the global optimization of expensive-to-evaluate functions.* Journal of Global Optimization, 2009. **44**(4): p. 509-534.

(494a) Optimal Design of Reverse Osmosis Based Water Treatment Systems

Ramkumar Karuppiah, Corporate Strategic Research, ExxonMobil Research and Engineering, Annandale, NJ, Scott J. Bury, Engineering Sciences, The Dow Chemical Company, Midland, MI, Adriana Vazquez, Supply Chain Technology Center, The Dow Chemical Company, Freeport, TX and Gregg Poppe, Dow Water Solutions TS&D, The Dow Chemical Company, Edina, MN

In this work we address the problem of optimal design of reverse osmosis based water treatment systems. A superstructure optimization method is proposed to solve the problem, where the superstructure for a reverse osmosis is structurally enhanced with additional features. We formulate the problem as mixed-integer nonlinear program (MINLP) which is solved to yield optimal results. A case study on desalination is considered in this work, and the numerical results obtained using our approach are validated using a commercial simulation tool. We further extend the problem by considering the effects of degradation of membrane performance over time and solve it by representing the problem as a two-stage stochastic program.

- -

(494b) Addressing the Operational Challenges in the Manufacture of Advanced Materials and Performance Products

John Wassick, Dow Chemical Company, Midland, MI, Scott J. Bury, Engineering Sciences, The Dow Chemical Company, Midland, MI, Naoko Akiya, Engineering & Process Science R&D, The Dow Chemical Company, Freeport, TX, Jeff Ferrio, The Dow Chemical Company and Anshul Agarwal, Department of Chemical Engineering, Carnegie Mellon University, Pittsburgh, PA

The challenges of supplying advanced materials and performance products are very different from those associated with commodity chemicals. Yet these challenges are well addressed by enterprise-wide optimization techniques. The aim of this talk is to describe the operational challenges in the manufacture of advanced materials and performance products and to show where and how process systems concepts can be used to address those challenges. Several challenges and their solutions will be reviewed. The first is the design of a formulated product through numerical optimization. The goal here is to more quickly and more completely identify candidate products that meet new customer requirements. We will then review the problem of crew scheduling in a job shop where the challenge of scheduling technicians on machines is key to maximizing throughput of the system. This will be followed by discussion of a generic approach to discrete event simulation that accelerates model development. The simulator, having a fixed model structure, relies on capturing the details of the simulated system in the data base supporting the model. Finally the logistical problem of optimally loading a semi-trailer with various packages is examined.

Synthesis of Two-Dimensional Reactor Network Based on Transfer Fundamentals

Da-yu Jiang[1], Hong-Guang Dong[2]

(1) ACRE Coking And Refractory Engineering Consulting Corporation, Dalian, 116085, PRC (2)School of Chemical Engineering , Dalian University of Technology, Dalian, 116024, PRC

Abstract

A two-dimensional model, which focuses on fluid flow, heat-mass transfer, and reaction kinetics both in axial and radial directions, is developed to represent a practical reactor. And the basic model is employed to construct a state-space framework for reactor network synthesis. Total Annual Cost (TAC) as an objective is considered to describe the performance of reactor system.

Keywords: Reactor Network Synthesis; Two-Dimensional Reactor Model; State Space Framework; Mixed-Integer Dynamic Optimization (MIDO)

1. Introduction

In the past decades, most efforts on the reactor network are classified into two general categories: superstructure based optimization, and targeting based methods. Traditionally, synthesis of reactor network, whose objective function commonly optimize maximization of yield, or minimization of total volume, is usually based on PFR, CSTR, DSR, even CFR models (Lakshmanan & Biegler, 1996; Schweiger & Floudas 1999). Though, Peclet number is considered to describe a non-ideal reactor instead of those ideal models (Achenie & Biegler, 1986; Zhou & Manousiouthakis, 2006). Actually, Peclet number describes only back-mixing in axial direction, but axial and radial directions in reactors are influenced by each other. Therefore, a practical reactor model which contains velocity distribution, heat transform, and mass transform in both axial and radial directions is more significant on industrial applications. Lakshmanan and Biegler (1996) developed a general DSR model, however the superstructure contained no recycle reactors. Schweiger and Floudas (1999) used CFRs and CSTRs as the fundamental units to form a network which contains more possible structural alternatives, but actually streams in CFR model are not allowed to mix or split arbitrarily. Schweiger and Floudas (1999) also solved cases of non-isothermal reactions, whose results show that the temperature profiles changed with reactor volume are vague. In fact, the same volume has different sizes, such as a long-thin reactor or a short-thick one. Although Zhou and Manousiouthakis (2006) constructed a perfect network using

IDEAS framework, there are any energy balance equations in the reactor model, so the approach is only suitable to solve isothermal problems.

In this work, a two-dimensional model based on transfer fundamentals is proposed to approximate to a practical reactor, and then a state space framework (Bagajewicz & Manousiouthakis, 1992, 1998) composed by series two-dimensional models and distribution network (DN) is constructed for the synthesis of reactor networks. That state space superstructure aims at developing a systematic framework which contains all possible initial reactor networks, in which the two-dimensional reactor model makes sure that every reactor has its certain size. Moreover, heat transfer rate equation is considered as a constraint in a non-isothermal reactor system, because the same volume with variable parameters, such as length and diameter, will lead to different areas which work on heat transfer effect. Compared with describing Total Annual Cost (TAC) through traditional coefficient method (Zhou & Manousiouthakis, 2008), we also apply a more detailed objective function, which involves installed cost, catalyst cost, and utility cost, to obtain economical optimal design. Finally, two cases collaborated between academia and industries are illustrated to show the application of the proposed approach.

2. Two-dimensional model

In a practical reactor model, fluid flow, heat-mass transform, and reaction kinetics is considered simultaneously. In cylindrical coordinate, when only heat conduction exists, the differential heat transfer equation is:

$$\rho Cp(\frac{\partial T}{\partial t}+u_r\frac{\partial T}{\partial r}+u_\theta\frac{\partial T}{r\partial\theta}+u_z\frac{\partial T}{\partial z})=\lambda(\frac{1}{r}\frac{\partial}{\partial r}\left(r\frac{\partial T}{\partial r}\right)+\frac{\partial^2 T}{r^2\partial\theta^2}+\frac{\partial^2 T}{\partial z^2})-\sum R_q H_q \quad (1)$$

Similar, when only diffusion exists, the differential mass transfer equation is:

$$\frac{\partial C_j}{\partial t}+u_r\frac{\partial C_j}{\partial r}+u_\theta\frac{\partial C_j}{r\partial\theta}+u_z\frac{\partial C_j}{\partial z}=D(\frac{1}{r}\frac{\partial}{\partial r}\left(r\frac{\partial C_j}{\partial r}\right)+\frac{\partial^2 C_j}{r^2\partial\theta^2}+\frac{\partial^2 C_j}{\partial z^2})+\sum_j R_q \quad (2)$$

For steady-state laminar flow, temperature and concentration only change with space but not time, and velocity is along axial direction, i.e.:

$$\frac{\partial T}{\partial t}=0,\quad \frac{\partial C_j}{\partial t}=0,\quad u_r=u_\theta=0 \quad (3)$$

The framework proposed in section 3 allows neglecting back-mixing in axial direction which makes the problem solved easier,

$$\frac{\partial^2 T}{\partial z^2}=0, \frac{\partial^2 C_j}{\partial z^2}=0 \quad (4)$$

Under the assumption of constant density, heat conductivity and diffusion coefficients, simplified heat and mass transfer equations of a practical reactor model are:

$$\rho Cpu_z\frac{\partial T}{\partial z}=\lambda(\frac{\partial^2 T}{\partial r^2}+\frac{1}{r}\frac{\partial T}{\partial r})-\sum R_q H_q \quad (5)$$

$$u_z\frac{\partial C_j}{\partial z}=D(\frac{\partial^2 C_j}{\partial r^2}+\frac{1}{r}\frac{\partial C_j}{\partial r})+\sum_j R_q \quad (6)$$

Specific heat capacity of mixed flow in (5) is changed with composition as

$$Cp = \frac{\sum_j C_j Cp_j}{\sum_j C_j} \tag{7}$$

Velocity profile on radial dispersion in (5) and (6) is:

$$u_z = u_{\max}[1-(\frac{2r}{d})^2] \tag{8}$$

Here, r is the radius of the reactor. u_{max} depends on both of reactor numbers k and discretized elements nfe:

$$u_{k,nfe} = \frac{4F_{k,nfe}}{\pi d_k^2 C_{k,nfe}} \tag{9}$$

Kinetics of reactions in (3) and (4) is:

$$R_q = k_{0q} \times \exp(-E_q / RT) \times \prod C_j^{m_{qj}} \tag{10}$$

Here, m_{qj} is the reaction order of the jth component in the qth reaction.

For non-isothermal reactions, when a minimal volume reactor achieves the objective of concentration, but heat transfer area is not enough sometimes. So designing a reactor not only requires reaching yield, but also needs to satisfy heat transfer rate equation:

$$K \times 2\pi rL \times \Delta t_m \geq Q \tag{11}$$

A practical reactor model is completed with boundary conditions:

$$z = 0,\ T = T_0,\ C_j = C_{j0} \tag{12}$$

$$r = 0,\ \frac{\partial T}{\partial r} = 0,\ \frac{\partial C_j}{\partial r} = 0 \tag{13}$$

$$r = a,\ \frac{\partial T}{\partial r} = \frac{Q}{2\pi a \times L \times \lambda},\ \frac{\partial C_j}{\partial r} = 0 \tag{14}$$

where constant heat flux $\frac{Q}{2\pi a * L * \lambda}$ is applied on tube wall to control temperature.

In this two dimensional model, partial differential equations (PDEs) are used to describe the relationship between axial and radial direction in a practical reactor.

3. State-space framework

A state-space framework is constructed to synthesize the two-dimensional reactor network. The distribution network (DN) contains all possible mixing, splitting, recycling and bypassing. The process operator (OP) accomplishes every unit operation. In this SS framework, CSTR, DSR and CFR are approximated obviously. For example, a two-dimensional reactor with recycling is equivalent to back-mixing, when the recycle ratio approaches ∞, it approximates a Continuous Stirred Tank Reactor (CSTR).

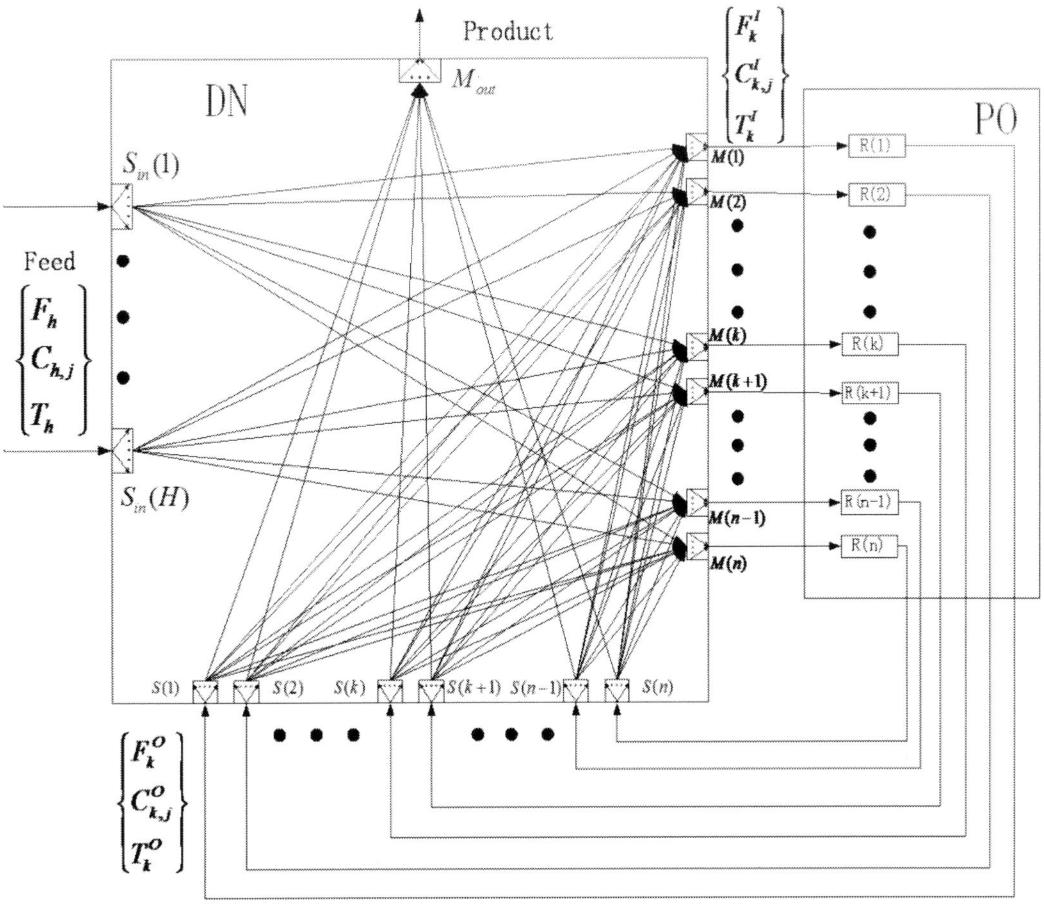

Fig 1. State-Space Framework for Reactor Network Synthesis

Fig.1 illustrates a state-space framework with J components, H feeds, N inlet and N outlet streams of reactors. The objective function containing fixed cost of reactors, catalyst cost and utility cost is written as

$$\text{Cost}_{\text{total}} = \sum_k (a + b \times (2\pi r_k L_k)^c + \text{Price}_{\text{catalyst}} \times (\pi r_k^2 L_k) + \text{Price}_{\text{utility}} \times (Q_k / Cp_{\text{utility}} / \Delta t) \times \text{time}_{\text{operating}}) \quad (15)$$

The mathematical formulations of DN are described as follows:

Material balance

$$F_k^I = \sum_h x_{h,k} F_h + \sum_i y_{i,k} F_i^O \quad (16)$$

$$F_{k,j}^{I} = \sum_{h} x_{h,k} F_{h,j} + \sum_{i} y_{i,k} F_{i,j}^{O} \quad (17)$$

$$F_{k,j}^{I} = \frac{C_{k,j}^{I}}{\sum_{j} C_{k,j}^{I}} F_{k}^{I} \quad (18)$$

$$F_{h,j} = \frac{C_{h,j}}{\sum_{j} C_{h,j}} F_{h} \quad (19)$$

$$F_{i,j}^{O} = \frac{C_{i,j}^{O}}{\sum_{j} C_{i,j}^{O}} F_{i}^{O} \quad (20)$$

Energy balance

$$\sum_{h} (T_{k}^{I} - T_{h}) \sum_{j} x_{h,k} F_{h,j} Cp_{j} = \sum_{i} (T_{i}^{O} - T_{k}^{I}) \sum_{j} y_{i,k} F_{i,j}^{O} Cp_{j} \quad (21)$$

Normalization

$$\sum_{k} x_{h,k} = 1 \quad (22)$$

$$\sum_{k} y_{i,k} \leq 1 \quad (23)$$

Here, $x_{h,k}$ is the distribution coefficient from $S_{in}(h)$ to $M(h)$, and $y_{i,k}$ is the distribution coefficient from $S(i)$ to $M(h)$.

The concentration of product and by-products is better to meet demands

$$\frac{C_{\text{prodcut}}}{\sum_{j} C_{j}} \geq \text{Yield} \quad (24)$$

$$\frac{C_{\text{by-prodcut}}}{\sum_{j} C_{j}} \leq \text{Limited value} \quad (25)$$

The existence of each reactor depends on integer variables

$$F_{k}^{I} \leq \text{Binary variables} \times F^{\infty} \quad (26)$$

Moreover, ideal gas equation is considered in gas system

$$p_{j} V = F_{j} \text{R} T \quad (27)$$

So the synthesis problem is Mixed-Integer Dynamic Optimization (MIDO).

4. Solution Methodology

This MIDO involves partial differential equations (PDEs) as well as mixed-integer problem. Finite element collocation was proposed by Cuthrell and Biegler (1987) for the full discretization of ordinary differential equations (ODEs). While Schweiger and Floudas (1999) used control parametrization approach to solve ODEs. The Shrink-wrap algorithm proposed by Zhou and Manousiouthakis (2006) is not suitable for non-isothermal condition or optimal control problem.

The Crank-Nicholson method is implicit and numerically stable, especially for heat and diffusion equations (Thomas, 1995). So the PDEs are discretized into algebraic equations through the Crank-Nicholson method in this work,. Then the MINLP is easily rewritten as a disjunctive program with a consequent simplification of NLP sub-problems. The liquid reactor system, which is epoxidation of propylene by hydrogen peroxide to produce propylene-oxide (HPPO), is solved by IPOPT version 3.8 stable in AMPL environment. The vapor-only reactor system, which is Alkylation of toluene by methanol to produce p-xylene, is solved in GAMS environment which shows the feasibility of the proposed approach.

5. Case study

5.1 Liquid system

Epoxidation of propylene by hydrogen peroxide to propylene-oxide (PO), whose by-products are propylene glycol methyl ether (PG) and oxygen, is a new technique.

$$\text{Propylene} + \text{Hydrogen peroxide} \xrightarrow{k_1} \text{PO} + \text{Water}$$

$$\text{PO} + \text{Methanol} \xrightarrow{k_2} \text{PG}$$

$$\text{Hydrogen peroxide} \xrightarrow{k_3} 0.5\ \text{Oxygen} + \text{Water}$$

The reaction rates are

$$R_1 = k_1 C_{propylene}^{0.165} C_{Hydr}^{0.63}$$

$$R_2 = k_2 C_{Meth}^{0.66}$$

$$R_3 = k_3 C_{Hydrogen}^{0.082}$$

The kinetic and thermodynamic data are shown in Table 1. Avoiding inactivation of catalyst, the bounds on the temperature are limited between 313 and 333 K, moreover the feeds are (1) 4.22 L/h 30wt % hydrogen peroxide (HP) and half of methanol (M), (2) 5.32 L/h propylene (PL) and the other methanol. The density is constant because reactions are under high pressure. In the objective function, fixed cost a=$50, purchased cost b=$150/m^2, exponent c=1, price of catalyst is $800/m^3, price of utility is $0.1875/t, temperature of utility changed 10K, annual operating time is 7200h.

Table 1. Parameters for reactions

reaction	k_0	E (K J/mol)	H (K J/mol)
1	143	22.83	-223.806
2	239000	54.05	-109.26
3	1.72×10^7	63.90	-105.708

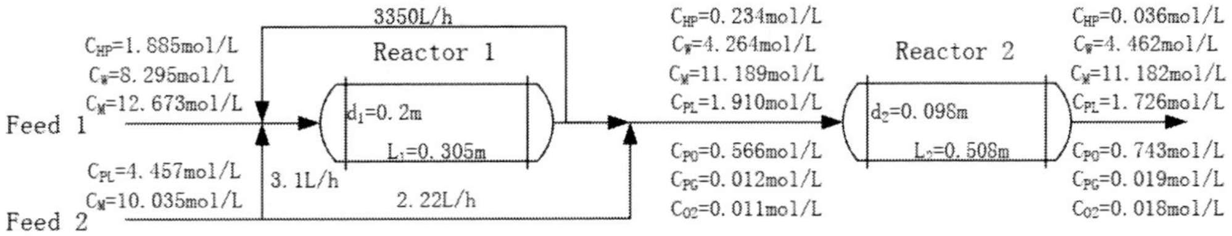

Fig. 2 Optimal reactor network of propylene-oxide system

In Fig. 2, feed (2) is split in two streams entrancing different reactors, and reactor 1 has a large recycle stream which performs the same as a CSTR. Total annual cost is $208.397. The optimal operating temperature of reactor 1 is 315 K, and the temperature profile in the center of reactor 2 is 315-333 K along the length.

5.2 Vapor system

Alkylation of toluene by methanol to produce P-Xylene is a vapor-only reactor system, which means both mole flow and volume flow are changed with length of a reactor.

$$\text{Toluene} + \text{Methanol} \xrightarrow{k_1} \text{P-Xylene} + \text{Water}$$
$$\text{Methanol} \xrightarrow{k_2} 0.5\ \text{Ethylene} + \text{Water}$$
$$\text{Toluene} \xrightarrow{k_3} 0.5\ \text{Benzene} + 0.5\ \text{P-Xylene}$$
$$\text{P-Xylene} \xrightarrow{k_4} 0.5\ \text{Ethylene} + \text{Toluene}$$
$$\text{P-Xylene} \xrightarrow{k_5} 0.5\ \text{M-Xylene} + 0.5\ \text{O-Xylene}$$

A reaction kinetic of the system was proposed by Steol (1993). The reaction rates are described as

$$R_1 = k_1 \times P_{\text{Toluene}} \times P_{\text{Methanol}}$$
$$R_2 = k_2 \times P_{\text{Methanol}}^2$$
$$R_3 = k_3 \times P_{\text{Toluene}}$$
$$R_4 = k_4 \times P_{\text{P-Xylene}}$$
$$R_5 = k_5 \times P_{\text{P-Xylene}}$$

The kinetic and thermodynamic data are shown in Table 2. Avoiding inactivation of catalyst, the feeds are both 713 K： (1) 6.667 mol/min toluene and 10 mol/min nitrogen, (2) 3.333 mol/min methanol and 10 mol/min nitrogen. In the objective function, fixed cost a=$50, purchased cost b=$150/m^2, exponent c=1, price of catalyst is $800/m^3, price of utility is $0.1875/t, temperature of utility changed 10K, annual operating time is 7200h.

Table 2 Kinetic and thermodynamic parameters in alkylation system

reaction	k_0	E (K J/mol)	H (K J/mol)
1	384.5 mol/g.h.atm^2	60.4	-69.965
2	1999 mol/g.h.atm^2	62.4	-11.82
3	250.7 mol/g.h.atm	66.6	0.253
4	4.110 mol/g.h.atm	26.4	58.153
5	165.8 mol/g.h.atm	54.2	0.978

In Fig. 3, feed (2) is split in two streams entrancing different reactors, in which 52.8% molar fraction enters into Reactor 1. Reactor 2 has a little recycle stream which returns to reactor 1. Total annual cost is $212.505. For example, the temperature profile in the center of reactor 1 is 713-712 K along the length, and is 712-707 K in reactor 2.

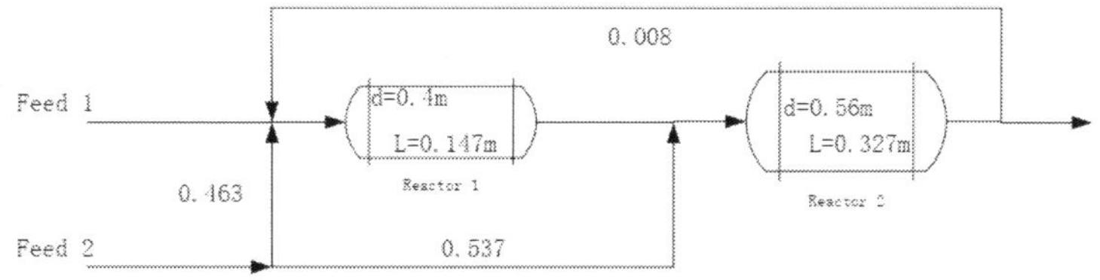

Fig. 3 Optimal reactor network of alkylation system

6. Conclusion

An axial-radial reactor model is established on fluid flow, heat-mass transfer and reaction kinetics. Then a state-space framework based on this two-dimensional model is constructed for synthesis of reactor networks. The Crank-Nicholson method is employed discretizing PDEs to transform this MIDO problem into a MINLP. Liquid and vapor reactor systems are both illustrated to achieve optimal decision variables simultaneously under minimal TAC, includes length, diameter, interconnections and operating conditions. Temperature and concentration distribution on axial and radial directions are easily attained at the same time through the approach proposed, which is demonstrated to be effective.

References

Biegler, L. T. Algorithmic Synthesis of Chemical Reactor Networks using Mathematical Programming. Ind. Eng. Chem. Fund., 1986, 25, 621.

Cuthrell J. E., Biegler L. T. On the Optimization of Differential-Algebraic Process Systems. AIChE J., 1987, 33, 1257.

Lakshmanan A., Biegler L. T. Synthesis of Optimal Chemical Reactor Networks, Ind. Eng. Chem. Res., 1996, 35, 4, 1344-1353.

Schweiger C. A., Floudas C. A. Optimization Framework for the Synthesis of Chemical Reactor Networks, Ind. Eng. Chem. Res., 1999,38, 3, 744-766.

Thomas J. W. Stability, Numerical Partial Differential Equations: Finite Difference Methods,, 1995, Springer-Verlag, 97-137.

Zhou W., Manousiouthakis V. I. Non-ideal Reactor Network Synthesis through IDEAS: Attainable Region Construction, Chemical Engineering Science, 2006, 61, 6936-6945.

Zhou W., Manousiouthakis V. I. Global capital/total annualized cost minimization of reactor network. Industrial and Engineering Chemistry Research, 2008, 47, 3771-3782.

Steol J. L., Uguina M. A., Valverde J. L., et al. Kinetics of Toluene Alkylation with Methanol over Mg-Modified ZSM－5: Ind. Eng. Chem. Res., 1993, 32(11): 2548－2554.

(494d) Efficient Computational Strategies for Multi-Objective Optimization of Multiperiod Waste Management in Chemical Sites

Elisabet Capon, Claude Rerat, Stavros Papadokonstantakis and Konrad Hungerbühler, Chemistry and Applied Biosciences, Institute for Chemical- and Bioengineering, Swiss Federal Institute of Technology, Zurich (ETHZ), Zurich, Switzerland

Waste minimization, material recovery and utilities rationalization have been traditionally dealt as integral parts at the design stage of process plants [1,2]. However, once the production process is established, the management of the generated wastes and its coordination with the production processes and utilities systems have a crucial role in the optimal plant resources allocation from an economic and environmental point of view [3].

In industrial practice, many facilities have centralized waste treatment plants, given that these plants provide a service for the whole chemical complex. Such centralized treatment services have a limited flexibility in their operating conditions, and must deal with a large variety of effluents, which differ in chemical features, mass flow and production time. Thus, the treatment plants have resource requirements, such as steam or electricity, and result in treated wastes which must comply with existing regulations, and may even have added value, such as recovered solvent or heat valuable streams. Hence, it is necessary to account for the production facilities and the waste management plants from an integrative perspective considering production planning, operation and resource management.

This work presents a multi-period waste management multi-objective optimization, considering economic and environmental issues. The specific behavior of the considered waste management treatments is included in the optimization problem as black-box models based on practical industrial practice [4,5,6] and computing utility requirements and emissions. To achieve more realistic solutions, the estimation of waste treatment costs and environmental impacts has been explicitly added to the assessment scheme, as well as the constraints of the operating conditions in the treatment units and the fulfillment of environmental regulations for water and air emissions. This system was applied to an industrial based case study and used to analyze the waste mixing potential.

Two main strategies are proposed to tackle the problem, namely a metaheuristic approach [7] and a rigorous mathematical problem formulation. The performance of both strategies is compared in terms of solution quality and computational complexity with the final aim of finding an efficient methodology for posing the heat and waste management integration problem.

References

[1] Barbosa-Povoa, A. P. A critical review on the design and retrofit of batch plants. *Computers & Chemical Engineering,* 2007, 31, 833-855

[2] Chakraborty, A.; Linninger, A. A. Plant-Wide Waste Management. 1. Synthesis and Multiobjective Design. *Industrial & Engineering Chemistry Research,* 2002, 41, 4591-4604

[3] Melnyk, S. A.; Sroufe, R. P.; Montabon, F. L.; Hinds, T. J. Green MRP: identifying the material and environmental impacts of production schedules. *International Journal of Production Research,* 2001, 39, 1559-1573

[4] Seyler, C.; Hofstetter, T. B. & Hungerbuhler, K. Life cycle inventory for thermal treatment of waste solvent from chemical industry: a multi-input allocation model. *Journal of Cleaner Production*, 2005, 13, 1211-1224

[5] Koehler, A.; Hellweg, S.; Recan, E.; Hungerbuehler, K. Input-dependent life-cycle inventory model of industrial wastewater-treatment processes in the chemical sector. *Environmental Science & Technology*, 2007, 41, 5515-5522

[6] Capello, C.; Hellweg, S.; Badertscher, B.; Hungerbuhler, K. Life-cycle inventory of waste solvent distillation: Statistical analysis of empirical data. *Environmental Science & Technology*, 2005, 39, 5885-5892

[7] Rerat, C.; Papadokonstantakis, S.; Hungerbuhler, K. Integrated liquid waste management in batch chemical industry based on multi-objective optimization. *Journal of Air and Waste Management Association*. Submitted for publication

FEEDSTOCK OPTIMIZATION AND SCHEDULING OF A POLYMER PLANT

Pablo A. Marchetti, Ignacio E. Grossmann
Department of Chemical Engineering, Carnegie Mellon University, Pittsburgh, PA

Wiley A. Bucey and Rita A. Majewski
Braskem America, Pittsburgh, PA

Polymer manufacturing is a continuous process where several campaigns of different product families are scheduled in a cyclic manner. The raw materials, usually referred as 'feedstock', are by-products of petroleum or natural gas such as ethylene and propylene, and their prices are inextricably linked to the prices of such commodities. The polymer market is extremely competitive and the quality expectation of customers is very high. Polymer plants usually operate at full capacity, seeking to improve the return on assets and to reduce supply chain costs[1]. In the last years, significant progress has been made in the development of new computational tools and models to solve problems in the context of Enterprise-wide Optimization[2]. Nonlinear programming formulations as the ones required to model the manufacturing processes of polymer production can now be accurately solved with reasonable computational effort. As an example, a multiperiod nonlinear programming optimization model for planning of production and operation of a real world multi-plant polymer facility as been developed by Jackson et al.[3]. They include detailed nonlinear empirical process models obtained from experimental results and actual plant operation data.

We propose in this work a nonlinear programming model for the selection of the optimal balance of feedstocks for manufacturing multiple grades of polymer product in a polypropylene plant. The feedstocks have different prices and propylene purities, and the process includes a reactor and a distillation column with various types of recycles. The model constraints consist of material balances around the system, lower and upper bounds on flow rates, constraints on compositions, and limits on the catalyst yield and flow. The main decision variables are the production rates of polypropylene for each product type, the flowrates and compositions of each stream, and the catalyst flow. The optimal operation balances the production rate with the costs of feedstocks, maximizing the plant throughput. The different products differ on how much catalyst is needed to produce a given production rate, and the objective function is driven by the economic tradeoff of selling prices versus feedstock costs.

The representation of the distillation process has been one of the major challenges. Approximate solutions for the performance of the distillation column are obtained using the aggregated group-based method of Kamath et al.[4] This method models a counter-current cascade of trays, without providing tray-by-tray details, and was chosen in order to maintain an adequate balance between solution accuracy and model size. Comparisons of the model with real plant data are presented.

Based on simplified and aggregated distillation models, single and multiple-product nonlinear programming feedstock optimization models have been developed. In the latter, a fixed time horizon is considered to produce a given schedule of polymer products. The optimal solution slows down the production rate for certain products to obtain a better separation on the distillation unit, thus reducing the production cost by using more refinery grade as feed to the column. Different time horizons were tested to measure the tradeoff of decreasing feedstock costs versus increasing the production rate. In addition, a "slack" product has been included to assess the benefits of having some extra production when the schedule finishes earlier. Overall, the results consistently showed the convenience of running the plant at its maximum flowrate whenever possible. A user-friendly computational tool that implements the proposed models to assist on purchase decisions by practicing engineers is also described.

References:

(1) Kadipasaoglu S., Captain J., James M. Polymer supply chain management. Int. J. Logistics Systems and Management. 2008; 4:233-253.

(2) Grossmann I. E. Enterprise-wide Optimization: A new frontier in Process Systems Engineering. AIChE Journal. 2005; 51:1846-1857.

(3) Jackson J. R., Hofmann J., Wassick J., Grossmann I. E. A nonlinear multiperiod process optimization model for production planning in multi-plant facilities. Proceedings FOCAPO 2003 (Eds. I. E. Grossmann and C. M. McDonald), pp. 281-284 (2003).

(4) Kamath R. S., Grossmann I. E., Biegler L. T. Aggregate models based on improved group methods for simulation and optimization of distillation systems. Computers and Chemical Engineering. 2010; 34:1312-1319.

(494f) Real-Time Scheduling of Batch Chemical Processes Via Multi-Agent Systems

Yunfei Chu[1], John Wassick[2] and Fengqi You[1], (1)Department of Chemical and Biological Engineering, Northwestern University, Evanston, IL, (2)Dow Chemical Company, Midland, MI

Scheduling is a crucial decision-making in batch processes [1]. Optimization methods provide a systematic approach to the scheduling problem and the optimality of the solution is guaranteed [2, 3]. However, due to the combinatorial nature of the resulting mixed-integer programming (MIP) problems, the computational complexity is still a main challenge inspite of the significant advances in the optimization theories and algorithms as well as computational power in recent years. Due to these limitation, optimization methods are often applied to the short-term scheduling problem under a set of assumptions which are generally made to control the problem complexity even if they compromise the resulting schedule. For example, the number of time points is often confined to a small value to reduce the size of the problem while sacrificing the resolution in the time domain. Applications of the optimization-based methods to the large-scale, long-term, or real-time scheduling problems are rare.

To circumvent the computational complexity in the MIP based approach, various alternatives have been presented. One alternative that has received wide attention is multi-agent based modeling. A major difference between the agent-based approach and others is that it models a process from the bottom up and decisions in the system are distributed across many intelligent agents [4]. This feature equips the agent-based approach with the power of providing a fast and efficient solution which is applicable to real time rescheduling and thus respond to unanticipated events and uncertain processing times. As a promising alternative for scheduling problems, a number of agent-based techniques have been proposed (see examples [5-11]).

However, most agent-based techniques deal with the sequential scheduling problems while applications to the more complex network problems are scarce. In a general network representation, the task agent interacts with not only the equipment agent but also the state agent and the storage agent. The network representation also adds more constraints to the system which in turn require stronger interactions among agents. The complexity of the network structure imposes more difficulties on the implementation of the agent-based approach.

The goal of this work is to develop a real-time scheduling approach based on the multi-agent systems for general batch processes. A novel scheduling algorithm is developed that provides a good balance between computational efficiency and solution quality. Due to the bottom-up feature in the agent-based model, the developed scheduling method can quickly adapt to a changing environment. The high computational efficiency and good solution quality allow the method to be implemented online to rapidly respond to the uncertainties in the process, e.g. the rush order, the equipment breakdown and the batch failure.

We first propose an agent architecture based the general resource-task network (RTN) representation. This allows the proposed approach to be applicable to a wide range of batch scheduling problems. Second, to improve the optimality of the solution returned by the agent based approach, we develop a simulation based scheduling algorithm via a two-level agent system as an alternative to a single level agent system. The simulation is conducted by an inner agent based system at the request of the outer agent system responsible for the schedule. As a result, an embedded agent based system is created. The inner agent system copies the current environment of the outer agent system when a scheduling decision needs to be made in the outer agent system. The simulation by the inner agent system returns the predicted objective function value for proposed decisions so that the outer agent system can make the final selection based on the predicted objective function value.

Comparison with the MIP approaches based on discrete time model model is performed through a real-world case study. The results show that agent-based approach can return good quality solution with significantly shorter computational time. Therefore, it can be applied to a long term scheduling problem and can also be more easily implemented online to reschedule the process under the uncertainties. Though the computational time is considerably less than the optimization based method, the solution quality is comparable to the optimal solution.

References:

[1] C. A. Mendez, J. Cerda, I. E. Grossmann, I. Harjunkoski, and M. Fahl, "State-of-the-art review of optimization methods for short-term scheduling of batch processes," *Computers & Chemical Engineering,* vol. 30, pp. 913-946, May 2006.

[2] E. Kondili, C. C. Pantelides, and R. W. H. Sargent, "A general algorithm for short-term scheduling of batch operations—I. MILP formulation," *Computers & Chemical Engineering,* vol. 17, pp. 211-227, Feb 1993.

[3] C. C. Pantelides, "Unified frameworks for optimal process planning and scheduling," in *Foundations of computer-aided process operations*, ed New York: Cache publications, 1994, pp. 253-274.

[4] W. M. Shen, L. H. Wang, and Q. Hao, "Agent-based distributed manufacturing process planning and scheduling: A state-of-the-art survey," *Ieee Transactions on Systems Man and Cybernetics Part C-Applications and Reviews,* vol. 36, pp. 563-577, Jul 2006.

[5] P. I. Cowling, D. Ouelhadj, and S. Petrovic, "A multi-agent architecture for dynamic scheduling of steel hot rolling," *Journal of Intelligent Manufacturing,* vol. 14, pp. 457-470, Oct 2003.

[6] P. Gu, S. Balasubramanian, and D. H. Norrie, "Bidding-based process planning and scheduling in a multi-agent system," *Computers & Industrial Engineering,* vol. 32, pp. 477-496, Apr 1997.

[7] T. N. Wong, C. W. Leung, K. L. Mak, and R. Y. K. Fung, "Dynamic shopfloor scheduling in multi-agent manufacturing systems," *Expert Systems with Applications,* vol. 31, pp. 486-494, Oct 2006.

[8] A. Aldea, R. Banares-Alcantara, L. Jimenez, A. Moreno, J. Martinez, and D. Riano, "The scope of application of multi-agent systems in the process industry: three case studies," *Expert Systems with Applications,* vol. 26, pp. 39-47, Jan 2004.

[9] A. Yang, B. Braunschweig, E. S. Fraga, Z. Guessoum, W. Marquardt, O. Nadjemi, D. Paen, D. Pinol, P. Roux, S. Sama, M. Serra, and I. Stalker, "A multi-agent system to facilitate component-based process modeling and design," *Computers & Chemical Engineering,* vol. 32, pp. 2290-2305, Oct 2008.

[10] M. Metzger and G. Polakow, "A Survey on Applications of Agent Technology in Industrial Process Control," *Ieee Transactions on Industrial Informatics,* vol. 7, pp. 570-581, Nov 2011.

[11] N. Julka, R. Srinivasan, and I. Karimi, "Agent-based supply chain management-1: framework," *Computers & Chemical Engineering,* vol. 26, pp. 1755-1769, Dec 2002.

(494g) Investigation of Hydrogen Gas Safety for the Electrolysis Based Ballast Tank Treatment System

Bumjoon Cha[1], Hyungtae Cho[1], Iksang Yoo[2] and **Il Moon**[1], (1)Chemical and Biomolecular Engineering, Yonsei University, Seoul, South Korea, (2)99 Seorin-dong, Jongro-gu, SK Innovation, Seoul, South Korea

Ballast water has been identified as a serious risk to marine environment and a major cause of ecological disturbance due to the pathogens and harmful adventive species discharged through ballast water. The International Maritime Organization (IMO) promulgates the international guidelines containing regulations about technical standards and requirements for ballast water management in order to control adventive species in the discharges. A number of techniques are developed to diminish the aquatic invasive species in ballast water and sedimentations, electrolysis coupled treatment system shows advantages in disinfection efficiency for target species and simple installation and maintenance of system. Simultaneously, the explosion risk of hydrogen gas generated through electrolysis system is issued in terms of safety of the ship. The main aim of this study is to verify dynamic behavior of ballast water and concentrations of hydrogen gas in the ballast tank during ballasting process. Computational fluid dynamic (CFD) simulations were implemented to analyze multiphase behavior of hydrogen gas and ballast water within the ballast tank. Hydrogen gas concentration at vent pipe is also measured to verify CFD simulation model. Consequently, hydrogen gas is released into vent pipe normally at 200 ppm, and ballast water contains 0.3 ppm of hydrogen in average. Based on the analysis, we analyze that hydrogen explosion is highly unlikely in the ballast tank. Finally, a recommendation on operationally feasible ways for sustaining hydrogen safety against potential explosion is suggested.

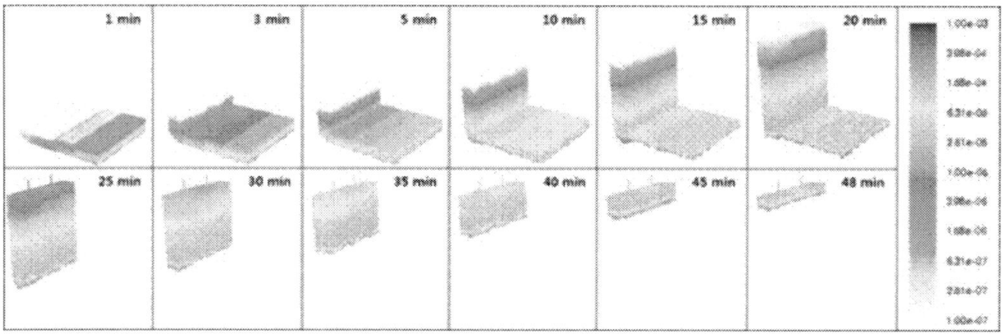

Figure SEQ Figure * ARABIC 1. Hydrogen gas concentration profile

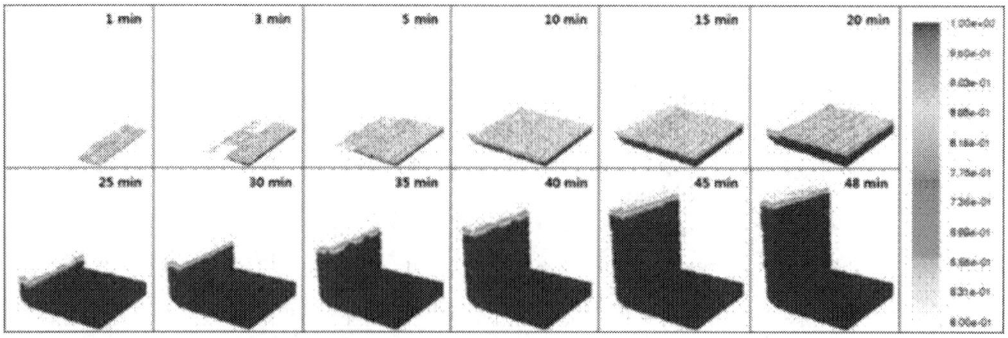

Figure SEQ Figure * ARABIC 2. Volume fraction of sea water

References

[1] M. M. Abu-Khader, O. Badran, and M. Attarakih, Ballast water treatment technologies: hydrocyclonic a viable option, *Clean technologies and Environmental Policy*, 13(2), 403-413 (2010).

[2] E. Tsolaki and E. Diamadopoulos, Technologies for ballast water treatment: a review, Chemical Technology and Biotechnology, 85(1), 19-32 (2009).

[3] V. A. Danilov and Il Moon, Gas management in flow field design using 3D DMFC model under high stoichiometric feed, *Korean Journal of Chemical Engineering*, 23(5), 753-760 (2006).

[4] E. Kim, K. Lee, J. Kim, Y. Lee, J. Park and I. Moon, Development of Korean hydrogen fueling station codes through risk analysis, International Journal of Hydrogen Energy, 36(20), 13122-13131 (2011).

(496a) Partial Control: Reuel Shinnar's Impact On Control of Complex Chemical Systems

Mayuresh V. Kothare, Chemical Engineering, Lehigh University, Bethlehem, PA

Many complex chemical processes having a large number of process variables and poorly understood models can be controlled reasonably well by controlling only a small subset of process variables using an equally small number of manipulated variables. This basic idea, termed "partial control", was originally proposed by Reuel Shinnar in 1981. Knowingly or unknowingly, this idea has been and continues to be applied to successfully control numerous complex industrial processes. We present a number of terms such as process variable dominance, modelable responses, practical degrees of freedom and sufficiency of partial control. We demonstrate the applicability of these concepts with several examples.

AIChE 2012 Annual Meeting

Session: In Memory of Professor Reuel Shinnar: A Major Contributor to Process Control.

From Stability to Performance-Driven Control Configurations for Multi Delay Systems Or The True Extension of Smith Predictor to Multi-Delay Cases

Prof. Zalman J. Palmor, *Faculty of Mechanical Engineering*

Technion –Israel Institute of Technology

Abstract

Most dead-time compensation (DTC) schemes developed to-date can be regarded as stability driven since their predictive elements are conventionally structured to eliminate delays from the closed-loop characteristic equation. In the single-delay case this approach happens to also agree well with a number of optimization-based design methods. This, however, is no longer true in the multiple-delay case, where stability driven DTCs might yield poor closed-loop performance. In fact, there appears to be no clear choice of the DTC configuration and conventional wisdom even suggests that in certain situations it may be advantageous to add artificial loop delays to control channels with shorter delays to equalize input dead times. In this talk, given in memory of my mentor for many years the late Professor Reuel Shinnar, I will show that DTC structures should be performance driven. To simplify the exposition, systems with multiple input delays will be considered and used to show how performance analysis can lead to a novel DTC configuration termed FASP (feedforward action Smith predictor). Alongside the conventional internal feedback, this configuration contains novel interchannel feedforward controllers and never adds artificial loop delays. The structure of FASP has an intuitive explanation and effectively prompts a paradigm shift in DTC for systems with multiple loop delays. It is shown that the FASP, like the (modified) Smith Predictor in the single delay cases, is inherent in many of the optimization-based design methods and thus may be considered as the true extension of the Smith Predictor to multiple delay processes. Furthermore, the development of the FASP enables to dispel many of the long standing myths associated with the Smith Predictor and its state of the art extensions. Application of the FASP to an upgraded version of the quadruple tank process with multiple delays laboratory set-up confirms the potential superiority of the FASP over existing state of the art dead-time compensators.

(496c) Nano-Scale Process Systems Engineering: Conceptual Design, Fabrication and Operation of Nano-Scale Processes

Gregory Stephanopoulos, Department of Chemical Engineering, Massachusetts Institute of Technology, Cambridge, MA

Process Systems Engineering (PSE) has effectively solved all major problems in simulation, design, control, diagnosis, scheduling and planning of large-scale chemical processes. As the focus of research moved in scale from m^3 to mm^3, "plants or labs on a chip" benefited from the accumulated PSE technologies, since the underlying phenomena could still be handled as effective continuous media. Current research, however, has pushed the scale of processing operations to a few nanometers. With the proposition of "nanoscale factories" as the next frontier, PSE must offer new theories and tools to realize the design, fabrication, simulation, operation and control of processing systems.

This presentation will address the foundational questions of nanoscale PSE that need to be tackled with new theories and computational methodologies. In particular, we will discuss: (1) How to synthesize the three networks of chemical reactions (process topology; energy production and dissipation system; monitoring and control), which compose the essential design of any nanoscale process, using new views on self-replicating reaction networks. (2) How to fabricate nanoscale structures with desired geometries, using guided self-assembly processes. (3) How to engineer monitoring and control systems, as integral parts of a processing topology, and how to deploy self-regulating control architectures.

(496d) Using Probaility Models to Represent Process Engineering Phenomena

Babatunde A. Ogunnaike, Department of Chemical and Biomolecular Engineering, University of Delaware, Newark, DE

This presentation highlights how probabilily models have been used successfully to represent randomly varying phenomena in process engineering. It is presented as an appropirate tribute to the late professor Reuel Shinnar one of hte earliest pioneers of probabilistic modeling in chemical engineering.

Using confocal microscopy and image analysis to quantify the size, shape, and location of integrin clusters, we observed that cells exhibit large variability in these integrin cluster properties. To describe these heterogeneous populations of clusters quantitatively, we identified appropriate probability models to characterize the size, shape, and location of integrin clusters in a population of adherent cells. We determined that integrin cluster sizes are lognormally distributed, integrin cluster eccentricities are Beta distributed, and the distances of integrin clusters from the cell center are Gamma distributed. We estimated the parameters corresponding to these probability models based on our measurement of integrin clusters in a population of cells, and the resulting best fit probability models describe the heterogeneous populations of clusters and provide the means to create an accurate mathematical description of spatially-localized integrin signaling compartments for use in computational models.

(496e) Novel Application of Control and Dynamic Systems Analysis to Coral Ecosystems

Francis J. Doyle III, Chemical Engineering, UCSB, Santa Barbara, CA

Inspired by Prof. Reuel Shinnar's creative work in such diverse application areas as crime control, this talk will address another unusual application for process systems engineers: the exquisite control of annual reproduction in spawning coral. The species under investigation has a well characterized spawning event that occurs once a year, on one evening for approximately one hour. Using a variety of mathematical models and systems analysis techniques, a novel hypothesis will be detailed to account for the precise timekeeping achieved in the face of stochastic environmental disturbances. A cascade network of controllers, at time scales of months (circa-annual), days (circa-lunar), and days (circa-diel) will be described. The model predictions will be supported by field data collected over the last 5 years in the pacific republic of Palau. The principles underlying this biological clock will be explored for relevance in chemical engineering problems, such as distributed control of sensors over networks. Ongoing work to unravel the gene regulatory circuitry will also be described.

(508a) Dynamic Modeling and Recipe Optimization of Semibatch Polymerization Processes

Yisu Nie[1], Lorenz Biegler[1], Carlos M. Villa[2] and John Wassick[3], (1)Chemical Engineering, Carnegie Mellon University, Pittsburgh, PA, (2)The Dow Chemical Company, Freeport, TX, (3)Dow Chemical Company, Midland, MI

Dynamic Modeling and Recipe Optimization of Semibatch Polymerization Processes

Y. Nie L.T. Biegler

Department of Chemical Engineering

Carnegie Mellon University

C.M. Villa J.M. Wassick

The Dow Chemical Company

Dynamic modeling and optimization techniques offer advantages for the design and improvement of process recipes. In this study, a mathematical model is first developed to describe the dynamic behavior of a polymerization process, based on first-principles including the heat and mass balances, reaction kinetics and vapor-liquid equilibrium (VLE). Next, a dynamic optimization strategy based on the direct transcription method is applied to optimize the process recipe, where the residence time is minimized, given a target product molecular weight as well as requirements on byproduct formation and reactor temperature.

The polymerization reaction can be carried out in conventional stirred- tank reactors with heat exchangers, following a semi-batch manner. The raw materials are first formed by mixing with catalysts in the reactor tank, and then monomer is continuously fed into the tank to form homopolymers. Each polymer chain undertakes the initiation, propagation, exchange and proton transfer reactions. There are two major considerations during the operation of the reactor. First, impurities are generated, which are undesired. Second, the unreacted monomer concentration should be limited to ensure safe operations in case of losing the cooling capacity. To deal with both issues, the monomer feed rate and the coolant flow rate need to be well controlled, and the concept of the adiabatic end temperature is introduced as a safety criteria with a specified upper limit.

The developed polymer process model is complicated by a large set of differential algebraic equations (DAEs) due to the population balances written with respect to chains of the same number of repeating units. Therefore, the optimization formulation is in the form of a dynamic optimization problem. The direct transcription method serves as an efficient tool for solving such problems. In this method, the DAEs are fully discretized into nonlinear algebraic equations at collocation points over finite elements and thus the dynamic optimization problem is translated into a large-scale nonlinear program (NLP) that can be handled by the off-the-shelf nonlinear solvers such as IPOPT. In this study, we model and solve the translated NLP in GAMS with IPOPT. From the simulation results, the batch time is significantly reduced compared to industrial data.

(508b) A Theoretical and Experimental Investigation On the Ring-Opening Polymerization of L,L-Lactide in the Presence of a Co-Initiator with Different Numbers of Hydroxyl Groups

Prokopis Pladis[1], Konstantina Karidi[1], Theofanis Mantourlias[2] and **Costas Kiparissides**[2], (1)Chemical Process Engineering Research Institute, Thessaloniki, Greece, (2)Department of Chemical Engineering, Aristotle University of Thessaloniki & Chemical Process Engineering Research Institute, Thessaloniki, Greece

Currently, there are two main manufacturing routes employed for the production of PLA. The first one involves the direct polycondensation of lactic acid. The main drawback of this technology is the requirement for continuous removal of produced water with the aid of a suitable solvent, under high vacuum and high temperature conditions. Moreover, the molecular weight of PLA produced by this process is in general low. The second production route employed by the industry is the ring-opening polymerization (ROP) of L,L-lactide that leads to the synthesis of high molecular weight polymers. The main advantage of this method is that by controlling the purity of the monomer, a wide range of molecular weights and polymer chain architectures can be obtained in the presence of a suitable co-initiator (i.e., mono-, bi- and multi-functional alcohols). Moreover, the fact that PLLA is produced in melt does not require the use of any environmentally hazardous solvents. After polymerization, the unreacted monomer is removed from the polymer melt, under vacuum, and is recycled back to the reactor.

Over the past twenty years, a great number of papers have been published on the synthesis of polylactide and several kinetic mechanisms have been proposed to describe the ring-opening polymerization of L,L-lactide [1-3]. The proposed kinetic mechanisms include a number of elementary reaction steps (i.e., chain initiation and propagation, different transesterification reactions, etc.). However, the most controversial reaction step regards the chain initiation rection. The two initiation mechanisms that have prevailed in the literature, are the "monomer activated mechanism" [4, 5] and the "alkoxide initiation mechanism" [6-8]. The former mechanism involves the polarization of the L,L-lactide carbonyl bond by interaction with stannous octoate and a proton, resulting in the formation of a carbocation and a cyclic intermediate that initiates the polymerization. In the latter case, stannous octoate reacts with the OH-bearing species to form an alkoxide, which is considered to be the actual initiating species.

In the open literature, there is a limited number of publications dealing with the mathematical modeling of the ROP of L,L-lactide. The first modelling attempts on the ROP of L,L-lactide, after the pioneering work of Eehnik [9], were reported by Paux et al. [10] and Mehta et al. [11, 12] who developed simplified mathematical models based on a cationic kinetic mechanism. The derived kinetic models included irreversible chain initiation, irreversible propagation, chain transfer to monomer and to impurities reactions. It was shown that model predictions were in good agreement with the experimental data on monomer conversion and molecular weight. The values of the kinetic parameters were also reported. Recently, Yu et al. [13] developed a kinetic model to describe the ROP of L,L-lactide. Experimental measurements on monomer conversion and number and weight average molecular weight were used to validate the mathematical model and estimate the unknown kinetic parameters. The polymerization experiments were carried out at 130 ^{0}C, in the presence of stannous octoate as initiator and 1-dodecanol as co-initiator, for different values of the initiator to co-initiator molar ratio.

The present study deals with the experimental and theoretical investigation of the ROP of L,L-lactide, in the presence of $Sn(Oct)_2$ and a co-initiator (i.e., 1,4-butanediol, glycerol, di-trimethylol-propane, polyglycidol) with different numbers of hydroxyl groups, for the production of high molecular weight PLLA. The polymerization of L,L-lactide was carried out at different monomer to co-initiator molar ratios and different temperatures. Particular attention was paid to the purification of all the reagents employed in order to control the impurities concentration that resulted in a very high reproducibility of the kinetic data. The produced polymers were fully characterized with respect to their molecular and thermal properties. In addition, a comprehensive mathematical model was developed based on a detailed kinetic mechanism of the ROP of L,L-lactide to predict the dynamic evolution of the monomer conversion and molecular weight developments (i.e., M_n, M_w, PD index of PLLA). The proposed kinetic mechanism for the ROP of L,L-lactide comprised a series of elementary reaction steps, including initiator activation, chain initiation, propagation, chain transfer to water and to octanoic acid, transesterification and chain scission reactions. In the presentation, the predictive capabilities of the mathematical model will be demonstrated by a direct comparison of model predictions with experimental measurements on monomer conversion as well as number and weight average molecular-weights of PLLA.

References

1. Mehta R, Kumar V, Bhunia H, Upadhyay SN. J. Macrom. Sci. Part C, Polym. Rev. 2005;45:325-349.

2. Gupta AP, Kumar V. Eur. Polym. J. 2007;43:4053-4074.

3. Jerome C, Lecomte P. Adv. Drug Deliv. Rev. 2008;60:1056-1076.

4. Kricheldorf HR, Kreiser-Saunders I, Boettcher C. Polymer 1995;36:1253-1259.

5. Schwach G, Coudane J, Engel R, Vert M. J. Polym. Sci. Part A: Polym. Chem. 1997;35:3431-3440.

6. Kowalski A, Duda A, Penczek S. Macromolecules 2000;33:7359-7370.

7. Majerska K, Duda A, Penczek S. Macromol. Rapid Commun. 2000;21:1327-1332.

8. Penczek S, Duda A, Kowalski A, Libiszowski J, Majerska K, Biela T. Macromol. Symp. 2000;157:61-70.

9. Eenink MJD. Synthesis of Biodegradable Polymers and Development of Biodegradable Hollow Fibers for the Controlled Release of Drugs. Ph.D. Thesis, Technische Hogeschool Twente Enschede, Enschede, The Netherlands 1987.

10. Puaux JP, Banu I, Nagy I, Bozga G. Macromol. Symp 2007;259:318-326.

11. Mehta R, Kumar V, Upadhyay S. N. Polym.-Plast. Technol. Eng. 2007;46:933-937.

12. Mehta R, Kumar V, Upadhyay S. N. Polym.-Plast. Technol. Eng. 2007;46:257-264.

13. Yu Y, Storti G, Morbidelli M. Macromolecules 2009;42:8187-8197.

(508c) Modelling of High-Pressure Ethylene Polymerization – From Reactor to Rheology

Thomas Herrmann[1], Markus Busch[2], Dieter Lilge[3], Gerd Mannebach[3] and Iakovos Vittorias[3], (1)Manufacturing Engineering Support EAI, Basell Polyolefine GmbH – LyondellBasell, Frankfurt, Germany, (2)Ernst-Berl-Institut, TU Darmstadt, Darmstadt, Germany, (3)Basell Polyolefine GmbH, a LyondellBasell company, R & D, Frankfurt, Germany

Developing a new polymer grade normally requires numerous experiments in pilot plants. Modeling studies can help to reduce the number of these expensive experiments and allow assessing the potential of different approaches.

For extrusion coating applications the processability of the polymers is of great importance. The processability is strongly coupled to the polymeric microstructure which in turn is governed by the reaction conditions during the polymerization. Not only pressure and temperature but also the reactor type (tubular or autoclave) have a strong influence on the polymer properties. Due to the differences in the polymer microstructure (e.g. molar mass distribution, branching density distribution, topology, etc.) for extrusion coating applications autoclave grades are predominantly used. However, the heat transfer and therefore as a direct consequence the conversion in autoclave reactors is inferior to tubular reactors. Thus it is desirable to use modeling techniques to assess potential routes to produce autoclave-like grades on tubular reactors.

As outlined above such a model has to be able to

- solve the heat balance for tubular reactors with counter-current cooling
- provide detailed information on the polymer microstructure
- capture and visualize the main differences between autoclave and tubular grades which are relevant for the processability in extrusion coating applications

Within the software Predici® [1] a model has been developed which meets these requirements. Based on a simplistic topological approach a characteristic function is derived which quantifies whether a polymer possesses a more comb-like or tree-like structure. It is found that tubular reactors produce more comb-like shaped polymers whereas autoclave reactors yield tree-like structures, which is documented in literature long time ago. However, the quantitative approach allows to assess whether modifications on the reaction conditions lead to autoclave-like grades from tubular reactors. The model clearly shows that by feeding a functionalized comonomer to the tubular reactor more tree-like polymers can be created.

A direct coupling to rheological quantities is highly desirable. Coupling the Predici simulation with Monte Carlo techniques yields a representative set of sample molecules for which the complete polymer architecture is known. This information is required to use the rheology models of the McLeish group [2] to predict shear viscosity functions. The computed shear viscosity functions of commercial tubular grades agree well with the experimental values.

[1] M. Wulkow, *Macromol. React. Eng.* **2008**, *2*, 461 – 494.

[2] D. J. Read, D. Auhl, C. Das, J. den Doelder, M. Kapnistos, I. Vittorias, T. C. B. McLeish, Science 2011, 333, 1871 – 1874.

(508d) How to Deal with Fast Process Dynamics of the Ethylene Solution Polymerization Process During Optimal Grade Transitions?

Karen Pontes[1], Marcelo Embiruçu[1], Inga Wolf[2] and Wolfgang Marquardt[2], (1)Industrial Engineering Postgraduate Program (PEI), Federal University of Bahia (UFBA), Salvador, Brazil, (2)Process Systems Engineering, RWTH Aachen University, Aachen, Germany

How to Deal with Fast Process Dynamics of the Ethylene Solution Polymerization Process During Optimal Grade Transitions?

Karen V. Pontes[1], Marcelo Embiruçu[1], Inga J. Wolf[2] and Wolfgang Marquardt[2]

[1]PEI – Industrial Engineering Postgraduate Program, Bahia Federal University

[2]AVT – Process Systems Engineering, RWTH Aachen

For highly non-linear processes with frequent product changes such as polymerization processes, operators' and engineers' experience may not suffice to operate the plant in an economically optimal way while maintaining product quality. In this context, real-time optimization strategies coupled with automatic process control offer opportunities for optimal process operation during the production of target polymer grades.

Recent advances in the solution of dynamic optimization (DO) problems have enabled their online solution, so that several approaches considering real-time optimization strategies for polymerization processes have been reported in the last years. The computational delay is usually lower than the grade transition duration, which may take minutes or even hours to complete. However, when approaching large-scale processes with fast dynamics, computational delay may hinder online optimization. This is the case for the ethylene polymerization in solution with catalyst in a series of stirred and tubular reactors, introduced by Braskem S.A. (Camaçari Petrochemical Pole - Brazil). A typical grade transition may take around 2-5 minutes due to high loads and small residence time. These fast dynamics are confirmed by a phenomenological mathematical model validated with industrial data. Bearing this in mind, this work seeks to investigate how real-time optimization strategies can be formulated to optimize the grade transition and production of this ethylene polymerization process.

Firstly, a two-layer dynamic real-time optimization (D-RTO) architecture, as formulated by Würth et al. (2011), is investigated which is then compared to a modification of the established two-layer real-time optimization (RTO) architecture (Marlin and Hrymak, 1997). The economic optimization at the upper-layer is trigged periodically to compensate the impact of slow disturbances (such as trends resulting from heat exchanger fouling or catalyst decay) on economically optimal operation and to compute trajectories for scheduled grade transitions. A linear time-varying MPC (Model-Predictive Controller) on the lower-layer tracks the reference trajectories at a higher sampling time to reject fast disturbances.

The D-RTO suggested here relies on a novel formulation of a two-stage DO problem for polymer grade transitions, considering solely economic goals in the objective function while the polymer properties are specified through constraints. Results show that a typical optimal grade transition problem, represented by a large-scale model made up of 140 differential and 2275 algebraic state variables can be efficiently solved in around 270 seconds. Furthermore, the two-layer architecture avoids the solution of the rigorous DO problem at every controller sampling time. Instead, it has to be solved less frequently. Other approaches, such as the one suggested by Zavala et al. (2006) combine the optimizer and the controller into a single-layer so that a DO problem has to be solved at every sampling time during the grade transition of low-density polyethylene tubular reactors. The DAE model used in the case study of Zavala et al. (2006), containing 294 differential and 64 algebraic state variables, is solved in around 351 seconds. Since grade transition in this case study is slow, a controller sampling time of 6 minutes can be used.

However, since the polymerization process investigated here presents very fast dynamics, a typical grade transition is completed already in around 2-5 minutes hindering the implementation of D-RTO in this industrial scenario. Attempting to ensure profitable process operation during the whole production cycle while targeting polymer properties, a modification of the established RTO approach is investigated. The steady-state optimization followed by the simulation of the optimal step response replaces the solution of the DO problem in the upper-layer of the D-RTO architecture when computing the reference trajectory.

Results show that the two-layer D-RTO allows higher profits and less off-spec production than the modified RTO, as expected. Important to mention, though, is that the RTO framework proposed here can already increase process profitability up to 30% compared to standard process operation, thus presenting itself as a promising alternative for implementation in industrial scenarios with less computational load. The closed-loop responses of both approaches are satisfactory since the underlying MPC can efficiently track the reference trajectories for the controlled and manipulated variables in the presence of fast disturbances.

Würth L, Hannemann R, Marquardt W. A two-layer architecture for economically optimal process control and operation. *J. Proc. Cont.* **21**, 311-321 (2011);

Marlin TE, Hrymak AN. Real-time operations optimization of continuous processes. In: *AIChE Symp. Ser.* **93**, 156-164 (1997);

Zavala VM, Laird CD, Biegler LT. Fast Implementations and Rigorous Models: Can both be accommodated in NMPC? In: *Int. J. Robust Nonlinear Control* 1–20 (2006).

(508e) Kinetics of Mechanisms of Chain Transfer to Polymer Reactions in Alkyl Acrylates: A Theoretical Study

Nazanin Moghadam[1], Masoud Soroush[1], Shi Liu[2], Andrew M. Rappe[2], Michael C. Grady[3] and Sriraj Srinivasan[4], (1)Department of Chemical and Biological Engineering, Drexel University, Philadelphia, PA, (2)Makineni Theoretical Laboratories, Department of Chemistry, University of Pennsylvania, Philadelphia, PA, (3)DuPont Experimental Station, Wilmington, DE, (4)Arkema, King of Prussia, PA

Environmental regulations [1,2] to reduce the volatile organic contents (VOCs) of coatings have been responsible for changes in the basic formulation of resins used in automobile coatings. High-temperature (>100 °C) polymerization has been used widely to produce high-molecular-weight high-solid-content resins. High-temperature free-radical polymerization is a complex process, which involves many competing secondary reactions that strongly influence the polymerization rate and polymer end-use properties [3]. Examples are *b*-scission, monomer self-initiation, and intra- and inter-molecular chain transfer reactions.

Previous studies [4-9] showed the successful use of computational quantum chemistry to better understand self-initiation and propagation reactions in polymerization systems. In our previous work [10], we studied the kinetics of a chain transfer to polymer (CTP) mechanism in polymerization of methyl acrylate using B3LYP/6-31G*. To the best of our knowledge, a comprehensive computational study of CTP reactions in polymerization of alkyl acrylates has not been reported as of yet.

This paper presents a comprehensive computational study of CTP reaction mechanisms in self-initiated high-temperature polymerization of alkyl acrylates. Different levels of theory (functionals and basis sets) have been applied to predict molecular structures, transition state geometries and barriers. The sensitivity of the levels of theory (B3LYP, X3LYP and M06-2X functionals, and 6-31G(*d*), 6-31G(*d,p*), 6-311G(*d*), and 6-311G(*d,p*) basis sets) to barriers and rate constants was evaluated. The effects of polymer chain-length and type of radical (*e.g.* tertiary *vs*. secondary radical) on the rate constant for chain transfer to polymer have been investigated. We have identified that the abstraction of hydrogen from a tertiary carbon atom in a dead polymer by a mono-radical is the most likely mechanism for CTP in ethyl acrylate (EA) and *n*-butyl acrylate (*n*-BA). We have found that that the mono-radicals generated via self-initiation have little influence on the capability of EA and *n*-BA live polymer chains to undergo CTP reactions. The effect of live polymer-chain length on the rate coefficients of chain transfer to polymer reactions was found to be insignificant.

References

[1] K. R. Schultz, Proc. Adv. Coat. Technol. Conf., Chicago, IL, 1992.

[2] VOC's Directive, EU Committee of the American Chamber of Commerce in Belgium, ASBL/VZw, Brussels, July 8, 1996.

[3] Cunningham, M. F.; Hutchinson, R. A. In Handbook of Radical Polymerization; Matyjaszewski, K., Davis, T. P., Eds.; John Wiley & Sons: Hoboken, NJ, 2002; pp 333-360.

[4] Nikitin, A.N.; Hutchinson, R.A.; Kalfas, G.A.; Richards, J.R.; Bruni, C.; The effect of Intramolecular Transfer to Polymer on Stationary Free-Radical Polymerization of Alkyl Acrylates, Macromolecular, 2009, 18, 247-258.

[5] Heuts, J.P.A.; Gilbert, R.G.; Radom, L.; A Prior Prediction of Propagation Rate Coefficients in Free Radical Polymerizations: Propagation of Ethylene, Macromolecules, 1995, 28, p. 8771-81.

[6] Khuong, K.S.; H. Jones, W.H.; Pryor, W.A.; Houk, K.N.; The Mechanism of Self-Initiated Thermal Solution of a Classic Problem, J. Am. Chem. Soc. 2005. 127, 1265-77.

[7] Yu, X.; Pfaendtner, J.; Broadbelt, L.J.; Ab Initio Study of Acrylate Polymerization Reactions: Methyl Methacrylate and Methyl Acrylate Propagation, J. Phys. Chem. A. 2008, 112, 6772-82.

[8] Srinivasan, S.; Lee, M.W.; Grady, M.C.; Soroush, M.; Rappe, A.M.; Computational Study of the Self-Initiation Mechanism in Thermal Polymerization of Methyl Acrylate, J. Phys. Chem. A 2009, 113, 10787-94.

[9] Srinivasan, S.; Lee, M.W.; Grady, M.C.; Soroush, M.; Rappe, A.M.; Self-Initiation Mechanism in Spontaneous Thermal Polymerization of Ethyl and n-Butyl Acrylate: A Theoretical Study, J. Phys. Chem. A 2010, 114, 7975-83.
[10] Moghadam, N.; Soroush, M.; Srinivasan, S.; Rappe, A. M.; Grady, M. C.; Chain Transfer to Polymer Reactions in Thermal Polymerization of Methyl Acrylate: Computational Study, Paper 300e, AIChE Annual Meeting 2011.

[10] Moghadam, N.; Soroush, M.; Srinivasan, S.; Rappe, A. M.; Grady, M. C.; Chain Transfer to Polymer Reactions in Thermal Polymerization of Methyl Acrylate: Computational Study, Paper 300e, AIChE Annual Meeting 2011.

(508f) Monitoring Emulsion Polymerization Processes by Means of Ultra-Sound Velocity Measurements

Heiko Brandt, Dominik Sühling and Sebastian Engell, Department of Chemical and Biochemical Engineering, Technische Universität Dortmund, Dortmund, Germany

Monitoring Emulsion Polymerization Processes

by Means of Ultra-Sound Velocity Measurements

Heiko Brandt, Dominik Sühling, Sebastian Engell, TU Dortmund, Germany

Introduction

Emulsion polymerization is an important industrial (semi-) batch process for the production of latices, which are used as adhesives or in paints. To drive the process to its economic optimum, the desired polymer properties must be achieved at an as high as possible reaction rate maintaining tight temperature control and respecting safety constraints. This can only be achieved by model based control, which requires the knowledge of the process state. In this work the usability of sound velocity measurements to determine the process state is analyzed.

In a homopolymerization process the measurement of the jacket and reactor temperature combined with a qualitatively good energy balance is sufficient to monitor the reaction progress using classical heat balance calorimetry. This can be combined with an open loop observer for the monomer hold up and is widely applied in polymerization processes, e.g. by (Gugliotta et al, 1994), (de Buruaga et al, 1997), or (Vicente et al, 2001). Today instead of calorimetry a state estimator is often applied as it offers the opportunity to combine information from different measurement devices, to filter the measurement noise and to estimate unknown process parameters at the same time. Examples are the application of an extended Kalman Filter by e.g. (Gesthuisen et al, 2004), who estimated the heat of reaction as well as the heat transfer coefficient, or a high gain observer used by (Sheibat-Othman and Othman, 2006), who estimated the overall reaction rate.

In a copolymerization process the measurement of the temperatures enables only to determine the overall reaction rate, i.e. the sum of all monomer reaction rates. But as the copolymer composition is an important polymer property, for its control it is necessary to determine the individual monomer consumptions. An individual measurement of the monomer consumption is possible e.g. by the application of online gas chromatography. This was successfully shown by (Dimitratos et al, 1989) and (Urretabizkaia et al, 1994). But it is not used much in industry due to the infrequently available and delayed data in addition to the necessary sampling device. Another direct measurement method for the individual monomer consumption is spectroscopy, often being (Near) Infrared Spectroscopy (N)IR or Raman Spectroscopy. This method offers not only the opportunity to determine the individual monomer hold up directly, but may also be used to measure e.g. the particle size distribution. Examples for the application of spectroscopy are given in (Elizalde et al, 2005) or (Santos et al, 2008). For a more frequent industrial use, the necessary effort to obtain a calibration model, the need of maintenance, and the high investment cost are problematic up to now.

Beside direct measurements, indirect measurements exist that can provide information about the monomer concentrations. Examples are densimetry and ultra-sound velocity measurements. The first one has been applied e.g. by (Canegallo et al, 1994), but it was claimed by (Vicente, 2001) to be not robust enough for further application. The usability of ultra-sound velocity measurements as a second independent measurement is analyzed in this work. As a first step towards the application to a copolymerization process, we investigate the monitoring of the reaction progress of a styrene homopolymerization reaction. Two different models for the computation of the ultra-sound velocity in aqueous dispersions are studied experimentally and compared.

Ultra-Sound Velocity

Ultra-sound velocity (c) is a simple and cheap measurement that is sensitive to changes in the different monomer concentrations due to the big difference in c of monomers and polymers. As the ultra-sound velocity of a medium is mainly characterized by its properties on the molecular level, detailed modelling of c becomes very complex. For liquids the sound velocity can be described according to (Urick, 1947) by the simple relation:

$$c = \frac{1}{\sqrt{\bar{\rho}\bar{\beta}}},$$

where $\bar{\rho}$ represents the effective density and $\bar{\beta}$ the effective adiabatic compressibility of the medium. According to (Siani et al, 1999), this equation remains valid for suspensions as long as the operation takes place in the Long-Wavelength Regime, meaning that the sound wavelength is much larger than the polymer particles. As the densities as well as the adiabatic compressibility are temperature dependent, the sound velocity is likewise sensitive to temperature changes. For the computation of the effective parameters $\bar{\rho}$ and $\bar{\beta}$, two models are compared in this work. (Mauntz, 2010) assumed them to be linear combinations with respect to the volume fractions φ_i of water (w), monomer (M) and polymer (Pol):

$$\bar{\rho} = \varphi_W \rho_W + \varphi_M \rho_M + \varphi_{Pol} \rho_{Pol}, \qquad \bar{\beta} = \varphi_W \beta_W + \varphi_M \beta_M + \varphi_{Pol} \beta_{Pol}.$$

(Siani et al, 1999) assumed the parameters to be linear combinations with respect to the volume fractions of the different phases involved being droplets (d), polymer particles (p) and water phase (w). They also included three adjustable parameters (j_{cr}, b, j_{MG}) in a compressibility model for the polymer particles to take care of nonlinear effects:

$$\bar{\rho} = \varphi_d \rho_d + \varphi_p \rho_p + \varphi_w \rho_w, \qquad \bar{\beta} = \varphi_d \beta_d + \varphi_p \beta_p + \varphi_w \beta_w,$$

$$\beta_p = \left(\varphi_{p,M} \beta_M + \varphi_{p,pol} \beta_{pol} + \Delta\beta_p\right) T,$$

$$T = \begin{cases} 1 & if\, \varphi_{p,M} \geq \varphi_{cr} \\ \left(\frac{1 - \varphi_{cr}}{1 - \varphi_{p,M}}\right)^b, & else \end{cases}, \qquad G = \frac{\varphi_{MG}^3 - 2\varphi_{MG} + 1}{\varphi_{MG}^3 - 2\varphi_{MG}^2 + \varphi_{MG}},$$

$$\Delta\beta_p = \varphi_{p,M}(1 - \varphi_{p,M})(\beta_{pol} - \beta_M)\left(1 - 2\varphi_{p,M} + \varphi_{p,M}(1 - \varphi_{p,M})(1 - G)\right),$$

with $j_{p,M}$ $j_{p,pol}$ as the volume fractions of monomer, respectively polymer in the particle phase. The adjustable parameters do not have a direct physical meaning. The idea of introducing T is to summarize effects due to a possible glass transition and the idea of introducing $\Delta\beta_p$ is to summarize all other nonlinear effects.

Experimental Validation

Independent of the chosen approach to model the ultra-sound velocity, the experimental procedure to determine its temperature dependency consists of three major steps:

· determine the ultrasound-temperature calibration curve for the monomer,

· determine the ultrasound-temperature calibration curve for different emulsifier-water mixtures and fit a data driven response surface model (e.g. a second order polynomial function),

· determine the ultrasound-temperature calibration curve of the final lattice with known composition.

With the temperature dependent densities of water, monomer and polymer taken from literature their compressibilities can be determined from the recorded calibration curves at each measurement point by means of Uricks law. While the model of (Mauntz, 2010) requires to fit the individual compressibility points to a temperature dependency curve, the model of (Siani etal, 1999) requires in addition the run of a complete polymerization process to estimate the adjustable parameters by solving an optimization problem:

$$\min_{\varphi_{cr}, b, \varphi_{MG}} J = \sum_i \left(c_{meas,i} - c_{calc,i}\right)^2.$$

For given measurements $c_{meas,i}$, the volume fraction of the monomers is computed by solving an optimization problem that minimizes the difference between the measured and calculated sound velocity using the conservation of mass and the known volumes fed up to this point in time:

$$\min_{\varphi_M} J = \sum_i \left(c_{meas,i} - c_{calc,i}\right)^2.$$

For some polymerizations both approaches work well (see Figure 1) while for others only the model of (Siani et al, 1999) represents the data well when the free parameters are readjusted to the recipe (see Figure 2).

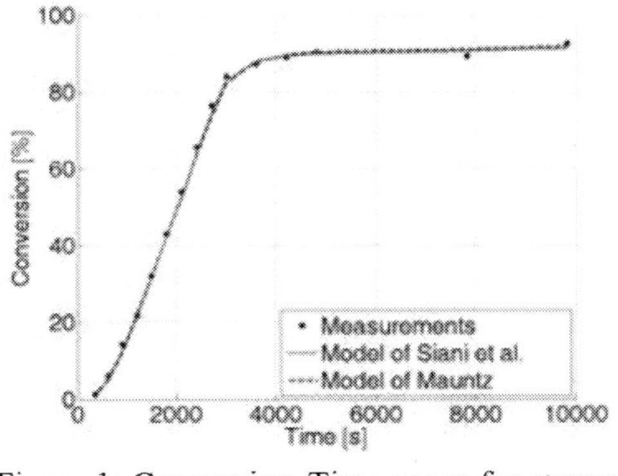

Figure 1: Conversion-Time curve for styrene polymerization 1 (low solid content)

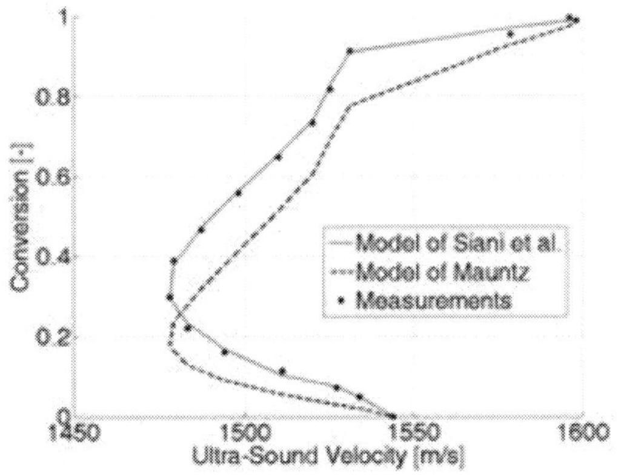

Figure 2: Conversion-Ultra-Sound Velocity curve for styrene polymerization 2 (high solid content)

We will demonstrate the use of ultra-sound velocity measurements in combination with calorimetry to monitor the reaction progress in an emulsion polymerization.

References

1. Gugliotta, L.M. et al, Estimation of conversion and copolymer composition in semicontinous emulsion polymerization using calorimetric data, Polymer, Vol. 36, No. 10, 1995

2. de Buruaga, I.S. et al., Nonlinear Control for Maximum Production Rate of Latexes of Well-Defined Polymer Composition, Ind. Eng. Chem. Res., Vol. 36, 1997

3. Vicente, M. et al, Simultaneous Control of Copolymer Composition and MWD in Emulsion Copolymerization, AIChe Journal, Vol. 47, No. 7, 2001

4. Gesthuisen, R. et al, Hierarchical Control Scheme for Time-Optimal Operation of Semibatch Emulsion Polymerizations, Ind. Eng. Chem. Res., Vol 43, 2004

5. Sheibat-Othman N. and Othman, S., Control of an Emulsion Polymerization Reactor, Ind. Eng. Chem. Res., Vol. 45, 2006

6. Dimitratos, J. et al, Control of Product Composition in Emulsion Copolymerization, Polymer Reaction Engineering, K.H. Reichert and W. Geiseler, V.C.H.: New York, p. 33, 1989

7. Urretabizkaia, A. etal, On-Line Terpolymer Composition Control in Semicontinous Emulsion Polymerization, AIChe J., Vol. 40, 1994

8. Canegallo, S., Composition Control in emulsion copolymerization. II. Application to binary and ternary systems, Journal of Applied Polymer Sciences, Vol. 54, 1994

9. Elizalde, O. et al. Monitoring Emulsion Polymerization Reactors: Calorimetry Versus Raman Spectroscopy, Ind. Eng. Chem. Res., Vol 44, 2005

10. Santos, J.C. et al, Comparison of techniques for the determination of conversion during suspension polymerization reactions, Brazilian Journal of Chemical Engineering, Vol. 25, 2008

11. Urick, R.J., A sound velocity method for determining the compressibility of finely divided substances, Journal of Applied Physics, Vol. 18, 1947

12. Siani, A. et al, Procedure for Calibrating an Ultrasonic Sensor for Online Monitoring of Conversion in Latex Reactors, J. Appl. Polym. Sci, Vol. 72, 1999

13. Mauntz, W., A Contribution to Observation and Time-Optimal Control of Emulsion Co-Polymerization Reactions, Dissertation, TU Dortmund, 2010

(508g) Frame Invariance in the Simulation of Single Screw Extruder

Florian Habla, Stephan Obermeier, Matthias Steib and Olaf Hinrichsen, Catalysis Research Center and Chemistry Department, Technische Universität München, München, Germany

Introduction

With a production volume of 280 million tons per year the polymer industry is one of the major fields in chemical industry. Polymer extrusion is probably the most important operation in polymer-processing being used for example for reactions and forming processes. Understanding and predicting the flow in an extruder is therefore an important task in view of defining and improving the final product quality.

Often the kinematics are created by keeping the screw stationary and rotating the barrel instead of rotating the screw, which is done mainly because this results in a more simplified modeling approach. The question of whether this simplification can be made and the obtained results are identical is still under examination, see for example Campbell et al. [1] and Rauwendaal et al. [2]. The conclusions are commonly drawn for simplifying geometries and conditions such as unwrapped channels, in which the curvature is neglected [1]. In this work we aim at clarifying this issue by doing a rigorous CFD-modeling and showing whether the results can be transferred into each other both in two and three dimensions. We therefore developed a technique, which rotates the whole mesh and subsequently corrects all variables according to the given rotation.

Transport of a passive scalar

The kneading and mixing behavior of the extruder is of major interest. The behavior can be determined by examining the transport of a passive scalar according to:

$$\frac{\partial c}{\partial t} + \nabla \cdot (Uc) = 0$$

where c is the concentration, U is the velocity and t is time. In Fig. 1 the mesh used for the two-dimensional analysis is given. Furthermore, the concentration fields when rotating the extruder or the barrel are shown, wherein the bottom half of the extruder was assigned with the passive scalar. It can be seen, that the concentration fields match each other perfectly and there is frame invariance. Furthermore, the results are in accordance with the DEM analysis of Conelly and Kokini [3].

Fig. 1: Two-dimensional mixing analysis. From left to right: Mesh used for analysis, results when rotating the extruder, results when rotating the barrel, results of Conelly and Kokini [3], right: This work.

Kinematics

Fig. 2 shows streamline plots for both cases. When rotating the extruder, the primary velocity field is obtained, wherefore if the barrel is rotated, the secondary flow field is revealed. Fig. 2 clearly proves that the velocity fields can be transferred into each other by $v_{re} = v_{rb} - \omega \times r$ (cf. the two figures on the right hand side).

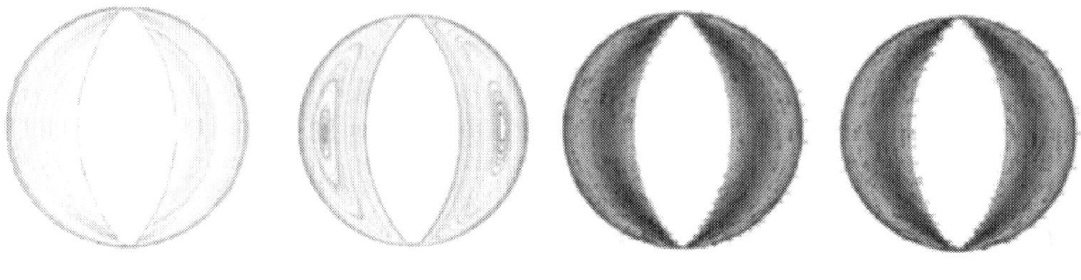

Fig. 2: Two-dimensional kinematics. From left to right: Streamlines for rotating the extruder, streamlines for rotating the barrel, velocity magnitude and vectors when rotating the extruder, transformed velocity magnitude and vectors when rotating the barrel by $v_{re} = v_{rb} - \omega \times r$.

We extended the simulations towards three dimensional simulations using a mesh of over 3,000,000 cells as being shown in Fig. 3a. The results shown in Fig. 3b prove that the results can be interchanged into each other even in three dimensions. The red line shows the velocity profile in the centerline when rotating the barrel. When doing the transformation $v_{re} = v_{rb} - \omega \times r$ the resulting velocity profile (red marks) almost perfectly matches the profile when rotating the extruder (back line) and almost no deviation can be perceived.

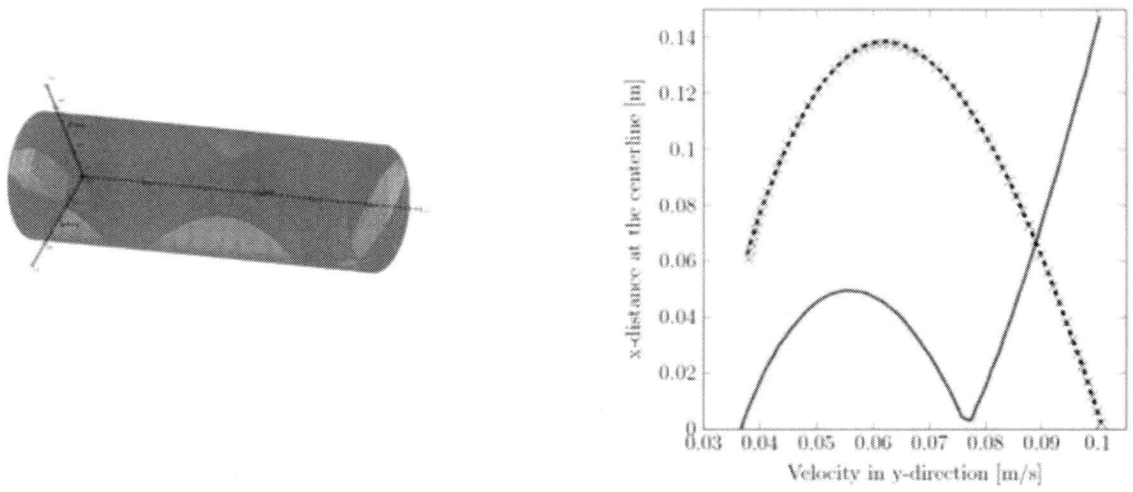

a) b)

Fig. 3: a) Mesh used for three-dimensional computations. b) Velocity profiles in the centerline: black: rotating screw, red line: rotating barrel, red marks: rotating barrel with transformed velocity by $v_{re} = v_{rb} - \omega \times r$.

Conclusions and future work

In this work we proved that the results for the velocity and a passive scalar are identical when rotating the screw and rotating the rod. This was done using a rigorous CFD modeling approach. One major advantage when rotating the barrel is the computational efficiency: The results are twice as fast as the simulations when rotating the screw. This is because each of the nodes needs to be updated at every time step according to the current position of the screw when rotating the screw, which results in a high computational cost. Currently we are evaluating whether frame invariance also applies to non-isothermal flows and here in particular to the heating term due to viscous dissipation.

References

[1] G. A. Campbell, M. A. Spalding and F. Carlson, Prediction of melt temperature rise in single-screw pump extruders, ANTEC 267 (2008).

[2] C. Rauwendall, T. A. Osswald, G. Tellez, P. J. Gramann, Flow Analysis in screw extruders – effect of kinematic conditions, Int. Polym. Proc. 4 (1998) 327-333.

[3] R. K. Conelly, J. L. Kokini, Examination of the mixing ability of single and twin screw mixers using 2D Finite Element Method simulation with particle tracking, J. Food Engr. 79(3) (2007) 956-969.

(516a) An Approach Considering Both Operation Stability and System's Hopf Bifurcations to Chemical Process Design

Hangzhou Wang[1], Tong Qiu[1], Bingzhen Chen[1], Jinsong Zhao[2] and Xiaorong He[1], (1)Department of Chemical Engineering, Tsinghua University, Beijing, China, (2)Institute of Process Systems Engineering, Department of Chemical Engineering, Tsinghua University, Beijing100084, China

In order to consider the inherent safety of chemical processes, researchers proposed many methods and strategies, many quantitative indices have been developed to describe the potential hazard and dangerous of different reaction routes and reactants. In authors' previous paper[1] an integrated quantitative index considering both speed to return to steady state point and capability of resistance to disturbances is introduced. In the other hand, oscillation behavior was observed in both experiment and mathematical simulation in some chemical processes[2-10], in which the concentration of feedstock and product oscillation occur under certain operation conditions. In order to design a more stable process, these potential oscillations should be avoided. And it is reported that the oscillation was caused by the existence of Hopf bifurcations in systems[2, 11]. Since all Hopf singularity points can be identified by using mathematic calculation under given operation condition range, consequently, the Hopf singularity point range can be determined. This paper proposed a design approach to chemical process design, considering both operation point stability and systems Hopf singularity points to improve the process stability and avoid potential oscillation in process. The detailed steps of this approach are described as follows:

(1) Solve the steady state solution of the dynamic model of chemical process.

(2) Judge the stability of these solutions and divide the solution curve into parts with different stabilities (stable and unstable).

(3) Calculate the integrated quantitative stability index of the process system

(4) Calculate Hopf points under the possible operation condition range.

(5) Formulate a multi-objective optimization problem for the process design in which the quantitative stability index together with the economic index are considered as objectives and the Hopf points ranges are involved as constraints,

(6) Obtain the optimal operation conditions.

The proposed approach is applied to a continuous fermentation process and a multi-objective optimization problem considering both economic and stability factors were conducted, and the Hopf points ranges are considered as constraints at the same time in this optimization problem. After calculation, a Pareto set is obtained. The optimization results provide information that is very useful for guiding process design and operation. From this study, conclusions can be drawn:

(1) Quantitative stability index is useful in multi-objective optimization problem when stability was considered as one of the objectives.

(2) Hopf singularity points range can be considered as constraints in optimal process design problems to avoid the potential oscillation in process.

References:

1. Wang, H., et al., *An integrated quantitative index of stable steady state points in chemical process design*, in

Proceedings of the 11th International Symposium on Process Systems Engineering, I.A. Karimi and R. Srinivasan, Editors. 2012, Elsevier B.V.: Singapore

2. Sridhar, L.N., *Elimination of Oscillations in Fermentation Processes.* AIChE Journal, 2011. **57**(9): p. 2397-2405.

3. Paz Astudillo, I.C. and C.A. Cardona Alzate, *Importance of stability study of continuous systems for ethanol production.* Journal of Biotechnology, 2011. **151**(1): p. 43-55.

4. Bai, F.W., W.A. Anderson, and M. Moo-Young, *Ethanol fermentation technologies from sugar and starch feedstocks.* Biotechnology Advances, 2008. **26**(1): p. 89-105.

5. Bai, F.W., et al., *Parameter oscillations in a very high gravity medium continuous ethanol fermentation and their attenuation on a multistage packed column bioreactor system.* Biotechnology and Bioengineering, 2004. **88**(5): p. 558-566.

6. McLellan, P.J., A.J. Daugulis, and J.H. Li, *The incidence of oscillatory behavior in the continuous fermentation of Zymomonas mobilis.* Biotechnology Progress, 1999. **15**(4): p. 667-680.

7. Daugulis, A.J., P.J. McLellan, and J.H. Li, *Experimental investigation and modeling of oscillatory behavior in the continuous culture of Zymomonas mobilis.* Biotechnology and Bioengineering, 1997. **56**(1): p. 99-105.

8. Sprenger, G.A., *Carbohydrate metabolism in Zymomonas mobilis: a catabolic highway with some scenic routes.* FEMS Microbiology Letters, 1996. **145**(3): p. 301-307.

9. Ghommidh, C., et al., *Oscillatory behaviour of Zymomonas in continuous cultures: A simple stochastic model.* Biotechnology Letters, 1989. **11**(9): p. 659-664.

10. Jöbses, I., et al., *Fermentation kinetics of Zymomonas mobilis at high ethanol concentrations: oscillations in continuous cultures.* Biotechnology and Bioengineering, 1986. **28**(6): p. 868-877.

11. Kuznetsov, Y.A., *Elements of applied bifurcation theory.* 1998, New York: Springer

---

* Corresponding author. Tel.: +86 10 62781499;fax: +86 10 62770304.

E-mail address: dcecbz@tsinghua.edu.cn (Bingzhen Chen)

(516b) Methodology for the Analysis and Design of Chemical Conversion Procesess Applied to the Methanol Synthesis

Paraskevas Klimantos and Magne Hillestad, Chemical Engineering, NTNU, Trondheim, Norway

A broad range of methods for process design appears in the literature which can be classified based on the chosen strategy for representing the design alternatives or the design space. These can range from methods that attempt to enumerate the alternatives in an explicit space, a coordinated search in the space of design decisions, evolutionary methods, superstructure optimization, targeting, problem abstraction, and combinations of these Biegler (1997). In the field of reactor design the methods that have made considerable contributions can be classified into attainable region methods or targeting, and optimisation methods involving either construction of superstructures that are formulated as MINLP problems or state space design representation that exploit analogies from dynamic optimization techniques.

Attainable region methods have their origin in the early work of Horn (1964), and is a representation of the attainable concentration space when reaction and mixing take place. The theoretical framework of the attainable region approach has been extensively further developed by Glasser et al. (1987) , Hildebrandt and Glasser (1990) , Feinberg and Hildebrandt (1997). Given a chemical reaction, the method constructs graphically the consecration space based on the fact the all points in its boundaries are represented by the convex hull of combination of ideal reactor (PFR and CSTR) trajectories. The constructed convex hull contains the entire attainable region in the concentration space and given the reaction kinetics one can predict upper bounds of the performance of reaction system. The attainable region can be considered as a fundamental contribution in the area of reactor design since it provides critical insight into the properties of reaction and mixing. However the applicability of the method has some limitations inherently imposed by the graphical approach considered restricting the designer to analyse problems in two or three dimensions.

Superstructure approaches were developed by several authors , Achenie and Biegler (1986) worked on an idea originated by Jackson (1968) and on the basis of one dimensional dispersion model they formulate the reactor synthesis problem as a NLP optimisation problem with split fractions, source points and sink points that determine the mixing pattern, this network has been extended also to non isothermal reactors. Kokossis and A. (1990) proposed a reactor representation that utilizes as a basis the continuous stirred tank reactor (CSTR) and approximates the plug flow reactor as a cascade of CSTRs. A reactor network superstructure is presented in which all possible structural configurations of interconnected reactors are embedded. Marcoulaki and Kokossis (1999) formulated the reactor synthesis problem in a same way and attempted to solve it with stochastic optimisation approaches such as simulated annealing. Many contributions can be found also in the area of determinist global optimisation of superstructure formulations Esposito and Floudas (2002). The combinatorial nature of the MINLP superstructure formulations result in a computational intractable problems and most of the aforementioned contributions focused on developing efficient optimisation algorithms and less focus has been given to the practical implementation in real industrial applications.

Within this contribution we present an alternative methodology for reactor design considered within the context of conceptual process design. The main idea is to represent basic operations, reaction, fluid mixing and heat transfer, and separation through a continues state space model describing the reaction path.A fluid element is tracked along the reaction path and optimal conditions are determined that satisfy a design objective subject to process and product quality constraints. By encapsulating fundamental phenomena in a composite model rich enough to resemble the feasible design space and conducting a systematic search we are given the possibility to identify regions where improved and innovative reactor configuration are located. We illustrate the methodology by considering the methanol synthesis loop,a process of industrial interest. A multistage case for methanol synthesis with interstage product removal is further considered and its economic feasibility is assessed.

Keywords: process, design, reactor design, optimization, methanol synthesis

References

Achenie, L. E. K., Biegler, L. T., 1986. Algorithmic synthesis of chemical reactor networks using mathematical programming. Industrial Engineering Chemistry Fundamentals 25 (4), 621–627.

Biegler, L.T., G. I. E. . W. A. W., 1997. Systematic methods of chemical process design. Prentice Hall., Upper Saddle River, NJ.

Esposito, W. R., Floudas, C. A., 2002. Deterministic global optimization in isothermal reactor network synthesis. Journal of Global Optimization 22, 59–95.

Feinberg, M., Hildebrandt, D., 1997. Optimal reactor design from a geometric viewpoint i. universal properties of the attainable region. Chemical Engineering Science 52 (10), 1637 – 1665.

Glasser, D., Crowe, C., Hildebrandt, D., 1987. A geometric approach to steady flow reactors: the attainable region and optimization in concentration space. Industrial Engineering Chemistry Research 26 (9), 1803–1810.

Hildebrandt, D., Glasser, D., 1990. The attainable region and optimal reactor structures. Chemical Engineering Science 45 (8), 2161 – 2168.

Hillestad, M., 2010. Systematic staging in chemical reactor design. Chemical Engineering Science 65 (10), 3301 – 3312.

Horn, G., 1964. Attainable region in chemical reaction technique. in The Third European Symposium on Chemical Reaction Eng. Pergamon, London.

Jackson, R., 1968. Optimization of chemical reactors with respect to flow configuration. Journal of Optimization Theory and Applications 2, 240–259.

Kokossis, A. C., A., F. C., 1990. Optimization of complex reactor networks i. isothermal operation. Chemical Engineering Science 45 (3), 595 – 614.

Marcoulaki, E. C., Kokossis, A. C., 1999. Scoping and screening complex reaction networks using stochastic optimization. AIChE Journal 45 (9), 1977–1991.

(516c) Innovative Cryogenic Air Separation Processes

Chao Fu, Department of Energy and Process Engineering, Norwegian University of Science and Technology, Trondheim, Norway and Truls Gundersen, Energy and Process Engineering, Norwegian University of Science and Technology, Trondheim, Norway

Cryogenic Air Separation Units (ASUs) are widely used for the high-volume supply of O_2, N_2 and rare gases in industrial processes such as steel plants, gasification plants, fertilizer plants and petrochemical plants. More recently, due to the increased concern about CO_2 emissions, a new large scale ASU application is the O_2 supply in CO_2-capturing power plants by oxy-combustion or pre-combustion. The cryogenic ASUs are causing considerable energy penalties related to CO_2 capture. New research activities on improving the performance of ASUs targeted for CO_2-capturing plants are essential.

Traditional cryogenic ASUs are based on the double-column distillation scheme as developed by Carl von Linde in 1910. In this scheme, air is crudely separated in a higher pressure (HP) column and then further separated in a lower pressure (LP) column. The high pressure N_2 from the top of the HP column can be condensed against the boiling O_2 at the bottom of the LP column. The condensed N_2 liquid is used as reflux for the two distillation columns. The double-column distillation scheme has solved the most considerable challenge in sub-ambient distillation processes, i.e. the production of reflux. However, the air feed has to be generally compressed to 400-600 kPa in order to be separated in the HP column. This causes unnecessary compression of O_2 in the air feed, thus wasting expensive compression work, in applications such as coal based power plants with oxy-combustion, where O_2 slightly above atmospheric pressure is sufficient.

An exergy analysis of a traditional double-column air distillation cycle indicates that the actual energy consumption (0.199 kWh/kgO_2) is around 4 times the theoretical minimum (around 0.05 kWh/kgO_2) when all unit operations are assumed reversible [1]. The compression of the air feed and the distillation system are responsible for the two largest exergy losses: 38.4% and 28.2% respectively. The irreversibility in distillation columns can be reduced by using "distributed reboiling". The energy consumption can be reduced by 10% when dual reboilers are used. For the air compression process, the energy consumption can be reduced by (i) improving the compressor efficiency, (ii) reducing the compression ratio, (iii) lowering the operating temperature, and (iv) reducing the mass flow through the compressor. This paper focuses on the last of these measures, i.e. ways to reduce the mass flow through the air compressor. A recuperative vapor recompression (RVRC) heat pump distillation scheme is proposed for the separation of air. As a result, only one column is used for air separation. In addition, the principle of "distributed reboiling" is applied to reduce irreversibilities in the distillation column.

In a simple RVRC heat pump distillation scheme, the air feed is slightly compressed to compensate for pressure losses along its flow path. The air is separated into N_2 and O_2 in a single distillation column. The reflux for the column is produced in the following way: the N_2 from the top of the column is heated to ambient temperature; a portion of the heated N_2 is compressed and cooled to its dew point after the compression heat is removed (by cooling water); the compressed and cooled N_2 is then condensed against the boiling O_2 in the bottom of the distillation column; the condensed N_2 liquid is then sub-cooled if necessary, expanded and finally sent back to the distillation column as reflux. In this case, the O_2 in the air feed is not significantly compressed. When this scheme is used for the production of O_2 (125 kPa) with a purity of 95 mole%, the energy consumption is calculated to be 0.184 kWh/kgO_2, a 7.5% reduction compared to a traditional double-column air distillation process [2, 3]. The energy consumption is reduced to 0.165 kWh/kgO_2 (17.1% reduction) when an intermediate RVRC heat pump is applied for distributing the reboiling. In addition, the energy consumption can be further reduced by efficient organization of the refrigeration cycle, e.g. using a portion of the air feed instead of the N_2 as the refrigerant.

The RVRC heat pump distillation scheme can also be applied for the production of high purity O_2. An argon side column can be added in this case. The energy consumption is calculated to be 0.232 kWh/kgO_2 in the case that the air feed is separated into the following products: O_2-99.4 mole%, N_2-99.5 mole% and argon-98.2 mole%. A 5.3% reduction in energy consumption has been achieved compared to a traditional three-column air separation cycle [4]. Since much more reflux is required in the high purity O_2 production process than in the low purity (95 mole%) case, the energy saving potential is expected to be less for high purity than for low purity O_2 production.

For the production of high pressure O_2, a simple way is to compress the gaseous O_2 (GOX) to the desired pressure in a low pressure O_2 production cycle. The pumped liquid O_2 (LOX) cycle is more commonly applied in recent years for the reason of safety and cost. In a pumped LOX cycle, the O_2 is produced in liquid phase and then pumped to the desired pressure. The pumped LOX is then vaporized against a portion of the air feed and heated to ambient temperature. When the RVRC heat pump distillation scheme is applied for producing high pressure O_2, a pumped LOX process can be similarly applied. The energy consumption is calculated to be 0.301 kWh/kgO_2 when O_2 (95 mole%) is produced at 4000 kPa, while the value is reported to be 0.313 kWh/kgO_2 when the double-column distillation scheme is used as the basis [5].

In conclusion, a recuperative vapor recompression heat pump distillation scheme is proposed to substitute the traditional double-column distillation scheme in cryogenic air separation processes. When the scheme is applied for the production of O_2 with low purity and low pressure, the energy consumption can be substantially reduced by avoiding the compression of O_2. The scheme can also be applied for producing O_2 with high purity and/or high pressure, however, the energy saving potential is smaller since the mass to be compressed is larger.

References

[1] Fu, C.; Gundersen, T. Using exergy analysis to reduce power consumption in air separation units for oxy-combustion processes. Energy 2012, in press.

[2] Fu, C.; Gundersen, T.; Eimer, D. Air separation. GB Patent, application number: GB1112988.9, 2011.

[3] Fu, C.; Gundersen, T. Using PSE to develop innovative cryogenic air separation processes. Proceedings of the 11th International Symposium on Process Systems Engineering, 15-19 July 2012, Singapore.

[4] Beysel, G. Enhanced cryogenic air separation: a proven process applied to oxyfuel. 1st International Oxyfuel Combustion Conference, 8-11 September 2009, Cottbus, Germany.

[5] Dillon, D. J.; White, V.; Allam, R. J.; Wall, R. A.; Gibbins, J. Oxy combustion processes for CO_2 capture from power plant; IEA GHG report 2005/9, 2005.

SIMULTANEOUS PROCESS AND RECIPE DESIGN OF AN ACRYLIC FIBER PRODUCTION SYSTEM

Marta Moreno-Benito[a], Kathrin Frankl[b], Wolfgang Marquardt[b], Antonio Espuña[a]

[a]*Chemical Engineering Department, Univesitat Politècnica de Catalunya, Av. Diagonal 647, E08028, Barcelona, Spain.*
[b]*Process System Engineering, RWTH Aachen University, Turmstraße 46, D-52056 Aachen, Germany*
E-mail: antonio.espuna@upc.edu

Abstract

Polymerization has been historically an area where process systems engineering tools have been widely applied by combining polymer science, chemistry, and technology, with process engineering principles [1]. A significant amount of work has been devoted to the modeling and simulation of polymerization reaction systems, which are characterized by complex interactions between productivity indicators (conversion, batch time, profit, etc.) and polymer properties (polydispersity, molecular weight distribution, etc.). As a result, the design, operation and control of such polymerization processes constitute challenging problems, where the choice of the trajectories of the reactor temperature and monomer feed rate is targeting at an optimum trade-off between productivity and polymer quality [1,2]. However, other downstream tasks, in addition to the polymerization reaction, also contribute to the optimization of production objectives, either economic or environmental, as shown in the solution of the scheduling problem for polymerization manufacturing facilities [3].

In this context, the aim of this work is to direct attention towards simultaneous process and recipe design of industrial-sized polymerization processes, understood as the selection and interconnection of process stages and the specification of their operating strategy. A model-based optimization approach is used to optimize simultaneously the process and the corresponding recipe to produce a polymer in single-product batch campaigns as part of the design of a polymerization plant. This way, a trade-off between process and recipe design can be achieved to improve production targets while maintaining polymer quality. Decisions related to the design and synthesis of the process include the selection of the type and number of process stages (also referred as tasks) and operations, the technological specification of processing units, the location of storage units, the recirculation of intermediate flows, the configuration of process stages in single, series or parallel operation, and the synchronization of tasks. Decisions referring to the operational design include the reference trajectories of the control variables, the batch times, material transfer synchronization between tasks, and batching decisions. Besides, equipment investment, processing costs, raw material costs, waste treatment costs, and economic impact of product quality are considered in the optimization objective function.

In model-based optimization approaches for process synthesis, all processing alternatives are first represented in a superstructure, which is later formulated into a mathematical program to be finally optimized [4]. Particularly, the modeling approach used in this work relies in the combination of mixed-logic modeling [5], previously applied to continuous process synthesis, and the use of multistage models [6] to represent the performance of each batch task. The approach is similar to the disjunctive multistage modeling by Oldenburg and Marquardt [7], developed in the context of individual batch unit optimization, and now extended to represent concurrent process stages and their synchronization.

The main steps are following described: The available information for each processing stage is first identified, such as appropriated technologies or practical control variables. Then, a superstructure is constructed, which gathers all potential synthesis alternatives. For that, state-equipment network (SEN) representation is used, assuming a direct relation between task and equipment. Logic variables and disjunctive equations are following defined to represent mathematically the qualitative information of such alternatives. Further available knowledge of the process that relates the various processing options can be incorporated into the model through logic propositions, which also limit the dimension of the combinatorial problem. Finally, multistage models are provided to represent each potential processing task, where material input and output stages are identified to allow the synchronization with other tasks. Dynamic, steady state, or linearly approximated models can be used, according to the impact of each process stage on the decision criterion.

The acrylic fiber process is presented as an example of the proposed modeling approach to solve the process and recipe design problem simultaneously. The production target is a fiber composed of 85% in weight of acrylonitrile (AN), and 15% of vinyl acetate (VA) in spun format. Hence, primary copolymer production and secondary spun production stages are required. For this purpose, the principal features of the manufacturing process are incorporated into the superstructure and the optimization model from related state-of-the-art literature. In this contribution, the complete process is included in the superstructure, although only the first part of the process is mathematically modeled and solved, obtaining the intermediate copolymer in either pellets shape or solution. The principal information featuring the complete process is composed of:

- the main stages in acrylic fiber production [3],
- the use of suspension and solution copolymerization technologies [8],
- the possibility of avoiding solvent separation after the solution copolymerization reaction [9,10],
- the operational information and dynamic models of copolymerization [11] and separation [7] processes,
- and the possibility of recovering and reusing the water used to wash the final product [8].

As a result, the following process stages or tasks are subject to inclusion in the process model: (1) copolymerization reaction, (2) recovery of unreacted monomer, solvent, and suspension medium after reaction, (3) washing and filtration, (4) repulping, (5) filtering, (6) wet spinning, (7) second washing and filtration, and (8) second recovery of solvent after spinning. For each process stage, several subtasks and configurations are allowed, which determine the required equipment pieces and their synchronization. All in all, nine alternative disjunctions are considered for plant and process synthesis in the example, like the selection of solution or suspension copolymerization technology, organic or aqueous solvent in solution polymerization, separation of solvent or direct transference of solved copolymer to repulping, etc. Each of these options leads to a particular cost profile in the objective function. For example, the use of organic solvent drastically increases the waste treatment costs. The process superstructure is presented in Figure 1, where all synthesis alternatives for the complete process are included, as well as the Booleans indicating if such alternative is selected (true value) or not (false value).

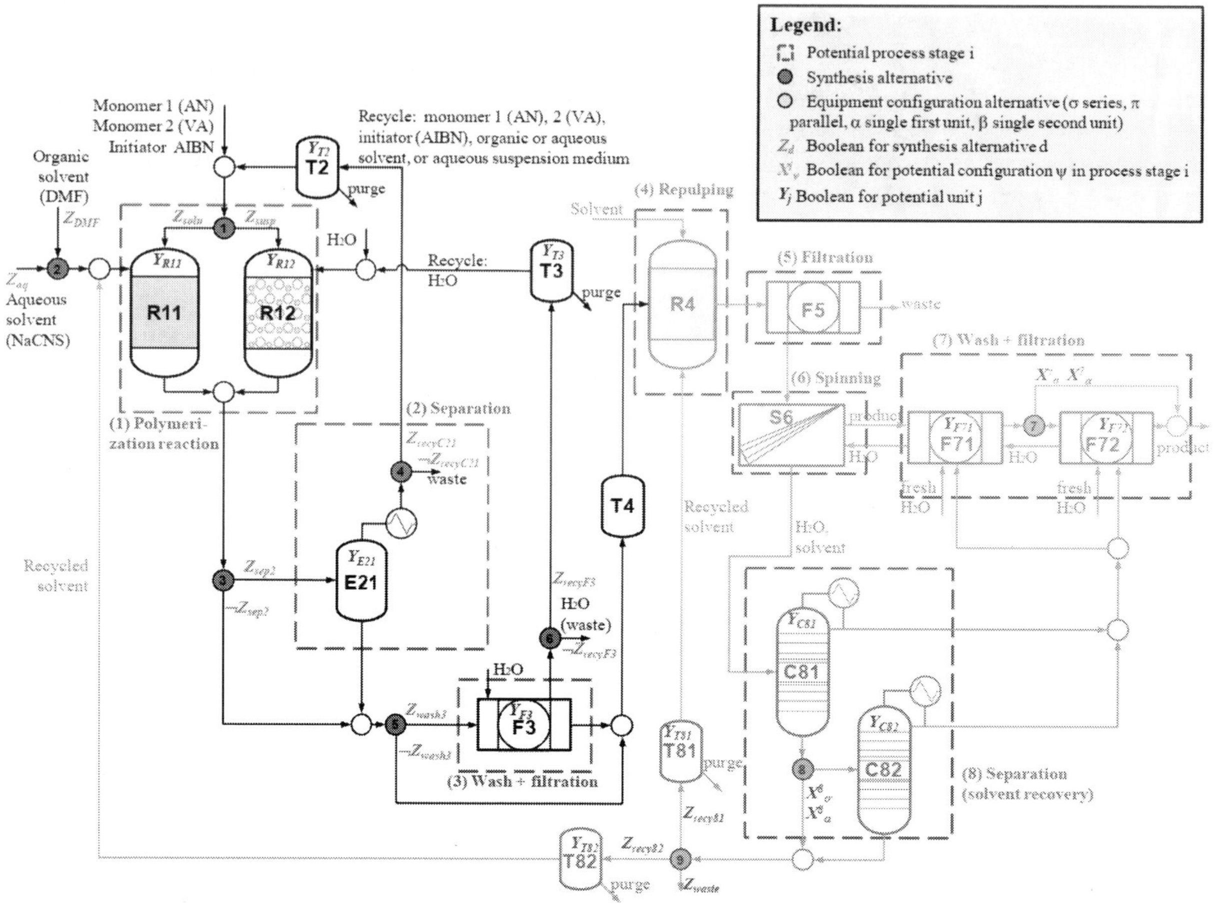

Figure 1: Superstructure of the acrylic fiber process. The equipment pieces are: R11 and R12 (solution and suspension polymerization reactors), E21 (evaporator and condenser), F3, F5, F71 and F72 (washing and filtering units), R4 (repulping unit), S6 (spinneret), C81 and C82 (distillation columns), and T2, T3, T4, T81 and T82 (buffer tanks). Secondary spun production stages, not solved in this contribution, are in grey.

Regarding processing conditions, the process stages whose control variables have a critical impact on cost function are: task 1 (copolymerization reaction), task 2 (recovery of unreacted monomer, solvent and suspension medium after reaction), and task 8 (recovery of the solvent after spinning and washing the final spun). For example, a large separation and reuse of solvent may partially mitigate its corresponding waste treatment cost, or the achievement of low conversions in the polymerization reaction may have an effect on higher processing costs in the subsequent distillation column. Hence, dynamic models are used to describe the tasks of these process stages, and their control variable trajectories and batch times are optimized. Particularly, monomer feed rate and temperature are the main dynamic variables used for reaction control, whereas separation processes are controlled through the dynamic adjustment of their respective heat input and reflux ratios. Quality restrictions are formulated for polydispersity and composition in order to keep the product properties in the reactor within a predefined quality range. In other process stages, which are not critically contributing to the objective function, steady-state assumptions or other approximations can be done in their corresponding tasks.

By modeling and optimizing the process and recipe design simultaneously, a potential improvement in the objective function should be met, with respect to the use of fixed process models and recipes. This result is expected due to the holistic evaluation of the decision criteria. In particular, the trade-offs between the various process stages within the complete process is considered in the optimization, as well as the interactions between synthesis and operational design degrees of freedom. Additionally, the optimization model permits the evaluation and comparison of processing alternatives according to multiple points of view, like processing, economic, or sustainable production policies.

Acknowledgements

This work has been developed during the research project SIGERA (DPI2012-37154-C02-01), funded by the Ministerio de Economía y Competitividad of Spain and the European Commission (ERD Funds). Additional financial support of the Aachen group from the EU FP 7 project F³ Factory (grant agreement no: 228867) and of the MegaCarbon project (NRW-EU Ziel 2-Programm 2007 – 2013, EFRE) is gratefully acknowledged.

References

1. R. Giudici, 2000. Latin American Applied Research, Vol. 30, pp. 351-356
2. M. Embiruçu, E.L. Lima, and J.C. Pinto, 1996. Polymer Engineering and Science, Vol. 36(4), pp. 433-447
3. E. Capón-García, A.D. Bojarski, A. Espuña, and L. Puigjaner, 2011. AIChE Journal, Vol. 57, pp. 2766-2782
4. I.E. Grossmann and G. Guillén-Gosálbez, 2010. Computers and Chemical Engineering, Vol. 34(9), pp. 1365-1376
5. R. Raman and I.E. Grossmann, 1994. Computers and Chemical Engineering, Vol. 18(7), pp. 563-578
6. V.S. Vassiliadis, R.W.H. Sargent and C.C. Pantelides, 1994. Industrial and Engineering Chemistry Research, Vol. 33(9), pp. 2111-2122
7. J. Oldenburg, W. Marquardt, D. Heinz, and D.B. Leineweber, 2003. AIChE Journal, Vol. 49(11), pp. 2900-2917
8. United States Environmental Protection Agency, 1995. http://www.epa.gov/ttn/chief/ap42/ch06/
9. M. Gol'dfein and B. Zyubin, 1990. Review Polymer Science, Vol. 32(11), pp.2145-2166
10. P. Bajaj, K. Sen, and S.H. and S.H. Bahrami, 1996. Journal of Applied Polymer Science, Vol. 59(10), pp. 1539-1550
11. D. Butala, K. Choi, and M. Fan, 1988. Computers and Chemical Engineering, Vol. 12(11), pp. 1115-1127

(516e) Optimal Design of Hydrocarbon Biorefinery Via Gasification Pathway: Multiobjective Optimization Coupled with Life Cycle Assessment

Belinda Wang, Berhane H. Gebreslassie and Fengqi You, Department of Chemical and Biological Engineering, Northwestern University, Evanston, IL

The amount of available energy is rapidly deceasing because of the large demand and consumption of fuel. Among all the energy alternatives, biomass-derived hydrocarbon fuels are most similar to fossil fuels. Concerns about global warming, national security and the diminishing supply of fossil fuels have being increasing. Therefore, the U.S. government proposed the Energy Independence and Security Act (EISA) of 2007 to encourage the development of biofuels. The EISA requires the total amount of biofuel production to increase to 36 billion gallons per year by 2022 from 4.7 billion gallons per year in 2007 [1]. Among all types of biofuels, EISA defines advanced biofuel as renewable fuel, other than ethanol derived from corn starch. The advanced biofuels would be more chemically similar to gasoline, jet fuel and diesel. Therefore, it would be more cost-effective because of the compatibility with the existing infrastructure. For example, renewable gasoline, renewable diesel, renewable jet fuel, cellulosic biobutanol, and algae-derived biofuels are infrastructure-compatible advanced biofuels. The production process would be more cost-effective and sustainable that utilizes the existing refining and distribution infrastructure.

In this work, we propose a superstructure for the optimization of hydrocarbon biorefinery via gasification pathways. The superstructure demonstrates the network configuration of the process which considers different process designs for the production of biofuels from corn stover. There are at least two options for selected technology blocks. In order to model the different alternatives, a binary variable is used to limit to only one selection when there are several technical options [2]. For gasification, we consider two types of gasifiers, high-temperature and low-temperature. Each type of gasifier would have different product distribution. Therefore, the sequential processing steps would be different. There are two ways to cool the raw syngas, which are direct quench and indirect quench. A significant amount of hydrogen is needed for hydrocracking and hydrotreating. Therefore, hydrogen can be produced through two pathways, internal production or steam methane reforming. Internal production uses the hydrogen within the syngas. While steam methane reforming uses additional natural gas to produce hydrogen. Then the conditioned syngas enters Fischer-Tropsch synthesis unit that uses cobalt, iron, or nickel based catalysts. Fischer-Tropsch liquids then can be refined to gasoline and diesel biofuel.

The properties of the feedstock are known, such as the moisture content and element compositions. The species in each stream and the reactions in each unit are also known from simulation results [3]. Therefore, component balances are performed over each unit and atomic balances are used in units with chemical reactions. Energy balances are also included as constraints. The heat consumption and generation rates are calculated using the total enthalpy differences between inlet and outlet flows. The capital costs are scaled based on the scaling stream and sizing factor. Other cost includes annual operational cost and feedstock cost. Operational cost consists of maintenance, steam, power, catalyst, and natural gas costs. The process makes profit from selling gasoline and diesel at market price. The net present value accounts for the investment and discounted annual cost and revenue.

We proposed a multi-objective mixed-integer nonlinear programming formation for the superstructure optimization of this hydrocarbon biorefinery under economic and environmental concerns. The bi-criteria optimization problem maximizes the net present value (NPV) and simultaneously minimizes the environmental impact. The environmental impact is measured using the global warming potential (GWP) metric following the recent development of the life cycle assessment (LCA) procedures. GWP is a relative measure of how much greenhouse gas emit during a process, and is expressed in terms of carbon dioxide equivalent. GWP is calculated over a specific time interval that has to be stated, commonly 20, 100 or 500 years. In this work, 100 year is adopted following Kyoto protocol. To obtain the optimal trade-off between the two contradicting objective functions, we use the epsilon constraint method. The resulting Pareto curve demonstrates the trade-off that exists between the NPV and GWP. The optimal solutions will form a Pareto-optimal curve which represents the set of optimal solutions of the two objectives. The curve separates the region of feasible and infeasible solutions. A case study is presented and the results show that the optimal choice is using high temperature gasification, cobalt catalyst in Fischer-Tropsch synthesis, direct quench and internal hydrogen production under both the economic and environmental criteria.

References

[1] *Energy Independence and Security Act of 2007,* U. S. Congress, 2007.

[2] P. Liu, E. N. Pistikopoulos, and Z. Li, "A Multi-Objective Optimization Approach to Polygeneration Energy Systems Design," *Aiche Journal,* vol. 56, pp. 1218-1234, May 2010.

[3] R. M. Swanson, A. Platon, J. A. Satrio, and R. C. Brown, "Techno-economic analysis of biomass-to-liquids production based on gasification," *Fuel,* vol. 89, pp. S2-S10, Nov 1 2010.

(516f) Process Analysis and Generic Optimization of the Synthesis of n-Butyl Acrylate in an Industrial Scale Reactive Distillation Column

Alexander Niesbach[1], Hanns Kuhlmann[1], Philip Lutze[2] and Andrzej Górak[1], (1)TU Dortmund University, Department of Biochemical and Chemical Engineering, Laboratory of Fluid Separations, D-44227 Dortmund, Germany, (2)Department of Biochemical and Chemical Engineering, Laboratory of Fluid Separations, TU Dortmund University, D-44227 Dortmund, Germany

Process analysis and generic optimization of the synthesis of *n*-butyl acrylate in an industrial scale reactive distillation column

Alexander Niesbach*, Hanns Kuhlmann, Philip Lutze, Andrzej Górak

Laboratory of Fluid Separations, TU Dortmund University, Germany

(*Corresponding Author's E-mail: **alexander.niesbach@bci.tu-dortmund.de**)

ABSTRACT

The development of innovative apparatuses and techniques has led to process intensification and achieving ecologic and economic improvements, such as decreasing energy consumption and increasing process efficiency.

Reactive distillation (RD) has already emerged as an excellent technology for process intensification exploiting the synergy effects of the combination of reaction and distillation phenomena at the same place and time. This combination allows overcoming limitations regarding thermodynamic and chemical equilibria which leads to an increased process yield. RD has been successfully studied mostly for esterifications with low carbon reactants.

Some of the reasons for this limited application to higher carbon numbers are the more complex chemistry as well as the risk for polymerization of some of the chemical compounds in the system. Besides, the identification of a near optimal set of operational as well as decisive variables for a complex RD system is difficult.

Therefore, within this project, the results of the experimental as well as model-based development of an RD process for a high carbon number esterification which is the heterogeneously catalyzed synthesis of *n*-butyl acrylate (BA) from acrylic acid (AA) and *n*-butanol (BuOH) is presented. The near optimal design is identified by a three step approach: 1) An indicator based feasibility analysis for identification if RD may be promising; 2) Experiments for rigorous rate-based model validation; 3) Optimization for identification of the near-optimal design. Furthermore, in step three, different solvers such as commercial solvers from ASPEN as well as an evolutionary algorithm have been checked for best performance.

For the design of the RD column, a theoretical and experimental feasibility analysis based on the properties of the system was done. At first the technical feasibility of the application of an RD column for the synthesis of *n*-butyl acrylate was proven by a theoretical study at the institute that was presented before (Keller et al, 2010).

In the second step, an experimental study was performed and the experimental results were used to validate a rate-based reactive distillation model implemented in the simulation environment Aspen Custom Modeler™, which was developed in a previous work (Klöker et al., 2005) and adapted to the present system.

Within this presentation, the third step will be presented, where the complex mixed-integer non-linear design problem needs to be solved. For this system, it turned out that an evolutionary algorithm showed best performance. Hence, the validated rate-based RD model was used and implemented into an evolutionary optimisation method to design an optimal RD column for the production of *n*-butyl acrylate. The production costs for *n*-butyl acrylate are used as objective function. For the column design several structural parameters (e.g. number of stages, feed position) as well as operational parameters (e.g. reflux ratio, distillate-to-feed ratio) have to be considered. Optimised variables have been found minimising the production cost of *n*-butyl acrylate for a chosen annual production of 20.000 t *n*-butyl acrylate in consideration of the required product purity. Several boundary conditions, like product purity and maximum temperature in the catalytic section, were taken into account. Hereby, a process design based on a heterogeneously catalysed reactive distillation has been developed and the overall production cost of *n*-butyl acrylate is reduced compared to the conventional process (Figure 1).

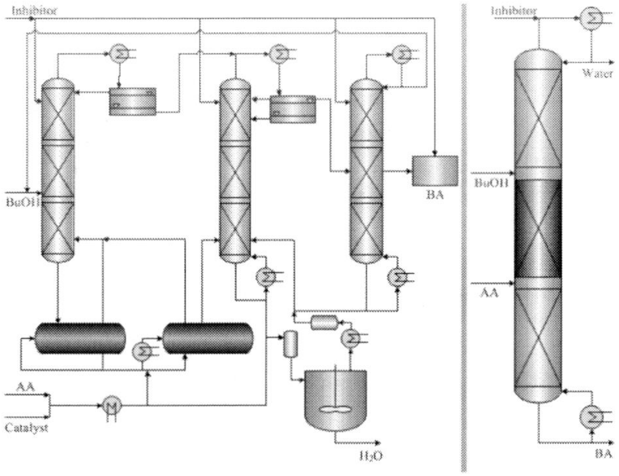

Figure 1: Comparison of conventional and intensified process for the production of *n*-butyl acrylate.

Acknowledgement: The research leading to these results has received funding from the European Community's Seventh Framework Programme (FP7/2007-2013) under grant agreement n° 228867,

F^3-Factory

References

1. Keller, T., T. Tretjak, S. Lacroix, A. Hoffmann and A. Górak, *Presentation at CHISA 2010.*, (2010), Prag, Czech Republic.

2. Klöker, M., E.Y. Kenig, A. Hoffmann, P. Kreis and A. Górak, *Chemical Engineering and Processing*, **44**, 617-629 (2005).

(530a) Forecasting Naphtha Crack Based On Multiple Regression and System Dynamics

Chaeeun Sung[1], Hweeung Kwon[1], Jinsuk Lee[2], Haesub Yoon[2] and Il Moon[1], (1)Chemical and Biomolecular Engineering, Yonsei University, Seoul, South Korea, (2)SamsungTotal Corporation, South Korea

Naphtha prices directly depend on crude oil prices since naphtha is produced by refining crude oil. Naphtha plays an important role as one of basic petrochemicals for downstream products. Uncertainty in forecasting naphtha price has increased due to expanding price variability of naphtha affected by various factors. Therefore, forecasting naphtha crack (price difference between naphtha and crude oil) is a major requirement for decision making and planning. Several studies about price forecasting have been published. In this paper, we considered two methods: multiple regression model [1] and system dynamics model [2] for forecasting the naphtha price. This study is concerned with the derivation of a set of major parameters affecting naphtha prices and identification of the most dominating factors. Naphtha price depends mainly on Asia demand and supply of naphtha as well as naphtha substitute, margin, global economy and operational rate of oil company. The data for these factors are collected from March 2007 to November 2011 except from October 2008 to March 2009 to avoid unusual economic effect, Lehman brothers collapse. In multiple regression model based on statistical approach used data from March 2007 to December 2010 (training period). This model has been verified by comparing naphtha crack trend of actual naphtha crack and predicted naphtha crack from January 2011 to November 2011(forecasting period). In forecasting, the trend of naphtha crack is more important than the accurate value of naphtha crack. Therefore, we focus on the direction of predicted naphtha crack.

Figure 1. shows actual naphtha crack and predicted naphtha crack in training period. From this period, *R2* of this model is 91.8%. This means that 91.8% of the variations in naphtha crack can be explained by the major factors. Figure 2. presents actual naphtha crack and predicted naphtha crack forecasting period. The trend of naphtha crack is more important than precise naphtha crack.

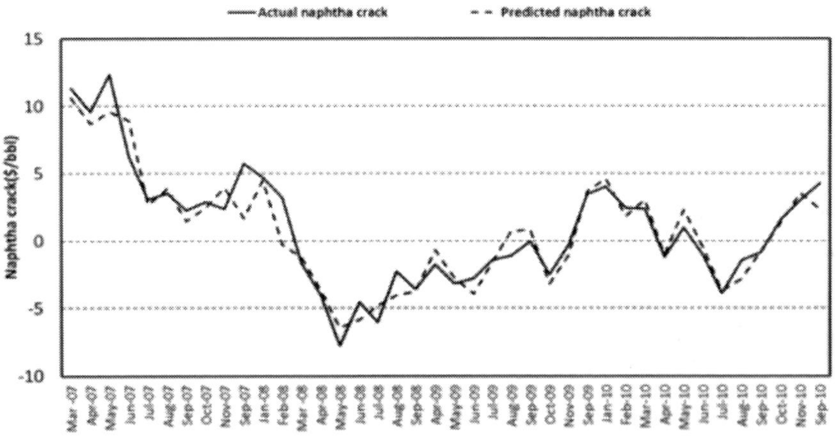

Figure 1. Time series plot of actual naphtha crack and predicted naphtha crack for training.

(From March 2007 to December 2010)

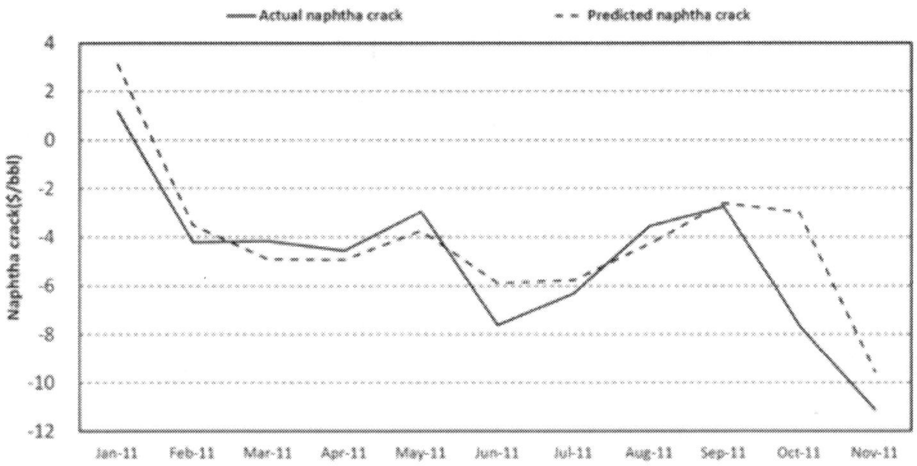

Figure 2. Time series plot of actual naphtha crack and predicted naphtha crack for forecasting.

(From January 2011 to November 2011)

Owing to 9 same predictions out of total 10 months in forecasting period, the percentage of correct predicted trend is 90%. Also, a model of forecasting naphtha crack is presented that is based on system dynamics. System dynamics is thinking model and simulation methodology. This model supports the study in complex systems. Therefore, this model is developed to support the changing petroleum markets. We draw a causal loop diagram for this model. This model forecasts naphtha crack and shows the relations among major factors. We presented two approaches for forecasting naphtha crack. Between two methods, we suggest higher percentage of correct predicted naphtha crack trend and the best approach in forecasting naphtha crack. The modeling approaches can be extended to forecast prices of other downstream chemicals such as LPG and NGL.

Keywords: Naphtha crack, Multiple regression, System dynamics, Forecasting

References

1. W. Zhang, H. Chen, M. Wang, 2010, A forecast model of agricultural and livestock products price, Mechanics and materials 20-23, 1109-1114

2. V. Karavezyris, K. Timpe, R. Marzi, 2002, Application of system dynamics and fuzzy logic to forecasting of municipal solid waste, Mathematics and Computers in Simulation, Vol. 60, 149-158

3. Kim, J., Lee, Y., Moon, .I., 2008, Optimization of a hydrogen supply chain under demand uncertainty, International Journal of Hydrogen Energy, Vol.33, Issue 18, 4715-4729

4. Kim, J., Moon, .I., 2008, Strategic design of hydrogen infrastructure considering cost and safety using multiobjective optimization, International Journal of Hydrogen Energy, Vol.33, Issue 21, 5887-5896

(530b) Data Management in the Processes for Enterprise Wide Optimization Using an Ontological Framework

Edrisi Muñoz[1,2], Elisabet Capon[1], José Miguel Laínez[3], Antonio Espuña[4] and Luis Puigjaner[1], (1)Chemical Engineering, Universitat Politecnica de Catalunya, Barcelona, Spain, (2)Software Engineering, Centro de Investigación en Matemáticas A.C., Guanajuato, Mexico, (3)School of Chemical Engineering, Purdue University, West Lafayette, IN, (4)Chemical Engineering Department, Univesitat Politècnica de Catalunya, Barcelona, Spain

Nowadays the process industries are very complex systems that involve complex decision-making. They require some advanced methods and/or tools in order to integrate and optimize information from different decision levels. In this sense, semantic-based decision support systems seem to offer an efficient tool for facilitating interoperability across multiple, heterogeneous systems. Moreover, enterprises consist of multiple businesses and process units working together at different time and space scales. The organization of the different scales and levels within such complex systems is crucial to understand, analyse, synchronize and improve their operations.

In order to deal with the process complexity, it is necessary to decouple the system across a hierarchy of appropriately chosen levels without disregarding the interrelationship that exists among them. For this purpose, we consider as basis of our analysis the supply chain (SC) concept which can be defined as the group of interlinked resources and activities required to create and deliver products and services to customers. Regarding the integration of different decision levels, an ontology (semantic structure) allows to coordinate the information exchange among the different modeling paradigms/conventions currently used for the different decisions to be taken. Its key role consists of capturing the relevant distinctions of the enterprise structure (specific domain) at the highest level of abstraction, embodying the results of academic research, and offering an operational method to put theory into practice [1].

This work focuses on the integration within the operational decision levels (control, scheduling and supply chain). In this work, an ontological framework is built as the mechanism for information and knowledge models sharing for multiple hierarchical levels and information systems. The potential of the general proposed semantic framework (model maintenance, usability and re-usability) is demonstrated in a case study dealing with an enterprise supply chain network design-planning problem. Further work is underway to unveil the full potential to implement a large-scale semantic web approach to support business processes decisions allowing the mining of quality data which will be translated into useful information by the decision maker. It is important to remark that the generality should be maintained since it is a basic principle in any ontological framework.

[1] Gruber, T. Ontology, Encyclopedia of Database Systems, Ling Liu and M. Tamerözsu, 2008. Springer-Verlag.

[2] Munoz, E., Espuna, A., Puigjaner, L. Towards an ontological infrastructure for chemical batch process management. "Computers & chemical engineering", 2010, 34 (5), p. 668-682.

[3] Brandl, D., Emerson, D., September 2003. Batch markup language batchml.

[4] International Society for Measurement and Control. (1995). Batch control. Part 1. Models and terminology. International Society for Measurement and Control.

[5] International Society for Measurement and Control. (2001). Data structures and guidelines for languages. International Society for Measurement and Control.

[6] International Society for Measurement and Control. (2003). Batch control. Part 3. General and site recipe models and representation. International Society for Measurement and Control.

[7] International Society for Measurement and Control. (2006). batch Control part 4. Lot production registers. International Society for Measurement and Control.

[8] International Society for Measurement and Control. (2007). Batch control part 5 automated equipment control models & terminology. International Society for Measurement and Control.

(530c) Dynamic Modeling, Data Reconciliation, and Simulation of Faults in a Sour Water Gas Shift Reactor

Parham Mobed[1], Jeevan Maddala[1], Richard Turton[2], Debangsu Bhattacharyya[2] and Raghunathan Rengaswamy[1], (1)Chemical Engineering, Texas Tech University, Lubbock, TX, (2)Department of Chemical Engg., West Virginia University, Morgantown, WV

For pre-combustion CO_2 capture in an integrated gasification combined cycle (IGCC) plant, efficient operation of the water gas shift reactors (WGSRs) is necessary. The operation of the WGSRs should ensure that the desired H_2/CO ratio at the inlet of the acid gas removal (AGR) unit is achieved and most of the COS content in the syngas is hydrolyzed in the WGSR process. Due to the typical operating conditions in a WGSR process for an IGCC plant, a number of faults can occur that can cause significant deviation from the operational targets and can result in considerable down-time. For example, the residual fly ash in the syngas can deposit on the catalyst reducing the number of active sites for reaction and decreasing the porosity of the catalyst bed. As the WGS reactions are exothermic, a temperature higher than that allowed by the catalyst manufacturer can cause micro-structural changes to the catalyst. A dynamic model can help in developing a better understanding of the faulty operation of the WGSR process and therefore, can be used for monitoring and development of fault-tolerant control.

With this motivation, a first-principles dynamic model of a sour shift reactor is developed in this work. The 1-D model comprises of mass, momentum, and energy balance equations. In order to reduce the computational time, the method of Thiele modulus and effectiveness factor is used in the modeling approach. The hydrolysis of CO is assumed to be a pseudo first-order reaction and the hydrolysis of COS is assumed to follow Eley-Rideal mechanism. The model is validated with the experimental data available in the open literature for an alkali-metal-impregnated catalyst. A gross error detection and reconciliation procedure is first performed on data. The kinetic parameters are then obtained from the reconciled data. The kinetic model is integrated into the conservation equations.

The validated model is used to study the effect of changes in the operating conditions such as the inlet flow rate composition and temperature of the syngas. In addition, the detailed model is used to study the dynamics of a number of key variables in response to various faults such as changes in the catalyst activity, changes in the surface area of the catalyst, and changes in the bed porosity. The presentation will also include a number of key observations that can be useful for process monitoring and fault diagnosis of the WGSR process.

(530d) Tuning Global Optimization Software Using Derivative-Free Optimization (DFO) Algorithms

Jianfeng Liu and Nick Sahinidis, Chemical Engineering, Carnegie Mellon University, Pittsburgh, PA

Options can have a significant impact on the performance of optimization solvers. Tuning is often necessary in order to identify values of options that improve solver performance in terms of execution time and solution quality. Despite extensive work in the development of optimization algorithms and software, no work has been done in tuning options for global optimization solvers. The objectives of this paper are twofold.

First, derivative-free optimization (DFO) algorithms are used to find optimal values for global optimization solver options. Tuning solver options can be regarded as an optimization problem. Our goal is to automatically `learn' a configuration of options which leads to the best solver performance in terms of execution time and solution quality. However, the relationship between options and solver performance is implicit and the performance function is not smooth. Therefore, we treat the global optimization solver as a black-box model, whose input is values for the options and output is a performance function. Common gradient-based optimization method cannot be used to solve this black-box optimization problem. DFO approaches, however, are suitable choices to solve the tuning problem, since they do not require derivative information for the objective function or constraints [1]. A set of 350 instances including problems in GlobalLib and MINLPLib are studied [2]. We show that tuning options can improve the performance of the global optimization solver BARON [3]. Performance improvements are observed for each problem in this test set collection but this requires using different options for each problem. Interestingly, we have been unable to find a single set of options that improves default BARON settings across the entire test library.

Second, a systematic comparison is made of different DFO methods in solving tuning problems. A total of 28 state-of-the-art DFO implementations are applied to find the optimal options of BARON. The algorithms are tested in terms of their ability to improve the performance of BARON. For the problems we studied, the DFO methods show remarkably different ability to solve tuning problems. Comparisons show that several DFO approaches are much better than others in terms of finding optimal options of BARON.

The main conclusion of this work is that users of global optimization software can resort to automatic machine learning techniques based on DFO solvers in order to identify optimal parameters for software options for their specific applications.

References cited

1. Rios, L. M. and Sahinidis, N. V., Derivative-free optimization: A review of algorithms and comparison of software implementations, http://thales.cheme.cmu.edu/dfo.
2. http://www.gamsworld.org/index.htm
3. Tawarmalani, M. and N. V. Sahinidis, A polyhedral branch-and-cut approach to global optimization, Mathematical Programming, Ser. B, 103, 225-249, 2005

(530e) An Intelligent Agent for On-Line Optimization of Gasoline Blending

César Pernalete, Refining and Industrialization Management, PDVSA Intevep, Los Teques, Venezuela and Jacinto Dávila, CESIMO, Universidad de Los Andes, Mérida, Venezuela

An intelligent agent for on-line optimization of gasoline blending

C. G. Pernalete [1,2], J. Dávila [2]

[1]Gerencia Técnica de Refinación. PDVSA Intevep. Venezuela.

[2] Centro de Simulación y Modelos (CESIMO). Universidad de los Andes. Venezuela

Abstract

Gasoline blending is a batch process where different intermediate products of the refining process are mixed in order to produce a final product to meet some quality specifications. The most important standard specifications for gasoline include: RVP (Reid Vapor Pressure), AKI (Anti knock index), RON (Research octane number), and D-86 distillation as leading properties. It is important to notice that each feedstock differs from the others on its properties and cost. Therefore, in order to reach a finished product that respects the aforementioned constraints and using the available volume of each feedstock, an optimization problem must be resolved, normally off-line, that is, previously and without intervention in the actual blending process in real time.

In this off-line optimization process two important assumptions are normally made: firstly, that the feedstock quality parameters are constant within a certain period of time for the blending and, secondly, that the models that describe the blending processes are linear or, at most, require only some non-linear adjustments. These assumptions, together with the possible heterogeneity of each feedstock inside the tank, caused by compounds separation because of density differences, often cause the production of a blend to end at some distance from the expected quality , generating high additional costs because of re-blend or quality giveaway: products with better quality than required.

In this work, it was designed and developed an optimizer that implements a novel strategy for executing real time optimization in gasoline blending processes. In order to account for the problems of process-model differences, the optimizer in this work uses blending models automatically obtained from machine learning experiments, specifically learning tasks with Support Vector Machines (SVM). For this purpose we took advantage of the large amount of operational blending data available in a Venezuelan refinery to produce different grades of gasoline.

The basic elements of any strategy for optimization of the gasoline blending process are: 1) the distributed control system (DCS), which executes the control actions on the plant, 2) the optimizer, which makes the calculations using current information from the plant, 3) the on-line analyzer, which determines on-line the feedstock and product properties, and 4) the tank information system (TIS). All of them must and have been considered in order to implement the solving application. All these operational constraints suffice to consider the on-line optimization system as a complex software application that must cleverly interact with other elements of the system during the optimization process. Considering that an intelligent agent is a computational system located in an environment and that is able to behave in an autonomous and flexible manner in such environment in order to reach its designed objectives, it was decided to develop the gasoline blending optimization system as an intelligent agent.

Summarizing, there is evidence that the modeling of blend properties for an specific refinery using support vector machines, is a promising technique as the average error in our experiments is always below that obtained by conventional methods. Also, we can show that the agent developed and tested in this work can deal efficiently with the calculation of the specific optimization problems. In our experiments, a solution was always found and the convergence time was in the neighborhood of one minute. The agent was also tested in simulations to evaluate its capacity to deal with perturbations in the product quality during the blending process and the results are enticing. Based on conservative estimated explained in the paper, we estimate that, by using this intelligent agent for the on-line optimization of gasoline blending in a medium capacity refinery of the National Refining Circuit of PDVSA, the system could generate savings in the order of 5MM$/year.

(534a) Exergy-Loss-Minimization Based Process Synthesis for Cascade Refrigerant System in Ethylene Plants

Ha Dinh, Jian Zhang and Qiang Xu, Dan F. Smith Department of Chemical Engineering, Lamar University, Beaumont, TX

Refrigeration system holds an important role in most chemical/petrochemical processes. The traditional cascade refrigeration system (CRS) used in ethylene plants includes multiple refrigerants and multiple temperature and pressure levels. Each component subsystem includes a compression section, a condensing section and several temperature levels with sub-coolers, expansion valves, evaporators and flash drums. In this study, at first, in-depth thermodynamic analysis for a single-loop refrigeration system is carried out to identify exergy loss in heat exchanger network with non-isothermal phase changes, especially at sub-cooler and evaporator in the refrigerant system. The relationship between compressor work consumption and exergy change from one pressure level to another is also comprehensively demonstrated. The achieved methodology is applied in a cascade refrigeration system used in ethylene plants to explore system energy loss below ambient temperature.

The procedure of process synthesis involves three consecutive steps: Thermodynamic analysis (for both single-loop and cascade refrigeration systems), Synthesis model development and solving, and solution validation and evaluation. Rigorous steady-state model of a propylene cycle in the cascade system is firstly constructed and employed to generate Exergy-Temperature (B-T) chart in order to analyze their thermodynamics and energy distributions. An exergy-based mixed-integer linear programming (MINLP) model is developed to minimize the overall exergy loss of the system, where multiple refrigerants with multiple recycling loops are simultaneously addressed and all of cooling demands are fulfilled. The optimal solution is subsequently examined by conducting rigorous simulations to check their feasibility and consistency. Economic evaluation is also carried out from the calculated capital cost of new equipments and annual operating cost. The efficacy of the developed methodology is demonstrated by a case study of a cascade propylene refrigeration system used in an ethylene plant.

(534b) Development of the Optimal Flowsheet of Isomerization of Light Gasoline Fractions

Nickolay V. Kuzichkin, Department of resource saving, Institute of Chem. Technology, Saint Penersburg, Russia

The two-tiered approach to the synthesis of the optimal structure of the process of isomerization of light gasoline fractions is proposed.

Problem optimal choosing of reactors and separation's system with a minimal use of auxiliary heating and cooling is solved at the low level, and on the top-level is carried synthesis of heat exchangers network.The volume of reactor, reflux ratio of the temperature and pressure in the distillation columns are the independent variables

The minimum amount of heat consumed by the auxiliary heating and cooling agents can be determined from the temperature - enthalpy diagram (TED).
The maximum possible heat recovery, and hence the minimal costs associated with the use of auxiliary fluids, obtained from TED are the lower boundary of the synthesis problem of heat exchange networks. Synthesis of heat exchangers network is executed by means of combinatoric method with aid estimated functions.

The developed method of choosing the optimal structure is universal and can be used to calculate construction and retrofit installations refinery

(534c) A Systematic Framework for Synthesis and Design of Multi-Scale Processing Networks Using Incremental-Based Solution Strategy

Alberto Quaglia[1], Maxim Shishmarev[2], Bent Sarup[3], Gürkan Sin[2] and Rafiqul Gani[1], (1)CAPEC, Department of Chemical and Biochemical Engineering, Technical University of Denmark, Kgs. Lyngby, Denmark, (2)Department of Chemical and Biochemical Engineering, Technical University of Denmark, Kgs. Lyngby, Denmark, (3)Vegetable Oil Technology Business Unit, Alfa Laval Copenhagen A/S, Søborg, Denmark

A systematic framework for synthesis and design of multi-scale processing networks using incremental-based solution strategy

A.Quaglia,[1] B. Sarup,[2] G. Sin,*[1] R. Gani[1]

[1]CAPEC, Department of Chemical and Biochemical Engineering, Technical University of Denmark, DK-2800 Kgs. Lyngby, Denmark; email: aq@kt.dtu.dk. *email:gsi@kt.dtu.dk.

[2]Vegetable Oil Technology Business Unit, Alfa Laval Copenhagen A/S, 2860, Denmark

For processing companies, the synthesis and design of optimal industrial operations includes all the strategic and tactical decisions such as the selection of raw materials and products portfolio, as well as the synthesis and design of the processing network and the optimization of the material flows through it.

Recent development in process system engineering (PSE) focused on a class of multi-scale models (from meso to mega-scale), formulating them as Mixed Integer Non Linear Programming (MINLP) problems (Grossmann, 2005). As many contributions in the literature demonstrate, problems of this class are certainly non-trivial, where typical factors contributing to the increase in problem complexity are number of process alternatives considered, size and complexity of the process models used to describe the process alternatives and presence of recycles, bypasses and parallel processing. Several methods and solution strategies (e.g. model reformulation, problem decomposition, variable initialization and bounding) have been successfully proposed for the solution of complex network synthesis problems. These methods and tools are often tailor-made for the specific problem to be solved, and therefore their selection and implementation is not trivial and requires expert judgment.

In this contribution, we propose a systematic framework for the formulation and solution of processing network synthesis problem, and illustrate its application through the solution of industrial case studies, developed in collaboration with Alfa Laval. The key characteristic of the framework is to manage the complexity of the multi-scale problem through a systematic approach, in order to allow simple, efficient and robust formulation and solution of the problem. In the first step of the framework, the user is guided through the formulation of the problem. A superstructure containing all possible alternatives is generated, the objective function is selected, and all relevant data and models are collected, reconciliated and systematized in a database structure. All available engineering and commercial insights are translated in logical constraints aiming at excluding redundant solutions from the search space, that is, the superstructure. In the second step the formulated MINLP problem is analyzed to identify and suggest the most appropriate solution strategy.

One of the strategies employed in the framework is the incremental decomposition-based solution, the complexity of the formulated problem is decomposed, and a series of sub-problems with incremental levels of complexity is automatically generated. The generated sub-problems are then solved in a sequence, where a lower level sub-problem solution provides initialization and bounds to the higher level, until the full problem solution is obtained. Moreover, sensitivity analysis is used at each step, in order to prioritize the use of complex (non linear) models to describe only processes and phenomena which have a large impact on the overall objective, so to obtain a good tradeoff between model complexity and quality of the solution.

The proposed framework is applied to the formulation and solution of industrial process synthesis problems, in the field of edible oil production and refinement, wastewater treatment and biorefinery.Through the presentation of these industrial case studies, the features of the proposed framework are highlighted, and the complexity related to industrial application of process synthesis (large number of process alternatives, complex and non-linear models, presence of recycles, bypass and parallel processes etc.) is discussed.

References

A.Quaglia, B.Sarup, G.Sin, R.Gani, 2012, Integrated business and engineering framework for synthesis and design of enterprise-wide processing networks, Computers & Chemical Engineering (38) 213-223.

I.E. Grossmann, 2005, Enterprise-wide Optimization: A New Frontier in Process Systems Engineering, AIChE Journal, 51, 1846-1857.

(534d) Design of a Reactive Distillation Column: Comparison Between Trays and Packings

Paola Bastidas[1], Julio C. Vargas[1], Alvaro Orjuela[2], Iván D. Gil[1] and Gerardo Rodriguez[1], (1)Departamento de Ingeniería Química y Ambiental, Universidad Nacional de Colombia, Bogotá, Colombia, (2)Departamento de Ingeniería Química y Ambiental, Universidad Nacional de Colombia, Bogota, Colombia

Design of a reactive distillation column: comparison between trays and packings

Paola Bastidas, Julio Vargas, Álvaro Orjuela, Iván Gil, Gerardo Rodríguez

Grupo de Procesos Químicos y Bioquímicos. Departamento de Ingeniería Química y Ambiental, Universidad Nacional de Colombia-Sede Bogotá, Cr. 30 45 – 03, Bogotá, Colombia

During the last decade, there has been an increasing application of reactive distillation technology in industry, due to the economic benefits that it offers. For example, the production rate and the conversion are improved by the constant removal of the generated products, and the capital costs are reduced since less units operations are required. This growing interest demands deeper study, in order to have more efficient processes.

The key to have efficient separation-reaction process is the proper design and selection of the hardware to be implemented inside the column. The current process simulators are powerful tools that allow comparing different types of internals, that is, to study the effect of the equipment geometry to achieve given separation and reaction grades. That is the case of the Aspen Plus software, where rigorous design of reactive distillation columns is possible through the non-equilibrium (NEQ) model, which has in to account the effect of heat and mass transfer velocities in the global and phase balances, and assumes that the equilibrium is reached only at the interphase.

In this work, the design of a reactive distillation column to produce isoamyl acetate is done with a NEQ model approach. Comparative studies between using trays and packings are developed through process simulations using Aspen Plus®software. In both cases, NRTL-HOC thermodynamic model is used to describe the phase equilibrium, and an equilibrium kinetic model using Amberlyst 35 as catalyst is implemented to describe the reaction behavior. A stream containing acetic acid, amyl alcohol, amyl acetate and water is feed in the lower zone of the reactive section. At the same time an excess of amyl alcohol is fed at the top of that section.

For the same column configuration, better results are obtained with packings than with trays. When using packings as internals, the isoamyl acetate purity is higher than with trays, as well as the global conversion. According to the global heat transfer rate, the feed streams vaporize when enter to the column, with a stronger tendency when using packings than when using trays. This change of energy in the reactive zone promotes the reaction in the desired direction, due to its endothermic nature. Because of the vaporization is weaker in the case of the tray column, the temperature profile goes down in this zone, making that one of the reactants be in the phase vapor rather than in the liquid vapor, which leads to a lower conversion than in the case of the packed column.

It was encountered also, that in the reactive zone were the higher tray efficiency and the lower HETP value, for the tray and packed column, respectively. That is, the reactive zone is the most efficient zone, mainly because that is the zone with bigger differences between the vapor and liquid temperatures.

A State-Space-Based MIDO Formulation for Batch Distillation Process Synthesis

Shan JIN, Hong-Guang DONG and Xin-Wen GUO, School of Chemical Engineering, Dalian University of Technology, Dalian, China

Introduction

Batch distillation is widely used in fine chemical and pharmaceutical industries, for the purification or recovery of high value-added products. The transient nature of the batch distillation process provides great flexibilities in both various operating policies and alternative complex configurations. Such unique characteristics help further explore the economic potential and process efficiency through the optimal synthesis of batch distillation process. However, the dynamic behavior and complicated economical trade-offs make a huge challenge for the synthesis problem.

Over the past decades, the optimal design and operation of batch distillation have received significant attentions (Logsdon, Diwekar and Biegler, 1990; Mujtaba and Macchietto, 1996; Sharif, Shah and Pantelides, 1998; Low and Sørensen, 2003). The previous works have provided some general insights on the comparative performances of various batch distillation systems and indicated that the interdependent process design and operating control variables should be optimized simultaneously. Furthermore, different configuration of batch distillation column was also taken into account as an optimization parameter with the method of a disjunctive model consisting a given set of equipment modules and the corresponding logical relationships provided by the design context (Oldenburg, Marquardt and Leineweber, 2003; Low and Sørensen, 2005).

With the development of manufacture and control in chemical process, more efficient and energy saving techniques are applied in batch distillation system. As a result, the number of possible batch distillation column configurations and complex operating strategies proposed in the literature continue to increase, such as the side withdrawals, the adaptive vapor recompression structures, thermally linked streams and other complex designs. So there is an urgent need to propose a completed superstructure of batch distillation system that encompasses all possible configurations and operating strategies to reduce the energy consumption and capital investment, and eventually achieve the profit maximization. The incompletion of the superstructure may preclude a series of design alternatives, where the optimal solution may actually lie.

In addition, because the batch distillation is a dynamic process, the separation difficulty also varies with the processing time, especially for multitask separation, the corresponding structure of the batch distillation system should be optimally changed in different time intervals. For example, two separation tasks with different relative volatilities, the optimal number of trays will certainly change during the different mixtures are processed. Even for a single separation duty, at the beginning it is relatively easier to get high purity of distillate, so less trays and reflux ratio are needed. As the process goes on, the separation difficulty is added, the number of trays and reflux ratio should be increased to accomplish the task. So there is plenty of room for the column configuration being optimized in different time intervals to make a better trade-off between capital and operating costs, which aims to maximize the total profit. Even more, due to the flexibility of batch distillation process, the temporary idle trays in certain time interval can even be used in other separation tasks, which can not only make the best of equipment resource, but also get extra profits. However, the constant vapor load and the fixed configuration on total time horizon are always adopted in most of previous works, and this limit may result in the compromise of the operation strategy and the design of configuration instead of the perfect match between each other in every time interval, which will increase the total cost.

In this work, we expect to propose a systematic method based on mixed integer dynamic optimization (MIDO) formulation for synthesizing the batch distillation process on both space and time dimension, which means that not only more novel advanced column configurations and operation strategies are taken into account, but also both of them can be optimized in every time interval. An improved superstructure based on state space representation (Bagajewicz and Manousiouthakis, 1992), a framework that takes all streams splitting and mixing possibilities into consideration, is constructed to achieve any potential optimal process. In the next section, the batch distillation process synthesis problem is presented including the problem definition, superstructure representation, and model formulation followed by objective function establishment. Finally, Binary and multicomponent separation case studies will be presented to illustrate the validity and advantages of the proposed approach.

The Batch Distillation Process Synthesis Problem

Problem Definition

The objective in this synthesis problem is to determine the most economical batch distillation process, which is capable of fulfilling all separation requirements. All the optimal parameters including configuration, design and operation strategy are achieved simultaneously. The problem is stated as follows:

Given: feed mixture to be separated with known batch size and composition; a set of product specifications (purities of key components in main products, amount, etc.); available production time.

Determine: For each time interval, the optimal configuration (position of the vessels, connections between trays and vessels), design (number of trays and vessels; vapor loads) and the operating strategy (liquid/vapor ratio, etc.).

To maximize: the profit.

Superstructure Representation

The state space based superstructure has been successfully used for the synthesis of continuous distillation process to reduce both the utility and capital costs (Zou et al., 2012), and we take the unique characteristics of batch distillation process (dynamic behavior and tray-vessel configuration) into account to develop the superstructure.

To be specific, the superstructure of batch distillation system is viewed as a system of two interconnected parts (see Figure 1). One is referred to as the distribution network (DN), in which all mixers, splitters and the connections between them are embedded. The other is the so-called process operator (PO), which is further divided into two kinds of equilibrium stages, i.e. the vessels and the trays, with intermediate condenser and/or reboiler. From the superstructure we can see that the initial feed stream can be charged into arbitrary number of vessels according to its property, such as the number of components and each composition in the feed stream. Then the streams out of the vessels are delivered to DN for distribution to arbitrary trays for further mass and heat exchanges, and similarly, the trays are fully connected to each other through splitters and mixers on DN and also have opportunities to distribute streams to arbitrary vessels. Finally, when the required purities are satisfied, each product can be collected in certain vessels or withdrawn from certain trays.

Under this framework, different stream-connecting structures between trays and vessels are used to generate all kinds of feasible configurations and operating strategies. Moreover all the connecting streams are fluctuating over time, as a result the connecting structures have time dimension, which means that the operation strategies and corresponding configurations can be both optimized in different time intervals.

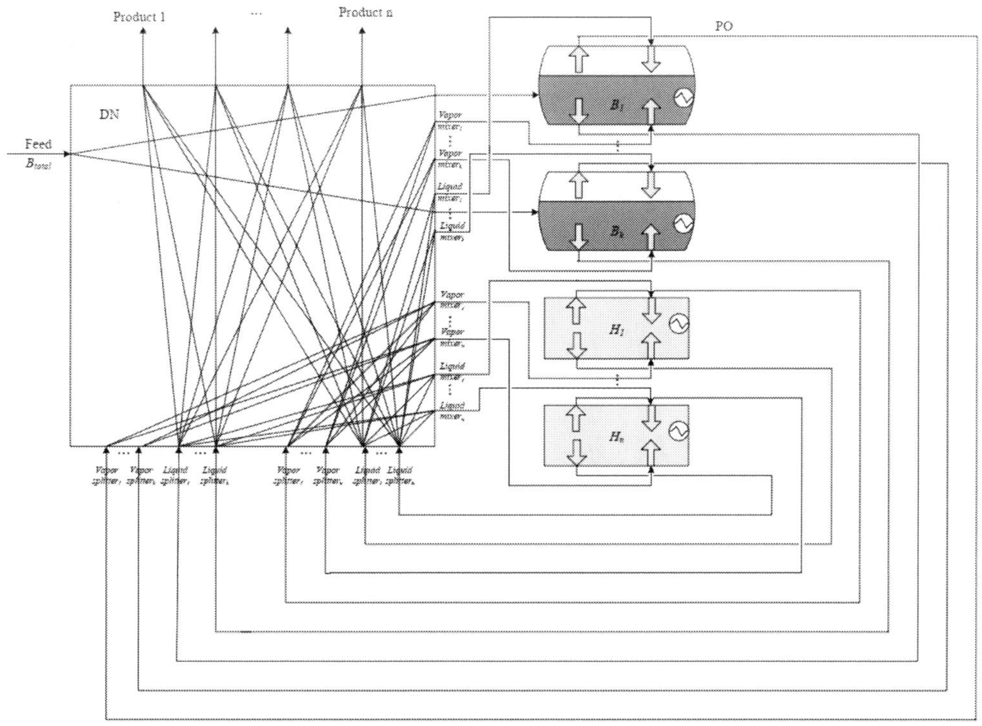

Figure 1. The improved state-space superstructure of batch distillation system

Model Formulation

In this work we have employed a detailed dynamic model considering Mass balance, Equilibrium relation, Summation check and Heat balance (MESH) equations for each of equilibrium stage, all mixers and splitters. Reasonable assumptions are also adopted here, such as constant molar liquid holdup on the trays, negligible vapor holdup, total condenser, constant pressure, perfect mixing and adiabatic trays. Thus the overall synthesis problem can be formulated as a mixed integer dynamic optimization (MIDO). The major mathematical model can be described by the following equations:

Total mass balance for feed:

$$B_{total} = \sum_{b=1}^{k} B_b \qquad \forall b \in NB \tag{1}$$

PO: representation of mass and heat exchanges process on equilibrium stage. (Take tray i as example, the formulations of vessels have the similar forms.)

Mass balance for component j of tray i:

$$\frac{d(H_i x_{i,j})}{dt} = L_{i,j}^{in} + V_{i,j}^{in} - V_{i,j}^{out} - L_{i,j}^{out} \qquad \forall i \in NT, \forall j \in C \qquad (2)$$

Where B_{total}, B_b, H_i, L, V, $x_{i,j}$ are the total feed, vessel holdup, tray holdup, liquid flowrate, vapor flowrate, and liquid composition, respectively.

Heat balance for tray i:

$$\frac{d(H_i h(H_i))}{dt} = L_i^{in} \times h\left(L_i^{in}\right) + V_i^{in} \times h\left(V_i^{in}\right) - V_i^{out} \times h\left(V_i^{out}\right) - L_i^{out} \times h\left(L_i^{out}\right) + Q_{R,i} + Q_{C,i} \qquad \forall i \in NT \qquad (3)$$

Where h represents the specific heat content, and it is the function of composition of the mixture and tray temperature. $Q_{R,i}$, $Q_{C,i}$ denote the intermediate reboiler and condenser heat duty.

DN: determination of stream-connecting structures. (For each time point)

Mass balance for component j of mixer i:

$$L_{i,j}^{in} = \sum_{i'=1}^{n}(L_{i',i} \times x_{i',j}) + \sum_{b=1}^{k}(L_{b,i} \times x_{b,j}) \quad \forall i,i' \in NT, \forall b \in NB, \forall j \in C \qquad (4)$$

$$V_{i,j}^{in} = \sum_{i'=1}^{n}(V_{i',i} \times y_{i',j}) + \sum_{b=1}^{k}(V_{b,i} \times y_{b,j}) \quad \forall i,i' \in NT, \forall b \in NB, \forall j \in C \qquad (5)$$

Mass balance for component j of splitter i:

$$L_{i,j}^{out} = \sum_{i'=1}^{n}(L_{i,i'} \times x_{i,j}) + \sum_{b=1}^{k}(L_{i,b} \times x_{i,j}) + D_i \times x_{i,j} \qquad \forall i,i' \in NT, \forall b \in NB, \forall j \in C \qquad (6)$$

$$V_{i,j}^{out} = \sum_{i'=1}^{n}(V_{i,i'} \times y_{i,j}) + \sum_{b=1}^{k}(V_{i,b} \times y_{i,j}) \qquad \forall i,i' \in NT, \forall b \in NB, \forall j \in C \qquad (7)$$

Where $L_{i',i}$, $V_{b,i}$, $L_{i,b}$, $V_{i,i'}$, D_i, $y_{i,j}$ are the flowrate of liquid stream from tray *i'* to tray *i*, the flowrate of vapor stream from vessel *b* to tray *i*, the flowrate of liquid stream from tray *i* to vessel *b*, the flowrate of vapor stream from tray *i* to tray *i'*, flowrate of product stream from tray *i* and vapor composition, respectively.

From the DN formulation, we can notice that $L_{i,b}$, $V_{i,b}$, $L_{b,i}$, $V_{b,i}$ are used to represent the connecting relations between trays and vessels, and $L_{i',i}$, $V_{i',i}$, $L_{i,i'}$, $V_{i,i'}$ express the connecting relations between different trays. Through the analysis of all these variables on different time intervals, we can get the connecting structures of the batch distillation system, which means that the optimal configurations and operation strategies are obtained on the total time horizon.

Objective Function Establishment

In this work, we formulate the objective function to contain the sales revenue, operating costs (including costs of cold and hot utilities) and equipment costs (including costs of distillation columns and heat exchangers) and we convert these costs to an hourly based. The objective function has the following form (Logsdon, Diwekar and Biegler, 1990):

$$\psi = \frac{DP_r - B_0 C_0}{t_f + t_s} - \frac{T.A.C.}{Hrs} \tag{8}$$

The function of $T.A.C.$ is adopted from Zou (2012) as follows:

$$T.A.C. = f(N_T, D_{tray,i}, A_{R,i}, A_{C,i}, A_{R,b}, A_{C,b}) + C_w(\sum_{i\in NT} Q_{C,b} + \sum_{i\in NT} Q_{C,i}) + C_S(\sum_{i\in NT} Q_{R,b} + \sum_{i\in NT} Q_{R,i}) \tag{9}$$

For complex columns systems, the cost of a column was normalized on a "per tray" basis in Zou (2012), this strategy is also adopted in this work in order to make the best of every operation unit in every time interval.

The function of the cost of single tray is based on Warren D. Seider (2004)

$$CT = (193.04 + 22.72 D_{tray,i} + 60.38 D_{tray,i}^{\ 2}) \times 1.2 \times 1.5 \tag{10}$$

Then the $f(N_T, D_{tray,i}, A_{R,i}, A_{C,i}, A_{R,b}, A_{C,b})$ can be written as:

Invest cost= $\sum_{i\in NT} (193.04 + 2.49 V_i^{out0.5} + 1.304 V_i^{out}) +$

$$500(\sum_{i\in NT} (\frac{Q_{R,i}}{7000(T_s - T_i)})^{0.8} + \sum_{i\in NT} (\frac{Q_{C,i}}{3000(T_i - T_w)})^{0.8} + \sum_{b\in NB} (\frac{Q_{R,b}}{7000(T_s - T_b)})^{0.8} + \sum_{b\in NB} (\frac{Q_{C,b}}{3000(T_b - T_w)})^{0.8}) \tag{11}$$

Where V_i^{out} denotes the vapor load of tray i, T_i, T_b, T_s, T_w represent the temperature of tray i, vessel b, heating and cooling utilities respectively.

The mathematical model of complex batch distillation system has been formulated as a mixed integer dynamic optimization (MIDO) problem with a set of differential-algebraic equations (DAE). Then the integer variables are eliminated by means of continuous variables directly. To be specific, the binary variables assigned to each tray and heat exchanger with a value of "0–1", denoting the logic of existence or absence, can be also indicated with the value of nonzero or zero of the continuous variables respectively, like the vapor flowrate of certain tray or the heat duty of certain exchanger. The central difference method is used to discretize the problem into nonlinear program (NLP) in this work. The NLP model is solved using the CONOPT solver in GAMS environment with the random initial values and perturbation strategy to increase global search ability.

Case Studies

In following sections, two synthesis case studies involving binary and multicomponent separation process are presented.

Example 1

The case of binary mixture separation of benzene and toluene is studied. In this example, the setup time and hold up of the trays are neglected, and we expect to synthesize the optimal batch distillation process to satisfy the separation requirements. The given data is presented in Table 1.

Table 1. Given data for example 1

Column Specifications	
Upper bounds on the number of stages, N_T	11
Upper bounds on the vapor boilup rate, V_i^{out} *(kmol/h)*	2
Operation pressure, *P (kPa)*	100
Available annual production time, *Hrs (h/year)*	8000
Utility available	
Cooling water: 300K *($/GJ)*	0.7
Steam: 421K *($/GJ)*	3
Initial charge	
Feed, B_b *(kmol)*	8
Feed composition, Benzene/Toluene $x_{b,1}/x_{b,2}$ *(mole fraction)*	0.58/0.42
Product specification	
Lower bounds on average Product purity of benzene x_D *(mole fraction)*	0.90
Product yield of benzene per batch, *D (kmol)*	4
Top product price P_{r1} *($/kmol)*	34
Bottom product price P_{r2} *($/kmol)*	34
Raw material cost C_0 *($/kmol)*	2

Case 1 permanent configuration model

First, permanent configuration model is employed, which means that the configuration of batch distillation column is unchangeable in total time horizon. The configuration, batch time, vapor flow, and distillate rate of the batch distillation system are optimized simultaneously to maximize the profit (See Table 2).

The result shows that under this scenario, the profit realized by regular column is about 61% more than that by inverted column, and this is mainly because the batch time of the regular column is much shorter than that of the inverted column, as a result the product revenue achieved by the regular column is greater than that of the inverted column.

Table 2. Summary of the result for permanent configuration model

Configuration	t_b	*D*	Nt	*OC*	*CC*	*P*
	h	*kmol/b*		*$/h*	*$/h*	*$/h*
regular	5	4	11	0.213	0.486	50.501
inverse	8	4	11	0.214	0.488	31.298

Case 2 changeable configuration model

Then the changeable configuration model, meaning that the configuration of batch distillation column can optimally change in each time interval, is adopted to illustrate its advantage. In this case, we consider the condition of fixed product demand and strict product quality specifications to avoid unlimited production of the products (Mujtaba and Macchietto, 2004). We assume that the batch time and the amount of distillate per batch are fixed with 6 hours and 4 kmol respectively, so the total revenue is fixed and the objective function is transformed into minimizing the total annual cost (TAC). The results are represented in figure 2 and Table 3.

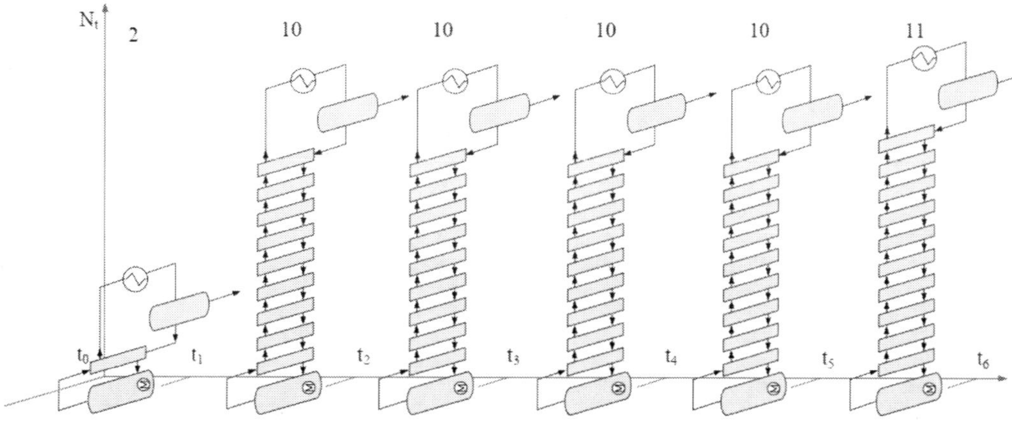

Figure 2. Optimal batch distillation process of changeable configuration model

Table 3. Summary of the result for changeable configuration model

Configuration	t_b	D	Nt	AOC	ACC	TAC
	h	$kmol/b$	Each time interval	$\$/yr$	$\$/yr$	$\$/yr$
permanent	6	4	11,11,11,11,11,11	1522.7	3854.7	5377.3
changeable	6	4	2,10,10,10,10,11	1408	3816	5224

The result shows that the number of trays is gradually increasing with the processing time in the changeable configuration model, which is in response to the increasing difficulty of the separation task as the process goes on. So there is a multistage optimization of both capital and operating costs in every time interval instead of compromising the operation strategy due to the fixed design of configuration in entire batch time. Thus, a 3% lower TAC is achieved in the changeable configuration model.

Example 2

In this section, the synthesis of optimal batch distillation process is investigated for the case of ternary mixture separation of benzene, toluene and ethylbenzene. In this example, the reflux of both condenser and reboiler is given changes to connect to arbitrary trays in different time intervals in order to adjust the liquid/vapor ratio on each equilibrium stage in real time. We expect to fully exploit the performance of the batch distillation system to synthesize an optimal process to maximize the amount of distillate of a specified concentration for a specified batch time. The given data is presented in Table 4.

Figure 3 shows that the liquid/vapor ratio on each equilibrium stage is optimized in each time interval to achieve the maximum amount of distillate. And the maximum amounts of each product are $D_{benzene}=4kmol$, $D_{toluene}=3.9kmol$ respectively, and the composition profile of each component in liquid holdup in condenser is represented in Figure 4.

Table 4. Given data for example 2

Column Specifications	
Number of stages, N_T	11
Upper bounds on the vapor boilup rate, V_i^{out} *(kmol/h)*	2
Operation pressure, P *(kPa)*	100
Tray holdup, *M(kmol)*	0.3
Condenser holdup, M_1*(kmol)*	0.5
Processing time, *Hrs* (*h*)	50
Initial charge	
Feed, B_b *(kmol)*	8
Feed composition, Benzene/Toluene/Ethylbenzene $x_{b,1}/x_{b,2}/x_{b,3}$ *(mole fraction)*	0.4/0.4/0.2
Product specification	
Lower bounds on average product purity of Benzene/Toluene $x_{D,1}/x_{D,2}$ *(mole fraction)*	0.95/0.80

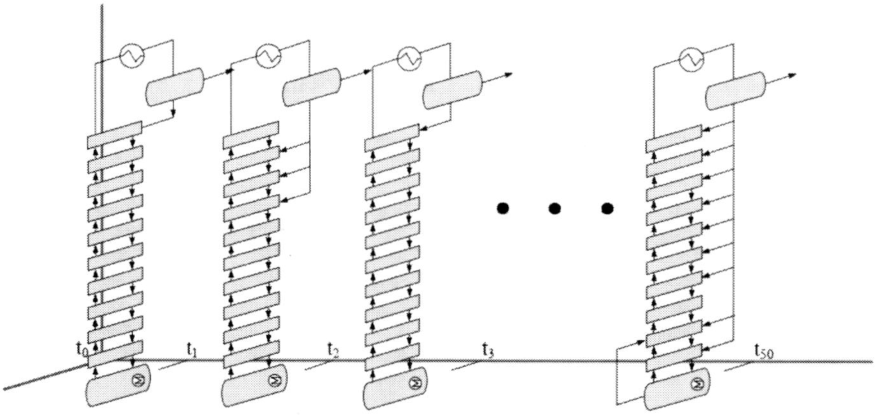

Figure 3. Optimal batch distillation process of ternary components separation task

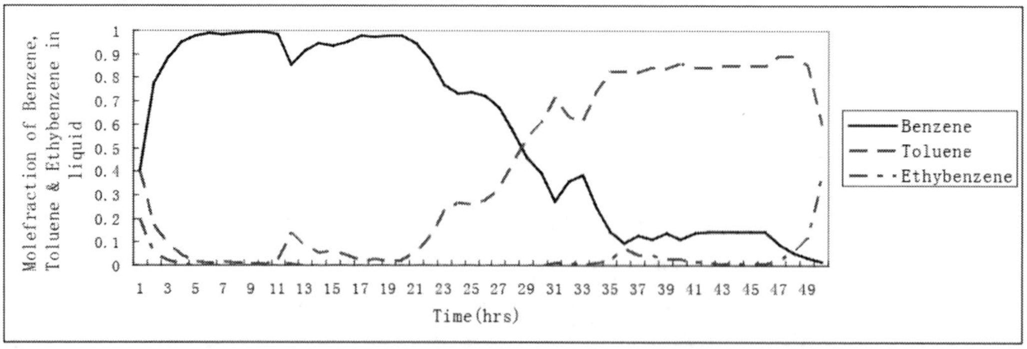

Figure 4. Composition profile in liquid holdup in condenser

Conclusion

In this work, a systematic method based on mixed integer dynamic optimization (MIDO) formulation for synthesizing the batch distillation process on both space and time dimension is proposed. An improved superstructure based on state space representation is introduced to encompass all potential optimal process.

The case studies indicate that under this framework, optimal configuration, design and operating policies of the batch distillation system are solved simultaneously. Moreover, the advantage of changeable configuration model has been highlighted. From the comparison analysis of the example results, not only energy saving but also capital investment reduction has been brought in our novel conceptual designs. Additionally, a novel operating strategy to fully exploit and adjust the performance of batch distillation system in real time is represented. Due to complexity of considering all potential optimal design and operating strategy of multicomponent separation, more work on optimizing multicomponent batch distillation system, including both design and operating strategy, will be implemented in the future.

Acknowledgment

This work is supported by the National Natural Science Foundation of China, under Grant no.21276029.

References

1. Logsdon JS, Diwekar UM, Biegler LT. On the simultaneous optimal design and operation of batch distillation columns. *Transactions of IChemE*. 1990; 68: 434-444.
2. Mujtaba IM, Macchietto S. Simultaneous optimization of design and operation of multicomponent batch distillation column-single and multiple separation duties. *J of Process Control*. 1996; 6: 27-36.
3. Sharif M, Shah N, Pantelides CC. On the design of multicomponent batch distillation columns. *Comp & Chem Eng*. 1998; S22: 69-76.
4. Low KH, Sørensen E. Simultaneous optimal design and operation of multivessel batch distillation. *AIChE J*. 2003; 49: 2564-2576.
5. Oldenburg J, Marquardt W, Heinz D, Leineweber DB. Mixed-logic dynamic optimization applied to batch distillation process design. *AIChE J*. 2003; 49: 2900-2917.
6. Low KH, Sørensen E. Simultaneous optimal configuration, design and operation of batch distillation. *AIChE J*. 2005; 51: 1700-1713.
7. Bagajewicz M, Manousiouthakis V. Mass/heat-exchange network representation of distillation networks. *AIChE J*. 1992; 38: 1769-1800.
8. Zou X, Cui YH, Dong HG, Wang JQ, Grossmann IE. Optimal design of complex distillation system for multicomponent zeotropic separations. *Chem Eng Sci*. 2012; 75: 133-143.
9. Seider W, Seader J, Lewin D. *Product and process design principles - synthesis, analysis and evaluation*. 2nd ed. New York: Wiley; 2004.
10. Miladi MM, Mujtaba IM. Optimization of design and operation policies of binary batch distillation with fixed product demand. *Comp & Chem Eng*. 2004; 28: 2377-2390.

(534f) Design and Synthesis of Energy Integrated Multipurpose Batch Plants Using a Robust Scheduling Platform

Esmael Reshid Seid and Thokozani Majozi, Chemical Engineering, University of Pretoria, Pretoria, South Africa

In recent years batch plants have attracted more attention due to their flexibility to produce different products using the same facility. Batch manufacturing is typically used in the pharmaceutical, polymer, food and speciality chemical industries, because it provides the necessary flexibility to accommodate various production requirements using the same processing facility. However, these very important features introduce extra degrees of complexity into the design, synthesis and operation of such plants. Utility requirements in these plants can be a major contributor to overall running costs. Heating and cooling are unavoidable aspects of many chemical processing facilities, with operations where heat is generated and others where heat is required. It is because of this occurrence that heat integration becomes a possibility for reducing utility requirements. Energy savings have often been neglected in the design and synthesis of batch processes in the past and so large percentage savings are possible.

The contribution in this paper is developing a comprehensive optimization technique for design and synthesis of heat integrated batch plants. In batch chemical processes it is highly imperative to optimize the schedule before design. The recent scheduling technique for multipurpose batch plants by Seid and Majozi (2012) is used as a platform. Computational studies are presented to illustrate the effectiveness of the proposed model. The result obtained shows that the utility design should be part of the same model for economic saving.

(534g) Global Optimization of Nonconvex MINLP Problems by Domain and Image Partitioning with Applications to Heat Exchanger Networks

Sung Young Kim, Debora Faria and Miguel Bagajewicz, Chemical Engineering, University of Oklahoma, Norman, OK

Global Optimization of Nonconvex MINLP Problems by Domain and Image Partitioning with Applications to Heat Exchanger Networks

Sung Young Kim, Debora Faria and *Miguel Bagajewicz

School of Chemical Engineering and Material Science, University of Oklahoma

100 East Boyd Street, T-335 – Norman, OK 73019-0628 USA

** Corresponding Author*

Abstract

We propose a new method to obtain the global optimum of full nonconvex MINLP problems. The method is based on partitioning the domain and image of nonconvex functions. The procedure we propose uses an MILP lower bound constructed using domain/image partitioning. Then a newly developed bound contraction procedure is applied and compared to branch and bound as well as branch and bound with bound contraction at each node. To illustrate the method we focus on the heat exchanger network stage-wise model. Results show a robust behaviour where the solution does not need initial values as local MINLP model like Dicopt need. In the presentation we will compare the performance of all options.

(544a) Experimental Design of Sample Plan to Enable Data-Based Decision Making

Swee-Teng Chin, Analytical Tech Center, Dow Chemical Company, Freeport, TX and **Leo H. Chiang**, The Dow Chemical Company, Freeport, TX

There are hundreds of analytical labs in The Dow Chemical Company. A main purpose of these labs is to enable process control and to ensure product quality consistency. Another purpose is to support special improvement projects. Lab operation needs to follow a sample plan (i.e., a document describing what, where, how, and how often to sample from the manufacturing process). A statistically sound sample plan is crucial across all businesses in Dow because insufficient samples can lead to wrong decision making and too many samples could be just a waste in time and cost.

This talk will discuss experimental design of sample plan to enable data-based decision making such as product variation investigation, accepting or rejecting incoming raw materials, and deciding analytical measurement improvement activities. Two industrial examples are given here: 1. Optimize a current sample plan to reduce number of samples and 2. Design a new sample plan to understand the physical properties of catalyst. In both of these projects, sound statistical methods were used that take into account both risk and costs. These are critical information that helps Dow to make important data-based decisions.

(544b) Multivariate Analysis of Process Data for Product Formulation Optimization in Property Cluster Space

Subin Hada[1], Robert Herring III[1], J. Colin Haser[1], Nishanth G. Chemmangattuvalappil[2] and Mario Richard Eden[1], (1)Department of Chemical Engineering, Auburn University, Auburn, AL, (2)Department of Chemical and Environmental Engineering, University of Nottingham - Malaysia, Semenyih, Malaysia

Mixture processes play an important role in today's manufacturing of value-added chemical products such as commodity chemicals, food, cosmetics, oil and pharmaceuticals. In classical non-mixture designs such as factorial and response surface designs, all the factors are orthogonal or independent. This means that it is possible to freely choose the level of a factor regardless of the other factors' levels. However, this freedom does not exist for mixture designs, because each component in a mixture is dependent upon the settings of the other component settings. In mixture design, the factors are interdependent and the effects of the factors on the responses are not separable. In addition, traditional mixture models are usually employed to investigate the relationships between the blend ratio matrix (**R**) and the final blend product property matrix (**Y**) only, given that the properties of the pure raw materials (**X**) and the process conditions used to manufacture them (**Z**) are already chosen. However, in the development of formulated products that meet target properties with minimum experiments and minimum total material cost, it is important to simultaneously account for all three degrees of freedom (**X**, **R**, and **Z**) available in the blending operations.

In this paper, projection methods like PCA and PLS are employed to extract and utilize the necessary information from all three degrees of freedom that are available to control the properties of any product manufactured in blending operation. This information is used to build models to simultaneously explore variations in the raw material properties data (**X**), their blend ratios (**R**), and process conditions (**Z**) and the resulting effect on the product quality (**Y**). These methods capture information about the variance and correlation structure within predictor and quality variables that were present during the production of previous product grades and that are important for consistency of the desired process conditions with past operating procedures. An integrated solution approach is employed that formulates two reverse problems, one for determination of design targets and another for matching of the design targets. The empirical models are used in a reverse problem formulation to ensure that a complete set of candidate materials and their mixtures are found which match the design target subject to the predictive power of the model. Mapping the design problem onto the lower dimensional cluster space utilizing property clustering algorithm facilitates visualization and simultaneous solution of the problem and also avoids combinatorial explosion when dealing with multi-component mixtures.

The method and concepts will be illustrated using a case study describing the design of polymer blending problems in a latent variable space.

(544c) Estimation of Permeability and Porosity for CO2 Sequestration in Heterogeneous Formations with Derivative-Free Optimization Based On Observed Data

Yan Zhang and Nick Sahinidis, Chemical Engineering, Carnegie Mellon University, Pittsburgh, PA

The modeling and simulation of CO2 sequestration in geological formations have received increased attention in recent years. Oftentimes, some of the parameters such as porosity and permeability in these models remain uncertain, as it is unrealistic to fully characterize the heterogeneity of geological formations. Therefore, to quantify the impact of such uncertainties on CO2 plume migration in heterogeneous formations, parameters such as porosity and permeability must be estimated from observed data. The methodologies of this kind of uncertainty quantification problem have been comparatively studied in oil production forecasts [1]. However, the optimization algorithms used for inverse modeling are mostly based on gradient methods which require the gradient of objective function. In numerical simulations of CO2 sequestration, to obtain such gradient information is computational prohibitive. To cope with the high computational requirements, we apply various derivative-free optimization (DFO) methods [2] to solve this inverse problem.

In our work, a 3D benchmark for CO2 injection into a synthetic heterogeneous (known porosity and permeability field) saline aquifer is modeled using TOUGH2 [3]. Selected model outputs, such as gas saturation map, from TOUGH2 simulations are perturbed with random Gaussian noise to generate a test data set. The data set is then used to test the ability of 28 DFO solvers to infer the original porosity and permeability field using Bayesian inference. The inferred porosity and permeability field are then used for uncertainty quantification of CO2 plume migration. Extensive computational results are reported to evaluate the ability of the 28 algorithms to estimate unknown porosities and permeabilities correctly from the observed data.

References cited:

1. Floris, F. J. T., Bush, M. D., Cuypers, M., Roggero, F. and Syversveen, A-R., Methods for quantifying the uncertainty of production forecasts: a comparative study, Petroleum Geoscience, 2001(7), S87-S96

2. Rios, L. M. and Sahinidis, N. V., Derivative-free optimization: A review of algorithms and comparison of software implementations, http://thales.cheme.cmu.edu/dfo.

3. Pruess, K. ECO2N: A TOUGH2 fluid property module for mixtures of water, NaCl, and CO2. Lawrence Berkeley National Laboratory: Berkeley, 2005.

(544d) A Comprehensive Evaluation of Different Variable Selection Methods for Soft Sensor Development

Zi Xiu Wang, Chemical Engineering, Auburn University, Auburn, AL, Q. Peter He, Chemical Engineering, Tuskegee University, Tuskegee, AL and Jin Wang, Auburn University, Auburn, AL

In recent years, rapid developments in technology facilitated the collection of vast amount of data from different industrial processes. The data has been utilized in many different areas, such as data-driven soft sensor development and process monitoring, to control and optimize the process. The performance of these data-driven schemes can be greatly improved by selecting only the vital variables that strongly affect the primary variables, rather than all the available process variables. Consequently, variable selection has been one of the most important practical concerns in data-driven approaches. By identifying the irrelevant and redundant variables, variable selection can improve the prediction performance, reduce the computational load and model complexity, obtain better insight into the nature of the process, and lower the cost of measurements [1-2].

A comprehensive evaluation of different variable selection methods for soft sensor development will be presented in this work. Among all the variable selection methods, seven algorithms are investigated. They are stepwise regression, PLS-BETA, PLS-VIP, UVE-PLS, PLS-SA, PCA-SA and GA as discussed below. Stepwise regression methods are often used for variable selection in linear regression [3]. The procedure is carried out in such a way that individual predictor/secondary variable is sequentially introduced into the model to observe its relation to the primary variables. Partial Least Squares (PLS) regression is a model parameter based algorithm. Both the regression coefficients estimated by PLS (PLS-BETA) and variable importance in projection (PLS-VIP) are discussed [4]. Another model parameter based method, called Uninformative Variable Elimination by PLS (UVE-PLS), is also related to regression coefficients. However, instead of looking at the regression coefficients only, the reliability of the coefficients is explored [5]. Variable selection algorithms based on sensitivity analysis (PLS-SA and PCA-SA) are also studied. In these approaches, the importance of variables is defined by their sensitivity, which is defined as the change in primary variables by varying the secondary variable in its allowable range [6-7]. Furthermore, properties of genetic algorithms (GA), which have been recently proposed for variable selection applications [8], are also investigated.

The algorithms of these variable selection methods and their characteristics will be presented. In addition, the strength and limitations when applied for soft sensor development are studied. The soft sensor prediction performance of models developed by these variable selection methods are compared using PLS.

A simple simulation case is used to investigate the properties of the selected variable selection methods. The dataset is generated to mimic the typical characteristics of process data, such as the magnitude of correlations between variables and the magnitude of signal to noise ratio [4]. In addition, the algorithms are applied to an industrial soft sensor case study. In both cases, independent test sets are used to provide fair comparison and analysis of different algorithms. The final performances are compared to demonstrate the advantages and disadvantages of the different methods in order to provide useful insights to practitioners in the field.

References

1. Andersen, C. M., & Bro, R. (2010). Variable selection in regression — a tutorial. *Journal of Chemometrics*, *24*(11-12), 728-737.
2. Reunanen, J. (2003). Overfitting in making comparisons between variable selection methods. The Journal of Machine Learning Research, 3, 1371-1382. Retrieved from http://dl.acm.org/citation.cfm?id=944978
3. Ma, M.-D., Ko, J.-W., Wang, S.-J., Wu, M.-F., Jang, S.-S., Shieh, S.-S., & Wong, D. S.-H. (2009). Development of adaptive soft sensor based on statistical identification of key variables. *Control Engineering Practice*, *17*(9), 1026-1034. Elsevier.

4. Chong, I.-G., & Jun, C.-H. (2005). Performance of some variable selection methods when multicollinearity is present. *Chemometrics and Intelligent Laboratory Systems*, *78*(1-2), 103-112.
5. Centner, V., Massart, D. L., de Noord, O. E., de Jong, S., Vandeginste, B. M., & Sterna, C. (1996). Elimination of uninformative variables for multivariate calibration. *Analytical chemistry*, *68*(21), 3851-3858.
6. Arciniegas, F. A., Embrechts, M., & Rueda, I. E. A. (2006). Variable Selection with Partial Least Squares Sensitivity Analysis: An Application to Currency Crises' Real Effects. *SSRN eLibrary*. Retrieved from http://papers.ssrn.com/sol3/papers.cfm?abstract_id=909508
7. Zamprogna, E., Barolo, M., & Seborg, D. E. (2005). Optimal selection of soft sensor inputs for batch distillation columns using principal component analysis. *Journal of Process Control*, *15*(1), 39-52.
8. Chiang, L. H., & Pell, R. J. (2004). Genetic algorithms combined with discriminant analysis for key variable identification. Analytical Sciences, 14, 143-155.

(544e) Use of Workflows in Knowledge Management

Girish Joglekar[1], Arun Giridhar[2], Linas Mockus[3] and Gintaras V. Reklaitis[2], (1)School of Chemical Engineering, Purdue University, West Lafayette, IN, (2)Chemical Engineering, Purdue University, West Lafayette, IN, (3)Purdue University, West Lafayette, IN

Within an organization, explicit knowledge creation is a very laborious and time consuming activity, accounting for a significant investment of resources. It also requires active participation from people from a wide variety of business functions, for example, management, IT, research, development and so on. The payoff in the investment is through better decision making and competitive advantage over competition on a sustained basis. As the decision making process gets more complex, it becomes imperative that the knowledge be managed such that it is easily discernible and accessible.

We have designed a novel framework for knowledge modeling which will be the foundation of a powerful knowledge management system. It builds on the common underlying theme between business workflows, scientific workflows, manufacturing recipes and experimental procedure. These four areas of application cover almost all of knowledge creation activity. We posit that the proposed framework will facilitate exchange of knowledge horizontally as well as vertically within an organization.

A workflow, the common theme mentioned above, captures the operational aspect of a work procedure: how tasks are structured, who performs them, what their relative order is, how they are synchronized, how information flows to support the tasks and how tasks are being tracked. A work procedure is developed with a particular end objective, such as, performing an experiment, running a simulation model, running a process or making a decision. Each work procedure uses previously generated information or knowledge, and/or creates new information or knowledge. By storing information in a semantically rich form, it can be processed by machines and used by agents involved in the work procedure. The work procedures, in turn, can invoke other work procedures, thus allowing a hierarchical representation of the entire knowledge creation process. Such a representation is not only crucial for integration of all functions, but also for long term use of knowledge.

In this paper, the key features of the framework are presented with the help of a research project to generate kinetic parameters for one of the reactions in the synthesis of a pharmaceutical.

(544f) Online Model-Based Experimental Design

Harvey Arellano-Garcia, Diana López and Tilman Barz, Chair of Process Dynamics and Operation, Berlin Institute of Technology, Berlin, Germany

In chemical processes, unknown model parameters have to be calibrated to data from dynamic experiments by solving nonlinear constrained parameter estimation problems minimizing a weighted sum of squares of the deviations between data and model responses subject to the model equations. For nonlinear models it further depends on the values of the parameter estimate. To obtain a parameter estimate with maximum reliability, an experimental design is computed minimizing a functional of the variance-covariance matrix subject to the model equations and constraints to experimental cost and operability. This results in intricate nonlinear non-standard constrained optimal control problems. Moreover, in order to cope with the problem of large uncertainties in the initial parameter values, we follow the idea of an adaptive experiment design. This idea is implemented via online estimation of parameter values as well as the generation of these optimal input actions. In other words, in each sampling instant, where measurements data are taken, current estimates of the parameters as well as the currently planned experiment design are then repeatedly updated.

In this work, an experimental verification of the developed approach will also be presented. The experimental results correspond to an ion-exchange high performance liquid chromatography column (HPLC) for protein separation, where transport and adsorption parameters and parameters for the calibration of a concentration sensor are determined. The Basis of the HPLC system is given by a Smartline Pump with Manager (Knauer GmbH, Berlin, Germany), which is used for the generation of time dependent feed concentrations, which are then supplied to the chromatography column. The HPLC system is fully automated by a commercial process control system Freelance 800F ABB (Ladenburg, Germany) and interfaced with the numeric programming environment Matlab®, where algorithms for parameter subset selection, parameter estimation, and the generation of optimal input designs are implemented.

Acknowledgement: The authors would like to thank the Knauer GmbH, Berlin-Germany for their support regarding the experimental case study.

(544g) Selection of Individualized Dosage Regimens by Using a Stochastic Optimization Approach

José Miguel Laínez, School of Chemical Engineering, Purdue University, West Lafayette, IN, Gary Blau, e-Enterprise Center, Purdue University, West Lafayette, IN, Seza Orcun, Purdue University, West Lafayette, IN and G. V. Reklaitis, Chemical Engineering, Purdue University, West Lafayette, IN

Specific techniques for individualizing and optimizing the dosage regimen and clinically monitoring each patient are necessary to offer her/him the proper doses of medicine promptly. This is of special concern for medicines that are costly or whose toxic effects are severe (e.g., oncological agents). The optimal dosage regimen for an individual is a combination of dose amount and dosing interval (i.e., time between doses) which minimize the risk (1-α) that the drug exposure deviates from the desired therapeutic window. The therapeutic window is defined as the range of drug exposure (e.g., blood concentration, area under the curve concentration-time) $[TW^L, TW^U]$, which is below a threshold defining an acceptable risk of toxic side effect TW^U and above a threshold defining a minimum acceptable level of therapeutic efficacy TW^L. Laínez et al. (2011) show how a dose feasible range can be obtained provided an acceptable risk level. Such an approach can be applied for those pharmacokinetic (PK) models for which the dose variable can be isolated. In this work, the dosage regimen optimization problem for general pharmacokinetic models (i.e., those described by differential-algebraic equations) is presented by using a scenario based stochastic NLP which optimizes a downside risk metric. The scenarios are derived from the posterior joint distribution of the individual's PK parameters which is obtained following an approximate Bayesian inference approach. A Smolyak rule is used for (i) the selection of the scenarios (i.e., combination of PK parameters) to be considered and (ii) the approximation of the downside risk metric. Two case studies, Gabapentin and Cyclophosphamide, are presented to emphasize the advantages and limitations of the proposed approach.

References

Laínez, J.M., G. Blau, L. Mockus, S. Orcun and G. Reklaitis, "Pharmacokinetic Based Design of Individualized Dosage Regimens Using a Bayesian Approach", Ind. Eng. Chem. Res, **50**, 5114-5130 (2011)

Acknowledgements

Support from the United States National Science Foundation (Grant NSF-CBET-0941302) is gratefully acknowledged.

(545a) Synthesis and Design of Processing Networks: Decision Making Under Uncertainty

Alberto Quaglia, CAPEC, Department of Chemical and Biochemical Engineering, Technical University of Denmark, Kgs. Lyngby, Denmark, Bent Sarup, Vegetable Oil Technology Business Unit, Alfa Laval Copenhagen A/S, Søborg, Denmark, Gürkan Sin, Department of Chemical and Biochemical Engineering, Technical University of Denmark, Kgs. Lyngby, Denmark and Rafiqul Gani, CAPEC, Department of Chemical and Biochemical Engineering, Technical University of Denmark (DTU), Kgs. Lyngby, Denmark

Synthesis and design of processing networks: decision making under uncertainty

A. Quaglia,[1] B. Sarup,[2] G. Sin,*[1] R. Gani[1]

[1]CAPEC, Department of Chemical and Biochemical Engineering, Technical University of Denmark, DK-2800 Kgs. Lyngby, Denmark; email: aq@kt.dtu.dk. *email:gsi@kt.dtu.dk.

[2]Vegetable Oil Technology Business Unit, Alfa Laval Copenhagen A/S, 2860, Denmark

The synthesis and design of processing networks is a complex decision-making problem, consisting of the selection of raw material, product portfolio and definition of the process technologies and the process configuration. As these decisions induce large capital investments, this problem has high impact on the profitability of process industries.

The solution of this important decision making problem is challenging. It is multidisciplinary and requires cross-functional collaboration of different corporate functions (e.g. marketing, engineering, finance). Moreover, it is largely based on estimations and forecasts of future data (i.e. market prices and sizes, raw material availability and compositions etc.), whose uncertainty has to be taken into account and properly managed in the decision-making process.

Even though several solution approaches have been proposed in the literature for this class of problems, the decision-making under uncertainty still poses a number of challenges (Sahinidis, 2004), especially when the complexity of real industrial cases is considered.

In this contribution, we propose a systematic framework for synthesis and design of processing networks under uncertainty. The framework is based on a previous work dealing with decision-making (Quaglia et al. 2012), which has been extended to include uncertainty.

The key novelty of the framework is to employ a systematic approach to manage the complexity of the resulting stochastic problem (e.g. large problems with high number of uncertain parameters). The framework employs a decomposition approach, guiding the user through some preliminary steps of uncertainty characterization and mapping, prior to the formulation of the stochastic problem. These steps aim at providing a deeper knowledge of the uncertain space and of its consequences on the optimal decisions. This information is then used in order to facilitate the solution of the stochastic problem.

To this goal, the framework employs different strategies including i) variable initialization ii) search space reduction via variable bounding iii) search space reduction via inactive constraints elimination. Moreover, the generated knowledge is also analyzed to systematically generate new configurations (not included in the search space of the original problem), to hedge against uncertainty.

In order to highlight the main features of the framework, case studies from the edible oil industry and wastewater treatment are formulated and solved. The optimal processing network under uncertainty is identified, the Value of Stochastic Solution (VSS) is calculated, and strategies to manage and hedge against uncertainty are suggested and evaluated.

References:

N.Sahinidis, 2012, Optimization under uncertainty: state-of-the-art and opportunities, Computers & Chemical Engineering (28) 971-983

A.Quaglia, B.Sarup, G.Sin, R.Gani, 2012, Integrated business and engineering framework for synthesis and design of enterprise-wide processing networks, Computers & Chemical Engineering (38) 213-223.

(545b) Optimization of a Carbon Dioxide-Assisted Nanoparticle Deposition Process Using Sequential Experimental Design with Adaptive Design Space

Michael J. Casciato[1], Sungil Kim[2], J.C. Lu[2], Dennis W. Hess[1] and Martha Grover[3], (1)School of Chemical & Biomolecular Engineering, Georgia Institute of Technology, Atlanta, GA, (2)School of Industrial & Systems Engineering, Georgia Institute of Technology, Atlanta, GA, (3)Georgia Tech, Atlanta, GA

Optimization of a carbon dioxide-assisted nanoparticle deposition process using sequential experimental design with adaptive design space

Michael J. Casciato*,†, Sungil Kim‡, J.C. Lu‡, Dennis W. Hess†, and Martha A. Grover†

†*School of Chemical & Biomolecular Engineering, Georgia Institute of Technology, Atlanta, GA 30332*

‡*H. Stewart Milton School of Industrial & Systems Engineering, Georgia Institute of Technology, Atlanta, GA 30332*

*E-mail: michael.casciato@chbe.gatech.edu

Abstract

A sequential design of experiments methodology with an adaptive design space is proposed and implemented to optimize a nanofabrication technique that possesses significant uncertainty in terms of design region and model structure along with an engineering tolerance requirement. This method is termed Layers of Experiment (LoE) with Adaptive Combined Design (ACD) and is developed specifically for advanced technology processes where there is little or no foreknowledge of the design region and/or model structure from either mechanistic understanding or empirical studies. This technique was applied to optimize an elevated pressure, elevated temperature carbon dioxide (epet-CO_2) nanoparticle deposition process where silver nanoparticles were deposited directly from an organometallic precursor in the fluid phase onto a silicon wafer surface.

A target mean nanoparticle size of 40nm was chosen, with surface enhanced Raman spectroscopy (SERS) as the motivating application. Using the LoE/ACD method, it was possible to find the process optimum for the epet-CO_2 process and build a reliable model using only 12 experiments conducted in two sequential layers. With temperature as the design variable, the first layer of experiments was conducted in the region [60°C, 150°C], while the second layer was conducted in the region [98°C, 128°C], indicating that the LoE/ACD algorithm reduced the size of the design region by 60°C while building a reliable model for statistical inference. In the first layer, a purely space-filling design was used since the model structure was unknown; in the second layer, the ACD approach yielded a design weighted between space-filling and D-optimal, favoring the D-optimal design. The optimized temperature for fabricating 40nm mean silver nanoparticles in this system was 117°C.

(545c) Bounding the Solutions of Chemical Kinetics Models Using ODEs with Linear Programs Embedded

Stuart Harwood, Process Systems Engineering Laboratory, Deptartment of Chemical Engineering, Massachusetts Institute of Technology, Cambridge, MA and Paul Barton, Process Systems Engineering Laboratory, Dept. of Chemical Engineering, Massachusetts Institute of Technology, Cambridge, MA

A method is presented for the computation of rigorous bounds enclosing all solutions of chemical kinetics models with uncertain parameters and/or controls. In many chemical processes, there are uncertain parameters that can lead to a wide range of possible process behaviors. Given a range of possible parameter or control values, one often wishes to determine the range of possible states that results, called the design envelope. It is similar in nature to related concepts such as the "reachable set". Rigorously bounding the design envelope is an important task in safety verification and global dynamic optimization of these processes.

This work will focus on bounding the design envelope of a batch reactor subject to uncertainty. The main assumption is that the parameters or controls have known upper and lower bounds. Mathematically, this problem reduces to bounding the set of possible solutions of an initial value problem in parametric ordinary differential equations (ODEs). Recent work has shown that tight component-wise upper and lower bounds can be computed by formulating a related ODE system which enforces physical information that might not be explicit in the model. Based on arguments involving differential inequalities, the resulting ODE system depends on the solutions of "embedded" parametric optimization problems. The physical information that is enforced takes the form of an invariant set, or a set which the states cannot leave at any time or for any parameter/control value. The invariant set restricts the feasible set of the optimization problems.

In chemical kinetics models, the invariant set is typically a convex polyhedron. This is because mass conservation arguments typically furnish upper and lower bounds that must hold for all time and parameter values. Meanwhile the states must obey affine relations due to stoichiometry. By taking affine relaxations of the objective functions of the embedded optimization problems, one obtains an ODE with linear programs (LPs) embedded.

A previously developed algorithm has been extended to handle further parametric dependence that occurs in the embedded LPs. This algorithm reformulates the system as differential algebraic equations, which leads to a robust and efficient way to solve numerically the resulting system of ODEs with LPs embedded. The resulting bounds are tight.

(545d) Offshore Oil and Gas Field Infrastructure Planning Under Complex Fiscal Rules and Endogenous Uncertainties

Vijay Gupta, Department of Chemical Engineering, Carnegie Mellon University, Pittsburgh, PA and Ignacio E. Grossmann, Chemical Engineering Department, Carnegie Mellon University, Pittsburgh, PA

The planning of offshore oil and gas field infrastructure is a very critical problem as it involves huge investments which may lead to large profits or losses. This problem has received significant attention in recent years given the new discoveries of large oil and gas reserves in the last decade around the world. The optimization of investment and operations planning of offshore oil and gas field infrastructure is traditionally modeled using the net present value (NPV) as the objective function, without considering the effect of fiscal rules that are associated to these development sites [1-2]. These rules determine the share of the oil company and host government in the total oil produced in a given year. Therefore, the models with simple NPV objective functions may yield solutions that are optimistic, which can in fact be suboptimal after considering the impact of fiscal terms.

The goal of this paper is to extend a mixed-integer nonlinear programming (MINLP) model for NPV-based oilfield development planning to include complex fiscal rules. In particular, we consider a recently proposed multi-field site strategic planning model [3] for offshore oil and gas fields as a basis to include generic fiscal rules with ring-fencing provisions. The proposed MINLP model is rather large as it includes many new discrete variables and additional constraints. We show that this model can be reduced to a variety of specific contracts models. Results on realistic instances show improved investment and operations decisions due to the explicit consideration of the fiscal terms during planning. Since the model becomes computationally very expensive to solve with the extension to fiscal rules, we provide several reformulation/approximation techniques and solution strategies that yield orders of magnitude reduction in the solution time [4].

In addition, uncertainty in the model parameters such as field size and its deliverability can have significant impact on investment and operations decisions. Therefore, we discuss the extension of the proposed model to handle endogenous uncertainties such as reservoir size and productivity using a framework based on two-stage and multi-sage stochastic programming approaches where timing of uncertainty realization depends on investment decisions. In particular, we first formulate the problem as a multistage stochastic programming model in the spirit of the work by Goel and Grossmann [5] and Tarhan et al. [6] in which special disjunctive constraints are considered as the conditional non-anticipativity constraints for endogenous uncertainties. Numerical results on several instances with uncertainty in the model parameters are reported as well as computational performance of the model with the proposed solution strategies [7]. In order to simplify the solution of this complex problem, we also consider a recursive two-stage programming model that is solved under a shrinking time horizon and present a comparison of the solution time and investment decisions with a multistage stochastic programming approach.

References

[1] Iyer, R. R.; Grossmann, I. E.; Vasantharajan, S.; Cullick, A. S. Optimal planning and scheduling offshore oilfield infrastructure investment and operations. *Ind. Eng. Chem. Res.* 1998, *37*, 1380–1397.

[2] van den Heever, S. A.; Grossmann, I. E. An iterative aggregation/disaggregation approach for the solution of a mixed integer nonlinear oilfield infrastructure planning model. *Ind. Eng. Chem. Res.* 2000, *39*, 1955–1971.

[3] Gupta V.; Grossmann I. E. An Efficient Multiperiod MINLP Model for Optimal Planning of Offshore Oil and Gas Field Infrastructure. *Ind. Eng. Chem. Res. 2012,* **DOI:** 10.1021/ie202959w

[4] Gupta V.; Grossmann I. E. Modeling and Computational Strategies for Optimal Development Planning of Offshore

Oilfields under Complex Fiscal Rules. *To be Submitted.*

[5] Goel, V.; Grossmann, I. E. A stochastic programming approach to planning of offshore gas field developments under uncertainty in reserves. *Comput. Chem. Eng.* 2004, *28* (8), 1409–1429.

[6] Tarhan, B.; Grossmann, I.E.; Goel, V. Stochastic programming approach for the planning of offshore oil or gas field infrastructure under decision- dependent uncertainty. *Ind. Eng. Chem. Res.* 2009, 48(6), 3078–3097.

[7] Gupta, V.; Grossmann, I. E. Solution strategies for multistage stochastic programming with endogenous uncertainties. *Comput. Chem. Eng.* 2011, 35, 2235–2247.

(545e) Nonconvex Generalized Benders Decomposition and Piecewise Convex Relaxations for Optimal Process Design and Operation Under Uncertainty

Xiang Li[1,2], Yang Chen[2] and Paul I. Barton[2], (1)Chemical Engineering, Queen's University, Kingston, ON, Canada, (2)Chemical Engineering, Massachusetts Institute of Technology, Cambridge, MA

Integrated process design and operation problems are often cast as mathematical programming problems for systematic solution. If a process involves uncertain factors that could impact the process performance significantly, the uncertainties need to be explicitly addressed in the mathematical programming, which results in a stochastic programming problem.

Recently, a nonconvex generalized Benders decomposition (NGBD) [1] has been developed to solve a class of stochastic nonconvex mixed-integer nonlinear programming (MINLP) problems arising from integrated process design and operation under uncertainty. By exploiting the decomposable structure of the problem, NGBD can solve the problem to global optimality finitely, and it exhibits dramatic computational advantages over traditional branch-and-bound based global optimization methods. As the convergence rate of NGBD is largely dependent on the tightness of the convex relaxations of the nonconvex functions, the efficiency of NGBD can be significantly improved by generating tighter convex relaxations.

It has been recognized in the process systems engineering literature that piecewise linear relaxation enables much tighter relaxation of bilinear programs and can expedite global optimization significantly in the branch-and-bound framework (e.g. [2] [3] [4] [5]). This paper generalizes the notion of piecewise relaxation to factorable functions in the context of McCormick relaxation [6] [7] and integrates the so-obtained piecewise McCormick relaxation technique into the NGBD algorithm to reduce the gap between the original problem and its convex relaxation. In addition, most subproblems in a NGBD iteration can be solved without exchanging information among them, and therefore they can be solved in parallel to reduce the solution time. The parallelization for NGBD will be briefly discussed in the paper as well.

Case studies of a pump network configuration problem [8] and an energy polygeneration problem [9] show that, while NGBD can solve problems that are intractable for a state-of-the-art global optimization solver (BARON [10]), integrating the proposed piecewise convex relaxation into NGBD helps to reduce the solution time by up to an order of magnitude. The case study results also show that parallel computation can reduce the NGBD solution time significantly.

References

[1] Li, X.; Tomasgard, A.; Barton, P. I. Nonconvex generalized Benders decomposition for stochastic separable mixed-integer nonlinear programs. Journal of Optimization Theory and Applications 2011, 151, 3, 425-454.

[2] Karuppiah, R.; Grossmann, I. E. Global optimization for the synthesis of integrated water systems in chemical processes. Computers and Chemical Engineering 2006, 30, 650–673.

[3] Meyer, C. A.; Floudas, C. A. Global optimization of a combinatorially complex generalized pooling problem. AIChE Journal 2006, 52, 1027–1037.

[4] Wicaksono, D. S.; Karimi, I. A. Piecewise MILP under- and overestimators for global optimization of bilinear programs. AIChE Journal 2008, 54, 991–1008.

[5] Gounaris, C. E.; Misener, R.; Floudas, C. Computational comparison of piecewise-linear relaxations for pooling problems. Industrial and Engineering Chemistry Research 2009, 48, 5742–5766.

[6] McCormick, G. P. Computability of global solutions to factorable nonconvex programs: Part I - Convex underestimating problems. Mathematical Programming 1976, 10, 147–175.

[7] Scott, J. K.; Stuber, M. D.; Barton, P. I. Generalized McCormick relaxations. Journal of Global Optimization 2011, 51, 4, 569-606.

[8] Westerlund, T.; Pettersson, F.; Grossmann, I. E. Optimization of pump configurations as a MINLP problem. Computers and Chemical Engineering 1994, 18, 845–858.

[9] Chen, Y.; Adams II, T. A.; Barton, P. I. Optimal design and operation of flexible energy polygeneration systems. Industrial and Engineering Chemistry Research 2011, 50, 4553–4566.

[10] Tawarmalani, M.; Sahinidis, N. V. Global optimization of mixed-integer nonlinear programs: A theoretical and computational study. Mathematical Programming 2004, 99, 563–591.

(545f) Optimal Design and Scheduling of Multipurpose Batch Plants Under Uncertainty

Dajun Yue and Fengqi You, Department of Chemical and Biological Engineering, Northwestern University, Evanston, IL

Due to its necessary flexibility, batch processing mode is predominant in the commercial manufacture of many chemicals of high value and low volume, or products which require complex synthesis procedures [1]. Meanwhile, the special characteristics of batch processes introduce considerable complexities to the systematic design of batch plants. It is known that efficient designs cannot be achieved without taking into account the detailed planning and scheduling. Hence, the integration of design and scheduling is desired [2]. Assuming the demand pattern is stable, a campaign mode operation is often employed for the design of batch plants [3]. In this work, rather than looking at a deterministic case, we assume that the batch plant is exposed to the uncertainty of demand variations from external markets and supply delays from external suppliers. In general, inventories can serve as buffers in dealing with uncertainties. Moreover, adding intermediate storages can increase the efficiency of equipment utilization, thus achieving more economical designs for batch plants. Therefore, when determining the optimal inventory levels for storage tanks of chemicals, we need to look at two aspects. One is the working inventory to satisfy the external demands as well as the internal consumptions. The other is the safety stock to hedge against the unexpected demand variations. A body of research closely related to this work is the multi-echelon stochastic inventory theory, which is originally developed for the supply chain management. By using the guaranteed service approach, one is allowed to model the inventory allocation across the entire system [4-6]. Applications of this approach were reported in studying continuous chemical process networks [7], but how to capture the stochastic nature of demand variations and supply delays within batch plants remains a challenge. While multiproduct plants are easier to model, most of the batch products are manufactured in multipurpose facilities [8]. Thus, in this work, we address the design of multipurpose batch plants and the first step is to establish a unified framework to account for the planning, scheduling and inventories.

By dividing the multipurpose batch plant system into macrostructure (connections between stages) and microstructure (a series of basic processing tasks within a stage) [3], we are able to derive a multi-stage network which can serve as the basis for the integration of planning, scheduling and stochastic inventory management for the design of batch plants. To approach this aggregated problem, we propose a single MINLP model that simultaneously determines the batch sizing, equipment assignments, production schedules, purchases and sales as well as the working and safety inventory profiles. This model exhibits multi-tradeoffs among decision variables from all problem levels, thus seamlessly integrate the planning, scheduling and stochastic inventory management. While this basic model contains multi-linear and concave terms, which would be computationally intractable for large-size problems, we reformulate the model to an MINLP with only square root and linear terms by exploiting the problem properties and using general linearization methods. In order to obtain global optimal solutions with modest computational times, we further develop a tailored branch-and-refine algorithm based on successive piecewise linear approximations. At last, examples are presented to illustrate the application of the model and the performance of the proposed algorithm.

References

[1] S. Papageorgaki and G. V. Reklaitis, "Optmal Design of Multipurpose Batch Plants. 1. Problem Formulation," *Industrial & Engineering Chemistry Research,* vol. 29, pp. 2054-2062, Oct 1990.

[2] A. P. Barbosapovoa and S. Macchietto, "Detailed Design of Multipurpose Batch Plants," *Computers & Chemical Engineering,* vol. 18, pp. 1013-1042, Nov-Dec 1994.

[3] N. Shah and C. C. Pantelides, "Optimal Long-Term Campaign Planning and Design of Batch Operations," *Industrial & Engineering Chemistry Research,* vol. 30, pp. 2308-2321, Oct 1991.

[4] S. C. Graves and S. P. Willems, "Optimizing the Supply Chain Configuration for New Products," *Management*

Science, vol. 51, pp. 1165-1180, Aug 2005.

[5] F. Q. You and I. E. Grossmann, "Integrated Multi-Echelon Supply Chain Design with Inventories Under Uncertainty: MINLP Models, Computational Strategies," *Aiche Journal,* vol. 56, pp. 419-440, Feb 2010.

[6] F. Q. You and I. E. Grossmann, "Mixed-Integer Nonlinear Programming Models and Algorithms for Large-Scale Supply Chain Design with Stochastic Inventory Management," *Industrial & Engineering Chemistry Research,* vol. 47, pp. 7802-7817, Oct 2008.

[7] F. Q. You and I. E. Grossmann, "Stochastic Inventory Management for Tactical Process Planning Under Uncertainties: MINLP Models and Algorithms," *AIChE Journal,* vol. 57, pp. 1250-1277, May 2011.

[8] D. W. T. Rippin, "Batch Process Systems Engineering: A Retrospective and Prospective Review," *Computers & Chemical Engineering,* vol. 17, pp. S1-S13, 1993.

(545g) Solution of Optimal Placement Problems in Water Networks with Intractably Large Scenario Spaces

Gabriel Hackebeil[1], Carl D. Laird[1], Sean Legg[1] and Jean-Paul Watson[2], (1)Chemical Engineering, Texas A&M University, College Station, TX, (2)Sandia National Laboratories, Albuquerque, NM

Design of contaminant detection and response systems in municipal water networks is crucial for protecting consumers from malicious or accidental contamination events. In order to minimize the health risks to the public, these systems must incorporate early warning detection systems to allow for adequate response measures in the event a contaminant is detected. Cleanup and control strategies should also be implemented to return the water system to an operational state with minimal disruption to service. The installation of chlorine booster stations for activation immediately following contaminant detection is one method to help reduce the impact of potential contamination scenarios. Optimal placement of sensors and booster stations within these municipal systems is an active area of research where a large number of methods have been proposed that employ linear or mixed-integer linear programming techniques.

We formulate these placement problems as mixed-integer linear stochastic programming problems. These problems must consider uncertainty in the location and time of these contamination events. Assuming contamination can occur from any node in the network and at any number of discrete time points can produce an intractably large scenario space when studying water networks with tens of thousands of nodes. In order to apply these solution techniques to actual city-scale design problems, the issue of finding representative scenario sets of a tractable size needs to be addressed.

We propose a method for finding provably good placements using random sampling from the entire scenario space. Using the method of Mak, Morton, and Wood (1999), we employ a sampling procedure that generates confidence intervals on the optimality gap when optimal placements are determined with a limited number of scenarios. In addition to variability in the objective function, we examine variability in the optimal placements. Using a number of distance metrics for comparing placements, we demonstrate near optimal placements using significantly smaller representative subsamples of the entire scenario space.

Sandia National Laboratories is a multi-program laboratory managed and operated by Sandia Corporation, a wholly owned subsidiary of Lockheed Martin Corporation, for the U.S. Department of Energy's National Nuclear Security Administration under contract DE-AC04-94AL85000.

(546a) An Integrated Methodology for Design of Tailor-Made Blended Products: Biofuels and Bio-Based Lubricants

Nor Alafiza Yunus, CAPEC, Department of Chemical and Biochemical Engineering, Technical University of Denmark, Kongens Lyngby, Denmark, Krist V. Gernaey, Department of Chemical and Biochemical Engineering, Technical University of Denmark (DTU), Kgs. Lyngby, Denmark, John M. Woodley, Department of Chemical and Biochemical Engineering, Technical University of Denmark, Lyngby, Denmark and Rafiqul Gani, CAPEC, Department of Chemical and Biochemical Engineering, Technical University of Denmark (DTU), Kgs. Lyngby, Denmark

The demand for bio-based products has been growing in recent years due to increasing concerns about the environment, resource depletion, and/or sustainability. Bio-based products can be designed by blending of conventional materials (e.g. fossil fuels) with other chemicals produced from renewable resources, such as biomass, reducing thereby the fossil fuel consumption and amount of pollutants released into the environment, and increasing the product safety. A tailor-made bio-based blended product is a mixture of several chemicals and designed to match specific product attributes. The most challenging task and critical step in tailor-made bio-product design is to identify and select the appropriate chemicals to be used as building blocks for the blend. At the initial stage, searching for the blend chemicals (additives) is uncertain and involves a very large search space. Therefore, it is important to have a method to systematically and efficiently reduce the search space within which the solution to the blending problem may be found. Once the candidate chemicals have been identified, their optimal compositions in the blend are determined, while still matching the target properties.

In this study, an integrated methodology for design of tailor-made bio-based blended products is presented. The methodology is able to find the composition of optimal blended bio-based products, and consists of two stages; The first stage is product design, where a computer-aided methodology is implemented to identify and evaluate the most promising blend candidates. Then, in the second stage, the blend attributes are validated experimentally or with rigorous simulation models. The first stage consists of four tasks. The first task is to define the design problem, where the product needs are identified, translated into target properties and given target values. In the second task, the required target property models are retrieved from a property model library developed specifically for bio-based blended products. The property model library stores the pure and mixture property models with their parameters. Also, a chemical database, which is one of the important tools in this integrated methodology, has been developed. This database stores the pure compounds which originate from renewable and non-renewable sources with their properties that may be used as building blocks in the blend design. In the third task, the blend problem is solved using a blend design algorithm to find the blends that best match the design targets. In this way, the amount of bio-based chemicals in the blend can be controlled and adjusted. This algorithm employs a decomposition based solution strategy to screen for the feasible blend candidates, thus reducing the search space in a systematic way. The result is a set of blends that match the desired targets, the compositions, values of the target properties and information about their miscibility. Finally, the mixture target property values are verified by rigorous models for the properties and mixtures that require it. Here, collection of data in the database helps in verifying the predicted blend properties.

The application of this systematic methodology is highlighted through case studies related to the design of sustainable and better performing blended gasoline and lubricants with bio-based chemicals. The objective of the case studies is to identify suitable bio-based chemicals as well as the blended products where they can be used.

(546b) Process Synthesis and Technoeconomic Evaluation of a Catalytic Biomass-to-Fuels Strategy

Sercan Murat Sen[1], David Martin Alonso[1], Stephanie G. Wettstein[1], Elif I. Gurbuz[1], Carlos A. Henao[2], James A. Dumesic[1] and Christos T. Maravelias[3], (1)Chemical and Biological Engineering, University of Wisconsin-Madison, Madison, WI, (2)Chemical and Biological Engineering, University of Wisconsin - Madison, Madison, WI, (3)Department of Chemical and Biological Engineering; DOE Great Lakes Bioenergy Research Center, University of Wisconsin-Madison, Madison, WI

Lignocellulosic biomass, an alternative for petroleum, is a sustainable source of carbon for producing chemicals and fuels in an environmentally friendly way. Conversion of lignocellulosic biomass to valuable intermediates is a critical step to develop effective biomass-to-biofuel strategies. Levulinic acid (LA) is one of these platform chemicals that can be produced from lignocellulosic biomass and transformed into liquid fuels, fuel additives and even other specialty chemicals. In this respect, we developed an LA-based catalytic strategy, in which an alkylphenol solvent is used for extraction of intermediates from the sulfuric acid solution, to convert lignocellulosic biomass into liquid hydrocarbon fuels. Then, we compared the alkylphenol strategy with a previously reported butyl acetate strategy in terms of cost-effectiveness.

In the butyl acetate strategy, first, the hemicellulose fraction of the biomass is solubilized and removed using dilute acid pretreatment. Pretreated biomass, which contains cellulose and lignin, is sent to a cellulose deconstruction unit, in which cellulose is converted to levulinic acid and formic acid with 55% yield in the presence of sulfuric acid, whereas the remaining cellulose is degraded to humins as a result of undesired polymerization reactions. After cellulose deconstruction, solid residues (lignin and humins) are filtered from the product stream, then mixed with the hemicellulose fraction that was removed by dilute acid pretreatment, and sent to a boiler/turbogenerator to produce heat and electricity covering the utility requirements of the process. Excess electricity is sold to the grid. Then, LA is converted to γ-valerolactone (GVL) over a RuRe/C catalyst in the presence of sulfuric acid. GVL is extracted from the sulfuric acid and GVL aqueous solution using butyl acetate solvent and sulfuric acid is recycled back to the cellulose deconstruction reactor. Finally, GVL is separated from butyl acetate by distillation and purified GVL is converted to butene and to butene oligomers.

The potential shortcoming of the butyl acetate strategy is the use of an expensive RuRe/C catalyst to prevent the negative effects of sulfuric acid in the catalytic processing. To address this shortcoming, an alkylphenol (2-sec-butyl phenol (SBP)) solvent is used for extraction of LA, which is produced in low concentration, from sulfuric acid solution. This strategy also allows for the aqueous phase containing sulfuric acid to be recycled back to the cellulose deconstruction reactor efficiently. Another advantage of alkylphenols is their high boiling points, so that extraction at elevated temperatures, i.e. at cellulose deconstruction temperature, is possible and leads to potential energy and equipment cost savings. The LA obtained from pretreated biomass is selectively reduced to GVL (while the FA is decomposed to H_2 and CO_2) over a RuSn/C catalyst in the presence of SBP solvent without hydrogenating the solvent. After the GVL production, SBP and GVL can be separated by distillation recovering GVL as a top product. The GVL would then be processed to liquid fuels. This sulfuric acid management strategy provides downstream catalytic processing at high yields in the absence of sulfuric acid and with a cheaper catalyst to convert LA into GVL.

After developing a model for the alkylphenol-based separation strategy, we investigated the economic impact of integrating this strategy with the original process. To determine the economic potential of these strategies, we carried out detailed process simulation studies (based on experimental results for all associated reactions) and capital/operational cost calculations. After optimizing the process configurations of the alkylphenol strategy, the minimum selling price (MSP) for butene oligomers is around $4.70 per gallon of gasoline equivalents (GGE). Also, we identified that the MSP is sensitive to changes in the catalyst lifetime. For a catalyst lifetime of less than 6 months, the alkylphenol strategy results in better MSP than that can be obtained from the butyl acetate strategy. Finally, a comparison of the alkylphenol strategy with a lignocellulosic ethanol production facility, which has an MSP equal to

$5.13/GGE, reveals the attractiveness of this LA-based catalytic approach.

(546c) Use of Octanol-Water Partition Coefficient and Toxicity Prediction in Design of Ionic Liquids for *in Situ* Fermentation Extraction

John White[1], Brock C. Roughton[2], Kyle V. Camarda[1] and Aaron M. Scurto[1], (1)Chemical and Petroleum Engineering, University of Kansas, Lawrence, KS, (2)Bioengineering Graduate Program, University of Kansas, Lawrence, KS

Fermentation is widely used to produce fine and bulk chemicals Due to toxicity of the product towards the producing microorganisms, overall product concentrations are often dilute. After fermentation, downstream separation is required which often has high energy inputs. *In situ* product removal with an immiscible solvent, such as certain ionic liquids (ILs),offers a mechanism by which product could be continually removed during production, reducing the effects of toxicity and product inhibition while also decreasing downstream separation demands. The IL solvent should have good partitioning of the product, but must have low toxicity to the microorganism to allow in situ processing. Environmental and health concerns should be addressed to ensure sustainability. Ionic liquids, consisting of cations and anions, are highly tunable towards fermentation extractions. For instance, they can be designed with hydrophobicity, low to no vapor pressure, low toxicity, etc. Hundreds of ionic liquids have been synthesized with up to 10^{14} possible combinations. The use of molecular descriptors within property prediction models provides an efficient method of searching the possible chemical space and selecting/designing candidate ionic liquids for a given extraction.

Predictive models for octanol-water partition coefficient and toxicity for various model organisms (*Escherichia coli*, *Vibrio Fischeri*, and rat) are presented. Experimental data collected from previously published literature and presented here have been used to develop the models. Ionic liquid descriptors were calculated using E-Dragon (http://www.vcclab.org/lab/edragon/). Descriptor selection was performed using an exhaustive search to find descriptors that provided a good fit to the experimental data (high R^2) while avoiding overfitting (represented by Mallow's C_p statistic). Using the selected descriptors, linear regression was used to generate the predictive property models. Leave-one-out cross-validation (LOOCV) was used to evaluate the predictive power of the models. Once the models were finalized, they were used in a computational molecular design (CMD) framework to optimally design ionic liquids for a given extraction. The optimization problem was formulated as a mixed-integer nonlinear program (MINLP) where a molecule structure is determined which minimizes the difference between target properties and the predicted properties of the solution molecule. A stochastic method, Tabu search, was used to identify locally optimal solutions to the optimization problem. The overall framework has been used previously by the authors in the design of ionic liquids for use in refrigeration. The target properties included low octanol-water partition coefficient, low *E. coli* toxicity, low *V. Fischeri* toxicity (indicative of aqueous environmental safety) and low rat toxicity (indicative of human toxicity). The results indicate that several designed ionic liquid candidates have desirable properties for solvent extraction from fermentation reactions. Further property prediction tools will allow more refined design of ionic liquids for more specified fermentation applications.

(546d) Bio-Based Gasoline Additives: Product and Chemistry Selection Through Network Generation and Optimization

W. Alex Marvin, Srinivas Rangarajan and Prodromos Daoutidis, Chemical Engineering and Materials Science, University of Minnesota, Minneapolis, MN

Several biomass-derived oxygenates, such as butanol and dimethylfuran (DMF), have been proposed as alternatives to ethanol for blending into current gasoline fuels. Each of these oxygenates can be synthesized from bio-based platform chemicals such as 5-hydroxymethylfurfural (HMF), acetone, ethanol, etc. in multiple ways. Consequently, there are several options for choosing the product (gasoline additive) and the synthesis route. We propose a strategy for optimal product and process chemistry selection by combining rule-based reaction network generation and mixed-integer linear programming (MILP).

Our strategy is composed of two steps. First, we use a reaction network generation tool, RING [1], developed in our group to construct a network of all possible reactions that are involved in converting bio-based platform compounds to potential gasoline additives. RING takes in as input reaction rules that determine the chemistries being explored. In this case, we include a broad spectrum of heterogeneous acid, base, and metal catalyzed chemistries as reaction rules.

Second, we formulate an MILP to determine the optimal vector of steady-state fluxes through the reaction network that minimizes catalyst use and reactor capital, and corresponds to an oxygenate blend that satisfies the requirements for gasoline additives. The objective function incorporates estimated reaction kinetics for each reaction based on the corresponding reaction rule. Due to the detailed chemistry of the reaction network being embedded into the MILP, the optimal vector determines (1) an oxygenate blend compatible with gasoline and (2) synthesis routes to produce each oxygenate from bio-based platform chemicals. Alternative synthesis routes and oxygenate blends are identified that have competitive objective functions.

This two-step strategy proposed here is generic and scalable for any product-cum-chemistry selection problem. The set of reaction rules, rate estimation, objective function and constraints formulation, and thermochemical analysis of the optimal synthesis routes will be presented and discussed.

[1] Rangarajan, S.; Bhan, A.; Daoutidis, P. Rule-based generation of thermochemical routes to biomass conversion, *Industrial & Engineering Chemistry Research* **2010**, *49* (21), 10459-10470

(546e) On the Systematic Synthesis Screening and Integration of Real-Life Biorefineries

Marinella Tsakalova, Athanassios Nikolakopoulos and **Antonis Kokossis**, School of Chemical Engineering, National Technical University of Athens, Athens, Greece

Biomass technologies have been inundated with new ideas and innovations the industrial world and the labs posing new challenges to chemical engineering. Individual bio-based paths, production of a single chemical product or fuel have gradually taken a significant position in markets. Given the plethora of degrees of freedom related to different feedstocks, paths and portfolio of products, the area seems to be chaotic. Opposing to the vast amount of options and possibilities, systems technology has on offer powerful methods to optimize designs and systematize improvements of biomass processing systems. Systematic screening leads to a holistic approach of the problem far from the applications that focus on individual processes or plants.

This paper introduces a methodology to address the systematic integration of biorefineries supporting decisions for selection and the integration of processing paths. The implementation takes the form of a synthesis model that is generalized with process and intermediates as appropriate to the case. For the estimation of cost the method uses regression models based on the thermodynamics of the process. The investment cost is quickly and roughly calculated based on energy losses resulting from the difference of the Lower Heating Value (LHV) of the inlet and outlet flows of a process.

The methodology is applied to 82 different chemistries that lead to the production of more than 80 intermediate and final products. The biomass is decomposed to three main streams of C5 sugars, C6 sugars and lignin. Those intermediates are the precursors for bio-based products and fuels such as polyamides, PVC, polypropylene, ethanol, aromatic polyols, resins, vanillin etc. that can be their petrochemical counterparts. The model produces a rank of products identifying the optimal integrated flows. The optimization model results to an optimal solution of coproduction of xylonic acid polyamide and lignin based castor oil. An impressive observation coming out from the results is that high market price of the product do not offer tops in the ranking lists. Several case studies are presented additionally to illustrate the wide framework of the methodology. The case studies concern the effect of the market demand and prices, the impact of the supply chain – transport cost, risk factors and LCA aspects.

Integrated Pathway Optimization for Biorefineries

Pascual Eduardo Murillo-Alvarado[a], Ismael Leonardo Raso[a], José María Ponce-Ortega[a], Medardo Serna-González[a], and Mahmoud El-Halwagi[b]

[a] *Chemical Engineering Department, Universidad Michoacana de San Nicolás de Hidalgo, Morelia, Mich., 58030, México.*

[b] *Chemical Engineering Department, Texas A&M University, 3122 College Station, TX, USA, 77843-3122.*

Abstract

This work proposes a new systematic approach to find the optimal pathway of a biorefinery considering simultaneous the selection of the products, feedstocks and processing steps, the model is based on a disjunctive programming formulation. The proposed model also allows selecting several products and feedstocks in the optimal solution. Also is possible considering the simultaneous maximization of the net profit and the minimization of the greenhouse gas emissions. The applicability of the proposed methodology is shown through a case study for the optimal pathway of a biorefinery and several scenarios have been analyzed. Results show that the combination of different feedstocks and products allows compensating simultaneously the economic and environmental objectives.

Introduction

Along of the years the main energy sources are the petrochemical fuels; however, the depletion of the proved oil reserves as well as the significant quantities of greenhouse gas (GHG) emissions associated to the use of these fossil fuels have promoted the searching of new and cleaner energy sources. For this reason, biofuels offer the potential to solve this problem. The biomass processing is performing in an industrial facility called biorefinery, where biomass is processed through physical, chemical or/and biological technologies to obtain biofuels in addition to other products as can be seen in **Figure 1**. However, several factors must be considered for the installation of a biorefinery like the selection of the feedstocks, the pre-processing stages, processing technologies and the manufactured products in addition to biofuels for any specific case. Different types of feedstocks can be considered, and the processing routes depend on the selected biomass because of the constituted chemical components is different, which may include lignocellulose, sugars, starch and lipids. According to the content of the main components is possible to determine the optimal processing routes for obtaining the desired products.

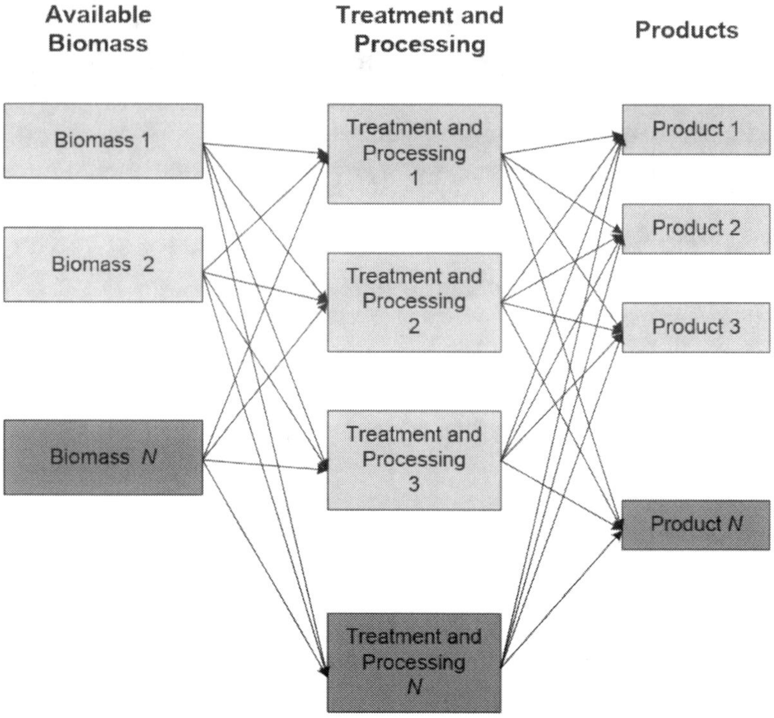

Figure 1. Biorefinery configurations.

In this context several works have been reported to highlight the importance of biorefineries; for example, Huber et al. (2006), Saxena et al. (2008, 2009) and Goyal et al. (2008) presented reviews of the processing of the biomass to obtain several products. However, these works did not present systematic procedures to obtain the optimal route of biomass processing. Recently, Pham and El-Halwagi, (2011) and Ponce-Ortega et al. (2012) presented a systematic approach and a formal optimization model based on disjunctive programming for the automatic optimization to determine the optimal pathway for a biorefinery configuration, respectively. Previous works did not consider the optimal selection of the type of feedstock and product, and the possibility to manufacture several types of products and feedstoks. For this reason, this paper proposes new mathematical programming model to take into account the selection of several feedstocks and products according with the availability and demand in the optimization of the pathway for a biorefinery. The proposed model considers simultaneously maximizing the total net profit and minimizing the associated net GHG emissions in the entire life cycle.

Superstructure Formulation

The problem here is to identify the interconnections between feedstocks and products through the available chemical routes. The approach first consists in the forward synthesis of biomass, where all potential intermediate products and their corresponding conversion technologies are identified. The second stage consists in backward generated product; where for a given final product, the required preceding compounds as well as their corresponding technologies are determined. The final step for the superstructure generation is the interception of two intermediate compounds obtained in the forward and backward steps identifying the set of conversion technologies required to connect these compounds. Using this approach, a superstructure like the one shown in **Figure 2** is obtained for each specific problem (i.e., given available feedstocks and desired products).

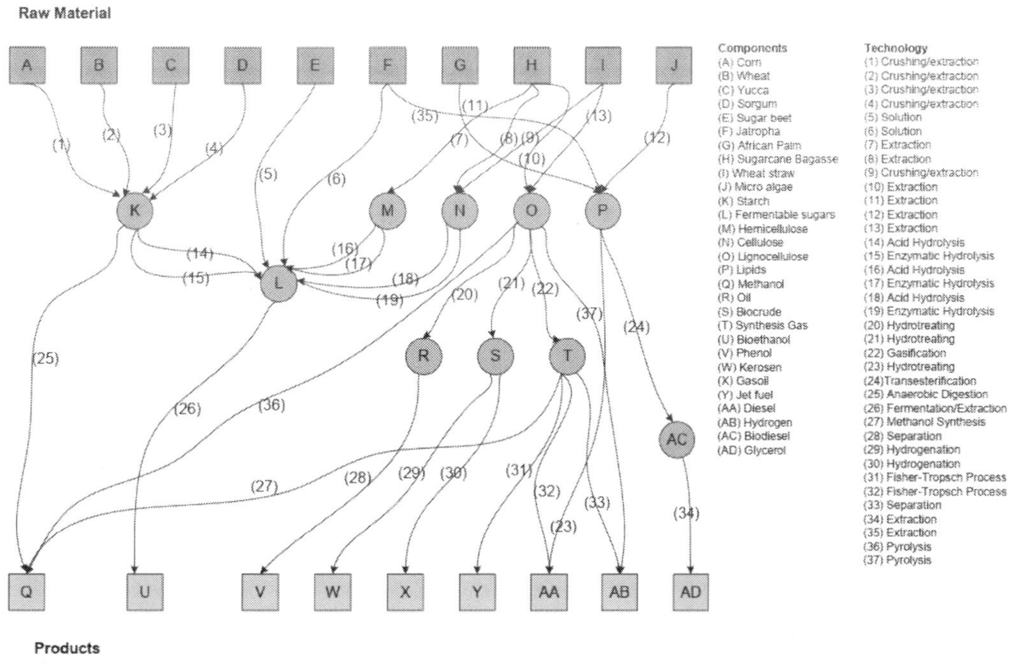

Figure 2. Example for the options of biorefinary pathway configuration.

Notice in previous figure that there are several possibilities for the biorefinery pathway, and that is not obvious to identify the pathway with the lowest cost, lowest GHG emissions and minimum processing steps; in addition, the solution with the minimum cost may not correspond to the solution with the lowest environmental impact or the minimum GHG emissions (and remember that this is one of the purposes for the use of biofuels).

Optimization formulation

Based on the generated superstructure, first the following scripts are defined, i represents any compound preceding the compound j in the superstructure. To yield any compound j, several preceding i can be used. The script t is used to represent any chemical route that connects any pair of compounds.

The following logical relationship states that when the compound j is selected to appear in the optimal pathway, then one technology t_j to produce it must be used. This is modeled using binary variables as follows:

$$y_j = \sum_{t_j} z_{j,t_j}, \ \forall j \in J, j \neq feedstock$$

When one technology t_j is used to produce j from i, then there is required an additional constraint to state that the preceding compound i must exist and this is modeled as follows:

$$\left(1 - z_{j,t_j}\right) + y_i \geq 1, \ \forall j \in J, j \neq feedstock, \forall t_j \in T_j$$

When the chemical route t_j is selected, then this technology has associated a conversion factor $\alpha_{i,t_j}\left(\mathbf{O}_{t_j}\right)$ to yield j, and the corresponding cost $Cost_j$ and environmental impact EI_j have to be determined properly. On the other hand, when the component j is not present in the optimal pathway, then the associated flowrate, cost and environmental impact must be zero. This situation is modeled through the following disjunctive formulation:

$$\begin{bmatrix} Y_j \\ M_{f_j}^{\min} \leq f_j \leq M_{f_j}^{\max} \\ Cost_j = M_{Cost_j} \\ EI_j = M_{EI_j} \end{bmatrix} \vee \begin{bmatrix} \neg Y_j \\ f_j = 0 \\ Cost_j = 0 \\ EI_j = 0 \end{bmatrix}, \ \forall j \in J, j = feedstock$$

$$\bigvee_{t_j \in T_j} \begin{bmatrix} Z_{j,t_j} \\ f_j = \alpha_{i,t_j}\left(\mathbf{O}_{t_j}\right) f_i, \ i \text{ preceds } j \text{ through } t_j \\ Cost_j = c_{t_j}\left(\mathbf{O}_{t_j}\right) f_j \\ EI_j = ei_j\left(\mathbf{O}_{t_j}\right) f_j \end{bmatrix}, \ \forall j \in J, j \neq feedstock$$

Previous disjunction can be reformulated as an algebraic problem as follows. First the Boolean variables Y_j and Z_{j,t_j} are reformulated as binary variables y_j and z_{j,t_j}. This means that when the Boolean variable is true, the associated binary variable must be one and the equations inside the disjunction must be applied; on the on the hand, when the Boolean variable is false, the associated binary variable must be zero and the relationships inside do not apply. Therefore, the objective function consists in the simultaneous maximization of the total profit and the minimization of the environmental impact taking into account the number of processing steps. This objective function is formulated as follows:

$$TotCost = H_Y \sum_{j \in J} Cost_j \qquad Sales = H_Y \sum_{j \in j_{products}} \text{Price}_j \times f_j$$

$$Profit = Sales - TotCost$$

The total environmental impact for the pathway is calculated as follows:

$$TotEI = \sum_{j \in J} EI_j$$

Results

To show the applicability of the proposed model, a case study is used to determine the optimal configuration of a biorefinery attending the economical and environmental aspects and considering the availability of feedstocks and demands of products of Mexico. The available feedstocks, unit costs and unit GHG emissions considered in this case study are showed in **Table 1**, which represents the most important.

Table 1. Data for the feedstocks for the case study.

Raw Material	Available (Kg/Day)	Cost (USD/kg)	Gas Emission (Kg)
Corn	80 000 000	0.0878	-0.3835
Wheat	35 000 000	0.0912	-0.4165
Yucca	1 000 000	0.0827	-0.2465
Sorghum	25 000 000	0.0914	-0.2191
Sugar beet	1 500 000	0.096	-0.2164
Jatropha	7 500 000	0.0931	-0.4273
African Palm	1 800 000	0.08205	-0.3561
Sugarcane Bagasse	27 500 000	0.0907	-0.4082
Wheat Straw	2 000 000	0.0769	-0.2794
Micro Algae	1 250 000	0.0825	-0.1878

The model considers the demand of products, price of sale and GHG emissions generated; which is shown in **Table 2**.

Table 2. Data for the products for the case study.

Products	Demand (Kg/Day)	Cost Sale (USD/kg)	Emission (kg CO_2/kg product)
Methanol	90 000	0.87	3.239
Bioethanol	150 000	0.523	1.375
Phenol	45 000	0.594	2.095
Kerosene	43 000	0.634	2.808
Gasoil	40 000	0.692	2.289
Jet Fuel	120 000	0.769	2.458
Biodiesel	175 000	0.79	2.346
Hydrogen	30 000	0.818	3.239
Glycerol	15 000	0.668	0.1

This case consists of 158 continuous variables, 66 binary variables and 409 constraints; and this model was coded in the GAMS software where the solver CPLEX was used to solve it. The optimal solution selects only one product (biodiesel), where 546,190 kg of micro algae per day to produce 175,000 kg of diesel per day are required. The optimal solution has a net profit of US$17,638,614.74/year. The associated net GHG emissions are 1.8139×10^8 kg CO_2 eq./year that represent the CO_2 captured in the growth of the micro algae, the CO_2 produced in the processing steps and the CO_2 produced in the combustion of the biodiesel.

For the case when is needed to satisfy the bioethanol demand (150,000 kg/day; this is called the scenario B), the optimal values for the net profit and the associated GHG emissions are 1,170,988.55 US$/year and 2.732x108 kg/year, respectively. **Figure 3** shows the Pareto curve for the objectives Profit versus GHG emissions for this scenario.

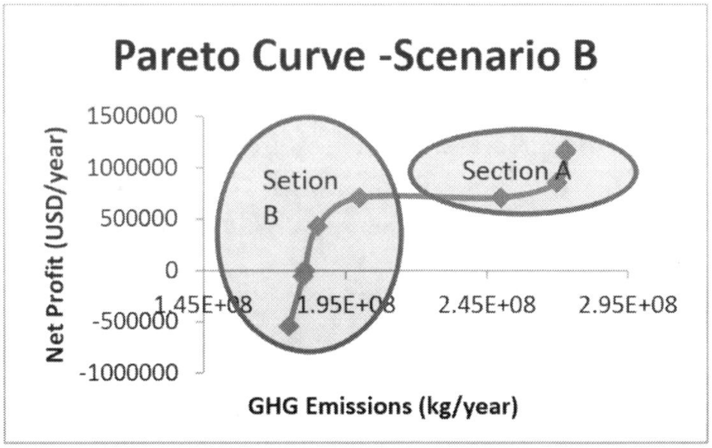

Figure 3. Pareto curve for Profit versus GHG emissions for scenario B.

The corresponding optimal pathway configuration is shown in **Figure 4**, where two sections are identified, one corresponding to high profits and high GHG emissions (section A) and the other one corresponding to negative profits and low GHG emissions (section B).

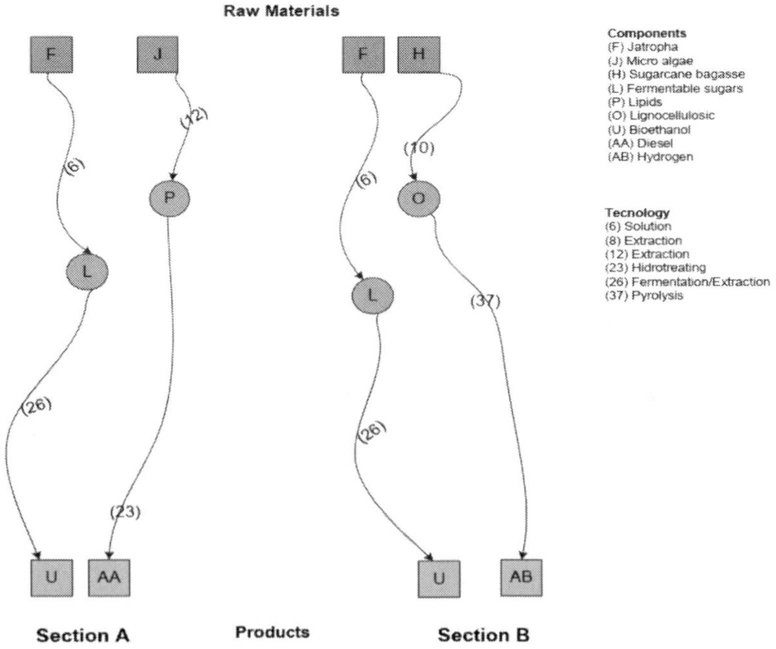

Figure 4. Optimal pathway for section A and B for scenario B.

Conclusions

This paper presents a systematic approach for the optimal pathway configuration for a biorefinery based on a distinctive programming formulation. The proposed model takes into account the simultaneous selection of the raw materials, products and processing steps. Also, the proposed model considers the simultaneous maximization of the net profit and the minimization for the associated GHG emissions taking into account the processing steps. The results obtained are presented through Pareto curves to identify the tradeoffs of the considered objectives. In addition, the proposed approach allows selecting several feedstocks and manufactured products. Finally, the proposed approach is applied to a case study of Mexico, and several scenarios have been identified to account the tradeoffs of the proposed objectives.

References

- Goyal H.B.; Seal D.; Saxena R.C. Bio-fuels from thermochemical conversion of renewable resources: A review. *Renewable and Sustainable Energy Reviews*. **2008**, 12 (2), 504-517.
- Huber G.W.; Iborra S.; Corma A. Synthesis of transportation fuels from biomass: Chemistry, catalysts, and engineering. *Chemical Reviews*. **2006**, 106 (9), 4044-4098.
- Pham V.; El-Halwagi M.M. Process synthesis and optimization of biorefinery configurations. *AIChE Journal*, **2011**, 58 (4), 1212-1221..
- Ponce-Ortega J. M.; Pham V.; El-Halwagi M.M.; El-Baz A. A disjunctive programming formulation for the optimal design of biorefinery configurations. *Industrial and Engineering Chemistry Research*, **2012**, 51(8): 3381-3400.
- Saxena, R.C.; Adhikari, D.K.; Goyal, H.B. Biomass-based energy fuel through biochemical routes: A review. *Renewable and Sustainable Energy Reviews*. **2009**, 13(1), 167-178.
- Saxena, R.C.; Seal, D.; Kumar, S.; Goyal, H.B. Thermo-chemical routes for hydrogen rich gas from biomass: A review. *Renewable and Sustainable Energy Reviews*. **2008**, 12 (3), 1909-1927.

(546g) Optimization of Biochemical Conversion Process of Lignocellulosic Biomass to Sugars by Integrating the Kinetic Models

Vasanth Natarajan, Department of Chemical and Biomolecular Engineering, National University of Singapore, Singapore, Singapore and Iftekar A. Karimi, Dept of Chemical & Biomolecular Engineering, National University of Singapore, Singapore, Singapore

Over the years, interest in new ways of producing biofuels is increasing rapidly in response to rising prices of gasoline, depletion of fossil fuels and the non-renewable nature of crude oil. Bioethanol, the first generation biofuel is a highly established renewable fuel today which comes from sugar or grain (starch). The production of second generation biofuel from cheaper, abundant lignocellulosic feedstocks (LCF) emerged as a potential alternative feedstock to highly competitive food based raw materials. For lignocellulosic biofuels to become a reality, research in the development of cost and energy effective biochemical conversion of biomass to sugars are highly needed. With a lot of improvements in the experimental studies of various pretreatment, saccharification and fermentation processes, a significant improvement from a modeling and optimization perspective is required.

In the literature, studies have been done on kinetic modeling [1], statistical and non-statistical techniques [2, 3] for process design and optimization. Until the present, most of the modeling based works focus on the optimization of size reduction process [4], pretreament process, saccharification and fermentation individually. For predicting the behaviour of the entire conversion system it is essential to integrate existing models for efficient process optimization [5]. In this work, we propose a model based optimization approach for the biochemical conversion of biomass to sugars by considering the overall conversion process. Our major objective of this work is to minimize the amount of energy required for the conversion process by determining the optimal design variables. Here, we integrate the kinetic models of dilute acid pretreatment, enzymatic hydrolysis and fermentation process. Also, we consider the physical (size reduction) and chemical aspects (diffusion models) of the biomass for a more detailed study of the above system. The above problem has been formulated in GAMS including the mass balance equations, kinetic equations, and the equations involved in size reduction and diffusion processes.

Finally, we present the results based on the minimal amount of energy consumed by the entire conversion process of lignocellulosic biomass to sugars. We have considered cornstover as our raw material due to its vast availability of experimental and theoretical data. This study shows the need for sophisticated models that considers the integrated models of the conversion process and also, the physical and chemical aspects of the biomass which is the novelty and the contribution of this work.

1. Esteghlalian, A., et al., *Modeling and optimization of the dilute-sulfuric-acid pretreatment of corn stover, poplar and switchgrass.* Bioresource Technology, 1997. 59(2-3): p. 129-136.

2. Ramirez, M.T. and E.S. Fraga, *Mathematical Modelling of Feed Pretreatment for Bioethanol Production*, 2009. p. 1299-1304.

3. Vargas Betancur, G.J. and N. Pereira Jr, *Sugar cane bagasse as feedstock for second generation ethanol production. Part I: Diluted acid pretreatment optimization.* Electronic Journal of Biotechnology, 2010. 13(3): p. 1-9.

4. Sayed Ali Hosseini, R.L., Sergei Kucherenko, and Nilay Shah, *Multiscale Moldeling of Hydrothermal Pretreatment: From Hemicellulose Hydrolysis to Biomass Size Optimization.* Energy and Fuels, 2010. 24: p. 8.

5. Ziyu Wang, J.X., and Jay J. Cheng, *Modeling Biochemical Conversion of Lignocellulosic Materials for Sugar Production: A Review.* BioResources, 2011. 6(4): p. 25.

(551a) IGCC Power Plant Dispatch Using Infinite-Horizon Economic MPC

Benjamin P. Omell, Chemical and Biological Engineering, Illinois Institute of Technology, Chicago, IL and Donald J. Chmielewski, Illinois Institute of Technology, Chicago, IL

The Integrated Gasification Combined Cycle (IGCC) possesses a number of benefits over traditional power generation plants, ranging from increased efficiency to flex-fuel and carbon capture opportunities. A lesser known benefit of the IGCC configuration is the ability to load track electricity market demands. The idea being that process modifications to enable dispatch capabilities will allow for a time-shift of power production away from periods of low energy value to periods of high value. The work begins with an illustration of Economic Model Predictive Control (EMPC) as a vehicle to exploit dispatch capabilities by pursuing directly the objective of maximizing revenue. However, implementation of EMPC can result in unexpected and at times pathological closed-loop behavior, including inventory creep and bang-bang actuation. To address these issues, an infinite-horizon version of EMPC is developed and shown avoid many of the performance issues observed in the finite-horizon version. The paper concludes with an in depth discussion of energy value forecasting and how the quality of forecasts can be incorporated into the design of the infinite-horizon EMPC controller.

(551b) Hedging Energy Commodities for Flexible Fuel Operations

Fanghui Fan[1], Fernando Garcia[2] and **Jeffrey Kantor**[2], (1)Chemical & Biomolecular Engineering, University of Notre Dame, Notre Dame, IN, (2)Chemical and Biomolecular Engineering, University of Notre Dame, Notre Dame, IN

Operators of flexible fuel utilities are subject to substantial risk due to stochastic price uncertainty in the commodity markets. To reduce financial risk and increase return, operators manage a portfolio consisting of physical or otherwise beneficial ownership of the underlying commodities [2, 3, 6]. The coordinated management of the portfolio and the flexible fuel operation is the hedging problem considered in this paper. The results of the paper can be used either by the process operator directly to reduce financial risk, or by an energy banker to price a an 'energy swap' to finance process operation.

The mathematical development is based on self-financing hedging portfolio consisting of the flexible fuel process, physical ownership of the underlying energy commodities, and a risk-free asset. The returns on portfolio assets include the stochastic yields on the underlying assets, convenience yields attributable to physical ownership of commodities, and the obligation to operate the process to meet known demand. Compared to conventional formulations available in the literature [1, 4, 5], the novel aspects of this formulation are the inclusion of a flexible fuel process with a first and second law consistent model, and the inclusion of convenience yield.

Following standard techniques, the feedback control strategy for the operation of a coordinated, optimal, risk-free hedge is found as the solution to a stochastic optimal control problem. The associated Hamilton-Jacobi-Bellman equation is solved using finite difference methods. Special consideration is given to the necessary boundary conditions, which appear to be novel. The solution is validated by an independent Monte-Carlo solution.

The control strategy consists of feedback rules for managing the hedging portfolio as a function of time-to-go and spot prices of the underlying commodities. Performance of the control strategy is demonstrated through extensive back-testing with real-world price data for coal and natural gas subject to trading and ownership constraints. The hedging error is quantified and discussed.

References

1. Fred Benth, Jurate Benth, and Steen Koekebakker. Stochastic Modeling of Electricity and Related Markets. World Scientific, 2008.
2. S.J. Deng and S.S. Oren. Electricity derivatives and risk management. Energy, 31(6-7):940–953, 2006.
3. M. Denton, A. Palmer, R. Masiello, and P. Skantze. Managing market risk in energy. Power Systems, IEEE Transactions on, 18(2):494–502, 2003.
4. Alexander Eydeland and Krzysztof Wolyniec. Energy and Power Risk Management: New Developments in Modeling, Pricing and Hedging. Wiley, 2003.
5. Hélyette Geman. Commodities and Commodity Derivatives: Modeling and Pricing for Agriculturals, Metals and Energy. Wiley, 2005.
6. Yanbo Jin and Philippe Jorion. Firm value and hedging: Evidence from u.s. oil and gas producers. Journal of Finance, 61(2):893–919, April 2006.

(551c) Optimal Rules for Central Bank Interest Rate Subject to Zero Lower Bound

Ajay Singh, Chemical and Biomolecular Engineering, University of Houston, Hoston, TX and M. Nikolaou, Chemical and Biomolecular Engineering, University of Houston, Houston, TX

Abstract

The celebrated Taylor rule (Taylor, 1993) is a feedback policy used to regulate the output gap and inflation of an economy by adjusting the central bank interest rate. The main issue addressed in this work is the effect of zero lower bound on the optimal interest rate determined by a central bank. We address this issue in a multi-parametric model predictive control (mpMPC) framework, which allows the derivation of explicit feedback rules even when inequality constraints are present. Application of this framework to a simple model of the US economy produced a number of Taylor-like rules (Singh and Nikolaou 2012), depending on the form and parameter values in the objective function employed by MPC. The results suggest that one from a small number of simple Taylor-like rules can be applied at each time, depending on the state of the economy. However, it was also shown that simply setting to zero negative interest rates produced by unconstrained Taylor rules is optimal in situations of negative output gap, as happened recently. Furthermore, it was observed, as has been noted elsewhere, that rules with inertia appear to better capture past decisions by the Federal Reserve Bank. Such rules have been systematically derived here by considering penalties on the rate of interest rate change in the MPC objective function. The proposed approach is illustrated through simulations on US economy data. A number of issues for future study are proposed.

Reference:

Taylor, J.B., 1993. Discretion versus policy rules in practice, Carnegie-Rochester Conference Series on Public Policy 39, North-Holland, pp. 195-214.

Singh, A. P. and M. Nikolaou (2012). "Optimal Rules for Central Bank Interest Rate Subject to Zero Lower Bound." Journal of Economic Dynamics and Control.(submitted)

(551d) Exploiting Price Fluctuations and Volatility Through Discontinuous Feedback Control

Christopher V. Rao, Chemical and Biomolecular Engineering, University of Illinois, Urbana Champaign, Urbana, IL

Many products of the chemical and petroleum industries are fungible goods. They are highly liquid and, in many case, actively traded on centralized exchanges. Examples include not only petroleum distillates, such as reformulated gasoline blend-stock for oxygen blending (RBOB) gasoline, but also basic plastics such as polyethylene and polypropylene. While this liquidity reduces market friction, it also subjects producers to rapid changes and potential shocks in market prices. Even in cases where the products themselves are not explicit commodities, the feedstocks are -- petroleum and natural gas being prime examples -- and further subject the producer to the whims of the market and potential dislocations in prices and spreads. While financial instruments such as future contracts and swaps can be used to hedge against these disruptive market forces, they may not always exist or require exceptionally large premiums due to limited volume and the absence of market makers.

We previously developed a toy model to explore how feedback control can be used to make chemical producers responsive to market forces through dynamic operating policies as opposed to through the use of financial instruments. A key result from this work is that these dynamic operating policies yield discontinuous control laws exhibiting hysteresis. In addition, we found that these control problems needed to be formulated as risk-sensitive Markov decision processes in order to obtain meaningful solutions.

In the present work, we extent this toy model to consider a larger class of problems involving price fluctuations and volatility. Using these models, we explore the factors that contribute to the resultant discontinuous control laws and derive analytic solutions to the governing dynamic programs. These results provide a basis for exploring more complex control problems that include the effects of market forces.

(551e) Integrating Dynamic Economic Optimization and Model Predictive Control for Optimal Operation of Nonlinear Process Systems

Matt Ellis[1], Mohsen Heidarinejad[2], Jinfeng Liu[3] and Panagiotis D. Christofides[3], (1)Chemical Engineering, UCLA, Los Angeles, CA, (2)Electric Engineering, University of California, Los Angeles, Los Angeles, CA, (3)Department of Chemical and Biomolecular Engineering, University of California, Los Angeles, Los Angeles, CA

Economic optimization of chemical processes has traditionally been addressed through real-time optimization (RTO). In general, an RTO system carries out economic optimization and computes optimal process operation steady-states using steady-state process models. These steady-states are used by the feedback control systems, typically designed via model predictive control methods, that force the process to operate on these steady-states. In recent years, numerous calls for the development of the so-called "smart manufacturing paradigm" have led to several attempts to integrate MPC and economic optimization of chemical processes to deal with variable demand, energy prices, variable feedstock and product transitions. In previous work [1], we developed a Lyapunov-based economic MPC (LEMPC) method capable of optimizing closed-loop performance with respect to general economic consideration for nonlinear systems. In the present work, we propose a conceptual framework for integrating dynamic economic optimization and MPC for optimal operation of nonlinear process systems. First, we introduce the proposed integrated framework. This optimal operation strategy consists of an LEMPC that preforms upper layer dynamic economic optimization to compute optimal process operating time-varying trajectories which are used by the lower-level control layer that may use conventional MPC schemes or even classical control. Such a framework takes advantage of the LEMPC ability to compute optimal process operation time-varying policies using dynamic process model instead of traditional steady-state models. Second, we prove practical stability of the proposed integrated dynamic economic optimization and control framework including an explicit characterization of the stability region of the closed-loop system. Third, we demonstrate through extensive simulations using realistic chemical process models that such an integrated control paradigm can both (1) achieve stability and (2) perform economically better than traditionally RTO systems using steady-state models. 1. Heidarinejad, M., J. Liu and P. D. Christofides, "Economic Model Predictive Control of Nonlinear Process Systems Using Lyapunov Techniques," AIChE J., 58, 855-870, 2012.

(551f) Flexible Manufacturing and Demand Response for Smart Grid Coordination

David Mendoza-Serrano, Chemical and Biological Engineering, Illinois Institute of Technology, Chicago, IL and Donald J. Chmielewski, Illinois Institute of Technology, Chicago, IL

One aspect of the smart grid is to use a real-time price structures within electricity markets as a vehicle to mitigate source-load inequities resulting from the uncertainty associated with renewable power sources (solar and wind). Under such price structures, manufacturing facilities with an ability to change energy consumption rates can expect to capture more profit while providing the societal benefit of improving grid reliability. Although facility managers qualitatively agree with the dual benefit of adopting smart grid operating policies, they lack quantitative methods to construct these policies as well as predict revenue gains.

In this paper we will illustrate the use of Economic Model Predictive Control (EMPC) as a vehicle to construct demand response policies for the flexible manufacturing application. In addition, we will propose a linear surrogate for the demand response policy, which will facilitate the development of a quantitative method to assess smart grid coordination opportunities. As an illustration of the approach a chemical processing plant will be investigated where process steam can be generated from a simple furnace or from an electric co-generation plant. In the second part of the example we will consider the additional opportunity of deferment of sub-operations based on supply-chain inventory levels.

(551g) System Identification and Distributed Control for Multi-Rate Sampling System

Quan Min Shao and Ali Cinar, Chemical and Biological Engineering, Illinois Institute of Technology, Chicago, IL

Chemical processes usually have many controlled variables, which may have different sampling times. For example, concentration measurements usually have much longer sampling time than temperature measurements. When implementing multiple-input-multiple-output (MIMO) control, traditional system identification (SI) and control techniques need to be extended to handle multi-rate sampling process.

Simply applying standard SI technique to the multi-rate system by using a single rate (the greatest common factor of different sampling times) would yield poor prediction results for the variables with large sampling times. A better approach for SI for multi-rate systems is by using *lifted model*, in which the system inputs and outputs with slower sampling rate are lifted to single rate, which will generate larger dimensions of inputs and outputs for the system models. The benefit of this approach is that all the variables are sharing the same sampling time, thus the traditional control techniques can be applied. Moreover, the faster sampling rate model contains more information than the slower rate model. However, this approach has limitations mainly on two aspects: first the dimensionality of the model will increase due to lifting, and the worst situation happens when some of the variables have much slower sampling rates compared to the fast rate; second, even slight inaccuracy of the lifted model may cause noticeable error of single-rate model when extracting single-rate model from the lifted model.

We propose a new approach to solve the control problem of multi-rate sampling systems that still utilizes the lifted model, but we leverage the advantages of distributed control techniques. The system outputs will be assigned to different subsystems based on their sampling times (in each subsystem, all the controlled variables have the same sampling time, but it is not necessary to have all the variables that have the same sampling time in a single subsystem). Then when identifying the lifted model of one subsystem, just input lifting (to single rate) is required, with the output unchanged, which would further reduce the dimensions of the subsystem models. Moreover, in order to develop distributed control, the influence from neighbor subsystem also needs to be considered. Thus the SI should also include the related inputs of neighbor subsystems as inputs to this subsystem. Subspace identification will be utilized as the SI method to obtain state space model for each subsystem.

After the distributed models under different sampling times are available, distributed model predictive control could be designed to deal with the control problem of multi-rate system. Only at the sampling time of each subsystem, the states of subsystems are updated. At each single-rate sampling time, first the input trajectories from neighbor subsystems are acquired, as constant inputs to this subsystem, then the objective function of the whole system is optimized with respect to its own inputs; more iterations of exchange of optimal trajectory with neighbor controllers and optimization could improve the global optimality. Finally the control sequences are updated to each subsystem. Asymptotic stability could also be guaranteed from the design of distributed model predictive control algorithm. This is important for this approach and needs to be proved carefully via Lyapunov stability.

The case study is the Tennessee Eastman challenge problem. The continuous variables are sampled every 1 minute, while there are also concentrations that are sampled every 6 or 15 minutes. Base control is implemented to ensure the system stability by a set of PI controllers. Then the distributed lifted models are built by subspace identification, and distributed control is used to control the system.

In summary, this work focuses on the control problem of multi-rate sampling systems. The concept of distributed control helps the decomposition of the multi-rate control problem, then standard SI could be conducted and traditional control algorithms can be applied to local controllers. Furthermore, distributed control can lower the dimension of the control problem, which could be challenging when the lifted model approach is utilized and centralized control is implemented.

(572a) Surrogate-Based Optimization of Simulated Energy Systems

Alison Cozad[1], Nick Sahinidis[1] and David C. Miller[2], (1)Chemical Engineering, Carnegie Mellon University, Pittsburgh, PA, (2)U.S. Department Of Energy, National Energy Technology Laboratory, Morgantown, WV

Standard optimization techniques fall short of being able to solve simulated systems or, more generally, black-box systems since these techniques require the use of first and second derivatives of algebraic objectives and constraints. In these cases, black-box or derivative-free optimization (DFO) solvers are used. However, when the design of a process involves discrete variables and complex decisions, the resulting problem topology is often too complex for these DFO methods to solve reliably.

To access standard algebraic optimization methods to solve simulated systems, we first build a set of algebraic surrogate models to describe the system. These models can be combined with design constraints and algebraic objectives or cost functions to formulate an algebraic optimization problem. This optimization model is to design all discrete and continuous process alternatives, technology choices, etc., while simultaneously solving for continuous operating conditions across the process network using a superstructure optimization framework.

In order to solve such a large superstructure model, we aim to build not only accurate but also simple surrogate models. A combination of machine learning, statistical, and optimization techniques are used to build low complexity algebraic models. These models are test, exploited, and improved using an adaptive sampling method we refer to as error maximization sampling. New simulation points are strategically chosen to help rule out the previous iterates' surrogate with the goal of increasing model accuracy using relatively few costly simulation.

By using low complexity, accurate surrogate models we are able to optimize energy systems by efficiently exploring the full set of process alternatives to design low cost, environmentally conscious energy systems.

(572b) Integrating the Carbon Capture Materials Database with the Process Simulation Tools of the Carbon Capture Simulation Initiative

Hosoo Kim[1], Maciej Haranczyk[2], Tom Epperly[3], Mahmoud Abouelnasr[4], Joseph A. Swisher[5], Kuldeep Jariwala[2], David Mebane[6], Berend Smit[5], Joel Kress[7] and David C. Miller[8], (1)Computational Science Division, National Energy Technology Laboratory, Morgantown, WV, (2)Lawrence Berkeley National Laboratory, (3)Lawrence Livermore National Lab, Livermore, CA, (4)Chemical and Biomolecular Engineering, University of California - Berkeley, Berkeley, CA, (5)Chemical and biomolecular Engineering, University of California-Berkeley, Berkeley, CA, (6)National Energy Technology Lab, Morgantown, WV, (7)Los Alamos National Lab, Los Alamos, NM, (8)U.S. Department Of Energy, National Energy Technology Laboratory, Morgantown, WV

Significant work is underway to identify new materials that could be effective for capturing CO_2 from anthropogenic sources, such as coal-fired power plants, because the current technologies are too expensive. Among the potential materials being studied are various classes solid sorbents, such supported amines, zeolites and metal organic frameworks (MOFs). Since millions of potential materials could be developed, it can be useful to use computational techniques to identify the most promising. One of the primary goals of the U.S. Department of Energy's Carbon Capture Simulation Initiative (CCSI) is the development of state-of-the-art computational toolset to enable promising concepts to be more quickly identified through rapid computational screening of devices and processes. The components of the CCSI Toolset, which enable initial screening, consist of a set of detailed process models coupled with a process synthesis and design framework to facilitate development of complete, integrated carbon capture processes using advanced optimization techniques including deterministic superstructure-based methods and derivative-free optimization (DFO). Recently these models and optimization framework have been linked with Carbon Capture Materials Database (CCMDB, www.carboncapturematerials.org), a comprehensive data collection on materials for carbon dioxide capture. The CCMDB has been developed by the Energy Frontier Research Center for Gas Separations Relevant to Clean Air Technologies, which is using computational chemistry methods to estimate thermophysical properties for crystalline porous materials such as zeolites and metal organic frameworks. The database currently contains data on approximately 200,000 materials. This presentation will describe how the properties from the CCMDB can be used within the CCSI models and optimization framework to screen the potential of new materials within the context of an optimized carbon capture process integrated with the other components of the energy generation system (i.e., power plant and compression system). Results from initial demonstration cases will also be discussed.

(572c) A Sub-Grid Model for an Array of Immersed Cylinders in Coarse-Grid Multiphase Flow Simulations of a Carbon Capture Device

Avik Sarkar[1], Xin Sun[1] and Sankaran Sundaresan[2], (1)Pacific Northwest National Laboratory, Richland, WA, (2)Chemical Engineering, Princeton University, Princeton, NJ

A post-combustion carbon-capture system utilizing a bubbling fluidized bed of sorbent particles is currently being developed as a part of the Carbon Capture and Simulation Initiative (CCSI) efforts. Adsorption of carbon dioxide (CO_2) by these amine based sorbent particles is exothermic and arrays of immersed cylindrical heat transfer tubes are often utilized to maintain the lower temperatures favorable for CO_2 capture. In multiphase computational fluid dynamics (CFD) simulations of the full-scale devices, which can be up to 10 m in size, approximately 10^3 cells are required in each dimension to accurately resolve the cylindrical tubes, which are only a few centimeters in diameter. Since the tubes cannot be resolved explicitly in CFD simulations, alternate methods to account for the influence of these immersed objects need to be developed.

In this work a sub-grid CFD model is developed to account for the influence of cylindrical arrays on flow dynamics in a multiphase flow. Highly resolved simulations of flow around a single cylinder are performed in a periodic domain, from which drag measurements are filtered to obtain the sub-grid model. This model may then be used in coarse-grid simulations where the cylinder array is treated as a uniform porous medium exerting an anisotropic drag on the gas-solid flow. The influence of these cylinders can appear as drag exerted on the gas and solid phases, or it may change the clustering behavior of particles thereby affecting the gas-particle drag indirectly. Both mechanisms are investigated to formulate a new sub-grid drag model for the gas-cylinder and solid-cylinder drag and to develop corrections for the existing gas-solid drag due to the presence of cylinders. Although this work addresses cylindrical rods in a carbon-capture device, the approach can be generalized to model other immersed obstruction geometries in industrial-scale multiphase flow simulations.

(572d) A Global Optimization Approach for the Short-Term Scheduling of Hydro Power Generation

Ricardo M. Lima, Energy Systems Modeling and Optimization Unit (UMOSE), LNEG - Laboratorio Nacional de Energia e Geologia, Lisboa, Portugal, Marian G. Marcovecchio, INGAR - CONICET - UTN, Santa Fe, Argentina and Augusto Novais, Energy Systems Modeling and Optimization Unit (UMOSE),, Laboratório Nacional de Energia e Geologia, I.P., Lisboa, Portugal

The paradigm of production and consumption of electricity has been changing due to increasing electricity demand, time variable electricity prices for industry, the growing penetration of renewable sources, and due to public policies regarding supply security and environmental impact. Hydro power is a widely used form of renewable energy with clear advantages in terms of no fuel consumption, and no direct waste or CO2 emissions. Furthermore, hydro plants have high flexibility to be re-scheduled and the capability to store energy using pumped hydro storage in coordination with wind power generation.

This paper addresses the optimization of the short-term scheduling of a cascade of hydro plants using detailed and accurate models, and a deterministic global optimization approach to deal with the non-convexities. Cascades are represented by integrated systems of hydro plants involving plants with reservoirs and plants without the capacity to store water, the so-called run-off-the-river plants. The need for this integrated and necessarily more complex approach derives from the strong interaction that arises between upstream and downstream plants caused by river connection and water releases. The scheduling of a hydro plant involves the determination of the start and duration of the operation of each turbine for the next day, with the time horizon divided into periods of one hour. The operation of each turbine is characterized by its output flow and the respective energy generated for each period of one hour that maximizes the operating profit of the cascade subject to the water management constraints of each reservoir and limits on the performance of the turbines.

Most of the scheduling models for hydro power generation proposed in the literature are based on Mixed Integer Linear Programming (MILP) models, where simplified expressions are used to calculate the power generated by the hydro plants and the forebay and tailrace levels, or where piecewise linear approximations are considered to replace nonlinear functions (Conejo et al. 2002, Borghetti et al. 2008). In addition, most models include a single turbine per plant. Recently, some authors have proposed more detailed models involving nonlinearities in the calculation of the power output, resulting in Mixed Integer NonLinear Programming (MINLP) models that are solved without guarantee of global optimality (Catalão et al. 2009, Diaz et al. 2011).

In this work hydro plants are considered, where the majority has a set of turbines in parallel linked to the same reservoir, but with different specifications in terms of maximum turbined flow and power output. A model is developed, which is based on a discrete time formulation with a time horizon of one day, divided in time periods of one hour, involving a) the respective time dependent constraints for mass balances of water, and power generated; b) the topological, reservoir and plant constraints; c) nonlinear functions to determine the power generated of each turbine; and d) polynomial functions to calculate the forebay and tailrace levels. Furthermore, the power output of each turbine is a function of three variables: 1) the flow rate of water discharged by the turbine; 2) the tailrace level of water downstream the plant; and 3) the forebay level of water in the reservoir. Therefore, the short-term scheduling results in a detailed MINLP problem, involving bi-linear terms and fourth degree polynomials.

In order to find the global optimum within a pre-specified tolerance, a Branch and Bound framework is proposed, which is based on the solution of an MILP overestimator model and of a MINLP model in each node of the tree. In the MILP model the bilinear terms are replaced by piecewise linear estimators defining their convex hull envelopes over partitions (McCormick, 1976, Sherali and Alameddine 1992, Bergamini et al. 2005), while the fourth degree polynomials are relaxed using valid over and under estimators. The formulation of the MILP model is improved by enforcing symmetry breaking constraints on the turbines with the same specifications that operate in parallel, and by a specific partition for the relaxations of the bilinear terms. This partition is constructed taking into consideration that one of the variables involved in the bilinear term is a semi-continuous variable, and therefore forces the minimum positive value of this variable to be a bound of the first region of the partition.

The MILP model provides a tight linear overestimation of the nonconvex region of the original MINLP problem and thus a valid upper bound of the objective function on each node of the Branch and Bound framework. The lower bound of the original MINLP problem is obtained by fixing the binary variables associated with the turbines at the value of the MILP solution if they are equal to one, leaving the variables equal to zero free, and solving a reduced MINLP problem. The larger difference between the non-linear term and its approximation in the MILP model is used to select the branching variable. The search stops once the pre-specified global optimality has been reached.

In order to assess the performance of the proposed framework, three case studies consisting on cascades with a different number of hydro plants and turbines with varied specifications are presented. The results show that the proposed MILP over estimator leads to a tighter upper bound, and then small gaps between the upper and lower bound can be achieved. In addition, the best solutions obtained represent an improved profit when compared with solutions obtained with local solvers for MINLP problems. The performance of the proposed framework is also generally superior to the one obtained with the available solvers for global optimization.

References

Bergamini, M. L., Aguirre, P., & Grossmann, I. E. (2005). Logic-based outer approximation for globally optimal synthesis of process networks. Computers and Chemical Engineering., 29, 1914–1933.

Borghetti, A. , D'Ambrosio C., Lodi A., Martello S., 2008, An MILP approach for short-term hydro scheduling and unit commitment with head-dependent reservoir, IEEE Trans. Power Syst., 23,3,1115-1124.

Catalão, J.P.S., Mariano, S.J.P.S., Mendes, V.M.F., and L.A.F.M. Ferreira, 2009, Scheduling of head-sensitive cascade hydro systems: A nonlimear approach, IEEE Trans. Power Syst., 24,1, 336-346.

Conejo, A.J. , Arroyo, J.M., Contreras J., Villamor, F.A., 2002, Self-Scheduling of a Hydro Producer in a Pool-Based Electricity Market. IEEE Trans. Power Syst., 17, 4, 1265-1272.

Díaz, F.J., Contreras, J., Munoz, J. I., Pozo D., 2011, Optimal scheduling of a price taker cascade reservoir system in a pool-based electricity market, IEEE Trans. Power Syst., 26,2,604-615.

McCormick, G. P. (1976). Computability of global solutions to factorable nonconvex programs. Part I. Convex underestimating problems. Mathematical Programming, 10, 146–175.

Sherali, H. D., & Alameddine, A. (1992). A new reformulation linearization technique for bilinear programming problems. Journal of Global Optimization, 2, 379–410.

(572e) Renewable Wind Power Prediction and Optimization Using Gaussian Mixture Copula Model and Bayesian Inference Based Local-Global Gaussian Process Regression Method

Mudassir Rashid and **Jie Yu**, Department of Chemical Engineering, McMaster University, Hamilton, ON, Canada

Renewable energy has attracted considerable attention recently because of the deterioration of the environment and the depletion of conventional energy resources. As a non-pollution renewable energy source for electricity generation, wind energy is increasingly attractive and has been well recognized as environmentally friendly, socially beneficial and economically competitive. The efficiency of wind turbines can be significantly improved if the wind farm operation is controlled and optimized through accurate wind speed forecasting for load requirements and maintenance scheduling. However, the intermittency of wind power generation and the stochastic, non-stationary and unpredictable nature of wind speed pose significant challenges as to the most efficient utilization of the wind energy source.

To overcome the limitations of conventional wind speed prediction methods, a Gaussian mixture copula model (GMCM) based approach is first applied to the wind speed data to identify the multi-seasonality and non-Gaussian component of diverse atmospheric conditions that influence wind speed. Then a Bayesian inference based local-global Gaussian process regression (GPR) approach is developed for wind speed prediction. Further, Gaussian process regression models within localized copula components are aggregated using the Bayesian inference based posterior probabilities into a global model for time series prediction of wind speed. The integrated GMCM-GPR approach is able to model the stochastic and non-stationary wind speed with appropriate handling of system uncertainty and multi-seasonality.

The proposed approach is demonstrated using real wind speed data from various locations in the US and compared against the auto-regressive integrated moving average (ARIMA) and support vector regression (SVR) methods. The comparison indicates that the GMCM-GPR approach provides more accurate and reliable predictions on wind speed than ARIMA and SVR methods. The intermittency and random fluctuations in wind speed that are difficult to characterize using regular dynamic modeling techniques are well captured by the presented approach.

(572f) Examining the Variability of Load, Wind, and Solar Power in the Regulation Timeframe

Bri-Mathias S. Hodge, Anthony Florita and Kirsten Orwig, National Renewable Energy Laboratory, Golden, CO

The integration of variable renewable power sources (wind and solar) has been a source of concern for power system operators because of the variability and uncertainty of the power output. However, power system operators are already used to accommodating variable and uncertain load. One way that the impacts of load variability and uncertainty are mitigated at is through the holding of reserves that can quickly increase or decrease generation to mirror changes in demand. At small timescales this is known as regulation reserve. Increasing the amount of regulation reserve needed for higher levels of wind and solar penetration implies that the wind and solar power changes that occur in the four second to five minute timescale are correlated. In this work we examine the distribution of changes in load, wind power, and solar power for different timescales in the regulation timeframe. Additionally, the correlation between changes in power output for individual wind turbines in a wind plant and solar arrays in a solar plant are examined. Power spectral density analysis has also been performed on the data to quantify the differences in power output character at different geographic scales.

(572g) Industrial Sectors and Integrated Solutions (ISIS) - A Techno–Economic and Multi-Pollutant Modeling Framework for Comprehensive Regulatory Analysis

Nabanita Modak[1], Gurbakhash S. Bhander[2], Jacky Rosati[3], Alex Macpherson[4] and Elineth Torres[4], (1)Office of Research and Development, Oak Ridge Institute of Science and Education/US EPA, Durham, NC, (2)Office of Research and Development, US Environmental Protection Agency, Durham, NC, (3)Office of Research and Development, US EPA, Durham, NC, (4)US EPA, Durham, NC

Industrial Sectors and Integrated solutions (ISIS) - A Techno–Economic and Multi-Pollutant Modeling Framework for Comprehensive Regulatory Analysis

Nabanita Modak[1], Gurbakhash Bhander[2], Jacky Rosati[2], Alex Macpherson[3], Elineth Torres[3]

[1]Oak Ridge Institute of Science and Education

[2]APPCD, NRMRL, Office of Research and Development, US EPA

[3]Office of Air Quality Planning and Standards, US EPA

In 2004, based on recommendations made by the National Research Council (NRC), the U.S. Environmental Protection Agency (EPA) decided to develop integrated assessment techniques to monitor and manage air quality for the protection of the environment. The NRC recommendations and the EPA response provided the foundation to develop “multi-pollutant and sector-based analyses and approaches” (EPA Documents, 2012) to manage air quality. However, an appropriate mathematical modeling framework was required to facilitate the required analyses to support this new approach to air quality management. After careful consideration of data needs, EPA decided to initiate development of the Industrial Sectors Integrated Solutions (ISIS) model.

ISIS is a multi-sector, multi-product, dynamic linear modeling framework that evaluates the economic and environmental impacts of emissions reduction strategies for multiple air pollutants. It is primarily a techno-economic model that considers several plant-level economic and technical factors such as the type of emissions units, production capacity, location, cost of production, and applicable emission controls. The model analyzes and evaluates both environmental and economic performances with and without policy constraints. Thus, it is essentially an optimization tool that helps policymakers identify cost-effective regulations for criteria and hazardous air pollutants that could potentially offer some attractive solutions to the industrial sectors.

ISIS was initially used to analyze holistic multi-pollutant options for the cement industry. Currently, EPA's Office of Research and Development (ORD) in collaboration with Office of Air Quality Planning and Standards (OAQPS) is expanding the ISIS modeling framework to include the pulp and paper and iron and steel sectors.

This paper discusses ISIS model's capabilities, basic framework, and modeling flexibilities. The paper also elucidates how the design and implementation of the model successfully address the substantial benefits and challenges of multi-pollutant and sector-based analyses and approaches. ISIS is coded in General Algebraic Modeling Systems (GAMS) language. The input data for the model include production costs, fuel type and intensity, emission intensity, elasticity of demand, discount rate, escalation rate, economic life of a plant, emission abatement technologies, recycled products, import, and export. All the data are organized into a Microsoft Excel and Access database. Based upon user specifications, the model extracts input data from the database and pre-processes it to obtain suitable input parameters for the model equations. Once the data have been pre-processed, ISIS uses a solver application (from GAMS) to determine the optimal levels of production, imports, exports and controls required to meet the market demand and emission constraints, while maximizing total surplus. From the illustration of the output data, users would be able to identify optimal emission reduction strategies that are cost-effective and protective of the environment.

Model Reduction in Multi-scale Simulation and Optimization

Yidong Lang, Lorenz T. Biegler*
Department of Chemical Engineering,
Carnegie Mellon University, Pittsburgh, PA 15213

David C. Miller,
U.S. Department Of Energy,
National Energy Technology Laboratory,
Morgantown, WV

AIChE 2012 Annual Meeting, 2012
October 28, Pittsburgh, PA

Motivations

- Challenges of describe energy process by complex, multi-scale models
- Reduced models (Reduced order models and surrogate models) balance the accuracy with computational cost.
- How to build the RM for multi-scale optimization
- The algorithm of optimization with RM
- How to evaluate the quality the optimal result with RM

Introduction

Trust-region method

$$\min_x f(x)$$

$$\min_x f^r(x_k + s) \quad s.t. \quad s \le \Delta_k$$

Acceptability

$$\rho_k = \frac{f(x_k) - \overset{f(x_k)}{f(x_k + s_k)}}{f^r(x_k) - f^r(x_k + s_k)} \qquad \Delta_{k+1} = \min(\gamma_{inc}\Delta_k, \Delta_{\max})$$

$$\Delta_{k+1} = \gamma \cdot \Delta_k$$

Recourse of ODM is unavoidable

Building $\mathcal{E}$-Exact reduced model

Off-line Procedures

- Determine variables and their ranges ---- Domain
- Design of experiments ----- Set of sampling points
- Run original detail model ---- Database
- Build reduced model ----- RM
- Verification and validation ----- $\mathcal{E}$-exact RM

On-line benefit:

No recourse of ODM during simulation and optimization

Fully Linear Property

Assumptions

- Continuously differentiable: $f(x)$ and $f^R(x)$ within $B=\{x:\|x-x_0\|\le\Delta\}$
- Lipschitz continuous $\|\nabla f(x_1)-\nabla f(x_2)|\le L_0\|x_1-x_2\|$

  $\|\nabla f^R(x_1)-\nabla f^R(x_2)|\le L_R\|x_1-x_2\|$
- Interpolation condition $f(x_i^s)=f^r(x_i^s),\quad i=1,n_s$
- B enclosure at least n+1 sampling points $n_s\ge n+1$

For any $x\in B$ errors are bounded

$$|f(x)-f^r(x)|\le\kappa_f\Delta^2$$

$$\|\nabla f(x)-\nabla f^r(x)|\le\kappa_g\Delta$$

$f(x)$ Original detail model (ODM) $f^R(x)$ ε-Exact reduced model

Solution with ε -Exact reduced model

$$\nabla f^R(\bar{x})\Rightarrow 0$$

KKT condition $\|\nabla f(\bar{x})-\nabla f^R(\bar{x})\|=\|\nabla f(\bar{x})\|\Rightarrow 0?$

How close to true $\|x^*-\bar{x}\|=?$

$$\|\nabla f(\bar{x})\|\le\kappa_{rg}\varepsilon^{1/2}$$

$$\kappa_{rg}=2\sqrt{n_x(\sigma+\sigma_r)}$$

$$\|x^*-\bar{x}\|\le\kappa_{rx}\varepsilon^{1/2}$$

$$\kappa_{rx}=\frac{(2\bar{\sigma})^{3/2}n_x^{1/2}}{\underline{\sigma}^2}$$

$$\min f(x) = x_1 \exp(-(x_1^2 + x_2^2) + 0.1(x_1^2 + 0.1x_2^2)$$
$$s.t. \qquad -1 \le x_1, x_2 \le 3$$

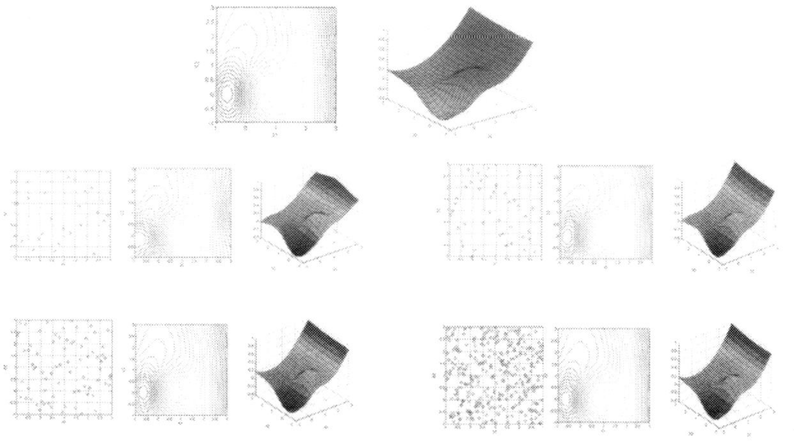

Performances with different sets

		Minimum(1e-3)	Mean	maximum
Nx	25	0.305220659659057	0.331071478967155	13.125077612243128
	45	0.050153816203088	0.053944741487930	1.447502480974255
	75	0.001916150304728	0.005771424034136	0.213013670785779
	300	0.000000392960082	0.000001231055892	0.000063629968679

		Objective	X1	X2
True		-0.383961517686256	-0.636127392935813	-0.000000857336992
Nx	25	-0.345313638078372	-0.885991340468813	-0.180555435248475
	45	-0.388611142796268	-0.674466888436212	-0.036661304974945
	75	-0.384091685139422	-0.635796362811184	0.000689867117202
	300	-0.383961629002243	-0.636128179573110	-0.000004742679934

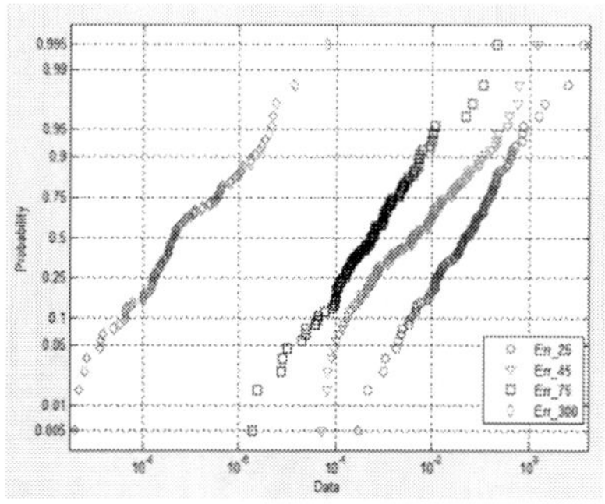

		$\bar{\Delta}$	ε	$\lVert \nabla f(\bar{x}) \rVert$	$\lVert x^* - \bar{x} \rVert$
Nx	25	0.6928	0.331071478967155	0.452642722230171	0.308272502916270
	45	0.5164	0.053944741487930	0.084103660376176	0.053048155539477
	75	0.4000	0.005771424034136	9.111840489579035e-004	7.671798391074405e-004
	300	0.2000	0.000001231055892	1.695752273876660e-006	8.337029715041763e-007

Randomness of LHD

Rosenbrock function

$$\min f(x) = (x_2 - x_1^2)^2 + (1 - x_1)^2$$
$$s.t. \qquad -1 \le x_1, x_2 \le 2$$

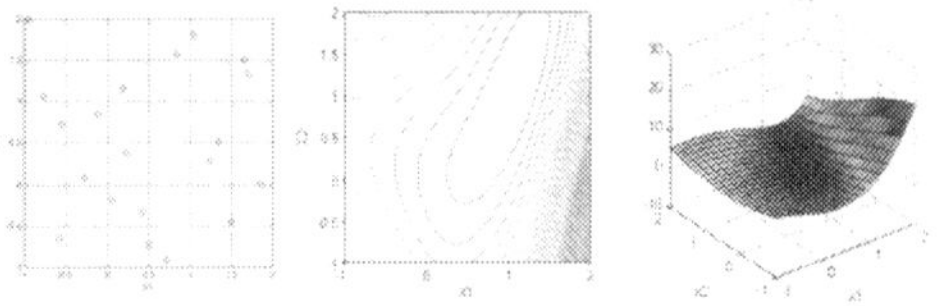

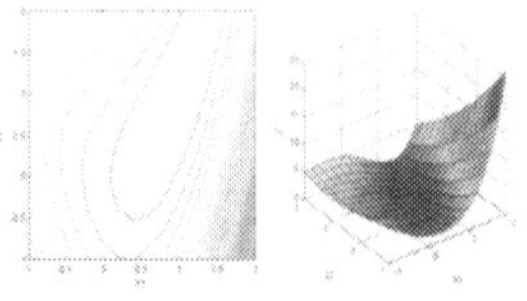

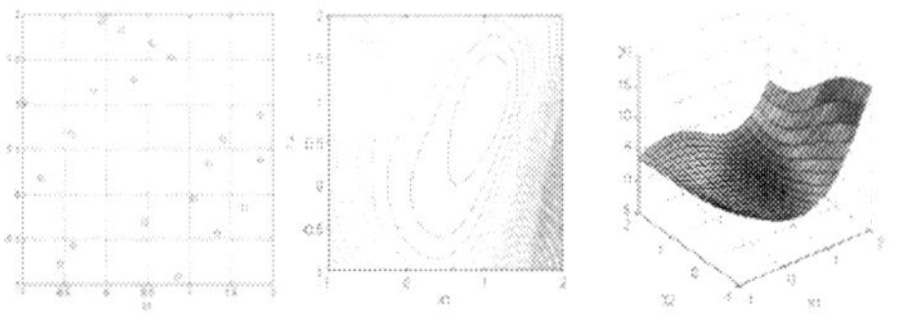

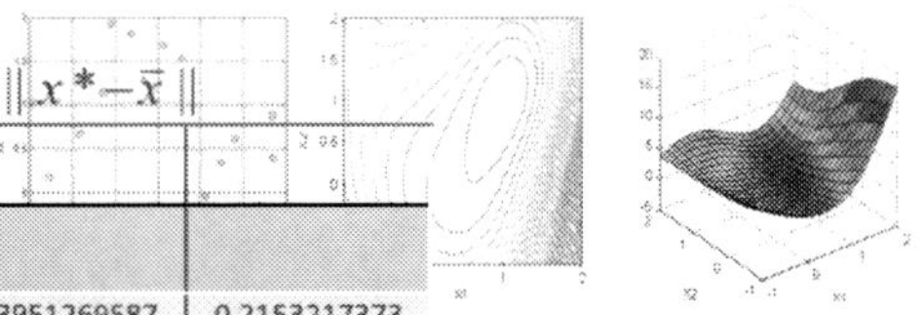

	X1	X2	Objective	$\|\nabla f(\bar{x})\|$	$\|x^* - \bar{x}\|$
True	1.00000000000	1.00000000000	0.00000000000		
LHD_1	1.10716717618	1.18675722495	-0.10930406390	0.3951269587	0.2153217373
LHD_2	1.01665842326	0.98976525241	-0.23645212712	0.2289955014	0.0195511641
LHD_3	1.00082903764	0.99789183742	3.860241e-004	0.0183556123	0.0022646742

Quality of space-filling

To guarantee RM to have fully linear property in the domain, it is demand that a squared ball centered any point in the domain enclosure at least n+1 sampled points

Higher quality of space filling is desired.

Translation propagation LHD is introduced

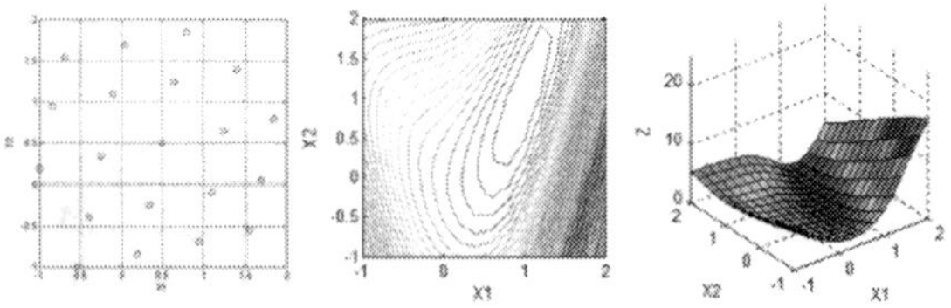

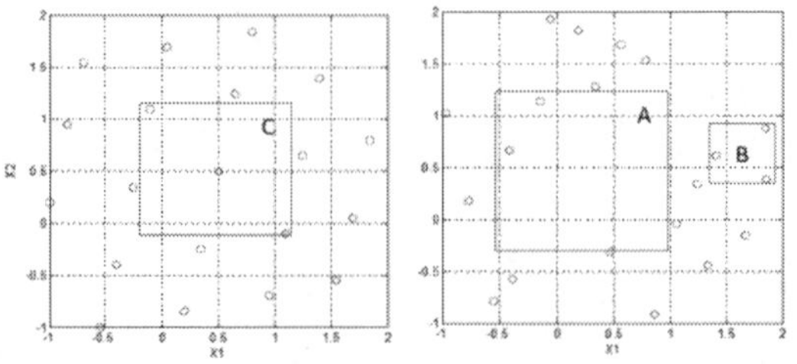

Two stage RM optimization

2-stage Algorithm

•Build RM1 within entire Ω1
•Solve NLP with RM1 to find x~.
•Form a neighborhood Ω2 around x~
•Build RM2 in Ω2.
•Solve new NLP to fine x*

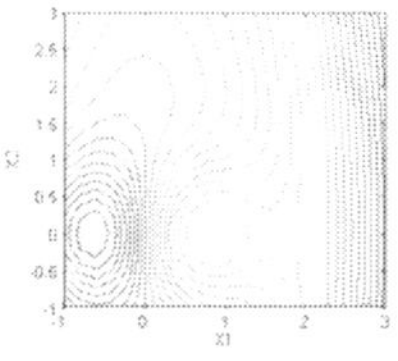

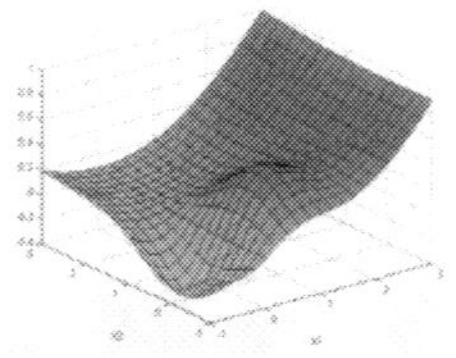

Stg	Nx	$\bar{\Delta}$	Minimum	Mean	maximum
1	25	0.6928	3.0522065965905E-004	0.3310714789671	13.125077612243
2	20	0.0775	3.9696247145388e-007	5.6403536627973e-005	8.38888715061664e-004

		X1	X2	Objective	$\lVert \nabla f(\bar{x}) \rVert$	$\lVert x^* - \bar{x} \rVert$
True		1.00000000000	1.00000000000	0.00000000000		
Stage	1	1.00170291094	1.004458261939	-0.31274748886	2.2461787E-003	4.7724213E-003
	2	1.00003013513	1.00005631658	-0.23645212712	7.6499096E-005	6.3872387E-005

Example 2: Rosenbrock Function

Challenge: very flat in the area of minimum

2-Stage algorithm solves it well

		X1	X2	Objective	$\lVert \nabla f(\bar{x}) \rVert$	$\lVert x^* - \bar{x} \rVert$
True		1.00000000000	1.00000000000	0.00000000000		
Stage	1	1.00170291094	1.004458261939	-0.31274748886	2.2461787E-003	4.7724213E-003
	2	1.00003013513	1.00005631658	5.0187520e-008	7.6499096E-005	6.3872387E-005

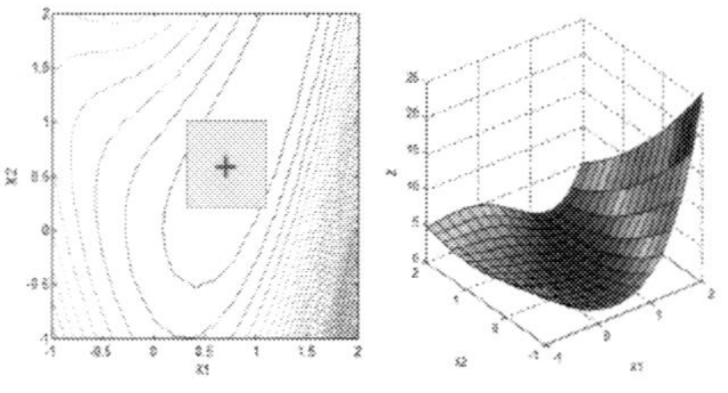

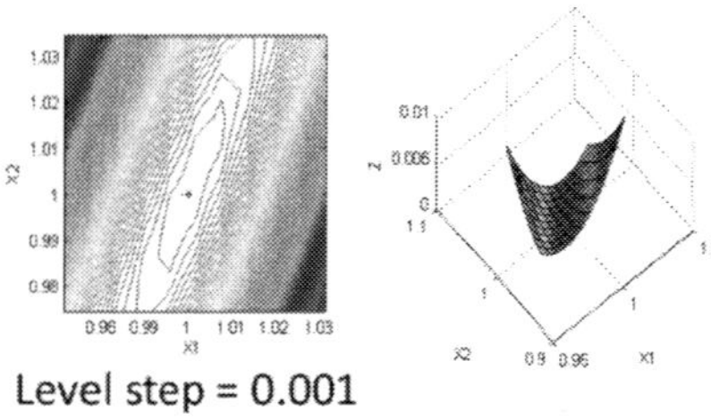

Level step = 0.001

Conclusion and further work

- To build a qualified RM, we proposed a set of procedures of development method.
- Theoretical derivations for constrained optimization offer support to the practices with RM
- The built RM maintains fully linear properties within the domain. --- its errors are bounded
- The solution with ε -Exact reduced model is close to its true which can be evaluated based on ε
- In future, this work will be expanded to flowsheet optimization for energy system.

Acknowledgements

Support

US Department of Energy (NETL)

Through project CCSI

(627a) Complementarity Formulations for the Nonlinear Model Predictive Control of Non-Smooth Systems

Ajit Gopalakrishnan and Lorenz Biegler, Chemical Engineering, Carnegie Mellon University, Pittsburgh, PA

When addressing the optimal control of systems governed by differential equations, an implicit assumption is that the differential equations that describe the evolution of these systems are continuous, and differentiable. However, dynamic process models are frequently subject to switches and other non-smooth elements. Examples include relief valves, check valves, phase changes, flow reversals, piecewise smooth correlations and so on. Such 'switching systems' fall under the umbrella of hybrid dynamic systems, with both continuous dynamics and discrete events embedded in them. The discrete events are autonomous to the system, and arise when the state trajectory intersects a switching surface (event trigger), causing the underlying differential equations to change (mode switch).

The focus of this work is on computing both open-loop and closed-loop optimal control solutions for such switching systems. We focus on the simultaneous method to transcribe the optimal control problem into an NLP, through a full discretization of states and controls. We represent the switching events through complementarity conditions, which retain the NLP nature of the optimal control problem without introducing binary variables. We modify the simultaneous method by introducing the switching time as a decision variable. This allows us to locate the switching event accurately at a finite element boundary resulting in an accurate computation of both the states and gradients of the optimal control problem. We also extend the proposed method to a Nonlinear Model Predictive Control framework, which involves a closed-loop solution of the above dynamic optimization problem. Several examples are presented to highlight the efficacy and computational performance of the suggested approach.

References:

[1] Baumrucker, B. T. and L.T. Biegler, "MPEC Strategies for Optimization of Hybrid Dynamic Systems," Journal of Process Control, 19, 1248-1256 (2009)

(627b) Optimal Design and Control of Batch Distillation Columns

Jonathan Moreno, Department of Chemical Engineering, Universidad de los Andes, Bogotá, Colombia and Jorge Mario Gómez, Chemical Engineering, Universidad de los Andes, Bogotá, Colombia

Batch distillation is maybe the oldest separation process used for separate liquid mixtures. It is used for the output in small and big scale of chemical products with high quality. The flexibility of the production arrangements can also cope with the fluctuations or rapid changes in demand. This operation is widely used for the production of fine chemicals and specialized products such as alcoholic beverages, essential oils, perfumes, pharmaceutical and petroleum products. [1]

The main advantages of batch distillation over a continuous distillation lie in the use of a single column as opposed to multiple columns and its flexible operation. Moreover, only one batch distillation column and only one sequence of operation are necessary to separate all the components in a mixture. [1]

Given the transient nature of this operation, to describe its behavior, a dynamic model is taken into account according to rigorous model (MESH) that includes energy and mass balance, equilibrium relationships and summation of composition. On the other hand, there are conventional and unconventional configurations, between the last ones are the inverted batch distillation column, the middle vessel batch distillation column, the multi-vessel batch distillation column, but for this work a conventional configuration is selected.

After implementation of the dynamic simulation without binaries variables done in *GAMS*, profiles obtained are compared with profiles extracted by simulation results in *Aspen Batch Distillation®*, in order to check the model used.

The optimization of batch distillation has been intensively studied in the literature, where overall objective is to determine the optimal strategy based on a given objective function and satisfy several constraints. In particular, the goal is to determine the optimal reflux policy to obtain a specified quality of product. In the literature are three categories, the problem of maximize distillate, the problem of minimize time and the problem of maximize profit. [2] In addition, an optimal design had been integrated with the optimal operation; where those issues are optimized simultaneously.

In this work a methodology of optimal design and control in batch distillation column is presented in order to maximize an economic objective function, which binaries variables are taken into account. That is done using a mixed integer dynamic optimization strategy [3], in special a MINLP solver in *GAMS*is used, which all continuous variables are discretized in order to obtain a simultaneous approach categorized as large-scale MINLP problem. [4] In addition, orthogonal collocation is implemented as discretization method, based on the work done by Hendro et al. [5] where they allude about the stability and accuracy of results so as improve to computational efficiency significantly.

Once the optimal policies are obtained, these are tested with post-optimal analysis which basically consists of a sensitivity analysis when profiles are manipulated. Finally, a case of study of hydrocarbons mixture is presented in order to illustrate the methodology presented in this work.

References:

[1] I. M. Mujtaba, Batch Distillation: Design and Operation, London, UK.: Imperial College Press, 2004.

[2] A. Arias Barreto, I. Rodriguez-Donnis, V. Gerbaud y X. Joulia, «Optimization of Heterogeneous Batch Extractive Distillation,» *Industrial & Engineering Chemistry Research,* vol. 50, 2011.

[3] B. Chachuat, A. B. Singer y P. I. Barto, «Global Mixed-Integer Dynamic Optimization,» *AIChE JOURNAL,* vol.

51, p. 2235–2253, 3 June 2005.

[4] L. T. Biegler, «An overview of simultaneous strategies for dynamic optimization,» *Chemical Engineering & Processing,* vol. 46, pp. 1043-1053, 2007.

[5] P. L. Hendro, P. Hoo y G. Wozny, «Efficient simulation of Batch distillation Process by Using Orthogonal Collocation,» *Chem. Eng. Technol.,* vol. 21, nº 11, pp. 853-862, 1998.

Derivation of the Optimal Control of an Extractive Distillation Unit in the Production Process of Fuel Grade Ethanol.

Manuel A. Ramos, Jorge M. Gómez*

Grupo de Diseño de Productos y Procesos, Departamento de Ingeniería Química, Universidad de los Andes, Carrera 1 No. 18ª-10, Bogotá, Colombia

Key words: Extractive Distillation, Optimal Control, Fuel Grade Ethanol, Orthogonal Collocation, Non Linear Programming.

1. Introduction

Recently, a research presented by Garcia-Herreros et al. [1] proposed the design and operating conditions that maximizes a profit function for the extractive distillation of fuel grade ethanol using glycerol as entrainer. That process is aimed to offer the greatest economic benefit in stationary conditions. However, to maintain the optimal operating conditions of the process time invariant parameters are required. In practice, distillation processes are subjected to disturbances that cannot be avoided and adversely affect normal operation.

In order to establish the usefulness of the proposed process for the industrial production of fuel grade ethanol, it is necessary to analyze its dynamic behavior and its controllability. The dynamic stability of the extractive distillation might represent a forceful advantage over the azeotropic distillation with benzene [2], the traditional method for the production of fuel grade ethanol. The azeotropic distillation has proved to have high parametric sensitivity and the presence of multiple steady states [3] - [4], which often implies low ethanol recovery [5].

The use of dynamic optimization for analyzing the control possibilities allows identifying the optimal process behavior and selecting the best control strategy based on quantitative criteria. It implies finding the variables profiles that minimize an objective function in a system subjected to disturbances. The optimal control problem is solved by discretizing the time domain through orthogonal collocation on finite elements [6] , which converts it into a large-scale, non-convex, Non-Linear Programming (NLP) problem.

2. Model of the Extraction Column

The separation of the ethanol-water mixture beyond the azeotropic composition is achieved by extractive distillation whose general representation is shown in Figure 1. The addition of an entrainer such as glycerol modifies the relative volatilities of the components present in the mixture. It allows obtaining fuel grade ethanol as distillate and a water-glycerol mixture as bottom product.

The process is modeled as a series of counter-current separation stages. At each stage, liquid and vapor flows get in contact in order to reach thermodynamic equilibrium, as shown in Figure 1. The mathematical model of the dynamic behavior of the distillation column is based on the MESH equations. The process is described by total and partial mass balances (M), thermodynamic equilibrium - equations (E), mole fraction summations (S) and energy balances (H).

*To whom correspondence should be addressed. E-mail: ma.ramos43@uniandes.edu.co

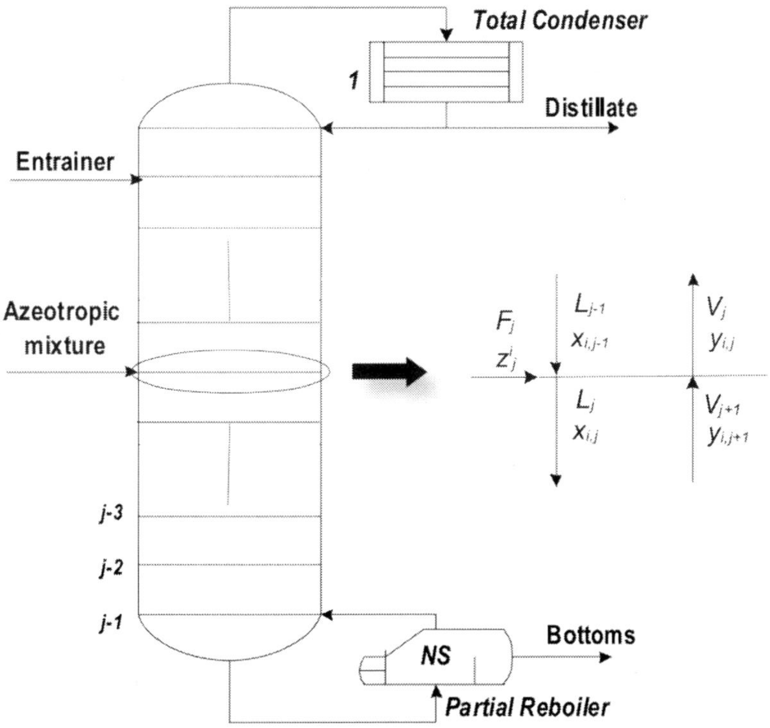

Figure 1. Extractive distillation column.

2.1. *Operating conditions*

The extraction column is made up by a total condenser (stage 1) and 18 theoretical stages, including a partial reboiler (stage 19). The azeotropic mixture is fed into stage 12 at a rate of 100 kmol/h and a temperature of 293.15 K. Pure glycerol is fed into stage 3 at a rate of 35 kmol/h and a temperature of 305 K. The extraction column operates at atmospheric pressure (101 kPa). These operating conditions are the same reported by Garcia – Herreros et al. [1]. The azeotropic mixture composition is subjected to two different disturbances in order to obtain the optimal control profiles for both scenarios.

2.2. *Discretization of the dynamic model.*

In order to accomplish the simulation and optimization of the dynamic model of the extractive distillation column, the DAE system is discretized. The state and control variable profiles are approximated into a family of orthogonal polynomials on finite elements [3], [6–8] ; it allows solving the complete formulation simultaneously as a NLP problem. The time horizon is divided into *i* finite elements. In this case, 3-point Radau collocation was selected because of its compatibility with NLP formulations and their high stability with truncation errors of order $O(h^{2K-1})$ [7]. This approach allows to make a *"direct transcription"* of the optimization problem and to solve it simultaneously in an Equation Oriented (EO) environment. Moreover, it maintains accuracy of the state and control profiles and allows the consideration of unstable systems, path constraints and singular controls [6]·[7]. It is important to remark that this approach leads to a large – scale sparse NLP formulation, that requires flexible decomposition strategies to solve the problem efficiently. The modeling was carried out in *GAMS 23.8,* and solved using *IPOPT* [9].

3. Formulation of the optimization problem.

3.1. Objective Function

The criterion for the derivation of the optimal control for the extractive distillation column taking into account the profit of the process had the aim of minimizing the off-spec products while maximizing the profit of the process. The objective function has therefore two main elements: the first one is the optimal control structure and the second one is the profit.

$$\min_{\substack{RR(t)\\ Q_R(t)}} \quad J=\int_0^{t_f}\begin{bmatrix}(\alpha_1(x_{d,eth}{}^{t}-x_{d,eth}{}^{sp})^2+\alpha_2(Q_R{}^{t}-Q_R{}^{sp})^2+\alpha_3(R_R{}^{t}-R_R{}^{sp})^2\\ -NP_t\end{bmatrix}dt \qquad \text{Eq. 1}$$

The first three terms of Eq. 1 stand for the optimal control problem. The aim of these three terms is to represent the quadratic error between the desired set point (e.g. $x_{d,eth}{}^{sp}$) of the variable and its nominal value (steady state value) in time (e.g. $x_{d,eth}{}^{t}$). The first term belongs to the controlled variable, the ethanol molar composition in distillate. This expresses the aim to minimize off-spec products during the time of perturbation. Terms two and three belong to control variables: the net heat duty of the reboiler and the molar reflux ratio of the condenser. These two terms are included into the objective function in order to make the solution as smooth as possible [3],[10],[11]. The terms $\alpha_1, \alpha_2, \alpha_3$ are weight parameters associated with ethanol molar composition, reboiler duty and molar reflux ratio. These parameters were calculated iteratively (and offline). The last term in the objective function is the net profit. The cost parameters of this term of the objective function are taken as market values, according to the ones reported by Garcia-Herreros et al. [1].

3.2. Optimization Variables

As it was said earlier, the variable to be controlled is the ethanol molar composition of distillate. The control variables are the reboiler duty and the molar reflux ratio of condenser. All variables have a trajectory through time that is represented by their values at each collocation point in each finite element. Therefore, the solution of the optimal control problem requires finding the trajectory of these variables. Initial values of the state and control variables are also considered decision variables of the optimization problem.

3.3. Constraints

The equality constraints of the optimal control problem are the collocation equations of the dynamic version of the MESH model.

As path operational constraints the derivation of the optimal control for the extractive distillation column has the following inequalities [1]:

- minimum purity of ethanol produced at each time step:

$$x_{d,eth}{}^{t} \geq 0.995, \ \forall t \in [0, t_f] \qquad \text{Eq. 2}$$

- maximum operating temperature of the column in order to avoid glycerol decomposition temperature:

$$T_s \leq 555K, \ \forall s \in [1, NS], \forall t \in [0, t_f] \qquad \text{Eq. 3}$$

3.4. Perturbations

Two different types of perturbations are taken into account: a sinusoidal one, and a step-like perturbation. Both are shown in Figure 2. It is important to note that the sinusoidal perturbation provides an ethanol molar composition above the azeotropic boundary (a condition that cannot occur in real world operation), because the present work wants to show optimal control in extreme operating conditions.

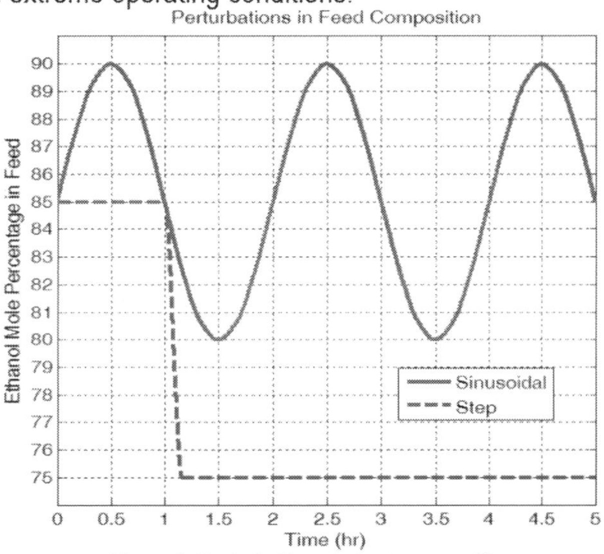

Figure 2. Perturbations in feed composition.

4. Solution Strategy.

The problem was modeled and solved in GAMS 23.8 on a quad core Intel i5 2.7 GHz CPU with 8 GB of RAM. IPOPT [9] was chosen as the NLP solution algorithm, because of its tested advantages for the solution of Large-scale NLP's and optimal control problems [7]·[9]. IPOPT was configured to run with MA57 and MA86 (w/MC77 ordering option) linear solvers in order to improve performance [12]. It was found that MA86 linear solver was by far faster than MA57, because it can run in parallel and is more recently developed [12]. Dramatic changes in solution times were observed with MA86, decreasing from 1320 CPUs (MA57) to 87 CPUs (MA86) in some cases. Table 1 shows the different CPU times observed with different linear solvers. The optimization problem contained 30792 variables and 30731 constraints (31 inequalities, 30700 equalities).

Table 1. Comparison of solution times with different linear solvers.

	MA57		MA86 (w/MC77)	
Optimization Problem	**Average CPUs**	**Average iterations**	**Average CPUs**	**Average iterations**
Step perturbation	1023	326	80	245
Sinusoidal perturbation	1329	338	87	269

For the initialization of the optimization problem, the steady-state values of the variables at every time step were chosen. For this purpose, three different initializations were used. Steady state values were obtained by carrying out steady state simulations with different values of reflux ratio and reboiler duty, obtaining temperature, molar composition (liquid and vapor) and liquid and vapor molar flow profiles. Running the NLP problem with 3 different initializations also contributes in finding different local optima [3].

5. Results and discussion.

The comparative profiles of the different perturbations are shown in figures Figure 3, Figure 4. It can be seen that ethanol composition in distillate is controlled by changing control variables. In both cases the objective function optimal value is negative, indicating the reliability of the overall controlled process.

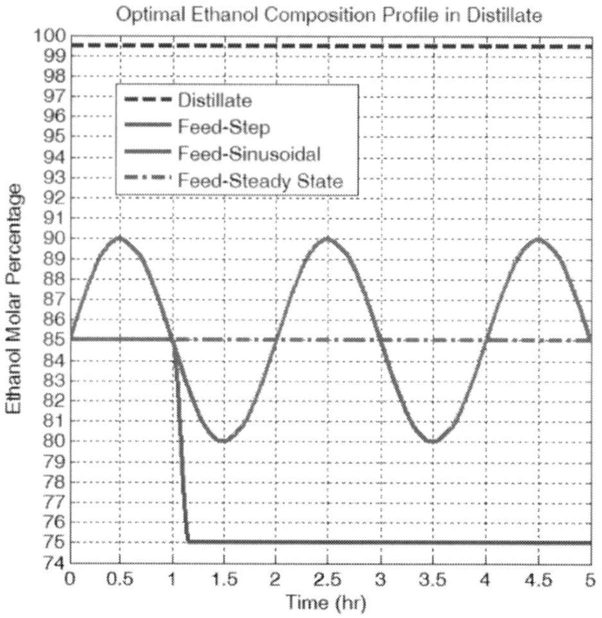

Figure 3. Optimal profile of ethanol composition in distillate.

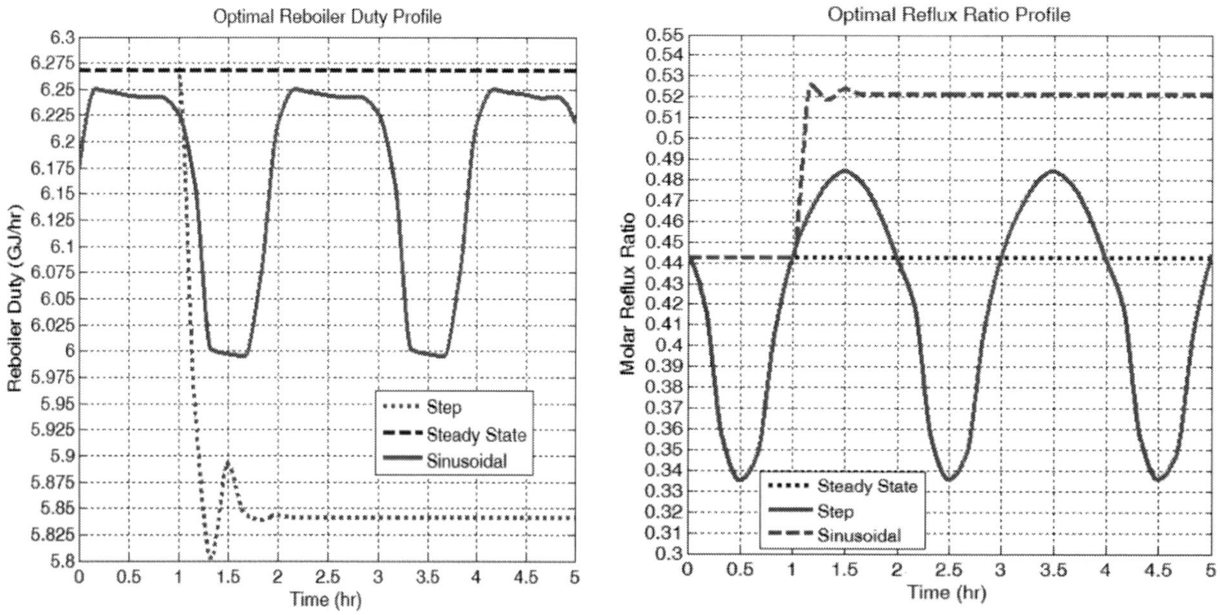

Figure 4. Optimal reboiler duty profile (left) and optimal reflux ratio profile (right).

The latter figures show the optimal behavior of the decision variables. As it can be observed, ethanol composition in distillate can be maintained constant despite the perturbation in feed. It is important to remark that control variables respond in a way consistent to perturbation in feed and manage to keep the system in the required operation point and to minimize off-spec product. It is important to remark that the reboiler duty profile in steady state differs from the sinusoidal perturbation case initial point. This result is consistent with the inclusion of the steady state variables into the optimization problem, on the framework of the multiplicity of the steady states of the extractive distillation process [4], [13]. The optimal point is reached changing the initial steady state value of reboiler duty. All unscaled objective function values (at the optimal point) are negative, indicating the economic reliability of the overall controlled process.

Table 2. Objective function value for the three optimization problems.

Optimization Problem	Complete Objective Function Value /5 hr of operation
Steady State	$-1.2733 \cdot 10^4$
Step perturbation	$-1.1494 \cdot 10^4$
Sinusoidal perturbation	$-1.2817 \cdot 10^4$

The optimization was performed with the three different initializations as said before in order to increase the probability of finding the global optimum within variable bounds. The results obtained with the three different initializations were exactly the same. This is likely to confirm that at least, if the global optimum has not been found, a very good local optimum has been found.

On the other hand, in order to analyze if there is an improvement on the overall profit with the composite objective function (Eq. 1), an optimization was carried out with a control - only objective function (terms 1, 2 and 3 of Eq. 1) with the sinusoidal perturbation. Table 3 shows that the overall profit (computed according to Eq. 1) is greater in the composite objective function.

Table 3. Comparison of the profit value in different optimization problems.

Objective function	Value (USD/5 hr of operation)
Control – only	9501.88
Composite (sinusoidal)	12845.12

It is also shown the improvement on the overall profit of the profit – control objective function over the control – only one.

6. Conclusions.

The derivation of the optimal control strategy for the extractive distillation of fuel grade ethanol was successfully addressed in this work. Using state of the art algorithms, the solution of this large scale problem can be obtained in matter of minutes. Our experience using different linear solvers with IPOPT shows that MA86 (w/mc77) performs very well and behaves well with the large – scale NLP derived from the optimal control problem.

It is also shown that through optimization it is possible to find the optimal control strategy of such a change-prone system as extractive distillation. The solution of the three different problems addressed in this research confirms this statement.

It is also remarkable that the controllability of the extractive distillation process was confirmed, showing that the designed process proposed by Garcia – Herreros et al. [1] is not only stable and operable in steady state operation but also subjected to dynamic disturbances.

This kind of optimization problem has multiple local minima due to non-convexities in the model, although IPOPT in most of the cases was able to find the local optima inside the feasible region. This is also confirmed by solving each problem with 3 different initializations. The results obtained show an economically profitable process even if the process itself is subjected to strong perturbations in its operation conditions, as the ones proposed in this research. As future work, it is proposed the simultaneous design and control of the extractive distillation column and the insertion of uncertainty parameters to the optimization problem.

References.

[1] P. García-Herreros, J. M. Gómez, I. D. Gil, and G. Rodríguez, "Optimization of the Design and Operation of an Extractive Distillation System for the Production of Fuel Grade Ethanol Using Glycerol as Entrainer," *Industrial & Engineering Chemistry Research*, vol. 50, pp. 3977–3985, 2011.

[2] Yasuki Kansha, Naoki Tsuru, Chihiro Fushimi, and Atsushi Tsutsumi, "New Design Methodology Based on Self-Heat Recuperation for Production by Azeotropic Distillation," *Energy Fuels*, vol. 24, no. 11, pp. 6099–6102, 2010.

[3] R. López-Negrete de la Fuente and A. Flores-Tlacuahuac, "Integrated Design and Control Using a Simultaneous Mixed-Integer Dynamic Optimization Approach," *Ind. Eng. Chem. Res.*, vol. 48, no. 4, pp. 1933–1943, 2009.

[4] K. Esbjerg, T. R. Andersen, D. Müller, W. Marquardt, and S. B. Jørgensen, "Multiple Steady States in Heterogeneous Azeotropic Distillation Sequences," *Ind. Eng. Chem. Res.*, vol. 37, no. 11, pp. 4434–4452, 1998.

[5] G. J. Prokopakis and W. D. Seider, "Feasible Specifications in Azeotropic Distillation," *AIChE Journal*, vol. 29, no. 1, pp. 49–60, Jan. 1983.

[6] J. E. Cuthrell and L. T. Biegler, "On the Optimization of Differential - Algebraic Systems of Equations in Chemical Engineering," Carnegie Mellon University, Carnegie Mellon University, 1986., Pittsburgh, PA, 1986.

[7] L. T. Biegler, *Nonlinear Programming: Concepts, Algorithms, and Applications to Chemical Processes*. Philadelphia: Society for Industrial and Applied Mathematics : Mathematical Optimization Society, 2010.

[8] U. Diwekar, *Batch Distillation: Simulation, Optimal Design, and Control*, Second ed. CRC Press, 2012.

[9] A. Wächter and L. T. Biegler, "On the Implementation of an Interior-Point Filter Line-Search Algorithm for Large-Scale Nonlinear Programming," *Mathematical Programming*, vol. 106, pp. 25–57, Apr. 2005.

[10] A. Ghaee, R. S.- Gharebagh, and N. Mostoufi, "Dynamic Optimization of the Benzene Extractive Distillation Unit," *Brazilian Journal of Chemical Engineering*, vol. 25, no. 04, pp. 765–776, 2008.

[11] M. Miranda, J. M. Reneaume, X. Meyer, M. Meyer, and F. Szigeti, "Integrating Process Design and Control: An Application of Optimal Control to Chemical Processes," *Chemical Engineering and Processing: Process Intensification*, vol. 47, no. 11, pp. 2004–2018, Oct. 2008.

[12] HSL, "A Collection of Fortran Codes for Large Scale Scientific Computation." 2011.

[13] M. Rovaglio and M. F. Doherty, "Dynamics of Heterogeneous Azeotropic Distillation Columns," *AIChE Journal*, vol. 36, no. 1, pp. 39–52, Jan. 1990.

(627d) Parametric Sensitivities of Dynamic Systems with Linear Programs Embedded

Kai Hoeffner, Process System Engineering Laboratory, Department of Chemical Engineering, Massachusetts Institute of Technology, Cambridge, MA, Stuart Harwood, Process Systems Engineering Laboratory, Deptartment of Chemical Engineering, Massachusetts Institute of Technology, Cambridge, MA and Paul I. Barton, Chemical Engineering, Massachusetts Institute of Technology, Cambridge, MA

Microorganisms are the main driving force in industrial processes such as biofuel production and pharmaceutical biotechnology. Hence, dynamic simulations of the microorganisms and their interactions with the bioreactor environment are important for design, optimization, and control of these processes. As a structured model for processes involving microorganisms, dynamic flux balance analysis (DFBA) combines the flux balance analysis model (Palsson, 2006) used in metabolic engineering with the dynamic mass balances of the reactor. The mathematical formulation is given by dynamic systems with linear programs (LPs) embedded. Sensitivity analysis for this class of systems is not only essential for simulation and optimization, but also of theoretical interest as they represent a class nonsmooth dynamic systems.

In this work, we develop first results on the nonsmooth parametric sensitivity analysis for dynamic systems with LPs embedded. Nonsmoothness is due to the nonsmooth dependence of the LP solution with respect to the problem data. In particular, the solution is a Lipschitz continuous function of the right-hand side of the equality constraints. Hence, the sensitivities are possibly set-valued, and this has to be accounted for during the integration and optimization.

The theoretical results are derived from parametric linear programming, sensitivity analysis of linear programs, and nonsmooth analysis. The dual problem of the embedded LP plays an important role since the dual solution set characterizes the subgradient of the LP solution map.

In the numerical implementation, the system is rewritten as a hybrid differential-algebraic equation (DAE) system such that the embedded LP only has to be solved at points along a trajectory where the optimal basis set, determined by the simplex algorithm, changes. A code, called DSL48LPR, has been developed which employs DAEPACK (Tolsma and Barton, 2000) component DSL48E, and is a numerical integrator for a dynamical system with LPs embedded, where the dynamic states determine the right-hand side of the equality constraints. The results are illustrated in a case study of ethanol production using genetically modified yeast and E. coli microorganisms.

Palsson, B., 2006. Systems Biology: Properties of Reconstructed Networks. Cambridge University Press, New York, NY.

Tolsma, J., Barton, P. I., 2000. DAEPACK: An open modeling environment for legacy models. Industrial & Engineering Chemistry Research 39 (6), 1826{1839.

(627e) Sensitivity-Based Output Feedback in Dynamic Optimization

Vinicius de Oliveira, Chemical Engineering, Norwegian University of Science and Technology, Trondheim, Norway, Sigurd Skogestad, Department of Chemical Engineering, Norwegian Univeristy of Science and Technology, Trondheim, Norway and Johannes Jäschke, Norwegian University of Science and Technology, Trondheim, Norway

We consider batch process optimization and robust implementation of optimal control policies. The dynamic optimization of such processes is in most cases model based, and therefore subject to uncertainties. This may lead to sub-optimal control trajectories with significant economic losses.

Our goal is the development of a simple strategy which guarantees near-optimal operation under all conditions. Here, 'under all conditions' means for the defined disturbances, plant changes and implementation errors.

The main idea is to apply sensitivity-based output feedback to update the control inputs using the deviation of the measurements (outputs) from the nominal optimal trajectory. In this approach, fast updates can be computed online without the need of reoptimization when disturbances occur. Similar idea has been presented in [1]. However, no systematic guidelines are given for the computation of such controller.

The use of sensitivity information to update nominal input has been recently used in the context of nonlinear model predictive control (see [2]) and dynamic real time optimization (see [3]). In both cases, however, the disturbances need to be measured or estimated.

We show here that an output feedback controller can be easily obtained from the optimal sensitivity of outputs and inputs provided there are enough independent measurements. This controller uses directly the process outputs and no state or disturbance estimation is required. The resulting closed-loop satisfies the necessary conditions of optimality to first order and is exact in the linear case. Interestingly, if the measurements outnumber the disturbances, there will be extra degrees of freedom that can be used to mitigate the effect of noise.

The proposed method was tested in two case studies with nonlinear dynamics, namely, a fed-batch reactor with uncertain kinetic parameters and a high-order continuous stirred-tank reactor where the initial states are disturbed. In both cases the controller was able to recover almost completely the economical loss due to small perturbations when compared to a nominal open-loop approach.

The method is simple and intuitive and yet has shown surprisingly good performance for relevant industrial processes.

References

[1] Z. K. Nagy and R. D. Braatz. 'Open-loop and closed-loop robust optimal control of batch processes using distributional and worst-case analysis". Journal of Process Control, vol.14, 411-422, 2004.

[2] V. M. Zavala and L. T. Biegler, "The advanced-step nmpc controller: Optimality, stability and robustness," Automatica, vol. 45, pp. 86-93, 2009.

[3] L. Wrth, R. Hannemann, and W. Marquardt, "Neighboring-extremal updates for nonlinear model-predictive control and dynamic real-time optimization," Journal of Process Control, vol. 19, pp. 1277 -1288, 2009.

(627f) Parallel Decomposition of Nonlinear Dynamic Optimization Problems

Jia Kang[1], Daniel Word[1], Johan Akesson[2] and Carl D. Laird[1], (1)Chemical Engineering, Texas A&M University, College Station, TX, (2)Department of Automatic Control, Lund University, Lund, Sweden

Problems with block structure arise in a number of areas, including dynamic optimization, parameter estimation, and nonlinear stochastic programming. While nonlinear programming has proven to be an effective tool for tackling these problems, the size of these problems can become prohibitively large for serial approaches on modern computer workstations due to both efficiency concerns and memory constraints. To address these issues, tools and algorithms have been developed to exploit the block structure of these problems and solve the NLP in parallel.

The simultaneous approach for solution of dynamic optimization problems discretizes all the variables and includes the discretized model as constraints in a large-scale optimization problem. These problems are very large, but they are sparse and inherently structured as a result of the discretization. The inherent structure of these problems allows for problem decomposition to solve the linear KKT systems in parallel. Interior-point methods provide a framework for efficient solution of these problems since the linear systems that are solved to find the step maintain a consistent structure from iteration to iteration.

The KKT system solved at each inner iteration of the interior-point method has a block structure with coupling induced by the continuity equations resulting from the discretization. Through the addition of coupling variables, a Schur-complement decomposition approach can be used that decouples individual KKT blocks associated with portions of the time horizon. The individual KKT blocks have a structure that is consistent with the overall structure of the dynamic optimization problem. The Schur-complement can be efficiently formed in parallel and then solved in serial to find the step in the coupling variables and multipliers. The remaining step variables can be solved in parallel. In this strategy, the size of the Schur-complement increases with the number of processors used. Therefore, there is a trade-off between parallel speedup and the time required to factorize and solve the Schur-complement in serial. The approach shows very favorable speedup for problems with few state variables relative to the number of algebraic variables.

This parallel algorithm is currently being integrated with the existing Modelica/Optimica modeling framework. In this presentation, we will show timing results for constructed dynamic optimization problems to show the parallel speedup as a function of the ratio of state to algebraic variables. In addition, we will discuss current research into the use of iterative linear solvers for solution of the Schur-complement system.

(629a) IDEAS Minimization of Entropy Generation for Isothermal/Isobaric/Constant Density Reactor Networks with Bounded Utility Consumption

Paul Ghougassian, Chemical Engineering, UCLA, Los Angeles, CA and Vasilios Manousiouthakis, Chemical & Biomolecular Engineering Department,, University of California Los Angeles, Los Angeles, Los Angeles, CA

In this paper, a methodology for the globally optimal synthesis of isothermal/isobaric/constant density reactor networks featuring minimum entropy generation and bounded utility consumption is presented. The proposed methodology employs the Infinite DimEnsionAl State-space (IDEAS) conceptual framework, which is shown to be applicable to the problem under consideration. The IDEAS framework considers all possible reactor units, and all possible mixing and splitting interconnections among them. The resulting mathematical formulation is an infinite-dimensional program with linear constraints and a possibly non-linear nonconvex objective function to be minimized. It will be shown mathematically that, under certain conditions, entropy generation and utility consumption are functions of only the inlet and outlet stream compositions and flowrates and do not depend on the reactor network structure, as long as there exists a network able to deliver the considered outlets from the known inlets. This theoretical result provides the foundation for a graphical method that can identify the global optimum of the considered optimization, by first quantifying the reactor network's Attainable Region (AR) and then depicting the behavior of the entropy generation and utility consumption functions within the AR. The proposed method is demonstrated on a case study. Finally, conclusions are drawn.

(629b) A New Approach to the Use of Exergy Analysis in Synthesis and Design - Focus On Low Temperature Processes

Truls Gundersen and Danahe Marmolejo-Correa, Energy and Process Engineering, Norwegian University of Science and Technology, Trondheim, Norway

A new Approach to the Use of Exergy Analysis in Synthesis and Design – Focus on Low Temperature Processes

Truls Gundersen[1], Danahe Marmolejo Correa[2]

Department of Energy and Process Engineering, Norwegian University of Science and Technology, Kolbjoern Hejes v 1B; NO-7491, Trondheim, Norway.

[1]corresponding author: truls.gundersen@ntnu.no;

[2]danahe.marmolejo@gmail.com.

Abstract

This paper presents a summary of some recent discoveries and developments related to the use of Exergy Analysis for the synthesis and design of industrial production plants with special emphasis on sub-ambient processes. The key elements of the paper are (1) the importance of decomposing certain exergy forms into their exergy components, (2) the importance of decomposing thermo-mechanical exergy at ambient temperature, (3) the systematic definition of exergy sources and exergy sinks for use in exergy efficiencies, (4) the establishment of linear Exergy diagrams using basic thermodynamic equations and simplifying assumptions, (5) the concept of targeting for Exergy similar to Energy targeting in Pinch Analysis, and (6) the use of Exergy as a design tool in conceptual process development. Since Exergy is defined as the maximum work that can be produced when a system reaches equilibrium with its environment through reversible processes, the concept of Exergy is particularly useful for sub-ambient processes. Here, external cooling is provided by refrigeration cycles, where the cooling duty is produced by a sequence of expansion and compression. This requires work or mechanical energy, which is pure exergy. The paper will use a very small example to illustrate the basic concepts, and the Reverse Brayton process for liquefaction of natural gas as an industrial case study.

Background

Despite a rich literature on Exergy (a term that was "coined" by Rant in 1956) in the last decades, and the fact that Exergy is closely related to the 2nd Law of Thermodynamics (often formulated by the use of the logical counterpart Entropy), the use of Exergy in the process industries has been rather limited. One contributing factor to this situation can have been the lack of standardization in the field of Exergy. Different symbols and different names have been used across disciplines (such as chemical versus mechanical engineering) and across continents (such as Europe versus the US). This lack of standardization also includes issues of classification, reference conditions and exergy efficiency definitions. Another contributing factor can have been the fact that Exergy considerations in quite a few cases are in conflict with economic considerations. In the process industries, the bottom line is always economy, expressed either as maximum annual profit or minimum total annualized cost. Since, however, there are considerable uncertainties in both operating cost and investment cost, using measures such as Exergy at least has the advantage of developing a "sound" design from a thermodynamic point of view. For sub-ambient process, exergy losses translate into compression work, which again translates directly into economy.

When using the Exergy concept for sub-ambient processes such as liquefaction of natural gas (LNG) and separation of air by cryogenic separation (ASU), more limitations have been encountered (Marmolejo-Correa and Gundersen, 2011) in the available literature on Exergy Analysis that limit the ability to properly evaluate the quality of process designs. Some of the proposed exergy efficiencies are unable to account for the fact that sub-ambient processes utilize the option to switch between two components of thermo-mechanical exergy, i.e. the temperature based part and the pressure based part (Marmolejo-Correa and Gundersen, 2012a). By expansion in a turbine below ambient temperature, pressure based exergy can be transformed into temperature based exergy (refrigeration or cold exergy) and work. Similarly, compression can be used to "store" work (i.e. exergy) in the form of pressure based exergy. A very important part of this picture is the fact that temperature based exergy for a material stream as well as the exergy of heat exhibits a remarkable behavior below ambient temperature. The exergy of heat is always positive, and when passing ambient temperature from above to below, there is a discontinuity at ambient temperature. Temperature based exergy for a material stream is also always positive, and even though there is no discontinuity at ambient temperature, the exergy value of the stream increases more rapidly per degree Celsius or Kelvin when the temperature is reduced below ambient compared to increasing the temperature above ambient.

This is why decomposition of exergy into different components and to decompose exergy evaluations at ambient temperature are important for understanding the behavior of certain process equipment as well as for the development of exergy efficiencies that properly measures the quality of process designs.

Contributions and Conclusions

The paper describes some modest contributions to the field of Exergy Analysis with emphasis on sub-ambient processes. The issues of standardization, classification and decomposition were discussed in the Background section. Based on insight developed by studying the behavior of exergy components and certain pieces of equipment above and below ambient temperature, a new general exergy efficiency definition called Exergy Transfer Effectiveness (ETE) has been developed (Marmolejo-Correa and Gundersen, 2012c) that focus on exergy change and exergy transformation in processes. Changes in exergy, supply or removal of exergy and transformation of exergy are carefully divided into Exergy Sources and Exergy Sinks. A set of rules has been developed to make a proper classification into sources and sinks.

Existing Exergy diagrams (Linnhoff and Dhole, 1992) for overall processes have been developed from temperature vs. enthalpy diagrams in Pinch Analysis (i.e. from Composite and Grand Composite Curves) by replacing temperature with Carnot factor as the y-axis (i.e. Exergy Composite Curves and the Exergy Grand Composite Curve). These diagrams have a number of shortcomings. First, in their construction, a considerable number of points have to be calculated due to the nonlinear shape of the curves (the Carnot factor is a non-linear function of temperature). Then, in the interpretation of the curves, targets for maximum exergy recovery, minimum exergy losses, etc. are not readily available from the curves. Exergy losses can be measured in a cumbersome way geometrically by measuring the area between the Exergy Composite Curves.

Using basic thermodynamic equations and simplifying assumptions, new Exergetic Temperatures have been proposed (Marmolejo-Correa and Gundersen, 2012b) that can be used to develop (piece-wise) linear Exergy Diagrams that are (1) easy to construct from the process stream data, and (2) provide easily and explicitly various targets for Exergy such as maximum exergy recovery, minimum exergy losses, minimum exergy requirement, and minimum exergy rejection. These diagrams enable the use of Exergy Analysis in early stages of process design, while currently it is used as a post-design measure of quality. More specifically, the paper will discuss its use in combination with the ExPAnD methodology (Aspelund et al., 2007, and Wechsung et al., 2011).

There is considerable scope for increased use of Exergy in analysis, synthesis and design of processing plants; however, there is a lack of expertise in industry and a lack of suitable software for application in industrial environments. The fact that Exergy Analysis must be used with care, since reducing exergy losses caused by irreversibilities often is in conflict with investment cost minimization, may also have contributed to its rather limited industrial use in the past.

References

ADDIN EN.REFLIST

Aspelund, A., Berstad, D.O., Gundersen, T. (2007). An extended pinch analysis and design procedure utilizing pressure based exergy for subambient cooling. *Applied Thermal Engineering*, **27**(16), 2633-2649.

Linnhoff, B., Dhole, V.R. (1992). Shaftwork targets for low-temperature process design. *Chem. Engng. Sci.*, **47**(8), 2081-2091.

Marmolejo-Correa, D., Gundersen, T. (2011). Low temperature process design: Challenges and approaches for using exergy efficiencies, in Pistikopoulos, E.N. et al. (editors). Proceedings of the 21st European Symposium on Computer Aided Process Engineering - ESCAPE 21, Thessaloniki, Greece, vol. 2, pp. 1909-13.

Marmolejo-Correa, D., Gundersen, T. (2012a). A comparison of exergy efficiency definitions with focus on low temperature processes. In review for *Energy*.

Marmolejo-Correa, D., Gundersen, T. (2012b). A new graphical representation of exergy applied to low temperature process design, in Karimi I.A., Srinivasan, R. (editors). Proceedings of the 11th International Symposium on Process Systems Engineering - PSE 2012, Singapore.

Marmolejo-Correa, D., Gundersen, T. (2012c). Exergy transfer effectiveness for low temperature processes. To be submitted to *International Journal of Thermodynamics*.

Rant, Z. (1956). Exergy, a new word for technical work potential. *Forsch. Ing. Wes.*, **22**, 36-37 (in German).

Wechsung, A., Aspelund, A., Gundersen, T. (2011). Synthesis of Heat Exchanger Networks at Sub-ambient Conditions with Compression and Expansion of Process Streams. *AIChE Journal*, **57**(8), 2090-2108.

(629c) Retrofit of Heat Exchanger Networks Using Superstructure Modeling and Global Optimization

Sung Young Kim, Debora Faria and Miguel Bagajewicz, Chemical Engineering, University of Oklahoma, Norman, OK

Retrofit of Heat Exchanger Networks using superstructure modeling and Global Optimization

Sung Young Kim, Debora Faria and *Miguel Bagajewicz

School of Chemical Engineering and Material Science, University of Oklahoma

100 East Boyd Street, T-335 – Norman, OK 73019-0628 USA

** Corresponding Author*

Abstract

In this paper, we present a robust global optimization method to design heat exchanger networks using superstructures and a recently developed method of domain and imgae partitioning of nonconvex functions. In previous work, we applied this method to the Synheat model. When using superstructures bilinear terms are also used, so we extend the bound contraction model to include these. The procedure we propose uses an MILP lower bound constructed using domain/image partitioning (Faria, Kim and Bagajewicz, 2012) and bilinear decomposition (Faria and Bagajewicz, 2011a,b). Then a newly developed bound contraction procedure is applied and compared to branch and bound as well as branch and bound with bound contraction at each node. We finally solve the retrofit problem, by proposing that certain heat exchangers in the superstructure exist, and new exchangers can be added.

1. D. Faria and M. Bagajewicz. A New Approach for Global Optimization of a Class of MINLP Problems with Applications to Water Management and Pooling Problems. AIChE Journal (2011a).
2. D. Faria and M. Bagajewicz. Global Optimization of Water Management Problems Using Linear Relaxations and Bound Contraction Methods. Industrial and Engineering Chemistry Research. 50, 7, 3738–3753 (2011b).
3. D. Faria, S.Y. Kim and M. Bagajewicz. Global Optimization of Nonconvex MINLP Problems by Domain and Image Partitioning : Applications to Heat Exchanger Network. Computers and Chemical Engineering, Submitted (2012)

A NEW APPROACH TO HEAT EXCHANGER NETWORK SYNTHESIS PROBLEM

Abstract ID# 279571

Işıl KİRKİZOĞLU[1], *Sinan GÜREL, Haldun SÜRAL*

Industrial Engineering Department, Middle East Technical University, 06800, Ankara, TURKEY

Introduction

Heat exchanger networks (HENs) bring several fluid streams into their desired temperatures by using available heat in the process. Since water and energy that are wasted to produce utility streams cause environmental and economic costs, processing a fluid to be heated with the one to be cooled is better than producing utility streams for efficient usage of energy.

In this study, we present a new solution approach for the HEN synthesis problem. Our aim is to provide a complete network design coupled with a detailed equipment design for heat exchangers (HEs). The approach can generate all possible HE alternatives instead of structuring the HEN at the very beginning as usually done in the literature (see Yee and Grossmann, 1990). Alternatives are generated systematically by discretizing the inlet and outlet temperatures for hot and cold stream pairs due to heat load equalities for a given solution quality. A mixed-integer nonlinear programming (MINLP) model is solved to design each HE alternative in detail. The model decides on the number of shell-and-tube passes and tubes, diameter of shell, inner/outer tube diameters, and tube pattern while minimizing HE area or the total construction and operation cost.

The Yee-Grossmann approach (Yee and Grossmann, 1990), based on developing a MINLP formulation for a proposed superstructure of the design of network, dominates the HEN design literature since 1990 (see Furman and Sahinidis, 2002; Morar and Agachi, 2010; and references therein). Related studies that keep the superstructure mostly focus on reviewing viability of the formulation assumptions, improving the methodology for solving the MINLP, and/or extending the framework. Mizutani *et al.* (2003), an example of those who extend the Yee-Grossmann approach, combine detailed HE design with the HEN superstructure. Our study differentiates from previous studies in the literature because we produce a detailed design for every heat exchange network alternative and evaluate possible alternatives on a shortest-path network problem whose each path represents a HEN design alternative. Considering all thermo-physical and transport properties, the heat transfer coefficients are calculated for every stream at the environment temperature and pressure. One can also

[1] Correspondence address: isil.kirkizoglu@metu.edu.tr

control the solution quality by deciding on the minimum temperature difference and heat load between inlet and outlet streams as the main parameters.

Problem Formulation and Solution Approach

The HEN synthesis problem is to achieve maximum heat transfer between streams and to use utility streams minimally. The decisions are:

- Which cold and hot streams meet in each HE?
- How much heat will transfer each HE?
- What should be the design properties for each HE?
- How many HEs should be in the network?
- How should HEs be arranged in the network?

The main idea behind the solution approach is to generate all possible HE alternatives, each of which can be represented by a node on a network. We call a stream 'hot (cold) stream' if that would be cooled (heated) during the process. 'System inlet (target) temperature' refers to the temperature of a stream at the beginning (after being processed) in the system. Start and end nodes of the network represent the initial and target status of the streams, respectively. Arcs between node pairs are generated by considering the inlet and outlet temperatures and energy requirements of the streams and have associated costs. The minimum-cost path on the network gives the final HEN design (Figure-1).

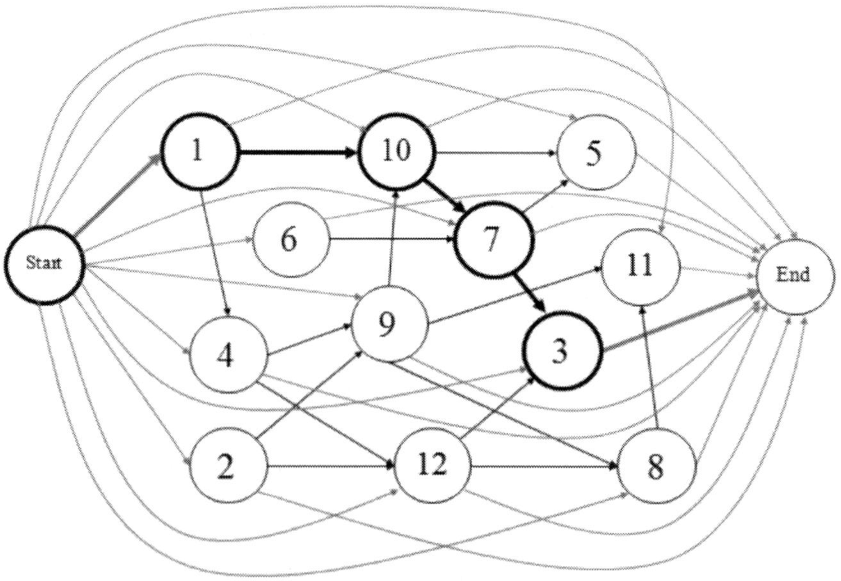

Figure-1. Representation of the network and the selected HEN shown in bold.

Note that nodes (i.e. alternatives HE designs) are generated for all hot and cold stream pairs and for every step change of their temperatures. For example, let us consider the example given in Yee and Grossmann (1990). There are two hot (H1, H2) and two cold (C1, C2) streams and their system inlet and target temperatures are given. If the minimum heat load difference ΔQ is selected as 300 kW, by discretization, the number of steps (β) between system inlet and target temperatures is set as represented in Table-1. After that, from the step heat loads, the step temperature differences (α) are calculated accordingly. Table-2 shows some possible alternatives of HEs for this example.

For every pair of hot and cold streams (i.e., alternative HE), we introduce a design error, γ, that defines the solution quality. Let us consider the example given in Table-1. Every step of C1 corresponds to 287.5 kW where each step of the other streams is 300 kW. For a HE alternative of H1 and C1 with one step change, the design error will be $\gamma = (300 - 287.5)/300 = 4.167\%$, which implies that the energy balance in the HE is established with this amount of an error. Hence, γ reflects the quality of the solution; where lower γ values imply better solution quality.

At each node on the graph, we keep the temperature information for all streams so that we can keep track of them along the network. Figure-2 shows what data is stored at a node if there are m hot streams (i=1,...,m) and n cold streams (j=1,...,n).

Table-1. Data for a HEN synthesis problem instance

	System inlet (°C)	Target (°C)	Fcp (kW/°C)	Total T diff. (°C)	Total Q diff. (kW)	ΔQ (kW)	A (°C)	B	$\Delta Q - \Delta Q^{max}$ (kW)
H1	170	60	30	110	3300	300	10	11	0
H2	150	30	15	120	1800	300	20	6	0
C1	20	135	20	115	2300	287.5	14.375	8	12.5
C2	80	140	40	60	2400	300	7.5	8	0

Table-2. Several HE alternatives generated for the example at Table-1.

Hot stream	Cold stream	Hot inlet	Cold inlet	Hot outlet	Cold outlet
H1	C1	11	6	5	0
H1	C1	9	4	7	2
H1	C2	8	5	3	0
H2	C1	6	8	0	2
H2	C2	4	7	1	4

The solution approach consists of 5 main steps summarized below:

a) ***Node Generation***: Generate all necessary nodes by pairing all possible discretized inlet and outlet temperatures for every stream in the system. The discretization of temperatures would be performed as explained above.

	Inlet temperatures of hot streams					Outlet temperatures of hot streams				
Hot	$T_1^{h(in)}$	...	$T_i^{h(in)}$	...	$T_I^{h(in)}$	$T_1^{h(out)}$	...	$T_i^{h(out)}$	...	$T_I^{h(out)}$
i										
	Inlet temperatures of cold streams					Outlet temperatures of cold streams				
Cold	$T_1^{c(in)}$	...	$T_j^{c(in)}$	...	$T_J^{c(in)}$	$T_1^{c(out)}$	...	$T_j^{c(out)}$	...	$T_J^{c(out)}$
j										

Figure-2. Node content of the network.

b) ***Eliminating Infeasible Nodes:*** Calculate the thermo-physical properties for every processed stream for each HE. Note that at a node (HE), the inlet temperature of a processed hot stream should be higher than its outlet temperature, and inverse is also true for a processed cold stream. For each unprocessed stream, the outlet temperature should be equal to its inlet temperature. Accordingly, eliminate the node if the temperature feasibility and energy balance is not satisfied. The heat load required by the cold fluid should be equal to the heat load generated by the hot fluid.

c) ***Individual HE Design:*** For every node in the network, solve a cost-minimizing MINLP model for the detailed design of shell-and-tube heat exchangers with TEMA standards. The MINLP model consists of necessary design (Geankoplis, 2003) and cost equations for each HE (i.e. node). If the model returns infeasible, eliminate that HE alternative.

d) ***Connecting Nodes***

i. *Dummy nodes:* There are two dummy nodes (Start, End) in the network. Every node v should be directly connected to the start and end dummy nodes. If the inlet (outlet) temperatures of streams are not at their system inlet (target) temperatures, they are connected to the start (end) node by using hot/cold utilities (such as cold water for cooling and superheated steam for heating). The utility usage and HE costs are associated to the arc between nodes and dummy nodes. If the streams are at the system inlet temperature, the cost of an arc from the starting node to that node should be 0. Similarly, when the streams reach their target temperatures, the cost of an arc from that node to the ending node is zero.

ii. *Connecting HE Nodes:* For any two nodes v and w, if outlet temperatures of all streams in node v is equal to the inlet temperatures of all streams in node w, arc (v,w) is generated. The cost associated to that arc is the capital cost of HE in node w.

e) ***Shortest Path:*** Solve the shortest-path problem on the whole network to find the best HEN with detailed design.

Computational Results and Discussion

For implementation of our solution approach we use MATLAB. For solving the MINLP models, MATLAB calls GAMS/BARON solver. We solve the shortest-path problem by using the Dijkstra's algorithm. Tests are done with two different PC settings: Intel® Core™ 2 Duo, 1.30 GHz, 3.50 GB RAM and Intel® Core™ i5-2400, 3.10 GHz, 6.0 GB RAM.

Some preliminary experiments are performed for determining the solution parameters and their behavior. As expected, it is seen that better results are obtained when the step heat load difference is decreased (i.e., the number of steps is increased) for each stream. But generally, increasing number of steps results in computational difficulties due to technical limitations related with increasing CPU time and memory requirements. Consequently, the number of steps for each stream should be arranged by taking this trade-off into account.

To test the solution approach, 12 test instances from the literature are solved. As it is explained earlier, our solution approach includes detailed design for every HE in the network. However, the test instances available in literature that solve for detailed HE designs in a network setting are very limited. Therefore, we also tested our approach on problems without detailed design, as well.

The approach we suggest is designed to handle all the physical properties of all streams, namely specific heats, thermal conductivities, viscosities and densities, at their environment temperature. In order to be able to identify and use all these properties at the substance's related temperature level, the composition of each stream (i.e., molar or mass fractions of substances within the stream) should be known. However, the composition of streams is usually not provided for the test instances in the literature. Therefore, we use the average values given for the physical properties. In addition, for the case that no detailed design is included, the overall heat transfer coefficients are also provided for the streams, which are used as an input to the model, rather than a decision variable.

Unfortunately, every instance has a different cost function (both capital investment and utility costs), overall heat transfer coefficient calculations (for instances that do not consider detailed

design of every HE), and utility environments. These changes are implemented for every instance. More experiments are done for some instances in order to find better solutions. Such experiments are developed by decreasing heat load difference ΔQ.
In Table-3, the comparison of our results that are obtained with the current best solution setup (discretization configuration) with the results in literature is given. The properties of the instances are also shown. The number of hot and cold streams explains the problem size. The stated notes show which features are allowed in the literature for the given instance.

The HENS problem has a multiobjective aspect. The heat exchange area, hence the capital cost of HEs, and the utility usage are tried to be minimized. This is why different measures can be compared to show the performance of the solution approach. For the comparison of results, the performance measures are the number of process and utility HEs and their areas, total heat exchange area, operating cost (cost of utility usage), total capital cost and the total cost. Percentage improvements are given for these measures. Positive values show the enhancement in results.

According to the results given in Table-3, our solution approach performs better for the cases where a detailed design is done. Heat exchange areas are mostly improved. Increase in usage of utilities are due to the discretization effect. Of course these results can be improved by increasing the number of steps for each stream, thus a greater portion of the solution space is explored. Note that stream splitting and placing utility HEs between process HEs as done by Ponce-Ortega et al. (2010) are not allowed in our approach. Even so, it is also seen from the results that the performance of the solution approach is comparable for the instances where detailed design is not considered.

Conclusion

In this study, we suggest a new solution approach to the heat exchanger network problem with detailed equipment design. By its nature, it is a complex and hard-to solve problem. Our approach solves the problem using discretization. It provides us to easily change and add features to the equipment design part without a need to change the entire HEN solution procedure.

Our approach is flexible and successfully finds the required number of heat exchangers and their connections. Note that the decision maker can change the parameter ΔQ to manage the HEN synthesis achieved at the end. As many solutions for HEN are created, decision makers can select any other design among these solutions according to his/her multiple preferences.

References

Furman, K.C. and Sahinidis, N.V. (2002). A critical review and annotated bibliography for heat exchanger network synthesis in the 20th century. *Ind. Eng. Chem. Res.*, 41(10), 2335-2370.

Geankoplis, C.J. (2003). Transport Processes and Separation Process Principles (4th ed.). New Jersey: Prentice Hall.

Gupta, A. and Ghosh, P. (2010). "A randomized algorithm for the efficient synthesis of heat exchanger networks". *Comput. Chem. Eng.*, 34(10), 1632-1639.

Huang, S.Y. and Chang, C.T. (2012). "An improved mathematical programming model for generating optimal heat-exchanger network (HEN) designs". *Ind. Eng. Chem. Res.*, 51(8), 3508-3515.

Mizutani, F.T., Pessoa, F.L.P., Queiroz, E.M., Hauan, S. and Grossmann, I.E. (2003), "Mathematical programming model for heat exchanger network synthesis including detailed heat-exchanger designs. 2. Network synthesis", *Ind. Eng. Chem. Res.*, 42(17), 4019-4027.

Morar, M. and Agachi, P.S. (2010). Review: Important contributions in development and improvement of the heat integration techniques. *Comput. Chem. Eng.*, 34(8), 1171-1179.

Ponce-Ortega, J.M., Serna-Gonzalez, M. and Jimenez-Gutierrez, A. (2007). "Heat exchanger network synthesis including detailed heat exchanger design using genetic algorithms". *Ind. Eng. Chem. Res.*, 46(25), 8767-8780.

Ponce-Ortega, J.M., Serna-Gonzalez, M. and Jimenez-Gutierrez, A. (2010). "Synthesis of heat exchanger networks with optimal placement of multiple utilities". *Ind. Eng. Chem. Res.*, 49(6), 2849-2856.

Turton, R., Bailie, R.C., Whiting, W.B. and Shaeiwitz, J.A. (2003). Analysis, Synthesis, and Design of Chemical Processes (2nd ed.). New Jersey: Prentice Hall.

Yee, T.F. and Grossmann, I.E. (1990). Simultaneous optimization models for heat integration – II. Heat exchanger network synthesis. *Comput. Chem. Eng.*, 14(10), 1165-1184.

Table-3. Computational results – % Improvement

Instance	Size	Notes	Current best solution setting			# of Process HEs	# of Utility HEs	Total Process HE Area (%)	Total Utility HE Area (%)	Total Area (%)	Total Utility (kW)	Total Capital Cost (%)	Utility Operating Cost (%)	Total Annual Cost (%)
			Hot streams	Cold streams	ΔQ (kW)									
HC-E1[a](R1)	2H2C	SS	14-22	18-10	200	same	same	-	-	8.90	-4.24	4.58	-9.47	-2.25
HC-E2[a] (R1)	1H2C	SS	70	26-40	30	same	1	-	-	13.46	-300.0	-3.62	-750.00	-40.70
HC-E3[a] (R1)	1H2C	SS	22	18-4	90	same	2	-	-	17.11	-	-8.89	-	-17.49
HC-E1[a] (R2)*	2H2C	SS	14-22	18-10	200	same	-1	-	-	8.90	-3.63	5.85	-8.38	-1.05
HC-E2[a] (R2)*	1H2C	SS	70	26-40	30	same	same	-	-	2.18	-3.45	10.03	-3.91	6.26%
HC-E3[a] (R2)*	1H2C	SS	22	18-4	90	same	same	-	-	-2.48	13.46	0.81	14.10	1.92%
YG-E1[b]	2H2C	-	33-18	23-24	100	same	1	-165.3	-6.26	-	62.33	-71.81	39.33	-13.98
YG-E2[b]	2H2C	-	12-16	18-15	150	-1	same	46.01	-107.24	-	-110.7	19.37	-87.95	6.92%
YG-E3[b]	5H1C	SS	9-3-5-3-6	55	120	same	1	41.01	-16.61	-	-21.17	14.35	-12.68	-9.88
PO-E2[c]	2H1C	MU - SS	50-47	75	16	-1	-1	-5.80	22.86	-	8.19	43.18	-124.80	-3.77
PO-E3[c]	2H3C	MU - SS	9-14	12-6-6	1150	2	-1	6.50	2.31	-	-6.45	-0.50	-21.86	-11.56
MG-E1[d]	2H2C	DD	4-10	24-5	100	same	same	67.24	0.00	-	0.00	10.12	0.00	0.59
4S1[e]	2H2C	DD	4-8	9-4	300	-3	same	65.08	-	-	-32.74	87.83	-46.24	10.51
Chp13-E2[f]	3H3C	-	12-2-5	5-8-6	100	-2	same	11.41	-4.96	-	-26.67	19.33	-3.81	13.14
5SP[g]	3H2C	SS	10-2-7	5-15	19	-2	2	-	-	-	-39.70	7.74	-38.63	-6.55

[a]: Huang & Chang, 2012
[b]: Yee & Grossmann, 1990
[c]: Ponce-Ortega et al., 2010
[d]: Mizutani et al., 2003
[e]: Ponce-Ortega et al., 2007
[f]: Turton et al., 2003
[g]: Gupta & Ghosh, 2010

*Result by using Yee and Grossmann (1990) methodology

SS: Stream splitting allowed
DD: Detailed design
MU: Multiutility + intermediate utility HE

(629e) Analysis of Different Sampling Methods for the Simulation-Based Optimization Approach to Model Feedstock Development for Chemical Process Industry

Ismail Fahmi, Chemical Engineering, The University of Tulsa, Tulsa, OK and Selen Cremaschi, Department of Chemical Engineering, University of Tulsa, Tulsa, OK

Analysis of Different Sampling Methods for the Simulation-based Optimization Approach to Model Feedstock Development for Chemical Process Industry

Ismail Fahmi, Selen Cremaschi

Department of Chemical Engineering, The University of Tulsa, 800 South Tucker Drive, Tulsa, Oklahoma 74104, USA

10C04: Energy and Sustainability in Operations

Abstract

Chemical Process Industry (CPI) has been highly dependent on the fossil-based materials as the feedstock. As the reserve for this non-renewable resource depletes, alternative resources need to be discovered. Among many options, biomass is considered very promising because it is renewable, locally available, and abundant. However, incorporating biomass as a feedstock for CPI requires significant amounts of investments for both research and development (R&D) and production capacity expansions. How these investments will shape the evolution of the biomass to commodity chemicals (BTCC) system should be investigated. Moreover, the uncertain decision-dependent endogenous technology evolutions of the BTCC system complicate the analysis. Fahmi and Cremaschi [1] have presented a prototype simulation-based optimization (SIMOPT) approach to study the feedstock development for CPI. The SIMOPT framework explained in [1] uses the concept of generating multiple unique timelines to represent various controlled evolution of the system under uncertainties. The SIMOPT framework [1] uses deterministic optimization with stochastic simulation in an integrated fashion to study the impact of uncertainties on the system performance. Deterministic optimization is used to produce the optimum decision set for the system at its current state. The optimization formulation for the BTCC investment planning used in the SIMOPT framework is explained in detail in [2]. Stochastic simulation is used to predict the system performance under uncertainties. In a single SIMOPT run, the simulation proceeds through the time steps based on the decision set determined by the deterministic optimization. Because of the accumulated differences between the expected and the realized behavior of the system, the system performance under the original decisions may no longer be applicable or feasible at one point in time. This condition is called a trigger event. In case of a trigger event, the simulation is halted and deterministic optimization is recalled to determine a new set of optimal decisions for the remainder of the timeline. The control is passed back to simulation along with the new set of decisions. The iteration between the deterministic optimization and stochastic simulation continues until the final time step. After the completion of the final step, a unique controlled evolution of the system performance under uncertainties (i.e., one unique timeline) is generated. Multiple unique timelines are produced because of different realizations of the uncertain parameters. To have a thorough analysis of the BTCC system performance accounting for different uncertainties, a statistically significant number of timelines are required, i.e. the quality of the results depends on the quality of the uncertain variables' space coverage. This can simply be done by gathering a virtually infinite number of timelines. However, such an approach would be prohibitively expensive in computational costs. Therefore, the minimum number of timelines that will achieve preset statistical significance in the results should be determined. In this work, a systematic analysis using different sampling methods to cover the uncertain parameters space of the BTCC investment planning problem is presented. The sampling methods considered in this work are Monte Carlo Sampling, Latin Hypercube Sampling, Univariate Dimension Reduction, Hammersley Sampling, and Halton Sampling. The performance metrics used for the comparison of these methods are the stabilization of four statistical moments (i.e., mean, variance, skewness, and kurtosis) of the objective and the overall objective distribution shape using the Smirnov test. A BTCC system of ethylene production is presented as the case study.

Keywords: BTCC investment planning, simulation-based optimization, sampling methods comparison

References:

[1] I. Fahmi and S. Cremaschi, "A Prototype Simulation-based Optimization Approach to Model Feedstock Development for Chemical Process Industry," *22nd European Symposium on Computer Aided Process Engineering,* 2012.

[2] I. Fahmi and S. Cremaschi, "Stage-gate Representation of Feedstock Development for Chemical Process Industry," *Foundations of Computer-Aided Process Operations 2012,* vol. , 2011.

(629f) Strategic Model with Transmission Network Design for the Energy Supply Chain Optimization in Petroleum Industries

Ariel Uribe[1], Ivan Ordoñez Sr.[1], Raul Rodriguez Peña Sr.[2] and Omar J. Guerra[3], (1)Colombian Petroleum Institute, ECOPETROL S.A., Piedecuesta, Colombia, (2)Colombian Petroleum Institute (ICP), ECOPETROL S.A, Piedecuesta, Colombia, (3)Colombian Petroleum Institute (ICP), ECOPETROL S.A., Piedecuesta, Colombia

Strategic model with transmission network design for the energy supply chain optimization in petroleum industries

Nowadays, energy management has become an increasingly important issue in each consumption sector (commercial, industrial, residential and transportation sectors). It is basically because energy plays an important role in every human activity and the need for more environment friendly energy resources (less greenhouse gas (GHG) emissions) like biomass, sun and wind power, etc has increased. In the particular case of oil and gas supply chain (production, transformation and transportation), the consumption of energy is estimated to approximately 10 % of gross oil and gas production (IPIECA 2007). About 90 % (48 % natural gas and 42 % oil) of the primary energy used by this industry takes the form of oil and gas, as supplies are available on site and are typically the cheapest source of energy. Some of the oil and gas used directly by the industry is transformed into electricity and heat, especially in the case of oil refineries and other large facilities. In total, about 10 % of the electricity and heat consumed by the industry is supplied from the grid (IPIECA 2007).

In this work, a strategic model for the energy supply chain design and optimization in oil industries is developed. The model considers GHG emissions, renewable energy sources energy efficiency, generation based on Gas, electricity purchase in a highly dependable market of the hydropower (64%), reliability and security constraints in the electric energy supply. The model was applied for the energy planning in an oil field production facility (GEC) at ECOPETROL. This facility include: Apiay field (4 active production fields and 48 oil wells), Castilla field (3 active fields and 126 oil wells), Chichimene field (2 active fields and 56 oil wells), Libertad field (3 active fields and 5 oil wells), Suria field (9 active field and 44 oil wells).

Due to an increased in oil production at GEC, an expansion plan in the energy supply system at GEC should be consider, this expansion plan include energy generation, transmission, distribution and final users. In the current situation at GEC, most of the energy requirements are provided by self-generation with two thermal plants (TermoOcoa (47 Mw) and TermoSuria (47 Mw)). The remaining of the energy requirements is purchased from an external energy supply system SIN (Interconnected National System) at 115 kV. The energy transmission take place at 115 kV with two distribution systems 115/34,5 kV – CDC and CDS. The distribution system for oil fields, facilities and oil Wells is at 34,5 kV.

The analysis is performed for a planning horizon of 5 -10 years, in the analyzed scenarios has an estimated growth of the demand from 120 Mw in 2012 to almost 220Mw in 2015, this huge growth is driven by an intensive campaign of development of all areas at GEC, which is estimated to grow from a production of 147 BOPD in 2012 to more than 300 kbopd in the next years, with an associated production of water slightly above from 400 kbwpd. For the energy supply the combining of different sources was analyzed: SIN and Self-generation (gas, oil, deasphalted, and other renewable alternative sources), in order to ensure energy supply required with an adequate reliability of the system so timely and cost-competitive, thus ensuring the expected volumes.

[3] IPIECA. 2007. "The global oil and gas industry association for environmental and social issues." (http://www.ipieca.org/sites/default/files/publications/Saving_Energy.pdf).

(647a) Design of a Parameter and State Estimation Method for Detecting Local Concentration On the Surface of a Carbon-Nanotube Based Sensor

Hong Jang, Biomolecular and Chemical Engineering, Korea Advanced Institute of Sciencen and Technology, Daejeon, South Korea, Kwang-Ki K. Kim, Department of Chemical Engineering, MIT, Cambridge, MA, Jay H. Lee, Chemical and Biomolecular Engineering, Korea Advanced Institute of Science and Technology, Daejeon, South Korea and Richard D. Braatz, Department of Chemical Engineering, Massachusetts Institute of Technology, Cambridge, MA

Carbon nanotube-based sensors have been developed by using the indirect detection of the adsorption and desorption of surrounding target molecules with light or electronic property changes. The adsorption and desorption phenomena can be modeled by the chemical master equation (CME), which is a system of stochastic ordinary differential equations (ODEs) describing the time evolution of the probability of each possible adsorption state. By discretizing the ODEs in CME with respect to time, standard linear parameter-varying (LPV) state-space system with Gaussian noises is obtained. In this work, an estimation algorithm with two optimization steps for estimating the sets of adsorption and desorption rate constant parameters and the probability states is proposed. The first step is the least-squares (LS) optimization for estimating the set of rate constant parameters, which changes the LPV system into a linear time-invariant (LTI) system. The second step is the full-information batch estimation method for estimating the probability distribution among the possible states in the LTI system. By iterating between the two optimization steps, both the parameter set and the probability states are converged to optimum values, which should closely track the true parameter and state values. The proposed combined parameter/state estimation method is shown to be more practical compared to the previous analytical method which has many limitations in terms of application to real systems.

(647b) Parameter Estimation in Stochastic Kinetic Models

Rishi Srivastava, University of Wisconsin-Madison, Madison, WI and James B. Rawlings, Chemical and Biological Engineering, University of Wisconsin-Madison, Madison, WI

In deterministic settings, we can use a chemical kinetic description in terms of continuous ODE to model a chemically reacting system. However, there are situations where the deterministic description of the chemically reacting systems in terms of continuous ODE model is inaccurate and unsatisfactory. One such situation occurs when some reactant, intermediate, or product species is present in a small concentration (e.g. only $10-1000$ molecules in the whole reactor). Another situation occurs inside a cell or tissue where several proteins/RNAs/DNAs are present in small amount (e.g. $1-100$ molecules). For these situations researchers have developed alternative representation for the chemically reacting system known as stochastic chemical kinetics (SCK). This alternative representation considers reactions as micro scale interaction between molecules. Unlike the continuous ODE model where in a small time instant all reactions occur with their respective rates, the SCK representation considers the probabilities of different reactions within a small amount of time. One of the salient features of SCK description is that in the limit of populations of all species tending to very large values, it converges to continuous ODE model representation.

Even though the SCK description is an accurate way to model a system with small population of some species, estimation of parameters for this description remains still difficult. The reason for the difficulty of the estimation problem is primarily due to two reasons. First, the estimation problem either requires the solution of infinite dimensional chemical master equation governing the probabilistic evolution of the system or it requires too many stochastic simulations of the realization of the system. Second, it is difficult to get accurate gradients and Hessians of the objective function for which we only have an estimator.

In this talk, we present reasons for accommodating measurement noise in experimental data. Then we present a new expression for experimental data likelihood.Next, we show utility of this new likelihood and methods of sensitivity estimation of this likelihood. Finally, we present methods of optimizing this new likelihood expression.

(647c) High Temperature Microreactors with Inherent Dynamic Robustness

Richard Pattison, Chemical Engineering, University of Texas at Austin, Austin, TX and Michael Baldea, University of Texas

High Temperature Microreactors with Inherent Dynamic Robustness

Richard Pattison and Michael Baldea

Department of Chemical Engineering

The University of Texas at Austin, 1 University Station C0400, Austin, TX 78712

email: mbaldea@che.utexas.edu

Hydrogen is produced industrially from light hydrocarbons (notably methane) via steam reforming. In recent years, shale gas production and landfill gas production has caused methane capture to become wide spread and highly decentralized. As a result, there has been a technological push for localized processing of methane for hydrogen production [1]. Catalytic plate microreactors are a promising alternative for low cost, localized production of hydrogen via methane steam reforming. Millimeter channel heights and micron level catalyst thicknesses minimize transport limitations, resulting in a process that is controlled by reaction rates, rather than transport rates. Microreactors also afford the benefit of a greatly simplified scale-up: production rates can be increased by augmenting the number of units. These benefits notwithstanding, the widespread use of microreactors has been hindered by a number of design and operation challenges.

In hydrogen production microreactors, (endothermic) methane steam reforming is supported by the (exothermic) catalytic combustion of methane; the reactions occur in a layered structure with alternating microchannels. *Counter-current*flow designs are advantageous due to simplified reactant distribution and potentially lower feed temperatures. However, in this case, consumption of the reactants in the combustion and reforming channels occurs in opposite spatial directions, and it is difficult to synchronize the respective rates of heat generation and consumption. Uneven heat fluxes can result in reactor extinction or localized "hot spots" that can damage the catalysts. Design modifications have been proposed for making the counter-current design feasible including distributed feed [2], multiple flow passes [3], and an offset catalyst distribution [4].

From an operational perspective, microreactors must be robust and capable of handling frequent demand fluctuations. For example, a rapid drop in reforming flow rate can result in temperature excursions detrimental to the catalysts along the reactor length. The implementation of feedback control mechanisms based on distributed temperature measurement devices and distributed actuators is limited for practical reasons (sensors and actuators cannot be implemented or simply do not exist at such small scales). Consequently, control is limited to a boundary control approach (i.e. manipulating feed flow rates) [5].

In this work, we propose an innovative approach for reactor temperature control based on the use of Phase Change Materials (PCMs). PCMs absorb energy in the form of latent heat when the environment reaches the material melting temperature. Confining a properly selected PCM layer between wall plates separating the reforming and combustion channels thus constitutes a potential means for distributed temperature control: the PCM melting process absorbs thermal energy at constant temperature, preventing the formation of hotspots. Metallic PCMs are also typically highly conductive, having the added benefit of rapidly dissipating thermal energy away from the hotspot. If the thickness of the PCM layer is properly selected, the melting dynamics are fast and represent an effective means for rejecting fast, transient fluctuations in the operation of the reactor. However, the temperature control effect provided by the melting process is limited in time by the quantity of PCM (once the material is melted, the temperature of the system will continue to rise).

In light of the above, we introduce a two-tiered control framework, consisting of i) a PCM layer that acts as a *distributed*local controller to mitigate the thermal effects of fast disturbances, e.g., fluctuations in hydrogen demand, and ii) a supervisory model-based output feedback controller aimed at rejecting the effect of slow, persistent disturbances by manipulating the flow rate to the combustion channel.

We also introduce a novel method for selecting the optimal geometry of the PCM layer (which amounts to "tuning" the fast component in the above hierarchical control framework). The proposed method relies on concepts from nonlinear system identification (notably, the use of multi-level binary sequences) and a recently developed time relaxation-based algorithm to determine the optimal thickness of the PCM layer in the presence of stochastic production disturbances.

The effectiveness of the proposed control scheme is demonstrated via simulations on a 2D first-principles microreactor model. The simulation results show the proposed control structure has excellent performance, completely avoiding the formation of hotspots in the reactor in the presence of significant disturbances in the production rate.

References

[1] N. Muradov, F. Smith, A. T-Raissi. Hydrogen production by catalytic processing of renewable methane-rich gases *Int J Hydrogen Energy*, 33: 2023–2035, 2008.

[2] G. Kolios, A. Gritsch, A. Morillo, U. Tuttlies, J. Bernnat, F. Opferkuch, and G. Eigenberger. Heat integrated reactor concepts for catalytic reforming and automotive exhaust purification. *Appl. Catal. B: Environmental*, 70(1-4):16-30, 2007.

[3] K. Venkataraman, JMRedenius, and LD Schmidt. Millisecond catalytic wall reactors: dehydrogenation of ethane 1. *Chem. Eng. Sci.*,57(13):2335-2343, 2002.

[4] M. Zanfir, M. Baldea, and P. Daoutidis. Optimizing the catalyst distribution for counter-current methane steam reforming in plate reactors. *AIChE J.*, 57(9):2518-2528, 2011.

[5] M. Mendorf, H. Nachtrodt, A. Mescher, A. Ghaini, and D. W. Agar. Design and Control Techniques for the Numbering-up of Capillary Microreactors with Uniform Multiphase Flow Distribution. *Ind. Eng. Chem. Res.*, 49(21):10908-10916, 2010.

(647d) Modeling the Polymeric Microstructure of Ldpe with a Novel Hybrid Simulation Approach

Eric Neuhaus[1], Markus Busch[1], Thomas Herrmann[2], Dieter Lilge[3], Gerd Mannebach[3] and Iakovos Vittorias[3], (1)Ernst-Berl-Institut, TU Darmstadt, Darmstadt, Germany, (2)Ernst-Berl-Institut für Technische und Makromolekulare Chemie, TU Darmstadt, Darmstadt, Germany, (3)Basell Polyolefine GmbH, a LyondellBasell company, R & D, Frankfurt, Germany

A novel hybrid simulation approach is developed, which combines the advantages of deterministic and stochastic modeling for complex polymerization networks. The first field of application is the description of industrial LDPE tubular reactors with peroxide or oxygen initiation.

This hybrid concept is motivated by the fact that the well established modeling techniques such as deterministic simulation handles with respect to microstructural properties mean values (e.g. averaged long and short chain branch densities), but is not able to generate individual macromolecule samples with exact polymeric topology. This type of information can only be derived from the stochastic Monte-Carlo simulation. It plays an important role for the determination of characteristics the control the (application) behavior of the polymeric species. A parameter of interest is e.g. the contraction factor g, which is the ratio of the mean square radius of gyration of a branched and a linear molecule with same molar mass. Furthermore, the access of the exact topology of an exemplary macromolecule allows the description of the topological β-scission, which has a significant effect on the chain-length distribution. Unfortunately, a pure stochastic simulation with sufficient accuracy to microstructural properties, heat and pressure balances requires extreme large molecule ensembles and therefore enormous computational demand. Thus the hybrid concept is based on the idea to use the Monte-Carlo approach as an optional add-on for the computation of the polymeric microstructure based on deterministic simulation of complex reactor operation.

The combined simulation approach is implemented as follows:

The deterministic simulation tool Predici is used to compute the effective reaction rates for each elemental reaction connected to the polymerization of ethylene as a function of the position inside the tubular reactor. In this step also the speed of flow and the probability function for the formation of radicals with the chain-length one are determined. In a second step this data set is used in the stochastic algorithm to simulate a certain number of distinct macromolecules with completely known microstructure one after another. Thus there is no necessity to handle large molecule ensembles. The resulting topology array is used as input for a three dimensional random walker, which computes the contraction factor g of the individual molecule without any model assumption like for example Zimm-Stockmayer-Theory.

The results of this hybrid simulation technique show good agreement with experimental chain-length distributions and contraction factors. Furthermore, the topological correct description of the β-scission in the stochastic approach allows a back coupling and improvement of the deterministic simulation. Applying it without any modification to autoclave reactors tests the validity of the demonstrated concept.

(647e) Multi-Scale Model Assisted Design of Fermentation Strategies for Efficient Microbial Production of Polyhydroxyalkanoates (PHAs)

Christos Chatzidoukas, Department of Chemical Engineering, Aristotle University of Thessaloniki, Thessaloniki, Greece and **Costas Kiparissides**, Department of Chemical Engineering, Aristotle University of Thessaloniki & Chemical Process Engineering Research Institute, Thessaloniki, Greece

Biocatalysis, despite the advantages regarding sustainability and selectivity of product formation, is still not viewed as a first-line alternative, but only as a last resort when other synthetic schemes have failed. The limited production of polymers using non-conventional methods (i.e., use of non-fossil resources and biocatalysts instead of chemical/metal catalysts) is a typical example that highlights the reluctance of the chemical industry to turn to biocatalysis. Thus, biopolymers, either are biodegradable and/or biobased, despite the fact that are broadly recognised for driving the evolution of plastics and contribute significantly to a sustainable society, they have apparently a niche market, considering that they represent only the 1% of the total polymer materials with a global production level of about 250 million tons.

A member of the biopolymers family that is drawing considerable attention is the polyhydroxyalkanoates (PHAs), which are completely biodegradable polymers produced by a wide variety of bacteria through the fermentation of sugars, lipids, alkanes, alkenes and alkanoic acids. PHAs hold the third position in the global ranking of biodegradable polymers with respect to their annual consumption, after the starch-polymers and polylactic acid and they stand out because of their thermoplastic and elastomeric properties and a wide range of physical and mechanical properties which are comparable to the respective one of commercial polymers like polyethylene and polypropylene. Despite the great potential of PHAs to act as alternatives to common plastics derived from petrol, their introduction to the world-wide polymer market is inconsistent with their advantageous characteristics. This is due to the fact that since several biotechnological production strategies of PHAs have barely passed the test of economic viability. The reasoning behind this is the lack of a knowledge-based methodology for the bioproduction of PHAs that will enable the control of production efficiency and product quality. Apparently, little effort has been made over the development of methods for the control of the molecular properties and the optimal fermentation process operation for the commercially efficient production of PHAs with tailored molecular properties, which are important determinants of their respective mechanical properties and end-uses.

It is known that in bacterial cultures, the accumulation of PHAs in the cells strongly depends on the culture nutritional conditions and aeration, affecting accordingly the followed pathways in the metabolic network of the cells. Thus, it is evident that there is inherited complexity over the synthesis of PHAs in microbial cell factories where a number of interactive biochemical phenomena of different length and time scale occurring simultaneously. Therefore in order to deal with such a problem, it is required extensive knowledge with respect to:

i.) the synthesis rates of a number of enzymes and metabolites in the metabolic reaction network and the polymerization kinetic mechanism occurring in the cytoplasm of each individual cell, reflecting micro-scale phenomena;

ii.) the growth, division and death of the cell population, the cell-cell interactions and the swelling of the cells by the accumulation of PHA granules, reflecting meso-scale phenomena and

iii.) the fermentation substrates consumption, the increasing culture density affecting the mixing pattern in the culture and the heat and mass (nutrients, oxygen and CO_2) transfer phenomena in the culture medium, reflecting macro-scale phenomena.

It is understood that the above described cellular activities, that is the substrates consumption and their catabolism by the cells, the cells growth and respiration, the synthesis of products as well as the cells division and death are highly interactive. Hence, the use of individual sub-models focusing on the above phenomena independently can only give results of low accuracy. This nature of the problem calls for an integrated model that combines the individual sub-models in order to ensure bottom-up and top-down flow of information (i.e., from micro-scale to macro-scale model and reversely) resulting in a multi-scale modeling approach of the studied problem. It is the scope o the present work primarily to analyze the model requirements in response to the various problem dimensions. Then a combined metabolic/polymerization/macroscopic model will be developed, accounting for a highly structured biomass, for the accurate simulation and prediction of the culture growth (lag-, exponential- and stationary phase), the PHA production rate and the molecular properties of PHA and thus, for the efficient control and operation of a PHA microbial production process.

(647f) Evaluating Trade-Offs in the Operation of a Semi-Batch Polymerization Reactor

Fady Assassa[1,2], Filip Logist[3], Jan Van Impe[3] and Wolfgang Marquardt[1], (1)Process Systems Engineering, RWTH Aachen University, Aachen, Germany, (2)German Research School for Simulation Sciences GmbH, Jülich, Germany, (3)Chemical Engineering, Katholieke Universiteit Leuven, Leuven, Belgium

Evaluating trade-offs in the operation of a semi-batch polymerization reactor

F. Assassa[a,c], F. Logist[b], J. Van Impe[b], W. Marquardt[a]

[a] *AVT – Process Systems Engineering, Aachener Verfahrenstechnik, RWTH Aachen University, Turmstraße 46, D-52056 Aachen, Germany*

[b] *BioTeC & OPTEC, Department of Chemical Engineering, Katholieke Universiteit Leuven, W. de Croylaan 46, B-3001 Leuven, Belgium*

[c] *German Research School for Simulation Sciences GmbH, 52425 Jülich, Germany*

{Filip.Logist,Jan.VanImpe}@cit.kuleuven.be

{Fady.Assassa,Wolfgang.Marquardt}@avt.rwth-aachen.de

Abstract. Optimizing the design and operation of polymerization reactors often involves multiple and conflicting objectives. In the current study, the trade-off between economic profit and energy consumption is evaluated for the operation of a semi-batch reactor, which produces a co-polymer of styrene and butyl-acrylate. A set of trade-off or so-called Pareto optimal solutions is generated based on a dynamic reactor model. Consequently, we have to solve a multi-objective dynamic optimization problem. A scalarization method (normal boundary intersection) is employed to tackle the multi-objective aspect. This method converts the original multi-objective optimal control problem into a series of single-objective optimal control problems, which are each solved with direct single shooting (adaptive control vector parameterization). For this series of optimization problems, it can be assumed that neighboring values of the scalarization parameter result in neighboring solutions. We exploit this feature by building an accurate approximation for the solution of the first single-objective optimization by grid adaptation and structure detection. The detected structure is repeatedly updated along the Pareto set. The resulting Pareto set as well as the resulting profiles for controls and states are discussed.

Case study.

Process. The case study deals with a semi-batch reactor, which produces a co-polymer of styrene and butyl-acrylate in solution with pentyl acetate. Di-cumyl peroxide is chosen as the initiator. Both monomers and the diluted initiator are fed separately. Their feed flow rates as well as the coolant inlet temperature are time-varying control variables.

Model. Hartwich [1] developed a model of the reactor from kinetic models of [2, 3] including reaction rates, polymer distribution moment equations and mass balances. The energy balance including the cooling jacket has been added. This gives rise to a model with 199 algebraic and 23 differential states.

Decision variables. The feed flow rates of the monomers and the diluted initiator as well as the coolant inlet temperature are decision variables of the optimal control problem. The optimization algorithm tries to minimize the objective function by varying these control variables.

Constraints. Feasible and safe batch operation is ensured by specifying bounds along the batch duration for the reactor content, the reactor temperature set-point, the reactor temperature and the accumulation of monomer. To guarantee production, lower bounds are imposed on the reactor content, the conversion, the amount of polymer and the reactor temperature at the end of the batch. Moreover, pre-specified product quality bounds on, e.g., the weight average molecular weight and the polydispersity, have to be fulfilled at the end time of the batch. The batch duration time is fixed.

Objectives. Two conflicting objectives are evaluated: (i) the profit, i.e., the return due to the product minus the cost for monomers and initiator, and (ii) the energy consumption, i.e., the energy needed to heat up the reactor.

Optimization procedure.

The approach for computing the Pareto set of the multi-objective dynamic optimization problem is similar to [4] and combines (i) the normal boundary intersection (NBI) method [5] for scalarization of the objectives with (ii) control grid adaptation and structure detection techniques [6-8] to efficiently solve the resulting dynamic optimization problem. The approach was implemented in the software package DyOS.

Normal boundary intersection (NBI). Das and Dennis [5] have proposed NBI, a geometrically intuitive approach, to mitigate the drawbacks of the weighted sum approach for scalarization of multiple objectives. NBI first builds a plane in the objective space which contains all convex combinations of the individual minima, i.e., the convex hull of individual minima (CHIM), and then constructs (quasi-)normal lines to this plane. The rationale is that the intersection between the (quasi-)normal from any point on the CHIM, and the boundary of the feasible objective space closest to the utopia point (i.e., the point which contains the minima of the individual objectives) is expected to be Pareto optimal. Hereto, the multi-objective optimization problem is reformulated as to maximize the distance from a point on the CHIM along the quasi-normal through this point, without violating the original constraints.

Grid adaptation and structure detection. The scalarized multi-objective dynamic optimization problem is converted into a nonlinear programming problem (NLP) by adaptive control vector parameterization developed by Schlegel and Marquardt [6]. Here, the control profiles are successively refined employing a wavelet-based analysis of the optimal solution obtained in the preceding optimization step. Furthermore, a control structure detection algorithm [7-8] was used to improve computational performance of the optimal solution.

Results.

Results have been obtained for some points along the Pareto set. Figure 1 shows an exemplarily Pareto set for (i) the negative value of the scaled profit vs. (ii) the scaled energy consumption with the scalarization vector as $\mathbf{w} = [1\text{-}A, A]^T$ in which A varies from 0 to 1 in steps of 0.1. In general, the process is profitable but requires energy supplied from outside. More profit can be made at the expense of consuming more energy. Hence, an engineer can use this information to select a suitable operation according to the current economic situation. Grid adaptation techniques, structure detection and multi-stage reformulations allow a lean control parameterization, which is updated going along the Pareto set.

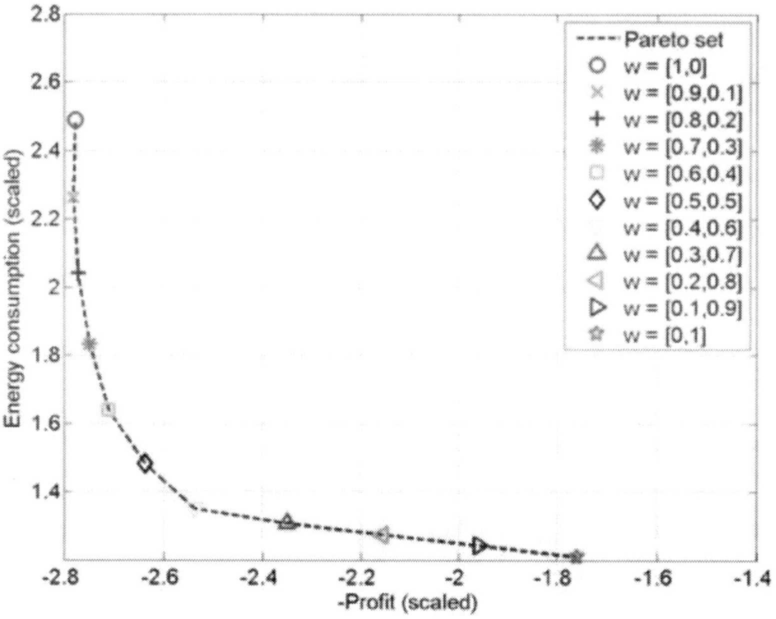

Figure 1: Sampled Pareto set of -profit vs. energy consumption with 11 points. For increasing A-values in w = [1-A,A]T optimizations proceed along the set from left to right.

Acknowledgements. Work supported in part by Projects OPTEC (Center-of-Excellence Optimization in Engineering) PFV/10/002 and SCORES4CHEM KP/09/005 of the KULeuven. FL acknowledges the FWO travel grant. JVI holds the chair Safety Engineering sponsored by the Belgian chemistry and life sciences federation essenscia.

This work has been partially funded by the German Research School for Simulation Sciences.

References

1. A. Hartwich (2012), PhD Thesis, RWTH Aachen University, AVT Prozesstechnik

2. A.E. Hamielec, J.F. MacGregor and A. Penlidis (1987). Multicomponent free-radical polymerization in batch, semibatch and continous reactors. Macromolecular symposia, 10:521–570.

3. J. Gao and A. Penlidis (2002). Mathematical modeling and computer simulator/database for emulsion polymerizations. Progress in Polymer Science, 27:403–535.

4. F. Logist, F. Assassa, J. Van Impe and W. Marquardt (2012), Exploiting grid adaptation and structure detection in multi-objective dynamic optimization problems, Accepted for oral presentation in the 22nd European Symposium on Computer-Aided Process Engineering (ESCAPE 22), Londen (UK), June 17-20 2012.

5. I. Das and J. Dennis (1998). Normal-Boundary Intersection: A new method for generating the Pareto surface in nonlinear multicriteria optimization problems. SIAM Journal on Optimization 8:631–657.

6. M. Schlegel, K. Stockmann, T. Binder, and W. Marquardt (2005). Dynamic optimization using adaptive control vector parameterization. Computers & Chemical Engineering, 29: 1731–1751.

7. M. Schlegel and W. Marquardt (2006). Detection and exploitation of the control switching structure in the solution of dynamic optimization problems. Journal of Process Control, 16:275–290.

8. M. Schlegel and W. Marquardt (2006). Adaptive switching structure detection for the solution of dynamic optimization problems. Industrial and Engineering Chemistry Research, 45:8083–8094.

(647g) Improving Safety and Performance of Polymer Processes by On-Line Applications

Klaus-Dieter Hungenberg, Polymer Processing and Engineering, BASF SE, Ludwigshafen, Germany

Technology leadership is based on two pillars – disruptive innovations leading to new superior processes and incremental process optimization of existing processes to maintain their superiority.

This lecture will shed light on one important factor for this incremental process optimization, i.e. the application of modern on-line methods to typical batch and semi-batch polymerization processes.

The basis for process optimization is the knowledge of the state of the process, which can be gained by the application of on-line sensors. Hard sensors like spectroscopic methods give direct access to polymer properties or monomer concentrations, however they are expensive and in many cases need a high effort for maintenance, whereas soft sensors from fusion of simple but robust sensors need a model behind to link information of e.g. temperatures, flows, pressure, torque to process or product information.

Based on the knowledge of the process state, a model-based safety concept is used to ensure the safety of the process. Such a concept avoids mechanical safety devices.

Once, process safety is assured, non-linear model predictive control technology can be used to optimize process efficiency and product quality, e.g. by feed time reduction in a semi-batch process and controlling the resulting polymer structure.

Though there are a lot of achievements over time, there still remain problems and challenges which are emphasized.

(649a) Microscopic Modeling of a Two-Species Thin Film Deposition Process Arising in Transparent Conducting Oxide Layer Manufacturing

Jianqiao Huang[1], Gerassimos Orkoulas[2] and Panagiotis D. Christofides[1], (1)Department of Chemical and Biomolecular Engineering, University of California, Los Angeles, Los Angeles, CA, (2)Chemical Engineering, University of California, Los Angeles, Los Angeles, CA

Thin film solar cells constitute an important and growing component of the overall solar cell market owing to their potential of
yielding improved light conversion efficiencies. The Transparent Conducting Oxide (TCO) layer, which typically consists of zinc oxide (ZnO) and aluminum (Al), is an important component of thin film solar cells and has a crucial influence on the performance of thin film solar cell systems. In addition to investigating the performance with respect to light conversion efficiency and long-term stability of an array of materials, thin film solar cell technology stands to benefit from fundamental microscopic modeling and optimal thin film manufacturing (deposition) control strategies that produce thin films with desired light trapping properties. This work focuses on the modeling of a large-scale two-species thin film deposition process and its application in the manufacturing process of TCO layers.
Aggregate surface roughness and slope are introduced to characterize the properties of TCO thin film surface properties. Both kinetic Monte Carlo (kMC) models and stochastic partial differential equation (SPDE) models are introduced to describe the film surface morphology dynamics. In the kMC models, a solid-on-solid square lattice is used to simulate the process and different growth mechanisms are utilized for each component, zinc oxide (ZnO) and aluminum (Al). Specifically, a deposition/migration mechanism is used for ZnO and a random deposition with surface relaxation mechanism is used for Al. In SPDE models, an Edwards-Wilkinson type equation is used to predict the process dynamics, which can be used in future work to design a feedback controller. Extensive simulation results will be presented to elucidate the dynamic behavior of surface morphology and pave the way for model-based feedback control.

(649b) Current-Driven Morphological Evolution of Monolayer-Thick Epitaxial Islands On Elastic Substrates

Dwaipayan Dasgupta, Georgios I. Sfyris and Dimitrios Maroudas, Chemical Engineering, University of Massachusetts Amherst, Amhest, MA

The driven assembly of confined quantum structures, such as monolayer-thick islands grown epitaxially on elastic substrates, of either the same or different material than that of the islands, is of particular importance to nanoelectronics and nanofabrication technologies. A typical driving force for the morphological evolution of these islands is an externally applied electric field; this evolution gives rise to morphological patterns, the stability and control of which may have significant impact on nanofabrication approaches. For coherently strained heteroepitaxial islands, it is particularly interesting to study their response to the action of the electric field simultaneously with the induced misfit strain.

In this presentation, we report theoretical and computational results on the current-driven morphological response of single-layer epitaxial islands on elastic substrates based on a two-dimensional fully nonlinear model, according to which mass transport due to curvature-driven diffusion, stress-driven diffusion, and electromigration is limited only to the island circumference. The model also accounts for diffusional anisotropy for such "periphery diffusion" along the island's edge. Within this mass transport regime, we have carried out a theoretical analysis of the morphological stability of such isolated islands and of the current-driven migration dynamics of morphologically stable islands; the theory predicts the dependence of the stable island migration speed on the island size and on the heteroepitaxial system parameters, such as strain due to lattice mismatch. We have also developed self-consistent dynamical simulators of the driven morphological evolution of such islands, combining front tracking methods with analytical expressions derived for the electric-field component tangential to the island's circumference and the elastic strain energy density.

Prior to analyzing the current-driven motion of stable heteroepitaxial islands, we modeled systematically such motion for single-layer homoepitaxial islands in order to compare with published experimental results on Ag(111) surfaces and assess the predictive capabilities of the model. The comparisons demonstrate an excellent quantitative agreement between our modeling predictions and the experimental findings for both the stable island morphology and the dependence of the current-driven island migration speed on island size. This agreement provides strong validation for our model.

Results are reported of a detailed simulation study that determined the dependence of the island migration speed, v_m, on island size, R_s given by the square root of the island's area, for stable steady states in the current-driven island motion on <110>-, <100>-, and <111>-oriented surfaces of single-crystalline face-centered cubic metals. We found that as the island size increases, the stable island morphology changes from round to faceted; the faceted island morphology is determined by the surface orientation through its effect on the diffusional anisotropy. In both homoepitaxial and heteroepitaxial islands, we found the island migration speed to be directly proportional to the inverse of the island size over much of the island size range examined starting from very small islands. In all cases, we found that there exists a critical island size, which, if exceeded, causes the $v_m(1/R_s)$ dependence to deviate from linearity: the dependence becomes nonlinear with a sublinear trend. For <110>-oriented surfaces, further increase of the island size beyond a second, larger critical size leads to morphological instability. For these {110} surfaces, the nonlinearity in the $v_m(1/R_s)$ dependence at larger-than-critical island sizes is due to an island morphological transition, where one of the island's facets that forms along a direction perpendicular to the electric field at smaller-than-critical island sizes splits into two such facets at steady state. For {100} surfaces, the source of the nonlinearity in the $v_m(1/R_s)$ dependence at larger-than-critical island sizes is a dynamical transition in the island's motion: the island's center-of-mass trajectory deviates from the direction of the applied electric field, a straight trajectory characteristic of all the stable islands at steady state for smaller-than-critical island sizes on {100} surfaces.

Furthermore, we have conducted a systematic parametric study to investigate thermal and size effects on the island's driven motion and morphological evolution, as well as effects on the island's driven dynamical response of elastic properties of the substrate and island materials (such as elastic moduli and misfit strain). We have predicted a variety of stable asymptotic states in the driven dynamical response of such single-layer islands and characterized in detail the resulting morphological patterns. We have also developed a universal scaling theory for the dependence of the island's migration speed on the island size that is validated by our simulation results for both homoepitaxial and heteroepitaxial islands over the entire parameter space explored. Our work motivates experimental measurements that can be compared directly with our modeling predictions and sets the stage for examination of collective phenomena, such as island coalescence.

(649c) Physiochemical Properties of Functional Polymer / Solid Carbon Surface Composites Via Molecular Simulation

Pil Seung Chung, Robert Smith, Sesha Hari Vemuri, Lorenz T. Biegler and Myung S. Jhon, Chemical Engineering, Carnegie Mellon University, Pittsburgh, PA

The discovery of new materials has introduced a fundamental paradigm shift in nanotechnology in numerous systems including micro/nano electromechanical systems (MEMS/NEMS) and efforts are being directed towards their applications by converging various functionalities. For example, graphene has recently attracted scientific/technological interest due to its exotic electric & thermo-mechanical properties, elasticity, low surface energy, and consequently low friction coefficient. Due to these superior physiochemical properties, graphene has recently been considered as the most promising candidate for high endurance materials applicable in MEMS/NEMS. However, at the nano-scale dimensions of these devices, the molecular mechanisms govern the system properties, and stringent confinement of the systems induced by composites of nano materials further complicate the experimental measurement of the physiochemical properties. Therefore, it is essential to develop a computational approach enabling molecular simulation for the interfacial phenomena of functional polymers and solid surfaces via controlling molecular architectures and external operation conditions of the integrated system.

In this study, we investigated the physiochemical properties of nano composites by combining various types of functional polymers and solid carbon surfaces (e.g., diamond, diamond-like carbon (DLC), and graphene) via large scale coarse-grained bead-spring molecular dynamics (MD) based on the endgroup-surface potential energy parameters determined by atomistic level theory [1, 2]. Intermolecular force fields between functional endgroups in polymers and carbon surfaces were atomistically estimated from the equilibrium geometries of truncated functional polymers and the models of the surface materials. By using MD, polymeric film conformations were characterized by anisotropic radii of gyration, where the perpendicular component determines the thickness of a monolayer film, and functional bead density defines the adhesion distribution of functional endgroups and the internal structure (e.g., layering and clustering) of the multilayer films. The dynamic responses were also investigated by calculating the self-diffusion coefficient of a tagged molecule. Various structures of functional polymers were utilized by tuning the polymer properties such as the functionalities (endgroup position and strength), molecular weight, and number of branches, where the star-like polymers with n branches (n=3 or 4) contain additional functional endgroups on the branches [3]. In summary, we focused on investigating the key parameters of nano composites governing the physiochemical properties by considering the functionality, the structures of polymers, and their interactions with the various carbon surfaces.

References

1. M.S. Jhon, R. Smith, S.H. Vemuri, P.S. Chung, D. Kim, and L.T. Biegler, "Hierarchical multiscale modeling method for head/disk interface," *IEEE Trans. Magn.*, **47**, 87-93 (2011).
2. R. Smith, P.S. Chung, J.A. Steckel, M.S. Jhon, and L.T. Biegler, "Force field parameter estimation of functional perfluoropolyether lubricants," *J. Appl. Phys.*, **109**, 7B728 (2011).
3. D. Shirakawa and K. Ohnishi, "A study on design and analysis of new lubricant for near contact recording," *IEEE Trans. Magn.*, **44**, 3710 (2008).

(649d) Integration of Polymer Electrolyte Membrane Fuel Cell Subcomponent Models Using Reduced Order Methodology

Robert L. Smith[1], Parag Jain[2], Lorenz T. Biegler[1] and Myung S. Jhon[1], (1)Chemical Engineering, Carnegie Mellon University, Pittsburgh, PA, (2)UOP LLC - A Honeywell Company, Des Plaines, IL

The polymer electrolyte membrane fuel cell (PEMFC) is a green energy device that utilizes hydrogen fuel to convert chemical energy to electrical energy with a higher efficiency than the internal combustion engine producing only water and waste heat. This device is composed of several subcomponents on both the anode and cathode sides including the gas channel (GC) where fuel enters, the gas diffusion layer (GDL) which creates a uniform distribution of fuel, the catalyst layer (CL) where the electrochemical reaction occurs, and the PEM through which the protons are conducted.

These subcomponents contain multiphysical phenomena and can be simulated by models on various scales ranging from phenomenological/continuum models to ab initio/molecular scale descriptions. In a multiscale/ multi-physical device such as the PEMFC, the integration between models must be performed within and across subcomponents. Integration of models within subcomponents requires an innovative methodology to coarse grain high resolution descriptions by reducing complexity while preserving the essential physics. The reduced order method (ROM) can serve as an excellent tool to integrate models. Linking across subcomponents requires determining states of two adjacent subcomponents such that continuity of states and fluxes at the interacting boundary are satisfied.

We examine two different procedures for integrating models for various scenarios. In the first application of the horizontal procedure, model accuracy is preserved by linking high resolution, complex models for two subcomponents. However, computational expense often makes this approach intractable. In an alternative horizontal approach, two reduced, coarse-grained models for each subcomponent can be to a reduction in complexity but also in model accuracy. In a second procedure, model cross-linking is used to link one high resolution subcomponent model to a coarse-grained, lower resolution model with an adjacent subcomponent model. This procedure provides a balance between accuracy and computational expense and is one of the most accessible approaches for model integration.

We demonstrate the best procedure by integrating simple GDL and CL subcomponent models with an interactive boundary via a cross-linking procedure. [1], [2] A ROM for the CL is constructed using principal component analysis and Kriging mapping.[3] We maximize PEMFC current density by solving the CL model with the GDL boundary as an input. Our methodology based on simple models will be generalized from 2-D to 3-D as well as integrated with the membrane subcomponent. This general methodology can be used for other energy areas dealing with multiphysics and multiscale phenomena.

(649e) Modeling Densities and Equilibria of Ice and Gas Hydrate Phases

Angelo Lucia, University of Rhode Island, Kingston, RI, Heath Henley, Chemical Engineering, University of RHode Island, Kingston, RI and Edward Thomas, Chemical Engineering, University of Rhode Island, Kingston, RI

The general warming of land and oceans around the world has generated interest in understanding long term environmental impacts associated with the melting of ice sheets and thawing of permafrost regions. In this talk extensions of the multi-scale Gibbs-Helmholtz Constrained (GHC) equation of state to modeling the phase densities and equilibrium behavior of hexagonal ice and gas hydrates are described. A number of numerical results for mixtures of N_2, O_2, CH_4, CO_2, water, and NaCl for permafrost conditions are presented to illustrate the predictive capabilities of the multi-scale GHC equation. In particular, it is shown that the GHC equation correctly predicts

1) The density of 1h ice and methane hydrate to within 1%.

2) The melting curve for hexagonal ice.

3) The heat of fusion of 1h ice at high pressure.

4) The hydrate-gas phase co-existence curve.

5) Various phase equilibrium involving ice and hydrate phases.

We also show that the GHC equation approach can be readily incorporated into subsurface flow simulation programs like FEHM to predict the behavior of permafrost and other reservoirs where ice and/or hydrates are present. Many geometric illustrations are used to elucidate key concepts.

(654a) Constrained Dynamic Programming of Hybrid Systems and Explicit Model Predictive Control – a Multi-Parametric Programming Approach

Pedro Rivotti[1], Martina Wittmann-Hohlbein[1] and Efstratios N. Pistikopoulos[2], (1)Department of Chemical Engineering, Centre for Process Systems Engineering, Imperial College, London, United Kingdom, (2)Chemical Engineering, Imperial College London, Centre for Process Systems Engineering, London, United Kingdom

In this work, we consider the solution of constrained dynamic programming problems involving hybrid linear models, such as piece-wise affine systems, where binary variables are introduced to describe the switches between different affine dynamics of the system. Linear or quadratic objective functions are considered, resulting in mixed integer linear (MILP) or quadratic (MIQP) formulations.
First, we present a multi-parametric (mp) programming approach to solve such hybrid linear constrained problems that involves: (i) reformulation of the original problem as an mp-MILP or mp-MIQP; (ii) decomposition of the problem into a set of sub-problems using dynamic programming; (iii) sequentially solving the resulting set of sub-problems of smaller dimensionality using suitable mp-MILP or mp-MIQP algorithms; (iv) obtaining the enclosure of parametric solutions as a piece-wise affine function of the vector of parameters.
Based on this, we then show how, by recasting model predictive control problems for hybrid systems within a dynamic programming formulation, the explicit solution of the hybrid MPC problem can be efficiently computed. By considering the states of the system to be controlled as parameters, the explicit control law is obtained as a piece-wise affine function of the states, thus reducing the computational burden involved in the online implementation of such controller to simple function evaluations.
Finally, we show how this approach can be extended to account for the presence of different types of uncertainty in the model equations. The work here presented outlines a general framework towards the design of explicit robust controllers for hybrid systems.

(654b) Advanced-Multi-Step Nonlinear Model Predictive Control

Xue Yang[1], Gilvan A. G. Fischer[2] and Lorenz T. Biegler[1], (1)Chemical Engineering, Carnegie Mellon University, Pittsburgh, PA, (2)Henrique Lage Refinery, Petróleo Brasileiro S.A. – Petrobras, São José dos Campos, Brazil

Nonlinear Model Predictive Control (NMPC) has gained wide attention through the application of dynamic optimization. It has the ability to handle variable bounds and deal with multi-input-multi-output systems. However, NMPC requires that the optimization problem be solved within one sampling time. Also, computational delay may lead to deterioration of controller performance and system stability. Here we propose an advanced-multi-step NMPC (amsNMPC) method with negligible computational delay. It is based on nonlinear programming (NLP) and NLP sensitivity.

Two variants are developed: the serial approach and the parallel approach. The basic ideas of the two approaches are identical:

- Background: the controller predicts the state multiple steps ahead in the future, uses the prediction as an initial value and solves an NLP problem to get predictions of manipulated variables.
- On-line: as state measurements are obtained, the controller uses NLP sensitivity to update predicted manipulated variables.

The main difference between these two approaches is that the NLP problems are solved at different frequencies and the Karush-Kuhn-Tucker (KKT) matrices used for update are formed in different ways. The *serial approach* applies one processor, and solves the NLP problem over a number of sampling times. The KKT matrix from the NLP solution is updated at every sampling time in order to update the manipulated variables, until the next KKT matrix is achieved. On the other hand, the *parallel approach* solves an NLP problem every sampling time. It applies multiple processors to assure that a free processor is always available for a new NLP problem, even if the solution of the previous one has not finished. Since we solve an NLP problem at every sampling time, we would use the KKT matrix at the solution only once and no KKT matrix update is needed. Both approaches can be applied to optimization problems, which take longer than one sampling time to solve. As both of them use predictions of states as initial values of the NLP problem, the optimization problem can be solved in advance without knowing the real states, and the computational delay can be avoided. In this presentation we detail the application of Lyapunov stability theory to prove the nominal stability and robust stability of the amsNMPC method. In addition, a large-scale first principles distillation column problem is used as an example to demonstrate the performance of the amsNMPC, and to compare it with competing strategies.

(654c) Composite Fast-Slow MPC Design for Nonlinear Singularly Perturbed Systems

Xianzhong Chen[1], Jinfeng Liu[2], Mohsen Heidarinejad[3] and Panagiotis D. Christofides[2], (1)Chemical and Biomolecular Engineering, Univ. of California, Los Angeles, Los Angeles, CA, (2)Department of Chemical and Biomolecular Engineering, University of California, Los Angeles, Los Angeles, CA, (3)Electric Engineering, University of California, Los Angeles, Los Angeles, CA

Time-scale multiplicity arises in numerous chemical processes and industrial plants due to the strong coupling of physico-chemical phenomena, like slow and fast reactions, occurring at different time-scales. Also, the dynamics of control actuation and measurement sensing systems very often induces a fast-dynamics layer in the closed-loop system. Conventionally, the analysis and controller design problems for multiple-time-scale systems are formulated through taking advantage of the mathematical framework of singular perturbations [1]. Model predictive control (MPC) has been widely employed in industrial process control applications due to its ability to satisfy state and input constraints by minimizing a meaningful objective function and taking advantage of system model to predict future evolution of the system over a predefined horizon while applying its solution in a receding horizon manner.

In this work, we focus on MPC of nonlinear singularly perturbed systems in standard form where the separation between the fast and slow state variables is explicit. Specifically, a composite control system comprised of a ``fast" MPC acting to regulate the fast dynamics and a ``slow" MPC acting to regulate the slow dynamics is designed. The composite MPC system uses multirate sampling of the plant state measurements, i.e., fast sampling of the fast state variables is used in the fast MPC and slow-sampling of the slow state variables is used in the slow MPC as well as in the fast MPC. Both fast and slow MPCs take advantage of their corresponding Lyapunov-based controllers to characterize closed-loop system stability region [2]. Using singular perturbation theory, the stability and optimality of the closed-loop nonlinear singularly perturbed system are analyzed. The proposed control scheme does not require exchange of information between the two MPC layers, and thus, it can be classified as decentralized in nature. The theoretical results are demonstrated through a nonlinear chemical process example.

[1] P. D. Christofides, P. Daoutidis. "Feedback control of two-time-scale nonlinear systems". International Journal of Control. Vol. 63, 965-994, 1996.
[2]P. D. Christofides, J. Liu, and D. Munoz de la Pena. "Networked and Distributed Predictive Control: Methods and Nonlinear Process Network Applications". Advances in Induatrial Control Series. Springer-Verlag, London, England, 2011.

(654d) Towards Model Predictive Dual Control

Tor Aksel N. Heirung[1], B. Erik Ydstie[2] and Bjarne Foss[1], (1)Department of Engineering Cybernetics, Norwegian University of Science and Technology, Trondheim, Norway, (2)Department of Chemical Engineering, Carnegie Mellon University, Pittsburgh, PA

We present a novel approach to optimal adaptive control that involves closed-loop system identification with minimal deterioration of output regulation. The control algorithm excites the system in order to improve the parameter estimates, but does not disturb the system when the estimates are judged good enough. Without requiring persistent excitation, the method seeks a balance between excitation for identification and output regulation.

The algorithm is designed such that standard optimization codes can be used in the implementation, avoiding the use of techniques like stochastic dynamic programming, which often lead to complicated control designs. We base our method on model predictive control (MPC) and formulate a controller that can be realized with small changes to an existing MPC framework.

The current implementation estimates parameters online using a recursive least squares algorithm. At each sampling time, an optimal input is computed based on the current estimate, leading to a certainty-equivalence type controller. The computation of the input involves solving a nonlinear programming (NLP) problem, minimizing a finite time-horizon objective function whose terms include future input and output magnitudes, future rates of change in input, as well as a measure of future parameter estimate error variances. The constraints include the model with its estimated parameters, bounds on future inputs and outputs, and a modified least squares algorithm implemented as a set of equality constraints. Adding the estimation equations as constraints informs the NLP of how the input sequence will affect the future parameter estimate quality. This leads to a controller with dual features (after Feldbaum [1]).

We illustrate the method on an example system with unknown input gain. The algorithm intelligently excites the system only when the parameter estimate is uncertain. We also show how a slight increase in system excitation by the controller can greatly enhance parameter identification with minimal effect on the system output.

[1] Feldbaum, A. A. Dual control theory. I-IV. Automation and Remote Control, 1961.

Model Adequacy for Real-Time Optimization

Dominique Bonvin

Laboratoire d'Automatique , Ecole Polytechnique Fédérale de Lausanne
EPFL – Station 9, 1015 Lausanne, Switzerland
dominique.bonvin@epfl.ch

Optimization is important in science and engineering as a way of finding "optimal" situations, designs or operating conditions. Optimization is typically performed on the basis of a mathematical model of the process under investigation. In practice, optimization is complicated by the presence of uncertainty in the form of plant-model mismatch and unknown disturbances. Without uncertainty, one could use the model at hand, optimize it numerically off-line and implement the optimal inputs in an open-loop fashion. However, because of uncertainty, the inputs need to be re-computed or adjusted in real-time based on measurements. This is the field of real-time optimization, which is labeled RTO for static optimization problems [1]. A standard way of implementing RTO is via the so-called two-step approach of repeated parameter estimation and optimization. In this scheme, measurements are used to adapt the model parameters, and the updated model is used for optimization. The fixed point of this iterative scheme is an important issue.

The quality of the model is often overlooked in real-time optimization. The model is typically inaccurate due to lack of time for detailed modeling or shear complexity. More surprising is the fact that a model can predict the plant outputs well but be inadequate to push the plant to its optimum, which means that the model does not adequately represent the optimality conditions of the plant. The problem of model selection in the two-step RTO approach has been discussed in [2]. If the model is structurally correct and the parameters are identifiable, convergence to the plant optimum can be achieved in a single iteration. However, in the presence of plant-model mismatch, whether the scheme converges, or to which point it does converge, becomes anyone's guess. This is due to the fact that the objective for parameter adaptation might be unrelated to the cost and constraint values and gradients that drive optimality in the optimization problem. Hence, minimizing the mean-square error of the plant outputs may not help in our quest for feasibility and optimality. Convergence under plant-model mismatch has been addressed in [3] and [4], where it has been shown that optimal operation is reached if model adaptation leads to matched KKT conditions for the model and the plant.

This contribution addresses the convergence of two-step RTO schemes in the presence of structural plant-model mismatch. We propose to investigate the parameter estimation and optimization problems in the light of the second-order sufficient conditions of optimality to show that, in general, there are too few degrees of freedom to be able to reach plant optimality. A possible solution to this problem consists in reconciling the objectives of the parameter estimation and optimization problems along the lines of "modeling for optimization" [5,6]. One could for example use as "measured outputs" in the parameter estimation problem estimates of the necessary conditions of optimality for the optimization problem. These issues will be investigated both theoretically and via the simulation of chemical processes at steady state.

References

[1] Marlin, T.E. and Hrymak, A.N. (1997). "Real-Time Operations Optimization of Continuous Processes", In AIChE Symposium Series - CPC-V. Vol. 93. 156–164.

[2] Forbes, J.F. and Marlin, T.E. (1996). "Design cost: A Systematic Approach to Technology Selection for Model-Based Real-Time Optimization Systems", Comp. Chem. Eng. 20, 717–734.

[3] Biegler, L.T., Grossmann, I.E. and Westerberg, A.W. (1985). "A Note on Approximation Techniques Used for Process Optimization", Comp. Chem. Eng. 9, 201–206.

[4] Forbes, J.F., Marlin, T. E. and MacGregor, J.F. (1994). "Model Adequacy Requirements for Optimizing Plant Operations", Comp. Chem. Eng. 18(6), 497–510.

[5] Srinivasan, B. and Bonvin, D. (2002). "Interplay Between Identification and Optimization in Run-to-Run Optimization Schemes", ACC, Anchorage, 2174–2179.

[6] Bonvin, D. and Srinivasan, B. (2012). "On the Role of the Necessary Conditions of Optimality in Structuring Dynamic Real-Time Optimization Schemes", Comp. Chem. Eng. (in press), http://dx.doi.org/10.1016/j.compchemeng.2012.07.012

(654f) Multi-Tiered Dynamic Optimization of Partial Shutdowns with Constrained Ukf-Based State and Parameter Estimation

Zhiwen Chong and Christopher L. E. Swartz, Chemical Engineering, McMaster University, Hamilton, ON, Canada

Process units are shut down from time to time either for maintenance or due to equipment failure. Shutdowns in intermediate units constitute a disruption in the processing chain, which can adversely affect the plant's ability to continue operation. They frequently also have a negative effect on the operating economics of a plant due to the attendant loss of production capacity.

Our research focuses on the development of systematic strategies and formulations for the optimal operation of plants under partial shutdown conditions. A partial shutdown is a type of circumscribed plant unit shutdown that permits the rest of the plant to continue operating to some degree. The goal of a partial shutdown strategy is to manipulate the available degrees-of-freedom in a plant, during and after a shutdown, such that production is restored in a cost optimal fashion while meeting all safety and operational constraints. This can be accomplished through adjustments of production rates, recycles and buffer levels.

In order to compute these adjustments, we solve a differential-algebraic equation (DAE) based dynamic optimization problem containing a model of the plant. The DAE system is discretized using orthogonal collocation on finite elements, which results in a large-scale optimization problem. An economics-based nonlinear model predictive control (MPC) control structure was used to mitigate effects of model uncertainty and disturbances. Within this broad framework, we adopt a multi-tiered dynamic optimization approach that allows us to prioritize multiple competing objectives (economics, product quality, move suppression, etc.) and specify the trade-offs from one tier to the next. In this approach, separate optimization problems are solved sequentially, with the preceding tiers providing information to the latter tiers. The optimal value of the objective function in a current tier is enforced as a constraint in the subsequent tier. An attractive feature of the multi-tiered optimization approach is that unlike a weighted-penalty approach, it permits one to specify, for example, the exact amount of the economics one is willing to sacrifice (in dollar terms) in order to obtain say, smoother input trajectories. In contrast, the specification of weights in a weighted penalty approach may lead to unpredictable trade-offs in the final solution.

A further aspect of this work is the investigation of the use of nonlinear state and parameter estimation algorithms to moderate the adverse effects of plant-model mismatch. The estimation algorithms are based on novel configurations of the constrained Unscented Kalman Filter (UKF), a nonlinear derivative-free observer that received considerable attention in recent years. The UKF was used to perform state and parameter estimation of large-scale DAE models. Constraints on the state and parameter estimates are enforced through a simple projection method that is based on solving a quadratic program (QP). States, parameters, as well as disturbance biases are estimated using this scheme.

Plant-model mismatch can often render terminal/restoration constraints infeasible. To address this, we propose a dynamic feasibility tier (in our multi-tiered optimization problem) that ensures that terminal constraints and parameter estimates from a one-step estimator are feasible throughout the model prediction horizon in the control optimization problem. This represents a practical method for ensuring the feasibility of terminal constraints and parameter values obtained from a state/parameter estimator whose prediction horizon is shorter than that of the control optimization problem.

The strategies in this work are illustrated through simulation case studies based on Kraft pulp mill. Conclusions are drawn and avenues for further work are identified.

(654g) Quasi-Decentralized Model Predictive Control of Process Systems Using Adaptive Sampling

Ye Hu, Department of Chemical Engineering and Materials Science, University of California, Davis, Davis, CA and Nael H. El-Farra, Department of Chemical Engineering & Materials Science, University of California, Davis, Davis, CA

As one of the few systematic methods suited for handling constraints, multivariable interactions and optimization requirements, Model Predictive Control (MPC) has been the subject of considerable interest in process control research. Numerous MPC formulations and methods have already been developed over the past few decades to address different kinds of problems arising from process control practice. Despite this progress, a close examination of the literature shows that the majority of existing MPC formulations are often developed on the basis of the conventional feedback control paradigm where the sensors are assumed to collect the measurements and transmit them to the controller at a fixed sampling/communication rate. With the increasing complexity of the process/controller interface brought about by, for example, the incorporation of wireless sensor networks in control systems, constraints on the information collection, processing and transmission are encountered more frequently. In some applications, it is also advantageous for the control system to have the ability to respond to changes in operating conditions by adaptively changing the rate of measurement collection and transmission so that the control system performance can be optimized under limited resources. These considerations require re-examination of the applicability of existing MPC methods developed within the conventional feedback control paradigm.

In this work, we focus on predictive control of large-scale dynamical systems consisting of interconnected subsystems, and develop a quasi-decentralized MPC framework using an adaptive sampling strategy. The framework aims to enforce closed-loop stability and achieve the desired performance specifications in each subsystem with minimal measurement sampling and communication between the constituent subsystems. Specifically, a Lyapunov-based model predictive controller is initially synthesized for each subsystem to enforce closed-loop stability within a well-characterized stability region for a fixed measurement sampling rate. The controller minimizes a local performance index over a finite horizon subject to constraints on the dynamics of the prediction models, bounds on the states and inputs, as well as a stability constraint that essentially enforces a prescribed decay rate for the local Lyapunov function over a sampling interval. The Lyapunov stability constraint is used to characterize the minimum time required for the local state to escape from the stability region (starting from a given initial state), and this characterization is then used as the basis for devising a sampling strategy that can adaptively adjust the sampling and communication rates based on the evolution of the local Lyapunov functions. The key idea is to implement the control sequence generated when the optimization problem is solved for as long as possible until a new measurement is sampled when either the minimum escape time or the end of the prediction horizon is reached. The new measurement is then used to update the states of the locally embedded prediction models, re-solve the optimization problem and assess whether the sampling rate should be maintained or changed based on a comparison with the last available measurement. The implementation of this strategy requires only local monitoring of the state within each subsystem to determine when the flow of information between the constituent subsystems can be suspended or restored. Switched system techniques are used to analyze closed-loop stability and derive precise conditions for the implementation of the proposed adaptive sampling policy. Finally, the theoretical results are illustrated using a reactor-separator process example.

(687a) Tutorial Overview On the Modeling and Control of Power Transmission Networks

Donald J. Chmielewski, Illinois Institute of Technology, Chicago, IL

A large-scale electric power system is a rather complicated interconnection of many systems. At one end there are generation units that convert available energy sources (either fossil, nuclear or renewable) into electric energy. This generated power must then be transported to load centers via high voltage AC transmission lines. The received energy is then delivered to individual customers via a distribution network. During operation, one of the highest priorities is the availability of high quality power to all consumers at all times. Such reliability is difficult to achieve, because of two challenges: (a) Consumer demand is continually changing and cannot be controlled by the power provider, and (b) With current storage technology, electric energy is virtually impossible to store in large quantities. Consequently, a provider must track consumer demand to send dispatch commands to the generators. Furthermore, these dispatch commands must be aware of the power flow limitations imposed by the transmission and distribution network. Failure to implement a suitable dispatch policy will result in poor power quality (erratic voltage levels) and in the worst case power outages.

Poor power quality is addressed by the implementation of a power management strategy, which considers small fluctuations in power demand over small time-scales (seconds to minutes). Fundamental to power management is the concept of spinning reserves. The idea is that a generator can quickly increase or decrease power output within a reasonably sized window of power conditions. Power outages can be prevented by energy management, which considers large demand changes over large time-scales. Fundamental to energy management is the fairly predictable nature of consumer demand. The basic idea is that generation facilities can be scheduled to provide power within specific time periods.

As renewable sources, such as wind and solar, are introduced to the system, additional burden will be placed on the transmission network. Renewable sources tend to be located great distances from load centers and will likely require significant expansion of the transmission hardware. In addition, renewable sources are inherently intermittent and non-dispatchable. With capacity factors of 30 to 40%, power output from renewable sources will vary from zero to three times the average. Short term variations in power output (due to wind gusts and small clouds) are also expected. It is further noted that the power production gaps created by renewable sources must be made up by the remaining dispatchable sources.

In this paper we will present a tutorial overview of the modeling and control system methods used in the operation of power transmission networks. In particular, effort will be made to highlight parallels between these methods and those used in the chemical process industry.

(687b) Modeling and Model Predictive Control Strategies for Building Energy Management

Cara Touretzky, Chemical Engineering, University of Texas at Austin, Austin, TX and Michael Baldea, University of Texas

Modeling and Model Predictive Control Strategies for Building Energy Management

Cara Rose Touretzky and Michael Baldea

Department of Chemical Engineering

The University of Texas at Austin, 1 University Station C0400, Austin, TX 78712

Email: CaraTour@utexas.edu, MBaldea@che.utexas.edu

Residential and commercial buildings account for over 40% of primary energy consumption in the U.S., with heating and cooling systems accounting for the majority of this figure [1]. While novel materials, new architectural designs, and the availability of alternative energy sources have contributed to reducing building electricity consumption, there still remains a significant amount of unrealized energy and cost savings that can be captured through the optimal control and management of day-to-day building energy use. Making efficient use of energy resources in the presence of fluctuations in weather, occupancy and cost involves complex decisions using both continuous and discrete decision variables, the latter corresponding to the on/off state of various pieces of equipment, e.g., heaters and chillers. The inherent complexity of the building energy management process is thus a natural candidate for optimization [2, 3].

The practical implementation of an optimization-based control and energy management approach is, however, hindered by the considerable size and complexity of the mathematical models representing the dynamics of a building; such models do not lend themselves easily to use in online calculations. An additional complication originates in the building dynamics: the significant discrepancy in the thermal mass of the air circulated in by the Heating Ventilation and Air Conditioning (HVAC) and the thermal mass of the building itself, along with the presence of energy recovery devices (e.g., air preheating using the exhaust air), give rise to a dynamic behavior that evolves over multiple time scales [2]. In this context, the evolution of local (e.g., room-level) temperatures is fast, while the total energy stored in the building evolves over a much longer time horizon. In turn, this multiple time scale behavior is at the origin of the stiffness of the mathematical models of buildings. Using such for controller design is impractical: the resulting controllers are almost invariably difficult to tune and sensitive to modeling errors and noise.

Motivated by the above, we propose a novel model reduction and energy management framework for buildings. We demonstrate that, owing to energy recovery features such as air preheating, air recirculation and thermal energy storage, the dynamic behavior of a building is similar to that of an integrated chemical process. Based on this analogy, we expand on our recent results [4] to develop a method for deriving a reduced-order, nonlinear model that describes the dominant, slow dynamics of a building, reflecting the influence of the large thermal mass of the system. Subsequently, we introduce a set of techniques that enhance model portability, i.e., facilitate a fast, optimal identification of the parameters of the model from building operation data, and the deployment of the model in specific building applications. The resulting models are low-dimensional and non-stiff, and thus well suited for control applications. Using the nonlinear model predictive control (NMPC) paradigm, we develop a new approach to building energy management based on the aforementioned reduced-order models. The stability properties of the proposed NMPC scheme are proved theoretically. Finally, we provide a set of experimental results obtained at the University of Texas at Austin Thermal Façade Laboratory, demonstrating the effectiveness of the proposed building energy management scheme.

References:

[1] *Buildings Energy Data Book Table 1.1.3*, U.S. Department of Energy (2011)

[2] Ma, Y., Anderson, G., Borrelli, F. *A Distributed Predictive Control Approach to Building Temperature Regulation*. American Control Conference (2011)

[3] Wang, S., Ma, Z. *Supervisory and Optimal Control of Building HVAC Systems: A Review*. HVAC&R Research (2008)

[4] Baldea, M., Dautidis, P. *Dynamics and Nonlinear Control of Integrated Process Systems*, Cambridge University Press (2012)

(687c) Constraint Handling in Inventory Control

Timothy P. McFarland, Chemical Engineering, Carnegie Mellon University, Pittsburgh, PA and B. Erik Ydstie, Department of Chemical Engineering, Carnegie Mellon University, Pittsburgh, PA

Control systems for chemical processes must address restrictions imposed by the equipment and safety constraints. Constraints may be enforced by alarming and manual intervention or through logic control. While enforcement of process constraints avoids potentially dangerous operation, the closed-loop stability properties of the process will change. By including constraints within the control system, the process may be safely steered away to retain the desired closed-loop stability. Model predictive control includes constraints in the calculation of control inputs. However, the handling of constraints by local control methods, such as inventory control, in a large chemical process network needs to addressed

In previous work, we have introduced a formalized approach for inventory control of chemical process networks using extensive variables : internal energy, volume, molar inventories. However, the inventory control approach has lacked a consistent method to handling process and input constraints. Most process constraints are in terms of intensive variables such as temperature and pressure rather than the extensive variables. The constraints must be converted to the extensive state-space and incorporated into the inventory control law.

In the current work, we propose constraint transformations from the intensive to extensive state-space for a number of example problems. The extensive variable constraints can be calculated from thermodynamic state relationships. These constraints are typically nonlinear functions of physical properties, intensive variables, and active intensive variable constraints. This approach trades nonlinear dynamics with constant constraints in the intensive space for nearly linear dynamics and nonlinear constraints in the extensive variable space. The extensive variable constraints are used to directly calculate setpoints for a particular active set of constraints. The process control computer system determines the active set and uses the appropriate setpoint relationship.

Even when the process constraints are included in the control system, active input constraints may still cause many problems for inventory control. Passive systems theory has been used to show stability of the extensive variable inventory control laws. The control laws require feed-forward terms relating to production and process flows. If the inputs become actively constrained, certain states in the chemical process become uncontrollable and the closed-loop system may no longer be passive. In the current work, we address input constraints by placing further restrictions on the setpoints of the controlled variables. Motivating examples are used to demonstrate the issues in dealing with steady state process constraints and input constraints in the inventory control framework.

(687d) Reduction and Control of Tightly-Integrated Plants Using Graph-Theoretic Methods

Seongmin Heo[1], Sujit S. Jogwar[2] and Prodromos Daoutidis[1], (1)Chemical Engineering and Materials Science, University of Minnesota, Minneapolis, MN, (2)Praxair, Inc., Tonawanda, NY

Energy integration is ubiquitous in modern chemical plants. Even though it leads to significant cost savings by reducing external utility consumption, integrated process networks become harder to be operated and controlled, especially in the context of transitions between different operating conditions. To this end, the control of large scale, integrated plants has been an active area of research for the past several years.

In our previous work, we have identified two different prototype structures with energy integration: networks with large energy recycle and networks with large energy throughput. It has been shown that each prototype network exhibits unique two-time scale dynamics, and model reduction has been done to obtain the description of the dynamics of the networks in each time scale. However, in the case of complex tightly-integrated plants, networks are complex, comprising of interconnections of these prototype networks. The resulting multi-loop structure has a potential to exhibit multi-time scale dynamics.

One possible approach for the analysis of these networks is detailed mathematical analysis with a successive application of singular perturbations. However, such an analysis becomes cumbersome with increasing size and complexity of a process network. An alternative is a graph theory based approach which is a generic and scalable analysis framework.

To this end, we have developed a graph-theoretic algorithm which can be used to analyze complex process networks with segregation of the energy flows. Specifically, the developed algorithm uses information about the structure and order of magnitude of the different energy flows involved in a network to automatically generate information on i) the time scales exhibited by the network, ii) control objectives to be pursued in each time scale and manipulated variables to enforce them, and iii) form of the reduced order models in each time scale.

In this work, we demonstrate the application of the developed algorithm to a benchmark chemical plant example, the toluene HDA process. Specifically, we focus on the reaction part of a design of the process which contains one feed-effluent heat exchanger (FEHE). First, we show that the process exhibits three-time scale dynamics using both numerical simulation and the developed graph theory algorithm. Specifically, the enthalpies of the process units associated with the inner recycle loop (recycle through the FEHE) are evolving in the fast time scale, the enthalpies of the process units associated with the outer recycle loop (gas recycle) as well as the total enthalpy of the inner recycle loop are evolving in the intermediate time scale and the total network enthalpy is evolving in the slow time scale. We propose a hierarchical control structure based on the results given by the algorithm. Specifically, we employ a nonlinear (input/output linearizing) controller as the slow time scale controller to control the variables at the exit of the network such as the benzene mole fraction of the liquid outlet of the separator. We use simple PI controllers as the intermediate and the fast time scale controllers to control the variables associated with the inner recycle loop such as the inlet temperatures of the furnace and the reactor. Simulation studies illustrate the effectiveness of the proposed control structure.

(687e) Quasi-Decentralized Control of Networked Process Systems with Discrete and Delayed Communication

Nael H. El-Farra, Department of Chemical Engineering & Materials Science, University of California, Davis, Davis, CA and Wei Wan, University of California, Davis, Davis, CA

The fundamental and practical challenges associated with the control of networked process systems have been the focus of significant research activities in process control over the past two decades and have motivated many research studies in this area. Examples include results on traditional centralized and decentralized control, distributed model predictive control, passivity-based control, agent-based systems, and singular perturbation formulations. In addition to these works, research has recently begun to address communication issues (e.g., resource constraints, real-time scheduling constraints, etc.) in the plant-wide control problem, leading to the design of resource-aware networked plant-wide control systems with explicitly-characterized stability and performance properties [1,2]. These efforts have been motivated by the increased reliance in the process industries on sensor and control systems that are accessed over shared communication networks instead of dedicated links, as well as the growing calls for expanding the traditional process control and operations paradigm in the direction of smart plant operations.

An important consequence of the tight integration between the different process subsystems (through inventory recycle) is the need for a similarly tight integration of the information flow between the local control systems. A balance must be maintained, however, between the desired control quality and the extent of information transfer. On the one hand, the need to account for the interactions between the different units and minimize the propagation of process upsets demands increased levels of local data collection and broad dissemination throughout the plant. On the other hand, the costs associated with sensor deployment and communication, together with the difficulty of obtaining frequent or timely measurements for certain process variables, impose constraints on the sensing and communication.

An approach that reconciles these conflicting objectives is model-based quasi-decentralized control, which is a distributed control strategy in which the local control systems rely on predictive models of the plant units to compute the local control action at times when communication is suspended, and communicate only at discrete times (to reduce communication costs) to provide updates to the models' states. In this approach, it was assumed that the communication delays were negligible and that the data transmitted by each subsystem was immediately available to the target control systems. In distributed control architectures, however, various forms of delays in measurement transmission are inevitable. These include communication delays (e.g., due to varying battery levels in a wireless sensor network) and processing delays (e.g., due to the use of computationally sophisticated data processing techniques). Such delays may degrade the closed-loop performance and even cause instability if not taken into consideration in the plant-wide control system design.

This paper focuses on networked control of multi-unit distributed plants subject to discrete and delayed communication between the local control systems. A model-based quasi-decentralized networked control structure that enforces closed-loop stability with minimal data transfer, while simultaneously taking into account the different communication delays associated with the transfer of information between the constituent subsystems, is developed. In this structure, each control system includes: (1) a set of predictive models each of which predicts the evolution of the state variables of a given neighboring unit when measurements are unavailable and is updated when communication is restored, and (2) a set of propagation models each of which uses the delayed measurements from a given unit, together with the past values of the estimated local control action, to generate an estimate of the current state which is then used to update the state of the corresponding model. A combined discrete-continuous system formulation of the networked closed-loop plant is developed and analyzed, leading to an explicit characterization of the closed-loop stability properties in terms of the update period, the communication delay sizes, the degree of plant-model mismatch, and the selection of controller design parameters. Finally, the theoretical results are illustrated using a reactor-separator plant example.

References:

[1] Sun, Y. and N. H. El-Farra, ``Resource-Aware Quasi-Decentralized Control of Networked Process Systems over Wireless Sensor Networks," *Chemical Engineering Science*, **69**, 93-106, 2012.

[2] Sun, Y. and N. H. El-Farra, ``Quasi-decentralized Model-Based Networked Control of Process Systems," *Computers and Chemical Engineering*, **32**, 2016-2029, 2008.

Functional Observers for Nonlinear Systems

Costas Kravaris

Department of Chemical Engineering, University of Patras, 26500 Patras, Greece

Paper No. 687f, 2012 AIChE Annual Meeting, Pittsburgh, PA

Abstract - In this work, the problem of designing functional observers is formulated for general nonlinear systems, for a single output to be estimated and a single measured output. Notions of functional observer linearization are also formulated, the objective being to achieve exactly linear error dynamics in transformed coordinates, with prescribed rate of decay of the error. Necessary and sufficient conditions for the existence of a lower-order functional observer with linear dynamics and linear output map are derived.

I. INTRODUCTION

THE problem of estimating a function of the state vector, without the need of estimating the entire state vector, arises in many applications. The design of output feedback controllers based on a state feedback design is a classical example, where it is only the state feedback function that needs to be estimated and not the entire state vector. Another important class of applications is related to the design of inferential control systems, where one output is measured and a different output, which is unmeasured, needs to be regulated to set point.

These applications motivate the development of functional observers, the aim being a reduction of dimensionality relative to a full-state observer. The notion of a functional observer was first defined in Luenberger's pioneer work on observers for linear multivariable systems ([17],[18]). Luenberger proved that it is feasible to construct a functional observer with number of states equal to observability index minus one.

The basic theory of linear functional observers can be found in standard linear systems texts, e.g. [4]. In recent years, there has been a renewed interest in linear functional observers ([5],[6],[10],[11],[21],[22]), the goal being to find the smallest possible order of the linear functional observer, which, under certain conditions, could be less than observability index minus one.

For nonlinear systems, there have been significant developments in the theory of full-state observers, with a variety of methods and approaches. In particular, in the context of exact linearization methods ([1],[8],[9],[12-16]), Luenberger theory for full-state observers has been extended to nonlinear systems in a direct and analogous manner. The goal of the present work is to develop a direct generalization of Luenberger's functional observers to nonlinear systems.

In the present work, we will consider unforced nonlinear systems of the form

$$\begin{aligned} \frac{dx}{dt} &= f(x) \\ y &= h(x) \\ z &= q(x) \end{aligned} \tag{1}$$

where:

$x \in \mathbb{R}^n$ is the system state

$y \in \mathbb{R}$ is the measured output

$z \in \mathbb{R}$ is the output to be estimated

and $f:\mathbb{R}^n \to \mathbb{R}^n$, $h:\mathbb{R}^n \to \mathbb{R}$, $q:\mathbb{R}^n \to \mathbb{R}$ are smooth nonlinear functions. The objective is to construct a functional observer of order $\nu < n$, which generates an estimate of the output z, driven by the continuous measurement of the output y.

Section II will define the notion of functional observer for a system of the form (1) in a completely analogous manner to Luenberger's definition for linear systems. Section III will provide two examples where a low-order functional observer can be constructed guided by physical intuition. These will motivate our line of attack for the functional observer design problem, outlined in Section IV. Section V will define notions of exact linearization for the functional observer problem, in the same vein as they have been defined for full-state observers in the literature. Section VI will develop necessary and sufficient conditions for the solution of the total linearization problem, as well as a simple formula for the resulting functional observer. In Section VII, the results of Section VI will be specialized to linear time-invariant systems, leading to simple and easy-to-check conditions for the design of lower-order functional observers.

II. Definition of a Functional Observer for a Nonlinear Dynamic System

In complete analogy to Luenberger's construction for the linear case, we seek for a mapping

$$\xi = \theta(x) = \begin{bmatrix} \theta_1(x) \\ \vdots \\ \theta_\nu(x) \end{bmatrix} \tag{2}$$

from $\mathbb{R}^n$ to $\mathbb{R}^\nu$, to immerse system (1) to a ν-th order system ($\nu < n$), with input y and output z:

$$\begin{aligned} \frac{d\xi}{dt} &= \varphi(\xi, y) \\ z &= \omega(\xi, y) \end{aligned} \tag{3}$$

where $\varphi : \mathbb{R}^\nu \times \mathbb{R} \to \mathbb{R}^\nu$, $\omega : \mathbb{R}^\nu \times \mathbb{R} \to \mathbb{R}$, the aim being that system (3), when driven by the measured output y of (1), can be utilized to calculate the unmeasured output z of (1).

But in order for system (1) to be mapped to system (3) under the mapping $\theta(x)$, the following relations have to hold:

$$\frac{\partial \theta}{\partial x}(x) f(x) = \varphi\big(\theta(x), h(x)\big) \tag{4}$$

$$q(x) = \omega\big(\theta(x), h(x)\big) \tag{5}$$

The foregoing considerations lead to the following definition of a functional observer:

Definition 1: *Given a dynamic system*

$$\begin{aligned} \frac{dx}{dt} &= f(x) \\ y &= h(x) \\ z &= q(x) \end{aligned} \tag{1}$$

where $f : \mathbb{R}^n \to \mathbb{R}^n$, $h : \mathbb{R}^n \to \mathbb{R}$, $q : \mathbb{R}^n \to \mathbb{R}$, y *is the measured output and* z *is the output to be estimated, the system*

$$\begin{aligned} \frac{d\hat{\xi}}{dt} &= \varphi(\hat{\xi}, y) \\ \hat{z} &= \omega(\hat{\xi}, y) \end{aligned} \tag{6}$$

where $\varphi : \mathbb{R}^\nu \times \mathbb{R} \to \mathbb{R}^\nu$, $\omega : \mathbb{R}^\nu \times \mathbb{R} \to \mathbb{R}$ $(\nu < n)$, *is called a functional observer for (1), if in the series connection*

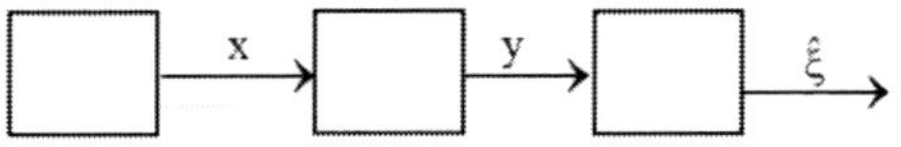

the overall dynamics

$$\begin{aligned} \frac{dx}{dt} &= f(x) \\ \frac{d\hat{\xi}}{dt} &= \varphi\big(\hat{\xi}, h(x)\big) \end{aligned} \tag{7}$$

possesses an invariant manifold $\hat{\xi} = \theta(x)$ *with the property that* $q(x) = \omega\big(\theta(x), h(x)\big)$.

In the above definition, the requirement that $\hat{\xi} = \theta(x)$ is an invariant manifold of (7), i.e. that

$\hat{\xi}(0) = \theta(x(0)) \Rightarrow \hat{\xi}(t) = \theta(x(t)) \quad \forall t > 0$, translates to $\left[\frac{\partial \theta}{\partial x}(x)\right] f(x) = \varphi\big(\theta(x), h(x)\big)$,

which is condition (4) stated earlier.

If the functional observer (6) is initialized consistently with the system (1), i.e. if $\hat{\xi}(0) = \theta(x(0))$, then $\hat{\xi}(t) = \theta(x(t))$, and therefore $\hat{z}(t) = \omega\big(\hat{\xi}(t), y(t)\big) = \omega\big(\theta(x(t)), h(x(t))\big) = q(x(t)) \quad \forall t > 0$, which means that the functional observer will be able to exactly reproduce z(t).

In the presence of initialization errors, additional stability requirements will need to be imposed on the $\hat{\xi}$-dynamics, for the estimate $\hat{z}(t)$ to asymptotically converge to $z(t)$.

At this point, it is important to examine the special case of a linear system, where $f(x) = Fx$, $h(x) = Hx$, $q(x) = Qx$ with F, H, Q being $n \times n$, $1 \times n$, $1 \times n$ matrices respectively, and a linear mapping $\theta(x) = Tx$ is considered. Definition 1 tells us that for a linear time-invariant system

$$\begin{aligned} \frac{dx}{dt} &= Fx \\ y &= Hx \\ z &= Qx \end{aligned} \tag{8}$$

the system

$$\begin{aligned} \frac{d\hat{\xi}}{dt} &= A\hat{\xi} + By \\ \hat{z} &= C\hat{\xi} + Dy \end{aligned} \tag{9}$$

will be a functional observer if the following conditions are met:

$$TF = AT + BH \tag{10}$$
$$Q = CT + DH \tag{11}$$

for some $\nu \times n$ matrix T. These are exactly Luenberger's conditions for a functional observer for a linear time-invariant system ([18]).

III. Two Motivating Examples

A. Non-isothermal Continuous Stirred Tank Reactor

Consider a Continuous Stirred Tank Reactor (CSTR), where consecutive irreversible exothermic reactions $A \rightarrow B \rightarrow C$ take place. Assuming that the reactor is well-mixed and has constant volume, that the feed is pure A, and that both reactions have first-order kinetics with Arrhenius dependence of the reaction rate constants on temperature, the dynamics of the reactor follows the equations ([7]):

$$\begin{aligned} \frac{dc_A}{dt} &= \frac{F}{V}(c_{A_{in}} - c_A) - k_{01} e^{-\frac{E_1}{RT}} c_A \\ \frac{dc_B}{dt} &= -\frac{F}{V} c_B + k_{01} e^{-\frac{E_1}{RT}} c_A - k_{02} e^{-\frac{E_2}{RT}} c_B \\ \frac{dT}{dt} &= \frac{F}{V}(T_{in} - T) - \frac{UA}{\rho c_p V}(T - T_J) + \frac{(-\Delta H)_{R_1}}{\rho c_p} k_{01} e^{-\frac{E_1}{RT}} c_A + \frac{(-\Delta H)_{R_2}}{\rho c_p} k_{02} e^{-\frac{E_2}{RT}} c_B \end{aligned} \tag{12}$$

where c_A, c_B are the concentrations of species A and B respectively, T is the temperature, $c_{A_{in}}$ is the concentration of species A in the inlet stream, T_J is the cooling jacket temperature, V is the reactor volume, F is the flow rate of inlet and outlet streams, E_1, E_2, k_{01}, k_{02} are the activation energies and pre-exponential factors of the reaction rates, $(-\Delta H)_{R_1}$, $(-\Delta H)_{R_2}$ are the heats of reaction, U, A are the heat transfer coefficient and heat transfer area respectively and ρ, c_p are the density and heat capacity respectively.

Suppose now that the measured output is the temperature

$$y = T \tag{13}$$

and that we want to monitor the conversion fraction of the reactant

$$z = \frac{c_{A_{in}} - c_A}{c_{A_{in}}} \tag{14}$$

By inspection of the dynamic model (12) of the reactor, we see immediately that the first equation of (12) could serve as an observer for c_A and then, z can be calculated from (14). Thus, we obtain the following functional observer:

$$\frac{d\hat{c}_A}{dt} = \frac{F}{V}(c_{A_{in}} - \hat{c}_A) - k_{01}e^{-\frac{E_1}{RT}}\hat{c}_A$$
$$\hat{z} = \frac{c_{A_{in}} - \hat{c}_A}{c_{A_{in}}} \tag{15}$$

This functional observer has been generated through the map $\xi = \theta(c_A, c_B, T) = c_A$ that has all the required properties to generate a functional observer.

However, in order for (15) to be of practical use, it must be a stable system, so that initialization error will decay. It is not difficult to see that this is indeed the case:
The error dynamics of (15) follows

$$\frac{d(\hat{c}_A - c_A)}{dt} = -\left(\frac{F}{V} + k_{01}e^{-\frac{E_1}{RT}}\right)(\hat{c}_A - c_A) \tag{16}$$

from which, because of the strict positivity of the quantity in parenthesis, it follows that the error asymptotically approaches zero.

B. Anaerobic Digester

Anaerobic digestion is a complex biochemical process, in which organic compounds are mineralised to biogas (a useful energy product), consisting primarily of methane and carbon dioxide. Anaerobic digestion of a soluble substrate can be modeled as a two-step process: the organic soluble substrate is first converted to a volatile fatty acid mixture by acidogenic bacteria and then the acids are utilized by methanogenic bacteria to form the biogas ([2]). Assuming that the anaerobic digestion takes place in a CSTR and that the feed only contains soluble substrate (no biomass and no volatile fatty acids), the mathematical description of the anaerobic digester dynamics takes the form ([3],[20]):

$$\begin{aligned}
\frac{dX_1}{dt} &= -DX_1 + \mu_1(S_1)X_1 \\
\frac{dS_1}{dt} &= D(S_0 - S_1) - \frac{1}{Y_1}\mu_1(S_1)X_1 \\
\frac{dX_2}{dt} &= -DX_2 + \mu_2(S_2)X_2 \\
\frac{dS_2}{dt} &= -DS_2 + c_{12}\frac{1}{Y_1}\mu_1(S_1)X_1 - \frac{1}{Y_2}\mu_2(S_2)X_2
\end{aligned} \tag{17}$$

where S_1 and S_2 are the concentrations of the soluble organic substrate and volatile fatty acids respectively, X_1 and X_2 are the concentrations of acidogenic and methanogenic biomass respectively, $\mu_1(S_1) = \frac{\mu_{max1} S_1}{K_{s1} + S_1}$ and $\mu_2(S_2) = \frac{\mu_{max2} S_2}{K_{s2} + S_2}$ are the specific growth rates of acidogenic and methanogenic bacteria respectively, with μ_{max1}, μ_{max2} the corresponding maximum specific growth rates and K_{s1}, K_{s2} the saturation constants, Y_1 and Y_2 are the biomass yield coefficients, c_{12} is the stoichiometric coefficient of conversion of S_1 to S_2, S_0 is the concentration of organic substrate in the feed, and D is the dilution rate.

Suppose now that the measured output is the methane production rate

$$y = Y_m\mu_2(S_2)X_2 \tag{18}$$

where Y_m is the yield coefficient for methane production, and that we want to monitor the conversion fraction of the organic substrate

$$z = \frac{S_0 - S_1}{S_0} \tag{19}$$

From the dynamic model (17) and the measured output equation (18), it is not difficult to see that the quantities

$$\begin{aligned}
\xi_1 &= \theta_1(X_1, S_1, X_2, S_2) = X_2 \\
\xi_2 &= \theta_2(X_1, S_1, X_2, S_2) = c_{12}(S_0 - S_1) - S_2
\end{aligned} \tag{20}$$

follow the differential equations

$$\frac{d\xi_1}{dt} = -D\xi_1 + \frac{1}{Y_m} y$$
$$\frac{d\xi_2}{dt} = -D\xi_2 + \frac{1}{Y_2 Y_m} y \qquad (21)$$

Moreover, the output z to be estimated can be expressed as a function of ξ_1, ξ_2, as follows:

$$z = \frac{1}{c_{12}S_0}\left(\mu_2^{-1}\left(\frac{y}{Y_m\xi_1}\right)+\xi_2\right) = \frac{1}{c_{12}S_0}\left(\frac{K_{s2}y}{Y_m\mu_{max\,2}\,\xi_1 - y}+\xi_2\right)$$

Thus, the dynamic system

$$\frac{d\hat{\xi}_1}{dt} = -D\hat{\xi}_1 + \frac{1}{Y_m} y$$
$$\frac{d\hat{\xi}_2}{dt} = -D\hat{\xi}_2 - \frac{1}{Y_2 Y_m} y \qquad (22)$$
$$\hat{z} = \frac{1}{c_{12}S_0}\left(\frac{K_{s2}y}{Y_m\mu_{max\,2}\,\hat{\xi}_1 - y}+\hat{\xi}_2\right)$$

is a functional observer for system (17) - (19).

Here the functional observer (22) has linear dynamics. As a result, its error dynamics is also linear:

$$\frac{d(\hat{\xi}_1 - \xi_1)}{dt} = -D(\hat{\xi}_1 - \xi_1)$$
$$\frac{d(\hat{\xi}_2 - \xi_2)}{dt} = -D(\hat{\xi}_2 - \xi_2) \qquad (23)$$

and, since $D > 0$, the error asymptotically converges to zero.

IV. Designing Lower – Order Functional Observers

For the design of a functional observer, one must be able to find a continuously differentiable map $\theta(x) = \begin{bmatrix} \theta_1(x) \\ \vdots \\ \theta_\nu(x) \end{bmatrix}$ to satisfy conditions (4) and (5), i.e. such that $\frac{\partial \theta_i}{\partial x}(x)f(x),\ i = 1,\cdots,\nu$ is a function of $\theta_1(x), \cdots, \theta_\nu(x), h(x)$ and $q(x)$ is a function of $\theta_1(x), \cdots, \theta_\nu(x), h(x)$.

However, such scalar functions $\theta_1(x), \cdots, \theta_\nu(x)$ may not exist, if ν is too small.

Moreover, even when they do exist, there is an additional very important requirement:

Since $\frac{\partial \theta}{\partial x}(x)f(x) = \varphi(\theta(x), h(x))$ will define the right hand side of the functional observer's dynamics, it must be such that the functional observer's dynamics is stable and the decay of the error is sufficiently rapid.

All the above requirements can be satisfied when $\begin{cases} \frac{dx}{dt} = f(x) \\ y = h(x) \end{cases}$ is a linearly observable system and $\nu = n-1$:

Available design methods for reduced order state observers (e.g. observer error linearization methods [9]) will generate a functional observer of order $\nu = n-1$, by only modifying the output map of the observer (so that the estimate of z is the observer output instead of the entire state vector).

The question is whether a lower order $\nu < n-1$ would be feasible and how to go about designing the functional observer. This is not an easy question because we will be trying to impose too many requirements at the same time.

In the examples of the previous Section, lower-order functional observers were successfully constructed. The first step involved identifying functions $\theta_1(x), \cdots, \theta_\nu(x)$ such that $\frac{\partial \theta_i}{\partial x}(x)f(x), i = 1,...,\nu$ and $q(x)$ can be expressed as functions of

$\theta_1(x), \cdots, \theta_v(x), h(x)$. The second step was to check stability of the error dynamics. This approach was successful because the selection of $\theta_1(x), \cdots, \theta_v(x)$ was directed by physical intuition.

Alternatively, one could try to follow the opposite path. As a first step, try to enforce stability: given some desirable dynamics for the observer $\frac{d\hat{\xi}}{dt} = \varphi(\hat{\xi}, y)$, with the φ - function so as to guarantee stability and rapid decay of the error, find $\theta(x) = \begin{bmatrix} \theta_1(x) \\ \vdots \\ \theta_v(x) \end{bmatrix}$ to satisfy $\frac{\partial \theta}{\partial x}(x)f(x) = \varphi(\theta(x), h(x))$ (4). The second step will then be to check if q(x) can be expressed as a function of $\theta(x)$ and $h(x)$ (condition (5)).

V. Exact Linearization of a Functional Observer

Along the second line of attack of the functional observer design problem, the most natural φ - function to work with is the linear one:

$$\varphi(\xi, y) = A\xi + By$$

or, slightly more generally,

$$\varphi(\xi, y) = A\xi + \beta(y)$$

It will then be the eigenvalues of the matrix A that will determine stability of the functional observer and the rate of decay of the error.

If we can find a continuously differentiable map $\theta(x)$ to satisfy the corresponding condition (4), i.e. to be a solution of the linear partial differential equation

$$\frac{\partial \theta}{\partial x}(x)f(x) = A\theta(x) + \beta(h(x))$$

for some Hurwitz matrix A, and if in addition $\theta(x)$ satisfies condition (5), i.e. that q(x) can be expressed as a function of $\theta(x)$ and $h(x)$, then we have a stable functional observer with linear dynamics:

$$\frac{d\hat{\xi}}{dt} = A\hat{\xi} + \beta(y)$$
$$\hat{z} = \omega(\hat{\xi}, y)$$

This leads to the Functional Observer Linearization Problem.

Note that stricter linearity requirements could be imposed, in particular, that $\beta(\cdot)$ and $\omega(\cdot,\cdot)$ be linear as well. This leads to the Functional Observer Total Linearization Problem.

A. Functional Observer Linearization Problem

Given a system of the form (1), find a functional observer of the form

$$\frac{d\hat{\xi}}{dt} = A\hat{\xi} + \beta(y), \qquad \hat{z} = \omega(\hat{\xi}, y) \tag{24}$$

where A *is a* $v \times v$ *matrix with stable eigenvalues and* $\beta: \mathbb{R} \times \mathbb{R}^v \to \mathbb{R}^v$, $\omega: \mathbb{R}^v \times \mathbb{R} \to \mathbb{R}$.

Equivalently, find a continuously differentiable mapping $\theta: \mathbb{R}^n \to \mathbb{R}^v$ *such that*

$$\left[\frac{\partial \theta}{\partial x}(x)\right] f(x) = A\theta(x) + \beta(h(x)) \tag{25}$$

and

$$q(x) = \omega(\theta(x), h(x)) \tag{26}$$

Assuming that the above problem can be solved, the resulting error dynamics will be linear

$$\frac{d}{dt}\left(\hat{\xi}-\theta(x)\right)=A\left(\hat{\xi}-\theta(x)\right)$$
$$\hat{z}-z=\omega\left(\hat{\xi},h(x)\right)-\omega\left(\theta(x),h(x)\right) \tag{27}$$

from which, for A having stable eigenvalues, the effect of the initial error $\hat{\xi}(0)-\theta(x(0))$ will die out, $\hat{\xi}(t)$ will approach $\theta(x(t))$ asymptotically, therefore $\hat{z}(t)$ will approach $z(t)$.

B. Functional Observer Total Linearization Problem

Given a system of the form (1), find a functional observer of the form

$$\frac{d\hat{\xi}}{dt}=A\hat{\xi}+By$$
$$\hat{z}=C\hat{\xi}+Dy \tag{28}$$

where A, B, C, D *are* $\nu\times\nu$, $\nu\times 1$, $1\times\nu$, 1×1 *matrices respectively, with* A *having stable eigenvalues. Equivalently, find a continuously differentiable mapping* $\theta:\mathbb{R}^n\to\mathbb{R}^\nu$ *such that*

$$\left[\frac{\partial\theta}{\partial x}(x)\right]f(x)=A\theta(x)+Bh(x) \tag{29}$$

and

$$q(x)=C\theta(x)+Dh(x) \tag{30}$$

Assuming that the above problem can be solved, the resulting error dynamics will be linear

$$\frac{d}{dt}\left(\hat{\xi}-\theta(x)\right)=A\left(\hat{\xi}-\theta(x)\right)$$
$$\hat{z}-z=C\left(\hat{\xi}-\theta(x)\right) \tag{31}$$

from which $\hat{z}(t)-z(t)=Ce^{At}\left(\hat{\xi}(0)-\theta(x(0))\right)$.

With the matrix A having stable eigenvalues, the effect of the initialization error $\hat{\xi}(0)-\theta(x(0))$ will die out, and $\hat{z}(t)$ will approach $z(t)$ asymptotically.

In order to solve either version of the Functional Observer Linearization Problem, it is natural to first try to solve the system of partial differential equations (25) or (29) given some small-size matrix A with fast enough eigenvalues, and then check to see if q(x) can be expressed as a function of θ(x) and h(x) according to (26) or (30). If it can, we have a functional observer with linear error dynamics - if not, we can try a different matrix A with different eigenvalues and/or larger size, up until we can satisfy both conditions.

The above approach will be feasible as long as the system of partial differential equations (25) or (29) is solvable. This is a singular system of partial differential equations because their principal part vanishes at the equilibrium point, and therefore standard existence/uniqueness results do not apply. However, we are covered by Lyapunov's Auxiliary Theorem for singular partial differential equations [19], as long as the nonlinear system (1) is analytic and in the Poincaré domain. The following Proposition is an immediate consequence of Lyapunov's Auxiliary Theorem. (See also [8].)

<u>Proposition 1</u>: *Let* $f:\mathbb{R}^n\to\mathbb{R}^n$, $h:\mathbb{R}^n\to\mathbb{R}$, $\beta:\mathbb{R}\to\mathbb{R}^\nu$ *be real analytic functions with* $f(0)=0$, $h(0)=0$, $\beta(0)=0$ *and denote by* $\sigma(F)$ *the set of eigenvalues of the matrix* $F=\frac{\partial f}{\partial x}(0)$. *Also, let A be a* $\nu\times\nu$ *matrix. Suppose:*

1. *All the eigenvalues of* A *are non-resonant with* $\sigma(F)$, *i.e. no eigenvalue* λ_j *of* A *is of the form* $\lambda_j=\sum_{i=1}^{n}m_i\kappa_i$, *with* $\kappa_i\in\sigma(F)$ *and* m_i *nonnegative integers, not all zero.*
2. *0 does not lie in the convex hull of* $\sigma(F)$.

Then the system of first-order partial differential equations

$$\frac{\partial\theta}{\partial x}(x)f(x)=A\theta(x)+\beta(h(x))$$

with initial condition $\theta(0)=0$, *admits a unique solution in a neighborhood of* $x=0$.

VI. Solvability Conditions for the Functional Observer Total Linearization Problem

The trial-and-error approach outlined in the previous section is in principle feasible, but it is far from being practical due to the computational effort involved in solving (25) or (29), which will be multiplied by the number of trials until (26) or (30) is satisfied.

To be able to develop a practical approach for designing functional observers, it would be helpful to develop criteria to check if for a given set of ν eigenvalues, there exists a functional observer whose error dynamics is governed by these eigenvalues. This will be done in the present Section for the Functional Observer Total Linearization Problem.

Before we proceed to the main result, we will need the following Lemma, which is a Corollary of Proposition1:

Lemma 1: *Let* $f:\mathbb{R}^n\to\mathbb{R}^n$, $h:\mathbb{R}^n\to\mathbb{R}$, $\gamma:\mathbb{R}\to\mathbb{R}$ *be real analytic functions with* $f(0)=0$, $h(0)=0$, $\gamma(0)=0$ *and denote by* $\sigma(F)$ *the set of eigenvalues of* $F=\frac{\partial f}{\partial x}(0)$ *and by* L_f *the Lie derivative operator* $L_f=\sum_{i=1}^{n} f_i(x)\frac{\partial}{\partial x_i}$.

Also, let $P(\lambda)=\lambda^\nu+\alpha_1\lambda^{\nu-1}+\cdots+\alpha_{\nu-1}\lambda+\alpha_\nu$ *be a given ν-th degree polynomial. Suppose:*

1. *All the roots of* $P(\lambda)$ *are non-resonant with* $\sigma(F)$, *i.e. no root* λ_j *of* $P(\lambda)$ *is of the form* $\lambda_j=\sum_{i=1}^{n} m_i\kappa_i$, *with* $\kappa_i\in\sigma(F)$ *and* m_i *nonnegative integers, not all 0.*
2. *0 does not lie in the convex hull of* $\sigma(F)$.

Then the scalar ν-th order partial differential equation

$$L_f^\nu \upsilon(x)+\alpha_1 L_f^{\nu-1}\upsilon(x)+\cdots+\alpha_{\nu-1}L_f\upsilon(x)+\alpha_\nu\upsilon(x)=\gamma(x)$$

with initial condition $\upsilon(0)=0$, *admits a unique solution* $\upsilon(x)$ *in a neighborhood of* $x=0$.

Proof of Lemma 1: Under the stated assumptions, the system of partial differential equations

$$\frac{\partial\theta}{\partial x}(x)f(x)=\begin{bmatrix}0 & 1 & \cdots & 0\\ 0 & 0 & \cdots & 0\\ \vdots & \vdots & & \vdots\\ 0 & 0 & \cdots & 1\\ -\alpha_\nu & -\alpha_{\nu-1} & \cdots & -\alpha_1\end{bmatrix}\theta(x)+\begin{bmatrix}0\\0\\ \vdots\\0\\1\end{bmatrix}\gamma(x) \tag{32}$$

admits a unique solution, as a consequence of Proposition 1. Moreover, setting $\theta_1(x)=\upsilon(x)$, it easily follows that

$$\begin{aligned}\theta_2(x)&=L_f\upsilon(x)\\ \theta_3(x)&=L_f^2\upsilon(x)\\ &\vdots\\ \theta_\nu(x)&=L_f^{\nu-1}\upsilon(x)\end{aligned},$$

and that $\upsilon(x)$ satisfies

$$L_f^\nu \upsilon(x)+\alpha_1 L_f^{\nu-1}\upsilon(x)+\cdots+\alpha_{\nu-1}L_f\upsilon(x)+\alpha_\nu\upsilon(x)=\gamma(x) \tag{33}$$

Finally, because of the uniqueness of the solution of the system of partial differential equations (32), it follows that the solution of (33) is also unique. □

We can now establish the main result:

Proposition 2: *Under the assumptions of Proposition 1, for a nonlinear system of the form* (1), *there exists a functional observer of the form*

$$\begin{aligned}\frac{d\hat{\xi}}{dt}&=A\hat{\xi}+By\\ \hat{z}&=C\hat{\xi}+Dy\end{aligned} \tag{28}$$

with (A,B) controllable pair and the eigenvalues of A being the roots of a given polynomial $\lambda^\nu+\alpha_1\lambda^{\nu-1}+\cdots+\alpha_{\nu-1}\lambda+\alpha_\nu$, *if and only if* $L_f^\nu q(x)+\alpha_1 L_f^{\nu-1}q(x)+\cdots+\alpha_{\nu-1}L_f q(x)+\alpha_\nu q(x)$ *is a* $\mathbb{R}$*-linear combination of* $h(x), L_f h(x),\ldots,L_f^\nu h(x)$.

Proof of Proposition 2: Without loss of generality, consider the observer dynamics in controllable canonical form

$$\frac{d\hat{\xi}}{dt}=\begin{bmatrix}0 & 1 & \cdots & 0\\ 0 & 0 & \cdots & 0\\ \vdots & \vdots & & \vdots\\ 0 & 0 & \cdots & 1\\ -\alpha_v & -\alpha_{v-1} & \cdots & -\alpha_1\end{bmatrix}\hat{\xi}+\begin{bmatrix}0\\0\\ \vdots\\0\\1\end{bmatrix}y \tag{34}$$

Suppose now that (34) is a functional observer for the nonlinear system (1). Then, condition (29) implies that there exists a mapping $\theta(x)$ such that

$$\frac{\partial\theta}{\partial x}(x)f(x)=\begin{bmatrix}0 & 1 & \cdots & 0\\ 0 & 0 & \cdots & 0\\ \vdots & \vdots & & \vdots\\ 0 & 0 & \cdots & 1\\ -\alpha_v & -\alpha_{v-1} & \cdots & -\alpha_1\end{bmatrix}\theta(x)+\begin{bmatrix}0\\0\\ \vdots\\0\\1\end{bmatrix}h(x) \tag{35}$$

or, written component-wise,

$$\begin{aligned}
&L_f\theta_1(x)=\theta_2(x)\\
&L_f\theta_2(x)=\theta_3(x)\\
&\quad\vdots\\
&L_f\theta_{v-1}(x)=\theta_v(x)\\
&L_f\theta_v(x)=-\alpha_v\theta_1(x)-\alpha_{v-1}\theta_2(x)-\cdots-\alpha_1\theta_v(x)+h(x)
\end{aligned}$$

from which, it follows that

$$\begin{aligned}
&L_f^v\theta_1(x)+\alpha_1L_f^{v-1}\theta_1(x)+\cdots+\alpha_v\theta_1(x)=h(x)\\
&L_f^v\theta_2(x)+\alpha_1L_f^{v-1}\theta_2(x)+\cdots+\alpha_v\theta_2(x)=L_fh(x)\\
&\quad\vdots\\
&L_f^v\theta_v(x)+\alpha_1L_f^{v-1}\theta_v(x)+\cdots+\alpha_v\theta_v(x)=L_f^{v-1}h(x)
\end{aligned} \tag{36}$$

At the same time, the functions $\theta_1(x),\cdots,\theta_v(x)$ must satisfy condition (30), i.e.

$$q(x)=C_1\theta_1(x)+C_2\theta_2(x)+\cdots+C_v\theta_v(x)+Dh(x)$$

for some $C_1,C_2,\cdots,C_v,D\in\mathbb{R}$.

Then, applying the operator $L_f^v+\alpha_1L_f^{v-1}+\cdots+\alpha_{v-1}L_f+\alpha_vI$ on both sides of the above relation and using (36),

$$\begin{aligned}
&L_f^vq(x)+\alpha_1L_f^{v-1}q(x)+\cdots+\alpha_{v-1}L_fq(x)+\alpha_vq(x)\\
&=C_1h(x)+C_2L_fh(x)+\cdots+C_vL_f^{v-1}h(x)+D(L_f^vh(x)+\alpha_1L_f^{v-1}h(x)+\cdots+\alpha_{v-1}L_fh(x)+\alpha_vh(x))
\end{aligned}$$

or

$$\begin{aligned}
&L_f^vq(x)+\alpha_1L_f^{v-1}q(x)+\cdots+\alpha_{v-1}L_fq(x)+\alpha_vq(x)\\
&=DL_f^vh(x)+(\alpha_1D+C_v)L_f^{v-1}h(x)+\cdots+(\alpha_{v-1}D+C_2)L_fh(x)+(\alpha_vD+C_1)h(x)
\end{aligned}$$

Hence, $L_f^vq(x)+\alpha_1L_f^{v-1}q(x)+\cdots+\alpha_{v-1}L_fq(x)+\alpha_vq(x)$ is $\mathbb{R}$-linear combination of $h(x),L_fh(x),\ldots,L_f^vh(x)$.

Conversely, suppose that $L_f^vq(x)+\alpha_1L_f^{v-1}q(x)+\cdots+\alpha_{v-1}L_fq(x)+\alpha_vq(x)$ is $\mathbb{R}$-linear combination of $h(x),L_fh(x),\ldots,L_f^vh(x)$, i.e. there exist $\beta_0,\beta_1,\cdots,\beta_{v-1},\beta_v\in\mathbb{R}$ such that

$$L_f^vq(x)+\alpha_1L_f^{v-1}q(x)+\cdots+\alpha_{v-1}L_fq(x)+\alpha_vq(x)=\beta_0L_f^vh(x)+\beta_1L_f^{v-1}h(x)+\cdots+\beta_{v-1}L_fh(x)+\beta_vh(x)$$

Then

$$\begin{aligned}
&L_f^vq(x)+\alpha_1L_f^{v-1}q(x)+\cdots+\alpha_{v-1}L_fq(x)+\alpha_vq(x)\\
&=\beta_0(L_f^vh(x)+\alpha_1L_f^{v-1}h(x)+\cdots+\alpha_{v-1}L_fh(x)+\alpha_vh(x))+(\beta_v-\beta_0\alpha_v)h(x)+\cdots+(\beta_1-\beta_0\alpha_1)L_f^{v-1}h(x)
\end{aligned} \tag{37}$$

Now let $\theta(x)=\begin{bmatrix}\theta_1(x)\\ \vdots \\ \theta_v(x)\end{bmatrix}$ be the solution of (35). Then, the functions $\theta_1(x),\cdots,\theta_v(x)$ will satisfy (36), and consequently, the function $p(x)=\beta_0 h(x)+(\beta_v-\beta_0\alpha_{v-1})\theta_1(x)+\cdots+(\beta_1-\beta_0\alpha_1)\theta_v(x)$

will satisfy

$$\begin{aligned}&L_f^v p(x)+\alpha_1 L_f^{v-1}p(x)+\cdots+\alpha_{v-1}L_f p(x)+\alpha_v p(x)\\ &=\beta_0(L_f^v h(x)+\alpha_1 L_f^{v-1}h(x)+\cdots+\alpha_{v-1}L_f h(x)+\alpha_v h(x))+(\beta_v-\beta_0\alpha_v)h(x)+\cdots+(\beta_1-\beta_0\alpha_1)L_f^{v-1}h(x)\end{aligned}$$

From the above and (37), Lemma 1 implies that $p(x)=q(x)$, therefore

$$q(x)=\beta_0 h(x)+(\beta_v-\beta_0\alpha_{v-1})\theta_1(x)+\cdots+(\beta_1-\beta_0\alpha_1)\theta_v(x)$$

The conclusion is that both conditions for the solution of the Functional Observer Total Linearization Problem are satisfied and

$$\begin{aligned}\frac{d\hat{\xi}}{dt}&=\begin{bmatrix}0&1&\cdots&0\\0&0&\cdots&0\\ \vdots&\vdots&&\vdots\\0&0&\cdots&1\\-\alpha_v&-\alpha_{v-1}&\cdots&-\alpha_1\end{bmatrix}\hat{\xi}+\begin{bmatrix}0\\0\\ \vdots\\0\\1\end{bmatrix}y\\ \hat{z}&=\left[(\beta_v-\beta_0\alpha_v)\ \ (\beta_{v-1}-\beta_0\alpha_{v-1})\ \ \cdots\ \ (\beta_1-\beta_0\alpha_1)\right]\xi+\beta_0 y\end{aligned}\tag{38}$$

is a functional observer.

This completes the proof.

VII. Lower-Order Functional Observers for Linear Systems

The results of the previous section can now be specialized to linear time-invariant systems. The following is a Corollary to Proposition 2.

Proposition 3: *For a linear time-invariant system of the form*

$$\begin{aligned}\frac{dx}{dt}&=Fx\\ y&=Hx\\ z&=Qx\end{aligned}\tag{8}$$

there exists a functional observer of the form

$$\begin{aligned}\frac{d\hat{\xi}}{dt}&=A\hat{\xi}+By\\ \hat{z}&=C\hat{\xi}+Dy\end{aligned}\tag{28}$$

with (A,B) *controllable pair and the eigenvalues of* A *being the roots of a given polynomial* $\lambda^v+\alpha_1\lambda^{v-1}+\cdots+\alpha_{v-1}\lambda+\alpha_v$, *if and only if*

$$(QF^v+\alpha_1 QF^{v-1}+\cdots+\alpha_{v-1}QF+\alpha_v Q)\in\operatorname{span}\left\{H,HF,\ldots,HF^v\right\}.$$

The above Proposition provides a simple and easy-to-check feasibility criterion for a lower-order functional observer with a pre-specified set of eigenvalues governing the error dynamics.

VIII. Conclusion

The present work has developed a direct generalization of Luenberger's functional observers to nonlinear systems. It has formulated notions of exact linearization for functional observer design and has derived a specific criterion for total linearization, including a simple formula for the resulting functional observer.

REFERENCES

[1] V. Andrieu and L.Praly, "On the existence of a Kazantzis-Kravaris/ Luenberger observer," *SIAM J. Control and Optimization,* Vol. 45, pp. 432-456, 2006.

[2] J. E. Bailey and D. F. Ollis, *Biochemical Engineering Fundamentals,* 2nd ed. New York: McGraw-Hill, 1986.

[3] G. Bastin and D. Dochain, *On-Line Estimation and Adaptive Control of Bioreactors.* Amsterdam: Elsevier, 1990.

[4] C. T. Chen, *Linear System Theory and Design.* New York: Holt, Rinehart and Winston, 1984.

[5] M. Darouach, "Existence and design of functional observers for linear systems," *IEEE Trans. Automat. Contr.*, Vol. 45, No.5, pp. 940-943, 2000.

[6] T. L. Fernando, H.M. Trinh, and L.Jennings, "Functional observability and the design of minimum order linear functional observers," *IEEE Trans. Automat. Contr.*, Vol. 55, No.5, pp. 1268-1273, 2010.

[7] H. S. Fogler, *Elements of Chemical Reaction Engineering.* Englewood Cliffs: Prentice Hall, 1992.

[8] N. Kazantzis and C. Kravaris, "Nonlinear observer design using Lyapunov's auxiliary theorem," *Systems & Control Letters,* Vol. 34, pp. 241-247, 1998.

[9] N. Kazantzis and C. Kravaris, "Nonlinear observer design for process monitoring," *Ind. Eng. Chem. Res.*, Vol. 39, pp. 408-419, 2000.

[10] S. K. Korovin, I. S. Medvedev, and V. V. Fomichev, "Minimum dimension of a functional observer with a given convergence rate," *Doklady Mathematics*, Vol. 78, No.3, pp. 940-943, 2008.

[11] S. K. Korovin, I. S. Medvedev, and V. V. Fomichev, "Minimum-order functional observers," *Computational Mathematics and Modeling*, Vol. 21, No.3, pp. 275-296, 2010.

[12] G. Kreisselmeier and R. Engel, "Nonlinear observers for autonomous Lipschitz continuous systems," *IEEE Trans. Automat. Contr.*, Vol. 48, No.3, pp. 451-464, 2003.

[13] A. J. Krener and A. Isidori, "Linearization by output injection and nonlinear observers," *Systems & Control Letters*, 3, 1983, pp. 47-52.

[14] A. J. Krener and W. Respondek, "Nonlinear observers with linearizable error dynamics," *SIAM J. Control and Optimization,* Vol. 23, pp.197-216, 1985.

[15] A. J. Krener and M. Xiao, "Nonlinear observer design in the Siegel domain," *SIAM J. Control and Optimization*, Vol. 41, No. 3, pp. 932-953, 2002. See also "Erratum: Nonlinear observer design in the Siegel domain," *SIAM J. Control and Optimization*, Vol. 43, No.1, pp. 377-378, 2004.

[16] A. J. Krener and M. Xiao, "Nonlinear observer design for smooth systems," in *Chaos in Automatic Control*, W. Perruquetti and J-P Barbot, Eds., Marcel Dekker, 2005, pp. 411-422.

[17] D. G. Luenberger, "Observers for multivariable systems," *IEEE Trans. Automat. Contr.,* Vol. AC-11, No.2, pp.190-197, 1966.

[18] D. G. Luenberger, "An introduction to observers," *IEEE Trans. Automat. Contr.,* Vol. AC-16, No.6, pp.596-602, 1971.

[19] A. M. Lyapunov, *The General Problem of the Stability of Motion.* (English Translation by A. T. Fuller) London: Taylor & Francis, 1992.

[20] M. Perrier and D. Dochain, "Evaluation of Control Strategies for Anaerobic Digestion Processes," *International Journal on Adaptive Control and Signal Processing,* Vol. 7, pp. 309-321, 1993.

[21] H. Trinh, T. Fernando, and S. Nahavandi, "Partial-state observers for nonlinear systems," *IEEE Trans. Automat. Contr.*, Vol. 51, No.11, pp. 1808-1812, 2006.

[22] C.-C. Tsui, "What is the minimum function observer order?," *J. Franklin Inst.*, Vol. 335B, No.4, pp. 623-628, 1998.

(687g) Distributed MPC Design for Nonlinear Two-Time-Scale Process Networks

Xianzhong Chen[1], Mohsen Heidarinejad[2], Jinfeng Liu[3] and Panagiotis D. Christofides[3], (1)Chemical and Biomolecular Engineering, Univ. of California, Los Angeles, Los Angeles, CA, (2)Electric Engineering, University of California, Los Angeles, Los Angeles, CA, (3)Department of Chemical and Biomolecular Engineering, University of California, Los Angeles, Los Angeles, CA

Chemical processes and plants are characterized by nonlinear behavior and strong coupling of physico-chemical phenomena occurring at different time-scales. Tradionally, singular perturbation theory is used as a framework for modeling, analysis, order reduction and controller design for nonlinear two-time-scale processes [1]. Rapid advance of wireless communication has inspired augmentation of traditional point-to-point and wired local control systems with networked control systems. Model predictive control (MPC) is an appropriate framework to deal with the design and coordination of networked control systems because of its ability to account for process/controller interactions in the calculation of the control actions. MPC is an online optimization-based approach, which takes advantage of a system model to compute a future manipulated input trajectory by minimizing a typically quadratic cost function involving penalties on the system state and control action. Typically MPC is studied within a centralized architecture; however, when the number of manipulated inputs increases, it may not be computationally efficient to obtain control inputs in a centralized manner. To reduce the dimensionality of centralized MPC as well as computational complexity, different manipulated inputs may be obtained through a number of distributed MPCs in a coordinated manner. Depending on the application and certain specifications, sequential or iterative distributed schemes are employed to obtain manipulated inputs [2]. Furthermore, accounting for two-time-scale process network behavior may lead to further simplifications on the distributed MPC structure.

In this work, we focus on distributed MPC of nonlinear singularly perturbed systems in standard form where the separation between the fast and slow state variables is explicit; such models arise naturally in the context of chemical process networks. Specifically, a composite control system comprised of distributed ''fast'' MPCs acting to regulate the fast dynamics and distributed ''slow'' MPCs acting to regulate the slow dynamics is designed. The composite distributed MPC system uses multirate sampling of the plant state measurements, i.e., fast sampling of the fast state variables is used in the distributed fast MPC and slow-sampling of the slow state variables is used in the distributed slow MPC as well as in the fast MPC. Both fast and slow distributed MPCs take advantage of their corresponding Lyapunov-based controllers in an iterative or sequential manner to characterize closed-loop system stability region [3]. Using singular perturbation theory, the stability and optimality of the closed-loop nonlinear singularly perturbed system are analyzed. The proposed distributed control scheme does not require communication between the fast and slow layers, and thus, it can be classified as decentralized in terms of interaction between two layer while at each layer it implements a distributed control scheme. A chemical process network example which exhibits two-time-scale behavior is used to demonstrate the structure and implementation of the distributed fast–slow MPC architecture in a practical setting. Extensive simulations are carried out to assess the performance and computational efficiency of the fast–slow MPC system.

[1] P. D. Christofides, P. Daoutidis. "Feedback control of two-time-scale nonlinear systems". International Journal of Control, Vol. 63, Pages 965–994, 1996.

[2] J. Liu, X. Chen, D. Munoz de la Pena, P. D. Christofides . "Sequential and iterative architectures for distributed model predictive control of nonlinear process systems". AIChE Journal, Vol. 56, Pages 2137–2149, 2010.

[3] P. D. Christofides, J. Liu, and D. Munoz de la Pena. "Networked and Distributed Predictive Control: Methods and Nonlinear Process Network Applications". Advances in Induatrial Control Series. Springer-Verlag, London, England, 2011.

(698a) Non-Catalytic Oxidation of Carbon Black and Diesel Engine Soot Samples - Kinetics and Structure-Activity Relationships

Hom Sharma[1], Lakshitha Pahalagedara[2], Ameya Joshi[3], Steven Suib[2] and Ashish B. Mhadeshwar[1], (1)Department of Chemical, Materials, and Biomolecular Engineering, University of Connecticut, Storrs, CT, (2)Department of Chemistry, University of Connecticut, Storrs, CT, (3)Modeling and Simulation, Corning Incorporated, Corning, NY

More than 13 million [[1]] diesel vehicles are responsible for transportation of 94% of the goods in US [[2]] and use 4 million barrels of diesel per day [[3]]. Diesel cars contribute up to 53% of the passenger cars sold in the European market [[4]]. Despite the efficiency, durability, low operating cost, and reliability of diesel engines, Particulate Matter (PM) emissions remain responsible for various human health and environmental problems [[5]]. Diesel Particulate Filters (DPFs) are the most popular aftertreatment technology to meet the stringent PM emissions standards [[6], [7]]. However, periodic regeneration of DPF through efficient oxidation of PM or soot remains a major challenge, especially due to the complex nature of soot. Structure and oxidation activity of diesel soot is attributed to the combined effects of fuel, lubricating oil, operating conditions, and engine type [[8]]. A wide range of kinetic parameters have been reported for soot oxidation, e.g., activation energy lies in the range of 40-300 kJ/mol [[9]]. Current understanding about the soot morphology and its quantitative impact on the oxidation activity is also limited. In this work, we have investigated the non-catalytic oxidation kinetics of 14 carbon black samples including diesel engine soot using Thermo Gravimetric Analysis and various characterization methods such as SEM, TEM, XRD, BET, and Raman (see Figure below for an example). Effect of various parameters, such as partial pressure of oxygen, ramp rate, flow rate, is also investigated in the Temperature Programmed Oxidation (TPO) experiments. Based on this comprehensive study – carried out for the first time – we will present insights regarding how the soot morphology affects its oxidation activity and the corresponding structure-activity correlations.

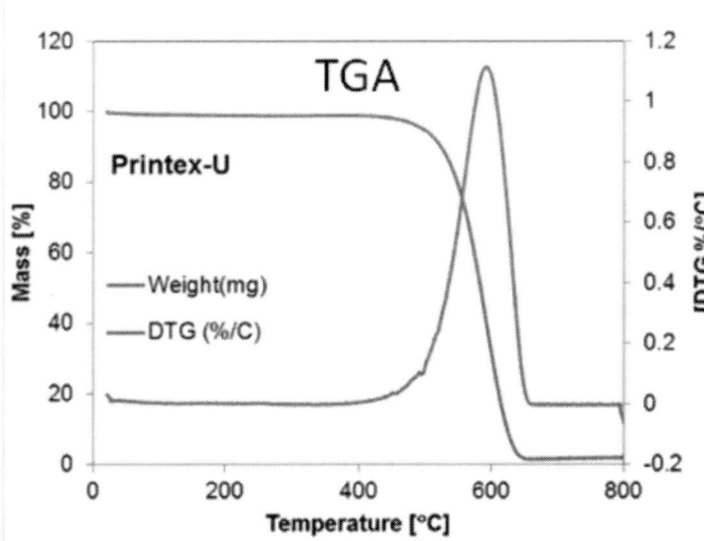

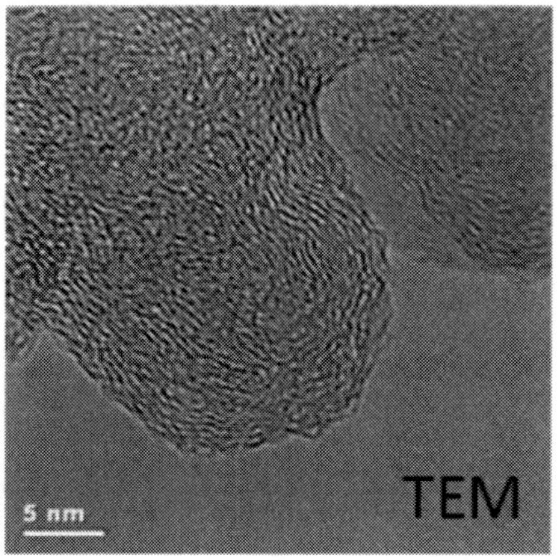

References:

[[1]] Schneider, C.G. and Hill, L.B., "Diesel and Health in America: The Lingering Threat", http://www.catf.us/resources/publications/files/Diesel_Health_in_America.pdf, 5 Feb. 2005.

[[2]] "Diesel Fuel Explained, Use of Diesel", http://www.eia.gov/energyexplained/index.cfm?page=diesel_use

[[3]] "Petroleum watch", http://energyalmanac.ca.gov/petroleum/petroleum_watch/2011-09-16_Petroleum_Watch.pdf, 16 Sep. 2011.

[[4]] Bensaid, S., Caroca, C.J., Russo, N., and Fino, D., "Detailed investigation of non-catalytic DPF regeneration", *The Canadian Journal of Chemical Engineering*, *89*(2), 401-407 (2011).

[[5]] Prasad, R. and Venkateswara Rao B. "A Review on Diesel Soot Emission, its Effect and Control", *Bulletin of Chemical Reaction Engineering and Catalysis*, 5(2) 69-86 (2010).

[[6]] "National Ambient Air Quality Standards (NAAQS)", http://epa.gov/air/criteria.html, 8 Nov. 2011.

[[7]] "Emissions Standards", http://www.dieselnet.com/standards/eu/hd.php, 6 Feb. 2012.

[[8]] Yezerets, A., Currier, N.W., Kim, D.H., Eadler, H., Epling, W.S., and Peden, C.H.F., "Differential kinetic analysis of diesel particulate matter (soot) oxidation by oxygen using a step–response technique", *Applied Catalysis B: Environmental*, *61*(1-2), 120-129 (2005).

[[9]] Kalogirou, M. and Samaras, Z., "Soot oxidation kinetics from TG experiments", *Journal of Thermal Analysis and Calorimetry*, *99*(3), 1005-1010 (2010).

(698b) Green Power Production From Natural Gas

Abdulrahman Albassam, University of California, Los Angeles, Los Angeles, CA and **Vasilios Manousiouthakis**, Chemical & Biomolecular Engineering Department,, University of California Los Angeles, Los Angeles, Los Angeles, CA

Natural gas accounts for 22% of the electrical power production in the US, a figure which means a massive carbon dioxide emission to the atmosphere which results in a notable environmental impact. In an effort to avoid such emissions we are proposing an idea in which existing technologies could be used in a unique design for the cogeneration of electrical power and economically valuable chemicals that would be produces via further processing of the carbon dioxide resulting from the natural gas processing. The design proposed might offer a slight decline in electrical power output compared to existing power plants but further economical analysis provides interesting findings in support of the new design in which the chemicals produced as co product could be marketable as a new source of revenue. The idea proposed discusses a sustainable system of producing power and formic acid via natural gas and provides flow sheet simulation through uni sim.

(698c) Construction of Attainable Region with Volumetric Constraints

Zayna Alhusseini, chemcial and biomolecular Engineering, UCLA, Los Angeles, CA and **Vasilios Manousiouthakis**, Chemical & Biomolecular Engineering Department,, University of California Los Angeles, Los Angeles, Los Angeles, CA

Since reactor design is always under challenge for chemical engineers, cost factors in the design process must also be considered. One way of considering the cost of a reactor is my constructing the AR with an economic consideration incorporated into the calculation. The reactor cost can be correlated to volumetric capacity. Constructing the AR under the constrain of total network volume can provide a measure of cost control.

In this work, the AR problem is investigated, within the Infinite DimEnsionAl State-space IDEAS framework, for a network of reactor with known RTD and isothermal reactor networks. Constructing of the AR of a Segregated Laminar Flow Reactor (SLFR) to produce the AR for the network subject to volumetric constraints. The IDEAS conceptual framework is realized through solution of a series of finite dimensional linear programs the resulting mathematical formulation is an infinite dimensional program with linear constraints and a separable concave objective function. The proposed methodology is illustrated using the Trambouze reaction kinetics.

(698d) Evaluation of Cathode Materials for Benthic Microbial Fuel Cells

Lewis Hsu[1], Adriane Wotawa-Bergen[2], Jeffery Kagan[2], Richard Bell[2], Ken Richter[2], Bart Chadwick[2] and Y. Meriah Arias-Thode[2], (1)Energy and Environmental Sciences Division, SPAWAR Systems Center Pacific, San Diego, CA, (2)SPAWAR Systems Center Pacific

Microbial fuel cells are devices which are able to directly produce electrical energy by the consumption of biodegradable compounds. Major research thrusts in this area focus largely on simultaneous energy production and clean water production from wastewater streams rich in organic material. A different approach to utilizing this technology is the application of benthic microbial fuel cells. These devices seek to harvest energy from the organic material contained in the ocean environment to power sensors or communication devices.

One of the major challenges to implementing these devices is the use of appropriate materials for system construction. The study presented here looks at different possible materials for use at the cathode of the fuel cell to perform oxygen reduction. The performance of several carbon-based electrodes (graphite, carbon cloth, and graphite fiber brushes) will be shown. Determinations of optimal anode:cathode size ratios and system loadings were also determined. Influence (beneficial or detrimental) of biological growth on these materials will also be discussed.

(698e) Reactive Absorption and Kinetics of Nitric Oxide (NO) Removal by Aqueous Persulfate Activated by Temperature and Iron(II) Ions

Yusuf G. Adewuyi, Chemical and Bioengineering, North Carolina A&T State University, Greensboro, NC and Nana Y Sakyi, Chemical and Bioengineering, North Carolina Agricultural and Technical State University, Greensboro, NC

The combustion of fossil fuel (e.g., in coal-fired plants) releases a large amount of NO_X (mainly NO and NO_2) and SO_2 into the atmosphere. Both NO_X and SO_2 contribute to the formation of ground-level ozone, urban photochemical smog, acid rain, eutrophication, respiratory and cardiovascular diseases. The efficient, cost-effective and environmentally friendlier removal of NO_X from flue gases poses a considerable industrial problem. Available control technologies such as Selective Catalytic Reduction (SCR), used for high NO_X removal has very high capital cost and suffer from various problems including disposal of spent toxic catalysts. Selective Noncatalytic Reduction (SNCR, urea injection) is highly temperature dependent and SNCR, ammonia injection (Thermal $DeNO_X$) has moderately high capital cost and ammonia handling problems. Low-NO_X burners (LNB) also has moderately high capital cost. Aqueous scrubbing with chemical oxidants provides alternative treatment techniques. While a number of chemicals such hydrogen peroxide, peroxymonosulfate (oxone) and sodium chlorite have been used, there are continuing need for reagents that are inexpensive, efficient, and environmentally benign. Sodium persulfate (or peroxydisulfate) is one such compound. The peroxydisulfate anion ($S_2O_8^{2-}$) is a strong oxidizing agent and because it is stable prior to activation, it is becoming increasingly popular in the in situ chemical oxidation. In this work, the absorption and kinetics of nitric oxide (NO) oxidation by aqueous solutions of sodium persulfate ($Na_2S_2O_8$) activated by temperature with and without Fe^{2+} were investigated via experimental and modeling studies in a bubble column reactor operated in the semibatch mode. The effects of persulfate (0.01−0.2M), Fe^{2+} (0−0.1M), and gas-phase NO (500−1000ppm) concentrations, and temperatures (23−90°C) on fractional conversion of NO were investigated. Initial runs conducted at 50°C showed 0.01M Fe^{2+} as the optimum for NO absorption for the different persulfate concentrations studied. At 0.1M $Na_2S_2O_8$ and 0.01M Fe^{2+} concentrations, fractional conversions of NO (percent of inlet NO removed) were 52.1% and 63.4% respectively, for temperature-alone and combined temperature-Fe^{2+} activation. In general, the NO conversion at all temperatures in or without the presence of Fe^{2+} increased sharply with persulfate concentration up to about 0.1M before leveling off. Increased temperature led to increased conversions of NO at all persulfate levels. At 0.1 M persulfate concentration, conversions up to 69% and 92% were observed at 70°C and 90°C, respectively. However, in the presence of 0.01M Fe^{2+} concentrations, conversions of up to 79% and approximately 100% were observed at 70°C and 90°C, respectively. The results indicate that Fe^{2+} activation further improved NO removal by about 10% at all temperatures. Solution pH and chloride concentration both were found to affect NO absorption at the higher temperatures but the effect of initial NO concentration was found to be marginal. The results also demonstrate the synergistic and antagonistic effects of Fe^{2+}, which is dependent on the persulfate-Fe^{2+} ratio. In addition to mechanistic reaction pathways proposed, models were developed to correlate the experimental data and estimate kinetic rate constants, reaction activation energies, and mass-transfer parameters for the NO_X-persulfate-iron (II) reactions. The results of this work and current evaluation of the effects of SO_2 and other flue gas components will be presented.

References:

1. Adewuyi, Y.G.; Sakyi, N.Y. Kinetic and Mechanistic Investigations of Nitric Oxide (NO) Removal by Aqueous Persulfate Activated by Temperature and Iron (II) Ions. Environmental Science and Technology. 2012. In review.

2. Khan, N.E.; Adewuyi, Y.G. Absorption and Oxidation of Nitric Oxide (NO) by Aqueous Solutions of Sodium Persulfate in a Bubble Column Reactor. Industrial & Engineering Chemistry Research. 2010, 49 8749-8760.

3. Adewuyi, Y.G.; Owusu, S.O. The Aqueous Absorption and Oxidation of Nitric Oxide with Oxone for the Treatment of Tail Gases: Process Feasibility, Stoichiometry, Reaction Pathways and Absorption Rate. Industrial & Engineering Chemistry Research, 2003,42, 4084-4100.

4. Adewuyi, Y.G., Khan, N.E. Modeling the Ultrasonic Cavitation-Enhanced Removal of Nitrogen Oxide (NO) in a Bubble Column Reactor. AICHE J. 2011 http://onlinelibrary.wiley.com/doi/10.1002/aic.12751/pdf

(698f) Challenges of Lowering Aromatic Content in KSA Gasoline

Nasiru M. Tukur and Saeed S. Al-Alloush, P&CSD, Saudi Aramco, Dhahran, Saudi Arabia

The number of registered vehicles in the Kingdom of Saudi Arabia (KSA) has increased more than 45% during the last 10 years. This has been attributed to high population growth (CAGR of 3.1% from 2004-10) and rising per capita income of the Kingdom (CAGR of 3.5% during 2000-08). Furthermore, road transport fuel outlook has shown an annual average growth rate of 4.7% reaching about 1.9 million B/Doe in 2030. The anticipated traffic growth is expected to present a key challenge to maintaining and improving air quality.

Mobile sources have since been identified as the primary contributors of NOx, VOC, PM and CO emissions in KSA – back in 1998 when field measurements were conducted by King Fahd University of Petroleum & Minerals (KFUPM). Improvements in fuel quality will reduce air pollutants from vehicles and have a direct positive impact on human health and the environment. Saudi Aramco took the proactive initiative to further improve air quality by developing a clean fuel quality roadmap to achieve the following goals:

· improve the quality of the environment to provide clean air to protect human health for current and future generations;

· maintain a competitive position in global markets for continuing fuel export; and

· ensure fuel compatibility with future automotive engine technology.

The critical elements of the clean transportation fuels (CTF) program are a reduction in sulfur content in the gasoline and diesel products to 10 ppm, reduction in the gasoline benzene to 1 vol.% and a maximum gasoline aromatic content of 35 vol.%. These reductions are to be implemented at different stages and vary by refinery. As a consequence, CTF projects were initiated that will install new processing units at the company's domestic refineries to meet the new gasoline and diesel products requirements.

This presentation will focus on the challenges on lowering the aromatic content at the various refineries. The projects at Ras Tanura, Jazan and Rabigh II will install new units that will extract the aromatics from reformate pool in an aromatic complex to produce Benzene, Toluene and mixed xylenes (BTX). The benzene and toluene which can be used as solvents or industrial feedstock will be used to support new petrochemical projects in the Kingdom. Paraxylene will be exported for use in the growing worldwide polyester-based clothing industry. YASREF (Yanbu Aramco Sinopec Refining Company) is planning to extract Benzene only in Benzene Extraction Unit, while Riyadh Refinery is to incorporate a Benzene Saturation Unit to remove the benzene precursors.

(698g) Extent and Kinetics of Hexavalent Chromium Removal by Selected Chelating Agents

Jude Ighere and Ramesh Chawla, Chemical Engineering, Howard University, Washington, DC

Improper storage and disposal practices have rendered hundreds of sites contaminated with hazardous substances. Hexavalent chromium (Cr(VI)) which is toxic and probable carcinogenic is listed as one of the common contaminants. In this study, three chelating agents - disodium dihydrogen ethylenediaminetetraacetate (Na_2H_2EDTA), ethylenediaminetetraacetic acid (H_4EDTA) and Nitrilotriacetic acid (NTA) were used to remove hexavalent chromium from soil obtained from a non-chromium contaminated hazardous waste site in Chelsea, Michigan. Batch experiments were performed to determine the degree and rate of chelation of Cr(VI) in Chelsea soil. In the chelation reaction, the oxidation state of Cr(VI) does not change, rather it is rendered non-toxic by forming an inactive complex with the chelating agent, which claws the metal ion within its structure. Average chelation complex (chelating agent-Cr (VI) ion) was 23% for Na_2H_2EDTA, 25% for H_4EDTA and 30% for NTA, over a 36 hour-time period (equilibrium) at room temperature. All experiments were conducted at pH of 5, which yielded the highest rate, and is feasible for field operation. There was an increase of 20-50% in the degree of chelation with rising temperature from 15C to 65C. Kinetic analysis using differential method indicated that the reaction rate was first order with respect to Cr(VI) for Na_2H_2EDTA and second order in the case of H_4EDTA and NTA.

(708a) Role of Surface Chemistry in Sub-Einstein Rheology of Nano-Composites

Moulik Ranka, Chemical and Biomolecular Engineering, University of Illinois, Urbana-Champaign, Urbana, IL and C.F. Zukoski, Department of Chemical & Biomolecular Engineering, University of Illinois at Urbana-Champaign, Urbana, IL

Sub-Einstein and negative intrinsic viscosities resulting from the addition of nano-fillers to polymer melts approximately 2-10 nm diameter have been reported. Here we explore the intrinsic viscosity of polyhedral oligomeric silisequioxane (POSS) particles of approximately 1.6-2.5 nm diameters as a function of polymer molecular weight. These nanoparticles consist of cubes of 8 corner silicon atoms edge bridged with oxygen atoms. A variety of functional groups are attached to the corner silicon atoms. Three different POSS particles are explored: POSS with 8 tethered alkyl chains (butyl and octyl) and POSS with 8 tethered polyethylene glycol, PEG, chains of degree of polymerization, n ~ 13 segments. The tethered particles are dissolved in a homopolymer matrix (polydimethyl siloxane, PDMS, for octyl-POSS and butyl-POSS and PEG for PEG-POSS). The POSS intrinsic viscosity drops with increasing polymer matrix molecular weight from approximately 2.5 (as predicted by Einstein for no slip boundary conditions) to negative values. We discuss these results in light of previous studies of the intrinsic viscosity of nanoparticles in polymer melts with special attention to particle-polymer segment surface interactions that result in particle-polymer miscibility and ideas based on the ability of nanoparticles to alter polymer relaxation rates. We develop a hypothesis that negative intrinsic viscosities are associated with weak particle-polymer segment surface interactions while at the same time developing mechanisms that ensure particles remain soluble in the polymer melt.

(708b) Recent Advancements for Simulating Long Glass Fiber Composites in Injection Molding Applications

Kevin J. Meyer[1], John Hofmann[2] and Donald Baird[1], (1)Chemical Engineering, Virginia Tech, Blacksburg, VA, (2)Virginia Tech, Blacksburg, VA

Recent Advancements for Simulating Long Glass Fiber Suspensions in Injection Molding Applications

Kevin J. Meyer, John T. Hofmann, and Donald G. Baird

The properties of fiber-filled composites are strongly influenced by the orientation of the fibers within the composite structure. A popular way to reinforce composite materials cheaply is through the introduction of both short glass fibers (SGF) and long glass fibers (LGF) into the polymer matrix. This allows current industrial molding techniques to be utilized with little or no modification while significantly increasing the molded part's properties. Traditionally the Folgar-Tucker model, with the addition of a delay parameter, has been employed to predict fiber orientation in injection molded parts and has attained wide success when the fibers are short . In suspensions where the glass fibers can be considered long bending of fibers has been seen as a consequence of complex flow fields which arise in injection molding. The modified Folgar-Tucker model becomes less accurate as the flexibility of fibers increases. Hence, the "Bead-Rod" model, where the semi-flexible fiber is represented as two rods connected by a hinge, is proposed as an alternative to the modified Folgar-Tucker model. The Folgar-Tucker and Bead-Rod models are used to predict fiber orientation in center-gated and end-gated test geometries for LGF systems. Predictions for both models are compared to experimentally obtained data of long fiber orientation. The Bead-Rod model is observed to agree more closely with experimentally observed LGF orientation than the modified Folgar-Tucker model at a number of sampling points through the mold cavity.

(708c) Vinyl Ester Resin Crosslinking Using Molecular Dynamic Simulations

Changwoon Jang[1], Thomas E. Lacy[1], Steven R. Gwaltney[2], Charles U. Pittman Jr.[2] and **Hossein Toghiani**[3], (1)Department of Aerospace Engineering, Mississippi State University, Mississippi State, MS, (2)Department of Chemistry, Mississippi State University, Mississippi State, MS, (3)Dave C. Swalm School of Chemical Engineering, Mississippi State University, Mississippi State, MS

Molecular dynamics simulations on an atomistic scale were performed to create a series of crosslinked vinyl ester resins, where the crosslink density was varied at a constant degree of conversion. Thermodynamic and mechanical properties were obtained as a function of the crosslink density. In this study, a new algorithm was developed to generate the crosslinking network by incorporating the correct regioselectivity (head-to-tail chain propagation), and simulated the actual monomer reactivity ratios during the cure. These features had never been achieved in previous simulations. Most previous crosslinking simulations have been performed on epoxy resins, which undergo step-growth type polymerization. Vinyl esters, in contrast, cure by free radical chain polymerization. In this study, vinyl ester crosslinking networks with conversions up to 98% were successfully achieved. Volume shrinkage, glass transition temperature, and Young's modulus of the equilibrated structures were calculated, where the resin's crosslink densities were varied. The isotropic Young's modulus was compared with available experimental data. This newly developed method holds great promise for generating other realistic thermoset and thermoplastic polymer systems containing two or more type of monomers.

(708d) Multiscale Modeling of Nanoparticle Suspensions: Rheology, Drying and Bulk Properties

Dan S. Bolintineanu[1], Jeremy Lechman[2], Gary S. Grest[3] and P. Randall Schunk[1], (1)Sandia National Laboratories, Albuquerque, NM, (2)Sandia National Laboratories, New Mexico, Albuquerque, NM, (3)Sandia National Laboratories, New Mexico

Colloidal suspensions have garnered significant interest due to their widespread use in various nanomanufacturing technologies. We present a multiscale approach to modeling the drying of colloidal suspensions and prediction of the bulk properties of the resulting composite materials. We discuss two relatively recent discrete element modeling techniques that are used to coarse-grain the solvent: stochastic rotation dynamics (SRD) and fast lubrication dynamics (FLD). We briefly compare their effectiveness in capturing the correct inter-particle hydrodynamic interactions. We then introduce novel approaches to simulate drying using these techniques, which we apply to the densification of suspensions of spherical, polydisperse particles with realistic physical properties. Finally, we use a continuum mechanics/finite element modeling approach to predict a number of important material properties of the resulting microstructure. In this way, we are able to model the effects of particle properties as well as drying parameters on the resulting microstructure and ultimately on the physical properties of interest in industrial applications.

(708e) Quantification of Layered Silicates Dispersion in Polymer Nanocomposites

Qian Gou[1], Mark D. Wetzel[2] and Babatunde A. Ogunnaike[1], (1)Department of Chemical Engineering, University of Delaware, Newark, DE, (2)Experimental Station, E323/227, E. I. du Pont de Nemours and Co., Inc., Wilmington, DE

Over the past few decades, polymer nanocomoposites have drawn increasing interest from both academia and industry because their properties (e.g., mechanical, permeability, and gas barrier properties) are superior to those of pristine polymers and conventional modified composites. These property enhancements arise mainly as a result of excellent dispersion of the nanofillers (clays such as layered silicates; nanotubes; etc.) in the polymer matrix; if the fillers are poorly dispersed, the advantage is lost. For example, in polymer nanocomposites employing clays as nanofillers, poor dispersion, which manifests as agglomerates of clays, leads to unreinforced composites, resulting in little or no property improvements. Therefore, the key to producing high performance nanocomposites is to ensure excellent dispersion of clay particles in the polymer matrix, which in turn requires an effective method for quantifying the degree of dispersion in the manufactured product. Unfortunately, most currently available methods, which employ techniques such as X-ray diffraction (XRD), transmission election microscopy (TEM), rheological measurements, atomic force microscopy (AFM), and nuclear magnetic resonance (NMR), only provide *qualitative* assessments of clay dispersion, which is inadequate for quality assurance and for on-line control of dispersion. Recently proposed methods employ TEM image analysis to quantify clay dispersion in the form of particle size distribution [1], or particle density [2]. However, what these techniques provide, even though quantitative, are surrogate metrics with *no* clear and direct indication of how the measured values can be used to distinguish good dispersions from poor ones.

To address this problem, we have developed a quantitative method involving a probability model-based analysis of TEM images, which introduces an explicit dispersion parameter δ for quantitatively distinguishing among various possible degrees of dispersion. In this presentation, we will discuss the details of our approach (which is based on using the gamma probability model to describe particle length distribution in polymer nanocomposites), show how the probability model parameters combine to produce the dispersion parameter, δ, and establish from literature data, an explicit empirical expression linking δ directly to physical dispersion states such as exfoliated dispersion (good), less exfoliated dispersion with several stacked platelets (average), and a combination of exfoliated particles and intercalated stacks (poor).

References:

1. Shah, R. K.; Kim, D. H.; Paul, D. R., Morphology and properties of nanocomposites formed from ethylene/methacrylic acid copolymers and organoclays. *Polymer* **2007,**48, (4), 1047-1057.

2. Fornes, T. D.; Yoon, P. J.; Keskkula, H.; Paul, D. R., Nylon 6 nanocomposites: the effect of matrix molecular weight. *Polymer* **2001,** 42, (25), 9929-9940.

(713a) Development of a Parallel Predictive Control Algorithm for Complex Processes with Multiscale Objectives

Antonios Armaou and **Stevan D. Wilson**, Chemical Engineering, The Pennsylvania State University, University Park, PA

For many advanced materials and biological processes of interest, performance is measured both with respect to product yield and quality. The elusive latter term is usually a measure of certain aspects of product microstructure that needs to be within strict limits. Sensor information at the microstructure scale is impractical or infeasible, necessitating multiscale process modeling to provide soft sensors for controller synthesis.

Multiscale models are traditionally used to quantify process evolution across all relevant length scales and characterize product behavior within current computational limitations. Such models, however, pose significant challenges for controller design due to a) the unavailability of closed form equations of the molecular-level description components and b) the computational demands of the resulting system that prevent direct implementation inside the control structure. An industrially relevant example for this is the thin film deposition processes widely used by the microelectronics and solar energy industries (such as the production of photovoltaic systems). Motivated by the complex and nonlinear process dynamics, strict quality requirements, high product value, and high production costs, a significant amount of research has been applied to this problem. While earlier work focused on designing structures that allow the use of standard feedback controllers by directly using simulations in lieu of measurements, more recent results have been focused on combinations of model-based controllers with observers. These structures are based on models derived by using system identification methods. Unfortunately, the use of predictive controllers adds an additional computational burden due to mathematical complexity.

To address both the computationally demanding issues of process modeling and predictive control, we investigated the application of parallel computing. Based on our previous efforts, we developed a computationally tractable online system identification scheme based on a combination of subspace and fuzzy system identification methods. This scheme allows for the derivation of stochastic differential equations where both the expectation and drift terms are identified. This model was incorporated into a predictive control scheme that drives the process towards desired objectives. By employing optimization algorithms that account for the nonlinear and uncertain nature of the underlying dynamics, the predictive controller was capable of efficiently identifying near optimal manipulated variable profiles. The change in computing resources available in real time impacted both the design and the algorithms used for both the model and controller. Discussion of each algorithm and the total system are presented along with results for the process under control.

(713b) Self-Optimizing Control Using the Sensitivity Features of SIpopt

Johannes Jäschke, Norwegian University of Science and Technology, Trondheim, Norway, Lorenz T. Biegler, Chemical Engineering, Carnegie Mellon University, Pittsburgh, PA and Sigurd Skogestad, Chemical Engineering, Norwegian University of Science and Technology, Trondheim, Norway

Self-optimizing control is a systematic framework for designing a control structure. The idea is to systematically find sets of controlled variables, which when controlled at constant setpoints, give optimal or near optimal operation in economic terms [1]. Thus, a self-optimizing control structure is required to have the two main properties:

1. The control structure optimizes process operation in terms of an economic cost
2. The control structure should be easy to implement (constant setpoints for the controlled variables)

Over the last decade, a framework has been developed to find and evaluate candidate sets of controlled variables. The methods are based on a linearization around the nominal optimal operating point, and yield sets of controlled variables, which minimize the economic loss in a vicinity of the nominal operating point [2].

However, one of the challenges when designing a self-optimizing control structure for large-scale problems is to find the optimal solution and the required sensitivity information. In particular, the sensitivity with respect to disturbances, the gain matrices and the reduced Hessian at the nominal solution are required for applying the tools from self-optimizing control theory. To the authors' knowledge, in all published case studies, the optimal sensitivity, the gain matrices and the reduced Hessian where estimated "manually", by re-solving the optimization problem and using a finite difference approximation. This is quite tedious and prone to errors.

In a related research area there has been progress on this topic. In nonlinear model predictive control and dynamic real-time optimization, the sensitivity of the optimal solution has been used to find fast approximations to the optimal solution, e.g. [3]. Some modern nonlinear programming algorithms contain features, which can calculate the required sensitivities at very small additional cost [4].

The combination of these NLP sensitivity calculation features and the well established framework of self-optimizing control is the topic of this paper. We show how the data which is needed for the methods from self-optimizing control, can be obtained with the powerful NLP sensitivity calculation features included in the software sIPOPT [4]. This software reuses the matrix factorizations from the NLP solver to obtain the sensitivities of the KKT solution at minimal additional computational cost. Moreover, reduced Hessian information can be obtained efficiently.

Using sIPOPT, the required sensitivity information is obtained with a very high degree of accuracy, and can be readily used for developing a self-optimizing control structure. This facilitates the task of designing a good control structure significantly, and avoids the cumbersome "manual" sensitivity calculations. We present some case studies including optimization and controlled variable selection of distillation columns and a network of heat exchangers for preheating the feed of a crude oil unit in a refinery.

[1] S. Skogestad "Plantwide control: the search for the self-optimizing control structure", J. Proc. Control, 10, 487-507 (2000)

[2] V. Alstad, S. Skogestad and E.S. Hori, "Optimal measurement combinations as controlled variables", J. Proc. Control, 19, 138-148 (2009)

[3] R. Huang, V. M. Zavala, L.T. Biegler, "Advanced step nonlinear model predictive control for air separation units" J. Proc. Control, 19, 678-685 (2009)

[4] H. Pirnay, R. Lopez-Negrete, and L. T. Biegler. "Optimal sensitivity based on IPOPT". submitted for publication, (2011).

(713c) Nonlinear Model Predictive Control of a Thermal Oxidizer System

Reza Asgharzadeh Shishavan and John D. Hedengren, Chemical Engineering, Brigham Young University, Provo, UT

Nonlinear Model Predictive Control of a Thermal Oxidizer System

RezaAsgharzadehShishavan and John D. Hedengren Department of Chemical Engineering Brigham Young UniversityEnvironmental regulations restrictthe allowed amount of pollutant gases emitted to the surrounding environment. Thermal Oxidizers (TO)are of paramount importance indestroying organic pollutant compounds in waste streamsof low heating value. This capability has increasedthe use of TOs in many chemical plants which in turnhas raised the importance of developing more precise and robustdynamiccontrol systems for these reactors.In this work a thermal oxidizer was modeled dynamically using both empirical and first principles models. Next,Advanced Process Monitoring (APM) software[1] which is available online and has found applications in many downstream,upstream,and chemical industries [2-3]was utilized to implementthe nonlinear Model Predictive Control (MPC) algorithm as the control strategy on the system.Nonlinear MPC is based on determining the best manipulated variable control action using optimization techniques and a nonlinear model of the process. In estimation mode, APMattempts to optimally estimate the true state of the dynamic system, given a real-time stream of measurements and a model of the physical process.MPC has been applied to numerous systems[4] but none of these studies have focused on implementing nonlinearMPC on a TO system. The complex nature of the combustion reactions have made it challenging to model and control the thermal oxidation. Part of the control challenge is due to the combustion reactions that are very sensitive to changes in temperature. For example, a100°C increase in temperature can make the reaction rateconsiderablyfaster than its original value and drastically changes the characteristics of the process response.Because of the inherent complexity of dynamic control systemsit is not feasible to use a more detailed combustion model with all potential chemical species and at the same time get a fast control decision. Likewise, using a detailed model can even lead to ahigher deviation of the predicted results from the measured ones as complexity does not always correlate to accuracy. Therefore, arelatively simple but practically demonstrated combustion mechanism was used as the basis for the model equations.This kinetic mechanism is restricted to the species of carbon monoxide, carbon dioxide, nitrogen, oxygen,water, and intermediate radicals.As mentioned before, the TO model used in this work takes advantage of both first principles and empirical equations. The first principles model consists of material, species, and energy balances generally shown as

Rate of mass or energy accumulation in the system = rate of mass or energy entering the system -- rate of mass or energy leaving the system

Because the Reynolds number in the TO chamber was found to be higher than 10,000, it can be assumed that there is turbulence inside the TO and the species are completely mixed in the radial direction[5]. Thusthe momentum balance is of much less importance than the material and energy balances.Moreover, average outlet concentration is of higher interest in industrial systems which excludes the need for detailed species profiles in the reactor. A momentum balance can be taken into account in design of TOs in order to increase the mixing and is of secondary importance for a control- relevant model.An empirical reaction rate introduced by Howard et al.[6]was used for modeling the combustion reaction. The applicability of this equation has been demonstrated for the post flame combustion region which is suitable for the TO case.The simplified kinetic mechanism was determined after review of numerous experimental studies for a wide temperature range (840-2360 K). This reaction equation has also been considered as a valid expression for determining CO combustion rate in a TO by Cooper et al [7].A schematic of the TO including

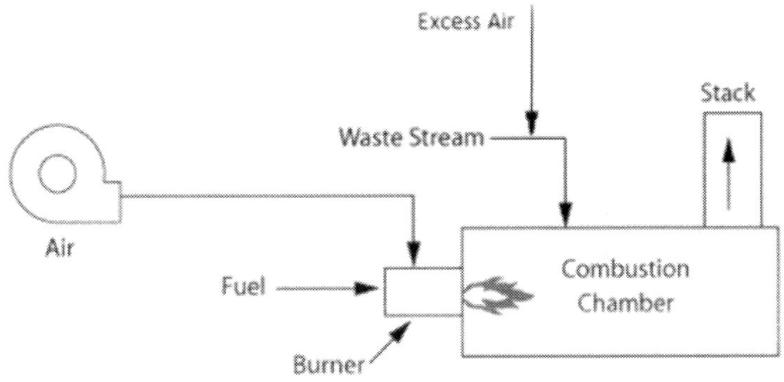

manipulated and control variables is shown in Figure 1.

Figure 1.Schematic of the modeled thermal oxidizer. The fuel (often natural gas) is combusted to provide thermal energy to react the waste stream (low heating value) into acceptable waste products.

Excess air and the burner fuel flow rates were selected as manipulated variables while exit CO and O2 concentrations were assignedas controlled variables.Typical values of 0.1% and 0.01% were selected as upper and lower CO exit concentration limits while 3 and 10 % were used for lower and upper limits forO2exit concentrations, respectively. Allowable CO exit concentrations are determined by environmental regulations while anO2 lower limit of 3% was selected to insure excess oxygen and to avoid a reducing environment inside the TO[5].The partial differential algebraic equation (PDAE) model of the thermal oxidizer system was first discretized in space and then in time. For the spatial discretization, the chamber was modeled as successive mixed flow reactors in which outlet of each CSTR reactor serves as the input for the next one. First, steady state modeling was performed and the result was used to initializethe dynamic control application.Economiceffects were included in the optimization by adding a cost on the burner fuel flow rate. Fuel cost drives the controller to the maximum allowable CO and minimum O2exit concentrations. This behavior can be explained by considering that the lower CO exit concentration reflects the need for lower conversion which in turn requires less burner fuel flow rate. Moreover, lower oxygen exit concentration means less excess air is required to be mixed with the waste gas entering the TO which is equivalent toneedingless burner fuel to heat the incoming gas to the desired temperature. Figure 2 show the trends of manipulated variables and controlled variables respectively.

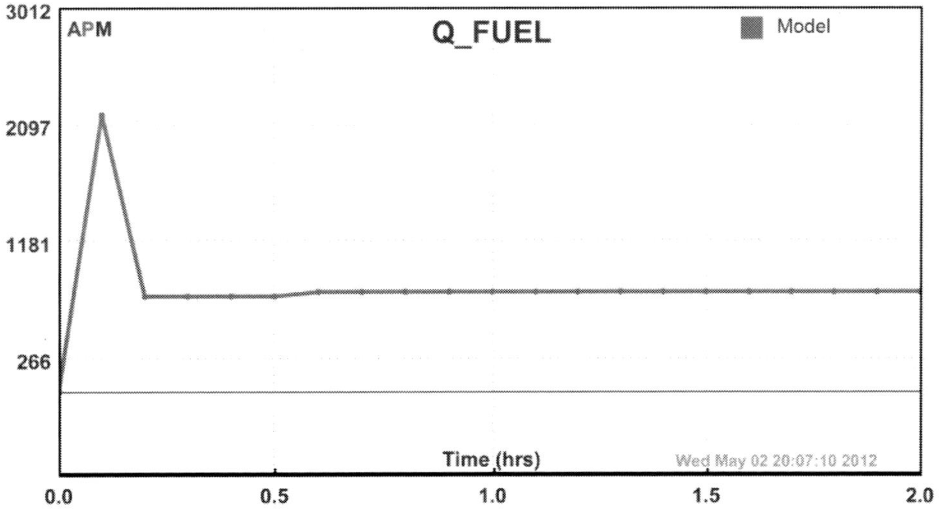

Figure 2.A: Trend of

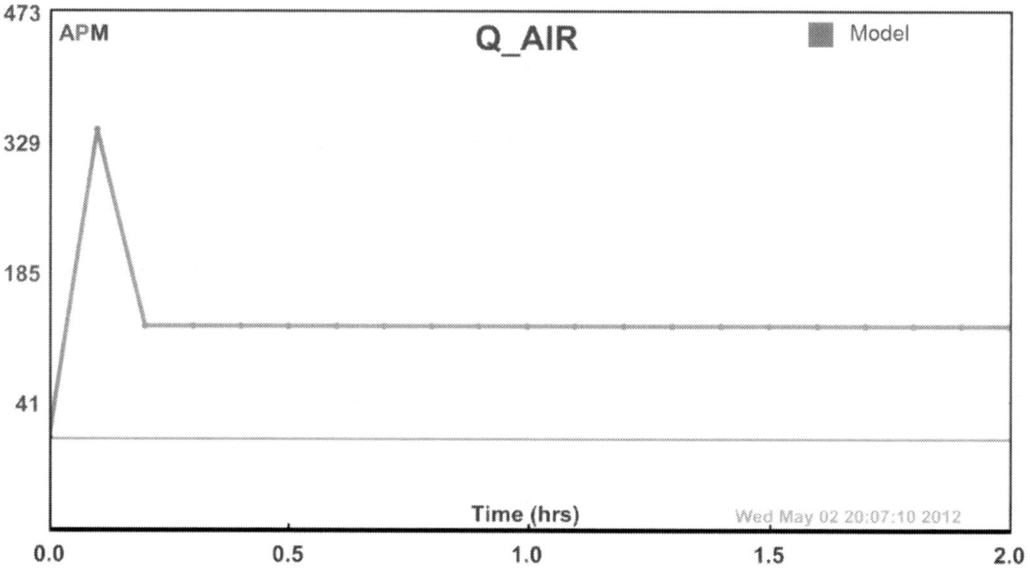

fuel flow rate change
Figure 2 B: Trend of excess flow rate change

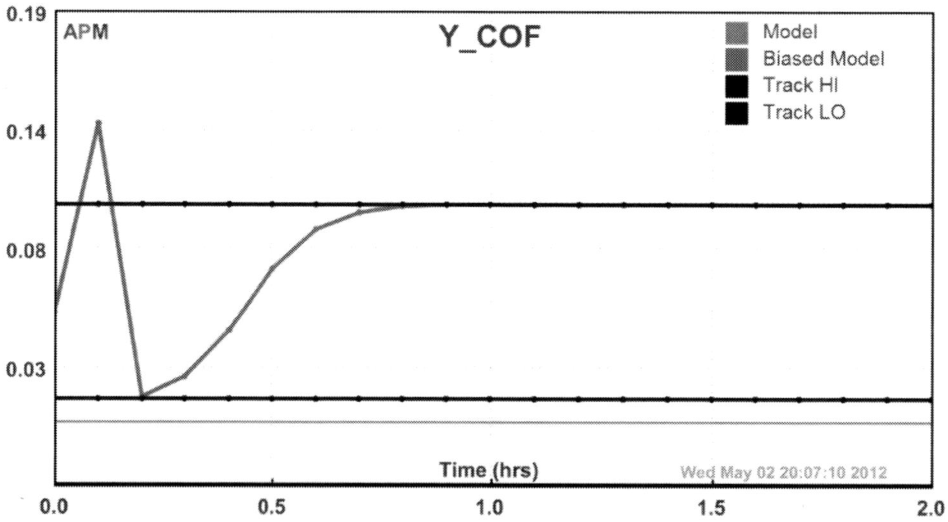

Figure 3 A:Trend of

carbon monoxide exit concentration changeFi

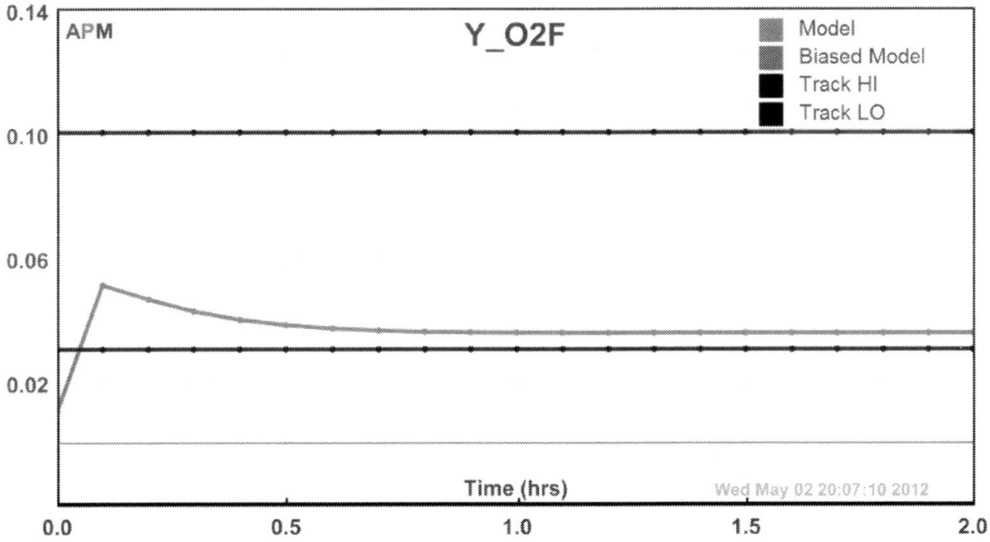

Figure 3 B: Trend of Oxygen exit concentration changeTo check the validity of the simulation result,an attempt was made in comparing the steady state modeling results with empirical statistical equation results for COpresentedbyLee et al.[7]. in several temperature and reaction times.Lee et al. developed their model after conducting numerous experiments and reported this to be a precise method. Results showed good agreement between the current model results and the statistical analysis.Table 1 shows a comparison between the temperature required for 99% and 99.9% destruction efficiencies at the same residence times from both works.Table 1. Comparison of Temperature required for destruction of CO to desired levels

99% destruction	99.9% destruction	
Resuts from Lee [7]	1058 K	1075 K
Results from current work	1051 K	1141 K

Simulation results showed thatthe nonlinear behavior of the TO can be completely captured by the controller. Therefore, nonlinear MPC can be considered as an option of choice for controlling the TOs in practice. Nonlinear MPC control systems are more precise in adjusting the controlled variable to its set- point and can use less control actions to reach this goal compared to traditional PID or linear MPC controllers.

References1. Hedengren, J., and Brower, D., *Advanced process monitoring of flow assurance with fiber optics*, in *In AIChE Spring Meeting*. 2012.2. Hedengren, J., *Optimization and Analytics in the Oil and Gas Industry*. Springer-Verlag BerlinHeidelberg, ch. Advanced Process Monitoring , p. submitted., 2012.3. Brower, D., et al. in *The 31st International Conference on Ocean, Offshore and Arctic Engineering*

OMAE 2012. 2012. Rio de Janeiro, Brazil.

4. Scattolini, R., *Architectures for distributed and hierarchical Model Predictive Control - A review.*Journal of Process Control, 2009. **19**(5): p. 723-731.5. Lewandowski, D.A., *Design of Thermal Oxidation Systems for Volatile Organic Compounds*. 1999, Boca Raton, Fla: CRC Press.6. J.B. Howard, G.C.W., D.H. Fine, *Kinetics of carbon monoxide oxidation in postflame gases.*Symposium (International) on Combustion, 1973. **14**(1): p. 975--986.7. Cooper CD, A.F.,

Air pollution control: a design approach. 1994, Prospect Heights (IL).

(713d) Discrete Mechanics Optimal Control (DMOC) and Model Predictive Control (MPC) Synthesis for 3D Spatial Actuation

Yu Yang and Stevan Dubljevic, Chemical and Materials Engineering, University of Alberta, Edmonton, AB, Canada

The common problem of the optimal controller synthesis for the process and mechanical robotic system is explored in this work. Namely, the optimal actuator motion control law and actuation policy for the reaction-diffusion process is considered. The common in process control practice realization admits the transport-reaction system coupled with servo-robotic mechanical device. For example, the welding arm scans over the specified domain in attempt to achieve the desired speed of welding.

In the considered model, the system consists of the four-degree robotic arm modeled by the discrete Lagrange-d'Alembert principle and a reaction-diffusion process governed by the parabolic partial differential equation (PDE). To address the optimal operational problem for the finite dimensional mechanics coupling with infinite-dimensional representation of reaction-diffusion process in different time scale, a two-layer control scheme is proposed in this research. The regulation of the robotic arm is performed in the lower layer by the discrete mechanics optimal control (DMOC) framework [1], which preserves the exact dynamics of the actuator even after the state discretization, to minimize the energy consumption of the arm's motion and meet the tight constraints of the velocity and position. An optimization problem is solved in real time to chose the control forces in the dual of tangent bundle *(TQ*)* and generate the optimal trajectory over the configuration space *(Q)*. For the higher level controller, synthesis benefits from the linear operator theory, spectrum decomposition and eigenfunction expansion [2], such that the parabolic PDE for the reaction-diffusion process is decomposed by the standard spectral technique and converted to a set of ordinary differential equations (ODEs). Then the modal model predictive control (MMPC) [3] is employed on this reduced model to optimize the actuation rate as well as guarantee the satisfactions of both the states and inputs constraints. The coordinated law of MMPC and DMOC for this two-layer controllers is established to achieve the overall processing optimality. Finally, the simulation study is conducted for the robotic application in the welding process of the 3D object [4], which evaluates and demonstrates the effectiveness of the control framework presented in this research.

References

[1] J. Mardson and M. West, Discrete mechanics and variational integrators, Acta Numerica, vol. 10, pp.357-514, 2001.

[2] R. F, Curtain and H. J. Zwart, An Introduction to Infinite-Dimensional Linear Systems Theory, Springer, 1995.

[3] S. Dubljevic, N. H. El-Farra, P. Mhaskar and P. D. Christofides, Predictive control of parabolic PDEs with state and control constraints, International Journal of Robust and Nonlinear Control, vol. 14, pp.133-156,2004.

[4] R. M. Murry, Z. Li, and S. S. Sastry, A Mathematical Introduction to Robotic Manipulation,CRC Press, 1994.

2012 AIChE Annual Meeting
Pittsburgh, PA, USA
October 28 – November 2, 2012

APPLICATION OF LINEAR MULTIPLE MODEL PREDICTIVE CONTROL (MMPC) FRAMEWORK TOWARDS DYNAMIC MAXIMIZATION OF OXYGEN YIELD IN AN ELEVATED-PRESSURE AIR SEPARATION UNIT

Priyadarshi Mahapatra, Advanced Virtual Energy Simulation Training and Research (AVESTAR™) Center
National Energy Technology Laboratory, U.S. Department of Energy
URS Corporation, 3592 Collins Ferry Road (Suite 160), Morgantown, WV 26505, USA
priyadarshi.mahapatra@netl.doe.gov, Tel: 304-285-4063, Fax: 304-285-0903

Stephen E. Zitney, AVESTAR™ Center
National Energy Technology Laboratory, U.S. Department of Energy, Morgantown, WV 26507, USA
stephen.zitney@netl.doe.gov, Tel: 304-285-1379, Fax: 304-285-0903

B. Wayne Bequette, Department of Chemical & Biological Engineering
Rensselaer Polytechnic Institute, Troy, NY 12180
bequette@rpi.edu, Tel: 518-276-6683, Fax: 518-276-4030

Abstract

In a typical air separation unit (ASU) utilizing either a simple gaseous oxygen (GOX) cycle or a pumped liquid oxygen (PLOX) cycle, the flowrate of liquid nitrogen (LN2) stream connecting high-pressure and low-pressure ASU columns plays an important role in the total oxygen yield. It has been observed that this yield reaches a maximum at a certain optimal flowrate of LN2 stream. At nominal full-load operation, the flowrate of LN2 stream is maintained near this optimum value, whereas at part-load conditions this flowrate is typically modified in proportion with the load-change (oxygen demand) through a ratio/feed-forward controller. Due to nonlinearity in the entire ASU process, the ratio-modified LN2 flowrate does not guarantee an optimal oxygen yield at part-load conditions. This is further exacerbated when process disturbances in form of "cold-box" heat-leaks enter the system. To address this problem of dynamically maximizing the oxygen yield while the ASU undergoes a load-change and/or a process disturbance, a multiple model predictive control (MMPC) algorithm is proposed. This approach has been used in previous studies to handle large ramp-rates of oxygen demand posed by the gasifier in an IGCC plant. In this study, the proposed algorithm uses linear step-response "blackbox" models surrounding the operating points corresponding to maximum oxygen yield points at different loads. It has been shown that at any operating point of the ASU, the MMPC algorithm, through model-weight calculation based on plant measurements, naturally and continuously selects the dominant model(s) corresponding to the current plant state, while making control-move decisions that approach the maximum oxygen yield point. This dynamically facilitates less energy consumption in form of compressed feed-air compared to a simple ratio control during load-swings. In addition, since a linear optimization problem is solved at each time step, the approach involves much less computational cost compared to a first-principle based nonlinear MPC.

Introduction

The cryogenic systems of air separation have the capability to deliver the largest capacities for products at a moderate to high purity level, compared to non-cryogenic based systems such as pressure-swing adsorption (PSA) and membrane technologies employed at lower end of production scale and product purities. The elevated-pressure cryogenic air separation units (ASU) have found application in integrated gasification combined cycle (IGCC) power plants where a typical oxygen molar-purity of 95% is required [1]. Furthermore, in such plants, the high oxygen delivery pressure (~1000 psi) required by the gasifier and the partial integration of an ASU with the combustion turbine, where GT-compressed air is available at high pressures (between 200–250 psi), makes the elevated pressure ASU a promising source of oxygen. Recently, an EP-ASU has been utilized as the primary oxygen source within the Advanced Virtual Energy Simulation Training and Research center at National Energy Technology Center (Morgantown) which features a high-fidelity, real-time dynamic simulator of an IGCC with CO2 capture process [1]. A typical downside of high-pressure operation is decreased separation efficiency, making the operating cost of such units escalate, in comparison to a low-pressure plant. This cost is by-far compensated with the decreased compressor work during oxygen and nitrogen compression.

It has been shown in previous work [3] that the liquid nitrogen distillate stream from HP-column (stream 'LN2'), which also serves as a liquid reflux to the LP-column, plays a significant role in the total oxygen yield. In another study by the authors, the MMPC approach was used to capture dynamic nonlinearities while controlling oxygen purities during rapid ramping operation of the ASU [4]. This operation was done with the abovementioned liquid nitrogen distillate flowrate at either a fixed ratio/feedforward to the oxygen demand or as an augmented feedback loop to the distillate stream nitrogen purity, an approach which is generally utilized by various industries as seen in many patent literatures. This approach closely approaches the optimal yield condition at or near steady state design conditions (assuming the plant is optimally designed at steady state) but does not guarantee an optimal oxygen yield at part-load conditions.

Steady State Design and Dynamic Model

The steady-state and dynamic flowsheet design including plant configuration, operating pressures, flowrates, optimization of oxygen recovery, condenser-reboiler heat-integration, "neat" operation and many other intricacies involved in modeling the ASU process have been provided in a recent paper by the authors [3]. In addition, some of the interesting problems encountered during operation including refrigeration imbalance and snowball effect have also been highlighted. It must be realized by the readers that EP-ASU is a perfect example of process complexity and serves as a good test-bed for various operability and controllability studies. Building the regulatory control layer in this process is, in itself, a daunting task. In fact numerous patents have been claimed which are centered at developing various lower-level controllers. Many sources also highlight the inclusion of a liquid oxygen storage tank for fast load-changes, especially during ramp-up conditions, although maintaining such a volumetric space at cryogenic conditions is a challenging and capital-cost intensive task.

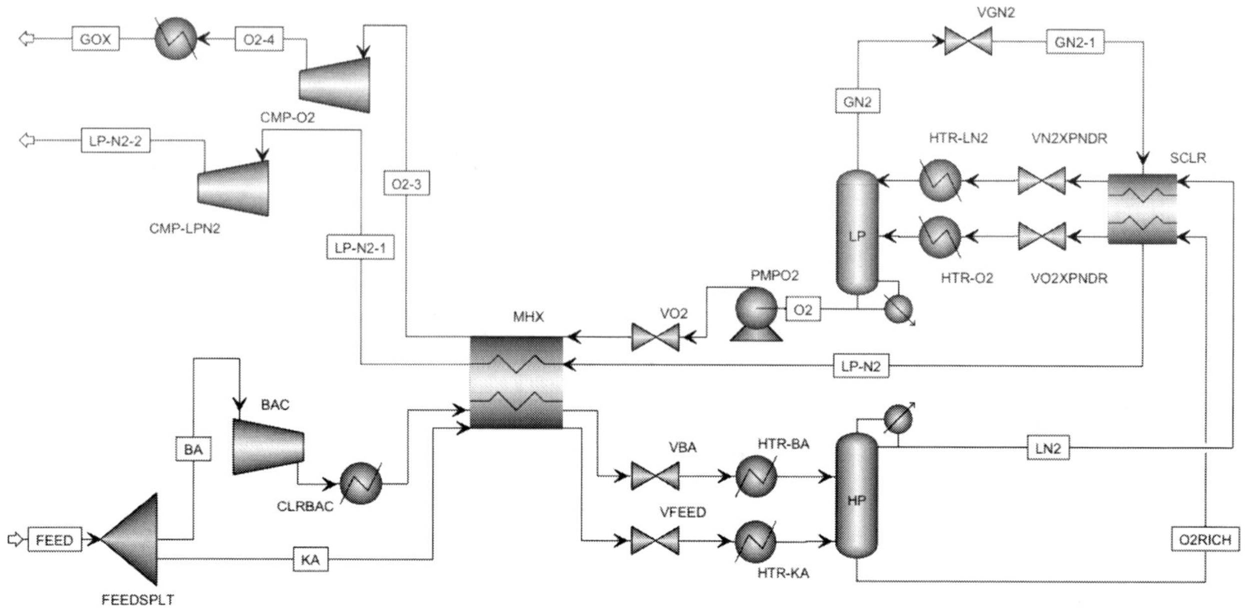

Figure 1. ASU process flowsheet as seen in Aspen Plus®

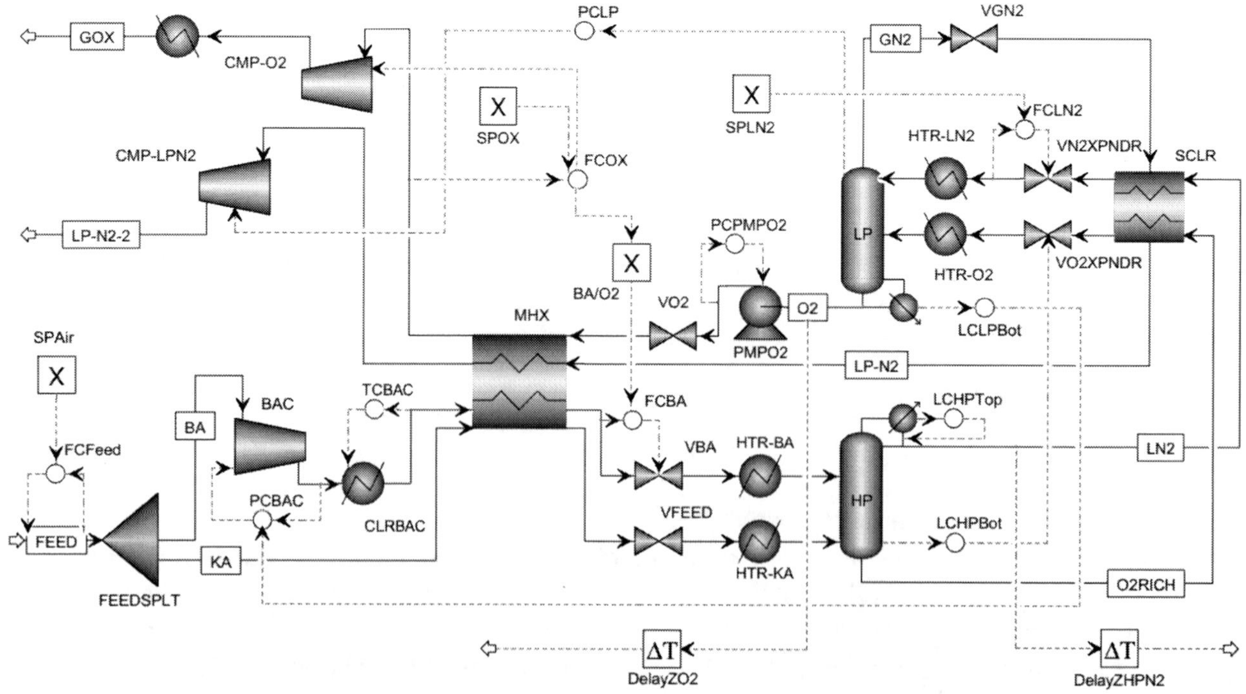

Figure 2. EP-ASU flowsheet in Aspen Plus Dynamics® showing regulatory controllers and various IO variables exposed for supervisory control layer

A schematic of steady-state flowsheet modeled in Aspen Plus® V7.3 is given in Figure 1. The dynamic flowsheet modeled in Aspen Plus Dynamics® V7.3 has been shown in Figure 2. Majority of the steady-state and dynamic flowsheet configurations used in the current study are identical to the previous work by the authors [3]. The only exceptions to such configurations are the absence of supervisory composition control loops in the current design. Instead the input blocks marked by 'SPAir', 'SPLN2' and 'SPOX' are left open. These blocks receive input "setpoint" signals via the Simulink-APD bridge from the MATLAB environment where the supervisory control calculations are being done. Similarly output signals from the delay blocks ('DelayZO2' and 'DelayHPN2') are being sent to MATLAB as "measured" variables (commonly also known as control variables, CV or process variables, PV).

Problem Definition

In general, an efficient plant design takes into account various optimization calculations at the design stage and tries to attain the plant objective in the most effective way. In purview of the EP-ASU for IGCC, the ultimate design objective is to minimize the operating cost of the cold-box. Predominant factors which contribute to the operating cost are the compression cost of main air compressors (MAC), oxygen and nitrogen compressors. For a given oxygen demand, purity and delivery pressures the operating costs involved in the product-side are do not fluctuate significantly, whereas those on the feed-side i.e. total flowrate of feed-air (and hence the energy consumed in MAC) serves as a good candidate for optimization.

During steady state optimization, rather than minimizing the feed-flow at fixed oxygen purity, an alternate yet equivalent approach is adapted for maximizing the oxygen yield (and hence minimizing the operating costs). This is achieved by maximizing the oxygen purity at a fixed feed-flow. It was identified that liquid-nitrogen stream (Stream 'LN2' in Figure 1) connecting HP and LP columns, which serves as a distillate for the HP-column and a liquid reflux to the LP-column, has a significant effect on oxygen yield. This has been shown in Figure 3 (see black curve marked with circles), where a clear maxima is observed at a certain liquid-nitrogen flowrate (~27100 lbmol/hr). On the left hand side (LHS) of this maximum, the oxygen purity gets degraded due to low LP-column reflux leading to lost oxygen in the waste nitrogen stream (Stream 'GN2') whereas on the right hand side (RHS), due to high HP-column distillate extraction, the purification within the HP-column is negatively affected which gets transferred to the overall oxygen separation.

In terms of process dynamics, when the system is operated around the optimal flowrate, an *input-multiplicity* behavior is observed. This is clearly visible in Figure 4 where 5% step increase and decrease in the 'LN2' flowrate is simulated. Both the open-loop transients settle down to lower oxygen purity values from nominal purity of 0.955. This gives both a negative and positive process gain around the steady-state operating point depending on which direction the process moves. This poses a challenging control problem wherein closed-loop process stability cannot be ensured using a single model / single controller-gain for all possible uncertainties or operating regimes around the nominal operating point.

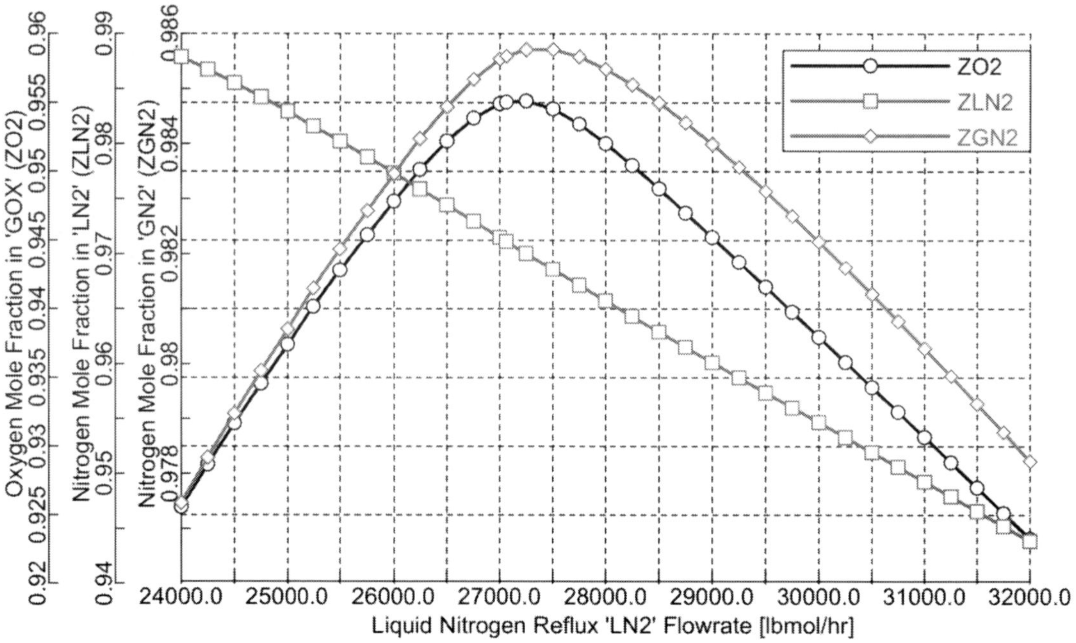

Figure 3. Sensitivity plots showing effect of liquid-nitrogen reflux 'LN2' flowrate towards oxygen product purity, liquid nitrogen 'LN2' purity and gaseous nitrogen 'GN2' purity at fixed oxygen withdrawal rate (14511.1 lbmol/hr) and total feed air flowrate (69000 lbmol/hr)

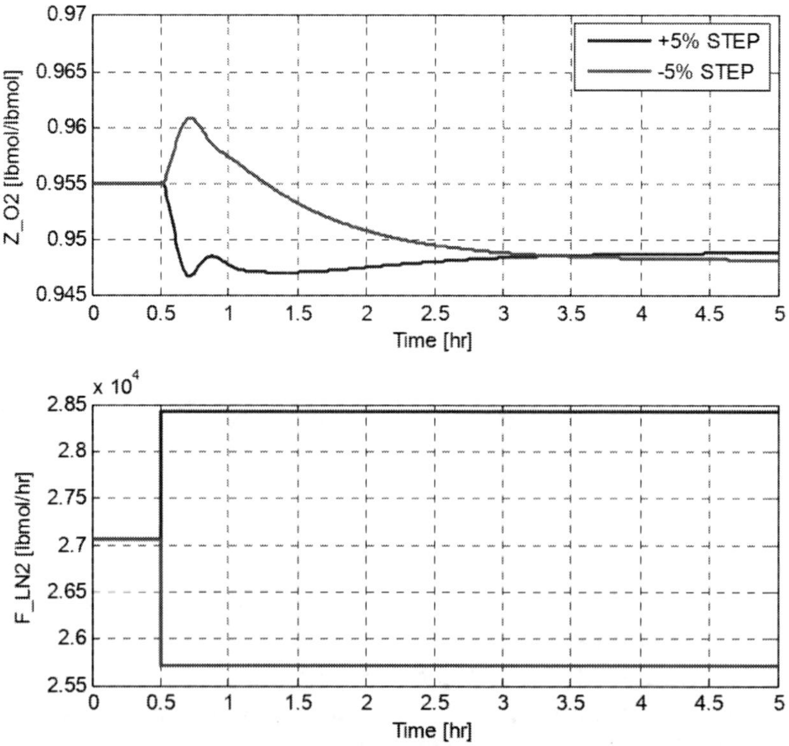

Figure 4. Transient responses for 5% STEP changes in liquid nitrogen flowrate (F_LN2) for oxygen demand of 14511.1 lbmol/hr and nominal value of F_O2/F_airASU. It can be seen that oxygen purity dynamically decreases to new steady states for F_LN2 changes in either direction.

It can also be seen from Figure 3 that purity of the abovementioned liquid nitrogen stream denoted by 'ZLN2' monotonically decreases with increasing flowrate ('F_LN2') of the same stream. This is fairly intuitive since with increase in HP-column distillate flowrate, the reflux to this column decreases, leading to a decrease in nitrogen (component having the least molecular wt.) purity. Dynamically this implies that F_LN2 is a strong candidate for controlling ZLN2. This approach has been used by the authors in their previous studies where for a multiloop PID based control, ZLN2 has indeed been tied with F_LN2; and oxygen purity ('ZO2') has been controlled using the total feed air ('F_airASU'). In context of an IGCC power plant, the waste 'gaseous' nitrogen is generally utilized as a gas turbine diluents or as a purge gas during system startup. Hence, meeting the nitrogen purity levels up to a considerable accuracy is not a strict system requirement. Therefore, various other configurations for multiloop control, where ZLN2 is not actively controlled, were also investigated in search of a control design which would promote faster load-following and/or disturbance rejection.

It must be emphasized that in this study the objective is not to control the oxygen and nitrogen purities at nominal values but to dynamically maximize the oxygen yield at the face of a fixed given feed-air flowrate (F_airASU) and oxygen demand (F_O2). As mentioned earlier, this is conceptually equivalent to the case where oxygen purity and demand are kept fixed whereas the feed-air flowrate is minimized (leading to a significant decrease in MAC compression cost). From a problem formulation and implementation perspective, these two cases are dissimilar and scope has been limited to the former in this paper. The following section discusses the multiple model predictive control (MMPC) formulation used for such dynamic maximization of oxygen purity solely based on linear models.

MMPC Formulation

The multiple model predictive control strategy [5] is based on the use of *n* models in the model bank that have the general state-space form given in Eq. (1).

$$\begin{aligned} {}^i\hat{x}_{k|k-1} &= {}^i\Phi\, {}^i\hat{x}_{k-1|k-1} + {}^i\Gamma^u u_{k-1} + {}^i\Gamma^l l_{k-1} + {}^i\Gamma^d\, {}^i\hat{d}_{k-1|k-1} \\ {}^i\hat{y}_{k|k-1} &= {}^iC\, {}^i\hat{x}_{k|k-1} + {}^iD^u u_{k-1} + {}^iD^l l_{k-1} + {}^iD^d\, {}^i\hat{d}_{k-1|k-1} \end{aligned} \quad (1)$$

where, given the current time step *k*, the following notations are used:

u_{k-1} : input (calculated by the controller) between time-step *k*-1 and *k*

l_{k-1} : measured and modeled disturbance between time-step *k*-1 and *k*

$\hat{d}_{k-1|k-1}$: unmeasured (estimated) but modeled disturbance between time-step *k*-1 and *k* for model *i*

$\hat{x}_{k|k-1}$: process states at time-step *k* given the plant measurements at time-step *k*-1(predicted) for model *i*

$\hat{x}_{k|k}$: process states at time-step *k* given the plant measurements at time-step *k* (corrected) for model *i*

$\hat{y}_{k|k-1}$: measured outputs at time-step k given the plant measurements at time-step k-1 (predicted) for model i

$\hat{y}_{k|k}$: measured outputs at time-step k given the plant measurements at time-step k (corrected) for model i

$\Phi, \Gamma^u, \Gamma^l, \Gamma^d, C, D^u, D^l, D^d$ are various state-space parameters for i^{th} model. The left superscript i denotes the model number, with i ranging from 1 to n models. Although the plant being controlled in practice is highly nonlinear (as demonstrated in Figure 4), the models in Eq.(1) are all linear. Each linear model is chosen to represent a discrete subspace of the overall nonlinear operating space. When all n models are linearly combined, the resulting bank of linear models spans the entire nonlinear operating space. As Figure 5 shows, the models within the model bank are updated in parallel with the plant.

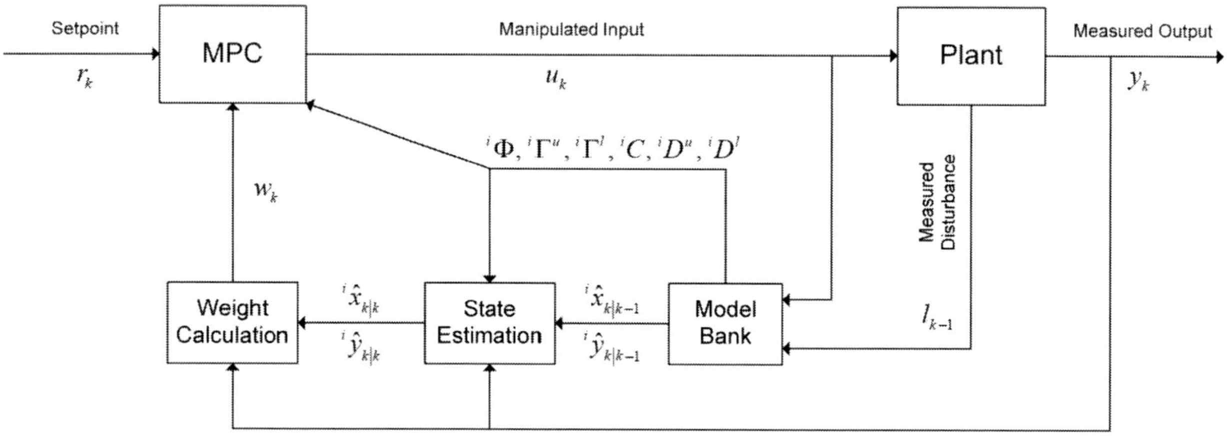

Figure 5. Block diagram showing the generic multiple model predictive control (MMPC) formulation

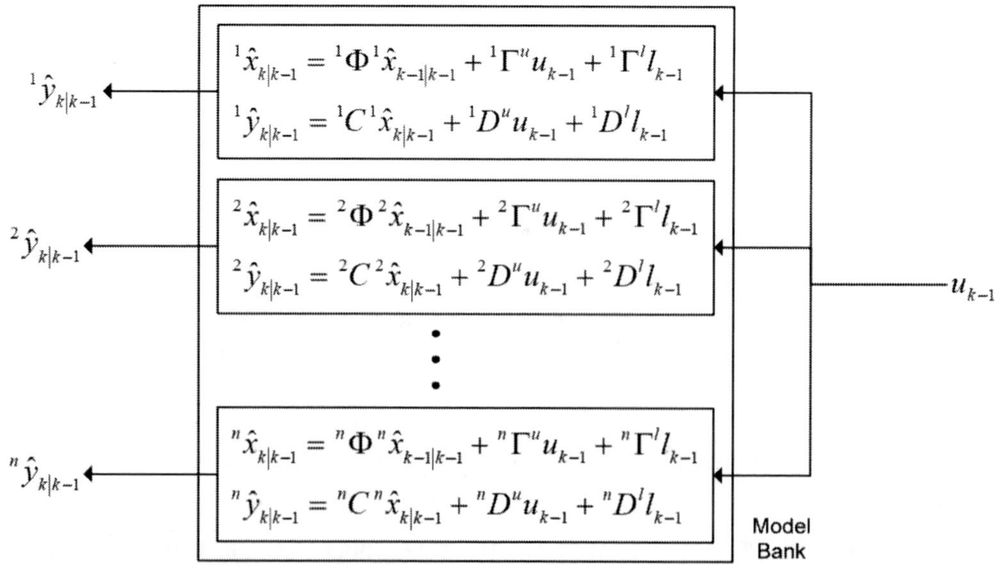

Figure 6. Schematic showing 'insides' of the model bank

The state-estimation involves a correction step, which modifies/corrects the plant states based on plant measurements at current time-step. For achieving this, an appended state formulation is used. The following equations list the prediction and correction steps involved in additive output formulation used in this study.

Prediction Step:

$$\begin{bmatrix} {}^i\hat{x}_{k|k-1} \\ {}^i\hat{d}_{k|k-1} \end{bmatrix} = \underbrace{\begin{bmatrix} {}^i\Phi & 0 \\ 0 & I \end{bmatrix}}_{{}^i\Phi^a} \underbrace{\begin{bmatrix} {}^i\hat{x}_{k-1|k-1} \\ {}^i\hat{d}_{k-1|k-1} \end{bmatrix}}_{{}^i\hat{x}^a_{k-1|k-1}} + \underbrace{\begin{bmatrix} {}^i\Gamma^u & {}^i\Gamma^l \\ 0 & 0 \end{bmatrix}}_{{}^i\Gamma^a} \begin{bmatrix} u_{k-1} \\ l_{k-1} \end{bmatrix}$$

$${}^i\hat{y}_{k|k-1} = \underbrace{\begin{bmatrix} {}^iC & I \end{bmatrix}}_{{}^iC^a} \begin{bmatrix} {}^i\hat{x}_{k|k-1} \\ {}^i\hat{d}_{k|k-1} \end{bmatrix} + \underbrace{\begin{bmatrix} {}^iD^u & {}^iD^l \end{bmatrix}}_{{}^iD^a} \begin{bmatrix} u_{k-1} \\ l_{k-1} \end{bmatrix} \tag{2}$$

Correction Step:

$$\begin{bmatrix} {}^i\hat{x}_{k|k} \\ {}^i\hat{d}_{k|k} \end{bmatrix} = \begin{bmatrix} {}^i\hat{x}_{k|k-1} \\ {}^i\hat{d}_{k|k-1} \end{bmatrix} + \underbrace{\begin{bmatrix} 0 \\ I \end{bmatrix}}_{{}^iL_k} \left(y_k - {}^i\hat{y}_{k|k-1} \right)$$

$${}^i\hat{y}_{k|k} = {}^iC^a \begin{bmatrix} {}^i\hat{x}_{k|k} \\ {}^i\hat{d}_{k|k} \end{bmatrix} + {}^iD^a \begin{bmatrix} u_{k-1} \\ l_{k-1} \end{bmatrix} \tag{3}$$

The model residuals are calculated as given in Eq. 4 and thereafter the current Bayesian probability ρ_k is calculated recursively using actual-plant output as given in Eq. (5).

$${}^i\varepsilon_k = y_k - {}^i\hat{y}_{k|k} \tag{4}$$

$${}^i\rho_k = \frac{\exp\left(-\frac{1}{2}\,{}^i\varepsilon_k^T\,{}^i\Lambda\,{}^i\varepsilon_k\right){}^i\rho_{k-1}}{\sum_{j=1}^{n} \exp\left(-\frac{1}{2}\,{}^j\varepsilon_k^T\,{}^j\Lambda\,{}^j\varepsilon_k\right){}^j\rho_{k-1}} \qquad {}^i\rho_k = \begin{cases} {}^i\rho_k & {}^i\rho_k \geq \delta \\ \delta & {}^i\rho_k < \delta \end{cases} \tag{5}$$

where, ${}^i\Lambda = \begin{bmatrix} {}^i\lambda_1 & & 0 \\ & \ddots & \\ 0 & & {}^i\lambda_{ny} \end{bmatrix}$ gives the model-residual covariance matrix. For brevity, each of the diagonal elements is kept identical. Here, δ represents a small non-zero value used to prevent model-probabilities reaching zero during recursive calculations thereby ensuring that all models remain active during weight-calculations.

The normalized probabilities (model weights) w_k are calculated in Eq. (6) after which the average linear model output vector ($\bar{y}$) is computed in Eq. (7).

$$
{}^{i}w_k = \begin{cases} \dfrac{{}^{i}\rho_k}{\sum_{j=1}^{n} {}^{j}\rho_k} & {}^{i}\rho_k \geq \delta \\ 0 & {}^{i}\rho_k < \delta \end{cases} \tag{6}
$$

$$
\bar{y}_{k+j|k} = \sum_{i=1}^{n} {}^{i}w_k \, {}^{i}\hat{y}_{k+j|k} \tag{7}
$$

This average output vector is used while formulating the constrained MPC objective function (J) as shown in Eq. (8).

$$
\min J = \sum_{j=1}^{P} \left(r_k - \bar{y}_{k+j|k}\right)^T W^y \left(r_k - \bar{y}_{k+j|k}\right) + \sum_{j=0}^{M-1} \Delta u_{k+j}^T W^u \Delta u_{k+j} \tag{8}
$$

where, $u_{\min} \leq u_{k+j} \leq u_{\max}$, $\Delta u_{\min} \leq \Delta u_{k+j} \leq \Delta u_{\max}$ and $y_{\min} \leq \bar{y}_{k+j|k} \leq y_{\max}$ are the imposed constraint conditions, r_k is the setpoint vector at current time-step, W_u and W_y are the input and output weighting matrices respectively, P is the prediction horizon and M is the control horizon.

System Identification

As discussed earlier, flowrate of liquid nitrogen (F_LN2), product oxygen (F_O2) and feed air (F_airASU) play an important role in the oxygen purity (ZO2). Oxygen purity monotonically increases with feed-air flowrate (not shown) and inclusion of this variable in the optimization formulation will only lead to unrealistically high feed-air flowrate giving ultra-high purity oxygen (ZO2 $\approx$ 1). Hence, for solving the current dynamic maximization problem the flowrate of feed-air needs to be fixed with respect to the oxygen flowrate. Here, a fixed ratio of these quantities based on steady-state values has been imposed as given in Eq. (9).

$$
\frac{\text{F\_airASU SP}}{\text{F\_O2 SP}} = 4.755 \tag{9}
$$

where, SP following the variable name denotes the setpoint value sent to the regulatory flow-controllers (signals from supervisory layer) which control the corresponding variable. F_airASU SP and F_O2 SP correspond to signals from 'SPAir' and 'SPOX' (shown in Figure 2) respectively.

Before moving to control implementation, the model parameters for all the control-models in the model-bank have to be determined. In essence, the state-space time-invariant matrices $\Phi, \Gamma^u, \Gamma^l, \Gamma^d, C, D^u, D^l, D^d$ shown in Eq. (1) have to be defined for each model. In this study, two models have been specified. One of them spans the left hand side (LHS) of the maximum point (Figure 3) and has a positive process gain. In contrast, the other model captures the right hand side (RHS) dynamics and possesses a negative process gain. Starting from the optimal steady-state design conditions, the LHS-model (Model-1) is obtained by first moving to a new operating point on the LHS. This is attained by keeping the oxygen demand (F_O2 SP) unchanged while providing a 10% step decrease in liquid nitrogen (F_LN2 SP). The process is brought to a new steady state by simulating the process for a long time (10 hr). Thereafter, multiple simultaneous random step changes are triggered on both the inputs (F_LN2 SP and F_O2 SP) at regular intervals (until significant settling is observed). The output "plant" signals are fitted to a 4th order subspace-based "n4sid" model (termed as "control-model") using System Identification Toolbox in MATLAB™. A similar approach is used to obtain the RHS-model (Model-2) as well by providing a 10% step increase in nitrogen flowrate to obtain a new steady state.

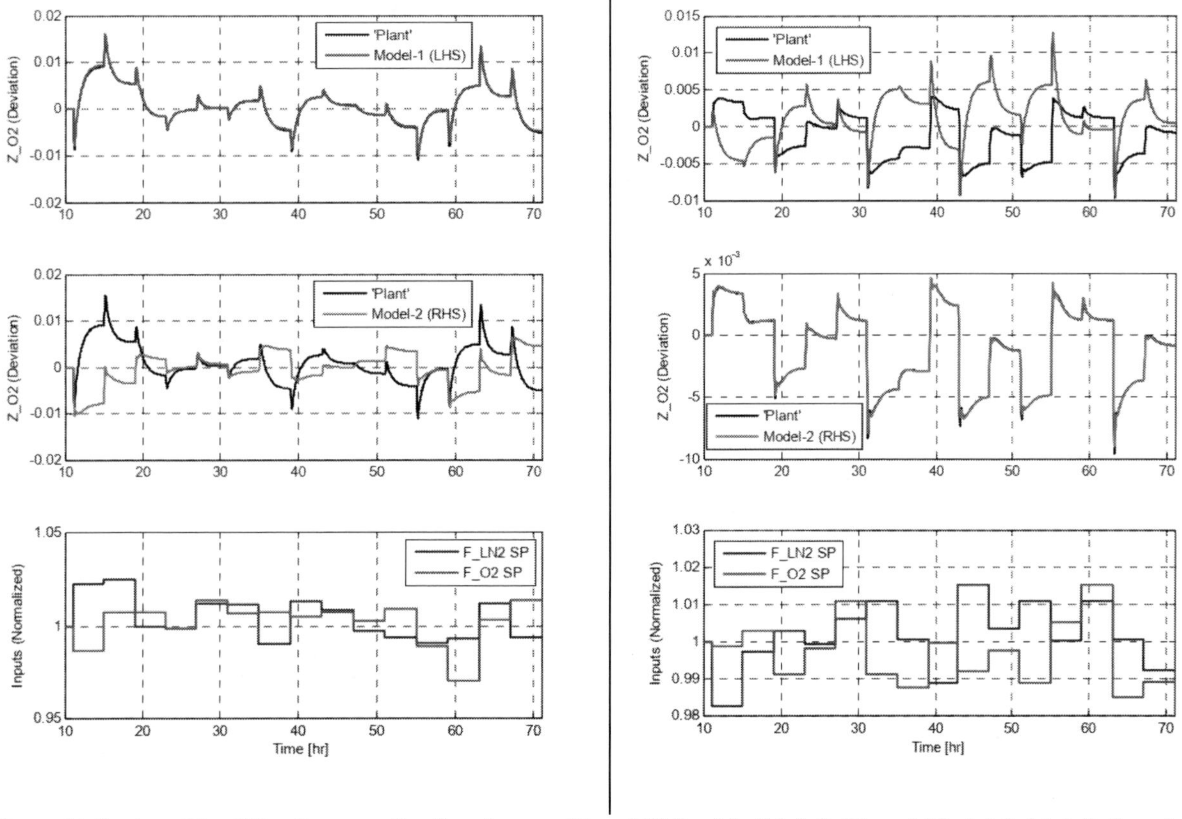

Figure 7. System Identification results showing quality of fit for Model-1 (left) and Model-2 (right). Input excitations have been shown at the bottom (on each side). Plant outputs are shown in black.

Identification (model-fit) results are shown in Figure 7 for LHS and RHS operating points. The resulting fits show that both the LHS (Model-1) and RHS (Model-2) models capture the plant responses with reasonable accuracy. For comparing and contrasting the behavior of both models, the outputs from the "other" model are also shown (middle plot for LHS and top plot for RHS). It is clearly visible that both models have gains (and even transients) in opposite direction. It must also be noted that the fixed-ratio criterion given in Eq. (9) ensures, to a certain extent, that for all possible combination of input variables the model does not crossover to the other regime. This is consistent with the plots.

Results & Discussion

A summary of simulation results obtained for the unconstrained MMPC control design using the two identified models is shown in Figure 8. Starting with a RHS operating point (ZO2 = 0.95 and F_LN2 = 28260 lbmol/hr), a pseudo-setpoint of high oxygen-purity value (0.956) is provided. This setpoint can take any value higher than expected maximum oxygen purity. The control algorithm tries to reach this SP by manipulating F_LN2 in a direction adherent to increase in purity but eventually the plant crosses-over to the other regime where further manipulating F_LN2 in the same direction starts bringing down the purity levels. The MMPC algorithm through weighing calculation start prioritizing the "other" model based on decreasing plant purity and the direction of F_LN2 actuation changes. This implies that the plant does not reach the given pseudo-setpoint value (since it is physically not possible to do so for given inlet flowrates), but operates very close to this maximum value. In Figure 8, the optimal plots for F_LN2 and Z_O2 have been shown. Clearly, the manipulated inputs seem to converge to a point where minor wobbling around optimal point (red-dotted line) can be observed.

Extensions and Future Work

In the current study, the demand for oxygen (and hence the feed-air) is maintained at a fixed value. In future work, the use of this strategy during load changes including ramp operations will be investigated. It is anticipated that feed-air flowrate will be a better candidate for regulating oxygen purities during fast load changes compared liquid nitrogen flowrate. In addition, if model crossovers occur while the plant is ramping, it might lead to sudden dip in purity level before the MMPC algorithm switches the model. In such situations, a two input (F_airASU, F_LN2) – two output (ZO2, ZLN2) single-model MPC implementation to meet oxygen purity during fast ramp-changes, should be prioritized. Once the transients settled down, the MMPC algorithm may be triggered for maximizing the purity around the new steady-state.

Another implementation for run-time optimization of the ASU may be done using feed-air flowrate minimization approach (at constant oxygen purity) in contrast to the current approach. Here, constant oscillations in feed-air flowrate may be observed as the liquid nitrogen fluctuates to maximize oxygen purity. Although such dynamic oscillations may at first discourage the implementation of proposed algorithms, it must be realized that over a period of time the difference in required feed-air prevent huge costs involved in expensive air compression.

Acknowledgement

The authors would like to acknowledge AVESTAR™ center at the National Energy Technology Laboratory, Morgantown, for insights on ASU modeling and IGCC related processes.

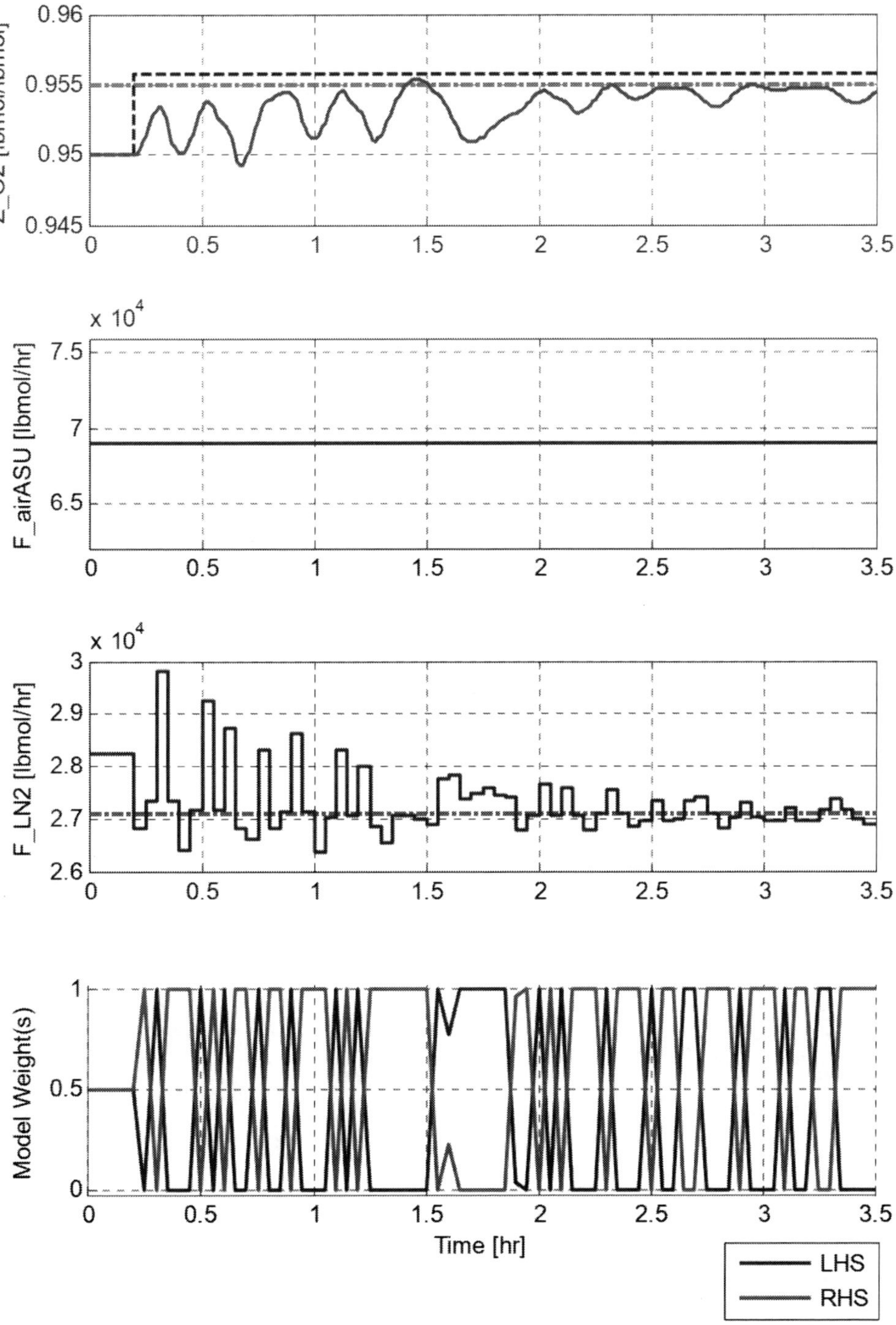

Figure 8. Simulation results showing dynamic maximization of oxygen purity (Z_O2) for fixed oxygen demand (14511.1 lbmol/hr) and feed-air flowrate (69000 lbmol/hr). Optimum operation is shown in dotted-red. Pseudo-SP for oxygen purity is given in dotted-black. MMPC parameters: P = 100, M = 20, $\lambda_{ii} = 10^8$.

References

[1] U.S. Department of Energy. Cost and Performance Baseline for Fossil Energy Plants, Volume 1: Bituminous Coal and Natural Gas to Electricity. Technical Report DOE/NETL-1281, National Energy Technology Laboratory, August 2007.

[2] S.E. Zitney, E.A. Liese, P. Mahapatra, R. Turton, D. Bhattacharyya, and G. Provost. Advanced Virtual Energy Simulation, Training, and Research: IGCC with CO_2 Capture Power Plant, *Proc. of the 28th Annual International Pittsburgh Coal Conference*, Pittsburgh, PA, September 12-15, 2011.

[3] P. Mahapatra and B.W. Bequette. Design and Control of an Elevated-Pressure Air Separation Unit for IGCC Power Plants in a Process Simulator Environment, *Industrial & Engineering Chemistry Research, special issue on Process Systems Engineering in Energy* (submitted, 2011).

[4] P. Mahapatra, B.W. Bequette, and S.E. Zitney. Multiple Model Predictive Control of Air Separation Unit as part of IGCC Power Plant during Rapid Load Changes, presented at the 2011 *AIChE Annual Meeting, Minneapolis, MN*, October 16-21, 2011

[5] M. Kurre-Kinsey and B.W. Bequette. Multiple Model Predictive Control Strategy for Disturbance Rejection, *Industrial & Engineering Chemistry Research*, 49(17): 7983–7989, 2010

(713f) Integrated Continuous-Flow System for Automated Online Optimization of Chemical Synthesis

Nikolay Zaborenko[1], Marc Champagne[2], Wei-Ming Sun[3], Michael T. Miller[4], Todd D. Maloney[4] and Edward Sheldon[5], (1)Chemical Product Research and Development, Eli Lilly and Company, Indianapolis, IN, (2)Analytical R&D, Eli Lilly, Indianapolis, IN, (3)Chemical Product Research and Development, Eli Lilly &Co., Inc., Indianapolis, IN, (4)Analytical Technologies Design and Development, Eli Lilly & Co., Indianapolis, IN, (5)Analytical Technologies Design & Development , Eli Lilly & Co., Indianapolis, IN

A great deal of experimental effort and cost is expended in chemical research and development towards mapping reaction response surface areas and determining optimal operating conditions. Frequently, such experimentation is performed by synthetic chemists in flask (batch) reactors, leading to overall large material consumption and time expenditure. Automated integration of continuous-flow micro-scale reactors with fluid delivery, reactor temperature control, and analytical equipment, combined with a decision-making algorithm, has been previously demonstrated in an academic setting to be a very powerful tool for real-time optimization with minimal material usage or operator involvement.[1] Herein, we present the application of this concept to a pharmaceutical R&D laboratory system. DeltaV was used as the communication framework to integrate and communicate with a network of robust high-pressure syringe pumps, a recirculating heater/chiller, and an on-line PAT HPLC, allowing widespread remote user interfacing. MATLAB was used as the system engine, coordinating all of the devices and applying the process algorithm. The system was used to apply Dantzig's simplex algorithm to a reaction step in an actual investigational pharmaceutical route. The reaction was performed in a micro-scale tube reactor, with the algorithm automatically optimizing product yield and selectivity by varying residence time, temperature, and equivalents of two components. The system design is fully modular, allowing for a wide variety of optimization or model investigation algorithms, reactor configurations, and target fitness functions.

---

[1] McMullen, J. P. 2010. Automated microreactor system for reaction development and online optimization of chemical processes. Ph. D. thesis, Massachusetts Institute of Technology, Cambridge, MA.

A New Extended Prediction Self-Adaptive Control (EPSAC) Strategy for Batch Control

Qing Lin Su[1], Martin Wijaya Hermanto[2], Richard D. Braatz[3], Min Sen Chiu[1]

[1]Department of Chemical and Biomolecular Engineering, National University of Singapore, 4 Engineering Drive 4, Singapore 117576

[2]Institute of Chemical and Engineering Sciences, 1 Pesek Road, Jurong Island, Singapore, 627833

[3]Department of Chemical Engineering, Massachusetts Institute of Technology, 77 Massachusetts Avenue, Cambridge, MA, 02139, USA

Abstract

It is well documented that linear model predictive control (MPC) techniques are predominantly used in industrial applications[1]. Driven by the stringent specifications on product quality, tighter environmental regulation of effluent streams, and higher competition in the process industries, the development of nonlinear model predictive control (NMPC) techniques is of interest to both the academic and industrial sectors. The main benefit of NMPC lies in its capability to handle nonlinearities and time-varying characteristics inherent in process dynamics while performing real-time dynamic optimization with constraints and bounds on both system states and manipulated variables[2,3]. Toward this end, various methods have been employed to approximate the process nonlinearity in the various NMPC design methods reported in the literature, including successive linearization[4], neural networks[5], multiple local models[6,7], piecewise linearization[8], and hybrid models[9].

Among the aforementioned NMPC design methods, the extended prediction self-adaptive control (EPSAC) algorithm[10] adopted a distinct approach by approximating process variables through iterative optimization around the pre-specified base input trajectory and corresponding base output trajectory. In this manner, the outputs in the prediction horizon are obtained as the sum of a base term and a deviation term. The former is computed from the nonlinear process model based on current value of input variable given by the predefined base input trajectory, while the latter obtained from finite step response or impulse response models, from which a quadratic programming (QP) problem is formulated and can be solved iteratively[11]. Though successful applications of EPSAC have been reported[10-13], one potential drawback of the previous EPSAC algorithms is the incorporation of finite step response or impulse models in the formulation of the control algorithm. Since model parameters are obtained by introducing a step change to the current input value specified by the base input trajectory, those model parameters obtained further away from the current time become less accurate due to process nonlinearity, leading to inevitable modeling error that degrades the achievable control performance. This shortcoming may become even worse when the EPSAC is applied in batch process control, where the objective is often to control the product quality at batch end. As a large number of step response parameters are needed at the beginning of the batch run to predict the future process outputs for the remaining batch, this eventually leads to inaccurate predicted outputs and poor control performance. Furthermore, another aim of this paper is to formulate the EPSAC algorithm to use state-space models due to its inherent flexibility to represent stable, integrating, and unstable processes[1-3, 14].

Motivated by the aforementioned discussion, a new EPSAC algorithm based on the Just-in-Time Learning (JITL) method[15] is developed in this work. The JITL method is a data-based modeling methodology that approximates a nonlinear system with a set of local models valid in the relevant operating regimes. In the proposed JITL-based EPSAC design, the deviation term is obtained by a set of local state-space models identified by the JITL method along the base trajectories. In simulation results of case studies for batch control of end-product quality, including an evaporative crystallizer[16] and a bioreactor[17], the proposed EPSAC algorithm provides better closed-loop performance than its previous counterpart[11].

References

[1] Qin, S.J., Badgwell, T.A. A survey of industrial model predictive control technology. *Control Engineering Practice*. 2003; 7(11): 733 - 764.

[2] Manenti, F. Considerations on nonlinear model predictive control techniques. *Computers and Chemical Engineering*. 2011; 11(35): 2491 - 2509.

[3] Darby, M.L., Nikolaou, M., MPC: Current practice and challenges. *Control Engineering Practice*. 2012; 4(20): 328 - 342.

[4] Lee, J.H., Ricker, N.L. Extended kalman filter based nonlinear model predictive control. *Industrial & Engineering Chemistry Research*. 1994; 6(33): 1530 - 1541.

[5] Peng, H., Nakano, K., Shioya, H. Nonlinear predictive control using neural nets-based local linearization ARX model-stability and industrial application. *IEEE Transactions on Control Systems Technology*. 2007; 1(15): 130 - 143.

[6] Dharaskar, K.P., Gupta, Y.P. Predictive control of nonlinear processes using interpolated models. *Chem. Eng. Res. Design*. 2000; 4(78): 573 - 580.

[7] García-Nieto, S., Martínez, M., Blasco, X., Sanchis, J. Nonlinear predictive control based on local model networks for air management in diesel engines. *Control Engineering Practice*. 2008; 12(16): 1399 - 1413.

[8] Cervantes, A.L., Agamennoni, O.E., Figueroa, J.L. A nonlinear model predictive control system based on Wiener piecewise linear models. *Journal of Process Control*. 2003; 7(13): 655 - 666.

[9] Hermanto, M.W., Braatz, R.D., Chiu, M.S. Integrated batch-to-batch and nonlinear model predictive control for polymorphic transformation in pharmaceutical crystallization. *AIChE Journal*. 2011; 4(57): 1008 - 1019.

[10] De Keyser, R.M.C., Cauwenberghe, A.R.V. Extended prediction self-adaptive control. In *IFAC Symposium on Identification and System Parameter Estimation*. 1985; 1255 - 1260.

[11] Hermanto, M.W., Chiu, M.S., Braatz, R.D. Nonlinear model predictive control for the polymorphic transformation of L-glutamic acid crystals. *AIChE J.* 2009; 10(55): 2631 - 2645.

[12] Gálvez-Carrillo, M., De Keyser, R., Ionescu, C., Nonlinear predictive control with dead-time compensator: Application to a solar power plant. *Solar Energy*. 2009; 5(83): 743 - 752.

[13] Niño, J., De Keyser, R., Syafiie, S., Ionescu, C., Struys, M., EPSAC-controlled anesthesia with online gain adaptation. *International Journal of Adaptive Control and Signal Processing*. 2009; (23): 455 - 471.

[14] Froisy, J.B. Model predictive control building a bridge between theory and practice. *Computers and Chemical Engineering*. 2006; 10-12(30): 1426 - 1435.

[15] Cheng, C., Chiu, M.S. A new data-based methodology for nonlinear process modeling. *Chemical Engineering Science*. 2004; 13(59): 2801 - 2810.

[16] Mesbah, A., Huesman, A.E.M., Kramer, H.J.M., Nagy, Z.K., Van de Hof, P.M.J. Real-time control of a semi-industrial fed-batch evaporative crystallizer using different direct optimization strategies. *AIChE Journal*. 2011; 6(57): 1557 - 1569.

[17] Peters, N., Guay, M., Dehaan, D. Real-time dynamic optimization of non-linear batch systems. *Canadian Journal of Chemical Engineering*. 2006; 3(84): 338 - 348.

(723a) Supply Chain Management Optimization Using a Hybrid Simulation Based Optimization Approach

Fnu Nihar and Marianthi G. Ierapetritou, Department of Chemical and Biochemical Engineering, Rutgers, The State University of New Jersey, Piscataway, NJ

Supply Chain Management Optimization using a Hybrid Simulation based Optimization Approach

Nihar, Marianthi Ierapetritou

Department of Chemical and Biochemical Engineering, Rutgers University, Piscataway, NJ

A supply chain is a network of suppliers, production facilities, warehouses and markets designed to acquire raw materials, manufacture and store and distribute products among the markets. The entire process is driven by the demand generated at the markets. Since there are different entities involved and they work towards their individual interests, the optimum performance of the whole network is often not achieved. It has been shown that greater efficiency and reduced costs can be achieved through proper coordination among the entities in terms of material, financial and information flow [1, 2].

In order to gain competitive edge in a global economy, companies aim at optimizing their overall supply chain. In a typical supply chain, there are a large number of entities. The interacting behavior of these entities among themselves is complex. This makes the problem of optimization large and complex. Mathematical modeling has been very widely used to formulate these problems. Several mathematical models use mixed integer linear programming (MIP) and mixed integer non-linear programming to solve the SC optimization problems [3, 4]. Stochastic models have also been developed for supply chain networks in uncertain environments [5]. However, these approaches lack realism since they are not able to capture the complex relations among the different entities comprising the network. Simulation models are another category of models that have been used to solve such problems. They help mimic complex relations among the different entities of the supply chain. Agent based models, a class of simulation models, treat the different entities as autonomous agents, each having its own set of behavioral rules. It has been shown that using this approach, it is possible to govern the interaction among the different entities of the whole network and thus reach at improved solutions [6]. Compared to the optimization based methods, limited work has been done using this approach.

Analytic and simulation models can be regarded as the two extremes of a range of mathematical models available for modeling [7]. Hybrid simulation/analytic models which are combinations of these two types of models enable us to take advantages of both of them and can be very useful [8, 9]. Such models are useful for stochastic optimization as well since they can mimic real systems including their stochastic and nonlinear behaviors. In the recent years, such models have been used to explore the supply chain domain. It is an active area of research and rapid progress is being made.

In this work, we demonstrate the use of a hybrid modeling approach to solve a small scale supply chain management problem. The problem has been formulated as a mathematical model minimizing the overall cost incurred over a number of planning periods. The objective of using the hybrid approach is to overcome the computational complexity involved in solving the large scale mixed integer nonlinear problem (MINLP) and to obtain a solution which represents supply chain reality more closely. The proposed framework involves the integration of the solution strategies of separate optimization and simulation models. The optimization model solves the planning and scheduling problem while the simulation model is used to verify the quality of the solution obtained by comparing it with real world scenarios. An iterative solution procedure has been used such that there is communication between the two models in terms of some decision variables in every cycle. The optimization model of the problem has been developed in GAMS while the simulation model has been developed using the JAVA Repast tool. Various scenarios involving different number of planning periods and different sets of deterministic demand have been used to investigate the proposed framework. The effectiveness of the approach has been shown in terms of the computational effort required and the quality of the solution achieved.

Considering the increasing environmental concerns and associated legislations these days, the proposed framework has been extended to account for environmental impact through a multi-objective optimization approach. In addition to the economic objective which minimizes the overall operation cost of the supply chain, an environmental objective which minimizes the environmental cost associated with the manufacturing and transportation processes of the network is also considered. The consideration of an environmental objective at the stage of supply chain management is critical in order to make a holistic assessment. Incorporating the environmental aspects at a later stage might lead to decreasing the environmental impact locally while causing an overall increase in the environmental costs. The simulation model is able to assess the various environmental aspects of the supply chain and decide whether the solution provided by the optimization model meets the environmental requirements or not. The presence of the two objective functions helps maintain the environmental performance of the supply chain while minimizing the overall cost.

References:

1. Stadtler, H., *Supply chain management and advanced planning--basics, overview and challenges.* European Journal of Operational Research, 2005. **163**(3): p. 575-588.

2. Varma, V.A., et al., *Enterprise-wide modeling & optimization--An overview of emerging research challenges and opportunities.* Computers & Chemical Engineering, 2007. **31**(5-6): p. 692-711.

3. Li, Z. and M.G. Ierapetritou, *Production planning and scheduling integration through augmented Lagrangian optimization.* Computers & Chemical Engineering, 2010. **34**(6): p. 996-1006.

4. Timpe, C.H. and J. Kallrath, *Optimal planning in large multi-site production networks.* European Journal of Operational Research, 2000. **126**(2): p. 422-435.

5. Petrovic, D., R. Roy, and R. Petrovic, *Modelling and simulation of a supply chain in an uncertain environment.* European Journal of Operational Research, 1998. **109**(2): p. 299-309.

6. Behdani, B., et al., *Agent based model for performance analysis of a global chemical supply chain during normal and abnormal situations*, in *Computer Aided Chemical Engineering*. 2009, Elsevier. p. 979-984.

7. Shanthikumar, J.G. and R.G. Sargent, *A Unifying View of Hybrid Simulation/Analytic Models and Modeling.* Operations Research, 1983. **31**(6): p. 1030-1052.

8. Lee, Y.H., S.H. Kim, and C. Moon, *Production-distribution planning in supply chain using a hybrid approach.* Production Planning & Control, 2002. **13**(1): p. 35-46.

9. Young Hae, L. and K. Sook Han. *Optimal production-distribution planning in supply chain management using a hybrid simulation-analytic approach*. in *Simulation Conference, 2000. Proceedings. Winter*. 2000.

(723b) Multiple Time Grids in Discrete-Time Models for Scheduling and Supply Chain Planning

Christos Maravelias[1], Sara Zenner[1] and Kaushik Subramanian[2], (1)University of Wisconsin, Madison, WI, (2)University of Wisconsin - Madison, Madison, WI

Multiple Time Grids in Discrete-time Models for Scheduling and Supply Chain Planning

Sara Zenner, Kaushik Subramanian, and Christos T. Maravelias

Department of Chemical and Biological Engineering, University of Wisconsin – Madison

1415 Engineering Dr., Madison, WI 53706, USA

The modeling of time plays a key role in the formulation of mixed-integer programming (MIP) models for scheduling, production planning, and operational supply chain planning problems. It affects the size of the model, the computational requirements, and the quality of the solution. While many types of time grids have been proposed in the literature for continuous-time models, discrete-time models have always been thought of as having a single grid with equally spaced time points. However, the ability to incorporate different grids into a single model is useful in many situations. For example, processes in different facilities that interact with each other may use different grids. Also, tasks in the same facility (executed on the same or different units) may have very different processing times. Finally, externally imposed events, such as due times or raw material deliveries, may not occur at a point of the grid used to represent the processing tasks.

To model all the aforementioned situations, existing approaches either introduce approximations or employ a very fine grid, typically based on the largest common factor among all time scales of interest, for the entire scheduling/planning horizon. Unfortunately, having a large number of time points leads to models that may be too large to solve in a reasonable time.

We present methods to formulate discrete-time multi-grid models that allow different tasks, units, or materials to have their own time grid. The length of the planning periods of the grids do not have to be multiples of the same factor. In addition, each grid may have periods of (known but) unequal length. Since a discrete-time approach is used, the multiple time grids are easily "synchronized" because all time points are known beforehand. The proposed methods lead to substantial reductions in the size of the formulations and thus the computational requirements. In addition, they can yield better solutions than formulations that use approximations. Finally, we show how to carefully select the different time grids, so that we obtain solutions that are comparable with the ones obtained using a single fine grid to represent the processing of all tasks and occurrence of events.

(723c) Robust Decision Making for Hybrid Process Supply Chain Systems Via Model Predictive Control

Richard Mastragostino and **Christopher L. E. Swartz**, Chemical Engineering, McMaster University, Hamilton, ON, Canada

A supply chain (SC) is a network of facilities that performs the functions of raw material procurement, raw material transformation into intermediate and finished products, and distribution of products to customers. In a chemical process supply chain (PSC), such as in the petrochemical or pharmaceutical industry, manufacturing is a major component. Supply chain management (SCM) and supply chain optimization (SCO) are concerned with the coordinated optimization of business processes, including strategic network design, purchasing, production, transportation and distribution in order to bring greater net value to the customer, at minimum overall cost. Poor SCM can contribute to undesirable performance such as the bullwhip effect, defined as the amplification in demand variability observed when moving up the supply chain from retailers to suppliers. Applications for SCM that integrate classical control technology have been motivated by the need for effective quantitative approaches to mitigate inefficiencies in the supply chain. These applications generally apply feedback laws for maintaining inventory positions and satisfying demand [Sarimveis et al., 2008]. However, the limitations associated with classical control technology, such as the inability to explicitly consider interactions or delays in the supply chain, can be averted by applying an advanced control method like model predictive control (MPC). As a result, MPC has been successfully applied for SCM in the literature [e.g., Wang and Rivera, 2008].

The supply chain model is a vital component of an MPC-based strategy, which must describe adequately the dynamic behavior of the system. The simplest model captures inventory dynamics based on the quantity of material entering and leaving the echelon at each time period; however a more sophisticated model addresses the hybrid continuous-discrete nature of the supply chain. In a supply chain the discrete nature arises from disjunctive logical conditions governing decision-making (e.g. production scheduling), generally giving rise to a mixed-integer-programming model. This has motivated MPC-based formulations for SCM of hybrid supply chain systems [e.g., Perea-Lopez et al., 2003; Mestan et al., 2006]. Uncertainty is a relevant issue in SCO, and if not addressed, can contribute to poor decision making. An approach to capture uncertainty explicitly at each execution of the model predictive controller is provided through robust MPC. With robust MPC, the model used for control, and unmeasured disturbances are considered uncertain; therefore, the future closed loop behavior is uncertain because the control action depends on the uncertainty propagation. An "open loop" approach to robust MPC, which neglects the partial compensation for uncertainty by feedback may lead to overly conservative control action. Previous work on robust MPC has attempted to approximate the future closed loop behavior of the system [e.g., Kothare et al., 1996; Li and Marlin, 2009].

In this paper we present an optimization-based decision making tool for the control of a hybrid supply chain system, via a robust MPC strategy. The proposed formulation captures uncertainty in model parameters and demand by stochastic programming, accommodates hybrid process systems with production decisions governed by logical conditions, and addresses multiple supply chain performance metrics, within an integrated optimization framework. In this work, a scenario-based approach is used to capture uncertainty through a finite number of discrete realizations of uncertain parameters. Furthermore, a nuance in this work is the application of a stochastic forecasting model to generate demand scenarios. Multi-objective optimization is useful for addressing simultaneously conflicting supply chain performance metrics such as customer service and economics. A Pareto frontier is generated to illustrate the trade-off existing between these metrics. Furthermore, this paper proposes an approach to reduce the conservativeness of "open loop" decision making under uncertainty, by approximating the closed loop prediction of uncertainty propagation with two-stage stochastic programming. The formulation is applied to the SCM of a multi-product supply chain. Through simulation, the robust framework is shown to provide a substantial reduction in the occurrence of back orders when compared with a nominal MPC implementation. Additionally, the "closed loop" approach provides an equivalent level of customer service at a reduced operating cost, when compared with the "open-loop" approach to robust MPC.

This paper will present the following: (i) a mathematical formulation of the dynamic supply chain model describing the hybrid system, (ii) details of the "open loop" and approximated "closed loop" approach to robust MPC applied to SCM, and (iii) results of a case study where the formulation is applied to the control of a multi-product supply chain. Additionally, key challenges and future avenues for research are identified.

References

Kothare, M., Balakrishnan, V., and Morari, M. (1996). Robust constrained model predictive control using linear matrix inequalities. Automatica, 32, 1361–1379.

Li, X. and Marlin, T. (2009). Robust supply chain performance via model predictive control. Comput. Chem. Eng., 33, 2134–2143.

Mestan, E., Turkay, M., and Arkun, Y. (2006). Optimization of operations in supply chain systems using hybrid systems approach and model predictive control. Ind. Eng. Chem. Res., 45, 6493–6503.

Perea-Lopez, E., Ydstie, B., and Grossmann, I. (2003). A model predictive control strategy for supply chain optimization. Comput. Chem. Eng., 27, 1201–1218.

Sarimveis, H., Patrinos, P., Tarantilis, C., and Kiranoudis, C. (2008). Dynamic modeling and control of supply chain systems: A review. Comput. Oper. Res., 35, 3530–3561.

Wang, W. and Rivera, D. (2008). Model predictive control for tactical decision-making in semiconductor manufacturing supply chain management. IEEE Transactions on Control Systems Technology, 16(5), 841 –855.

(723d) Design of Supply Chains with Random Facility Disruptions

Pablo Garcia-Herreros[1], John Wassick[2] and Ignacio E. Grossmann[1], (1)Chemical Engineering Department, Carnegie Mellon University, Pittsburgh, PA, (2)Dow Chemical Company, Midland, MI

The impact of supply chain performance on companies' competitiveness is widely recognized in the process industries. The formulation of the design and management of supply chains as an optimization problem allows selecting the alternatives that offer the best decisions based on a quantitative basis. However, the performance of any given supply chain is subject to uncertainties derived from product demand, operating costs, and reliability of facilities.

The design and management of supply chains under demand uncertainty has received significant attention over the past years. However, it is only recently that the possibility of disruptions in supply chains has been considered in the formulation of design problems [1]. Such disruptions, localized in specific facilities, can have a catastrophic impact on the entire supply chain as it might propagate through the system. In order to protect against the undesirable effects of such events, disruptions must be anticipated in the formulation of the design problem. In this work, the optimization of the supply chain design under random facility disruptions is formulated within the framework of multistage stochastic programming with recourse.

We consider the supply of a single commodity from a fixed set of production facilities, passing through distribution centers, and serving a fixed set of customers with deterministic demand. Lead times are considered between facilities. The objective is to select among the set of candidate distribution centers and design their capacity in order to minimize the investment cost and the expected operational cost of satisfying the customers demand. We develop a multi-period formulation where the first-stage variables correspond to the selection of distribution centers and their capacities; the recourse variables that are selected in the different time periods correspond to the inventory level at each distribution center, the amounts transported from production facilities to distribution centers, the amounts transported from distribution centers to customers, and penalties for unsatisfied demand.

At any instant in time, one or more disruptions can occur in the distribution centers according to given probabilities. The combination of disruption in a time instant is called a *discrete state*of the system [2]. The transition of the system among discrete states through the different time periods can be modeled as a time-homogeneous Discrete-Time Markov Chain (DTMC).

Although the discrete nature of disruptions in the system determine a finite number of possible discrete states, the multi-period formulation with multi-stage recourse variables implies an exponential growth in the number of scenarios with respect to the time periods. In order to overcome this difficulty, a constant inventory level is maintained throughout all time periods. This policy eliminates the dependence of the system state from its history, avoiding the explosion in dimensionality that is caused by the multi-stage inventory recourse. The resulting problem is then equivalent to a two-stage stochastic programming problem where the scenarios correspond to the possible discrete states of the facilities.

In order to illustrate the application of the proposed model we present several example problems. Simulation results on the predicted supply chain designs show the economic advantages of the stochastic programming approach over the design based on deterministic supply when scenarios with random disruptions are considered. Furthermore, the periodic review of inventory is found to be particularly well suited for the design stage of supply chains. It is also shown that inflexibility on the inventory management offers a conservative estimate of the supply chain's performance.

References

[1] Qi, L.; Shen, Z.-J. M. & Snyder, L. V. The Effect of Supply Disruptions on Supply Chain Design Decisions. *Transportation Science,* 2010, *44*, 274-289.

[2] Terrazas-Moreno, S.; Grossmann, I. E.; Wassick, J. M. & Bury, S. J. Optimal design of reliable integrated chemical production sites. *Computers & Chemical Engineering,* 2010*, 34*, 1919 – 1936.

(723e) Optimal Supply Chain Redesign in the Electric Motor Industry

Maria Analia Rodriguez[1], **Ignacio E. Grossmann**[2], Aldo Vecchietti[1] and Iiro Harjunkoski[3], (1)INGAR, Santa Fe, Argentina, (2)Chemical Engineering Department, Carnegie Mellon University, Pittsburgh, PA, (3)Industrial Software and Applications, ABB AG, Corporate Research Germany, Ladenburg, Germany

Optimal supply chain redesign in the electric motor industry

Maria Analia Rodriguez, Aldo R. Vecchietti, Iiro Harjunkoski and Ignacio E. Grossmann

The integration of supply chain redesign and tactical decisions such as defining inventory levels and how supply chain nodes are connected is a challenging problem that can greatly impact companies' economy. Rising transport costs are key factors in decisions about both where to place the assets (factories and distribution centers) and how much inventory to store. In addition, managing inventory has become a major target in order to simultaneously reduce costs and improve customer service in today's increasingly competitive business environment (Daskin, Coullard and Shen, 2002). For that reason, over the last few years, there has been an increasing interest in developing enterprise-wide optimization (EWO) models to solve problems that are broad in scope and integrate several decision levels (Grossmann, 2005).

In the case of the electric motors industry, the relevance of this problem is given by some key issues. On the one hand, electric motors are expensive products, so keeping them in inventory means that a significant capital cannot be used for other purposes. On the other hand, a motor malfunction may block the entire production of a plant and therefore, obtaining a spare motor as soon as possible is crucial.

Another special characteristic of this industry is given by the type of product. Most works from the literature, assume that the products are only moved forward in the supply chain and only the demand of new products is considered. In this case, the situation is more complex. As usual, demand can be originated by new customers or new investments at customer sites but, in addition, motors that are already in use at customer sites can fail. When that happens, clients require a motor in order to replace the one that failed. An important decision in this context is whether to replace failing motors with new units or with repaired products. An efficient inventory management of new and used motors in the supply chain warehouses is another challenge of this problem.

Customer plants typically have tens or more different type of motors in their production processes, and also identical motors can be used for a variety of purposes. According to the type of motor and application, the criticality of a given unit can be very different so the time a customer allows that a motor is out of service until another one replaces it, is case dependent. If the time requirement is very tight, it might be necessary to have some motors in stock at customer sites.

Taking into account that motor demand is uncertain and depends on motors failure at customer plants, a responsive supply chain can only be guaranteed when an effective inventory management, as well as an appropriate distribution and storage structure are planned together. Furthermore, demand uncertainty might also have a relevant influence on warehouses capacity requirement. In that sense, if the plan for storage capacity does not consider demand uncertainty, it might be infeasible to provide the motors as required.

You and Grossmann (2010) propose an optimization model to design a multi-echelon supply chain and the associated inventory systems under demand uncertainty in the chemical industry. The original model is an MINLP with a non-convex objective function so they develop a spatial decomposition algorithm to obtain near global optimal solutions with reasonable computational expense. The supply chain involves one product, design decisions consider the installation of new distribution centers, but no expansions or elimination of installed warehouses are considered. In addition, the model assumes one planning period so investment costs are annualized and capacity constraints are not analyzed. Our approach extends this previous work introducing new considerations regarding the particular industrial context, complexities from the model point of view and novel concepts that were not considered before.

We develop an optimization model to redesign the supply chain of the electric motors industry under demand uncertainty from strategic and tactical perspectives in a planning horizon consisting of multiple periods. Long term decisions involve new installations, expansions and elimination of warehouses handling multiple products. It is also considered which warehouses should be used as repair work-shops in order to store, repair and deliver used motors to customers. Tactical decisions include deciding inventory levels (safety stock and expected inventory) for each type of motor in distribution centers and customer plants, as well as the connection links between the supply chain nodes. Capacity constraints are also considered when planning inventory levels. At the tactical level it is analyzed how demand of failing motors is satisfied, and whether to use new or used motors.

The uncertain demand is addressed by defining the optimal amount of safety stock that guarantees certain service level at a customer plant. In addition, the risk-pooling effect described by Eppen (1979) is taken into account when defining inventory levels in distribution centers and customer zones. One novel consideration is given by inclusion of lost sales costs in the objective function, which was extended from the work by Parker and Little (2006). Due to the nonlinear and large size nature of the original formulation, piece-wise linearization and lagrangean relaxation algorithms are applied to obtain the optimal solution.

References

Daskin, M., Coullard, C. and Shen, Z-J. An Inventory-Location Model: Formulation, Solution Algorithm and Computational Results. Annals of Operations Research, 2002, 110, 83–106.

Eppen, G. D. Effect of centralization on expected costs in a multi-location newsboy problem. Management Science, 1979, 25, 498 – 501.

Grossmann, I.E. Challenges in the New Millennium: Product Discovery and Design, Enterprise and Supply Chain Optimization, Global Life Cycle Assessment. Computers and Chemical Engineering, 2005, 29, 29-39.

Parker, L. L. and Little, A. D. Economical reorder quantities and reorder points with uncertain demand. Naval Research Logistics Quarterly, 2006, 11, 351–358.

You, F. and Grossmann, I. E. Integrated Multi-Echelon Supply Chain Design with Inventories Under Uncertainty: MINLP Models, Computational Strategies. AIChE Journal, 2010, 56, 2, 419 – 440.

AN INTEGRATED PRODUCTION AND DISTRIBUTION MODEL FOR THE OPTIMAL OPERATION OF INDUSTRIAL GAS SUPPLY-CHAINS

Pablo A. Marchetti, Ignacio E. Grossmann
Department of Chemical Engineering, Carnegie Mellon University, Pittsburgh, PA

Tong Li, Jeffrey Arbogast, and Jean André
Process Control & Logistics, Air Liquide, Newark, DE

In this work, we address the problem of determining optimal operational level decisions for the coordinated production and distribution of industrial gas supply-chains. In this industry, cryogenic air separation processes are used to produce oxygen, nitrogen, and argon both as gaseous and liquid products. Air separation units consume large amounts of electricity, mainly due to the operation of the compressors used at different stages of the process. Depending on the equipment configuration used, alternative operation modes with different efficiencies and energy requirements are available at each plant. Thus, the main driver of the production cost is the cost of electricity, which prices fluctuate during the day either on an hourly or on a peak/off-peak basis, depending on the pricing scheme adopted. Gaseous and liquid customers of industrial gases are usually served by pipeline or truck delivery, respectively, and the transportation cost needed for truck deliveries is the main variable cost of the distribution side. Both the frequency of deliveries to a given customer and the selection of routes supplying product to multiple customers should be considered in order to reduce this cost. Besides, distribution capacity depends not only on the availability of trucks, trailers, and drivers but also on the inventory levels of liquid products at each plant. Thus, at any given time the connection between production and truck-distribution is given by the available amount of liquid products stored at the plants. The replenishment of storage tanks in customer locations must be secured by an appropriate distribution schedule.

On the production side, previous work addressing the minimization of production costs in air separation plants has been proposed by Ierapetritou et al.[1] and Karwan and Keblis[2]. Ierapetritou et al.[1] developed a two-stage stochastic formulation where uncertainty in the power prices is considered within a given portion of the optimization horizon. Karwan and Keblis[2] developed a mixed integer mathematical formulation embedded in a rolling horizon procedure to minimize the cost of running an air separation unit under a real time pricing (RTP) scheme. They also conducted simulation studies to assess the robustness of the production plans obtained. On the distribution side, several publications tackle the estimation of the delivery scheduling costs for different purposes through the use of truck routing heuristics or continuous approximations. In one of the most recent papers, You et al.[3] developed a mixed-integer linear programming model to integrate long term planning decisions of sizing storage tanks at customer locations with the estimation of the delivery costs at the operational level. In order to solve large-scale instances, they develop a first approach based on a two-level decomposition strategy and the second one based on a continuous approximation method. In a recent effort to model both production and distribution within a holistic approach, Glankwamdee et al.[4] developed a simplified linear production and distribution planning model for industrial gas supply chains. In order to account for uncertainty, they extended this formulation both via a minmax model and a two-stage stochastic program, and tested the effectiveness of the proposed methods using simulation. However, only time-aggregated planning decisions were considered and neither plant mode selection nor vehicle routing details are included in the model.

The main goal of this contribution is to assess the benefits of optimal coordination for production-distribution in an industrial gases supply-chain, with the same level of accuracy on both sides. A mixed-integer linear programming (MILP) formulation minimizing the overall cost of production and distribution over a limited time horizon (7 to 14 days) is presented. The problem involves the production of gaseous and liquid products at multiple plants, and the distribution of liquid products to multiple customers through alternative routes using a limited number of trucks. On the distribution side, multiple depots are considered, and trucks at a given depot can deliver product from multiple sources. Besides, in order to ensure customer storage replenishments, product can be purchased from an alternative source. As the number of sources and customers increase, the selection of the alternative routes to be included in the formulation becomes a critical issue, and special methods to choose the routes are evaluated. The proposed methodology has been successfully tested on small and medium-size examples with a limited number of products, plants, customers, depots, trucks, and routes. Clusters of customers are considered for the larger problem instances. Comparison of different levels of coordination on the production-distribution supply-chain shows that significant benefits can be obtained with a higher coordination among all plants/depots in order to fulfill the shared customer demands.

References:

(1) Ierapetritou M. G., Wu D., Vin J., Sweeney P., Chigirinskiy, M. Cost minimization in an energy-intensive plant using mathematical programming approaches. Industrial and Engineering Chemistry Research. 2002; 41:5262–5277.

(2) Karwan M. H., Keblis M. F. Operations planning with real time pricing of a primary input. Computers and Operations Research. 2007; 34:848–867.

(3) You F., Pinto J. M., Capón E., Grossmann I. E., Arora N., Megan L. Optimal distribution-inventory planning of industrial gases. I. Fast computational strategies for large-scale problems. Industrial and Engineering Chemistry Research. 2011; 50:2910–2927.

(4) Glankwamdee W., Linderoth J., Shen J., Connard P., Hutton J. Combining optimization and simulation for strategic and operational industrial gas production and distribution. Computers and Chemical Engineering. 2008; 32:2536–2546.

(723g) Optimizing Tank Fleet in Chemical Supply Chains

Meng Sha, Department of Chemical and Biomolecular Engineering, National Unviersity of Singapore, Singapore, Singapore and Rajagopalan Srinivasan, Department of Chemical and Biomolecular Engineering, National University of Singapore, Singapore, Singapore

Tank fleet flows constitute an important and expensive part of chemical supply chains as the storage and transportation of chemical products are strictly controlled under safety and environmental regulations. A typical tank fleet flow starts from manufacturer which fills up a tank fleet with chemical and transports it to distribution center. Distribution center then delivers it to customer based on existing orders. Customer holds the tank fleet till it consumes all the chemical products, and returns the empty tank fleet to distribution center or manufacturer. Shortage of chemical tank fleets on the manufacturer side would force the plant to shut down, resulting in huge financial losses and customer service level decreases. In the meantime, the lead time for enterprises to get new tank fleets after order placement is very long. Therefore in the reality, chemical enterprises always prefer to hold a huge amount of expensive tank fleets and keep purchasing new ones from suppliers. As a result, it provides large opportunities for cost reductions in chemical supply chains through efficient and effective management of tank fleets.

To address the tank fleet management problem, we developed an agent-based discrete event simulation model of supply chain. This model contains a chemical enterprise, multiple customers, and market. The chemical enterprise comprises multiple plants, multiple warehouses, order coordinator, replenishment coordinator and logistics department. These supply chain entities are functioning on certain rules and policies, and make decisions based on their own interests. And they communicate and collaborate with each other in various activities including end product sale, order placement, order assignment, inventory management and replenishment planning, which construct the tank fleet flows of the supply chain. We started from a base case with simple policies and investigated the usage of tank fleets on entities and transportation under different scenarios. Afterwards we employed various policies and rules into plants, warehouse and logistics department, and evaluated the respective tank fleets movement performance through simulation-optimization techniques. The emphasis of our work is on the development of efficient and effective chemical tank fleet management policies.

(725a) Use of Aggregation Prediction As a Function of Protein and Excipient Properties for Lyophilized Formulation Design

Brock C. Roughton[1], Lavanya K. Iyer[2], Haider S. Tarar[3], Anthony I. Pokphanh[3], Elizabeth M. Topp[2] and Kyle V. Camarda[3], (1)Bioengineering Graduate Program, University of Kansas, Lawrence, KS, (2)Department of Industrial and Physical Pharmacy, Purdue University, West Lafayette, IN, (3)Department of Chemical and Petroleum Engineering, University of Kansas, Lawrence, KS

Protein stability is a major concern during development of a protein drug product, with physical and chemical instability playing major roles. Lyophilization, or freeze-drying, is often used to help stabilize protein formulations. However, protein aggregation may still occur at significant levels in lyophilized formulations. Protein aggregation is undesirable, resulting in product deactivation and an increased potential for immunogenic response in patients. Predicting the aggregation propensity of lyophilized proteins is desirable for aiding formulation decisions, such as whether lyophilization should be pursued and whether stabilizing additives would be required. Such additive molecules are referred to as excipients and often comprise the bulk of a drug product. If lyophilization of a protein is necessary, prediction of the effect of added excipients on aggregation would provide a useful tool for formulation design. Several excipients, such as sucrose and glycine, are commonly used to prevent aggregation in lyophilized formulations but are not effective for all proteins and exhibit different protective effects from protein to protein.

A predictive model for aggregation of lyophilized proteins as a function of protein descriptors and excipient molecular descriptors is presented. The protein descriptors are comprised of physical/structural descriptors as well as computationally determined descriptors developed in previous literature for predicting aggregation based on primary or tertiary structure. Connectivity indices describing atomic and bonding configurations were used as molecular descriptors for the excipients. The protein and excipient descriptors were used to correlate percent native protein retained after lyophilization and reconstitution, as determined experimentally by size-exclusion chromatography (SEC). Losses in percent native protein correspond to protein aggregation. Selection of descriptors was performed using an exhaustive search method to provide sufficient fit while avoiding overfitting, as evaluated by R^2 and the Akaikie Information Criterion (AIC). The predictive power of the resulting correlation was evaluated using leave-one-out cross-validation (LOOCV). Initial results showed that use of protein descriptors alone was sufficient to predict aggregation on a formulation-by-formulation basis but was not able to predict aggregation across formulations. Fifteen proteins were considered under sucrose, buffer, urea, and glycine formulations.

From the set of proteins used in the initial work, four proteins were selected based on diversity of descriptor values. An exhaustive search was performed to identify the four proteins that best spanned the descriptor space, specifically for the descriptors that were selected for the sucrose and glycine formulations. The four selected proteins were lyophilized with 15 different excipients, resulting in 60 unique formulations. The excipients spanned the structural space between glycine and sucrose. After lyophilization and reconstitution, SEC analysis was used to quantify protein lost to aggregation. Combination of protein and excipient descriptors was successful in predicting aggregation across all 60 formulations through use of a single correlation. The resulting correlation was used in a computational molecular design (CMD) framework to optimally design an excipient for a given protein that minimized aggregation. A stochastic optimization method was used to solve for optimal excipient molecular structures and prediction intervals were calculated for the solutions at a 95% confidence level. The correlation was also used to search for proteins that would show minimal aggregation for a given excipient commonly used in lyophilized formulations. Use of the predictive model provides a basis for rational lyophilized formulation design, using either approved molecules or optimally-designed excipients for a given protein.

Analysis of high-throughput multiparametric flow cytometry data to identify cellular phenotypes underlying alcohol mediated aberrant differentiation of embryonic stem cells

Rajanikanth Vadigepalli, Eleni Anni, Joshua Ogony
Department of Pathology, Anatomy and Cell Biology
Thomas Jefferson University
Philadelphia, USA
Rajanikanth.Vadigepalli@jefferson.edu

Arun Tangirala, Sreekanth Rajagopalan
Department of Chemical Engineering
Indian Institute of Technology Madras
Chennai, India
sreekanthr@smail.iitm.ac.in

I. Introduction

Flow cytometry (FCM) is an established high-throughput research and clinical tool. Yet, analysis of data is often limited to extracting a small subset of information, which may not meet requirements such as re-engineering biological processes. The current work presents a pipeline to uncover biological information from single-cell multiplex data using available analysis tools, serving as a backdrop in stem cell research.

Prenatal exposure to alcohol by embryo through consumption by mother during pregnancy may lead to abnormalities in the form of craniofacial malformations and cognitive deficits in the newborn infant. The severity of these disabilities comprises a wide range of disorders termed FASD or Fetal Alcohol Spectrum Disorder. Though the exact time point of interference of ethanol during gestation is indeterminable due to experimental constraints (even *in vitro* studies are limited to few days in mouse embryonic stem (ES) cells to extract good viable samples from long passages of cell cultures), our previous studies[1] have indicated interference of ethanol during early stages of differentiation itself. We have studied ES cells which are capable of self-renewal (pluripotency), and can differentiate into various fetal cell lines, controlled by a group of transcription factors (TFs). The three primary germ layers are i) endoderm, ii) ectoderm and iii) mesoderm, and the principal regulatory network controlling differentiation of ES cells is known to comprise of three TFs - Sox-2, Oct-4 and Nanog. While down-regulation of these core TFs trigger differentiation, the balance in expressions of these TFs determine cell fate and state of pluripotency. Thomson *et al.*[2] have shown that the balance between Sox-2 and Oct-4 determines the germ layers mesoderm (ME) and neuroectoderm (NE) - while higher expression of Sox-2 triggers progenitor TFs towards NE fate and represses those determining ME fate, the *vice versa* occurs with higher expression of Oct-4. In our studies, we have forced the differentiation towards NE with all-trans retinoic acid (RA) treatment over 6 days.

The consolidated data analysis indicates i) that ethanol delays down-regulation of core TFs and differentiation, ii) presence of multiple subpopulations in RA and RA + ET mediated ES differentiation, iii) ethanol affects TFs and phenotypic distribution in an ethanol dose-dependent manner. The findings point to an early interference of ethanol in ES differentiation networks that alter cell commitment against normal development of NE lineage towards FASD in mice and humans.

II. Experimental Method

Undifferentiated E14Tg2a.4 cell-line mouse cells were plated on 0.1% gelatin-coated plates, in an ES cell medium supplemented with 15% fetal bovine serum, 1000 U/ml Leukemia Inhibitory Factor (LIF), and 0.1 mM β-mercaptoethanol. The control samples for normal course of differentiation were treated with 10nM RA (to drive differentiation towards NE) in an adherent monolayer[3]. Another set of plates passaged from these control samples were supplemented with 100mM ethanol (ET/EtOH) treatment. Also, three biological replicates were cultured identically to compare the quality of results.

The permeabilized cells samples were then stained with a cocktail of monoclonal antibodies along with their isotypes (to measure background signal or non-specific binding of anti-bodies) for the TFs - Sox-2 (APC), Oct-4 (PerCP-Cy5.5), Nanog (PE). Along with these antibodies, a Live/Dead viability dye (APC-Cy7) was used to subgate live and dead populations. The entire experimental procedure was carried out in two phases - one to study time points D0 (or ES), D2 and D4, and the other to study D0, D4 and D6. The FCM data were obtained from BD LSRII, analazyed in FlowJo to gate Live cells and exported for further analysis.

III. Computational Approach

A. *Data Preprocessing*

Data were preprocessed using a linear-logarithmic transformation. Two choice of transformations - i) Logicle Transformation (proposed by Parks *et al.*[4]) and ii) Arcsinh Transformation (or arc-sine-hyperbolic) were compared with a criterion that the final transformed distribution of ES cells is normally distributed since ES cells are a lump of undifferentiated and mono-characteristic cells. At the second stage of preprocessing, the data were corrected for apparent increase in fluorescent intensities due to cell sizes through morphology correction proposed by Knijnenburg *et al.*[5] Finally, the data was normalized in each of the fluorescent channels across time points.

B. *Results from Data Analysis*

1) Ethanol Delays Differentiation: To study the variation of transcription factors across days and between treatment (RA+ET) and control (RA alone), MFI or Mean Fluorescence Intensities of the sample conditions were computed, and subjected to Kruskal-Wallis test[6] (rank-order based analysis of mean and is non-parametric) rather than traditional T-test which requires t-distributed TFs (not the case in FCM data, in general.) Figure 1a shows the bar plots of mean TF levels across days, along with a box plot of median TF levels with 5% and 95% quantile whiskers.

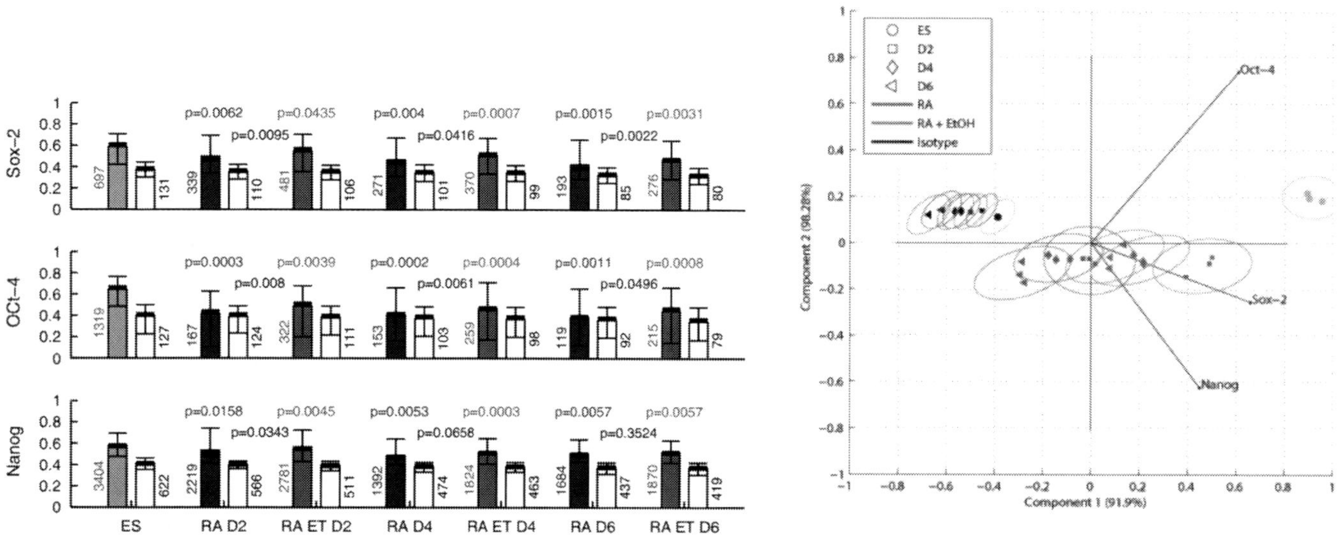

(a) MFI Comparison

(b) PCA of MFIs

Fig. 1: **(a)**: Mean Fluorescence Intensities: Ethanol delays down-regulation of Sox-2 and Oct-4. p-values represent probability of means being same at 0.05 significance levels from Kruskal-Wallis test. High value (0.3524) between RA D6 and RA+ET D6 Nanog level shows that means are the same, indicating similar down-regulation in Nanog under both the conditions, while low p-values (0.0022 and 0.0496) in Sox-2 and Oct-4 represent withholding of down-regulation by ethanol. Colored p-values represent test results between ES and respective condition, while black p-values represent that between RA and RA + ET samples on a particular day. White bars represent isotype levels, and numbers by the side of bars represent raw scale MFI.
(b): PCA of MFI of all samples from the three biological replicates indicates that ethanol delays down-regulation in TFs. In normal course of differentiation in RA, the ES samples move along the first principal component towards the isotypes. The RA + ET treated samples lie between ES and differentiated samples in RA. The ellipses represent one standard deviation of respective sample populations.

The raw scale values to the side of the bar shows that Oct-4 is down-regulated to the order of 11 folds (in raw scale) in the normal course of NE-directed differentiation while that in presence of ethanol is about 6 folds. The down-regulation in Sox-2 between normal course of differentiation and ethanol-mediated differentiation is 2.5 and 3.5 respectively. These inferences point to a delay in down-regulation of TFs during differentiation in presence of ethanol. This phenomenon is verified by a simple PCA (Principal Component Analysis) of the MFI levels. here, FSC and SSC were left out as they correlated poorly with the TF values. Also, their apparent intensity effects were corrected as previously mentioned.

Figure 1b shows the first two components of the PCA with the MFI scores of the three biological replicates along with one standard deviation in expressions, indicated by interval ellipses. The line-up of the MFIs along the first principal component, which accounts for 91% variance, is dominated by Sox-2 (0.6572) and Oct-4 (0.6062) with Nanog loading value at 0.4478. This shows that much of the variations can be explained by Sox-2 and Oct-4. The delay in down-regulation of these TFs is indicated by the scatter of ethanol MFIs between ES and RA samples.

2) Identification of Phenotypes: Typically FCM data can have upto 17 fluorescent parameters, in which case MFI or PCA becomes hard to interpret. Here, we build Minimum Spanning Tree (MST) of clusters of cells, similar to SPADE developed by Bendall and Qiu[7]. Rather than a faithful down-sampling, we have used Self-Organization Maps (SOM) or Kohonen Maps[8] to ensure guaranteed representation of minor subpopulations. Figure 2 represents the Euclidean distance based MST of control samples including all 6 parameters (FSC,SSC,Sox-2,Oct-4,Nanog,Live-Dead). Here, the cells are further segregated into based on isotype levels to estimate positives and negatives in TFs. The progression of highlighted cells in bottom strips, from ES to D4, shows that phenotypic subpopulations exist and branch out from ES. A clear progression with differentiation (on D4) can be seen towards the isotypes (black marks). Whereas with treatment, from Figure 3, cluster nodes or subpopulations do not branch out completely towards isotypes. The overlap of ES and D4 nodes indicate the dealy in differentiation, or even termination without expressing required phenotypic variation (since the right TF strips indicate down-regulation in TFs on D4 with treatment as well.)

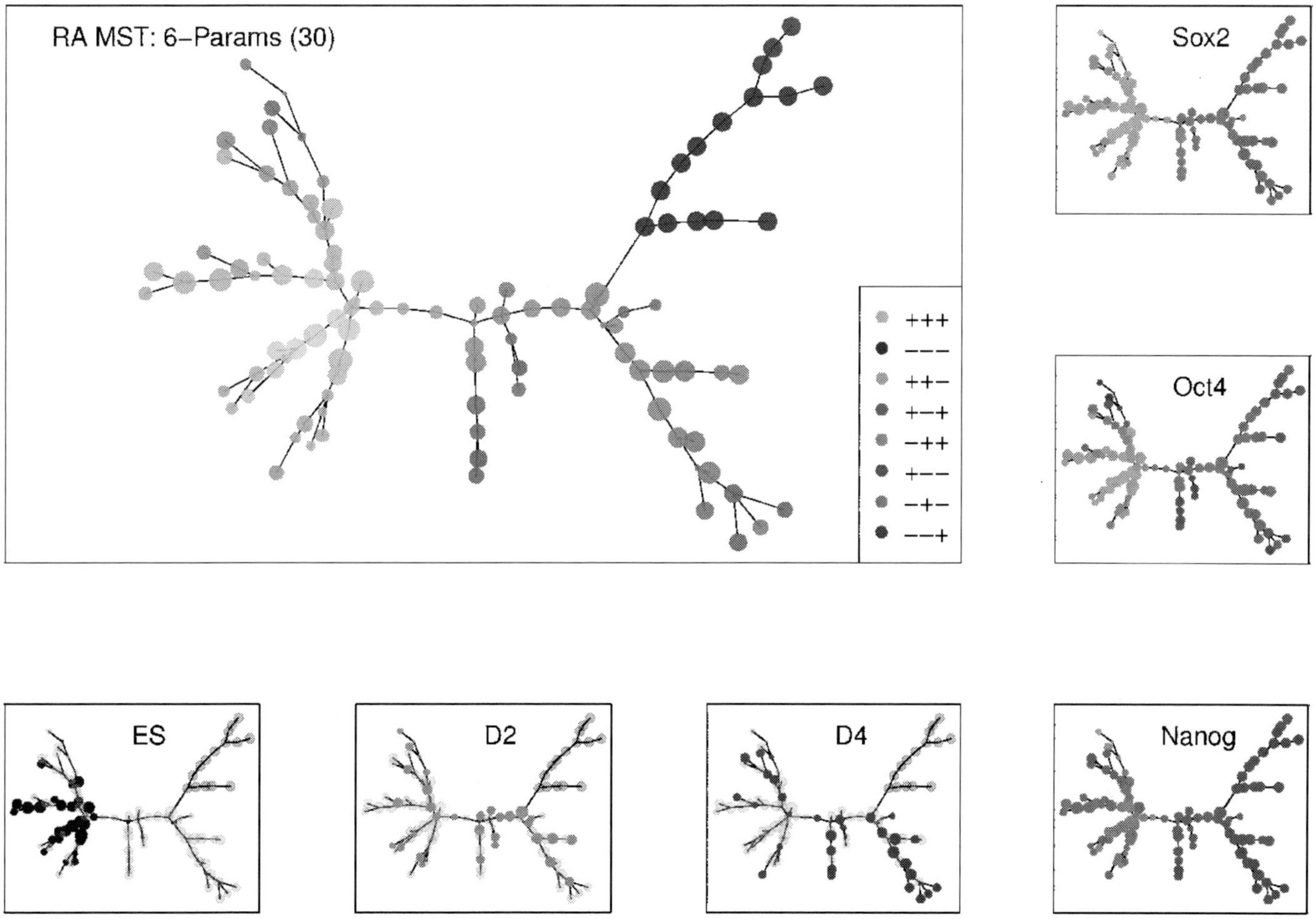

Fig. 2: MST of Control Samples: Down-sampled cells are clustered into 30 clusters through k-means, and their center-to-center Euclidean distances making use of all 6 parameters are measured to build the adjacency matrix for MST. Thus, closely related subpopulations or clusters are closer to eachother. The TFs - Sox-2, Oct-4 and Nanog are coded with colors R, G, B respectively, ranging from light (low expression) to dark (high expression). They are then mixed to color the clusters in the main plot, and node size represents % representation of the respective subpopulation in the overall sample. Black nodes represent isotypes. The bottom strip of figures highlight only those cell populations for respective days, and right strip of figures represent individual TF expression levels for the nodes.

The major phenotypes are identified through clustering of the data, and further visualization through a novel 3-D density plots. Here the appropriate number of clusters are identified through spectral clustering and clustered as t-skewed mixture model from EMMIX package[9]. The workflow of the algorithm is shown in Figure 4.

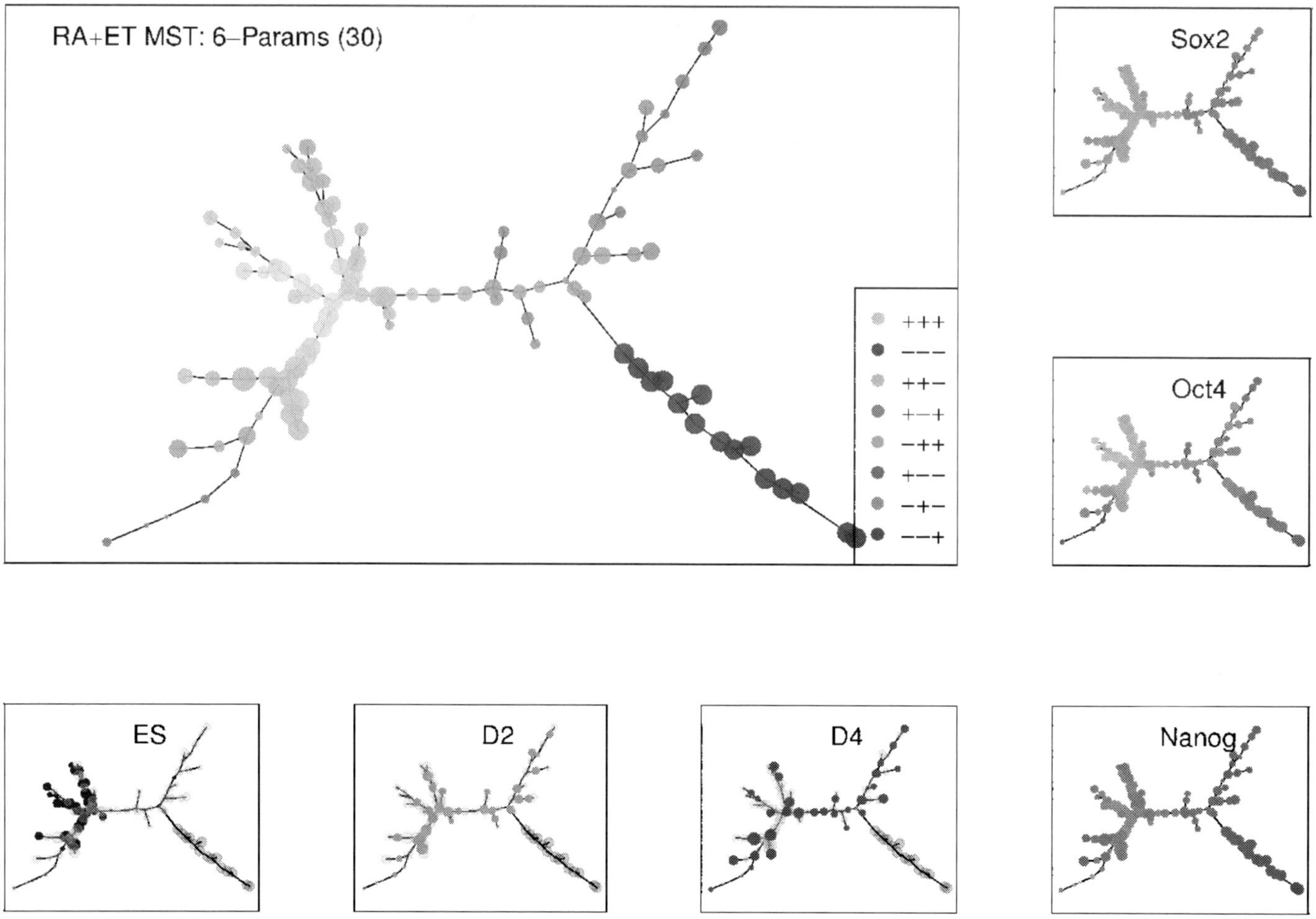

Fig. 3: MST of Treatment Samples: Follows similar plotting methodology as control samples. While control samples clearly distinguish cells from ES-D4, separated from eachother, it is not the case in treatment. A major portion of D4 cells overlap with D2 and even ES cells. At the same time, comparison between TF levels of treatment and control shows that differentiation has progressed, yet the cells have not phenotypically differentiated towards NE (as seen from control MST). Shows the alteration in lineage commitment of ES cells on EtOH treatment.

In terms of phenotypic features, RA condition samples show (refer Figure 5) a unique Sox-2(+)Nanog(+) subpopulation on day 2 that subsequently gets down-regulated over days. After 6 days of differentiation this subpopulation contributes to about a third of the differentiated cells. However, this particular population, is not well-expressed in RA + ET condition. This population is of interest due to its high Sox-2 and Nanog expression which may serve as an intermediate for NE differentiation. A direct interference of ethanol on this population by suppression may affect signalling of NE progenitors.

3) Effect of Ethanol Concentration: Finally, effect of ethanol on TFs and phenotype distributions were identified through three sets of experiments - 25mM, 50mM and 100mM EtOh concentration at two time points - ES and D3. A summary of the results are depicted in Figure 6. It indicates - A) distribution of triple positives and triple negatives deviate from control with treatment dosage, B) changes (down-regulation) in TFs of triple positive cells increase with dosage, C) the ration Oct-4/Sox-2 increases with concentration, indicating more differentiation towards ME and away from NE, D) further verifies the increasing Oct4+ cells and decreasing Oct4- cells.

IV. Discussion and Conclusion

The analysis involved an effective inferential of FCM data that would have been otherwise missed with simple MFI calculations, and 2D color density visualizations. The computational approach introduced a novel 3D density visualization, effective for the current dataset with 3 TFs, and a SOM based density down-sampling that captures most of the cell phenotypes. The use of MST and further clustering of down-sampled samples revealed dominant cell subpopulations that were expressed

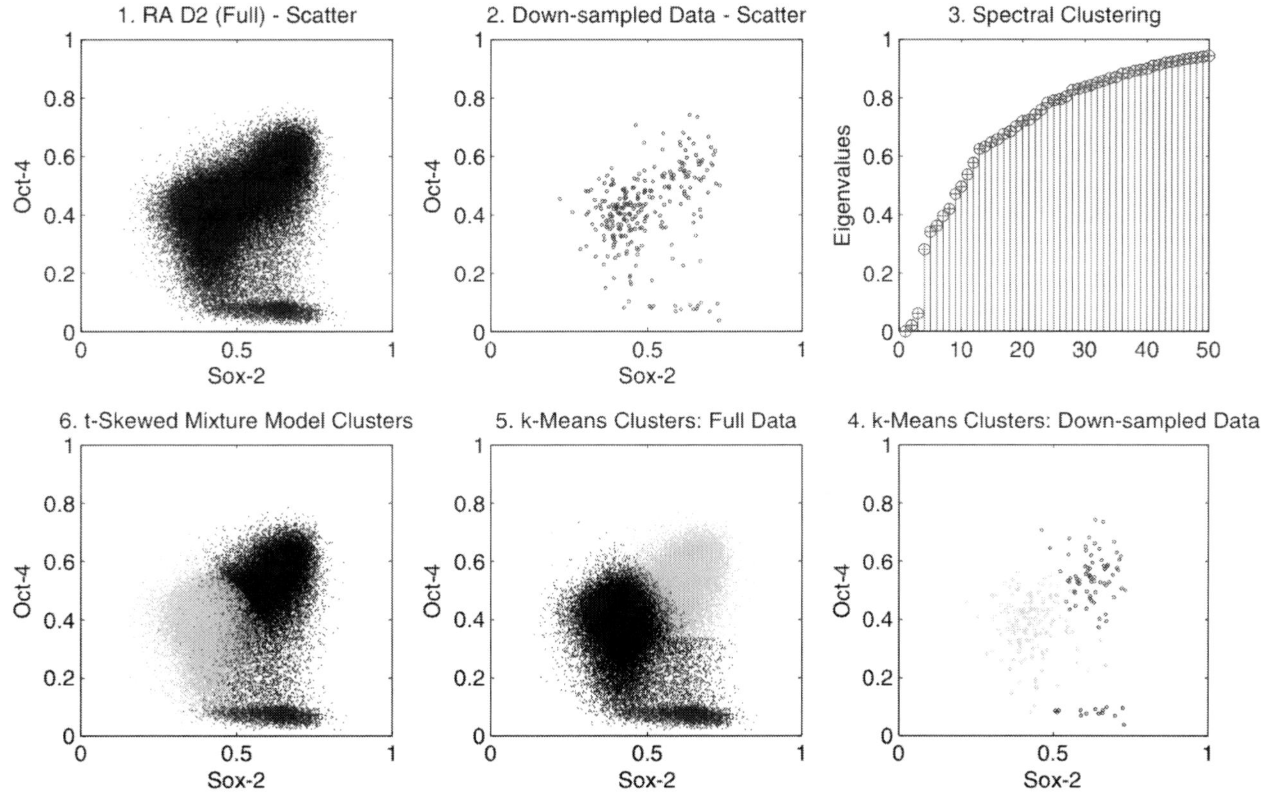

Fig. 4: Clustering Workflow: 1 - Scatter distribution of RA D2 cells in Sox-2 - Oct-4 plane; 2 - Down-sampling of data based on density from SOM cluster grid; 3 - Spectral decomposition of down-sampled data Laplacian indicates 3 clusters in the graph; 4 - initial k-Means on down-sampled data provides approximate center locations; 5 - k-Means clustering of complete data set determines cluster memberships for all points along with final cluster centers. But boundaries are hard and artificial; 6 - t-skewed mixture modeling with initialized settings from k-Means result provides better cluster memberships.

differently in control and treatment samples. The highlighted findings strongly suggests a deviation from normal differentiation pathway, leading to higher expression of Oct-4+ cells that have been shown previously to suppress NE cell lineage. All these factors contribute to a change in cell commitment leading to FASD.

ACKNOWLEDGMENT

REFERENCES

[1] R. Vadigepalli, J. Ogony, and H. Anni, "Ethanol effects on transcription factor network regulating stem cell differentiation," in *IEEE International Conference on Bioinformatics and Bioengineering*.

[2] M. Thomson, S. J. Liu, L. Zou, Z. Smith, A. Meisnner, and S. Ramanathan, "Pluripotency factors in embryonic stem cell regulate differentiation into germ layers," *Cell*, vol. 145, no. 6, pp. 875–889, 2011.

[3] J. Lu, P. L. H. Gao, B. Fang, Z. Geng, P. Zheng, and H. Song, "All-trans retinoic acid promotes neural lineage entry by pluripotent embryonic stem cells via multiple pathways," *BMC Cell Biology*, vol. 10, no. 57, 2009.

[4] D. Parks, M. Roederer, and W. Moore, "A new logicle display method avoids deceptive effects of logarithmic scaling for low signals and compensated data," *Cytometry Part A*, vol. 69A, no. 6, pp. 541–551, 2006.

[5] T. Knijnenburg, O. Roda, Y. Wan, G. Nolan, J. Aitchison, and I. Shmulevich, "A regression model approach to enable cell morphology correction in high-throughput flow cytometry," *Molecular Systems Biology*, vol. 7, no. 531, 2011.

[6] W. Kruskal and A. Wallis, "Use of ranks in one-criterion variance of analysis," *Journal of American Statistical Association*, vol. 47, no. 260, pp. 583–621, 1952.

[7] S. C. B. et al., "Single-cell mass cytometry of differential immune and drug responses across a human hematopoietic continuum," *Science*, vol. 332, pp. 687–696, 2011.

[8] T. Kohonen, "Self-organized formation of topologically correct feature maps," *Biological Cybernetics*, vol. 43, no. 1, pp. 59–69, 1982.

[9] G. McLachlan and D. Peel, "The emmix algorithm for the fitting of normal and t-components," *Journal of Statistical Software*, vol. 4, no. 2, 1999.

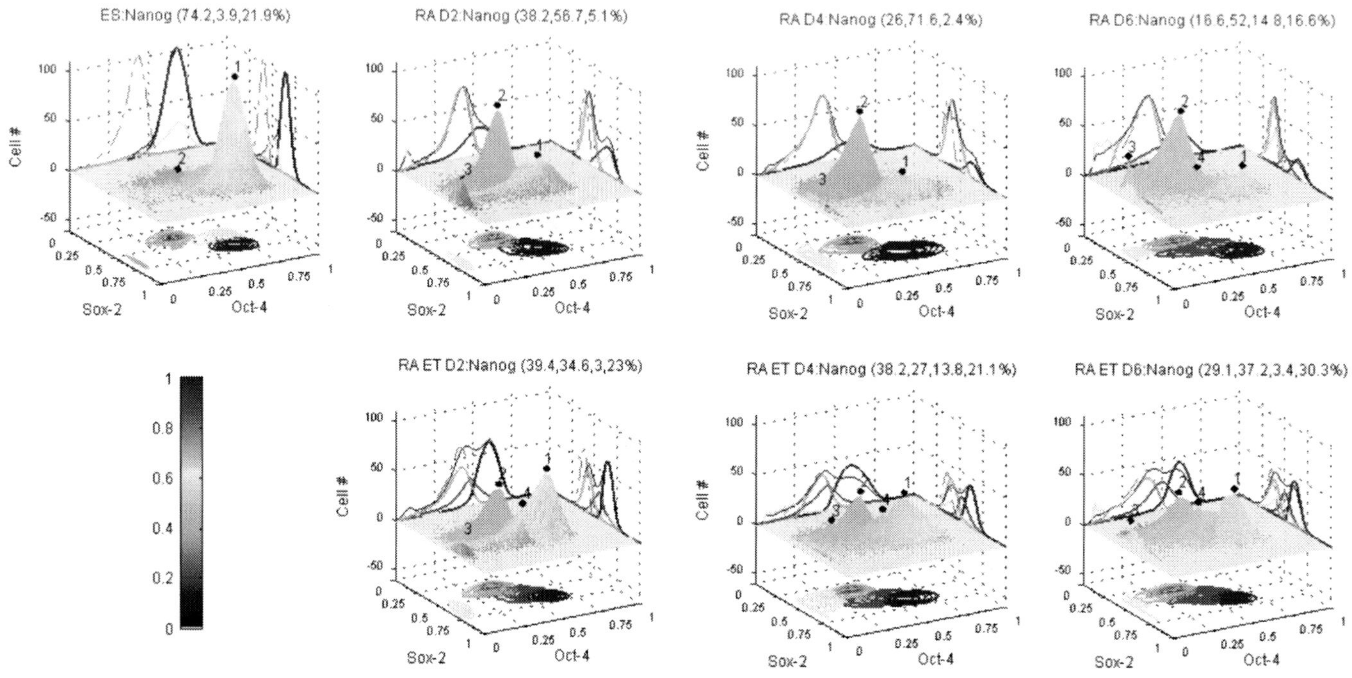

(a) ES - D2

(b) D4 - D6

Fig. 5: Three dimensional density plots indicating multiple phenotypes. Shows i) bivariate density of cells in Sox-2 and Oct-4 plane, with texture overlay of average Nanog values over the bins, ii) univariate density plots Sox-2 and Oct-4 for the entire sample along with those of individual subpopulations in the background, iii) contours of subpopulations along with isotype controls in black. Values in title shows percentage cells in each subpopulation. Smaller peaks near isotype levels in RA + ET samples indicate either a delay in differentiation in ethanol conditions or higher proliferation rate in undifferentiated cells. Plot also indicates a unique Sox-2(+)Nanog(+) population that is suppressed in EtOH.

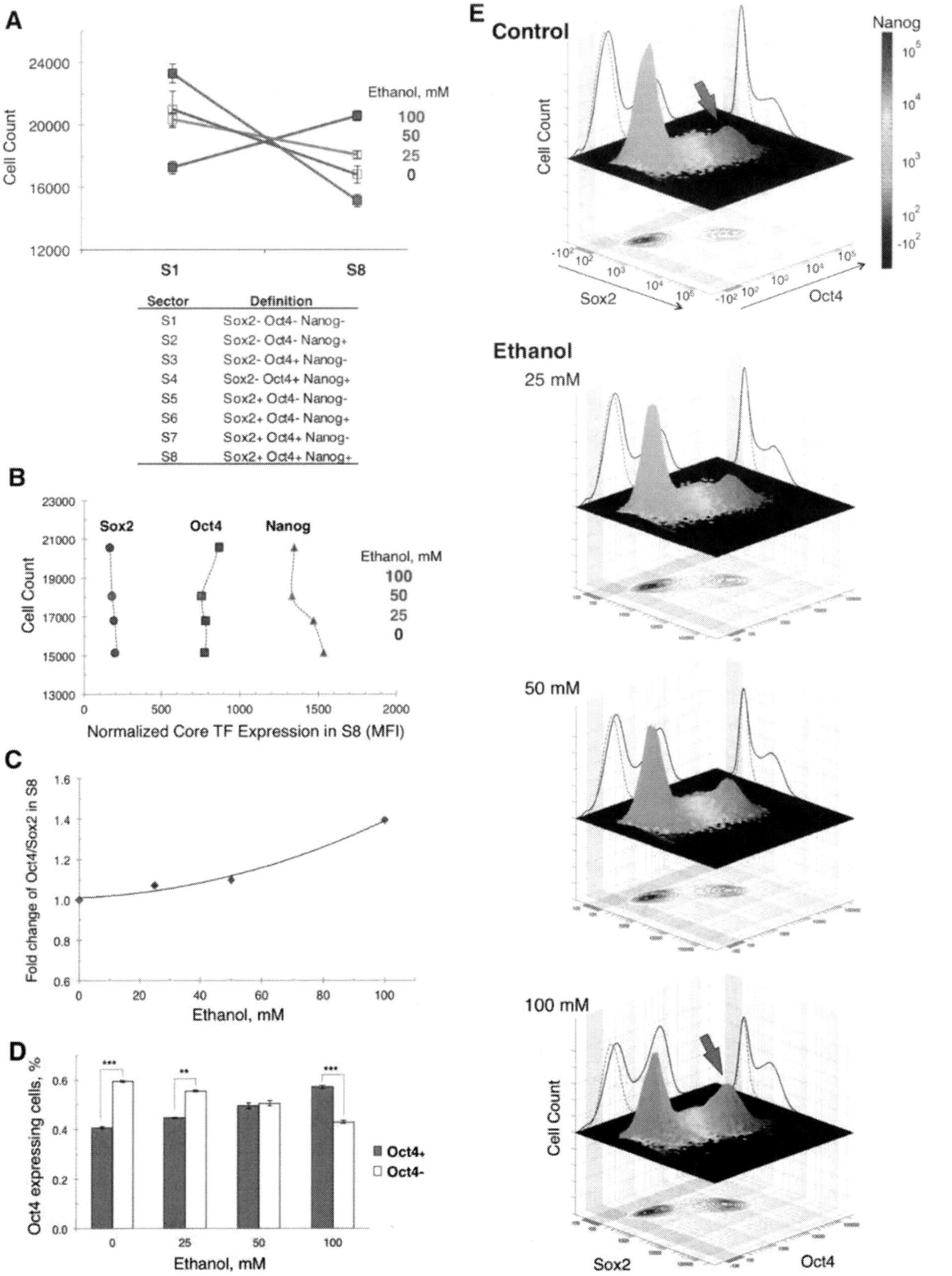

Fig. 6: Summary of effect of EtOH concentration: **A**: Shows distribution of Sector-8 (+++ or triple positives) and Sector-1 (— or triple negatives) cells with concentration. **B**: Normalized (across treatments) TF expression levels in original raw scale. **C**: The Oct-4/Sox-2 ratio or fold change with concentration. **D**: Oct4+ and Oct4- cell counts with concentration. **E**: Summary of all inferences of effects with dosage.

(725c) A Quantitative Assessment of the Izhikevich Neuron Model Against Experimental Data

Gautam Kumar[1], William E. Schiesser[2] and Mayuresh V. Kothare[1], (1)Chemical Engineering, Lehigh University, Bethlehem, PA, (2)ChE and ESE, Lehigh University and University of Pennsylvania, Bethlehem, PA

Neurons communicate with each other in a network by generating and transferring action potentials. The exact phenomenon through which these action potentials encode the information contained in presynaptic activities is still unknown. To understand this phenomenon of neural coding using rigorous mathematical analysis is an important problem in computational and theoretical neuroscience. One of the challenges in attempting such analysis is to develop computationally efficient mathematical models of neurons which can predict action potentials observed in experimental conditions, both qualitatively and quantitatively. Recent efforts towards developing single neuron models show the capability of mathematically simplified low-order models in capturing broad range of qualitative behavior of biological neurons. The obvious benefits of using these low-order models in the analysis of network of neurons for neural coding are their mathematical simplicity and existence of small number of model parameters which can easily be extracted from experimental data. Other benefits are in implementing neural prostheses where these models can be simulated at low costs. In this work, we analyze the quantitative predictive capability of one such model, **the Izhikevich single neuron model** [1], using the benchmark set of experimental data available from "**The quantitative single-neuron modeling competition**" [2,3,4]. Our analysis reveals the prospective applicability of this model in aforementioned applications. In addition, it may serve as a benchmark comparison of the performance of this model against other existing models.

The dynamical behavior of the Izhikevich single neuron model is expressed by two coupled non-linear ordinary differential equations with state-based reset conditions. Existence of these reset conditions shows discontinuities in the membrane potential trajectory predicted by the model. To make a quantitative assessment of the model, a part of the data can be used to estimate unknown model parameters and the rest of the data can be used to validate the model predictability against the experimental data. To estimate unknown parameters of the model, we introduce stochasticity in the model and take the well-known approach of maximum likelihood estimation (MLE). The objective here is to maximize the joint probability distribution of the occurrence of an experimentally observed sequence of action potentials over these unknown parameters. For computing the joint probability distribution, we derive the two dimensional Fokker-Planck equation [5], a convective-diffusive (hyperbolic-parabolic) partial differential equation, using the stochastic Izhikevich model. For computational efficiency, we approximate the two dimensional Fokker-Planck equation with a one dimensional Fokker-Planck equation coupled with two ordinary differential equations. We solve the Fokker-Planck equation at the time of occurrence of action potentials, in a given sequence of action potentials, by implementing the **method of lines with** a **5-point upwind scheme** [6]. Finally, estimation of parameters is performed by computing and maximizing the joint probability distribution of the first passage times (the time of occurrence of action potentials) using the MATLAB optimization toolbox.

In the presentation, we will first show the validation of the MLE approach used in estimating model parameters against simulated data sets. For this, we simulate the stochastic Izhikevich model with given parameters and generate 50 realizations of spike trains, each consisting of 300 action potentials. We estimate 9 model parameters of the Izhikevich model using the first 200 action potential timings for 100 realizations. Average values of parameters over these 50 realizations are then used to predict the next 100 action potential timings in all realizations. This establishes the validation and efficacy of the MLE approach used in estimating model parameters. Next, we show the efficacy of the model predictability against experimental data available from the competition. For this, we use two-third of the data (200 action potentials) in each data set of available 13 data sets from the competition to estimate 9 model parameters of the Izhikevich model. Average values of parameters over 13 data sets are then used to predict timings of the rest of the one-third of the data in each data set. Finally, we demonstrate the efficacy of the model in predicting experimentally observed action potentials by making the benchmark comparison stated in the competition.

1. Izhikevich, E. M. Dynamical systems in neuroscience. The MIT Press, 2007.
2. Jolivet et.al. The quantitative single-neuron modeling competition. Biol Cybern, vol. 99, 417-426, 2008.
3. http://www.incf.org/community/competitions/spike-time-prediction/2009 : Quantitative single-neuron modeling 2009.
4. Gerstner et.al. How good are neuron models? Science, vol. 326, no. 5951, 379-380, 2009.
5. Gihman and Skorohod. Stochastic differential equations. Springer, 1972.
6. Schiesser and Griffiths. A compendium of partial differential equation models. Cambridge University Press, 2009.

A DYNAMIC PHYSIOLOGY BASED PHARMACOKINETIC MODEL FOR ASSESSING LIFELONG INTERNAL DOSE

Dimosthenis A. Sarigiannis, Spyros P. Karakitsios

Aristotle University of Thessaloniki, Department of Chemical Engineering, Thessaloniki 54124, Greece

Introduction

A generic dynamic lifetime PBPK model, capable to describe xenobiotics ADME (Absorption, Distribution, Metabolism, Elimination) processes was developed to support chemical risk assessment. All physiologic parameters are given by functions of time describing the several lifetime stages (from embryo conception through mature life), including the differences of enzymatic activity, depended on the ontogenesis of the regulating genes. In addition, the organ/blood partition coefficients are given by generic formulas taking into account the lipid composition of the tissue and the K_{ow} of the toxic compound in consideration. The generic model was applied to Bisphenol A, describing also the ADME procedure of its metabolites, namely and BPA-glucuronide. Because BPA-glucuronide is the dominant metabolite, it is excreted totally in urine, thus the overall model allows as to link external exposure to urine biomonitoring data.

Infant exposure to BPA and the potential adverse health effects is one of the hottest public health issues. The main controversy regards the toxicokinetic behavior of BPA; although BPA glucuronidation acting as detoxification mechanism is complete and fast, due to the reduced metabolic capacity of infants-neonates, there is still ample opportunity of internal exposure [1, 2]. Existing biomonitoring studies regarding BPA exposure cannot provide substantial support, because most of them track only urinary metabolites (BPA-glu), hence are useful only for assessing the overall daily dose. Due to the fast metabolism of BPA to BPA-glu, the extent of red blood cells binding (fraction of 0.95) and the limitations of the analytical techniques, the related biomonitoring studies either fail to detect free-BPA in the plasma, or when they detect it, the observed values are considered a result of background contamination from labware and indoor dust [3], an opinion which is under strong contestation [4]. Contradiction is further amplified by the toxicity testing results. BPA was found to produce adverse neurodevelopmental effects in rats given an oral dose considered as environmentally relevant, but the results were under serious criticism by the regulatory authorities [5, 6] mainly based on the validity of the applied methods (not GLP compliant). Additional arguments included the relevance of bioavailability for the same normalized oral dose among rodents and humans, due to substantial differences in BPA excretion route (humans only via urine, rodents feces and urine due to hepatobilic recirculation) and rate of elimination and excretion (elimination half life time 5.3 h and 10.5h for humans and rodents respectively).

Preliminary efforts to develop PBPK models of BPA were made by Shin et al. [7] and Teeguarden et al. [8], but the most comprehensive model with whole body compartments which clearly modeled metabolic scaling from adults to neonates-infants was developed by Edginton and Ritter [1]. The core finding of the study was that BPA plasma concentrations could be approximately 11 times greater in newborns than in adults exposed to the same weight-normalized dose, indicating the need for further human biomonitoring data in order to fully optimize the plasma clearance rate. The latest published BPA PBTK model [9], implemented sulfation as an additional metabolic pathway, considering that sulfation activity is well expressed in newborns and activity is at least as high as in adults and might be even greater. Besides the importance of the additional clearance pathway, the developed model was much more simplified than the one developed by Edginton and Ritter, with significantly fewer compartments, without BPA binding to red blood cells implementation, and did not deal at all with the conjugated metabolites. The estimated steady-state free plasma BPA concentrations in neonates – infants were lower than the ones estimated by Edginton and Ritter, due to the addition of sulfation. As a result, the comprehensive PBPK model presented in this study is the only one addressing exposure during pregnancy, lactation period and as well as exposure through multiple routes.

Methodology

The overall modeling system was developed in asclXtreme (AEgis® technologies), allowing us to derive dynamic estimations. Ours is practically a double PBPK model coupling the mother-fetus interaction (Figure 1).

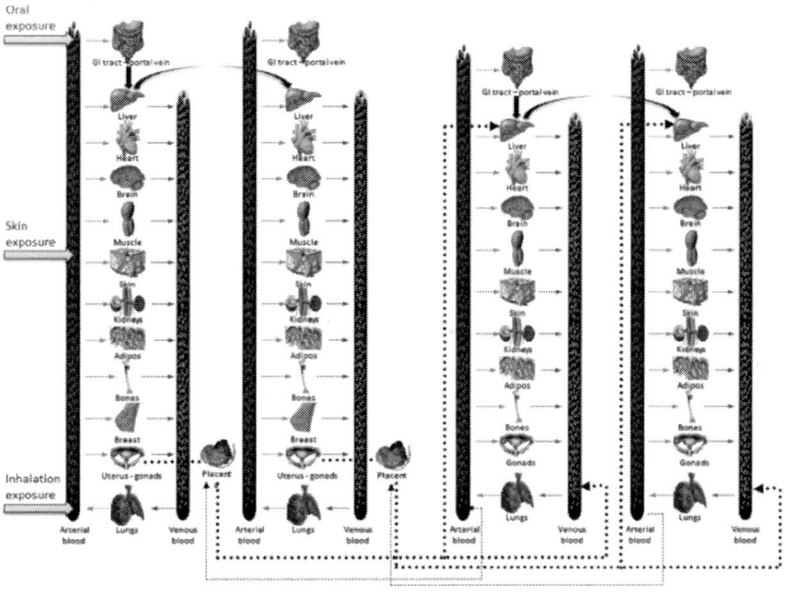

Figure 1: Visual representation of the Mother-Fetus generic PBPK model

As such, the first complex describing mother physiology includes additionally the breast and the uterus compartments. The second complex describes the fetus and consequently the infant development.

For the several tissue compartments of the model, ADME procedure regarding mass conservation is given by the following formula:

$$V_i \frac{dC_{ij}}{dt} = Q_i(CA_j - CV_{ij}) - Metab_{ij} - E\lim_{ij} + Absorp_{ij} - \Pr Binding_{ij} \qquad (1)$$

where V_i represents the volume of tissue group i, Q_i is the blood flow rate to tissue group i, CA_j is the concentration of chemical j in arterial blood, and C_{ij} and CV_{ij} are the concentrations of chemical j in tissue group i and in the effluent venous blood from tissue i, respectively. $Metab_{ij}$ is the rate of metabolism for chemical j in tissue group i; liver, being the principal organ for metabolism would have significant metabolism and, with some exception, usually $Metab_{ij}$ is equal to zero in other tissue groups. $Elim_{ij}$ represents the rate of elimination from tissue group i (e.g., biliary excretion from the liver), $Absorp_{ij}$ represents uptake of the chemical from dosing (e.g., oral dosing), and $PrBinding_{ij}$ represents protein binding of the chemical in the tissue.

The parameters related to organ volumes (**V**) and blood flows (**Q**) were taken from the International Commission on Radiological Protection report [10] and fitted to time (**t**) (in Datafit

9.0 software) in order to derive continuous time depended non lineal polynomial formulas in the form of:

$$V = a \cdot t^b + c \cdot t^d + e \cdot t + f \tag{2}$$

The blood/tissue partition coefficients are contaminant specific and are estimated by the tissue lipids content and the octanol/water partition coefficient of the contaminant by the following formula [11]:

$$P_{tissue/blood} = \frac{K_{ow} \cdot Fl_{tissue} + Fw_{tissue}}{K_{ow} \cdot Fl_{blood} + Fw_{blood}} \tag{3}$$

Regarding uterus, beside the assumption of equal diffusion flow from uterus to placenta and vice-versa during pregnancy, uterus behaves like other organs:

$$\begin{aligned} V_{\text{int\_uterus}} \frac{dC_{\text{int\_uterus}}}{dt} &= Q_{uterus} \cdot (1 - HCT) \cdot \left(C_{pls\_art} - C_{\text{int\_uterus}}\right) \\ &- K_{d\_uterus\_pla} \cdot \left(C_{placenta} - C_{uterus\_M}\right) \\ &- PS_{rbc\_uterus} \cdot f_u \cdot \left(C_{\text{int\_uterus}} - \frac{C_{rbc\_uterus}}{K_{rbc}}\right) \\ &- PS_{cell\_uterus} \cdot f_u \cdot \left(C_{\text{int\_uterus}} - \frac{C_{cell\_uterus}}{K_{rbc}}\right) \end{aligned}$$

For the placenta, the corresponding formulation is based on a proper modification of an existing mother-fetus human PBTK model [12]:

$$\begin{aligned} V_{\text{int\_placenta}} \frac{dC_{\text{int\_placenta}}}{dt} &= K_{d\_uterus\_pla} \cdot \left(C_{\text{int\_placenta}} - C_{\text{int\_uterus}}\right) \\ &- K_{d\_pla\_amniot} \cdot \left(C_{\text{int\_placenta}} - C_{amniot} \frac{P_{placenta}}{P_{amniot}}\right) \\ &+ Q_{placenta\_fetus} \cdot f_u \cdot \left(C_{\text{int\_art\_fetus}} - \frac{C_{\text{int\_placenta}}}{P_{placenta}}\right) \\ &+ K_{glu\_deconj} \cdot f_u \cdot Q_{placenta\_fetus} \cdot C_{\text{int\_placenta\_glu}} \\ &- PS_{rbc\_placenta} \cdot f_u \cdot \left(C_{\text{int\_placenta}} - \frac{C_{rbc\_placenta}}{K_{rbc}}\right) \\ &- PS_{cell\_placenta} \cdot f_u \cdot \left(C_{\text{int\_placenta}} - \frac{C_{cell\_placenta}}{K_{rbc}}\right) \end{aligned}$$

The first two parts of the formula describe the exchange between uterus and amniotic fluid respectively, the third one describes the placenta-fetus exchange and the fourth the deconjugation of mother BPA-Glu to BPA in the tissue. Latest ex *in-vivo* experiments have shown that BPA freely diffuses across the human placenta [13]. De-conjugation kinetics was determined empirically, based on experimental findings [13-15] where concentration in the placenta was found to be almost twice the one in maternal blood.

The breast compartment was important to describe the process of lactation during the first three postnatal months. Excretion via lactation is described as an output from the mammary tissue compartment through a partitioning process between mammary tissue and milk, and milk withdrawal by suckling, as described for PCBs in rats [16] and further adopted for humans [17]. BPA plasma exchange is described as follows:

$$V \frac{dC_{\_breast}}{dt} = PS_{\_cell\_breast} \cdot f_u \cdot \left(C_{\_int\_breast} - \frac{C_{\_breast}}{K_{\_breast}}\right) - L_{excr} \tag{4}$$

and the related excretion via lactation

$$L_{excr} = Q_{\_milk} \cdot \frac{C_{\_breast}}{K_{\_breast}} \cdot P_{\_milk/blood} \tag{5}$$

Tissue:blood and milk:blood partition coefficients are estimated using the Poulin and Krishnan [11] approach based on tissue water and lipid composition [18]:

$$P_{\_milk/blood} = \frac{K_{ow} \cdot Fl_{\_tissue} + Fw_{\_tissue}}{K_{ow} \cdot Fl_{\_blood} + Fw_{\_blood}} \tag{6}$$

where K_{ow} represents n-octanol:water partition coefficient, and ***Fl*** and ***Fw*** stand for the lipid and water fraction, respectively, for either tissue or blood. Through lactation procedure, both free BPA and BPA-glu are excreted, in a proportion related to their free-plasma concentration and K_{ow}. For estimating the total amount of BPA to which the infant is orally exposed, the sum of BPA and BPA-Glu in breast milk needs to be taken into account.

The inhalation route of exposure was added to the model, in order to assess differences in BPA bioavailability with regard to the route of exposure. Lung free plasma BPA concentration is described by:

$$\begin{aligned} V_{\text{int}\_lung} \cdot \frac{dC_{\text{int}\_lung}}{dt} &= Q_{lung} \cdot (1 - HCT) \cdot \left(C_{pls\_ven} - C_{\text{int}\_lung}\right) \\ &- PS_{rbc\_lung} \cdot f_u \cdot \left(C_{\text{int}\_lung} - \frac{C_{rbc\_lung}}{K_{rbc}}\right) \\ &- PS_{cell\_lung} \cdot f_u \cdot \left(C_{\text{int}\_lung} - \frac{C_{cell\_lung}}{K_{rbc}}\right) \\ &+ Q_{vent} \cdot C_{amb\_air} \cdot P_{air} \\ &- Q_{vent} \cdot f_u \cdot \left(\frac{C_{\text{int}\_lung}}{P_{air}} \cdot (1 - V_{ds}) + C_{amb\_air} \cdot V_{ds}\right) \end{aligned} \tag{7}$$

Dermal absorption of BPA was done considering a diffusion process, described with a single lumped compartment [19]. BPA permeability is calculated based on the octanol-water (K_{ow}) partition coefficient and the molecular weight, according to the following function [20]:

$$\log K_p = -2.80 + 0.66 \log K_{ow} \quad 0.0056 MW \tag{8}$$

The most crucial parameters of the overall PBPK model are glucuronidation and sulfation capacities throughout the several developmental stages. In the simulations of Edginton and Ritter [1], it was assumed that glucuronidation capacity in neonates is 5% of the one in an adult, increasing to 30% at the age of three months, and reaching full capacity at the age of one year, based on previous findings with regard to a physiologically based approach for scaling to children [21]. This assumption is kept as such. In addition, it was assumed that sulfation is fully developed already during the fetal period and that it is equal to the 15% of the adult glucuronidation [9].

Results

For assessing fetal exposure, we assumed that the maternal overall daily intake is equal to 1.5 μg/kg-bw as estimated by EFSA. In this case, free plasma BPA in maternal blood is almost 0.006-0.007 μg/L, which is higher than what expected for a non-pregnant woman (0.005 μg/L). Placental concentration is 0.0013 μg/L and the corresponding fetal concentration is 0.004-0.005 μg/L. Although placenta has almost twice the concentration of free BPA compared to the other maternal tissue compartments, fetal concentration is lower because the whole fetus acts as a metabolic compartment to the system, due to the existing sulfation capacity. If sulfation was negligible, fetus would be an additional non-metabolizing compartment to the overall fetus –

placental system, and the steady state concentration would be slightly higher than the maternal one yet lower than the placental concentration (around 0.007-0.008 μg/L). Thus, free plasma BPA in the fetus is sensitive to the extent that sulfation develops, as well as to the effect of β–glucuronidase in the placenta and the corresponding BPA deconjugation. If the effect of β–glucuronidase was negligible, then the maternal concentration should be 0.0045 μg/L, which is slightly lower than the one of a non-pregnant woman. Placental concentration would be even lower (0.004 μg/L) and the corresponding fetal exposure would be 0.0015 μg/L. However, this hypothesis is not supported by relevant experimental data.

For the other developmental stages, the median values derived are in the same order of magnitude but generally lower compared to the ones derived by Edginton and Ritter [1], especially for the first postnatal period (till 3 months). Almost after the second month, glucuronidation exceeds the activity of sulfation and the differences tend to decrease, while no differences are observed after the first year of development. The largest difference between our results and the ones of Edginton and Ritter [1] is observed in of neonates, where by assuming the existence of sulfation as an additional clearance pathway, the overall clearance rate is four times higher. The internal dose variability between the highest and the lowest exposed population - accounting for inter-individual variability - is up to a factor of two. To realize the magnitude of internal dose, the corresponding steady state estimation for adults exposed to EFSA TDI of 50 μg/kg-bw derives an average free BPA plasma concentration of 0.16 μg/L. This concentration is almost five times higher than the ones derived for bottle fed neonates and twice as high as the concentration estimated in intensive care hosted premature bottle-fed neonates. The latter represent the worst-case realistic exposure scenarios, with regard to the immature metabolic capacity. For comparison, if a neonate was exposed to EFSA TDI, the corresponding steady state concentration would be 0.45 μg/L, which is almost three times higher than the one of an adult.

Considering that the software implements dynamic simulation through time allowing to track internal dose profile either on a daily basis depending on the nutritional schedule and/or other routes of exposure, or to simulate continuously the internal dose through life time from the moment of conception (even if for the first three months internal dose cannot be determined). Based on the average exposure scenarios as indicated by EFSA, the results of the corresponding run (from gestation to two post-natal years) are presented in Figure 2.

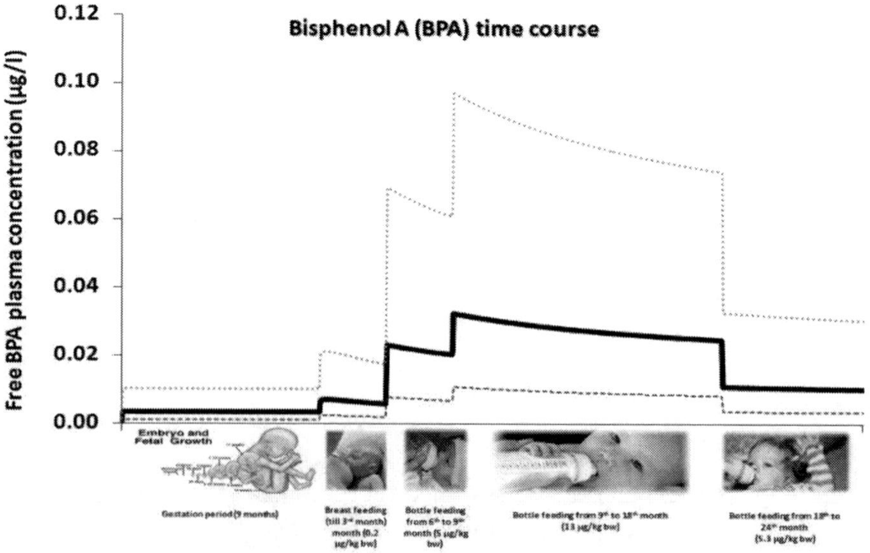

Figure 2: Continuous assessment of free plasma BPA from the developing fetus and infant until the age of two

The bold line in the middle indicates the steady-state concentration of free plasma BPA through these stages as derived by this run, the thump area around this line indicates the range of uncertainty for each developmental stage, originated either by the uncertainty/variability on intrinsic clearance capacity, or by uncertainty in the exposure scenarios, compiling the results of the Monte Carlo analysis when the exposure scenarios change. This kind of graph gives as an "overview" assessment of the overall steady state free plasma BPA variation through the developmental stages.

In a recent epidemiological study [22], inhalation exposure to BPA of employees in plastics factories in China was linked to sexual male dysfunction. The study triggered our interest in assessing BPA bioavailability under a different route of exposure. Indoor air exposure to BPA had a median value of 5 $\mu g/m^3$, which in a worst case scenario may results to an intake of almost 30 μg for the 8h working shift. Considering the lack of adequate data for fully validating an inhalation route of administration, we ran an exposure scenario according to which workers are exposed to a constant concentration of BPA, which in turn is fully absorbed in the lungs and transferred to systemic circulation. In this case, BPA transformation is limited by the liver blood flow rate and the steady state concentration in blood comes after 5-6 hours, at a concentration equal to 0.0022 μg/L. For comparison, if BPA was administered at the same rate orally (in terms of mass/time), the steady state concentration of free plasma BPA would be six times lower.

On the contrary, when BPA is administered dermally the corresponding steady state concentration is almost two times higher than oral, but steady state is achieved in about three to four days. Thus, due to the slow rate of absorption, the lack of first pass metabolism is less important. This is further reflected in the daily worst case exposure scenario of additional uptake of 71 μg of BPA during a 10 hours shift of a cashier from dermal contact with thermal paper that is used in cash receipts commonly [23]. In this case, the diurnal peak concentrations are higher only by 50% than the exposure levels corresponding to the typical EFSA exposure scenario. However, the contribution of this exposure route to the AUC is more significant, since the slow rate of absorption contributes to a continuous presence of BPA in plasma within the day.

Discussion and conclusions

The biologically effective dose of the fetus during gestation is highly linked to the one in maternal blood, due to the high lipophilicity and the relatively small molecular weight of BPA that allows free diffusion across the placenta. Fetal BED is increased through placental BPA-Glu and BPA-Sulf deconjugation by β -glucuronidase and arylsulfatace-C respectively, enzymes that are present in high concentrations in the placenta. Free plasma BPA in fetal blood may rise proportionally to maternal plasma after exposure, but it decreases at a slower pace since half of the intake is given intravenously due to the complexity of the uterus-placenta-fetus exchange system and erythrocyte binding. To the moment, there are no adequate pharmacokinetic data to allow the full quantification of the effect that β -glucuronidase has on BPA-GLU deconjugation and, consequently, fetal exposure. Sensitivity analysis (where the input parameter value is changed by 1% and the relative change in outcome is calculated when all other input parameters are fixed) performed on our model showed that this parameter is determining for placental concentration; it affects fetal exposure with a sensitivity coefficient of 0.67, which is higher than the contribution of fetal intrinsic clearance (sensitivity coefficient of 0.4), verifying the dependence on the de-conjugation activities [2]. With regard to the relative importance of the dissociation of the two de-conjugation activities, the effect of arylsulfatace-C is found to be minimal. Sulfation is considered as an alternative metabolic pathway for BPA when glucuronidation is not fully developed, and as such it is important only in fetuses and infants. Thus, the overall amount of BPA–Sulf subjected to possible de-conjucation in the placenta is mainly the one produced by the fetus, which is considered negligible compared to the amount of BPA–Glu circulating in the placenta, coming from maternal clearance. Thus, β-glucuronidase is expected to have a significantly higher importance.

Mother BPA-Glu bioavailability is also very important in the case of breast-fed infants. Transfer of BPA through milk is not sufficient enough to explain exposure of breast fed infants and the overall BPA exposure through breast feeding can only be explained by BPA-Glu cleaved in the gastrointestinal tract. Even when the worst-case scenario is taken into account, breast-fed infants seem to be significantly less exposed compared to bottle-fed infants and neonates.

From the methodological point of view, the generic PBPK model developed in this study successfully addressed the internal exposure assessment of a compound subjected to wide scientific and regulatory deliberation. The complexity of BPA exposure scenarios employing multiple pathways and routes, as well as the differences of enzymatic activity within the developmental stages, demanded the assessment of specific needs and the answer to specific questions [24]. The identification of the parameters affecting internal exposure might provide us guidance for applying the necessary protective measures, without implementing unnecessary conservatism in exposure scenarios management. Clearly BPA is a contaminant that needs to be banned from plastic material related to baby care, while for occupational groups protection measures are needed for specific exposure pathways: higher ventilation is required in BPA processing sites and use of gloves for people handling BPA containing paper receipts. In addition, the model gives the capability to reconstruct exposure by urine biomarker data. Hence, the overall uptake is estimated, but based on the knowledge of the different routes contribution, actual internal dose is estimated. This feed-forward procedure is an additional asset of a generic PBPK model, ensuring better data (environmental and biomonitoring) exploitation and interpretation.

References

1. Edginton A.N. and Ritter L. (2009) *Environmental Health Perspectives*. 117(4): 645-652
2. Ginsberg G. and Rice D.C. (2009) *Environmental Health Perspectives*. 117(11): 1639-1643
3. Dekant W. and Völkel W. (2008) *Toxicology and Applied Pharmacology*. 228(1): 114-134
4. Vandenberg L.N., Chahoud I., Heindel J.J., Padmanabhan V., Paumgartten F.J.R., and Schoenfelder G. (2010) *Environmental Health Perspectives*. 118: 1055-1070
5. EFSA.(2006) Opinion of the Scientific Panel on Food Additives, Flavourings, Processing Aids and Materials in Contact with Food on a Request from the Commission related to 2,2-bis(4-hydroxyphenyl)propane (bisphenol A) European Food Safety Agency
6. EFSA.(2008) Toxicokinetics of Bisphenol A - Scientific Opinion of the Panel on Food additives, Flavourings, Processing aids and Materials in Contact with Food European Food Safety Agency
7. Shin B.S., Kim C.H., Jun Y.S., Kim D.H., Lee B.M., Yoon C.H., Park E.H., Lee B.M., Yoon C.H., Park E.H., Lee K.C., Han S.Y., Park K.L., Kim H.S., and Yoo S.D. (2004) *Journal of Toxicology and Environmental Health - Part A*. 67(23-24): 1971-1985
8. Teeguarden J.G., Waechter Jr J.M., Clewell III H.J., Covington T.R., and Barton H.A. (2005) *Toxicological Sciences*. 85(2): 823-838
9. Mielke H. and Gundert-Remy U. (2009) *Toxicology Letters*. 190(1): 32-40
10. ICPR.(2002) Basic anatomical and physiological data for use in radiological protection: reference values, in The International Commission on Radiological Protection, J. Valentin, Editor.
11. Poulin P. and Krishnan K. (1996) *Toxicology and Applied Pharmacology*. 136(1): 126-130
12. Beaudouin R., Micallef S., and Brochot C. (2010) *Regulatory Toxicology and Pharmacology*. 57(1): 103-116
13. Balakrishnan B., Henare K., Thorstensen E.B., Ponnampalam A.P., and Mitchell M.D. (2010) *American Journal of Obstetrics and Gynecology*. 202(4): 393.e1-393.e7
14. Schonfelder G., Wittfoht W., Hopp H., Talsness C.E., Paul M., and Chahoud I. (2002) *Environmental Health Perspectives*. 110(11): A703-A707
15. Takahashi O. and Oishi S. (2000) *Environmental Health Perspectives*. 108(10): 931-935
16. Lee S.K., Ou Y.C., Andersen M.E., and Yang R.S.H. (2007) *Archives of Toxicology*. 81(2): 101-111
17. Verner M.A., Charbonneau M., Lopez-Carrillo L., and Haddad S. (2008) *Environmental Health Perspectives*. 116(7): 886-892
18. Price K., Haddad S., and Krishnan K. (2003) *Journal of Toxicology and Environmental Health - Part A*. 66(5): 417-433
19. McCarley K.D. and Bunge A.L. (2001) *Journal of Pharmaceutical Sciences*. 90(11): 1699-1719
20. USEPA.(2004) Risk assessment guidance for Superfund - Volume I: Human health evaluation manual (Part E, Supplemental guidance for dermal risk assessment).
21. Edginton A.N., Schmitt W., Voith B., and Willmann S. (2006) *Clinical Pharmacokinetics*. 45(7): 683-704
22. Li D., Zhou Z., Qing D., He Y., Wu T., Miao M., Wang J., Weng X., Ferber J.R., Herrinton L.J., Zhu Q., Gao E., Checkoway H., and Yuan W. (2009) *Human Reproduction*. 25(2): 519-527

23. Biedermann S., Tschudin P., and Grob K. (2010) *Analytical and Bioanalytical Chemistry*. 398(1): 571-576
24. Vandenberg L.N., Maffini M.V., Sonnenschein C., Rubin B.S., and Soto A.M. (2009) *Endocrine Reviews*. 30(1): 75-95

NEW CAPABILITIES FOR LARGE-SCALE MODELS IN COMPUTATIONAL BIOLOGY

Casey S. Abbott[1], Eric L. Haseltine[2], R. Abraham Martin[1], and John D. Hedengren[1]
[1]Brigham Young University, Provo, Utah
[2]Vertex Pharmaceuticals, Cambridge, MA

Introduction

Advances in biomedical research have lead to an increase of experimental data to be interpreted in the context of reaction pathways, molecular transport, and population dynamics. Kinetic modeling is one way employed to interpret this data and is used in the pharmaceutical industry in developing clinical trials for new medications [1]. Kinetic modeling's role will only increase as large pharmaceutical companies look to scale back and be more focused, spending less on R&D while expecting more results.

One collection of kinetic models is built on the System Biology Markup Language (SBML), which includes hundreds of contributions. Many models in this collection have detailed reaction metabolic pathways that describe biological systems, including cause and effect relationships in the human body. While simulations of these biological systems have been successfully applied for many years, the alignment to available measurements continues to be a challenge. The best available solution techniques continue to limit the size of the reconciliation of models and measurements to small and medium size problems. This limits the usefulness of the models due to the many assumptions and simplifications that are required in order for the optimizer to be able to perform the parameter estimation.

Recent developments may have changed this situation: an optimization technique known as the simultaneous approach has shown promise in efficiently optimizing large models (thousands of variables and parameters) [2]. In this method, the model and optimization problem are solved simultaneously, as opposed to the traditional approach of solving the differential and algebraic equation (DAE) model sequentially. In the sequential approach, each iteration of the optimization requires the solution of the DAE model. Much of the recent development for the simultaneous approach has occurred in the petrochemical industry, where on-line process control applications require optimization of nonlinear models with many decision variables in the span of minutes. The applications drive petrochemical processes to produce more from existing processing units, stay within safe operating conditions, and minimize environmental emissions. Using the same solution techniques, major advances in the biological systems area are possible. One software that takes advantage of the simultaneous approach is APMonitor (APM).

Results

To verify that APM can accurately simulate biological models it was used to simulate a toy model and the results were compared to literature values. The model used is a basic model describing the HIV virus over thirty days with nine parameters, three variables and three differential equations [3]. APM was successfully able to replicate the results published in the literature. APM simulation was also verified with a model that describes the dynamics of HIV infection of CD4+ T cells [4]. This model is a little larger with nine parameters, four variables, and five DAE. This model was obtained from the BioModels Database and was manually converted to a format that could be used by APM. It was found that APM could accurately simulate this model and match values from literature and simulations in MATLAB.

Once it was shown that APM could simulate biological models accurately the next step was to verify the parameter estimation capabilities of APM. This was done using a HIV model similar to those mentioned above. In order to perform the parameter estimation the objective function was set to minimize the absolute error between the model and synthetic data. There was measurement noise of plus or minus 0.5 log order added to the synthetic data. All six parameters values were changed to see if APM could find the correct parameter values from several different starting points. It was found that APM was able to accurately find the correct parameter values. Figure 1 shows the concentration of HIV viruses from the synthetic data and the predicted model values with the estimated parameters. As seen in the figure APM was able to correctly find the parameters that allowed the model to fit the synthetic data. With this same model it was shown that APM allows for parallel processing, allowing multiple parameter estimations to be run simultaneously.

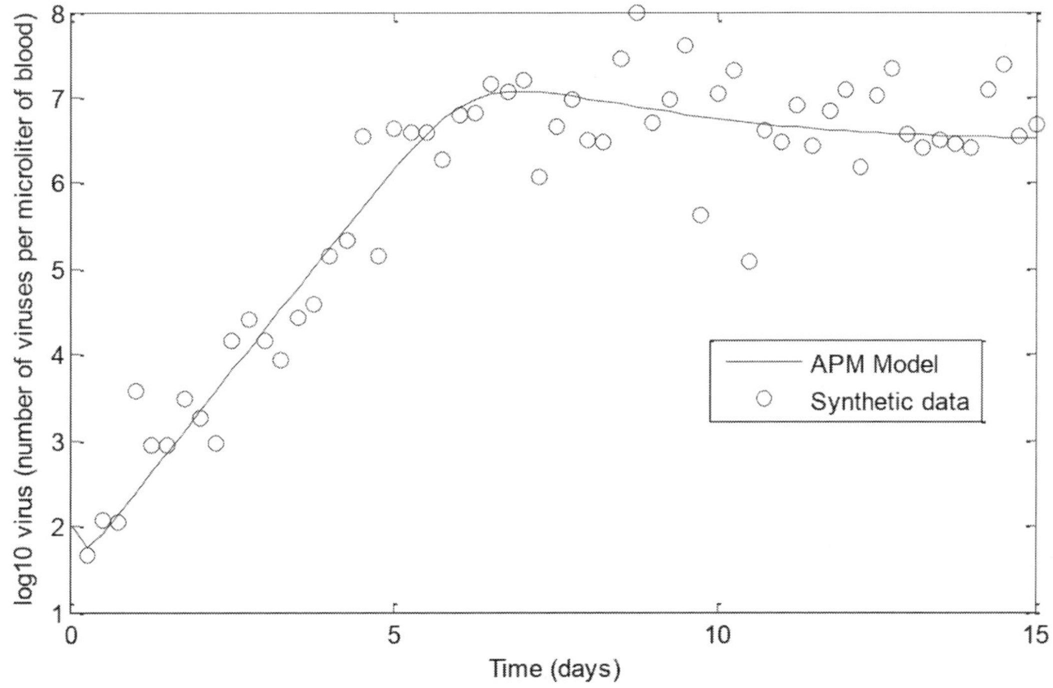

Figure 1: Shows the fit of APM model with estimated parameters to the synthetic data

Before the capability of APM's parameter estimation could be applied to large-scale biological models it was necessary to create an automatic conversion from the Systems Biology Markup Language (SBML) to a version usable by APM. This not only eliminates the human error in the conversion process, but it also allows for the quick evaluation of many publicly available models.

This conversion tool was used to automatically convert a model that describes the ErbB signaling pathways [5]. This model shall be referred to as the ErbB model. It is a large model with 225 parameters, 504 variables and 1331 DAEs. In the article the authors estimated 75 of the initial conditions and rate constants out of the 229 identified by the sensitivity analysis. This was accomplished through simulated annealing which required 100 annealing runs and 24 hours on a 100-node cluster computer on average to obtain just one good fit. APM is currently able to simulate the model and properly shows the dynamics found in the literature. It is believed that values do not match the literature values due to the limitations found in the conversion tool as it does not properly handle piecewise functions. Even if the literature values are not replicated exactly, parameter estimation can still be preformed. This is the next step in the project. The objective function will minimize the absolute error between the model and two measurable concentrations. Instead of using simulated annealing to estimate the parameters, a multi-start optimization will be used. To accomplish this, the parameter values will be randomly varied from 2.5 log order from the prior value. Many of these runs will be conducted and the results will be compared to those found using simulated annealing. If it is found that the design space is too flat or if there are too many optimums other optimizing techniques will be considered such as simulated annealing or a genetic algorithm.

In addition to the HIV and ErbB model parameter estimation, a benchmark test of Nonlinear Programming (NLP) solvers was conducted on the test set of over 400 curated models in the BioModels database. The test compares the performance of APOPT, BPOPT, IPOPT, SNOPT, and MINOS on this class of problems that is characterized by challenges unique to computation biology. This benchmark study points to a number of characteristics of interior point and active set methods for solving large-scale and sparse systems of differential and algebraic equations.

References

1. Adiwijaya, B. S. Herrmann, E. Hare, B. Kieffer, T. Lin, C. Kwong, A. D. Garg, Randle, J. C. R. Sarrazin, C. Zeuzem, S. and Caron, P. R. (2010), "A Multi-Variant, viral dynamic model of genotype 1 HCV to assess the in vivo evolution of protease-inhibitor resistant variants," *PLoS Comput. Biol.*, 6(4):e1000745.

2. Biegler, L. T. (2007), "An overview of simultaneous strategies for dynamic optimization," *Chemical Engineering and Processing: Process Intensification*, 46(11) pp. 1043 – 1053.

3. Nowak, M. and May, R. (2000), Virus Dynamics Mathematical Principles of Immunology and Virology. Oxford, New York: Oxford University Press.

4. Perelson, A. S. Kirschner, D. E. De Boer, R. (1993), "Dynamics of HIV infection of CD4 + T cells," *Math Biosci,* March, pp. 81-125.

5. Chen, William W. Schoeberl, Birgit Jasper, Paul J. Niepel, Ulrik B. Lauffenburger, Douglas A. and Sorger, Peter K. (2009), "Input-output behavior of ErbB signaling pathways as revealed by a mass action model trainded against dynamic data," *Molecular Systems Biology,* 5, pp. 239.

(744a) Modeling Stress-Induced Hormone Effects On Glucose-Insulin Dynamics in Critically Ill Patients

Thang Ho[1], Gilles Clermont[2], Balaji Yegneswaran[2] and Robert S. Parker[1], (1)Chemical and Petroleum Engineering, University of Pittsburgh, Pittsburgh, PA, (2)Department of Critical Care Medicine, University of Pittsburgh, Pittsburgh, PA

In the U.S., 5 million patients are admitted annually to intensive care units (ICUs) [1]. Critical care patients commonly suffer from stress hyperglycemia, resulting from a decrease in insulin dependent glucose uptake or/and relative lack of insulin secretion. Clinical evidence has shown that targeted glucose control could help reduce morbidity and mortality rate in ICUs [2]. However, the complex interaction of factors impacting patient glucose dynamics is challenging, where stress-induced hormone levels, e.g., epinephrine, cortisol, and glucagon, play pivotal roles in maintaining glucose homeostasis in ICU patients. Thus in order to efficiently and successfully control ICU patient's glucose level, it is necessary to exam the impact of these hormones on glucose-insulin dynamics. The focus of this work is to synthesize a patient-tailorable ICU model of glucose-insulin interactions including stress hormones. This involves the integration of epinephrine and cortisol models into the extended minimal model [3] to address changes in glucose-insulin dynamics resulting from trauma, stress, and inflammation. As an inflammation surrogate, we use lipopolysaccharide (LPS) challenge.

We synthesized a composite model starting from the extended minimal model (EMM) of Roy [3], which is more relevant to diabetic patients than ICU patients. The model's main structure includes the dynamics of free fatty acid and elevated metabolism, in addition to glucose-insulin interactions [3]. To capture the dynamics of pancreatic insulin secretion, a key characteristic of non-diabetic ICU patients, a three-state pancreas model was constructed from the biological mechanism of pancreatic insulin release [4,5]. The non-diabetic EMM successfully captures both glucose and insulin data from intravenous glucose tolerance tests. Two-compartment pharmacokinetic models are incorporated to capture the dynamics of epinephrine and cortisol in the body upon exogenous infusion [6, 7]. The steady state responses from the simulation are in agreement with 9 different epinephrine infusion studies; dynamic fit is also consistent with transient response data. Endogenous epinephrine and cortisol productions are calibrated against LPS challenge.

Elevated plasma epinephrine and cortisol concentrations decrease pancreatic insulin secretion and increase insulin resistance (decrease in insulin dependent glucose uptake). Epinephrine also increases hepatic glucose production. By incorporating these stress-induced hormonal effects, the resulting patient model provides a superior characterization of the glycemic response of trauma in ICU patients. Furthermore, this composite model can be used within a model-based control system to provide decision support to clinicians working to maintain targeted glucose control without hypoglycemia.

Reference:

1. Angus, D.C., Kelly, M.A., Schmitz, R.J., White, A., Popovich, J. (2000) "Current and Projected Workforce Requirements for Care of the Critically Ill and Patients With Pulmonary Disease: Can We Meet the Requirements of an Aging Population?" *Journal of the American Medical Association.* 284(21):2762-2770
2. Van den Berghe, G. et al (2001) "Intensive Insulin Therapy in Critically Ill Patients." *New England Journal of Medicine*.345: 1359-1367
3. Roy, A. (2008). Dynamic Modeling of Free Fatty Acid, Glucose, and Insulin During Rest and Exercise in Insulin Dependent Diabetes Mellitus Patients. Ph.D dissertation, University of Pittsburgh.
4. Bergman, R.N., Phillips, L.S., Cobelli, C. (1981) "Physiological Evaluation of Factors controlling Glucose Tolerance in Man". *J. Clin. Invest.* 68:1456-1467
5. Pedersen, M.G., Toffolo, G.M., Cobelli, C. (2010). "Cellular Modeling: Insight into Oral Minimal Models of Insulin Secretion". *Am. J. Phys. Endocrinol Metab.* 298: E597-E601

6. Rosen S.G., Linares O.A., Sanfield J.A., Zech L.A., Lizzio V.P., Halter J.B.(1989) "Epinephrinekinetics in humans: Radiotracer methodology". *J Clin Endocrinol Metab.* 69(4):753-61
7. Mager, M.E, Jusko, W.J.(2002) "Quantitative Structure-Pharmacokinetic/ Pharmacodynamic Relationships of Corticosteroids in Man". *J. of Pharmaceutical Science.* 91(11):2441-2451

(744b) Bayesian Inference Based Aggregated Gaussian Process Models for Identification of Ovarian Cancer Subtypes and Prediction of Survival Rates of Cancer Therapies

Mudassir Rashid and Jie Yu, Department of Chemical Engineering, McMaster University, Hamilton, ON, Canada

Cancer genome research effort aims to provide a comprehensive molecular characterization to accelerate the understanding of cancer biology and the discovery of new therapeutic targets. The immense wealth of multidimensional genomic data provides a new paradigm for developing new therapeutic solutions to improve the lives of patients.

The standard treatment for high-grade serous ovarian cancer is aggressive surgery followed by platinum-taxane chemotherapy. After therapy, platinum resistant cancer can occasionally reoccur resulting in a low overall five-year survival rate. The comprehensive measurements of genomic and epigenomic abnormalities on high-grade serous ovarian cancer samples are essential to identify the molecular abnormalities that may affect the outcome of different treatments. Moreover, accurate identification of the ovarian cancer subtypes can provide an opportunity to improve specialized cancer therapies. To that end, a genomic identification method based on Bayesian inference and Gaussian process classification is developed by using the gene expression data to identify the subtypes of cancer and predict the survival rates of cancer therapies. The proposed approach uses aggregate models developed through the integration of local Gaussian process classification models using the posterior probabilities estimated from Bayesian inference. Unlike the current standard analysis that involves separate clustering of different genomic data types followed by a manual integration of the cluster assignments, the new approach results in various localized models that can be amalgamated within the Bayesian framework. The application of the proposed approach to ovarian cancer patient samples demonstrates the significantly improved prediction accuracy and identification performance over the traditional techniques.

(744c) A NEW Framework for Online Optimization of Recombinant Protein Production in FED-Batch Fermentation Processes

Zheng Li, Department of Chemical Engineering, Texas Tech University, Lubbock, TX and **M. Nazmul Karim**, Chemical Engineering Department, Texas Tech University, Lubbock, TX

Online optimization of protein production is a challenging problem in the pharmaceutical industry. A new framework for online optimization of protein production in fed-batch fermentation processes is proposed in this study. This framework consists of trajectory planning and nonlinear model predictive control (NMPC) to realize online optimization of fed-batch fermentation. To avoid the effects of measurement noise and model-process mismatch, the proposed framework also includes dual Unscented Kalman Filter (UKF) to estimate both process states and model parameters, simultaneously. An online optimization of tissue-plasminogen activator (tPA) protein expression in mammalian cell culture is studied by using the proposed framework. In this case study, a modified Monod type kinetic model is applied to precisely predict the states of bioprocess including the nutrients consumption rates, cell numbers in all state of the cells (live, dead. lysed), and product formation rates. With the fed-batch mammalian cell culture model, the trajectory states are determined by solving a nonlinear optimization problem for maximizing the protein productivity. The calculated trajectory states are taken as reference states in NMPC. To avoid control degradation due to model uncertainties and perturbation during the real-time fermentation, a dual estimation strategy based on Unscented Kalman Filter is developed to estimate the actual process states and model parameters at each sampling time. To study the influence of the updated model parameters on the protein productivity in the fed-batch fermentation, variance-based global sensitivity analysis is performed to estimate the importance of the model parameters. The updated model parameters with their total-effect indices are used to evaluate the extent of the process-model mismatch. If the value of inner products of the vector of model parameters and the vector of their corresponding total-effect indices is over a threshold, the trajectory planning is executed again to update the trajectory states of the fermentation. Through the trajectory planning and NMPC algorithm, the optimal feeding rates and concentrations are determined for maximizing the productivity of tPA-I and tPA-II proteins during fermentation. The simulation result of protein production shows 12.64% improvement compared with the batch fermentation. The preliminary results show that under conditions of noisy measurement and structured plant-model mismatch, the dual estimation algorithm can precisely estimate the actual states.

DESIGN OF A MODULAR SAFETY SYSTEM FOR THE ARTIFICIAL PANCREAS: THE HEALTH MONITORING SYSTEM

Rebecca A. Harvey[1,2], Eyal Dassau[1,2], Howard Zisser[1,2], Dale E. Seborg[1,2], Lois Jovanovič[1,2], and Francis J. Doyle III[1,2]

[1]*University of California Santa Barbara, Santa Barbara, CA*
[2]*Sansum Diabetes Research Institute, Santa Barbara, CA*

Introduction

Type 1 diabetes mellitus (T1DM) is a disease in which blood glucose (BG) cannot be regulated by the body due to the auto-immune destruction of insulin-producing beta cells in the pancreas. Control of this disease may be achieved in a variety of ways, but is essential, because poor control can result in serious complications, both long- and short-term. Automatic control has become realizable recently, due to vast improvements in technology for sensing of BG and administration of insulin.[1,2] The ultimate goal in automatic control of T1DM is the artificial pancreas device system (APDS), a combination of continuous glucose monitors (CGMs), insulin delivery devices, and the control system that executes insulin delivery and monitors the system for safety. The APDS will be a high-risk device, due to its automatic delivery of a crucial hormone that is both necessary for life and dangerous in high doses.[3,4] Like any other high-risk system, the APDS must contain multiple safety layers to ensure the health of the user and the proper condition of the device.

The primary safety concern in the APDS is hypoglycemia (BG<70 mg/dL), or low BG, which can lead to coma or death of severe and prolonged. Hypoglycemia is the most dangerous short-term complication of intensive treatment of T1DM. Other major health concerns include hyperglycemia (BG>180 mg/dL) and missed meals. In addition, the system performance must be monitored, with such issues as pump or sensor error and communication problems, both among devices and in telemedicine.

There are several differences in the design of the safety system when compared to a traditional system, due to the constraints of a biomedical system. The lack of redundant sensors due to space constraints is a major concern, particularly due to the frequent noise encountered. In addition, only glucose can easily be measured in situ; insulin levels can only be measured by assay. One of the most notable disturbances is that the system itself has a human fully incorporated, who may intervene in an unforeseen manner.

Methods

The Health Monitoring System (HMS) has been designed as a process monitoring and alert module that can be executed in real time in parallel to any controller for the APDS (see Figure 1).

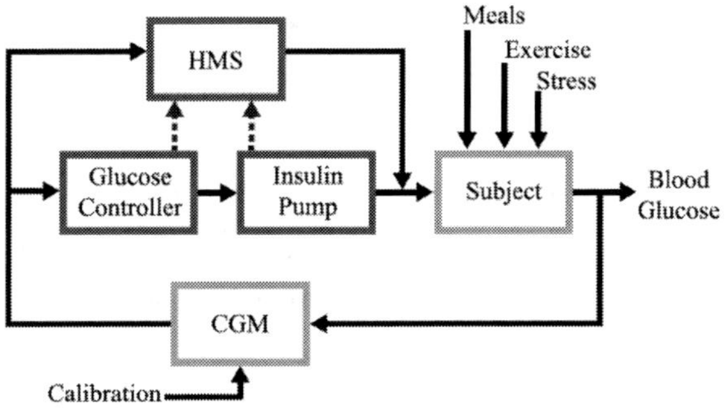

Figure 1: *APDS structure with parallel controller and safety system. Gray dashed lines are optional lines of communication.*

The HMS has been designed to easily accommodate several safety modules, including hypoglycemia, hyperglycemia, and missed meal detection; pump and sensor error detection; and communication monitoring. The primary module of the HMS, the Low Glucose Predictor (LGP), was designed first to predict and prevent hypoglycemia and is presented here.

LGP Design

The LGP was designed using retrospective, ambulatory CGM data from a large patient cohort. It was then tested in silico using the FDA-accepted UVA/Padova metabolic simulator.[5] There are three major modules in the LGP. The first module is a pre-processing module to prepare CGM data for prediction by filtering the data for noise and drift, and detecting sensor calibration. The handling of missing data comprises the second module. The third module is the core algorithm, where predictions of hypoglycemia are made by extrapolating an estimation of the rate of change of the CGM data. See Figure 2 for a flowchart of the LGP module.

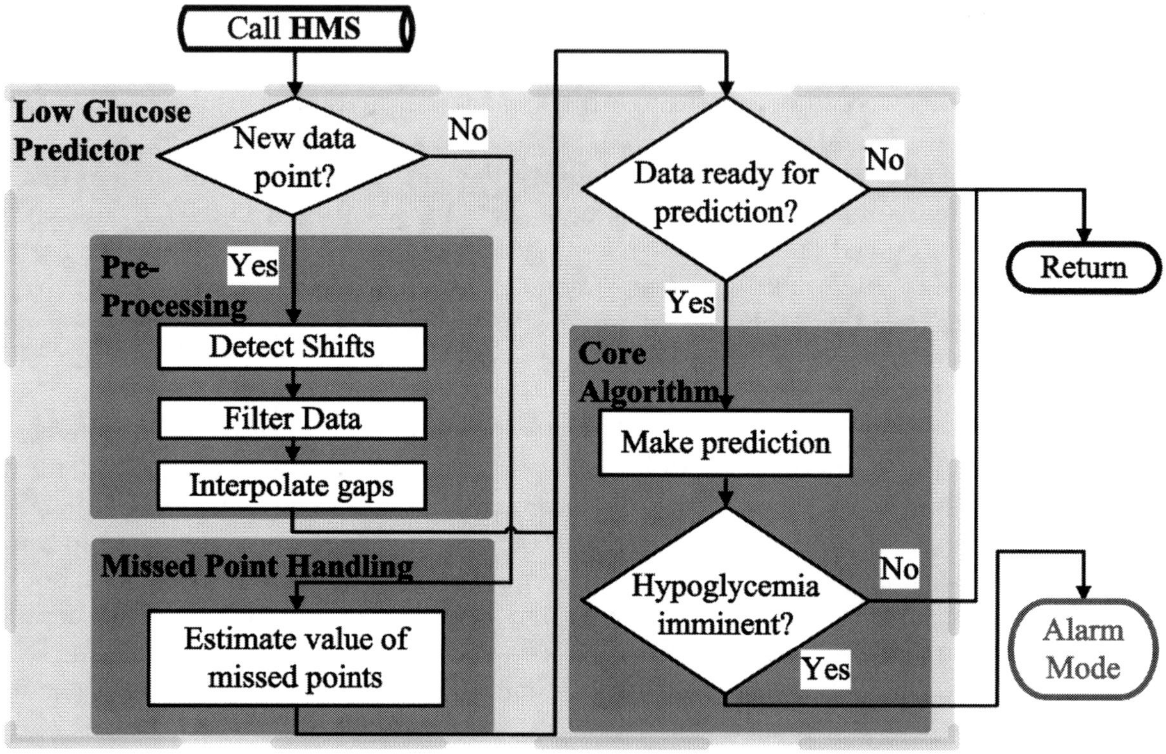

***Figure 2:** LGP algorithm flowchart*

The major sub-modules in the LGP are the filter and the prediction. The noisy CGM data are filtered using physiologically-based parameters to ensure that data reflect the glucose value accurately. The algorithm filters the data using a noise spike filter,[6]

$$G_{F,NS}(k)=\begin{cases} G_m(k) & if\,|G_m(k)-G_F(k-1)|\leq \Delta G \\ G_F(k-1)-\Delta G\, if\,(G_F(k-1)-G_m(k))>\Delta G \\ G_F(k-1)+\Delta G\, if\,(G_m(k)-G_F(k-1))>\Delta G \end{cases} \tag{1}$$

where k is the sampling instant, $G_F(k\text{-}1)$ is the previous filtered value, $G_{F,NS}(k)$ is the filtered value resulting from the noise-spike filter, $G_m(k\text{-}1)$ is the measurement, and ΔG is the maximum allowable change from one instant to the next.[6] The high frequency noise is dampened using a low pass filter[6]:

$$G_F(k)=\frac{\Delta t}{\tau_F+\Delta t}G_{F,NS}(k)+\left(1-\frac{\Delta t}{\tau_F+\Delta t}\right)G_F(k-1) \tag{2}$$

where $G_F(k)$ is the filtered value, Δt is the sampling period, and τ_F is the filter time constant.[6] An estimate of the rate of change is calculated in the core algorithm section, using filtered data in the following equation, the derivative of the Lagrange interpolation polynomial:

$$\begin{aligned} G'_F(j) \cong & \frac{t(j)-t(j-1)}{(t(j-2)-t(j-1))(t(j-2)-t(j))} G_F(j-2) + \\ & \frac{t(j)-t(j-2)}{(t(j-1)-t(j-2))(t(j-1)-t(j))} G_F(j-1) + \\ & \frac{2t(j)-t(j-2)-t(j-1)}{(t(j)-t(j-1))(t(j)-t(j-2))} G_F(j) \end{aligned} \quad (3)$$

where $G'_F(j)$ is the estimated rate of change at point j, t is time, G_F is the filtered glucose value, $j=k-SAR+1:k$, and SAR is the number of successive alarms required for an alert to be engaged.[7] The rate of change calculation is extrapolated linearly through the hypoglycemia threshold, TH, to decide if hypoglycemia will occur within the prediction horizon, PH.

The LGP outputs the prediction and current measurement to the HMS alarm module, which determines if an alarm must be issued (see *Figure 3*). The alarm mode module issues audible and visual alerts and sends warning short and multimedia messages (SMS and MMS), seen in *Figure 4*. The automatic, redundant alert system of SMS and MMS messages is the key safety feature of the HMS. These messages are sent to the physician in charge or other primary contact with a profile of the current trend and prediction for the upcoming 15 minutes.[8] The HMS alarms are acted upon by administration of 16g of rescue carbohydrates.

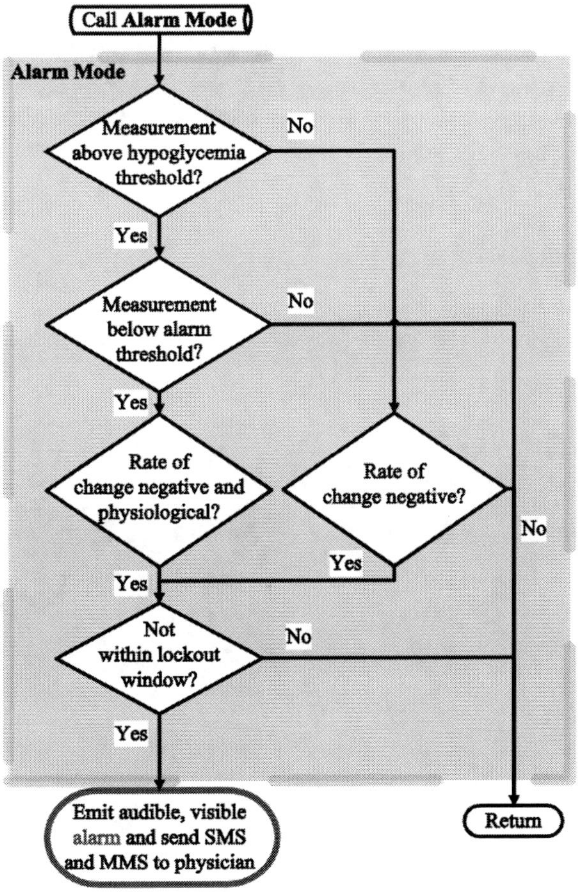

Figure 3: *Alarm mode module flow chart.*

Clinical Evaluation

The HMS was evaluated clinically in parallel to the zone-MPC controller[9,10] via the Artificial Pancreas System (APS©). The zone-MPC controller was designed at UCSB/SDRI to maintain BG in a predefined range, increasing or decreasing insulin delivery accordingly, when the CGM predictions by the internal model violate the boundaries of the zone. The zone-MPC is an updated implementation of MPC for diabetes control; use of MPC in this context was first published by Parker et al.[11,12] The Dexcom® SEVEN® PLUS (San Diego, CA) CGM was used for subcutaneous glucose sensing, and insulin delivery was effected using the Animas® OneTouch® Ping® (West Chester, PA). Reference IV blood samples were taken and assayed using the YSI 2300 STAT Plus (Yellow Springs, OH) approximately every 30 minutes. The protocol included automatic closed-loop control from approximately 5PM followed by an unannounced 50g CHO dinner meal, overnight control, an unannounced 40g CHO breakfast meal, unannounced 16g CHO snack (pre-exercise) if the reference glucose was < 120 mg/dl, 30 minute mild exercise and 16g CHO snack 3h post exercise.

Results and Discussion

The HMS was recently validated in clinic in conjunction with a zone-model predictive control (zone-MPC) controller via the Artificial Pancreas System (APS©), both designed at UCSB.[9,10,13,14] The ability of the HMS to alarm and emit SMS and MMS push notifications was validated (see Figure 4 for a screenshot from the attending physician's phone).

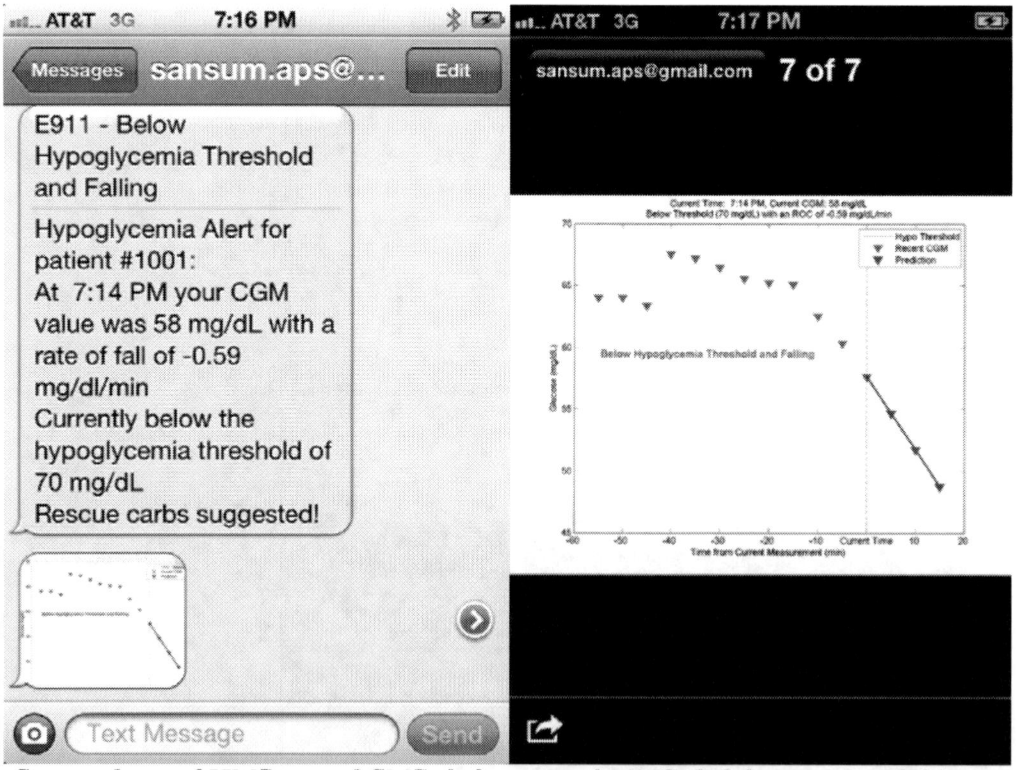

Figure 4: *Screenshots of HMS-issued SMS (left top) and MMS (left bottom), with blow-up of MMS (right) from a clinical trial.*

During the first clinical trial, the subject overdosed ~3U of insulin prior to her arrival to the closed-loop session (not per protocol). Even with insulin reduction by the controller and four carbohydrate supplements after HMS alerts, hypoglycemia was unavoidable but manageable due to the HMS, confirming that a safety system to detect adverse events is an essential part of the APDS.

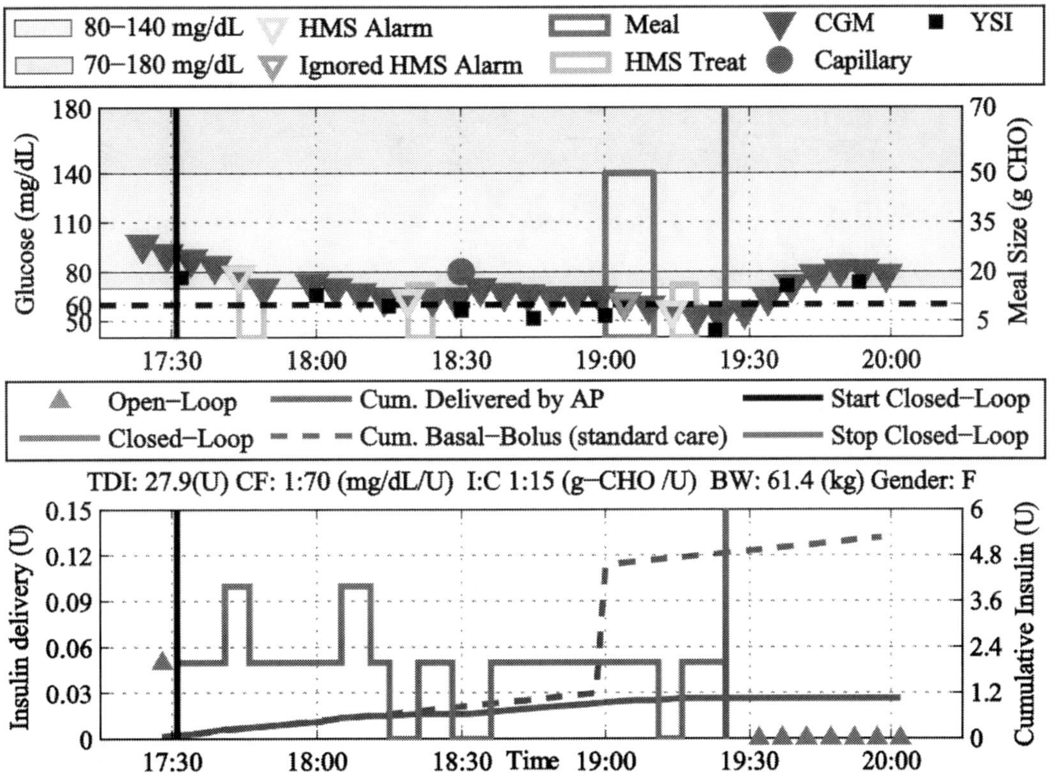

Figure 5: *Clinical case study using the HMS in which the trial was terminated early due to low blood glucose (per protocol). The start and end of closed loop are shown as black and red bars. On the top plot, CGM, YSI, and fingerstick values are shown as blue triangles, black squares, and red dots, respectively. The HMS alarms are plotted on top of the CGM values as orange triangles for ignored alarms and light blue for accepted alarms. Meals and HMS treatments are shown in purple and green, respectively. On the bottom plot, insulin delivery is shown as a green line (automatic) and blue triangles (manual) with Cumulative delivery in blue and the standard care in purple (dotted).*

A total of 12 full-length closed-loop sessions were conducted with the HMS/zone-MPC system (mean time in closed-loop 23.5 ± 0.7 hours). During closed loop, the time spent under 70mg/dL (considered hypoglycemia by the American Diabetes Association)[15] was only 2% by CGM and 1.4% by YSI. This result is a significant improvement over that found in ambulatory data (approximately 10-14%).[7,16] A total of 46 HMS alarms were sounded (mean 4 ± 2 alarms per subject), at an average CGM value of 82 ± 11 mg/dL, with a recovery of 10 mg/dL within 30 minutes on average. The CGM values, aligned from the time of the HMS alarm, are given in Figure 6, showing that alarms were sounded as CGM values were nearing the hypoglycemia region, causing recovery into the normoglycemia zone (70-180 mg/dL). The low standard deviation at the onset of the alarm suggests a high specificity of the safety system and confirmation of the necessity of the alarm. The recovery period of about 0.3-0.4 hours is in agreement with the clinical expectation of recovery in 15 minutes with 15-20g of rescue carbohydrates, depending on the rate of fall and amount of insulin on board.[17]

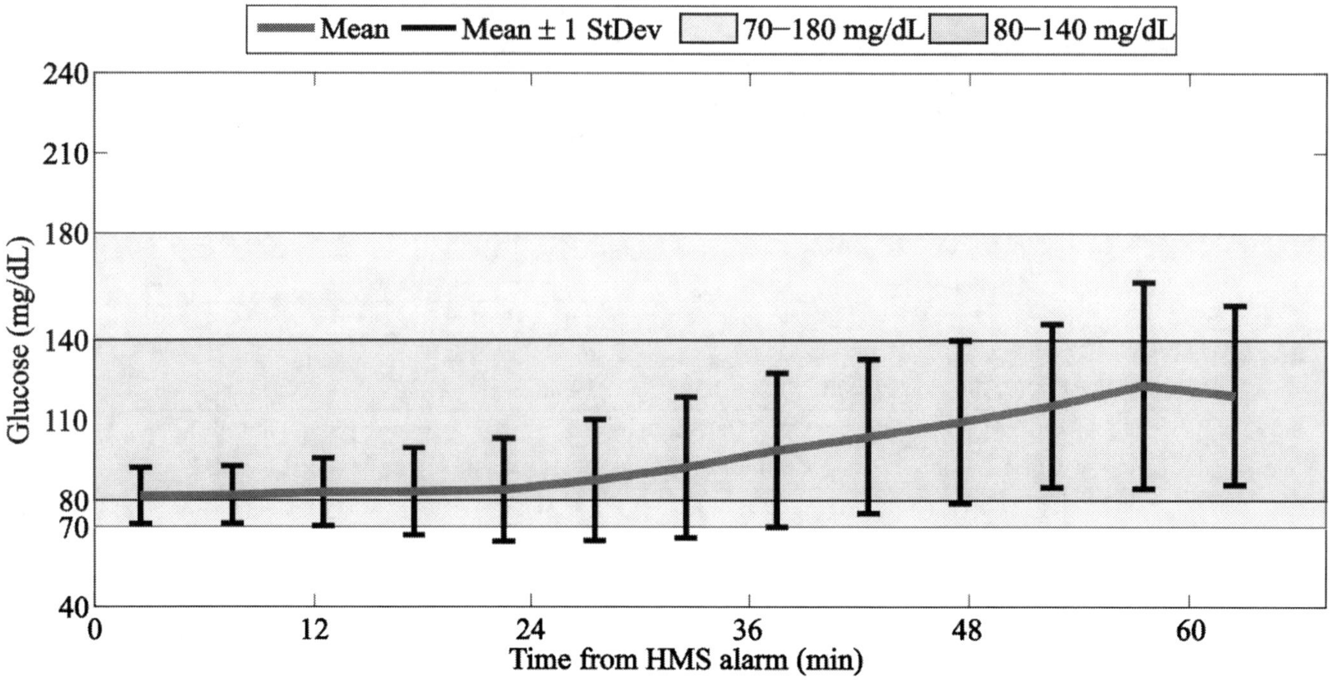

Figure 6: *CGM data from 12 clinical subjects with 46 HMS alarms, aligned from the time the alarm sounded. The average CGM value is in red, with standard deviation given in black brackets. The zone used for the zone-MPC controller is in the green filled area, with normoglycemia in the blue area.*

Conclusions

The HMS has been designed as a safety system in parallel to APDS that alerts for and mitigates adverse events. This design ensures the safety of the user and adds robustness to the overall system without interfering with the primary controller. The ability of the HMS to be an effective alert system that provides a safety layer to the APDS controller has been clearly demonstrated in clinic, detecting events that were unavoidable even with controller braking. With use of the HMS, adverse events that are not handled by the controller are readily detected and mitigated. The HMS has the ability to ensure safety through the implementation of escalating alarms; alarming locally initially, increasing the scope by alarming globally (SMS/MMS) to an emergency contact, followed by a health care provider or call center if no response is detected.

References

1. Zisser, H.C., T.S. Bailey, S. Schwartz, R.E. Ratner, J. Wise, "Accuracy of the SEVEN(R) Continuous Glucose Monitoring System: Comparison with Frequently Sampled Venous Glucose Measurements", *J Diabetes Sci Technol*, 3 (5) pp. 1146-54 (2009).
2. Keenan, D.B., B. Grosman, H.W. Clark, A. Roy, S.A. Weinzimer, R.V. Shah, J.J. Mastrototaro, "Continuous glucose monitoring considerations for the development of a closed-loop artificial pancreas system", *J Diabetes Sci Technol*, 5 (6) pp. 1327-36 (2011).
3. Tanenberg, R.J., C.A. Newton, A.J. Drake, "Confirmation of hypoglycemia in the "dead-in-bed" syndrome, as captured by a retrospective continuous glucose monitoring system", *Endocr. Pract.*, 16 (2) pp. 244-8 (2010).

4. Klonoff, D.C., C.L. Zimliki, L.A. Stevens, P. Beaston, A. Pinkos, S.Y. Choe, G. Arreaza-Rubin, W. Heetderks, "Innovations in technology for the treatment of diabetes: clinical development of the artificial pancreas (an autonomous system)", *J Diabetes Sci Technol*, 5 (3) pp. 804-26 (2011).
5. Kovatchev, B.P., C. Breton, C. Dalla Man, C. Cobelli, "In Silico Preclinical Trials: A Proof of Concept in Closed-Loop Control of Type 1 Diabetes", *J Diabetes Sci Technol*, 3 pp. 44-55 (2009).
6. Seborg , D.E., T.F. Edgar, D.A. Mellichamp, F.J. Doyle III, *Process Dynamics and Control*, 3rd ed., John Wiley & Sons, Hoboken, NJ, 2011.
7. Harvey, R.A., E. Dassau, H.C. Zisser, W. Bevier, D.E. Seborg, L. Jovanovic, F.J. Doyle III, "Clinically relevant hypoglycemia prediction metrics for event mitigation", *Diabetes Technol. Ther.*, 14 (8) pp. 719-27 (2012).
8. Dassau, E., L. Jovanovič, F.J. Doyle III, H. Zisser, "Enhanced 911/GPS Wizard: a Telemedicine application for the Prevention of Severe Hypoglycemia - Monitor, Alert and Locate", *J Diabetes Sci Technol*, 3 pp. 1501-06 (2009).
9. Zisser, H.C., E. Dassau, W. Bevier, R.A. Harvey, L. Jovanovič, F.J. Doyle III, Clinical Evaluation of a Fully-Automated Artificial Pancreas using Zone-Model Predictive Control with Health Monitoring System, in: *American Diabetes Association 72nd Scientific Sessions*, Philadelphia, PA, 2012.
10. Grosman, B., E. Dassau, H.C. Zisser, L. Jovanovič, F.J. Doyle III, "Zone model predictive control: a strategy to minimize hyper- and hypoglycemic events", *J Diabetes Sci Technol*, 4 (4) pp. 961-75 (2010).
11. Parker, R.S., F.J. Doyle III, J.E. Harting, N.A. Peppas, Model predictive control for infusion pump insulin delivery, in: *18th Annual International Conference of the IEEE Engineering in Medicine and Biology Society*, 1996, pp. 1822-3.
12. Parker, R.S., F.J. Doyle III, N.A. Peppas, "A model-based algorithm for blood glucose control in Type 1 diabetic patients", *IEEE Trans. Biomed. Eng.*, 46 (2) pp. 148-57 (1999).
13. Dassau, E., H. Zisser, C.C. Palerm, B.A. Buckingham, L. Jovanovič, F.J. Doyle III, "Modular Artificial β-Cell System: A Prototype for Clinical Research ", *J Diabetes Sci Technol*, 2 (5) pp. 863-72 (2008).
14. van Heusden, K., E. Dassau, H.C. Zisser, D.E. Seborg, F.J. Doyle III, "Control-relevant models for glucose control using a priori patient characteristics", *IEEE Trans. Biomed. Eng.*, 59 (7) pp. 1839-49 (2012).
15. Workgroup on Hypoglycemia, A.D.A., "Defining and reporting hypoglycemia in diabetes: a report from the American Diabetes Association Workgroup on Hypoglycemia", *Diabetes Care*, 28 (5) pp. 1245-9 (2005).
16. Harvey, R.A., E. Dassau, H. Zisser, D.E. Seborg , L. Jovanovič, F.J. Doyle III, "Design of the Health Monitoring System for the Artificial Pancreas: Low Glucose Prediction Module", *J Diabetes Sci Technol*, (2012).
17. American Diabetes Association, "Standards of Medical Care in Diabetes—2012", *Diabetes Care*, 35 (Supplement 1) pp. S11-S63 (2012).

(744e) Optimal Control for Predicting Drug Dosage in Superovulation Stage of in Vitro Fertilization

Kirti Maheshkumar Yenkie, Bioengineering, University of Illinois, Chicago, Chicago, IL, Urmila Diwekar, Vishwamitra Research Institute, Center for Uncertain Systems: Tools for Optimization and Management, Clarendon Hills, IL and Vibha Bhalerao, Jijamata Hospital, Nanded, India, Nanded, India

In vitro fertilization (IVF) is one of the most highly pursued assisted reproductive technology worldwide. The IVF procedure is divided into four stages: Superovulation, Egg-retrieval, Insemination/Fertilization, Embryo transfer. Superovulation is the most crucial stage in IVF, since it involves external injection of hormones to stimulate development and maturation of multiple oocytes. The maximum amount of effort and money for IVF procedure goes into superovulation. Although numerous advancements have been made in IVF procedures, medication quality, etc little attention has been given to modifying the protocols based on a predictive model.

A model for the follicle growth dynamics and number as a function of the injected hormones and patient characteristics is developed. The modeling basics are adapted from batch crystallization moment model, since moments are representatives of specific properties like number, shape and size of the particles under consideration. Based on this model, the dosage of the hormones to stimulate multiple ovulation or follicle growth is predicted by using the theory of optimal control. The objective of successful superovulation is to obtain maximum number of mature oocytes/follicles within a particular size range. Using the mathematical model involving follicle growth dynamics and the optimal control theory, optimal dose and frequency of medication is predicted for obtaining the desired result. The model will be modified to consider the sources of uncertainty due to patient's age, previous medical history, suitability of medicine and protocol used. The optimal drug delivery regime predicted in the presence of uncertainty will be compared to the current dosage regime predicted.

Thus, a phenomenon currently based on trial and error will get a supportive basis to start with. This will aid as a predictive tool for medical professionals and provide them with a specific dosage strategy for a patient. This will bring down the probability of failure, decrease cost of complex monitoring and excess medication. Thus, it will decrease the overall cost of IVF treatment for the patient as well as the physician.

(744f) Dynamic Treatment Strategies for Methanol Intoxication Using Ethanol As Alcohol Dehydrogenase Inhibitor in Human Beings

Ruben D. Vargas, Angel A. Galvis, Jonathan Moreno, Jorge M. Gomez and Watson L. Vargas, Department of Chemical Engineering, Universidad de los Andes, Bogotá, Colombia

Methanol poisoning is a fatal public health problem that affects population in both developed and developing countries particularly as alcohol consumption increases in most countries around the world. Several sources of poisoning are related to this issue (e.g. adulterated beverages, inappropriate distillation processes, methylated spirits, accidental exposure to this organic solvent, suicide attempt cases, among others). The inlet routes to the systemic metabolism are usually ingestion, inhalation or absorption through the skin. Several isolated cases of mass intoxications have taken place worldwide in recent years, Nicaragua (788 cases, 2006); Colombia (31 cases, 1989 and 88 cases, 2004); Jordan (17 cases 2006), Estonia (154 cases, 2001) [1-6]. Nevertheless, the total of cases per year in each country might be higher: Colombia (136 cases, 2007 and 250 cases, 2008); United States (2283 possible cases, 2007) [7].

Methanol toxicity arises due to two main accepted mechanisms: i) central nervous system depression and ii) biotransformation of methanol into formaldehyde and then to formic acid. Therefore, current treatments approaches are related to alcohol dehydrogenase (ADH) inhibition, since this enzyme is the one responsible for methanol metabolism to formaldehyde (the toxic metabolite). In emergency settings and with the purpose of treating patients for Methanol poisoning, inhibition is carried out using intravenous injection of "pure" ethanol or 4-methylpirazole (fomepizole) because these compounds may compete for the active sites on the ADH enzyme.

Several *in-silico* models for alcohols and glycols have been developed and the physiologically based pharmacokinetic (PBPK) ones are the most common because these enhance the fitting with experimental data keeping the physiological meaning of each differential equation [7-8]. Nevertheless, in the state of art there are few models that include non-reactive elimination pathways or dynamic control strategies in order to provide the most accurate doses for the inhibition kinetics of ADH in a physiological media.

In this study a multi-compartment physiologically based pharmacokinetic model including non-reactive elimination pathways, and a dynamic optimization technique is implemented in order to minimize the formaldehyde production on the alcohol dehydrogenase ADH enzyme using intra-venous ethanol as inhibitor. This technique is contrasted with a PID control strategy designed to stabilize the vein formaldehyde concentration under critical levels, the controller parameters are tuned up via nonlinear optimization techniques.

The proposed PBPK model includes separation between venous and arterial blood pools, and the main organs for methanol, ethanol and its metabolites biodistribution have been modeled in detail (i.e. Liver, lungs, kidneys, muscles and fat, stomach, gastrointestinal tract and skin), and the reactive unit (liver) is modeled as series of CSTR reactors with a reversible Michaelis-Menten kinetics. Additionally, the parameters for this model have been adjusted and validated using experimental data from the literature.

Keywords: Methanol intoxication, enzymatic inhibition, dynamic optimization, PID control.

References:

[1] Pan American Health Organization, Methanol Poisoning in Leon, Nicaragua, Report, September, (2006).

[2] Bennett I.L., Cary F.H., Mitchell G.L., et al., Acute methyl alcohol poisoning: A review based on experiences in an outbreak of 323 cases. Medicine 32:431-463, (1953).

[3] Kane R.L., Talbert W., Harlan J., et al., A methanol poisoning outbreak in Kentucky. A clinical epidemiologic study In Archives of Environmental Health, 17(1):119-29. (1968).

[4] Paasma R., Hovda K.E., Tikkerberi A., Jacobsen D., Methanol mass poisoning in Estonia: outbreak in 154 patients. Clin Toxicol (Phila). 45(2):152-7 (2007).

[5] Balaguer J., Intoxicación por Metanol. Toxicología Clínica. Bataller Editorial, Published in Spain p. 60-64 (2004).

[6] Colombian National Health Institute, Methanol monitoring and poisoning control (2010). (*Original language:* Protocolo de vigilancia y control de intoxicaciones por metanol, Instituto nacional de salud de Colombia (2010)).

[7] BrentJ. Fomepizole for Ethylene Glycol and Methanol Poisoning. *The new england journal of medicine* 360; 21 may 21, 2009

(744g) Revisiting the Foundations of Cybernetic Modeling Using Systems Engineering Methods

Aravinda Mandli and Jayant Modak, Department of Chemical Engineering, Indian Institute of Science, Bangalore, India

The cybernetic modeling framework for the growth of microorganisms provides an elegant methodology to account for unknown regulatory interactions using principles of optimal control theory[1, 2]. In this work, we revisit the various assumptions used in the cybernetic modeling framework. We first propose a generalized Monod model incorporating control variables, formulate various optimal control problems involving the generalized Monod model and solve them analytically to show that the long-term objective of maximizing cellular concentration at a fixed final time is equivalent to maximizing instantaneous growth rate during the growth of microorganisms in a single/two substrate batch environment. We then show that reformulation of the state and adjoint equations of the above optimal control problems result in generalized governing dynamic equations of the cybernetic model. We illustrate the influence of various constraints in the cybernetic model on the optimal growth behavior of microorganisms by solving several dynamic optimization problems using genetic algorithms. Finally, we will present a methodology for inferring the values of control variables for enzyme induction $u_i$of cybernetic models from experimental data.

References:

1. Ramkrishna D, Song HS: Dynamic models of metabolism: Review of the cybernetic approach. *AIChE J* 2012, 58(4):986-997.

2. Young JD, Ramkrishna D: On the matching and proportional laws of cybernetic models. *Biotechnol Prog* 2007, 23(1):83-99.

(747a) Control of Transparent Conducting Oxide Surface Morphology for Improved Light Trapping Using Wafer Grating and Model Predictive Control

Jianqiao Huang[1], Gerassimos Orkoulas[2] and Panagiotis D. Christofides[1], (1)Department of Chemical and Biomolecular Engineering, University of California, Los Angeles, Los Angeles, CA, (2)Chemical Engineering, University of California, Los Angeles, Los Angeles, CA

Thin-film silicon solar cells are currently among the most important and widely used solar cell systems and their share of the overall solar cell market is steadily increasing. Transparent Conducting Oxide (TCO) layers are an important component of thin film solar cell system. It has been demonstrated by many researches that the surface morphology, which is characterized by aggregate surface roughness and slope, has a crucial influence on the light trapping process and the conversion efficiency of thin film solar cell systems. Thus, it is very important to control the TCO surface morphology properties at desired values during the deposition process to improve the performance of thin film solar cells. This work focuses on the design of a feedback control system using both wafer grating and a model predictive controller to shape the surface morphology for improve solar cell performance. Specifically, a sinusoidal grated wafer is used for TCO thin film deposition process and a spatially distributed deposition rate profile is used to precisely control the TCO surface morphology during the deposition process. In this work, an Edwards-Wilkinson (EW) type equation is used to predict the process dynamics and the analytical solution of EW equation is derived as the basis of controller design. The model parameters of EW equations are identified by fitting the closed-form equation to open loop simulation data obtained from large-lattice kinetic Monte Carlo simulations. Finally, the controller is applied to a large scale kinetic Monte Carlo model of a TCO deposition process and the results demonstrate that the controller successfully regulates the surface morphology to desired values that lead to improved light trapping and solar cell performance.

(747b) Shape Control of Protein Crystals

Sangil Kwon, Chemical Engineering, UCLA, Los Angeles, CA, Michael Nayhouse, Chemical and Biomolecular Engineering, UCLA, Los Angeles, CA, Gerassimos Orkoulas, Chemical Engineering, University of California, Los Angeles, Los Angeles, CA and Panagiotis D. Christofides, Department of Chemical and Biomolecular Engineering, University of California, Los Angeles, Los Angeles, CA

Proteins play a key role as therapeutics in a number of diseases and protein crystallization is a central activity in the pharmaceutical industry. Specifically, the production of highly-ordered, high-quality protein crystals through batch crystallization processes is vital in devising proteins for therapeutic purposes. However, despite extensive experimental and computational work on understanding protein structure and function, there remains a lack of knowledge to model and control batch protein crystallizers.

Motivated by the above considerations the present work focuses on the simulation and control of globular proteins. Since the crystal growth of globular proteins is a non-equilibrium process, it will be simulated using kinetic Monte Carlo (kMC) methods. As is common practice in simulations of crystal growth, the solid-on-solid model will be adopted. In the solid-on-solid approximation, particles are deposited on the growing crystal without voids or overhangs and the resulting crystal is highly compacted. The implementation of the kinetic Monte Carlo methodology requires knowledge of the binding energies, the impingement rate, and the crystallization driving force. In previous simulations, a range of values was assigned to the previous parameters until satisfactory agreement with experiments was obtained for the protein under consideration. In the kMC simulations, molecular attachment, detachment, and migration events are considered and growth rates on the (101) and (110) faces can be obtained.Depending on the relative attachment energy of the crystal faces and assuming that the independentcrystal faces that are likely to appear during the growth are (101) and (110) faces, kMC simulations reproduce the experimentally observed cross-over behavior in the crystal growth rates between the two faces of the protein crystal.

This work focuses on control of the growth rate ratio between the two independent faces in single protein crystallization process using the temperature as the manipulated control input. A 3-D plot of steady-state relative growth rates versus solute concentration and temperature is derived to describe the evolution of the crystal growth accounting for the effect of solute concentration variations and temperature changes in batch systems from supersaturated solutions. The developed 3-D plot is then used as the basis for the design of a model predictive control algorithm that includes penalty on the deviation of crystals from its desired crystal shapes, specifically, subject to limit on the rate of change of the temperature of the batch crystallization.The calculated temperatures are used in the kMC models to achieve optimal crystal shape and to operate until the next sampling time.

The shape of the crystal is determined by the two faces and their relative face growth rates. Therefore it is possible to control the evolution of crystal shapes by controlling the ratio between the two face growth rates. Based on needs, more elongated to the direction of either horizon or vertical crystal can be obtained. Extensive simulation results covering the various aspects discussed above will be presented.

(747c) Multi-Rate Sampled-Data Control of Spatially Distributed Process Systems

Zhiyuan Yao, Department of Chemical Engineering and Materials Science, University of California, Davis, Davis, CA and Nael H. El-Farra, Department of Chemical Engineering & Materials Science, University of California, Davis, Davis, CA

With the wide-spread use of digital sensors and controllers in the control of continuous-time process systems over the past few decades, sampled-data control systems have been the focus of considerable research interest. The discrete availability and transfer of information (e.g., sensor measurements, control signals) in these systems typically stems from the intrinsic limitations on the data collection, transmission and processing capabilities of the measurement and control devices, but can also arise due to resource constraints on the communication medium interfacing the process with the control system. It is now well understood that these limitations may have a detrimental influence on the stability and performance properties of the closed-loop system, and that sampling effects need to be taken into consideration in the design of the control system.

These considerations have motivated considerable research work on the analysis and design of sampled-data control systems, with the overwhelming majority of existing studies focused on lumped parameter systems modeled by systems of ordinary differential equations. Many important engineering processes and systems, however, are characterized by significant spatial variations due to the presence of strong diffusive, convective and phase dispersion phenomena and are naturally described by partial differential equations. Compared with the extensive body and long history of research work on control of distributed parameter systems, the design of sampled-data control systems for spatially distributed processes has received only limited attention. Recently, we introduced in [1],[2] a framework for sampled-data control of distributed processes modeled by parabolic partial differential equations (PDEs) with measured outputs that are sampled over a resource-limited communication medium at discrete time instances. The main idea was to include an approximate finite-dimensional model of the infinite-dimensional system to provide the controller with estimates of the dominant slow states to compensate for the discrete availability of the sensor measurements, and to execute model updates using the actual measurements at each sampling time. In these studies, the controller design and analysis were carried out under the assumption that all the measurement sensors have the same fixed sampling rate and are forced to transmit their data to the controller concurrently. In many practical situations, however, limitations on the measurement capabilities of different sensors may result in a significant gap between the sampling rates, and in such cases a synchronized sampling mechanism may not be the best choice. For example, in chemical reactors, the composition and density measurements typically need several minutes of analysis, while the temperature can be measured at a relatively fast rate. Moreover, the importance of the measurement collected is another factor that can trigger the use of multi-rate sampling. It is reasonable to apply a fast sampling rate to the sensors placed at certain critical locations in the process (e.g., where frequent monitoring and tight control are required), while reducing the sampling rates of the other sensors in order to reduce cost and optimize energy resource consumption (e.g., battery life).

In this contribution, we present a methodology for the design of model-based output feedback controllers for spatially distributed systems modeled by highly-dissipative partial differential equations (PDEs) with multiple measured outputs that are sampled at different sampling rates. Initially, an approximate finite-dimensional system that captures the dominant dynamics of the infinite-dimensional system is obtained and used to design an observer-based output feedback controller. Due to the lack of continuous measurements, an inter-sample model predictor is included in the controller and used to provide the observer with estimates of the unavailable outputs. The model predictions are then updated and corrected at each time that a measurement becomes available. Owing to the different sampling rates of the available measurement sensors, the model update is performed using different outputs, or combinations of outputs, at each update time. A hybrid system formulation that captures the model update pattern dictated by the range of sensor sampling rates is used to analyze the stability properties of the sampled-data finite-dimensional closed-loop system and derive a necessary and sufficient condition for closed-loop stability. The condition is used to explicitly characterize the interdependence between the different sampling rates, the size of the model uncertainty, the controller and observer design parameters, and the spatial locations of the control actuators and measurement sensors. Singular perturbation techniques that exploit the two time-scale behavior of the spectrum of the differential operator are used to derive conditions for closed-loop stability of the sampled-data closed-loop infinite-dimensional system. Finally, the theoretical results are applied to the control of a low-density polyethylene tubular reactor.

References:

[1] Yao, Z. and N. H. El-Farra, ``On Model-Based Networked Control of Nonlinear Spatially Distributed Process Systems," *Proceedings of the 18th Mediterranean Conference on Control and Automation*, pp. 898-903, Marrakech, Morocco, 2010.

[2] Yao, Z. and N. H. El-Farra, ``Robust Fault Detection and Reconfigurable Control of Uncertain Sampled-Data Distributed Processes," *Proceedings of 50th IEEE Conference on Decision and Control*, pp. 4925-4930, Orlando, FL, 2011.

Output feedback control of transport reaction processes using adaptive proper orthogonal decomposition

Davood Babaei Pourkargar and **Antonios Armaou**

Most of the processes in the chemical process industry necessitate the consideration of complex transport phenomena often coupled with chemical reactions. Examples include reactive distillation in chemical processing, lithographic processes such as plasma etching in semiconductor manufacturing, and tin float bath process during glass production.

These processes' control problem is nontrivial due to the spatially distributed nature of the associated system dynamics. This problem can be addressed through model reduction, where the controller synthesis is based on finite dimensional approximations to the original infinite dimensional partial differential equations (PDEs) system [1]. The computation of basis functions that are subsequently utilized to obtain finite dimensional ordinary differential equation (ODE) models is one of the important parts of this design [2-6]. A common approach to this task is to use Karhunen-Loeve expansion combined with the method of snapshots, then using the method of weighted residuals to construct the finite ODE. KLE along with the method of snapshots has been used extensively in model reduction, optimization and control of distributed processes [7-14].

Our current research efforts focus on the recursive computation of eigenfunctions as additional data from the process becomes available to a) reduce computational requirements, & b) circumvent the issue of a priori availability of a sufficiently large ensemble of PDE solution data. Initially, an ensemble of eigenfunctions is constructed based on a relatively small number of snapshots. The dominant eigenspace is then utilized to compute the empirical eigenfunctions required for model reduction. This dominant eigenspace is recomputed with the addition of each snapshot with possible increase or decrease in its dimensionality [12], [13].

In this paper, our recent results on control of distributed parameter systems employing a new approach in adaptive proper orthogonal decomposition (APOD) methodology will be presented. In regular APOD approaches, the ensemble of the snapshots that is used to update the basis functions only possesses the most recent snapshots. This procedure could possibly lead to loss of important profiles as the process evolves. A move to consider the most independent snapshots instead of newest ones leads to design robust controllers with larger region of accuracy which is important when dealing with model uncertainty and fast evolving processes. Furthermore, a restricted number of sensors is usually available to deduce process behavior.

Through the construction of an accurate reduced-order model (ROM) recursive revisions based on the updated eigenfunctions, static output feedback controllers could be designed using the ROM and continuous measurements available from sensors [13]. To circumvent the lack of on-line measurements of the state profiles due to a restricted number of sensors, observers are designed based on the ROM.

The combination of an APOD methodology variant with a dynamic observer that is stable in the Lyapunov sense, allows the synthesis of nonlinear controller thus resulting in a computationally-efficient alternative to using large-dimensional ROMs with global validity for observer & controller pair design. Specifically, a robust Lyapunov-based controller synthesis approach is used to design a static controller [15] which is then combined with the dynamic observer. The observer/controller components are then recursively redesigned whenever the reduced order model is revised to retain relevancy and enforce closed-loop stability.

The proposed controller synthesis method was illustrated on two open-loop distributed systems. First, a typical diffusion reaction process that exhibits nonlinear dynamic behavior has been studied. We considered an elementary exothermic reaction taking place on a thin catalytic rod [1], [12]. The temperature of the rod is adjusted by means of limited actuators located along the length of the rod. The spatial profile of the temperature of the rod is described by a parabolic PDE. Second, the integral form of the Kuramoto-Sivashinksy equation (KSE) was considered. KSE can adequately describe incipient instabilities arising in a variety of physicochemical systems including falling liquid film, unstable flame fronts and interfacial instabilities between two viscous fluids [2], [3], [13]. The proposed controller structure successfully stabilizes the systems at a spatially uniform steady-state profile even though only limited information of the state is known.

References

[1] Christofides PD. Nonlinear and Robust Control of PDE Systems. New York: Birkhauser, 2000.
[2] Armaou, A.; Christodes, P. D. Feedback Control of the KuramotoSivashinsky Equation. Physica D 2000, 137, 49-61.
[3] Christodes, P. D.; Armaou, A. Global Stabilization of the KuramotoSivashinsky Equation Via Distributed Output Feedback Control. Syst. Control Lett. 2000, 39, 283-294.
[4] Armaou, A.; Christodes, P. D. Plasma-Enhanced Chemical Vapor Deposition: Modeling and Control. Chem. Eng. Sci. 1999, 54, 3305-3314.
[5] Armaou, A.; Christodes, P. D. Nonlinear Feedback Control of Parabolic PDE Systems with Time-Dependent Spatial Domains. J. Math. Anal. Appl. 1999, 239, 124-157.
[6] Armaou, A.; Christodes, P. D. Finite-Dimensional Control of Nonlinear Parabolic PDE Systems with Time-Dependent Spatial Domains Using Empirical Eigenfunctions. Int. J. Appl. Math. Comput. Sci. 2001, 11, 287-317

[7] Armaou A, Christofides PD. Dynamic optimization of dissipative PDE systems using nonlinear order reduction. Chem Eng Sci. 2002; 57:5083-5114.
[8] Balsa-Canto E, Alonso AA, Banga JR. Reduced-order models for nonlinear distributed process systems and their application in dynamic optimization. Ind Eng Chem Res. 2004;43:3353-3363.
[9] Shvartsman SY, Kevrekidis IG. Nonlinear model reduction for control of distributed parameter systems: a computer assisted study. AICHE J. 1998;44:1579-1595.
[10] Alonso AA, Frouzakis CE, Kevrekidis IG. Optimal sensor placement for state reconstruction of distributed process systems. AICHE J. 2004;50:1438-1452.
[11] Armaou, A. Output Feedback Control Of Dissipative Distributed Processes via Microscopic Simulations. Comput. Chem. Eng. 2005, 29, 771-782.
[12] Varshney, A.; Pitchaiah, S.; Armaou, A. Feedback control of dissipative distributed parameter systems using adaptive model reduction. AIChE J. 2008, 55, 906-918.
[13] Pitchaiah, S.; Armaou, A.; Output Feedback Control of Distributed Parameter Systems Using Adaptive Proper Orthogonal Decomposition, Ind. Eng. Chem. Res. 2010, 49, 10496-10509.
[14] Sirovich, L. Turbulence and the dynamics of coherent structures: part I, II and III. Q. Appl. Math. 1987, XLV, 561-590
[15] Sontag, E. A universal construction of artsteins theorem on nonlinearstabilization, Sys. & Contr. Lett., 13 (1989), pp. 117–123.

(747e) Control of Unstable Heat Equation On Two-Dimensional Time-Varying Domain Using Empirical Eigenfunctions

Mojtaba Izadi, Chemical & Materials Engineering, University of Alberta, Edmonton, AB, Canada and Stevan Dubljevic, Chemical and Materials Engineering, University of Alberta, Edmonton, AB, Canada

Many industrial transport-reaction nonlinear processes involve the change in the shape of the material and domain of analysis as a result of phase change, chemical reaction, external forces, and mass transfer. The mathematical models of such processes are obtained from conservation laws, such as mass, momentum and/or energy, and usually form nonlinear partial differential equations (PDEs).

The approach of modal decomposition, which implies that the dominant behavior of PDE systems can be approximately described by finite-dimensional systems, is extensively used for the controller synthesis of PDE systems. A well-known methodology in the extraction of eigenfunctions of nonlinear PDEs is the use of Karhunen-Loeve (KL) decomposition on an ensemble of solutions obtained from numerical or experimental resolution of the system. These modes, known as empirical eigenfunctions, are used in the derivation of accurate nonlinear reduced-order approximations of many diffusion-reaction systems and fluid flows.

Compared to the extensive research efforts on the order-reduction of distributed parameter systems modelled by PDEs, there are only few studies to address model-reduction and control of PDE systems with spatially time-varying domain. Assuming that the evolution of domain is known a priori, which can be measured in many processes, KL decomposition cannot be directly applied to the solutions of PDEs with time-varying domain. Armaou and Christofides used a mathematical transformation to represent the nonlinear PDE on an appropriate time-invariant domain and applied KL decomposition to obtain the set of eigenfunctions on the fixed domain [1,2]. In the study of the internal combustion engine flows by Fogleman et al., the velocity fields are stretched in one dimension to obtain data on a fixed grid such that the divergence of the original velocity field (continuity) is preserved [3]. Following these contributions, one way to deal with the aforementioned problem is to map the set of the solutions on a time-invariant domain and then apply KL decomposition, however, different mappings could change the energy content of the solutions. The idea here is to map the solutions of the PDE system on a fixed reference geometry while preserving the invariance of physical properties (energy) of the solutions.

To find the control law to stabilize the unstable steady-state of a nonlinear heat-equation on a two-dimensional spatially time-varying domain, we first map a set of the solutions of the nonlinear PDE describing the system behavior to a reference configuration while preserving the invariant property of thermal energy. A basis can be found by using the KL decomposition on the mapped solutions, and by applying the inverse mapping, a set of time-varying empirical eigenfunctions are obtained that capture the most energy of the system. Subsequently, the empirical eigenfunctions are used as a basis for Galerkin method to derive the reduced-order ODE model that accurately captures the dominant dynamics of the PDE system. Finally, the ODE system is used for the synthesis of nonlinear controller.

References:

1. A. Armaou, P. D. Christofides, Nonlinear Feedback Control of Parabolic Partial Differential Equation Systems with Time-dependent Spatial Domains, J. Math. Anal. Appl. 239 (1), 1999, 124-157.

2. A. Armaou, P. D. Christofides, Finite-Dimensional Control of Nonlinear Parabolic PDE Systems With Time-Dependent Spatial Domains Using Empirical Eigenfunctions, Int. J. Appl. Math. Comput. Sci. 11 (2), 2001, 287-317.

3. M. Fogleman, J. Lumley, D. Rempfer, D. Haworth, Application of the Proper Orthogonal Decomposition to Datasets of Internal Combustion Engine Flows, J. Turbul. 5, 2004, 023.

(747f) A Sufficient and Necessary Condition for Global Convergence of Continuous Crystallizers

Juan Du and B. Erik Ydstie, Department of Chemical Engineering, Carnegie Mellon University, Pittsburgh, PA

A Sufficient and Necessary Condition for global convergence

of Continuous Crystallizers

Juan Du and B. Erik Ydstie, Department of Chemical Engineering,

Carnegie Mellon University, Pittsburgh, PA

The main contribution of this work is three-fold. First a sufficient and necessary condition for continuous crystallization process is derived to guarantee the global exponential convergence of any trajectories. All of the possible trajectories converge to one nominal trajectory given that the process starts with different initial conditions or it is under temporary perturbation, due to change in boundary flows or external forces. The priori knowledge of an attractor is not required to analyze the convergence behavior. We use contraction theory to conclude the incremental stability, i.e. stability of the system trajectory with respect to each other. Incremental stability is a stronger form of stability than uniform global exponential stability with respect to origin, derived by traditional Lyapunov method. We use virtual displacement to measure the distance between any two trajectories. The nonlinear dynamics of the process is described in an exact differential form.

Secondly, we develop a control structure to make the system follow a reference trajectory. The closed-loop convergence analysis provides us guidelines to select process measurement and manipulated variables. Inventory control is applied to regulate the process dynamics. Two control structures are proposed based on the incremental stability condition. One is to use inlet concentration of mother liquid to control the degree of supersaturation. The other is to manipulate crystal withdrawal rate to keep total mass of crystals constant. We use contraction theory to prove their validity. Furthermore numerical experiments demonstrate the effectiveness of the control strategy.

The last but not least, we construct a generalized thermodynamic metric to measure the distance between two trajectories of the crystallization process. The metric goes to zero as the two trajectories approach each other. The metric is characterized by the thermodynamic extensive variable and their conjugated variables, i.e. intensive variable in the tangent differential form. It is directly related to the availability function defined in classic thermodynamics. In addition we show that the thermodynamic metric is convex which facilitates the generalization of current results.

(773a) Biorefinery Location and Technology Selection Through Supply Chain Optimization

W. Alex Marvin, Lanny D. Schmidt and Prodromos Daoutidis, Chemical Engineering and Materials Science, University of Minnesota, Minneapolis, MN

The current U.S. biofuel industry is dominated by ethanol (mainly produced from corn grain starch) and biodiesel (mainly produced from soybean oil and recycled cooking oils), but cellulosic biofuels (based on non-food resources such as agricultural waste, grasses and woody crops) are predicted to play a large role in the future. The U.S. EPA predicts that 8.7 million gallons of cellulosic biofuel will be produced this year, while government Renewable Fuel Standard (RFS2) mandates require the use of 16 billion gallons of cellulosic biofuel in motor-vehicle fuel by 2022. The expansion of the cellulosic biofuel industry to meet these mandates will impose significant logistical challenges at each stage of the biofuel supply chain (biomass production, harvesting, storage, processing and transportation, and biofuel distribution).

Supply chain optimization (SCO) can be applied to mitigate the risks associated with biomass logistics, and to determine efficient biomass-to-biofuel supply chain configurations. For these reasons, biofuel SCO has become an area of active research (e.g., see the review in [1]). This study differs from previous literature by (i) focusing on facility profitability through the inclusion of detailed taxation and capital depreciation, (ii) considering multiple competing biomass processing technologies producing a range of biofuels, (iii) including existing biofuel facilities, (iv) studying a large multi-state region, and (v) including each RFS2 fuel classification (renewable fuel, advanced biofuel and cellulosic biofuel) to analyze competition for shared feedstocks.

Specifically, this study proposes a mixed integer linear program to solve the SCO problem of locating and sizing biomass processing facilities in the Midwestern United States (a 12-state region) to meet the RFS2 biofuel mandates for 2015. A set of seven biomass processing technologies, each expected to be commercialized in the near-term, are considered for construction in the region. Existing corn ethanol facilities are included, and compete for biomass resources. The spatial distribution and farmgate cost of eight types of biomass is accessed from a recently updated U.S. Department of Energy database [2]. A detailed cash flow analysis that includes capital depreciation and taxation is embedded into the model formulation to give insights into the minimum biofuel selling price for each installed facility. Equilibrium market cost for each of the RFS2 fuel classifications, which is directly related to the Renewable Identification Number market price, is determined through sensitivity analysis of the delivered biofuel price.

[1] An, H.; Wilhelm, W. E.; Searcy, S.W. *Biomass Bioenergy* **2011**, 35, 3763–3774.

[2] U.S. Department of Energy, U.S. Billion-Ton Update: Biomass Supply for a Bioenergy and Bioproducts Industry, 2011.

(773b) Incorporating Water Resources, Electricity Supply, and Allocation of Sequestered CO2 in the Nationwide Hybrid Energy Supply Chain

Josephine A. Elia, Richard C. Baliban and Christodoulos A. Floudas, Chemical and Biological Engineering, Princeton University, Princeton, NJ

Previously developed frameworks on the nationwide energy supply chain analysis for transportation fuels based on hybrid coal, biomass, and natural gas to liquids (CBGTL) processes are expanded to incorporate water supply chain, electricity resources, and feasible allocation of sequestered CO_2 [1-9]. The CBGTL plants are optimized using from an expanded superstructure that includes syngas conversion to liquid fuel products via both Fischer-Tropsch reactions and methanol intermediate [6-8]. The fuels can be upgraded in a standard upgrading section similar to the operation of petroleum refineries, or over zeolite catalysts. Multiple carbon conversion levels (i.e., 40%, 50%, and 60%) are imposed to represent varying degrees of feedstock utilization and electricity production for the single plant, giving a total of 162 distinct configurations of CBGTL facilities that may exist in the supply chain. Each of the topology is optimized with simultaneous heat, power, and water integration approaches.

Water, electricity, and CO_2 sequestration considerations are incorporated both on the single plant level and on the supply chain level. On the individual plant, freshwater consumption and wastewater discharge are minimized via the water integration approaches. On the supply chain level, United States data on water availabilities are incorporated in the model constraints and the transportation cost to deliver water to the CBGTL plants is included in the objective function. The three carbon conversion levels give rise to plant topologies that are net producers (lower conversion levels) and net consumers of electricity (higher conversion levels). In the supply chain analysis, combinations of these plants are used to achieve a target of total electricity consumption in the United States such that the grid electricity expansion does not exceed 10% of the current capacity (i.e., 102.5 GW). Finally, in a single CBGTL plant, CO_2 is either vented, recycled, or captured for sequestration to fulfill a minimum of 50% emissions reduction from petroleum-based processes. The pipeline investment, transportation and injection costs of the captured CO_2 are incorporated in the supply chain analysis.

Case studies are presented to illustrate the performance of the optimal supply chain network for multiple levels of petroleum fuel replacement (i.e., 50%, 75%, and 100%) for the United States. A case study using the locations of current energy infrastructure (i.e., coal mines and oil refineries) is also included to examine the trade offs between facility location selections and the overall cost of fuel production.

[1] C.A. Floudas, J. A. Elia, R. C. Baliban (2012) Hybrid and Single Feedstock Energy Processes for Liquid Transportation Fuels: A Critical Review. Comp. Chem. Eng. 41:24-51.

[2] R. C. Baliban, J. A. Elia, C. A. Floudas (2010) Toward novel biomass, coal, and natural gas processes for satisfying current transportation fuel demands, 1: Process alternatives, gasification modeling, process simulation, and economic analysis. Ind. Eng. Chem. Res. 49:7343-7370.

[3] J. A. Elia, R. C. Baliban, C. A. Floudas (2010) Toward novel biomass, coal, and natural gas processes for satisfying current transportation fuel demands, 2: Simultaneous heat and power integration. Ind. Eng. Chem. Res. 49:7371-7388.

[4] R. C. Baliban, J. A. Elia, C. A. Floudas (2011) Optimization framework for the simultaneous process synthesis, heat and power integration of a thermochemical hybrid biomass, coal, and natural gas facility. Comp. Chem. Eng. 35:1647-1690.

[5] J. A. Elia, R. C. Baliban, X. Xiao, C. A. Floudas (2011) Optimal energy supply network determination and life cycle analysis for hybrid coal, biomass, and natural gas to liquid (CBGTL) plants using carbon-based hydrogen production. Comp. Chem. Eng. 35:1399-1430.

[6] R. C. Baliban, J. A. Elia, V. Weekman, C. A. Floudas (2012) Process synthesis of hybrid coal, biomass, and natural gas to liquids via Fischer-Tropsch synthesis, ZSM-5 catalytic conversion, methanol synthesis, methanol-to-gasoline, and methanol-to-olefins/distillate technologies. Submitted for publication.

[7] R. C. Baliban, J. A. Elia, C. A. Floudas (2012) Simultaneous process synthesis, heat, power, and water integration of thermochemical hybrid biomass, coal, and natural gas facilities. Comp. Chem. Eng. 37:297-327.

[8] R. C. Baliban, J. A. Elia, R. Misener, C. A. Floudas (2012) Global Optimization of a MINLP Process Synthesis Model for Thermochemical Based Conversion of Hybrid Coal, Biomass, and Natural Gas to Liquid Fuels. Comp. Chem. Eng. In press. DOI:10.1016/j.compchemeng.2012.03.008

[9] J. A. Elia, R. C. Baliban, C. A. Floudas (2012) Nationwide Supply Chain Analysis for Hybrid Feedstock Energy Processes with Significant CO2 Emissions Reduction. AIChE Journal. Submitted for publication.

(773c) Multiscale Strategic Planning Model for the Design of Integrated Ethanol and Gasoline Supply Chain

Federico Andersen, Chemical Engineering, Planta Piloto de Ingenieria Quimica (PLAPIQUI), CONICET-Universidad Nacional del Sur , Bahia Blanca, Argentina, Maria Soledad Diaz, Process Systems Engineering, Planta Piloto de Ingenieria Quimica (PLAPIQUI), CONICET - Universidad Nacional del Sur, Bahia Blanca, Argentina and Ignacio E. Grossmann, Chemical Engineering Department, Carnegie Mellon University, Pittsburgh, PA

Multiscale Strategic Planning Model for the Design of Integrated Ethanol and Gasoline Supply Chain

Federico Andersen[1], Soledad Diaz[1] and Ignacio Grossmann[2],

(1) Chemical Engineering, Planta Piloto de Ingenieria Quimica (PLAPIQUI), Universidad Nacional del Sur - CONICET, Bahia Blanca, Argentina

(2) Dept. of Chemical Engineering, Carnegie Mellon University, Pittsburgh, PA 15213

Ethanol continues to receive attention as a major energy source from biomass that can help to address the increasing needs of liquid fuels that are predicted over the next few decades. Research and industrial efforts have made of ethanol as one of the safest (both economically and environmentally) alternatives to conventional fuels, at least in the short and medium term. However several challenges remain, which have not yet been fully solved and developed. One of the main concerns is related to the implementation of higher blends of ethanol with gasoline and the utilization of alternative production processes for lignocellulosic ethanol. These concerns are based on the RFS Schedule under the Energy Independence and Security Act of 2007, which points to the utilization of 21 of 36 billion gallons of advanced biofuel by 2022. The widely used raw material to produce E10 within the US is corn, but due to increasing demands, alternative raw materials such as lignocellulosic material must be considered for the production of fuel grade ethanol. Facing the need to start the commercialization of higher ethanol blends, new challenges with respect to bioethanol production can be identified. Therefore in this transition from E10 to E30 it would be necessary not only to consider the addition of new raw materials, but also to allow for the distribution of different ethanol blends at the same time in the same gas station.

Because of the need to provide the market with multiple fuel blends, the business of gas stations has become a multi-blend market, where the demand of each blend depends on the driver's preference and legal requirements. This preference is mainly related to the price of the blend and the miles per gallon of fuel that can be driven, which is also different among vehicles. This feature requires an accurate forecast of demand for every blend and poses a new challenge to estimate and satisfy this demand with minimal cost, opening up a new market for multi-product gas stations, where blender pumps are used to obtain the desired blend mixing E85 with E10 at the gas station itself. Since most of the current gasoline distribution takes place through single-product gas stations and the investments to retrofit all of them into multi-product would be huge, we have modeled the coexistence of both of these types of gas stations and the prospect to move towards the multi-product ones in a gradual way, considering the possibility to enhance or reduce the total quantity of gas stations.

In this work we formulate the integrated ethanol and gasoline supply chain, taking into account the stages from the harvesting sites, through production sites for ethanol, petroleum refineries, blending stages and up to the retail of different blends in gas stations. The general problem can be stated as follows. Given a superstructure that combines all these stages and different means of transportation which connect the nodes, the integrated ethanol and gasoline supply chain can be represented with a multiperiod model and a time horizon of 20 years. Given also an initial capacity for ethanol plants and gasoline distribution center and data of type and quantity of existing gas stations; and given a forecast of demand for different blends over the entire time horizon, the main goal is to determine several major decisions in order to minimize cost. Some of these decisions involve: whether to install or not small; medium or large ethanol plants and gasoline distribution centers (taking into account economy of scale for the investments), and timing both for retrofits of different types of gas stations that involve blending pumps. The costs components included in the analysis are: investment capital cost (where economy of scale has been considered); raw materials; production; transportation; storage and distribution cost.

Since the industries and players of the bioethanol supply chain are widely spread over the US, the coordination over different geographical locations is a key feature for the optimization of the supply chain. Another key feature is based on different time scales required in the model. This fact arises from the inherent difference between the lifetime of investments in ethanol plants, which is on the order of years, and the replenishment of the gas stations that occurs every week or every two weeks. Because of both features the need to cope with two different formulations of the model can be identified. Each formulation contains a different level of detail of some stages, especially in the downstream supply chain (at gas stations level).

Within the more general formulation we have considered an aggregated strategic planning model in order to identify the regions of the US where more investments are needed and the optimal configuration of the network. Within the second formulation, even though the main modeling effort has been devoted to the operation of gas stations and the selection of blending pumps, the global characteristic of a supply chain model is also being reflected. Hence, this multiscale strategic planning model contains a high level of details about gas stations in comparison to other stages of the supply chain.

To illustrate the application of the proposed MILP models, we have considered over a geographical region a set of feasible retrofits between the single-product and multi-product gas stations, and decrease of capacity in the period that they are being retrofitted. Also we have included the possibility to satisfy high ethanol fuel demand with lower ethanol fuels.

Integer variables are associated to active gas stations; required retrofits; new gas stations and those that should be dropped off. Due to the size of the multiscale strategic planning model, we have applied a decomposition scheme to efficiently solve the problem using the Lagrangean decomposition strategy.

1- Datta et al., 2011. Ethanol – the primary renewable liquid fuel. J Chem Technol Biotechnol 86: 473 – 480.

2- Martin M., Grossmann I. 2011. Systematic synthesis of sustainable biorefineries: A review.

3- Neiro S., Pinto. J. 2004. A general modeling framework for the operational planning of petroleum supply chains. Computers and Chemical Engineering 28, 871-896

4- Rusell D., Ruamsook K., Thomchick E. 2009. Ethanol and the Petroleum Supply Chain of the Future: Five Strategic Priorities of Integration. Transportation Journal. 48-1

5- You F., Grossmann I. 2008. Mixed-Integer Nonlinear Programming Models and Algorithms for Large-Scale Supply Chain Design with Stochastic Inventory Management. Ind. Eng. Chem Res., 47, 7802-7817

6- You F., Wang B., 2011. Life Cycle Optimization of Biomass-to-Liquids Supply Chains with Distributed-Centralized Processing Networks. Industrial & Engineering Chemistry Research. 50(17): 10102-10127

(773d) Optimisation of Bioelectricity Supply Chains

Ozlem Akgul, UCL, London, United Kingdom, Niall Mac Dowell, Centre for Process Systems Engineering, Chemical Engineering Department, Imperial College London, London, United Kingdom, Nilay Shah, Centre for Process Systems Engineering, Department of Chemical Engineering, Imperial College London, London, United Kingdom and Lazaros G. Papageorgiou, Centre for Process Systems Engineering, UCL (University College London), London, United Kingdom

There has been a growing interest in the development and use of renewable energy technologies in the recent years to tackle major problems such as global climate change, depleting fossil fuel resources and increasing oil prices. Among the renewable energy technologies, bioenergy is considered to be a promising near-term solution. Bioenergy is obtained from the conversion of a variety of biomass feedstock including wood, dedicated energy crops, food crops and biomass waste. This diversity of types of biomass offers a significant advantage in terms of contributing to energy security through a diversified energy mix. Bioenergy can be used for different purposes such as dispatchable electricity generation, high grade heat and production of liquid biofuels. Mandates and targets have been set around the globe to promote the use of renewable energy. The UK is committed to supplying 15% of its energy demand from renewable resources and reduce its greenhouse gas emissions by at least 34% by 2020. Moreover, when combined with CO_2 capture and storage, the co-firing of biomass with fossil fuels leads to CO_2 negative energy. This is important in the context of mitigating anthropogenic CO_2 emissions, and eventually reducing atmospheric CO_2 concentration. Therefore, bioenergy is expected to play an important role as a part of the renewable energy mix to meet these long term renewables and emissions reductions targets.

A bioenergy supply chain is a multi-echelon network consisting of biomass cultivation sites, bioenergy production facilities and demand centres. Application of supply chain optimisation to such systems means consideration of all these nodes in the chain as well as transport of biomass and bioenergy between these nodes. This work presents a spatially-explicit optimisation framework for a bioenergy supply chain network based on a "neighbourhood flow" approach. Linear models of co-fired coal-based power plants with amine-based CO_2 capture have been derived from correlations of detailed simulations of the whole system using ASPEN plus.

The model applicability is highlighted with a case study of bioelectricity generation in the UK. The bioelectricity demand is determined based on a combination UK renewables and carbon mitigation targets. Availabilities of different types of woody biomass in the UK are accounted for in the generation of bioelectricity. Similarly, land-use-change (LUC) emissions are also incorporated in the carbon balance. Biomass imports are considered as a possible option to meet the domestic bioenergy demand

The model aims to minimise the total cost and/or total emissions of the whole network by optimising a variety of variables such as the material flows within the network, biomass supply and bioenergy production amounts in each region as well as the extent of biomass co-firing and CO_2 capture at a given power station.

(773e) Multi-Objective Optimization of a Biorefinery's Supply Network

Lidija Čuček[1], Mariano Martin[2,3], Ignacio E. Grossmann[4] and Zdravko Kravanja[1], (1)University of Maribor, Faculty of Chemistry and Chemical Engineering, Maribor, Slovenia, (2)Department of Chemical Engineering, University of Salamanca, Spain, (3)Carnegie Mellon University, Department of Chemical Engineering, Pittsburgh, PA, (4)Chemical Engineering Department, Carnegie Mellon University, Pittsburgh, PA

Abstract

Biomass is one of the renewable energy sources having increasing importance (Rentizelas et al., 2009), and from which heat, electricity and biofuels can be generated. Nowadays, the main concerns include the increasing shortages of energy and water resources, increasing prices, the demand and dependency on fossil fuels, as well as growing environmental concerns, global climate change being, in general, the greatest environmental threat, have led to the consideration of alternative and renewable energy sources. Developing clean and renewable energy resources ranks as one of the greatest challenges facing mankind over the medium- to long-term (Mata et al., 2011). In the short-term, in the case of the transportation sector, only biofuels from biomass provide an alternative that can be implemented because of its high density, compatibility with current automobile engines, and existing fuels' distribution infrastructures. Thus, bioethanol and biodiesel have become the more promising alternatives (Martín and Grossmann, 2012).

However, the competiveness of biomass as an energy source faces some challenges, and strongly depends on the biomass supply chain (Yu et al., 2009). Biomass is usually seasonally and locally available, has low energy density and high moisture content, degrades during storage, and requires extensive infrastructure for harvesting, transportation, storage, and processing (Lam et al., 2010). The first generation of biofuels, based on starch, sugar and oil crops brought about an ethical trade-off, since the raw material could also be used for food, resulting in an increase in food prices and, eventually, supply risks. Consequently, a second generation has been proposed as a solution, which is still at an early stage of development, not to mention the so-called third generation. Therefore, for the biofuel industry to be competitive with petroleum-based fuels, as well as providing further benefits, a more profitable, better integrated and more sustainable biorefinery supply network design is crucial for attracting investment in the production of more sustainable biofuels.

This contribution presents a synthesis of regional renewable biomass and bioenergy supply chains, based on a Mixed-Integer Linear Programming (MILP) approach. This method addresses the main challenges presented by biomass and water resources, the distributive and varied availabilities regarding location and time. The production processes from different sources of biomass include first, second, and third generations of biofuels such as bioethanol, biodiesel, hydrogen, and FT-diesel (Martín and Grossmann, 2012). The aim was to maximize the economically viable utilization of resources, by accounting for the competition between fuels and food production. Biorefinery supply networks are typically comprised of four-layer supply network superstructure, including the agricultural (layer 1 – L1), collection and preprocessing (layer 2 – L2), core processing (layer 3 – L3), and usage (layer 4 – L4) layers, including transportation between these layers. A region is divided into zones (Lam et al., 2010) for optimizing conversion operations and transportation flows. An MILP model for the synthesis of regional networks for the supply of energy and bioproducts (Čuček et al., 2010; Lam et al., 2011) was extended in order to be applied within an optimally-integrated biorefinery's supply network, so that the final products from one process could now become the raw materials for another process; e.g., bioethanol is a product from biochemical, thermochemical or thermo-biochemical processes, and could become a raw material for the production of biodiesel or hydrogen, a product from lignocellulosic materials that may be needed for FT synthesis. The multi-objective optimization of a heat-integrated biorefinery's supply network was performed, with maximization of the economic performance and minimization of the environmental impact, evaluated by the carbon and energy footprints. The objective of the synthesis was to determine economically-efficient and environmentally-benign solutions using the optimal selection of raw materials, technologies, intermediate, and final product flows.

Keywords: Biomass, Bioenergy, Biorefinery's supply network, Multi-objective optimization, Carbon footprint

References

Čuček L., Lam H.L., Klemeš J.J., Varbanov P.S., Kravanja Z., 2010, Synthesis of regional networks for the supply of energy and bioproducts, Clean Technologies and Environmental Policy, 12(6), 635-645

Lam H.L., Varbanov P.S., Klemeš J.J., 2010, Optimisation of regional energy supply chains utilising renewables: P-graph approach, Computers & Chemical Engineering, 34, 782-792

Lam H.L., Klemeš J.J., Kravanja Z., 2011, Model-size reduction techniques for large-scale biomass production and supply networks, Energy, 36(8), 4599-4608

Martín M., Grossmann I.E., 2012, Systematic synthesis of sustainable biorefineries: A review, Industrial & Engineering Chemistry Research, submitted for publication

Mata T.M., Martins A.A., Sikdar S.K., Costa C.A.V., 2011, Sustainability considerations of biodiesel based on supply chain analysis, Clean Technologies and Environmental Policy, 13, 655–671

Rentizelas A.A., Tolis A.J., Tatsiopoulos I.P., 2009, Logistic issues of biomass: The storage problem and the multi-biomass supply chain, Renewable and Sustainable Energy Reviews, 13(4), 887-894

Yu Y., Bartle J., Li C.Z., Wu H., 2009, Mallee biomass as a key bioenergy source in Western Australia: Importance of biomass supply chain, Energy Fuels, 26(6), 3290-3299

(773f) Multiobjective Stochastic Programming Models and Algorithms for Robust Design and Optimization of Biofuels Supply Chains

Yuan Yao, Berhane H. Gebreslassie and Fengqi You, Department of Chemical and Biological Engineering, Northwestern University, Evanston, IL

The growing concern about limited nonrenewable resources and severe pollution generated by traditional energy production triggers research on more sustainable energy resources. Hydrocarbon biofuels converted from biomass have been considered as promising alternatives.[1] Hydrocarbon biofuels can be used without significant changes of the current fuel distribution and utilization, and they provide vehicle performance similar to or better than their conventional counterparts.[2] Currently, several mathematical programming models, most of which are deterministic approaches, have been proposed to address the optimal design of hydrocarbon biorefinery supply chains.[3, 4] However, the final supply chain design is subject to various sources of uncertainty, such as seasonal and geographical fluctuation of biomass supplies, variability in biofuel demands due to unstable economic situations, population growth, the variability of feedstock purchase price and selling price of the hydrocarbon biofuels, and unexpected events. Deterministic model fails to consider these uncertain factors in the final solution. Thus it is of great importance to develop a comprehensive programming model which is robust from the economic performance perspective under the presence of uncertainties.[5] Stochastic programming model is an appropriate approach that optimizes the expected value of the objective function and obtains a solution performing well on average. When uncertainties are involved in, a critical issue is the existence of potentially high financial risk. Risk management should be taken into consideration in the final design of hydrocarbon biorefinery supply chain to reduce the impacts of unfavorable invent of uncertain parameters.[6] However, the standard stochastic programming methods usually do not provide control mechanisms on the unfavorable outcomes to implement risk management.[7] How to develop such a stochastic model that takes uncertainties into consideration while conducts financial risk management is a challenge. Another problem is that as the scenarios increase the resulting large scale model tends to be computationally expensive, how to solve it in a reasonable computational time needs to be addressed.

In this paper, a stochastic mixed-integer linear programming (MILP) model that predicts the optimal network design, technology selection, capital investment, production operations, and logistics management decisions under hydrocarbon biofuels demand and biomass supply uncertainties is developed. The design problem is formulated as a multi-objective optimization problem where the expected annualized cost of the supply chain network and a specific metric for financial risk are two objectives. The financial risk is measured by conditional value-at-risk and downside risk. A two-stage stochastic programming approach is used in formulation of the model.[6] In order to overcome the computational prohibition of the resulting MILP, an efficient decomposition approach is complemented, which is based on the multi-cut L-shaped method developed by You and Grossmann.[8] Finally, the modeling framework proposed and solution algorithm is illustrated by four cases studies in the State of Illinois. Optimal design of hydrocarbon biorefinery supply chain in deterministic case, stochastic cases with 4 scenarios, 100 scenarios and 1000 scenarios are presented. The comparison between deterministic and stochastic solutions, the conditional value-at-risk and downside risk, standard L-shaped method and multi-cut L-shaped method are discussed.

References

[1] F. Q. You, L. Tao, D. J. Graziano, and S. W. Snyder, "Optimal design of sustainable cellulosic biofuel supply chains: Multiobjective optimization coupled with life cycle assessment and input-output analysis," *AIChE Journal,* vol. 58, pp. 1157-1180, 2012.

[2] F. Q. You and B. Wang, "Life Cycle Optimization of Biomass-to-Liquid Supply Chains with Distributed-Centralized Processing Networks," *Industrial & Engineering Chemistry Research,* vol. 50, pp. 10102-10127, 2011.

[3] A. Dunnett, C. Adjiman, and N. Shah, "A spatially explicit whole-system model of the lignocellulosic bioethanol supply chain: an assessment of decentralised processing potential," *Biotechnology for Biofuels,* vol. 1, pp. 1-17, 2008.

[4] B. Aksoy, H. Cullinan, D. Webster, K. Gue, S. Sukumaran, M. Eden, and N. Sammons, "Woody biomass and mill waste utilization opportunities in Alabama: Transportation cost minimization, optimum facility location, economic feasibility, and impact," *Environmental Progress & Sustainable Energy,* vol. 30, pp. 720-732, 2011.

[5] J. Kim, M. J. Realff, and J. H. Lee, "Optimal design and global sensitivity analysis of biomass supply chain networks for biofuels under uncertainty," *Computers & Chemical Engineering,* vol. 35, pp. 1738-1751, 2011.

[6] F. Q. You, J. M. Wassick, and I. E. Grossmann, "Risk Management for a Global Supply Chain Planning Under Uncertainty: Models and Algorithms," *AIChE Journal,* vol. 55, pp. 931-946, 2009.

[7] B. H. Gebreslassie, Y. Yao, and F. Q. You, "Design under Uncertainty of Hydrocarbon Biorefinery Supply Chains: Multiobjective Stochastic Programming Models, Decomposition Algorithm and A Comparison between CVaR and Downside Risk," *Submited to AIChE Journal* 2012.

[8] F. Q. You and I. E. Grossmann, "Multicut Benders decomposition algorithm for process supply chain planning under uncertainty," *Ann Oper Res,* 2011.

AUTHOR INDEX

AUTHOR INDEX

AUTHOR INDEX

AUTHOR INDEX

AUTHOR INDEX

AUTHOR INDEX

AUTHOR INDEX

AUTHOR INDEX

AUTHOR INDEX

AUTHOR INDEX